8th IGRSM International Conference and Exhibition on Geospatial and Remote Sensing (IGRSM 2016)

IOP Conference Series: Earth and Environmental Science
Volume 37

Kuala Lumpur, Malaysia
13 - 14 April 2016

ISBN: 978-1-5108-2616-8
ISSN: 1755-1307

Printed from e-media with permission by:

Curran Associates, Inc.
57 Morehouse Lane
Red Hook, NY 12571

Some format issues inherent in the e-media version may also appear in this print version.

Printed by Curran Associates, Inc. (2016)

For permission requests, please contact the Institute of Physics
at the address below.

Institute of Physics
Dirac House, Temple Back
Bristol BS1 6BE UK

Phone: 44 1 17 929 7481
Fax: 44 1 17 920 0979

techtracking@iop.org

Additional copies of this publication are available from:

Curran Associates, Inc.
57 Morehouse Lane
Red Hook, NY 12571 USA
Phone: 845-758-0400
Fax: 845-758-2633
Email: curran@proceedings.com
Web: www.proceedings.com

Table of contents

Volume 37

8th IGRSM International Conference and Exhibition on Geospatial & Remote Sensing (IGRSM 2016)
13–14 April 2016, Kuala Lumpur, Malaysia

Accepted papers received: 2 June 2016
Published online: 1 July 2016

Preface

Papers

Technology Trends

Forestry

Agriculture

Infrastructure & Urban Planning

Land Use / Land Cover Mapping

Disaster Management

Environmental Monitoring

IOP Conf. Series: Earth and Environmental Science **37** (2016) 011001 doi:10.1088/1755-1315/37/1/011001

PREFACE

This proceedings consists of the peer-reviewed papers from the *8th IGRSM International Conference and Exhibition on Geospatial & Remote Sensing & GIS* (IGRSM 2016), which was organised by the Institution of Geospatial and Remote Sensing Malaysia (IGRSM) on 13–14 April 2016 at the Berjaya Times Square Hotel, Kuala Lumpur, Malaysia. The conference, with the theme *Geospatial on the Go,* aimed to facilitate the dissemination of knowledge, and the sharing of expertise and experience in geospatial sciences, in all aspects of its applications. It also aimed to build linkages between industry and local and international professionals in this field.

The co-organisers of the conference were Universiti Putra Malaysia (UPM), Science & Technology Research Institute for Defence (STRIDE), National Space Agency (ANGKASA), Universiti Teknologi Malaysia (UTM), Universiti Utara Malaysia (UUM), Universiti Teknologi Petronas (UTP), Malaysian Rubber Board (LGM), Forestry Department Peninsular Malaysia (JPSM), National Hydraulic Research Institute of Malaysia (NAHRIM), Malaysian Centre for Geospatial Data Infrastructure (MaCGDI), Malaysian Civil Defence Department (JPAM), Department of Agriculture (DOA), Ornet Sdn. Bhd. and Smart Digital International Sdn. Bhd. The conference's silver sponsor was Antaragrafik Systems Sdn. Bhd.

Some of the highlights of the conference included:

- A pre-conference talk on *Impact of Geographical Factors on Cross-Country Movement During Crisis Situations* by Assoc. Prof. Dr. Marian Rybansky, Department of Military Geography and Meteorology, University of Defence, Czech Republic
- Officiation by YB Datuk Seri Panglima Madius Tangau, Minister of Science, Technology & Innovation
- Keynote presentations by:
    - Hjh. Norizam Che Noh, Director, MaCGDI, Malaysia: *Addressing Current Challenges and Formulating Future Strategies for the National Geospatial Community*
    - Prof. Dr. Yukihiro Takahashi, Director, Space Mission Center, Hokkaido University, Japan: *Next-Generation Remote Sensing With Micro-Satellite*
    - Assoc. Prof. Dr. Reza Ehsani, Citrus Research & Education Center (CREC), University of Florida, US: *The Rise of Small UAVs: Applications, Opportunities and Challenges*
- Oral and poster presentations from 88 speakers, from both Malaysia (55) and abroad (33), covering areas of technology trends, infrastructure and urban planning, land use / land cover mapping, disaster management, and environmental monitoring
- 2016 Annual General Meeting (AGM) of IGRSM, which included the award of certificates of election to the first 74 members who are eligible to use the title Geospatialist (Gs), comprising of 3 Distinguished Fellows, 4 Fellows and 67 Professional Members

- Special sessions on:
    - Hexagon Ignite by Hexagon Geospatial and Antaragrafik Systems Sdn. Bhd.
    - Industry Outreach, with presentations by Ground Data Solutions R&D Sdn. Bhd., Oracle Corporation, Alam Sekitar Malaysia Sdn. Bhd. (ASMA), National Space Agency (ANGKASA), Hexagon Geospatial and Antaragrafik Systems Sdn. Bhd., Digital Globe, and Geospatial Media & Communications.

Two awards were presented during the conference's closing ceremony:

- **Best Paper Award:** Gs. Dr. Hamdan Omar, Forest Research Institute Malaysia (FRIM): Malaysia: *Quantifying Rate of Deforestation and* CO_2 *Emission in Peninsular Malaysia Using Palsar Imageries*
- **Best Student Paper Award:** Maher Aburas, University Putra Malaysia (UPM), Malaysia: *Landscape Analysis of Urban Growth Patterns in Seremban, Malaysia Using Spatio-Temporal Data*

The success of the IGRSM 2016 was due to commitments of authors, keynote speakers, session chairpersons, the organising and technical programme committees, student volunteers from Universiti Putra Malaysia (UPM), and many others of various roles. We would like to thank all exhibitors and contributors, namely Antaragrafik Systems Sdn. Bhd., Ground Data Solutions R&D Sdn. Bhd.; Bumitouch PLMC Sdn. Bhd.; E.J. Motiwalla; Geospatial Media and Communications Sdn. Bhd.; MaCGDI; Geoscience & Digital Earth Centre (INSTeG), UTM; Remote Sensing of Environment and Disaster Laboratory (RSED), University of Tokyo; Space Mission Center, Hokkaido University; and Global Navigation Satellite System (GNSS) Simulation Laboratory, STRIDE.

Assoc. Prof. Gs. Dr. Abdul Rashid Mohamed Shariff

Chairman
8th IGRSM International Conference on Geospatial & Remote Sensing (IGRSM 2016)

President
Institution of Geospatial and Remote Sensing Malaysia (IGRSM), 2014-2016

IOP Conf. Series: Earth and Environmental Science **37** (2016) 011002 doi:10.1088/1755-1315/37/1/011002

Organising Committee

Chairman	Assoc. Prof. Gs. Dr. Abdul Rashid Mohamed Shariff
Technical Programme	Gs. Dr. Dinesh Sathyamoorthy
Finance	Normalina Jamaluddin Gs. Kumar Veliayudam
Local Arrangement	Gs. Dr. Ahmad Fikri Abdullah Gs. Dr. Farrah Melissa Muharam Siti Sarah Emran
Sponsorship and Exhibition	Tn. Hj. Gs. Khairuddin Abd Rahman Gs. Mohd Hafiz Mohd Hazir
Publicity and Website	Gs. Samsuddin Ahmad Gs. Dr. Ruzinoor Che Mat Gs. Mohd Zakri Tarmidi Siti Nooradzah Adam

Steering Committee Members

Gs. Wan Abdul Hamid Shukri, Forestry Department Peninsular Malaysia (JPSM)
Gs. Khairul Anam Musa, National Hydraulic Research Institute of Malaysia (NAHRIM)
Zuhaidah Nordin, Malaysian Centre for Geospatial Data Infrastructure (MaCGDI)
Maj. (PA) Syed Azam Shah, Malaysian Civil Defence Department (JPAM)
Noor Azawani Wahap , National Space Agency (ANGKASA)
Tn. Hj. Ismail Hj. Hashim, Department of Agriculture (DOA)
Raja Aqma Raja Taihairan, Department of Agriculture (DOA)
Ezuan Suhaidy Suhaimi, Department of Agriculture (DOA)

International Program Committee

Dr Ramin Nourqolipour, Department of Survey, Iran
Assoc Prof Dr Yassine Charabi, Sultan Qaboos University, Oman
Assoc Prof Dr Naohiko Kohtake, Keio University, Japan
Assoc Prof Dr Wataru Takeuchi, University of Tokyo, Japan
Dr Ebrahim Jahanshiri, University of Nottingham
Dr Khosro Khorramnia, Ariss Smart Agriculture Solutions
Dr. Bheshem Ramlal, University of The West Indies, Trinidad and Tobago
Dr. Raid Al-Tahir, University of New Brunswick, Canada

Technical Programme Committee

Chairpersons:

Gs. Dr. Dinesh Sathyamoorthy, Science & Technology Research Institute for Defence (STRIDE), Ministry of Defence, Malaysia

Assoc. Prof. Gs. Dr. Abdul Rashid Mohamed Shariff, Universiti Putra Malaysia (UPM), Malaysia

Members:

Prof. Gs. Dr. Mazlan Hashim, Universiti Teknologi Malaysia (UTM), Malaysia
Prof. Dr. Li Jing, Beijing Normal University, China
Prof. Dr. Iyyanki Muralikrishna, Defence Research and Development Organisation (DRDO), India
Prof. Dr. Ismat Mohamed El Hassan, King Saud University, Saudi Arabia
Prof. Dr. Ewe Hong Tat, Universiti Tunku Abdul Rahman (UTAR), Malaysia
Prof. Dr. Christine Pohl, University of Osnabrueck, Germany
Prof. Dr. Mahender Kotha, Goa University, India
Prof. Dr. Paolo Gamba, University of Pavia, Italy
Assoc. Prof. Gs. Dr. Abdul Nasir Matori, Universiti Teknologi Petronas (UTP), Malaysia
Assoc. Prof. Gs. Dr. Khiruddin Abdullah, Universiti Sains Malaysia (USM), Malaysia
Assoc. Prof. Gs. Dr. Lim Hwee San, Universiti Sains Malaysia (USM), Malaysia
Assoc. Prof. Gs. Ir. Dr. Chan Yee Kit, Multimedia University (MMU), Malaysia
Assoc. Prof. Dr. Francis Harvey, University of Minnesota, USA
Assoc. Prof. Dr. Mustafa Din Subari, Universiti Teknologi Malaysia (UTM), Malaysia
Assoc. Prof. Dr. Marian Rybansky, University of Defence, Czech Republic
Assoc. Prof. Dr. Anuar Ahmad, Universiti Teknologi Malaysia (UTM), Malaysia
Assoc. Prof. Dr. Kasturi Devi Kanniah, Universiti Teknologi Malaysia (UTM), Malaysia
Assoc. Prof. Dr. Biswajeet Pradhan, Universiti Putra Malaysia (UPM), Malaysia
Assoc. Prof. Dr. Zulkiflee Abd Latif, Universiti Teknologi MARA (UiTM), Malaysia
Assoc. Prof. Dr. Raid Al-Tahir, University of New Brunswick, Canada
Assoc. Prof. Dr. Benny Peter, Universiti Teknologi Malaysia (UTM), Malaysia
Assoc. Prof. Dr. Siti Khairunniza Bejo, Universiti Putra Malaysia (UPM), Malaysia
Gs. Dr. Norzailawati Mohd Nor, International Islamic University Malaysia (IIUM), Malaysia
Gs. Dr. Ruzinoor Che Mat, Universiti Utara Malaysia (UUM), Malaysia
Gs. Dr. Ahmad Fikri Abdullah, Universiti Putra Malaysia (UPM), Malaysia
Gs. Dr. Farrah Melissa Muharam, Universiti Putra Malaysia (UPM), Malaysia
Dr. Josée Lévesque, Defence Research and Development Canada (DRDC), Canada
Dr. Lim Tien Sze, Multimedia University (MMU), Malaysia
Dr. Soni Darmawan, University of Tokyo, Japan
Dr. Lawal Billa, University of Nottingham, Malaysia Campus
Dr. Khamarrul Azahari Razak, Universiti Teknologi Malaysia (UTM), Malaysia

IOP Conf. Series: Earth and Environmental Science **37** (2016) 011004 doi:10.1088/1755-1315/37/1/011004

Peer review statement

All papers published in this volume of *IOP Conference Series: Earth and Environmental Science* have been peer reviewed through processes administered by the proceedings Editors. Reviews were conducted by expert referees to the professional and scientific standards expected of a proceedings journal published by IOP Publishing.

Genetic algorithm for investigating flight MH370 in Indian Ocean using remotely sensed data

***Maged Marghany, Shattri Mansor, Abdul Rashid Bin Mohamed Shariff**

Geospatial Information Science Research Centre, Faculty of Engineering
University Putra Malaysia, 43400 UPM, Serdang, Selangor, Malaysia
*Email: magedupm@hotmail.com

Abstract. This study utilized Genetic algorithm (GA) for automatic detection and simulation trajectory movements of flight MH370 debris. In doing so, the Ocean Surface Topography Mission(OSTM) on the Jason- 2 satellite have been used within 1 and half year covers data to simulate the pattern of Flight MH370 debris movements across the southern Indian Ocean. Further, multi-objectives evolutionary algorithm also used to discriminate uncertainty of flight MH370 imagined and detection. The study shows that the ocean surface current speed is 0.5 m/s. This current patterns have developed a large anticlockwise gyre over a water depth of 8,000 m. The multi-objectives evolutionary algorithm suggested that objects are existed on satellite data are not flight MH370 debris. In addition, multi-objectives evolutionary algorithm suggested that the difficulties to acquire the exact location of flight MH370 due to complicated hydrodynamic movements across the southern Indian Ocean.

1. Introduction

Since March 8 2014 and despite, twelve countries have joined in for the search-and rescue efforts of missing flight of MH370, it is so durable to find a plane like MH370 [1]. By the way, China deployed 10 high-resolution satellites to scurry the South China Sea, Digital Globe opened its crowdsourcing platform Tomnod and Airbus Defence and Space mobilized its five satellites to find some leads [2]. We live in an age and era where technology tracks our every little movement. Why then it is so difficult to find a plane like MH370? The Malaysian Airlines flight went missing on March 8 2014 and every new piece of information seems to shroud the flight's disappearance in more mystery. Theories ranged from hijacking to disruption to a conceivable suicide by one of the pilots. Finally, citing satellite-data analysis by British firm Inmarsat, has concluded that the flight, which vanished while flying to Beijing from Kuala Lumpur, had crashed thousands of miles away in the southern Indian Ocean (Figure 1) and none of the passengers or crew on board survived [2].

According to Grady [3], leveraging a combination of satellite and terrestrial assets, news networks continue to dig-deep into their portfolio of communications options. Among those options are their pre-existing full time, loss of satellite capacity, or contribution feeds. Nearly all major news networks leverage these contribution feeds to help deliver studio-to the studio feeds, move content from one corner of the globe to another, or supplement other terrestrial-based pathways.

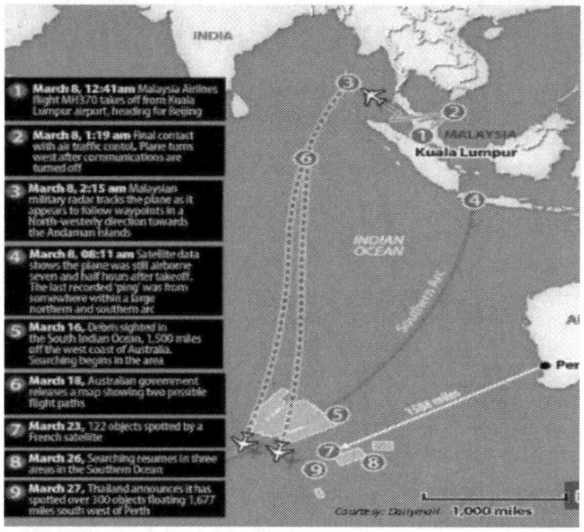

Figure 1. MH370 suspected location in the southern Indian Ocean.

Aireon will enable airlines and ANSPs to more accurately track the location of aircraft, something that, as evidenced by the recent MH370 disappearance, is very much needed. By providing near real-time data transmission via satellite, aircraft can be tracked within a few miles of its location at any time, no matter where it is on earth. In light of the disappearance of flight MH370, technological advancements in flight tracking have come under some serious scrutiny. How will Aireon play a role in flight tracking of commercial aircraft? How does it differ from the available current tracking technologies. The rummage for the flight MH370 has been a sobering reminder of the vastness and uncharted mystery of our planet's oceans. Conversely, the race against the clock to find the airliner's black box has also provided something of a showcase for the technologies that are enhancing our ability to operate in an environment often considered to be every bit as challenging as outer space [4].

The main question can be raised up what appropriate sensors can be used to monitor and detect flight MH370 debris? The high-resolution sensors either on board of satellite or airborne can detect and identify the flight MH370 debris. Even HF ground radar can detect any foreign objects moving in the coastal zone. This also is required the standard methods of object automatic detection by using high resolution microwave satellite data with 1 m as in the spot mode of both RADARSAT-2 SAR, TerraSAR-X satellite data. The RADARSAT-2 SAR satellite has a synthetic aperture Radar (SAR) with multiple polarization modes, including a fully polarimetric mode in which HH, HV, VV and VH polarized data are acquired. Its highest resolution is 1 m in Spotlight mode (3 m in Ultra-Fine mode) with 100 m positional accuracy requirement. In addition, RADARDSAT-2 SAR Scan Narrow SCNB beam is its and a high revisit period of 7 days. Further, has nominal near and far resolutions of 7 m. If the length of the flight is 24 m, means it could clearly be detected in RADARDSAT-2 SAR Scan Narrow. This suggests that, as high cloud covers are dominated in the southern Indian Ocean, it is suggested to use airborne SAR sensors like uninhabited aerial vehicle Synthetic aperture radar (UAVSAR, by JPL, L-band) with a 22-km-wide ground swath at 22° to 65°.

Clearly, the lack of remote sensing and in situ data made the hunt of flight MH370 even more convoluted. Satellite altimeter measurements could provide accurate data of bathymetry, sea level, and ocean wave and ocean current movements. In this regard, the main objective of this work is to utilize the capability of Ocean Surface Topography Mission (OSTM) on the Jason-2 satellite with genetic algorithm (GA) to investigate the existing of the flight MH370 debris in the southern Indian Ocean under the impact of Indian Ocean circulation.

2. Study Area

Figure 2 shows seafloor topography in the Malaysia Airlines flight MH370 search area. Dashed lines approximate the search zone for sonar pings emitted by the flight data recorder and cockpit voice recorder popularly called black boxes. The first sonar contact (black circle) was reportedly made by a Chinese vessel on the east flank of Batavia Plateau (B), where the shallowest point in the area (S) is at an estimated depth of 1637 meters. The next reported sonar contact (red circle) was made by an Australian vessel on the north flank of Zenith Plateau (Z) (Figure 2). The deepest point in the area (D) lies in the Wallaby- Zenith Fracture Zone at an estimated depth of 7883 meters. The Wallaby Plateau (W) lies to the east of the Zenith Plateau. The shallowest point in the entire area shown here is on Broken Ridge (BR). Deep Sea Drilling Project (DSDP) site 256 is marked by a grey dot. The inset in the top left shows the area's location to the west of Australia. Seafloor depths are from the general bathymetric chart of the oceans [5].

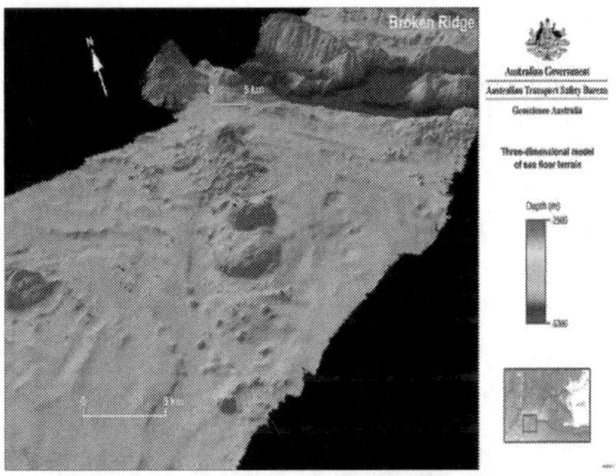

Figure 2. Bathymetry of Indian Ocean.

3. The publicly available data

According to Zweck [6], the only publicly available reliable data are: (i) the location of the INMARSAT 3-F1 satellite in a geostationary orbit at an elevation of 35,800 km over the point on the earth at 1.5°N, 64.5° E; (ii) the data released by the AAIB on March 25th which includes the times at which the satellite pinged the aircraft, together with the measured burst frequency offsets (i.e., the Doppler-shift frequencies) at those times; and (iii) a map released by the Malaysian government which shows that the angle between the 8:11 am arc and the satellite position was about 40°. The initial search area is accurately shown on a March 17th map on the web site of the Washington Post newspaper. This map also includes the approximate location of the satellite and four red arcs marked with the times 5:11am, 6:11am, 7:11am, and 8:11am. Although this map is qualitatively correct, the data it seeks to represent does not agree quantitatively with the three reliable data sets described above [6]. Since the methodology the Inmarsat team is used to deduce possible flight paths depends on accurate knowledge of the locations of the arcs, it is not currently possible for their calculations.

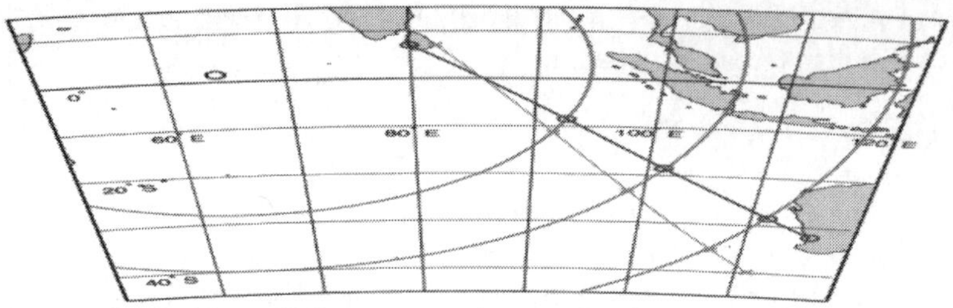

Figure 3. Map of path of a fictitious flight, XX123, from Colombo, Sri Lanka towards Perth Australia, departing at 12 noon and arriving in Perth at 7:48 pm [7].

In addition, in Figure 3, the three red arcs show the location of the plane at 3 pm, 5 pm, and 7 pm. Using only this information, the blue curve shows the 400-knot flight path. The magenta curve shows how the flight path that is deduced would change if the starting location were assumed to be 220 km due east of Colombo. A plane traveling along the magenta path would have a speed of 416 knots and be 970 km from Perth at 7:48 pm [7].

4. Genetic algorithm and debris of MH370

The genetic algorithm (GA) can work in two ways: (i) if a piece of debris is located in the southern Indian ocean, GA can predict, or backtrack, to the original location, or the possible crash site; and (ii) if a piece of debris is located by satellite, GA can decide either the debris belong to flight MH370 or not. The simulation of flight MH370 debris is done using GA. Following Anderson [8], GA encodes the parameter values associated with each candidate solution as a string, which usually in binary format. For each parameter, the number of bits provided must be sufficient to encode the full range of possible values associated with that parameter.

The string representing a solution is simply the concatenation of the sub-strings corresponding to the individual parameters; by analogy with biology, this string is referred to as a chromosome. Starting with an initial population of candidate solutions (i.e., chromosomes) constructed by means of a random number generator, a genetic algorithm iteratively applies three basic steps: (i) rank the members of the current population according to fitness, (ii) select superior members which will be used to breed the next generation, and (iii) apply operators on randomly-selected pairs of these members to mimic the transfer of genetic material to offspring that occurs during biological reproduction, thereby producing a new generation with statistically superior characteristics [8].

4.1 Multi-objective optimization based Pareto dominance

Following Anderson [9], the definition of the problem given above is in one sense incomplete – it does not specify the choice of norm for the space $\boldsymbol{Y}$. In a single objective optimization problem, the objective space is usually a subset of the real numbers and a solution $x_1 \in \boldsymbol{P}$ is better than another solution $x_2 \in \boldsymbol{P}$ if $y_1 < y_2$ where $y_1 = \mu(x_1)$ and $y_2 = \mu(x_2)$. In the case of a vector-valued objective function mapping, comparing solutions is more complex and one must endeavour to capture the essential priorities of the problem in the choice of norm. Herein lays the crucial distinction between single objective and multi-objective problems - whereas the former afford simple scalar measures of fitness that can be used to rank individual members of the design space, the latter are characterised by conflicts of interest among the competing objectives as measured by $\mu_i, i=1,m$ [9-11]. There are several ways to deal with this complication. Perhaps the simplest is to create a scalar figure of merit as a weighted sum of the separate objective measures,

$$\text{(i) minimise} \quad \mu_i = \sum_{i=1}^{m} \alpha_i \mu_i \qquad (1)$$

Another approach is to convert all but one of the objectives into constraints,

$$\text{(ii) minimise } \mu_j \text{ subject to } \mu_i \le z_i \ \forall i = 1, m; i \ne j \qquad (2)$$

While convenient, these methods shed little light on the nature of the trade-offs made. As there may be subtle, non-quantifiable considerations involved in site selection, such as risks to personnel or to equipment, a better approach is to map the trade-off surface so that the decision maker can execute judgment in making a final selection. To perform this mapping, it is not necessary to run (i) or (ii) above for a large number of parameter selections α_i, z_i and to inspect the outcomes. Instead, we can use an evolutionary stochastic optimisation algorithm to reveal the Pareto front, as described below.

Pareto optimality is based on the binary relation of dominance. A solution $\boldsymbol{x}_1 \in \boldsymbol{X}$ is said to be dominated by another solution $\boldsymbol{x}_2 \in \boldsymbol{X}$, written $\boldsymbol{x}_2 \prec \boldsymbol{x}_1$, if $\boldsymbol{x}_2$ is at least as good on all counts (objectives) and better on at least one, that is,

$$\mu_i(x_2) \le \mu_i(x_1) \ \forall i = 1, m \text{ and } \mu_j(x_2) < \mu_j(x_1) \text{ for some } j. \qquad (3)$$

With this relation, the Pareto set of optimal (non-dominated) solutions $\boldsymbol{P}^*$ will usually have multiple entries, associated with different trade-offs between the objectives. The image $\boldsymbol{Y}^* \subset \boldsymbol{Y}$ of the Pareto set $\boldsymbol{P}^* \subset \boldsymbol{P}$ is referred to as the Pareto front and knowledge of its shape greatly assists in choosing the best compromise solution [9]. The Genetic algorithm has implemented on the data of the Jason-2/Ocean Surface Topography Mission (OSTM), and QuikSCAT [12].

5. Results and discussion

Figure 4 shows the possible MH370 debris which acquired by Chinese satellite in 44° 57'S 90° 13'E in the southern Indian ocean with the length of 24 m. Further, Figure 5b, shows the corrected image generated by using a Genetic Algorithm. Figure 5b shows the segmented image generated by Genetic Algorithm which indicates that the bright object which claimed by Chinese satellite as MH370 debris is belong to aircraft carrier. The correlation between real aircraft carrier and one produced by genetic algorithm is 0.87 with standard errors of 0.002. This means that the optical remote sensing data either THEOS satellite or a China satellite are not able to determine precisely any debris belong to MH370 which considered as unusual uncertainty.

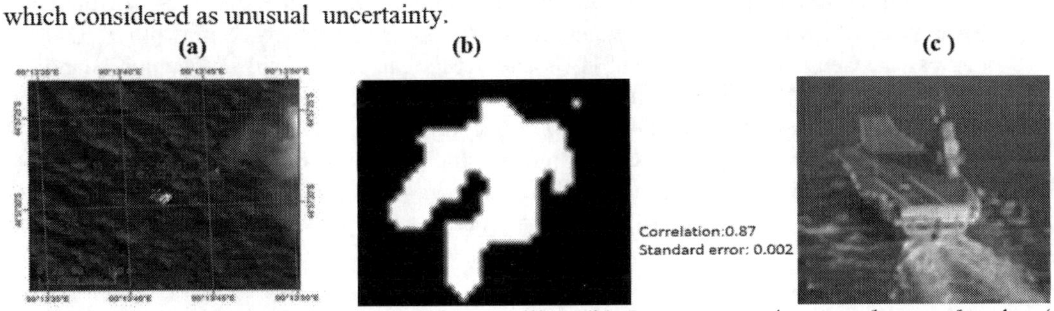

Figure 4. Flight MH370 debris in (a) China satellite , (b) GA segmentation results correlated to (c) aircraft carrier as example.

Figure 5 shows the simulated debris trajectory movements based on clues were given by Inmarsat-4 satellites. The debris moved from Australia across the Southern Indian Ocean since March 2014 to February 2016. This simulation took 7 hours to be completed using GA. Figure 6e shows that debris flow in anti-clockwise direction with root mean square error of current velocity of 10 cm/sec which is coincided with the Southern Indian current movement.

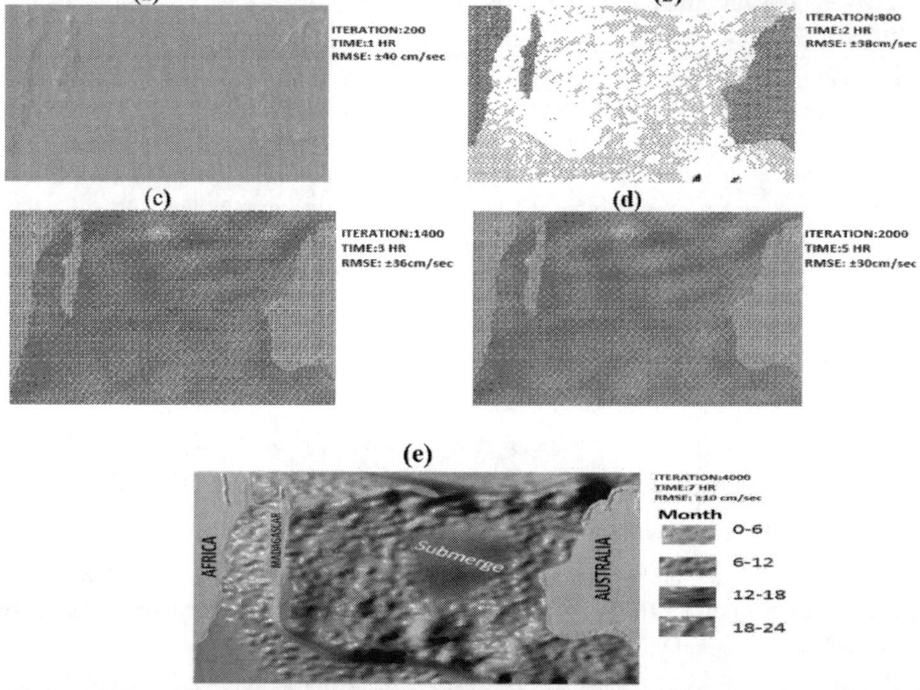

Figure 5. Genetic algorithm iterations of current movement and possible flight MH370 debris (a) 200,(b)800,(c)1,400, (d)2,000 and 4,000.

However, the element of the debris would not have floated for several months at the water's surface but would have drifted underwater a few meters deep. In fact, the Antarctic Circumpolar Current (ACC) will cause instabilities for the debris trajectory movements. In this regard, the MH370 debris could transport westward and spin in a large scale counter-clockwise eddies rotation and drifted westward to the African east i.e. Mozambique and Madagascar coastal waters. During this dynamic instability, either debris are more buoyant than water, in which case they float, or they are less buoyant, in which case they sink. Therefore, the turbulent movements with 50 km/ day of the large southern Indian gyre with width of 100 km would cause the debris to submerged in depth of 3,000 m to 8,000 m across the Southern Indian Ocean [12]. The debris has been found in Reunion Island are not belong to MH370. In fact, the debris would sink under sea surface of 3000 water depth within less than few months as explained above. If there is no clue confirms the existence of debris either from remote sensing data or ground search across the Southern Indian Ocean, this means the MH370 have landed vertically through the ocean surface and broke down to several pieces through the water column due to huge hydrostatic pressure of more than 2,000 m water depth. This confirms the theory of Marghany [12].

6. Conclusions

This study has used optimization techniques of Genetic algorithm to investigate the impact of ocean surface current on flight MH370 debris. The southern Indian Ocean during the months of March-April has dominated by anticlockwise large gyre moving with maximum velocity of 0.5 m/s and slowly drifts westward. It means that flight MH370 debris can potentially travel up to 50 km/day with large eddies of a width of 100 km wide. The study shows that flight MH370 debris could not move to Africa within 24 months and with less than 2 months it would sink before washed up on Réunion Island. However, it can be said that the turbulent flow due to large Southern Indian gyre would make the debris submerged in deep water more than 3000 m across the Southern Indian Ocean. In conclusion, intelligent system based on multi-objectives Genetic algorithm can be used to investigate uncertainties in data and information. In conclusion, it is difficult to determine the MH370 debris in the Southern Indian Ocean because of sophisticated and turbulent current, which could drift the debris away to westward that required large-scale search areas.

References

[1] R Curtis 2014. The mainstream media's shameful coverage of missing flight MH370. The Bates Student. **17** 3.
[2] Asia News 2014 Geospatial World **9** 13.
[3] B Grady 2014 NSR Analysis: OU Or Contribution The Business of Pre-Planning For Breaking News. Sat magazine June 2014 60.
[4] J Excell 2014 Down deep. The Engineer, **296** 3.
[5] H F Walter and M M Karenm 2014 EOS, **95** 173.
[6] J Zweck 2014 How Satellite Engineers are Using Math to Deduce the Flight Path of the Missing Malaysian Airliner [*www.utdallas.edu/~zweck/MH370.pdf*].
[7] J Zweck 2014. How Did Inmarsat Deduce Possible Flight Paths for MH370? SIAM News http://www.siam.org/news/news.php?id=2151.
[8] S J Anderson 2014 In Ed. M. Marghany. www.intechopen.com/books/advanced-geoscience-***remote-sensing.***
[9] S J Anderson, PJ Edwards, P Marrone, and Y A Abramovich 2003. Investigations with SECAR - a bistatic HF surface wave radar, *Proceedings of IEEE International Conference on Radar, RADAR 2003*, Adelaide.
[10] S J Anderson 2013 IEEE Tran. on Geosc. and Rem. Sens., **51**, 1805.
[11] S J Anderson, Darces, M., Helier, M., and Payet, N. 2013 . Accelerated convergence of genetic algorithms for application to real-time inverse problems, Proceedings of the 4th Inverse Problems, Design and Optimization Syposium, IPDO-2013, Albi, France, 149-152.
[12] M.Marghany 2014. Developing genetic algorithm for surveying of MH370 flight in Indian Ocean using altimetry satellite data. 35th Asian conference of remote sensing, at Nay Pyi Taw, Mynamar, 27- 31 October 2014. a-a-r-s.org/acrs/administrator/components/com.../OS-081%20.pdf.

IOP Conf. Series: Earth and Environmental Science 37 (2016) 012002 doi:10.1088/1755-1315/37/1/012002

Online 3D terrain visualisation using Unity 3D game engine: A comparison of different contour intervals terrain data draped with UAV images

Mohd Hafiz Mahayudin and Ruzinoor Che Mat
School of Multimedia Technology and Communication,
Universiti Utara Malaysia,
06010 UUM Sintok, Kedah

E-mail: ruzinoor@gmail.com

Abstract. The main objective of this paper is to discuss on the effectiveness of visualising terrain draped with Unmanned Aerial Vehicle (UAV) images generated from different contour intervals using Unity 3D game engine in online environment. The study area that was tested in this project was oil palm plantation at Sintok, Kedah. The contour data used for this study are divided into three different intervals which are 1m, 3m and 5m. ArcGIS software were used to clip the contour data and also UAV images data to be similar size for the overlaying process. The Unity 3D game engine was used as the main platform for developing the system due to its capabilities which can be launch in different platform. The clipped contour data and UAV images data were process and exported into the web format using Unity 3D. Then process continue by publishing it into the web server for comparing the effectiveness of different 3D terrain data (contour data) draped with UAV images. The effectiveness is compared based on the data size, loading time (office and out-of-office hours), response time, visualisation quality, and frame per second (fps). The results were suggest which contour interval is better for developing an effective online 3D terrain visualisation draped with UAV images using Unity 3D game engine. It therefore benefits decision maker and planner related to this field decide on which contour is applicable for their task.

1. Introduction

In recent years, Geographical Information System (GIS) technology have advanced a great deal with the internet, this is due to terrain visualisation technology that uses the internet as the main platform for presentation such as Google earth, Marble and Bing Map3D has emerged rapidly thus increase the user expectation and demand for terrain visualisation [1]. These platforms are capable of producing a 3D visualisation of GIS data with more realism. Most of the terrain data is derived from the offline system before it being used inside online environments. This is due to that online system is very dependent on the offline system for visualising the terrain, for example if the terrain data and file size of 3D data too large, it will slow the loading and rendering time considerably.

This study aims to compare three different contour intervals (1m, 3m, 5m) terrain data that have been draped with UAV images, this study will utilize GIS data of palm oil plantation in Sintok, Kedah which consist of terrain data and UAV images. The study will utilize three different software's, which are ArcGIS from ESRI (http://www.esri.com), Global Mapper [2] and Unity 3D. The objective of this paper is to compare the effectiveness of visualising online 3D terrain draped with UAV images for different contour intervals data based on the data size, loading time (office and out-of-office hours), response time, visualisation quality, and fps.

2. Related works

Terrain visualisation has advanced a lot in these recent years. Thus, there are a lot of different platforms that GIS data is applied. Ruzinoor et al. [3] have done a good review on 3D terrain visualisation of GIS data starting from the earlier stage of the technology emerged. They also produce many related work regarding this research by utilising Virtual Reality Markup Technology (VRML) as main platform by performing different kind of experiment for online 3D terrain visualisation such as using different contour interval data [1], different GIS software's [4], different rendering technique [3, 5], different web servers [6], and different multi-sensor satellite images [7]. Besides that, Ruzinoor also study on how to utilized Unity 3D game engine in developing online 3D terrain visualisation for an oil palm plantation [8-10]. In another study by Shin et al. [11], Unity 3D game engine also used to visualize 3D terrain but their project focus more on sea navigation. In a study by Lv et al. [12], they visualize 3D multi-layer seabed terrain in Smartphone environment. Resch, Wohlfahrt, & Wosniok [13] studies the effectiveness of 3D seabed visualisation with time control in different platforms. Horowitz & Schultz [14] uses 3D printing technology to print digital terrain models of Mars surface. Wu, Deng, & Paul [15] studies on terrain visualisation using graphic card technology and openGL. Mazurek, Paces, & Filla [16] uses 3D visualisation technology with electronic flight instrument system to solve flight disorientation during flight for aircraft pilots. Finally, Zocco, Livatino, & De Paolis [17] studies on improving military operations by providing 3D visualisation of the area of military operations.

3. The implementation of online 3D terrain visualisation using Unity 3D game engine

Terrain visualisation is using the terrain data in a 3D view that would allow a much more detail view of the terrain surface and elevation. The process of analysis begins with contour data preparation, continue with UAV data preparation and finally Unity 3D implementation. The URL used to launch the applications is at "http://www.hafsaruz.net".

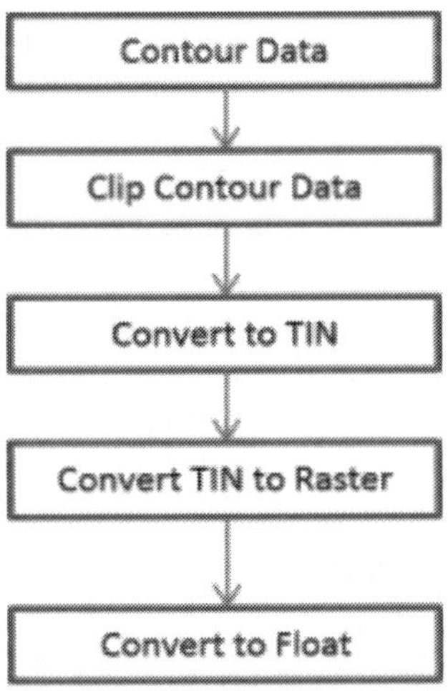

Figure 1: The flow of contour data preparation.

3.1 Contour data preparation

The contour data used for this study was gathered from palm oil plantation in Sintok, Kedah. The size of the study area was 2.08829 acre. The projection used was WGS 84 UTM Zone 47N. The contour data of the study area obtained from Digital Terrain Model (DTM) generated from UAV images that capture from the aerial imagery of the study area. As mentioned earlier, the contour data is divided into three different intervals which is 1m, 3m, and 5m. In order to generate different contour intervals from DTM, the Global Mapper software [2] by Blue Marble Geographic was used. This three different contour intervals contour data need to be clipped. These contour data are in vector format which contain the height value of the terrain based on their intervals. The contours data, however, do not used the same projections as the UAV images data and have to be changed into similar projections. ArcGIS define projection tool being utilized to set this two types of data into similar projections. In order to clip the contour data, the process begins by generating new SHP files which contain the size of the clipped area needed. The size of the clipping area is set to 712331.186567 top, 665056.150775 left, 712241.534425 bottom and 665150.415246 right. After that, each of the contour intervals data was clipped based on this clipped area. Figure 1 show the whole process of contour data preparation. When the contour data being clipped, its need to be convert into Triangular Irregular Network (TIN) format. After that, the data were converted into raster format. Finally, the data were converted into float file (.FLT and .HDR) using the conversion tools that available in ArcGIS.

Figure 2 shows the three contour data with different interval of the same area when the clipping process was done.

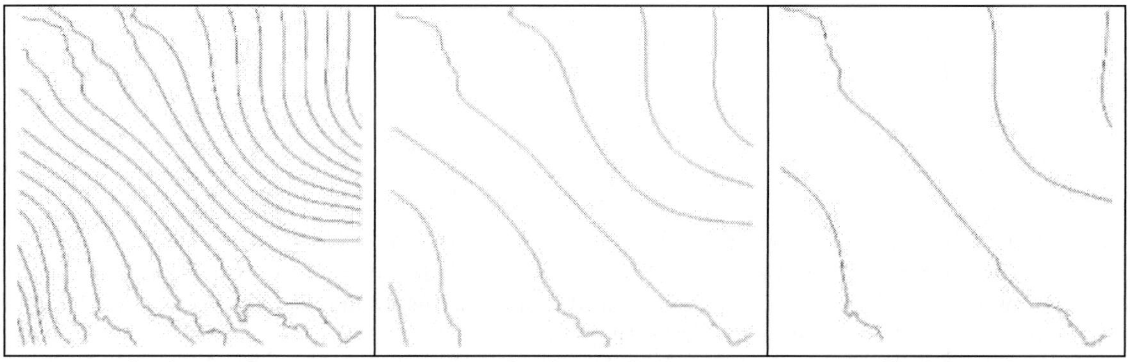

1m interval **3m interval** **5m interval**

Figure 2: Different contour interval data.

3.2 UAV data preparation

The UAV images data of the study area are collected by flying over the area using UAV at 318.309m altitude with Canon Powershot s100 (5.2 mm). These UAV images need to be clipped with the same size of the contour data clipped earlier. Again, the ArcGIS software was used to clip this data. The same size of the clipped area was used which is 712320.63 top, 664934.95 left, 712231.96 bottoms and 665044.44 rights. The process begins by opening the UAV images data. After that, open the clipped area SHP files. Then the UAV images were clipped based on the clipped area selected. Figure 3 show the UAV images data after clipped. This file then need to be export as JPG image in order to be used in Unity 3D game engine.

Figure 3: The UAV images data after clipped.

3.3 Unity 3D implementation

After finishing the contour data preparation and UAV images data preparation, the data is imported into Unity 3D. In Unity 3D, the terrain is generated based on terrain data that was clipped before. 3D object of the terrain is generated using unity 3D terrain engine and viewing functions is added such as zoom in and out. It is then published for web environments. The file is uploaded at web server with address link "http://www.hafsaruz.net". Figure 4 show the flow on how the terrain visualisation being published into web environment using Unity 3D game engine.

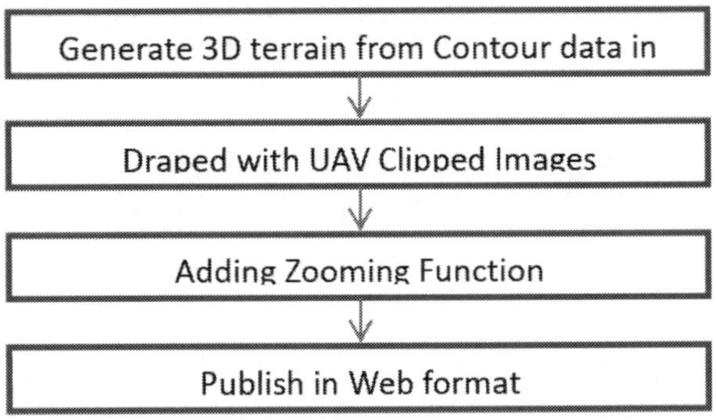

Figure 4: The flow of publishing terrain visualisation for web environment.

4. Results and discussion

The result of this study was three Unity 3D web player format that was created from different contour interval data which are 5m, 3m, and 1m. The output was measured based on data size, loading time (office and out-of-office hours), response time, visualisation quality, and fps. The first output is from 5m interval. Figure 5 show the image of this output. The results of this image show "Acceptable" of visualisation quality where the area inside the circle can be seen clearly but difficult to identify the slopes. Overall the area of oil palm plantation can be viewed clearly.

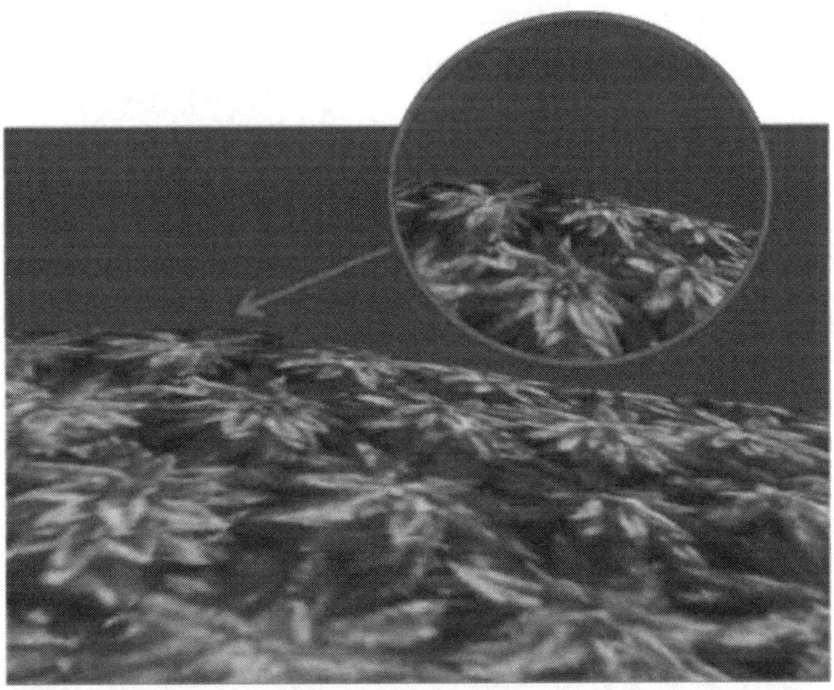

Figure 5 : The view of UAV images draped with 5m interval contour data.

The second result is from 3m interval output. Figure 6 show the image of this output. The results of this image show "Clear" of visualisation quality where the area inside the circle can be seen clearly but with slight slopes, overall, most of the area inside the oil palm plantation can be viewed clearly.

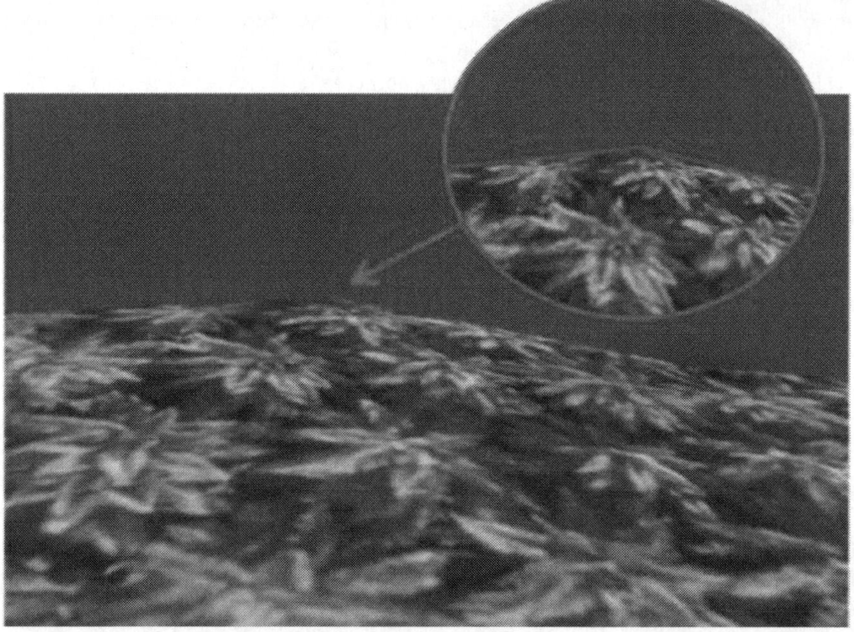

Figure 6: The view of UAV images draped with 3m interval contour data.

The final result is from 1m interval output. Figure 7 show the image of this output. The results of this image show "Very clear" of visualisation quality where the area inside the circle can be seen very clearly with a good view of slopes, overall, most of the area inside the oil palm plantation can be viewed clearly.

Figure 7: The view of UAV images draped with 1m interval contour data.

Table 1 shows the results produced from each contour intervals data. As mentioned earlier, there are five criteria to be measured which are data size, loading time (office and out-of-office hours), response time, visualisation quality, and fps. The results of data size for 1m data interval produce the bigger file size after published compared to 3m and 5m data interval. This is due to usually 1m data interval holds a lot of data compare to 3m and 5m data interval. The same size of UAV images was overlaid on each data interval. The visualisation qualities in these three outputs were already discussed earlier in this section.

Table 1: Result of each contours data.

Criteria	**Contour Intervals**					
	1m		**3m**		**5m**	
Data size	1.401 kb		1.399 kb		1.399 kb	
Loading time office hour (Off) and out-of-office hour (Out)	Off 3.2 sec	Out 2.0 sec	Off 1.2 sec	Out 1.3 sec	Off 1.0 sec	Out 0.9 sec
Response time	139 ms		109 ms		110 ms	
Visualisation quality	Very clear		Clear		Acceptable	
Frames per second (fps)	58.11 fps		57.54 fps		56.45 fps	

The other criteria to be measured was fps value where Comodo Dragon browser performance tool is used for the purpose of this testing. Testing was done by recording the web page in five seconds for different contour interval to show the value of fps. For 5m data interval 56.45 fps value was recorded while for 3m data interval 57.54 fps value was recorded and 1m data interval recorded fps value of 58.11. After that, the test continue with measuring the response time for each contour intervals data using Yslow by Marcel Duran [18]. In order to view the response time value, YSlow plugin are installed inside Mozilla Firefox web browser. The results of the response time acquired are 139 ms for 1m, 109 ms for 3m, and 110 ms for 5m. The response time of each contour intervals are slightly different from each other. Finally, the loading time was measured using tools from App.telemery name "page speed monitor" by Fabasoft group [19]. The results that was recorded are from two-time frames that are during office hour and out-of-office hour. The time taken for office-hour was between 8 am to 5 pm while the out-off-office hour was after 5 pm. The results recorded for office hour are 3.2 sec for 1m, while for 3m and 5m each recorded an average of 1sec. While the results for out-of-office hour is 2 sec for 1m, 1.3 sec for 3m and 0.9 sec for 5m is slightly faster compare during office hour.

5. Conclusions

In conclusion, it can be concluded that in order to develop online 3D terrain visualisation overlaid with UAV images by different contour intervals terrain data using Unity 3D game engine, there is some consideration that has to be looked upon. One of it is the accuracy of the data used. By converting the terrain data in multiple formats of data which is raster to vector in ArcGIS, the terrain data has extra features when generating terrain in Unity 3D game engine. Thus, the result shows slightly different from the expectation on 1m, 3m, and 5m contour intervals. As for 1m contour interval, it shows that the result as expected where 1m contour interval generated bigger data size compared to 3m contour interval, however, the data size of 5m contour interval is slightly bigger than 3m contour interval. This situation occurred, may be due to the extra features generated from the 5m contour interval. It can be concluded that the most appropriate and suitable data to be used for implementation of online 3D terrain visualisation draped with UAV images using Unity 3D game engine was 3m interval data. It giving promising results as same as 1m interval data which has less data size, less loading time and response time and less fps value compare to 1m interval data.

6. References

[1] Ruzinoor CM, Shariff ARM, Mahmud AR, Pradhan B, Rahim MSM. 2010. 3D Terrain Visualisation: A Comparison of Different Contour Intervals for Terrain Data Draped with Aerial Photograph *MRSS 6th Remote Sensing and GIS Conference and Exhibition (MRSS 2010)*; Kuala Lumpur, Malaysia

[2] Blue Marble Geographics. 2016 Global mapper. [cited 2015 30 December]; Available from: http://www.bluemarblegeo.com/products/global-mapper.php.

[3] Ruzinoor CM, Shariff ARM, Pradhan B, Ahmad MR, Rahim MSM. 2012 A review on 3D terrain visualisation of GIS data: techniques and software. *Geo-spatial Information Science*;**15**(2):105-15.

[4] Ruzinoor CM, Shariff ARM, Mahmud AR. 2009. Online 3D Terrain Visualisation: A Comparison of Three Different GIS Software. *International Conference on Information Management and Engineering (ICIME2009)*; Kuala Lumpur, Malaysia

[5] Ruzinoor CM, Shariff ARM, Mahmud AR, Pradhan B. 2011. 3D Terrain Visualisation for GIS: A Comparison of Different Technique. In: Manfred, editor. True-3D in Cartography: Autosteroscopic and Steric Visualisation of Geodata: Springer-Verlag; p.

[6] Ruzinoor CM, Shariff ARM, Mahmud AR. 2009. Online 3D Terrain Visualisation of GIS data: A Comparison between Three Different Web Servers *International Advanced Technology Congress (ATCi) 2009*; PWTC Kuala Lumpur, Malaysia.

[7] Ruzinoor CM, Shariff ARM, Pradhan B, Mahmud AR, Rahim MSM. 2013 An effective visualisation and comparison of online terrain draped with multi-sensor satellite images. *Arabian Journal of Geosciences*;**6**(12):4881-9.

[8] Ruzinoor CM, Shariff ARM, Zulkifli AN, Mohd Rahim MS, Mahayudin MH. 2015. Web Based 3D Terrain Visualisation Using Game Engine. *5th International Conference on Computing and Informatics (ICOCI 2015)*; Istanbul, Turkey

[9] Ruzinoor CM, Zulkifli AN, Nordin N, Mohd Yusof SA. 2013 A Review on Technique in Managing Oil Palm Plantation towards a Digitilized Online 3D Application. *Information Management and Business Review*;**5**(11):547-52.

[10] Ruzinoor CM, Zulkifli AN, Nordin N, Mohd Yusof SA. 2016 Online 3D Oil Palm Plantation Management Based on Game Engine A Conceptual Idea. *Jurnal Teknologi*;**78**(2-2).

[11] Shin I-S, Beirami M, Cho S-J, Yu Y-H. 2015 Development of 3D Terrain Visualisation for Navigation Simulation using a Unity 3D Development Tool. *Journal of the Korean Society of Marine Engineering*;**39**(5):570-6.

[12] Lv Z, Su T, Li X, Feng S. 2015. 3D visual analysis of seabed on smartphone. *8th IEEE Pacific Visualisation Symposium (PacificVis 2015)*; 14 to 17, 2015; Hangzhou, China

[13] Resch B, Wohlfahrt R, Wosniok C. 2014 Web-based 4D visualisation of marine geo-data using WebGL. *Cartography and Geographic Information Science*;**41**(3):235-47.

[14] Horowitz SS, Schultz PH. 2014 Printing space: Using 3D printing of digital terrain models in geosciences education and research. *Journal of Geoscience Education*;**62**(1):138-45.

[15] Wu J, Deng L, Paul A. 2015 3D Terrain Real-time Rendering Method Based on CUDA-OpenGL Interoperability. *IETE Technical Review*;**32**(6):471-8.

[16] Mazurek P, Paces P, Filla J. 2015. High-fidelity terrain landscape EFIS visualisation in comparative navigation to solve disorientation. *IEEE/AIAA 34th Digital Avionics Systems Conference (DASC), 2015* Prague IEEE.

[17] Zocco A, Livatino S, De Paolis LT. 2014. Stereoscopic-3D vision to improve situational awareness in military operations. Augmented and Virtual Reality: Springer International Publishing; p. 351-62.
[18] Duran M. 2015 Yslow. [cited 2015 30 December]; Available from: htp://yslow.org.
[19] Fabasoft Group. 2015 App.telemetry. [cited 2015 12 December]; Available from: http://www.apptelemetry.com/index_en.html.

A comparative accuracy of Google Earth height with MyGeoid, EGM96 and MSL

Noradila Rusli[1], Muhammad Faiz Pa`suya[2] and Noorfatekah Talib[3]

Center of Study for Surveying Science & Geomatics, Faculty of Architecture & Surveying, Universiti Teknologi MARA, 02600 Arau, Malaysia

Email: noradilarusli@perlis.uitm.edu.my

Abstract. This study aims to provide understanding on the accuracy of height derived from Google Earth ($H_{GoogleEarth}$) as compared to height obtained from Malaysian Geoid Model ($H_{MyGeoid}$), Mean Sea Level (H_{MSL}) and Earth Geoid Model 96 (H_{EGM96}). Total of 50 established points with height acquired from $H_{MyGeoid}$ and H_{MSL} were measured within UiTM (Universiti Teknologi MARA) Arau Campus. These points were also used to extract height from Google Earth and EGM96. Statistical results showed a good range of R^2 between $H_{GoogleEarth}$-$H_{MyGeoid}$, $H_{GoogleEarth}$-H_{MSL} and $H_{GoogleEarth}$-H_{EGM96} i.e. 0.823, 0.843 and 0.824 respectively. It shows $H_{GoogleEarth}$ strongly correlated with H_{MSL}.

1. Introduction

A study conducted by [1] has proved the usage of digital elevation model (DEM), obtained from Google Earth. They were able to delineate a watershed boundary with appropriate accuracy. The same researcher also proved that DEM from Google Earth correlate significantly with other free source height data which are SRTM90 and ASTER30 [2]. However, the accuracy of Google Earth height data as compared to real-site measurement height data remains unknown. The accuracy of DEM depends on the location that can be estimated using Ground Control Point (GCP) measured with the help of Global Navigation Satellite System (GNSS) on the ground survey. Among various methods of accuracy assessment, GPS (Global Positioning System) survey provides the best way to map features on terrain with high accuracy.

Therefore, the main objective of this study is to evaluate the accuracy of height extracted from the Google Earth by comparing the height with 50 GNSS Ground Control Point (GCP) and local Mean Sea Level derived from ground survey levelling in UiTM Arau, Perlis.

2. Method and data

The method and data acquisitions will be explains more in this section.

2.1. Study area

UiTM (Universiti Teknologi MARA) Arau branch located between 6°27'19.41"N, 100°16'19.06"E (upper left) and 6°26'47.21"N, 100°17'10.24"E (lower right) (Figure 1). Total 50 height points were marked within the campus known as cadastral reference marks (CRM). These points were measured on site using GPS instrument which later in this paper will be referred as $H_{MyGeoid}$. By using the same points (50 points) heights were extracted from Google Earth ($H_{GoogleEarth}$) and EGM96 (H_{EGM96}). In addition, height from mean sea level (H_{MSL}) was also measured at the same points which are taken as the datum height for this study.

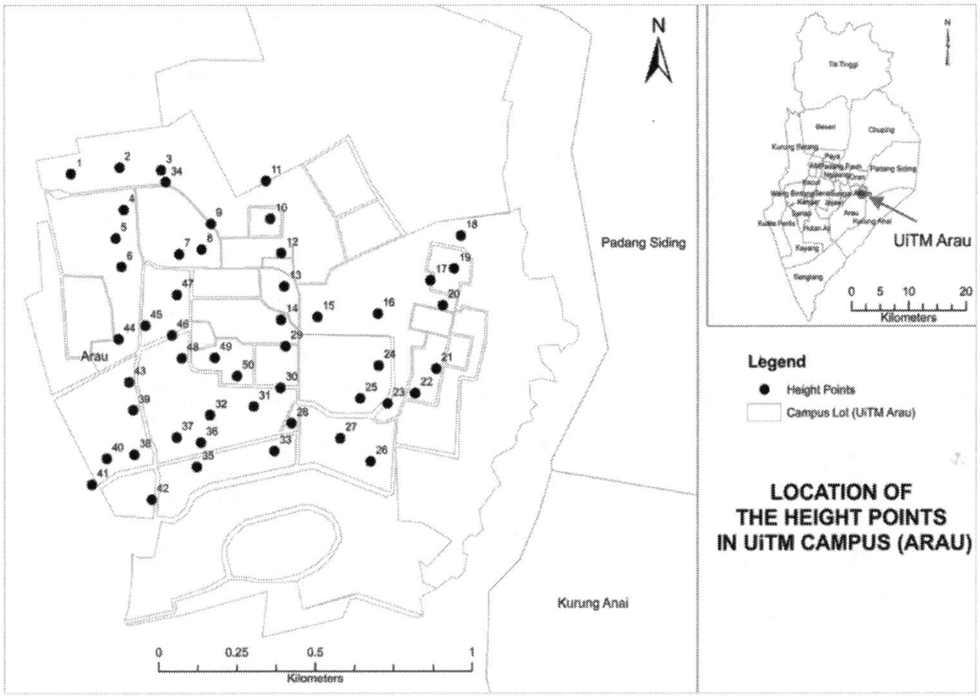

Figure 1. Study area and points location of elevation/height in UiTM Arau.

2.2. Source of data and processing
There are four (4) types of heights obtained in this study which are the heights from Google Earth, EGM96, MyGeoid and MSL. Both $H_{MyGeoid}$ and H_{MSL} were measured on site using GPS and levelling instruments respectively.

2.2.1. Height from Google Earth. The $H_{GoogleEarth}$ was derived from Google Earth using Terrain tool provided at http://www.zonums.com/gmaps/terrain.php. As declared by Zonum, the source of $H_{GoogleEarth}$ height was SRTM90 (Shuttle Radar Topographic Mission 90) produced by NASA (National Aeronautics and Space Administration). Subsequently, all of the 50 points from different heights ($H_{MyGeoid}$, $H_{GoogleEarth}$, H_{MSL} and H_{EGM96}) were analyzed to identify its correlation.

2.2.2. GPS surveys and processing. The 50 GPS GCP in this study has been observed using static GPS with two (2) hours observation to obtain accurate three-dimensional coordinates. GPS observations were carried out using Topcon GR5 field observation and the observation procedure was conducted in accordance with the regulation prescribed by the Department of Survey and Mapping Malaysia (DSMM). For the determination of GCP coordinates, all measurements were connected to the Arau and UUM Continuously Operating Reference Station (CORS) from MyRTKnet and the coordinates were established in the Geocentric Datum Malaysia 2000 (GDM 2000). In GPS data processing, all the raw GPS observations were processed using commercial software named Topcon Tool Software. Table 1 show the processing parameters used in this study.

Table 1. Processing Parameters. [3]

Items	Parameters
General procedure	Procedure provided in manufacturer manual must be followed
Datum	GDM2000
Elevation mask	15^0
Ephemerides	Short baseline of less than 30 km: Broadcast Long baseline: Precise
Baseline processing	RMSE less than two (2) cm
Quality	Maximum data rejection: less than 10% Ambiguity fixed solution
Adjustment	Least square adjustment should be used
Minimally constrained adjustment	One control station fixed in GDM2000 coordinate
Quality indicator	Pass chi-square test at 95% confidence region All baseline must pass local test
Over-constrained adjustment	At least two (2) control stations must be fixed in the final adjustment

2.2.3. Determination of orthometric height. Before height comparisons, the GPS GCP with height values of the DEM extracted from the Google Earth, the coordinate of GCP should be converted from GDM2000 to World Geodetic System (WGS84). Meanwhile, the ellipsoidal height with referenced to the WGS84 ellipsoid also should be converted to orthometric height with the EGM96 geopotential model using a well-known formula in Equation 1:

$$H = h - N \qquad \text{(Equation 1)}$$

where N is the geoid height from EGM96 (NEGM96), h is the ellipsoidal height regarding World Geodetic System (WGS84) and H is the orthometric height.

The geoid height from NEGM96 at each point can be computed online using the following link: http://earthinfo.nga.mil/GandG/wgs84/gravitymod/egm96/intpt.html. Besides that, the height from Google Earth has also been compared with orthometric height computed using ellipsoid height as referenced to GDM2000 and precise local geoid model, Malaysia Geoid Model MyGEOID provided by Department of Survey and Mapping Malaysia (DSMM).

2.2.4. Precise levelling survey. The accuracy of Google DEM was analyzed by comparing it with the height above local mean sea level. The height was based on the mean sea level (MSL) value from the tide gauge in Port Klang. In this study, Leica NA3003 digital level has been used for the execution of leveling works to determine local mean sea level value at 50 GPS GCP. A collimation test for levelling or two (2) -peg test has been performed to determine the level's collimation error. The minimally constrained adjustment was carried out by fixing Standard Benchmark (SBM) located at

UiTM Arau. The height of SBM 0121 is 45.022 metres above the mean sea level (MSL). The permissible discrepancy for this study is 3mm$\sqrt{km}$ for first order two (2)-way levelling.

3. Results and analysis

To analyze the accuracy of height obtained from Google Earth ($H_{GoogleEarth}$) with another source of height data ($H_{MyGeoid}$, H_{EGM96} and H_{MSL}), fifty points of height were overlapped with $H_{MyGeoid}$ and H_{MSL} at the same coordinates. The correlation results are plotted in Figure 2. $H_{GoogleEarth}$ was placed on Y-axis in each graph.

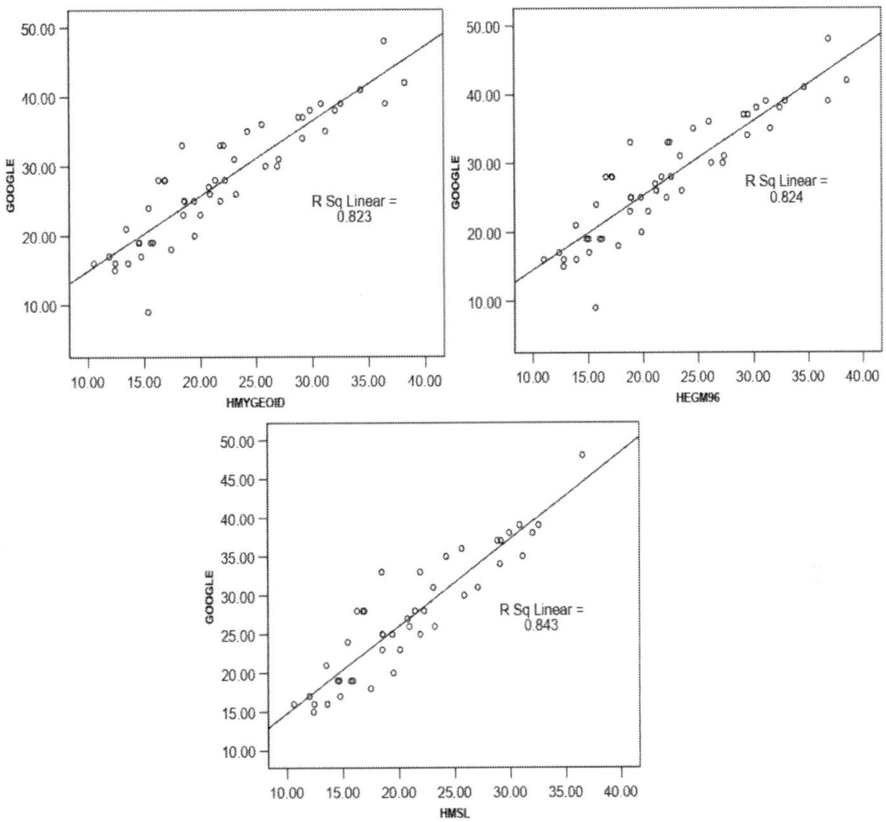

Figure 2. Results of R^2 for each comparison.

Statistically, $H_{GoogleEarth}$ was highly correlated with H_{MSL} data (0.843), H_{EGM96} (0.824) and $H_{MyGeoid}$ (0.823) accordingly. The results indicated that the $H_{GoogleEarth}$ accuracy is much closer to H_{MSL}. Figure 3 illustrates the location of $H_{GoogleEarth}$ among three (3) sources of height.

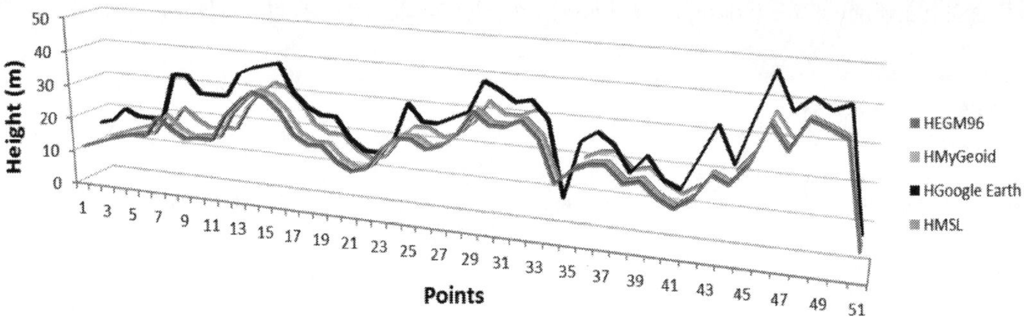

Figure 3. Location of $H_{GoogleEarth}$, $H_{MyGeoid}$, H_{EGM96} and H_{MSL} based on results.

Then, $H_{Google\ Earth}$ was further analyzed to see a range of differences with the three (3) sources of height data. $H_{Google\ Earth}$ was substracted from $H_{MyGeoid}$, H_{MSL} and H_{EGM96} to obtain the differences (Figure 4). The range of differences between $H_{Google\ Earth}$ and $H_{MyGeoid}$ is from 0.5 m (minimum) to 14.6 m (maximum), $H_{Google\ Earth}$ and H_{EGM96} is from 0.18 m (minimum) to 14.1 m (maximum) and $H_{Google\ Earth}$ and H_{MSL} is from 0.5 m (minimum) to 14.5 m (maximum).

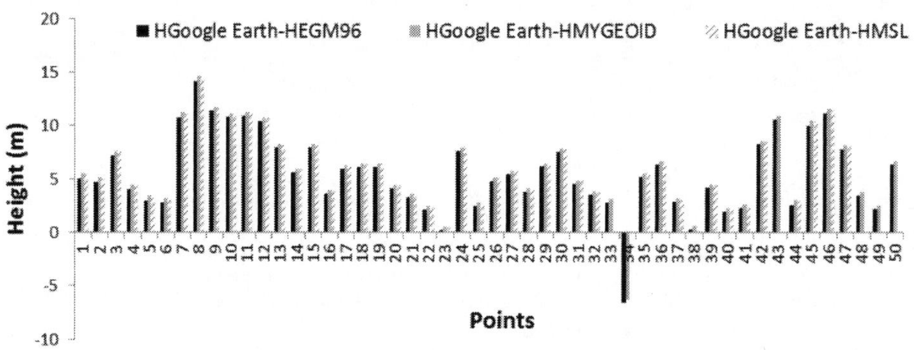

Figure 4. $H_{Google\ Earth}$ differences with $H_{MyGeoid}$, H_{EGM96} and H_{MSL}.

4. Conclusion

The results from the analysis of data indicate that the height from Google Earth ($H_{GoogleEarth}$) proves to be highly correlated with on ground levelling height data which are $H_{MyGeoid}$, and H_{MSL}. Therefore, these findings will hopefully assist researchers to obtain a free source of height data with appropriate accuracy without spending more money and facing tedious data collection and processing.

References

[1] N. Rusli and M. R. Majid, "Digital Elevation Model (DEM) Extraction from Google Earth: a Study in Sungai Muar Watershed," in *Geoinformation – Catalyst for planning, development and good governance-Applied Geoinformatics for Society and Environment 2012*, 2012, pp. 24–28.

[2] N. Rusli, M. R. Majid, and A. H. Din, "Google Earth-Derived Digital Elevation Model : A

Comparative Assessment against Aster and SRTM Data," *IOP Conf. Ser. Earth Environ. Sci.*, vol. 18, p. 012065, 2014.

[3] Jamil, H., "GNSS Heighting and Its Potential Use in Malaysia. In *GNSS Processing And Analysis*". Marrakech, Morocco: FIG Working Week, 2011.

[4] Rapp, R.H., "Use of potential coefficient models for geoid undulation determination using a spherical harmonic representation of the height anomaly/geoid undulation difference". Journal of Geodesy 71 (5), 282–289, 1996.

Assessment of geometric errors of Advanced Himawari-8 Imager (AHI) over one year operation

Wataru Takeuchi
Institute of Industrial Science, University of Tokyo
6-1, Komaba 4-chome, Meguro, Tokyo 153-8505, Japan

E-mail: wataru@iis.u-tokyo.ac.jp

Abstract. This paper presents an approach to check a geometric performance of Advanced Himawari-8 imager (AHI) and demonstrate and evaluate a new approach to ensure more geometric accurately focusing on visible imagery in 500 meters. A series of processing is supplemented by ground control points of shore lines, land mark locations and digital elevation model. Firstly, a template matching technique is conducted to find a best matching point by simply moving the center of AHI sub-image over each point in a reference image of shore lines and calculating the sum of products between the coefficients and the corresponding neighbourhood pixels in the area spanned by the filter mask. Secondly, ortho-rectification processing is carried out to compensate for the geodetical distortions with respect to the acquisition condition including viewing geometry and so on. As a result, an average of root mean square sum of residual errors with system correction and that of precise geometric correction are shown. Overall geometric accuracy is about 1 to 1.5 pixels from 2015 March to July and it also gradually decreased down to 0.2 to 0.8 from 2015 September to 2016 February. AHI is officially open to public for operational use as of July 1, 2015 and after that operation date geometric errors are reasonably satisfied within one pixels of errors.

1. Introduction

Advanced Himawari-8 Imager (AHI) is a Japanese weather satellite for geosynchronous with a three-axis stabilized spacecraft with a meteorological mission. The Japan Meteorological Agency (JMA) contracted for AHI as a successor to MTSAT and GMS series, in cooperation with the Civil Aviation Bureau (CAB), of the Ministry of Transport of Japan [1].

Geostationary satellites such as Meteosat, MTSAT and Geostationary Operational Environmental Satellite (GOES) were designed for meteorological applications. As a consequence, the onboard sensors have very low geometric, radiometric and spectral resolution. Despite these characteristics, during recent years, several studies have used geostationary satellite data for land surface remote sensing application. In most of them, these data have been used to retrieve geophysical parameters used as an input for models GMS data and are very effective to detect burned areas in different tropical environments as well [2].

AHI breaks through limitations of earlier three-axis stabilized GEO instruments with significant improvements in many areas, including spatial sampling, radiometric sensitivity, calibration and performance around local midnight. The process for the automatic detection of Landmarks and Earth Edges has been working operationally for the first image open to public via JMA. The ability to use these data as the basis to determine the satellite orbit and attitude has also been demonstrated [3]. In this sense, it is necessary to have a continuous accurate and timely determination of the satellite orbit and attitude in order to perform geometric correction of Earth images acquired by geostationary meteorological satellites [4].

The objective of this study is to check a geometric performance of AHI imagery and demonstrate and evaluate a new approach to ensure more geometric accuracy focusing on visible imagery in 500 meters from 2015 March to 2016 February.

2. Methodology

2.1. Framework and data used in this study

Figure 1 shows a framework of precise geometric correction of AHI. A time-series of AHI images from March 2015 to February 2016 at 03:00 (UTC) were collected for evaluation and band 3 (red, 0.64μm) with 500-meters footprint was calibrated to a top-of-atmosphere (TOA) reflectance for further processing.

A series of processing is supplemented by ground control points (GCPs) from MODIS MOD44W land-sea mask (LSM), land mark locations (LML) developed for GMS and digital elevation model from shuttle radar topographic mission (SRTM3). LMLs consists of a collection of land marks developed for a geometric correction of Japanese geostationary meteorological satellite (GMS) which was a predecessor of AHI [5]. It consists of 792 locations with latitude and longitude values and are expected to have a sufficient performance on AHI as well which has the same spatial coverage with MTSAT and GMS [6].

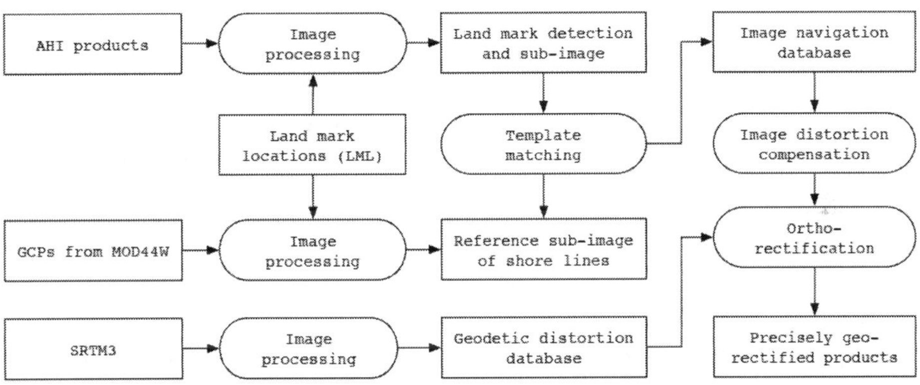

Figure 1. Flowchart of a precise geometric correction of AHI using a template matching technique.

2.2. Sub-image generation and template matching technique

Firstly, a AHI visible image was clipped out into 64 × 64 sub-images centered on 792 LML. Secondly, a suit of sub-images was digitized into a land-sea mask image using a discriminant analysis which enables a dynamic thresholding [7]. Thirdy, shore lines were extracted from LSM database and they were digitized into 128 × 128 sub-images centered on 792 LMLs used as a reference. Finally, a template matching technique was conducted to find a best matching point by simply moving the center of 64 × 64 sub-image over each point in a reference image and calculating the sum of products between the coefficients and the corresponding neighbourhood pixels in the area spanned by the filter mask [8].

2.3. Image distortion compensation and ortho-rectification
Image distortion compensation is conducted by an affine transformation shown as following formula;

$$\begin{pmatrix} \Delta x \\ \Delta y \end{pmatrix} = \begin{pmatrix} a_{11} & a_{12} \\ a_{21} & a_{22} \end{pmatrix} \begin{pmatrix} x \\ y \end{pmatrix} + \begin{pmatrix} b_1 \\ b_2 \end{pmatrix} \quad (1)$$

where x and y values are defined in image coordinates of AHI image and Δx and Δy values are measured by a template matching technique between a AHI sub-image and a reference sub-image derived from shore lines. Then a_{ij} and b_i are computed by a least-squres method over selected cloud-free sub-images by minimizing the sum of the residuals squared. Finally, orthorectification processing is carried out to compensate for the geodetical distortions with respect to the acquisition condition including viewing geometry, platform attitude, Earth rotation and of the relief effects (parallax). It is based on the intersection of viewing direction and the WGS-84 earth ellipsoid using SRTM3 digital elevation model with sufficient accuracy in 500-meters.

3. Results and discussions
Figure 3-(a) shows the number of fine pixels among 796 LSLs which are identified as cloud free and used to detect land mark and generate sub-image from Band 3. It ranges from 5 to 25 LSLs from 2015 March to 2016 February. The number of fine pixels are relatively lower from May to August because of rainy season in monsoon Asian region whereas 20 to 25 LSLs are available from September to October which guarantees higher geometric error compensation reliability. A series of template processing is carried out by using those fine LSLs for the further analysis.

Figure3-(b) shows geometric errors in east-west direction which is obtained from template matching results. Overall geometric accuracy is about 0.7 to 1.2 pixels from 2015 March to July and it gradually decreased down to 0.2 to 0.5 from 2015 September to 2016 February. In July and November, a relatively higher errors more than 2.0 pixels are often found because of less fine GCP's are available for processing as shown in Figure3-(a).

Figure3-(c) shows geometric errors in north-south direction which is obtained from template matching results. Overall geometric accuracy is about 0.2 to 0.5 pixels from 2015 March to July and it also gradually decreased down to -0.2 to 0.2 from 2015 September to 2016 February. In July and November, a relatively higher errors more than 2.0 pixels are often found like those in east-west direction because of less fine GCP's are available for processing as shown in Figure3-(a).

Figure3-(d) shows overall geometric errors which is obtained from root mean square errors from east-west and north-south directions. Overall geometric accuracy is about 1 to 1.5 pixels from 2015 March to July and it also gradually decreased down to 0.2 to 0.8 from 2015 September to 2016 February. AHI is officially open to public for operational use as of July 1, 2015 and after that operation date geometric errors are reasonably satisfied within one pixels of errors.

4. Conclusions and future works
This study was carried out to check a geometric performance of AHI imagery and demonstrate and evaluate a new approach to ensure it more accurately focusing on visible imagery in 500 meters. The ability to use these data as the basis to determine the satellite orbit and attitude has also been demonstrated. Overall geometric accuracy is about 1 to 1.5 pixels from 2015 March to July and it also gradually decreased down to 0.2 to 0.8 from 2015 September to 2016 February. AHI is officially open to public for operational use as of July 1, 2015 and after that operation date geometric errors are reasonably satisfied within one pixels of errors. Extensive modeling and analysis work over one year has shown that good correlation with the true orbit and attitude can be achieved.

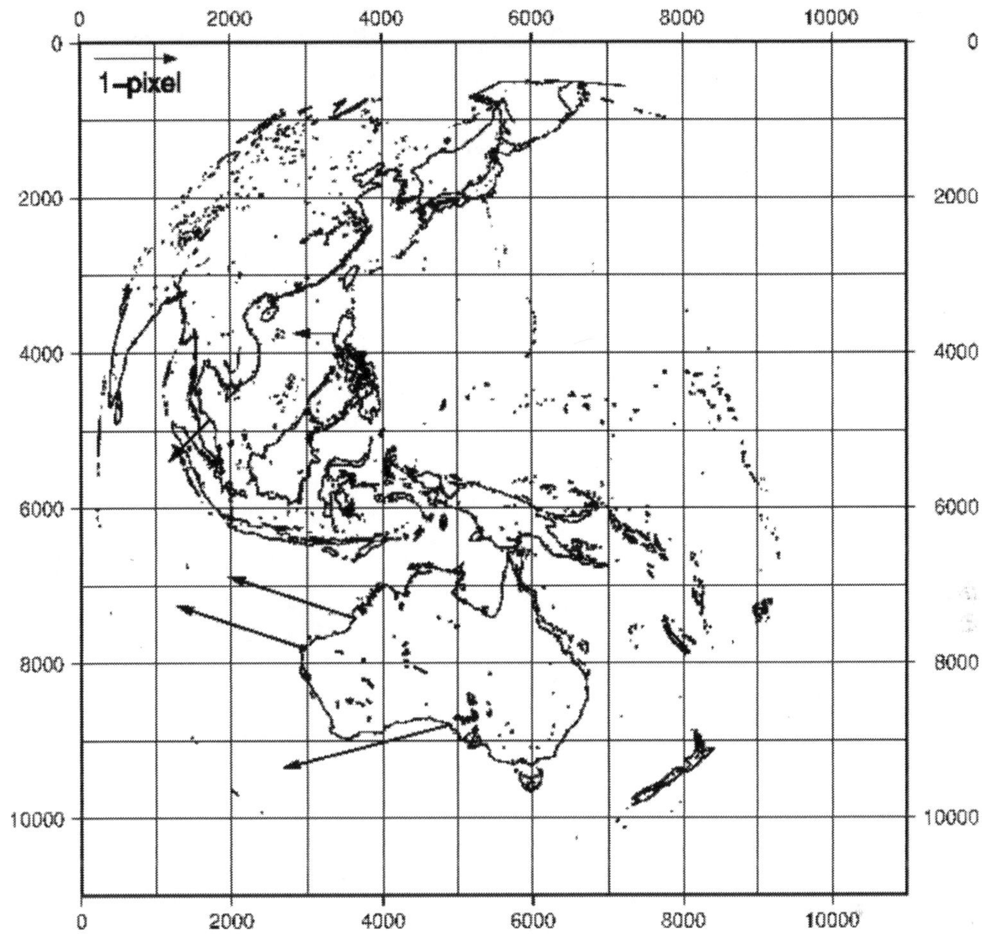

Figure 2. Error vector map of geometric correction on Dec. 18, 2014.

References

[1] Japan Meteorological Agency, 2003. JMA HRIT Mission Specific Implementation, Ver. 1.2.

[2] Boschetti, L., P. A. Brivio, J. M. Gregoire, 2003. The use of Meteosat and GMS imagery to detect burned areas in tropical environments. *Remote Sens. Environ.*, 85, 78?91.

[3] Milnes, M., 2006. Orbit/Attitude Determination for MTSAT through automated Landmark and Earth Edge Detection. *Proceedings of 25th ISTS (International Symposium on Space Technology and Science) and 19th ISSFD (International Symposium on Space Flight Dynamics)*, Kanazawa, Japan.

[4] Takeuchi, W., K. Oyoshi and S. Akatsuka, 2015. Assessment of geometric accuracy with Advanced Himawari-8 Imager (AHI). *Proceed. Int. symp. remote sens. (ISRS 2015)*, Tainan, Taiwan.

[5] Yasukawa, M. and M. Takagi, 2003. Geometric correction of GMS S-VISSR using elevation distortion compensation (in Japanese with English abstract). *Japanese journal of photogrammetry and remote sensing*, 42(6), 33-41.

[6] Takagi, M., 2004. Precise geometric correction for NOAA and GMS images considering elevation effects using GCP template matching and affine transform. *Proceedings of the SPIE*, 5238, 132-141.

[7] Kittler, J. and M. J. Duff, 1985. Image Processing System Architectures. *John Wiley & Sons, Inc., New York, USA.*

[8] Takeuchi, W., T. Nemoto, T. Kaneko and Y. Yasuoka, 2010. Development of MTSAT data processing, distribution and visualization system on WWW. *Asian Journal of Geoinformatics*, 10(3), 29-33.

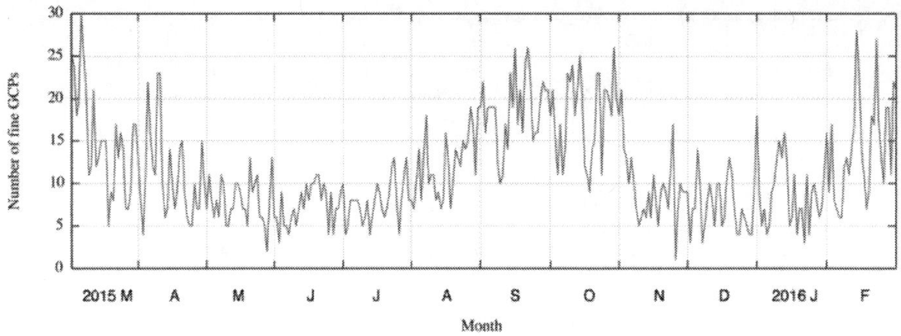

(a) The number of fine GCPs

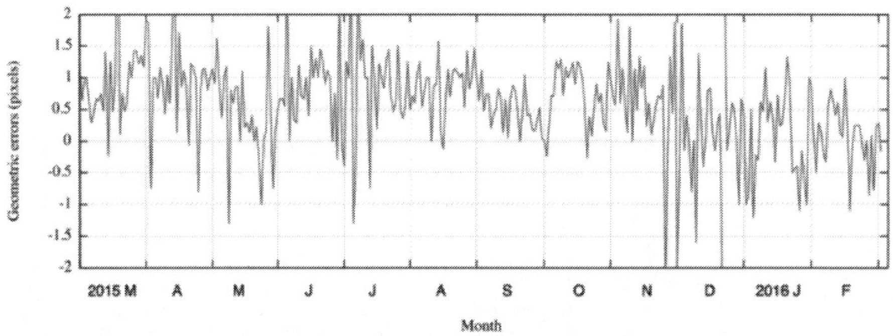

(b) Geometric errors in east-west direction

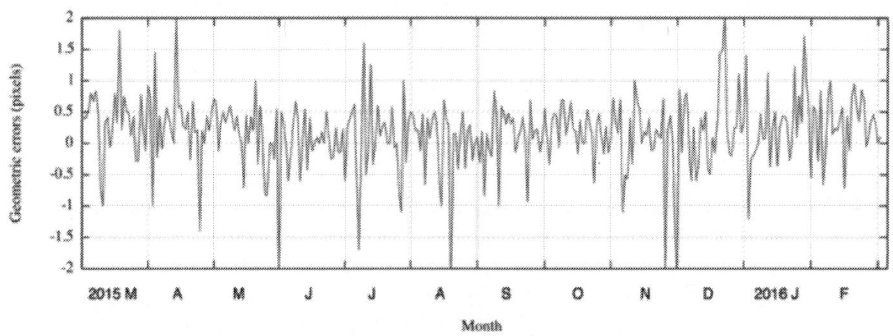

(c) Geometric errors in north-south direction

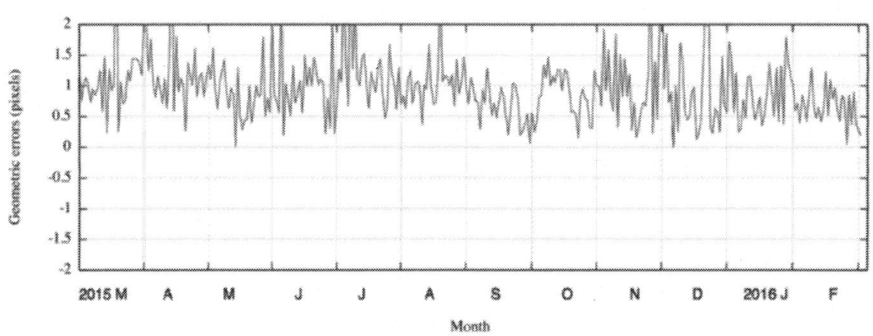

(d) Overall geometric errors

Figure 3. The number of fine GCPs selected for template matching and geometric errors.

Sediment-hosted/orogenic gold mineral systems exploration using PALSAR remote sensing data in Peninsular Malaysia

Amin Beiranvand Pour[1], Mazlan Hashim

Geoscience and Digital Earth Centre (Geo-DEC), Universiti Teknologi Malaysia

Abstract. The Bentong-Raub Suture Zone (BRSZ) is genetically related to the sediment-hosted/orogenic gold deposits associated with the major lineaments and form-lines in the Central Gold Belt of Peninsular Malaysia. In this investigation, the Phased Array type L-band Synthetic Aperture Radar (PALSAR) satellite remote sensing data were used to analyse major geological structures in Peninsular Malaysia and provide detailed characterization of lineaments and form-lines in the BRSZ, as well as its implications for sediment-hosted/orogenic gold exploration in tropical environments. The pervasive array of N-S faults in the study area and surrounding terrain is mainly linked to the N-S trending of the BRSZ Suture Zone. N-S striking lineaments are often cut by younger NE-SW and NW-SE-trending lineaments. Three generations of folding event have been discerned from remote sensing structural analysis. Gold mineralized trends lineaments are associated with the intersection of N-S, NE-SW, NNW-SSE and ESE-WNW faults and curvilinear features in shearing and alteration zones. Compressional tectonics structures such as NW-SE trending thrust, ENE-WSW oriented faults in mylonite and phyllite, recumbent folds and asymmetric anticlines in argillite are high potential zones for gold prospecting.

1. Introduction

Phased Array type L-band Synthetic Aperture Radar (PALSAR), put onboard the Advanced Land Observing Satellite (ALOS), and was launched on January 24, 2006 by an H-IIA rocket from Tanegashima Space Center. It was developed by Japanese Ministry of Economy, Trade and Industry (METI) as a joint effort with Japan Aerospace Exploration Agency (JAXA). PALSAR is an active microwave sensor for all-weather conditions observation and operable both daytime and nighttime [1]. It has L-band synthetic aperture radar with multi mode observation function (Fine mode, Direct downlink, ScanSar mode, and Polarimetric mode) of multi polarization configuration (HH, HV, VH, and VV), variable off-nadir angle (9.9 to 50.8 degrees), switching spatial resolution (10 m, 30 m, 100 m for Fine, Polarimetric, and ScanSar modes, respectively) and swath width observation (30 km, 70 km, and 250-350 km for Polarimetric, Fine and ScanSar modes, respectively) [1]. PALSAR data can be used in

[1] Corresponding Author. Tel: +607 -5530666; Fax: +607- 5531174; Email address: beiranvand.amin80@gmail.com; a.beiranvand@utm.my; mazlanhashim@utm.my.

specific fields, include (i) land area basin mapping (geological structural analysis of target areas); (ii) coastal area basin mapping; (iii) monitoring of environments and natural disasters; and (iv) research and development for the processing and application of multi polarimetric SAR data (geological structural analysis on the first stage of resource exploration).

Many gold mines and prospects in the Peninsular Malaysia are located in the Central Gold Belt (CGB). Gold mineralization in this belt is structurally controlled. The CGB is a highly potential region for prospecting gold exploration targets along the major lineament structures using remote sensing technology. To date, this gold belt has not investigated using recent generation of very high resolution satellite remote sensing imagery [2, 3]. This research presents a remote sensing approach for geological structure mapping in tropical environments. The objectives of this study is to map structural elements associated with sediment-hosted/orogenic gold deposits in the Central Gold Belt of the Peninsular Malaysia using the Phased Array type L-band Synthetic Aperture Radar (PALSAR) satellite remote sensing data at both regional and local scales.

2. Materials and methods

In this investigation, two fine Mode dual polarization Level 4.1 PALSAR scenes were obtained from the Earth and Remote Sensing Data Analysis Center (ERSDAC) Japan (http://gds.palsar.ersdac.jspacesystems.or.jp/e/) for the Central Gold Belt (CGB) of the Peninsular Malaysia. PALSAR fine observation scene covering many of the gold mining districts in Kuala Lipis region in the state of Pahang was selected for detailed analysis of structural features associated with known gold deposits. It covers the eastern part of the Bentong-Raub Suture Zone and the central part of the CGB. Panjom (101° 58′ 58″ E, 4° 08′ 27″ N), Buffalo reef (101° 47′ 11″ E, 4° 15′ 59″ N), Selinsing (101° 47′ 38″ E, 4° 14′ 57″ N), Rubber hill (101° 47′ 55″ E, 4° 14′ 38″ N), Kechau-Tui (101° 58′ 49″ E, 4° 16′ 27″ N) and Tersang (101° 51′ 96″ E, 4° 04′ 81″ N) mines are located in this PALSAR scene. Figure 1 shows the district-scale geological map of the selected study area.

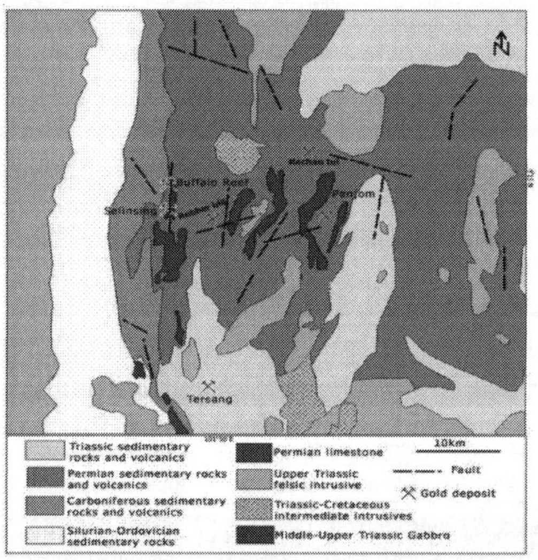

Figure 1. District-scale geological map of the gold deposits, Kuala Lipis, Pahang, Central Malaysia [2].

To facilitate tracing the structural patterns and investigate the relationship between structural setting and gold mineralization in the Central Gold Belt (CGB) of the Peninsular Malaysia, Level 4.1 PALSAR data required to be filtered for speckle reduction. Adaptive filters remove radar speckle from images without seriously affecting the spatial characteristics of the data [4]. Therefore, adaptive filtering was applied to the Level 4.1 PALSAR data. The data were processed using the ENVI (Environment for Visualizing Images) version 4.8 software package.
Directional filters were used to enhance specific linear trends in the Local Sigma resultant images. Four principal Directional filters: N-S, E-W, NE-SW, and NW-SE with a 7*7 kernel size were applied (Table 2). Filters were chosen to highlight the main lineament directions in the Central Gold Belt (CGB) of the Peninsular Malaysia. Directional filter angles were adjusted as N-S: 0°, E-W: 90°, NE-SW: 45°, and NW-SE: 135°. North (up) is zero degrees and the other angles are measured in the counterclockwise direction. 7*7 kernel matrix was selected to enhance semi-smooth and smooth/rough features. Image Add Back value was entered 40%. The Image Add Back value is the percentage of the original image that is included in the final output image. This part of the original image preserves the spatial context and is typically done to sharpen an image.

3. Results and discussion

Directional filtering was implemented to the PALSAR data for tracing structural elements in the selected spatial scene covering the Panjom, Buffalo reef, Selinsing, Rubber hill, Kechau-Tui and Tersang goldfields. Figure 2 shows the RGB results for N-S, NE-SW, and NW-SE (R: 0°, G: 45°, B: 135°) filtering directions applied to HV image. The above mentioned Directional filters have been selected for RGB color combination because NE-striking thrust, NS-striking normal faults and NW-striking strike-slip faults are the most important structural elements for gold targets exploration in the CGB [5]. Lineaments and form-lines are detected (Figure 2), including the long lineaments and short lineaments that form linked systems of long lineaments. The western and northern parts of the images exhibit longer and more lineaments than eastern part. Two major trends NS and NE are detected generally in the western part of the images. However, the central and eastern parts of the images are dominated by lineaments, which are elongated mainly to the NE and few to the NW (Figure 2). Lineaments mapped in northern and central part of the images express several fold systems as curvilinear structures. Lineaments associated with stream are interpreted to be fracture or fault controlled on the central north part. NS and NE-SW trending lineament systems are clearly extensive in the region. Most of longer lineaments are concentrated in the NS direction. NS-trending, normal-slip faults parallel to the Bentong-Raub Suture Zone trend are defined by an obvious, west facing fault escarpment. This NS trend is similar to the orientation of the Bentong-Raub Suture Zone (Figure 2). Penjom gold deposit is located along of the significant splay faults. The Kelau-Karak fault (normal) is one of the major faults running across the Penjom goldfield that control major plutonic emplacements. Localized distribution is caused by faulting and folding, and the Penjom thrust has an NE-SW strike and southerly dip within the deposit. Major gold mineralization took place within the footwall of this thrust. The Kelau-Karak fault and Penjom thrust are detected in the south-eastern part of the image (Figure 2). A regional view of the open-pit quarry of Penjom is shown in Figure 3. The Penjom thrust is the dominant feature controlling the distribution of ore at Penjom and strikes NE (35°) and dips to the southwest (30°-40°).

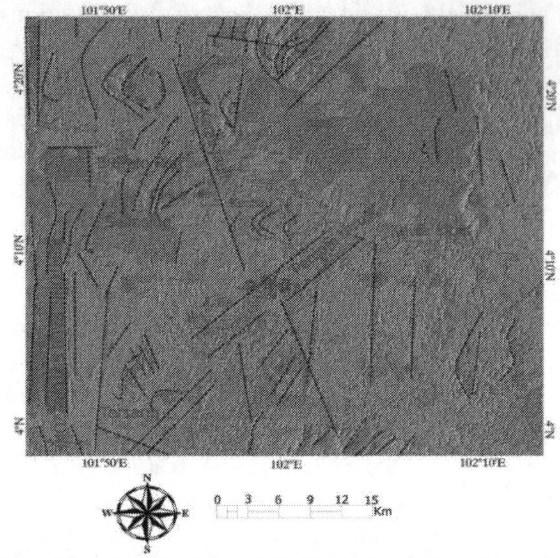

Figure 2. RGB image of N-S (0°), NE-SW (45°), and NW-SE (135°) directional filters covering gold mining districts in Kuala Lipis region, Pahang.

Figure 3. A regional view of the open-pit quarry of Penjom ore deposit.

Figure 3 shows selected spatial subset scene covering northern parts of the Bentong-Raub Suture Zone and CGB in the Kelantan state. RGB color combination was assigned for N-S (R: 0°), NE-SW (G: 45°), and NW-SE (B: 135°) directional filters. The HV polarization PALSAR image was selected to perform directional filtering. The N-S orientation of the Bentong-Raub Suture Zone as a regional strike is detected in the western part of the image (Figure 3). The structural lines with NE-SW and NW-SE directions are in accordance with the tectonic framework of the study area. Major N-S and NE-SW orientations and strike-slip faults with sharp boundaries are revealed in Figure 3. Several small faults and fractures with NE-SW direction are clearly visible in the south and central part of the image. However, few small faults with N-S and NW-SE directions are observable in the north and eastern part of the image. Curvilinear features and some curving faults with NE-SW and NW-SE trends are apparent in the image especially in the central part (Figure 3). These lineament and curvilinear structures have great potential for hosting many quartz-gold lodes. Many of major gold deposits have been found along N-S Bentong-Raub Suture Zone trend in the CGB. In primary stages of exploration plans for identifying potential areas in the north of the CGB a particular attention would be to N-S structural elements at a regional scale.

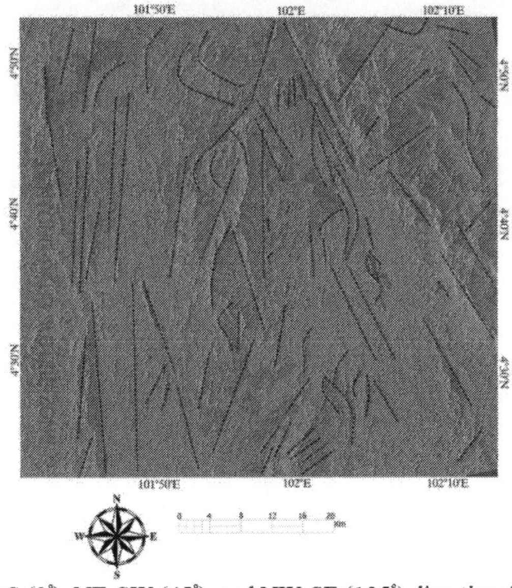

Figure 4. RGB image of N-S (0°), NE-SW (45°), and NW-SE (135°) directional filters covering the northern part of the CGB, Kelantan state.

4. Conclusions

Results of this investigation provide an exploration approach using PALSAR data to map structural elements associated with gold mineralization along the Bentong-Raub Suture Zone in the Central Gold Belt (CGB) of the Peninsular Malaysia. Structural investigation has shown sets of N-S, NE-SW, NNW-SSE and ESE-WNW mineralized trends associated with fault-related rocks and hydrothermally alteration zones. These main trends were intercepted by many shear or lateral fault zones. Exploration plan should also give more attention to narrow down target zones of alteration types and shear zones that show the presence of mylonite, cataclasite and felsic intrusive. The results of this investigation should assist more feasible mineral exploration plans in the CGB of the Peninsular Malaysia. The study presented here encourages further applications of PALSAR satellite remote sensing data for mapping regional and district structural elements associated with epithermal and polymetallic vein-type mineralization in tropical environments.

Acknowledgements

This study was conducted as a part of Fundamental Research Grant scheme (FRGS) (Vote no: R.J130000.7809.4F455), Ministry of Higher Education (MOHE) Malaysia. We are thankful to the Universiti Teknologi Malaysia for providing the facilities for this investigation.

References

[1] Igarashi T 2001. ALOS Mission requirement and sensor specifications. Advances in Space Research 28(1), 127-131.

[2] Pour AB, Hashim M, 2015. Structural mapping using PALSAR data in the Central Gold Belt Peninsular Malaysia, Ore Geology Reviews 64, 13-22.

[3] Pour AB, Hashim M, 2015. Integrating PALSAR and ASTER data for mineral deposits exploration in tropical environments: a case study from Central Belt, Peninsular Malaysia. International Journal of Image and Data Fusion 6(2), 170-188.

[4] Pour BA, Hashim M, Marghany M, 2014. Exploration of gold mineralization in a tropical region using Earth Observing-1 (EO1) and JERS-1 SAR data: a case study from Bau gold field, Sarawak, Malaysia. Arabian Journal of Geosciences 7(6), 2393-2406.

[5] Pour BA, Hashim M, 2014. Structural geology mapping using PALSAR data in the Bau gold mining district, Sarawak, Malaysia. Advances in Space Research, 54(4), 644-654.

A novel built-up spectral index developed by using multiobjective particle-swarm-optimization technique

Maher Ibrahim Sameen and Biswajeet Pradhan*

Department of Civil Engineering, Faculty of Engineering, University Putra Malaysia, 43400 Selangor, Malaysia

* Author to whom correspondence should be addressed; Tel: + 6(03)-8946-6383; Email: biswajeet24@gmail.com or biswajeet@lycos.com

Abstract. In this study, we propose a novel built-up spectral index which was developed by using particle-swarm-optimization (PSO) technique for Worldview-2 images. PSO was used to select the relevant bands from the eight (8) spectral bands of Worldview-2 image and then were used for index development. Multiobjective optimization was used to minimize the number of selected spectral bands and to maximize the classification accuracy. The results showed that the most important and relevant spectral bands among the eight (8) bands for built-up area extraction are band4 (yellow) and band7 (NIR1). Using those relevant spectral bands, the final spectral index was formulated by developing a normalized band ratio. The validation of the classification result using the proposed spectral index showed that our novel spectral index performs well compared to the existing WV-BI index. The accuracy assessment showed that the new proposed spectral index could extract built-up areas from Worldview-2 image with an area under curve (AUC) of (0.76) indicating the effectiveness of the developed spectral index. Further improvement could be done by using several datasets during the index development process to ensure the transferability of the index to other datasets and study areas.

1. Introduction

Recent advances in remote sensing and computing technology have created several new applications. There are numerous applications, which need information about urban features (i.e. buildings, paved roads) such as urban heat island [1], land use mapping [2], flood simulation and modelling [3-4], site suitability analysis [5] etc. Remote sensing data such as very high-resolution satellite images (e.g. Worldview-2) can be used to extract that information. One of the rapid and efficient method and data processing approach is the spectral indices developed and proposed by researchers to extract built-up areas from multispectral satellite images.

There are several studies proposed spectral indices for built-up area extraction from satellite images. Reference [6] proposed a new spectral index to extract built-up areas from Landsat OLI images. The spectral index developed by analyzing different band ratios using spectral and thermal bands. The developed index was compared with four previously used spectral indices; the Band Ratio for Built-up Area (BRBA), the Normalized Built-up Area Index (NBAI), the New Built-up Index (NBI) and the Normalized Built-up Index (NDBI). The results showed that the developed spectral index was superior as compared to ones of the other spectral indices. In another paper, reference [7] developed a technique to extract urban built-up land features from Landsat imagery using three indices, NDBI, Modified Normalized Difference Water Index (MNDWI), and Soil Adjusted Vegetation Index (SAVI). The results indicated that the urban built-up lands were extracted with an overall accuracy ranging from 91.5 to 98.5 percent. In addition, reference [8] proposed a new spectral index for delineating built-up areas from satellite images. The new index combined three existing indices to derive a new improved index for rapid extraction of built-up areas from satellite images. Furthermore, reference [9] presented a new technique for extracting built-up areas from panchromatic images using texture information. The results showed that the method could detect built-up areas from panchromatic images with an overall accuracy of 86.7%.

The literature review revealed that most of the studies developed spectral indices for rapid extraction of built-up areas from various types of satellite images based on specific image data. Therefore, it is required to develop spectral indices for other types of satellite images, which is not investigated in literature. In addition, investigation more advanced and robust methods are required to be involved in the development of spectral indices. Therefore, this study proposes a novel spectral built-up index for rapid built-up extraction from Worldview-2 images. Most of the literature presented above used spectral analysis for the development of spectral indices while in this study we use optimization techniques to develop such indices. Optimization techniques provide best solution for several problems including the selection of relevant wavelengths and determining the best combination of spectral bands for the development of final spectral indices.

1.1 Study area and image data

The study area of this research is a subset of Universiti Putra Malaysia (UPM) campus located in the state of Selangor (ranged from upper left longitude 3°00′14.48′N and latitude 101°42′14.71′E, lower right longitude 3°00′00.71′N and latitude 101°42′44.12′E, of WGS84 coordinate system) (Figure 1). The study area is a mix of urban and vegetation features. Buildings are vary in terms of roofing materials, shape, size and spectral signatures which is very important to take into account when developing new methods or algorithms for feature extraction.

An eight-band pan-sharpened multispectral Worldview-2 standard image (resolution 0.5 m) was utilized to be cloud-free and to consist of different land-cover classes. The image was collected in March 2009. The data set has 1.81m spatial resolution with eight channels: Coastal, Blue, Green, Yellow, Red, Red Edge, Near-Infrared1 (NIR1) and Near-Infrared2 (NIR2), with center wavelengths at 425, 480, 545, 605, 660, 725, 835, and 950 nm, respectively. The radiometric resolution of the data set is 16bit.

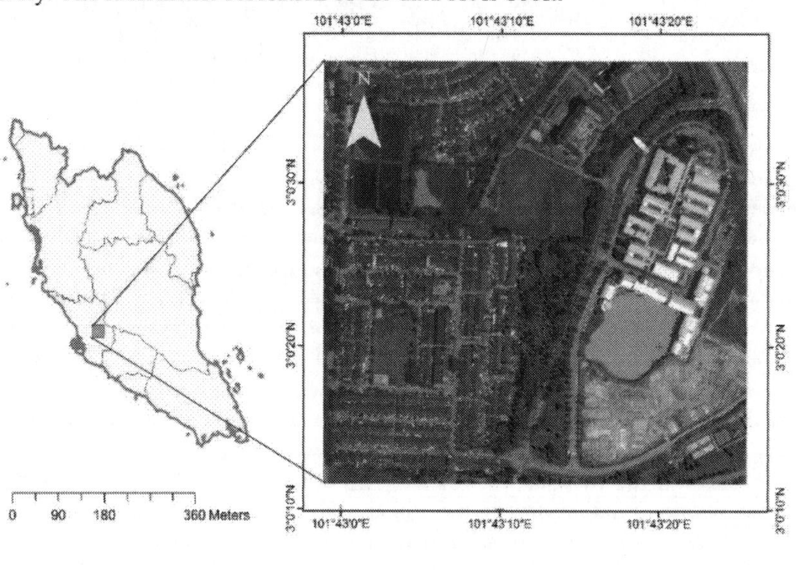

Figure 1. Location of the study area used to develop and validate the new proposed spectral index for built-up area extraction from worldview-2 image.

2. Methods

Figure 2 shows the steps followed to develop the new spectral built-up index based on Worldview-2 satellite image. First, we applied the standard preprocessing steps on the data to reduce the present noises in the dataset. Radiometric calibration was applied to convert the digital numbers of the satellite image pixels into physical units (radiance). Then the radiance values were converted into reflectance to get at surface reflectance signature for the objects found in the image subset. After that, IARR atmospheric correction was applied due to the lack of reference and field data.

The next step was the collection of endmembers of built-up objects as well as the other classes found in the Worldview-2 image. These endmembers were collected directly from the image for both built-up and other classes. For each class, 200 randomly selected samples were collected to ensure accurate index development. Once the training samples (end-members) are selected, the next step was to select the relevant (useful) spectral bands from the available eight (8) bands of Worldview-2 image. These relevant (useful) bands were assumed to best separate the classes and extract the built-up areas from the image. Particle swarm optimization (PSO) technique was used to select those relevant bands.

Once the relevant and useful spectral bands selected, the new spectral index can be developed. Band ratio is one of the most common spectral indices developed and widely used to extract features from satellite images. Therefore, we selected such type of index for the new built-up index aimed to be developed in this study. In order to find the best band ratio that could separate the other classes from the built-up class, an optimization technique should be applied. In this study, the same optimization technique (PSO) which was used for feature selection is used to find the best band ratio.

Finally, when the new spectral built-up index is developed, it was used to classify the Worldview-2 image and extract the built-up areas. Then the results were validated and compared with the existing spectral NDBI index to measure its accuracy and performance. The subsequent sections describe the details of those steps followed to develop the new index for built-up area extraction from Worldview-2 image.

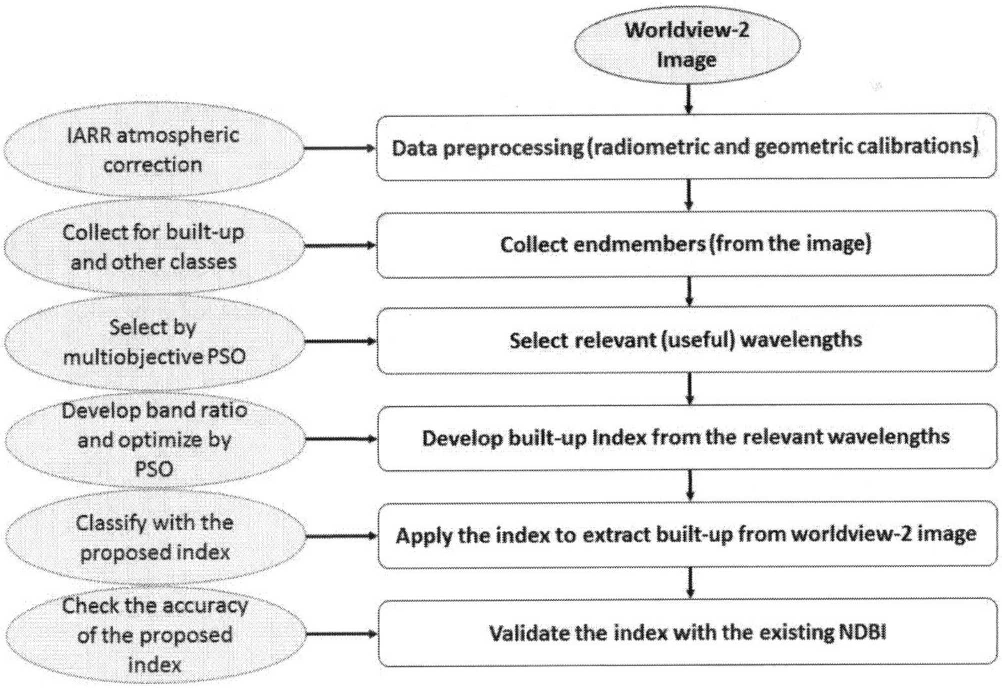

Figure 2. Overall methodology applied to develop the new spectral index for built-up area extraction from worldview-2 image.

2.1 Image processing

Original Worldview-2 satellite image contains variety types of noises and it is important to reduce those noises before the actual processing. First, the digital number of the image pixels were converted to radiance to reduce the sun illumination and some of the sensor related noises. After that, the radiance values of image pixels were converted into reflectance values to reduce the other types of noises such as atmospheric effects. In order to get the surface reflectance signatures, it was important to apply atmospheric correction models to reduce the effect of atmospheric layers on the object reflectance signatures. In this study, IARR model was used because this type of atmospheric correction requires no input data. After that, it was important to apply pan sharpening to the original image to improve the spatial resolution of the multispectral image by fusing the panchromatic band (0.5 cm) due to complexity of the urban environment.

2.2 Feature selection by multiobjective particle swarm optimization (PSO)

Particle Swarm Optimization (PSO) is one of the successful algorithms [10] reported in literature. PSO has been accepted to be effective in a wide variety of applications, being able to produce very good results at a very low computational cost [11]. A multi-objective extension, the Non-dominated Sorting Particle Swarm Optimizer (NSPSO), was proposed in [12] to cover multi-objective problems. The algorithm initializes the population and evaluates the initial particles using the fitness functions. The solutions are organized into fronts called Pareto fronts, and the non-dominated solutions construct the first Pareto front. Once the population is ranked, the nondominated solutions are identified and ranked using one niching method (e.g. crowding distance). The population is then evolved using velocities calculated with regards to the sorted list of the non-dominated solutions.

The fitness function is shown in Equation (1) which is used to minimize the classification error rate obtained by the selected features during the evolutionary training process and the number of selected features [13].

$$F = \begin{cases} w \times Train_{ER_s} + (1-w) \times Test_{ER_s} \\ \alpha \times \dfrac{\#Features}{\#All\ Features} + (1-\alpha) \times \dfrac{ER_s}{ER_{all}} \end{cases} \tag{1}$$

where w is a constant number (weight of classification error rate obtained from training data) and $w \in [0,1]$, $Train_{ER_s}$ is the classification error rate obtained by the selected feature subset and training subset data, $Test_{ER_s}$ is the classification error rate obtained by the selected feature subset and testing subset data, α is a constant number (weight of number of selected features), $\#Features$ stands for number of selected features, $\#All\ Features$ represents the total number of features available for classification, ER_s is the classification error rate obtained by the selected features, ER_{all} is the classification error rate obtained by all the available features. Error rate of classification results can be calculated using Equation (2) [14].

$$Error\ Rate\ (ER) = \frac{FP + FN}{TP + TN + FP + FN} \tag{2}$$

where TP, TN, FP, and FN stand for true positives, true negatives, false positives, and false negatives, respectively.

IOP Conf. Series: Earth and Environmental Science **37** (2016) 012006 doi:10.1088/1755-1315/37/1/012006

2.3 Index development

Once the relevant spectral bands are selected, the next step is to put them in a mathematical form to be used for rapid built-up area extraction. The most common practice technique for formulating such mathematical equations from a set of variables is to use band ratios which best classifies a set of testing data. In this study, trial-and-error approach is replaced with a method that is more efficient. PSO optimization is used to find the best band ratio from the relevant spectral bands that could classify the image data.

To find the best band ratio using the relevant spectral bands, an optimization problem were formulated as shown below,

Find optimal values of a, b, c, d in the Equation (3), which minimizes the function (f) presented in Equation (4)

$$BSI = \frac{a \times Band1 + b \times Band2}{c \times Band1 + d \times Band2} \quad (3)$$

where $Band1, Band2$ are the two most significant spectral bands selected by PSO.

$$f = Error\ Rate\ (ER) = \frac{FP + FN}{TP + TN + FP + FN} \quad (4)$$

The above optimization problem can be solved with any optimization techniques and in this research; PSO was used for its efficiency. The optimization is implemented in Matlab 2015 software and the best solutions were determined. The algorithm tries to find the best solution for the coefficient values in Equation (3) which minimizes the error rate to achieve the high the possible classification accuracy. Finally, the spectral index for built-up area extraction was produced.

3. Results and discussion

This section presents the results obtained from the data processing and optimization solutions applied as discussed above.

3.1 Results of feature selection

The worldview-2 image has eight (8) spectral bands; however, not all of them are useful for built-up area extraction. Therefore, PSO was applied to select the best bands that can achieve similar and better classification results than using all the available spectral bands. An optimization problem was formulated as discussed in the methods section and solved by using Matlab software.

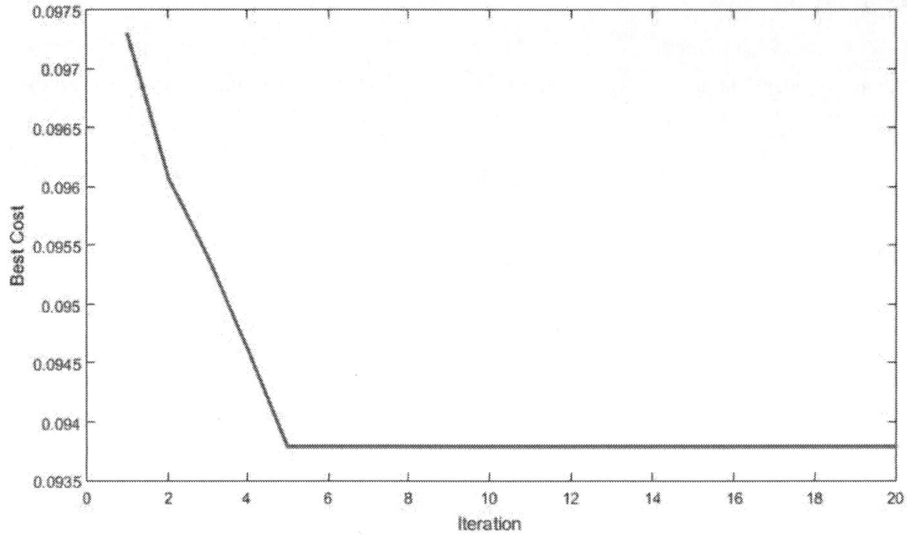

Figure 3. Best costs versus the number of iterations used in PSO.

In the first iteration, PSO found that the bands (7, 4) are the best bands for built-up area extraction based on the classification error rate values. Following the other 19 iterations, it was found that the same bands (7, 4) remained the best for built-up area extraction. These bands in worldview-2 image are Near Infrared 1 (NIR1, 770 - 895 nm), Yellow (585 - 625 nm).

3.2 Results of index development

Based on the best spectral bands selected by PSO, the second optimization problem was solved to find the best mathematical function which best classifies the image data. The second optimization was applied and the optimum values for the coefficients (a, b, c, d) were found. Table 1 shows the values for the coefficients in Equation (3) which formulate the final spectral index (BSI) developed for rapid built-up area extraction from worldview-2 images.

Table 1. Best values of coefficients in BSI found by PSO.

Coefficient	Best value found by PSO	Approximated value
a	1.33	1
b	-2.46	-2
c	1.07	1
d	1.77	2

The final spectral index (BSI) is formulated as presented in Equation (5) and Equation (6),

$$BSI = \frac{Band4 - 2 \times Band7}{Band4 + 2 \times Band7} \quad (5)$$

$$BSI = \frac{Yellow - 2 \times NIR1}{Yellow + 2 \times NIR1} \quad (6)$$

3.3 Results of image classification by the proposed index

The developed spectral index can be used in band math available in ENVI software to classify the Worldview-2 image into built-up and non-built-up classes. The following results were produced by applying the proposed and existing Worldview-2 spectral indices.

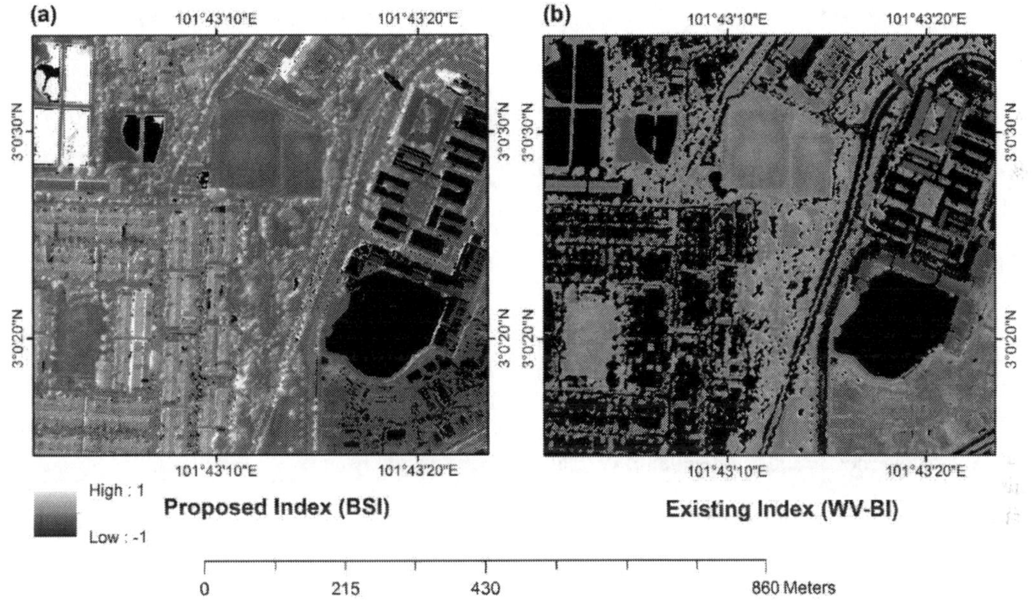

Figure 4. Results of the proposed (a) and existing (b) spectral indices of built-up area extraction applied in ENVIs band math.

The existing worldview-2 built-up area index based on band1 (coastal) and band6 (red edge) spectral bands, while in our approach, PSO found that band4 (yellow) and band7 (NIR1) are the most suitable bands to extract the built-up areas from the worldview-2 image. However, this study limited on using only one dataset for the development of spectral index for built-up area extraction and this could affect the transferability of the index for other study areas.

In order to extract the built-up areas from the above raster images, a threshold needs to be determined. In this study, we determined the suitable thresholds by visual interpretation of the images. The results obtained from the application of the selected thresholds are represented in Figure 5.

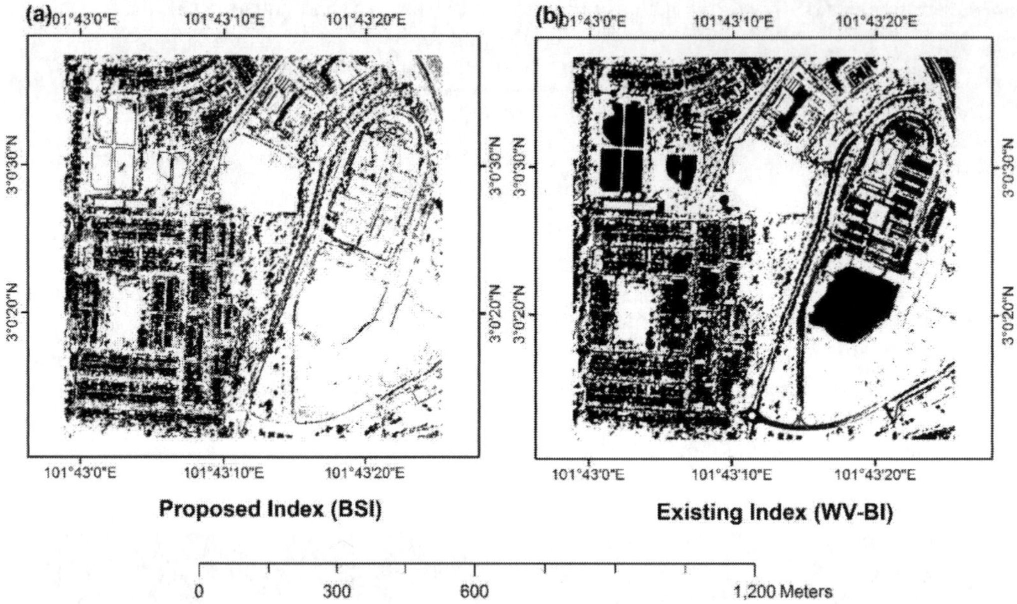

Figure 5. Built-up areas extracted from worldview-2 image using the new proposed and existing worldview indices.

3.4 Validation

Testing data was generated in ArcGIS software based on the original Worldview-2 image for validation purposes. Using the testing data and classified images, ROC curves were generated in Matlab software (Figure 6). After that, area under curve (AUC) was calculated for both ROC curves indicating the accuracy of the detected built-up areas. The AUC values were calculated for the proposed spectral index (BSI) and the existing index (WV-BI) and they found to be equal to 0.76 and 0.69 respectively.

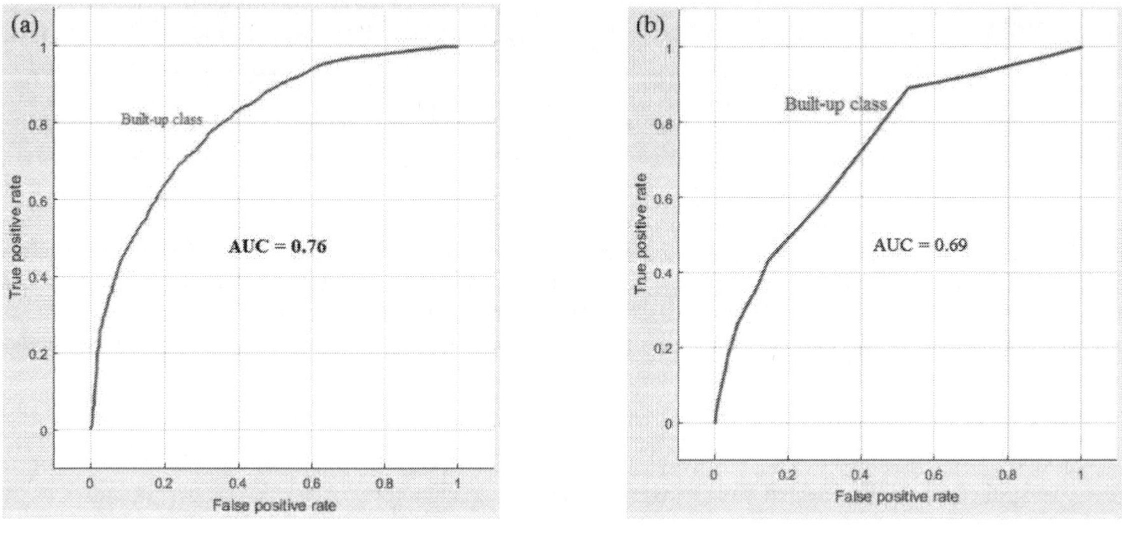

Figure 6. Area under curve and ROC curve of detected built-up areas, (a) the proposed spectral index (BSI), (b) the existing Worldview index (WV-BI).

Conclusion

This study proposed a new spectral built-up index for Worldview-2 data. The proposed index was developed by using multi-objective particle swarm optimization. The optimization technique applied selected the most two significant bands for built-up area extraction. Then it was applied in this research for the second time to develop the final index in mathematical form by using band ratio strategy. The two significant bands selected were band4 (yellow) and band7 (NIR1). The results of built-up area extraction were then compared with an existing built-up index proposed by Digital Globe called WV-BI. The accuracy assessment showed that our index outperform the existing spectral index quantitatively and qualitatively. However, the validation is only applied on the image that we used for the index development and it was tested for its transferability. Therefore, it is recommended for other researchers for further improvements on the index and develop more indices for various purposes using optimization techniques.

References

[1] Akbari, H., Cartalis, C., Kolokotsa, D., Muscio, A., Pisello, A. L., Rossi, F., . . . Zinzi, M. (2016). Local climate change and urban heat island mitigation techniques–the state of the art. Journal of Civil Engineering and Management, 22(1), 1-16.

[2] Chaudhary, S. K., Kumar, D., & Jain, M. (2016). Multi-Classifier Fusion for Land Use Land Cover Mapping in Jharia Coal Field Geostatistical and Geospatial Approaches for the Characterization of Natural Resources in the Environment (pp. 773-777): Springer.

[3] Kia, M. B., Pirasteh, S., Pradhan, B., Mahmud, A. R., Sulaiman, W. N. A., & Moradi, A. (2012). An artificial neural network model for flood simulation using GIS: Johor River Basin, Malaysia. Environmental Earth Sciences, 67(1), 251-264.

[4] Tehrany, M. S., Pradhan, B., & Jebur, M. N. (2013). Spatial prediction of flood susceptible areas using rule based decision tree (DT) and a novel ensemble bivariate and multivariate statistical models in GIS. Journal of Hydrology, 504, 69-79.

[5] Brewer, J., Ames, D. P., Solan, D., Lee, R., & Carlisle, J. (2015). Using GIS analytics and social preference data to evaluate utility-scale solar power site suitability. Renewable Energy, 81, 825-836.

[6] Bouzekri, S., Lasbet, A. A., & Lachehab, A. (2015). A New Spectral Index for Extraction of Built-Up Area Using Landsat-8 Data. Journal of the Indian Society of Remote Sensing, 43(4), 867-873.

[7] As-Syakur, A. R., Adnyana, I., Arthana, I. W., & Nuarsa, I. W. (2012). Enhanced Built-Up and Bareness Index (EBBI) for mapping built-up and bare land in an urban area. Remote Sensing, 4(10), 2957-2970.

[8] Xu, H. (2008). A new index for delineating built-up land features in satellite imagery. International Journal of Remote Sensing, 29(14), 4269-4276.

[9] Pesaresi, M., Gerhardinger, A., & Kayitakire, F. (2008). A robust built-up area presence index by anisotropic rotation-invariant textural measure. Selected Topics in Applied Earth Observations and Remote Sensing, IEEE Journal of, 1(3), 180-192.

[10] Eberhart, R. C., & Kennedy, J. (1995). A new optimizer using particle swarm theory. Paper presented at the Proceedings of the sixth international symposium on micro machine and human science.

[11] Kennedy, J., Kennedy, J. F., Eberhart, R. C., & Shi, Y. (2001). Swarm intelligence: Morgan Kaufmann.

[12] Li, X. (2003). A non-dominated sorting particle swarm optimizer for multiobjective optimization. Paper presented at the Genetic and Evolutionary Computation—GECCO 2003.

[13] Xue, B., Zhang, M., & Browne, W. N. (2013). Particle swarm optimization for feature selection in classification: A multi-objective approach. Cybernetics, IEEE Transactions on, 43(6), 1656-1671.

[14] Sheikhpour, R., Sarram, M. A., & Sheikhpour, R. (2016). Particle swarm optimization for bandwidth determination and feature selection of kernel density estimation based classifiers in diagnosis of breast cancer. Applied Soft Computing, 40, 113-131.

The temporal scale research of MODIS albedo product authenticity verification

Yongxing Cao[1], Zhihang Xue[1], Hui Cheng[2], Yajv Xiong[2], Yunping Chen[2], Ling Tong[2]

[1] Electric Power Research Institute of Sichuan, Chengdu, China
[2] University of Electronic Science and Technology of China, Chengdu, China

E-mail: xyj0943@163.com

Abstract. This study introduces a method that normalizes the inversed ETM+ albedo to the local solar noon albedo for the temporal scale of the MODIS albedo validation. Firstly, the statistical relation model between the surface albedo and the solar elevation angle was set up, and then deducing relationship between ETM+ albedo and the solar elevation angle, so the ETM+ albedo at local solar noon could be got. Secondly, the ground measurement albedo at the local solar noon was used to assess the inversed ETM+ albedo and the normalized albedo. The experiment results show that the method can effectively improve the accuracy of product certification.

1. Introduction

In remote sensing products, land surface albedo is one of the most important parameters to evaluate the earth's energy budget. Albedo is defined as the ratio of outgoing to incoming radiation at the Earth's surface. In order to understand and characterize the fluxes of energy at the Earth's surface, albedo must be known as accurately as possible [1].With the launch of the Moderate Resolution Imaging Spectrometer (MODIS) instrument aboard the NASA Terra platform in December 1999, a new capability for deriving moderate resolution estimates of broadband albedo was initiated [2]. Validating these land surface products is important because their accuracy is critical to the scientific community for various applications. Any feedback from the validation activity will also help improve the generation of these products [3].

The traditional validation method assesses the uncertainties of the albedo product by comparing it with field measurements [4-5]. However, the result of this method is suspicious because MODIS pixels cover larger areas than in situ measurements [6-7]. In order to reduce the effect of the scale mismatch, we use the ETM+ as the medial data in the validation procedure. But the ETM+ and MODIS are not synchronous at temporal scale. In the former validation the main concern is the spatial scale, The MODIS albedo products represent albedo at the local solar noon and the inversed ETM+ albedo products represent albedo the satellite passby. The abledo is changed in one day from dawn to dark as the solar elevation angle changed [8]. So normalizing the inversed ETM+ albedo to the local solar noon albedo is a necessary method to reduce the error in the validation procedure.

In this study the MODIS albedo productions are MCD43B3, the medium resolution image are ETM+ and the ground measurement albedo are from Surface Radiation Budget Network Goodwin Creek, MS station.

2. Methods

In general the Land surface albedo changes accompanying with the solar elevation [8]. The Land surface albedo are showing U-shaped distribution in one day. The Figure 1 shows the albedo variations and the solar elevation angle variations from 9:00 am to 6:00 pm in one day. There is a large different between the albedo at the local solar noon and albedo at dawn or dark. If the Landsat-7 satellite passby at the dawn or dark, the error will be great.

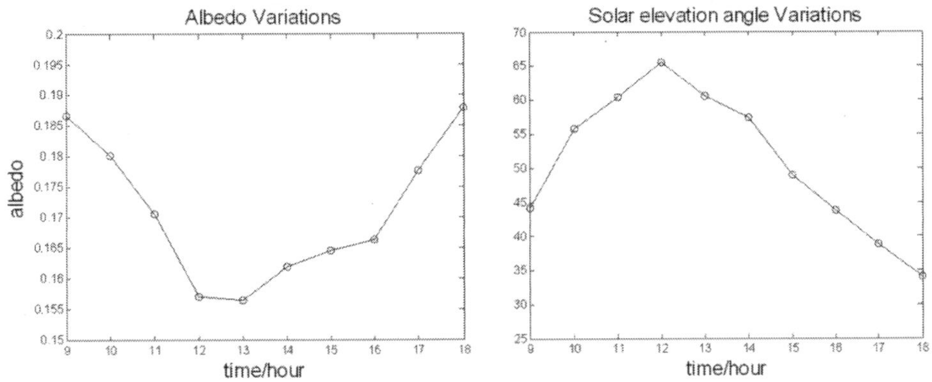

Figure 1. The albedo variations (left) and the solar elevation angle variations (right) in one day.

In this study we set up the statistical relation model between the surface albedo and the solar elevation angle, and then deducing relationship between ETM+ albedo and the solar elevation angle, so we can get the ETM+ albedo at local solar noon. The relationship between albedo and the solar elevation angles are usually given by Partridge. The Formula:

$$\alpha_s = \alpha_0 + (1-\alpha_0)e^{-k\varsigma} \tag{1}$$

There the α_s is the surface reflectance and the ς is the solar elevation angle, the α_0 is the surface reflectance of the height solar elevation angle (at the local noon), the k is the fitting coefficient. The formula is not suit for all conditions, so we simplify the formula to a new one, and give the right at noon solar elevation angle the right k. The formula:

$$\alpha_s = \alpha_0 e^{k/\xi} \tag{2}$$

Here we set up regression relation model in different at noon solar elevation angle. After we get the coefficient *k*, we deduce the albedo at local solar noon.

$$\alpha_o = \alpha_s e^{-k/\xi} \quad (3)$$

Here the α_s is the ETM+ albedo when the Landsat-7 pass-by, the ς is the solar elevation angle, the α_o is the albedo at local solar noon.

Table 1. Coefficient values under different angles in regression formula.

ς	k	Rlated coefficient	STD	Samples
ς<=35	3.329	0.832	0.010	43
35<ς=<55	4.247	0.857	0.006	37
55<ς=<75	5.832	0.846	0.012	45
ς>75	6.556	0.876	0.011	45

3. Results & Discussion

Finally, we use the ground measurement albedo at the local solar noon to assess the inversed ETM+ albedo and the normalized albedo. There we take 10 samples in different senson; the solar elevation angle will be cover the whole range. The Figure 2 is the result of the assessment.

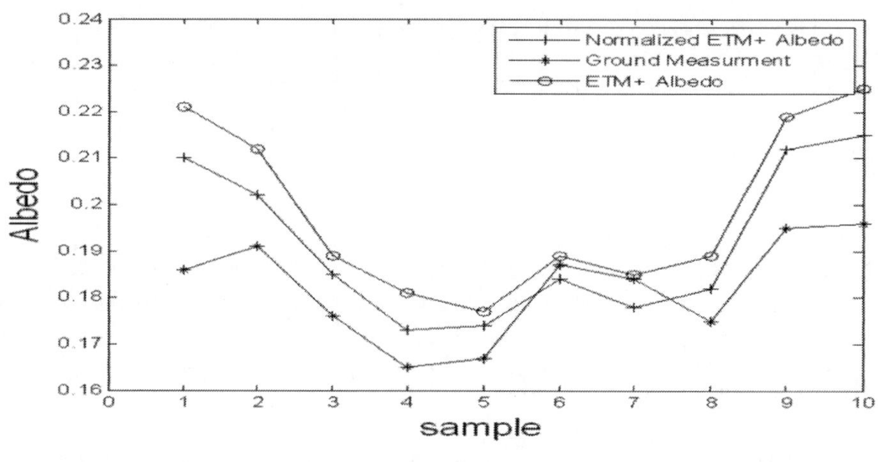

Figure 2. Comparison of normalized albedo and the original inversed albedo and the ground measurement albedo.

In the chart the normalized ETM+ albedo more close to the ground measurement albedo, the inverted albedo when the satellite passby are always large than the real at the local solar noon albedo. So only the normalized ETM+ albedo can represent the at local solar noon albedo, using this albedo will access the MODIS at local solar noon albedo production accurately.

The Figure 3 is the result of the comparison of normalized albedo and the original inversed albedo and the MODIS albedo.

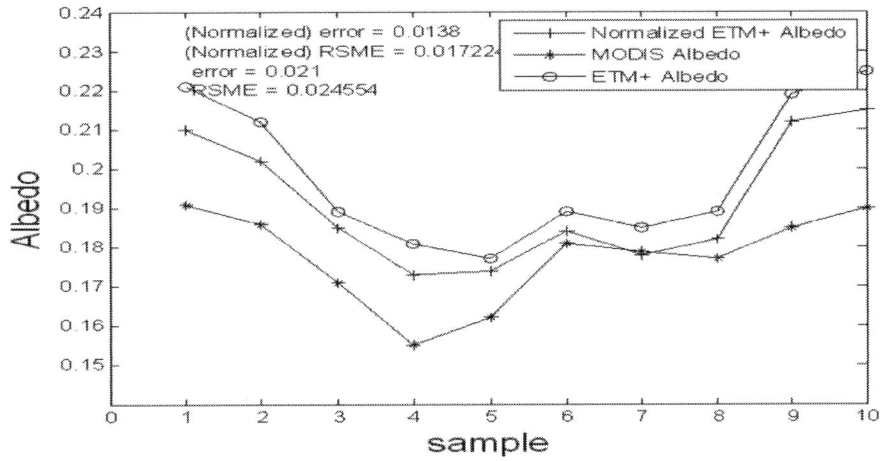

Figure 3. Comparison of normalized albedo and the original inversed albedo and the MODIS albedo.

Using the normalize albedo at local solar noon to access the MODIS albedo the error will be reducing. In this sample the error is reduce 0.0034, about 24% of the error.

4. Conclusions

In this paper, the major concern is the temporal scale of the MODIS albedo validation, According to the ground measurement albedo, we set up the regression model to normalize the ETM+ inversed albedo. The ETM+ albedo is normalized to the local solar noon albedo, it will be the same time as the MODIS albedo production. It will decrease the errors which come from the asynchronous of the two albedo production. The experiment results show that the method can effectively improve the accuracy of product certification.

References

[1] Sellers P J 1992 Remote sensing of the land surface for studies of global change, NASA/GSFC. International Satellite Land Surface Climatology Project Report, Greenbelt, MD, USA.

[2] Schaaf C B, GaoF, Strahler A H, Lucht W, LI X and Privette J 2001 Land surface albedo, nadir BRDF-adjusted reflectance and BRDF product from the Moderate Resolution Imaging Spectroradiometer (MODIS). Proceedings of the 11th Conference on Satellite Meteorology and Oceanography, AMS, Madison, WI, 15–18 October 2001 (Lawrence, KS: Allen Press), pp. 5–21.

[3] Liang S, Fang H, Chen M, Shuey C J, Walthall C, Daughtry C, Morisette J, Schaaf C, and Strahler A 2002 Validating MODIS land surface reflectance and albedo products: Methods and preliminary results, Remote Sens. Environ., 83, 149– 162.

[4] Wang Z, Schaaf C B, Chopping M J, et al 2012 Evaluation of moderate-resolution Imaging Spectroradiometer (MODIS) snow albedo product (MCD43A) over tundra. Remote Sens Environ, 117: 264–280

[5] Peng J J, Liu Q, Wen J G, et al 2014 Multi-scale validation strategy for satellite albedo products and its uncertainty analysis. Science China: Earth Sciences, 45: 66-82

[6] Román M O, Gatebe C K, Shuai Y, et al 2013 Use of in situ and Airborne Multiangle Data to Assess MODIS-and Landsat-Based Estimates of Directional Reflectance and Albedo. Ieee Trans Geosci Remote Sensing, 51: 1393-1404

[7] Wang Z, Schaaf C B, Strahler A H, et al 2014 Evaluation of MODIS albedo product (MCD43A) over grassland, agriculture and forest surface types during dormant and snow-covered periods. Remote Sens Environ, 140: 60-77

[8] Chen J W, Zuo H C, Wang Y, et al 2014 Parameterization scheme about albedo changing with solar altitude angle over different underlying surface in arid areas of Northwest China. Plateau Meteorology, 33(1): 80-88

Identification of topographic elements composition based on landform boundaries from radar interferometry segmentation (preliminary study on digital landform mapping)

Wirastuti Widyatmanti, Ikhsan Wicaksono, Prima Dinta Rahma Syam

Laboratory of Remote Sensing, Department of Geographic Information Science
Faculty of Geography, Universitas Gadjah Mada, Yogyakarta, Indonesia

Email: wwidyatmanti@ugm.ac.id

Abstract. Dense vegetation that covers most landscapes in Indonesia becomes a common limitation in mapping the landforms in tropical region. This paper aims to examine the use of radar interferometry for landform mapping in tropical region; to examine the application of segmentation method to develop landform type boundaries; and to identify the topographic elements composition for each type of landform. Using Idrisi® and "eCognition ®" softwares, toposhape analysis, segmentation and multi-spectral classification were applied to identify the composition of topographic elements i.e. the types of land-cover from Landsat 8, elevation, slope, relief intensity and curvatures from SRTM (DEM). Visual interpretation on DEM and land-cover fusion imagery was conducted to derive basic control maps of landform and land-cover. The result shows that in segmentation method, shape and compactness levels are essential in obtaining land-cover, elevation, and slope class units to determine the most accurate class borders of each element. Despite a complex procedure applied in determining landform classification, the combination of topographic elements segmentation result presents a distinct border of each landform class. The comparison between landform maps derived from segmentation process and visual interpretation method demonstrates slight dissimilarities, meaning that multi-stage segmentation approach can improve and provide more effective digital landform mapping method in tropical region. Topographic elements on each type of landforms show distinctive composition key containing the percentage of each curvature elements per area unit. Supported by GIS programming and modeling in the future, this finding is significant in reducing effort in landform mapping using visual interpretation method for a very large coverage but in detail scale level.

1. Introduction

The application of remote sensing in geomorphic mapping has been developing vastly, coinciding with the launching of radar interferometry and its ability to provide digital surface/elevation model (DSM/DEM) [1]. One of the important parameters in geomorphic mapping is land physical characteristics information derived from landform classification and mapping, both in regional or global level. Landform, as the main element in geomorphic mapping, is a natural object feature that divides the earth surface into one main spatial unit [2]. It comprises the information of relief, elevation and the common land activities that dominating the landscape processes. DEM provides more promising data to the researchers in increasing the quality of landform mapping, because it consists of elevation information with 3-dimentional relief features. Nowadays, landform classification using DSM/DEM is one of basic research domain in digital terrain analysis, including GIS landscape integrated studies. The common free access and reliable 3-dimentional relief data used in many mapping applications is Shuttle Radar Topography Mission (SRTM) (Figure 1). This imagery has horizontal and vertical spatial resolution of 90 m and 10-16 m respectively. However, it only provides a DSM features instead of DEM, because when a surface covered by very dense vegetation, it is only a small part of soil surface reflectance that is received by sensor [3]. Therefore, addition corrections of radiometric and geometric (vertical) are needed to obtain more accurate elevation values (DEM).

Figure 1. The 1 arc second global resolution SRTM raw data of Yogyakarta and surrounding areas.

Landform mapping using remote sensing imagery in Indonesia mostly involves visual interpretation technique to manually obtain landform features instead of digitally. Generally, visual image interpretations for land-use/land-cover and landform application utilize multi-spectral imageries from Landsat images combined with DEM from SRTM [4,5]. The use of DEM in applying digital technique is more focused on the modelling of land-cover changes, hydrology, natural disaster, and general land physiographic changing using multi-temporal data [6,7]. The feature elements of landforms are diversely complex in each type of classification, and the requirement of high experience and understanding of their characteristics is needed in interpreting them. Indonesian geographical condition located in tropical country where vegetation covering most of its earth surface, is a common issue in remote sensing studies, especially in gaining the accurate feature of landforms. This issue also causes the landform identification with visual interpretation could not be easily done by the amateurs. This current study is expected to provide a preliminary concept and procedure to map landform in a rapid but reliable way.

Landform digital mapping is a process to obtain landform information from digital data like DEM, satellite imagery and aerial photos [8]. The development of remote sensing technology provides digital data with higher spectral and spatial resolution. However, such consequences are inevitable that this more meticulous data will have more complex information which results in producing some technical issues, in this case in how to combine the enormous topographic elements information derived from pixel based image analysis. With preconceived assumption that topographic elements associates with segments as the cells groups which have equal values, then object based image analysis (OBIA) becomes the valuable approach for landform mapping. Derivation of land-cover and landform using OBIA have been so prominent in the last couple of years, because of its speciality in overcome this pixel analysis issues which is only calculating the image pixel values instead of considering geometric and contextual information [9;10]. Thus, OBIA have been well-proven for classifying landform from DEM [11; 12) because the object contextual model of the identified landform is more reliable compared to per-pixel traditional method [5]. The results of classification from previous studies only provide topographic elements in every mapping unit, such as information of elevation, slope, relied and curvature, but do not conclude their type of landform. On the other hand, there is a need to provide rapid digital landform identification to assist researchers in determining type of landform in a very large landscape such as in Indonesia such as the applications on land-suitability and planning which need a fine scale of landform map. Further study on how to develop a landform type criteria based on topographic elements information from multi-segmentation process will assist the researcher in generate a program language system for automatic landform type identification.

In general, OBIA includes two working steps, which are segmentation and classification. OBIA stands out as a promising technique among the diversity of digital mapping methods for landcover/landuse object classifications. OBIA has the ability to produce uniform objects and then associate those objects into contextual information on the higher framework level [4; 5]. OBIA application for land-cover mapping purpose, generally speaking, grows more rapidly because it is relatively simpler (but not effortless) than for landform mapping [5]. In OBIA, land form classification is based on expert judgment and knowledge-based on DEM, satellite image and/or over both combinations. The extended algorithm to reduce the uniform segments from the image segmentation result is explored through multi resolutions segmentation (MRS). The newest usage of OBIA for combined land-cover and landform mapping has just been applied for mapping the vegetation, canal, and mud flats using very high spatial resolution imageries [7, 8, 9]. Therefore, in this paper, the aims of the study are to examine the ability of radar interferometry imagery in mapping landforms using segmentation method, and to identify the topographic elements composition of each landform mapping unit derived from "toposhape" analysis.

2. Study area, conceptual basic, material and method

2.1 Study area

This study was conducted in Yogyakarta and surrounding areas where its topography characteristics are a combination of flat to mountainous landscape features. The volcano landform, on the north part of this area, is complemented by karst and structural landforms on the east and west parts respectively. This landform complexity assists this study to identify the terrain-shape of each type of landform using topographic composition that represents elevation, slope, basin, relief intensity, and curvature.

2.2 Conceptual basic

Landform is an object that being studied as a part of Geomorphology [11]. Landform is a terrain surface feature which naturally formed by a certain process. The landform itself however is limited and discontinued by nature. In landform mapping, the use of DEM has been recently common in gaining more accurate landform information. The available method consist of mere visual interpretation, and supported by multispectral fusion or combination of image and DEM, contour map, and photogrammetry or radar interferometry to provide better visual topography appearance compared to the 2-dimensional appearance [12, 13]. On the other hand, there are many aspects we can gain from the image with digital approach, although will bear the consequences of complexities and difficulties. Moreover, years of experience and flight hours dealing with multi-scale land form mapping, is critical prerequisite. This research will use the classification of landform with 1:250,000 scale [14] as basic segmentation of DEM and InSAR Image [14, 15, 16, 17]. This landform classification includes: volcanic, structural, fluvial, and karst using elemental landform approaches: elevation, slope, relief intensity and curvature (basin, valley and others slopes represented by "toposhape" fetures). Marine and aeolian landforms are excluded into consideration as the sample areas didn't meet the minimum requirements for this study.

2.3 Material and method

The main method used in this study is designed to facilitate the need of classifying landform using digital technique. It involves some digital image processing to gain landform boundaries and characteristics. It includes radar image processing, using ENVI 5.0, "toposhape" identification using Idrisi® and segmentation processes using eCogniton®. Three steps of digital analysis activities were conducted after image radiometric and geometric pre-processing (Figure 2). First step was conducting visual interpretation on Landsat 8 with several band composites to produce different landform type boundaries using multi-level landform mapping approach [18], a method that is available to assist user to gain more accurate landform interpretation from several remote sensing imageries. These landform boundaries become a foundation to separate information on the process of topographic elements identification.

Second was segmenting the DEM based on the elevation region into high- and low-land classes (Table 1), followed by similar step but was based on the slope and elevation classification [17] (Table 1). The third step was classifying "toposhape" identification (using Idrisi®) per unit area based on landform boundaries from step 2 that has been compared to landform boundary from visual interpretation. The detail topographic elements are provided in Table 2.

Table 1. Elevation and slope classification for segmentation purpose [17].

Class	Elevation – relative height (m)		Slope (%)	
1	<50	: lowlands	0-2	: flat or almost flat
2	50-200	: low hills	3-7	: gently sloping
3	200-500	: hills	8-13	: sloping
4	500-1000	: high hills	14-20	: moderately steep
5	>1000	: mountains	21-55	: steep
6		-	56-140	: very steep
7		-	>140	: extremely steep

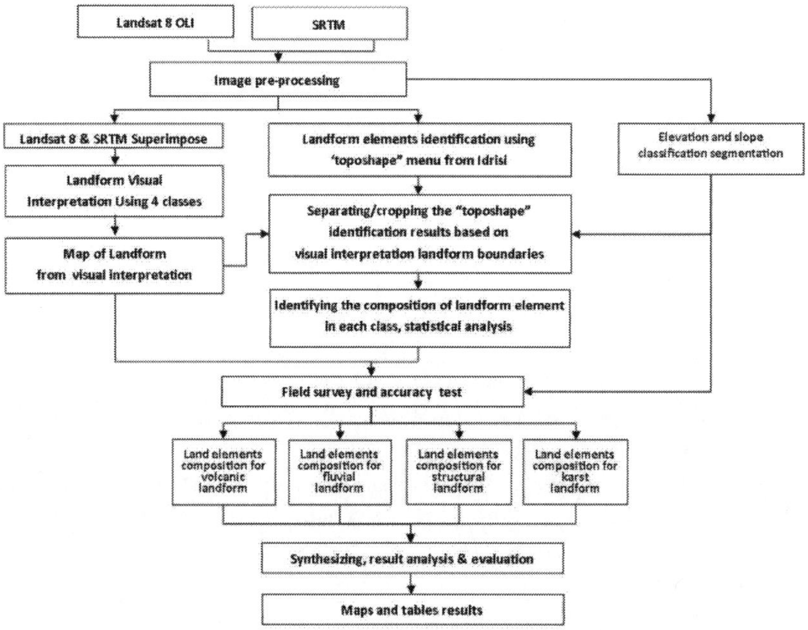

Figure 2. Research flowchart of land elements identification.

Table 2. Slope curvature topographic element identified using "toposhape" features in Idrisi®.

No.	Type of slope curvature	No.	Type of slope curvature
1.	Peak	7.	Saddle hillside
2.	Ridge	8.	Slope hillside
3.	Saddle	9.	Concave hillside
4.	Ravine	10.	Inflection hillside
5.	Pit	11.	Unknown hillside
6.	Convex hillside		

3. Result and discussion

The first result provides landform boundaries generated using visual interpretation methods. Multi-level landform interpretation analysis [18], supported by geological and topographic maps, and DEM developed from SRTM, was applied to identify more accurate landform boundaries visually. The segmentation process for SRTM (Figure 2) in this study was utilizing eCognition® software. The segmentation process applied multi-resolution segmentation algorithm method which reduced the average heterogeneity of the objects in the image locally, on the available resolution. In this method, the user can apply various scale parameters to obtain the best object unit in accordance with the classification and resolution. The final classification produced landform boundaries based on the best scale parameters on segmentation process.

The identification of landform boundaries from several scale parameters in segmentation processes shows various details of boundaries. This study use range of scale parameter from 50-250 with 50 unit increments. The lowest value of scale parameter 50 was chosen according to the possible map scale that can be generate from SRTM imagery. Figure 3 shows the image results from segmentation with scale parameter of 50 and 200, indicating that the boundaries generating from scale parameter 50 is finer compared to its of 200. The observation applied on segmentation image result with other different scale parameter presents similar result that the smaller value of scale parameter the finer boundaries of landform is produced. For this reason, the scale parameter of '50' segmentation image is chosen as the main image result from segmentation process.

The landform boundaries results from segmentation method and visual interpretation were compared to examine their accuracy. This study uses an assumption that landform boundaries obtained from visual interpretation is more accurate than the one from digital processing since the visual interpretation involves expertise judgment and several afore mentioned supportive data to determine the boundaries.

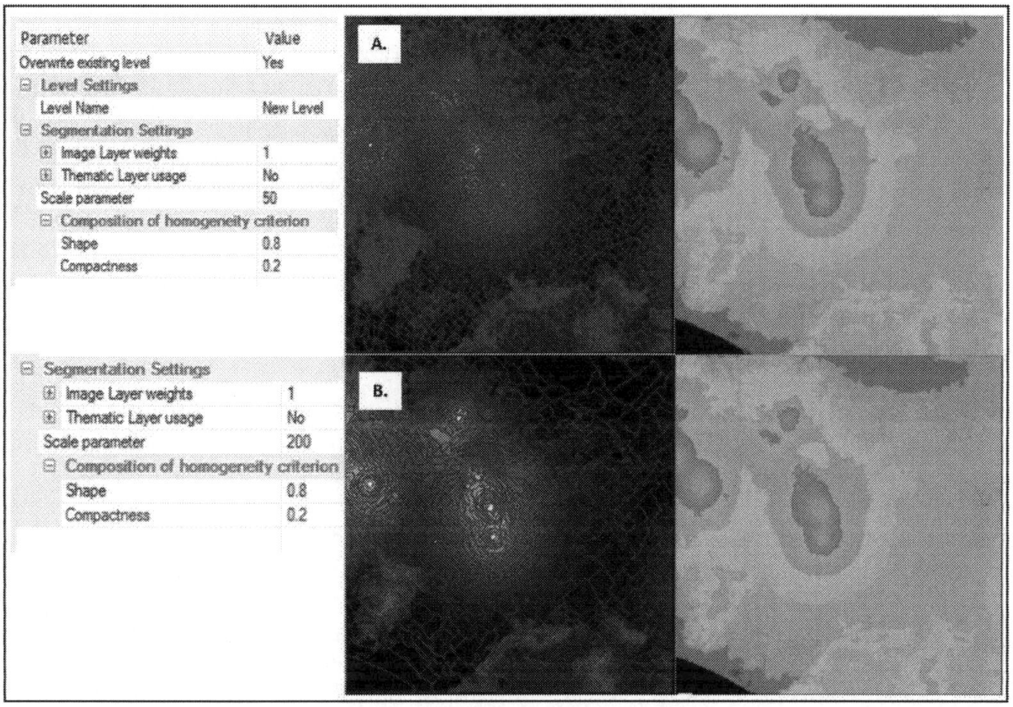

Figure 3. Landform classification imagery from segmentation method in various scale parameters. Figure A and B showing segmentation results and their classification using scale parameter 50 and 200 respectively.

Subsequently, these examined landform unit boundaries were overlain with "toposhape" analysis result to reveal the composition of the topographic elements of each landform unit represented by type of slope curvature (Table 2). Field survey was conducted based on the temporary result of classification and landform. The result shows that dividing the area based on elevation and slope from segmentation process to assist land elements identification does not significantly contribute to its composition diversity. Therefore, the composition identification was done directly using landform boundaries.

IOP Conf. Series: Earth and Environmental Science **37** (2016) 012008 doi:10.1088/1755-1315/37/1/012008

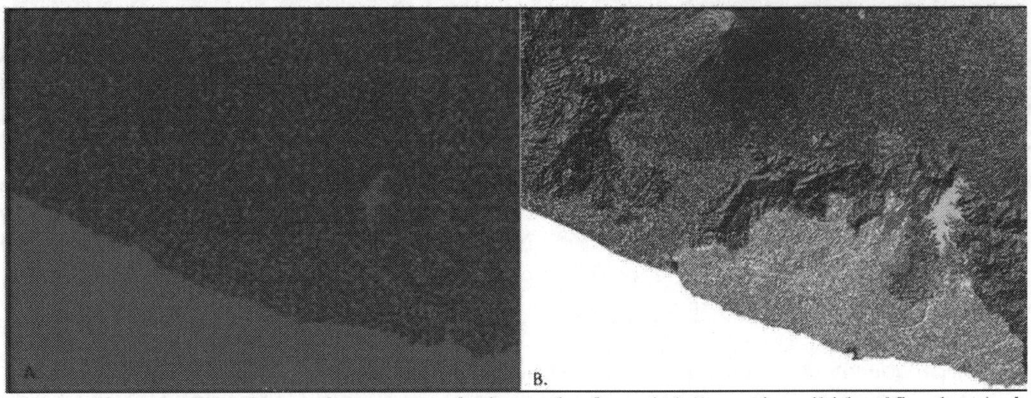

Figure 4. Topographic element imagery analysis results from (A) "toposhape" identification (using Idrisi®), and (B) landform visual interpretation on SRTM and Landsat, combined with the "toposhape" analysis result.

Figure 5 shows various results of topographic elements represented by number of peak, ridge, saddle, ravine, pit, convex hillside, saddle hillside, slope hillside, concave hillside, inflection hillside, and unknown hillside. Those numbers of topographic elements from "toposhape" analysis are displayed using graphs to perceive their pattern, distribution and importantly their composition in each landform unit which are presented in percentage (%). Largely, the graphs shows different pattern on volcanic, structural, fluvial and karst landform type area units. Field survey was conducted randomly to validate the accuracy of the slope, elevation, relief and "toposhape" analysis result in the every single landform type. The obstacle found in the field commonly dealt with the vegetation and build-up area land-covers that conceal the original shape of these covered land. In this case, the expertise on landform analysis is needed to confirm the accurate type of "toposhape".

On the volcanic landform (Figure 5.A), the dominant topographic elements are ravine (>17%) and inflection hillside (> 20 %). Those compositions are showing the dominant of valley and inflection points that commonly occurred in young volcanic areas. Concave and saddle hillside (both are below 10% of the composition) complement the features of valleys and slopes. The structural landform composition shows three dominants land elements which are ridge (13-15%), ravine (15%), convex hillside (14%), saddle hillside (16%), concave hillside (13%), and inflection hillside (17) (Figure 5.B). The various elements dominance on structural landform presents the roughness of landscape textures seen on the satellite imagery. High slope, erosion and dissected valley on different resistant layers have created the diversity of this structural's topographic elements. Conversely, the fluvial topographic elements composition (Figure 5.C) shows highest number of flat "toposhape" element type which is more than 5%. This percentage is considered high compared to ones on other landforms which are below 1%. However, the flat "toposhape" in fluvial landform was expected to have much higher percentage (> 5). Observing that Yogyakarta flood plains are dissected by many rivers which have steep and large valleys on its middle part, the low percentage of flat "toposhape" actually is able to present specific characteristics on this fluvial landform. In spite of this, another study on topographic elements of various fluvial landforms needs to be explored specifically.

Karst landform "toposhape" composition (Figure 5.D) shows high percentage on ridge and ravine topographic elements (>10%). The karst landform in Gunung Kidul district is characterized by cone karst with sinusoidal hills mixed together with enclosed star-shape depression and linked valleys [11]. These ridges elements mostly represent the high dissected cone shape karst landscape that during dry season (time of this SRTM imagery recorded) emerge observably because the vegetation leaves are withering. On the other side, on the cone areas where the vegetation are not classified as withering type, the appearance of cone karst is concealed by vegetation of which the grow direction is shaping the cone as a sharper ridge. The abundant depression and interconnected valley are identified by high percentage of ravine topographic element. Intensive solution processes followed by erosion during rainy season produce deeper valley that lead to more depression formation.

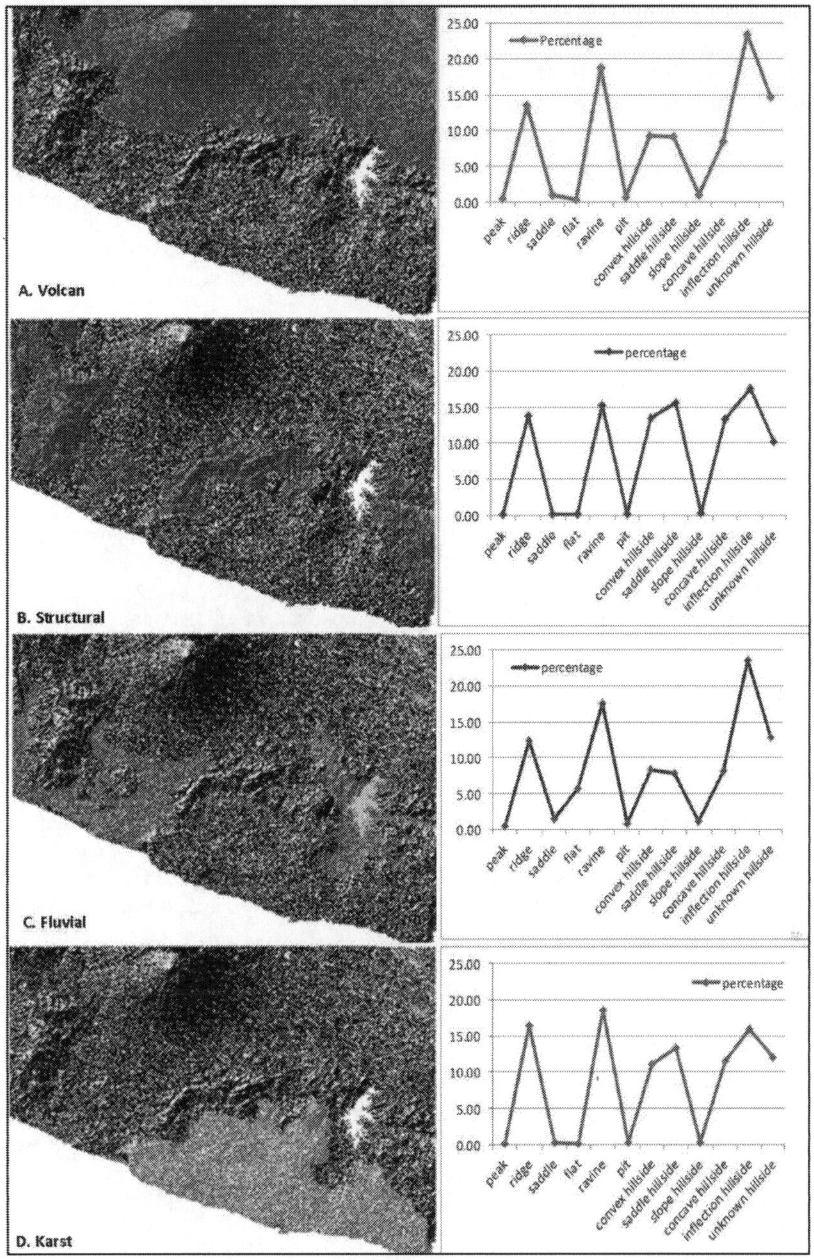

Figure 5. The boundaries of (A) volcanic, (B) structural, (C) fluvial and (D) karst landforms based on visual interpretation and their each composition of topographic elements presented in a line graph showing the percentage of each elements number compared to the total.

4. Conclusion

This preliminary study on identifying topographic element composition on each landform class provides important conclusion on how this study can lead into a better result. More sample areas of each type of landform are obviously needed to ensure that the elements compositions are exclusively consistent in every type of landform. General understanding on geomorphological processes occurred on land formation are the main factors in determining the validation of "toposhape" analysis result. For example, the topographic element composition for fluvial landform in Yogyakarta that have steep valley and large riverbank is not always applicable on others fluvial plains. Further analysis of "toposhape" identification has to be applied on other fluvial landforms, i.e. Semarang or Surabaya (estuarine and delta) which have different characteristics with the ones in Yogyakarta.

The same landform type from different areas does not guarantee that similar topographic element composition could be occurred among them. However, these possible differences compositions have a good chance to emerge a new sub-landform classification. Further study in applying segmentation process in digital landform classification is also worth to adopt to examine its delineation and boundaries accuracies. In summary, this current study result is important as one of foundations for the development of landform digital mapping in tropical country.

References

[1] Gercek, D.; Toprak, V.; Stribl, J. (2011) Object-based classification of landforms based on their local geometry and geomorphometric context. *International Journal of Geographic Information Science*. Vo. 25, pp 1011-1023.

[2] Lu, Z. and O. Kwoun (2007). Radarsat-1 and ERS Interferometric Analysis over Southeastern Coastal Louisiana: Implication for Mapping Water-Level Changes Beneath Swamp Forests, IEEE Transactions on Geoscience and Remote Sensing, submitted.

[3] Norheim, R.A., Queija, V.R., Haugerud, R.A., 2002. Comparison of LiDAR and INSAR DEMs with dense ground control. In: Proceedings of the ESRI 2002 User Conference. ESRI, San Diego, 9 pp

[4] Drăguţ, L. & Eisank, C. (2011). Object representations at multiple scales from digital elevation models. *Geomorphology* 129, 183–189.

[5] Blaschke, T., (2010). Object based image analysis for remote sensing. *ISPRS Journal of Photogrammetry and Remote Sensing*, 65, 2–16.

[6] Drăguţ, L. & Blaschke, T. (2006). Automated classification of topographic elements using object-based image analysis. Geomorphology 81, 330–344.

[7] Van Zuidam, R. (1985). Aerial photo-interpretation in terrain analysis and geomorphic mapping. International Institute for Aerospace Survey and Earth Science (ITC). Smits Publication. The Hague.

[8] Ahnert, F. (1988). Introduction to geomorphology. *Earth Surface Processes and Landforms*. Vol. 24: 4. Pp 379. London.

[9] Camargo, F., C. Almeida, T. Florenzano, C. Heipke, R. Feitosa, and G. Costa (2011). ASTER/Terra imagery and a multilevel semantic network for semi-automated classification of landforms in a subtropical area. *Photogramm. Eng. Remote Sens.*, 77 (6), 619–629.

[10] Haryono, E. & Day, M. (2004). Landform differentiation within the Gunung Kidul Kegelkarst, Java, Indonesia. *Journal of Cave and Karst Studies*. V. 66, no. 2, pp 62-69.

[11] Kim, M., Warner, T.A., Maffen, M., Atkinson, D.S. (2011). Multi-scale GEOBIA with very high spatial resolution digital aerial imagery: scale, texture and image object. *International Journal of Remote Sensing*, Vol. 32, 10, pp 2825-2850.

[12] Li, S., Romanovsky, V., Lovick, J., Wang, Z., & Peterson, R. (2003). Application of satellite SAR imagery in mapping the active layer of Arctic permafrost. Final Report NAG5-8614. : NASA Centre for Aerospace Information.

[13] Liu, L., Zhang, T., & Wahr, J. (2010). InSAR measurements of surface deformation over permafrost on the North Slope of Alaska. Journal of Geophysical Research, 115, F03023, Doi:10.1029/2009JF001547.

[14] MacMillan, R.A., Shary, P.A., (2009). Chapter 9 landforms and topographic elements in geomorphometry. In: Hengl, T., Reuter, H.I. (Eds.), Geomorphometry—Concepts, Software, Applications. : Developments in Soil Science, vol. 33. Elsevier, Amsterdam, pp. 227–254.

[15] Sandwell, D. and M. Wei (2006). ALOS Interferometry. http://topex.ucsd.edu/alos/.

[16] Suharsono, P. (1988). Landform identification and satellite imae imagery for geomorphology. Laboratorium Module. PUSPICS. Yogyakarta. Indonesia

[17] Takeuchi, S., & Yamada, S. (2002). Comparison of InSAR capability for land subsidence detection between C-band and L-band SAR. Proceedings of IGARSS 2002, 24–28 June, Toronto, Canada (pp. 2379–2381). New York, NY: IEEE.

[18] Van Asselen, S. & Seijmonsbergen, A.C. (2006). Expert-driven semi-automated geomorphological mapping for mountainous area using a laser DTM. Geomorphology, 78 (3-4), pp 309-320.

[19] Widyatmanti, W. (2015). Hydro-geomorphic controls on the development and distribution of acid sulfate soils in Central Java, Indonesia. *Dissertation*. University of New South Wales Sydney, Australia.

Clustering of landforms using self-organizing maps (SOM) in the west of Fars province

Marzieh Mokarram[1] and Dinesh Sathyamoorthy[2]

[1] Department of Range and Watershed Management, College of Agriculture and Natural Resources of Darab, Shiraz University, Iran.
[2] Science & Technology Research Institute for Defence (STRIDE), Ministry of Defence, Malaysia

Email: m.mokarram@shirazu.ac.ir

Abstract. The aim of this study is to cluster landforms in the west of the Fars province, Iran using self-organizing maps (SOM). In SOM, according to qualitative data, the clustering tendencies of landforms were investigated using six morphometric parameters, which were slope, profile, plan, elevation, curvature and aspect. First, topographic position index (TPI) was used to prepare the landform classification map. The results of SOM showed that there were five classes for landform classification in the study area. Cluster 5 corresponds to high slope, high elevation but with different of concavity and convexity that consist of ridge landforms. Cluster 3 corresponds to flat areas, possibly plantation areas, in medium elevation and almost flat terrain. Clusters 1, 2 and 4 correspond to channels with different slope conditions.

1. Introduction

Landforms are the result of geomorphologic processes that occur on the earth's surface. The term landform is used for morphometrically homogeneous land-surface regions due to the effects of common geological and geomorphological processes [1]. As this concept of landforms is an idealized one, it then follows that the closer the studied landform conforms to its definition, the greater the accuracy of the obtained model. A subdiscipline of geomorphology is geomorphometry, which provides quantitative and qualitative description and measurement of landforms [1, 2, 3, 4], and is based on the analysis of variations in elevation as a function of distance. A basic underlying principle of geomorphometry is digital elevation models (DEMs). A number of studies have employed DEMs for the automated mapping of landforms [5, 6, 7, 8, 9]. Derivation of landforms can be carried out using various approaches, including classification of terrain parameters [10, 11], filter techniques [12], cluster analysis [13] and multivariate statistics [14]. A common focus of the study of landforms is to consider them as formed by small and simple elements which are topologically and structurally related. A more complete description of landform may be achieved by using complex terrain parameters calculated from simple ones, e.g., the topographic wetness index [15], stream power index [16], aggradation and degradation indices [17], thresholds [10], automated classification using object-based image analysis multivariate descriptive statistics [10, 18], double ternary diagram classification [1], discriminant analysis [19], fuzzy logic and unsupervised classification [6, 14, 20, 21, 22] and neural networks [23].

Self-organizing map (SOM) is an unsupervised and nonparametric artificial neural network (ANN) algorithm that clusters high dimensional input vectors into low dimensional (usually two-dimensional) output map that preserves the topology of the input data [23]. Preserving topology means that the SOM map preserves the relations between the input neighboring points in the output space [24]. Hosokawat and Hoshit (2001) used SOM to generate a damage distribution map in Kobe city in Japan that corresponds with the actual damage recorded following the 1995 earthquake [25]. Ehsani and Quiel (2008) employed SOM and Shuttle Radar Topography Mission (SRTM) data to characterize yardangs in the Lut desert, Iran [23]. The results demonstrate that SOM is a very efficient tool for analyzing aeolian landforms in hyper-arid environments that provides very useful information for terrain feature analysis in remote regions. Ferentinou and Sakellariou (2010) applied SOM in order to rate slope stability controlling variables in natural slopes, while Ferentinou et al. (2010) used SOM to classify marine sediments [26]. Mokarram et al. (2014) used SOM to study the relationships between geomorphological features of alluvial fans and their drainage basins. The results of the analysis showed that several morphologically different fan types were recognized based on their geomorphological characteristics in the study area [22].

The west of Fars, Iran has the highest points in the Fasa city, with a rich wildlife preserve, known as Kharmankooh, which is a hunting prohibited region. The area has a variety of landforms that it is suitable for the wildlife. As different animals live in different landforms, clustering of landforms can be used to predict the type of wildlife in the study area. To this end, the aim of this study is to cluster the landforms using SOM based on morphometric characteristics.

2. Material and methods

2.1. Study area

The study area is located in west of Fars province, southwest Iran, which is shown in Figure 1. It located between 29° 00′ to 29° 25′ northern latitude and 53° 19′ to 53° 54′ eastern longitude (Figure 1). Six morphometric parameters were analyzed; slope, profile, plan, elevation, curvature and aspect (Figure 2 and Table 1).

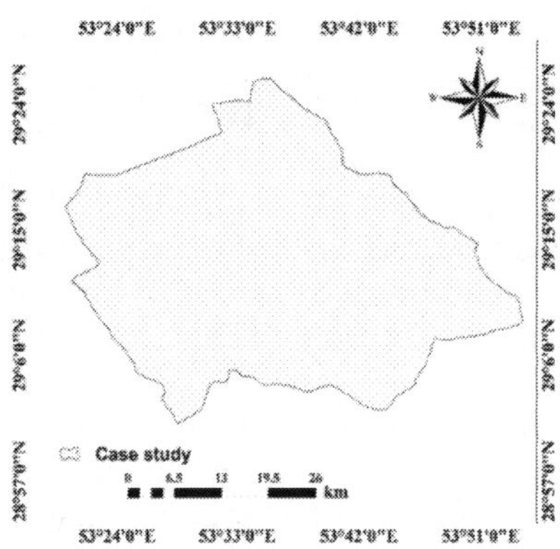

Figure 1. Location of the study area.

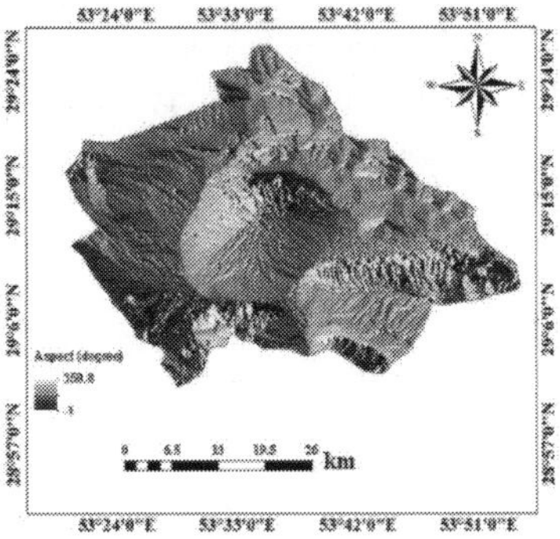

Aspect

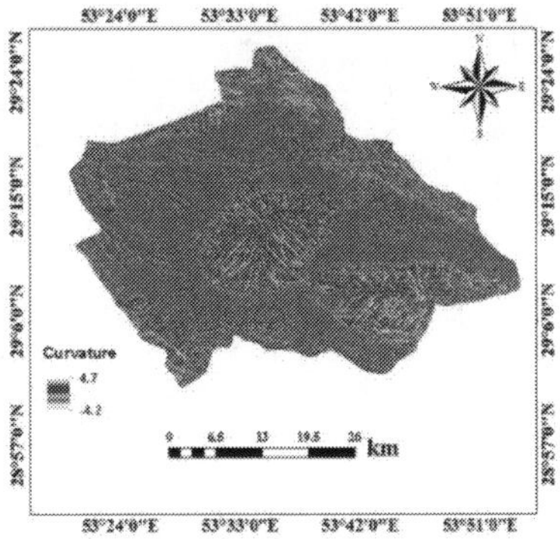

Curvature

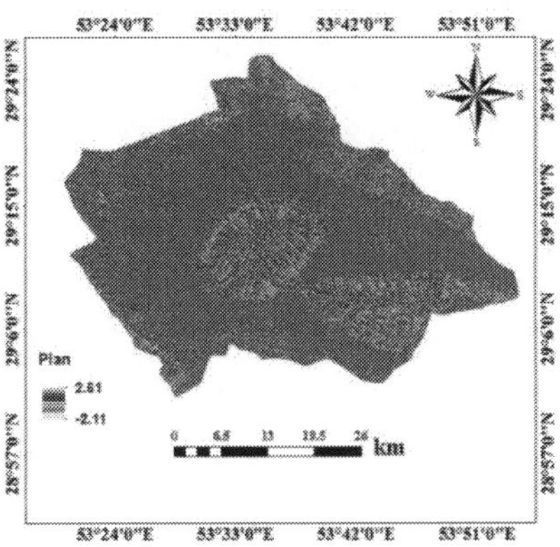

Plan

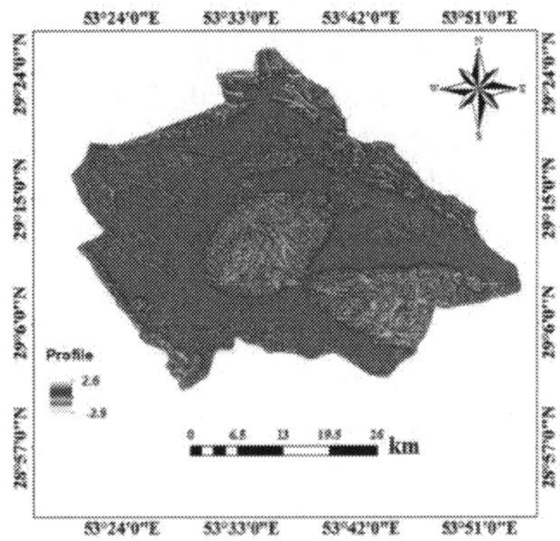

Profile

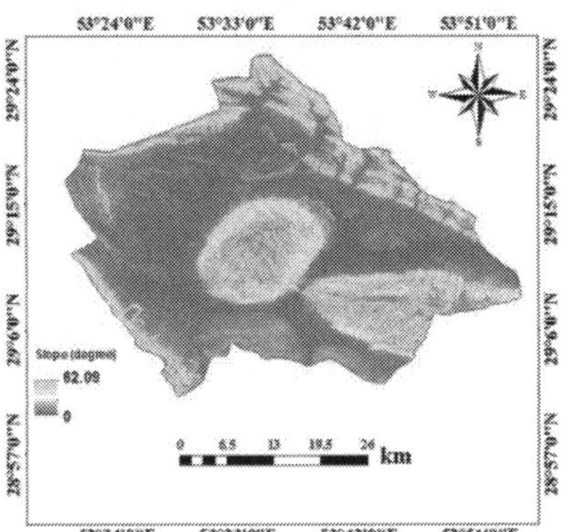

Slope

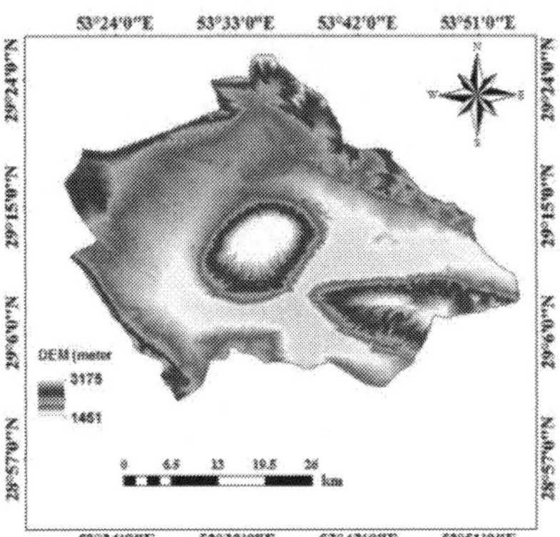

Elevation

Figure 2. Morphometric parameters of the study area.

IOP Conf. Series: Earth and Environmental Science **37** (2016) 012009 doi:10.1088/1755-1315/37/1/012009

Table 1. Morphometric parameters measured for the determination of landform classification.

Parameter	Min	Max	Average	STDVE
Slope (°)	0.38	42.12	9.94	12.11
Profile (1/m)	-0.40	0.44	0.01	0.14
Plan (1/m)	-0.48	0.66	0.01	0.20
Elevation (m)	1484.00	2805.00	1898.58	345.93
Curvature (1/m)	-0.48	0.98	0.01	0.26
Aspect (°)	7.70	358.15	169.42	111.06

2.2 Landform classification using topographic position index (TPI)

TPI [27] compares the elevation of each cell in a DEM to the mean elevation of a specified neighborhood around that cell. Positive TPI values represent locations that are higher than the average of their surroundings, as defined by the neighborhood (ridges). On the other hand, negative TPI values represent locations that are lower than their surroundings (valleys). TPI values near zero are either flat areas (where the slope is near zero) or areas of constant slope (where the slope of the point is significantly greater than zero) [27].

TPI (Eq. (1)) compares the elevation of each cell in a DEM to the mean elevation of a specified neighborhood around that cell. Mean elevation is subtracted from the elevation value at the center [27]:

$$TPI = Z_0 - \sum_{n-1} Z_n / n \quad (1)$$

where;

Z_0= elevation of the model point under evaluation

Z_n= elevation of grid

n= the total number of surrounding points employed in the evaluation.

Combining TPI at small and large scales allows a variety of nested landforms to be distinguished Table 2 and Figure 3. The exact breakpoints among classes can be manually chosen to optimize the classification for a particular landscape. As in slope position classifications, additional topographic metrics, such as variances of elevation, slope, or aspect within the neighborhoods, may help delineate landforms more accurately [27].

Table 2. Landform classification based on TPI [27].

Classes	**Description**
Canyons, deeply incised streams	Small Neighborhood:$z_o \leq$-1 Large Neighborhood:$z_n \leq$-1
Midslope drainages, shallow valleys	Small Neighborhood:$z_o \leq$-1 Large Neighborhood: -1 $< z_n <$ 1
upland drainages, headwaters	Small Neighborhood:$z_o \leq$-1 Large Neighborhood:$z_o \geq$1
U-shaped valleys	Small Neighborhood: -1 $< z_o <$ 1 Large Neighborhood:$z_o \leq$-1
Plains small	Neighborhood: -1 $< z_o <$ 1 Large Neighborhood: -1 $< z_o <$ 1 Slope $\leq$5°
Open slopes	Small Neighborhood: -1 $< z_o <$ 1 Large Neighborhood: -1 $< z_n <$ 1

IOP Conf. Series: Earth and Environmental Science **37** (2016) 012009 doi:10.1088/1755-1315/37/1/012009

	Slope > 5°
Upper slopes, mesas	Small Neighborhood: $-1 < z_o < 1$ Large Neighborhood: $z_o \geq 1$
Local ridges/hills in valleys	Small Neighborhood: $z_o \geq 1$ Large Neighborhood: $z_o \leq -1$
Midslope ridges, small hills in plains	Small Neighborhood: $z_o \geq 1$ Large Neighborhood: $-1 < z_o < 1$
Mountain tops, high ridges	Small Neighborhood: $z_o \geq 1$ Large Neighborhood: $z_o \geq 1$

The classes of canyons, deeply incised streams, midslope and upland drainages, and shallow valleys tend to have strongly negative plane form curvature values. On the other hand, local ridges / hills in valleys, midslope ridges, small hills in plains and mountain tops, and high ridges have strongly positive plane form curvature values.

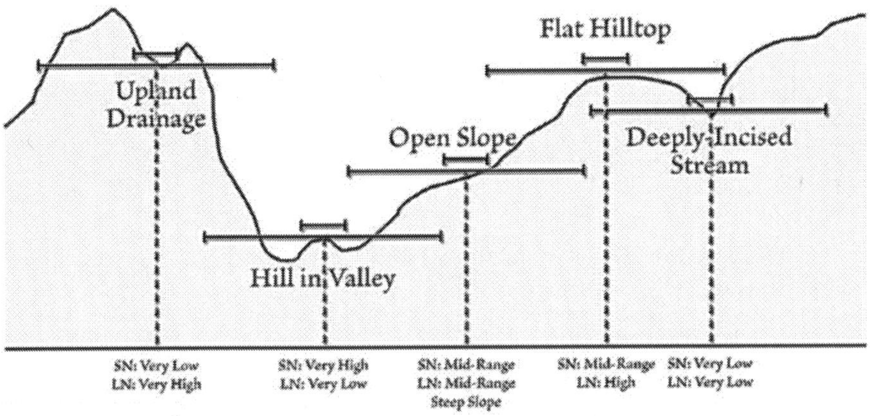

SN: Small neighborhood
LN: large neighborhood
Figure 3. Position of each class on topographic lines [28].

2.3. Self-organizing map (SOM)

SOMs are unsupervised ANNs formed from neurons located on a regular, two-dimensional regular planar array grid (Figure 4). In fact SOM is based on unsupervised learning, which means that no human intervention is needed during the learning and little needs to be known about the characteristics of the input data. SOM offers a solution to apply a number of visualizations linked together [29]. When several visualizations are linked together, scanning through them is very efficient because they are interpreted in a similar way [30]. The U-matrix produced from SOM visualizes distances between neighboring map units and thus shows the cluster structure of the map. Samples within the same cluster will be the most similar according to the variables considered [30].

The SOM algorithm consists of two individual stages: the competitive and cooperative stages. In the competitive stage, the best matching neuron is selected, while in the cooperative stage, the weights of the winner are adapted as well as those of its immediate lattice neighbors [31]. Further explanation for each of the steps is as follows:

1. Competitive stage:

Let A be a lattice of N neurons with weight vectors $w_i = [w_{ij}] \in R^d, W = (w_1, \ldots, w_N)$. All the neurons receive the same input vector $v = [U_1 \ldots U_d] \in V \subset R^d$. For each input v, we select the neuron with the smallest Euclidean distance (''winner-takes-all'', WTA) [32]:

$$i^* = \arg\min_i \|w_i - v\| \quad (2)$$

where w_i is neuron weights and v is input vector.

2. Cooperative stage

The weight update rule in incremental mode is as follows [32]:

$$\Delta w_i = \eta \Lambda(I, i^*, \sigma\Lambda(t))(v - w_i), \forall_i \in A \quad (3)$$

where Λ is the neighborhood function, which is a scalar-valued function of the lattice coordinates of neurons i and i^*, r_i and $r_i{}^*$, mostly a Gaussian:

$$\Lambda(i, i^*) = \exp(-\|r_i - r_{i^*}\|2 / 2\sigma\Lambda 2) \quad (4)$$

with range $\sigma\Lambda$ (i.e., the standard deviation). The positions r_i are usually taken to be the nodes of a discrete lattice with a regular topology [32].

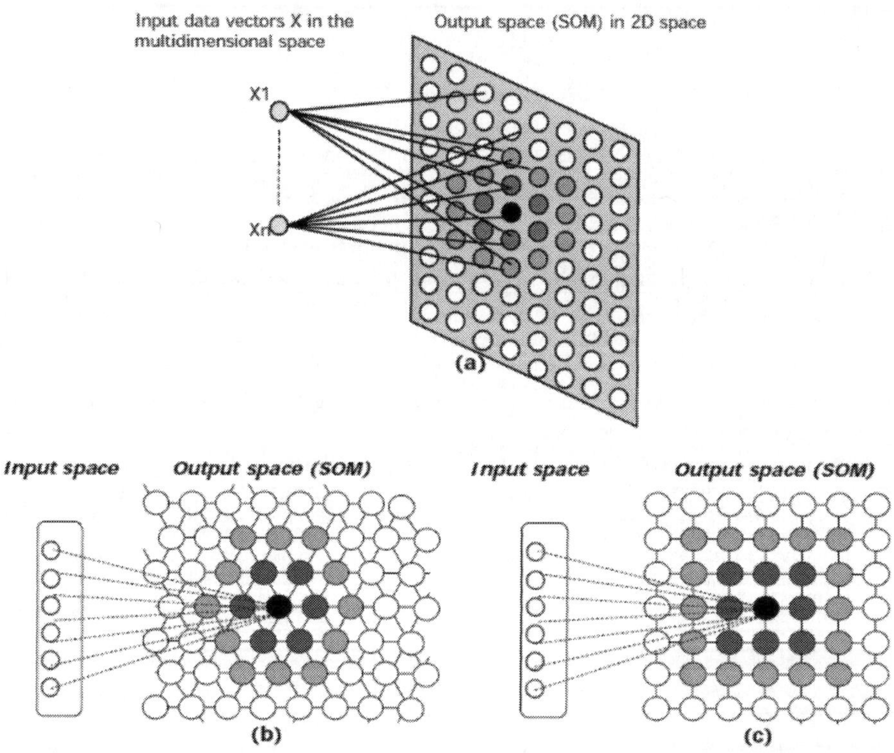

Figure 4. The structure of a SOM network: (a) Selection of a node and adaptation of neighboring nodes to the input data. The SOM grid can be (b) hexagonal or (c) rectangular. The black object indicates that the node that was selected as the best match for the input pattern [33].

3. Results and discussion

3.1. Landform classification

The TPI maps generated using small and large neighborhoods are shown in Figures 5. TPI is between -94 to 77 and -203 to 260 for 5 and 15 cells respectively (Figure 5). The landform maps generated based on the TPI values are shown in Figure 6.

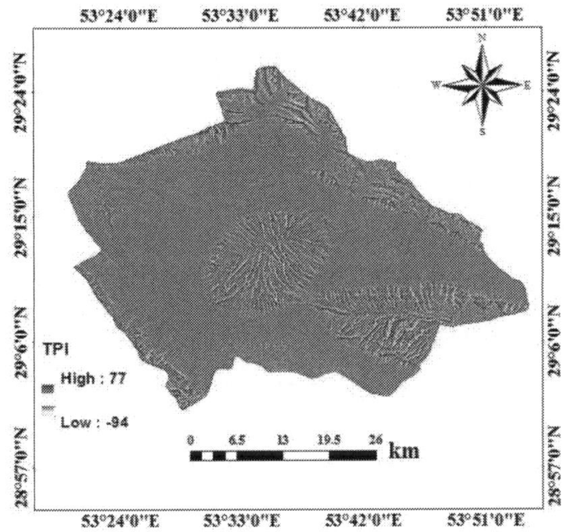

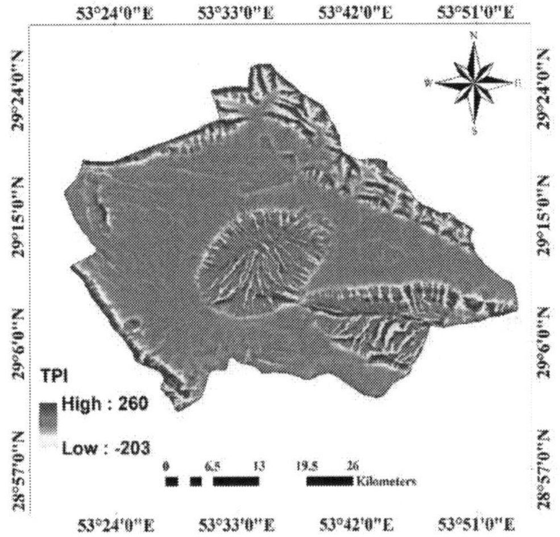

(a) (b)

Figure 5. TPI maps generated using (a) small (5 cell) and (b) large (15 cell) neighborhood.

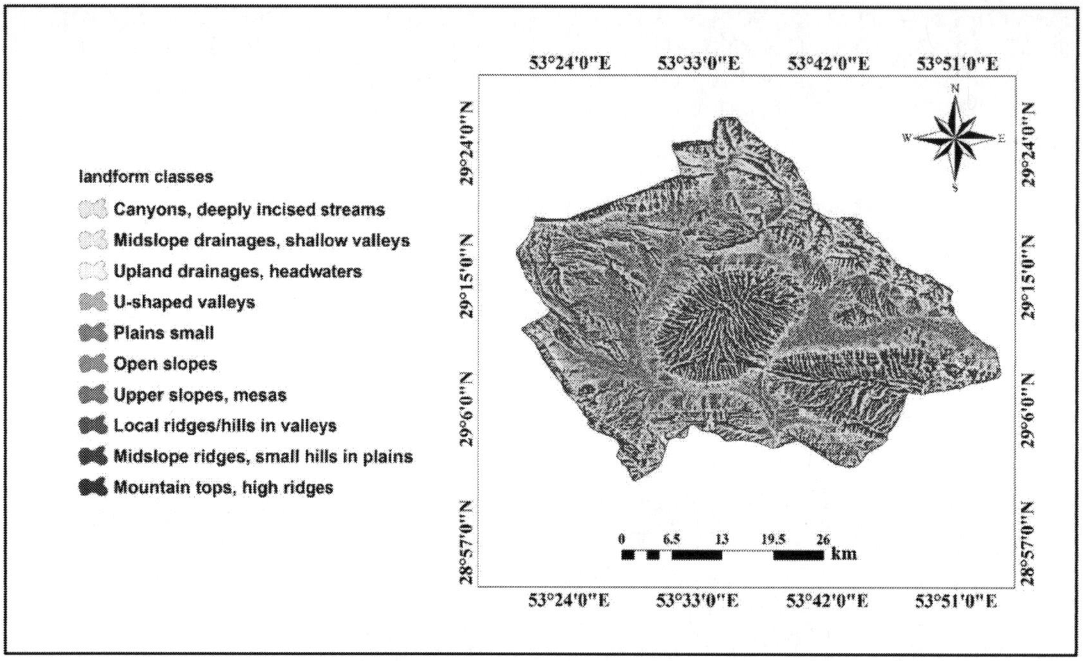

Figure 6. Landform classification using the TPI method.

3.2. SOM for landform classification

SOM was applied for the study area to describe the landform classification. The visualizations in Figures 7 and 8 (for landform classification) consist of 16 hexagonal grids, with the U-matrix in the upper left, along with the six component layers (one layer for each morphometric parameter examined in this study). As previously mentioned, the clustering of landform classification used the morphometric parameters of slope, profile, plan, elevation, curvature, aspect (Figure 2).

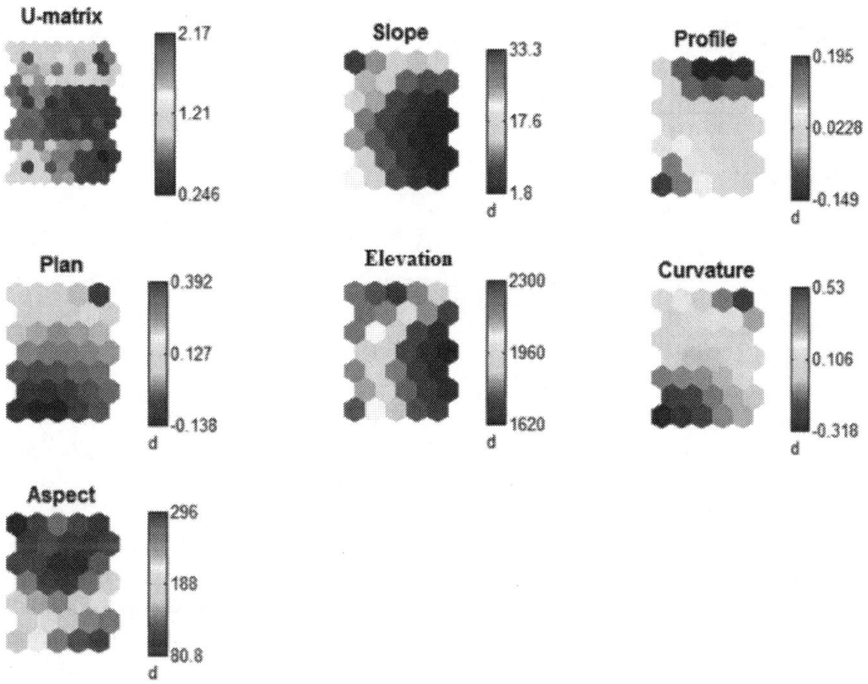

Figure 7. SOM visualization through U-matrix (top left) and the six component layers for landform classification.

According to Figure 7, the six figures are linked by position: in each figure, the hexagon in a certain position corresponds to the same map unit. The legend for each of the hexagons shows the degree of color compared to each other. In the SOM method, similar colors show the direct relationship between the parameters. It can be seen that the elevation and slope are closely related to each other. In addition, profile and curvature are closely related to each other. The other parameters are not related together.

One the other hand, in the U-matrix, it is easy to see that the bottom rows of the SOM form an almost clear cluster (class A), as shown in the labels in Figure 8(c). The other classes that consist of C, D, E and H form the other clusters (Figure 8c).

As was shown in Figure 8, the numbers written in the hexagons are data that are absorbed by each of the nodes in the neural network [34]. According to Figure 8, the maximum number of hexagons was 5, indicating that the maximum data in these places is 5. In addition, the minimum number of hexagons is 0, indicating that in these places, there is no data. According to Figure 8, the principal component (PC) projection showed that the study data had high density with good distribution. Finally, using the label map (Figure 8c), the study data was classified into five classes for landforms.

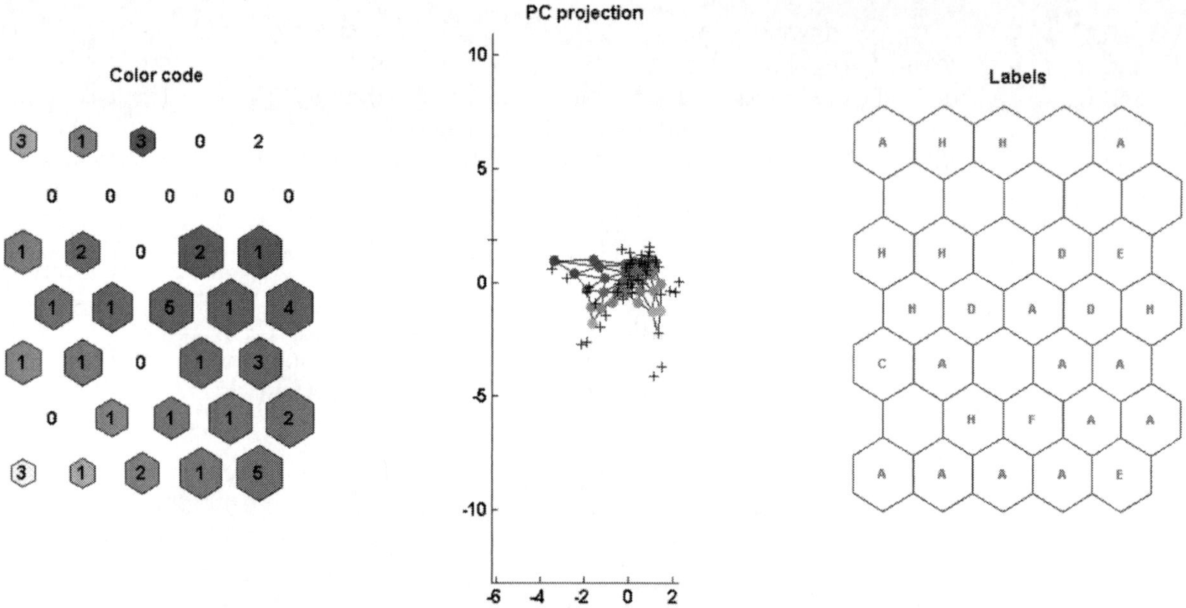

(a) (b) (c)

Figure 8. Different visualizations of the clusters obtained from the classification of the morphological variation through SOM: (a) Color code. (b) Principal component projection. (c) Label map with the names of the landform classification.

The characteristics of each group determined by the label map are provided in Table 3. It seems that the five clusters correspond to different terrain forms. From the component planes, the features differentiating the clusters can be seen. In this table, the categorized map units and the corresponding morphometric features are summarized. Cluster 5 corresponds to high slope and elevation, but with different of concavity and convexity that consist of ridge landforms. Cluster 3 corresponds to flat areas, possibly plantation areas, with medium elevation and almost flat terrain. Clusters 1, 2 and 4 correspond to channels with different slope conditions.

Table 3. Characteristics of the clusters from the SOM for the landform classification.

Group	Parameters	Slope (o)	Profile (1/m)	Plan(1/m)	Elevation (m)	Curvature(1/m)	Aspect (o)
Cluster 1(A)	Min	Low (<5)	-	-	Medium (1500-2500)	-	North
	Max	High (>23)	+	+	Medium (1500-2500)	+	North
Cluster 4 (C)	Min	Medium (5-16)	-	-	Medium (1500-2500)	-	North
	Max	Medium to high (16-23)	+	+	High (>2500)	+	North
Cluster 2 (D)	Min	Low (<5)	-	-	Medium (1500-2500)	-	East
	Max	Medium (5-16)	+	+	Medium (1500-2500)	+	North
Cluster 3 (E)	Min	Low (<5)	-	-	Low (<1500)	-	East
	Max	Low (<5)	+	+	Medium (1500-2500)	+	Northwest
Cluster 5 (H)	Min	Low (<5)	-	-	Medium (1500-2500)	-	Northeast
	Max	High (>23)	+	+	High (>2500)	+	Northwest

4. Conclusion

The purpose of the study was to determine the effectiveness of SOM as a clustering tool for landform classification. In SOM, according to qualitative data, the clustering tendencies of the landforms were investigated using six morphometric parameters (slope, curvature, aspect, elevation, plan and profile). The U- matrix showed that some of the data are closely related to each other, such as elevation and slope. In addition, considering that PC projection represents the amount of data relationship with each other, PC projection was used to determine the study's data had high density. The results showed that the data had high density and had correlation with each other. Finally, using the labels in the SOM method, five classes for the landforms were detected. Cluster 5 corresponds to high slope and elevation but with different of concavity and convexity that consist of ridge landforms. Cluster 3 corresponds to flat areas, possibly planation areas, with medium elevation and almost flat terrain. Clusters 1, 2 and 4 correspond to channels with different slope conditions. Based on the clusters, we can predict that there are five types of main wildlife in the study area.

References

[1] Crevenna AB, Rodrı´guez VT, Sorani V, Frame D and Ortiz MA 2005 Geomorphometric analysis for characterizing landforms in Morelos State, Mexico. Geomorphology, 67, 407–422.

[2] Dikau R, Brabb EE, Mark RK and Pike RJ 1995 Morphometric landform analysis of New Mexico. Zeitschrift für Geomorphologie, N.F., Suppl.-Bd., 101, 109-126.

[3] Pike RJ 2002 A Bibliography of Terrain Modeling (Geomorphometry), the Quantitative Representation of Topography-Supplement 4.0. Open-File Report 02-465, U.S. Geological Survey.

[4] Dehn M, Grtner H and Dikau R 2001 Principles of semantic modeling of landform structures. Computers & Geosciences, 27 (8), 1005– 1010.

[5] Macmillan R, Pettapiece W, Nolan S and Goddard T 2000. A generic procedure automatically segmenting landforms into landform elements using DEMs, heuristic rules and fuzzy logic. Fuzzy sets and Systems, 113, 81–109.

[6] Burrough PA, van Gaans PFM and MacMillan RA 2000 High resolution landform classification using fuzzy-k means. Fuzzy Sets and Systems, 113, 37-52.

[7] Meybeck M, Green P and Vorosmarty CJ 2001. A New Typology for Mountains and Other Relief Classes: An Application to Global Continental Water Resources and Population Distribution, Mount. Research & Development, 21, 34 - 45.

[8] Schmidt J and Hewitt A 2004 Fuzzy land element classification from DTMs based on geometry and terrain position. Geoderma, 121, 243–256.

[9] Saadat H, Bonnell R, Sharifi F, Mehuys G, Namdar M and Ale-Ebrahim S 2008 Landform classification from a digital elevation model and satellite imagery. Geomorphology, 100, 453–464.

[10] Dikau R 1989 The application of a digital relief model to landform analysis in geomorphology. In: Raper, J. (Ed.), Three-dimensional Applications in Geographical Information Systems. Taylor & Francis, London, pp. 51–77.

[11] Dikau R, Brabb EE, Mark RK and Pike RJ 1995 Morphometric landform analysis of New Mexico. Zeitschrift für Geomorphologie, N.F., Suppl.-Bd., 101, 109-126.

[12] Sulebak JR, Etzelmqller B and Sollid JL 1997 Landscape regionalization by automatic classification of landform elements. Norsk Geografisk Tidsskrift, 51 (1), 35– 45.

[13] Etzelmüller B and Sulebak JR 2000 Developments in the use of digital elevation models in periglacial geomorphology and glaciology. Physische Geographie, 41.

[14] Adediran AO, Parcharidis I, Poscolieri M and Pavlopoulos K 2004 Computer-assisted discrimination of morphological units on north-central Crete (Greece) by applying multivariate statistics to local relief gradients. Geomorphology, 58, 357–370.

[15] Moore ID and Nieber JL 1989 Landscape assessment of soil erosion and non-point source pollution. Journal Minnesota Academy of Science, 55, 18– 24.

[16] Moore ID, Gessler PE, Nielsen GA and Peterson GA 1993a Soil attribute prediction using terrain analysis. Soil Science Society of America, 57, 443– 452.

[17] Moore ID, Turner AK, Wilson JP, Jenson SK and Band LE 1993b GIS and Land–Surface–Subsurface Process Modeling. In: Goodchild, M.F., Parks, B.O., Steyaert, L.T. (Eds.), Environmental Modeling with GIS. Oxford University Press, New York, pp. 196– 230.

[18] Evans IS 1979 An integrated system of terrain analysis and slope mapping. Final report on grant DA-ERO-591-73-60040, University of Durham, England, 192pp.

[19] Giles PT 1998 Geomorphological signatures: classification of aggregated slope unit objects from digital elevation and remote sensing data. Earth Surface Processes and Landforms, 23, 581-594.

[20] Mokarram M, Roshan G and Negahban S. 2015. Landform classification using topography position index (case study: salt dome of Korsia-Darab plain, Iran). Model. Earth Syst. Environ. (2015) 1:40

[21] Mokarram M and Sathyamoorthy D. 2015. Modeling the relationship between elevation, aspect and spatial distribution of vegetation in the Darab Mountain, Iran using remote sensing data. Model. Earth Syst. Environ. (2015) 1:30

[22] Mokarram M, Seif A and Sathyamoorthy D. 2014. Use of morphometric analysis and self-organizing maps for alluvial fan classification: case study on Ostorankooh altitudes, Iran. IOP Conference Series: Earth and Environmental Science.

[23] Ehsani AH and Quiel F 2008 Application of Self Organizing Map and SRTM data to characterize yardangs in the Lut desert, Iran. Remote Sensing of Environment 112 (2008) 3284–3294.

[24] Kohonen, T 2001 Self-organizing maps, 3rd ed. Springer-Verlag.

[25] Hosokawat M and Hoshit T 2001 Landform Classification Method Using Self-Organizing Map and its application to Earthquake Damage Evaluation. 0-7803-7031-7/01/$10.00 (C) 2001 IEEE

[26] Ferentinou M, Hasiotis T and Sakellariou M 2010 Clustering of geotechnical properties ofmarine sediments through self – organizing maps: An example from the ZakynthosCanyon–Valley system, Greece. In: Submarine Mass Movements and their consequencesIV , Mosher, D., Shipp, C., Moscardelli, Chaytor J, Baxter C, Lee H and Urgeles R, (Ed.), Advances in Natural and Technological Hazards Research, v. 28, 43-54,Springer, The Netherlands.

[27] Weiss A 2001 Topographic Positions and Landforms Analysis (Conference Poster). ESRI International User Conference. San Diego, CA, pp. 9-13.

[28] Jenness J. 2006 Topographic Position Index (tpi_jen.avx) extension for ArcView 3.x, v. 1.3a. Jenness Enterprises, http://www.jennessent.com/arcview/tpi.htm, (last accessed on February 2010).

[29] Buza A, Mcdonald JA, Michalak J and Stuetzle W 1991 Interactive data visualization using focusing and linking. Proceedings of IEEE conference of visualization, 156-63.

[30] Dhubkarya DC, Nagariya D and Kapoor R 2010 Implementation Of A Radial Basis Function Using VHDL. Global Journal of Computer Science and Technology, 16-19.

[31] Kohonen T (1995). Self - Organizing Maps. Springer, Berlin.Lambeck, K., 1996. Sea-level change and shore-line evolution in Aegean Greece since UpperPalaeolithic Time. Antiquity 70, 588-611.

[32] Hulle MMV 2012 Self-organizing Maps. G. Rozenberg et al. (eds.), Handbook of Natural Computing, DOI 10.1007/978-3-540-92910-9_19, Springer-Verlag Berlin Heidelberg 2012.

[33] Dykes J 2005 Exploring Geovisualization. Elsevier.

[34] Venna J and Kaski S 2001. Neighborhood preservation in nonlinear projection methods: An experimental study. Lecture Notes in Computer Science 2130: 485-491.

Unveiling topographical changes using LiDAR mapping capability: case study of Belaga in Sarawak, East-Malaysia

T R Ganendra, N M Khan, W J Razak, Y Kouame, E T Mobarakeh

Ground Data Solutions R&D Sdn Bhd, Lot 6, Jalan 51A/223, 46100 Petaling Jaya, Selangor Darul Ehsan, Malaysia

E-mail: admin@grounddatasolutions.com

Abstract. The use of Light Detection and Ranging (LiDAR) remote sensing technology to scan and map landscapes has proven to be one of the most popular techniques to accurately map topography. Thus, LiDAR technology is the ultimate method of unveiling the surface feature under dense vegetation, and, this paper intends to emphasize the diverse techniques that can be utilized to elucidate topographical changes over the study area, using multi-temporal airborne full waveform LiDAR datasets collected in 2012 and 2014. Full waveform LiDAR data offers access to an almost unlimited number of returns per shot, which enables the user to explore in detail topographical changes, such as vegetation growth measurement. The study also found out topography changes at the study area due to earthwork activities contributing to soil consolidation, soil erosion and runoff, requiring cautious monitoring. The implications of this study not only concurs with numerous investigations undertaken by prominent researchers to improve decision making, but also corroborates once again that investigations employing multi-temporal LiDAR data to unveil topography changes in vegetated terrains, produce more detailed and accurate results than most other remote sensing data.

1. Introduction

The relentless reshaping and changing of the earth surface that results from nature and human activities triggers topographic variations, causing hazard, environment and ecological problems, land and farming degradation, etc. Several studies have emphasized the suitability of remote sensing techniques to map topography [1] [2] [3] [4]. However, this earth surface mapping using remote sensing techniques has not always yielded the expected outcome; either some features were difficult to detect [5], or low spatial resolution was obstructing the detection and identification of features reliably [6]. In densely vegetated terrain; the mapping via spectral signatures of the surface features below the vegetation may be unreliable [7]. In particular, the effective mapping of features such as hills, creeks, valley and rivers may be impossible because exposure of the surface beneath the vegetation canopy may be limited [7] [8] [9].

Unlike other remote sensing techniques, LiDAR remote sensing technology is unique in the field of topography change mapping, as it is able to measure features under vegetation, by scanning the target with laser pulses and recording backscattered information highly effectively and accurately [10][11][12].

In this study, the change detection was measured using LiDAR datasets collected in 2012 & 2014 to produce meaningful outputs depicting the topographic changes in Belaga. However, the mapping of topography under dense vegetation still remains a daunting task, because of the dense forest canopy. Figure 1 and Figure 2 below show the overlap areas of two projects in Belaga, Sarawak collected in 2012 and 2014. This study area mostly comprises of dense jungle.

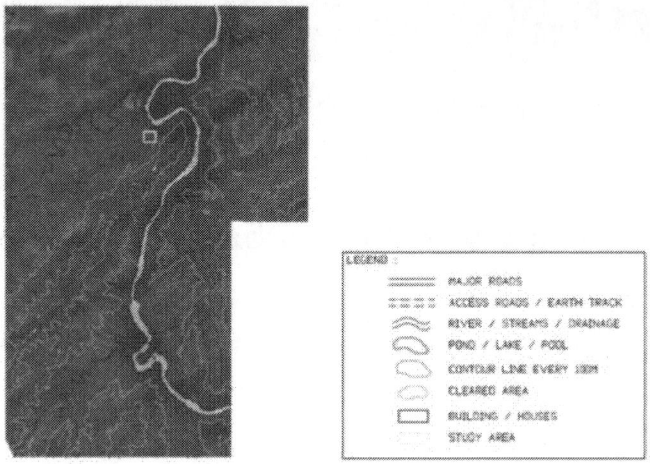

Figure 1. Overlap area which depicts parts of the northern section of Belaga.

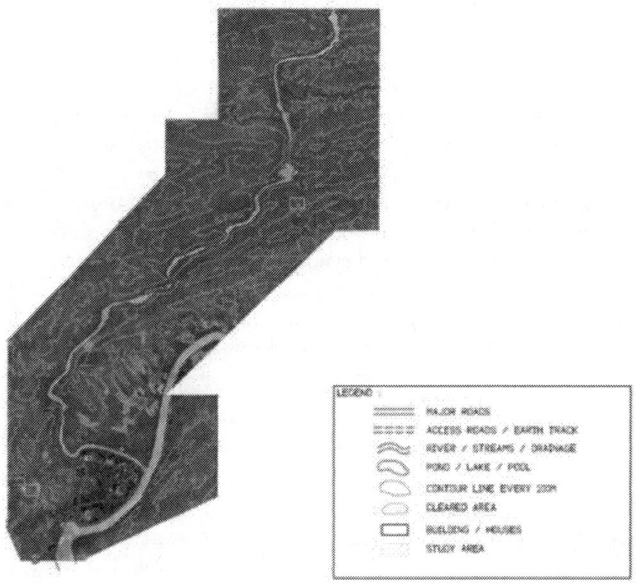

Figure 2. Overlap area which depicts the southern section of Belaga.

2. Materials and methodology

2.1. Datasets

The two sets of spatial data used in this study are airborne full waveform LiDAR and imagery data captured on 12th February 2012 and 8th September 2014. According to the American Society of Remote Sensing Society (ASPRS) [14], when using LiDAR-derived elevation data, it is essential to specify the accuracy expected for all final products being delivered. Vertical accuracy is assessed in the z dimension only and different methods are used for non-vegetated and vegetated terrain [15]. LiDAR data vertical accuracy should always be reviewed in terms of the potential harm that could be done to the public health and safety [14]. Examples for ten common vertical accuracy classes and other quality criteria for digital elevation data can be obtained from ASPRS's Positional Accuracy Standards for Digital Geospatial Data [15].

To determine the accuracy of different datasets, the elevation points of the both datasets were compared with established control points. The standard deviation, $RMSE_z$ and accuracy at 95% confidence level are computed based on equation 1, 2 and 3 respectively. Table 1 lists the properties of the data used in this study.

$$\textit{Standard Deviation} = \sqrt{\frac{\sum_{i=1}^{n}(z_i-\acute{z})^2}{n-1}} \tag{1}$$

$$RMSE_z = \sqrt{\frac{\sum_{i=1}^{n}\left(z_{i\,(surveyed)} - z_{i\,(LiDAR)}\right)^2}{n}} \tag{2}$$

$$\textit{Vertical Accuracy at 95\% Confidence Level} = RMSE_z \; X \; 1.96 \tag{3}$$

where z_i is the ith error in the specified direction, $\acute{z}$ is the mean error in the specified direction, n is the number of checkpoints tested and i is an integer ranging from 1 to n.

Table 1. Properties of the LiDAR and imagery datasets.

Properties	**2012**	**2014**
Scanner Type	Riegl LMS-Q560	Riegl LMS-Q560
Forward Speed	100 kph / 62 mph	100 kph / 62 mph
Flying Height Above Ground Level (AGL)	600 m	600 m
Max AGL Setting	850 m	850 m
Scanning Angle	52.9°	45°
Swath Width	597 m	349 m
Vertical Accuracy at 95% Confidence Level	0.096 m	0.057 m
Average All Point Spacing	0.2 m	0.3 m
All Point Density	11 points/m^2	8 points/m^2
Average Ground Point Spacing	1.5 m	1.8 m
Ground Point Density	0.4 points/m^2	0.2 points/m^2
Imagery Raw Pixel Size	0.1 m	0.1 m

2.2. Methodology

The multi-temporal airborne full waveform LiDAR data was captured using a Riegl LMS-Q560 scanner. The data was acquired in two different years on 12th February 2012 and on 8th September 2014 in Belaga, Sarawak. The flying heights of the sensor were on average 600 m AGL, which produced a laser swath width of 349 m for 45° and 597 m for 52.9° field of view setting for laser. The flightlines were designed with a 30% overlap between adjacent flightlines for the camera.

TerraSolid and ArcGIS software were used to conduct the data processing and analysis. First, the data from the laser, IMU and GPS sensors were processed to produce a raw point cloud. LiDAR point clouds were generated using the ITRF00 datum and UTM Zone49 map projection. Then, the points were classified into ground and non-ground points [16]. This included detailed ground editing to correct the errors in the automatic classification processes to improve the ground model. Then, the ground points were used to generate a DTM. Still imagery was collected simultaneously with LiDAR data with a 60% forward overlap and 30% side overlap to ensure sufficient imagery coverage.

Orthophoto mosaics were generated using trajectory data produced from GPS and IMU data, DTM keypoints and the imagery with a 0.1 m resolution. Statistical and geostatistical analysis were performed as described next.

3. Results and analysis

3.1. Validation of LiDAR dataset
Multi-temporal LiDAR datasets were captured in 2012 and 2014 in Belaga, Sarawak. These two sets of data were processed using the ASPRS standard of vertical accuracy as mentioned in the section *2.1.* The vertical accuracy of each LiDAR dataset in table 1 shows that the data is sufficiently accurate to be used in this study. However, to ensure agreement between both datasets, and to confirm that any variation in the datasets is due to topographical changes and not attributable to data production issues, the datasets were validated using cross sections of point clouds in built-up areas. Some buildings can be considered as static structures for verification. Figure 3 shows an example for the point cloud cross sections for both datasets along a building that matched in both datasets. The point cloud of 2012 (red) and point cloud 2014 (blue) match each other in the ground areas around and on the roofs of buildings which were confirmed to be unchanged in these 2 years.

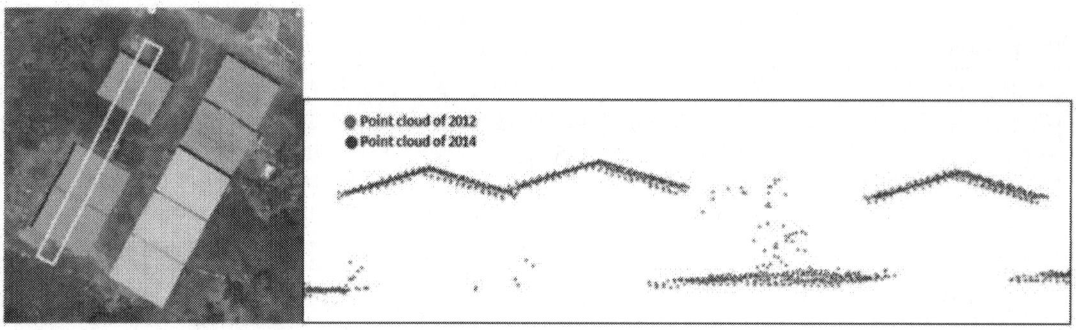

(a) (b)

Figure 3. (a) Cross section of buildings in the study area and (b) Point cloud of the buildings and surrounding ground for 2012 in red and 2014 in blue.

3.2 Growth measurement with multi-temporal LiDAR data

3.1.1. Canopy height model from LiDAR data. Canopy height models (CHM) derived from LiDAR data are widely recognised as an excellent tool for understanding and managing forest ecosystems. Here, a CHM was generated by subtracting the digital surface model of the first return (DSM) with the digital terrain model (DTM) of the ground points. These CHM were used to illustrate canopy changes of the selected area, which was completely heavily vegetated (figure 4). The black line indicates the cross-section profile that was used throughout this analysis. Figure 5 (a) and (b) illustrate CHMs for 2012 and 2014 respectively.

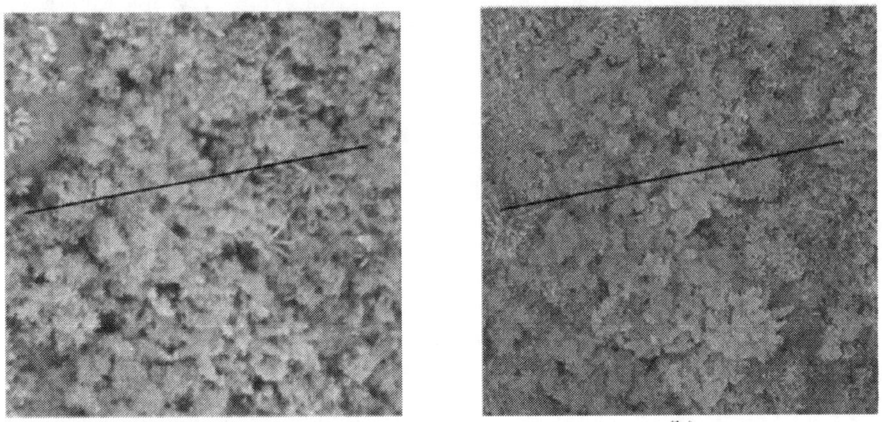

(a) (b)

Figure 4. Orthophoto of CHM study area. (a) 2012 and (b) 2014.

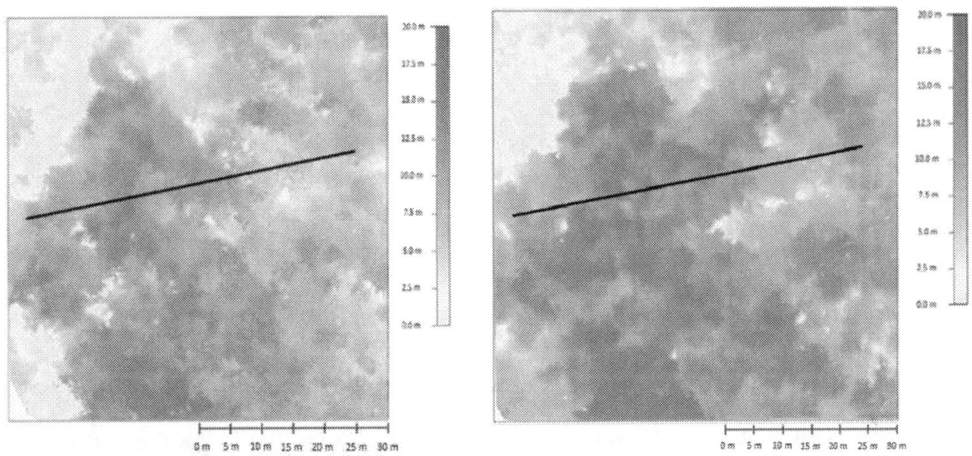

(a) (b)

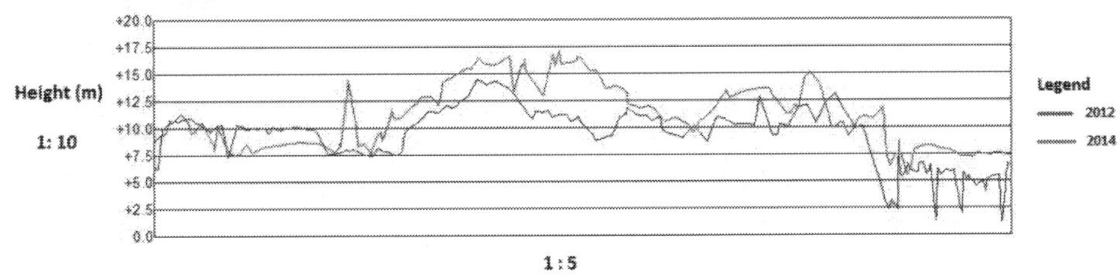

(c)

Figure 5. CHM. (a) 2012, (b) 2014 and (c) Cross-section profile of 2012 and 2014.

Based on the CHM of both years, the areas in red show higher tree heights, while yellow areas indicate areas of grass or low-lying shrubs. Figure 5 (c) shows the cross-section profile for both CHMs, indicating increases of canopy height along the cross section line. However, decreases of canopy height were also detected along the cross section. Therefore, by analyzing the differences between both CHMs (figure 6), the areas which experienced decreases (green) and increases (yellow to red) of canopy height can be identified easily. Areas digitized in blue show a few locations with decrease in canopy height, probably due to fallen trees or parts of trees. Orthophoto images of these blue areas of both years, figure 6 (b) and (c) shows the appearance of "black holes" due to some sort of disappearance of vegetation. The areas in red show high increase in canopy probably due to neighbouring higher trees increasing their canopy sideways. This can particularly be seen with areas digitized in magenta. Therefore, the use of LiDAR multi-temporal data is able to provide indicative CHM changes.

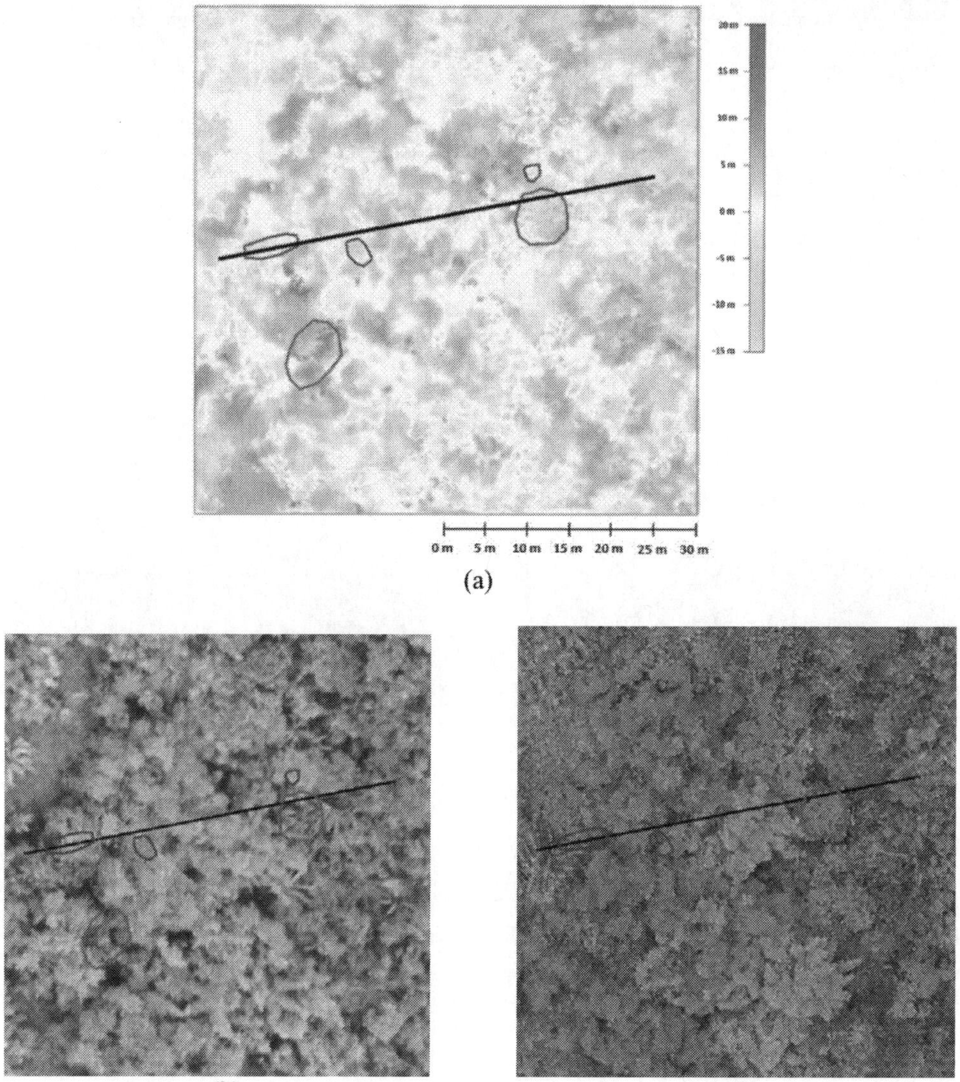

(a)

(b) (c)

Figure 6. The changes in canopy height. (a) CHM difference, (b) Orthophoto image of 2012, (c) Orthophoto image of 2014.

3.1.2. Individual tree height growth. The LiDAR technology is well suited for acquisition of highly detailed forest structure information, and can provide more detailed information relating to the individual tree growth. Individual tree height and canopy diameter growth can be measured by taking cross sections of the tree point cloud profile. Randomly, five sample individual trees were identified to measure their growth as shown in figure 7.

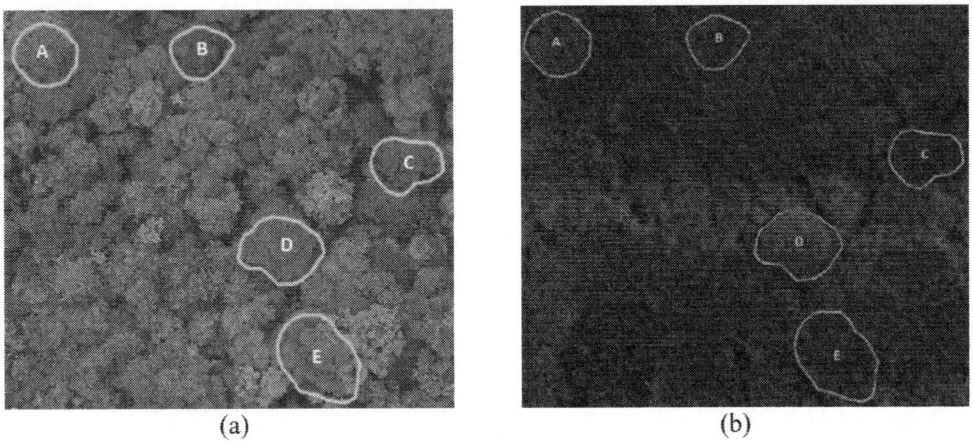

Figure 7. Sample of individual tree. (a) Orthophoto image and (b) Point cloud.

Figure 8 shows the measurement of individual tree heights from five sample trees from the study area. The red and blue points indicate the point cloud in 2012, 2014. The height was measured from ground to the top of the tree crown for both years, using their respective point clouds. From the measurement, all five sample trees showed growth over the two years estimated to be ranging from approximately 0.61 to 1.35m. The canopy diameter of the sample trees can also be seen to have increased in the two years. It can be inferred that trees A, B, & C were faster growing than D & E within this time interval.

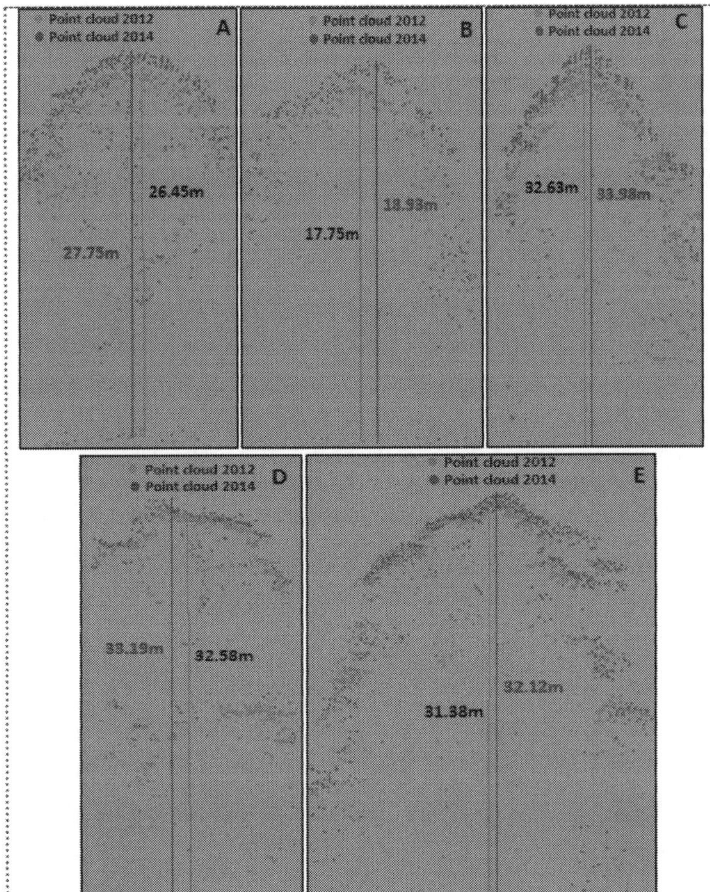

Figure 8. The height of each individual tree was measured from point cloud cross-sectional profile.

3.3 Comparison of area elevation changes

Figure 9 shows the respective orthophoto images and DEM for 2012 and 2014. Visual inspection of these orthophoto images and the generated DEM shows that there were significant changes in the DEM surface during these two years.

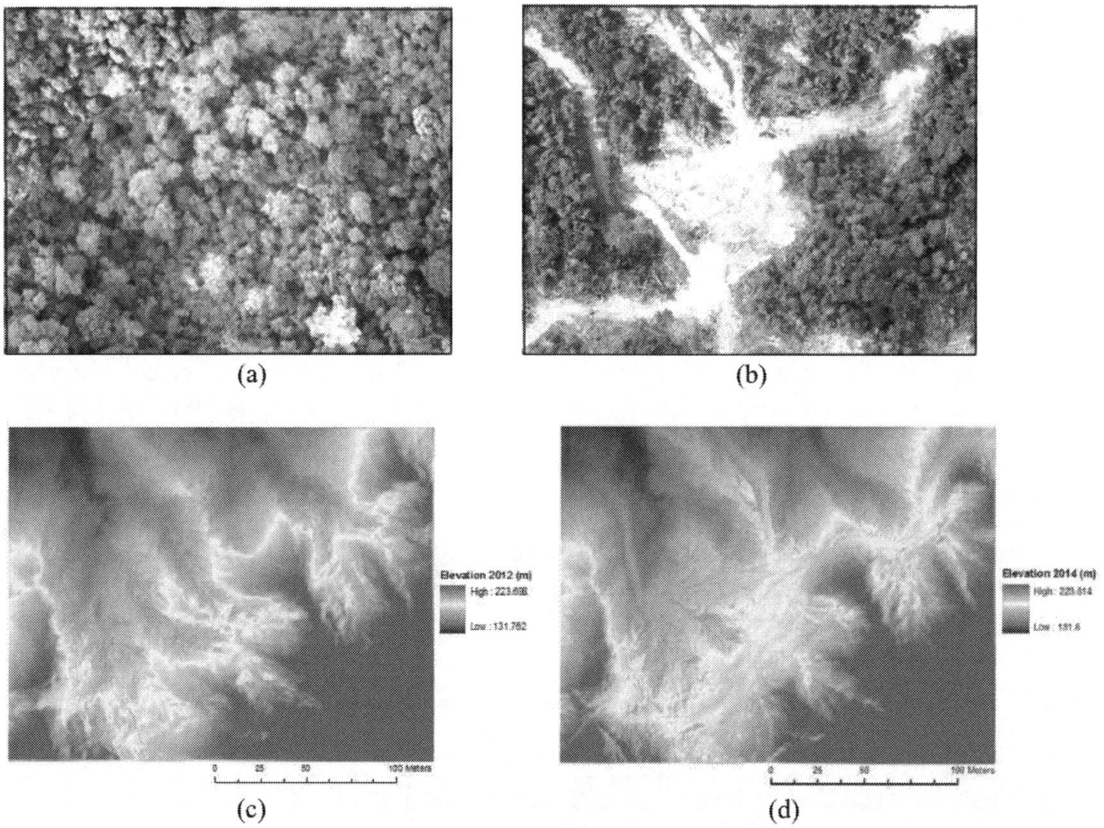

Figure 9. Study area for elevation changes, with the DEM generated from the LiDAR data. (a) Orthophoto image in 2012, (b) Orthophoto image in 2014, (c) DEM of 2012 data and (d) DEM of 2014 data.

Most of the changes are believed to be due to earthworks activities causing movement, compaction and expansion of soil. Soil erosion would be of particular concern. These earthworks activities can be analyzed by monitoring changes in the terrain. Raster data is used in this surface analysis, which is one of the more efficient methods to identify areas that have been significantly cut and filled. This analysis is based on changes in the surface elevation between 2012 and 2014 surfaces, which will identify the areas that have been affected by track construction, such that there is surface material loss, compaction, gain or expansion.

Figure 10 shows the area elevation changes, which are determined by comparing the difference between the 2012 and 2014 raster grids. The areas in blue indicate significant soil gain or expansion between 1 to 5.72 m. The areas in red and orange indicate significant soil loss or compaction between -1 to -6.07 m. The green and yellow areas show little elevation change, and cover approximately 60 percent of the study area.

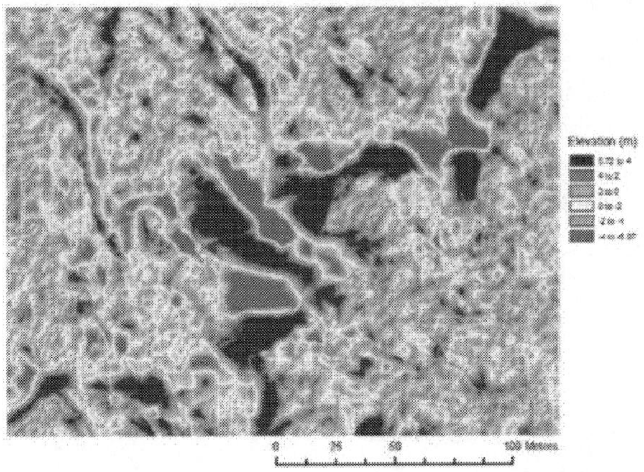

Figure 10. Interpolated surfaces of area elevation changes.

3.4 LiDAR for earthwork monitoring

LiDAR technology offers highly accurate elevation data with highly detailed and clear imagery allowing for identification of features and vegetation. These are useful for monitoring an area expected to undergo elevation changes initiated by earthwork activities. The changes can be analyzed using cross sections of the point cloud and visualization of the orthophoto mosaic images, as shown in figure 11. The mosaic image figure 11 (a) shows the area underwent construction during the data acquisition in 2012; while in figure 11 (b), the road construction seems to be complete. Figure 11 (c) shows the cross section of the road using both sets of data, which clearly indicate that in two years, terrain changes have occurred. In particular, there has been excavation work to lower ground levels by approximately 3 m for the road section shown.

(a)

(b)

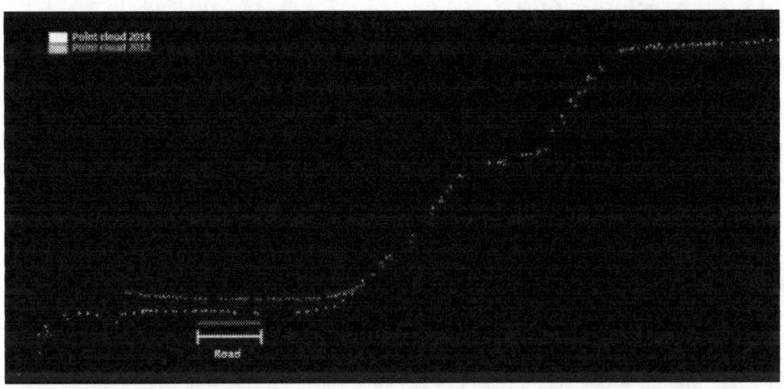

(c)

Figure 11. Images of road construction (a) 2012, (b) 2014, (c) Cross section of point cloud in 2012 (green) and 2014 (white).

3.5 Earthwork calculation

In most preliminary project planning and design, LiDAR derived DEM and digital imagery is necessary to identify favourable alignments, evaluate alternative routes and create final design drawings that optimize alignments and grades for the selected alternative. The elevations of the same area (figure 11) for both years can be seen in figure 12 (a) and (b). There were minor changes in the elevation between 3.5 to -3.5 m as represented in figure 12 (c).

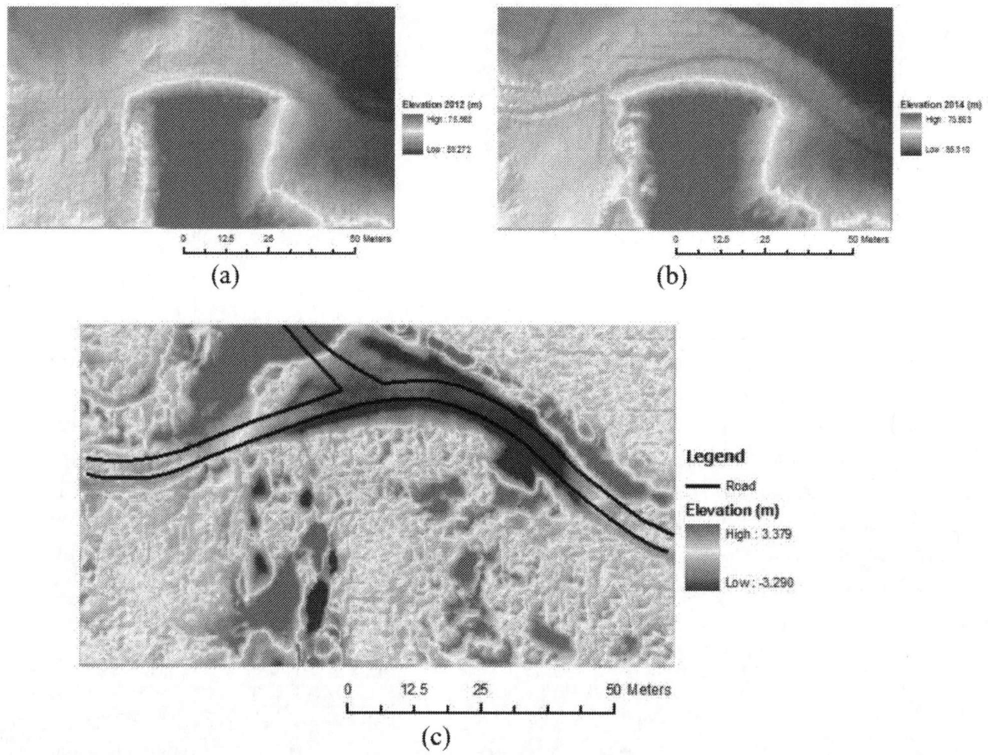

Figure 12. Elevation changes. (a) Elevation during road construction in 2012, (b) Elevation in 2014, (c) Elevation changes over 2 years.

GIS can also estimate the cut and fill that is required for the desired area. By using the same area (120 m) of the road, this study produced an estimate of cut and fill of the road segment. The road segment was divided into 12 sections of road, 3 me wide by 10 m long (figure 13). Table 2 shows the estimate of cut and fill for each section.

Figure 13. 12 section of road.

Table 2. Estimate of cut and fill.

Section	Soil Volume (m^3)		
	Cut	Fill	Total material disposed
1	0.9726	1.5848	- 0.6122
2	0.0000	10.3208	- 10.3208
3	0.0000	17.3907	- 17.3907
4	4.9899	4.9836	0.0063
5	21.5851	0.0000	21.5851
6	27.3640	0.0000	27.3640
7	24.5926	0.0000	24.5926
8	28.4032	0.0000	28.4032
9	31.8958	0.0000	31.8958
10	28.3945	0.0000	28.3945
11	4.8144	9.6253	- 4.8109
12	0.1337	3.3214	- 3.1877
Total	173.1458	47.2266	125.9192

Thus, for this 120 m road section, a total of 125.9192 m^3 of soil is estimated to have been disposed of. However, while ArcGIS software has been used for these analyses, past experience has shown that software specific to the analysis required, can more efficiently and accurately use the LiDAR data for the purpose of the analysis. Examples of such software include MX Road for Road Design and PLSCADD for Transmission Line Design.

4. Conclusion

When using multi-temporal full waveform datasets to do change analysis, it is important to ensure agreement between both datasets to ensure any change is due to topography changes and not mapping accuracy issues. Static structures, such as unchanged buildings were used to confirm agreement between datasets and showed no difference between 2012 and 2014 datasets.

There is a wide range of analyses that can be performed using multi-temporal LiDAR data, but this study focused on a few topographical changes that have occurred in the study area in two years. For densely vegetated areas, LiDAR was able to demonstrate its unique ability to analyze vegetation growth and elevation through tropical rainforest canopy. Individual tree growths were measured and the canopy height model (CHM) generated for each year and the changes between the two datasets were compared. The study also analyzed the elevation changes of both datasets, due to erosion and earthwork activities which can provide estimates of cut and fill for earthworks calculations. Thus, this paper demonstrates some of the useful change detection analyses that can be performed for multi-temporal full waveform LiDAR data, and more analyses should be explored in future.

References

[1] Ono T, Hattori S, Hasegawa H, and Akamatsu S 2000 *Int. Archives of Photogrammetry and Remote Sensing* **33** B3 672-77

[2] Shaker A and Shi W Z 2003 Polynomial models as a tool for mapping high-resolution satellite imagery *European Remote Sensing Conf. (SPIE)* Barcelona

[3] Shaker A, Nichol J E, Wong M S, 2007 Topographic mapping from small satellites: a case study of chris/proba data. *High-Resolution Earth Imaging for Geospatial Information, ISPRS Hannover Workshop* Germany

[4] Li A K and Batchvarova T 2008 Topographic mapping and terrain modeling from multi-

IOP Conf. Series: Earth and Environmental Science **37** (2016) 012010 doi:10.1088/1755-1315/37/1/012010

sensor satellite imagery *Proc. of the ISPRS Congress of the Int. Society for Photogrammetry and Remote Sensing ISPRS Annals* Beijing **311** 809-14
[5] Leatherdale, J D 1979 The practical contribution of space imagery to topographical mapping *New Technology for Mapping* 425-44
[6] Petrie G 1978 The status of topographic mapping from space imagery *Remote Sensing and National Mapping, Remote Sensing Society* 1-16
[7] Crippen R E and Blom R G 2001 Unveiling the lithology of vegetated terrains in remotely sensed imagery *Photogrammetric Engineering and Remote Sensing* **67** (8) 935-46
[8] Siegal B S and Goetz A F 1977 Effect of vegetation on rock and soil type discrimination
[9] Murphy R J and Wadge G 1994 The effects of vegetation on the ability to map soils using imaging spectrometer data *Remote Sensing* **15** (1) 63-86
[10] Lim K, Treitz P, Wulder M, St-Onge B and Flood M 2003 LiDAR remote sensing of forest structure *Progress in Physical Geography* **27** (1) 88-106
[11] Hsiao K H, Liu J K, Yu M F and Tseng, Y H 2004 Change detection of landslide terrains using ground-based LiDAR data *XXth ISPRS Congress* **7** (5) 5
[12] Yang X ed 2008 Remote sensing and geospatial technologies for coastal ecosystem assessment and management *Springer Science & Business Media*
[13] Wan Mohd, Abdullah and Hashim 2014 Accuracy comparison of digital elevation models generated from different sources *FIG Conf. 2014*
[14] Flood M 2004 ASPRS Guidelines vertical accuracy reporting for LiDAR data
[15] ASPRS 2005 ASPRS positional accuracy standards for digital geospatial data Edition 1, Version 1.0 **81** (3) A1-A26
[16] Li Z, Zhu C and Gold C 2004 Digital terrain modelling: principles and methodology CRC Press

Development of SPIES (Space Intelligent Eyeing System) for smart vehicle tracing and tracking

Suzanah Abdullah[1], Muhammad Ariffin Osoman[2], Chua Guan Liyong[2], Mohd Zulfadhli Mohd Noor[2], Ikhwan Mohamed[1]

[1]Faculty of Architecture, Planning and Surveying, Universiti Teknologi MARA Seri Iskandar Campus, Seri Iskandar,32610, Perak, Malaysia,
[2]Geoinfo Services Sdn Bhd, No 31, Jalan Bandar 2, 53100 Taman Melawati, Kuala Lumpur, Malaysia,

Email: suzan156@perak.uitm.edu.my

Abstract. SPIES or Space-based Intelligent Eyeing System is an intelligent technology which can be utilized for various applications such as gathering spatial information of features on Earth, tracking system for the movement of an object, tracing system to trace the history information, monitoring driving behavior, security and alarm system as an observer in real time and many more. SPIES as will be developed and supplied modularly will encourage the usage based on needs and affordability of users. SPIES are a complete system with camera, GSM, GPS/GNSS and G-Sensor modules with intelligent function and capabilities. Mainly the camera is used to capture pictures and video and sometimes with audio of an event. Its usage is not limited to normal use for nostalgic purpose but can be used as a reference for security and material of evidence when an undesirable event such as crime occurs. When integrated with space based technology of the Global Navigational Satellite System (GNSS), photos and videos can be recorded together with positioning information. A product of the integration of these technologies when integrated with Information, Communication and Technology (ICT) and Geographic Information System (GIS) will produce innovation in the form of information gathering methods in still picture or video with positioning information that can be conveyed in real time via the web to display location on the map hence creating an intelligent eyeing system based on space technology. The importance of providing global positioning information is a challenge but overcome by SPIES even in areas without GNSS signal reception for the purpose of continuous tracking and tracing capability

1. Introduction

The integration of GNSS technology with other technology and sensors allows accurate geo-locational acquisition of remotely sensed imagery by imaging sensors mounted on space borne satellites, ortho-rectified photo by camera mounted on airborne fixed-wing and rotary-wing platforms, and now, videos and still pictures by camera mounted on terrestrial-based vehicle. The output of terrestrial-based vehicle allows measurement to be made from the video and still pictures besides animation that can always be viewed to enable limitless site revisit. Furthermore, the integration of the above technologies and geo-spatial technologies allows geo-locational data to be integrated and utilized through web-based and desk top for tracking applications and others customized viewing, tracking and tracing related applications

The integration of GNSS technology with other technology and sensors allows accurate geo-locational acquisition of remotely sensed imagery by imaging sensors mounted on space borne satellites, ortho-rectified photo by camera mounted on airborne fixed-wing and rotary-wing platforms, and now, videos and still pictures by camera mounted on terrestrial-based vehicle. The output of terrestrial-based vehicle allows measurement to be made from the video and still pictures besides animation that can always be viewed to enable limitless site revisit.

The integration of the above technologies and geo-spatial technologies allows geo-locational data to be integrated and utilized through web-based and desk top for tracking applications and other customized viewing, tracking and tracing related applications. In terrestrial application, one of the challenges of using the GPS is positional data cannot be collected in the absence of GNSS services. As all the above mentioned technologies are capable of recording the visual, time and position, hence, the synchronization of the output along with other technology that allow position can be determined in the absence of GNSS services, will benefit not only the tracking application of a moving vehicle but also with tracing capability of its surrounding event. The purpose of the proposal is to develop a space-based integrated system for real-time monitoring or off-real-time movement of vehicle and its environment using the GNSS services with track and visual tracing capability even in the absence of GNSS services. The benefit of the product is to trace the cause of accident and as a mobile CCTV, it acts as eye-witness of an event.

1.1. Problem statement

Nowadays, many types of accidents cases that occur on the road often hear in the chest of press such as television, radio, internet and many more. Accidents regularly associated with unexpected situations such as human negligence or damage of the vehicle itself. Accident will occur anywhere without know when the accident will happen. It is difficult to describe how the accident happened especially accident cases without proof.

This study will assist in identifying the cause of an accident that happens, whether from human negligence or due to the condition of the vehicle itself. The method how to identify before the accidents occur also can be shown in this project. This study will help to identify how an accident occurred. The recorder system installed on the vehicle can be helpful to find the cause of the accident. This system is used to record an accident that occurred either before or after it periodically.

In terrestrial application, one of the challenges of using the GPS is positional data cannot be collected in the absence of GNSS services. As all the above mentioned technologies are capable of recording the visual, time and position, hence, the synchronization of the output along with other technology that allow position can be determined in the absence of GNSS services, will benefit not only the tracking application of a moving vehicle but also with tracing capability of its surrounding event.

1.2. Objective

The main objective of this study is to propose a new approach deep tackling vehicle user security problem especially to improving their behavior on way of driving that more well-mannered and safe. Meanwhile, the objectives of the study are as follows:

a) To develop a space-based integrated video system for monitoring real-time and archived movement of vehicle using the GNSS services.

b) To record incident with visual and tracking capability even in the absence of GNSS services.

c) To develop a prototype as a mobile CCTV, it acts as an eye-witness of an incident for itself and other vehicles.

2. Literature review

In this section involve with the explanation that based on elaboration of conceptual system tracing and tracking. Since the introduction of GPS Navigation systems in the marketplace, consumers and businesses have been coming up with innovative ways to use the technology in their everyday life. GPS Navigation and Tracking Systems keep us away from getting lost when we are in strange locations, they monitor children when they are away from home, keep track of business vehicles and can even let us know where a philandering partner is all the times (Khraisat et al., 2011). Mobile application is the most of the convenience way for everybody to access information. Combination of Voice, video, GPS, SMS, accelerometer and WIFI will introduce most intelligence gadget in a recent technology.

Mobile phones have made significant improvements from providing voice communications to advanced features such as camera, GPS, WI-FI, SMS, voice recognition, Internet surfing, and touch screen (Huei, 2012). The global Positioning System which also known as GPS is a satellite-based navigation system designed to help navigate on the earth, in the air and under water. It provides users with positioning, navigation and timing services(Al-Khedher 2012; RAKESH 2014). The GPS system was created by the Department of Defense of the United States and nowadays is still owed and operated by the United States. Originally, it was used exclusively by the U.S. military, but in 1983 an order allowed anyone to use the system.

The system was decalared fully operational in 1995(Abbott and Powell 1999). In recent times, other Global Navigation Satellite Systems (GNSS) have been either launched or being developed, such as the Russian GLONASS system, the European Union Galileo satellite navigation system, the Chinese Beidou and Indian satellite navigation system(Marron Monteserin 2014). Today GPS-based navigation systems can be found in motor vehicles, farming and mining equipment and variety of other land-based vehicles.

3. Methodology

In this section will explain the methods to be implemented in this research from the beginning until the end of the study. Methodology is an important thing to explain the appropriate method used in this project to achieve the objectives of the study. This chapter also important to arrange the process of analysis in order to making the good result based on the objective. The study was carried out by identifying issues for study, so it is necessary to plan the project to be implemented. The process of collecting information or data will also be described in this chapter to ensure that data required depends on the objectives of the study. Methodology is the systematic, theoretical analysis of the method applied to a field of study. It comprises the theoretical analysis of methods and principles associated with the branch of knowledge. Figure below shows the flow chart of methodology process.

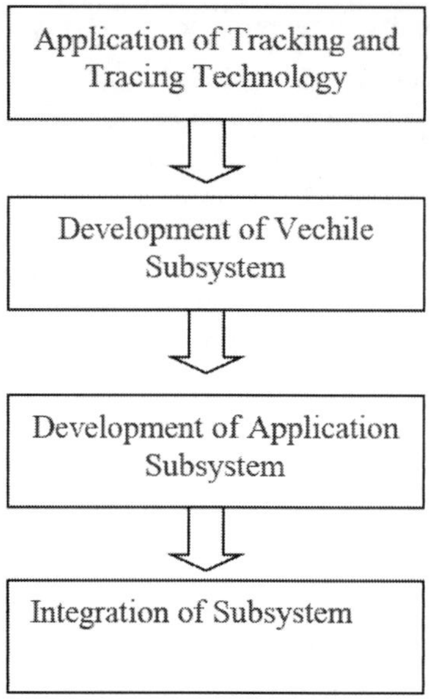

Figure 1. Process of Methodology.

3.1 Application of tracking and tracing technology

The preliminary part of the project is to study the availability of terrestrial imaging-tracking-tracing technologies and applications. The study shall improvise the existing system if available and prioritize applications in the results will be used for the final synthesis of the prototype. The protocol interoperability between sensors to be used for the hardware proposed and applications to convert the data captured in real-time and off-real-time application will be studied to enable data transfer.

3.2 Development of vehicle subsystem

The vehicle subsystem will be developed as a device capable of imaging-tracking-tracing and time synchronization with and without GNSS Services. The output of the device is real-time and off-real-time or recorded data. The imaging capability of the device all be able to capture the environment in video mode and in still picture with time and positioning synchronization to enable tracing of event when applicable. The vehicle sub-system is a system to collect data. The system is hardware to be mounted on to the vehicle comprises the integration of GNSS receiver to receive the positioning information from the GNSS services; Video recorder to record video in the field of view of the camera along the route of the moving vehicle. The video can be record the front or rear view and even both of the moving vehicles depending on the position and number of the hardware to be mounted; Gyroscope which is a rotation sensor, continuously calculate the position of the vehicle by using a previously known position without the need for external.

3.3 Development of application subsystem

The development of the system is divided into two applications: web-based and desktop. The application sub-system is a system to process the data collected into information. The web application receives the GNSS data (position) from the vehicle sub-system in real-time and off-real-time to monitor the movement and environmental vie (still picture) captured of the vehicle using map viewer. The desktop application retrieves the recorded data (video embedded with GNSS positional data or still picture) in the vehicle sub-system to further process and analysis for various view-track-trace related off-real-time applications.

3.4 Integration of subsystem

The completed vehicle and applications subsystems will be integrated as a prototype. Verification of its performance will be assessed upon its accuracy for imaging, positioning and time synchronization for real-time data capturing, recorded data capturing and tracking with and without GNSS services.

4. Expected Outcome

The result will be achieved in the implementation of this project are as below:

a) This prototype able to help track vehicle either real time or off real time.

b) This prototype also integrate with semi HD camera which is able to record incident either in video mode or still picture.

c) In the absence of satellite services, this prototype also developed with Gyro sensor which is able to continue calculation the position of movement from the last position stored by GNSS services.

d) Act as a mobile CCTV, it can be used as an eye-witness of an accident for itself and other vehicle.

e) Developed with system to view the movement of vehicle real time and also playback the recorded video in offline mode.

Conclusion

The project fulfils the first objective by successfully developing an intelligent space based integrated video system for monitoring real time and archived movement of vehicle using the GNSS services at prototype level. To fulfil the second objective of recording incident with visual and tracking capability even in the absence of GNSS services, the system exploits the data measured and recorded by the Gyroscope sensor by continuing the position of movement from the last known position stored by GNSS services.

References

[1] Al-Kherher, M. A. ((2012)). "Hybrid GPS-GSM localization of automobile tracking system. " arXiv preprint arXiv:1201.2630.

[2] Alharaki, O. O., F. S. Alaieri, et al. (2010). "The integration of gps navigator device with vehicles tracking system for rental cars firms." International Journal of Computer Science and Information Security **8**(6): 47-51.

[3] Chadil, Chadil, N., A. Russameesawang, et al. (2008). Real-time tracking management system using GPS, GPRS and Google earth. Electrical Engineering/Electronics, Computer, Telecommunications and Information Technology, 2008. ECTI-CON 2008. 5th International Conference on, IEEE.

[4] Forssell, U., P. Hall, et al. (2002). "Map-aided positioning system." Proceedings, International Federation of Automotive Engineering Societies.

[5] Kodavati, B., V. Raju, et al. (2011). "GSM and GPS based vehicle location and tracking system." International Journal of Engineering Research and Applications (IJERA) ISSN:

2248-9622.
[6] Mansell, J. P. and W. M. Riley (1993). Vehicle tracking and security system, Google Patents.

Variable methods to estimate the ionospheric horizontal gradient

Karthigesu Nagarajoo
Department of Electronics Communications, Faculty of Electrical and Electronics Engineering, UTHM, Batu Pahat, Johor.
Email: karthi@uthm.edu.my

Abstract. DGPS or differential Global Positioning System is a system where the range error at a reference station (after eliminating the error due to its' clock, hardware delay and multipath) will be eliminated from the range measurement at the user, which view the same satellite, presuming that the satellites path to both the reference station and the user experience common errors due to the ionosphere, clock errors etc. In this assumption, the error due to the ionospheric refraction is assumed to be the same for the two closely spaced paths (such as a baseline length between reference station and the user of 10km as used in simulations throughout this paper, unless otherwise stated) and thus the presence of ionospheric horizontal gradient is ignored. If a user's path is exposed to a drastically large ionosphere gradient, the large difference of ionosphere delays between the reference station and the user can result in significant position error for the user. Several examples of extremely large ionosphere gradients that could cause the significant user errors have been observed. The ionospheric horizontal gradient could be obtained instead from the gradient of the Total Electron Content, TEC observed from a number of received GPS satellites at one or more reference stations or based on empirical models updated with real time data. To investigate the former, in this work, the dual frequency method has been used to obtain both South-North and East-West gradients by using four different receiving stations separated in those directions. In addition, observation data from Navy Ionospheric Monitoring System (NIMS) receivers and the TEC contour map from Rutherford Appleton Laboratory (RAL) UK have also been used in order to define the magnitude and direction of the gradient.

1. Introduction

DGPS is a system where the range error at a reference station (after eliminating the error due to its' clock, hardware delay and multipath) will be eliminated from the range measurement at the user, which view the same satellite, presuming that the satellites path to both the reference station and the user experience common errors due to the ionosphere, clock errors etc. In this assumption, the error due to the ionospheric refraction is assumed to be the same for the two closely spaced paths and thus the presence of ionospheric horizontal gradient is ignored. If a user's path is exposed to a drastically large ionosphere gradient, the large difference of ionosphere delays between the reference station and the user can result in significant position error for the user. Several examples of extremely large ionosphere gradients that could cause the significant user errors have been observed. As an example, Datta-Barua *et al.* [1] had found an ionosphere gradient as large as 316mm/km which is more than 60 times the one-sigma nominal gradient (one-sigma nominal gradient is at 3~5mm/km) [2-3].

The ionospheric horizontal gradient could be obtained instead from the gradient of the TEC observed from a number of received GPS satellites at one or more reference stations or based on empirical models updated with real time data. To investigate the former, as an example, the dual frequency method has been used to obtain both South-North and East-West gradients by using four different receiving stations separated in those directions. In addition, observation data from NIMS receivers and the TEC contour map from RAL UK have also been used in order to define the magnitude and direction of the gradient.

2. Determining the magnitude of the gradient from GPS data

There are a number of ways in which an approximate value for the gradient could be determined to input into the empirical formula using either a *3D* ionosphere model such as IRI or real-time measurements or a model updated with real-time data.

2.1 Gradient from four GPS observation stations

The dual frequency model has been used to obtain the slant TEC (*sTEC*) from four different receiving stations separated in both South-North and West-East directions. Then, by using appropriate simultaneous equations, both the South-North and West-East gradients in units δTEC/°*lat* and *δTEC/°longi* respectively can be obtained. The latitudinal gradient (g_1) is considered positive in the South-North direction and the longitudinal gradient (g_2) is considered positive in the West-East direction.

Figure 1 below shows the four GPS stations that have been used in this gradient determination method. Those stations are being operated by the UK Ordnance Survey. They are Church Lawford Met at (52.36°N, 1.33°W) as station **A**, Leeds at (53.8°N, 1.66°W) as station **B**, Lincoln (53.25°N, 0.52°W) as station **C** and Nottingham (52.96°N, 1.2°W) as station **D.**

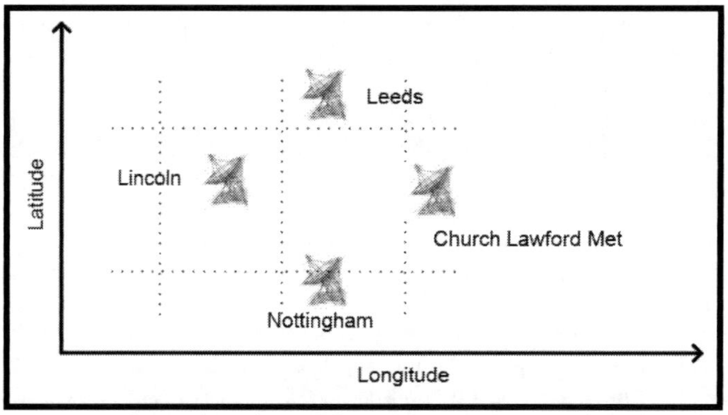

Figure 1. Approximate location of four GPS receiver stations in UK used to determine the magnitude and direction of the ionospheric horizontal gradient.

The simultaneous equations that were used to obtain g_1 and g_2 are as follows;

$$TEC_{sB} = TEC_{sA} + g_1(x_B - x_A) + g_2(y_B - y_A) \quad (1)$$

$$TEC_{sC} = TEC_{sA} + g_1(x_C - x_A) + g_2(y_C - y_A) \quad (2)$$

$$TEC_{sD} = TEC_{sA} + g_1(x_D - x_A) + g_2(y_D - y_A) \quad (3)$$

A, **B**, **C** and **D** stand for the respective reference stations. x and y represent the latitude and longitude respectively and g_1 and g_2 represent the gradient of TEC in latitude and longitude respectively.

To obtain *sTEC* from every station for a period of time, the dual frequency model was used;

$$TEC_{slant} = \frac{1}{40.3}\left(\frac{f_1^2 \bullet f_2^2}{f_1^2 - f_2^2}\right)(P_2 - P_1) \quad (4)$$

where f_1 is 1557.42MHz (L_1), f_2 is1227.6MHz (L_2), P_1 and P_2 are GPS pseudorange measurements on L_1 and L_2 respectively.

2.2 Gradient from NIMS observation data

Another method that have been tried to obtain the magnitude of the gradient was to use the NIMS observation data. Since NIMS satellites are polar orbiting satellites, their observation data should be able to give the magnitude of the horizontal gradient in the latitudinal direction.

The Navy Ionospheric Monitoring System, briefly known as NIMS satellites (formerly known as NNSS), are polar orbiting satellites at an altitude of 1100km above the Earth [4]. It is a constellation of 6 satellites. These navigation satellites were used by the US Navy before GPS became available. NIMS satellites transmit in the form of dual frequencies at 150 and 400 MHz. The *sTEC* along each path of the ray was obtained by a chain of ground-based NIMS receiver. Since the satellites are in low-Earth polar orbit (LEO), they move latitudinally across the sky to provide the scan of the ionosphere. Although NIMS could measure only the ionisation below the altitude of the satellite's orbit of about 1100km, compare with GPS which can measure up to 20200km altitude, at least during the day and for solar maximum, the TEC derived from NIMS well represents the total TEC to higher altitude.

Pryse *et al.* [5] obtained the TEC data from 3 NIMS Earth receiver stations in UK. These stations were Aberystwyth (52.42°N, 4.06°W), Hawick (55.42°N, 2.80°W) and Reay (58.57°N, 3.76°W). To obtain the *sTEC*, they processed the observations from individual satellites that pass over these stations by using standard, well documented techniques to produce the images of electron density in the height versus latitude plane. Then, vertical integration over the electron density was done to obtain the vertical TEC (*vTEC*), as a function of latitude for each satellite pass. A set of parameters was defined in the plot of *vTEC* versus latitude in order to represent the position and shape of electron density troughs. The parameters were chosen so that both the location of the trough and the form of the latitudinal gradients could be characterized. A schematic diagram showing these parameters is shown in figure 2.

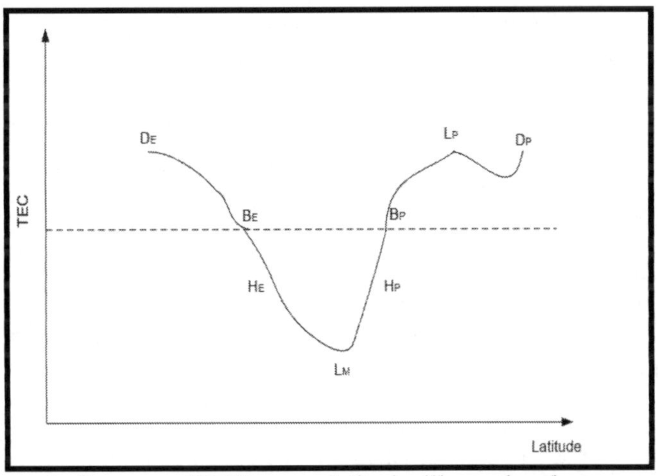

Figure 2. Schematic diagram of TEC versus latitude, illustrating the parameters defined to characterize the location and shape of the trough [5].

D_E and D_P are the limits of the usable data range, where E and P refer to the equatoward and poleward extremes respectively. The average *vTEC* was calculated for the entire range of the data between the extreme limits and was used to define the locations of two 'breakpoints' on the equatorward, B_E and poleward, B_P sides of the trough minimum. H_E and H_P are the halfway points between the minimum and equivalent breakpoint. Data on the poleward maximum, L_P, were used to define the poleward gradient of the trough wall.

The *vTEC* in between all these points have been compared with the *vTEC* from IRI at the same time, day and location. The comparison was done in order to see if there is any correlation in the value of *vTEC* from NIMS and IRI. If there is correlation, it will be helpful to obtain the magnitude of the ionospheric gradient and eventually will be helpful to improve the accuracy of location in DGPS. The work than furthered to define the magnitude of the gradient in DGPS by using the tomographic images of the ionosphere from RAL UK's TEC map.

2.3 Gradient from RAL UK's TEC Map

The Rutherford Appleton Laboratory, RAL UK provides the real-time *vTEC* profile (ionospheric tomographic images or contour of TEC) for the whole of Europe. The contour maps are updated every 10 minutes in order to see if there is any change in the structure of the TEC [6]. The contour was obtained by using the daily GPS Observation data provided by the International GPS Service for Geodynamics (*IGS*). The *vTEC* was computed by using the method developed by [7]. The ionospheric pierce point (*IPP*) was fixed at 400km. However, for this work, the TEC contour was taken only for the region of the UK.

Tomography is a technique that has been developed from a medical diagnostics tool to become an imaging technique for many applications. Tomographic imaging of the ionised atmosphere is one of the most recent advances in the use of TEC measurements. By using this method, the measured *sTEC* from a number of satellites can be inverted in a reconstruction algorithm to produce the two-dimensional spatial distribution of the electron concentration throughout the region of the ray-path intersections at a fixed IPP of 400km.

Several authors have reviewed the application of computerised reconstruction of radio tomographic imaging of the ionosphere. Leitinger [8] has explained the basis and the limitations of the ionospheric application. Kunitsyn *et al.* [9] have also derived a number of theoretical formulations regarding ionospheric tomography. Pryse [10] has analyzed the experimental results from the application of radio tomography to ionospheric imaging. All the development work carried out has demonstrated that radio tomography is really reliable in providing images of large scale spatial structures in a wide area of the ionosphere from a limited number of ground stations. Yizengaw *et al.* [11] has also expressed the view that tomographic imaging technique has the potential to provide a more accurate description of the average electron density of the ionosphere. Structures such as troughs, traveling ionospheric disturbances (*TIDs*), polar cap patches and horizontal gradients are some of the ionospheric features suitable for study by this technique.

3. Results and analysis

3.1 GPS observation stations

By using the simultaneous Equations 1, 2 and 3 and the *sTEC* values from Equation 4, the value of g_1 and g_2 were obtained for the same period of time. The *sTEC* can be seen in figure 3, whereas the g_1 and g_2 can be seen in figure 4.

It can be seen that the gradients g_1 and g_2 change quite rapidly (~every minute) and some temporal averaging seems to be needed. Though the temporal averaging can be done, the drastic variation of g_1 and g_2 does not seems to be helpful to give better gradients (g_1 and g_2) solution. Without a better g_1 and g_2, the idea of improving the position of a user in the vicinity of the four stations (figure 1) by inputting the effect of latitudinal and longitudinal gradient (g_1 and g_2 respectively), cannot be done.

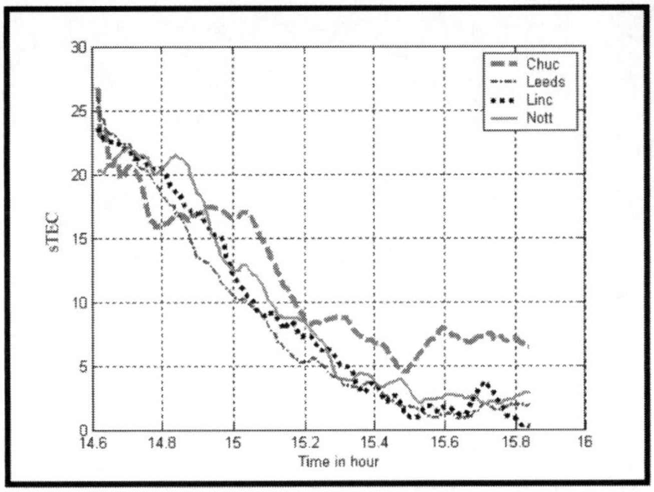

Figure 3. The *sTEC* at four different stations for a 75 minute daytime period.

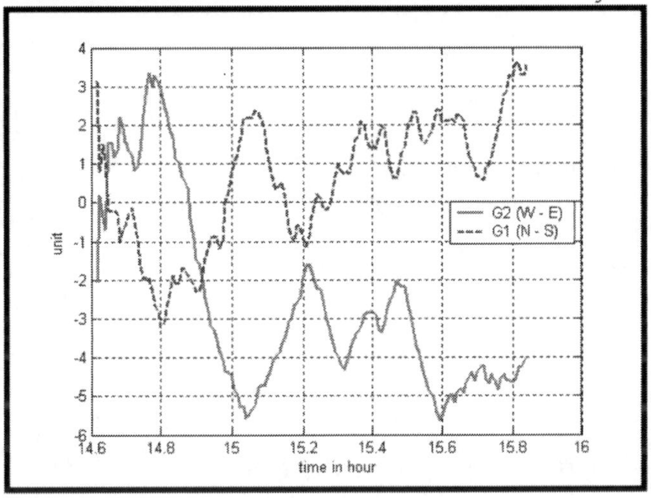

Figure 4. The value of g_1 and g_2 for the same duration of time as for figure 2.

3.2 NIMS observation stations

In order to determine this correlation, the difference between *vTEC* from NIMS and IRI was obtained on a few days in the months of September 2002, January 2003 and June 2003. Some of the findings are shown in figures 5, 6, 7 and 8 below.

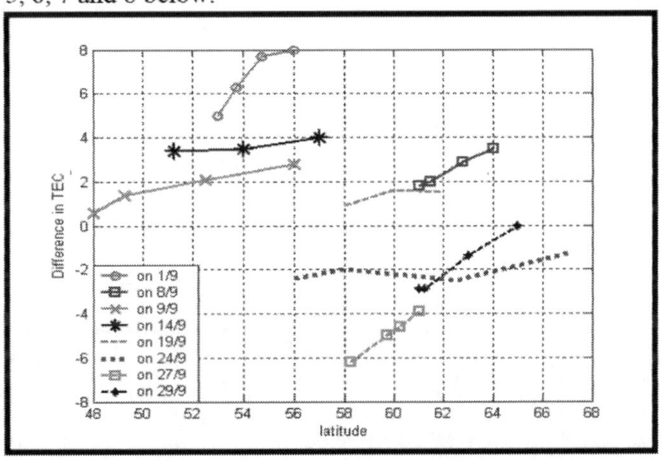

Figure 5. Difference in *vTEC* for NIMS and IRI at early morning in September, 2002.

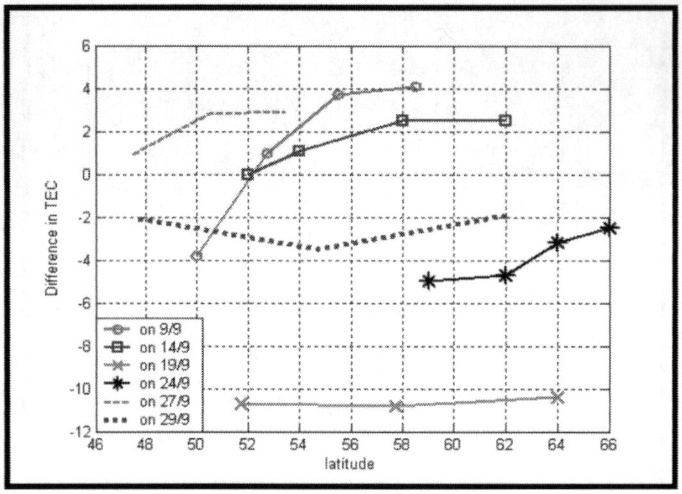

Figure 6. Difference in *vTEC* for NIMS and IRI at noon in September, 2002.

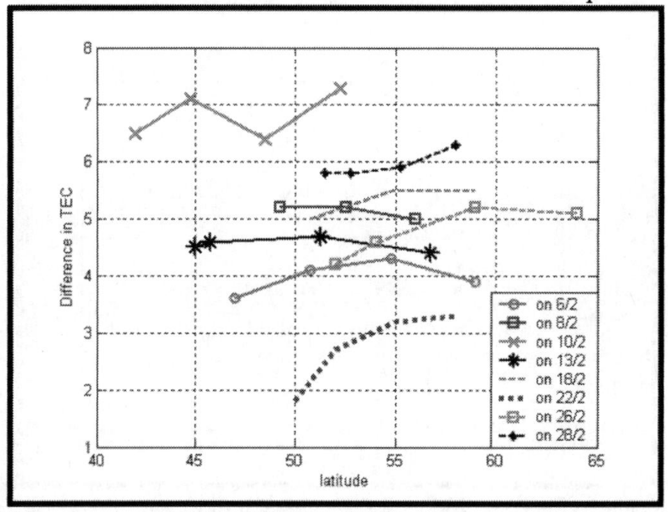

Figure 7. Difference in *vTEC* for NIMS and IRI at early morning in February, 2002.

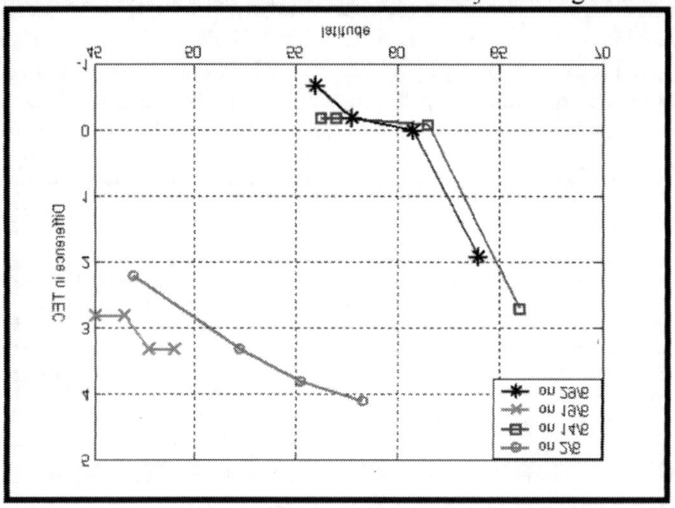

Figure 8. Difference in *vTEC* for NIMS and IRI at early morning in June, 2002.

From all those figures above, it can be seen that there is no correlation at all between *vTEC* from NIMS and IRI. Other than that, the *vTEC* was also not available from NIMS for each and every day

IOP Conf. Series: Earth and Environmental Science **37** (2016) 012012 doi:10.1088/1755-1315/37/1/012012

for the same time and same location (same range of latitude). This is the reason the plot of difference between *vTEC* from NIMS and IRI on different days varies at different locations of latitude. Further, NIMS *vTEC* data could only show the variation of TEC in latitude and not in longitude.

To further the analysis, comparison was also done between the variation of *vTEC* derived from NIMS and IRI with the K_p index. The K_p index is derived from the standardized K index from 13 magnetic observatories. It was found out that the correlation between the TEC from real observation data and IRI is good for higher *K* indices [12-13].

K_p indices were obtained from NGDC. Figures 9(a), 9(b) and 9(c) below show the K_p indices on the days the comparison was done between the *vTEC* from NIMS and IRI.

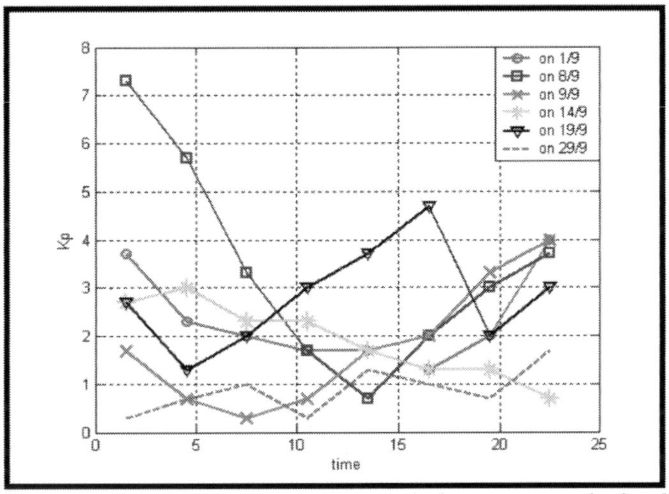

Figure 9(a). K_p indices at every 3 hours for a 24 hour period on some particular days of the month in September 2002.

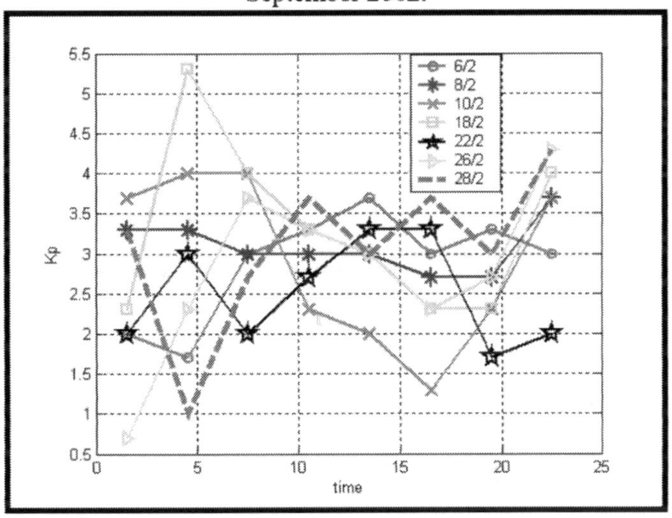

Figure 9(b). K_p indices at every 3 hours for a 24 hour period on some particular days of the month in February 2003.

Analysis has been carried out to see if the K_p index does give any indication of the variation between the *vTEC* from NIMS and IRI. As an example, from figure 5 above, the variation of the *vTEC* from NIMS and IRI from 6 to 9*am* on 14 September 2002 is increasing as latitude increases. When we see the K_p index from Figure 9(a) for that duration of time on the same day, the index decreases. It shows that the variation of *vTEC* between NIMS and IRI does not correlate with the variation of the K_p index. The same goes for the K_p indices for other NIMS and IRI differences. However, another analysis has been done to investigate the variation of K_p indices itself for 24 hours a day for seven consecutive days.

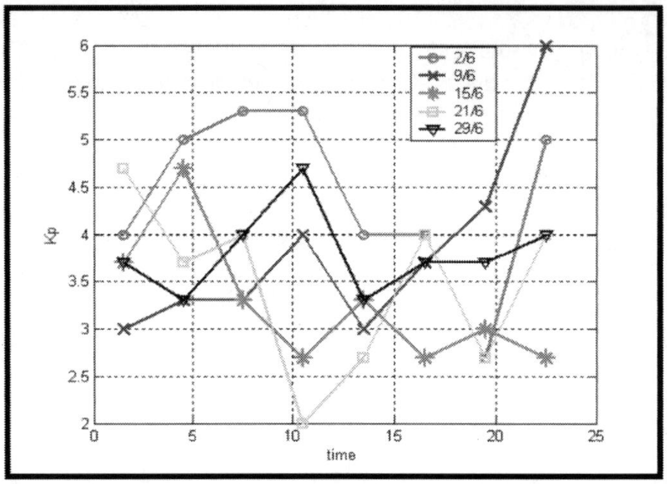

Figure 9(c). K_p index indices at every 3 hours for a 24 hour period on some particular days of the month in June 2003.

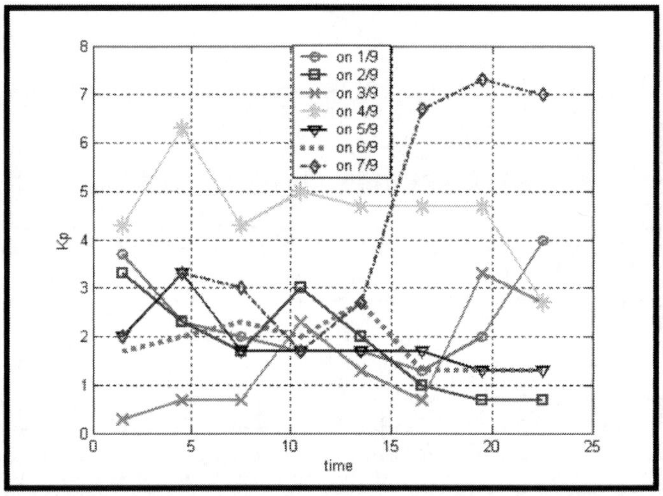

Figure 10(a). K_p indices for seven consecutive days on September 2002.

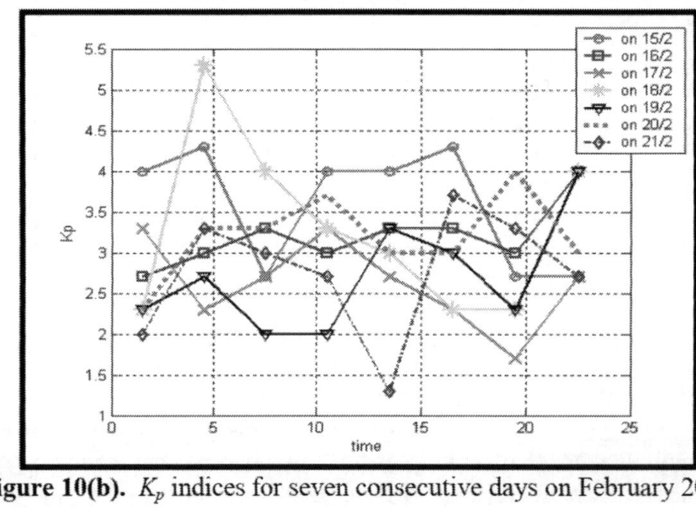

Figure 10(b). K_p indices for seven consecutive days on February 2003.

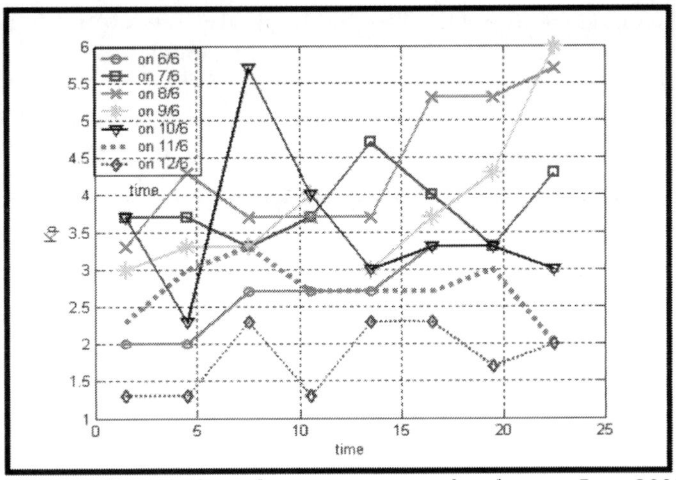

Figure 10(c). K_p indices for seven consecutive days on June 2003.

As was mentioned above, the plots of the K_p indices for seven consecutive days have been done in order to see the coherence between the indices from day 1 and 2, day 2 and 3 and so on and so forth. Unfortunately, from all those figures above, 10(a), 10(b) and 10(c), the correlation of the K_p index on consecutive days is very poor. This poor correlation of the K_p index cannot be used to explain the variation of *vTEC* between NIMS and IRI.

3.3 RAL's UK TEC

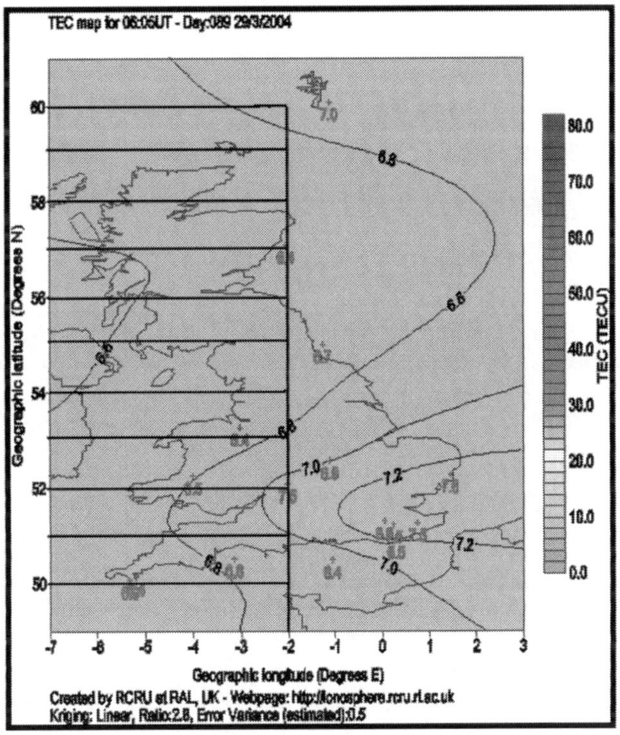

Figure 11. Contours of the TEC map on 29 March 2004 at 06:05UT (RAL, UK).

IOP Conf. Series: Earth and Environmental Science **37** (2016) 012012 doi:10.1088/1755-1315/37/1/012012

Figure 11 shows the contour map of TEC from RAL UK that was obtained on 29 March 2004 at 06:05UT. In order to know the *vTEC* at every degree of latitude, the contours of TEC will be interpolated linearly at every 1° of latitude at the longitude of 358°E. By doing so, the *vTEC* that is obtained at each degree of latitude can be compared with the *vTEC* from IRI at 400km. Contours of TEC were taken at dawn because the correlation of this *vTEC* with IRI is normally good at this time.

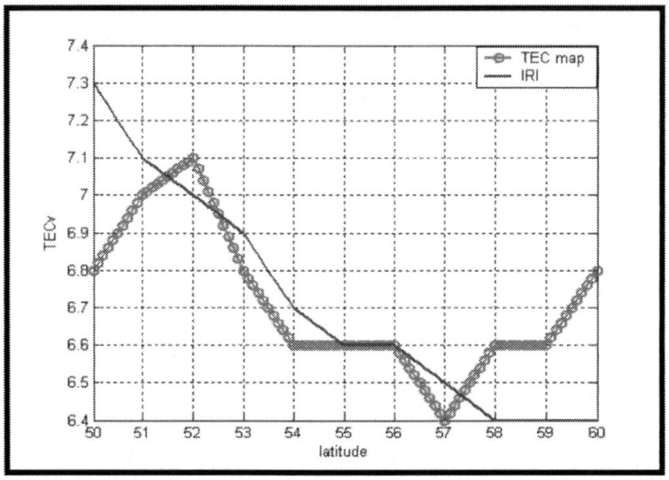

Figure 12. *vTEC* from RAL contour map (after linear interpolation) and IRI at dawn.

As expected, the *vTEC* from the contours correlates very well with the *vTEC* from IRI at dawn. This has shown that we can obtain the magnitude of gradient from IRI for this particular time and day. The numerical differentiation then was done to obtain the magnitude of the gradient (in unit $dTEC/dLat$) as shown in figure 13.

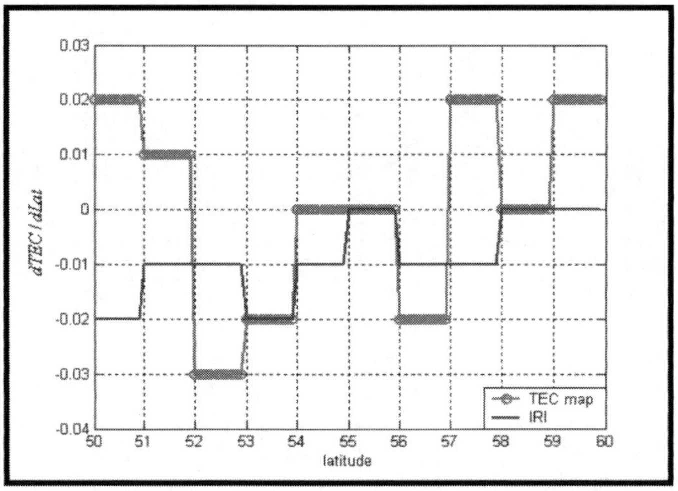

Figure 13. Gradient of TEC from RAL's contour map and IRI at dawn.

The differentiation ($dTEC/dLat$) also shows that the magnitude of the gradient does agree well between the vTEC from contour map and IRI values. Another contour map of TEC was also chosen to do the same analysis, but at dusk rather than dawn.

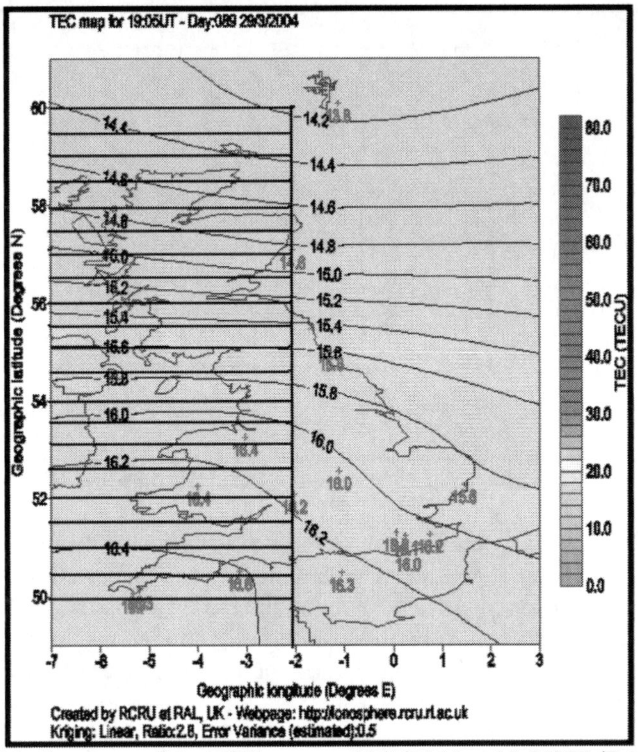

Figure 14. Contours of the TEC map on 29/3/2004 at 19:05UT (RAL, UK).

Figure 14 shows the variation of vTEC is much larger than the variation of *vTEC* at dawn (figure 13). So, in order to get the better *vTEC* observation data, the interpolation was done at every 0.5° of latitude. Using the same method as before, the *vTEC* from the contour map at every 0.5° is compared with the *vTEC* from IRI, as shown in figure 15 below.

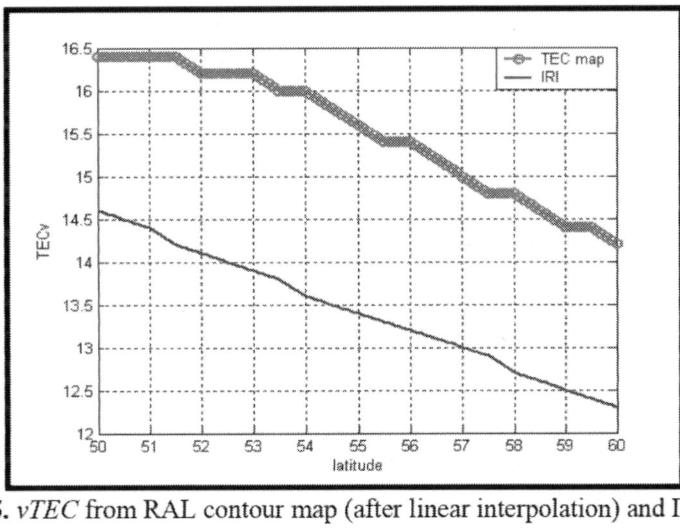

Figure 15. *vTEC* from RAL contour map (after linear interpolation) and IRI at dusk.

Again, it shows that the *vTEC* from the RAL UK's TEC map and IRI correlate very well. This will make us easier to obtain the magnitude of the gradient. The plot of $dTEC/dLat$ is shown below.

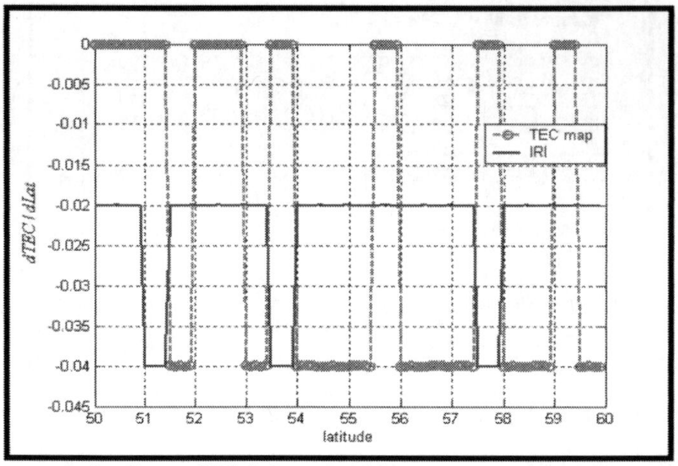

Figure 16. Gradient of TEC from RAL's contour map and IRI at dusk.

It shows that IRI really correlates well with the contour of TEC from RAL for the UK region, both at dawn and dusk. Both are giving about the same magnitude of gradient. However, figures 13 and 16 show that the magnitude of the gradient appears to be the same for a range of 1° of latitude, which is about 111km in baseline length. This is because, in region like UK, which lies in geographical mid-latitude of Northern Hemisphere, the horizontal variation of electron density latitudinally is generally not as large as in the equatorial region, where a significant latitudinal variation of electron density can occur for a short a distance as 10km of baseline length. Nevertheless, since the magnitude of the gradient can be obtained from RAL UK's TEC map, this can be used to show the improvement in the final position of a user station in DGPS.

4. Conclusion

In order to make a correction in practice, the horizontal electron density gradient needs to be known. It can be estimated from the gradient in *sTEC* observed from the received GPS satellites (since it has also been shown that the ratio of *sTEC* gives the ratio of electron density at the IPP) or based on an empirical model updated with real-time data, or both. As described above, a number of methods have been used to obtain the magnitude and direction of the gradient from real data. However, all of these methods have their own advantageous and disadvantages.

References

[1] Datta-Barua S, Walter T, Pullen S, Luo M, Blanch J and Enge P January 2002 Using WAAS ionospheric data to estimate LAAS short baseline gradients *Proceedings of the ION 2002 NTM, Anaheim, CA., pp. 523-530.*

[2] McGraw G A, Brenner M, Pullen S and Van Dierendonck A J September 2000 Development of the LAAS accuracy models *Proceedings of ION GPS 2000, Salt Lake City, UT., pp. 1-12.*

[3] Pullen S December, 14 2000 *Summary of ionosphere impact on PT 1 LAAS: Performance and mitigation options* (Stanford University, Dept. of Aeronautics)

[4] Materassi M, Mitchell C N and Spencer P S J 2003 Ionospheric imaging of the northern crest of the equatorial anomaly *Journal of Atmospheric and Solar-Terrestrial Physics, vol. 65, pp. 1393-1400.*

[5] Pryse S E, Kersley L, Malan D and Bishop G J 2006 Parameterization of the main ionospheric trough in the European sector *Radio Science, vol. 41.*

[6] Cander Lj R and Ciraolo L 2002 First step towards specification of plasmaspheric-ionospheric conditions over Europe on-line *Acta Geod. Geophys. Hung., vol. 37, pp. 153-161.*

[7] Ciraolo L and Spalla P 1997 Comparison of ionospheric total electron content from the Navy Navigation Satellite System and the GPS *Radio Science, vol. 32, no. 3, pp. 1071-1080.*

[8] Leitinger R 1996 *Tomography, in Modern Ionospheric Science – 50 Years of Ionospheric Research in Lindau* edited by Kohl H, Ruster R and Schlegel K (EGS, Katlenburg-Lindau, ProduServ GmbH Verlagsservice, Berlin), pp. 346-370.

[9] Kunitsyn V E and Tereshchenko E D 2003 Ionospheric Tomography *Springler-Verlag*.

[10] Pryse S E 2003 Radio tomography, a new experimental technique *Survey of Geophysics, vol. 24, pp. 1-38*.

[11] Yizengaw E, Moldwin M B, Dyson P L and Essex E A 2006 Using tomography of GPS TEC to routinely determine ionospheric average electron density profiles *Journal of Atmospheric and Solar-Terrestrial Physics, doi: 10.1016/j.jastp.2006.07.023*

[12] Jakowski N, Leitinger R and Ciraolo L 2004 Behaviour of large scale structures of the electron content as a key parameter for range errors in GNSS applications *Annals of Geophysics, Supplement to vol. 47, no. 2/3, pp. 1031-1047*.

[13] Mosert M, Gende M, Brunini C, Ezquer R and Altadill D 2007 Comparisons of IRI TEC predictions with GPS and digisonde measurements at Ebro *Journal of Advances in Space Research, vol. 39, pp. 841-847*.

[14] Leitinger R 1999 *Ionospheric tomography, in Review of Radio Science, 1996-1999* edited by Stone W.R. (Oxford University Press, Oxford, U.K.), 1999.

Acknowledgement
The author is grateful to Universiti Tun Hussein Onn Malaysia for providing the fund to attend and present this paper at this conference.

Evaluating the Effect of Global Positioning System (GPS) Satellite Clock Error via GPS Simulation

Dinesh Sathyamoorthy*, Shalini Shafii, Zainal Fitry M Amin, Asmariah Jusoh and Siti Zainun Ali
Science & Technology Research Institute for Defence (STRIDE), Ministry of Defence, Malaysia

*Email: dinesh.sathyamoorthy@stride.gov.my

Abstract. This study is aimed at evaluating the effect of Global Positioning System (GPS) satellite clock error using GPS simulation. Two conditions of tests are used; Case 1: All the GPS satellites have clock errors within the normal range of 0 to 7 ns, corresponding to pseudorange error range of 0 to 2.1 m; Case 2: One GPS satellite suffers from critical failure, resulting in clock error in the pseudorange of up to 1 km. It is found that increase of GPS satellite clock error causes increase of average positional error due to increase of pseudorange error in the GPS satellite signals, which results in increasing error in the coordinates computed by the GPS receiver. Varying average positional error patterns are observed for the each of the readings. This is due to the GPS satellite constellation being dynamic, causing varying GPS satellite geometry over location and time, resulting in GPS accuracy being location / time dependent. For Case 1, in general, the highest average positional error values are observed for readings with the highest PDOP values, while the lowest average positional error values are observed for readings with the lowest PDOP values. For Case 2, no correlation is observed between the average positional error values and PDOP, indicating that the error generated is random.

1. Introduction

Fundamental to the operation of GNSS is the one-way ranging that depends on satellite clock predictability. Even though the clocks in GNSS satellites are very accurate, they drift slightly, resulting in small errors which affect GNSS accuracy. This drift can be in the order of 9 to 18 ns per day and introduces a slow ramp type error in the transmitted signal. This error is difficult to detect because its signature resembles the typical relative motion between a GNSS satellite and receiver. The GNSS control segment continually monitors the satellite clocks and corrects any drift that is found. However, these corrections are based on observations and may not indicate the clock's current state, leaving residual error in the range of up to 7 ns [1-5]. On occasion, the satellite clocks behave unpredictably and produce errors that grow significantly before the operators can spot it and mark it as unhealthy. For example, on 1 January 2004, the clock on GPS satellite SV-23 drifted for approximately 3 h by a pseudo range error rate of 70.6 m/s before the command centre marked it unhealthy, by which time the pseudo range error had grown from 0 to 285 km [6]. A similar clock failure occurred for GPS satellite SV-22 on 28 July 2001, where its clock drifted for 90 min, leading to pseudo range error of up to 200 km [7].

This study is aimed at evaluating the effect of Global Positioning System (GPS) satellite clock error on GPS performance. It will be conducted using GPS simulation, which will allow the tests to be held with various repeatable conditions, as defined by the authors. As the tests are conducted in controlled laboratory environments, they will not be inhibited by unintended signal interferences and obstructions [8-10]. In previous studies, GPS simulation was used to evaluate the vulnerabilities of GPS to radio frequency interference (RFI) [11, 12] and multipath [13, 14].

2. Methodology

The apparatus used in the study are an Aeroflex GPSG-1000 GPS simulator [15], a notebook running GPS Diagnostics v1.05 [16] and a Garmin GPSmap 60CSx handheld GPS receiver [17]. The GPS receiver employs the GPS L1 coarse acquisition (C/A) signal, which is an unencrypted civilian GPS signal widely used by various GPS receivers. The signal has a fundamental frequency of 1,575.42 MHz and a code structure which modulates the signal over a 2 MHz bandwidth [2, 18, 19]. The study is conducted in STRIDE's mini-anechoic chamber [20] to avoid external interference signals and unintended multipath errors. The test setup employed is as shown in Fig. 1. Simulated GPS signals are generated using the GPS simulator and transmitted via the coupler. The following assumptions are made for the tests conducted:

- No ionospheric or troposheric delays
- Zero unintended clock and ephemeris error
- No obstructions or multipath
- No interference signals.

The tests are conducted for coordinated universal time (UTC) times of 0000, 0300, 0600 and 0900 for the following coordinates:

- N 2° 58', E 101° 48', 0 m (Kajang, Selangor, Malaysia)
- N 39° 45', W 105° 00', 0 m (Denver, Colorado, USA)
- S 16° 55', E 145° 46', 0 m (Cairns, Queensland, Australia)
- S 51° 37', W 69° 12', 0 m (Rio Gallegos, Argentina).

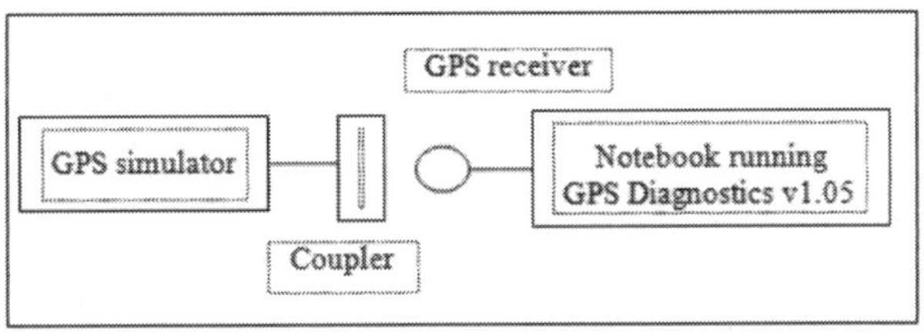

Figure 1. The test setup employed.

The almanac data for the periods is downloaded from the US Coast Guard's web site [21] and imported into the GPS simulator. The GPS signal power level is set at -130 dBm.

Trimble Planning [22] is used to estimate GPS satellite coverage at the test areas for the period of the study in terms of position dilution of precision (PDOP) (Table 1), which represents the effect of GPS satellite geometry on 3D positioning precision. A PDOP value of 1 is associated with an ideal arrangement of the satellite constellation. To ensure high-precision GPS positioning, a PDOP value of 5 or less is usually recommended. In practice, the actual PDOP value is usually much less than 5, with a typical average value in the neighbourhood of 2 [2, 18, 23].

The GPS simulator does not provide specific GPS satellite clock error simulation. However, it does allow for selection of pseudorange errors for the GPS satellites. For this study, GPS satellite clock error is simulated using the pseudorange error function, with 1 ns of clock error representing a pseudorange error of 0.3 m [1-5]. Two conditions of tests are used:

Table 1. PDOP of GPS coverage at the test areas for the period of the tests.

Location	Time	PDOP
Kajang	0000	1.33
	0300	1.46
	0600	1.24
	0900	1.72
Denver	0000	1.39
	0300	2.22
	0600	1.26
	0900	1.43
Cairns	0000	1.52
	0300	1.67
	0600	1.44
	0900	1.60
Rio Gallegos	0000	1.75
	0300	1.25
	0600	1.78
	0900	1.66

- **Case 1:** All the GPS satellites have clock errors within the normal range of 0 to 7 ns, corresponding to pseudorange error range of 0 to 2.1 m.
- **Case 2:** One GPS satellite, with the highest elevation (Table 2), suffers from critical failure, resulting in clock error in the pseudorange of up to 1 km (the maximum pseudorange error provided by the GPS simulator).

For each reading, the coordinates computed by the GPS receiver are recorded for a period of 15 min, and the values of average horizontal, vertical and overall errors are calculated.

Table 2. GPS satellites (SV) with the highest elevation at the start of each test period. These satellites are used to simulate critical failure that causes large clock error.

Location	Time	SV	Elevation	Azimuth
Kajang	0000	16	65.77	-133.38
	0300	3	86.43	-3.87
	0600	1	65.12	103.50
	0900	2	64.45	-37.06
Denver	0000	1	85.27	-35.07
	0300	29	76.79	-48.33
	0600	21	79.92	97.25
	0900	14	72.36	19.84
Cairns	0000	14	87.27	-163.54
	0300	31	63.11	119.28
	0600	11	57.69	-78.83
	0900	20	82.00	37.34
Rio Gallegos	0000	26	72.89	-107.89
	0300	24	73.81	-108.52
	0600	25	79.51	-125.26
	0900	21	63.00	-19.45

IOP Conf. Series: Earth and Environmental Science **37** (2016) 012013 doi:10.1088/1755-1315/37/1/012013

3 Results & Discussion

As observed in Fig. 2-9, increase of GPS satellite clock error causes increase of average positional error. This is due to increase of pseudorange error in the GPS satellite signals, which results in increasing error in the coordinates computed by the GPS receiver. It is observed that the maximum overall error caused by satellite clock error is in the range of 1.42 to 2.40 m for Case 1, and 929.36 to 1,393.98 m for Case 2. For Case 2, the overall errors caused are constrained by the limitation of the pseudorange error function (1 km) provided by the GPS simulator. In comparison, the critical failures suffered by GPS satellites SV-22 in 2001 and SV-23 in 2004 caused pseudorange errors of up to 200 and 285 km respectively [6, 7].

It is observed that for Case 1, for all the readings, the values of vertical error are larger than horizontal error, as GPS receivers can only track satellites above the horizon, resulting in the vertical solution being less precise than the horizontal solution [2, 18, 19]. For Case 2, vertical error is initially larger than horizontal error, but with increasing GPS satellite clock error, horizontal error becomes larger than vertical error. This is as the GPS satellite with simulated critical error has high elevation, causing horizontal error to increase at a higher rate as compared to vertical error. Critical failure simulation applied to GPS satellites with lower elevations would result in higher rate of vertical error increase and lower rate of horizontal error increase.

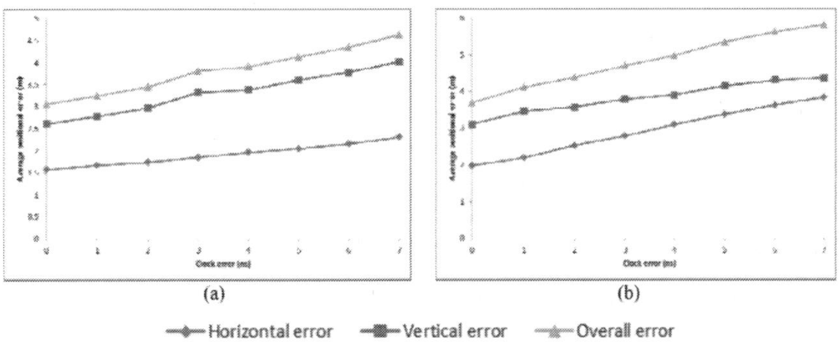

Figure 2: Recorded average positional error values for Case 1 at Kajang for periods of: (a) Lowest error: 0600 (b) Highest error: 0900.

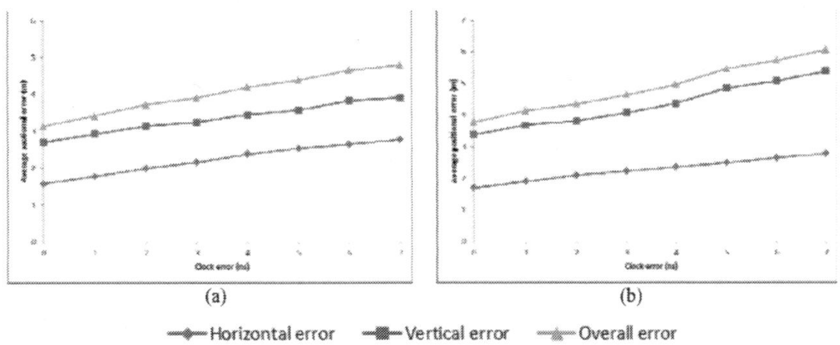

Figure 3: Recorded average positional error values for Case 1 at Denver for periods of: (a) Lowest error: 0600 (b) Highest error: 0300.

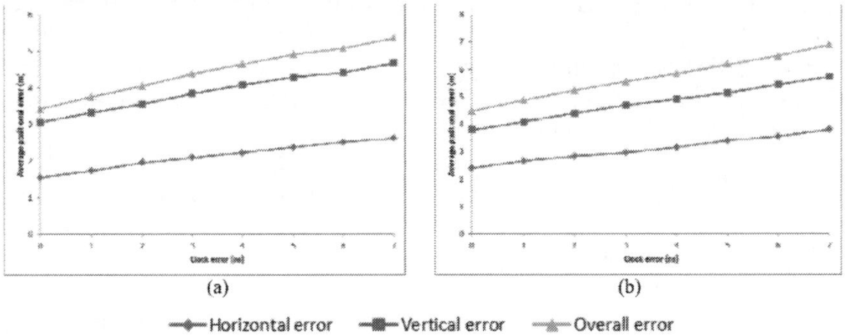

Figure 4: Recorded average positional error values for Case 1 at Cairns for periods of: (a) Lowest error: 0600 (b) Highest error: 0300.

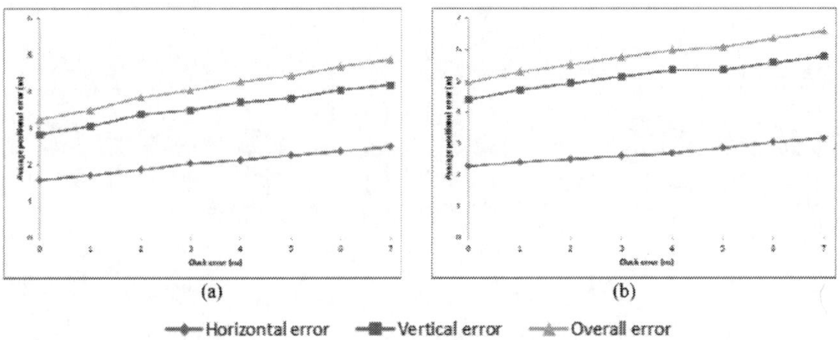

Figure 5: Recorded average positional error values for Case 1 at Rio Gallegos for periods of: (a) Lowest error: 0300 (b) Highest error: 0600.

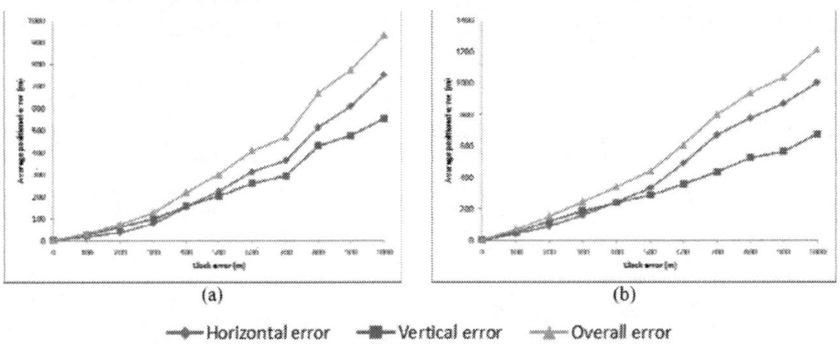

Figure 6: Recorded average positional error values for Case 2 at Kajang for periods of: (a) Lowest error: 0900 (b) Highest error: 0600.

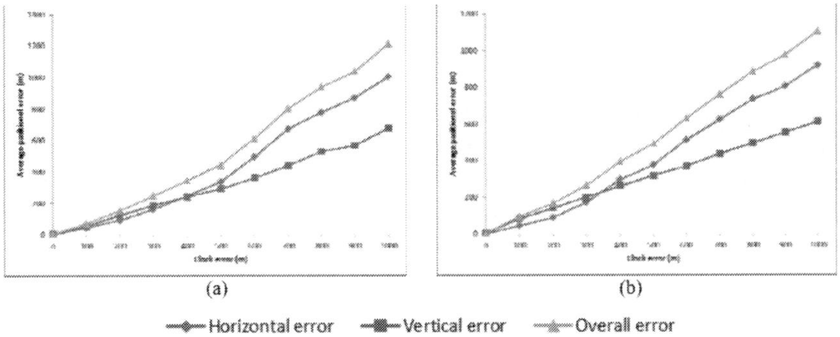

Figure 7: Recorded average positional error values for Case 2 at Denver for periods of: (a) Lowest error: 0600 (b) Highest error: 0000.

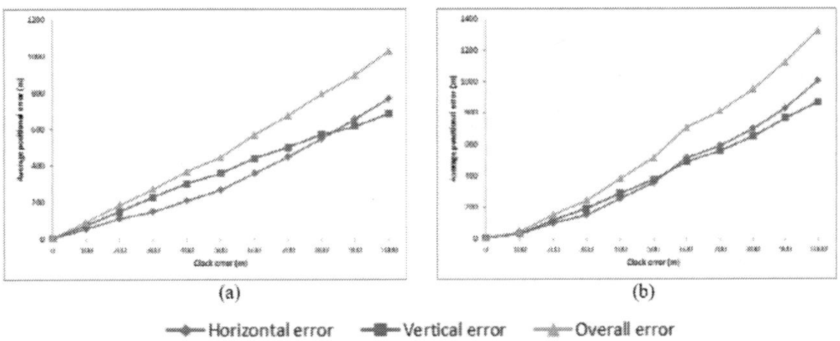

Figure 8: Recorded average positional error values for Case 2 at Cairns for periods of: (a) Lowest error: 0000 (b) Highest error: 0600.

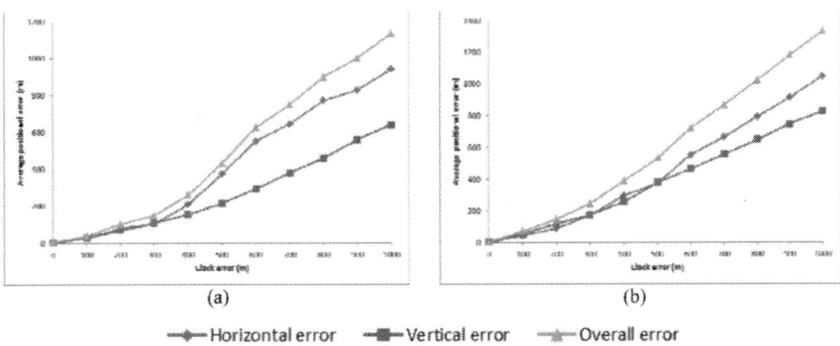

Figure 9: Recorded average positional error values for Case 1 at Rio Gallegos for periods of: (a) Lowest error: 0600 (b) Highest error: 0000.

Varying average positional error patterns are observed for the each of the readings. This is due to the GPS satellite constellation being dynamic, causing varying GPS satellite geometry over location and time, resulting in GPS accuracy being location / time dependent [2, 18, 23]. For Case 1, in general, the highest average positional error values are observed for readings with the highest PDOP values (Kajang at 0900, Denver at 0300, Cairns at 0300 and Rio Gallegos at 0600), while the lowest average positional error values are observed for readings with the lowest PDOP values (Kajang at 0600,

Denver at 0600, Cairns at 0600 and Rio Gallegos at 0300). For Case 2, no correlation is observed between the average positional error values and PDOP, indicating that the error generated is random.

The tests conducted in this study employed GPS signal power level of-130 dBm. Usage of lower GPS signal power levels would result in reduced carrier-to-noise density (C/N_0) levels, which is the ratio of received GPS signal power level to noise density. Lower C/N_0 levels would result in increased data bit error rate when extracting navigation data from GPS signals, and hence, increased carrier and code tracking loop jitter. This, in turn, results in more noisy range measurements and thus, higher rates of increase of positional error values [2,18, 19, 24]DOD, 2001; Kaplan & Hegarty, 2006; Petovello, 2009; USACE, 2011).

4. Conclusion

Based on the results of this study, it was found that increase of GPS satellite clock error caused increase of average positional error due to increase of pseudorange error in the GPS satellite signals, which resulted in increasing error in the coordinates computed by the GPS receiver. Varying average positional error patterns were observed for the each of the readings. This is due to the GPS satellite constellation being dynamic, causing varying GPS satellite geometry over location and time, resulting in GPS accuracy being location / time dependent. For Case 1 (all the GPS satellites have clock errors within the normal range of 0 to 7 ns), in general, the highest average positional error values were observed for readings with the highest PDOP values, while the lowest average positional error values were observed for readings with the lowest PDOP values. For Case 2 (one GPS satellite suffers from critical failure, resulting in clock error in the pseudorange of up to 1 km), no correlation was observed between the average positional error values and PDOP, indicating that the error generated was random.

References

[1] Olynik MC 2002 *Temporal Characteristics of GPS Error Sources and Their Impact on Relative Positioning*. Master's thesis (Alberta: The University of Calgary Calgary).
[2] Kaplan ED and Hegarty CJ 2002 *Understanding GPS: Principles and Applications.* (Massachusetts : Artech House, Norwood)
[3] Worley S, 2014 *GPS Errors & Estimating Your Receiver's Accuracy* Available online at: http://edu-observatory.org/gps/gps_accuracy.html (Last access date: 31 July 2014).
[4] Royal Academy of Engineering (RAE) 2011 *Global Navigation Space Systems: Reliance and Vulnerabilities* (London: Royal Academy of Engineering (RAE))
[5] Bidikar B, Rao GS, Ganesh L, and Kumar MS 2014 *Positioning* **5** 22-26
[6] Eastlack H 2004 *SVN-23/PRN-23 Integrity Failure of 1 January 2004.* (US Air Force, Washington, D.C.: Second Space Operations Squadron).
[7] Lavraks JW 2005 *An Overview of Civil GPS Monitoring* (Manassas, Virginia: Institute of Navigation (ION)).
[8] Aloi DN, Alsliety M and Akos DM 2011 *IEEE T. Instrum. Meas.* **56** 11-24.
[9] Kou Y and Zhang H 2011 *Inside GNSS* **6** 52-61
[10] Pozzobon O, Sarto C, Chiara AD, Pozzobon A, Gamba G, Crisci, M and Ioannides R 2013 *Inside GNSS* **8** 45-53.
[11] Dinesh, S, Mohd Faudzi, M. and Zainal Fitry 2012 MA *Defence. Sci. J.* **62** 338-347.
[12] Dinesh S, Mohd Faudzi M, Rafidah M, Nor Irza Shakhira B, Siti Robiah A, Shalini S, Aliah I, Lim BT, Zainal Fitry MA, Mohd Rizal AK and Mohd Hasrol Hisam, MY 2014 *ASM Sci. J.* **8**, 11-20.

[13] Dinesh S, Shalini S, Zainal Fitry MA and Siti Zainun A 2013 *Defence S&T Tech. Bull.* **6** 130-140.
[14] Dinesh S, Shalini S, Zainal Fitry MA, Siti Zainun A, Siti Robiah A, Mohd Idris I and Mohd Hasrol Hisam MY *Adv. Mil. Tech.* **9** 81-95.
[15] Aeroflex 2010 *Avionics GPSG-1000 GPS / Galileo Portable Positional Simulator* (Aeroflex Inc.: Plainview, New York).
[16] CNET 2004 *GPSDiag 1.0* Available online at: http://download.cnet.com/GPSDiag/3000-2130_4-4951103.html (Last access date: 31 January 2010).
[17] Garmin 2007 *GPSmap 60CSx Owner's Manual* (Olathe, Kansas: Garmin International Inc.).
[18] Department of Defence (DOD) 2001 *Global Positioning System Standard Positioning Service Performance Standard, Command, Control, Communications, and Intelligence* (Washington D.C.: Department of Defence (DOD))
[19] US Army Corps of Engineers (USACE) 2011 *Engineer Manual EM 1110-1-1003: NAVSTAR Global Positioning System Surveying* (Washington D.C., US Army Corps of Engineers (USACE)).
[20] Kamarulzaman M 2010 *Technical Specification for STRIDE's Mini-Anechoic Chamber* (Malaysia: Science & Technology Research Institute for Defence (STRIDE), Ministry of Defence).
[21] US Coast Guard (USCG) 2014 *GPS NANUs, Almanacs, & Ops Advisories* Available online at: http://www.navcen.uscg.gov/?pageName=gpsAlmanacs (Last access date: 31 July 2014).
[22] Trimble 2014 *Trimble's Planning Software* Available online at: http://www.trimble.com/planningsoftware.shtml (Last access date: 31 July 2014).
[23] Dinesh S, Wan Mustafa WH, Mohd Faudzi M, Kamarulzaman M, Hasniza H, Nor Irza Shakhira B, Siti Robiah A, Shalini S, Jamilah J, Aliah I, Lim BT, Zainal Fitry MA, Mohd Rizal AK, Azlina B and Mohd Hasrol HMY 2010 *Defence S&T Tech. Bull.* **3** 100-118.
[24] Petovello M *Inside GNSS* **4** 20-29.

Exploratory study on Marine SDI implementation in Malaysia

Zakri Tarmidi[1,3] **Abdul Rashid Mohd Shariff**[1], **Ahmad Rodzi Mahmud**[1], **Zelina Zaiton Ibrahim**[2], **Abdul Halim Hamzah**[4]

[1]Faculty of Engineering, Universiti Putra Malaysia, 43400 UPM Serdang, Malaysia

[2]Department of Environmental Management, Faculty of Environmental Studies, Universiti Putra Malaysia, 43400 UPM Serdang, Malaysia

[3]Department of Geoinformation, Faculty of Geoinformation and Real Estate, Universiti Teknologi Malaysia, 81310, Johor Bahru, Malaysia.

[4]Land Administration and Development Groups (LAnDS), Department of Real Estate,Faculty of Geoinformation and Real Estate, Universiti Teknologi Malaysia, 81310, Johor Bahru, Malaysia.

Corresponding Authors: zakritarmidi@gmail.com

Abstract. This paper discusses the explanatory study of the implementation of spatial data sharing between Malaysia's marine organisations. The survey method was selected with questionnaire as an instrument for data collection and analysis. The aim of the questionnaire was to determine the critical factors in enabling marine spatial data sharing in Malaysia, and the relationship between these indicators. A questionnaire was sent to 48 marine and coastal organisations in Malaysia, with 84.4% of respondents answering the questionnaire. The respondents selected were people who involved directly with GIS application in the organisations. The results show there are three main issues in implementing spatial data sharing; (1) GIS planning and implementation in the organisation, (2) spatial data sharing knowledge and implementation in the organisation and (3) collaboration to enable spatial data sharing within and between organisations. To improve GIS implementation, spatial data sharing implementation and collaboration in enabling spatial data sharing, a conceptual collaboration model was proposed with components of marine GIS strategic planning, spatial data sharing strategies and collaboration strategy.

1. Introduction

A shoreline and marine area contain marine resources and producing economic growth in the country, especially in Malaysia [1, 2]. Malaysia's marine area consist of five areas (Figure 1); Strait of Malacca, Gulf of Thailand, South China Sea, Sulu Sea and Sulawesi Sea. Malaysia's marine area is approximately 574,000km square kilometres, and its management and responsibilities are divided between local, state and federal government [3].

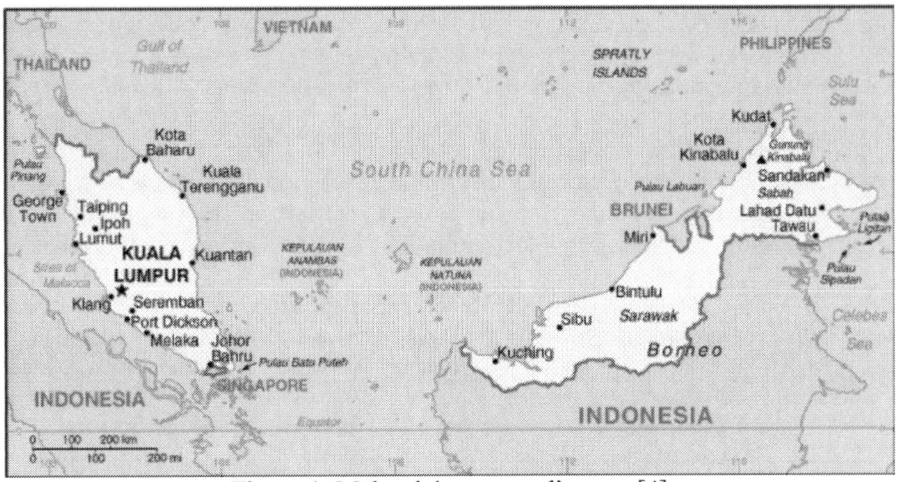

Figure 1: Malaysia's surrounding sea [4].

In managing marine activities, it is important to have good governance with assistant from marine Spatial Information System (GIS), to ensure sustainable planning and development of marine area [2, 3, 5]. Current practice shows lack of vital spatial information, spatial data not covering and concerning coastal dynamic information and the biological diversity is not completed [3]. Besides that, the developments of GIS for marine and coastal areas were autonomous, heterogeneous and distributed between organisations, and need for spatial data sharing between marine organisations for sustainable marine management [6, 7].

Marine GIS was used for spatial data viewing, textual and spatial record, planning, monitoring, forecasting, navigating, managing transportation and facilitate communication [8, 9]. To improve marine GIS development, marine spatial data infrastructure (SDI) was introduced to facilitate and coordinate the spatial data exchange and sharing between stakeholders in the marine spatial community [10-12]. Marine SDI is a framework that comprising a system of information, products and enabling technologies that are critical to sustainable development and management of coastal, marine and freshwater areas, and will improve spatial data quality and availability, reduce spatial data duplication and redundancies, and make the spatial data interoperability [13, 14]. Marine SDI consist of five components; spatial data, policies, access network and people [12]. Besides that, marine SDI involves cooperation between marine organisations, and required partnership engagement which included engagement management, information management and capacity building [15].

This paper focuses on a study on the degree of GIS implementation practice, issues and limitation and partnership engagement in enabling marine spatial data sharing in Malaysia's marine organisations.

2. Research methodology

This study used survey method, using questionnaire as an instrument for data collection and analysis. The aim of the questionnaire is to identify critical factors in implementing spatial data sharing in Malaysia's marine organisations. The objectives of the questionnaire are; (1) to identify the level of GIS implementation in the organisations, (2) to identify critical factors in implementing GIS and (3) relationship between these critical factors. The questionnaire consists of three parts; (a) information on respondent's background, (b) information on GIS implementation in the organisation and (c) the level of spatial data sharing implementation. The questions using a Likert Scale to measure the extent of agreement describe by each item. The scale ranged from 1 to 5, where 1 = strongly disagree, 2 = disagree, 3 = fair, 4 = agree, and 5 = strongly agree. The process of developing the questionnaire also included a pilot survey, where is used to modify and eliminate the number of variables. The feedbacks were analysed and few minor modifications were made, especially in the format. After the modification, the questionnaire was sent to respondents.

For reliability test, Cronbach's Alpha was conducted to measure the internal consistency of the research instrument. The generally agreed value of the lower limit for Cronbach's alpha is 0.70 [16]. The analysis was performed separately for the items of each factor, the summaries of the reliability analysis given in Table 1. All items show the results higher than 0.70 therefore it is reliable.

Table 1: Reliability test results.

Description	No of items	Alpha values	Items for deletion	Alpha if item deleted
GIS				
1. Geospatial Data	11	0.949	-	0.951
2. Technologies	7	0.958	-	0.961
3. Human Resources	11	0.922	-	0.929
SDI				
4. Data	6	0.974	-	0.975
5. Data 2	6	0.983	-	0.983
6. Data exchange	7	0.955	-	0.958
Collaboration				
7. Within organisation	4	0.994	-	0.994
8. Between organisation	7	0.934	-	0.938

3. Results

Analysis of the questionnaire was divided into three main categories; respondent's background, GIS implementation in the organisation, and implementation of spatial data sharing.

3.1. Respondent's background

Respondent's background focusing on respondent's experience in using GIS in the organisations. Table 2 shows the general background of the respondents such as user type, number of years using GIS and GIS function being used by respondents.

Table 2: Background of the respondents in using GIS.

	n	%
Background of respondents in using GIS		
Less than one year	3	7.9
One to two years	1	2.6
Two to three years	1	2.6
More than three years	33	86.8
Respondent's GIS user type		
Data User	14	36.8
Data Provider	1	2.6
Both (Data User and Provider)	23	60.5
Respondent's GIS functionality		
View information	6	15.8
Collect data	1	2.6
Analyze information	7	18.4
Integrated with other system	3	7.9
All above	21	55.3

Results from table 2 show that, the respondents were mostly using GIS for more than three years, most of the respondents were both data user and data provider, and using most of GIS functions. From these results, it can conclude that the respondents have knowledge of GIS background, and qualify to give an opinion on GIS.

3.2. GIS implementation in the organisation

GIS implementation discusses about the respondent's knowledge and experience in handling GIS, and the important aspect of GIS that need to have in the organisations. Analysis was based on three main components of GIS; data, personnel and software, hardware and network. Figure 2 shows the mean for GIS personnel understanding of the importance of spatial data components. Figure 3 shows the mean of organisations or people factors in succeeding spatial information system implementation in the organisations.

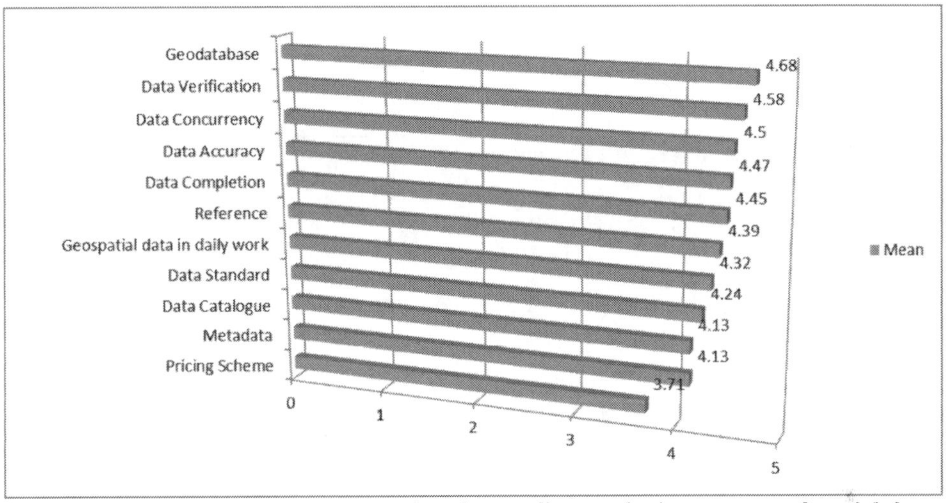

Figure 2: View of GIS personnel understanding on the importance of spatial data component.

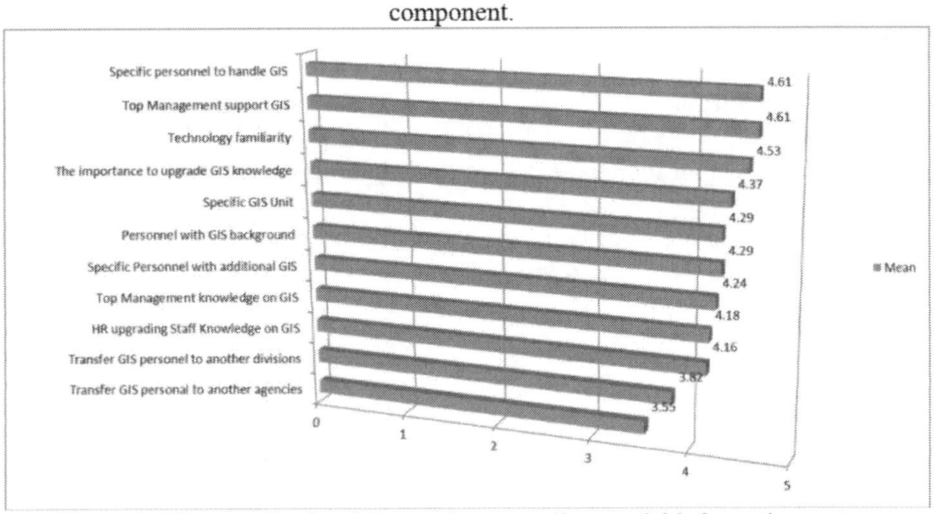

Figure 3: The organisational factors in succeeding spatial information system implementation in the organisations.

From the results, it shows most of the respondents are aware of the GIS components. The mean of the answer were mostly above 4, which indicated the respondents agreed with the importance of each component.

3.3. Spatial data sharing implementation

Implementation of spatial data sharing discussed on the knowledge of respondents in spatial data sharing, limitation in implementing spatial data sharing correlation between knowledge of GIS and SDI in the organisational implementation of spatial data sharing. For descriptive statistical analysis, three analyses were done. The first analysis is to understand the knowledge on cooperation on spatial data exchange in the organisation. The second analysis is to understand the cooperation for GIS implementation in the organisation, and the third analysis is to understand the opinion on collaboration in enabling spatial data sharing between marine's organisations. Table 3 shows the summary of the descriptive analysis

Table 3: Summary of descriptive analysis.

Factors	**Mean Score**
The knowledge on cooperation on spatial data exchange in the organisation	
Give data to other divisions to assist other divisions	4.05
Geospatial data pricing	4.08
Get data from other unit/division to assist processing	4.08
Get spatial data from other agencies	4.24
Give spatial data with other agencies	4.26
Integrate system with other divisions	4.39
The cooperation for GIS implementation in the organisation	
Geospatial data collection	4.39
Geospatial data upgrading	4.39
Cooperation on developing GIS	4.42
Cooperation on upgrading GIS	4.42
The opinion on collaboration in enabling spatial data sharing between marine's organisations	
Frequent meeting facilitating geospatial data protection	4.05
Frequent meeting facilitating geospatial data sharing	4.16
Lead organisation	4.21
Geospatial data development committee	4.26
Geospatial data sharing committee	4.26
Formal collaboration for spatial data sharing	4.55
Collaboration to improve GIS	4.66

To analyze correlation from the questionnaire, the inferential analysis was selected were Spearman's Rho analysis to analyze the correlation from the Likert scale question and Pearson Chi-Square to analyze the correlation of nominal data. The hypothesis were constructed based on three main criteria; the knowledge and implementation of GIS in the organisation, the knowledge and implementation of spatial data sharing in the organisation, and the collaborative process in enabling spatial data sharing. To understand the correlation between the three main criteria, these hypotheses have been listed:

- There is correlation between duration using GIS in the organisation with personnel knowledge in GIS
- There is a correlation between GIS user in the organisation with the personnel level of knowledge in GIS
- There is a correlation between GIS knowledge on spatial data with GIS technologies
- There is a correlation between spatial data sharing implementation with knowledge about GIS
- There is correlation between knowledge about spatial data sharing with knowledge, spatial data sharing implementation

For the first hypothesis, Spearman's rho analysis was used to determine the relationship between duration using GIS and knowledge on GIS. There was a positive correlation, which was statistically significant, $r = -0.373$, $p < .05$. For the second hypothesis, A Pearson Chi-Square test was used to determine whether there was significant correlation between types of GIS user in the organisation with the personnel level of knowledge in GIS. There was no significant correlation between types of GIS user with a level of knowledge on GIS, $X^2 = 69.450$, $DF = 38$, $p > 0.05$. The third hypothesis using Spearman's Rho analysis, to determine the relationship between respondent's knowledge on GIS data with knowledge of GIS technologies and GIS management. There was a positive correlation, which was statistically significant between respondent's knowledge on GIS data with knowledge of GIS technologies, $r = 0.665$, $p < 0.05$. There was also a positive correlation between respondent's knowledge of GIS technologies with GIS institutional management, $r = 0.414$, $p < 0.05$.

The fourth hypothesis was used Spearman's Rho analysis to determine the relationship between respondent's knowledge on GIS with the implementation of spatial data sharing. There was a positive correlation with statistically significant, $r = 0.671$, $p < 0.05$, which indicate that to successfully implement spatial data sharing, the knowledge on GIS in important. The last hypothesis was analyzed using Pearson Chi-square analysis, to determine the correlation between respondent's knowledge of spatial data sharing with the spatial data sharing implementation, where there was also a significant correlation between the two components, $X^2 = 176.263$, $DF = 38$, $p > 0.05$.

From the inferential statistical analysis, the findings show;

- There is a correlation between personal knowledge of spatial information system component in the spatial information system implementation in the organisation.
- There is a correlation between personal knowledge of spatial information systems with the implementation of spatial data sharing in the organisation.
- There is a correlation between spatial data sharing with spatial information system implementation.
- There is a correlation between cooperation in the organisation with spatial data sharing implementation.
- There is a correlation between collaboration with other organisation with spatial data sharing implementation.

4. Discussion

From the analysis, it shows that several issues and opportunities need to be managed to improve GIS implementation and to enable spatial data sharing in marine administration and management in Malaysia. It can be summarized as;

- Spatial data need for planning for standardized spatial data collection, storage and distribution with proper metadata to simplify data sharing process

- Spatial data also need to be verified, have information on its concurrency, accuracy and level of completion.
- A proper GIS hardware, software and access network are essential to facilitate in spatial data view and analysis.
- In organisational aspect, specific GIS personnel and GIS unit were needed to handle GIS, and GIS personnel should not be transferred to another unit or organisation.
- Knowledge on GIS should be improve for GIS personnel and top management in the organisation
- Awareness on the importance of integration of GIS and other system related should be increased
- Knowledge of spatial data sharing need to improve for GIS personnel and organisation's top management
- Cooperation in the same organisation was needed in developing GIS in an organisation before upgrading GIS data and functionality
- Formal collaboration between organisations with proper lead organisation, the committee and frequent meeting are needed to enable spatial data sharing

From the findings, the results then were grouped into three main groups: issues in GIS planning and development, issues in enabling spatial data sharing, and issues in collaboration between organisations to enable spatial data sharing. To improve marine spatial data sharing in Malaysia's organisations, several strategies need to be applied, based on the three issues, marine GIS strategic planning, marine spatial data sharing strategies and collaboration strategies as shown in figure 4.

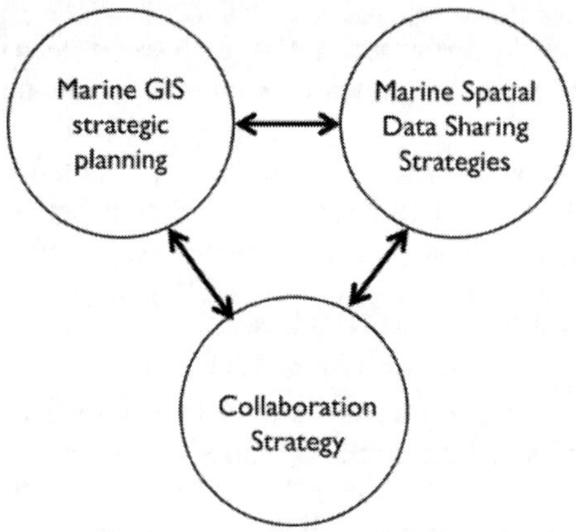

Figure 4: The key components to improve marine spatial data sharing in Malaysia's organisations.

5. Conclusion

This study has identified the critical factors in implementing marine spatial data sharing in Malaysia's environment, and also the relationship between the factors. This study is significant in improving the theory of marine spatial data sharing implementation, which should include three main components as shown in figure 4. This study was hoped to give a significant contribution to marine spatial data sharing implementation, hence improve the marine SDI development research.

6. Acknowledgment

Our utmost gratitude to the Geospatial Information Science Research Group (GISRC), Universiti Putra Malaysia (UPM), Malaysian Centre for Geospatial Data Infrastructure (MaCGDI), and the anonymous reviewers of the earlier related topic of this paper for their encouragements, comments and suggestions.

References

[1] T.-E. Chua, S. R. Bernad, and M. San, "Coastal and Ocean Governance of the Seas of East Asia: Towards an Era of New Regional Cooperation and Partnerships," presented at the Tropical Coasts, 2003.

[2] A. Rajabifard, A. Binns, and I. P. Williamson, "Administering the marine environment - The spatial dimension," *Journal of Spatial Science,* vol. 50, pp. 69-78, 2005.

[3] A. H. Saharuddin, "National ocean policy - New opportunities for Malaysian ocean development," *Marine Policy,* vol. 25, pp. 427-436, 2001.

[4] S. Navigator. (2006). *Malaysia*. Available: http://www.solarnavigator.net/geography/malaysia.htm

[5] S. Nichols, D. Monahan, and M. Sutherland, "Good Governance of Canada's Offshore and Coastal zone: Towards an Understanding of the Marine Boundary Issues," *Geomatica,* vol. 54, pp. 415-424, 2000.

[6] Z. Tarmidi, A. R. M. Shariff, A. R. Mahmud, and Z. Z. Ibrahim, "The Important of Information Integration in Marine Management: A Review," *World Applied Sciences Journal,* vol. 22, pp. 870-876, 2013.

[7] I. Masser, A. Rajabifard, and I. Williamson, "Spatially enabling governments through SDI implementation," *International Journal of Geographical Information Science,* vol. 22, pp. 5-20, 2008.

[8] L. Strain, A. Binns, A. Rajabifard, and I. Williamson, "Spatially Administering the Marine Environment," in *SSC 2005 Spatial Intelligence, Innovation and Praxis: The national biennial Conference of the Spatial Sciences Institute*, Melbourne: Spatial Sciences Institute, 2005.

[9] S. Vaez, "Building a Seamless SDI Model for Land and Marine Environments," PhD, Centre of Spatial Data Infrastructure and Land Administration, The University of Melbourne, Melbourne, 2010.

[10] A. Rajabifard, "Diffusion of Regional Spatial Data Infrastructures: with particular reference to Asia and the Pacific," PhD, Centre of Spatial Data Infrastructure and Land Administration, University of Melbourne, Melbourne, 2002.

[11] D. Coleman and J. McLaughlin, "Defining Global Geospatial Data Infrastructure(GGDI): Components, Stakeholders and Interfaces," *Geomatica,* vol. 52, pp. 129-143, 1998.

[12] L. Strain, "An SDI Model to Include the Marine Environment," Master of Geomatics Engineering Department of Geomatics, University of Melbourne, Melbourne, 2006.

[13] A. Meiner, "Spatial data management priorities for assessment of Europe's coasts and seas," *Journal of Coastal Conservation,* pp. 1-7, 2011.

[14] Z. Nedovic-Budic and J. K. Pinto, "Understanding interorganizational GIS activities: A conceptual framework," *URISA-WASHINGTON DC-,* vol. 11, pp. 53-64, 1999.

[15] M. Z. Tarmidi, A. R. Mohamed Shariff, Z. Z. Ibrahim, A. R. Mahmud, and A. H. Hamzah, "Issues and Challenges in Managing Malaysia's Marine Spatial Information Sharing," in *XXV International Federation of Surveyors Congress 2014*, 2014.

[16] A. Field, *Discovering statistics using IBM SPSS statistics*: Sage, 2013.

In search of the Malaysian Extended Continental Shelf

N A Z Yahaya[1]*, T A Musa[1], K M Omar[1], A H M Din[1,2], N M Abdullah[1], A H Othman[1] and M I A Wahab[3]

[1]Geomatic Innovation Research Group (GnG), Faculty of Geoinformation and Real Estate, Universiti Teknologi Malaysia, 81310 Johor Bahru, Johor, Malaysia.
[2]Geoscience and Digital Earth Centre (INSTEG), Universiti Teknologi Malaysia, 81310 Johor Bahru, Johor, Malaysia.
[3]Quality Engineering Department, Universiti Kuala Lumpur, Malaysian Institute of Industrial Technology, Persiaran Sinaran Ilmu, 81750 Bandar Seri Alam, Johor Bahru, Johor, Malaysia,
*nooranim@utm.my

Abstract. Over the years, the sovereignty proclamation of Coastal States for their extended continental shelf has been a crucial matter. The declaration and extension of a continental shelf under the United Nations Convention on the Law of the Sea provide significant potential for many developing nations in economics, trades, resource exploitation, communication and security. Hence, the application of satellite altimeter, as one of the solutions for collecting bathymetry data to define the approximate limits of the continental shelf, is reviewed. This paper also discusses the possible significance or contribution of space-derived bathymetry, i.e. the seafloor topography, either independently or harmoniously with different datasets, to meet the element of the Article 76 of UNCLOS.

1. The importance of continental shelf

Over the years, the sovereignty proclamation of countries for their extended continental shelf has been a crucial matter. The declaration and extension of a continental shelf under the 1982 United Nations Convention on the Law of the Sea (hereinafter referred to as UNCLOS) for many developing nations provide significant potential for many developing nations in economics, trades, resource exploitations, communication and security.

A coastal or archipelagic State (hereinafter referred to as Coastal State) has sovereign rights and exclusive jurisdiction over its maritime zones for exploring and exploiting the natural resources in the marine environment which is determined in accordance with customary international law as reflected in the UNCLOS. The maritime zones recognised under UNCLOS, as illustrated in Figure 1, are measured seawards from the baseline, which includes the internal waters, the 12 nautical miles (NM) Territorial Sea, the 24 NM Contiguous Zone, the 200 NM Exclusive Economic Zone (EEZ), along with the continental shelf of 200-350 NM (where applicable), the High Seas and the Area [1].

As the highlighted maritime zones in this paper, the continental shelf is a crucial marine area which is regarded as an oasis in the ocean for plants and animals due to the abundance of sunlight, shallow waters, and nutrient-packed sediments that washes in from rivers, wave actions, and in some areas, upwelling [2]. In addition to the vast amount of marine life, people have depended on the continental shelf for thousands of years to provide up to 90 percent of the fisheries production in the world and other resources such as natural gas, oil and mineral deposits. The natural resources of the continental shelf consist of minerals and other non-living resources of the seabed and subsoil together with living organisms belonging to sedentary species, i.e. organisms which, at the harvestable stage, are either immobile on or under the seabed or unable to move except in constant physical contact with the seabed or the subsoil [3]. A Coastal State is, among other things, entitled to explore and exploit the exclusive rights within the continental shelf area for its economic gain.

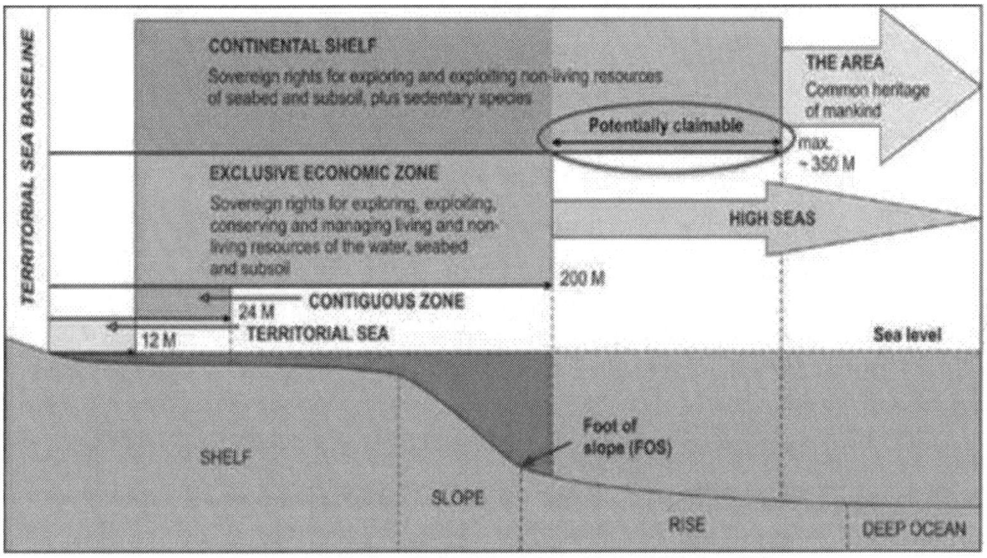

Figure 1. Marine Zones based on United Nation [3].

The Continental Shelf (Article 76 Part VI of the 1982 UNCLOS) is defined as "the natural prolongation of the land territory to the outer edge continental margin, or 200 NM from the coastal state's baseline, whichever is greater". Therefore, a State's continental shelf may exceed 200 NM until the natural prolongation (the land mass of the coastal State) ends (see Figure 2). However, it may never exceed 350 NM from the baseline; or it may never exceed 100 nautical miles beyond the 2,500 meter isobaths (the line connecting the depth of 2,500 meters). It is noted that the continental shelf of a coastal State comprises the seabed and subsoil of the submarine areas: the slope and the rise; it does not include the deep ocean floor with its oceanic ridges or the subsoil thereof.

The declaration and extension of a Continental Shelf under UNCLOS for many developing nations provide a significant economic potential. In order for a Coastal State to exercise its sovereign rights over a particular seabed area, it has to define the outer limits of the extended continental shelf (ECS). A submission containing the outer limits of its continental shelf has to be deposited to the Commission on the Limits of the Continental Shelf (CLCS, hereinafter referred to as the Commission) through the Secretary-General of the United Nations in order to carry out the rights. The final and binding outer limits of the ECS should then be established in accordance with the Commission's recommendations.

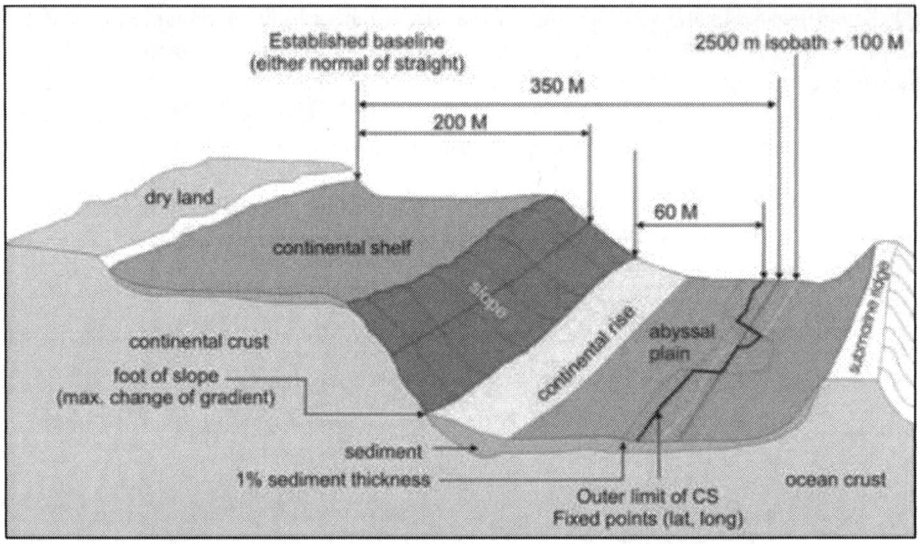

Figure 2. The Continental Shelf defined by UNCLOS [4].

2. The continental shelf submission's impact to Malaysia

In general, all Coastal States have the ability to define the outer limits of their territorial waters around the margins of the South China Sea without reference to opposite States. The People's Republic of China and the Socialist Republic of Vietnam have wide-ranging claims over the South China Sea, which are largely undefined, while the other countries, i.e. the Republic of Philippines, Malaysia, Brunei and Indonesia, have claimed their contiguous sea zones and continental shelf areas.

The main factor that motivates the Coastal States to submit their claims for ECS is the exclusive sovereign rights to explore and exploit the resources in the area. Being claimed as "the second Persian Gulf", South China Sea yields about 11 billion barrels of oil and 190 trillion cubic feet of natural gas in an unexplored area [5, 6]. Due to economic factors and conflicting resources in the South China Sea, the circumstances initiated a dispute involving complicated issues relating to UNCLOS, which does not offer clear guidelines in this situation [7, 8]. As a result, certain quandaries are faced, such as [7]:

i. Highly emotional territorial claims in a region of rising nationalism.
ii. Risks of accidental conflict that could escalate.
iii. Risks to freedom of navigation in EEZs.
iv. Disputes over the interpretation and applicability of international law, notably the UNCLOS.

On May 6, 2009, Malaysia lodged its joint submission with the Socialist Republic of Vietnam to the Commission, in accordance with Article 76, paragraph 8, of UNCLOS, regarding information on the limits of the continental shelf beyond 200 NM from the baselines, of which the breadth of the territorial sea measures in respect of the southern part of the South China Sea. Conforming to the 1979 New Map of Malaysia (*Peta Baru Malaysia*), Malaysia has the potential to extend its continental shelf beyond 200 NM (but not exceeding 350 NM from the breadth of the territorial sea) for the maritime area in Sarawak and West Sabah.

However, as soon as the executive summary of the submission is made public on the Commission's website, other neighbouring States had a chance to respond. The People's Republic of China and the Republic of Philippines submitted a protest note through its Notes of Verbale to the UN in response to the joint submission. The main reason for the People's Republic of China's protest to the joint submission is, as it covered the Spratly Islands in the South China Sea, of which its sovereignty is disputed by several surrounding States since ages ago. In the light of the fact, the Commission decided to defer further consideration of the submission until such time as the submission is next in line for consideration [9]. Despite the protests, issues and disputes from neighbouring States, Malaysia continues its Malaysian Continental Shelf Project (MSCP) to determine our extended Continental Shelf and gather as much data as possible to defend the claim.

3. Advancement of satellite altimeter technology

Since the Seasat mission in 1978, global marine gravity anomalies of different accuracies and spatial resolutions have been derived from satellite altimeter missions [10]. By using the gravimetric geodesy concept, this space technique provides an economic tool to investigate the first view of the ocean floor structures in many remote areas of the earth where shipborne techniques are not feasible. Moreover, altimetry data have provided the Geoscience community a revolutionary and dramatically improving view of the global marine gravity field and geoid [11]. The main principle of satellite altimeter is to measure the instantaneous shape of the ocean at the nadir with good precision [12]. A pulse-limited radar is used to measure the altitude of the satellite above the closest sea surface point which, to a first approximation, is a measure of geoid height, or gravitational potential, as shown in Figure 3.

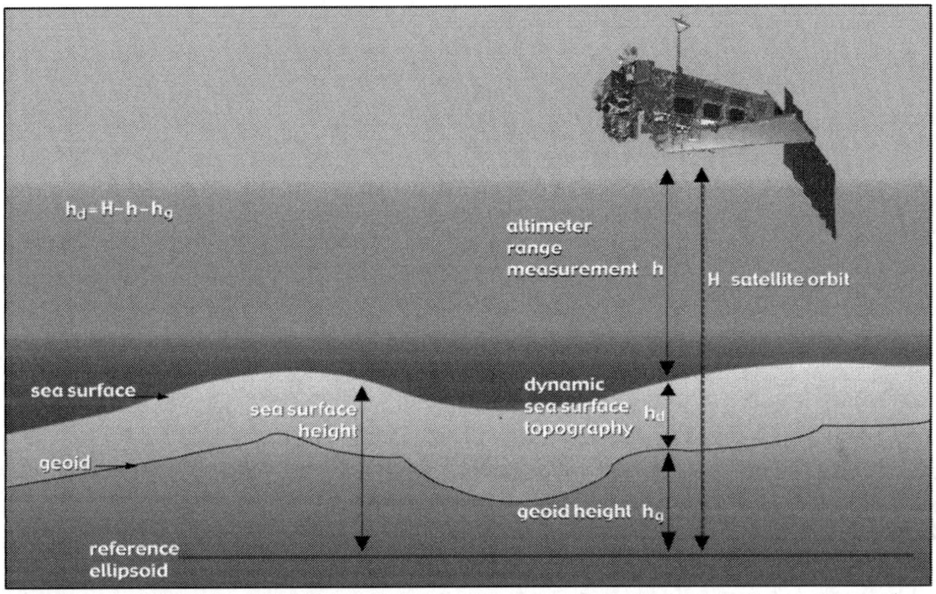

Figure 3. Principle of Satellite Altimeter (courtesy from European Space Agency).

4. Locating the Malaysia's continental shelf

Locating the continental shelf is a crucial component for a Coastal State's effort to lay claim to its juridical Continental Shelf under Article 76 of the UNCLOS. The Article 76 defines the 2500 m isobaths as "a line connecting the depth of 2,500 metres", but does not specify the surface that depth is measured from 2500 m isobaths fringe continents and islands, and lie on the flanks of mid-ocean ridges [13].

The primary step to derive data from satellite altimeter is obtaining the marine geoid for the recovery of marine gravity anomalies. The basic observations are the altimeter tracks which can be used to derive the sea surface height, the marine geoid, and the deflections of the vertical. The equation from Inverse Stokes or Vening-Meinesz can be applied to derive the marine gravity anomalies [14, 15, 16]. The gravity anomalies, then, can be used to determine the bathymetry of the seafloor. The variations of gravity anomaly are highly correlated with the seafloor topography [17, 18, 19, 20]. Figure 4 shows the graphical flow from the radar altimeter data used for the determination of the relevant parameters to determine the bathymetry, i.e. Marine Geoid, Gravity Anomalies and the final product, Bathymetry.

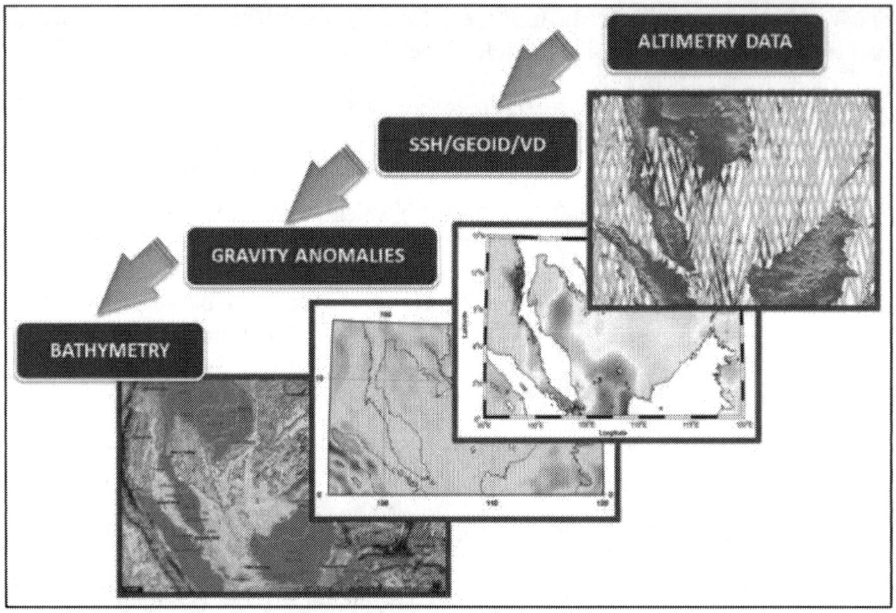

Figure 4. Information flow from the altimetry data, Marine Geoid, Gravity Anomalies and the final product, Bathymetry.

5. Summary

With technological advancements throughout the years, the demands from the worldwide population grew and there emerged a problem. Over-exploitation of the maritime resources is immensely felt towards the middle of the 20th century and many nations started feeling the need to ensure protection of their marine resources and maritime zones. Existing space-derived bathymetry can clearly be used for conducting a "desktop study" in the early stages of determining whether a Coastal State has a juridical Continental Shelf that is worth claiming, and for developing a plan of action in the likely area to be included and what data may be needed to substantiate it [21].

Furthermore, the data could be used as supportive evidence to strengthen our argument at the United Nation level. The data used in the early planning stages are not necessarily included with that submitted to the Comission, but the findings from the stage will determine the role of altimeter. Table 1 comprise the modified list of the mapping tasks required to substantiate the delineation of a Continental Shelf under Article 76 (UNCLOS) and the applicability of altimeter to each of the tasks [13].

Moreover, the space-derived bathymetry data are able to be employed as additional and/or validation of existing shipborne bathymetry data. In essence, recent advances in altimeter should be constantly reviewed to balance the paucity of seabed data in interpretation and determination of the extended continental shelf for Coastal States. To sum up, the cost-effective technique from the technology of satellite altimeter is capable to be a solution for mapping globally uniform bathymetry.

Table 1. The modified mapping tasks for establishing a juridical Continental Shelf and the applicability of altimeter [13].

Tasks	Applicability of Altimetry
Map baselines (shoreline points)	None
Map 2500 m depth contour	Applies and meets the standard
Map the base of the slope	Applies
Map the foot of the slope or prove it does not exist	Possibly
Optional -decide whether to use the morphological Foot of the Slope or "evidence to the contrary"	Possibly
Map sediment thickness seawards of the foot of the slope	Possibly
Determine whether isolated elevations are "natural prolongations"	Possibly
Resolve the "ridge exception"	Possibly
Create lines at distances of 60, 100, 200 and 350 nm	Not needed
Prepare a submission including "charts, maps and diagrams"	Could be used

Acknowledgement
The authors would like to express their sincere appreciation to Universiti Teknologi Malaysia (UTM) and the Ministry of Education (MOE) Malaysia for the Fundamental Research Grant Scheme (FRGS) Vot No. 4F227 and Research University Grant Scheme (GUP) Vot No. 12H99 given to complete this research project.

References
[1] United Nations, UN 1997 *The 1982 United Nations Convention On The Law Of The Sea* (United Nation, New York) ed 97 vol 10
[2] Marinebio 1998 *Continental Shelves-Marinebio.Org* (Marinebio Conservation Society) Retrieved on January 24, 2014, from http://Marinebio.Org/Oceans/Continental-Shelves.Asp
[3] United Nation 1993 *The Law Of The Sea: Defination Of Continental Shelf* (United Nation, New York) ed 93 vol 16
[4] United Nations 2012 *The Definition Of The Continental Shelf And Criteria For The Establishment Of Its Outer Limits* (Division for Ocean Affair and the Law of the Sea) Retrived on May 8, 2013, from http://www.un.org/depts/los/clcs_new/continental_shelf_description.htm
[5] Energy Information Administration, EIA 2013 *International Energy Statistics, South China Sea Regional Analysis Brief* (US International energy data and analysis) Retrieved on May 13, 2015 from https://www.eia.gov/beta/international/regions-topics.cfm?RegionTopicID=SCS
[6] Kaplan, R D 2014 *Asia's Cauldron: The South China Sea And The End Of A Stable Pacific* (New York: Random House Publishing Group)

[7] Bader, J A, Lieberthal, K G and Mcdevitt, M M 2014 *Keeping The South China Sea In Perspective.* (Australian Policy Online) Retrieved on September 15, 2014 from http://Apo.Org.Au/Node/41197

[8] Buszynski, L 2013 *National Security College Occasional Paper (*Australian National University, Australia)

[9] Najhan Bin Md Said 2013 *Seminar On Ocean Data Management Maritime Centre* (Putrajaya, Malaysia)

[10] Hwang, C, Kao, E and Parsons, B 1998 *Geophysical Journal International* **34** 449–460

[11] Mcadoo, D 2006 *15 Years Of Progress In Radar Altimetry Symp.* (Venice, Italy)

[12] Majumdar, T J and Bhattacharyya, R 2005 *Current Science* **89** 1754–58

[13] Monahan, D 2004 *Oceanography* **17** 75–82

[14] Hofmann-Wellenhof, B and Moritz H 2005 *Physical Geodesy* (Springer-Verlag, Vienna)

[15] Hwang, C 1998 *J. Geodesy* **72** 304–312

[16] Sandwell, D T 1992 *Geophys. J. Int.* **109** 437–448

[17] Smith,W. H. F., and D. T. Sandwell 1994 *Journal of Geophysical Research-Solid Earth* **99** 21803–21824.

[18] Smith,W. H. F., and D. T. Sandwell 2004 *Oceanography* **17** 8–23

[19] Baudry, N., and S. Calmant 1991 *Geophysical Research Letters* **18** 1143–1146

[20] Ramillien, G., and A. Cazenave 1997 *Journal of Geodynamics* **23** 129–149

[21] Lamont, I 1999 *International Conference On Technical Aspects Of Maritime Boundary Delineation And Delimitation* (International Hydrographic Bureau, Monaco)

Assessment of Gravity Field and Steady State Ocean Circulation Explorer (GOCE) geoid model using GPS levelling over Sabah and Sarawak

A H Othman[1]*, K M Omar[1], A H M Din[1,2], Z A M Som[1], N A Z Yahaya[1], and M F Pa'suya[3]

[1]Geomatic Innovation Research Group (GnG), Faculty of Geoinformation and Real Estate, Universiti Teknologi Malaysia, 81310 Johor Bahru, Johor, Malaysia.
[2]Geoscience and Digital Earth Centre (INSTEG), Universiti Teknologi Malaysia, 81310 Johor Bahru, Johor, Malaysia.
[3]Faculty of Architecture, Planning and Surveying, Department of Surveying Science and Geomatics, Universiti Teknologi Mara Perlis, Arau, Perlis, Malaysia.

*Email: amirhamzah_othman@yahoo.com

Abstract. The GOCE satellite mission has significantly contributed to various applications such as solid earth physics, oceanography and geodesy. Some substantial applications of geodesy are to improve the gravity field knowledge and the precise geoid modelling towards realising global height unification. This paper aims to evaluate GOCE geoid model based on the recent GOCE Global Geopotential Model (GGM), as well as EGM2008, using GPS levelling data over East Malaysia, i.e. Sabah and Sarawak. The satellite GGMs selected in this study are the GOCE GGM models which include GOCE04S, TIM_R5 and SPW_R4, and the EGM2008 model. To assess these models, the geoid heights from these GGMs are compared to the local geometric geoid height. The GGM geoid heights was derived using EGMLAB1 software and the geometric geoid height was computed by available GPS levelling information obtained from the Department Survey and Mapping Malaysia. Generally, the GOCE models performed better than EGM2008 over East Malaysia and the best fit GOCE model for this region is the TIM_R5 model. The TIM_R5 GOCE model demonstrated the lowest R.M.S. of $\pm$16.5 cm over Sarawak, comparatively. For further improvement, this model should be combined with the local gravity data for optimum geoid modelling over East Malaysia.

1. Background

Geoid is a fundamental surface in geoscience which is defined as an equipotential surface of the earth's gravity field associated with some average of sea level that is commonly adopted as the height reference datum [1]. The geoid model is strongly dependent on the quality and quantity of the gravity data utilised for the solution [2]. One of the primary data sources in geoid determination is from the Global Geopotential Model (GGM). GGM is one of the cheapest and reliable methods to derive geoid height, and it contributes to the long-wavelength parts of the geoid model. GGM is an important component in developing high resolution geoid models over local regions. For example, Malaysia used the GGM01C model [3]. In addition, GGM provides a reference field for local gravimetric geoid. The GGM model is used to derive parameters such as geoid height, gravity anomaly and deflection of vertical components. Nonetheless, the geoid height is the primary interest of engineers, surveyors and others, especially for obtaining the orthometric height from GPS.

The importance of the geoid has increased substantially due to the widespread use of Global Navigation Satellite Systems (GNSS) for positioning and navigation. GNSS, unlike traditional surveying instruments, has the ability to provide three-dimensional coordinates (latitude, longitude and height) anywhere in the world, at any time irrespective of the weather. However, the GNSS-determined heights, i.e. ellipsoidal heights, are geometrical heights. These cannot be used in surveying and engineering projects of which orthometric heights are required. Hence, there is a need for geoid determination, since it is the reference surface for orthometric heights. Thus, by combining the geoid and GNSS observations, the ellipsoidal heights can be converted to the physically meaningful orthometric heights [4].

The recent GGM can be accessed via http://icgem.gfzpotsdam.de/iCGEM. A GGM comprises a set of fully normalized spherical harmonic coefficients to model the long wavelength features of the external earth gravity field. Users of GGM should perform their own accuracy and precision verifications, such as comparing the GGM-derived gravity field quantities with local data [2]. There are a few classes of GGM such as satellite-only GGM, combined GGM, and tailored GGM. For this paper, we focus on satellite-only GGM which is derived from satellite GOCE.

The GOCE satellite gravity mission was launched in 2009 and ended in 2013, whereby the collected data lead to significant improvements on the long-wavelength aspects of the earth's gravity field. The advantage of satellite gravity missions is that it provides a homogenous and near global coverage of gravity field information [5]. However, usually, the error estimates for such GGMs are not too optimistic and presented as global averages, and thus does not necessarily represent the performance of the GGM over a particular region [2].

Sabah and Sarawak has a rough topography and many rivers that cause conventional levelling difficult to be carried out [6]. Hence, as an alternative, GPS levelling technique is employed for conducting levelling over rough and mountainous areas. GPS levelling requires a precise geoid model in order to transform geometric height to physical height, i.e. orthometric height. GPS leveling technique has been used widely; for example, to monitor dam subsidence due to water or natural gas removal, earth crustal movements, and as height control across water bodies for bridge construction [2]. Since the release of the GOCE GGM model, there has been no specific study conducted over Sabah and Sarawak related to the performance of GOCE GGM model over the region. Therefore, this paper aims to determine the best fit GOCE model over the East Malaysian region.

2. Data and method

2.1 Global Geopotential Model

GGM is the most essential component in modern geoid determination which significantly contributes to its long wavelength information. Currently, various GGMs have been derived and made available freely for the scientific community which can be downloaded from the official website of the International Centre for Global Earth Models (http://icgem.gfz-postdam.de/ICGEM). From this website, three GOCE GGM models were selected: JYY_GOCE04S, GO_CONS_GCF_2_TIM_R5, and GO_CONS_GCF_2_SPW_R4 model, which were released in 2014. The other GGM model selected is EGM2008 which is a known high resolution global geoid model. These GOCE GGM models and the EGM2008 model were chosen to test how accurate the GOCE models are compared to the EGM2008 over East Malaysia, with the use of local GPS levelling data as the benchmark. The list of GGM data used is summarised in Table 1.

Table 1. List of selected Global Geopotential Models in this study and their descriptions.

Model	Year	Data	Reference
EGM2008	2008	Satellite(Grace), Gravity data, Altimetery data	Palvis et al.,2008
JYY_GOCE04S	2014	Satellite(GOCE)	Yi et al.,2013
GO_CONS_GCF_2_TIM_R5	2014	Satellite(GOCE)	Brockman et al., 2014
GO_CONS_GCF_2_SPW_R4	2014	Satellite(GOCE)	Gatti et al., 2014

Based on Table 1, the models, year, data and reference are listed, whereby further information can be acquired via the ICGEM website (http://icgem.gfz-postdam.de/ICGEM).

GGMs were then evaluated by comparing geoid heights from GPS levelling with the GGM geoid heights computed using EGMlab through MATLAB. For EGM2008, 360 maximum degree order were used to minimise the time required in computing the geoid heights. The descriptive statistics of the differences between GGM-derived and GPS levelling geoid heights are reported section 3 of this paper.

2.2 GPS levelling data

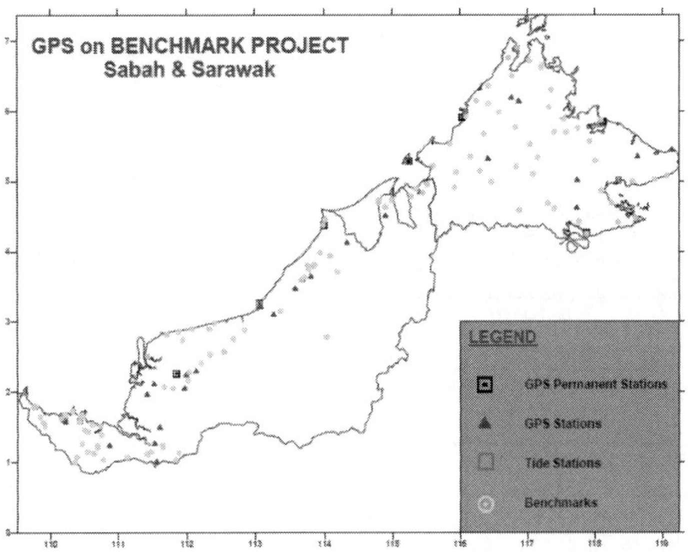

Figure 1. GPS levelling data on collocated benchmarks in 2004/2005 over East Malaysia.

The GPS levelling data was collected by the Department Survey and Mapping Malaysia in 2004/2005 and made available for this study (see Figure 1). The GPS levelling data on collocated benchmarks used in this study are basically on the Standard benchmarks and 2nd class benchmarks of the

Department of Survey and Mapping Malaysia. For this study, 26 collocated GPS levelling benchmarks in Sabah and 30 collocated GPS levelling benchmarks were selected.

2.3 Methodology
GPS levelling data has become one of the standard tools for validating and evaluating global and local gravimetric geoid models. Nowadays, the ellipsoidal height precision is about 1-2 cm; hence, GPS levelling is used as an external measure of the accuracy of geoid model. This quality evaluation of GGM became a routine procedure before computing any high resolution local gravimetric geoid model. Therefore, the statistical summary of the comparison between the GGMs and the GPS levelling data will be the result that will indicate which GGM model best fits over East Malaysia.

Based on Figure 2, the geoid height derived from GGM is computed using EGMLab [1]. The geometric geoid is determined from GPS levelling data on collocated benchmarks over Sabah and Sarawak from the Department of Survey and Mapping Malaysia. The geometric geoid is assumed as the most probable value. The comparison between GGM geoid height and geometric geoid height is obtained by the following formula;

$$\Delta N = N^{GGM} - Ngeometric \quad (1)$$

Where;
N^{GGM}: Geoid height derived from GGM
Ngeometric: Geoid height derived from GPS levelling
Ngeometric is obtained from ellipsoidal height, h_{GPS}, and orthometric height of benchmarks, H_{MSL}
The Root Mean Square (R.M.S.) of the differences, ΔN, will be obtained and the lowest R.M.S. will indicate the most suitable GGM over East Malaysia.

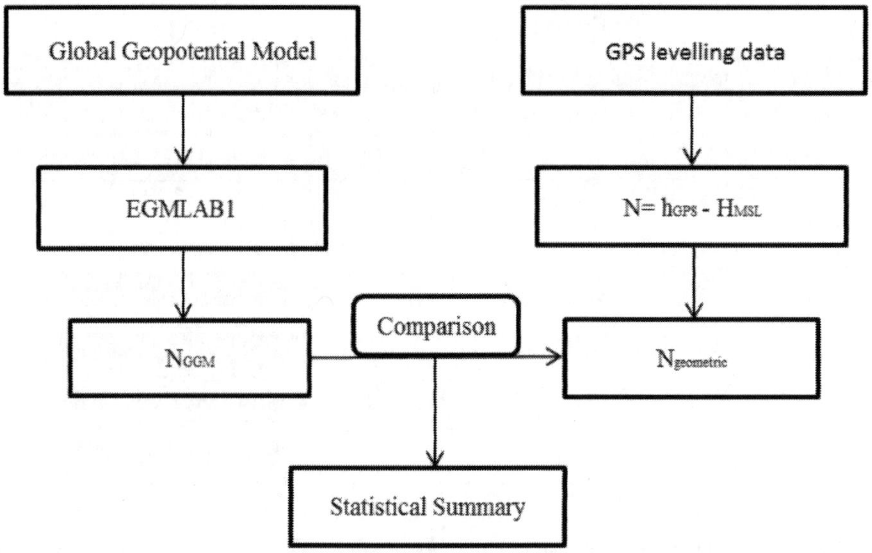

Figure 2. Flowchart of the geoid height assessment.

3. Results and analysis

The evaluation of GGM is important because the global geoid model will contribute to the quality of the high resolution local geoid model. The findings will be useful for users of GPS levelling. In this paper, the results are divided into two assessments, one for each state: Sabah and Sarawak.

3.1 Assessment of GGM in Sabah

Figure 3 shows the geoid heights for the selected GGMs and geometric geoid height at 26 GPS levelling benchmarks over Sabah. TIMR5 follows closely to the geometric geoid values. The EGM2008 is closer to the geometric geoid height values based on the average difference of 1.937 m. However, based on the R.M.S., EGM2008 has the highest R.M.S. of 0.378 m which indicates that the model does not best fit the region. The models from GOCE show better fitting, whereby TIMR5 has a R.M.S. of ±0.271 m. The difference in R.M.S. of TIMR5 and EGM2008 is about 10 cm which demonstrates that TIMR5 has a significant improvement compared to the other GGMs over Sabah.

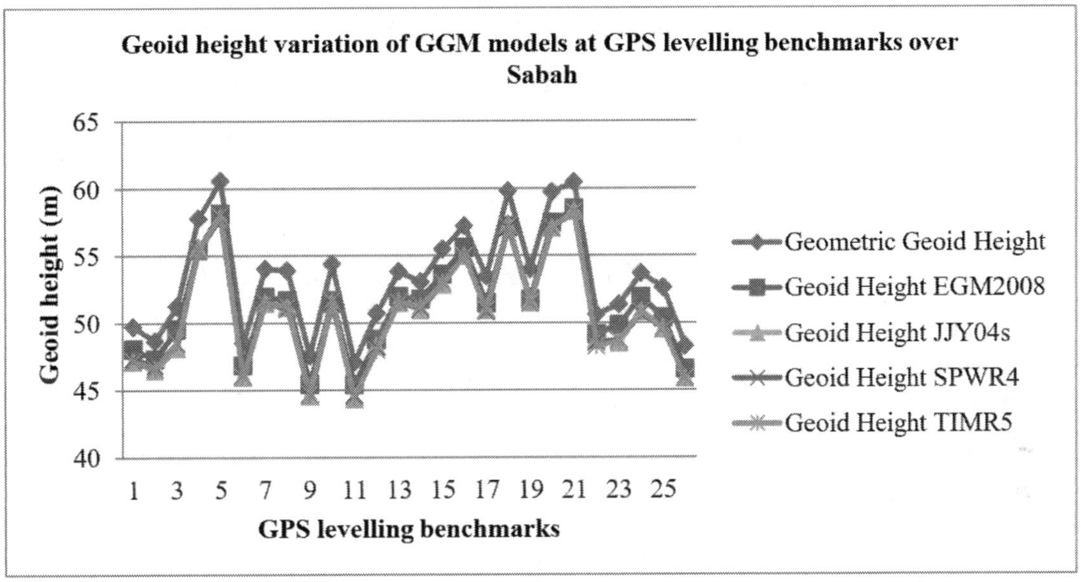

Figure 3. Geoid height variation of GGM models at GPS levelling benchmarks over Sabah.

Table 2 shows the numerical statistics for the geoid height evaluation of the GOCE GGM models and EGM2008 using 26 Sabah GPS levelling data. Based on the R.M.S. values, the best model is GO_CONS_GCF_2_TIM_R5. Therefore, GGMs from the GOCE satellite are better when compared to EGM2008. The lowest R.M.S. is ± 27.1 cm from the TIM_R5 GOCE model.

Table 2. Statistic of geoid heights from GGM compared to geometric geoid height over Sabah.

GGM	Min (m)	Max (m)	Average Difference (m)	R.M.S. (m)
EGM2008	1.269	2.729	1.937	0.378
JJY04S	1.874	3.089	2.534	0.318
SPWR4	1.820	3.022	2.514	0.311
TIMR5	1.787	3.079	2.517	0.271

3.2 Assessment of GGM in Sarawak

Based on Table 3, the best fit GGM for the Sarawak region is TIMR5 (GO_CONS_GCF_2_TIM_R5) as well. All GOCE GGM models show a good performance based on the R.M.S. values compared to EGM2008. Hence, GOCE GGM models significantly contribute to the improvement of geoid height. The difference of R.M.S. between EGM2008 and TIMR5 is about ±10 cm which proves that the GOCE GGM models have impact in GNSS heighting and height unification. It could be sufficient for engineering and most scientific applications, but for precise applications, a high resolution geoid model is needed. For further research, an investigation on the impact of GOCE GGM models in various scientific applications, such as in oceanography, i.e. in ocean circulation and sea level rise, needs to be carried out. We suggest that, for future geoid computations, further GGM evaluations need to be carried out and the contribution of GOCE models should be addressed.

Table 3. Statistic of geoid heights from GGM compared to geometric geoid height over Sarawak.

GGM	Min (m)	Max (m)	Average Difference (m)	R.M.S. (m)
EGM2008	1.878	2.792	2.179	0.254
JJY04S	2.131	3.169	2.776	0.244
SPWR4	2.569	3.208	2.793	0.213
TIMR5	2.377	3.230	2.762	0.165

4. Conclusion

The assessment of the GOCE geoid models based on recent GOCE Global Geopotential Model (GGM) and EGM2008 has been performed using GPS levelling data over Sabah and Sarawak. Three GOCE GGM 2014 models, which are GOCE04S, TIM_R5, and SPW_R4, and EGM2008 are evaluated using local GPS levelling data as the benchmark. Based on the results, the best fitting GOCE GGM model is the GO_CONS_GCF_2_TIM_R5 and the lowest is the JYY_GOCE04S. The R.M.S for the best fit TIMR5 model over Sabah and Sarawak is ± 27.1 cm and ±16.5 cm, respectively. In general, the GOCE model is better than EGM2008 over this study area. The R.M.S. difference between the best fit GOCE model, i.e. TIMR5 model, and EGM2008 is about ± 10 cm. In conclusion, the GOCE satellite improves the longwave of gravity field and geoid height. As a suggestion, for better results in future, the GOCE model should be combined with the local gravity data in order to develop a high resolution geoid model over an area.

5. Acknowledgments

The authors wish to acknowledge the Department Survey and Mapping Malaysia for the GPS levelling data, especially to Dr. Azhari, Mr. Soeb Nordin, and Mr. Zulkifli. We would like to express our appreciation to ICGEM for giving us free access to the global geopotential models. We would also like to acknowledge Dr. Ramin Kiamehr for sharing his EGMLab software, which is available from Mr. Zainal Abidin Mat Som. Furthermore, we are grateful for the assistance and encouragement of our colleagues from the Geomatic Innovation Research Group, FGHT, UTM. Last but not least, the authors would like to extend their sincerest appreciation to Ministry of Education (MOE) Malaysia and Universiti Teknologi Malaysia for funding this research under Research University Grant (Vot number: Q.J130000.2527.12H99).

6. References

[1] Sanso F and Sideris M. G 2013 *Geoid Determination Theory and Method* (Springer-Verlag Berlin Heildelberg)

[2] Ramin K 2010 *Practical Concept in Geoid Modelling with Geophysical and Geodynamical Interpretations* (Lambert Academic Publishing)

[3] Nordin A. F, Abu S, Hua C. L and Nordin S 2005 *Malaysia Precise Geoid (MyGEOID)* (Coordinates) Vol 1 issue 4

[4] Sjoberg L, Gidudu A, and Ssagendo R 2015 *The Uganda Gravimetric Geoid Model 2014 Computed by The KTH Method* J. Geod. Sci 2015 5:35-46

[5] Rummel R 2012 *Height Unification using GOCE* J. Geod. Sci 2012 2(4) 355-362

[6] Amir H. O, Kamaludin M O, Rusli O, Zainal Abidin M Som, Yusuf O 2013 U*nification of vertical datum in Sabah and Sarawak.* Journal Teknologi UTM

Pre- analysis assessment of Sea Surface Temperature (SST) products in the region of Malaysian coastal water

M A H Aziz[1*], K M. Omar[1], A H M Din[1, 2],M N M Reba[2]
[1]Geomatic Innovation Research Group (GnG), Faculty of Geoinformation and Real Estate, Universiti Teknologi Malaysia, 81310 Johor Bahru, Johor, Malaysia.
[2]Geoscience and Digital Earth Centre (INSTEG), Universiti Teknologi Malaysia, 81310 Johor Bahru, Johor, Malaysia.

Email: muhammadamirulhazwan@gmail.com

Abstract. This paper presents the pre-analysis of validation between the acquisition satellite data and in situ data. To carry out this assessment, Sea Surface Temperature (SST) data are acquired to be regressed with SST In situ. With the launch of the Moderate Resolution Imaging Spectroradiometer (MODIS) satellite with a sensor on the Terra spacecraft, data sets of the global distribution of sea surface temperature are retrieved, and need to be validated and analyzed. Radar Altimeter Database System (RADS) also has an archived data of Optimal Interpolation SST (OISST) that can be retrieved based on satellite track of altimeter. The aim of this paper is to present inter-comparison study between pixel based (MODIS SST) and point based (RADS SST). The value of root mean square error (rmse) is computed to see the performance of the data product. It is an assessment and evaluation to see the performance for both data. The objective of this paper is to evaluate Malaysian coastal area through validation with in situ data. To achieve the objective, we perform pre-analysis study of the MODIS products and RADS SST to see the performance of both data in terms of spatial value during seasonal changes. However, the scope of this analysis covers only on the spatial MODIS pixel value and the OISST point value during the southwest monsoon daytime. From the result, RADS SST/RADS show higher root mean square error (rmse) at 0.731/0.677 (before calibration) and 0.6951/0.476 (after calibration). From the rmse result, we could deduce that the RADS SST has random error arising from the fact that the interpolated points are based on the track.

1. Introduction

The Southern South China Seas (SSCS) is connected with the Sulu Sea in the east of Mindoro and Balabac straits with average depth of 60 metre. The SSCS has complex geographical setting that influences the sea surface current circulation. The sea surface current circulation at the SSCS is controlled by monsoon. Weaker southwesterly summer monsoon winds prevail from April to August causing great circulation coastal jet off the Vietname[6]. However, there is still lack of understanding of water characteristics during seasonal changes. There are previous studies on the water characteristic in the Malaysia coastal area through intensive hydrography observation but it is limited to certain area and time [8]. The understanding about the physical characteristics such as current circulation, temperature/salinity profile is limited to the observation carried out. Furthermore previous study by [12], provide some understanding of the SSCS in small area which does not include Malaysian coastal area.

With the availability of satellite derived sea surface temperature (SST), we have improve our understanding the characteristics of surface water. The satellite data covered large area and high resolution depending on the product of SST. For Moderate Resolution Imaging Spectroradiometer(MODIS)SST product, the data have been established since 2000 until present with improvement of the payloads sensor on the platform. MODIS is freely available at the web based oceancolor. SST studies using MODIS in Malaysia has been done by [8] with interest on the surface circulation monitoring. MODIS have sensor on board of TERRA and AQUA, the data is available in pixel based sensor. However, the Malaysian coastal area is less covered by the MODDIS daily resolution data, having SST spatial characteristics information about Malaysian coast limited only to the availability of the imageries data.

With various improvement such as data impairment techniques, SST is available on a weekly basis with a large scale area. Data undergoes interpolation process to fill gap and such interpolation technique requires error control. Interestingly, NOAA provides data interpolation known as Optimal interpolated Sea Surface Temperature (OISST) using AVHRR sensor with controlled error. This data has passed through the process of interpolation to obtain weekly resolution [11] and also has its resolution stretcher modified as discussed in version 2 [14]. The data is also known as the Reynolds SST, named after Richard W. Reynolds that produce these products for National Centre Environmental Prediction(NCEP). OISST can be used to study climate change such as El Nino; the El Nino anomalies values can be calculated by subtracting weekly OISST to monthly OISST. The spatial resolution of the data is 1° (111km) but has been resampled at 0.25° for OISST version 2. In Comparing the OISST product to MODIS, MODIS product data can also be found on a weekly basis but do not cover the whole region of Malaysia. However, MODIS has better spatial resolution of 9km. The degree of resolution is able to interpret the sea in more detail which can be used for monitoring current [8].

The issue concerning data inter-comparison of SST product over the region of Malaysia is still less studied. Previous studies have focused more on areas of extreme variability or areas that has big changes of SST value[15]. Studies for areas of low variability such as SSCS is still lacking, especially in the region of Malaysian coastal water. The reason for the comparison is to show credibility before the product data is used for operations at sea. Inter-comparison of SST product by the Global High Resolution Sea Surface Temperature (GHRSST) to ensure that data can be used continuously by scientific community[3]. According to Global Data Assimilation Experiment(GODAE)[9], the data has to be check regularly to perform consistence analysis. It is an assessment and evaluation to see the effectiveness of technicality based on the performance analysis of the data based. The objective of this paper is to evaluate Malaysian coastal area through validation with insitu data. To achieve the objective, we perform pre- analysis study of the MODIS products and OISST to see the performance of both data in terms of spatial value during seasonal changes. However, the scope of this analysis covers only on the spatial MODIS pixel based value and the OISST point based value during the southwest monsoon daytime. The next session will be detailing on the selection of data analysis.

2. Material and method

2.1 In situ observation

The data sets from 1992-2008 has been selected as primary data for benchmarking. The data consists of temperature profile from the observation of GTSPP project. GTSPP is an abbreviation for Global Temperature Salinity Profile Project. This pilot project has been initiated by the NOAA agency. The observation is obtained from many platforms of fixed platform, drifting to several instrumentation such as Expandable Bathy Thermograph (XBT), Conductivity Temperature Depth(CTD) cast, moored bouy and Thermosalinographs. The data have passed through multiple steps of quality control programs through 'QCed' and 'Mquest'[7]. In coastal Peninsular Malaysia, there are over 909 point/stations (figure 1) spanning from 1992-2006 and are well distributed. The data can be access through GTSSP website that provides in-situ primary data around the world [4]. However, only 14 selected stations have been utilised in this paper for southwest monsoon.

2.2Satellite Data

2.2.1 Optimal Sea Surface Temperature

The SST used in this study is the Optimum Interpolation Sea Surface Temperature (OISST) which was retrieved from the RADS database; the data was optimized by the altimeter track. SST Database from RADS has never been tested, thus, there is the question of how the track of altimeter may affect the variability OISST data. The SST data is set to epoch of altimeter track. For the analysis, we use RADS SST to represent the OISST from RADS . The selection of data is matched with the time of in situ with latency of 6 hour max. The data selected is between March to April and September to October during the transition seasonal changes.

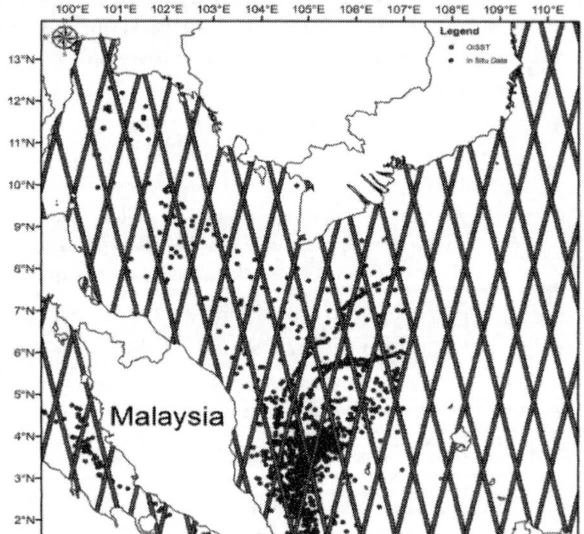

Figure 1 .Map of SST at region of Malaysia. Red line indicates of RADS SST and purple dots is the distribution of In situ SST.

2.2.2 Level 3 Binned MODIS Data

Level 3 MODIS Binned image in figure 2 shows the area of Malaysian coastal area,which is bounded by latitude 1° 15' to 14° N and longitude 97° 30' to 109° 15'E. It is a SST image data and retrieved from the oceancolor web based. The SST are available at a global scale,providing an image represent the water surface temperature at 9km resolution. MODIS data present an equalpixel based water surface and the data is an archived product from the NASA Goddard Space Flight Center Ocean Data Processing System.

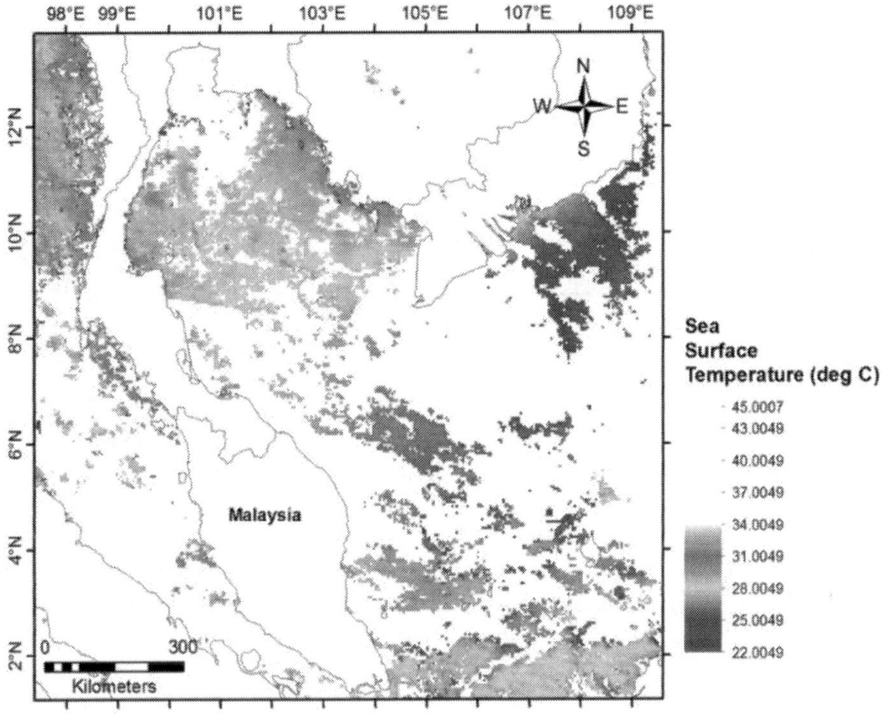

Figure 2. Level 3 data of SST data from MODIS satellite.

Table 1. Description of SST data product.

Name	Temporal Resolution	Spatial Resolution(km)	Period	Original data
RADS SST(OISST v2)	Weekly	30	1981-present	AVHRR AVHRR+AMSR-E
MODIS	Daily(daytime/nightime)	9	2000-present	TERRA/MODIS

2.3 Data processing

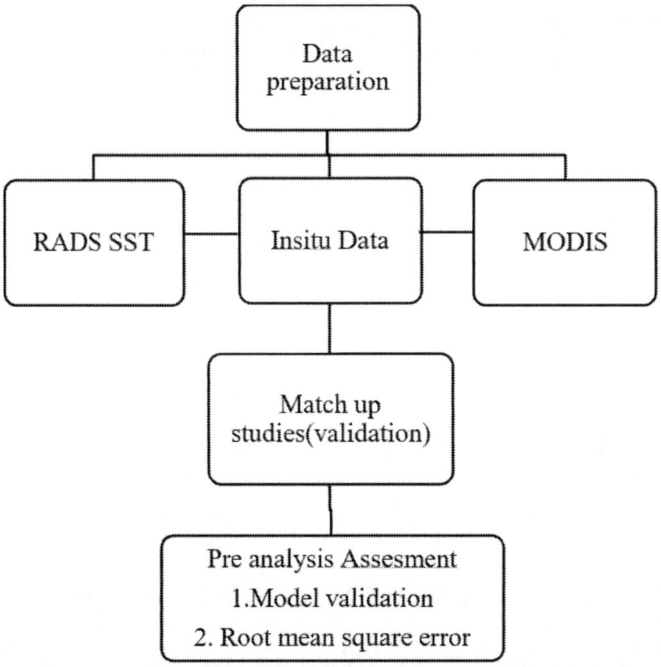

Figure 3. Flowchart of data validation between RADS SST/MODIS SST and In situ SST. From the validation, the equation from the regression is used to calibrate the data. The regression equation is tested with root mean square error(rmse).

Validation of satellite data collected may have some problem with observation data in situ. This is because the epochs of the satellite passes may not correspond with the exact time of observation at the ocean. Thus, The temporal resolution of the satellite pass over the sampling site does not always match exactly with the observation data, for instance, in the mid-latitudes, the drifter experience a difference in time ranging from an hour to 10 hours. While the typical average time between fixes are around 2-3 h [6], the latency of satellite track position can cause water circulation at the upper ocean with strong inertia and strong tidal activity near the continental margins [4]. Usually, the uptake of data for match up study can be up 6 hour latency under consideration to resolve inertial motion[3]. For the validation part, the latency of epochs for the validation is limited to 6 hours.

In order to clarify the characteristic of each SST product, the RADS SST represent point based SST of OISST v2 data optimize by the track of satellite altimeter (figure 1), while MODIS is the pixel based SST data which represents the sensor of TERRA on board sensor. The advantage of the OISST is that the product has been used by several researchers and the data has been converted to 0.25 daily resolution using Reynolds optimum interpolation. Based on the objectives of this paper, the comparison of the SST products, through validation of SST with the in situ data was conducted in order to reveal the relative difference of spatial characteristic of both data at the same period. Considering the temporal differences, the data has been selected during the season of southwest monsoon. Thus, the only independent variable to be tested is the spatial characteristic. The spatial characteristics by all means are the pixel based data (MODIS data) and the point based data (RADS SST). However, the validation was performed only based on the relative differences between the two SST products. For the first assessment, we would like to evaluate the quantitative differences of the SST products. The next section is to test the model of SST product based on its linear equation.

3. Result and Discussion

The quantitative difference between the two SST products is depicted in figure 4. However, the accuracies of these SST products are still unknown; thus, the evaluation of the two SST products was performed by analysing the bias of each data products from the in-situ data. The histogram indicates the bias at each selected point.

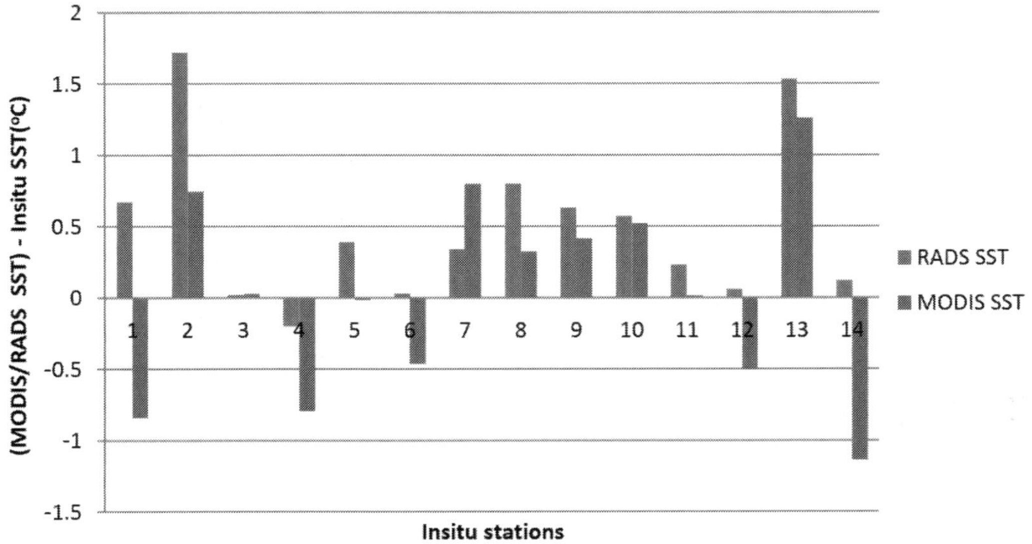

Figure 4. Histogram plot of sea surface temperature biases between MODIS/RADS SST data in situ (GTSPP).

Table 2. Root Mean Square Error(RMSE) of MODIS SST and RADS SST(before calibration).

SST product	RMSE
MODIS SST	0.677
RADS SST	0.731

The red bars represent MODIS data which presents data with huge biases. Nevertheless, the root mean square error (rmse) gave a contrasting result (that is, the rmse of the MODIS is smaller than that of the RADS SST). This large rmse of the RADS SST data could be said to originate from the interpolation method of Reynolds Optimum method. The same inter-comparison of the Reynolds product has been conducted by [14]. Based on Iwasaki, the Reynolds OI SST which has 1 week resolution has spatial differences up to 1.2°C. Furthermore temporal resolution of the Reynolds OI SST is 1 week which may be one of the causes of the large RMSE. RADS SST data has weekly resolution data of optimum interpolation and also optimized by the altimeter track. The altimeter track gives the value of the SST as point based which may also contain error.

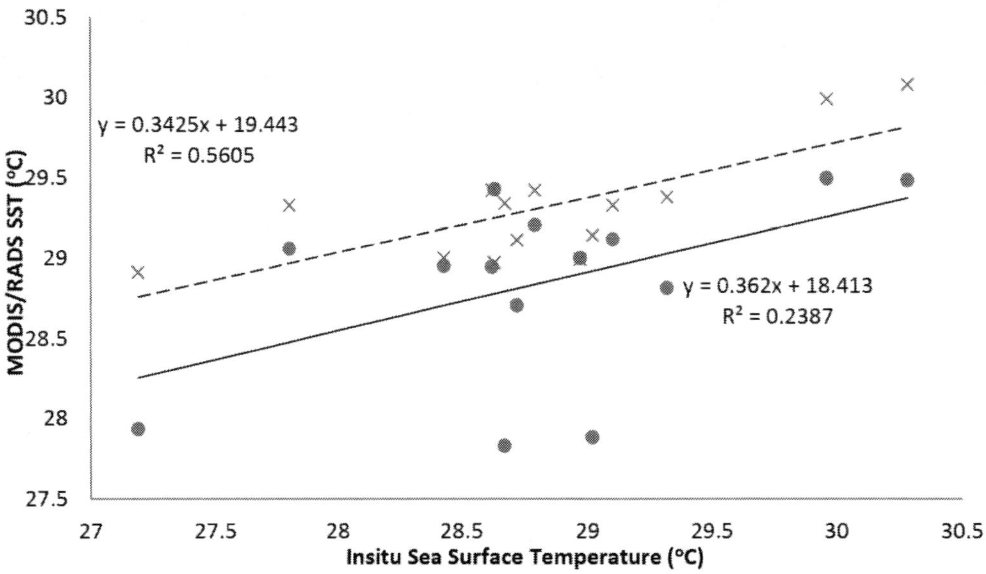

Figure 5. Regression plot of sea surface temperature between two models of MODIS/RADS SST data and in situ (GTSPP). Dashed line indicates regression fit using RADS SST and the solid line is for MODIS data.

In this section, we try to validate the data of both products with the in-situ to see the relationship. For this assessment, is to evaluate the performance of the linear model of the both product. The reason of this model performance assessment is to evaluate the data performance after calibration. The calibration is based on the linear equation (mx + c). The data performance will indicates of reliability of the data. The root mean square been compute to see the error value after the calibration.
The SST models were plotted based on its same condition. Both of the data has been selected during transitional of southwest monsoon seasons. There are some important differences that we consider of its spatial difference between the MODIS/RADS SST may not have the same difference distance. Figure 3 demonstrates the regression plots for validation of the two models in respect of MODIS SST and RADS SST; this is to examine the relationship between these two data.

Table 3. Root Mean Square Error(RMSE) of MODIS SST and RADS SST(after calibration)

Model	RMSE
MODIS SST	0.4762
RADS SST	0.6951

The R^2 statistics for the model was computed to assess the goodness of fit of the regression curves. The OISST model shows consistency of data near the line with moderate R^2 of 0.5605, while MODIS showed inconsistency of data away from the line with poor R^2 of 0.2387. From the comparison of both models, the data plots of RADS SST shows better relations of R^2 with the in-situ. This suggests that the data has moderate relation in term of the identical value. However, after the data calibration, it was revealed that the RADS SST model contains larger error than MODIS SST (see Table 2).

As shown on table 2, the rmse of RADS SST is higher before the calibration; the continuous data contains higher errors after calibration. Both tables have indicated that RADS SST inhibits error from the interpolation data over the track of satellite altimeter.

4.Conclusion

This paper has described the spatial characteristics between RADS SST and MODIS SST. The aim of this study was to carry out a comparison of SST data between two global models in order to extract the correct information about the data reliability. The first assessment based on figure 4 was performed to determine the biases of each data selected. As demonstrated by numerous researchers, Calibration/Validation (Cal/Val) of satellite data is crucial to obtain accurate information of the models' validity.This paper describes the spatial characteristics between RADS SST and MODIS SST. The aim of the study is to perform inter-comparison between pixel based (MODIS SST) and point based (RADS SST). From the above analysis, RADS SST shows higher root mean square error (rmse) of 0.731(before calibration) and 0.6951(after calibration). From the rmse result, we could deduce that the RADS SST has random error arising from the fact that the interpolated points are based on the track.

Acknowledgment
The authors would like to extend their sincerest appreciation to Universiti Teknologi Malaysia for funding this research under Research University Grant (Vote No. 12H99) and the eSciece fund (Vote No. 4S074) from the Ministry of Sciences, Technology and Innovation Malaysia. The authors wish to thank the Global Temperature and Salinity Profile Programme for providing the in-situ temperature profile data over the Malaysian region.

References

[1] Banzon, V. F., & Reynolds, R. W. (2013). Use of WindSat to Extend a Microwave-Based Daily Optimum Interpolation Sea Surface Temperature Time Series. *Journal of Climate*, 2557–2562.
[2] Cornell, J. A., & Berger, R. D. (1986). Factors that Influence the Value of the Coefficient of Determination. *Florida Agricultural Experiment Stations Journal Series Paper 6709*, 65-70.
[3] Dash, P., Ignatov, A., Kihai, Y., & Sapper, J. (2009). *The SST Quality Monitor (SQUAM)*. National Environmental Satellite, Data, and Information Service (NESDIS).
[4] December 2014. U.S. Department of Commerce, National Oceanic and Atmospheric Administration, National Oceanographic Data Center, Silver Spring, Maryland, 20910. 8 August 2015, http://www.nodc.noaa.gov/GTSPP/.
[5] Emery, W. J., & Thomson, R. E. (2001). Data Analysis Methods In Physical Oceanography. Amsterdam, The Netherland: Elsevier Sciences B.V.
[6] Gan, J., & Qu, T. (2008). Coastal jet separation and associated flow variabilityin the southwest South China Sea. *Deep Research Part 1*, 1-19.
[7] Kurian, N. (2010). *Quality control using QCed and Mquest: A Comparison*. Goa, India: National Institute of Oceanography.
[8] Mohd Akhir, M. F., Zakaria, N. Z., & Tangang, F. (2014). Intermonsoon Variation of Physical Characteristics and Current Circulation along the East Coast of Peninsular Malaysia. *International Journal of Oceanography*, 9.
[9] Provost, C. L., Bell, M., Chassignet, E., Cummings, J., Fukumori, I., Hurlburt, H., et al. (n.d.). *Assessment and testing of the GODAE products*. Toulouse: LEGOS/CNSR.
[10] Reynolds, R. W. (2009). *What's New in Version 2* . National Centers For Environmental Information(NCEP).
[11] Thomson,R.E.,P.H. LeBlond ,and A.B Rabinovich 1997:Oceanic odyssey of a satellite-tracked drifter:Nort Pacific variability delineated by a single trajectory,f.Oceanogr.,53,81-87
[12] W. Fang, Z. Guo, and Y. Huang, "Observational study of the circulation in the southern South China sea," Chinese Science Bulletin,vol.43,no.11,pp.898–905,1998.

[13] Richard W. Reynolds, Thomas M. Smith, Chunying Liu, Dudley B. Chelton, Kenneth S. Casey, and Michael G. Schlax, 2007: Daily High-Resolution-Blended Analyses for Sea Surface Temperature. *J. Climate*, **20**, 5473–5496.

[14] S, I., M, K., & H, T. (2008). Inter-comparison and evaluation of global sea surface temperature products. *International Journal of Remote Sensing*, 6263–6280.

Wave power potential in Malaysian territorial waters

Nor Asmida Mohd Nasir[1] **and Khairul Nizam Abdul Maulud**[1,2]

[1] Department of Civil and Structure Engineering, Universiti Kebangsaan Malaysia
[2] Earth Observation Center, Institute of Climate Change, Universiti Kebangsaan Malaysia

Email:norasmida@siswa.ukm.edu.my

Abstract. Up until today, Malaysia has used renewable energy technology such as biomass, solar and hydro energy for power generation and co-generation in palm oil industries and also for the generation of electricity, yet, we are still far behind other countries which have started to optimize waves for similar production. Wave power is a renewable energy (RE) transported by ocean waves. It is very eco-friendly and is easily reachable. This paper presents an assessment of wave power potential in Malaysian territorial waters including waters of Sabah and Sarawak. In this research, data from Malaysia Meteorology Department (MetMalaysia) is used and is supported by a satellite imaginary obtained from National Aeronautics and Space Administration (NASA) and Malaysia Remote Sensing Agency (ARSM) within the time range of the year 1992 until 2007. There were two types of analyses conducted which were mask analysis and comparative analysis. Mask analysis of a research area is the analysis conducted to filter restricted and sensitive areas. Meanwhile, comparative analysis is an analysis conducted to determine the most potential area for wave power generation. Four comparative analyses which have been carried out were wave power analysis, comparative analysis of wave energy power with the sea topography, hot-spot area analysis and comparative analysis of wave energy with the wind speed. These four analyses underwent clipping processes using Geographic Information System (GIS) to obtain the final result. At the end of this research, the most suitable area to develop a wave energy converter was found, which is in the waters of Terengganu and Sarawak. Besides that, it was concluded that the average potential energy that can be generated in Malaysian territorial waters is between 2.8kW/m to 8.6kW/m.

1. Introduction

Development of a country does not only depend on a sustainable development of energy resources, but also on its hygiene, safety and the ability of renewable energy (RE) itself. There is no doubt in the future that our world would still need oil for technological development and innovation since it is still dependent on it. However, dependence on certain energy sources cannot ensure sustainable development of a country. Therefore, RE is the best alternative which has fewer effects and impacts on the environment [1]. In the 1970s and the early 1980s, research and development on wave energy have been done to address the problem of world oil crisis. Since then, many countries have developed wave power for the generation of electricity, water desalination and also for pumping water into reservoirs.

Studies on the exploration of waves to be used as a new source of energy have newly taken the interest of researchers from all over the world. Malaysia is also one of the countries that are investing significantly in this research area, optimizing the production from the ocean [2]. Wave energy is related to solar energy where the wind generated by thermal heat reacts with the surface of the sea water producing energy to generate waves. The amount of energy produced and the size of waves depend on the wind speed, the duration of the wind blowing, and the distance of the wind to the waves [3]. Tenaga Nasional Berhad (TNB) has started to look at RE to meet the amount of energy needed in the future. Through projections that were supposed to be undertaken, TNB was expected to start generating RE of 2,013 GWh in 2012. This would increase to 5,385 GWh in 2015 and they expect to generate 11,244 GWh of RE by 2020 [4, 5].

Currently, the use of geographic information system (GIS) is very important especially when dealing with geospatial information and problems on earth. The concept and technology of GIS itself allow us to collect and manage spatial data and to also observe the correlation between the data [6, 7, 8, 9, 10, 11].

This study concerning spatial analysis is to determine suitable locations to generate waves in Malaysia. GIS analysis which was carried out included a variety of algorithms and parameters used to produce the best results in determining the location of optimal wave energy. In GIS, the calculation may be made either based on space or non-space data. The quantification of power delivered by a wave moving across the sea level has been measured and was used in this study for the South West region using the Seapower system [12].

2. Methodology

Wave data used in this research was provided by the Malaysia Meteorology Department (MetMalaysia) and is supported by satellite imaginary obtained from the National Aeronautics and Space Administration (NASA) and Malaysia Remote Sensing Agency (ARSM) within the time range of the year 1992 until 2007. In 2006, a comparison between the data from MetMalaysia and satellite images was made [13] with the results showing that the data from MetMalaysia can be applied to the study of wave energy in Malaysia. The process to identify the best location of wave energy is shown in Figure 1.

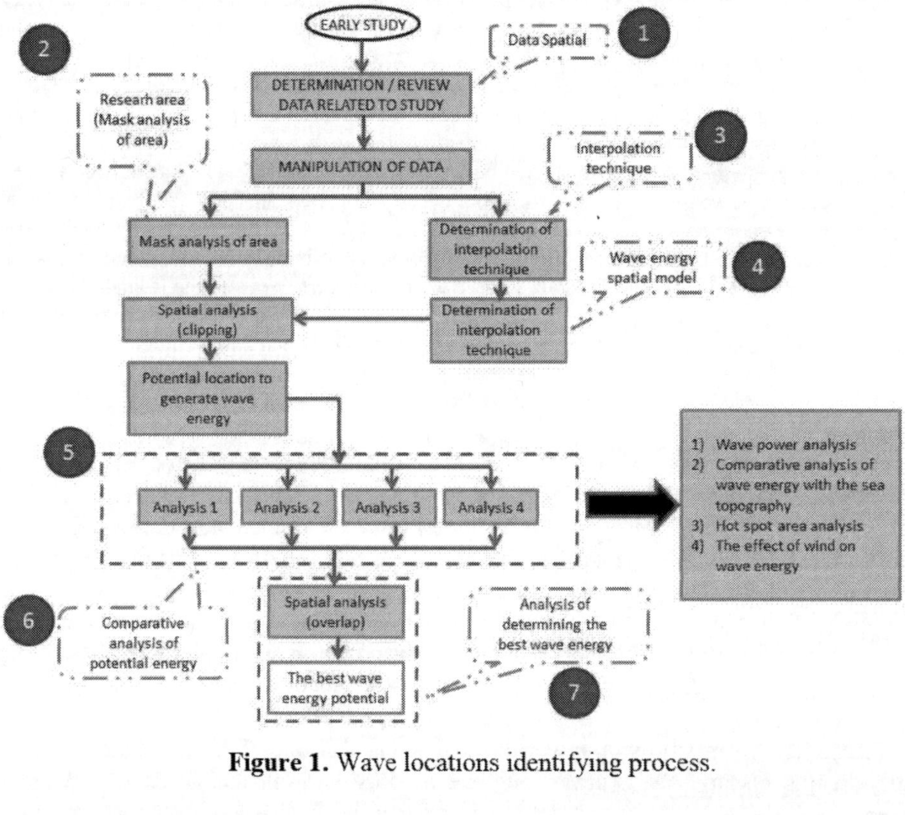

Figure 1. Wave locations identifying process.

2.1. Mask analysis of the research area

Mask analysis is an early process to identify suitable locations for the generation of wave power. In this analysis, all the sensitive and protected areas have been identified as they play an important role as a filter to determine which areas are the most ideal for generating waves. Factors that were used to obtain areas free from being sensitive and are protected are the Malaysian border, the bathymetry between 30 m to 250 m, 12 nautical miles from the coastline, oil and gas pipe routes, marine parks [14], oil and gas areas, underground cables and also ports [15,16]. Figure 2 show the map generated by using geographic information system (GIS) after considering the sensitive and protected area. The total area of Malaysian waters is 571,996.037 km^2. After the mask analysis has been carried out, the total area of research area (non-protected and sensitive areas) is 273,146.154 km^2. It shows that 48% of Malaysian waters are suitable for wave energy generation.

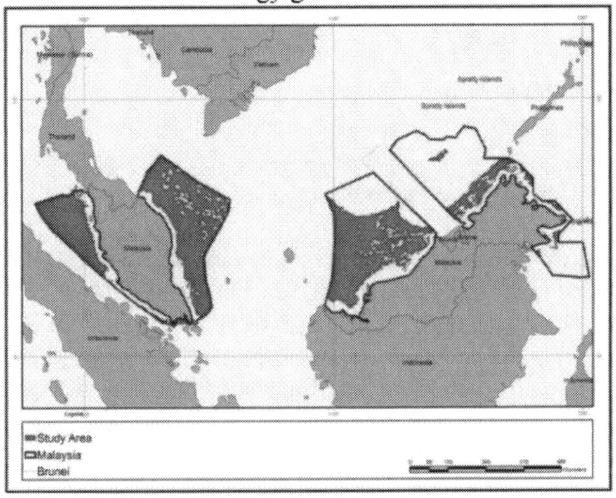

Figure 2. Map generated by mask analysis.

2.2. Comparative analysis

Comparative analysis is a method that has been accomplished to strengthen and evaluate the results obtained. It is conducted to compare between two or more groups of data samples. This analysis can prove the validity and accuracy of the results based on the criteria used and determines the pattern change in a set of data series compiled by factors such as year, month and hour. In this research, the data was collected from the year 1992 to 2007. Four analyses which have been carried out were:

1. Wave power analysis.
2. Comparative analysis of wave power with the sea topography.
3. Hot-spot area analysis.
4. The effects of the wind on wave energy.

2.2.1. Wave power analysis

Wave power can be generated when waves, currents and wind integrate into a steady rate. Wave power analysis is divided into two, which are the highest power and lowest power. This study provides an overview regarding the potential of wave energy that can be generated. From the results of the analysis to determine the highest and lowest power, the ability to generate wave energy is good. This is because there are areas with high values for the last 16 years. Process to get the maximum and minimum power was based on calculation of data from year 1992 to 2007. The total area of maximum or minimum power every year will divide with the total research area to get the percentage of maximum and minimum power. The wave power values have been computed using the expression [17,18,19]:

$$P = \frac{\rho g^2}{64\pi} H^2 mo.Te \approx (0.5 \frac{kW}{m^3 s}) H^2 mo.Te \tag{1}$$

Where P is the energy flux in watts per meter of crest (kW/m), H^2mo is the significant wave height, Te the wave energy period, ρ the water density and the g the acceleration by gravity.

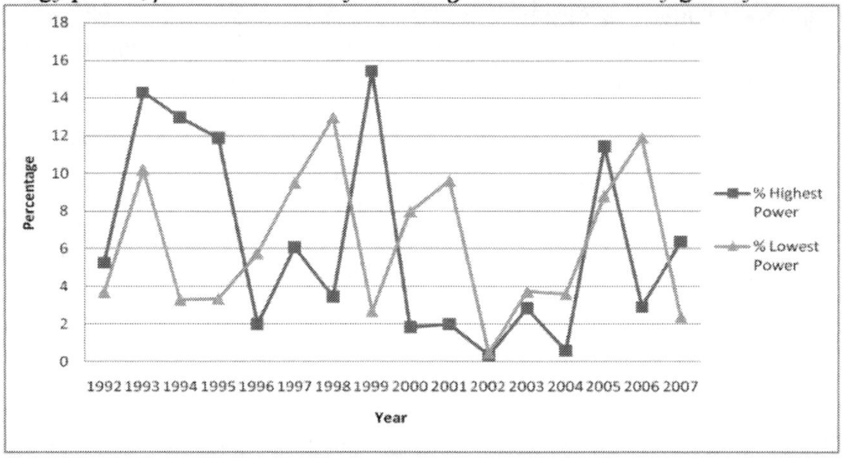

Figure 3. Results of the highest and lowest power generated from the year 1992 to 2007.

Figure 3 shows the results of analysis of the highest and lowest power generated from the year 1992 to 2007. The difference in value for each year depends on a variety of matters, including climate change, natural disasters, wind, changes in the sea temperature and other factors that affect the formation of waves. Based on the analysis, it is shown that the year 1993 and 1999 recorded the highest generation of wave energy while in 2002 and 2004, the rate of energy generated was at its lowest. The pattern of these changes indicates that the number of data capable of generating wave energy varies every year. From the lowest position analysis, the year 2002 gives the least percentage value, but in 1998 and 2006, it is shown that the percentage of the lowest position is high. In this analysis, the year 1993, 1999 and 2005 demonstrate the ability to generate wave energy which is high. This indicates that every six years, the value or rate of wave energy in Malaysian waters is high. However, this finding must be considered carefully to observe the significance with other factors such as tidal power.

2.2.2. Comparative analysis of wave power with the sea topography

A comparative analysis of the power of wave energy and the ocean topography provides a comparison between powers of wave energy with the characteristics of the ocean depth. In this analysis, the results of the ocean surface topography are overlaid with the wave energy potential.

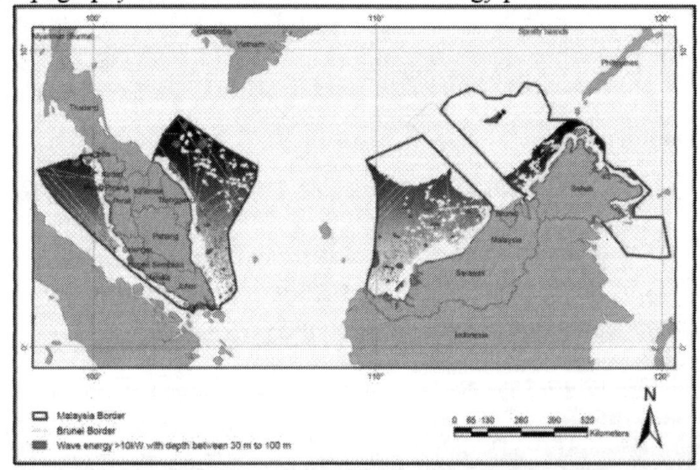

Figure 4. Wave power more than 10 kW with a depth of 30 m to 100 m.

Figure 4 shows the areas with a depth of 30 to 100 m which have the energy of more than 10 kW. Based on previous research conducted by other researchers, it is established that the ideal power to generate wave energy is 10 kW [20,21]. The analysis indicates that the areas that meet the analysis requirements are scattered. The results are not only focused on the South China Sea, but it also can be seen in the Straits of Malacca, though the area being considerably smaller compared to the South China Sea.

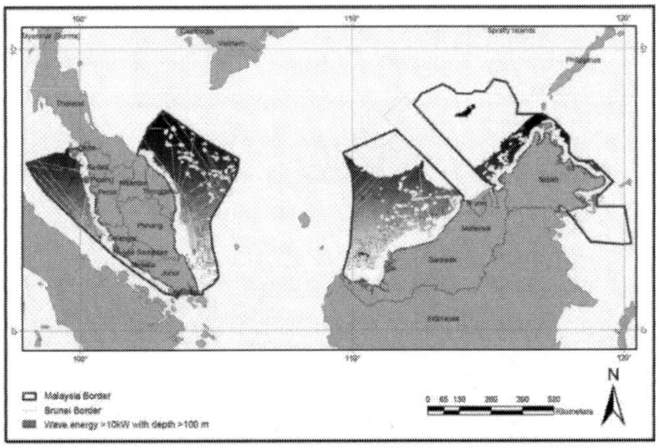

Figure 5. Wave power of more than 10 kW with depth of 100 m to 250 m.

Based on Figure 5, the areas which generate more than 10 kW are considerably limited, focusing more in the waters of Sabah and Sarawak. There is no spotted area around the Peninsular since the maximum water depth at Peninsular Malaysia is only about 90 m.

Based on this analysis, it clearly shows that the depth of the ocean plays a crucial role in determining the wave energy power. A depth of about 30 m to 100 m is able to generate more energy than a depth of 100 m to 250 m for more than 10 kW/m of energy generated.

2.2.3. Hot-spot area analysis

This analysis is to examine whether the potential wave energy produced has an influence on its surroundings. Figure 6 shows the hot-spot wave energy potential in Malaysia. The point indicating a wave energy that has a high value (positive) is the best area for generating wave energy. In reference to Figure 6, there are six potential locations, which are Kelantan, Terengganu, Pahang, Johor, Sarawak and Sabah. Based on this analysis, it shows that the South China Sea has the highest potential to generate wave energy compared to the Straits of Malacca. The position of the sea plays a crucial role since the Straits of Malacca is placed in between Peninsular Malaysia and Sumatera, Indonesia.

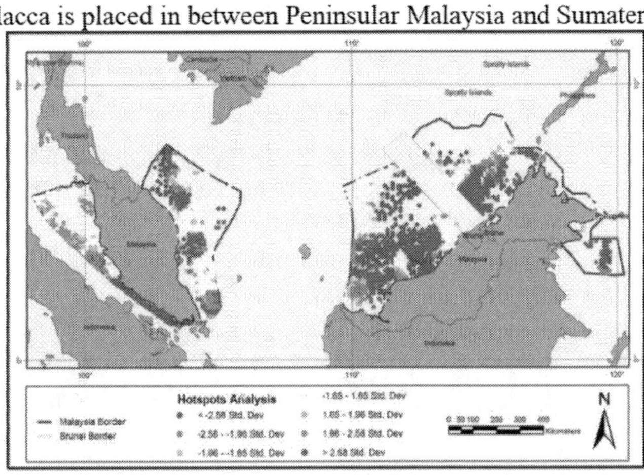

Figure 6. Hot-spot wave energy potential from the year 1992 to 2007.

IOP Conf. Series: Earth and Environmental Science **37** (2016) 012018 doi:10.1088/1755-1315/37/1/012018

2.2.4. The effect of the wind on wave energy

There is no doubt that the wind is a major factor in the occurrence of waves in the ocean. The data from 1992 to 2007 illustrates the significant link between wind and wave power. Figure 7 shows the locations where the significant link between wind and wave power occurs. The location is described as a significant area with the wind of more than 6 m/s (average wind speed), and the wave energy potential exceeds 10 kW/m. It can be concluded from these findings that strong wind can create valuable wave energy potential. This is because the wind is able to produce high and long waves. Therefore, this analysis shows that Kelantan, Terengganu, Pahang, Sarawak and Kedah are suitable locations to generate wave energy.

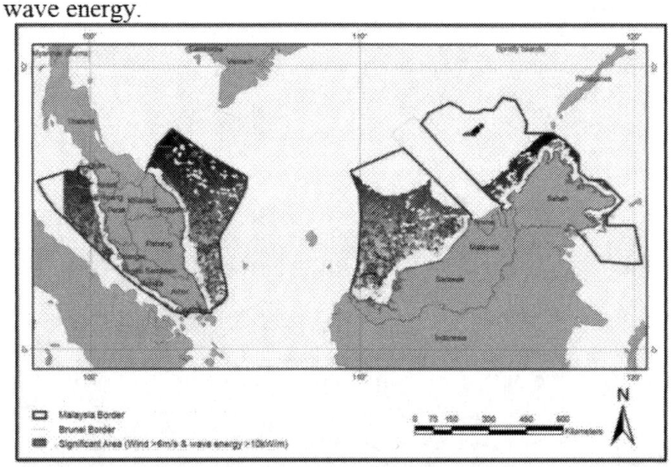

Figure 7. Suitable location to generate wave energy based on winds (>6 m/s).

3. The best locations for the generation of wave energy in Malaysia

The ability to locate the best location for the generation of wave energy requires extensive analysis of factors that needed to be carried out. The results of the comparative analysis conducted indicate the best location that can be explored and studied. There are six locations identified that have the best potential to generate wave energy. The total potential area to generate wave energy is about 59,227,711.71 m2. The details and the location of the potential place are shown in Table 1.

Table 1. Best location of wave energy generation.

No.	Latitude	Longitude	Location	Nearest Island	Total Area (m^2)
1.	5.3116	103.6067	Terengganu waters	Kapas Island (39km)	125 763.04
2.	4.9079	103.8294	Terengganu waters	Nyirih Island (19km) Tenggol Island (19km)	13 488.48
3.	4.8795	103.7162	Terengganu waters	Nyirih Island (5km) Tenggol Island (4km)	23 132 490.44
4.	4.7711	103.7228	Terengganu waters	Nyirih Island (9km) Tenggol Island (4km)	3 142 711.11
5.	4.6009	104.0952	Terengganu waters	Nyirih Island (51km) Tenggol Island (47km) Kerteh port (68km)	25 937 356.83
6.	5.1929	112.4087	Sarawak waters	Miri port (188km)	6 875 901.81
				Total Area (m^2) :	**59 227 711.71**

From Table 1, the best locations to generate wave energy are at the waters of Terengganu and Sarawak. The locations with the largest area that can generate wave energy are at location three and five. The most practical location being the third location since it is only 5 km and 4 km away from Nyirih Island and Tenggol Island, respectively. This location is the nearest location to the island, with a total area of 23,132,490.44 m^2. Currently, the electric power at Tenggol Island and Nyirih Island is generated using a generator to obtain electricity. The generator is powered by using diesel and it is not environmentally friendly. Since Tenggol Island is one of the most beautiful diving gateway islands in the south of Terengganu, a detailed study needs to be conducted to identify the level of commercialization of wave energy in that area to maximize the use of the energy.

4. Conclusion

This study shows that the average annual wave energy that can be generated in Malaysian territorial waters is between 2.8 kW/m to 8.6 kW/m. This shows that the rate of wave energy generated in Malaysia is comparable with other Asian countries such as China, Japan and India [22]. The value of the wave energy generated is the value without equipment that can help to increase wave energy power. Four detailed analyses have been carried out using the GIS models, and they have been proven to be helpful in strengthening the evidence of wave energy generation capability in Malaysia. In reference to the analyses that have been carried out, the waters of Terengganu and Sarawak have the highest potential to develop waves with a wave energy converter (WEC).

References

[1] Kaltschmitt, M., Streicher, W., and Wiese, A. (2007). *Renewable Energy – Technology, Economics and Environment*. New York: Springer.

[2] Lim, X. L. and Lan, W. H. (2014). Public acceptance of marine renewable energy in Malaysia. *Energy Policy*, Volume 65, 16-26.

[3] World Energy Council (WEC). (2010). *2010 Survey of Energy Resources*. London, UK: World Energy Council.

[4] Tenaga Nasional Berhad (TNB). (2011). *Laporan Tahunan 2011*. Kuala Lumpur: Tenaga Nasional Berhad.

[5] Tenaga Nasional Berhad (TNB). (2012). *Laporan Tahunan 2012*. Kuala Lumpur: Tenaga Nasional Berhad.

[6] Clarke, K. C. (1995). *Analytical and computer cartography*. Ed 2. Englewood Cliffs, New Jersey: Prentice Hall.

[7] Star, J., and Estes, J. (1990). *Geographic information system: An introduction*. New Jersey: Prentice Hall.

[8] Duecker, K. J. (1987). Geographic information systems and computer-aided mapping. *J. Am. Plann. Assoc.* 53: 383-390.

[9] Dangermond, J. (1988). Introduction and overview of GIS. Paper presented at Geographic Information Systems Seminar: *Data sharing- myth or reality*. Ontario: Ministry of Nature Resources.

[10] Andrew, L. and Lovett, I. (2008). *GIS for Environment Decision-Making*. New York, USA: Taylor and Francis.

[11] Johnson, L. E. (2009). *Geographic Information Systems in Water Resources Engineering*. London: CRC Press.

[12] Tucker, M. J.; and Pitt, E. G. (2001). Wave in Ocean Engineering: Elsevier Ocean Engineering Book Series, Ed 5. Kidlington, UK: Elsevier.

[13] Omar Yaakob, Tengku Mohd Ariff Tengku Ab Rashid and Mohamad Afifi Abdul Mukti. (2006). Prospects for Ocean Energy in Malaysia. *Proceedings of International Conference on Energy and Environment 2006 (ICEE 2006)*, Uniten, Selangor, page 62-68.

[14] Jabatan Taman Laut Malaysia. (2010). Pelan Strategik Jabatan Taman Laut Malaysia 2011-2015.Putrajaya. http://www.dmpm.nre.gov.my/files/BUKU%20PELAN20

STRATEGIK%20JTLM%202011-2015.pdf /http://www.dmpm.nre.gov.my/apa-itu-taman-laut.html?uweb=jtl

[15] Prest, R., Daniell, T., and Ostendorf, B. (2007). Using GIS to evaluate the impact of exclusion zones on the connection cost of wave energy to the electricity grid. *Energy Policy*. Volume 35(9): 4516-4528.

[16] Nobre, A., Pacheco, M., Jorge, R., Lopes, M. F. P. and Gato, L. M. C. (2009). Geo-spatial multi-criteria analysis for wave energy conversion system development. *International Journal of Renewable Energy*. 34:97-111.

[17] Iglesias, G. and Carballo, R. (2010). Wave energy resource in Estaca de Bares area (Spain). *Renewable Energy*, 35:1574-1584.

[18] Vicinanza, D., Cappietti, L., Ferrante, V. and Contestabile, P. (2011). Estimation of the wave energy in the Italian offshore. *Journal of Coastal Research*, SI 64 (Proceedings of the 11th International Coastal Symposium), 613-617. ISSN 0749-0280

[19] Rusu, E. and Onea, F. (2016). Estimation of the wave energy conversion efficiency in the Atlantic Ocean close to the European islands. *Renewable Energy*, 85:687-703.

[20] Spauling, M. l., Grilli, A., Damon, C., and Fugate, G. (2010). Application of technology development index and principle component analysis and cluster methods to ocean renewable energy facility siting. *Marine Technology Society Journal,* 44: 8-23.

[21] Kim, C. K., Toft, J. E., Papenfus, M., Verutes, G., Guerry, A. D., Ruckelshaus, M. H., Arkema, K. K., Guannel, G., Wood, S. A., Bernhardt, J. R., Tallis, H., Plummer, M. L., Halpern, B. S., Pinsky, M. L., Beck, M. W., Chan, F., Chan, K., Levin P. S., Polasky, S. (2012). Catching the right wave: evaluating wave energy resources and potential compatibility with existing marine and coastal uses. *PLoS One* 7(11): e47598 doi:0.1371/journal.pone.0047598

[22] Muzathik, A. M., Wan Nik, W. B., Ibrahim, M. Z. and Samo, K. B. (2010). Wave Energy Potential of Peninsular Malaysia. *ARPN Journal of Engineering and Applied Sciences.* 5:11-23.

Questionnaire-based person trip visualization and its integration to quantitative measurements in Myanmar

S Kimijiama[1] and M Nagai[2]

[1]School of Engineering and Technology, Asian Institute of Technology, Pathumthani, Thailand
[2]Center for Spatial Information Science, The University of Tokyo, Tokyo, Japan

E-mail: st114307@ait.asia

Abstract. With telecommunication development in Myanmar, person trip survey is supposed to shift from conversational questionnaire to GPS survey. Integration of both historical questionnaire data to GPS survey and visualizing them are very important to evaluate chronological trip changes with socio-economic and environmental events. The objectives of this paper are to: (a) visualize questionnaire-based person trip data, (b) compare the errors between questionnaire and GPS data sets with respect to sex and age and (c) assess the trip behaviour in time-series. Totally, 345 individual respondents were selected through random stratification to assess person trip using a questionnaire and GPS survey for each. Conversion of trip information such as a destination from the questionnaires was conducted by using GIS. The results show that errors between the two data sets in the number of trips, total trip distance and total trip duration are 25.5%, 33.2% and 37.2%, respectively. The smaller errors are found among working-age females mainly employed with the project-related activities generated by foreign investment. The trip distant was yearly increased. The study concluded that visualization of questionnaire-based person trip data and integrating them to current quantitative measurements are very useful to explore historical trip changes and understand impacts from socio-economic events.

1. Introduction

The economic transition from an isolated country to opening up to the global economy is creating opportunities to develop a high potential for economic growth for Myanmar. This is mainly due to the country's abundant natural resources, surplus labour and locational advantage, due to the geographical proximity of economic giants such as the People's Republic of China and India. [1] has predicted Myanmar's Gross Domestic Product (GDP) to grow by 8.3% and 8.2% in 2015 and 2016, respectively, through both government and foreign investments in the infrastructure, manufacturing and services sectors. For instance, in the telecommunications sector, mobile phone penetration is expected to increase from only about 20% now to 50% of the population in 2017. Similarly, mobile phone connectivity is expected to increase from 12% in 2014 to 70% by 2017 [2]. As the level of use and number of mobile phones users are increasing, analysing individual trip behaviour using mobile phone data can be a powerful alternative tool to capture a large volume of data and visualize the dynamics of individual trip behaviour.

In developing countries, socio-economic factors determine local life styles, including trip behaviour. Thus, evaluation of local impacts associated with socio-economic factors from a person trip perspective is very important and Southeast Asia is not an exception. However, even though person trips are known to be affected by factors such as increased investment and improved infrastructure, changes in trip patterns cannot be traced, mainly because of the absence of established methods, the lack of secondary data and its unreliability. In the field of social science, traditional methodologies to analyse a person trip largely depend on questionnaire surveys and travel diaries. However, integrating questionnaire-based data to quantitative measurements and visualizing such trips data in time-series can be a very helpful tool to assess the diverse individual impacts resulting from rural socio-economic changes. Therefore, the objectives of this study are to: (a) visualize questionnaire-based person trip data and validate it by applying GPS loggers and (b) compare the errors between the questionnaire survey and GPS data sets with respect to sex and age group and (c) assess the change of trip behaviour. This research is expected to contribute to the visualization of questionnaire-based survey data and integrate it with quantitative measurements in order to trace dynamic individual trip behaviour by social attributes, in time-series.

2. Data

In this study, two datasets were utilized. First, a questionnaire-based person trip was prepared. A field survey was conducted with the questionnaire and a GPS device to obtain information, both for a personal profile and one-day trip as well as the house location. Second, 38 GPS loggers were distributed the respondents, two days before conducting the questionnaire survey. The trip data obtained from the questionnaires was validated by the GPS logs data. 345 individual data sets were collected from the questionnaire and GPS loggers and trip behaviour was analysed.

2.1. Study area

The study area is located in Dawei, Tanuntharyi Region, Myanmar. This region is developing its industry and infrastructure to create local employment opportunities and ultimately to support national economic growth through the Dawei Development Project which was established in 2008. The project includes development of the Dawei deep seaports, an industrial estate called Dawei special economic zone (SEZ) and road and rail links to Thailand. The Dawei SEZ law is also established to enhance the Dawei Development Project in order to be a hub for the trade and transportation in the Southeast Asian region [3].

Villages in the SEZ mainly depend on agricultural activities such as plantation and paddy cultivation; however, the project development has led to the reconstruction of villagers' trip patterns due to the creation of employment opportunities and the development of roads. The study purposively selected rural villages dependent on agriculture and located near the Italia–Thai Development Public Company camp, which is the initial contractor with the Myanmar government for the Dawei Development Project (Figure 1.)

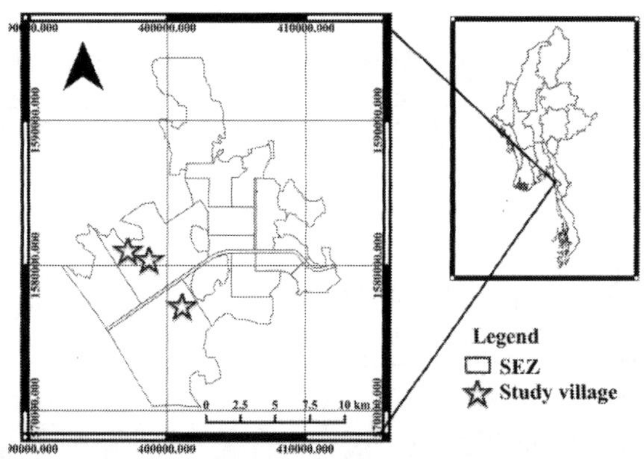

Figure 1. Special economic zone in Dawei.

2.2. Field survey data

A field survey was conducted in 2015 to collect information both for person trip behaviour and personal profiles. Stratified random sampling by sex and age was employed to understand trip characteristics. Non-spatial personal attributes such as age, sex, marital status, education level, household status, occupation and monthly household income were collected through a pre-tested questionnaire conducted earlier in 2014. Spatial information such as trip origin, destination, direction, distance, mode and duration in 2005, 2010 and 2015 was also collected. The study also employed formal and informal interviews with key informants such as village heads, using checklists. The data collection also involved direct field observation. The local language was used for communication with local people, assisted by a translator.

2.3. GPS log records

Wearable GPS devices such as " i-gotU USB Travel & Sports Logger – GT-600" were used to log trip data and validate the data obtained from the questionnaires. This device is lightweight (37g) and small (44 x 41.5 x 14 mm) with an automatic motion detector [4] which can be worn on the waist or clipped to clothes. The device recorded 24-hour trips with a 5-second interval using the motion detection mode. A maximum of 38 devices were distributed at one time to a total of 345 respondents aged 16 years plus. Both questionnaire and GPS log data sets are available from the 345 samples.

3. Methodology

3.1. Overall methodology

The following Figure 2 is the overall methodology employed in this study. The methodology focused on five major steps in order to achieve the main objective of integration of two data sets. First, questionnaire-based, non-spatial one-day trip data was manually converted to spatiotemporal data with the use of the respondents' house location collected by the GPS device and visualized. Second, trip parameters such as stay points, moving segments, number of trip, trip distance and trip duration were extracted, both from the questionnaire and GPS data. Third, errors of two data sets were calculated by the selected parameters such as number of trip, trip distance and trip duration. Fourth, errors of the selected parameters were further analysed by social parameters such as sex and age. Fifth, the trip behaviour in 2005, 2010 and 2015 was analysed.

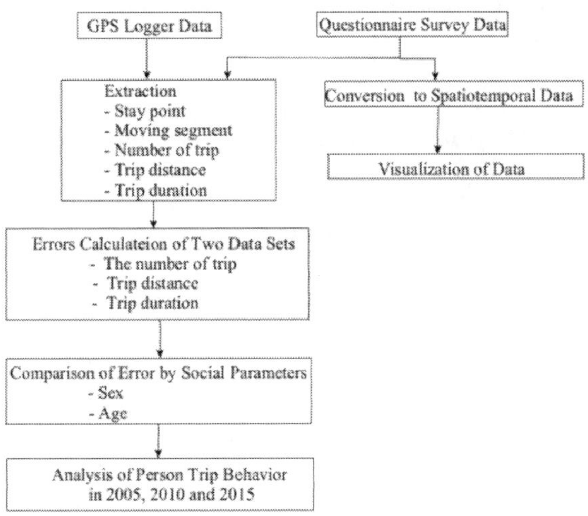

Figure 2. Overall methodology.

3.2. Process of questionnaire survey data

3.2.1. Conversion of questionnaire survey data to spatiotemporal data. Details of person trip obtained through interviews are ambiguous as, in rural areas, there are no significant landmarks such as prominent shopping centres and buildings as in big cities. In order to obtain a clearer picture of the trip behaviour of local people, trip information such as trip origin, destination, distance, direction and time in 2005, 2010 and 2015 was collected through the questionnaire survey. This information was manually converted to spatiotemporal information by utilizing Geographic Information Systems (GIS) and visualized in a moving format. The destination is determined using the original house location data recorded by GPS device, distance and direction. All spatiotemporal data are listed in a timeline and saved in a comma-separated values (csv) format. Simultaneously, non-spatial attributes such as age, sex, marital status, education level, household status, occupation and household income, were also integrated with the file.

3.2.2. Extraction of moving segments from questionnaire data. The total number of trip, the total trip distance and trip duration were manually extracted from the questionnaires and listed in excel format. In this study, a single trip is defined as staring from a location to a destination, such as from home to workplace. The total number of trip, distance and duration were summed for further data validation.

3.3. Process of GPS logger data

3.3.1. Stay point extraction from GPS logger data. After the trip data had been obtained with GPS loggers, spatiotemporal data such as time, latitude and longitude were extracted from the devices. The break up of the trip segment was acquired to find the stay points. In this study, stay point extraction with outlier detection and removal technique utilized by [5] were employed using the following equation (1):

$$Distance\ (p_{start}, p_{end}) < D_{threh}\ and\ TimeDiff\ (p_{start}, p_{end}) > T_{threh} \quad (1)$$

Where the parameters D_{threh}, considerable maximum distance as a stay point, and T_{threh}, minimum time spending at a same place, are adjustable. In this study, a stay point is detected if T_{threh} > 20 minutes and D_{threh} <= 300 metres. Based on the calculation, stay points are extracted and are listed by start-time, end-time, duration, distance in metres, average speed in kilometres/hour and the total number of stay points. Additionally outlier detection and noise removal technique were applied by using standard deviation (σ).

3.3.2. Moving segment extraction from GPS logger data. Once the stay points are extracted from the GPS data, the moving segments can be extracted. The extracted parameters in this study is total trip distance per trip in metre, duration per trip in minutes, starting time, ending time, average speed kilometres/hour and total points. These parameters were calculated by utilizing Java language. Furthermore, the selected parameters such as the number of trip, total trip distance and total trip duration from the GPS data were summarized by utilizing the PostgreSQL.

3.4. Error calculation of two data sets

After processing the two data sets, the following equation (2) was employed to calculate the error between the two data sets by the number of trip, trip distance and trip duration.

$$Relative\ Change\ (x, y) = Absolute\ difference / Max\ (x, y)*100 = |\triangle| / Max\ (x, y)*100 \quad (2)$$

Where x is the data from the questionnaires and y is the data from the GPS loggers. In this process, trips made out of villages such as in Yangon and Thailand, and unfixed trips such as daily employment at various places within or outside the villages, were excluded.

3.5. Comparison of errors by social parameters
The errors were further analyzed by social parameters such as sex and age. The age groups were categorized as (A) <20, (B) 21-30, (C) 31-40, (D) 41-50, (E) 51-60 and (F) 60< for this analysis. The difference from the average error was also calculated by the categories.

3.6. Analysis of person trip behaviour
Based on the level of error, the average trip behaviour in 2005 and 2010 and 2015 obtained from the questionnaires was further analysed in time-series.

4. Result and discussion
4.1. Conversion and visualization of person trip data
For visualization of person trip data from the questionnaire, the questionnaire-based data was converted (Figure 3) to the spatiotemporal data and visualized together with GPS log data (Figure 4). This person trip can be also displayed according to attributes. This visualization of person trip data provides more useful trip characteristics by attributes than text-based information in questionnaires. Illustrative visualization is easy to understand and more information can be obtained for a better understanding of the underlying tendency behind the data [6].

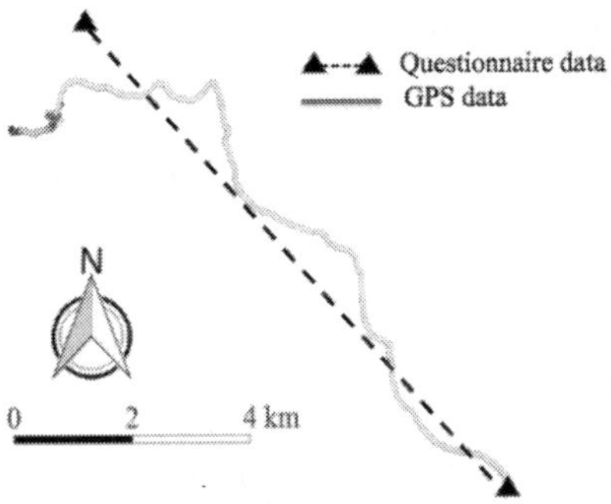

Figure 3. Comparison of questionnaire and GPS data.

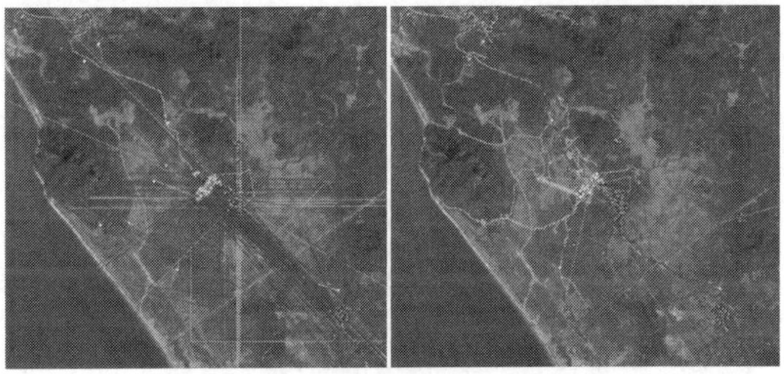

Figure 4. Comparison of visualized questionnaire and GPS data.

4.2. Comparison of two data sets
To validate the questionnaire-based data, the average errors between two data sets were calculated. The 238 data sets from a total of 331 samples are considered in this analysis. This reveals a total of 543 trips, with a total 1,936.3 km trip distance and 7,457.0 minutes of trip duration from the questionnaires while data from the GPS loggers shows a total of 403 trips, 2,302.0 km trip distance and 7403.6 minutes of trip duration. The result shows the average errors between the two data sets as 25.5% (A), 33.2% (B) and 37.2%, respectively (Table 1).

Table 1. Average errors of two data sets.

	Average differences (%)
(A) No. of Person Trip	25.5
(B) Trip Distance	33.2
(C) Trip Time	37.2

Both human and data processing errors are found to have significantly impacted the average errors within each of the above three parameters. Human error from the questionnaires, for example, involves under-reporting of short trips and exact trip routes. Because villagers did not report stays of more than 20 minutes on the way to a destination, this introduced errors in analysing GPS log data.

Additionally, errors from the GPS data processing are largely caused because of short trips of less than 300 metres, and short stays of more than 20 minutes.

4.3. Average error of person trip by sex and age

The level of errors by sex and age and differences from the average error are summarized in the Table 2 and Table 3, respectively. The notable error in the number of trip in male age group (A) and female age group (B) are found 17.6% and -13.4% more than the average rate of 25.5%, respectively. The average number of trip in the male age group (A) shows the highest time (5.5) in the questionnaires; however that of in GPS log shows only approximately half (2.5). This probably because stay points between each trip would be shorter than 30 minutes, thus some of trips are integrated to the previous trip during processing the GPS data in this age group. While in the female age group (B) shows lowest error probably resulting from a large engagement in project-related employments, including as office staff and laboratory workers at the project camp, travelling by motorbike to the workplace. As trip that they made were only two between home and workplace, thus the potential errors during processing data are minimized. On the other hand, for the average error of trip distance, the notable errors are found in the female age group (D) as 15.6%, while female age group (B) shows the lowest error (-15.3%) of the average. The female age group (D) shows the highest error of 15.6% as respondents in this group tend to be forgetful the trip that they made. Indeed, 27.3% of respondents show more than 88.3% of error. This tendency significantly affected on the female trip duration of the age group (D). For the female age group (B), this group show the lowest error among the whole group and this can be explained due to their working pattern. On the other hand, the average error of trip duration, the male age group (A) shows the lowest error (-15.1%) probably as 50% of respondents of this group made longer trips than average of total (50.9 minute). While, female age group (C) and (D) show notable errors of 10.2% and 13.1%, respectively. For the group (C), 77.8% of trip was made less than average total time, thus short trip resulted in the higher error.

Table 2. Error of trip behaviour by sex and age.

	No. of Trip		Distance		Trip duration	
Age	M	F	M	F	M	F
(A) <20	43.0	25.2	32.0	41.6	22.6	36.8
(B) 21-30	27.5	12.0	32.8	18.2	41.1	36.5
(C) 31-40	27.7	20.4	36.9	28.4	29.2	47.9
(D) 41-50	22.4	32.1	39.3	49.2	40.2	50.8
(E) 51-60	24.1	29.2	32.8	31.6	35.7	32.3
(F) 61<	19.9	22.2	30.5	25.2	41.9	31.7
Avg.	25.5		33.2		37.2	

Note: M and F stand for male and female, respectively.
Unit is percentage.

Table 3. Differences from the average errors.

	No. of Trip		Distance		Trip duration	
Age	M	F	M	F	M	F
(A) <20	17.6	-0.2	-1.5	8.4	-15.1	0.4
(B) 21-30	2.1	-13.4	-0.8	-15.3	3.4	-1.1
(C) 31-40	2.3	-5.0	3.4	-5.2	-8.5	10.2
(D) 41-50	-3.0	6.7	5.8	15.6	2.5	13.1
(E) 51-60	-1.3	3.8	-0.8	-2.0	-2.0	-5.3
(F) 61<	-5.4	-3.2	-3.0	-8.4	4.2	-5.9

Note: M and F stand for male and female, respectively.
Unit is percentage.

4.4. Change in person trip

Despite of errors revealed in the previous section, the level of errors are acceptable for analysing the trip data in 2015, 2010 and 2005 obtained from the questionnaire survey. Thus, trends of parameters in 2005, 2010 and 2015 were compared and the results of yearly change are described in Table 4. The results show that row (A) does not show so much difference over a period of 10 years. Row (B) increases yearly and the changes are 1.9 (2005-2010), 2.1 (2010-2015) and 4.1 (2005-2015) times. Considering no significant change in (A), the increase in (B) is probably caused by the increased trip distance per trip. Row (C) increased in 2010; however, again decreased to similar duration to 2005 in 2015. Based on the trend of (A), (B) and (C), it can be said that the main change in trip behaviour is the increase of the trip distance and this can result from the change in trip mode such as walking, travelling by motorbike, car or other modes. It can be confirmed that as the trip distance increases, the trip mode also changes. Indeed, the motorbike mode increased 5.9 times from 2005 (9.2 %) to 2015 (54.0 %).

Table 4. Average number of person trip, trip distance and trip duration by year.

	2005	2010	2015
(A) No. of Person Trip	2.3	2.1	3.0
(B) Trip Distance (km)	2.3	4.1	9.2
(C) Trip Duration (minutes)	40.6	58.4	46.0

5. Conclusion

Person trip survey is shifting from traditional questionnaire survey based to qualitative measurements with the advance of telecommunication technology. Integration of chronological questionnaire-based trip data to GPS survey and visualizing it in time-series can be very important to evaluate chronological trip changes with consideration to socio-economic and environmental events. In this study, questionnaire-based trip data were converted to spatiotemporal data and visualized. The average error between the questionnaire and GPS data sets in the average number of person trips, trip distance and trip duration were 25.5%, 33.2% and 37.2%, respectively. Furthermore, examination of error in the trip behaviour with respect to sex and age, found smaller errors among working age females mainly employed with the project-related activities. Furthermore, questionnaire-based trip behaviour made in 2005 and 2010 were further assessed and yearly increase of trip distance was found. Thus, visualizing questionnaire-based person trip data helps in tracing old trip behaviour, and the visualized trip data can be integrated to measurements by mobile phones and GPS devices. Furthermore, such assessment with respect to social parameters in time-series will improve understanding of the dynamic impacts from socio-economic events.

Reference

[1] Asian Development Bank, "Myanmar: Economy." [Online]. Available: http://www.adb.org/countries/myanmar/economy. [Accessed: 19-Sep-2015].

[2] World Bank, "Myanmar : Empowering People for Inclusive Growth," 2015.

[3] A. Min and T. Kudo, "Newly Emerging Industrial Development Nodes in Myanmar : Ports , Roads , Industrial Zones along Economic Corridors," *Emerg. Econ. Corridors Mekong Reg. (BRC Res. Rep. No.8)*, no. 8, pp. 187–230, 2012.

[4] Mobile Action, "MobileAction Technology - i-gotU Official Site." [Online]. Available: http://www.i-gotu.com/. [Accessed: 17-Sep-2015].

[5] A. Witayangkurn, T. Horanont, N. Ono, S. Yoshihide, and R. Shibasaki, "Trip Reconstruction and Transportation Mode Extraction on Low Data Rate GPS Data from Mobile Phone," 2013, no. June 2015.

[6] G. Andrienko and S. Augustin, "Visual Analytics Tools for Analysis of Movement Data."

Temporal geospatial analysis of secondary school students' examination performance

ND Nik Abd Kadir[1] and NA Adnan[2]

[1] Centre of Studies Surveying Science and Geomatics, Faculty of Architecture, Planning and Surveying,
Universiti Teknologi MARA, 40450 Shah Alam, Malaysia.

Email: nikdahlia@gmail.com

Abstract. Malaysia's Ministry of Education has improved the organization of the data to have the geographical information system (GIS) school database. However, no further analysis is done using geospatial analysis tool. Mapping has emerged as a communication tool and becomes effective way to publish the digital and statistical data such as school performance results. The objective of this study is to analyse secondary school student performance of science and mathematics scores of the Sijil Pelajaran Malaysia Examination result in the year 2010 to 2014 for the Kelantan's state schools with the aid of GIS software and geospatial analysis. The school performance according to school grade point average (GPA) from Grade A to Grade G were interpolated and mapped and query analysis using geospatial tools able to be done. This study will be beneficial to the education sector to analyse student performance not only in Kelantan but to the whole Malaysia and this will be a good method to publish in map towards better planning and decision making to prepare young Malaysians for the challenges of education system and performance.

1. Introduction

Education plays an important role in human capital development and is associated with prosperity and individual opportunity for a better life [1]. In order to achieve better student performance, the authority related to education sector realizes that high students' quality are needed to lead the country in the future. Unfortunately, quality is not easy to measure and define because of the complicated process increases due to different opinion of the interested parties with respect to individual quality attributes [2]. For example, the leader of the economic world, USA realizes that the importance of quality jobs depend directly on the ability of the education system that emphasizes producing of an excellent student in science and mathematics to join the industries. In addition, teaching and learning styles of mathematics and science should be improvised and become a priority to the education policy making agenda [3]. For that reason, this study focuses on the science and mathematics subjects to analyze secondary school performance with the aid of geospatial techniques. In mapping perspective, Malaysia Ministry of Education has improved the organization of the data to have the geospatial school database format. However, no further studied was done and the student performance results were not properly presented in graphical manners. The map has acted as a communication tool is not useful when the data are not published in the right way using the appropriate analysis data tool. To date, the portal of Ministry Malaysia Education is less efficient in presenting school performance information.

Sijil Pelajaran Malaysia (SPM) is a national examination organized by the Malaysian Examinations Board for student taken normally when their age is at 17 years old. SPM is a compulsory certificate for student to further the study to the next step level. Students would have a choice whether they want to continue the study to university, matriculation, overseas university, polytechnic, college or continue to take the STPM examination depend on the examination result they have. The analysis of school performance is not new and the researcher efforts to determine associations between school performance and catchment area was done by [4]. The study utilizing statistical techniques of Geographically Weighted Regression (GWR) to determine the spatial variation of the Mathematics school performance [5] was done.

The aim of this study is to find the potential used of geospatial techniques for school student examination performance in science and mathematics score in the year 2010 to 2014. The objectives of this study are to analyse school student performance of science and mathematics subject score for the SPM examination result using geospatial techniques and to produce GPA scoring map for these subjects from the year 2010 to 2014.

2. Methods and Materials

2.1. Area of study

The Kelantan state was chosen as a study area. The data involved GIS layers such as the boundary of Kelantan state and territories, school data (shape format) which provided by the Cartography Section of the Malaysia Survey and Mapping Department (JUPEM) and the SPM examination result (excel and .doc format) derived from the Department of Kelantan Education. All the GIS layer data use the Kertau Rectified Skew Orthomorphic (RSO) coordinate system. Only two SPM's subjects' performance which is the Science and Mathematics Grade Point Average (GPA) score were used for further analysis in this study. Figure 1 shows the study area of Kelantan includes the territory of Tumpat, Kota Bharu, Pasir Mas, Bachok, Pasir Puteh, Tanah Merah, Machang, Jeli, Kuala Krai and Gua Musang with the schools location point.

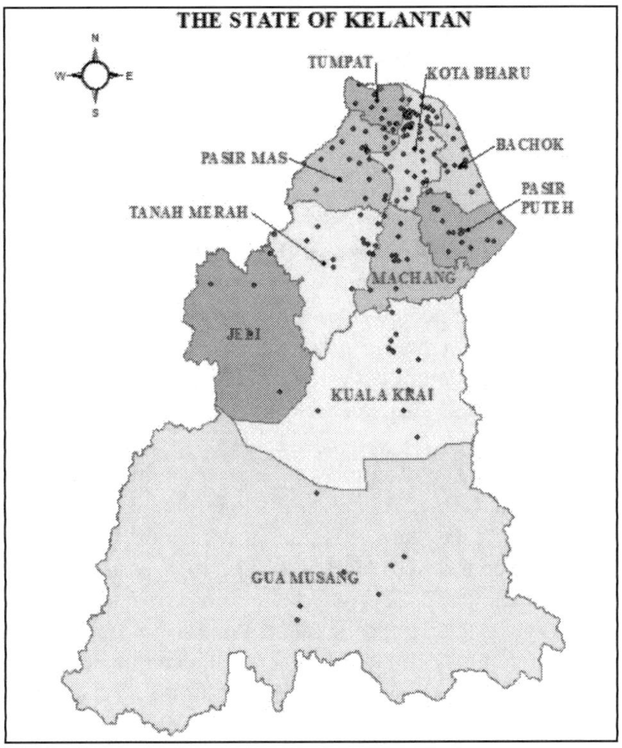

Figure 1. Study area of Kelantan and the location of schools

Table 1 shows the subject Grade System that has been used by the Malaysia Ministry of Education since 2009. The grade value is used as a reference for the analysis of Mathematics and Science subjects.

Table 1. SPM Grades System

Grades	Grade Value	Interpretations
A+	0	Super Distinction
A	1	High Distinction
A-	2	Distinction
B+	3	Super Credit
B	4	High Credit
C+	5	Upper Credit
C	6	Credit
D	7	Upper Pass
E	8	Pass
G	9	Fail

There are three category data in this study, first is the Science and Mathematics which are in word and excel format, second is the school point location in the ArcGIS format of shape and third is the softcopy base map data. The attribute data contains the name of school and the Grade Point Average (GPA) data. The GPA range value divided into 1 to 9 according to SPM system Grade. The relational database concept was used to relate and join excel and shape format. This study utilizing the inverse distance weighted (IDW) as an interpolation method to map the trend surface of Grade Point Average (GPA) of Science and Mathematics as a variable of this project. The IDW method is based on the supposition that the neighbouring point contributes more interpolation than the farther one. The IDW method states that impact of known data point is inversely proportion to the distance from unknown to know data point [6].

3. Result and discussion

3.1. Geospatial science and mathematics scoring performance pattern distribution analysis

The IDW interpolation method is used to derive spatial distribution patterns of SPM school performance for the years 2010 to 2014 from the GPA score data. Figure 2 and Figure 3 show the results for the spatial distribution of the Science and Mathematics GPA score for the year 2010 to 2014 in Kelantan state. The result indicated interesting trends of GPA for the Kelantan schools. The IDW interpolation map of GPA for the Mathematics score consist of nine (9) GPA values as proposed by the Malaysia Ministry of Education grade system (Table 1) known as a GPA value of 0 – 2 for the grades of A+ to A-, 2-4 for the grades of A- to B, 4-6 for the grades of B to C and 6-8 for the grades of C to G.

Overall, the result indicated that a high number of students achieved the super and high distinction for the grade value 0-2 consist of A+, A, A- for both subject in year 2010 to 2014 in the Kota Bharu territory. The Kota Bharu is the capital state of Kelantan which has the 48 out of 144 selected schools comprising boarding school, excellent regular school and also some of satisfactory regular school. However, in the year 2014 the results of GPA show declining trend when the GPA color for top score performance is reduced obviously for Science subject. Furthermore, the spatial distribution map also indicated that many territories in the Kelantan state have achieved top score and also super credit (B+) and high credit (B) for student performance as shown by the Pasir Mas, Pasir Puteh, Bachok, Tanah Merah, Machang and Tumpat. From the generated spatial distribution map the Jeli territory has resulted the highest GPA score as compared to the others territory for the science and mathematics SPM examination results. This is because the Jeli territory has the lowest number of schools located only five (5) primary schools with one school of science category school, thus make the result dominated by high performance color category (Figures 2 and 3). The Gua Musang territory presented the lowest GPA performance result with upper credit (C+), credit (C), upper pass (D) and pass (E) for both SPM subjects.

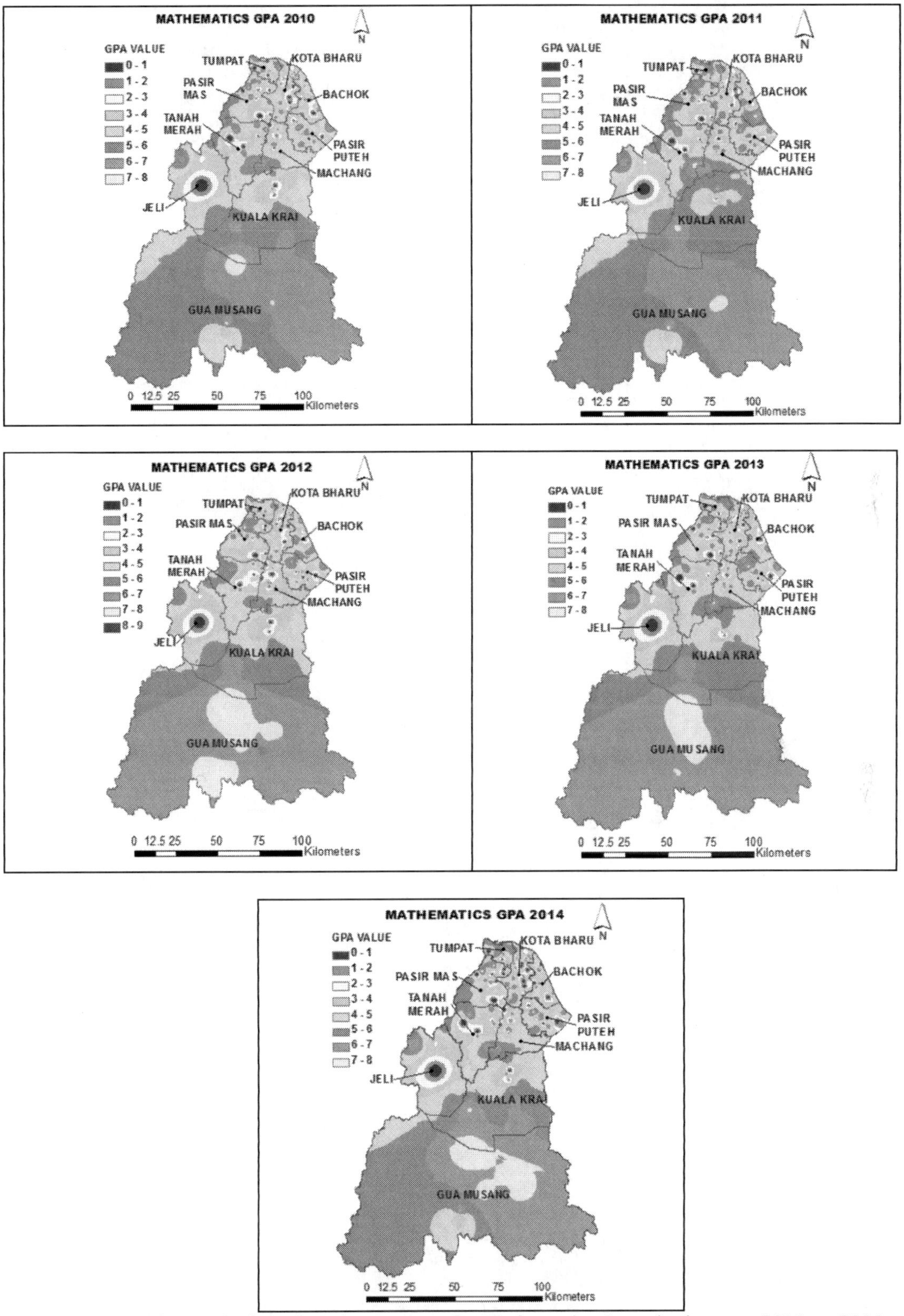

Figure 2. The trend of Grade Point Average for Mathematics Score in the years 2010 to 2014

IOP Conf. Series: Earth and Environmental Science **37** (2016) 012020 doi:10.1088/1755-1315/37/1/012020

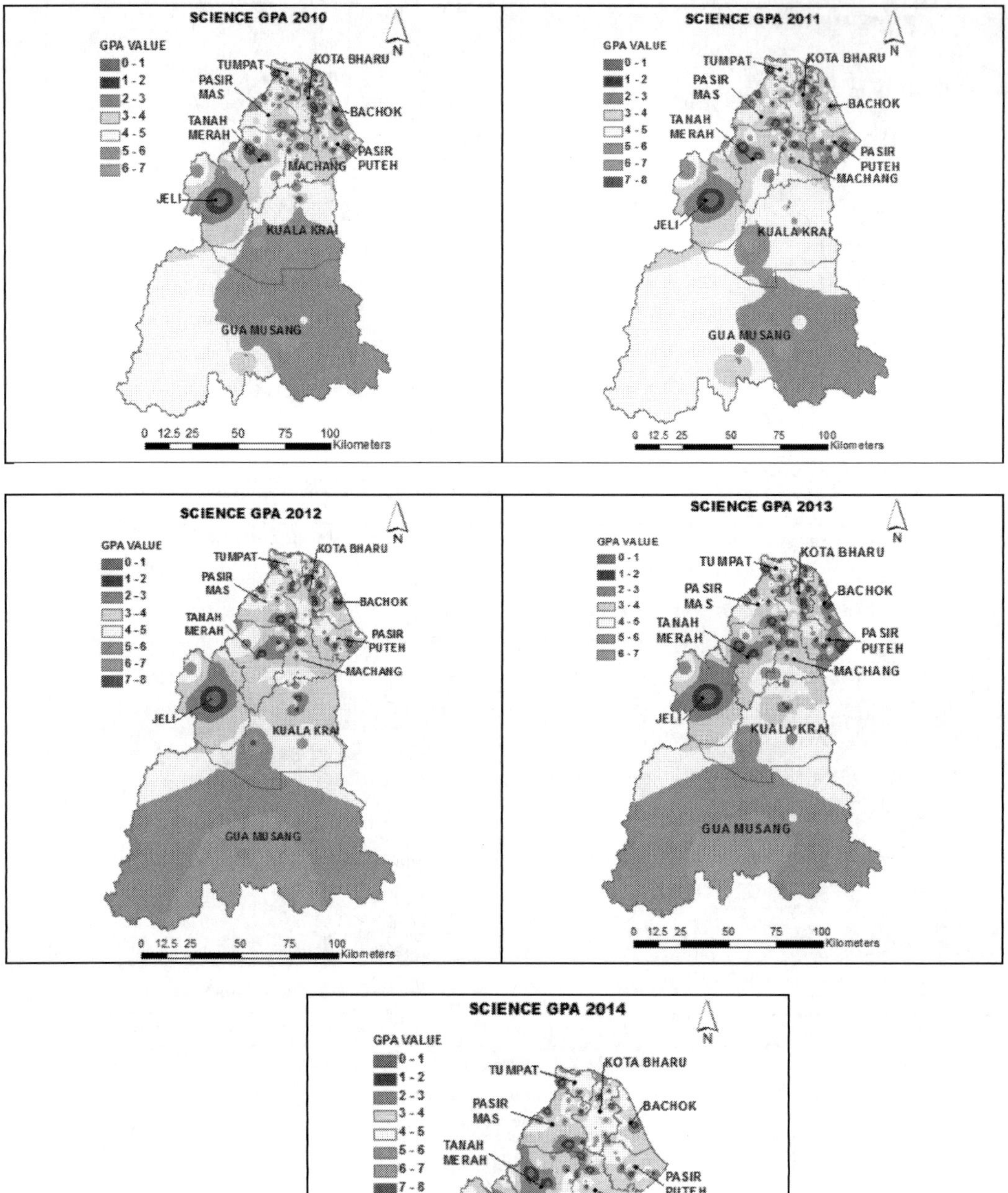

Figure 3. The trend of Grade Point Average for Science Score in the years 2010 to 2014.

Meanwhile, for the SPM 2014 GPA result, [7] stated that the result shows decline in several attainment, but it still on acceptable GPA and not significantly different from the previous three (3) years prior to 2013. The results can be considered as satisfactory results for candidates who sat for SPM 2014.

4. Conclusion

The used of GIS in education analysis particularly on Science and Mathematics score from the years of 2010 to 2014 is successfully done to achieve the objective of this study. The use of the IDW tool of interpolation is a highly potential approach to produce spatial distribution patterns of school's performance trend. Regarding the student performance result, the result indicated that in the overall, the Kelantan has good school performance of the SPM examination from 2010 to 2015 for every territory except for the Gua Musang with average to low performance result. This study able to provide effective ways of interpreting school performance results statistically and graphically not only to education management policy makers, but also benefited to school teachers as well as parents.

Acknowledgements

Acknowledgements for Department of Survey and Mapping Malaysia, Ministry of Education Malaysia and Department of Kelantan Education.

References

[1] Anonymous 2005 Keeping America competitive: Five strategies to improve mathematics and science education, reported by education commission of the states, http://www.ecs.org/clearinghouse.

[2] Battle, J., & Lewis, M. 2002 The increasing significance of class: The relative effects of race and socioeconomic status on academic achievement. Journal of Poverty, 6(2), 21-35

[3] Blevins, B. M. 2009 Effects of socioeconomic status on academic performance in Missouri public schools. Retrieved from http://gradworks.umi.com/3372318.pdf

[4] Brown, P., Marsden, J., Batey, P. & Hirsch® eld, A. 1998 Relationships between pupil performance and social conditions: a GIS-based analysis using geodemographics. Paper presented at the conference `Investigating Locational Data', Lancaster University, 9± 10 July.

[5] Brunsdon, C., Fotheringham, A.S. & Charlton, M.E. 1996 Geographically weighted regression: a method for exploring spatial nonstationarity. *Geographical Analysis*, 28, 281± 298.

[6] M. A. Azpurua and K. D.Ramos, 2010 "A comparison of spatial interpolation methods for estimation of average electromagnetic field magnitude," Progress In Electromagnetics Research M, vol. **14**, pp. 135–145.

[7] M Yassin 2015 Newspaper Statement by Ministry of Education, SPM Examination Result 2014 on 03 MAC 2014 http://www.moe.gov.my/v/kenyataan-media view?id=4671&page=1&keyword=& accessed on 25/5/2015.

Human leptospirosis distribution pattern analysis in Hulu Langat, Selangor

Zuhafiza Zulkifli[1], Abdul Rashid Mohamed Shariff[1], Zakri M. Tarmidi[12]

[1]Faculty of Engineering, Universiti Putra Malaysia, 43400 UPM Serdang, Malaysia

[2]Department of Geoinformation, Faculty of Geoinformation and Real Estate, Universiti Teknologi Malaysia, 81310, Johor Bahru, Malaysia.

Corresponding Authors: zakritarmidi@gmail.com

Abstract. This paper discussed the distribution pattern of human leptospirosis in the Hulu Langat District, Selangor, Malaysia. The data used in this study is leptospirosis cases' report, and spatial boundaries. Leptospirosis cases, data were collected from Health Office of Hulu Langat and spatial boundaries, including lot and district boundaries was collected from the Department of Mapping and Surveying Malaysia (JUPEM). A total of 599 leptospirosis cases were reported in 2013, and this data was mapped based on the addresses provided in the leptospirosis cases' report. This study uses three statistical methods to analyze the distribution pattern; Moran's I, average nearest neighborhood (ANN) and kernel density estimation. The analysis was used to determine the spatial distribution and the average distance of leptospirosis cases and located the hotspot locations. Using Moran's I analysis, results indicated the cases were random, with a value of -0.202816 which show negative spatial autocorrelation exist among leptospirosis cases. The ANN analysis result, indicated the cases are in cluster pattern, with value of the average nearest neighbor ratio is -21.80. And results also show the hotspots are has been identified and mapped in the Hulu Langat District.

1. Introduction

Ministries of Health, under the Planning and Development Divisions has reported the increasing number of leptospirosis cases from 263 in 2000 to 1418 in 2009, as shown in Table 1 [1]. To reduce the disease spreading, prevention plan of case morbidity and mortality of leptospirosis disease was needed, and control plan using Prevention and Control of Infectious Diseases Act 1988 leptospirosis has been gazetted as a notifiable disease on 9 December 2010.

The locations of leptospirosis disease were needed to answer the questions related to leptospirosis cases such as where the case happened. When the case happened? How the case happened? What are the action that have been taken to prevent this case and other questions have to be answered as soon as possible in order to make other people in the area affected felt they are in a safe condition [1]. This question can be answered using spatial pattern distribution, where the locations of leptospirosis can be collected, stored and analyze to determine the disease spreading pattern, hotspot area and prevention plan to reduce the cases.

This paper focuses on the spatial analysis to determine the distribution pattern of leptospirosis cases in the Hulu Langat District, Selangor. The main aim of the study is to facilitate Health Department of Malaysia in utilizing modern technology to analyze and act on identifying environmental risk factors for human leptospirosis in order to arrange strategies for disease control and prevention of case morbidity and mortality.

2. Background

Leptospirosis is a bacterial infection caused by a species of pathogenic Leptospira genus called Spirochetes, which can be found in fresh water contaminated by animal urine [2-4]. This disease is known worldwide as the most common re-emerging zoonotic disease, and infectious with a broad range of clinical manifestations, ranging from mild flu-like illness to very severe disease with hemorrhagic manifestations and multiorgan failures [1, 5, 6]. Leptospirosis also known as weil disease, icterohemorrhagic fever, swineherd's disease, rice-field fever, cane-cutter fever, swamp fever, mud fever, hemorrhagic jaundice Stuttgart disease and canicola fever [3]. Annually, approximately 500,000 severe cases occur worldwide, with a mortality rate of 5–20% [7].

In Malaysia, the first fatal case was reported by Fletcher in 1925, where he able to identify the disease from twenty one patient [1]. From 2004 to 2009, the leptospirosis case report is increasing and causes the significant number of deaths, and it is shown to be a common cause of acute febrile illness in Malaysia [1, 2]. Table 1 shows the leptospirosis cases reported in Malaysia from 2004 until 2009.

Table 1: Leptospirosis cases in Malaysia from 2004 to 2009 [1].

N	State	Year											
		2004		2005		2006		2007		2008		2009	
		Cases	Death	Cases	Death	Cases	Death	Cases	Death	Cases	Death	Cases	Death
1	**Perak**	29	4	71	4	93	3	149	3	289	16	280	19
2	**Selangor**	16	5	20	3	37	0	93	0	97	2	208	7
3	**Pahang**	29	1	24	0	51	3	184	3	198	7	184	5
4	**Kelantan**	15	1	38	1	17	1	81	1	180	4	138	4
5	**Terengganu**	7	1	17	0	42	1	55	3	107	4	126	9
6	**Kedah**	15	1	27	0	31	0	28	1	52	2	106	9
7	**N. Sembilan**	27	1	41	0	24	0	49	0	59	0	91	0
8	**Sarawak**	32	2	42	2	37	3	46	1	58	5	70	4
9	**Johor**	30	1	29	6	31	2	115	2	87	3	59	3
10	**WP KL**	31	0	20	1	27	1	31	0	45	0	54	0
11	**Sabah**	13	1	12	1	19	0	41	2	34	2	35	1
12	**P. Pinang**	9	0	25	1	28	0	37	0	25	2	32	0
13	**Melaka**	7	1	9	1	79	2	32	1	25	0	20	1
14	**WP Putrajaya**	1	0	1	0	3	0	3	0	1	0	14	0
15	**Perlis**	2	1	2	0	5	0	5	0	6	0	1	0
16	**WP Labuan**	0	1	0	0	0	0	0	0	0	0	0	0
	Total	**263**	**20**	**378**	**20**	**527**	**22**	**949**	**22**	**1263**	**47**	**1418**	**62**

To help in reducing the leptospirosis cases, it is important to map the area of leptospirosis cases to implement the prevention and control program [9, 13]. Using spatial analysis, the distribution pattern can be mapped, and the trend of the disease spreading can be predicted and control program can be implemented in this area.

A research study by Soares (2010) shown the leptospirosis incidence and lethality rates correlated with the population's socioeconomic conditions, independent of the period in Sao Paulo, Southeastern Brazil [3]. The variables were analyzed using the global and local Moran indices to identify the spatial pattern (dispersed, clustered or random). Spearman's correlation coefficient was used to test the associations between variables with clustered spatial patterns.

Another study by Yimsamran and Dasananda, (2011) used to apply Geoinformatics technology and spatial statistic to analyze the leptospirosis cases in Thailand during 2003 until 2009 [4]. Exploratory spatial data analysis (ESDA) was done to demonstrate the geographical distribution of the disease, and Kulldorff's scan statistic was used to detect significant area using spatial clusters, temporal clusters and spatiotemporal clusters of leptospirosis in Thailand provinces. Result from this study shown the most highly transmission of leptospirosis is in the northeast area of Thailand, which will need closer surveillance and monitoring.

3. Research methodology

This study involves three phases; background study, data acquisition and data analysis. The first phase was to identify the research topics, scope and objectives, where the study has identified the leptospirosis spreading cases locations, based on the epidemiology of leptospirosis cases, factors for emergence of leptospirosis, prevention and control, and spatial distribution pattern of leptospirosis. The second phase is data collection, data mining process, database design and development and database integration. Data was collected in the Hulu Langat District, Selangor Malaysia, and data mining was used to locate the leptospirosis cases, and make it spatially enable for the data analysis process. Figure 1 show the study area for this research. The third phase was to analyze the pattern distribution of leptospirosis cases; Global Moran's I was applied to determine the spatial autocorrelation based on both feature locations and values, Average Nearest Neighbors was applied to analyze the leptospirosis spatial distribution and Kernel density estimation was applied to locate the hot spot location for leptospirosis cases.

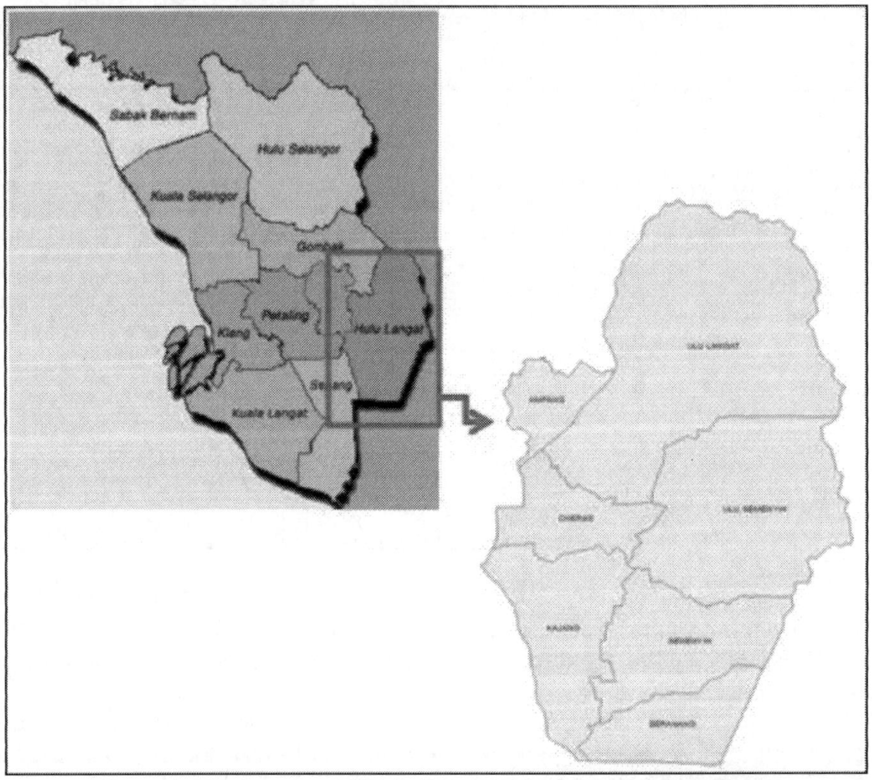

Figure 1: The location of Hulu Langat District in Selangor State.

4. Results

To obtain the results in this study, the analysis was conducted into descriptive and interferential analysis. The descriptive analysis purpose is to obtain the statistical result of leptospirosis cases based on time (month) and location. The inferential analysis was conducted to identify the distribution pattern of the leptospirosis cases, whether it's cluster, random or dispersed.

4.1. Descriptive analysis

The first analysis was to identify the distribution pattern of leptospirosis cases based on time and location. The data used was recorded in 2013, in the Hulu Langat District, which consists of seven mukim, including Ampang, Beranang, Cheras, Kajang, Semenyih, Ulu Langat and Ulu Semenyih. Leptospirosis cases in the Hulu Langat district show there is 599 cases was reported in 2013, with Ampang has highest leptospirosis cases (316), followed by Kajang (115), Cheras (55), Ulu Langat (46), Semenyih (44), Beranang (20) and Ulu Semenyih (3). From the analysis, it shows the distribution of human leptospirosis is random in Ulu Semenyih, clustered pattern in Ampang, Cheras, Kajang and certain area in Ulu Langat, Semenyih and Beranang. Figure 2 shows the leptospirosis distribution pattern of the Hulu Langat District. From the analysis, it is also shown the leptospirosis cases highest number in July (77), October (64) and November (62) as shown in figure 3.

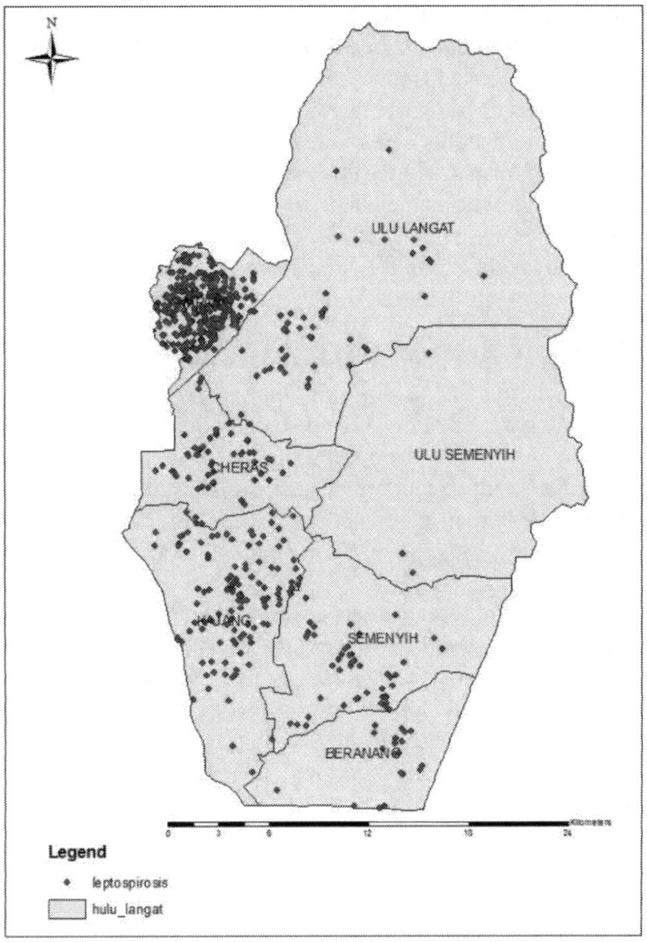

Figure 2: Distribution map of human leptospirosis cases in Hulu Langat District (2013).

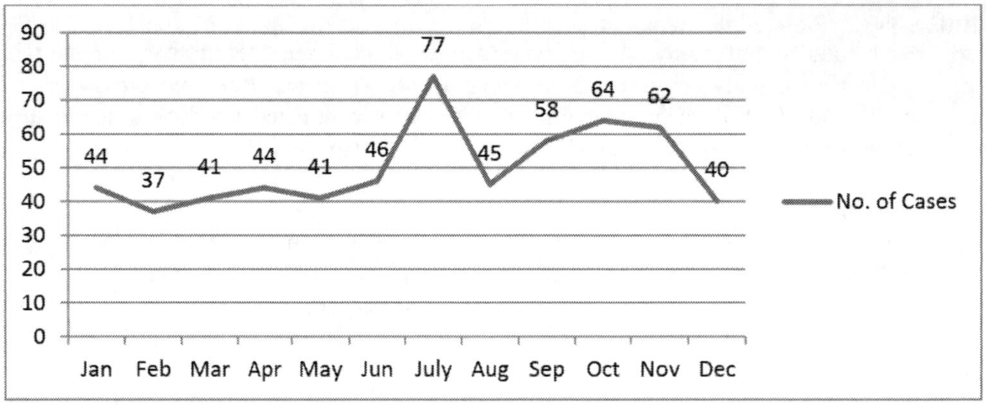

Figure 3: The graph of human leptospirosis cases in Hulu Langat district in 2013 based on Month.

4.2. Inferential analysis

The aim of the inferential analysis is to identify the distribution pattern of leptospirosis cases in the Hulu Langat District. To get the results, three analysis methods were conducted; Global Moran's I, Average Nearest Neighbor, and Kernel Density. Global Moran's I analysis was selected to identify the spatial correlation of the leptospirosis cases area; either it's dispersed, random or clustered. Global Moran's I were based on locations and data values, with simultaneous analysis. The null hypothesis for Global Moran's I was the distribution values were randomly distributed among features in the dataset. The results show there were a negative spatial correlation exist among leptospirosis cases within the Hulu Langat District, with result of cases is -0.202816, while z-score is -0.184058 with a p value of 0.853968. From the result, the null hypothesis was accepted; it shows the spatial distribution of the leptospirosis cases in the Hulu Langat District is spatially random.

The second analysis to analyze the leptospirosis cases spatial distribution using Average Nearest Neighbors analysis. The null hypothesis for this analysis is the leptospirosis cases are randomly distributed in the Hulu Langat District. The result shows the average nearest neighbor ratio is 0.534353 with p: 0.0000. With the value of z-score of -21.80, there is less than 1% likelihood that this clustered pattern could be the result of random chance, so the null hypothesis is rejected, and from the result, it can conclude the leptospirosis cases pattern in the Hulu Langat District is exhibiting a cluster pattern.

The last analysis is to locate the leptospirosis cases hot spot area in the Hulu Langat District, using Kernel density estimation analysis and the result of this analysis is the density map of the leptospirosis cases. Figure 4 shows the leptospirosis density map, with the dark red color area is the maximum leptospirosis incidence density, and the lighter color is the area with lower cases of leptospirosis. From the result, the most affected area was located in Ampang, with hot spot area including Taman Kosas, Kampung Tasik Tambahan, Jalan Pandan Indah and Taman Cempaka. Besides that, Sungai Chua in Kajang also detected as one of the high density area of leptospirosis cases.

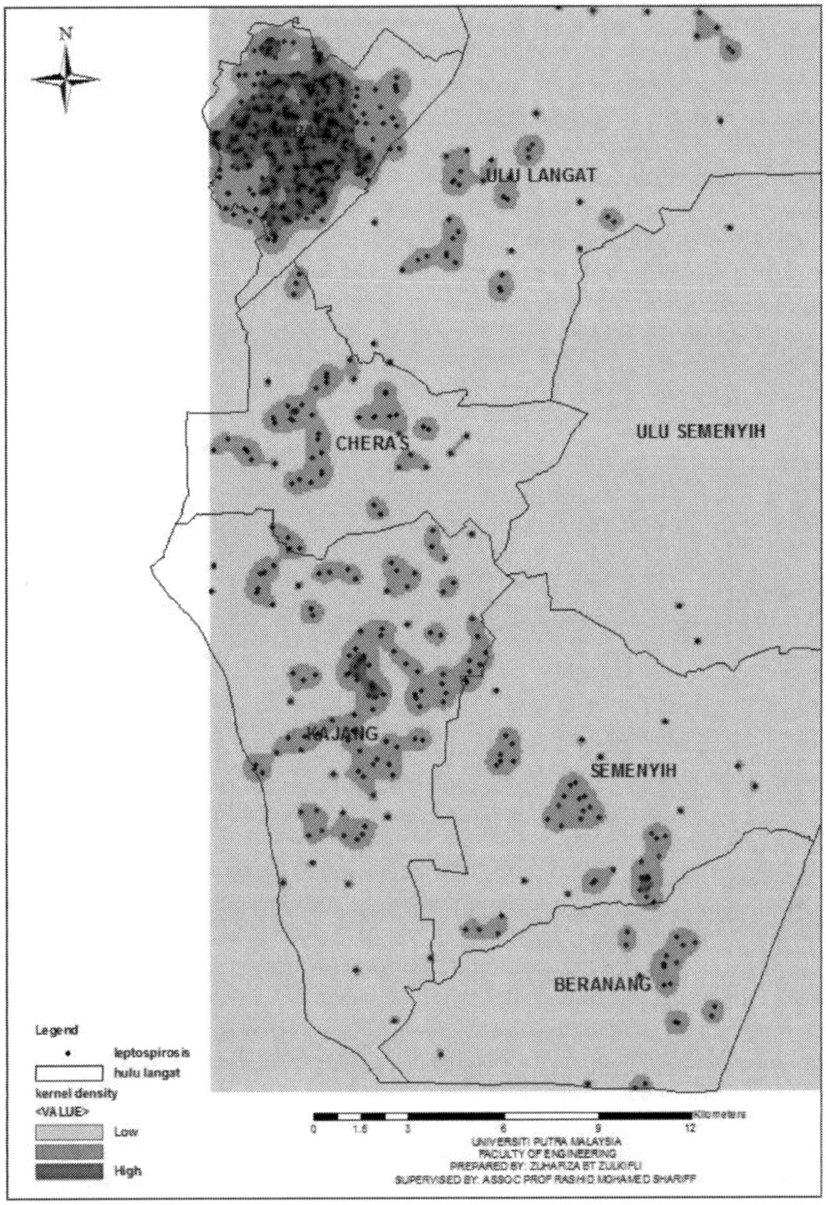

Figure 4: The hotspot map of human leptospirosis cases in Hulu Langat district in 2013.

5. Discussion

Several analyses were conducted to identify the distribution pattern of leptospirosis cases in the Hulu Langat district, including descriptive and inferential analysis. Analysis based on location show the distribution pattern of leptospirosis cases is clustered in Ampang, Kajang, Cheras and Semenyih, and random in Beranang, Ulu Langat and Ulu Semenyih. Analysis based on month, show the highest number of cases is on July 2013; with mukim Ampang contributes the highest number of leptospirosis cases.

From the inferential analysis, from Moran's I analysis, it indicated the negative spatial autocorrelation exist among leptospirosis cases within the district, and thus the null hypothesis is accepted where the leptospirosis cases in the district distribute spatially random. However, result from

ANN analysis showed the average nearest neighbor ratio is less than 1 (0.534353 with p = 0.00), which conclude the leptospirosis cases in the Hulu Langat district is exhibiting a cluster pattern, where it indicates the each case happened were near between each other, and most the hot spot locations of leptospirosis cases were located in a residential area in Ampang and Kajang.

To improve the results of the analysis, a recommendation for future research is; to include data of leptospirosis cases for five or more years, integrating the analysis of leptospirosis cases with other disease cases, especially dengue, to identify the relationship between different diseases, and included other spatial data, such as land use, demographic data, and water body, to see the relationship between these data with increasing number of leptospirosis cases. Besides that, a larger area of leptospirosis cases can also be added, to see other district pattern, and how to plan a prevention action in the Health Department of Malaysia.

6. Conclusion

This study has identified the distribution pattern of leptospirosis cases in the Hulu Langat District, Selangor Malaysia. This study is significant in identify the distribution pattern, and it can help Malaysia's Health Department to plan for preventing the spreading of the disease.

7. Acknowledgment

A special thanks to Pejabat Kesihatan Daerah Hulu Langat and member of Faculty of Engineering, Universiti Putra Malaysia.

References

[1] M. Ministry of Health. (2011). *Guidelines for the Diagnosis, Management, Prevention and Control of Leptospirosis in Malaysia*. Available: http://www.jknselangor.moh.gov.my/documents/pdf/sharingDoc/kawalan_penyakit_berjangkit/guidelines/Leptospirosis.pdf.

[2] C. H. Chang, M. Riazi, M. H. Yunus, S. Osman, and R. Noordin, "Limited diagnostic value of two commercial rapid tests for acute leptospirosis detection in Malaysia," *Diagnostic Microbiology and Infectious Disease,* vol. 80, pp. 278-281, 12// 2014.

[3] T. S. M. Soares, M. d. R. D. d. Latorre, G. Z. Laporta, and M. R. Buzzar, "Spatial and seasonal analysis on leptospirosis in the municipality of São Paulo, Southeastern Brazil, 1998 to 2006," *Revista de saude publica,* vol. 44, pp. 283-291, 2010.

[4] S. Yimsamran and S. Dasananda, "Pattern of Leptospirosis Incidence in Thailand: A Spatiotemporal Analysis," *Suranaree Journal of Science & Technology,* vol. 18, 2011.

IOP Conf. Series: Earth and Environmental Science 37 (2016) 012022 doi:10.1088/1755-1315/37/1/012022

Reclamation of mosquito breeding sites using Landsat-8 remote sensing data: A case study of Birnin Kebbi, Nigeria

Danboyi Joseph Amusuk, Mazlan Hashim[1], Amin Beiranvand Pour

Geoscience and Digital Earth Centre (Geo-DEC), Universiti Teknologi Malaysia

Abstract

It is believed by recent releases of World Health Organization (WHO) that more than half of the world's population (3.2 billion) live in areas that are at risk of malaria transmission. Although increased efforts are dramatically reducing the malaria burden in some places where the rate of new cases indicates a fall by 37% globally and 60% death rate. Unfortunately, the sub-Saharan Africa still shares 89 % of malaria and 91% of malaria deaths. Essentially, attacking the causative vectors and reclamation of the vector breeding sites could be remarkable for the rolling back the malaria epidemic project. Consequently, it is essential to explore the possibility of using recent Landsat-8 data remote sensing data and applications of Geographic Information System (GIS) technique in contributing to the realization of this objective. This investigation used for identifying mosquito breeding habitat (Derelict Ponds) zones the application of supervised classification of the Landsat-8 image in conjunction with GIS layering which allowed identification of high risk prone regions for mosquito breeding habitat. The methodology delineated 10 spatial locations of the Derelict Ponds (DP) spread around the Birnin Kebbi urban environment. Moreover, the results combined with comparative analysis of the link between warm climatic (temperature and rainfall data) conditions and Malaria prevalence that is associated with urban poverty. This study indicates that the application of Landsat-8 data and GIS techniques can be a useful tool for planning and management of environmental health and mapping of hot spot environmental problem areas.

1. Introduction

Rolling back Malaria from the depressed ground surfaces that emanated from human activities such as earth mining in the Birnin Kebbi environment. The capital of Kebbi state is not to neglect the pre-imaginable mosquito habitats existing and facing treatment or curative measures only but to find the interface between the host and the vectors (Derelict Ponds). This should form a critical audit of the environment by quantifying the applicability of preventive and the curative control measures. These ponds fill with water during rainy seasons bringing to people high cases of plasmodium falciparum within periods of time. It is

[1] Corresponding Author. Tel: +607 -5530666; Fax: +607- 5531174; Email address: mazlanhashim@utm.my.

one of the major medical challenges in the region. The medical centres within the urban environment records high percentage of malaria cases annually especially during the rainy seasons that occur between the months of June and October (rainy season).

It is evident that the malaria risk is connected to a wide variety of derelict mining ponds that existed in the environment. The mining ponds can be found everywhere in the state, but many of them have been abandoned due to lack of proper planning. Some of them are used as illegal refuse dump sites. The objective of the paper is to use Landsat-8 imagery and GIS techniques to delineate the positions of these mosquito breeding habitats (Derelict Ponds), around the study area and the level of their abandonment whose consequence bring about the malaria incidences.

2. Methodology

Birnin Kebbi city is located in north-western Nigeria (Figure 1). It is the capital city of Kebbi State and headquarters of the Gwandu Emirate. As of 2007 the city had an estimated population of 125,594. The geology of Kebbi State is dominated by two formations of the pre Cambrian Basement Complex in the south and south east and young sedimentary rocks in the north. The Basement Complex region is composed of very old volcanic and metamorphic rocks such as granites, schists, gneisses, quartzites and migmatites. In addition, there are meta-sediments such as phyllites and meta-conglomerates. The sedimentary region consists of rocks of the Gwandu, Illo and Rima groups whose ages range from the Cretaceous to the Eocene. The Gwandu group consists of massive clay grits inter-bedded with sandstone while the Illo and Rima groups consist of pebbly grits, sand stones and clays, mudstones and siltstones respectively. Minerals that can be found in the state include quartz, kaolin, piotolitic bauxite, clay, and potassium, silica sand and salt [1].

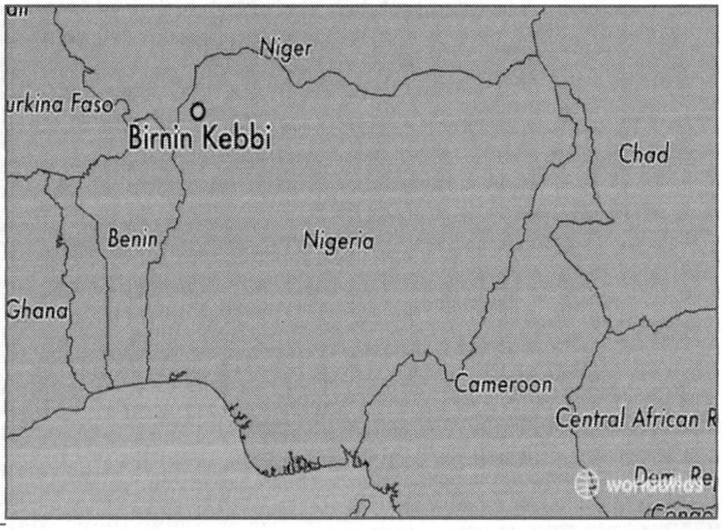

Figure 1. Location of Birnin Kebbi city in the north-western Nigeria.

In this study, Landsat-8 data of the north-western Nigeria were used for identification of Derelict Ponds (DP) location. A cloud-free level 1T (terrain corrected) Landsat-8 image were obtained through the U.S. Geological Survey Earth Resources Observation and Science Center (EROS) (http://earthexplorer.usgs.gov). It was acquired on May 3, 2014 for the study area. The image map projection is UTM zone 40 North (Polar Stereographic for Antractica) using the WGS-84 datum [2, 3]. Malaria data and GPS ground surveys were also utilized for this investigation. The data acquired for this study is from the annual returns of the Sir Yahaya Memorial Specialist Hospital for the cases of malaria and typhoid fever for a period of nine years (2006 to 2014), which shows the annual number of patients, treated of malaria and typhoid fever (Table 1). GPS ground surveys were carried out at the ten locations of the ponds obtaining the latitude and Longitudes of the ponds where obtained. The areas of these ponds were calculated from the coordinates obtained from the GPS observations.

Table 1: Malaria and Typhoid cases during 2006 to 2013.

YEAR	PERIOD	MALE	FEMALE	DISCHARGE	DEATH
2006	Jan-Dec	191	237	392	36
2007	Jan-Dec	202	225	383	44
2008	Jan-Dec	205	234	398	35
2009	Jan-Dec	181	219	378	32
2010	Jan-Dec	227	264	447	44
2011	Jan-Dec	207	238	405	40
2012	Jan-Dec	199	230	390	48
2013	Jan-Dec	194	239	387	46
2014	Jan-Dec	214	251	423	44

This work was carried out using ENVI and ArcGIS environments to derive information about the locations and features of ponds. The Landsat 8 image covering the study area was enhanced in the ENVI environment by contrast stretching method, this method of image processing enable transformation of the multispectral image data into an image display that reveals the background of information of each pixel of the image as well as layer stacking process. The stacked layered image was exported to ArcGIS environment for further processing. The image data was added as input raster into ArcGIS then the RGB combinations was done according to the choices of bands that will enhance the identification of details. The image Supervised classification technique was undertaken and it is a task for extracting information classes from a multi band raster image. The supervised classification procedure was adopted to categorize all pixels in a digital image to be one of the several classes or themes and come up with a thematic map. The spatial pattern within the data for each pixel was used as the numerical basis for categorization [4].

3. Result and discussion

Figure 2 shows supervised classified image map for Birnin Kebbi city derived from Landsat-8. Variety of different information was extracted for the study area. Derelict ponds are appeared as red color in north-east, center and western segment of the scene. Most of the derelict ponds are located in residential (build up) areas. However, some of them are associated with hilly areas that appeared as yellow color. Several small derelict ponds were detected in residential area as red pixel especially western part of the image map. Rice husk appears light green color in the north part of the image. Residential (build up) region manifested as dark blue and bare land as gray pixels, which are distributed in central and southern segments of the image. Water body and river are depicted in light mauve hue.

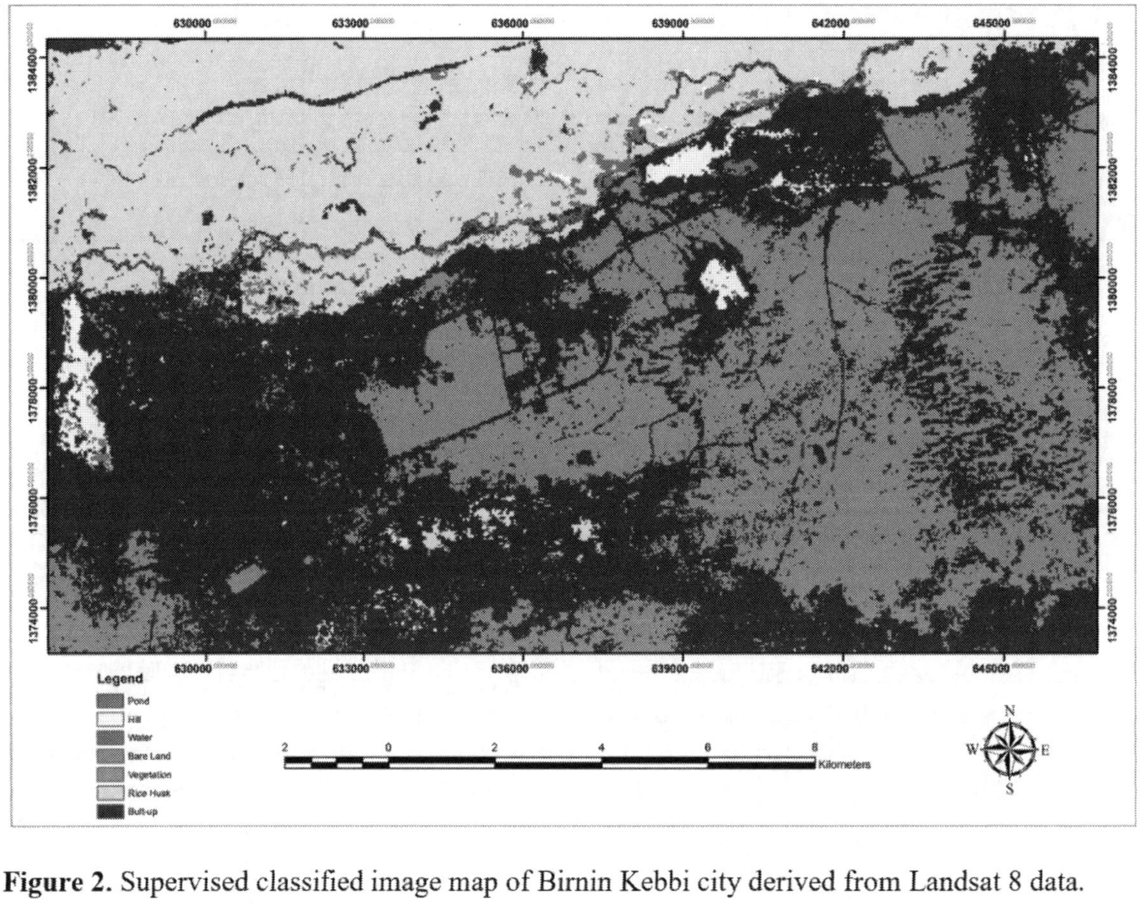

Figure 2. Supervised classified image map of Birnin Kebbi city derived from Landsat 8 data.

Table 2 shows the location and size of 10 large derelict ponds in Birnin Kebbi city, which were obtained by GPS ground surveys. The identified derelict ponds using Landsat 8 data were coincided with field survey results.

Table 2: Location and size of derelict ponds in the in Birnin Kebbi city.

S/N	NAME	AREA M^2	RECLAIMED?	ANY DRAINAGE?
1.	Tudun Wada Pond	251.001	No	No
2.	Presidential lodge pond	518.735	No	No
3.	Rafin Atiku Pond	278.890	Sand filled	Yes
4.	Illela Yari Pond	348.175	No	No
5.	New Cemetery Road Pond	139.445	No	No
6.	Ya Yara Market Pond	334.668	No	No
7.	Atiku Villa Pond	139.445	No	Yes
8	Dukku hill pond	250.230	No	No
9	Ambursa pond 1	390.460	No	No
10	Ambursa Pond 2	337.120	NO	No

Figure 3 and 4 show derelict ponds, which they filled with vegetation and solid waste dumped, which are suitable place for mosquito habitats. These ponds fill with water during rainy seasons bringing to people high cases of plasmodium falciparum within periods of time. Malaria and Typhoid cases during rainy seasons bloom in residential region around the derelict ponds in Birnin Kebbi city every year [5]. Figure 5 shows the affected population by the Malaria and Typhoid during rainy seasons in 2006 to 2014, which leaded to death in some cases.

Figure 3: A pond covered by water Lilly.

Figure 4. Solid waste dumped in one of the ponds.

IOP Conf. Series: Earth and Environmental Science **37** (2016) 012022 doi:10.1088/1755-1315/37/1/012022

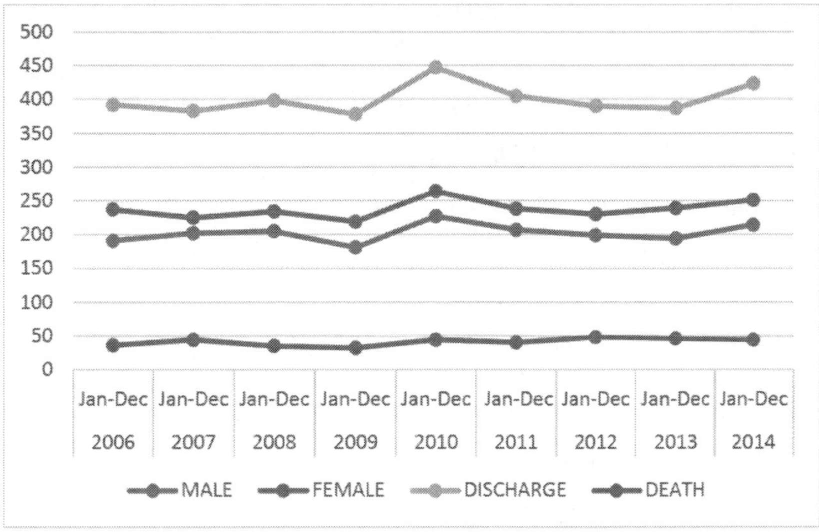

Figure 5. Malaria and Typhoid cases during rainy seasons in 2006 to 2014 for Birnin Kebbi city.

Land used map has been produced for Birnin Kebbi city using Landsat-8 and GIS techniques (Figure 6). Most of derelict ponds with different sizes and shapes are depicted on the map.

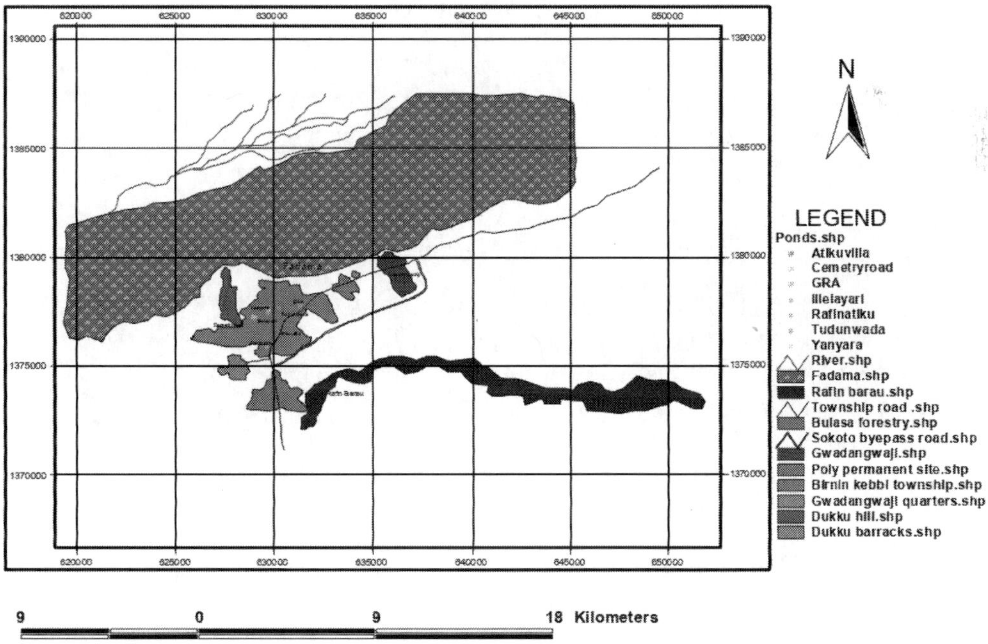

Figure 6. Land used map of Birnin Kebbi city.

Although, there is no climatic recording stations in Birnin Kebbi, but the town falls within the Theison polygon whose recording stations is situated at Sokoto. The average maximum temperature ranges between 30^{0}c and 41^{0}c, while the average minimum ranges between 15^{0}c and 26^{0}c. Average rainfall is 740mm annually, and spreads over a rainy season of about 120 days (Figure 7). A larger water deficit occurs due to a high potential of evaporation level during the hot season. At these periods, the ponds get filled up with from rain to overflow. The absence of drains to discharge the water away makes some of the derelict ponds to contain water all of the year. These places are suitable for mosquito breeding, which is a continuous exercise throughout the year.

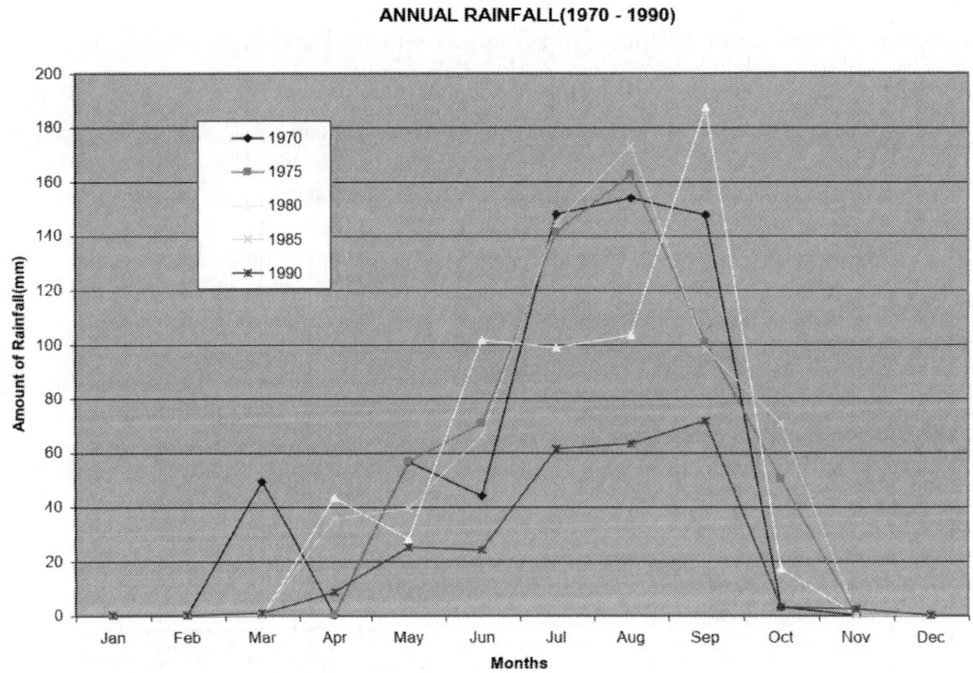

Figure 7. Annual Rainfall variability during 1970 to 1990 for Birnin Kebbi region.

4. Conclusion

Results of this study indicate that the combination of satellite remote sensing data from Landsat-8, GIS techniques and field survey is very useful tool for identification of derelict ponds as mosquito habitats in Birnin Kebbi city. Several problems are associated with derelict ponds, including the occupation of unnecessary space that is not beneficial to citizens except serving as breeding mosquito habitat, reducing the availability of the land that could be consider for the growing population and endangering tourism potentials. Reclamation is the best option to these ponds. The exercise of reclamation is usually aimed at improving water or land resources for agriculture and other purposes through the use of drainage or irrigations, restoration of mine workings and other construction activities. The reclamation work should be started immediately after the mining is completed, through planting of trees and grasses. In this type of procedures lovely forested hills, plains or valleys are the result which later becomes economical to the government since people visit such places as tourism potentials site in Birnin Kebbi city.

Acknowledgements

This study was conducted as a part of Fundamental Research Grant scheme (FRGS) (Vote no: R.J130000.7809.4F455), Ministry of Higher Education (MOHE) Malaysia. We are thankful to the Universiti Teknologi Malaysia for providing the facilities for this investigation. I wish to acknowledge Deacon Sunday Jaye of the Sir Yahaya Memorial Specialist Hospital Birnin kebbi for compiling the up data of malaria data for this study.

References

[1] Obajea MA 2013. The Sokoto Basin of Northwestern Nigeria: A Preliminary Assessment of the Hydrocarbon. Petroleum Development Technology Journal, 3(2).

[2] Pour AB, Hashim M, 2015. Hydrothermal alteration mapping from Landsat-8 data, Sar Cheshmeh copper mining district, south-eastern Islamic Republic of Iran.Science Direct Journal 155-160

[3] Roy E 2013. Turbid wakes associated with offshore wind turbines observed with Landsat 8. Remote Sensing of Environment, 145(2014), 105-115.

[4] Lillesand TM and Kiefer RW 1994. Remote Sensing and image interpretation, third edition. John Wiley & Sons, inc, printed in United States of America.9-12; 531-536;542-562

[5] Payne N and Hahn S 1995. Understanding your health 4th edition. Printed in the United States chp 18, pp446 – 475.

Spatial service delivery system for smart licensing & enforcement management

N A Wahap[1*], N M Ismail[2], N M Nor[1,3], N Ahmad[1], M F Omar[1], A A A Termizi[1], D Zainal[1], N M Noordin[2] and S Mansor[3]

[1] Agensi Angkasa Negara (ANGKASA), Pusat Angkasa Negara, 42700 Banting, Selangor, Malaysia
[2] Geoinfo Services Sdn. Bhd. 31, Jalan Bandar 2, Taman Melawati, 53100 Kuala Lumpur, Wilayah Persekutuan Kuala Lumpur, Malaysia
[3] Geospatial Information Science Research Centre (GISRC), Fakulti Kejuruteraan, Universiti Putra Malaysia, 43400 Serdang, Selangor, Malaysia

E-mail: azawani@angkasa.gov.my

Abstract. Spatial information has introduced a new sense of urgency for a better understanding of the public needs in term of what, when and where they need services and through which devices, platform or physical locations they need them. The objective of this project is to value-add existing license management process for business premises which comes under the responsibility of Local Authority (PBT). Manipulation of geospatial and tracing technology via mobile platform allows enforcement officers to work in real-time, use a standardized system, improve service delivery, and optimize operation management. This paper will augment the scope and capabilities of proposed concept namely, Smart Licensing/Enforcement Management (SLEm). It will review the current licensing and enforcement practice of selected PBT in comparison to the enhanced method. As a result, the new enhanced system is expected to offer a total solution for licensing/enforcement management whilst increasing efficiency and transparency for smart city management and governance.

1. Smart city administration and service delivery

Spatial information has introduced a new sense of urgency for a better understanding of the public needs in term of what, when and where they need services and through which devices, platform or physical locations they need them. It has grown in popularity among government agencies to have an existing system to be incorporated with the Global Positioning Systems (GPS) and Geographic Information Systems (GIS), which allows them to visualize and map the complex spatial relationship. Generally, the use of technology in the advanced delivery system at all levels will relate to tracking and tracing capability to induce mobility and efficient administration of a city [1]. Sensors for tracking and tracing are tools to determine the current and past locations and extensive information of a unique item or property.

Radio-frequency identification (RFID) is synonymous with track-and-trace solutions and has a critical role to play in supply chains [2]. RFID is a code-carrying technology and can be used in place of a barcode to enable non-line of sight-reading. Deployment of RFID was earlier inhibited by cost limitations but the usage is now increasing. RFID has been used for maintenance applications, such as an escalator maintenance called London Underground Step Tracking System developed by CoreRFID to protect passengers and reduces the risk of travel disruption [3]. The benefits of RFID includes of less contact, non-line-of-sight identification of objects over long distances, bulk or mass identification, insensitivity of the transponder against moisture, dirt and abrasion; possibility to store information directly at the item, and the possibility to write new data to the chip of the transponder [4].

An excellent city administration provides efficient service delivery to its citizens and stimulates commercial sector in the era of a service-based economy [5]. A smart government service is informed about its city's condition and is able to reach its citizens effectively. The Smart Licensing/Enforcement Management (SLEm) application objective is to value-add existing license management process for business premises which comes under the responsibility of Local Authority (PBT).

By manipulating geospatial technology, SLEm can improve manual processes of enforcement activity by utilizing satellite-related technology with the integration of various sensors and mobile platform. This application allows enforcement officers to work in real-time, use a standardized process which can simultaneously improve service delivery and optimize operation cost and time.

This application was tailored to the requirement of the selected PBT and in parallel to the project objective. By adopting Information and Communications Technology (ICT), SLEm offers a smart functional service delivery system by reducing waste of resources, hence resulting in cost and energy savings.

2. Smart Licensing and Enforcement management (SLEm) system approach

In order to achieve the objective, a user requirements study was conducted to identify a fitting smart concept for selected PBT. Figure 1 summarised the methodology stages of user requirements study for this project.

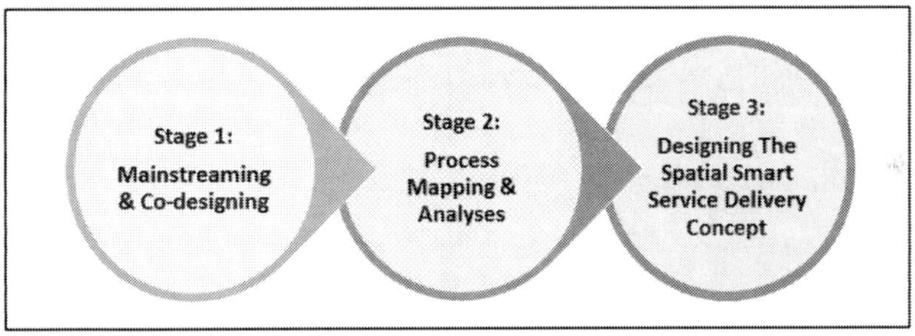

Figure 1. Methodology stages of user requirement study.

Series of meetings and site visits were carried out and questionnaires were distributed during stage one, to identify PBT's current services and to understand the technology readiness, governance and organizational factors within PBT. A process mapping and analyses was conducted in stage two, to identify the requirement of devices, software and data for development of the track-and-trace module for each PBT. Process of designing the conceptual workflow and spatial smart service delivery system were done in stage three.

2.1. User Requirement Analyses

Based on the assessment made from the user requirements study, a theme of a smart conceptual model was developed including the interoperability of systems and data-sharing platform. From the analyses, there are four main services provided by the selected six PBTs. The four services are the licensing and enforcement management, waste management, amenities and facilities management and public complaint. Table 1 shows the summary of PBT descriptions and findings.

Table 1. PBT descriptions and findings.

PBT	Theme/Focus	Functional Smart Related Project	Future Plans	Conceptual Smart Services
Johor Bahru Municipal Council (MBJB)	A City of International Standard by 2020	• Licensing to cater ISO 14001 for Low Carbon Society (under IRDA) • Waste Management • APK System • e-COMPLAINT • Tax System	• Environment Management • Crime/disaster prevention and Social Management (crime hot spot).	• Smart Complaint • Smart Licensing/Enforcement Management
Melaka Historic City Council (MBMB)	• Focus: Green Technology City • Theme: Melaka Historic City	• Real-time traffic control: CCTV/Real-Time Detection • Speed Monitoring • Traffic Information: LED MITCH • 1Map Melaka: Mineral piping BKI • Environment: Better Air Quality (BAQ) • Air Pollutant Index • Scheduled Waste, Domestic Trash: PPSPPA	• Traffic Management System • Environment Management • Crime/disaster prevention.	• Smart Complaint • Smart Amenities/Facilities Management
Sepang Municipal Council (MPS)	Green Environment, Green Energy, Bandar Selamat	• OSC Online • Compound System • Assessment Tax System • Sistem Penerimaan Hasil • E-Complaint • EIS & GIS Data Integration • GIS Data Management (offline – Map Info) Planning Department provides info to other departments. • Crime Mapping (collaboration with PDRM)	• GIS Based Waste Management System • GIS Based Complaint Management System for SJLR	• Smart Complaint • Smart Waste Management
Langkawi Municipal Council (MPL)	Langkawi City of Tourism	• GIS Data Management (offline – Map Info): not integrated and standalone. • e-PBT • OSC Online • i-Plan (Land & Property)	• Environment Management: Waste Management • Crime/Disaster Prevention: Safe City • Facilities Management (Licensing): Mobile Workforce Management System.	• Smart Complaint • Smart Amenities/Facilities Management
Pekan District Council (MDP)	Development Towards Vision 2020	• i-Plan	• Facilities Management – GIS Based Licensing Monitoring System • Location based e-Aduan.	• Smart Complaint • Smart Licensing/Enforcement Management
Kuala Langat District Council (MDKL)	-	• Sistem Aduan Rakyat Selangor (STARS)	• Location based e-Aduan • Contractor Management System.	• Smart Complaint • Smart Waste Management

2.2. Licensing activity and workflow

Based on the assessment, Johor Bahru Municipal Council (MBJB) was elected to be the case study for this paper. The prototype development started by reviewing the current work practice of the Licensing and Enforcement Department for Business and Advertisement Premise Licensing in MBJB, followed by the development of enhanced workflow and the conceptual design of the track-and-trace module. Finally, a general system architecture of SLEm was designed based on the hardware and software requirement and database design infrastructure. Figure 2 shows the current process workflow for licensing and enforcement activity of MBJB.

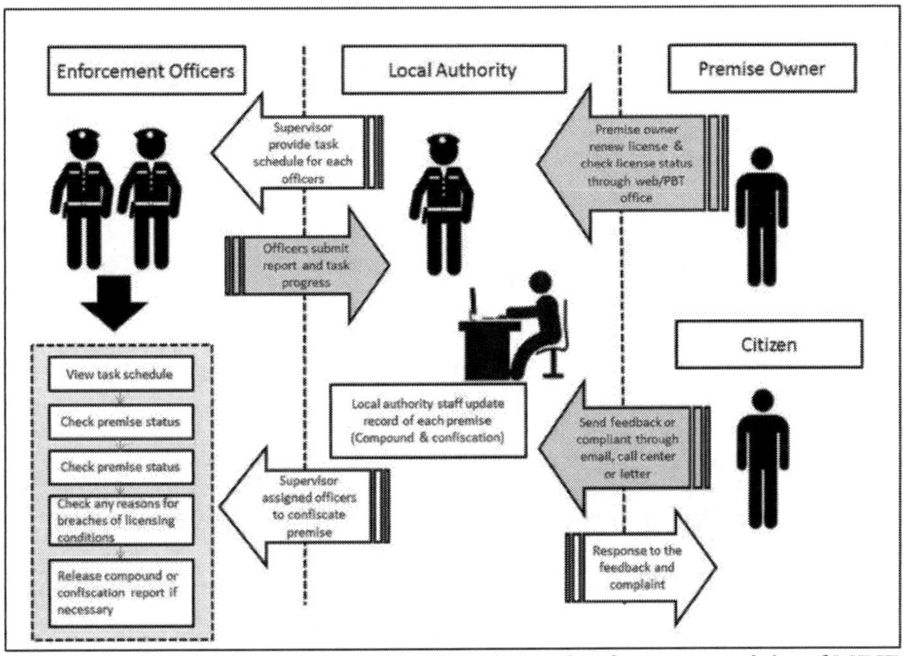

Figure 2. Current process workflow for licensing and enforcement activity of MBJB.

The workflow requires the enforcement officers to do an inspection in the premises of their task with a printed hardcopy of registered premises. On the site, enforcement officers will check premises license status and the license compliance with the registered license. This workflow allows enforcement officers to issue a compound letter and confiscation report if needed. All findings will be reported and submitted to the main office afterwards. The desk officers will update the system based on the reports from the enforcement officers.

3. Conceptual workflow & design

SLEm was designed with the consideration of having licensing and enforcement activity by using mobile devices with embedded Location Based Services (LBS) and RFID tracing capability. It promotes an efficient solution which provides a total mobility and real-time connectivity between the enforcement officers and the main office.

3.1. Conceptual workflow

This conceptual workflow was developed to tackle issues highlighted by MBJB regarding premise licensing and enforcement activity which fulfill the needs:

i. to monitor license status (premise license registration and renewal and its compliance with the registered license)
ii. to monitor licensing and enforcement officers on site
iii. to have an online reporting and submission hence avoiding missing reports
iv. to manage complaint effectively

By using SLEm, the enforcement officers can track and locate premises of his task by using map tracking. All information of premises will be stored in a spatial database so it can be presented in web-GIS applications. Therefore, enforcement officers can easily locate premises location as the premises were already been marked on the map. Figure 3 shows the conceptual flow of SLEm Application.

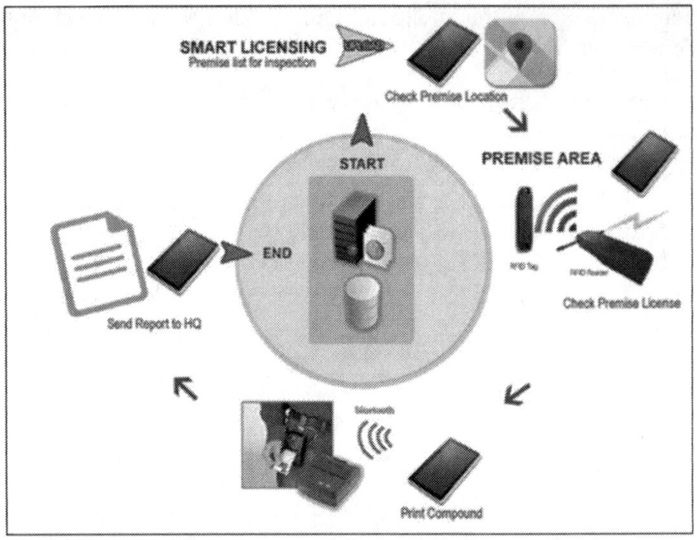

Figure 3. The conceptual flow of SLEm Application.

3.2. Conceptual system design

The system prototype comprises of the following components: Users, GPS service, operation, GSM/GPRS, SQL database, application server and Google map server. The user were citizens/premise owners, staffs (enforcement officers) and local authority (supervisor, management team). The sensor and satellite technology were incorporated in the track-and-trace module to locate and track the enforcement officers and licensed premises. Figure 4 demonstrates the conceptual system design for SLEm application.

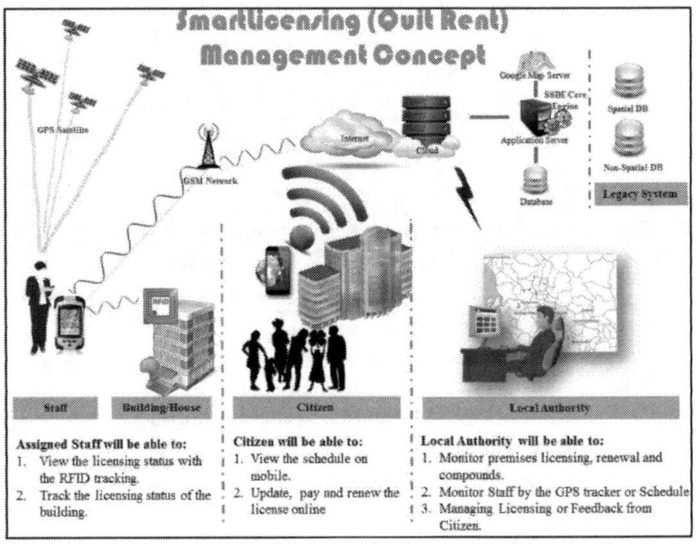

Figure 4. The conceptual system design for SLEm Application.

i. Local Authority (PBT): Local authority is responsible for management of licensing, monitoring staff, and managing complaint and feedback from citizen
ii. Staff (Enforcement officers): Staff is local authority staffs who is responsible for monitoring premise licensing status in his authority

iii. Premise Owner: Premise owner is the building owner or license holder that registered under local authority
iv. Citizen: Citizen refers to public citizen that resides in the city

4. Spatial service delivery elements

The most important aspect of SLEm was the integration of the tracking and tracing technology in a mobile platform. SLEm presented an end-to-end concept in a comprehensible spatial representation of a map viewing via the web-GIS application of its three major components which are not being emphasized by other developers as observed in the study. Table 2 summarises the key element to support SLEm Application.

Table 2. Summary of the key element of SLEm Application.

Actor	Spatial Element	Instrument/Sensor	Smarter Element
Management Supervisor	Map: Geographic interface for viewing, analysing, and managing the complaint. Location Based Services (LBS)	PC, Laptop	• Multiple task delegation • Real-time monitoring map • Analyses and report
Premise Owner Citizen	Map: Geographic interface for Location Based Services (LBS)	PC, Laptop, Smart Phone, Tab	• Online license renewal • Online license status • Online suggestion, complaint and feedback list
Enforcement Officers	Map: Geographic interface for viewing the problem. Location Based Services (LBS) for Navigation: urgent corrective complaint about navigation	Smart Phone, Tab, RFID A-GPS, M-GNSS	• Locating premise location on map • Tracing premise information • Automation of complaint notification • Generating & submitting of report to PBT

5. Conclusions

The objective of SLEm is to value-add existing license management process for business premises which comes under the responsibility of PBT. In this study, MBJB licensing and enforcement activity was reviewed and enhanced. By manipulating geospatial technology and RFID sensors via mobile devices, SLEm application has provided the PBT with a platform to work in mobile and real-time mode to improve the service delivery by optimizing operation time and cost. The new spatial smart service delivery system is expected to offer a total solution for licensing/enforcement management whilst increasing efficiency and transparency for smart city management and governance. By focusing on automation, the application is limiting human errors in data collection and reporting hence monitoring and decision making is made easy for the local authority.

Acknowledgement

The authors would like to thank to Ministry of Science, Technology, and Innovation Malaysia (MOSTI) for funding the project through TechnoFund and Johor Bahru Municipal Council (MBJB) for the support given in providing information of the existing licensing and enforcement management and platform to conduct the integration of modules and verification for the prototype.

References

[1] Nor NM and Wahap NA 2014 IOP *Conf. Ser.: Earth Environ. Sci.* **18** 012168

[2] Sarac A, Absi N and Dauzere-peres S 2009 A literature review on the impact of RFID technologies on supply chain management Working paper ENSM-SE CMP WP 2009/2.

[3] CoreRFID Ltd. 2008 Maintenance and Engineering: case study of the london underground escalator maintenance project Warrington: Author.

[4] Schultman F and Sunke N 2008 Life cycle information of buildings supported by RFID technologies [power point slide] Germany: University of Siegen.

[5] Evaluating the impact of local e-services The smart cities research Brief No.15 Retrieved January 28, 2016 from http://www.smartcities.info/files/Smart_Cities_Research_Brief_15_Evaluating%20the%20impact%20of%20local%20e-services.pdf

IOP Conf. Series: Earth and Environmental Science 37 (2016) 012024 doi:10.1088/1755-1315/37/1/012024

Daylighting estimation and analysis in residential apartment building: GIS based approach

Mahesh B Sonawane[1] and Sumedh Y Mhaske[2]

[1]Research Scholar, Civil & Environmental Engineering Department, Veermata Jijabai Technological Institute, Mumbai – 400019, India.
[2]Associate Professor, Civil & Environmental Engineering Department, Veermata Jijabai Technological Institute, Mumbai – 400019, India.

E-mail: er.mahesh99@gmail.com
symhaske@vjti.org.in

Abstract. The openings in the building envelope have a great influence on daylighting in the internal area of the building spaces. The amount of opening area, its orientation, outside obstruction & positioning of building affects the inside illumination. Most of the energy consumption occurs during the building's operational phase for heating, cooling & lighting purposes. This paper aims to provide a simplified analytical and GIS based approach to evaluate the potential of daylight inside the room under clear sky conditions. The work evaluates the intensity of internal illumination in residential apartment building from available outside external illumination.

1. Introduction

Daylight is nature's free gift which is free, non-depleting and most valuable resource received from the sun [1]. It is considered as an integral part in building design for residential and commercial buildings. It is essential particularly in multistoried apartment buildings since lower stories receive less illumination as compared to upper stories. Before invention of electricity, candles, kerosene lamps and gas lamps were used in the night period for illumination purpose but the lighting quality was very poor. Today as reduction and conservation of energy has become an important issue, daylighting is considered as an important aspect in the building design [2]. Daylight consists of two parts, direct solar illumination and sky radiation (diffuse light scattered by earth's atmosphere) [3]. For daylighting design, direct solar illumination is not considered since it creates discomfort, in the form of heat and glare, and only sky radiation is considered for contributing to the illumination of interiors of building in the day period [4].

The lighting constitutes about 20 % of total energy consumption in a fully air conditioned office building. In India, 15 % of the total electricity generated is used for lighting in various sectors. The good lighting design incorporates room surface brightness, glare reduction, uniform distribution, good lamp coloration and it also improves architecture as well as performance of building. The lighting energy can be reduced by use of efficient lamps, control devices, proper daylight utilization, and lighter finishes of ceilings, walls and furnishings [5]. Over the period of time there has been advancement in the illumination efficiency of the lamps, modern buildings demand high level of illumination particularly jewelers shops, five star hotels. The quality and quantity of light depends upon task to be performed, in most of the situations the quality and quantity of light becomes important [6].

In India, external available horizontal illumination assumed for design purposes is considered as 8000 lux. Clear design sky is considered as the basis of daylighting design by the *IS* code. The provision of large size windows does not increase the illumination in the same proportion [7]. Since the sky varies so much in its brightness from hour to hour, location and from season to season, it is difficult to predict daylight, no simple solution can be given. The lightness of the room surface is considered as an important factor since the illumination level inside a room will be much more with light color than dark color [4].

The indoor daylight depends upon size, location of window, room size, interior finish and external obstruction present outside the building. The calculation of daylight factors requires estimation of sky component, external reflected component and internal reflected component for clear design sky condition which is accepted as standard outdoor condition [8]. Daylight intensity varies with earth rotation, geographic location, atmospheric cover, season and time of the day. There is need to integrate daylight and artificial lighting energy system, installation of energy conserving dimmable lighting system provides energy savings [9].

In previous works, daylighting performance studies were carried out in an industrial building in China by experimental measurements and numerical simulations; Ecotect and Rediance software tools were used for simulation purpose. They presented an experimental method for the measurement of daylighting illuminance distribution in large space buildings [12]. An algorithm is prepared to compute the maps of solar irradiation, optimal angles for high resolution digital elevation models suitable for computer systems. Researchers developed a tool to compute the map of maximum solar irradiation, required for areas with stable climatic conditions [13]. Computer simulation studies using 'Relux' software had been carried between natural light, artificial light and related illumination in commercial building in Mumbai city. Comparisons were done with fluorescent luminaires without daylight harvesting and one with 'LED' fixtures with daylight harvesting using daylight dimming sensors. Energy and operating costs is reduced by 80 % with the latter scheme for over 25 years period [14]. The work evaluated a prototype dynamic integrated shading and light –redirecting system to optimize daylight conditions. Daylighting performance simulation studies were carried on this system. Daylight has been improved with the redirecting glass lamella shading system, another advantage was that it did not cause additional glare compared to the reference shading system. Due to higher penetration of daylight, the use of electricity for lighting was reduced up to 80% depending on the daylight-linked lighting control [15].

The researchers presented a luminous efficacy model for global horizontal irradiance to derive values of outdoor global horizontal illuminance data from local weather station data. It involves clearness factor as the main influencing variable, other variables i.e 'Humidity ratio and solar altitude' are incorporated in model formulation. The work shows the relationship between model predictions and the measured data [16]. With the help of daylight modeling techniques based on solar radiation data, hourly daylight has been estimated for horizontal and sloping surfaces for New Delhi (India) city. They developed a model to estimate interior illuminance and validated the same using experimental hourly inside illuminance data of an existing skylight integrated vault roof mud house [17].

An extensive review on daylight illuminance data measurements, determinations, and prediction model developments and lighting energy reductions due to daylighting schemes have been presented. For creating reliable daylight databases, long-term data measurement is required [18]. Externally reflected component due to neighboring buildings is the main source of natural light as it reduces the daylight levels inside the residential flats. The lower floors of multistoried buildings get insufficient and poor quality light. Hence these floors require more artificial light in day period too [19].

The intensity and distribution of natural light is not constant and varies with the geographic location. Therefore, the objective of this work is to present an effective GIS based approach to estimate illumination inside the room, the work also presents the daylight analysis of residential apartment building. Gramm ++ software tool is used in the work to get various areas of the floor, perimeter of floor plan, orientation and daylight analysis [20].

2. Methodology

In Pune city, most of the residential apartment buildings usually have many small rooms with exposure in only one or two directions. The aim of apartment building planning is to provide adequate light levels even if only one window is provided in a wall, duct is often provided to admit light and achieve ventilation. The methodology presented in this paper is to obtain the illumination intensity inside the room in residential apartment building, worked out at center of the room from the opening area of wall (fig. 3). The methodology is shown in figure 1. In Gramm ++ (GIS) tool, the floor plan of building is registered with its ground coordinates. The latitude and longitude (ground coordinates) of the building are 18.569° N and 73.768° E. This georeference plan is converted into digital format by digitizing the outer and inner walls of the building. This digitized plan is used to work out the perimeter of the floor plan, various internal areas of the building spaces. The geo registered and the GIS based vector analysis plans are shown in figure 4 and 5 respectively.

These illumination levels are determined at horizontal working plane of 60 cm from the floor level (fig.4). The external illumination (daylight) available outside the building is measured by 'Lux' meter simultaneously at four exterior sides of the building during three times of the day i.e at 7.00 am, 12.00 noon and at 5.00 pm for 'clear sky' condition. The *SC* for various values of *l/d* and *h/d* ratios of all the openings of the building are worked out using *IS* Code: 2440-1975 and mathematical computational tool program [11].

The percentage illumination inside the various rooms is obtained with respect to outside available daylight. The internal finishes of walls, ceil and floors are considered, the *IRC* factor of the surfaces has been considered to be 0.7 % as per the *IS* code and added in the total illumination.

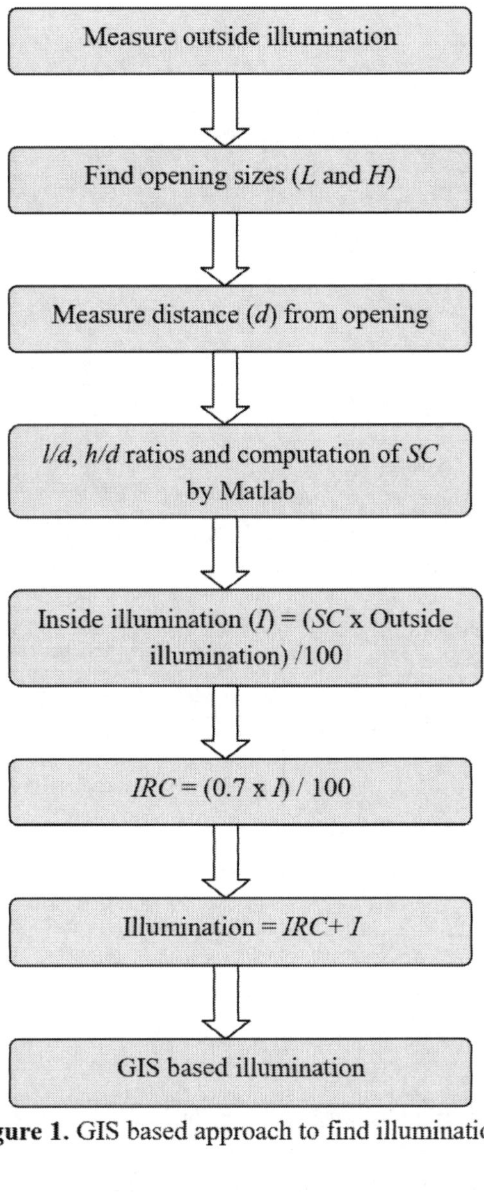

Figure 1. GIS based approach to find illumination.

IOP Conf. Series: Earth and Environmental Science **37** (2016) 012024 doi:10.1088/1755-1315/37/1/012024

2.1 Sky component
In India, the table of sky components had been developed and referred in the present work for the calculation purpose. The table gives various values of sky components on a horizontal plane due to a vertical opening (window) illuminated by a clear design sky of uniform luminance. The table shows that the *l/d* and *h/d* ratios larger than 2.0 up to infinity, the increase in the sky component is not significant. The illumination from a square window is much more than a wide window for the same height but much lower compared to a tall window [7].

2.2 Case study
The residential building chosen for the case study is situated in Pune city. The floor plan (fig.4) consists of two *BHK* and three *BHK* flats [10]. There are total four flats planned at this floor. Each flat is opened at three faces. The key plan of the buildings is shown in figure 2. The work presents the daylight analysis of first floor flats of *A1* building. The building is located at extreme left side of the project, it is symmetrical about vertical axis; north direction points upward.

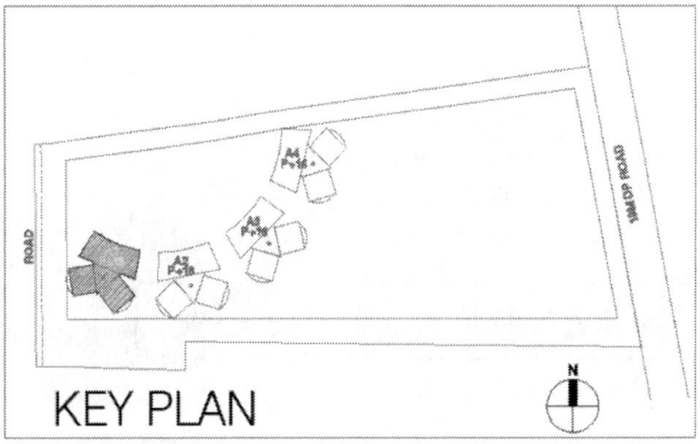

Figure 2. Key Plan of the residential building project (Source: www. nandanbuildcon.com).

The position of buildings in the project are shown in figure 2, four buildings (*A1*, *A2*, *A3* and *A4*) are planned along a curve. All the buildings are designed for parking and sixteen floors (i.e $p + 16$).

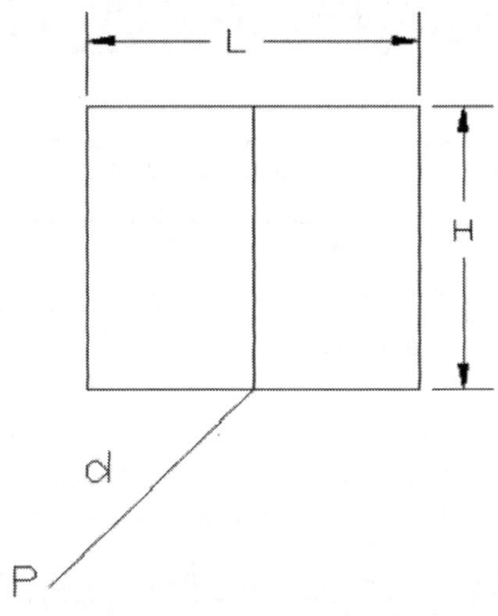

Figure 3. Point at which the internal illumination is worked out.

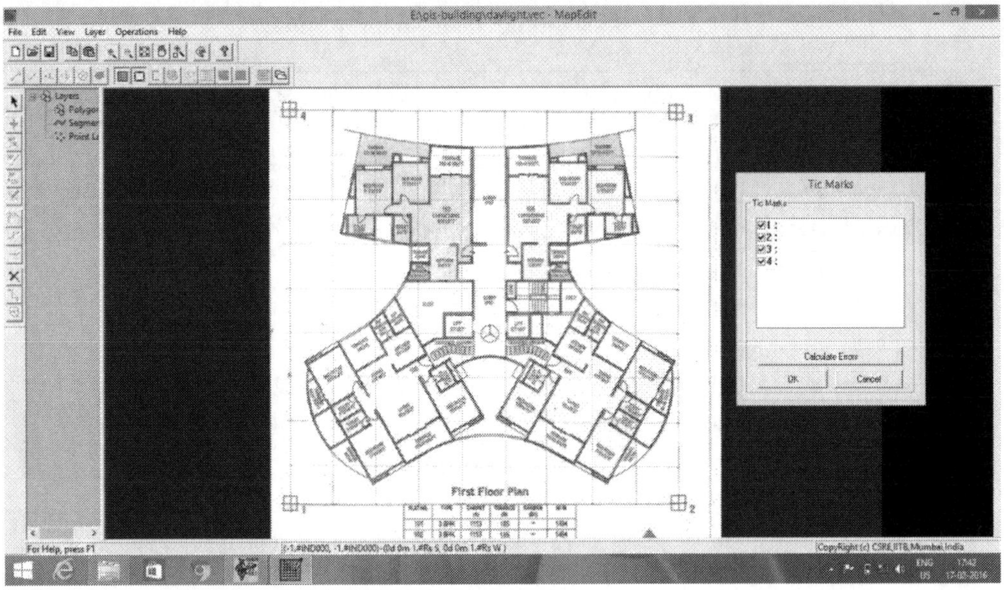

Figure 4. Georeference building floor plan.

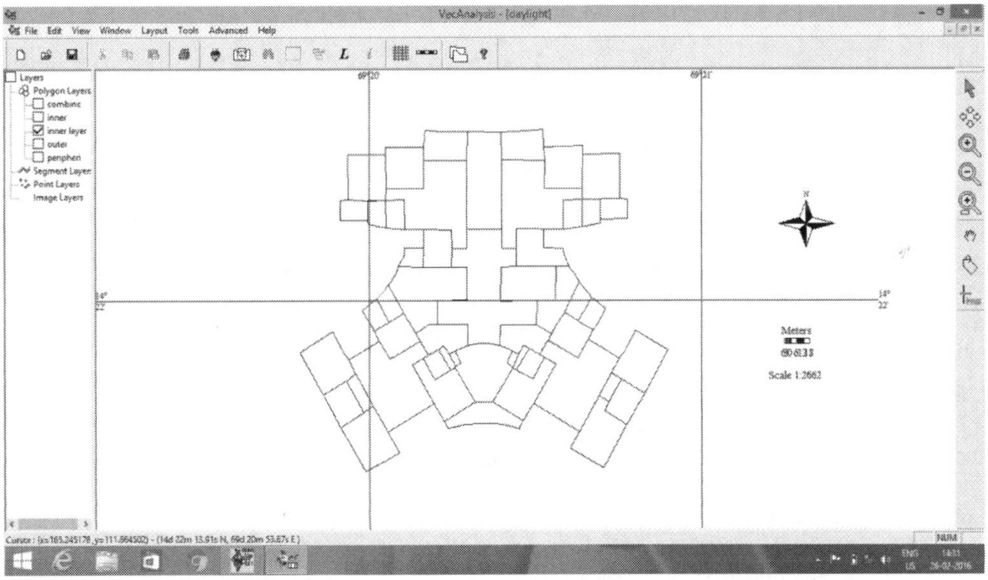

Figure 5. GIS based vector analysis view of floor plan.

Table 1. Schedule of door and window openings.

Door sizes			Window sizes		
Type	**'W' mm**	**'H' mm**	**Type**	**'W' mm**	**'H' mm**
D	3000	2270	*W*	3000	1800
Da	3750	2270	*Wa*	3750	1800
Db	2250	2270	*Wb*	2250	1800
D1	1800	2270	*W1*	1800	2100
D2	1060	2270	*W2*	1500	2100
D3	900	2270	*W3*	750	1200
D4	800	2270	*W4*	900	2100
D5	750	2270	*W5*	1800	1200

Table 2. Observed external illumination levels (Lux) outside building '*A1*'.

Time Face	7.00 am	12.00 noon	5.00 pm
North face	3940	12670	3908
East face	3378	9862	3323
South face	3694	15580	4182
West face	4730	11710	4085

The above table represents the illumination levels, measured with Lux meter, outside the chosen building (*A1*) on a sunny day. The illumination intensity is more at west direction than that of East direction because of presence of adjoining building (*A2*) in the eastern face of the building. The illumination intensity at south direction is more than that of North except morning time (i.e 7.00 a m).

3. Results and discussions

The daylight intensities are worked out in each room; the results of illumination of the corresponding openings are shown in table 3. It presents the illumination in *NW* and *NE* flats; table 4 shows the illumination in *SW* and *SE* flats.

Table 3. Daylight illumination (lux) of building *A1*, floor 1. (Orientation –*NW* and *NE*).

		Flat 1 (103) Orientation – *NW* (illumination in lux)			Flat 2 (104) Orientation – *NE* (illumination in lux)		
Rooms	O	7:00 a.m.	12:00 noon	5:00 pm	7:00 a.m.	12:00 noon	5:00 pm
BR1	*D1*	373.35	1212.93	374.09	377.15	1212.84	374.09
	W2	460.00	1140.00	397.37	328.58	960.00	323.24
BR2	*W1*	415.00	1334.55	411.63	415.00	1334.55	411.63
L D	*D*	315.79	1015.58	313.24	315.79	1015.58	313.24
K	*D5*	384.19	950.432	331.55	274.17	800.44	269.70
	W3	303.69	751.84	262.27	216.88	633.20	212.35

In this work, the analytical procedure is linked to GIS and further results of illumination are worked out. The results show that the maximum level of daylight is observed in *BR 2* (table3) and minimum in Kitchen room. Kitchen room of *NE* flat receives daylight from the window provided in eastern direction and through dry terrace but the kitchen room of *NW* flat receives daylight from the window and dry terrace located at western direction; still illumination in *NW* flat is more than *NE* due to the presence of another building at eastern side. Opening *W1* of *B R* 2 is admitting maximum illumination as compared to other openings; *W3* in kitchen room receives least light. The illumination through opening *W1* of *BR 2* in both the flats (1 and 2) is same as it is placed adjacent to Living- Dining room.

All the rooms of both flat receive more illumination from *NW* direction than *NE* direction. Openings *D*, *D1* and *W1* are receiving same illumination in both the flats (*NW* and *NE*) whereas variation is observed in other openings. At 7.00 am and 5.00 pm intensity of illumination is almost same and it is three times more at 12.00 noon. Room wise distribution of illumination (*NW* and *NE* orientation) is presented in figure 6a to figure 6d.

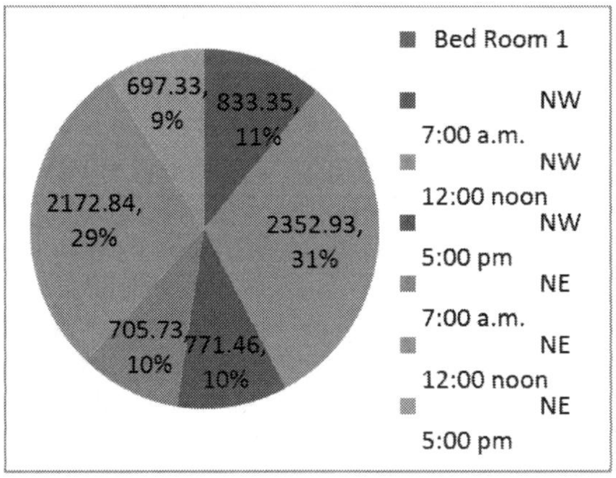

(a) Bed Room 1.

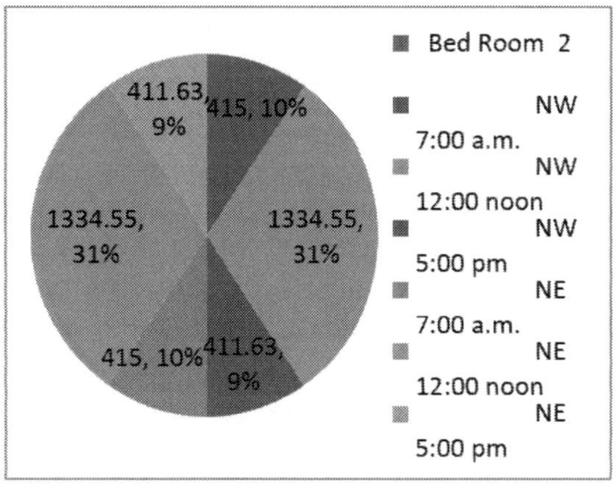

(b) Bed Room 2.

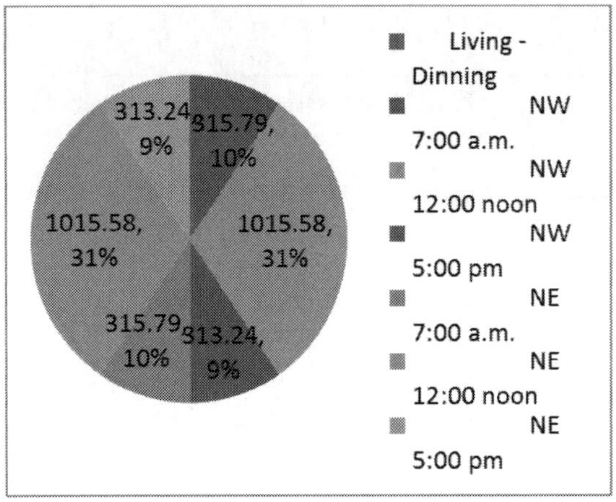

(c) Living -Dinning.

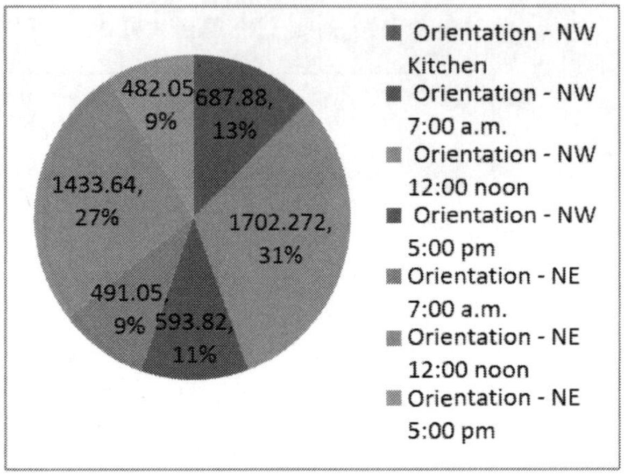

(d) Kitchen.

Figure 6. Percentage distribution of daylight for *NW* and *NE* flats.

Table 4. Daylight illumination (lux) of building *A1*, floor 1. (Orientation –*SW* and *SE*).

		Flat 3 (102) Orientation – *SW* (illumination in lux)			Flat 4 (101) Orientation – *SE* (illumination in lux)		
Rooms	O	7:00 a.m.	12:00 noon	5:00 pm	7:00 a.m.	12:00 noon	5:00 pm
BR 1	*W1*	303.82	1281.40	343.95	303.82	1281.40	343.95
	W4	281.29	696.40	242.93	200.83	586.50	197.62
BR 2	*W1*	389.02	963.11	335.97	277.74	811.20	273.30
	D3	289.68	717.17	250.18	206.82	603.99	203.51
BR 3	*W1*	401.41	1693.01	454.44	401.41	1693.01	454.44
L D	*Da*	479.81	2023.69	543.20	479.81	2023.69	543.20
	Wb	846.54	2095.78	731.10	604.39	1765.21	594.72
K	*W3*	200.24	495.74	172.93	142.96	417.13	140.67
	D5	278.69	689.96	240.69	198.97	581.137	195.79

As per the floor plan (fig. 4), southern flats (flat 1 and flat 2) are planned along a curve hence more space is available between the flats, resulting more light from this direction. In southern flats (table 4), maximum illumination is observed in living – dinning room, opening *Wb* and *Da* is receiving greater daylight than all other openings. At 7.00 am, 12.00 noon and 5.00 pm the intensity through *Da* is same in both the flats, but variation is observed in *Wb*, south west direction is receiving more illumination than south east, the variation is due to the fact that the daylight is obstructed because of terrace of above flat. These openings (*Wb* and *Da*) are provided in opposite walls hence more illumination is observed in this room. The least illumination is observed in kitchen room, admitting daylight only from one direction, opening *W3* is showing lowest illumination intensity whereas opening *D5* is receiving more daylight. The illumination through opening *W1* of *BR1* is more than that of *BR2* though location of *W1* is same. Room wise distribution of illumination (*SW* and *SE* orientation) is presented in figure 7a to figure 7e.

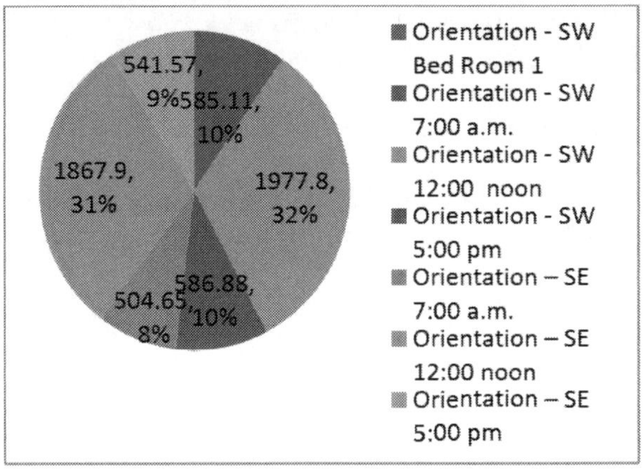

(a) Bed Room 1.

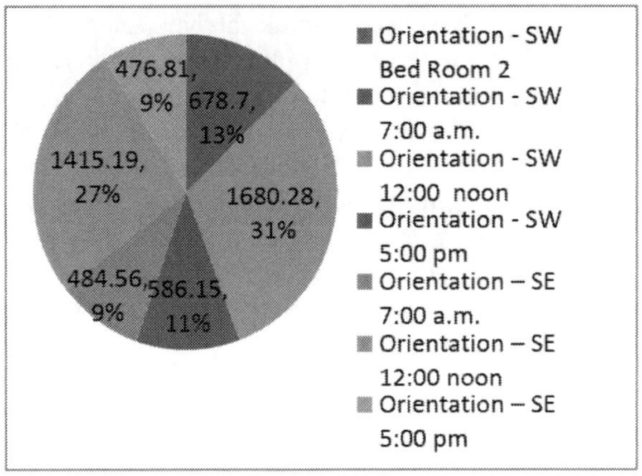

(b) Bed Room 2.

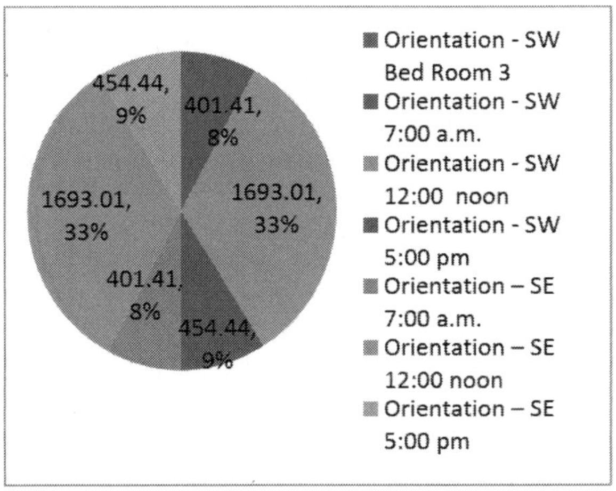

(c) Bed Room 3.

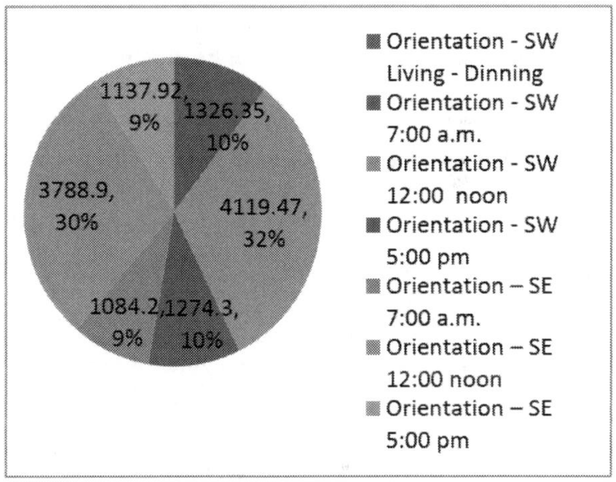

d) Living –Dinning.

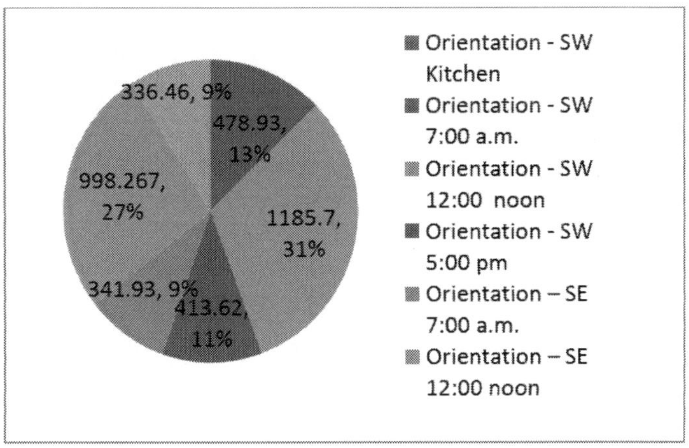

(e) Kitchen.

Figure 7. Percentage distribution of daylight for *SW* and *SE* flats.

In northern flats, *NW* flats admit more daylight than *NE* flats; and in southern flats, *SW* flats receive more daylight than *SE* flats. In general, better distribution of daylight intensity is observed in all the flats. The results of the above analysis are compared with the standard values of *IS* code. The allowable level of illumination as per *IS* Code of practice are for 1) Kitchen room = 200 lux 2) Bedroom = 200 lux and 3) Living/Dining = 300 lux.

In living-dinning room of northern flats, the illumination (315.79 lux and 313.24 lux) is just meeting the *IS* code requirement at 7.00 am and 5.00 pm, hence requires minimum artificial illumination at these timings. Similarly, in kitchen room of north east flat, the illumination at 7.00 am (216.88 lux) and 5.00 pm (212.35 lux) are just meeting the *IS* code criteria hence it also requires artificial illumination.

Kitchen room of south west flat is also receiving less daylight, at 7.00 am (200.24 lux), but at 5.00 pm it is just 172.93 lux which less than *IS* code value, in south east flat kitchen room is admitting 142.96 lux at 7.00 am and 140.67 lux at 5.00 pm which is far below than *IS* code standards hence use of artificial illumination is must. If combined effect of both the openings is considered in kitchen room, the illumination intensity is more than *IS* code standards. GIS tool is utilized to get the attribute data of various internal areas, perimeter and the associated coordinates of the floor plan.

4. Summary

This work summarizes the following aspects from the daylighting analysis of apartment building; following conclusions can be drawn.

- In this work, all the flats receive good daylight as per the existing orientation of the building.
- Analysis is done only for three timings of the day on a sunny day; for other timings and sky conditions, the work can be carried.
- During 7:00 am and 5:00 pm, for all the flats, the daylight which enters the kitchen room is below the Indian Standard Code values, hence artificial lighting is required to be provided to compensate the deficiency.
- The procedure could be useful to the engineers and planners to estimate daylight.
- The illumination intensity depends upon the distance from the window; closer the window more will be the illumination, weather shade provided over openings reduces daylight to large extent.
- Daylight depends upon the color of the surface finishes; light yellow color gives more illumination due to reflection.
- Daylight analysis is an important strategy in energy efficient building designs; software tools could play major role.
- Daylighting requires an integrated design approach about the building orientation, climate of the site and placement of building openings.
- Geographic Information System is to be considered as an effective tool in daylight assessment in the residential buildings.

Notation Following symbols and abbreviations are used in this paper:

L = Length of window;
H = Height of window/opening;
W = Width of opening;
O = Openings;
l/d = Length to depth ratio;
h/d = Height to depth ratio;
d = Depth of room from window face;
I = Illumination;
IRC = Internal Reflected Component;
IS = Indian Standard;
SC = Sky Component;
BHK = Bed, Hall, Kitchen;
BR = Bed Room;
L D = Living - Dinning;
K = Kitchen;

NW = North West;
NE = North East;
SW = South West;
SE = South East;

Acknowledgements
Authors are thankful to Prof S L Kolhatkar and Ar. Priya Sonawane for providing valuable help to carry out the research work.

References

[1] Hopkinson R G, Petherbridge P and Longmore J 1966 Daylighting, Heinemann, London
[2] Harvey J Bryan 1979 A simplified procedure for calculating the effects of daylight from clear skies Lawrence Berkeley Laboratory, University of California Berkeley, California
[3] Joshi M, Sawhney R L, Buddhi D 2007 Estimation of luminous efficacy of daylight and exterior illuminance for composite climate of Indore city in Mid-Western India, *Renew Energy* **32** pp 1363-78
[4] IS: 2440-1975, *Guide for day lighting of Buildings*– Second Revision, Bureau of Indian Standard, New Delhi, 1989, pp 8-9
[5] Manual on norms and standards for environment clearance of large construction projects, Ministry of Environment and Forests, Government of India, 15 Sept, 2006, pp 191-192, 207
[6] Bureau of Energy Efficiency 2011 *Building Lighting Design*, version 2.0, March 2011, USAID ECO-III Project, Ministry of Power, Government of India
[7] D Ajitha Simha 1985 *Building Environment* Tata McGraw-Hill Publication, New Delhi, India, chapter 6 pp168- 171
[8] SP: 41 1987 *Handbook on Functional Requirements of Buildings* (Other than Industrial Buildings), Part 1-4, Bureau of Indian Standards, New Delhi, India, 81-7061–011-7
[9] Anil Ahuja 1997 *Integrated M/E Design, Building system Engineering*, International Thomson Publishing pp 89-90 New York, U S A
[10] http://www.nandanbuildcon.com/completed-projects/nandan-prospera/image-gallery, downloaded on December 29, 2015
[11] MATLAB 2008 MATLAB 7.6.0, Release 2008a. The MathWorks, Inc
[12] Yuanyi C, Junjie L, Jingjing P, Xiaodong C, Qingyan C, Yi J 2014 Experimental and simulation study on the performance of daylighting in an industrial building and its energy saving potential *Energy and Buildings* **73** pp 184- 91
[13] Tabik S, Villegas A, Zapata E L and Romero L F 2012 A Fast GIS-tool to Compute the Maximum Solar Energy on Very Large Terrains *Int. Conf. on Computational Science* ICCS 2012, Procedia Computer Science 9, pp 364 –72
[14] Shailesh K R and Tanuja S Raikar 2010 Application of RELUX Software in Simulation and Analysis of Energy Efficient Lighting Scheme *International Journal of Computer Applications* (0975 – 8887), **9 (7)** pp 24-35
[15] David Appelfeld and Svend Svendsen 2013 Performance of a daylight- redirecting glass-shading system, *Energy and Buildings* **64** pp 309-16
[16] Sokol Dervishi and Ardeshir Mahdavi 2013 A simple model for the derivation of illuminance values from global solar radiation data, *Building Simulation* **6** pp 379-83
[17] M Jamil Ahmad and G N Tiwari 2010 Estimation of luminous efficacy of daylight and Illuminance for composite climate, *International Journal of Energy and Environment* **1(2)** pp 257-76
[18] Danny H W Li 2010 A review of daylight illuminance determinations and energy Implications, *Applied Energy* **87** pp 2109-18
[19] Li D H W, Wong S L, Tsang, Gary H W and Cheung 2006 A study of daylighting performance and energy use in heavily obstructed residential buildings via computer simulation techniques, *Energy and Buildings* **38** pp 1343- 48
[20] GRAMM ++, Version 2.0, GIS Package development and Applications, CSRE, IIT, Bombay, Mumbai

IOP Conf. Series: Earth and Environmental Science **37** (2016) 012025 doi:10.1088/1755-1315/37/1/012025

Accuracy assessment of NOGGIN Plus and MALÅ RAMAC X3M single channel ground penetrating RADAR (GPR) for underground utility mapping

Mas Sazali Hashim, Saiful Nizam Saip, Nurfauziah Hani, Biswajeet Pradhan* and Saleh Abdullahi[1]

[1]Department of Civil Engineering, Faculty of Engineering, University Putra Malaysia, 43400, UPM, Serdang, Selangor, Malaysia

Corresponding Address: Biswajeet Pradhan, Tel: +603-89466383,
Fax. +603-86567129; Email: biswajeet24@gmail.com or biswajeet@lycos.com

Abstract. Ground Penetrating Radar (GPR) becomes a popular device in investigation of the underground utilities in recent years. GPR analyses the type and position of utility objects. However, the performance accuracy of GPR models is an important issue that should be considered. This study conducts the accuracy analysis between two models of single channel GPR; NOGGIN PLUS and MALÅ RAMAC X3M, by focusing on the basic principles of single channel GPR, accuracy analysis and calibration methods implemented on GPR. The survey work has been performed to identify the most accurate instrument to detect underground utility objects. In addition, data analysis was carried out to compare between two models of single channel GPR. This study provides proper guidelines and assists surveyors to select the suitable instruments regarding on applications especially on utility mapping in terms of accuracy.

1 Introduction

There are several approaches have been proposed for land utility surveying and other manmade subsurface structures [1]. In terms of utility survey, utilities are some kind of buried objects under the ground that can be measured, identified and then determined by using the instruments [2-4]. Basically, the product of utility survey is utility maps of for examples; gas pipes, metal and plastic pipes, sewage, electricity cables, telephone cable and pipes etc. There are two basic approaches of utility mapping; active systems that are connected to the utilities, and passive systems that work without connection [1, 5].

Regarding Malaysian perspectives, utility surveying was not specially supported by any authorizing organizations [6]. Recently, utility survey is sheltered under the Department of Survey and Mapping Malaysia (DSMM) which located in Utility Section. Nowadays with the rising of GIS and remote sensing technology, there are various kinds of instruments or tools exist in market place. The utilities sector is complex and changing rapidly in terms of instrumental developments [6]. Different instruments have their own functions and specific usage. More recent developments, that been used for acquiring utilities data are Pipe and Cable Locator (PCL) and Ground Penetrating Radar (GPR) [7-9]. The usage of GPR is applied in many applications including environmental application, engineering application, earth science application and more other applications [10-13]. In engineering application,

GPR is not only focusing on surveying, it has been used in many other applications such as constructions, unbound pavement structures, wall condition, etc [14]. In utility mapping application, GPR can provide three dimensional (3D) image of surface which depends on the specification of the model [15]. Malaysia has experienced the growth of utility survey with increasing number of utilities survey instruments.

As a related study conducted in Malaysia Najua S. [6] performed a comparison analysis between PCL and GPR; build up utility database and three dimensional modelling of utility mapping. Jol H. [16] expanded the usage of GPR as the utility instrument based on the level of accuracy between two different models of GPR. The rising of technology of utility instruments makes a difficult situation for surveyors and other users in ascertaining suitable instruments, used to detect underground objects with high accuracy [17]. In response of this issue, the current comprehensive study have been conducted in three objectives; to get information on basic principle of single channel GPR; to analyze the accuracy between two different models of single channel GPR; and identifying GPR calibration method. This study will significantly improve and ease the approaches of utility data collection.

2 Study area and material

The location of study area was at GPR, Pipe and Cable Locator, Test Base, Jabatn Ukur dan Pemetaan Malaysia (JUPEM) Kuala Lumpur. This site was selected to fulfil the objectives of the study, whereby existing underground utilities such as water pipes, electric cables, and telecommunication cables were buried underground.

The instruments used were single channel NOGGIN PLUS and MALÅ RAMAC X3M Ground Penetrating Radar (GPR) provided by Utility Section, DSMM. Data processing and interpretation was processed using software provided by Utility Section, DSMM.

3 Methodology

3.1 Survey design

Survey design is the first requirement for most of the study implementation. It can be assumed as planning process in initial steps before conducting survey work, especially, where measurement lines were designed. In this study, survey design was divided in two categories; first, measurement lines were designed in paper; and second, design was input in the instruments itself before the actual survey work begin. Moreover, survey design were determine based on the area of survey. The survey line was designed based on Y line only, forward, with 1 meter station interval. The area was designed with (6×6) metres as shown in Figure 1.

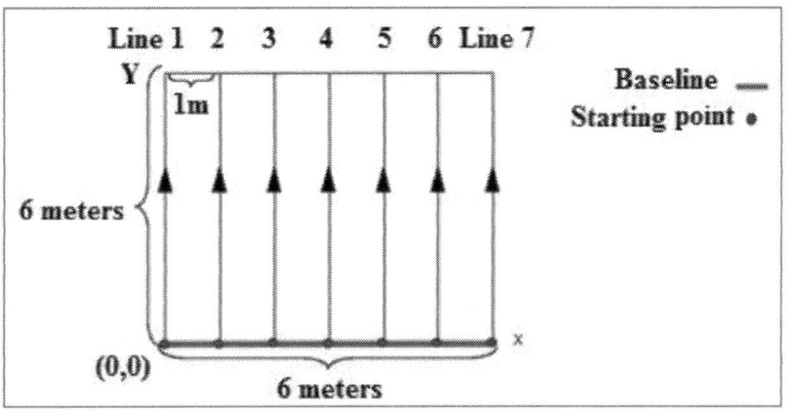

Figure 1. Survey design.

3.2 Survey planning
Survey planning is the method on how the survey will be started and ended. Firstly, before the actual survey starts, it is necessary to do reconnaissance work. Reconnaissance work is needed to explore the area of study and also to figure out various kinds of underground utilities involve at the survey site [18]. This method requires determining any obstacles that maybe encounter during survey work so that it can be ran smoothly without long time consumption. The second method was control survey with aim of establishing a control point at the survey area and to get the position of the survey site. Control survey is important part of this work to identify the location coordinates of the underground utilities. This control survey was performed using GPS instrument. Furthermore, the last method is detection survey or underground utility survey by using GPR.

3.3 Field work
This study requires two field work surveys as control survey and underground utility survey. A control survey requires to determine the position of the survey area. This is important in order to get accurate data positioning during data collection. GPS instrument, TOPCON GR-3 receiver with FC-200 controller was used for this process. Real Time Kinematic (RTK) GPS technique was used to obtain the coordinate of the unknown points. Before the survey was carried out, the receiver was set up at unknown point as shown in Figure 2. Then receiver was turn on to connect with Geodesy Section of DSMM to get correction data from Malaysia Active GPS System (MASS) station with radio link communication.

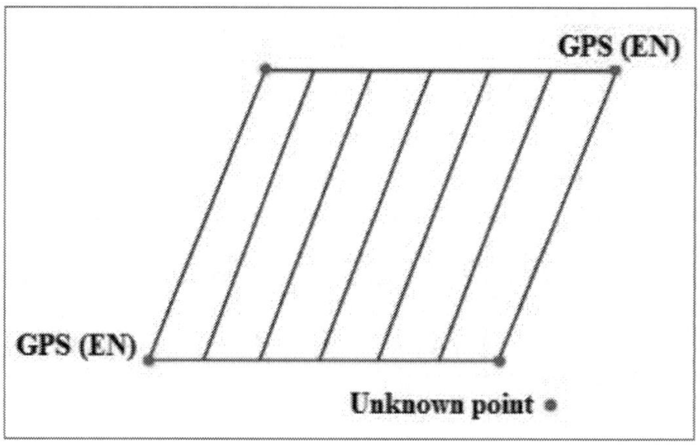

Figure 2. Control survey.

3.4 Underground utility survey
Underground utility survey was started after the completion of control survey. The purpose of using GPR is to detect the destructive and non-destructive underground utilities. Two different models of single channel GPR were used; NOGGIN PLUS and MALÅ RAMAC X3M, to evaluate the level of accuracy and image quality.

3.5 Data processing
Noggin data was processed using two software; EKKO_View and EKKO_Mapper. In addition, Ground Vision and Object Mapper software were used for MALÅ RAMAC X3M data view. The function of the software is for data entry, data analyse and summarization of the data measurements in automated format. Moreover, data processing involves recording data measurements, analysing,

sorting summarising, data storing etc. Typically, data processing manipulates the raw data into information in order to produce outputs.

The flowchart of the methodology is shown in Figure 3.

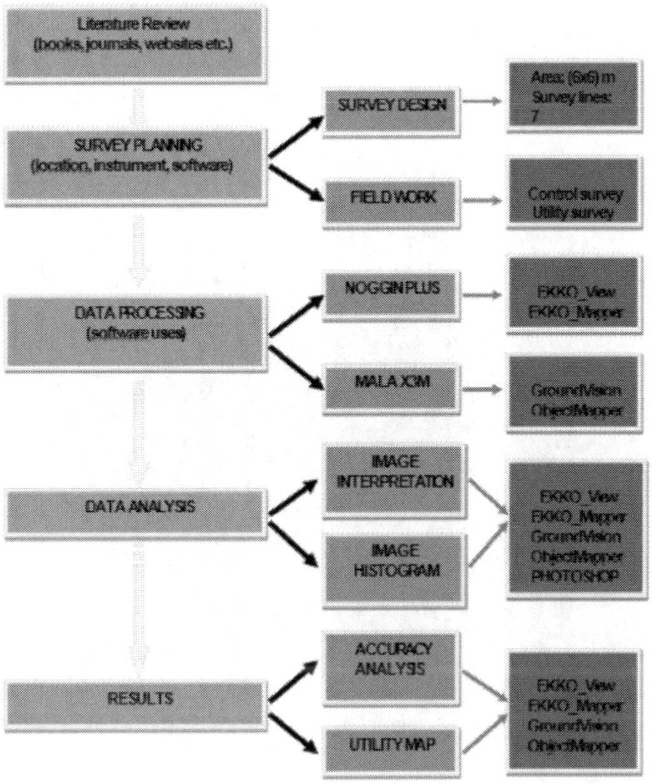

Figure 3. Flow chart of methodology.

4 Results and discussion

Image interpretation was done on GPR images for the purposes of identifying utility objects and their properties. The principle of image interpretation is according to the elements of the image such as colour, shadow, resolution, texture, pattern and depth. Image interpretation was done by comparing images collected by two GPR models (NOGGIN PLUS and MALÅ RAMAC X3M). NOGGIN images had clearer appearance compared to MALÅ images due to better image resolution. Based on images analysis in Adobe Photoshop Elements 8 software, NOGGIN model produces colour image while MALÅ default images were gray scale. This investigation was done to determine the depth and images pattern from detection survey. By using antenna of 250 MHz for both instruments, the survey was done in seven lines, but only the profile of line 4 and line 5 were analysed. The time interval and depth analysis were applied on these two profile lines. Same velocity (100 m/µsec) were applied on both instruments to determine the depth of water pipe and electric cable.

Figure 4 shows the interpretation image of line 4 which was taken by NOGGIN model. In this image, the blue hyperbola shape indicates water pipe underground utility while the red hyperbola shape indicates electric cable. The curvature of the dashed white line, represent hyperbola shape, indicates the interference of objects. Depth and time intervals are shown in Table 1. From survey, the detection of depth 1.3 m was taken in duration of 24 ns for water pipe while for electric cable the depth 1.35 m was noted in 25 ns.

The survey was repeated for line 5 in same place. Line 5 is the connection of line 4 to observe the relation of the utility objects. From NOGGIN image, Figure 5 shows the different colour lines which blue lines illustrate water pipe, red lines illustrate electric cable and white lines indicate interference of objects in survey line 5. Y-axis shows the time interval and depth while the X-axis shows the distances. Table 2 explains Figure 5 in terms of depth and time interval. A depth of 1.3 m for water pipe was taken within 23 ns time interval while depth of 1.35 m of electric cable was noted in 25 ns time intervals.

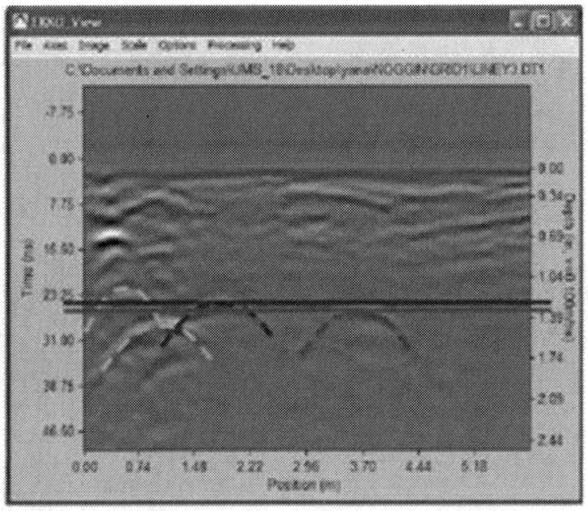

Figure 4. NOGGIN line 4.

Table 1. Depth vs time NOGGIN line 4.

LINE 4	Type	Time (ns)	Depth (m)
▬▬▬	Water pipe	24	1.3
▬▬▬	Electric cable	25	1.35

IOP Conf. Series: Earth and Environmental Science **37** (2016) 012025 doi:10.1088/1755-1315/37/1/012025

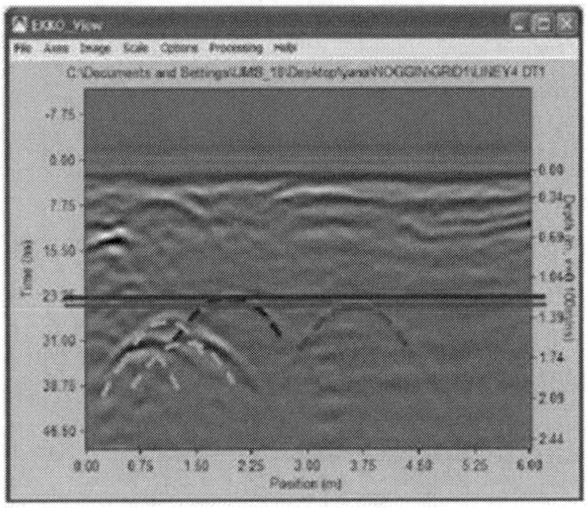

Figure 5. NOGGIN line 5.

Table 2. Depth vs time NOGGIN line 5.

LINE 4	Type	Time (ns)	Depth (m)
	Water pipe	23	1.3
	Electric cable	25	1.35

Figure 6 shows the image of line 4 which is taken by MALÅ. This image shows water pipe and electric cable by different colour of lines. Table 3 shows depth and time interval for this stage. For water pipe, a depth 1.4 m taken in 26 ns and for electric cable a depth 1.5 m was illustrious in 27 ns.

Figure 7 shows the MALÅ image with the blue and red lines for water pipe and electric cable respectively. Instead of those lines, it has white line which indicates the noise of objects. Based on the scale reading from Figure 7, Table 4 explains a depth of 1.3 m was detected in 26 ns for water pipe while depth 1.4 m was detected in 27 ns time interval for electric cable.

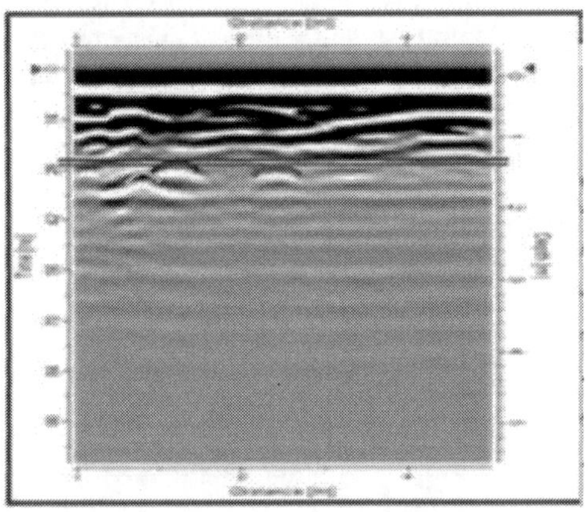

Figure 6. MALÅ line 4.

Table 3. Depth vs time MALÅ line 4.

LINE 4	**Type**	**Time (ns)**	**Depth (m)**
	Water pipe	26	1.4
	Electric cable	27	1.5

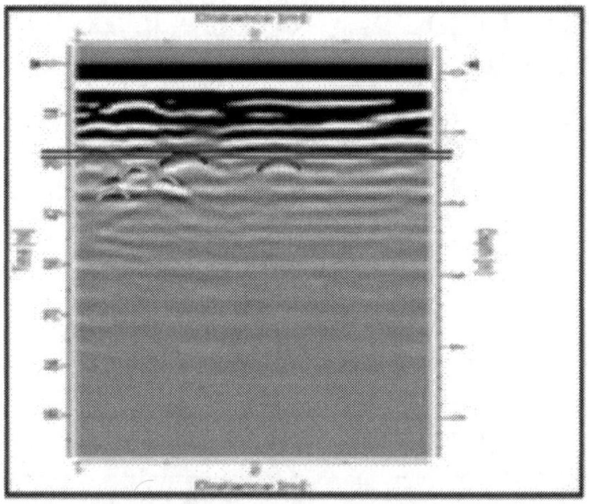

Figure 7. MALÅ line 5.

Table 4. Depth vs time MALÅ line 5.

LINE 4	**Type**	**Time (ns)**	**Depth (m)**
	Water pipe	26	1.3
	Electric cable	27	1.4

The results show that, there are a few differences in terms of depth measured from both instruments with the original depth from the previous measurement. The original depth of water pipe was 1.5 m while electric cable had depth of 1.4 m. The average value of utilities measured between line 4 and 5 have been counted, water pipe had depth of 1.3 m with NOGGIN model and 1.35 m for MALÅ model. Electric cable depth average was 1.35 m and 1.45 m for NOGGIN and MALÅ respectively. From the value compared in this analysis, GPR indicates accuracy up to cm-level.

Table 5 shows the comparison value of depth measured with the original value while Figure 8 represents the comparison value of Table 5 in bar graph.

Table 5. Measured depth and original depth.

	Water pipe	Electric cable
	Depth (m)	
Original	1.5	1.4
NOGGIN	1.3	1.35
MALÅ	1.35	1.45

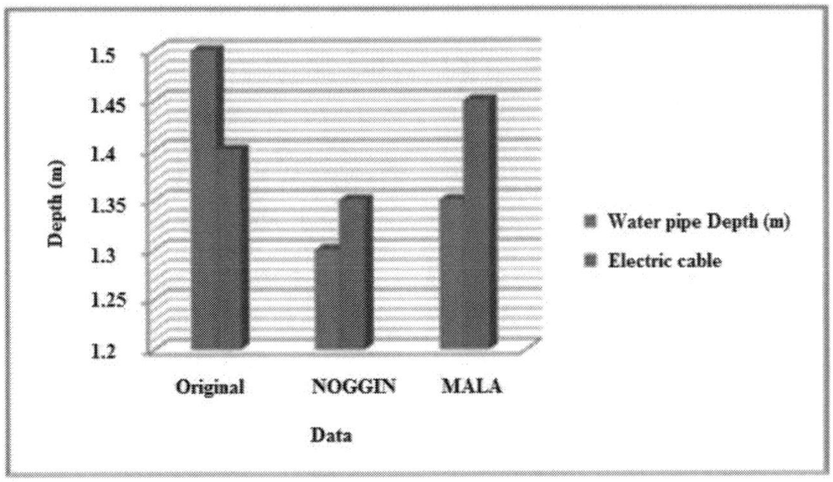

Figure 8. Data comparison.

Table 6 shows the difference between NOGGIN and MALÅ which counted from the average value of depth. Difference depth of 0.2 m for water pipe and depth of 0.05 m for electric cable are the difference between NOGGIN and original depth. On the other hand, different values of depth 0.15 m for water pipe and depth of 0.05 m for electric cables are from MALÅ model. It shows that, GPR instrument can provide cm-level accuracy in assessment between measurement values and original values.

Table 6. Value differences from the average counted values.

	Water pipe	Electric cable
	Depth (m)	
NOGGIN	0.2	0.05
MALÅ	0.15	0.05

5 Conclusion and recommendations

GPR has potential capabilities in various applications especially in utility mapping field. This technology is widely applied in the last decades to discover the subsurface phenomenon. As it is well known, GPR can give better accuracy compared to other instruments such as Pipe Cable Locator (PCL) in terms of producing high resolution image. This study can assist the surveyors and other users to make the selection between two models of single channel GPR either NOGGIN PLUS or MALÅ RAMAC X3M based on the output results. In order to determine which models can give better accuracy, the profiles along two lines have been analyzed. Along the same lines, the survey was repeated in two times by using both GPR models. The accuracies of models were different based on depth of utility objects and other properties. In general, the utilization of GPR improves and eases the data processing and outputs interpretation. GPR can also minimize the time consumption during data collection. Furthermore, the rising technology of survey instruments makes GPR as the best solution for utility survey mapping application regarding time-consumption and reliability of the results.

In order to introduce GPR to user, for further study it is recommended to study on:

i. Digital image processing for better interpretation of GPR image
ii. The 3D utility mapping application from the usage of GPR
iii. Enhance the calibration technique of GPR through upgrading the existing test base with respect to the accuracy of instruments

References

[1] Young G N and K L Alft, *Utility mapping and data distribution system and method*, V.M. Company, Editor 2014, Google Patents, US8775083 B2: USA.
[2] Daniels D J, *Ground penetrating radar*. Vol. 1. 2004: Iet.
[3] Wittenberg E and L A Prosser 2011, Ordering errors, objections and invariance in utility survey responses. *Applied health economics and health policy*. **9**(4): p. 225-241.
[4] Norta A, L Carlson, and R Yangarber 2010, Utility survey of ontology tools. *Department of Computer Science, Department of Linguistics-University of Helsinki, Finland*. **62**: p. 63-65.
[5] Luce R D, *Utility of gains and losses: Measurement-theoretical and experimental approaches*2014: Psychology Press.
[6] Najua S, *Three Dimensional Utilities Modelling Using Ground Penetrating Radar Multichannel Antenna*, in *Department of Geomatic Engineering,* 2010, University Teknologi Malaysia: Johor.
[7] Knight R 2001, Ground penetrating radar for environmental applications. *Ann Rev Earth Planet Sci*. **29**(1): p. 229-255.
[8] Linford N, P Linford, L Martin, and A Payne 2010, Stepped frequency ground-penetrating radar survey with a multi-element array antenna: Results from field application on archaeological sites. *Archaeological Prospection*. **17**(3): p. 187-198.
[9] Jaw S W and M Hashim 2013, Locational accuracy of underground utility mapping using ground penetrating radar. *Tunnelling and Underground Space Technology*. **35**: p. 20-29.
[10] Bristow C S, A Switzer, and D Kennedy 2013, Ground penetrating radar. *Methods, Treatise of Geomorphology. Maryland Heights, San Diego: Academic*: p. 183-194.
[11] Minet J, P Bogaert, M Vanclooster, and S Lambot 2012, Validation of ground penetrating radar full-waveform inversion for field scale soil moisture mapping. *Journal of Hydrology*. **424**: p. 112-123.
[12] Minet J, A Wahyudi, P Bogaert, M Vanclooster, and S Lambot 2011, Mapping shallow soil moisture profiles at the field scale using full-waveform inversion of ground penetrating radar data. *Geoderma*. **161**(3): p. 225-237.
[13] Selim E I, A A Basheer, G Elqady, and M A Hafez 2014, Shallow Seismic Refraction, Two-Dimensional Electrical Resistivity Imaging, and Ground Penetrating Radar for Imaging the Ancient Monuments at the Western Shore of Old Luxor City, Egypt. *Archaeological Discovery*. **2014**.
[14] Maser K R, *Ground penetrating radar surveys to characterize pavement layer thickness variations at GPS sites*1994: Strategic Highway Research Program, National Research Council.
[15] Böniger U and J Tronicke 2010, Improving the interpretability of 3D GPR data using target–specific attributes: application to tomb detection. *J Archaeol Sci*. **37**(4): p. 672-679.
[16] Jol H M, *Ground penetrating radar theory and applications*2008: Elsevier.
[17] Fisher E, G A McMechan, and A P Annan 1992, Acquisition and processing of wide-aperture ground-penetrating radar data. *Geophysics*. **57**(3): p. 495-504.
[18] Baiden G, Y Bissiri, S Luoma, and G Henrich 2014, Mapping utility infrastructure via underground GPS positioning with autonomous telerobotics. *Tunnelling and Underground Space Technology*. **39**: p. 6-14.

A tool for assessing ecological status of forest ecosystem

Abd Rahman Kassim[1], Muhammad Afizzul Misman[2], Mohd Azahari Faidi[2] and Hamdan Omar[2]

Head[1] and Research Officers[2]
Geoinformation Programme, Forestry & Environment Division, Forest Research Institute Malaysia, Kepong 52109, Selangor, Malaysia

Email: rahmank@frim.gov.my

Abstract. Managers and policy makers are beginning to appreciate the value of ecological monitoring of artificially regenerated forest especially in urban areas. With the advent of more advance technology in precision forestry, high resolution remotely sensed data e.g. hyperspectral and LiDAR are becoming available for rapid and precise assessment of the forest condition. An assessment of ecological status of forest ecosystem was developed and tested using FRIM campus forest stand. The forest consisted of three major blocks; the old growth artificially regenerated native species forests, naturally regenerated forest and recent planted forest for commercial timber and other forest products. Our aim is to assess the ecological status and its proximity to the mature old growth artificially regenerated stand. We used airborne LiDAR, orthophoto and thirty field sampling quadrats of 20x20m for ground verification. The parameter assessments were grouped into four broad categories: a. forest community level-composition, structures, function; landscape structures-road network and forest edges. A metric of parameters and rating criteria was introduced as indicators of the forest ecological status. We applied multi-criteria assessment to categorize the ecological status of the forest stand. The paper demonstrates the application of the assessment approach using FRIM campus forest as its first case study. Its potential application to both artificially and naturally regenerated forest in the variety of Malaysian landscape is discussed

1. Introduction

Managers and policy makers are beginning to appreciate the value of ecological monitoring of artificially regenerated forest especially in urban areas. With the advent of more advance technology in precision forestry, high resolution remotely sensed data e.g. hyperspectral, airborne light detection and ranging (LiDAR) are becoming available for rapid and precise assessment of the forest condition.

Foresters usually measure the status of the forest based on ground sampling through various scales of forest samplings from stand level, landscape level and national level. The sampling results provide indications of the forest ecological status in terms of its stocking density, structure and composition. For continuous forest inventory plots, the forester may be able to generate the forest dynamics status such as growth and mortality. With the advent of remote sensing technology, more forestry application has now embarked on forest assessment combined with field inventory over large areas. The remote sensing application may include the use of optimal and radar image as well as (LiDAR) to assess forest area. The application of LiDAR can provide detailed three-dimensional structure of forest canopies and has been widely used to characterize forest cover and structure [6]. In Malaysia, the application of LiDAR is still at research scale [4] and its potential application to support national forest inventories should be explored. Some the most important research areas where the LiDAR can play significant roles are canopy and tree height estimation, forest structure, and biomass and volume estimation [4]. The application of LiDAR may contribute to the assessment of the ecological status of a forest ecosystem. Knowing the ecological status or its integrity provides a useful framework for ecologically based monitoring and valuable information for assessing ecosystem condition and management effectiveness by Geraldine *et. al* [2]. The approach can be based on spatial and non-spatial information. In this presentation, we will demonstrate the application of non-spatial output in terms of tables of metrics and its ratings (see Geraldine *et. al* [2]). We also combine the metrics to generate a single index that describe the ecological status of the stand using Multi–Criteria Analysis Shell for Spatial Decision Support (MCAS-S) [5].

An assessment of the ecological status of forest ecosystem was developed and tested using FRIM campus artificially and naturally regenerated forest stands. The forest consisted of three major blocks; the old growth artificially regenerated native species forests, naturally regenerated forest and more recent planted forest for commercial timber and other forest products. In the past, the assessment of forest carbon had been conducted by Hamdan *et. al* [3] using remotely sensed L-Band SAR data. With the availability of LiDAR data for the year 2007 and 2013, we extended the assessment of FRIM stands in terms of carbon and other metrics of ecosystem structure and function.

Our aim is to demonstrate the assessment of the ecological status and its proximity to the mature old growth artificially regenerated stand using LiDAR data and FRIM's campus forest stands as its first case study. The study site has a long history of artificially regenerated forest stand of various species mixture planting of both native and exotics species. Its potential application to both artificially and naturally regenerated forest in variety of Malaysian landscape is briefly discussed.

2. FRIM's Campus Forest Stand as example

2.1. Site description

Major plantation researches began in the Forest Research Institute Campus in the 1920's and more than 100 species, both natives and exotics, were planted in 1920's and 30's (Watson 1935 cited in Francis Ng [1]). The planting was successful, however no scaling up of efforts towards industrial tree plantation levels were embarked. As a result, these areas serve as demonstration sites rather than production sites. Over the years, since 1970's, several plantation trials of exotic species such as pines, *Eucalyptus spp, Araucaria spp and Acacia spp* were established to support wood based industry. The establishment of all these plantations trials together with existing native forest within the FRIM campus and its close proximity to the permanent reserve forest makes it a unique setting for natural succession of flora and fauna species into the area. Some of the planted forest stands have grown to a fully stocked stand, perhaps approaching the condition of the pristine climax forest.

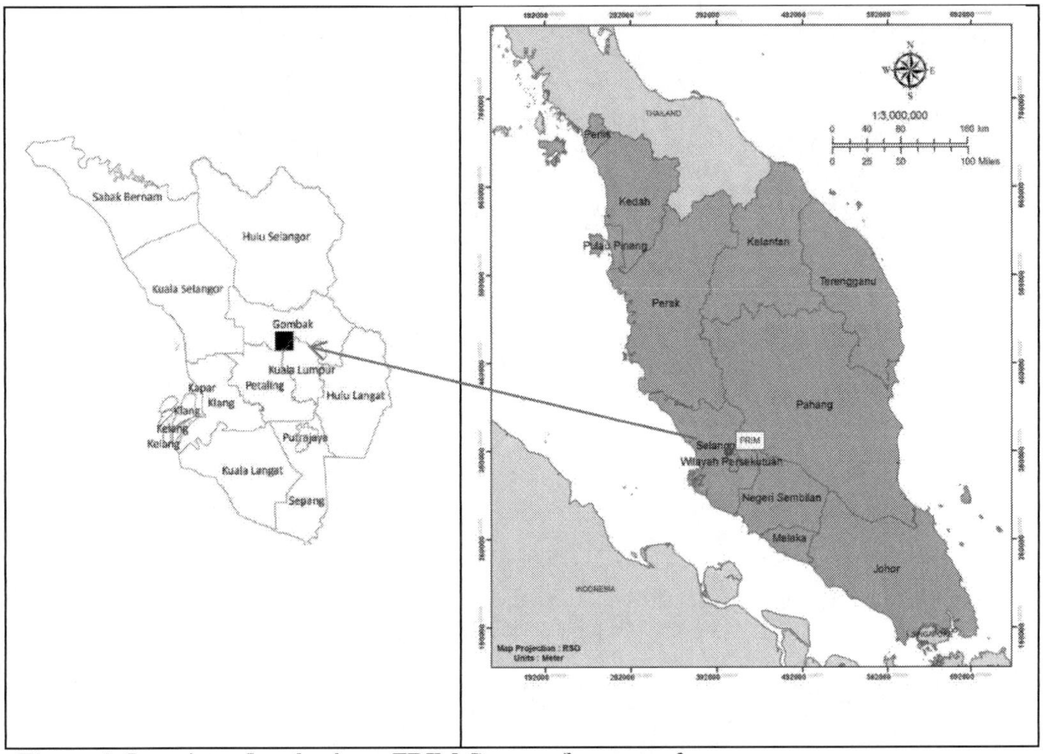

Figure 1. Location of study site at FRIM Campus forest stand.

2.2. Assessment approach

We used airborne LiDAR data set taken in 2007 and 2013 and 30 field sampling quadrats of 20x20m for ground verification to assess the metrics of ecological status.

2.2.1. LiDAR Processing

LiDAR data was used to produce several layers of ecological information. Generally, canopy height model (CHM) was generated from LiDAR data and this information was used to generate other layers for the assessment of the ecological status. CHM gave the information of the height of canopy trees from the ground. CHM was generated by substracting digital elevation model (DEM) with the digital terrain model (DTM) as shown in the Figure 2.

CHM was used as one of the layer and several other layers were generated from this information. This includes the estimation of biomass and volume of the forested area. The biomass and volume of the inventory plots established were correlated with the average CHM's values of each plot to get the function of both parameters as shown in Figure 3. Biomass and volume were then estimated using the functions established.

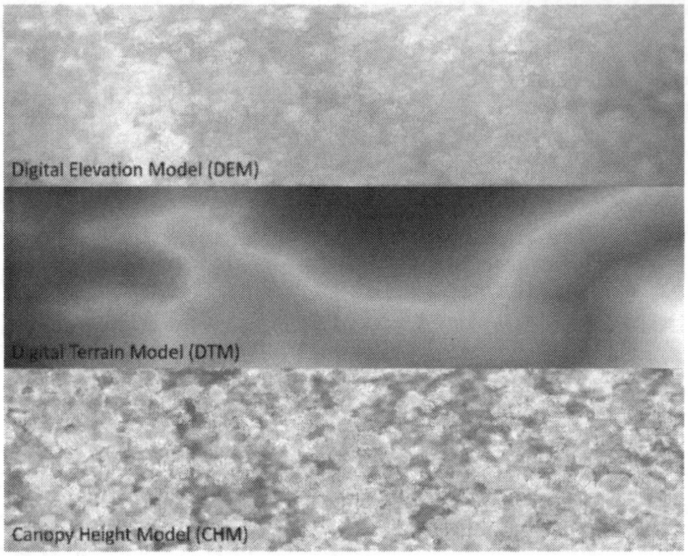

Figure 2. Generation of Canopy Height Model (CHM) from Digital Elevation Model (DEM) and Digital Terrain Model (DTM)

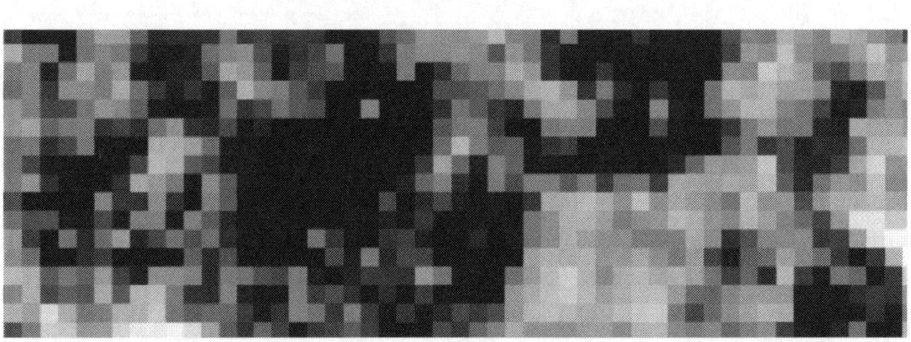

Figure 3. Example of biomass estimation using average CHM's values at 20 x 20 m resolution.

Other layers that were generated from CHM are height growth, volume and biomass changes, and mortality. These layers were produced by processing two sets of LiDAR data that were acquired from different dates. Height growth of the canopy tree was produced by calculating the difference of both CHMs. The same method was also applied with the volume and biomass changes layers. As for the mortality layer, it was generated by detecting canopy gaps or dead trees by using CHM's data of both years (Figure 4).

2.2.2. Developments of metrics and indicators

We take into account the component of forest ecosystem in describing the ecological status of the stand i.e. composition, structure and function. Beside the three aspects, we also included the landscape structures to indicate the threat to the stand especially in area potentially expose to disturbance due to its proximity to forest edges. Table 1 showed the metric parameter and its rating criteria.

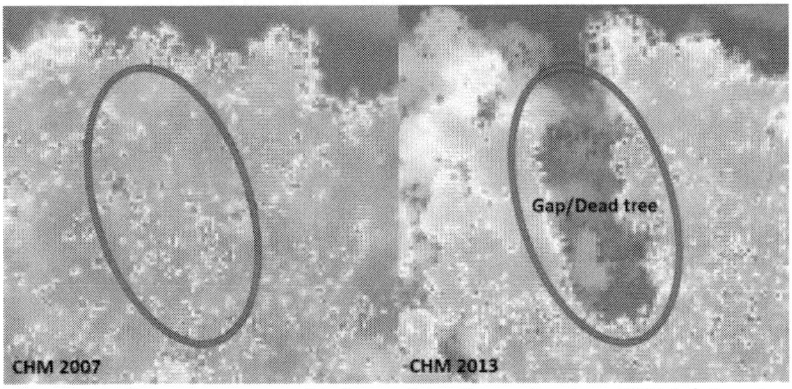

Figure 4. Canopy gaps or dead tree detection using two CHM data sets.

We used the "traffic light" color code to indicate the status of the ratings, namely, the status of the forest as good (Green), moderate (Yellow) and poor (Red). The Green color reflects the stand as in good condition and performing as expected; the Yellow color indicates that it is relatively good but can be further improved; while Red color indicate that the stand is in the stage of significant concern and may require human intervention to improve its ecological status

Table 1. A metric of structure, composition and function developed for FRIM campus site with its rating value.

Metric type	Metrics	Rating		
		Good	**Moderate**	**Poor**
Structures	Volume (m^3ha^{-1})	>300	100-300	<100
	Biomass (Mg^3ha^{-1})	>400	200-400	<200
	Canopy layers	Multi-layer	Mono-layer >20m	Mono-layer <20 m
	Top height (m)	>40	20-40	<20
	Height structural indices	TBA	TBA	TBA
	Light intensity (%)	TBA	TBA	TBA
Composition	Regeneration abundance	TBA	TBA	TBA
	Indicator exotic species	TBA	TBA	TBA
	Diversity indices	TBA	TBA	TBA
Function	Volume change (m3/ha)	>4	1-4	<1
	Biomass change (Mg/ha)	>10	3-10	<3
	Relative gap size	TBA	TBA	TBA
	Height growth (m)	>0.7	0.3-0.7	<0.3
Landscape	Relative disturbance	0	0-30	>30

Note: TBA means to be arranged

2.2.3. Ecological status based combine metrics

We applied multi-criteria assessment system (MCAS) to combine the ecological status of the forest stand. Using the spatial information of each metric, we combined each metric into clusters based on structure, function and composition of forest ecosystem to define the ecological status of the stand. In addition, we included the landscape structures which examine the edge effects within and at the boundary of FRIM's campus forest stand. MCAS is a technique that allows for the measurement and aggregation of the performance of alternatives or options, involving a variety of both qualitative and quantitative dimensions. The final output will be subjected to consultation with relevant experts.

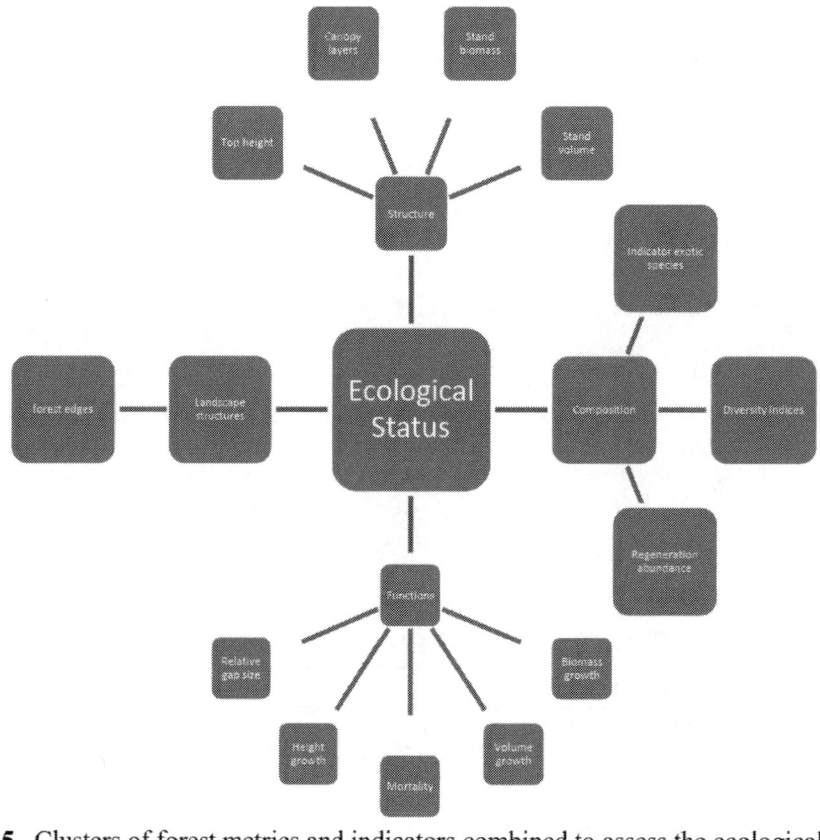

Figure 5. Clusters of forest metrics and indicators combined to assess the ecological status of the stand. Note that we have yet to establish the layers for composition component.

3. Results and Discussions

We presented the results in two parts: (a) Metrics and its rating output (b) Ecological status based on combined clusters of metrics and indicators.

3.1. Metrics and rating results

Here, we demonstrate the application of the ecological assessment based on the metrics and its indicators performance of three different sites in FRIM campus (a) Field 9a: mature old growth artificially regenerated Kapur Forest (b) Field 12 C: mature old growth of artificially regenerated mixed dipterocarp forest (c) Field 43: naturally regenerated forest. Kapur (*Dryobalanops aromatica*) is one of the major timber species in Malaysia. The stand consisted of almost pure crop of Kapus trees and form a two layer vertical structure i.e. the main canopy and the understory layer. Field 43 is a natural forest but probably in disturbed condition and remained less productive with stand volume growth is less than 1 m^3ha^{-1}.probably due to its past disturbance. Several actions are being taken by FRIM's Forestry and Environment Division to enrich the forest stand through enrichment planting of high quality timber trees.

Table 2. Metrics and its rating based on the traffic light colors for three field in FRIM campus.

Metric type	**Metrics**	**Field 43**	**Field 9A**	**Field 12C**
Structures	Volume (m3/ha)			
	Biomass (m3/ha)			
	Canopy layers			
	Top height (m)			
	Height structural indices	TBA	TBA	TBA
	Light intensity (%)	TBA	TBA	TBA
Composition	Regeneration abundance	TBA	TBA	TBA
	Indicator pioneer species	TBA	TBA	TBA
	Diversity indices	TBA	TBA	TBA
Function	Volume change (m3/ha)			
	Biomass change (Mg/ha)			
	Relative gap size			
	Height growth (m)			
Landscape	Relative disturbance			

TBA: to be arranged

3.2. Ecological status based on combined clusters of metrics
The application of MCAS allows us to generate a value based on combination of metrics and its parameters. We assumed that the old growth from multi species artificially regenerated stand can be a good indicator of productive forest resembling the pristine native forest condition. A closeness of the stand to the old growth from multi species artificially regenerated stand would be the stand that has high in volume, biomass, tall height, multi structure, low light intensity, dominance of pioneers or invasive species etc. In the assessment, we included the available structure and function metrics (Table 1 and 2) to produce the expected that reflect variation in ecological status indices within the FRIM campus forest stand. The embedded box in the graph illustrates the ecological status index for the final output, and its respective structure and function cluster metrics.

3.3. General remarks
We have demonstrated the assessment of ecological status for 3 stand management units or known as Field within FRIM campus based on the LiDAR data and field ground sampling. We take into considerations the structures and function of the artificially regenerated forest ecosystem of FRIM campus. We are still lacking with the assessment on the composition component.

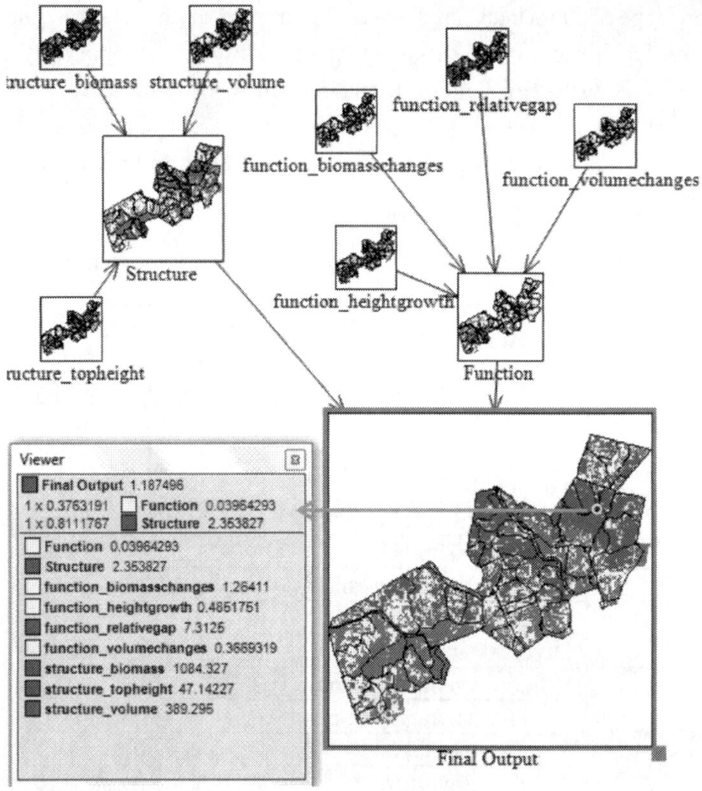

Figure 6. Example of GIS layer output on ecological status of forest stand based on combined metrics.

To address a complete assessment of the stand requires further effort to examine several other parameters: biodiversity indices, structural indices, and light intensity. Currently our assessment is based on LiDAR data and field sampling plot and development of statistical model to predict stand biomass and volume. Although the predictive model is perhaps sufficient for assessment of the parameters mentioned above, predicting biodiversity may requires different approach in sampling of the sites to include the variations across the campus forest. Stratification based on planting species mixture and age of stand may serve as a good criteria to examine the biodiversity variation status of various site. Height generated from LiDAR data may also be used as one criterion to relate with biodiversity variation. In general, a multi-structure forest stand consisting of the several ecological guilds have often been associated with diverse flora and fauna.

Many new forests of native and exotic species in the country are being established as parks especially in urban areas (e.g. Putrajaya) and FRIM's forest stand can therefore be a benchmark for other artificially regenerated forest. However the suitability of the tool application for forested areas of naturally regenerated stand has yet to be established. Effort has been initiated to establish a benchmark for ecological status assessment for primary and disturbed lowland forest in Sg Menyala Forest Reserve, Negeri Sembilan.

4. Conclusions

The forest stand consisted of naturally and artificially regenerated stand of FRIM campus had provide us the opportunity to assess the ecological status of forest stands due to its known long history and background of the stand. The differences in the species mixtures demonstrate its long-term impact on the timber volume and biomass as well as the forest structure and may explained variation in biodiversity. Further research is required to establish the relationship of biodiversity indices from ground sampling and spectral images to upscale the value over the entire FRIM campus.

LiDAR data provides us the opportunity to address metrics and its rating of ecological status of a forest stand. The application of MCAS-S provides us with the opportunity to develop an indices based on combine metrics that may reflect the ecological status of the forest stand. However, care should be taken in assigning the combine metrics so that it will reflect the expected results.

5. Acknowledgement

We would like to thanks all staff of the Geoinformation Programme of FRIM for their contribution in data collection and ground verification. We extend our gratitude to other colleagues of FRIM who have shared their views on the approach to assess ecological status of forest ecosystem at FRIM campus. Thanks to Mrs Rodziah Hashim for preparing location map. Lastly to our Director General of FRIM and other senior management officers who have been supportive of this study. The study is currently funded under the 11th Malaysian Plan, Ministry of Natural Resources and Environment entitled "Penyelidikan indeks integriti ekosistem (IIE) untuk pengurusan hutan tropika negara secara mapan".

References

[1] Francis SP 2010 100 Years of Tropical Forest Research *The Story of the Forest Research Institute Malaysia* FRIM Special Publications.

[2] Geraldine LT Faber-Langendoen D Brian RM Shriver WG Gibbs JP 2009 Monitoring and evaluating the ecological integrity of forest ecosystems *Frontiers in Ecology and the Environment* 7(6):308–316.

[3] Hamdan O Aziz KH Abd Rahman K 2011 Remotely sensed L-band SAR data for tropical forest biomass estimation *Journal of Tropical Forest Science* 23(3): 318-327.

[4] Ismail MH Manaf MS 2011 The potential of LiDAR application in Malaysia *International Journal of Remote Sensing Applications* 1(1):1–5.

[5] MCAS-S development partnership 2014 *Multi-Criteria Analysis Shell for Spatial Decision Support MCAS-S version 3.1 User guide* ABARES Canberra.

[6] Zald HSJ Wulder MA White JC Hilker T Hermosilla T Hobart GW Coops NC 2016 Integrating Landsat pixel composites and change metrics with LiDAR plots to predictively map forest structure and aboveground biomass in Saskatchewan, Canada *Remote Sensing of Environment* 176:188–201.

Comparison results of forest cover mapping of Peninsular Malaysia using geospatial technology

Wan Abdul Hamid Shukri B. Wan Abd Rahman
Forestry Department Peninsular Malaysia

E-mail : wanamid@forestry.gov.my

Abstract. Climate change and global warming transpire due to several factors. Among them is deforestation which occur mostly in developing countries including Malaysia where forested areas are converted to other land use for tangible economic returns and to a smaller extent, as subsistence for local communities. As a cause for concern, efforts have been taken by the World Resource Institute (WRI) and World Wildlife Fund (WWF) to monitor forest loss using geospatial technology – interpreting time-based remote sensing imageries and producing statistics of forested areas lost since 2001. In Peninsular Malaysia, the Forestry Department of Peninsular Malaysia(FDPM) has conducted forest cover mapping for the region using the same technology since 2011, producing GIS maps for 2009-2010,2011-2012,2013-2014 and 2015. This paper focuses on the comparative study of the results generated from WRI,WWF and FDPM interpretations between 2010 and 2015, the methodologies used, the similarities and differences, challenges and recommendations for future enhancement of forest cover mapping technique.

1.Introduction

Deforestation has been identified as one of the many contributing factors to climate change[5]. Keeping forests intact throughout the world has become a major focus and concern for scientists and environmentalists, as well as governments, as their deterioration in terms of size and quality may produce unwanted domino effects to the stability of the earth. While developed countries have generally ceased from opening their remaining forests due to their economic prosperity, developing countries with natural forests continue to exploit these land banks as an important source to spur their economic growth towards achieving similar developed status. This came into the limelight in 1992, when the United Nations Conference on Environment and Development (UNCED) was held in Rio De Janeiro, resulting in the manufacturing of agreements and guidelines for sustainable development that stressed on sustainable forest management [9].

For Malaysia, this comes in the form of submission of remaining forest area status reports at regional and global levels on specified periodical basis such as through the Country Report for the Convention on Biodiversity (CBD), Forest Resource Assessment (FRA), Monitoring, Assessment and Reporting (MAR), ASEAN Senior Officers of Forestry (ASOF) and ASEAN Ministers on Agriculture and Forestry (AMAF). The report can also be used as base information for the country if it decides to pursue the carbon stocking scheme such as the Reducing Emissions from Deforestation and Forest Degradation (REDD) which may provide monetary incentives through conservation and rehabilitation efforts of the country's forests[10].

The availability of remote sensing temporal satellite images in combination with GIS have opened a new dimension in forest cover change detection where official statistical information of particular areas of forest extent derived from conventional methods can be scrutinised and verified. For almost the past two decades, geospatial technology has been used by the government authorities such as the Forestry Department of Peninsular Malaysia (FDPM) in Malaysia for the planning, managing and monitoring of its forest area and world-based non-governmental environmental organisations such as World Resource Institute (WRI) and World Wildlife Fund (WWF) for their work including tracking the deforestation trends in temperate and tropical countries with still large traits of forest including Malaysia as part of the check and balance process in safeguarding the remaining world's forests. In this paper, a comparison of forest cover loss results from 2001 to 2015 produced by WRI,WWF and FDPM , wherever possible, is made to assess its accuracy. The FDPM's geospatial data used in this paper are research data, as it is not used as an official data in any publication by the organisation.

2. Forest, forest cover and Permanent Reserved Forest

The definition by the United Nation's Food and Agriculture Organisation (FAO) [1] of "Forest" is 'a land area not more than 0.5 ha with a tree canopy cover of more than 10% which is not primarily under agricultural or other specified non-forest use, and in young forests and regions where tree growth is climatically suppressed, the trees should be capable of reaching a height of 5m *in situ*.' The definition is generally accepted, including for Malaysia. The Forestry Department Peninsular Malaysia (FDPM) has published their forested area statistics for Peninsular Malaysia since the establishment of the entity during the colonial days. The initial base information of the forest areas was extracted from the topography maps of the Land Survey Department and the Standard Sheet maps of the Land Offices. Eventually the maps and statistic information were updated by the State Forestry Departments (SFDs) in form of deductions and additions on the statistical information whenever a forest area is approved for conversion to non-forest use or development or degazettement of a Permanent Reserved Forest (PRF), or a gazettement is made for a new Permanent Reserved Forest(PRF). The conventional method involved the use of manual measurement and calculations of charted areas on hardcopy maps to come up with the forest area figures. The importance of an area on the map tallying in size with its stated area statistic is important to maintain the integrity of the data, and this can be made possible through the use of computerised methods. In 2011, FDPM embarked on a pilot forest cover mapping project for Peninsular Malaysia using geospatial technology aiming to compare the acreage size of an area polygon with the ones generated traditionally. The initial definition for forest cover used was: 'Land covering an area of more than 0.5 ha with trees having height of more than 5 meters at maturity and with canopy coverage of more than 30% or trees enable to reach the thresholds mentioned *in situ* derived from satellite images. Excludes land dominated for agriculture and urban purposes but include water bodies.'As the project progresses, FDPM later redefined forest cover as 'Land covering an area of more than 0.5 ha with trees having more than 5 meters at maturity and with canopy coverage of more than 30% except in high altitude areas. Include forests in islands and Permanent Reserved Forests (PRFs) under Sustainable Forest Management but exclude water bodies such as natural or man-made lakes, major rivers and land dominated for agriculture and urban purposes.'

WRI assumes "trees" as forests and defined it as 'all vegetation taller than 5m in height, and may take the form of natural forests or plantations across a range of canopy densities'(Hansen et.al [7]) which may include agricultural plantation such as oil palm. Through their Global Forest Watch online system they have managed to produce statistics of "forest" cover loss for every state in Peninsular Malaysia beginning from the year 2001 to 2014. The data has been improved for the year 2013 and 2014 where they have succeeded in deducting non-forest tree plantation areas through their recent modification on the geospatial analysis.

Meanwhile WWF performed geospatial analysis on forest cover loss in Malaysia for the year 2001 to 2015(January to June) where they defined forest cover based on the pixel values of the processed MODIS images used.Values of 0 to 40 are interpreted as forests, while values of 41 to the maximum value of 255 are considered as deforestation[9].

Permanent Reserved Forest (PRF) is defined as 'any land constituted or deemed to have been constituted a permanent reserved forest under the National Forestry Act, 1984'[6]. A PRF is basically a forest, but may be a land without forest cover and legally gazetted administratively as one, while a land with forest cover may not be a PRF, as in the case of the Taman Negara and Krau Game Reserve.

3.Methodology

The comparing analysis with FDPM's official data and research geospatial data is made exclusively based on the area size statistics provided by WRI and WWF.

3.1 Forestry Department Peninsular Malaysia

FDPM uses an integration of remote sensing and Geographical Information System (GIS) procedures to accomplish the task. Software used are ERDAS Imagine, ArcGIS and ArcView 3.x. For 2010 to 2014, SPOT 5 pan-sharpened 2.5 m spatial resolution satellite images obtained from Remote Sensing Agency Malaysia (ARSM) covering 68 scenes for Peninsular Malaysia were used (with the least cloud cover (less than 20% cloud cover). Whenever needed, additional ground control points (GCP) were added to improve the accurate positioning of the images on the actual map during geo-referencing[8]. The type of images were replaced by Landsat 8 and SPOT 6/7 for forest cover mapping beginning the year 2015. Initial Unsupervised classification (UC) was conducted on several imagery scenes. Each classes were identified using relevant GIS layers and ground-truthing of vague spots before Supervised Classification (SC) was performed to improve the results. However the classification output of SC was only slightly better than that of UC. The SC raster output was converted to GIS shapefile vector format. The vector layer was then overlayed with the same imagery scene to check for classification inconsistencies and to enable correction/editing. Once again, relevant GIS layers were used for support and as reference during editing. The re-digitising and correcting exercise proved to be cumbersome tasks thus the exercise was abandoned and not duplicated for the remaining scenes. Ghorbani [4] found that visual interpretation still produce higher accuracy than digital image classification for Landsat TM and Google Earth images. Previously, FDPM has been using visual interpretation technique to extract vegetation features from Landsat TM images (FDPM [2]). Nevertheless, the elements for visual interpretation for SPOT 5 imagery remains the same – tone, colour,shape, size, pattern, texture, shadow, height /depth, location and association. This is supported by overlay analysis with GIS layers Land Use Map produced by the Department of Agriculture (DOA), National Forest Inventory Map and topography map to enable the early creation of the rough forest cover layer in vector shapefile format. On-screen digitising was conducted to delineate the forest feature (based on the forest cover definition used) with on and off comparison made to the other related layers (where relevant, the same delineated boundary of the other layer is used to maintain boundary consistency). The forest delineation was further re-classified to separate the four (4) main forest types; inland forest, peat swamp forest, mangrove forest and forest plantation, again referring to the related layers for guidance. The digitised data underwent the topology process in ArcGIS to eliminate data irregularities such as sliver polygons. Validation exercises were carried out for the Forest Cover geospatial data produced in the previous year through hands-on geospatial verification workshops the following year conducted with geospatial and field personnel of the 12 SFDs in attendance. Uncertain/vague/grey areas were marked and recorded. On the ground inspections with further aerial surveillance using helicopters on several areas were implemented to enhance accuracy. The verified data (through the officially signed acceptance letter of the respective SFDs) were to be endorsed for usage by FDPM. The verified data from each states were merged into one official Peninsular Malaysia forest cover geospatial map data. The methodology used is as shown in **Appendix A.** The forest cover geospatial mapping was conducted every two years starting from the year 2009 but shortened to annually beginning 2015. For the purpose of this comparison analysis , the FDPM forest cover loss figures during each period are divided into the number of years involved so that an average figure can be derived to represent a particular year size acreage.

3.2 World Resource Institute

The forest cover loss data set for Peninsular Malaysia inserted in the Global Forest Watch online system by WRI for the year 2001 to 2014 was produced through collaboration efforts with organisations such as University of Maryland, United States Geological Survey (USGS),Google and NASA[12]. They are generated from processed and analysed Landsat 5, Landsat 7 and Landsat 8 images. The clear land surface observations in the satellite images were assembled and a supervised learning algorithm was applied to identify per pixel tree cover loss. The analysis method has been modified in numerous ways, including new data for target year, reprocessed data for the year 2011 and 2012, and improved modelling and calibration (Hansen et al.[7]). The modifications improved forest change detection for 2011-2014 including better detection of smallholder agriculture in tropical forests, selective loss and short cycle plantations [13]. Further explanation of the methodology used are as in http://earthenginepartners.appspot.com/science-2013-global-forest.

3.3 World Wide Fund

WWF commissioned the Netherland-based remote-sensing agency, SarVision to undertake the processing of MODIS satellite images, processed into 7-band annual cloud free composites at 250m resolution [11] for spatial analyses of deforestation in Malaysia. The analysis done by WWF for forest cover loss are for the periods between July 2001-June 2005,July 2005June 2010,July2010-June 2013 and July 2014-June 2015. The processing encompassed the following procedures: i) generation of MODIS data, ii) classification of forest cover, iii) selection of images with appropriate pixel values for deforestation analyses, and iv) forest cover index (FCI) and map validation. The technical details of each procedure are as given in **Appendix B**. WWF additionally implemented deforestation analysis according to the respective states and PRFs of Peninsular Malaysia. For the purpose of this comparison analysis, the WWF forest cover loss figures during each period are divided into the number of years involved so that an average size figure can be derived for each particular year. The period between July to June of each period are assumed as the period between January to December of each year.

4. Results & Discussion

4.1 Forest cover loss 2001 to 2010

Table 1 : Forest cover loss (in '000 hectares) 2001-2010.

Source	Year ('000 hectares)										
	2001	2002	2003	2004	2005	2006	2007	2008	2009	2010	Total
WRI	169.4	179.1	77.0	156.1	180.7	144.6	167.3	120.5	224.3	190.1	1,609.1
WWF	264.9	264.9	264.9	264.9	264.9	104.2	104.2	104.2	104.2	104.2	1,845.5
FDPM Annual Report [a]	54.8	32.1	13.6	12.8	-21.3	-25.1	72.0	-3.0	-29.5	9.5	115.9

[a] Forest area loss.

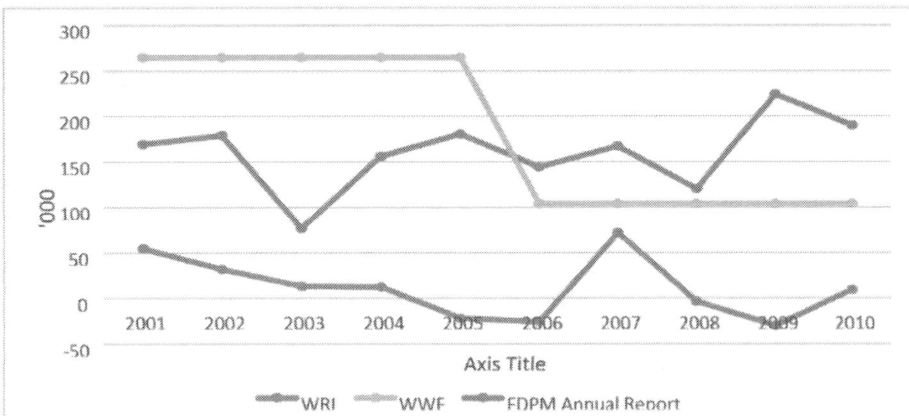

Figure 1 : Forest cover loss (in '000 hectares) 2001-2010.

Forest cover loss statistics based on "current year minus year before" calculations of forest area figures published in the FDPM Annual Reports as shown in Table 1 and Figure 1 are merely used as "partial" comparison of geospatial technology derivation by WRI and WWF versus conventional reporting by FDPM as it did not produce geospatial forest cover mapping data before the year 2010. The forest area published in the FDPM Annual Report is not based on satellite imagery interpretation but from conventional statistical reporting collected from the SFDs. The forest gain in 2005, 2006,2008 and 2009 reported are figures based on area additions through the inclusion of new areas gazetted as PRF and inclusion of forest areas identified through formal area applications received and ground surveys done by the SFDs on newly identified forest areas and areas not recognised as forest before(due to its immaturity or insufficient composition to identify it as one). Geospatial forest cover loss figures of WRI[12] and WWF showed higher forest cover loss of 1,388.4% and 1,592.3% respectively in comparison to FDPM's conventional derived forest area loss of only 115,900ha between 2001 to 2010. WWF's overall figure of 1,845,500 ha for the period is 14.7% higher than WRI's figure of 1,609,100 ha. The period between 2001 to 2005 showed WWF's period total size of 1,324.5 ha being 73.8% higher than WRI's figure of only 762.3 ha. In contrast, for the period 2006 to 2010, WWF's period total size of 521 ha was 38.5% lower when compared against WRI's 846.8 ha.

4.2 Forest cover loss 2011-2015

Table 2 : Forest cover loss (in '000 hectares) 2011-2015.

Source	Year ('000 hectares)					
	2011	2012	2013	2014	2015	Total
WRI	203.3	255.1	88.8[b]	200.7[b]	-	747.9
WWF	139.0	139.0	139.0	74.5	74.5	566.0
FDPM Geospatial	94.2	94.2	56.2	56.2	72.7	373.5
FDPM Annual Report [c]	57.3	18.5	-42.6	27.9	-	61.1

[b] Excluding agricultural tree plantations.
[c] Forest area loss.

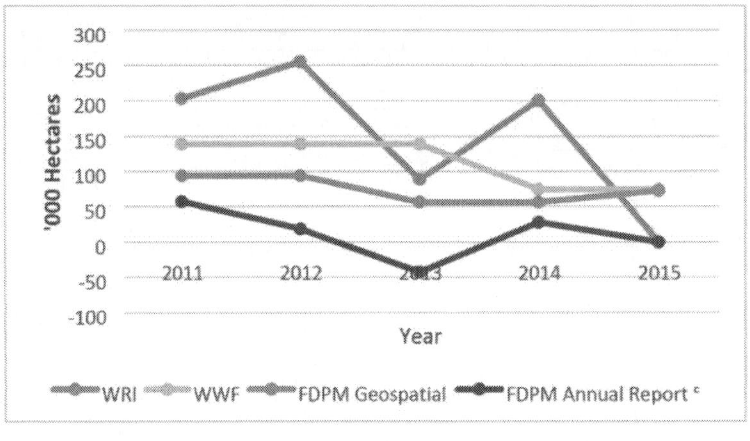

Figure 2 : Forest cover loss (in '000 hectares) 2011-2015.
Note : No data for WRI and FDPM Annual Report for the year 2015.
[c] Forest area loss.

The trend of forest cover loss analysis remained high in WRI[12] and WWF data as compared to FDPM's dataset of both geospatial and annual report as illustrated in **Table 2** and **Figure 2**. WRI's improved analysis

[13] where non-forest that includes agricultural tree crops are eliminated from the original forest cover loss figures for the year 2013 and 2014 still yield higher forest cover loss when compared to WWF(169.7%) and FDPM's geospatial(257.1%)figures for the year 2014. WWF and FDPM's geospatial results for the year 2014 and 2015 have smaller variations, where while WWF's data are larger in size, the percentages differ slightly at 24.5% for 2014 and 2.4% for 2015. FDPM have produced only bi-annual forest cover geospatial data for Peninsular Malaysia for 2009-2010,2011-2012 and 2013-2014 using SPOT 5 images due to the difficulty in obtaining cloud-free images over a one year period[8]. Since the FDPM's unavailability of SPOT 5 images from April, 2015, Landsat 8 images were mostly used for 2015 mapping exercise with several SPOT 6/7 images available for refinement. FDPM's geospatial forest cover loss figure is 5 times higher than FDPM's forest area loss data due to the difference between forest cover and forest area definitions where figure for the former used normally refers to only formal data. However the high variations between the two will be thoroughly investigated by FDPM to narrow the difference in the future.

4.3 Forest cover loss in PRFs 2011-2015

Table 3 : Forest cover loss in PRFs (in '000 hectares) 2011-2015.

Source	Year ('000 hectares)					
	2011	2012	2013	2014	2015	Total
WWF	35.3	35.3	35.3	29.5	29.5[d]	164.9
FDPM Geospatial	34.5	34.5	24.4	24.4	42.7	160.5

[d]Up to June 2015.

No analysis was conducted by WRI for this comparison. There are less variations between the figures of both entities where WWF's data was higher than FDPM's for most years except for 2013 and 2015 as indicated in **Table 3**. The different month period of both data to represent a particular year and the average figure used may be the reason for the variation shown in the year 2013. While WWF consistently applied the MODIS data, FDPM used Landsat 8 for 2015, replacing SPOT 5 which were used in previous forest cover mapping exercises. FDPM's geospatial data displayed higher figure compared to WWF's since FDPM's data was generated from available images dated throughout 2015 while WWF's 2015 actual image sources was from July 2014 to June 2015. The loss of forest cover within PRFs in a sizeable manner occurs when a large area is clear felled for the establishment/replanting of a forest plantation or other non-forest development such as for utilities/infrastructure constructions (e.g. dam, transmission line or highway/road network). The close figures between the two for 2011,2012 and 2014 indicated that images with low spatial resolution used by WWF can produce accurate results with the correct digital image processing technique as compared to visual interpretation using higher spatial resolution images. The analysis focusing on smaller specified areas (such as the PRF) may also be a contributing factor.

5.Conclusion

Conserving and protecting forest cover has never been so important as in these past few decades in order to sustain social, economic and environmental stability of a country. Malaysia has pledged to maintain at least 50% of its forest intact during the Rio Summit and the pledge has been renewed from time to time (the latest by the Malaysian Prime Minister during the opening ceremony of the Inter-governmental Plenary Session of Biodiversity and Ecosystem Services (IPBES 2016) in Kuala Lumpur on the 22nd February 2016)[14]. According to the Global Forest Resources Assessment Report 2015 [1] , Malaysia still has 68% of its land covered with forests where most of them are distributed in Sabah and Sarawak. The use of geospatial technology, especially remote sensing, has made forest cover monitoring possible in producing a better depiction of the status of the resource. The challenge in having cloud cover free images can be overcome in the future, through the use of radar, airplanes or drones mounted with appropriate sensors and cameras to complement the cloud-covered, "cataract" scenes. Forests are dynamic vegetation liable to grow back over time if left undisturbed thus forest gain is possible besides reforestation and afforestation. A forest plantation is managed in a way that it will be 100% felled at maturity and thus the forest cover will indicate a sizeable cleared area if an image taken on that date is used during analysis. This is in contrast to the "before" image of the same area with the forest or forest plantation still standing. Thus the date of the images used influenced the forest cover loss size output of the year in analysis. The use of different image from different sensor with different spatial resolution are additional factors to consider. Higher spatial resolution images is assumed to produce more accurate delineation. As FDPM has revert to using Landsat 8 beginning 2015 with a lower spatial resolution of 15 m as compared to using SPOT 5 before this, the accuracy consistency may no longer be maintained. However SarVision's (on behalf of WWF) use of MODIS images with lower spatial resolution of 250 m [11] but using different processing and analysis method to generate size figures with less variation in comparison to FDPM for the total period 2011-2015 for PRFs need to be explored further to determine its merit. These are the few issues which need to be resolved before the forest cover data through remote sensing interpretation and GIS can be used and published by FDPM in addition to the existing forest area statistic data in the near future.

6.References

[1] FAO 2015 *Global forest resources assessment 2015 Desk reference* (www.fao.org/publications)
[2] FDPM 1994 *Penginterpretasian visual imageri-imageri Landsat TM dalam bentuk "False Colour Composite" (FCC) dan "Georeferenced Image Map" (GIM)* Internal circulation.
[3] FDPM 2000-2014 *Annual reports 2000- 2014* Forestry Department Peninsular Malaysia.
[4] Ghorbani A and Pakravan M 2013 *Land use mapping using visual vs. digital image interpretation of TM and Google Earth derived imagery in Shrivan-Darasi Watershed (Northwest of Iran)* European Journal of Experimental Biology, 2013, **3(1)**:576-582
[5] Gorte R W and Sheikh P A 2010 *Deforestation and climate change*.Congressional Research Service.
[6] Government of Malaysia 1993 *.National Forestry Act 1984 (Amended 1993) (Act 313).*
[7] Hansen M C, Potapov P V, Moore R, Hancher M, Turubanova S A,Tyukavina A, Thau D, Stehman S V ,Goetz S J, Loveland T R, Kommareddy A, Egorov A, Chini L, Justice C O and Townshend J R G 2013 *High-resolution global maps of 21st century forest cover change*. Science **342**(15 November):850-53.Data available from :earthenginepartners.appspot.com/ science-global-forest.
[8] Shukri W A H W A R, Shahrulnizam K, Mohd Basri M 2014 *Forest cover mapping using geospatial technology for Peninsular Malaysia*.Paper presented at 17th Malaysian Forestry Conference,Sabah on 11 November 2014.
[9] United Nations 1992 *A/CONF.151/26 (Vol.III) forest principles*. Report of the United Nations Conference on Environment and Development.
[10] United Nations(2016) *About REDD+*. UN-REDD Programme Sheet.
[11] WWF Malaysia and WWF Netherlands 2015 *Analysis of Deforestation in Malaysia 2001-2015*.
[12] www.globalforestwatch.org/country/MYS Last access date on 28 February 2016.
[13] www.globalforestwatch.org/2016/01/when-tree-cover-loss-is-really-forest-loss-plantation-mapsimprove-forest-monitoring Last access date on 28 February 2016.
[14] Witnessed by the author through official attendance on 22 February 2016.

IOP Conf. Series: Earth and Environmental Science **37** (2016) 012027 doi:10.1088/1755-1315/37/1/012027

Appendix A Forest cover mapping methodology for FDPM

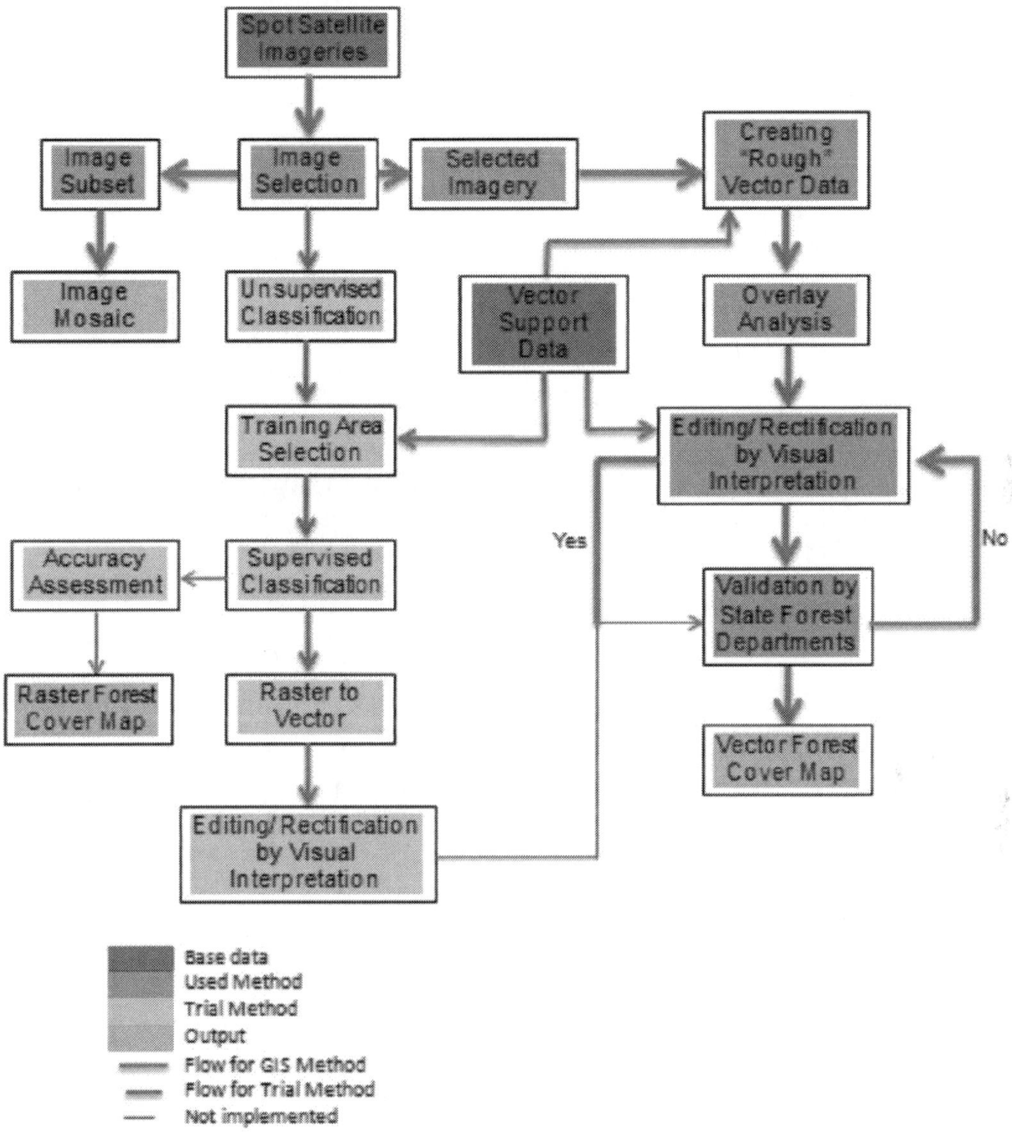

Appendix B Forest cover mapping methodology for WWF

Generation of MODIS satellite images – Based on published science-based materials, SarVision developed a fully automated system that received the MODIS daily surface reflectance products from Aqua and Terra satellites as data inputs. Calibration, radiometric, geometric, and topographic terrain corrections were performed on the automated satellite images. Computer analyses of automated cloud and haze removal and annual compositing techniques were used to select images with the best pixels, generating the MODIS annual composite images with the minimal amount of cloud cover as required. The targeted cloud cover was 0%. All daily images were processed into
7-band annual cloud-free composites at 250m resolution, and analyzed with the relevant statistics.

Classification of forest cover— Forest covers were classified based on two MODIS yearly cloud free images for each period of change, and where applicable, images of two consecutive years were used. A forest index was then calculated for each image using the threshold of the spectral information based on an empirical relationship established by SarVision using knowledge of the land-use practice in the region. The resulting maps gave an indication of the forest quality. A map of forest cover loss index was also generated to show the extent of deforestation or forest degradation. The final deforestation index was dependent on both the quality of the original vegetation in year 0, and the absolute difference with year 1. Small corrections were performed on the indices, where applicable, upon manual inspection.

Selection of images with appropriate pixel values for deforestation analysis – In analyzing the MODIS images for forest cover index, pixel values of 1 – 255 denoted the increasing level of detected forest cover loss, and the pixel value would become zero when no forest cover loss was detected. Images with pixel values of 1 – 40 (loss of forest cover below 15%) were indicative of slight or no significant loss of forest cover or they may be related to seasonal changes in vegetation. In comparison, images with pixel values of 41 to 255 (loss of forest cover exceeding 15%) were indicative of substantial deforestation, and these changes in forest cover are not likely to be reversible in the future. Hence, only MODIS images of pixel values exceeding 40 were used to assess the areas of deforestation, in which spatial analyses were performed using the Tabulate Area calculation for MODIS forest cover index from year 2001 to 2015.

Forest Cover Index (FCI) and FCI map validation – SarVision validated the FCI map as close to the reference date of 30th June 2013 as possible; the FCI map was produced against Very High Resolution (VHR) data available on GoogleEarth. The FCI map for Peninsular Malaysia showed an overall accuracy of 96% and all reliability figures were well above 90% accuracy, allowing us to discriminate dense, tropical forest from other land-cover or land-use types.

Quantifying rate of deforestation and CO_2 emission in Peninsular Malaysia using Palsar imageries

O Hamdan, K Abd Rahman and M Samsudin
Forest Research Institute Malaysia, 52109 FRIM, Kepong, Selangor, Malaysia

Email: hamdanomar@frim.gov.my

Abstract. Increasing human population and the rapid grows of Malaysia's economy are often associated with various environmental disturbances which have been contributing to depletion of natural resources and climate change. The need for more spaces for numerous land development activities has made the existing forests suffer deforestation. The study was carried out in Peninsular Malaysia, which currently has about 5.9 million ha of forests. Phased array type L-band SAR (Palsar) and Palsar-2 images over the years 2010 and 2015, respectively were used to identify forest cover and deforestation occurrences resulted from various conversion of forests to other land uses. Forests have been identified from horizontal-vertical (HV) polarization and then classified into three major categories, which are inland, peat swamp and mangrove. Pixel subtraction technique was used to determine areas that have been changing from forests to other land uses. Forest areas have been found declined from about 6.1 million ha in year 2010 to some 5.9 million ha in 2015 due to conversion of forests to other land uses. Causes of deforestation have been identified and the amount of carbon dioxide (CO_2) that has been emitted due to the deforestation activity has been determined in this study. Oil palm and rubber plantations expansion has been found the most prominent factor that caused deforestation in Peninsular Malaysia, especially in the states of Pahang, Terengganu, Johor and Kelantan. The rate of deforestation in the period was at 0.66 % yr^{-1}, which amounted a total of about 200,225 ha over the five years. Carbon loss was estimated at about 30.2 million Mg C, which has resulted in CO_2 emission accounted at about 110.6 million Mg CO_2. The rate of CO_2 emission that has been resulted from deforestation was estimated at 22.1 million Mg CO_2 yr^{-1}. The study found that the use of a series of Palsar and Palsar-2 images are the most appropriate sensors to be used for monitoring of deforestation over the Peninsular Malaysia region.

1. Introduction

In Malaysia, forest is defined as land spanning more than 0.5 ha with trees higher than 5 m and a canopy cover of more than 30 %, or trees able to reach these thresholds in situ. It does not include land that is predominantly under agricultural or urban land use [1]. Major forest types in Malaysia are lowland dipterocarp, hill dipterocarp, upper hill dipterocarp, montane, ericaceous, peat swamp and mangrove forests. In addition, there also smaller areas of freshwater swamp forest, heath forest, forest on limestone and forest on quartz ridges. Currently there are about 5.89 million ha of forest occur in Peninsular Malaysia, which covers about 44.7% of its land area. Out of this, 4.92 million ha falls under Permanent Reserved Forest (PRF), 0.58 million ha is Totally Protected Area (TPA) and the remaining 0.39 million ha belongs to state/alienated land [2].

Forests play an important role as a substantial coastal carbon sink. It is interesting to note that plant biomass in the ocean and coastal areas comprise of only 0.05% of the total plant biomass on land, get it cycles a comparable amount of carbon each year. Forests account for a large percentage of carbon fixing of about 80% of plant biomass carbon and 40% of the soil carbon [3]. Carbon sequestration results from the difference between photosynthetic carbon fixation and ecosystem respiration in the forest ecosystem [4] and overall it results in the determination of the net ecosystem carbon exchange [5].

The largest single source of greenhouse gases (GHG) emission in recorded history is from conversion of forest to other land use types [6]. This conversion has direct on-site effects resulting in degradation of vegetation, loss of biodiversity, destruction of property and occasional loss of life, while off-site impacts include carbon emissions and thus global warming and climate change. Deforestation in tropical region is considered as the second largest source of anthropogenic greenhouse gas emissions [7] and is expected to remain a major emission source for the foreseeable future. Deforestation is defined as the felling of trees without subsequent replanting or regeneration, i.e. the conversion of previously forested land to agricultural, urban or other land uses, which do not have a substantial tree canopy cover <10% [1]. Statistically, about 1 to 2 billion Mg of carbon per year which is equivalent to 15 - 25% of annual global greenhouse gases (GHG) emissions arises from tropical deforestation and forest degradation [8,9].

In Malaysia the conversion of existing forest to non-forest areas has increased the GHG emissions and contributed to the global climate change. Loss of forests have resulted from rapid clearing and conversion of land for agricultural purposes, and the forest had decreased by 1.2 million hectare by 1990 [1]. Half of the forests in the Peninsular Malaysia was cleared in the late 1980s [10], which decreased the total forest area to 57% of the original area by 2002 [11]. This has resulted in a serious land cover challenge in the Peninsular Malaysia [12]. The rapid expansion of oil palm over the two decades has led to the transformation of large areas of forest and plantation landscapes throughout Southeast Asia and is believed to be one of the major sources of greenhouse gas (GHG) emissions linked to land use in the region [13].

Malaysia is now committed to the program of reducing emissions from deforestation and degradation, and forest conservation (REDD+) under the United Nations Framework Convention on Climate Change (UNFCCC). The main aim of the program is to stabilize the GHG concentrations in the atmosphere at a level that would prevent dangerous anthropogenic interference with the climate system. However the amount and rate of deforestation, and to some extend emission resulted from the deforestation are yet to account. Identifying the drivers of deforestation is therefore essential as the first step to provide information for the decision-making processes and to establish land use policies in a way to combat land use change and be able to control any potential drivers that might contribute the carbon emission [7].

Given the current gap in knowledge and understanding of emission at national level, the research presented in this paper aims at providing methodology for estimating emission from deforestation with special reference to the use of synthetic aperture radar (SAR) satellite data. While national data on carbon emission have commonly not been available in the past, this research will generate new information for the country in-line with the REDD+ requirements.

2. Materials and methods

2.1. The study area

The study area covers entire region of Peninsular Malaysia, which encompasses an area of about 131,598 km^2. Peninsular Malaysia is located between latitudes 1°20' and 6°45' N and about longitudes 100' and 104'30' E. Practically the entire region is in the tropical moist forest life zone with annual rainfall in excess of 2000 mm fairly evenly distributed throughout the year. A mountain range runs along most of the peninsula with peaks >2000 m. More than 60% of the area is undulating lowland <300 m above sea level, about 35% is hilly with elevations of 300 to 1300 m, and the remainder is very mountainous Forests in this region comprises mainly inland dipterocarps, which consist of lowland, hill and upper hill, which are categorised based on land altitude, i.e. < 300, 300–750 and 750–1200 m respectively. Trees from the family Dipterocarpaceae are dominant in these forests. Other types of forests montane, peat swamp, and mangrove.

2.2. Satellite data

Satellite data that was used in data came from two satellites, which are Advanced Land Observing Satellite (Alos) and Alos-2. Alos-2 is the successor of the Alos, but the structure of the new satellite is quite different from its predecessor. Alos was launched in January 2006 and brought Phased Array type L-band Synthetic Aperture Rader (Palsar) on-board. After 5 years of observations it stopped transmitting in April 2011. Alos-2 was then launched on 24 May 2014 and bringing Palsar-2 sensor. Palsar-2 is currently operating and producing L-band SAR data, that has similar (with some advancements) characteristics with Palsar.

Both Palsar and Palsar-2 images datasets used in this study consisted of 1x1 degree mosaic tiles. A tile product consisted of two bands in horizontal-vertical (HV) and horizontal-horizontal (HH) polarisations at 25 m spatial spacing, geometrically and radiometrically corrected and normalised for topography. It also contains additional ancillary image data layers with information about acquisition dates, local incidence angle and a water- and no-data mask. The mosaic images were acquired from a number of individual scenes acquired in year 2010 from Palsar, and in year 2015 from Palsar-2. This 5-year interval data over Peninsular Malaysia is shown in Figure 1.

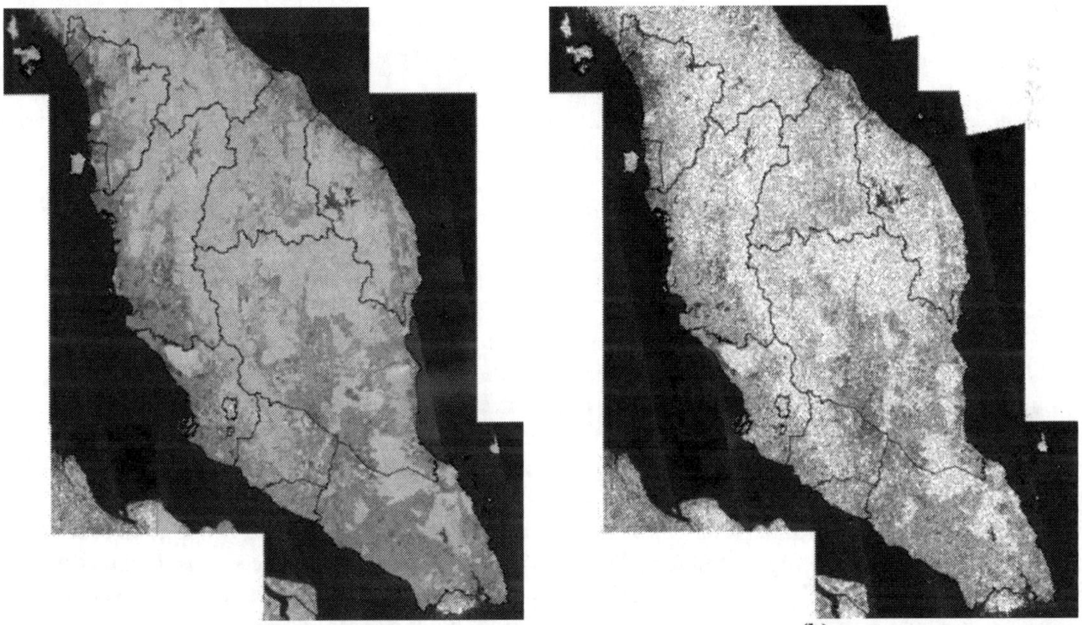

(a)   (b)

Figure 1. Mosaic images of Alos Palsar (a) and Alos-2 Palsar-2 (b) over the years 2010 and 2015, respectively.

2.3. Forest–non-forest classification

Forest–non-forest classification was performed on both mosaic images to delineate forests from other land uses. This process is critical to define the boundary of forests and to ensure that the estimated forest cover does not include other types of vegetation. The reason is that forests are often confused with rubber, teak and other timber tree plantations, which are common in Peninsular Malaysia and they appear almost identical on both HH and HV polarisations images. To minimise error associated with misclassification, image enhancement was applied to the images. Detailed explanation on the image enhancement technique can be found in [14]. The process was then finally produced a classified image that contained only forest areas for year 2010. This images were converted to vector format to be used for forest cover changes analysis.

2.4. Forest cover changes analysis
Changes of forest area is referred to as deforestation, which is human induced permanent conversion of forest land to non-forest, as all of the forest is cut and the land is cleared and used for other purposes. Temporary change in forests, like one rotation tree crop (up to 25 years) within forest reserves are not considered as deforestation and it falls under sustainable management of forests. Forest cover changes analysis has been conducted to determine the areas that have been converted from forest to other land uses.

Deforested areas were detected based on pixel differencing method. Mosaic images from both years were subtracted by using the following function;

$$\text{Change} = DN_{2010}\text{-}DN_{2015} \quad (1)$$

where *DN* denotes digital number of pixel from images over the years 2010 and year 2015, respectively.

This process produced another image that represent changes that have occurred between year 2010 and 2015. The vector layer of forest cover in 2010 was used as the base for changes analysis. Areas that have been changing from 2010 were detected and classified as deforestation. The vector layer was edited manually to produce another vector layer that represent forest areas of year 2015.

Land use categories that have been classified as deforestation were identified and were generalized into 6 major classes that consisted of oil palm plantation, rubber plantation, urban area, agriculture (other than oil palm and rubber plantations), water body and others (comprised mine and quarry, aquaculture, animal husbandry areas and cleared land) for the purpose of simplification. A ground truthing activity has been conducted at several locations to conform the current land use on the ground.

2.5. Carbon stock assessment
Carbon stock in the remaining forests have been assessed before carbon emission due to deforestation can be estimated. To complete this exercise, carbon stock in all forest types (i.e. inland, peat swamp and mangrove forests) in Peninsular Malaysia have been surveyed on the ground at a number of sample sites. The sampling design in this study was a modified sampling design according to the standard operating procedure (SOP) that has been developed by Winrock International [15], which follows the IPCC standards [16]. A cluster comprises four (4) subplots and the design is shown in Figure 2. The subplot was designed in circular with smaller nests inside. The biggest nest measures 20 m in radius, followed by the smaller nests measuring 12 and 4 m. The sizes of trees were measured according to the nest sizes, which is summarized in Table 1. Depending on the nest size, it indicates that not all stands were measured in a single subplot. In additional to these nests, there is another small nest measuring 2 m in radius, which is used to count saplings (i.e. trees measuring < 10 cm in diameter at breast height (dbh) and > 1.3 m in height). Clustering of subplots at each sampling unit is recommended for natural forest areas and areas that have been selectively logged. The sampling system is design in a way to make the data collection processes easier, faster but reliable and representative for a particular forest stratum.

A forest ecosystem normally has five terrestrial carbon pools, which are; (i) aboveground living biomass, (ii) belowground living biomass, (iii) deadwood, (iv) non-tree vegetation and litters, and (v) soil. However, one of the most significant carbon pools is aboveground biomass as it the easiest and the most practical pool to assess, while representative to an ecosystem. Aboveground biomass (AGB) comprises all the living aboveground vegetation, including stems, branches, twigs and leaves, deadwood, non-tree vegetation and litters. Allometric functions is the best way that above ground biomass can be estimated. In this study, regional allometric functions that are found in the literature (Table 2) were used to estimate the biomass of living trees in the inland, peat swamp and mangrove forests. A default factor of 0.47 was used to convert the biomass into carbon.

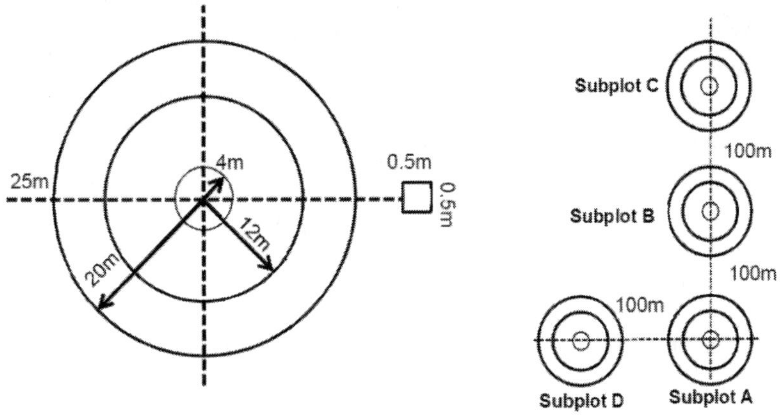

Figure 2. Layout of the sampling plot and sampling design of a cluster of sample

Table 1. Summary living trees measurement in a subplot.

Nest radius (m)	Size	Tree size, dbh (cm)
4	Small	≥ 10
12	Medium	≥ 20
20	Large	≥ 40

Table 2. Allometric functions that have been used for AGB estimations.

Forest type	Allometric function	Source
Inland forest	$AGB = [\exp(-1.803 - 0.976E + 0.976\, ln(\rho) + 2.673\, ln(D) - 0.0299\, [ln(D)]^2]$	[17]
Peat swamp forest	$AGB = 0.65*\exp(-1.239 + 1.98*ln(D) + 0.207*\, ln(D)^2 - 0.0281*ln(D)^2$	[18]
Mangrove forest	$AGB = 0.251\rho D^{2.46}$	[19]

AGB denotes aboveground biomass (kg/tree), *E* represents bioclimatic variable, ρ is wood specific gravity/ wood density, and *D* is dbh.

2.6. Estimation of carbon emission

There are two methods that can be used to estimate emission, which are the stock-difference method and the gain-loss method. Stock-difference requires two measurements in two different time periods. The estimated carbon stocks between the two periods divided by the time difference between the two time periods is the estimated carbon stock change per unit area. This method is better suited for deforestation. The gain-loss method, on the other hands, does not require two measurements; instead it simply measures the rate of carbon change in a given period, i.e. mean annual increment of carbon. Hence, this method is better suited for forest degradation where the stock change can be measured in two time periods through the expected change in carbon stock. Once the areal extents of deforestation is known, emission factor (*EF*) is then required. Emission factor is a measure of carbon loss that is caused by conversion of forest to other land uses. The carbon stock in forested areas was measured on the ground while carbon stock in the other landuse categories were adopted from literature to obtain the EF, as summarized in Table 3. Net emission from land use and land use changes can be estimated based on equations provided by [16], which can be expressed as

$$\Delta C = \frac{(Ct2 - Ct1)}{(t2 - t1)} \tag{2}$$

where ΔC is annual carbon stock change (Mg C yr^{-1}), C_{t1} and C_{t2} is carbon stock at time t_1 and t_2 (Mg C) respectively. However, whenever activity data (i.e. deforestation information) is available from changes analysis that has been conducted prior to the estimation of carbon emission, the following equation can be used,

$$Ct = \sum (AD \times EF) \tag{3}$$

where Ct is the carbon stock in time t, AD is the activity data or the area undergoing a specific type of land use change that emits carbon (measured in ha), and EF is emission factor, which is the total loss of the carbon stock per unit land area during the specific type of land use change (measured in Mg C ha^{-1}). Carbon emission can be expressed in terms of C loss or can be converted to CO_2 by multiplying with a factor of 3.67 which is the molecular weight of CO_2 per unit atomic weight of C.

Table 3. Emission factor for different types of land conversion.
(Summarized from [6])

Land use categories	Carbon stock (Mg C ha^{-1})	Conversion of forest to	Emission factor (Mg C ha^{-1})
Forest	186	Forest	0
Oil palm	36	Oil palm	150
Rubber	58	Rubber	128
Urban area	7	Urban area	179
Agriculture	54	Agriculture	132
Water body	0	Water body	186
Others	0	Others	186

3. Results and discussion

3.1. Forest distribution and extent

The study has classified the forests into three major types, which are inland, peat swamp and mangroves as indicated in Table 4. Inland dipterocarp forest is the major type of forest in Peninsular Malaysia, which accounted over 90% of the total forest cover. In addition to the identified major forest types, 'gelam' was also found within the study area. Gelam (*Melaleuca cajuputi*) is a monospecific trees which occurs in swamp forest behind beaches and mangroves. This forest – although in a small proportion – is found dominant, fringing behind the mangroves in the north eastern part of Terengganu, a state that resides in the East Coast of Peninsular Malaysia. However, due to difficulty in delineating this forest from inland forest, it was grouped into inland forest.

3.2. Deforestation in Peninsular Malaysia

The total land area in Peninsular Malaysia is estimated at about 131,598 km^2 with an overall forest area at 58.96 km^2 in the year 2015. Changes of forest cover in Peninsular Malaysia between 2010 and 2015, as resulted from deforestation are reported in Table 4. Factors of deforestation are generalized into six categories as summarized in Table 5. The total deforested area was found at 200,225 ha between years 2010 and 2015, which was accounted for deforestation at the rate of 40,045 ha yr^{-1}. Large piece of forest area in Peninsular Malaysia have been converted to oil palm and rubber at commercial scales, which occupied about 61% and 14%, respectively. Furthermore, the forest area has been replaced by urban area in terms of development, high density of human structures such as houses, commercial buildings, roads, and highways. Similarly, agricultural land was the share of the forests area that is converted to arable under permanent crops. The forest area were also converted to quarry and mine sites from which dimension stone, rock, construction aggregate, riprap, sand, gravel, or slate have been excavated from the ground. These activities resulted to open-pit mine from which minerals were extracted. Certain area of the forest were left idle that is they are not cut, burned, cropped, heavily grazed, cultivated, or otherwise disturbed. Other parts of the forest were converted to animal husbandry and farms for domestic supplies. Small patches of the clear land were also identified as scars of forest fires that occurred in the year 2014. A spatially distributed map indicating the deforestation that occurred between 2010 and 2015 was produced and depicted in Figure 3.

Table 4. Deforestation in Peninsular Malaysia between years 2010 and 2015.

Forest type	Forest cover 2010 (ha)	Forest cover 2015 (ha)	Deforestation (ha)	Rate of deforestation (ha yr^{-1})	Rate of deforestation (% yr^{-1})
Inland	5,690,816	5,525,034	165,782	33,156	0.58
Peat swamp	290,038	264,578	25,460	5,092	1.76
Mangrove	115,181	106,198	8,982	1,796	1.56
Total	6,096,035	5,895,810	200,225	40,045	0.66

Table 5. Categories of land uses that have been converted from forested areas.

No.	Land use category	Area (ha)	Percentage area (%)
1.	Water body	7,506	4
2.	Oil palm	122,137	61
3.	Rubber	28,031	14
4.	Agriculture	18,020	9
5.	Urban	12,013	6
6.	Others	12,516	6
Total		200,225	100

3.3. The assessed carbon stock

The assessment of carbon stock have been done in all major forest types found in Peninsular Malaysia. However, the calculation of EF used an average value of carbon stock from inland and peat swamp forests only. Mangrove forest was excluded from the estimation because its extent was found too small and if included in the estimation, will produce serious underestimate of carbon stock for all forests in entire study area. The average carbon stock of these forests was estimated at 186.46 Mg C ha^{-1}. Table 6 summarizes the assessed carbon stock for all forest types in different pools of aboveground carbon. Table 3 that was shown earlier used these values as a basis to calculate the *EF*.

Table 6. The estimated aboveground carbon stock at different forest types.

Forest type	Average carbon stock (Mg C ha^{-1})			
	Aboveground living	Deadwood	Non-tree vegetation and litters	Total carbon stock
Inland	174.49	4.92	1.29	180.70
Peat swamp	168.63	21.40	2.19	192.22
Mangrove	135.45	22.12	3.88	161.45

3.4. The estimated carbon emission

The loss of carbon was estimated by multiplying the extents of deforested areas (Table 5) with the emission factors for each corresponding land use category. Normally the rate of net emission is reported in yearly basis between two periods. The study found that the net CO_2 emission from deforestation in Peninsular Malaysia was at 110.6 million Mg CO_2 from year 2010 to 2015, with the rate of CO_2 emission at about 22.1 million Mg CO_2 yr^{-1}. Table 7 shows the estimated loss of carbon and CO_2 emission for the entire Peninsular Malaysia. It is obvious that the major contribution to the loss of carbon in the study area was from the expansion of oil palm and rubber plantations as compared to the other land uses, as shown in Figure 4.

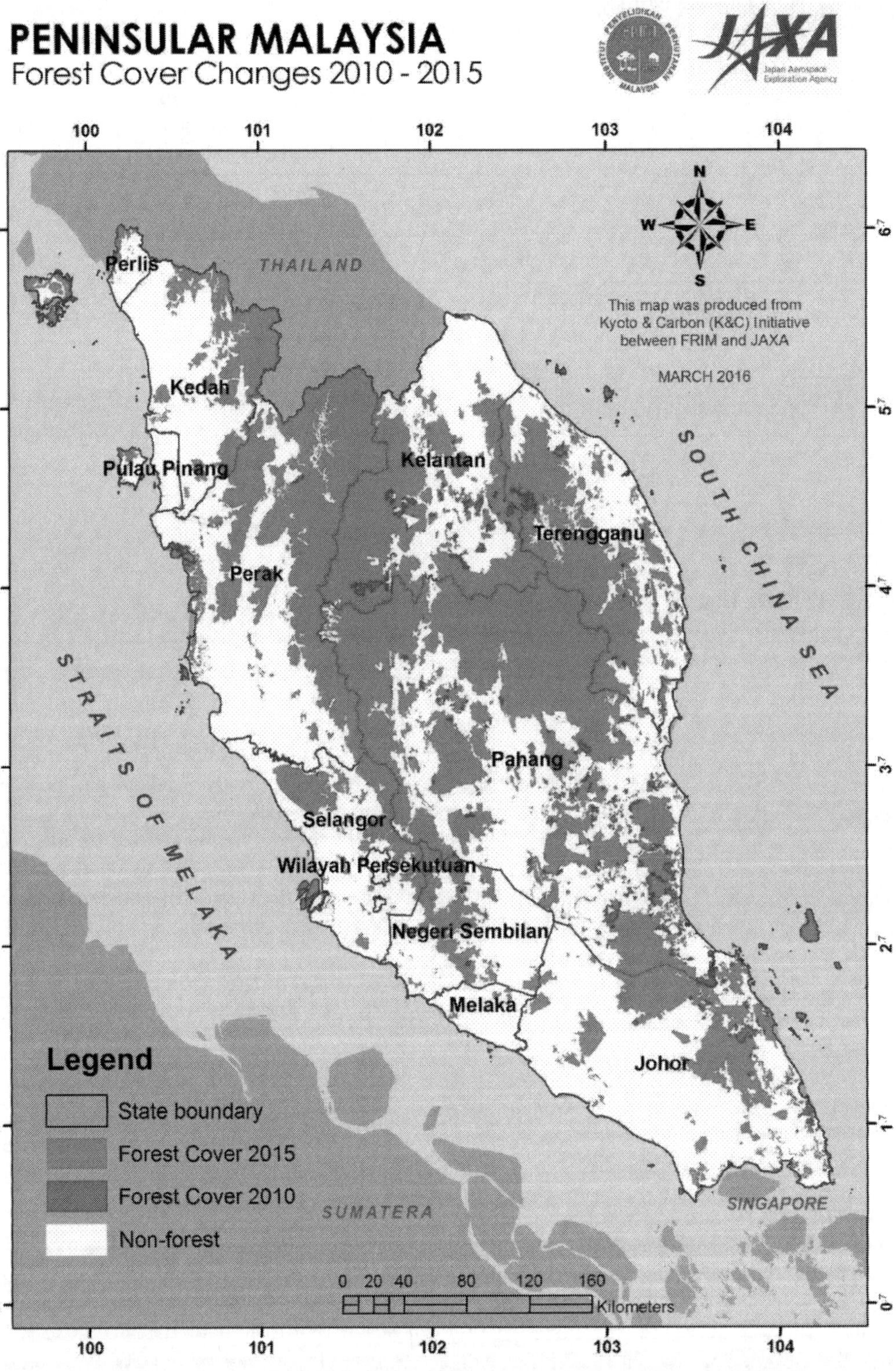

Figure 3. Map showing extents of changed forest cover in Peninsular Malaysia between years 2010 and 2015.

Table 7. Summary of carbon loss and emission of CO_2 resulted from deforestation between 2010 and 2015.

Land use categories	Area (ha)	Carbon loss (Mg C)	Emission (Mg CO_2)
Water body	7,506	1,396,116	5,119,092
Oil palm	122,137	18,320,555	67,175,367
Rubber	28,031	3,588,026	13,156,094
Agriculture	18,020	2,378,669	8,721,785
Urban	12,013	2,150,413	7,884,846
Others	12,516	2,328,062	8,536,228
Total	200,225	30,161,840	110,593,412

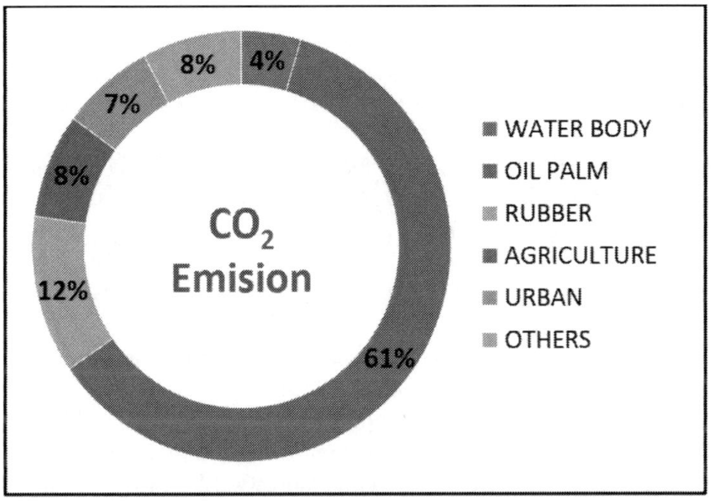

Figure 4. Composition of carbon emission due to conversion of forests to other land uses in Peninsular Malaysia between 2010 and 2015.

4. Conclusion

The study has provided methodology for estimating emission from deforestation with special reference to the use of Palsar data. The rate of deforestation was at 0.66 % yr^{-1}, which amounted a total of about 200,225 ha over the five years between 2010 and 2015. Carbon loss was estimated at 30,161,840 Mg C, which has resulted in CO_2 emission accounted at about 110,593,412 Mg CO_2. The rate of CO_2 emission that has been resulted from deforestation was estimated at 22,118,682 Mg CO_2 yr^{-1}.

Expansions of the palm oil and rubber plantations were found the common factor of deforestation in Peninsular Malaysia. The findings of this study will be an important indicator in assessing forest status and the trend of agricultural crops expansions in the country. It is also essential in providing activity data for carbon emission studies to make the REDD+ a reality. The information produced from this study is valuable for the policy makers to balance between development and environmental amenity. Although agricultural expansion has been determined as the key factor of deforestation in Peninsular Malaysia, the factors are actually vary regionally and change over time. The study also found that the use of a series of Palsar and Palsar-2 images are the most appropriate sensors to be used for monitoring of deforestation over the region.

The sampling design adapted for assessing carbon pools was appropriate to represent forests in Peninsular Malaysia. While national data on carbon assessment have commonly not been available in the past, this research has generated new information for the country in-line with the REDD+ requirements. However, some limitations have been identified from the study, which can be summarized as (i) generalization of net CO_2 emission from conversion, regardless the CO_2 intake during the growth of the agriculture crops, and (ii) the forest ecosystem was considered as one single stratum, regardless the variations of the status and management practices. Above all, to identify cost effective policies that will effectively reduce emission from agriculture expansions in Malaysia, REDD+ mechanism should depends on capacity to implement measures for effectively controlling deforestation and practicing good agriculture practices.

Acknowledgments
This work has been undertaken within the framework of the Japan Aerospace Exploration Agency (JAXA) Kyoto & Carbon (K&C) Initiative (Phase 4). Alos Palsar and Alos-2 Palsar-2 data were provided by JAXA EORC through RESTEC, Japan. Some information, especially primary data on forest biomass were adapted from the research funded by the International Tropical Timber Organization (ITTO) and the Government of Malaysia under the "Reducing emission from deforestation and forest degradation for enhancing ecosystem services" - (REDDES) thematic program. Thanks to the staff of Geoinformation Programme, FRIM who have involved in ground data collection and sampling.

References

[1] Food and Agriculture Organization of the United Nation – FAO. (2010). *Global Forest Resources Assessment 2010.* FAO Rome.
[2] Forestry Department Peninsular Malaysia – FDPM. (2015). *Annual Report 2014.* Kuala Lumpur.
[3] Dixon, R.K. (1994). Carbon pools and flux of global forest ecosystems. *Science,* **263**: 185–190.
[4] Gong, J., Ge, Z., An, R., Duan, Q., You, X. & Huang, Y. (2012). Soil respiration in poplar plantations in northern China at different forest ages. *Plant Soil,* **360**: 109–122.
[5] Valentini, R., Matteucci, G., Dolman, A.J., Schulze, E.D., Rebmann, C., Moors, E.J., Granier, A., Gross, P., Jensen, N.O, Pilegaard, K., Lindroth, A., Grelle, A., Bernhofer, C., Grünwald, T., Aubinet, M., Ceulemans, R., Kowalski, A.S. & Vesala, T. (2000). Respiration as the main determinant of European forests carbon balance. *Nature,* **404**: 861-865.
[6] Agus, F., Henson, E.I., Sahardjo, H.B., Harris, N., Noordwijk, V.M., & Killeen, J.T. (2014). *Review of emission factors for assessment of CO2 emission from land use change to oil palm in Southeast Asia.* Report from the technical panels of the 2nd Greenhouse Gas Working Group of the Roundtable on Sustainable Palm Oil (RSPO).
[7] IPCC (Intergovernmental Panel on Climate Change). (2007). *Climate Change 2007.* Cambridge University Press.
[8] Houghton, R., (2005). Tropical deforestation as a source of greenhouse gas emissions. Eds. Mutinho & Schwartzman, *Tropical deforestation and climate change.* Belém.
[9] Gibbs, H., Brown, K., Niles, S.J. & Foley, J.A. (2007). Monitoring and Estimating Tropical Forest Carbon Stocks: Making REDD a Reality. *Environ Res Lett,* **2**: No. 045023.
[10] Gillis, M. & Repetto, R. (1988). *Public Policies, the misuse of Forest Resource.* Cambridge University Press, Cambridge.
[11] Langner, A., Miettinen, J. & Siegert, F. (2007). Land cover change 2002–2005 in Borneo and the role of fire derived from MODIS imagery. *Glob Change Biol,* **13**: 2329–2340.
[12] Brookfield, H. & Byron, Y. (1990). Deforestation and Timber extraction in Borneo and the Malaysia Peninsular. *Glob Environ Chang,* **1**: 42-56.
[13] Wicke, B., Sikkema, R., Dornbury, V. & Faaij, A. (2011). Exploring land use changes and the role of palm oil production in Indonesia and Malaysia. *Land Use Policy,* **28**: 193–206.
[14] Hamdan, O., Mohd Hasmadi, I., Khali Aziz, H., Norizah, K. & Helmi Zulhaidi M.S. (2015). Determining L-Band Saturation Level for Aboveground Biomass Assessment of Dipterocarp Forests in Peninsular Malaysia. *J Trop For Sci,* **27**(3): 388 – 399.
[15] Walker, S.M., Pearson, T.R.H., Casarim, F.M., Harris, N., Petrova, S., Grais, A., Swails, E., Netzer, M., Goslee, K.M. & Brown, S. (2014). *Standard Operating Procedures for Terrestrial Carbon Measurement: Version 2014.* Winrock International.
[16] IPCC (Intergovernmental Panel on Climate Change). (2006) *Guidelines for National Greenhouse Gas Inventories,* Geneva, Switzerland.

[17] Chave, J. and co-authors. (2014). Improved allometric models to estimate the aboveground biomass of tropical trees. *Glob Change Biol*. gcb.12629: 14p.
[18] Chave, J. and co-authors. (2005). Tree allometry and improved estimation of carbon stocks and balance in tropical forests. *Oecologia*, **145**: 87–99.
[19] Komiyama, A., Ong, J.E. & Poungparn, S. (2007). Allometry, biomass, and productivity of mangrove forests: A review. *Aquat Bot,* **89**:128–137.

Deforestation alters rainfall: a myth or reality

M F Hanif, M R Mustafa*, A M Hashim and K W Yusof

Department of Civil and Environmental Engineering, Universiti Teknologi PETRONAS, 32610 Bandar Seri Iskandar, Perak Darul Ridzuan, Malaysia

*Email: raza.mustafa@petronas.com.my (Corresponding Author)

Abstract. To cope with the issue of food safety and human shelter, natural landscape has gone through a number of alterations. In the coming future, the expansion of urban land and agricultural farms will likely disrupt the natural environment. Researchers have claimed that land use change may become the most serious issue of the current century. Thus, it is necessary to understand the consequences of land use change on the climatic variables, e.g., rainfall. This study investigated the impact of deforestation on local rainfall. An integrated methodology was adopted to achieve the objectives. Above ground biomass was considered as the indicator of forest areas. Time series data of a Moderate Resolution Imaging Spectroradiometer (MODIS) sensor were obtained for the year of 2000, 2005, and 2010. Rainfall data were collected from the Department of Irrigation and Drainage, Malaysia. The MODIS time series data were classified and four major classes were developed based on the Normalised Difference Vegetation Index (NDVI) ranges. The results of the classification showed that water, and urban and agricultural lands have increased in their area by 2, 3, and 6%, respectively. On the other hand, the area of forest has decreased 10% collectively from 2000 to 2010. The results of NDVI and rainfall data were analysed by using a linear regression analysis. The results showed a significant relationship at a 90% confidence interval between rainfall and deforestation ($t = 1.92$, $p = 0.06$). The results of this study may provide information about the consequences of land use on the climate on the local scale.

1. Introduction

It has been documented that tropical deforestation plays a crucial role in the alteration of regional or global climate [1]. Moreover, the change in agricultural patterns and productions can alter the rainfall by intervening in the process of the heat energy flux and evapotranspiration of the earth surface. The conversion of forests to agricultural and urban lands disrupts the natural cycle of evapotranspiration. This consequently results in reduced rainfall due to the low humidity of the atmosphere [2].

In recent years, there has been an increasing amount of literature on the consequence of deforestation on the regional and global scales [3-6]. A key study examined the effect of deforestation in the Amazonian climate in which the researchers simulated the model for twenty years. The results of this study showed the 60% reduction in the rainfall [7]. To better understand the mechanism of deforestation and its impact on the global scale, a study found that the increase in land cover change, especially deforestation, by 8% leads to an increase in the global temperature by 0.02oC annually [8]. Another study over the Amazonian basin has highlighted the reduction in rainfall up to 1.8% during the deforestation in 2010 [9]. In 2013, a study found that the decrease in evapotranspiration due to the complete deforestation over the African continent can reduce the rainfall by 50 % [10]. Furthermore, a longitudinal study used the general circulation model (GCM) to investigate the impact of deforestation over the Asian region. Their results concluded that large-scale deforestation can lead to reduced precipitation [11].

Recently, investigators have examined the effects of deforestation on Southeast Asia, Africa, and Amazonian regions [12-15]. These studies have demonstrated that partial or complete deforestation of tropical forests can lead towards the increased global mean temperature. However, the variations in regional climate are influenced by many factors, such as vegetation, soil type, and land water relationship. Therefore, it is impossible to anticipate the changes in one region by extrapolating the changes from other regions [6].

The amount of published literature describing the role of describing the role of complete or partial deforestation in the Amazon forests is comparatively high from the African and Southeast Asian forests [1, 3, 9]. Moreover, it has been reported that the effect of land use change is more complex in the Asian tropical forests as compared to African and Amazonian forests [16]. The underlined reasons include the distribution of land and oceans, which is far more different from Amazonian forests. Moreover, the inclusion of monsoon circulations changes the situation in Asian tropical forests to a significant extent. In spite of these, only a few studies have investigated the effect of deforestation on Southeast Asian and African forests. A study has predicted a drier climate over the Southeast Asian deforested areas simulated by the GCMs [17]. Whereas, another study in Congo Basin, Africa has revealed that decreased evapotranspiration due to deforestation can reduce the rainfall up to 50 % over the entire basin [10].

Traditionally, climate change is believed to be a natural phenomenon but recent studies on global warming have heightened the need to consider the other factors as well. Over the past decade, researchers have agreed that the increase in temperature is due to anthropological changes, such as greenhouse gas (GHG) emissions and land use changes [18]. Whereas, in the context of Malaysia, the temperature and wind patterns of the Indian Ocean are amongst the most important factors to determine the rainfall distribution over Peninsular Malaysia. However, in this regard, a recent study by the Department of Irrigation and Drainage (DID) Malaysia, points out the two major factors for the increase in temperature and decline in rainfall in Southeast Asian forests, which are forest fires in Southeast Asian forests and rapid land use change [19].

The information regarding the changes in land use is of critical importance for the assessment of the relationship between natural and human-induced factors. The advancement of satellite remote sensing has made it easy to investigate the changes in land cover over the local or regional level. The use of Normalised Difference Vegetation Index (NDVI) values extracted from Moderate Resolution Imaging Spectroradiometer (MODIS) sensor, onboard NASA's Earth Observing System (EOS), Terra and Aqua satellites, for the assessment of forest cover, has been reported in several parts of world [20, 21]. The higher NDVI values indicate the presence of vegetation and vice versa. Therefore, the assessment of NDVI values extracted at different time span can provide the value of change [21].

Considering all of these evidences, it seems that much less is known about how deforestation really affects the precipitation at the local scale. Therefore, the objective of this study was to determine the association of precipitation and deforestation on a local scale. The methodological approach taken in this study is an integrated methodology based on the calculation of the rate of deforestation from 2001 to 2010 and quantifies the change with rainfall data using a simple linear regression model.

2. Methodology

2.1. Description of the study area

Perak is located in the western region of Peninsular Malaysia and covers an area of 21,035 km^2. Geographically it lies between 100°0'E to 102°0'E latitude and 3°30'N to 6°0'N longitude (Figure 1). The Perak river basin has a catchment area of about 14,908 km^2, which covers about 70% of the Perak state. The climate of the Perak river basin can be characterised by four seasons as of the whole Malaysia, from which two are monsoon and the other two are inter-monsoon seasons. The first monsoon season also called the southwest monsoon (SWM), comprises four months starting from May to August. Whereas, the northeast monsoon (NEM) occurs from November to February. Eastern parts of the Peninsula receive heavy rainfall during the northeast monsoon [22].

For the investigation of the rate of deforestation, the NDVI time series at 250 m × 250 m resolution at the time interval of 16 days were acquired for 2001, 2005, and 2010 from the MODIS sensor. Whereas, the selection of rain gauges was established based on (i) the land use changes in the study area as indicated in Figure 1 and Figure 3, and (ii) consistency of the data. Thus, daily rainfall data from 19 different rain gauges were acquired from the Department of Irrigation and Drainage for the period of 2000 to 2010.

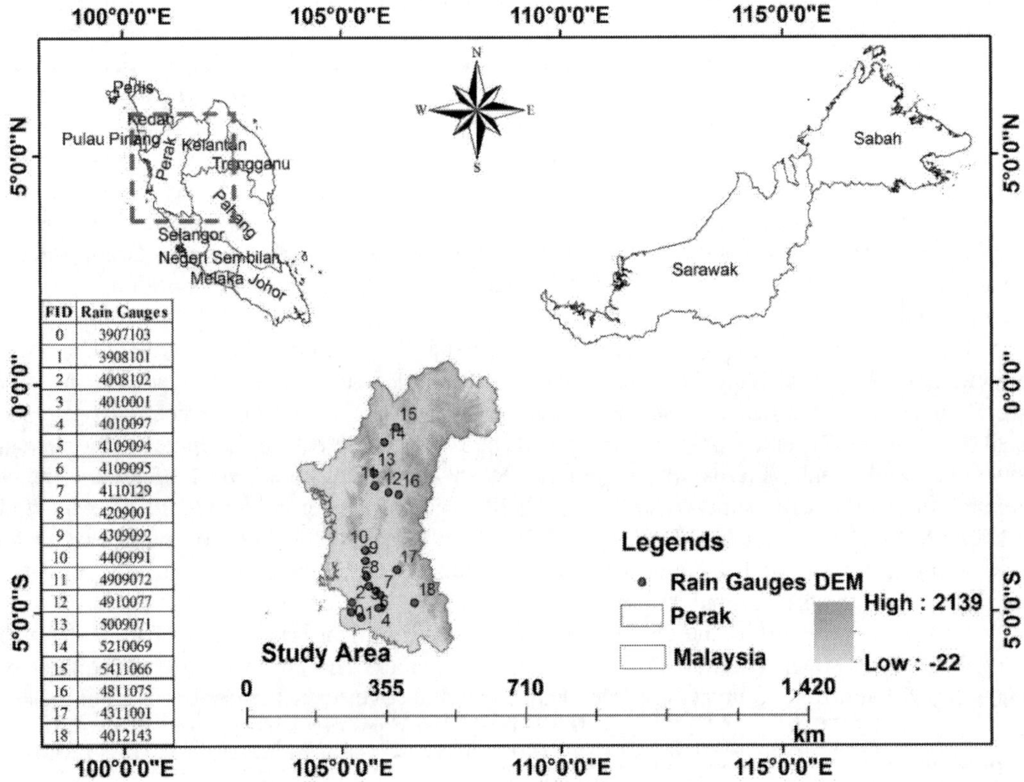

Figure 1: Description of the study area.

2.2. Data pre-processing

It was necessary to pre-process the MODIS data before land use classification to acquire better classification accuracy. The pre-processing included the extraction of the Normalised Difference Vegetation Index (NDVI) dataset from the composite datasets of MODIS product (MOD13Q1). After separating the NDVI dataset from the rest of the values, the next step was to re-project the dataset from the MODIS Integerised Sinusoidal Projection (ISN) to the World Geodetic System (WGS) 1984 North Zone 47 and then, sub-set the area of the Perak river basin.

2.3. Normalized Difference Vegetation Index (NDVI)

The use of the Normalised Difference Vegetation Index (NDVI) is a well-established approach for the assessment of above ground biomass. NDVI provides the information regarding the energy balance, i.e., energy received from the sun and emitted back by the Earth's objects. Concerning the plantation, it gives the information about the greenness or the health of the plants. Its values range from -1 to +1. In the practical scenario, it is considered that water bodies and barren land give the NDVI value less than 0.1. In contrast to this, the vegetation and forests depict the higher photosynthetic values and their NDVI ranges from 0.4 to 0.8, respectively [23]. NDVI can be explained as in equation (1):

$$NDVI=\frac{NIR-R}{NIR+R} \tag{1}$$

where the NIR represents the Near Infrared band of light spectrum and R represents the Red spectrum band.

2.4. *Classification of land cover*

The rate of deforestation was detected by subtracting the classified NDVI images over the entire Perak state. The time series information concluded the 12 images in each of the three years. All the images for each year were separately pre-processed and layer stacked. The time series analysis helped to analyse the trend of land use change in the study area. Classification of remotely sensed images can be categorised into two broad classes as pixel-based and object-based classification. Object-based image analysis (OBIE) has the ability to classify the images more accurately but it is only effective for high-resolution data as mentioned by [24]. This study used the moderate resolution data with a spatial resolution of 250 m × 250 m for the extraction of Normalized Difference Vegetation Index (NDVI) values. Therefore, the pixel-based hybrid classification method was used due to its appropriateness for coarse and moderate resolution data sets [25-27]. Another study in this context discovered that the accuracy of OBIE is much less as compare to Pixel based classification for MODIS images [28].

Due to above-mentioned limitation, the pixel-based hybrid classification was applied. In hybrid classification, first unsupervised classification was applied. Unsupervised classification works on the basis of spectral information from the different ground objects. It extracts the information of land covers depicting the same spectral signature and then combines them in one class [29]. For classification, a number of trials were performed to find the appropriate number of spectral classes for NDVI classification. It was observed that 20 classes were insufficient, as it was not providing the true classification of the NDVI images (i.e. forest and water zone were merged etc.). On the other hand, when 40 numbers of classes were used, then the overlapping behavior of classes was observed. Only 30 numbers of spectral classes were found optimum as the result showed the appropriate classification. The second step in hybrid classification was to combine the same spectral groups by analysing their spectral signatures. The four major land cover classes were developed based on the standards of Food and Agriculture Organisation (FAO) of United Nations [30]. The detail of combined classes is presented in Table 1. Moreover, the combined classes were then verified by using previously available ground truth information and past GPS surveys. In the end, we obtained the classified land cover maps for the years of 2000, 2005, and 2010.

Table 1: Level I and level II classification system

Level I	Class	Level II	Description
1	Water	11	Streams and Canals
		12	Lakes
		13	Reservoirs
2	Urban or Built-up Land	21	Residential
		22	Commercial and Services
		23	Industrial
		24	Transportation, Communications, and Utilities
		25	Industrial and Commercial Complexes
		26	Mixed Urban or Built-up Land
		27	Other Urban or Built-up Land
3	Agricultural Land	31	Cropland and Pasture
		32	Orchards, Nurseries, and Ornamental Horticultural Areas
		34	Confined Feeding Operations
		35	Other Agricultural Lands
4	Forest	41	Deciduous Forest Land
		42	Evergreen Forest Land
		43	Mixed Forest Land

2.5. Hypothesis development

Numerous studies have attempted to explain the effect of deforestation on the amount rainfall on the local scale as mentioned by [31], who stated that land use change plays an important role in altering the global temperature, which consequently alters the other climatic variables. Two hypotheses were drawn to analyse the effect of deforestation on local rainfall patterns.

H_o: Deforestation does not alter the rainfall on the local scale.

H_1: Deforestation does alter the rainfall on the local scale.

3. Results and discussion

3.1. NDVI classification

Figure 2 shows the results obtained from the NDVI classification. We also considered the NDVI value range given by [27] for the different objects in Malaysia. The classified images consisted of four classes, which were water, urban land, agricultural lands, and forests. The classified images were attributed in such a way that the value of the NDVI was more than 0.7 depicting the forest class, 0.4 to 0.69 presenting the agricultural lands and the NDVI values less than 0.4 were considered for the urban lands. Whereas, the negative values of the NDVI were considered as the water class. From the data in Figure 2, it is apparent that the water class shows negative or zero NDVI values for all years. The NDVI values of urban lands show the normal behaviour for the years of 2000 and 2005; but, for the year of 2010, some of the NDVI values reached up to 0.58. It is somewhat surprising, but it was due to the presence of some agricultural lands within the outskirts of the urban areas. Values for the agricultural and forest classes show the downward trend from 2000 to 2010. As with the deforestation, the value of above ground biomass also decreased which was explained by the NDVI values.

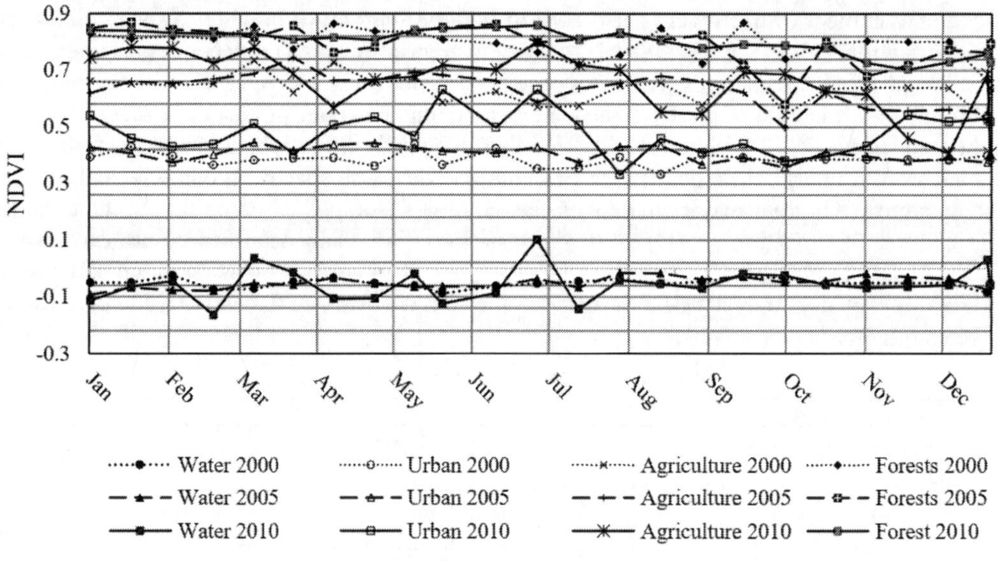

Figure 2: NDVI signature for all land cover classes.

3.2. Change in forest cover

The classified images for the years 2000, 2005, and 2010 are shown in Figure 3. Table 2 presents an overview of land use changes within the Perak river basin. In 2000, the area under the water class was 66,143 ha, which slightly increased in 2005 to 80,018 ha and reached up to 107,368 ha in 2010. These slight positive changes can be argued by the guidelines provided by the Department of Irrigation and Drainage in the manual of Urban Storm Water Management (MSMA) to build up the detention ponds in the urban areas for flood protection. The urban area was 112,110 ha in the year of 2000 and it slightly increased in 2005 and ended up at 161,152 ha in the year 2010. The behaviour of the agricultural class depicts rather different behaviour than the proceeding classes. It jumps up to 217,341 ha and then to 343,218 ha from 2000 to 2005 when the forests were cleared chiefly for the plantation of palm trees.

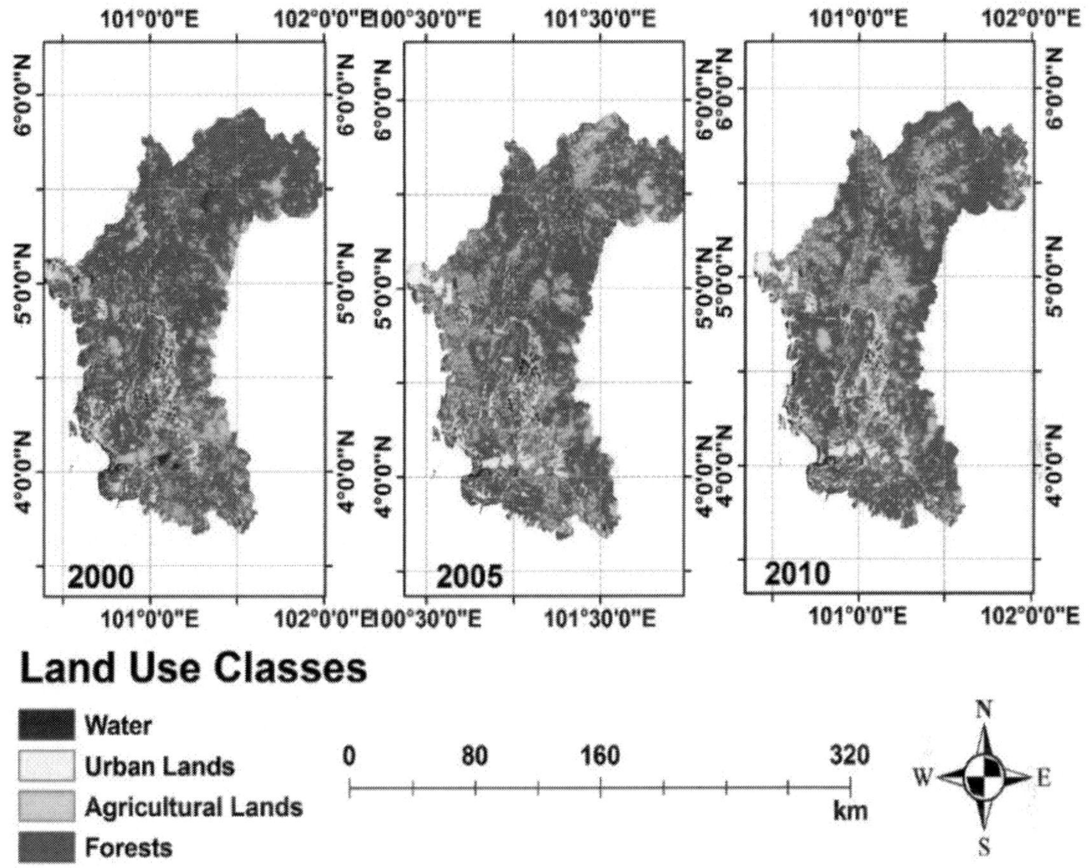

Figure 3: Land use classification for 2000, 2005, and 2010.

After 2005, there was no significant increase and it ended at 351,445 ha in the year of 2010. The forest class showed a significant decreasing trend. It was reduced from 1,663,179 ha to 1,507,383 ha from 2000 to 2010 (Table 1).

Table 2. Land use change (2000 - 2010).

Years	Classes Area (ha)				Percentage Change			
	Water	Urban	Agriculture	Forest	Water	Urban	Agriculture	Forests
2000	66143	112110	217341	1663179	3%	5%	11%	81%
2005	80018	139787	343218	1550850	4%	7%	16%	73%
2010	107368	161152	351445	1507383	5%	8%	17%	71%

3.3. Regression analysis between deforestation and rainfall

The regression analysis was used for the prediction of any significant relationship between the deforestation and rainfall. The details of the descriptive statics are given in Table 3. For the purpose of uniformity of both data sets, rainfall and deforestation information was used on a monthly basis. For three years, the total number of observation was 36. The mean value of the forest biomass was 0.8 and the mean value according to the logarithmic scale for the rainfall was 2.218. The standard deviation value for the forest biomass was 0.04 and for the rainfall, it was 0.18.

Table 3. Description of deforestation and rainfall data.

	N	Minimum	Maximum	Mean	Std. Deviation	Variance
Deforestation (Forest Biomass)	36	0.70	0.89	0.79	0.04	0.002
Rainfall	36	1.70	2.48	2.21	0.17	0.029

Table 4 reports the regression results for four different models which were the Southwest Monsoon model, Northeast Monsoon model, Inter-Monsoon model and a combined model. In the case of the Southwest Monsoon and Inter-Monsoonal seasons, the p-values 0.223 and 0.100 were found to be insignificant in the regression model, which implies that there was no relationship between the deforestation and the Southeast Monsoon and Inter-Monsoon rainfall. Opposite to that, the results of the Northeast Monsoon model reported that a unit increase in deforestation would increase the values of rainfall by 2.577 units. Therefore, the alternative hypothesis, in this case, is supported which stated that deforestation does alter the rainfall on the local scale. It is somewhat surprising, but these results are in accord with recent studies indicating that deforestation results in locally increased rainfall [32]. In contrast to the Northeast Monsoon model, the combined model for all seasons shows that a unit increase in deforestation will decrease the value of rainfall by 0.791 units. These results support the results from previous studies in this regard as the deforestation really affects the patterns of rainfall on the local scale [2].

Table 4. Regression analysis.

Season	Independent Variable	Dependent Variable	Linear Regression Model	Significance Value (P)
Southwest Monsoon	Deforestation	Rainfall	$R = 1.102 + 1.274_F + 0.46_{\Sigma t}$	0.223
Northeast Monsoon	Deforestation	Rainfall	$R = 4.229 + 2.557_F + 0.38_{\Sigma t}$	0.000
Inter-monsoon	Deforestation	Rainfall	$R = 1.196 + 1.358_F + 0.40_{\Sigma t}$	0.100
Accumulative	Deforestation	Rainfall	$R = 2.843 - 0.791_F + 0.34_{\Sigma t}$	0.000

Overall, these results indicate that deforestation has an impact on rainfall. The present study raises the possibility that a rapid decrease in forest areas may lead towards the drier climatic seasons. Similar results have been obtained from several other parts of the world [1, 33, 34]. Certainly, the study adopted the standard processing methods to quantify the impacts of deforestation on rainfall. However, most of the recent studies only rely on the complex climatic modelling [10, 11, 35], whereas, this study opted an observational approach in this regard as suggested by [1]. This study is a way forward to understand the impact of anthropogenic changes on regional climate. Moreover, this work will also serve as a base for future studies in term of town planning and land management on the regional or local scale. Furthermore, the results of this study are in line with the previous studies on the Southeast Asian forests which explain the significance of the study [17]. Despite these promising results, several questions remain unanswered at present. There is abundant room for further progress in determining the long-term effect of land use change on the climatic issues to understand the underlying phenomenon.

4. Conclusion

In this investigation, the aim was to assess the impact of deforestation on the rainfall distribution. The analysis of MODIS images reveals that the area under urban land is growing bigger and the area under agricultural and forest land is decreasing. The research has shown that the above ground biomass can be utilized for the indicator of deforestation. The lower value of biomass indicates the lower forest reserves and vice versa. This study has also identified the positive relationship of deforestation with the rainfall for the Northeast Monsoon season. This combination of findings provides some support for the conceptual premise that there exists a relation between deforestation and rainfall as has been shown by the regression analysis. This research will serve as a base for future studies and further research in order to understand the underlying phenomenon of complex climatic behaviour with land use change. More information on the relation of deforestation and rainfall would help us to establish a greater degree of accuracy on this matter.

Acknowledgement
The authors are also thankful to U.S. Geological Survey Department for providing us freely the temporal land use datasets for Perak, Malaysia. Moreover, we are also obliged to the Department of Irrigation and Drainage (DID) Malaysia for providing the rainfall data for Perak river basin. The first author is also thankful to Universiti Teknologi PETRONAS, Malaysia for providing financial support under graduate assistantship scheme.

References
[1] Lawrence D and Vandecar K 2015 Effects of tropical deforestation on climate and agriculture *Nature Clim. Change* **5** 27-36
[2] Spracklen D V, Arnold S R and Taylor C M 2012 Observations of increased tropical rainfall preceded by air passage over forests *Nature* **489** 282-5
[3] Bagley J E, Desai A R, Harding K J, Snyder P K and Foley J A 2014 Drought and deforestation: has land cover change influenced recent precipitation extremes in the Amazon? *Journal of Climate* **27** 345-61
[4] Davin E L and de Noblet-Ducoudré N 2010 Climatic impact of global-scale deforestation: Radiative versus nonradiative processes *Journal of Climate* **23** 97-112
[5] Werth D and Avissar R 2002 The local and global effects of Amazon deforestation *Journal of Geophysical Research: Atmospheres* **107** 8087
[6] Mahmood R, Pielke R A, Hubbard K G, Niyogi D, Dirmeyer P A, McAlpine C, Carleton A M, Hale R, Gameda S and Beltrán-Przekurat A 2014 Land cover changes and their biogeophysical effects on climate *International Journal of Climatology* **34** 929-53
[7] Nobre P, Malagutti M, Urbano D F, de Almeida R A and Giarolla E 2009 Amazon deforestation and climate change in a coupled model simulation *Journal of Climate* **22** 5686-97
[8] Lawrence P J and Chase T N 2010 Investigating the climate impacts of global land cover change in the community climate system model *International Journal of Climatology* **30** 2066-87
[9] Spracklen D and Garcia-Carreras L 2015 The impact of Amazonian deforestation on Amazon basin rainfall *Geophysical Research Letters* **42** 9546-52
[10] Nogherotto R, Coppola E, Giorgi F and Mariotti L 2013 Impact of Congo Basin deforestation on the African monsoon *Atmospheric Science Letters* **14** 45-51
[11] Dallmeyer A and Claussen M 2011 The influence of land cover change in the Asian monsoon region on present-day and mid-Holocene climate *Biogeosciences* **8** 1499-519
[12] Werth D and Avissar R 2002 The local and global effects of Amazon deforestation *Journal of Geophysical Research: Atmospheres (1984–2012)* **107** LBA 55-1-LBA -8
[13] Stibig H-J, Achard F, Carboni S, Rasi R and Miettinen J 2014 Change in tropical forest cover of Southeast Asia from 1990 to 2010 *Biogeosciences* **11** 247
[14] Fox J, Vogler J B, Sen O L, Giambelluca T W and Ziegler A D 2012 Simulating land-cover change in montane mainland southeast Asia *Environmental management* **49** 968-79
[15] Lorenz R and Pitman A J 2014 Effect of land-atmosphere coupling strength on impacts from Amazonian deforestation *Geophysical Research Letters* **41** 5987-95

[16] Mabuchi K, Sato Y and Kida H 2005 Climatic impact of vegetation change in the Asian tropical region. Part I: Case of the Northern Hemisphere summer *Journal of Climate* **18** 410-28

[17] Schneck R and Mosbrugger V 2011 Simulated climate effects of Southeast Asian deforestation: Regional processes and teleconnection mechanisms *Journal of Geophysical Research: Atmospheres (1984–2012)* **116**

[18] Loo Y Y, Billa L and Singh A 2014 Effect of climate change on seasonal monsoon in Asia and its impact on the variability of monsoon rainfall in Southeast Asia *Geoscience Frontiers*

[19] Section N W P D January 2009 Climate Change Scenarios for Malaysia 2001 - 2099.

[20] Eckert S, Hüsler F, Liniger H and Hodel E 2015 Trend analysis of MODIS NDVI time series for detecting land degradation and regeneration in Mongolia *Journal of Arid Environments* **113** 16-28

[21] Alphan H and Derse M A 2013 Change detection in Southern Turkey using normalized difference vegetation index (NDVI) *Journal of Environmental Engineering and Landscape Management* **21** 12-8

[22] Suhaila J and Jemain A A 2012 Spatial analysis of daily rainfall intensity and concentration index in Peninsular Malaysia *Theoretical and Applied Climatology* **108** 235-45

[23] Tovar C M 2012 El índice normalizado diferencial de la vegetación como indicador de la degradación del bosque *Unasylva: revista internacional de silvicultura e industrias forestales* 39-46

[24] Liu D and Xia F 2010 Assessing object-based classification: advantages and limitations *Remote Sensing Letters* **1** 187-94

[25] Jianya G, Haigang S, Guorui M and Qiming Z 2008 A review of multi-temporal remote sensing data change detection algorithms *The International Archives of the Photogrammetry, Remote Sensing and Spatial Information Sciences* **37** 757-62

[26] Coppin P, Jonckheere I, Nackaerts K, Muys B and Lambin E 2004 Review ArticleDigital change detection methods in ecosystem monitoring: a review *International Journal of Remote Sensing* **25** 1565-96

[27] Ghosh A, Mishra N S and Ghosh S 2011 Fuzzy clustering algorithms for unsupervised change detection in remote sensing images *Information Sciences* **181** 699-715

[28] Gao Y and Mas J F 2008 A comparison of the performance of pixel-based and object-based classifications over images with various spatial resolutions *Online Journal of Earth Sciences* **2** 27-35

[29] Tran P, Marincioni F and Shaw R 2010 Catastrophic flood and forest cover change in the Huong river basin, central Viet Nam: A gap between common perceptions and facts *Journal of Environmental Management* **91** 2186-200

[30] Carrão H, Gonçalves P and Caetano M 2008 Contribution of multispectral and multitemporal information from MODIS images to land cover classification *Remote Sensing of Environment* **112** 986-97

[31] Mahmood R, Quintanar A I, Conner G, Leeper R, Dobler S, Pielke R A, Beltran-Przekurat A, Hubbard K G, Niyogi D, Bonan G, Lawrence P, Chase T, McNider R, Wu Y, McAlpine C, Deo R, Etter A, Gameda S, Qian B, Carleton A, Adegoke J O, Vezhapparambu S, Asefi S, Nair U S, Sertel E, Legates D R, Hale R, Frauenfeld O W, Watts A, Shepherd M, Mitra C, Anantharaj V G, Fall S, Chang H-I, Lund R, Treviño A, Blanken P, Du J and Syktus J 2010 Impacts of Land Use/Land Cover Change on Climate and Future Research Priorities *Bulletin of the American Meteorological Society* **91** 37-46

[32] Butt N, de Oliveira P A and Costa M H 2011 Evidence that deforestation affects the onset of the rainy season in Rondonia, Brazil *Journal of Geophysical Research: Atmospheres (1984–2012)* **116**

[33] Comarazamy D E, González J E, Luvall J C, Rickman D L and Bornstein R D 2013 Climate impacts of land-cover and land-use changes in tropical islands under conditions of global climate change *Journal of Climate* **26** 1535-50

[34] Sen O L, Bozkurt D, Vogler J B, Fox J, Giambelluca T W and Ziegler A D 2013 Hydro-climatic effects of future land-cover/land-use change in montane mainland southeast Asia *Climatic change* **118** 213-26

[35] Garcia-Carreras L and Parker D 2011 How does local tropical deforestation affect rainfall? *Geophysical Research Letters* **38**

Aboveground biomass and carbon stocks modelling using non-linear regression model

Nurul Ain Mohd Zaki[1], Zulkiflee Abd Latif[1], Mohd Nazip Suratman[2] and Mohd Zainee Zainal[3]

[1]Applied Remote Sensing & Geospatial Research Group, Centre for Studies Surveying Science & Geomatics, Faculty of Architecture, Planning & Surveying, Universiti Teknologi MARA, 40450 Shah Alam, Selangor, Malaysia

Email: nurulainzaki@gmail.com; zabdlatif@gmail.com

[2]Faculty of Applied Sciences, Universiti Teknologi MARA, 40450 Shah Alam, Selangor, Malaysia

[3]Centre for Studies Surveying Science & Geomatics, Faculty of Architecture, Planning & Surveying, Universiti Teknologi MARA, Arau, 02600, Perlis, Malaysia

Abstract. Aboveground biomass (AGB) is an important source of uncertainty in the carbon estimation for the tropical forest due to the variation biodiversity of species and the complex structure of tropical rain forest. Nevertheless, the tropical rainforest holds the most extensive forest in the world with the vast diversity of tree with layered canopies. With the usage of optical sensor integrate with empirical models is a common way to assess the AGB. Using the regression, the linkage between remote sensing and a biophysical parameter of the forest may be made. Therefore, this paper exemplifies the accuracy of non-linear regression equation of quadratic function to estimate the AGB and carbon stocks for the tropical lowland *Dipterocarp* forest of Ayer Hitam forest reserve, Selangor. The main aim of this investigation is to obtain the relationship between biophysical parameter field plots with the remotely-sensed data using non-linear regression model. The result showed that there is a good relationship between crown projection area (CPA) and carbon stocks (CS) with Pearson Correlation ($p < 0.01$), the coefficient of correlation (r) is 0.671. The study concluded that the integration of Worldview-3 imagery with the canopy height model (CHM) raster based LiDAR were useful in order to quantify the AGB and carbon stocks for a larger sample area of the lowland *Dipterocarp* forest.

1. Introduction

Tropical forest plays a significant role in the climate change as a main carbon storehouse for the whole world. In the carbon sequestering photosynthesis process, tree and soils continuing to provide the oxygen for human being respiration cycle. Recent studies by [1], shows that there has been rapidly increasing in forest loss in most other parts of the tropics over the last decade, with South-East Asia displaying one of the highest rates of deforestation. It is confirmed in a recent forest cover research that shows the main cause of forest loss in South-East Asia is the conversion of the forest to cash crop plantation [2]. In the past few centuries, anthropogenic activities and natural consequences have increased the concentration of carbon dioxide including greenhouse gases (GHG) to the atmosphere [3]. The anthropogenic activities

such as degraded human-set fires, destructive timber logged, forest clear for agricultural and development purposes would affect the carbon cycle may lead to global warming affect [1].
There are several methods that can be used to measure aboveground biomass. Another method that usually used to estimate the aboveground biomass is using the method conversion of volume data to develop allometric equation. Tree stand allometric equations are developed by calculating the relationship between field measurement of the tree parameter such as the diameter of the trunk, tree height, and diameter at breast height (DBH), tree species, age, crown density, and also bioclimatic variable. Undeniable, field-based forest inventory remains vital in the carbon stocks estimation of forest structure [2] , [3]. Nevertheless, this can be improved by the use of remotely sensed data where it can reduce the massive field-based sampling if it comes to a huge area of forested land. Therefore, the objective of this research includes (i) investigating the relationship between carbon stocks and tree parameter (ii) developing a model of carbon stocks estimation using non-linear regression model develop from crown projection area (CPA) of the emergent and canopy layer and finally (iii) predicting the carbon estimation of the study area.

2. Material and methods

2.1. Study area

The research was conducted in a forest managed by University Putra Malaysia, Ayer Hitam Forest Reserved, Selangor State, Malaysia. The forest lies between Latitude 3°00'24.19"N, Longitude 101°38'25.24"E, the location of the study area is illustrated in Figure 1. The type of this forest is lowland *Dipterocarp* forest. It comprises of various species dominated by the family from *Dipterocarpaceae*, followed by family *Euphorbiaceae* as a second major family species. The average rainfall occurring on the average of 2178 mm annually while the humidity reaches 74% and the average temperature is 27.7 ° C minimum and 22.9 ° C maximum [5].

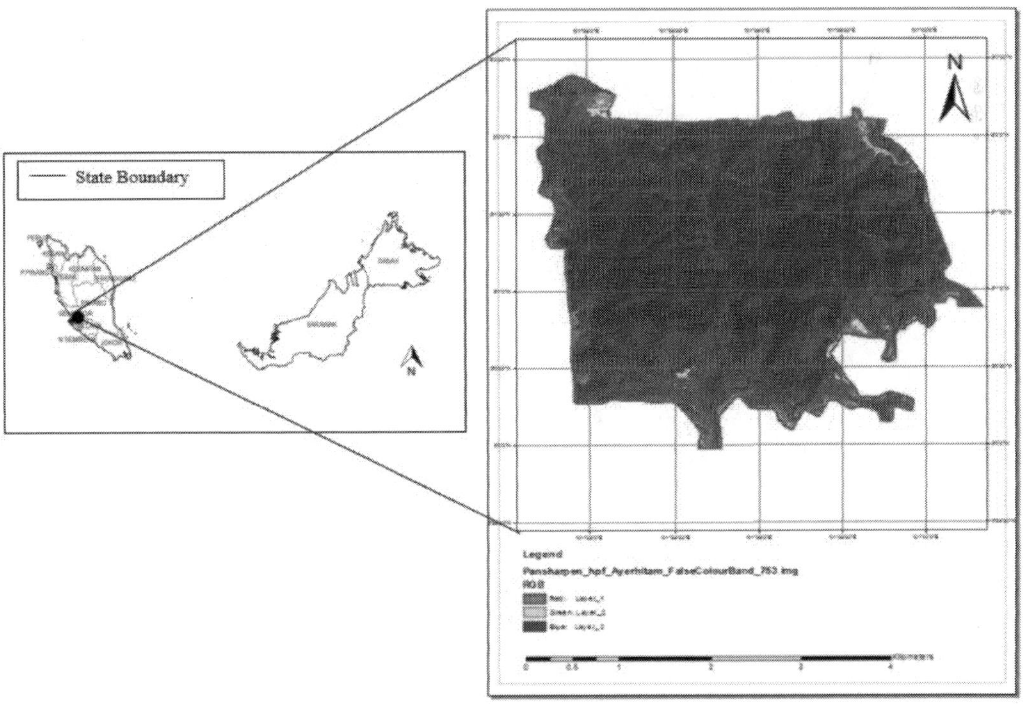

Figure 1. Location of Ayer Hitam Forest Reserve is shown in false colour band combination of 542 of WordView3 (WV-3).

IOP Conf. Series: Earth and Environmental Science **37** (2016) 012030 doi:10.1088/1755-1315/37/1/012030

2.2. Satellite image, LiDAR and supporting data

The light detection and ranging (LiDAR) data had been used in this dissertation were acquired in August 2013. The sensor used in this mission is by using LITEMAPPER-5600 that consist of RIEGL LMS-Q560 Laser Scanner. This equipment allows an airborne scanning laser to penetrates the distance between the aircraft and the ground and produce the precise digital surface models. The LiDAR system was mounted on board of aircraft, fixed with the real-time kinematic (RTK) survey that will provide the differential position of the aircraft and equip with the Inertial Measurement Unit (IMU). The altitude of the flying height was 1000 m above the ground level using the flying speed 90 knots. With the capacity of laser swath width of approximately, 1155m was a measure for every flight line, the pulse repetition frequency that had bet set during the LiDAR observation is 150 kHz (150, 000 pulses per second) in order to generate a number of return. The canopy height model (CHM) was created using LAStools plug-in software ArcGIS. Figure 2 show the CHM of the tree height in a different view.

Super-spectral high resolution of WV-3 (8 bands) 25 km^2 with the projection of UTM47 N and WGS84 datum, have been used in this study. The data had been obtained on 9th December 2014 that consists of a 31cm panchromatic band, 4 bands Very Near Infrared (VNIR) colours (blue, green, red and near infrared NIR-1) and 4 added VNIR colours which are coastal, yellow, red edge and NIR2. The spatial resolution was 0.30 m (panchromatic) and 1.2 m (multispectral). The classification was performed on a pan-sharpen WV-3 image using hyperspherical colour sharpening (HCS) techniques for better visualisation.

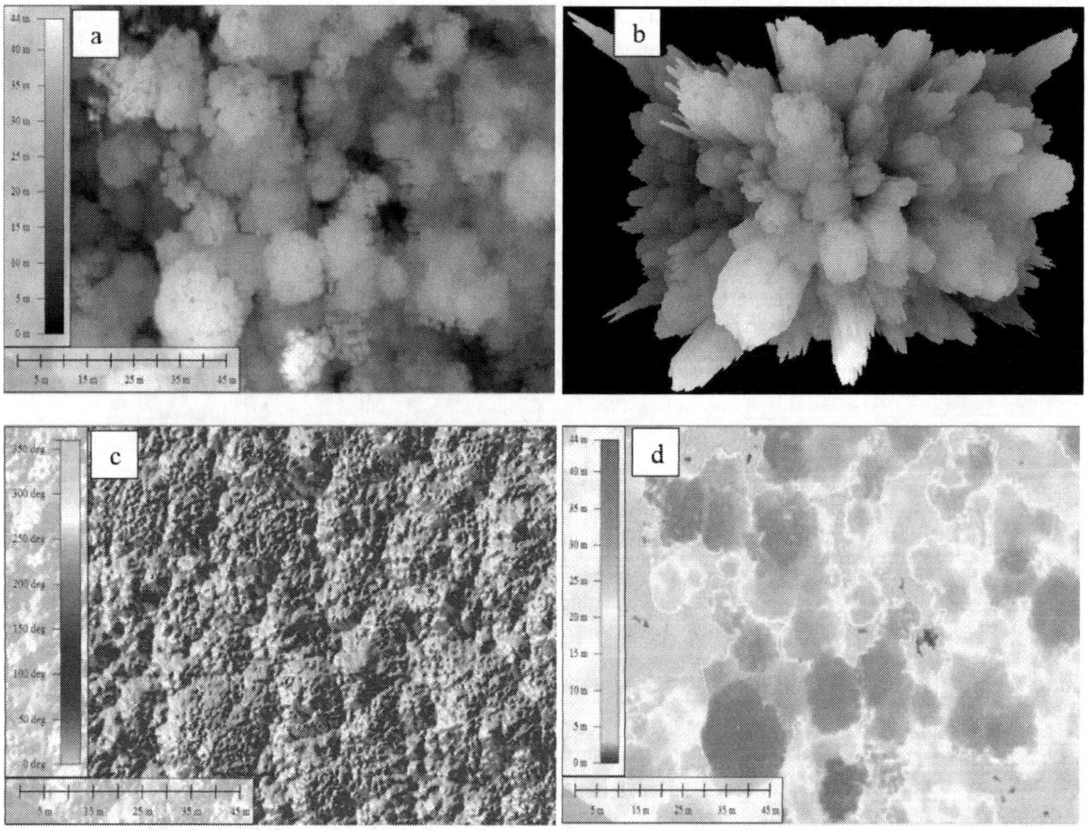

Figure 2. CHM derived from the digital surface model and the digital terrain model. **a)**. The top view of the CHM of the study area. **b)**. The 3D view from the top of the CHM. **c)**. The slope direction shade using CHM and finally **d)**. The colour ramp shade of the CHM.

2.3. *Field sampling plot*
Field sampling plot were conducted on May 2015 at 2-hectare rectangular plots that consist of 50 number of subplots (20 x 20 m) using stratification number of sampling. In each plot, every tree that has more and equal to 10 cm DBH were inventoried and recorded. The parameter that had been collected includes tree id, tree location, species, stem diameter at breast height, crown base height, crown diameter, tree height and leaf area index of the subplot area. The field plot and tree coordinates were tied by the traversing using total station from a known Global Positioning System (GPS) reference stations. The reference stations were observed using GPS using Topcon GTS with static L1/L2 and L5 bands.

$$AGB_{est} = \exp\left[\left[-1.803 - 0.976\,E + 0.976\ln\rho + 2.673\ln(D) - 0.0299\left[\ln(D)^2\right]\right]\right] \quad (1)$$

Where AGB = Aboveground biomass (kg / tree), ρ = wood density, E = Bioclimatic variables and D = DBH. The AGB and carbon stocks were calculated using allometric equation develop by Equation (1), [7] which incorporated the usage of wood density, height, DBH and bioclimatic variables as predictors. Despite height and DBH, wood density is one of the important predictors in the AGB calculation [8]. In addition, the AGB (kg) were converted to Mg ha^{-1} and converted to carbon stocks by applying the conversion factor 0.47 which represents 47% of the dry biomass assume to be carbon for all part of the tree that had been recommending by Intergovernmental Panel on Climate Change (IPCC) [9].

2.4. *Non-linear regression modelling*
The previous researcher had been used regression modelling [4], [5], and machine learning approaches [6], [7] in order to quantify the complexity of the AGB and carbon stocks for the tropical rain forest. The quadratic model of non-linear regression model were used to develop the model to estimate AGB using this mathematical function (2) and (3):

$$y = \alpha x^2 + \beta x + c \quad (2)$$

$$Carbon\ Stocks = \alpha x^2 + \beta x + c \quad (3)$$

2.5. *Validation of the statistical analysis*
The statistical analysis was performed by calculating the statistical parameter such as residual R^2, Root Mean Square Error (RMSE) of the absolute accuracy had been done. (Equation 4). The data was analysed in the scatter plots and box plots to identify the outliers and had been removed in order to get the robust model for regression. Out of 911 of tress, 183 were used to develop the model, while other 30% of the field sampling were used as a validation.

$$\text{RMSE} = \sqrt{\sum_{i=1}^{n} \frac{(yi - \widehat{yi})^2}{n}} \quad (4)$$

Where n is the number of samples, yi is the calculated value of carbon, $\widehat{yi}$ is the predicted carbon by the model of the response variables.

3. Results and Discussion

3.1. *Descriptive statistics*
The selection of the independent variables of the carbon stocks for tropical lowland was made through broad review of the literature on the subject matter[8]. Overall, the dominant tree species occurred in the area was *Dipterocarpaceae* spp, *Euphorbiaceaea* spp, *Burseraceae* spp, and *Sapotaceae* spp. The

most dominant species was *Endospermum diadenum* with 24.69% occurrence, *Hopea sulcata*, *Dipterocarpus verrucosus* and *Shorea macroptera* (Figure 3).

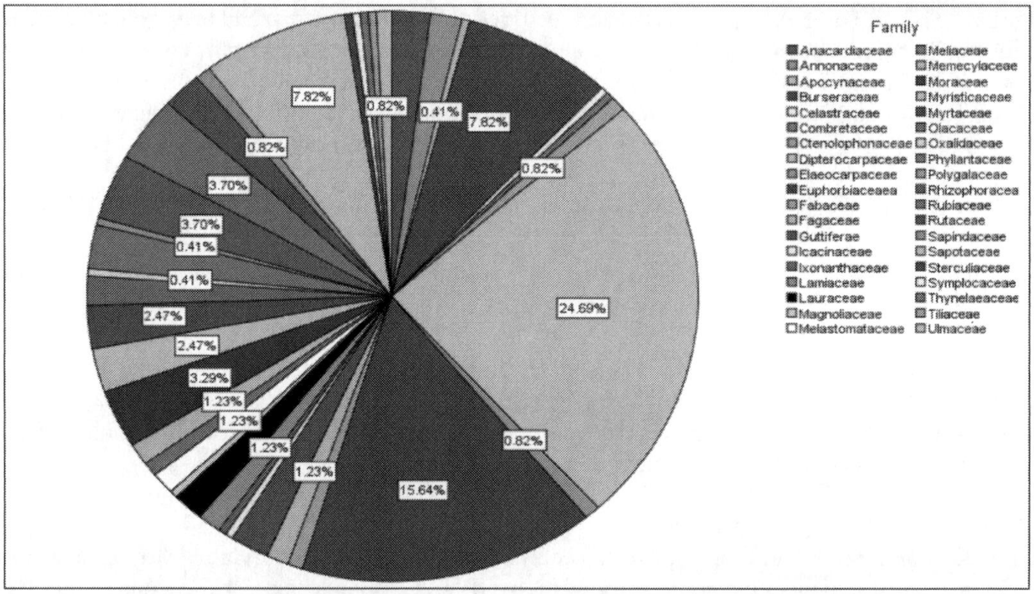

Figure 3. The occurrence of the different species of the study area.

Table 1. Summary statistics of variables collected from field sampling, LiDAR and satellite segmentation output and used for model development (n = 183) and validation dataset (n = 62), respectively.

Variable and unit	Model-building data set					Validation data set				
	No. of tree	Min[a]	Max[b]	Mean	SD[c]	No. of tree	Min[a]	Max[b]	Mean	SD[c]
Height from LiDAR	183	10.850	33.299	20.574	5.108	62	12.331	37.822	21.054	5.297
Crown Projection Area	183	7.514	117.26	31.094	19.517	62	10.034	214.28	29.456	28.789
Height from field	183	40.421	72.060	55.506	8.640	62	12.00	37	20.497	5.354
DBH	183	10	72.50	28.301	13.566	62	10.200	113.00	29.071	16.258
Aboveground Biomass	183	32	5325	697.54	901.42	62	46.00	17167	948.677	2276.47
Carbon Stocks (kg)	183	15	2503	327.81	423.71	62	22.00	8068.0	21.054	5.296
Carbon Stocks(Mg ha^{-1})	183	0.0075	1.2514	0.1639	0.2118	62	0.0108	4.03	0.2229	0.5349

(Min[a] = Minimum, Max[b] = Maximum and SD[c] = Standard Deviation)

3.2. Image segmentation for crown projection area (CPA) delineation

The image multispectral WV-3 and CHM derived from LiDAR were segmented using Multiresolution segmentation algorithm using eCognition software was used in order to obtain the appropriate segments. Then, the watershed transformation was done and split the tree cluster[3]. Morphology was applied to the segmented objects and give an appropriate figure to the trees. Finally, the undesired objects were removed and output the crown projection area of the study plots. The accuracy assessment used were calculated and tabulated using equation develop by [9].

3.3. Correlation analysis – Pearson correlations

In order to see the relationship between the variables which is diameter at the breast height (DBH), height derived from LiDAR (H_LDR), crown projection area (CPA), height derived from field (H_Field) and carbon stocks (CS), Pearson correlation had been done to measure the strength of correlation between independent and dependent variables of the carbon estimations.

Table 2. Pearson correlation of the variables

Variable		DBH	H_LDR	CPA	CS
DBH	PC	1	0.763**	0.718**	0.909**
	Sig.		0.000	0.000	0.000
	N	183	183	183	183
H_LDR	PC	0.763**	1	0.549**	0.709**
	Sig.	0.000		0.000	0.000
	N	183	183	183	183
CPA	PC	0.718**	0.549**	1	0.671
	Sig.	0.000	0.000		0.000
	N	183	183	183	183
H_Field	PC	0.763**	0.995**	0.556**	1
	Sig.	0.000	0.000	0.000	0.000
	N	183	183	183	183

**Correlation is significant at the 0.01 level (2-tailed)

(PC= Pearson Correlation, N = number of samples, H_LDR = tree height from LiDAR, H_Field = tree height from field, CPA = crown projection area, .Sig = significant level)

Based on Table 2, tabulated that the highest correlation strength exists between height derived from field sampling (H_Field) and height derived from LiDAR (H_LDR), with 0.995 with ($P < 0.01$). There is also a high positive correlation between carbon stocks (CS) and crown projection area (CPA) where the correlation between this two variables (0.671) with ($P < 0.01$) significant level. Next, the non-linear regression model was develop using CPA as independent variables and carbon stocks as dependent variables as the data yield in this result shown the strong evidence to develop a model.

3.4. Modelling AGB using non-linear regression models

The non-linear regression model was carried out using CPA and output the coefficient of determination of 0.67 which is ($R^2 = 0.45$). The correlation coefficient of the CPA and carbon was high but not strongly correlated which is (0.671) indicated there is a significant correlation between these two variables. According to ANOVA, the model is significant at 99% significance level. The details summarization of the regression of the developed model is tabulated in Table 3, 4 and 5.

Table 3. Model Summary[b]

R	R^2	Adj. R^2	Std. Error of the Estimate	Change Statistics					Durbin-Watson
				R^2 Change	F Change	df 1	df2	Sig.F Change	
0.673[a]	0.453	0.447	0.14751	0.453	74.589	2	180	0.000	1.866

a. Predictors: (Constant), CPA, CPA^2
b. Dependent Variable: Carbon Stocks (Mg ha^{-1})

Table 4. ANOVA[a]

Model	Sum of Squares	df	Mean Square	F	Sig
1	Regression	3.701	2	1.851	0.000[b]
	Residual	4.466	180	0.025	
	Total	8.167	182		

a. Dependent Variable: Carbon Stocks (Mg ha^{-1})
b. Predictors: (Constant), CPA, CPA^2

Table 5. Coefficients [a]

	Unstandardized Coefficients		Standardized Coefficients	t	Sig
	B	Std. Error	Beta		
Intercept	1.333	0.354		3.763	0.000
CPA	-2.166	0.487	-2.416	-4.443	0.000
CPA^2	0.919	0.165	3.030	5.574	0.000

a. Dependent Variable: Carbon Stocks (Mg ha^{-1})

Based on the summary obtained from the regression analysis (Table 4), a model was developed to estimate the carbon stocks of the study area (Equation 5 and 6) derived from the non-linear regression model using quadratic function (Equation 3).

$$ln\, Carbon\, Stocks = \beta 1 \ln\, x^2 + \beta 2 \ln x + \beta 0 \tag{5}$$

$$ln\, CS = 0.919 \ln\, CPA^2 - 2.166\, \ln CPA + 1.333 \tag{6}$$

Where *CS* is the carbon stocks in Mg Ha-1, *CPA* is the crown projection area, and $\beta 1$, $\beta 2$ and $\beta 0$ is the constant derived from the regression model. The variables CPA and carbon had been transforming and the natural logarithmic had been applying to the formula.

3.5. Validation of Non-Linear regression model

The Pearson product model of the non-linear regression using quadratic function were validated using a random selection of 30% of the dataset which is 62 tress from various species and tree height from out of the total of 245 total stands. The resulting coefficients of determination between carbon predicted from a non-linear model and the carbon observed from field dataset was (R^2=0.7463) (Figure 4).

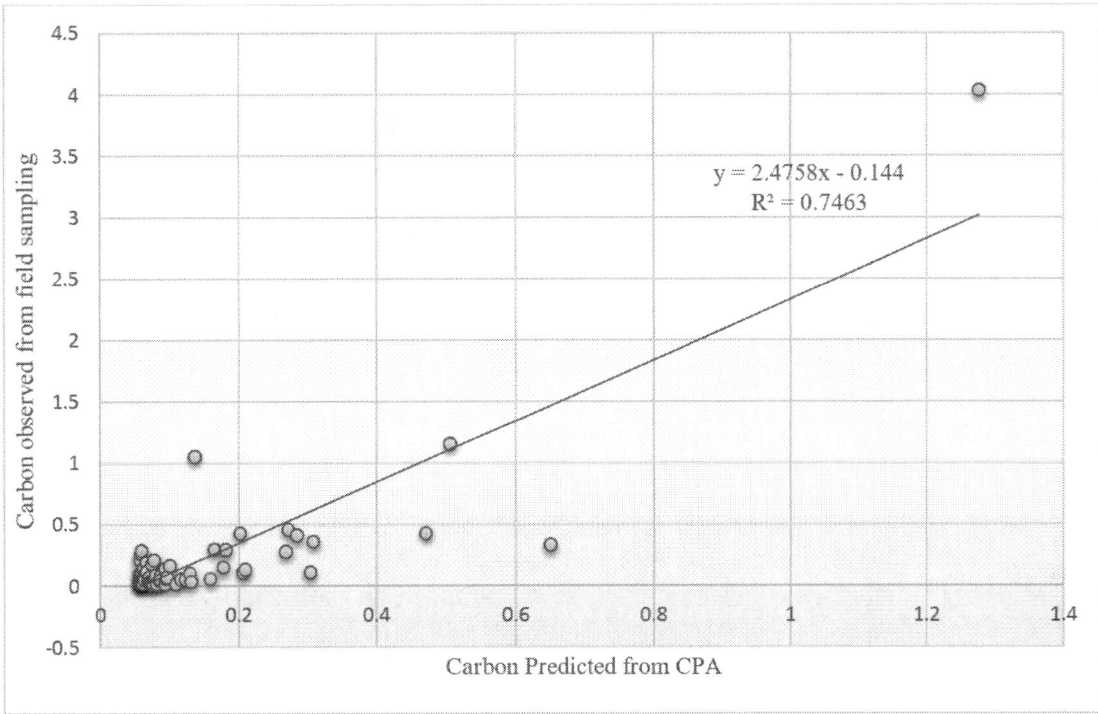

Figure 4. Scatter plots of non-linear regression model validation

The validation of the carbon predicted and carbon observed from field sampling were tabulated using scatter plots, and the coefficient of determination shows that the fit explain 74.63% of the total variation in the data about the average.

4. Conclusion

The relationship between carbon stocks with CPA obtained in this study showed a good relationship and impressive validation results. However the coefficient of determination R^2 of the regression model was not high ($R^2 = 0.453$), but there is an evidence that the correlation between CPA and carbon stocks was high which is 0.671. Due to the irregularities of the canopy of the tropical rainforest, the result can be considered as accepted based on comparison with the output of other research that had been done at other tropical rainforests for an example studies by [10] and [11].

References

[1] FAO, *Global Forest Resources Assessment 2015*, ISBN 978-9. 2015.
[2] Z. Latif and H. Zaqwan, "Deforestation and carbon loss estimation at tropical forest using multispectral remote sensing: Case study of Besul Tambahan Permanent Forest Reserve," *IEEE proceeding*, 2015.
[3] N. A. Mohd Zaki, Z. Abdul Latif, M. Z. Zainal, and K. Zainuddin, "Individual Tree Crown (ITC) Delineation Using Watershed Transformation Algorithm For Tropical Lowland Dipterocarp," *Proc. IEEE 2015*, vol. 2, pp. 237–242, 2015.
[4] S. Mebrat Tikabu, "Estimating Carbon Stock Using Very High Resolution Imagery and Terrestrial Laser Scanning In Tropical Rain Forest of Royal Belum, Malaysia," University of Twente, 2015.
[5] H. Omar, M. H. Ismail, H. Shafri, and K. Hamzah, "Estimating Biomass in Logged Tropical Forest Using L-Band SAR (PALSAR) Data and GIS Estimating Biomass in Logged Tropical Forest Using L-Band," *Sains Malaysiana*, no. October, 2015.

[6] Y. Zhu, K. Liu, L. Liu, S. Wang, and H. Liu, "Retrieval of Mangrove Aboveground Biomass at the Individual Species Level with WorldView-2 Images," *Remote Sens.*, vol. 7, pp. 12192–12214, 2015.

[7] M. E. J. Cutler, D. S. Boyd, G. M. Foody, and a. Vetrivel, "Estimating tropical forest biomass with a combination of SAR image texture and Landsat TM data: An assessment of predictions between regions," *ISPRS J. Photogramm. Remote Sens.*, vol. 70, pp. 66–77, 2012.

[8] Z. A. Latif, I. Zamri, and H. Omar, "Determination of tree species using Worldview-2 data," *2012 IEEE 8th Int. Colloq. Signal Process. its Appl.*, pp. 383–387, 2012.

[9] N. Clinton, A. Holt, J. Scarborough, L. Yan, and P. Gong, "Accuracy Assessment Measures for Object-based Image Segmentation Goodness," *Photogramm. Eng. Remote Sesing*, vol. 76, no. 3, pp. 289–299, 2010.

[10] P. R. Mbaabu, Y. A. Hussin, M. Weir, and H. Gilani, "Quantification of carbon stock to understand two different forest management regimes in Kayar Khola watershed, Chitwan, Nepal," *J. Indian Soc. Remote Sens.*, no. DoF 2012, 2014.

[11] Y. A. Hussin, H. Gilani, L. Van Leeuwen, M. S. R. Murthy, R. Shah, S. Baral, N. E. Tsendbazar, S. Shrestha, S. K. Shah, and F. M. Qamer, "Evaluation of object-based image analysis techniques on very high-resolution satellite image for biomass estimation in a watershed of hilly forest of Nepal," *Appl. Geomatics*, vol. 6, no. 1, pp. 59–68, 2014.

Acknowledgments

The authors would like to express their deepest gratitude to the Ministry of Higher Education (MOHE), Malaysia for financing the under Research Management Centre (RMC), (Ref:: RAGS/1/2014/STWN10/UITM//1) Universiti Teknologi MARA (UiTM), and also Faculty of Forestry, Universiti Putra Malaysia (UPM) for granting access to the study area. Thanks to Alhaffiz bin Abd Radzak (RS GIS Consultancy Sdn Bhd) for helping us in term of LiDAR pre-processing. Special thanks to a member of Ayer Hitam Rainforest Research Expedition 2015 (AHRRE 2015) who involve in data collection of forest inventories.

Vegetation structure determination using LIDAR data and the forest growth parameters

M Rybansky[1], M Brenova[1], J Cermak[2], J van Genderen[3] and Å Sivertun[4]

[1]University of Defence in Brno, Kounicova 65, 612 00 BRNO, Czech Republic,
[2]Mendel university Brno, Czech Republic,
[3]University of Twente, Netherlands,
[4]Swedish Defence University, Stockholm

Email: marian.rybansky@unob.cz

Abstract. The goal of this paper is to identify the main vegetation factors in the terrain, which are important for the analysis of forest structure. Such an analysis is important for forestry, rescue operations management during crises situations and disasters such as fires, storms, earthquakes and military analysis (transportation, cover, concealment, etc.). For the forest structure determination, both LIDAR and the forest growth prediction analysis were used. As main results, the vegetation height, tree spacing and stem diameters were determined

1. Introduction

The digital terrain model (DTM) represents the bare ground of the earth's surface without any objects such as plants and buildings. In contrast to a DTM, a Digital surface model (DSM) represents the earth's surface including all objects on it. The DTM does not change as frequently as does the DSM. The most important changes of the DSM are in forested areas due to the vegetation growth. Using LIDAR technology, see also [1-8] the canopy height model (CHM) is obtained by subtracting the DTM and the corresponding DSM. The crown DSM is calculated from the first pulse echo and DTM from the last pulse echo data. The main problem of the DSM and CHM data is the date of the airborne laser scanning.

This paper describes a method for calculating the forest structure changes using LIDAR data and the forest growth analysis results. The three main parameters are the mean height of vegetation, mean tree distances and thickness of mean tree trunk predicted using the relations between the canopy age and these parameters. The LIDAR DSM data (as a sample model) with an average density of 1 point per square meter was used to compare both data sources (vegetation growth curves values and laser data). The statistical calculations show that the tree growth curves increase relatively quickly from the beginning (young canopy), but becomes flattens out at the higher age.

That method was tested on a typical spruce forest stand. This spruce forest was chosen because it is the most common tree species in many countries located in central and northern Europe.

The results of this experiment show that this method is fully applicable for the DSM generated from LIDAR data. The method can be appropriately implemented as a relatively cheap up-dating tool in the GIS technology between two laser scanning campaigns of a territory. This method can also be refined using the growth curves of individual types of trees.

2. Canopy structure prediction methods

Vegetation growth used to be characterized by a description of the state and characteristic of changes resulting from growth over time. As the definition of word "growing" we can imagine the quantitative enlarging of volume of biomass of living organisms. The descriptions of state and characteristic changes coming up during forest vegetation development are made by use of enumeration survey values. Basic values of enumeration survey are e.g. number of trees per square unit (this is a basic value for computation of mean spacing), mean thickness, mean height, circular base, stock, growth figure etc.

2.1 Number of trees per square unit

Number of trees per square unit $N.\text{ha}^{-1}$ as a rule, tends to be regarded as value which is variable to a great extent. Generally, we can say that the development of this enumeration survey value depends on natural mortality and manmade operations. Basic factors which always affect this value are as follows:

- Age of vegetation.
- Woody plant species.
- Habitat.
- Technique of vegetation setup.

The number of trees per square unit is usually presented in relation to the age of the vegetation. We know that, with increasing age, the number of trees diminishes very quickly in the beginning, and later diminishes at a slower rate. We can express this fact by a simple analytic relation [9] and [10]:

$$N = \frac{B}{t^m} \tag{1}$$

where B and m are constants, t is an age.

The development of tree numbers depends also on woody plant species and habitat (see Fig. 1), more detailed see e.g. [9, 10, 11]. The spruce as the predominant tree type (in Central Europe) was selected.

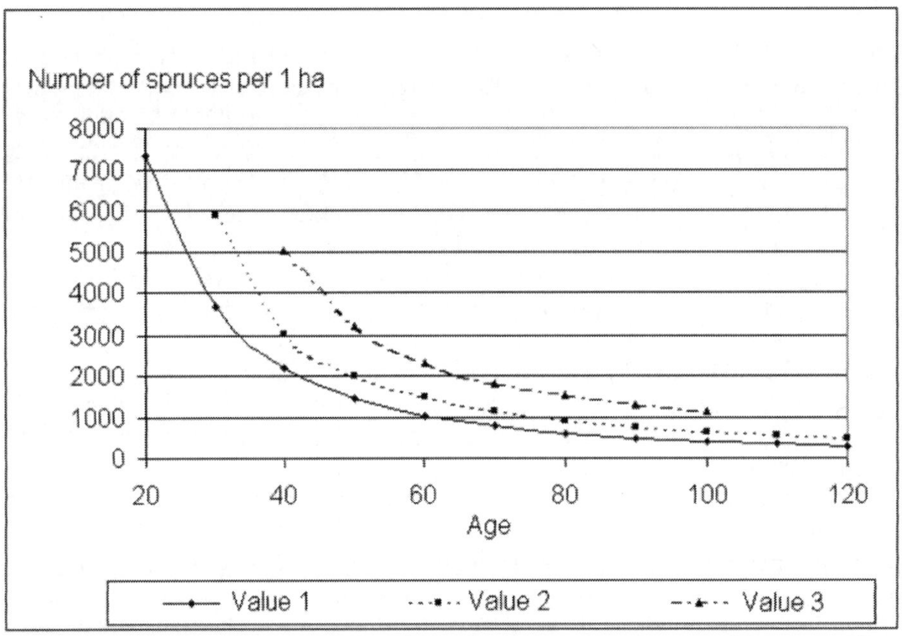

Figure 1. Number of spruce tree per ha in dependence on the age of the vegetation Values 1, 2 and 3 refer to the quality of the soil type.

The number of trees of the predominant species decreases rapidly in the period of youth, especially on fast-growing trees and light-demanding species and has considerably lesser value than on shade-loving woody plant species. At the same time, we have to remark that the growth of woody plants on fertile soils is more intensive. That is why the less fertile habitat has more trees than the rich fertile soil habitat at the same comparable age. Naturally, growing vegetation has in the early stage of development significantly greater number of trees per square unit than vegetation founded by plantation.

The development of tree number per square unit is presentable also by use of the other values. As an example, the relation of the evolution of tree numbers depends on:

- Mean tree spacing (MTS).
- Thickness of mean trunk.
- Mean height of growth.

The relations among the development of number of trees per square unit, thickness of mean trunk and mean tree spacing are considered as the closest one.

2.2 Mean tree spacing

The function of mean tree spacing is definable by number of trees for square spacing lay-out [9, 10]:

$$MTS \sim \sqrt{\frac{40{,}000}{\pi \, . \, N}} \, , \quad \text{then} \quad N \sim \frac{40{,}000}{\pi . MTS^2} \qquad (2)$$

where N – number of trees per ha.

The number of trees as the function of tree spacing a_3 among the three adjoining trees (see Fig. 2) is computable by mathematic calculation. Variation coefficients for a_3 take values which are usable to computation of necessary number of measuring.

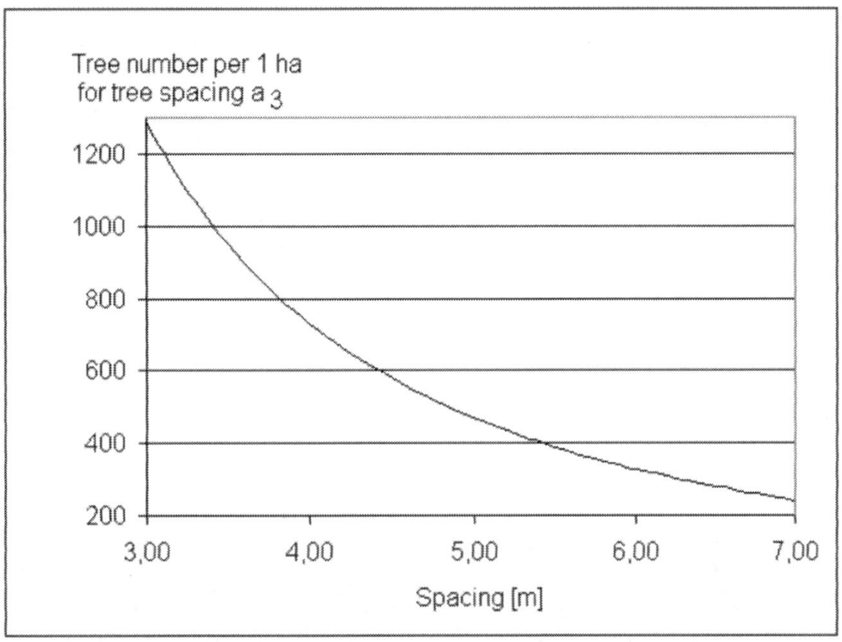

Figure 2. Interdependence of tree number on tree mean spacing.

2.3 Mean tree thickness

Mean thickness of vegetation D presents its thickness maturity, in most cases expressed as the thickness of mean trunk responding to the circular base of mean tree of vegetation given or, it is possible to define it as the weighed arithmetic mean and then it is the arithmetically determined mean thickness.

Mean thickness frequently tends to be related to the age of vegetation. These relations are presented by slightly concave curves closely approaching to a straight line. In the following, there exists a very important relation between mean thickness and quality of habitat. Generally speaking, we can say that with lowering quality of soil, the mean thickness also decreases.

Frequently the thickness maturity of vegetation is considered over distribution of tree frequency in height of 1.3m ($d_{1,3}$). This distribution can be presented by the use of known Gauss curve. Distribution of tree thicknesses is frequently a left-handed asymmetric one [9, 10, 11]. Relation [1] depicts the number of trees in dependence on stem diameter and mean spacing's:

$$N = B.x^{-k},\qquad (3)$$

where N is tree number, B, k are constants and x is mean trunk diameter or tree mean spacing (see Fig. 3).

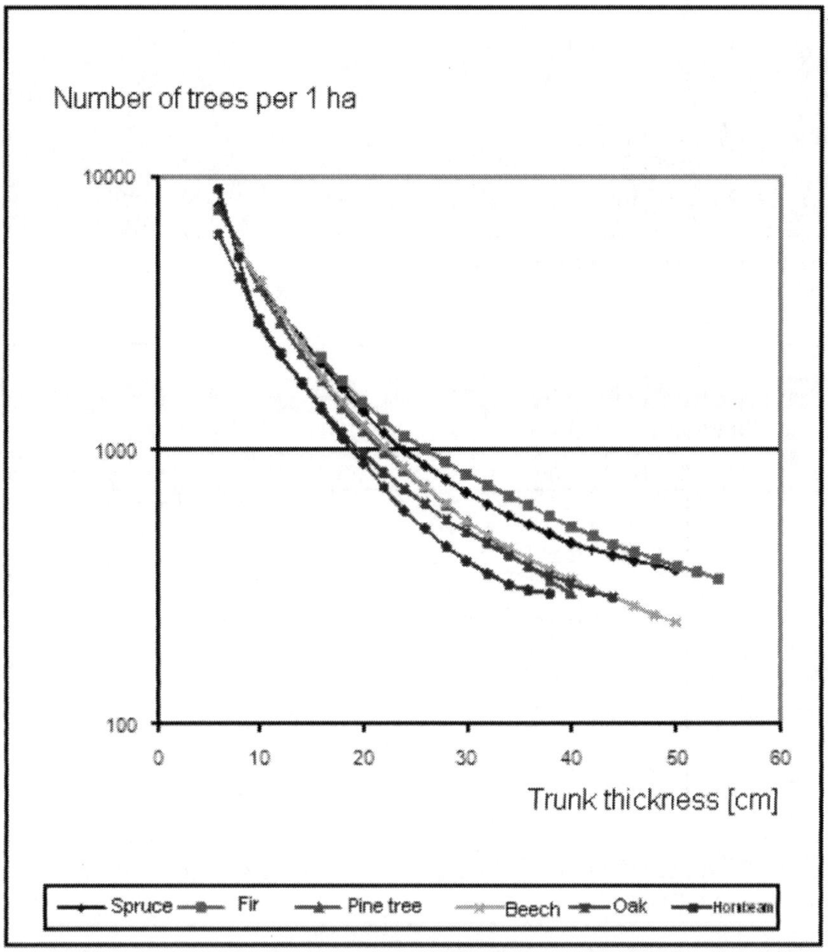

Figure 3. Number of trees per ha in dependence on trunk thickness.

2.4 Mean tree height

Mean height H tends to be considered for the rate of height maturity of vegetation. In practice, we understand this to be the height of mean trunk of woody plant species given.

In addition to mean height, the top height of vegetation is presented in [3, 5, 8, 11]. Its definition is not unified at the present time, but we can say that it is as a rule the mean height of a certain number of the thickest trees of vegetation given.

The height of mean trunk is mostly expressed in dependence on vegetation age. That way we can obtain so called curve of growing (developing). Generally, this curve rises from the beginning relatively quickly, but at the higher age flattens out. (see Fig. 4).

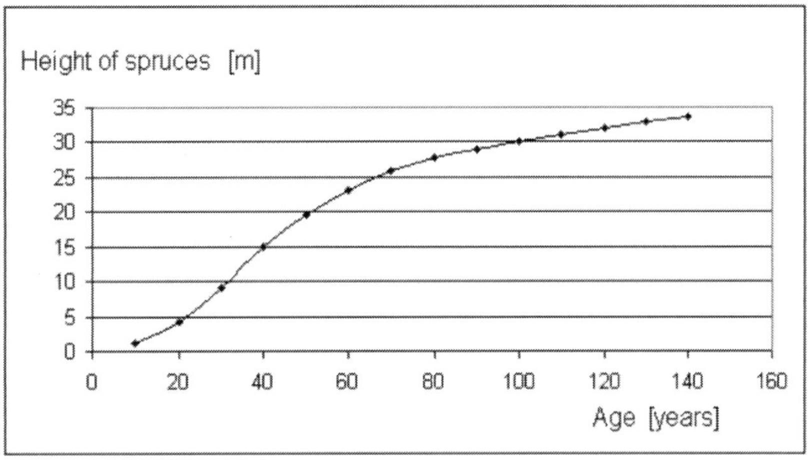

Figure 4. Process of height growth of a spruce forest.

Height growth of woody plants depends primarily on habitat quality, which is exploited at enumeration survey by use of its height and age. Generally, we can say that the better habitat of woody plant given, the better also the mean height of vegetation.

3. Results and discussion

The described methods were tested on a spruce forest stand, located in the south-eastern part of the Czech Republic, northeast of Brno. This spruce forest was chosen because it is the most common tree species in many countries located in central and northern Europe. The current age of this forest is about 40 years. To determine changes in the canopy height, distances among trees and tree diameters the series of aerial photographs were used. The photographs were acquired during the military topographic maps up-dating periods from 1973 to the present. In addition the DSM with an average density of 1 point per square meter was used to compare both data sources (aerial photographs and laser data). This DSM was generated from LIDAR data in 2013. Comparing the heights of corresponding trees on the aerial photographs of various ages, the statistical sets of the tree growth rate were obtained. These statistical data and LIDAR data (2013) were compared with the growth curve of the spruce forest (Fig. 4), which corresponds to a similar natural environment (soil quality, climate characteristics, geographic location, etc.). The results of the measurements in a representative spruce forest and statistical calculations show that the tree growth curves rises relatively quickly from the beginning (young canopy), but becomes more flattened at the higher age. The results of this experiment show – see Fig. 5, which this method is fully applicable for the DSM generated from LIDAR data.

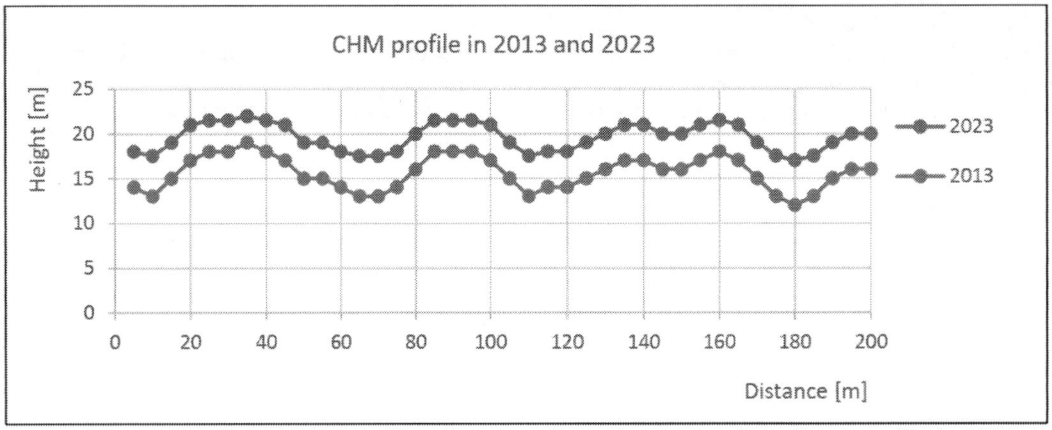

Figure 5. LIDAR CHM profile of a spruce forest in 2013 and predicted profile in 2023.

4. Conclusions

The method can be appropriately implemented as a relatively cheap up-dating tool in the GIS technology between two laser scanning campaigns of a territory. This method can also be refined using the growth curves of individual types of trees. It is also necessary to search the relationships between the natural environment factors and the specific canopy growth.

To determine the changes in the canopy height also the series of aerial photographs can be used. The photographs were acquired in the Czech Republic during the military topographic maps up-dating periods from 1973 to the present. In addition the DSM with an average density of 1 point per square meter can be used used to compare both data sources (aerial photographs and laser data).

References

[1] Hyyppä, J. at all. 2004 *Algorithms and methods of airborn laser scanning for forest measurement.* In International Archives of Photogrammetry and Remote Sensing. Vol. 36 – 8/W2, pp 82-89.

[2] Aschoff, T. and Spiecker, H. 2004 *Algorithms for the automatic detection of trees in lasser scanner data.* In International Archives of Photogrammetry and Remote Sensing.

[3] Gobakken, T. and Næsset, E. 2004 *Effects of forest growth on laser derived canopy metrics.* In International Archives of Photogrammetry and Remote Sensing. Vol. 36 – 8/W2, pp 224-227.

[4] Blair, J. B., Rabine, D. L., Hofton, M. A. 1999 *The Laser Vegetation Imaging Sensor: a medium-altitude, digitisation-only, airborne laser altimeter for mapping vegetation and topography.* In ISPRS Journal of Photogrammetry & Remote Sensing 54/1999, pp 115–122.

[5] Su, Y., Guo, Q., Fry, D. L., Collins, B. M., Jacob, M. K., Flanagan, P. Battles, J. J. 2015 *A Vegetation Mapping Strategy for Conifer Forests by Combining Airborne Lidar Data and Aerial Imagery.* Canadian Journal of Remote Sensing 12/2015, 53 pp. DOI: 10.1080/07038992.2016.1131114.

[6] Carson, W. W., Andersen, H. E., Reutebuch, S. E. and McGaughey, R. J. 2004 *Lidar Applications in Forestry – An Overview.* ASPRS Annual Conference Proceedings, May 2004, Denver, Colorado.

[7] Lim, K., Treitz, P., Wulder, M., St-Onge, B. and Flood, M. 2003 *LiDAR remote sensing of forest structure.* Progress in Physical Geography, 27; 88, DOI: 10.1191/0309133303pp360ra.

[8] Ahlberg, S., Söderman, U., Tolt, G. 2006 *High Resolution Environment Models from Sensor Data* In Defence Imagery Exploitation, London, UK.

[9] Simon, J., Kadavy, J. and Macku, J. 1998 *Forest economic adjusting (in Czech).* MZLÚ Brno.

[10] Simon, J., Horacek, P., Konas, P. and Rybansky, M. 2000 *Assessment of the critical stem diameter*

by-passed by a vehicle (in Czech). VA Brno. 21 pp.
[11] Rybansky, M. and Hyrs, F. 2001 *Cross-Country Movement analysis of the forest (in Czech)*. VA Brno, 102 pp.

Acknowledgement
This paper is a particular result of the defence research project DZRO K-210 and Specific research project 2016 at the department K-210 managed by the University of Defence in Brno.

GWR-PM - Spatial variation relationship analysis with Geographically Weighted Regression (GWR) – An application at Peninsular Malaysia

Jamhuri J[1], Azhar B M S[2], Puan C L[2], and Norizah K[3]

[1] Master student, Faculty of Forestry, Universiti Putra Malaysia, 43400 UPM Serdang, Selangor, Malaysia.
[2] Senior Lecturer, Deartment of Forest Management, Faculty of Forestry, Universiti Putra Malaysia, 43400 UPM Serdang, Selangor, Malaysia.
[3] Senior Lecturer, Department of Production, Faculty of Forestry, Universiti Putra Malaysia, 43400 UPM Serdang, Selangor, Malaysia.

norizah_k@upm.edu.my

Abstract. GWR-PM has been developed exclusively for decision makers in Peninsular Malaysia and the purpose is to provide them with additional flexibility in analysing spatial variation. While GWR extension analysis in ArcMap application has a universal coordinate system, GWR-PM is specifically designed with Peninsular Malaysia's coordinate system of Kertau RSO Malaya Meter. This paper presents the development of GWR-PM model by using a model builder, the application of which is to examine the forest fire risk at North Selangor Peat Swamp Forest. This model can be extended and improved by using ArcGIS language of phyton.

1. Introduction

A spatial pattern is a perceptual structure, placement, or arrangement of objects on earth. It includes the space in between those objects. Patterns may be recognised because of their arrangement, maybe in a line or by a clustering of points. Spatial patterns have been used by scientists, analysts and other professionals for planning, managing and estimation. Many researchers have used the traditionally regression method but there is a lack of information such as location and attribute data, besides weightage values not taken into account. Recently, with a new evolution of technology in spatial analysis, Geographically Weighted Regression (GWR) has been introduced to solve such problems related to spatial non-stationary. Geographically Weighted Regression (GWR) is a regression method capable of handling various relationships of variables in local spatial patterns for modelling, examining, monitoring and decision making [1]. The final outputs of spatial analysis using GWR provide users with spatial non-stationary, or in other words, variation of variables in local spatial patterns and a map of the spatial variation in relationships [1][2].

There are several studies using GWR in spatial regression which have been applied in forestry. For example, [3] used GWR to analyse the factors that drive the afforestation in Northern Vietnam. In this study, they used locations of timber demand as dependent variable and heterogeneity such as population, distance, industry and forest as independent variables. In the studies of [4], GWR was applied to investigate the impact of scales on prediction of rainfall uncertainty. This was carried out by modelling the relationship between vegetation and climate; they used vegetation as dependent variable and independent variables were precipitation and temperature. Besides that, [5] employed GWR analysis to estimate the net primary production (NPP) of the Chinese forest ecosystems. They used the Chinese forest NPP, derived from plot inventory data covering the entire country as dependent variable while independent variables were temperature, precipitation, altitude, and yearly Time-Integrated Normalised Difference Vegetation Index (TINDVI). GWR analysis is not limited to applications in wildlife studies. For example, [6] used GWR to investigate the effects of local spatial heterogeneity on multivariate relationships of white-tailed deer distribution, using land cover patch metrics and climate factors. The distribution data of deer were collected by using the numbers of pellets consumed at different places.

On the other hand, [7] used GWR to examine spatially non-stationary and scale-dependent relationships between urban landscape fragmentation and related factors. This study was intended to investigate the applicability of GWR in modelling the relationships between effective mesh size and related factors, and then examine their spatial non-stationarity and the scale-dependence. They used effective mesh size as dependent variable; distance to main roads, distance to district centres and slope were explanatory variables. [7] used an extension in ArcToolbox version ArcGIS 9.3 to run GWR analysis. In addition, a study by [8] was intended to improve the ecosystem service in Fuzhou City, China; this was done by providing geographic variations using GWR analysis to prioritise areas for conservation or construction, and design an ecological corridor. They used population density in Fuzhou City, China as dependent variable; proportion of urban and rural construction land, proportion of transportation land and proportion of forest land were explanatory variables used for GWR analysis in Arc GIS 10.0. ArcGIS provides such extension application under spatial statistics and helps to visualise and interpret spatial non-stationarity of dataset.

Dynamic changes in the Peninsular Malaysia forest ecosystem have brought new challenges to the process of decision making in forest management system. Multiple biotic and/or abiotic factors and additional side constraints such as uncertainty of their variability in space require a detailed explanation and regression analysis for spatially varying relationships. To date, there are several studies on spatial patterns in Malaysia such as land use planning, parcel management, census data, topographic mapping, environmental monitoring and natural resources management. An example of such studies was done by [9] in relation to a community's forest dependency and its effects on the forest resources and wildlife abundances in Sarawak, Malaysia. They used heterogeneity as variables to study the effects of the community's forest dependency on the forest resources and wildlife abundances in Sarawak. Other than that, [10] studied the pattern of the changes of the conservation policy of tigers in Malaysia. They used camera traps to collect visual data and the analysis was done by using spatial autocorrelation. However, both studies are limited to non-spatial display of data which cannot explain the relationships between some sets of variables and spatial locational variables; they may help to identify the nature of relationships that exist between variables by taking spatial location into consideration with GWR analyses, and the spatial variability between relationships across the study area can be detected [11] Incorporating the explanatory variables into the GWR command is however a problem to an amateur or beginner user of GWR analysis.

Hence, this paper presents an approach to develop a toolbox's extension of GWR for spatial non-stationary analyses with its related applications. A new extension of the GWR analysis development described in this paper provides users with an additional flexibility to choose input features of selected regions and pre-processing of all the spatial features before varying relationships analysis. Input feature tools with sequence help users to determine the right variables in relationships used in modelling and mapping with GWR analysis; this procedure has a special application to forest management in Peninsular Malaysia.

2. Model Development

The processes of capturing and managing the spatial data to be queried are an initial phase for spatial analysis in GIS. Decision makers are responsible for those requirements; identifying the issues and mitigating measures until the final analysis can be modelled and mapped. General requirements for the GWR model development in Peninsular Malaysia forests can be described as below:-

2.1. Define the problem - as spatial non-stationarity analysis.
2.2. Identify the issue - as locational or spatial variations
2.3. Mitigation measures - as explanatory variables which believed or expected to give effect on the variations to the issue highlighted.

Thus, the relationship between the significant problems, issues and mitigation measures can be illustrated in the Toolbox of ArcMap application as shown (Figure 1) below;

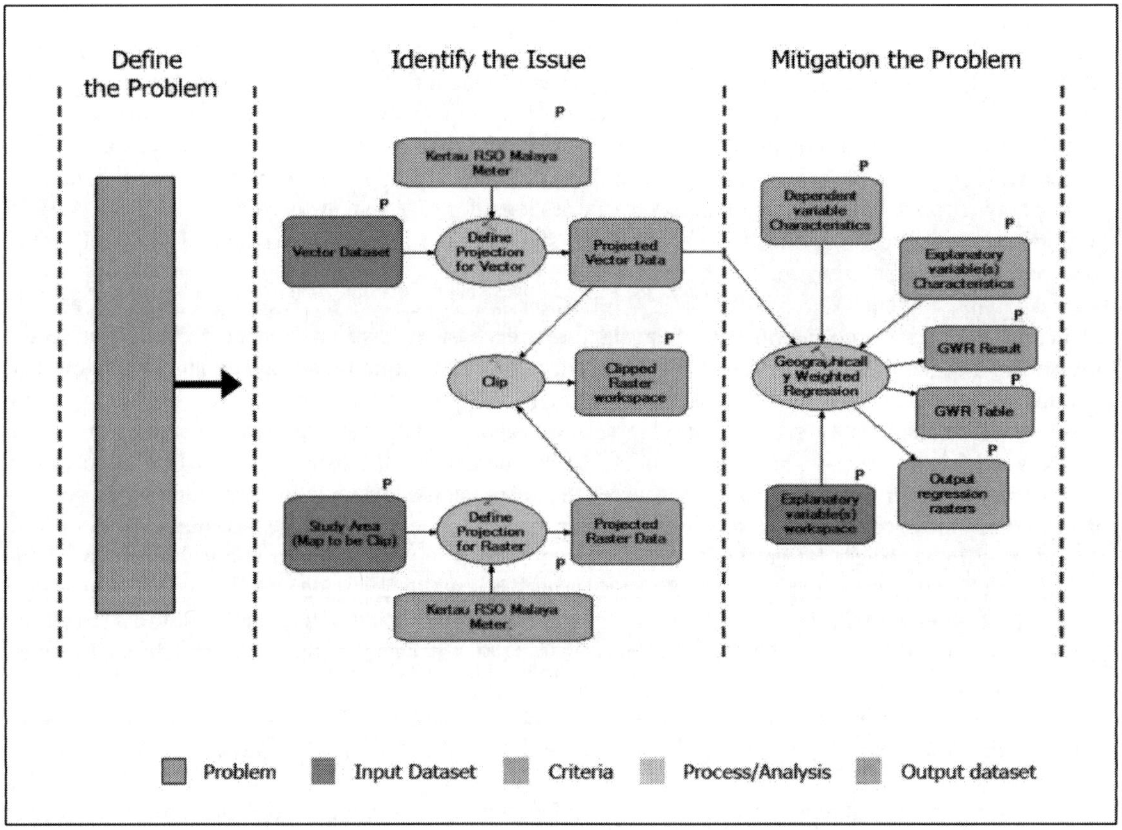

Figure 1. A simplified model of GWR analysis for Peninsular Malaysia.

2.1. Feature dataset of vector and raster data.

Feature datasets are used to present spatial variations to be analysed in GWR. The datasets consist of dependent and explanatory variables. Feature datasets are used because the location of the feature data will be used for GWR inquiry. Besides that, feature datasets have an attribute table that can be used to input dependent and explanatory variables. The raster data are used to overlay with the feature dataset for visual references.

2.2. Define projection
For the purpose of uniformity, feature datasets are referenced with the projected coordinate systems of rectified skew orthomorphic (RSO) Malaya Meters format. Malaysia uses RSO because it has a base geographic in 2 Dimensions (2D) under UTM zone 47N as projected Coordinate References System (CRS). The Kertau (RSO) / RSO Malaya (m) was revised on 14th August 2006. It was defined by information from Defence Geographic Centre (DFC) and used only for metrication of RSO grid.

2.3. Vector and/or Raster data analysis
Geoprocessing is a GIS operation used to manipulate spatial data. A typical geoprocessing operation takes an input dataset, performs an operation on that dataset, and returns the result of the operation as an output dataset. There are two types of geoprocessing in ArcGIS, which are geoprocessing and raster geoprocessing (Table 1). Geoprocessing process includes buffering, clipping, intersecting, union, merging and dissolving. On the other hand, raster geoprocessing includes raster clip (raster to vector clip), composite band, computing pan sharp weights, creating ortho-corrected raster dataset, creating pan-sharpened raster dataset and extracting sub-dataset. In this process, a raster dataset needs to be clipped with feature dataset to fit the area of study, as reducing the size speeds up ArcGIS processing. The spatial pattern will maximise production at the end point in every angle.

Table 1. Geoprocessing process available in ArcToolbox of ArcMap version 10.3 [12].

Geoprocessing	
Buffer	Creates buffer polygons around input features to a specified distance.
Clip	Clip is a process to cut out a piece of one feature class using one or more of the features in another feature class as a cookie cutter.
Intersect	Computes a geometric intersection of the input features. Features or portions of features which overlap in all layers and/or feature classes will be written to the output feature class.
Union	Computes a geometric union of the input features. All features and their attributes will be written to the output feature class.
Merge	Combines multiple input datasets of the same data type into a single, new output dataset. This tool can combine point, line, or polygon feature classes or tables.
Dissolve	Aggregates features based on specified attributes.
Raster Geoprocessing	
Clip	Cuts out a portion of a raster dataset, mosaic dataset, or image service layer.
Composite Band	Creates a single raster dataset from multiple bands
Compute Pan sharp Weights	Calculates an optimal set of pan-sharpened weights for new or custom sensor data.
Create Ortho Corrected Raster Dataset	Incorporates elevation data and image metadata to accurately line up imagery.
Create Pan-sharpened Raster Dataset	Fuses a high-resolution panchromatic raster dataset with a lower-resolution multiband raster dataset to create a red-green-blue (RGB) raster with the resolution of the panchromatic raster.
Extract Sub-dataset	Extracts raster datasets stored within a sub-dataset raster file.
Raster To DTED	Splits a raster dataset into separate files based on the DTED tiling structure.
Resample	Changes the spatial resolution of your raster dataset and sets rules for aggregating or interpolating values across the new pixel sizes.
Split Raster	Divides a raster dataset into smaller pieces, by tiles or features from a polygon.

3. GWR

Feature datasets are added as location data and are arranged as dependent variables and factors of influence to the dependent variables are fixed as explanatory variables. The result will be attributed with spatial pattern map, coefficient value and R2 value. The coefficient value describes to better understand of location sample either as a statistically significant hot spot or a statistically significant cold spot, while R2 value explains the significantly fraction between 0.0 and 1.0. Figure 2 shows the interface of GWR analysis developed by using the model builder in Arctoolbox of ArcMap version 10.3.

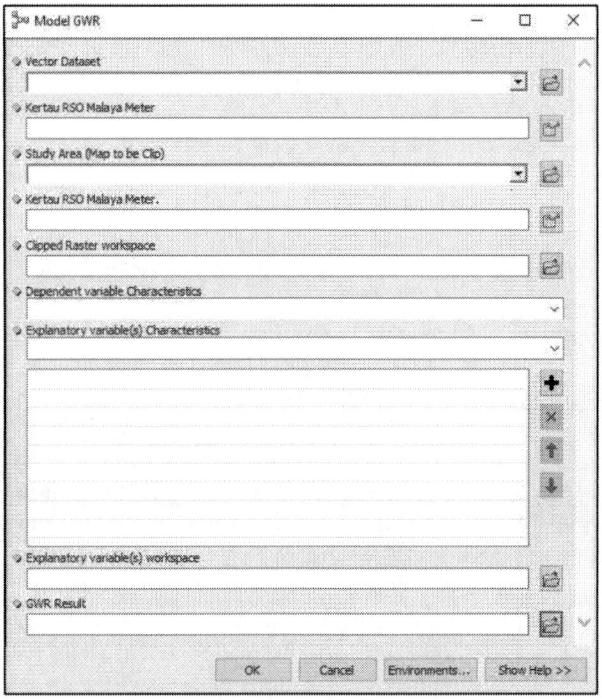

Figure 2: Interface of GWR modelling analysis.

4. Model applications

As a demonstration, we applied our model to Raja Musa Forest Reserve (RMFR), a part of North Selangor Peat Swamp Forest (NSPSF), to analyse the hotspot areas of high risk forest fire (Figure 3). The locations of the hotspot areas with frequent occurrences of fire were set as dependent variables; heterogeneity factors were taken into account as explanatory variables such as distance to road, distance to river, distance to oil palm, distance to paddy field, and distance to continuous forest. These data were collected by using the landscape metric measurement method in Google Earth Pro free version. Each variable was measured manually in each point of forest fire hotspot. Also, Landsat 8 image with 30 meter resolution of RMFR was used for Kernal mapping.

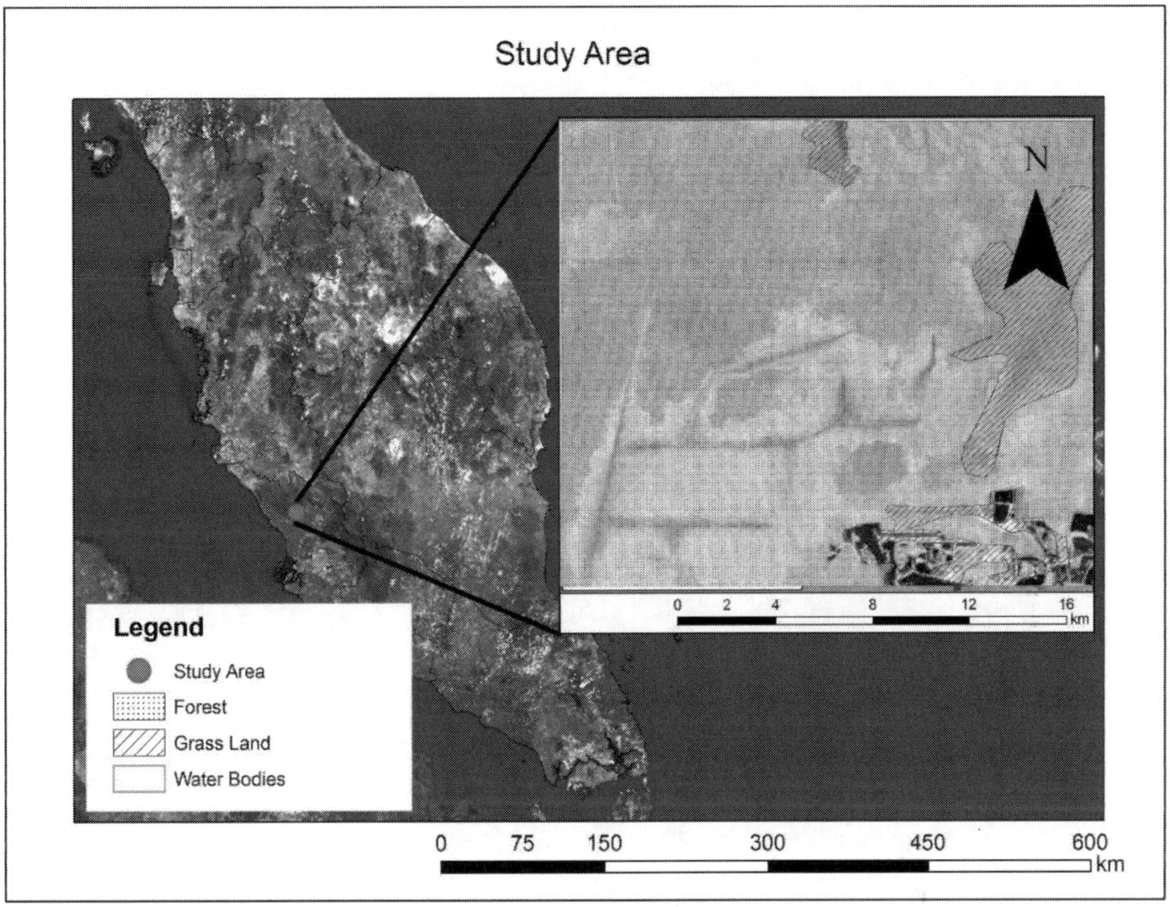

Figure 3. Application area located in the RMFR part of NSPSF

5. Results and discussion

The GWR analysis provides R^2 and coefficient raster to show the significantly of the model tested and the hotspot areas of the study sites. The results obtained from model application on hotspot areas of high risk forest fire in NSPSF show that 76.47% of R^2 is significantly related to heterogeneity variables which are distance to road (DTRD), distance to river (DTRV), distance to oil palm (DTOP), distance to paddy (DTPD), and distance to continuous forest (DTCF). Based on these results, all the variables are positively higher which is coefficient values (positive or negative) of hotspot is 4.25404. But, the highest coefficient value for single variable is distance to river DTRV which is the coefficient value of hotspot -9.94182 and followed by DTCF -9.25813, distance to paddy field DTPD 8.24629, distance to oil palm DTOP -7.71606 and lastly distance to road DTRD -0.000303741 (Figure 4). The result from the DTPD shows that, the closer paddy with the fire hotspot areas, the higher potential the area to get burn. This show the significant because paddy are easily to get burn especially in dry season (summer). Conversely, the result from DTRV show that the closer river with the fire hotspot areas, the lower potential of the area to get burn. This show the significant between fire hotspot areas and river because fire are less burning in wet area. In additionally, the same result to DTCF, DTOP and DTRD which are lower potential to burn when close and high potential of burning when far from fire hotspot areas. Forest and oil palm are dense area and ground level are covered by canopy which are help in maintain soil moisture and could prevent from fire burning while area with no canopy covered are directly exposed to sunlight heat which could make it dry and easily to get burn. Finally, distance to road not directly affect the fire because most of the road is surrounded by wild grass that grows in limited areas. Figure 5 below shows the hotspots of forest fire areas in RMFR in the year of 2015. In regard to this matter, the results show the ability of GWR coefficients to vary over spaces; GWR often produces better residual sum of squares, which is an indication of better model fit [13] [14].

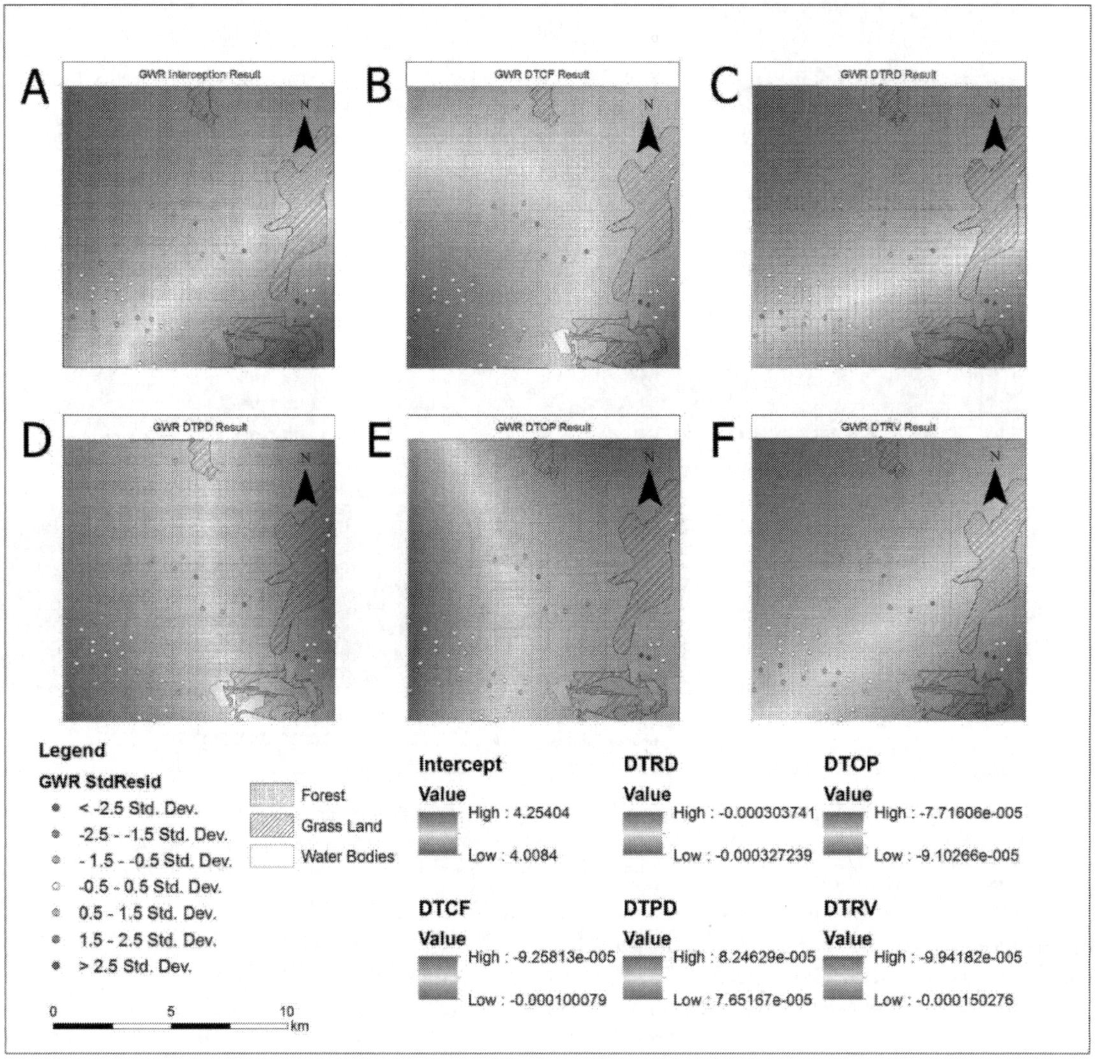

Figure 4. Result of the GWR analysis: (A) Interception of Variables, (B) distance to continoues forest (DTCF), (C) distance to road (DTRD), (D) distance to paddy field (DTPD), (E) distance to oil palm (DTOP), and (F) distance to river (DTRV).

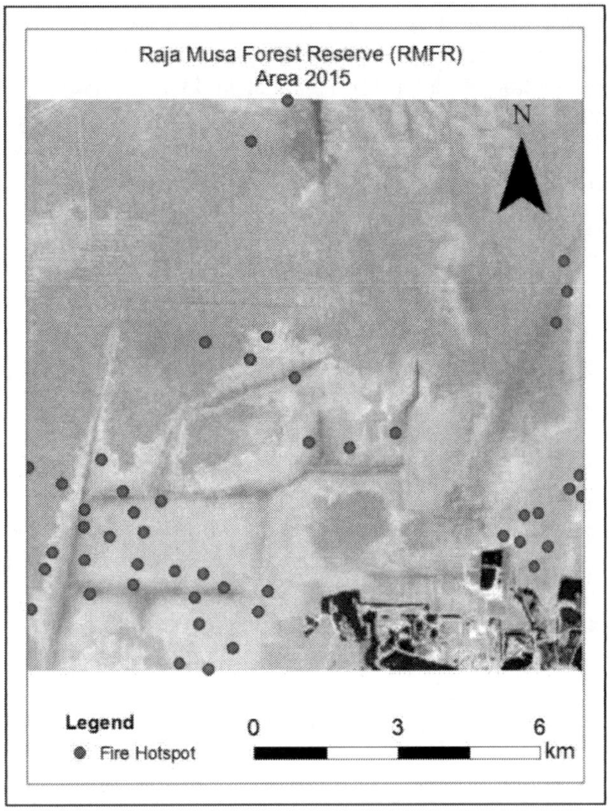

Figure 5. Map of Raja Musa Forest Reserve area.

6. Conclusion

As a conclusion, GWR provides better understanding in the form of kernel mapping such as in Figure 4 that displays the patterns of spatial map compared with other regressions which do not take location into account. Ignoring spatial variations under these conditions could produce results of biased estimation, misleading significant tests and sub-optimal predictions. Thus, GWR provides an effective alternative option by making use of spatial location as proxy variables. Therefore, GWR can achieve significantly improved results over those of other regression models. Furthermore, this model development fits well with GWR analysis, and this actually eases the analysis process. According to [15] more than 290 million hectares of land burned annually (between 79% and 91% of total global burned area). So, this modelling could lead to better management practices, especially in Forest Department. Currently, most of the practices in the department rely on elusive analysis; the modelling of this study actually improves the analysis by providing illustrations of spatial pattern maps that will be easy to understand.

7. References

[1] Fotheringham, A.S., Brunsdon, C., Charlton, M., 2002. *Geographically Weighted Regression: The Analysis of Spatially Varying Relationships*. Wiley, West Sussex.

[2] Luo, J., Wei, Y.H.D., 2009. *Modeling spatial variations of urban growth patterns in Chinese cities: the case of Nanjing*. Landscape and Urban Planning 91, 51– 64.

[3] Clement, F., Orange, D., Williams, M., Mulley, C., & Epprecht, M. (2009). *Drivers of afforestation in Northern Vietnam: Assessing local variations using geographically weighted regression.* Applied Geography, *29*(4), 561–576. http://doi.org/10.1016/j.apgeog.2009.01.003

[4] Propastin, P., Kappas, M., & Erasmi, S. (2008). *Application of Geographically Weighted Regression to Investigate the Impact of Scale on Prediction Uncertainty by Modelling Relationship between Vegetation and Climate.* International Journal of Spatial Data Infraestuctures Research, 3, 73–94. http://doi.org/10.2902/1725-0463.2008.03.art6

[5] Wang, Q., Ni, J., & Tenhunen, J. (2005). *Application of a geographically-weighted regression analysis to estimate net primary production of Chinese forest ecosystems.* Global Ecology and Biogeography, *14*(4), 379–393. http://doi.org/10.1111/j.1466-822X.2005.00153.x

[6] Shi, H., Laurent, E.J., LeBouton, J., Racevskis, L., Hall, K.R., Donovan, M., Doepker, R.V., Walters, M.B., b, Lupi, F., Liu, J., 2006. *Local spatial modeling of white- tailed deer distribution.* Ecological Modeling , 190, 171–189.

[7] Gao, J., & Li, S. (2011). *Detecting spatially non-stationary and scale-dependent relationships between urban landscape fragmentation and related factors using Geographically Weighted Regression.* Applied Geography, *31*(1), 292–302. http://doi.org/10.1016/j.apgeog.2010.06.003

[8] Hu, X., Hong, W., Qiu, R., Hong, T., Chen, C., & Wu, C. (2015). *Geographic variations of ecosystem service intensity in Fuzhou City, China.* Science of The Total Environment, 512-513, 215–226. http://doi.org/10.1016/j.scitotenv.2015.01.035

[9] Nelson, J., Muhammed, N., & Rashid, R. A. (2015). *Community's forest dependency and its effects towards the forest resources and wildlife abundances in Sarawak, Malaysia, 4509*(November). http://doi.org/10.1080/13504509.2015.1072117

[10] Rayan, D. M., & Linkie, M. (2015). *Conserving tigers in Malaysia: A science-driven approach for eliciting conservation policy change. Biological Conservation, 184,* 18–26. http://doi.org/10.1016/j.biocon.2014.12.024

[11] Shrestha, P. M. (2006). *Comparison of Ordinary Least Square Regression, Spatial Autoregression, and Geographically Weighted Regression for Modeling Forest Structural Attributes Using a Geographical Information System (GIS)/Remote Sensing (RS) Approach.* Master Thesis. Retrieved from http://people.ucalgary.ca/~mcdermid/Docs/Theses/Shrestha_2006.pdf

[12] ArcGIS (2006). *ArcGIS Desktop Help 9.3 - An overview of commonly used tools.* Retrieved June 27, 2006, from http://webhelp.esri.com/arcgisdesktop/9.3/index.cfm?TopicName=An_overview_of_commonly_used_tools

[13] Brunsdon, C., Fotheringham, a S., & Charlton, M. E. (1996). *Geographically Weighted Regression: A Method for Exploring Spatial Nonstationarity.* Geographical Analysis, 28(4), 281–298. http://doi.org/10.1111/j.1538-4632.1996.tb00936.x

[14] Zhang, L., Shi, H., 2004. *Local modeling of tree growth by geographically weighted regression.* Forest Science, 50, 225–244.

[15] Van Lierop, P., Lindquist, E., Sathyapala, S., & Franceschini, G. (2015). *Global forest area disturbance from fire, insect pests, diseases and severe weather events.* Forest Ecology and Management, 352, 78–88. http://doi.org/10.1016/j.foreco.2015.06.010

Evergreen broadleaf forest transition zone changes in Japan from 1961 to 2008 detected by aerial ortho-photos

Etsuko Nakazono[1], Nobuyuki Tanaka[2], Masatsugu Yasuda[3], Hiromu Daimaru[4], Wataru Takeuchi[1]

[1] Institute of Industrial Science, the University of Tokyo, Japan
[2] Tokyo University of Agriculture
[3] Japan Forest Technology Association
[4] Forestry and Forest Products Research Institute, Japan

Abstract. In order to detect the distribution change of evergreen broad-leaved trees (EBTs) in a old-growth forest on the transitional zone of cool-temperate and warm-temperate zones, we used the ortho-photo data conversed from the aerial photos. Comparing the crown map of EBTs in the 1-ha verification plot with the ground truth data of individual tree inventory, 14 out of 17 (82%) upper layer trees were found to be visually read on the aerial photo We chose two indices for detecting the distribution change of EBTs, crown number and total crown area. We made crown maps of the 20-ha plot based on ortho-photos in 1961, 1975, 1985, 2003, 2005 and 2008, and calculated crown number and total crown area for each photos. The crown number increased at a rate 0.18/year/ha from 1961 to 2000's, and total crown area also increased at a rate 0.21% for the 20-ha plot. The total crow area increase was highly probable because errors of area in ortho-photos were smaller than secular changes of the area.

1. Introduction

Global warming has become one of the most important environmental problems. The average global surface temperature has increased by about 0.8 degree Celsius since 1880. The past climate seems to causes vegetation changes. In this study, we focused the distribution changes of vegetation from the past to the present. There are some papers about the detection of the changes of distribution of vegetation (e.g., [1], [2], [3], [4]), but there are no papers to detect the distribution change of evergreen broad-leaved trees (EBTs) in the old-growth forest on the transitional zone of cool-temperate and warm-temperate zones with high accuracy. In this study we examined the method to detect the distribution change of EBTs with high accuracy.

In order to detect the distribution change of EBTs, we selected the aerial photos. In Japan, aerial photos were taken from the 1930's. From 1950's, Geographical Survey Institute has taken aerial photos all over Japan. And in the 1970's, color aerial photos are taken in all of Japan. So in Japan, the aerial photos are oldest remote sensing data. And the resolution is high enough to interpret crowns of EBTs ([5], [6]).

But in order to use the aerial photos as GIS data, it is necessary to convert the aerial photos into digital ortho-photos. But orthographically errors of the area are left, and changes of distribution of vegetation are occasionally small to ignore the errors.

We defined two indices for detecting the distribution change of EBTs: crown number, which has no connection with the area, and total crown area, which needs the investigation of the error to detect the changes.

2. Study area and data

In this study, we set up two plots in an old-growth forest on the transitional zone of cool-temperate and warm-temperate zones on the southern slope of Mt. Tsukuba in Japan. This forest has been protected as shrine forest of Tsukubasan Shrine. Deciduous broad-leaved trees are distributed over the high altitude with EBTs over the low altitude. One was the 1-ha verification plot, in which individual tree inventory was done in 2007, and the other one was 20-ha plot (actually 20.4 ha) for crown reading on ortho-photos. We selected aerial photos and ortho-photos, where were taken in the defoliation periods between winter and spring (table 1).

Table 1. Ortho and aerial photo data

Date of photography	Color/BW	Focal distance (mm)	Flight height	Pixel size (cm)	Single /Stereo	Ortho/Aerial
2008/5/21	color	152.90	2,429	20.0	Single image ortho	ortho only
2008/5/7	color	152.90	2,429	20.0	Single image ortho	ortho only
2005/11/18-28	color	100.00		8.0	Single image ortho	ortho only
2003/11/17	color	151.61	2,416	20.0	Single image ortho	ortho only
1986/11/6	color	151.43	1,640	20.0	Single image ortho	ortho and aerial photo
1975/1/21	color	152.69	1,580	18.0	Stereo matching	ortho and aerial photo
1961/2/1	BW	210.36	5,300	50.0	Stereo matching	ortho and aerial photo

3. Method

3.1. Making crown maps

First, we made a crown map of EBTs in the 1-ha verification plot, using ortho-photo taken in 2005. We determined whether every tree crown was evergreen broad-leaved or not by visually reading, and made the polygon for each tree crown. We also used the 2003 and 2008 ortho-photos as assistant data for visually reading. We compared the crown map with the ground truth data of individual tree inventory in the 1-ha verification plot to confirm the accuracy of visual reading.

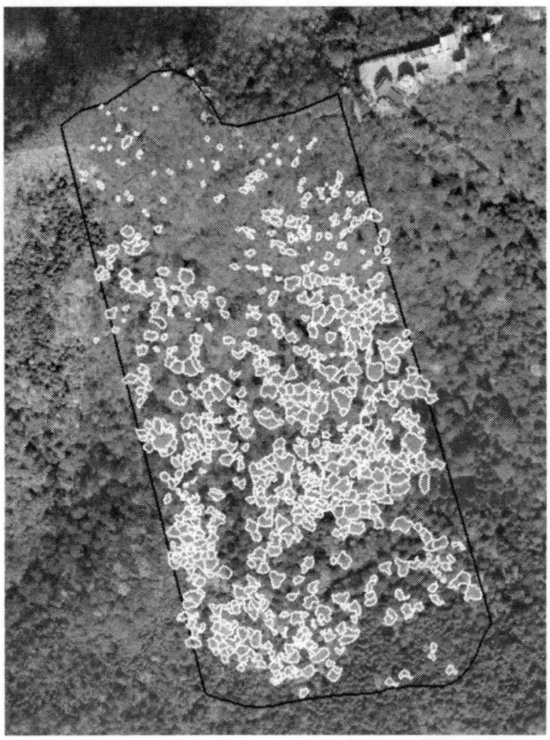

Fig. 1. The crown map of EBT in large plot (white line), using the 2005 ortho photo

Second, we made an EBT crown map of 20.4-ha plot based on ortho-photo taken in 2005 (Fig 1). We used 2003 and 2008 ortho-photos as assistant for visually reading. Next, we made the crown maps based on ortho-photo taken in 2003 and 2008 with reference of the crown map of 2005. We made crown maps based on ortho-photos taken in 1961, 1975 and 1986. In order to improve the accuracy of visual reading, we used the history data for each crown. We collected the presence-absence data for each crown in each photo, and used them to improve the accuracy.

3.2 Crown number and total crown area

We selected two indices, crown number and total crown area, in order to detect the distribution change of EBTs.

In order to use the aerial photos as GIS data, we converted the aerial photos into ortho-photos. There are errors of area because of orthographically errors, boundary lines errors and so on. It is too difficult to measure the crown area on the ground, and we can't know the true value of crown area. For this reason, we chose crown number as the index irrelevant to error of area. But it has another difficulty in separating EBT crowns, based on the shape and texture. Thus, to increase accuracy of this index, we identified each crown on ortho-photos of 2000's at first. Crown number of 2003, 2005, and 2008 was all same numbers. But it took a lot of time and we must repeat to adjust the boundary whenever ortho-photos were added. So for 3 data of 1900's, we calculated the crown number from crown history data (presence-absence data for each year of photography).

For determining crown number in ortho-photos in 1900's, we identified each crown in connection with that of ortho-photos in 2000's to make crown history data (presence-absence data for each year of photography).

The other index we chose was total crown area, sum of the crown polygons area, and it isn't necessary to identify the each crown. But in order to detect the distribution change by total crown area, we must consider errors of area. We assumed "errors of area" as follows:

1. From 2003 to 2008, the change of true total crown area is 0. Then "true total crown area" of 2000's is defined as mean of total crown area in 2003, 2005 and 2008.
2. The biggest value in three differences between true total crown area and each total crown area of 2000's is defined as the "error of area".
3. The "error of area" and secular changes between true total crown area and total crown area of 1900's is compared. If

"Error of area" < "distribution changes"

over the time, we detect the distribution change by total crown area.

4. Result and Discussions

4.1 Determination accuracy on visual reading of EBT crowns

Comparing the crown map of EBTs in the 1-ha verification plot with the ground truth data of individual tree inventory, 14 out of 17 (82%) upper layer trees were found to be visually read on the aerial photo (Table 2). None of the middle-layer and lower-layer trees were detected.

Table2. Accuracy on visual reading of evergreen broad-leaved crowns

	Upper	Middle	Lower
Ground truth	17	47	23
Visual reading	14	0	0

4.2 Change of crown number and total crown area.

Total crown number in the 20-ha plot increased almost lineally at a rate 0.18/year/ha from 1961 to 2000's (Fig.2 Top). That increased steadily in every elevation between 500m and 850m, especially in the high elevation, more than 750m (Fig. 2 Bottom).

4.3 Change of total crown area.

The differences (D) between true total crown area (M) and total crown area of each photos are shown in Table 3. The maximum difference of D/M in 2003, 2005 and 2008 was 4.2%, which was assumed to be the error of total crown area of otho-photos. D/M in 1961, 1975 and 1986 was between 21.08 and 40.22, which was much larger than the error of total crown area. This results suggested that total crown area increased from 1961 to 2000's steadily.

Table 3. Total crown area, "true total crown area", and "errors of area" in 20-ha plot

Year	Total crown area(m^2)	M-total crown area (D)	D/M (%)
2008	43959	646	1.45
2005	46480	-1876	4.20
2003	43376	1229	2.76
"True total crown area" (M)	44605		
1986	35201	9403	21.08
1975	30430	14175	31.78
1961	26663	17942	40.22

Total crown area increased linearly at a rate of 0.21% in the 20-ha plot, and total crown area of 2000's is 1.67 times larger than that of 1961 (Fig.2 top). Total crown area increased steadily in every elevation between 500m and 850m (Fig.2 bottom). The increase was the largest in middle elevation, between 600m and 650m.

These results that crown number and total crown area increased from 1961 to 2000's indicated that EBTs increased in the past 47 years.

5. Conclusion

We tested to detect the distribution change of EBTs in an old-growth forest on the transitional zone of cool-temperate and warm-temperate zones, using ortho-photo data, conversed from the aerial photos. Two indices, crown number and total crown area, showed the distribution change for 47 years. But both indices have strong and weak points. Crown number does not need to consider the errors of area, but needs to identify each EBT crown. On the contrary, total crown area does not need to identify each EBT crown, but needs to consider the errors of area.

On the southern slope of Mt. Tsukuba, the number and total area of EBT crowns increased from 1961 to 2008 steadily. These results need to ecologically estimate the relation to the fact that the average global surface temperature increased during the past 100 years.

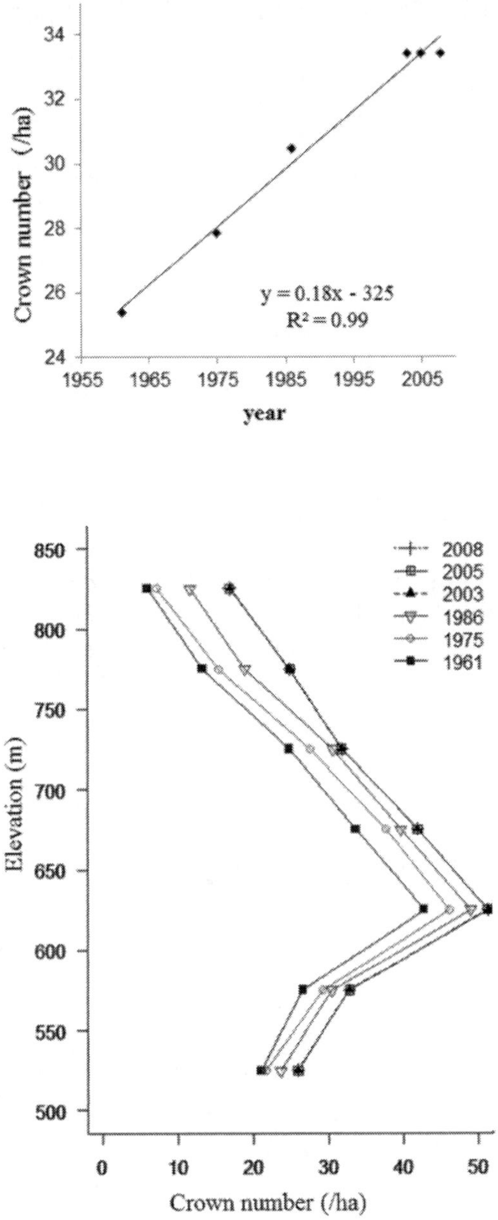

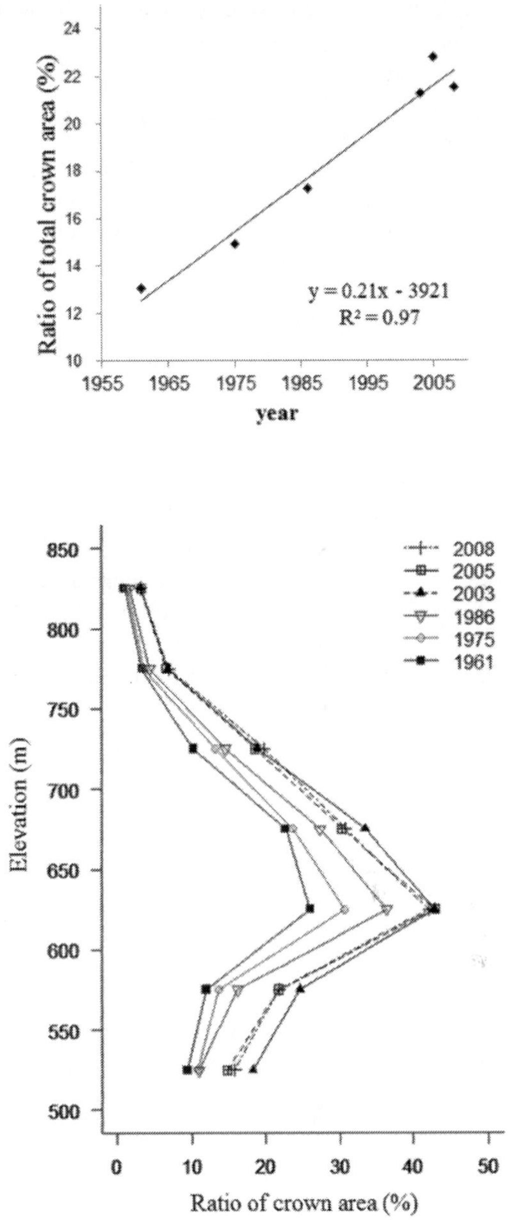

Fig 2: the change of crown number per ha in 20-ha plot.
Top: whole elevation,
Bottom: every 50m elevation

Fig 3: the change of ratio of total crown area in 20-ha plot
Top: whole elevation,
Bottom: every 50m elevation

Acknowledgement

This research was partly funded by Social Implementation Program on Climate Change Adaptation Technology
(SI-CAT) funded by MEXT.

Reference

[1] Beckage B., Osborne B., Gavin D.G., Pucko C. & Perkins T., 2008. A rapid upward shift of a forest ecotone during 40 years of warming in the Green Mountains of Vermont. *Proceedings of the National Academy of Sciences of the USA (PNAS)* 105, pp4197-4202.

[2] Peñuelas J., & Boada M., 2003. A global change-induced biome shift in the Montseny mountains (NE Spain). Global Change Biology 9, pp131-140.

[3] Shimazaki M., Sasaki T., Hikosaka K. & Nakashizuka T., 2011. Environmental dependence of population dynamics and height growth of a subalpine conifer across its vertical distribution: an approach using high-resolution aerial photographs. *Global Change Biology* 17(11), pp3431-3438.

[4] Yasuda M., Daimaru H. & Okitsu S., 2007. Detection of Alpine Moor Vegetation Change by Comparison of Orthonized Aerophotographs at Different Times. *Geographical Review of Japan* 80(13) pp842-856.

[5] Setojima M., Akamatu Y., Fukui Y., Imai Y., Asahiro K., Shigematu T., 2001. Change in the Color Tone of Representative Evergreen Broad-leaved Trees in Satoyama Coppice Forests based on Time Series Aerial Color Photographs collected in a Year. *Environmental information science. Extra, Papers on environmental information science* 15, pp31-36.

[6] Okuda M., Minowa Y., Takahara H., Ogura J., 2007. Range expansion of Castanopsis forests during the last 70 years in the Higashiyama hill area, Kyoto. Japanese journal of forest environment 49(1), pp19-26.

A review of uses of satellite imagery in monitoring mangrove forests

Rhyma Purnamasayangsukasih P[1], Norizah K[2], Ismail Adnan A M[2], and Shamsudin I[2]

[1] Faculty of Forestry, Universiti Putra Malaysia (UPM) Serdang, Selangor, Malaysia
[2] Department of Forest Management, Faculty of Forestry, Universiti Putra Malaysia (UPM) Serdang, Selangor, Malaysia

E-mail: norizah_k@upm.edu.my

Abstract: Satellite image could provide much information of earth surfaces in a large scale in a short time, thus saving time. With the evolution and development of sensors providing satellite image, resolution of object captured enhanced with advance image processing techniques. In forestry, satellite image has been widely used for resources management, planning, monitoring, predicting, etc. However, the uses of satellite image are reported to be moderate and sometimes poor for mangrove forests due to homogenous species existed in salty and inundation areas. Many researches had been carried out to improve the uses of satellite imagery of either optical or radar data for mangrove forests. This paper reviews the uses of satellite imagery data in mangrove with the main focus of the literature related to mangroves monitoring.

1. Introduction

In 1957, the world witnessed the launch of the first satellite into the orbit of the earth. Subsequently in the early 1960s, satellites associated with other purposes such as weather forecast, communication and commerce were introduced and successful launched. Since the 1970s, data acquired from these satellites have been successfully used for weather prediction, monitoring global environmental conditions, and geographical and geological applications [1]. Data captured from the satellite systems provide information on earth surfaces.

Satellite imageries are beneficial to numerous areas of work such as monitoring work as these imageries cover large areas, virtually any part of the earth, including inaccessible areas which were previously too remote or too dangerous to reach when using conventional aerial photography such as dense forest, peat swamp forest and mangrove forest. Through the usage of satellites, researchers are able to collect data unaffected by local air traffic constraints; they can analyse comparisons of land covers at different times, which is suitable for long-term studies. Research methods based on satellites are time-saving, cost-effective and enhance the possibility of classifying the vegetation through spectral and texture analyses; on the other hand, ground measurement methods are difficult, expensive, time-consuming and labour-intensive [2-6].

Previously, satellite images were expensive and the initial-stage satellite technology was used for military applications such as threat monitoring and assessment; this type of machine was known as reconnaissance satellite, for instance CORONA satellite [7]. The satellite was launched in 1960 and the operation ended in 1972 due to a leakage of confidential images on the internet; these pieces of sensitive information belonging to the United States of America (USA) were related to military security and defence. CORONA was developed primarily for area surveillance covering the regions of the Soviet Union, China and other parts of the world. The first civilian earth satellite, Landsat was launched in 1972. Landsat 1 to Landsat 5 were successfully launched in the absence of competitors until 1986 when the first commercial satellite, SPOT was introduced. Since the launching of commercial satellites, satellite images have gradually become less expensive and the areas of applications increase. Commercial satellites such as Ikonos and QuickBird provide global, accurate and high-resolution images to individuals, organisations and governments. Satellite imaging systems can be classified into radar and optical systems. Radars provide their own energy to illuminate an area of interest and measure the reflected signals while optical systems acquire the reflected electromagnetic waves of the sunlight and/or the infrared radiation emitted by objects on the ground [8]. Optical satellites are more frequently used than radar satellites. Rapid advancement of satellite technology causes wide usages of these data. In mangrove studies, the optical satellite with medium spatial resolution is normally used. Tables 1 and 2 show the common optical and radar satellites used in mangrove researches in the past 3 decades [9-11]. Table 3 shows their years of launchings and active applications until the current year.

Satellite images are available in different spatial resolutions, from low to high. Satellite images of low-resolution with ground pixels of more than 10 m are useful for applications such as environmental assessment, mapping, forestry management, disaster assessment and urban monitoring. In contrast, images of high-resolution data with ground pixel sizes of less than 5 m provide detailed information of small objects on the earth surface such as buildings, streets, river and trees which are useful for applications such as transportation network mapping, disaster management, urban planning and farming.

Satellite images also can be used to determine the elevation and topographical features of a particular land. Also, it is useful in viewing agricultural fields where farmers are able to monitor the health of their crops. In research, researchers or scientists can study environmental changes to forecast weathers or natural disasters that may happen in the future. Meanwhile, city planners are able to plan developments of new residential areas for communities. In transportation, satellite data can be used for traffic studies and facilitate the planning of new road networks. In disaster management, satellite images can be used to plan evacuation routes during earthquakes or fire events. Lastly, in Forestry, satellite images can provide detailed information about forest status such as species distribution, stand density, forest extend, operational monitoring, and discrimination species at dryland and wetland.

Table 1. Optical satellite [10].

Sensor / System	Platform	No. of band(s)	Spectral range	Spatial resolution	
				MSS	Pan.
High resolution sensor					
MSS (Multi Spectral sensor)	Landsat 1, 2, 3	4	B, G, R, NIR	~ 80 m	
TM (Thematic Mapper)	Landsat 5	6	B, G, R, NIR, SWIR	30 m	
ETM+ (Enhanced Thematic Mapper Plus)	Landsat 7	6	VNIR, SWIR	30 m	15 m
HVR (High Resolution Visibility)	SPOT (Satellite Pour l'Observation de la Terre) 1, 2, 3	3	G, R, NIR	20 m	10 m

HRVIR (High Resolution Visible and Infrared)	SPOT (Satellite Pour l'Observation de la Terre) 4	4	G, R, NIR, SWIR	20 m	10 m
HRG (High Resolution Geometrical)	SPOT (Satellite Pour l'Observation de la Terre) 5	4	G, R, NIR, SWIR	10 m (VNIR); 20 m (SWIR)	2.5 m
ASTER*	Terra	10	G, R, NIR; 6-SWIR	15 m (VNIR); 30 m (SWIR)	
IRS (Indian Remote Sensing Satellite) 1C, 1D		4	G, R, NIR, SWIR	23 m	5.8 m
ALI (Advance Land Imager)	EO-I (Earth Observing)	9	2-B, G, R, 2-NIR, 2-SWIR	30 m	15 m
Very high resolution sensor					
Quickbird		4	VNIR; Pan	2.4 m	0.6 m
IKONOS		4	VNIR; Pan	4 m	1 m
PRISM**	ALOS (Advanced Land Observation System)	1	Pan	N/A	2.5 m
WorldView-2		8	VNIR; Pan	<2 m ***	<0.5 m ***
GeoEye-1		4	VNIR; Pan	1.65 m	0.41 m
Other optical sensor					
GLAS (Geoscience Laser Altimeter System)	IceSAT (Ice, cloud and land elevation Satellite)	LiDAR, 2	Green (532 nm), NIR (1064 nm)	70 m footprint; 170 m spacing	
HYPERION	EO-I (Earth Observing)	Hyper-spectral: 220	400-2500 nm	30 m	

B = blue; **G** = green; **R** = red; **NIR** = near-infrared; **SWIR** = shortwave infrared; **V** = visible
*Advanced Spaceborne Thermal Emission and Reflectance Radiometer
**Panchromatic Remote-sensing Instrument for Stereo Mapping
***Maximum resolution limited by US government

Table 2. Radar satellite [10].

Sensor	Platform	Band(s)	Polarization(s)	Spatial resolution
SIR-C (Space-borne Imaging Radar)	Space Shuttle	C, L, X	HH, HV, VV	10-200 m
ERS-1 (European Remote-Sensing Satellite)	European Remote-Sensing Satellite	C	VV	25-100 m
JERS-1 (Japanese Earth Resource Satellite)	Japanese Earth Resource Satellite	L	HH	25-100 m
Radarsat-1		C	HH	8-100 m
Radarsat-2		C	HH, HV, VH, VV	3-100 m
ASAR (Advanced Synthetic Aperture Radar)	ENVISAT	C	HH, VH, HV, VV	25-150 m
PALSAR (Phased Array type L-band Synthetic Radar)	ALOS (Advanced Land Observation System)	L	HH, HV, VH, VV	10-100 m

Polarization indicated by transmit and receive polarizations, respectively (**H** = horizontal polarization; **V** = vertical polarization)

Table 3. Year of launch and active application of satellite until current year.

Satellite(s)	Year of Launch	Active Application
SPOT 1	1986	Inactive (2002)
SPOT 2	1990	Inactive (2009)
SPOT 3	1993	Inactive (1996)
SPOT 4	1998	Inactive (2013)
Landsat 1	1972	Inactive (1978)
Landsat 2	1975	Inactive (1982)
Landsat 3	1978	Inactive (1983)
Landsat 5	1984	Inactive (2011)
Landsat 7	1999	Active
IRS 1C	1995	Inactive (2004)
IRS 1D	1997	Inactive (2006)

Satellite(s)	Year of Launch	Active Application
ASTER	1999	Active
IKONOS	1999	Inactive (2015)
QuickBird	2001	Active
RADARSAT-1 SAR	1995	Inactive (2013)
ENVISAT ASAR	2002	Inactive (2012)
ERS-1 SAR	1991	Inactive (2000)
JERS-1	1992	Inactive (1998)
AIRSAR	1990	Inactive (2004)
ALOS PALSAR	2006	Inactive (2011)

2. Mangrove forest

Mangrove forests flourish in intertidal zones and coastlines between the sea and the land within tropical and subtropical regions of the world (Figure 1), approximately located 30° N and 30° S latitude [12-13]. They grow in harsh environmental conditions with high levels of salinity, high temperature, extreme tides, high rates of sedimentation and muddy anaerobic soils [13-15].

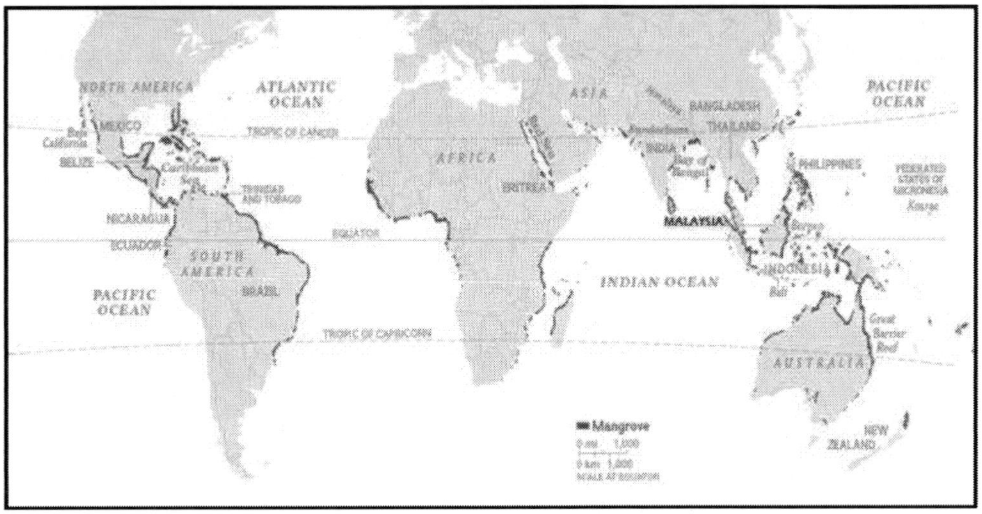

Figure 1. The world's mangrove distribution (Source: [13]).

Mangrove forests are among the most productive and biologically important ecosystems as they provide various goods and services to the society as well as benefits to coastal and marine systems; in fact, they are one of the most valuable ecosystems in the world [13][16-17]. [17] further comment that mangroves constitute less than 0.4% of the world's forests; they provide habitats for thousands of marine life, and serve the local communities with food, medicine, fuel and building materials. They are among the most carbon-rich forests in the tropics and important factors in mitigating the impact of climate change by sequestering carbon dioxide (CO_2) (the main greenhouse gas, apart from water vapour) from the atmosphere [18-19]. They also act as a security barrier in protecting the coastal areas from tidal waves, tsunamis and cyclones [15][20-22].

Even though they are important sources of goods and services, FAO (2007) reported that the total global area of mangroves decreases at about 1% per year. Mangrove forests experienced a significant worldwide loss estimated at about 35% to 36% in the period from 1980 to 2005; this rate of decline is faster than that of the tropical rain forest and coral reef [12][16][23-24]. Natural disasters such as increased sea level, cyclones, storm, lightning strikes, tsunami, and flood cause negative impacts on the growth of mangrove forests; on the other hand, human interferences such as overexploitation, conversion to aquaculture, urbanisation, developments related to agriculture, infrastructure and tourism bring about shrinking area of the mangrove forests [12][16][20][23][25-27]. [24] predict that the mangrove forest could be 100% vanished if the present rate of loss continues. As a consequence, the important goods and services provided by the ecosystem of mangrove forests will diminish [24].

3. Remotely sensed imagery data

Basically, there are two types of remotely sensed satellite images depending on the main source of energy: active and passive [28-29]. An active system controls its own electromagnetic radiation source, which is known as radar. A passive system uses natural energy from the sun as the source of electromagnetic radiation, which is known as optical system.

3.1 Radar satellite image

A radar imagery system or Radio Detection and Ranging is a method for the detection and ranging of earth surface features. Radar is an active remote sensing system which means that it provides its own source of energy to produce an image; it is commonly referred to as synthetic aperture radar (SAR). Radar carries its own electromagnetic radiation source, which is then directed to the surface and the energy reflected back from the surface is recorded [29]; this is expressed in Figure 2. Some examples of radar satellites are TerraSAR, RazakSat, ENVISAT and RADARSAT. The data are acquired either by day or by night. Due to the specific wavelength of radar, cloud covers can be penetrated without any adverse effect on the imagery. According to [30], radar has special properties that make it a viable alternative and/or partner to the optical remote sensing technique. For instance, microwave energy is capable of penetrating atmospheric conditions that render traditional space borne optical and multispectral systems useless [31]. Therefore, radar has the ability to produce images through rain, fog, hail, smoke, and, most importantly, clouds and it does not depend on daylight [28][32]. These characteristics hold enormous data collection potential in many countries around the world [32].

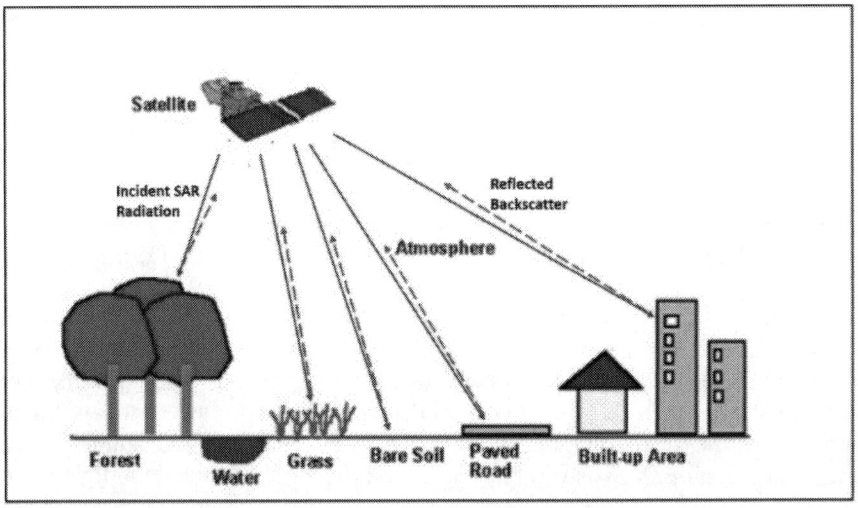

Figure 2. Principal of radar satellite (modified from [33]).

Advantages of radar satellite [8]:

- all weather capability (small sensitivity of clouds, light rain)
- day and night operation (independent of sun illumination)
- no effects of atmospheric constituents (multi-temporal analysis)
- sensitive to dielectric properties (water content , biomass, ice)
- sensitive to surface roughness (ocean wind speed)
- accurate measurements of distance (interferometry)
- sensitive to man-made objects
- sensitive to target structure (use of polarimetry)
- subsurface penetration

3.2 Optical satellite image

Optical imageries are acquired during the daylight since the satellite depends on the reflections of sunlight from objects on the Earth's surface in the absence of cloud cover [28][33]. Some examples of optical satellite systems include Landsat, SPOT, Ikonos, GeoEye and WorldView.

[33] describes an optical satellite as using visible, near infrared and short-wave infrared sensors to form images of the earth's surface by detecting the solar radiation reflected from targets on the ground (Figure 3). With different wavelengths, different materials reflect and absorb them differently. Thus, the targets can be differentiated by their spectral reflectance signatures in the remotely sensed images.

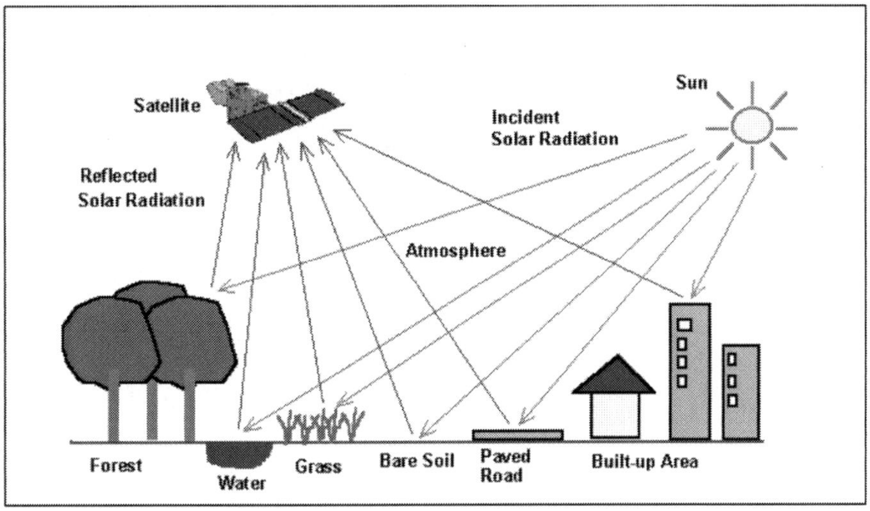

Figure 3. Principal of optical satellite [33].

Advantages of optical satellites [8]:

- specific wavelength sensitivity
- recent innovations with some satellite systems enable mission operators to more frequently update tasking plans to accommodate cloud-coverage forecasts
- manoeuvrable, with varying speed levels to move from one target to another
- several images can be collected during the same pass at the same latitude

4. The role and needs of satellite imagery data in mangroves monitoring

In the last 3 decades, a significant decline in mangrove forests has become a major environmental issue. The increased awareness of people, particularly environmentalists exerts a pressure on the need for conservation and restoration. As such, retrieving up-to-date information with regard to the extent of damage and condition of mangrove ecosystems is an essential aid to management as well as policy- and decision-making processes [9].

Remotely sensed satellite images have been used to characterise mangrove ecosystems for more than two decades; aerial photographs has gradually become obsolete because high resolution images are more advanced, and promise detailed and accurate data [9][34-35]. The benefits and limitations are listed in Tables 4, 5 and 6. With high resolution images, it is relatively easy to study the costal ecosystems. The embedded detailed information provides a synoptic view, multi-temporal, coverage and multi-spectral ability in the whole range of wavelengths, from visible to microwave. These advantages can effectively act as a tool par excellence in providing advance and reliable information on mangrove extent and status of its growth along the coastal areas [36], which would not be possible by using ground based techniques. The reflectance patterns of vegetation in visible and near infrared (NIR) spectral region give information on the conditions of vegetation covers [36].

In mangrove monitoring, satellite imagery data are able to deliver information such as land cover changes, disaster observation, species distribution, zonation pattern, etc. [37] state, a particular advantage of using optical data that mangroves are relatively distinct from non-mangrove areas, although confusion with adjoining tropical forests often leads to errors in the mapping of mangrove extent. [9] state that the interpretation of radar data over mangrove ecosystems is very complex because in radar data, the backscatter signal of mangrove ecosystems is influenced by the geometric properties of the stand such as canopy closure, canopy geometry, leaf structure, cell structure, stem structure, and the underlying surface component and its roughness: soil mudflats water; it is also affected by dielectric properties, which vary, depending on the soil moisture, plant moisture, and underlying water surfaces. The responses of these different conditions fluctuate, depending on the incidence wavelength, the polarisation of the radar beams, and the incidence angle of the radar waves.

Various image processing and analysis techniques have been developed to facilitate remotely sensed imagery interpretation and to extract as much information as possible from the images. Based on a review study by [9], several methods have been employed to classify remotely sensed data on mangroves which are visual interpretation, pixel based classification (supervised and unsupervised), object-based methods, vegetation indices, and leaf area index (LAI). According to [38], visual interpretation highly depends on an interpreter's ability to recognise and analyse several characters of a satellite image, for example shape, size and pattern. Supervised and unsupervised classifications are common pixel based techniques of satellite image processing. In these techniques, spectral properties of every pixel within the area of interest are analysed without taking into account the spatial or contextual information related to the pixel of interest while object-based method is based on information of a group of similar pixels [39]. A vegetation index is a number generated by some combinations of remote sensing bands and may have some relationships to the amount of vegetation in a given image pixel; and normalised difference vegetation index is most widely used to separate vegetation from non-vegetation [9].

In Malaysia, [17] used Landsat images to monitor and analyse the changes over a period of 25 years of mangrove areas in Iskandar Malaysia. Two techniques were applied: the maximum likelihood classification (MLC) was used to compare and support vector machine (SVM). As a result, 6 land use/land cover (LULC) have been classified, which are forest, oil palm, rubber, mangrove, urban, and water bodies for both techniques. The comparisons of accuracies are between 62% (for SVM) and 95% (for MLC) from the years 1989, 2000, 2005, 2007, 2009, 2013 and 2014. Thus, MLC achieved the highest overall classification accuracy compared to with SVM for all the seven years considered.

In recent years, unmanned aerial vehicles (UAVs) grow in popularity as a monitoring technique. UAV is an unmanned aerial vehicle or aerial vehicle without a pilot [40]; it is also known as drone. UAVs can fly using a manual remote controller, semi-automatic and automatic or combination of these methods. As an alternative way to acquire images from the space, UAVs can carry many types of sensors such as LiDAR, multispectral camera, or hyper spectral camera to capture images. As for viewing, sensors such as Synthetic Aperture Radar (SAR) can be applied to view through cloud and Daylight (RGB) Video Sensor to see what is going on at the ground. However, it all depends on the capability of UAV payload. The application of UAVs has widely spread and they are used in vegetation monitoring. For example, in a study conducted by [41], a high altitude long endurance (HALE) UAV, Pathfinder Plus was used to demonstrate this technique in monitoring a coffee plantation in Hawaii. The UAVs also have been applied in wetland observation, such as platform option of free-flying satellite, UAVs and International Space Station (ISS) for remote sensing assessment of the littoral zone; this work was carried out by [42]. He found that UAVs have much greater efficiency since they can be deployed as needed directly to the object areas and avoid poor weather conditions. Next, in a study on hazard detection and analysis at natural wetland of Taijiang National Parks, [43] utilised UAVs to collect ground truth data.

Table 4. The benefit and limitation of medium resolution of optical imagery data [9]

Medium-resolution imagery	Benefits	Limitation
Spectral resolution	Several multispectral bands, always including R, G, B; near-infrared; and oftentimes even mid-infrared; and thermal bands	Skilled trained personnel are required to best exploit the information content of the multiple bands (considering transformation, etc.)
Spatial resolution	Ideal for mapping on a large regional scale	Too coarse for local observation requiring in-depth species differentiation and parameterization
Temporal resolution	Frequent mapping (e. g., rainy season and dry season within 1 year; or repeated annual mapping) is possible	Repetition rate may be too low to record impact of extreme events (e. g., cyclones, floods, tsunami); furthermore, very weather dependent (clouds) = critical in subtropical and tropical regions
Costs	Depending on sensor, freely available (e. g., Landsat), very cost efficient (ASTER), or expensive (e. g., SPOT); but all are cost efficient compared with field surveys and airborne campaigns	Software for image processing need (common software, such as erdas, ENVI, and ArcGIS, have high licene fees), but usually not real limitation)
Long-term monitoring	Data availability over three decades	depending on the future duration of the systems and subsequent comparable sensors
Purposes	Inventory and status maps; change detection, such as assessment of impact damages; assessment of forestation and conservation success	For some species-oriented botany-focused studies, resolution may already be too coarse
Discrimination level	Mangrove-non-mangrove, density variations, condition status, mangrove zonation, in rare cases also species discrimination	High regional differences; classification result depends highly on the ecosystem conditions, such as biodiversity, heterogeneity of forest, adjacent targets; species identification is rarely possible
Methods	visual interpretation with on screen digitizing, pixel-based, object-based, and hybrid classification approaches; image transformation and analyses (PCA, TCT,HIS indices, etc)	To exploit the full potential of the data skilled analysts needed

Other	Data easy to access or order; best explored data type and , thus, most literature available; long-term monitoring granted	

Table 5. The benefit and limitation of high resolution of optical imagery data [9]

High-resolution imagery	Benefits	Limitation
Spectral resolution	Red-near-infrared spectral information with red-edge slope; usually panchromatic band allowing image fusion (pan-sharpening)	Relatively few spectral bands
Spatial resolution	High resolution (0.5-4 m range) for mapping on a local scale	Only small area is covered
Temporal resolution	Regular mapping is possible on demand	Weather dependent (clouds); cost intensive if repeated monitoring is required
Costs	Moderate costs for single acquisitions (depending on are)	Very high costs if repeated monitoring is requested. Also, high costs of object-oriented image processing software
Long-term monitoring	Theoretically possible, but usually not used because of expense. Sensors, such as Ikonos, QuickBird, etc, available since late 1990s/2000.	Depending on the future duration of the systems and subsequent comparable sensors. Only back to the late 1990s.
Purposes	Discrimination of mangrove species, spatial distribution and variability, health status, parameterization.	Single tree species discrimination usually not possible.
Discrimination level	Down to species communities; detailed parameterization	Regional differences; classification result depends highly on the ecosystem conditions, such as biodiversity, heterogeneity of forests, adjacent targets
Methods	Visual interpretation with on screen digitizing, pixel based, object-based, and hybrid-classification approaches.	Skilled analysis needed to exploit the full potential of the data.
Other	Valuable information source to support field survey and accuracy assessment. Easy to close the scale gap to *in situ* investigations	In some countries, data of the relevant sensors very difficult to purchase, few studies published based on the data type.

Table 6. The benefit and limitation of radar imagery data [9]

Radar Imagery	Benefits	Limitation
Spectral resolution	Active microwave radiation; delivers alternative information about surface structure; various wavelengths and polarizations are selectable	No spectral information
Spatial resolution	Varies	Usually low, except TerraSAR-X
Temporal resolution	High; weather independent	None
Cost	Many data types available at low cost in the context of science proposal	Restricted access to data (certain number of scene; also some data not sharable with certain developing countries)
Long-term monitoring	Good; long duration systems	None
Purposes	Mangrove extent, condition, canopy properties, deforestation, biomass estimation	No information derivable from spectra
Discrimination level	Age structure, forest parameters, biomass estimation	Vegetation forms without a prior knowledge; no separation among species
Methods	Analyses of the backscatter signal using advanced image-processing technique; very quantitative physics-based manner of image analysis	Extremely skilled analysis with experience in radar image processing needed
Other	Most promising results when SAR data combined with optical imagery	Relatively few studies have been conducted; special software or modules are needed for radar image processing.

5. Conclusion and prospective

In short, usages of satellite images for mangrove forests management are crucial in obtaining information. Satellite imagery data offer considerable advantages in mangrove studies and are useful to monitor mangrove ecosystems. Options to use either optical or radar satellite images depend on a study's needs. The output of a satellite imagery analysis can provide accurate information to facilitate conservation planning and policy-making as the technology is currently in an advanced stage. In the future, the quality of images processing can be improved further if radar and optical data are integrated. These images can be utilised to explore and monitor the development of mangrove forests in a more effective manner, as [44] aptly state that radar data complement optical data.

6. References

[1] Kuthadi, S. 2005. Detection of Objects from High Resolution Satellite Images. *Phd Dissertation,* University of Minnesota: USA. **67** pp.

[2] Mumby, P.J., green, E.P., Edwards, A.J. and Clark, C.D. 1999. The cost-effectiveness of remote sensing for tropical coastal resources assessment and management. *Journal of Environment Management,* **55**: 157-166.

[3] Dahdouh-Guebas, F. 2002. The use of remote sensing and GIS in the sustainable management of tropical coastal ecosystems. Environment, Development and *Sustainability,* **4**:93-112.

[4] Holland, D. and Marshall, P. 2003. Using high-resolution satellite imagery in a well mapped country. Proceeding of ISPRS-EARSel Joint Work on "High Resolution from Space", Hannover, Germany, October 2003.

[5] Held, A., Ticehurst, C., Lymburner, L. and Williams, N. 2003. High resolution mapping of tropical mangrove ecosystems using hyperspectral and radar remote sensing. *International Journal of Remote Sensing,* **24(13)**:2739-2759.

[6] Lee T.M. and Yeh H.C. 2009. Applying remote sensing technique to monitor shifting wetland vegetation: a case study of Danshui river estuary mangrove communities, Taiwan. *Ecological Engineering,* **35**:487-496.

[7] Ekblad, U. Earth Satellites and Detection of Air and Ground-based Activities. *Phd Dissertation.* Royal Institute of Technology: Stockholm, Sweden. **193** pp.

[8] Hennig, S. 2013. Exploring the benefits active and passive spaceborne system. from http://eijournal.com/print/articles/exploring-the-benefits-of-active-vs-passive-spaceborne-systems. Retrieved 25 January 2016

[9] Kuenzer, C., Bluemel, A., Gebhardt, S., Quoc, T.V. and Dech, S. 2011. Remote sensing of mangrove ecosystems: a review. *Remote Sensing,* **3**:878-928.

[10] Heumann, B.W. 2011. Satellite remote sensing of mangrove forests: recent, advance and future opportunities. *Progress in Physical Geograpgy,* **35(1)**:87-108.

[11] Zhang, Y. 2010. Ten years of technology advancement in remote sensing and research in the CRC-AGIP LAB in GGE. *Geomatica,* **64(2)**:173-189.

[12] FAO. 2007. *The World's Mangroves 1980–2005.* FAO Forestry Paper 153. FAO: Rome, Italy. **77** pp.

[13] Giri, C., Ochieng, E., Tieszen, L.L., Zhu, Z., Singh, A., Loveland, T., Masek, J. and Duke, N. 2011. Status and distribution of mangrove forests of the world using earth observation satellite data. *Global Ecology and Biogeography,* **20**:154–159.

[14] Spalding, M., Kainuma, M. and Collin, L. 2010. *World Atlas of Mangroves*. Earthscan, UK, **319** pp.

[15] Kathiresan, K. and Bingham, B.L. 2001. Biology of mangroves and mangroves ecosystem. *Advance in Marine Biology,* **40**:81-251.

[16] Valiela, I., Bowen, J.L. and York, J.K. 2001. Mangrove forests: One of the world's threatened major tropical environments. *Bioscience,* **51**:807–815.

[17] Kanniah, K.C., Sheikhi, A., Cracknell, A.P., Goh, H.C., Tan, K.P., Ho, C.S. and Fateen N.S. 2015. Satellite images for monitoring mangrove cover changes in a fast growing economic region in southern Peninsular Malaysia. *Remote Sensing,* **7**:14360-14385.

[18] Donato, D.C., Kauffman, J.B., Murdiyarso, D., Kurnianto, S., Stidham, M. and Kanninen, M. 2011. Mangroves among the most carbon-rich forests in the tropics. *Nature Geoscience,* **4**:293–297.

[19] Hamdan, O., Aziz, H.K. and Mohd Hasmadi, I. 2014. L-band ALOS PALSAR for biomass estimation of Matang mangroves, Malaysia. *Remote Sensing and Environment,* **155**: 69–78.

[20] Giri, C., Zhu, Z., Tieszen, L.L., Singh, A., Gillette, S. and Kelmelis, J.A. 2008. Mangrove forest distributions and dynamics (1975–2005) of the tsunami-affected region of Asia. *Journal of Biogeography*, **35**:519–528.

[21] Kamaruzaman, J. and Dahlan, T. 2008. Managing sustainable mangrove forests in Peninsular Malaysia. *Journal of Sustainable Development*, **1(1)**:88-96.
[22] United Nations Environment Programme (UNEP). 2014. *The Importance of Mangroves to People: A Call to Action*. Van Bochove, J., Sullivan, E. & Nakamura, T. (Eds). United Nations Environment Programme World Conservation Monitoring Centre, Cambridge. **128** pp.
[23] Alongi, D.M., 2002. Present state and future of the world's mangrove forests. *Environmental Conservation*, **29 (3)**:331–349.
[24] Duke, N.C., Meynecke, J.O., Dittmann, S., Ellison, A.M., Anger, K., Berger, U., Cannicci, S., Diele, K., Ewel, K.C., Field, C.D., Koedam, N., Lee, S.Y., Marchand, C., Nordhaus, I. and Dahdouh-Guebas, F. 2007. A world without mangroves? *Science*, **317(5834)**: 41–42.
[25] Simard, M., Zhang, K., Rivera-Monroy, V.H., Ross, M.S., Ruiz, P.L., Castaneda-Moya, E., Twilley, R.R. and Rodriguez, E. 2006. Mapping height and biomass of mangrove forests in Everglades national park with srtm elevation data. *Photogrammetric Engineering & Remote Sensing,* **72(3)**:299-311.
[26] Gilman, E., Ellison, J., Duke, N.C. and Field, C. 2008. Threats to mangroves from climate change and adaptation options: a review. *Aquatic Botany*, **89(2)**:237–250.
[27] Polidoro, B.A., Carpenter, K.E., Collins, L., Duke, N.C., Ellison, A.M., Ellison, J.C., Farnsworth, E.J., Fernando, E.S., Kathiresan, K. and Koedam, N.E. 2010. The loss of species: Mangrove extinction risk and geographic areas of global concern. *PLoS ONE,* **5(4)**:1-10.
[28] Musa, Z.N., Popescu, I. and Mynett, A. 2015. A review of applications of SAR, optical, altimetry and DEM data for surface water modelling, mapping and parameter data. *Hydrology and Earth System Science,* **19**:3755-3769.
[29] Al-Wassai, F.A. and Kalyankar, N.V. 2013. Major limitations of satellite images. *Journal of Global research in Computer Science,* **4(5)**:51-59.
[30] Foody, G.M., 1988. Crop classification from airborne synthetic aperture radar data. *International Journal of Remote Sensing,* **9(4)**:655-668.
[31] Elachi, C. 1988. *Spaceborne Radar Remote Sensing: Applications and Techniques.*New York: IEEE Press, 254 pp.
[32] Hack, B.N., Herold, N.D. and Bechdo, M.A. 2000. Radar and optical data integration for land-use/land-cover Mapping. *Photogrammetric Engineering and Remote Sensing*, **66(6)**:709-716.
[33] Liew, C.S. 2001. Principles of Remote Sensing. http://www.crisp.nus.edu.sg/~research/tutorial/rsmain.htm. Retrieved 1 January 2016.
[34] Vaiphasa, C., Skidmore, A. K., & de Boer, W. F. (2006). A post-classifier for mangrove mapping using ecological data. *ISPRS Journal of Photogrammetry & Remote Sensing,* **61**:1-10.
[35] Green, E. P., Clark, C. D., Mumby, P. J., Edwards, A. J., & Ellis, A. C. (1998). *Remote Sensing Techniques for Mangrove Mapping*. International Journal of Remote Sensing, **19 (5)**:935- 956.
[36] Ajay, P., Vijay, S., Mehmood, K., Jaydipsingh, K., Kalubarme, M.H., Pandya, C.H., Nischal, J. and Lomesh B. 2014. Mapping and monitoring mangrove in the costal districts of Gujerat state using remote sensing and geo-informatics. *Asian Journal of Geoinformatics,* **14(1)**:15-26.
[37] Lucas, R.M., Bunting, P., Clewley, D., Proisy, C., Viergever, K., Woodhouse, I., Carreiras, J., Rosenqvist, A., Accad, A. and Armston, J. 2009. Characterisation and Monitoring of Mangroves Using ALOS PALSAR Data. JAXA Earth Observation Research Center (EORC).
[38] Riebeek, H. 2013. *How to Interpret a Satellite Image: Five Tips and Strategies: Feature Articles. Earthobservatory.nasa.gov.* http://earthobservatory.nasa.gov/Features/ColorImage/. Retrieved 31 December 2015.
[39] Weih, R.C. and Riggan, N.D. 2010. Object-based classification vs. pixel-based classification: comparitive importance of multi-resolution imagery. *The International Archives of the Photogrammetry, Remote Sensing and Spatial Information Sciences,* **Vol. XXXVIII-4/C7**. Ghent, Belgium.
[40] Eisenbeiss, H. 2009. UAV Photogrammetry. *Phd Dissertation.* University of Technology Dresden: Germany. **237** pp.

[41] Herwitz, S.R., Johnson, L.F., Dunagan, S.E., Higgins, R.G., Sullivan, D.V., Zheng, J., Lobitz, B.M., Leung, J.G., Gallmeyer, B.A., Aoyagi, M., Slye, R.E. and Brass, J.A. 2004. Imaging from an unmanned aerial vehicle: agricultural surveillance and decision support. *Computers and Electronics in Agriculture*, **44**:49-61.

[42] Peterson, D.L., Brass, J.A., Smith, W.H., Langford, G., Wegener, S., Dunagan, S., Hammer, P. and Snook, K. 2003. Platform options of free-flying satellites, UAVs or the international space station for remote sensing assessment of the littoral zone. International Journal of Remote Sensing, **24(12)**: 2785–2804.

[43] Liu, C.C., Chen, Y.Y. and Chen, C.W. 2013. Application of multi-scale remote sensing imagery to detection and hazard analysis. Natural Hazard, **65**:2241-2251.

[44] Aschbacher, J., Tiangco, P., Giri, C.P., Ofren, R.S., Paudyal, D.R. and Ang,Y.K., 1995. Comparison of different sensors and analysis techniques for tropical mangrove forest mapping. In: editors (Eds.), Proc. Geoscience and Remote Sensing Symposium, IGARSS 95, Quantitative Remote Sensing for Science and Applications, Firenze, 10-14 July 1995, 2109-2111 pp.

Characterization and spatial distribution of mangrove forest types based on ALOS-PALSAR mosaic 25m-resolution in Southeast Asia

S Darmawan [1,2,3], W Takeuchi [1], E Nakazono [1], E Parwati [4], V T Dien [5], K S Oo [6], K Wikantika [3] and D K Sari [2]

[1] Institute of Industrial Science, University of Tokyo, Tokyo, Japan.
[2] Institut Teknologi Nasional (ITENAS), Bandung, Indonesia.
[3] Center for Remote Sensing, Institut Teknologi Bandung (ITB), Bandung, Indonesia.
[4] Indonesian National Institute of Aeronautics and Space (LAPAN), Jakarta, Indonesia.
[5] Forest Inventory Planning Institute (FIPI), Hanoi, Vietnam.
[6] Geospatial and Resource Center (MGRC), Yangon, Myanmar.

Email: soni@iis.u-tokyo.ac.jp and soni_darmawan@yahoo.com

Abstract. The objective of this research is to investigate characteristics of mangrove forest types and to identify spatial distribution of mangrove forest based on ALOS PALSAR mosaic 25m-resolution in Southeast Asia. Methodology consists of collecting of ALOS PALSAR image for overall Southeast Asia region, preprocessing include converting DN to NRCS and filtering, collecting regions of interest of mangrove forest in Southeast Asia, plotting, characterization and classification. Result on this research we found characteristics of mangrove forest on HH values around -10.88 dB to -6.65 dB and on HV value around -16.49 dB to -13.26 dB. On polarization of HH which the highest backscattering value is mangrove forest in Preak Piphot River Cambodia, Thái Thủy Thai Bình Vietnam, and Vạn Ninh tp. Móng Cái Quảng Ninh Vietnam whereas the lowest backscattering value is mangrove forest in Thailand area. On polarization of HV which the highest backscattering value is mangrove forest in Preak Piphot River Cambodia, Sorong and Teluk Bintuni Indonesia whereas the lowest backscattering value is mangrove forest in Subang Indonesia, Giao Thiện Giao Thuỷ Nam Định, Vietnam and Puyu Mueng Satun Thailand. Based on characterization, we create a rule criteria for classification of mangrove areas and non mangrove area. Finally we found spatial distribution of mangrove forest based on ALOS PALSAR 25m-resolution in Southeast Asia.

1. Introduction

1.1 Background

Mangrove forests can be found along ocean coastlines throughout the tropics area, and provide important products and services [1]. They also provide among the most intense coastal carbon sinks in the world and play a growing and central role in the global carbon cycles [2]. According to Daniel et al. [3] mangroves have five times larger numbers of total carbon storage per unit area basis on average than those typically observed in temperate, boreal and tropical terrestrial forests. High rates of trees and plant growth, coupled with anaerobic, water logged soil that slow decomposition, result in large long term carbon storage [4]. This suggests that mangroves play an important role in global climate change management. In order to gain and build a solid understanding of the global carbon budget and ultimately the effects of diminishing mangrove forests on climate change, it is crucial to obtain an assessment and quantification of the spatial distribution of mangrove forest. As most countries do not have sufficient information to include mangroves in their national reporting to the United Nations [4].

From a global perspective, Southeast Asia is well endowed as it supports the world's largest area of mangroves, originally extending over 6.8 million hectares and representing 34-42 percent of the world's total [1]. Southeast Asia's mangroves are the best developed and probably the most species-diverse in the world [5]. According to Giri et al. [6] the current estimate of mangrove forests of the world is less than half of what it once was and reductions suggest that 30–40% of coastal wetlands and 100% of mangrove forests could be lost in the next 100 years if the present rate of loss continues. Suggested the growing importance of mapping and monitoring of mangrove forest. However, conducting field survey for mangrove forest and its productivity in regional area is proven to be very difficult due to muddy soil condition, heavy weight of the wood [7], the vast area to cover, and tidal influences.

Remote sensing has been widely proven to be essential in monitoring and mapping of land use land cover including mangrove forest. According to Lucas et al. [8] the launch of the Japanese Space Exploration Agency's (JAXA) Advanced Land Observing Satellite (ALOS) Phased Arrayed L-band SAR (PALSAR) represented a milestone in the global observation, characterization, mapping and monitoring or mangroves largely, because these provide more information on the three dimensional structure and biomass of woody vegetation and the presence and extent of (primarily tidal) inundation. As data can be day or night regardless of weather conditions, mangroves can be observed more frequently, even in regions with prevalent cloud cover. The backscattering coefficient depends upon the interactions of microwaves of varying configurations with component of the mangrove (e.g. leaves, branches and trunks) of varying size, dimension, density, orientation and dielectrics constants (moisture contents) [8]. The longer L-band microwaves have a greater likelihood of penetrating the foliage and small branches of the upper canopies of the forest and interacting with woody trunk and larger branch components as well as the underlying surface [9,10]. The objective of this research is to investigate characteristics of mangrove forest types and to identify spatial distribution of mangrove forest based on ALOS PALSAR mosaic 25m-resolution in Southeast Asia. Methodology consists of collecting of ALOS PALSAR image for overall Southeast Asia region, preprocessing include converting DN to NRCS and filtering, collecting regions of interest of mangrove forest in Southeast Asia, plotting, characterization, classification and spatial distribution of mangrove forest.

1.2 Site description

Southeast Asia or Southeastern Asia is a subregion of Asia, consisting of the countries such as Myanmar, Thailand, Cambodia, Laos, Vietnam, Malaysia, Singapore, Brunei, Philippines, East Timor and Indonesia. According to Giesen et al. [1] Mangroves occur throughout Southeast Asia, from the Irrawaddy delta in Myanmar in the Northwest, through the more than 17 000 scattered islands of the Indonesian and Philippines archipelagos to Papua Island in the East, spanning a distance of more than 6000 kilometres from east to West and 3500 kilometres from North to South. We choice 46 samples, their location is representative of all mangrove forest in Southeast Asia. Distribution of sample site on study areas can be seen on figure 1.

2. Methodology

For this study, methodology consists of collecting of ALOS-PALSAR data and some references data such as land cover maps, Landsat data and information from Google earth, preprocessing focus on converting Digital Number to Normalize Radar Cross Sections (NRCS) and filtering. Delineation of samples of mangrove forest based on some references, characterization for knowing mean and standard deviation, profiling and statistics, and classification. The Methodology will follow on figure 2.

2.1 Data collection

ALOS PALSAR L-band with 1.27 GHz center frequency and 14 or 28 MHz bandwidths and dual polarization HH and HV developed by Japan Aerospace Exploration Agency (JAXA). The ALOS PALSAR was developed by JAXA and launched in January 2006. ALOS PALSAR data was collected

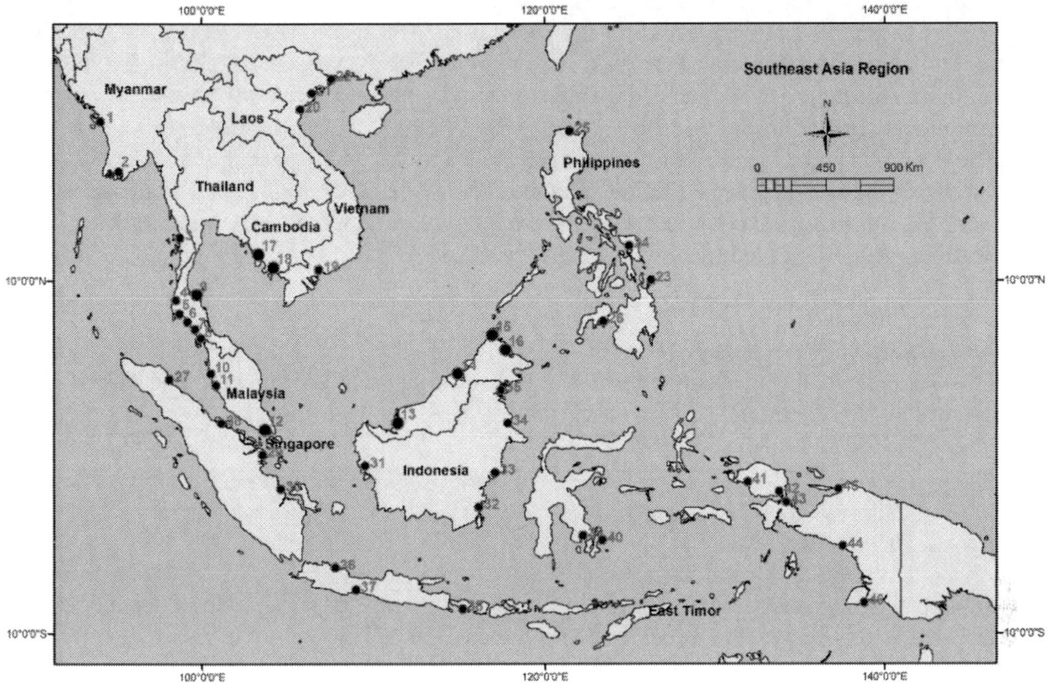

Figure 1. Study area and distribution of 46 samples.

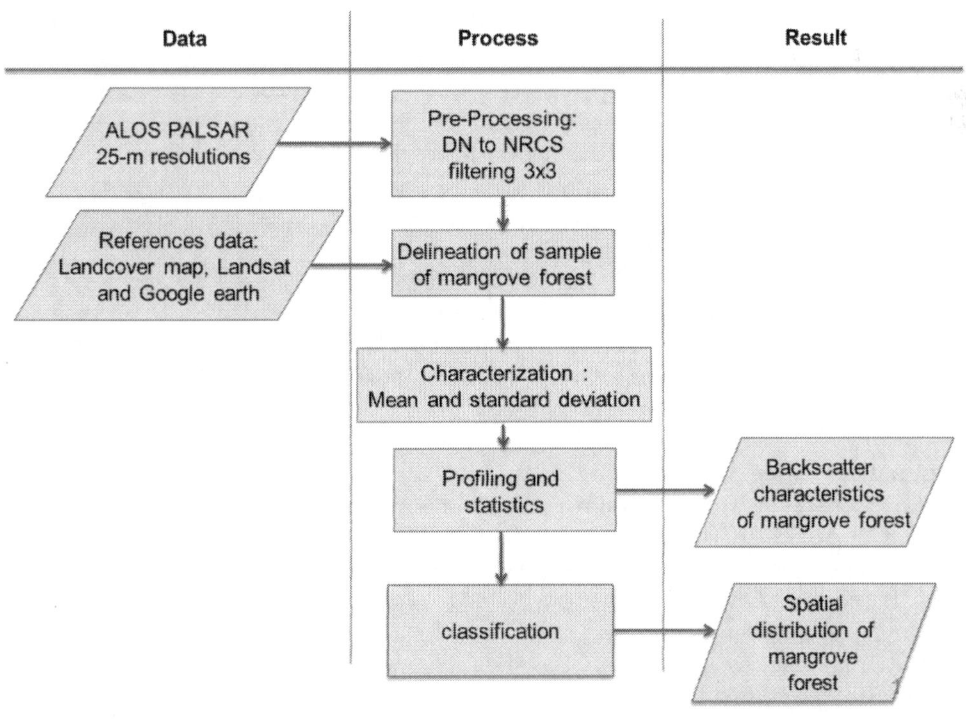

Figure 2. Methodology.

as part of Kyoto and Carbon Initiative project. We collected ALOS PALSAR from http://www.eorc.jaxa.jp/ALOS/en/palsar_fnf/fnf_index.htm. We used ALOS PALSAR mosaics on 2010 that total 60 tiles for covering Southeast Asia regions. All data were acquired in fine beam dual mode at a viewing angle of 34.4 and delivered in single-look complex (SLC) as the normalized backscattering coefficient in slant-range geometry by JAXA. PALSAR product is geometry and radiometric corrected and normalized for topography at 25 meter pixel spacing. PALSAR radiometric and geometric calibration is given in Shimada et al. [11]. We collected all scenes to cover Southeast Asia region. For secondary data we used land cover maps from each country and Landsat data also Google earth for supporting identification of mangrove area.

2.2 Pre-processing

All images were projected into geographical latitude and longitude coordinates. The digital number (DN) value of HH (DN_{HH}) and HV (DN_{HV}) backscatter intensities were converted to a normalized radar cross section (NRCS) in decibel (dB), (i.e., $\sigma°_{HH}$ and $\sigma°_{HV}$) by the following equations [12] :

$$\sigma°_{HH} = 10 \times \log_{10}(DN^2_{HH}) - CF \quad (1)$$

$$\sigma°_{HV} = 10 \times \log_{10}(DN^2_{HV}) - CF \quad (2)$$

Where, σ° is backscattering coefficient and CF is the calibration factor. The CF is dependent on the processing date, in this study CF is equal to -83 both for HH and HV.

All images were mosaicked into one continuing images to cover overall Southeast Asia region using the image analysis software. For visualization we used image composites with R=HH, G=HV and B=HH/HV. We also use Enhanced Lee filtering with windows size 3x3 to reduce speckle noise.

2.3 Characterization and profiling

We have collected samples of mangrove forest as Region of Interest (ROI) on ALOS PALSAR data based on references data. We collected 46 ROI of mangrove forest dispersion on Southeast Asia region. The size of ROI different for each mangrove forest. This condition because large of mangrove forest areas is different on each region. After that calculating of mean and standard deviation for each sample. For visualization we create profiling of mean and standard deviation for each sample.

2.4 Classification of mangrove forest area

Classification of mangrove forest based on backscattering characteristics of mangrove forest on each location and also based on topography data. We used rule based classification method based on characteristics of mangrove for classification of mangrove and non mangrove in study area. Parameter that was used on rule based classification is HH, HV and topography data.

3. Results and discussion

3.1 Backscatter characteristics of mangrove forest in Southeast Asia

We processed all ALOS-PALSAR image include mosaicking, converting DN to NRCS, filtering and then create image composites with R=HH, G=HV and B=HH/HV. The image cover overall Southeast Asia region. Based on ALOS-PALSAR composite and also land cover maps from each country, Landsat data and Google earth, we can identify mangrove forest. In the figure 3 we show dispersion of mangrove forest from ROI in Southeast Asia region. Based on references data we can identify mangrove forest and non-mangrove forest on study area.

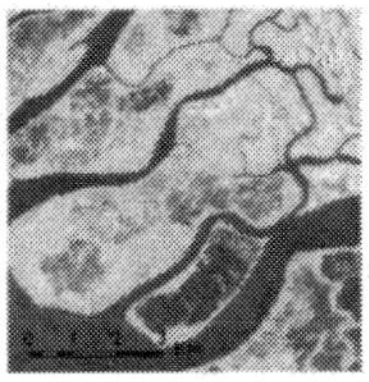
1. Lay Taung, Myanmar

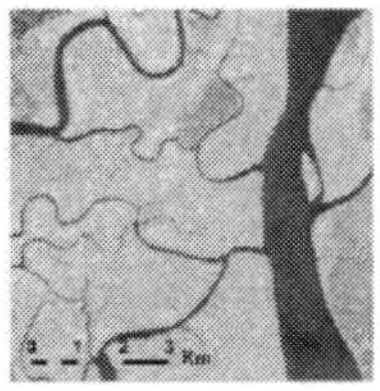
2. Myaungmya, Myanmar

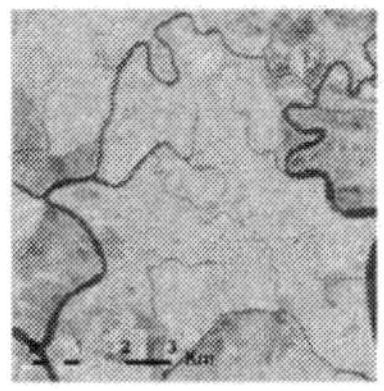
3. Myeik, Myanmar

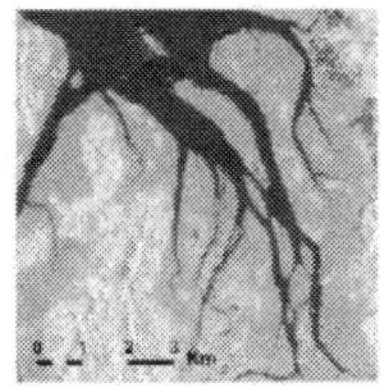
4. Shi Phangnga National Park, Thailand

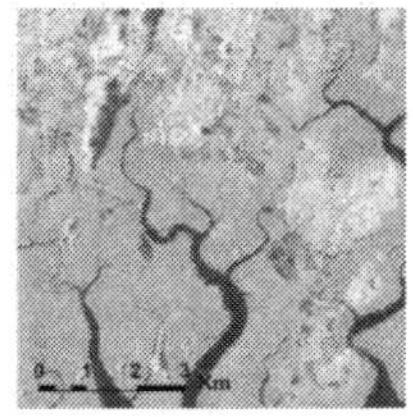
5. Bang Toei Mueang Phangnga, Thailand

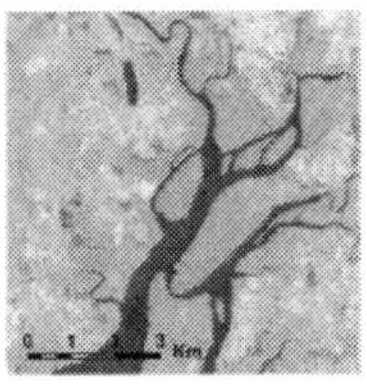
6. Khlong Khanan, Thailand

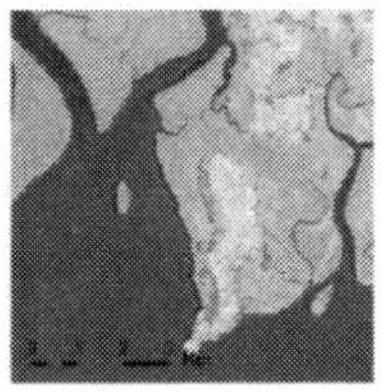
7. Suso Palian, Thailand

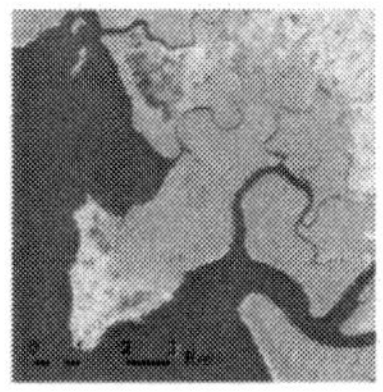
8. Puyu Mueng Satun, Thailand

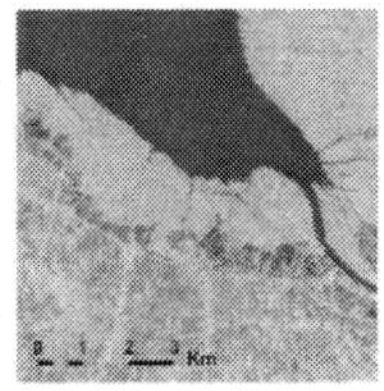
9. Pak Phanang Fangtawanok, Thailand

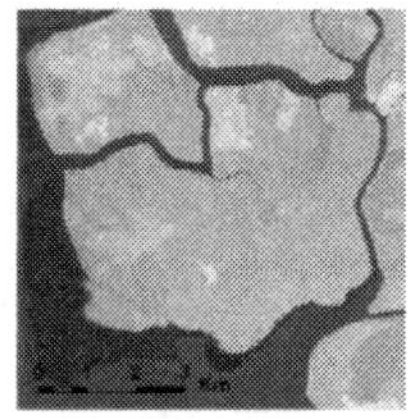
10. Jebong Perak, Malaysia

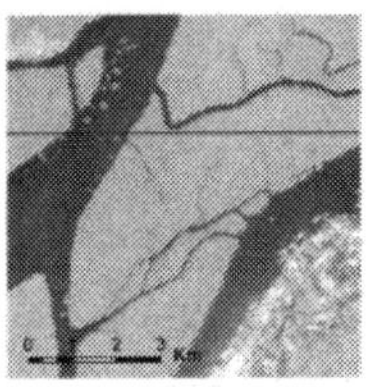
11. Pulau Kelang, Malaysia

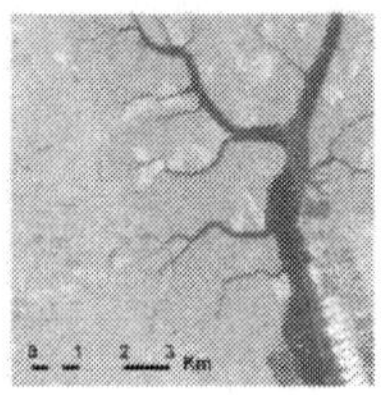
12. Serkat Johor, Malaysia

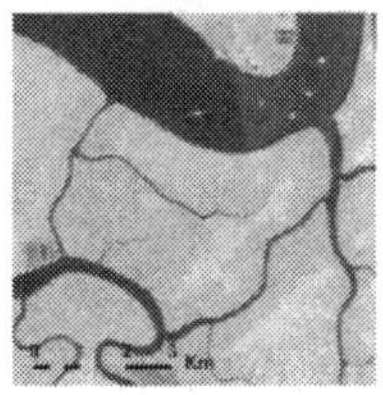
13. Sarikei Sarawak, Malaysia

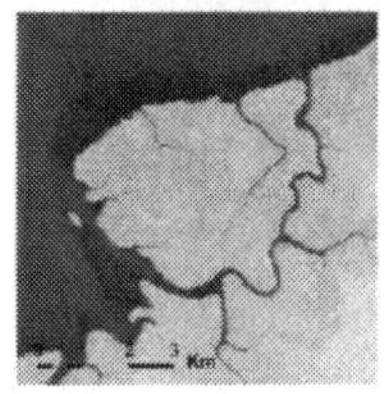
14. Pulau Selirong, Brunei

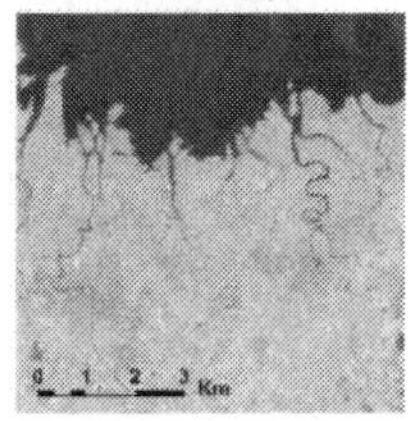
15. Kota Marudu Sabah, Malaysia

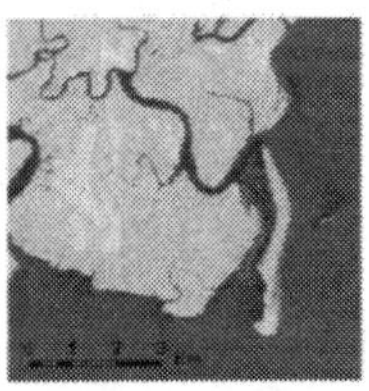
16. Beluran Sabah, Malaysia

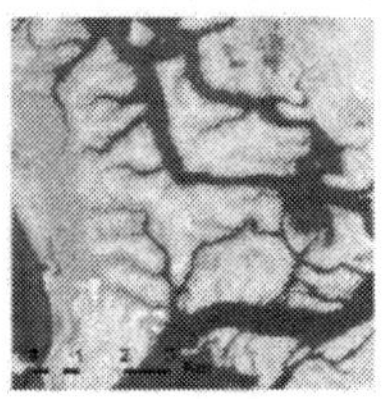
17. Peam Krasaop Wildlife Sanctuary, Cambodia

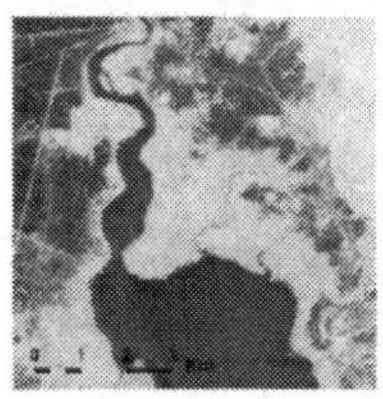
18. Preak Piphot River, Cambodia

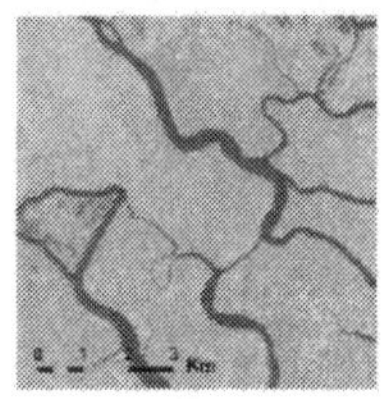
19. Thạnh An, Cần Giờ, Vietnam

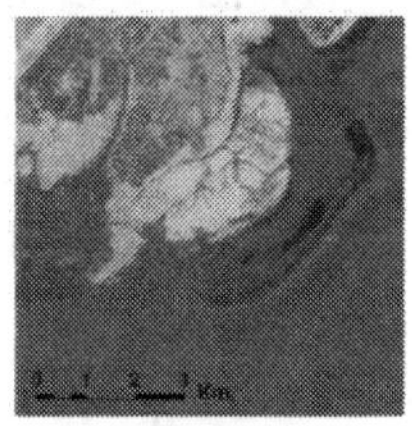
20. Giao Thiện, Giao Thuỷ Nam Định, Vietnam

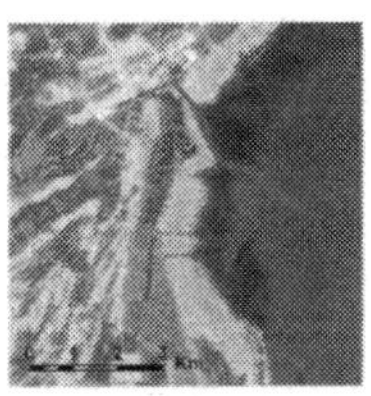
21. Thái Thủy Thai Binh, Vietnam

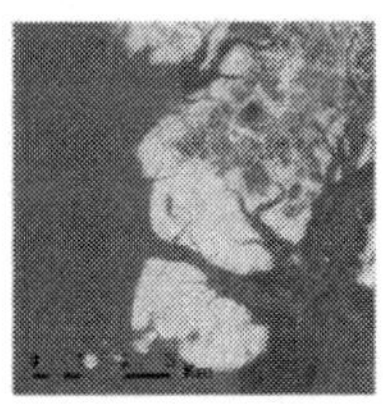
22. Vạn Ninh, tp. Móng Cái Quảng Ninh, Vietnam

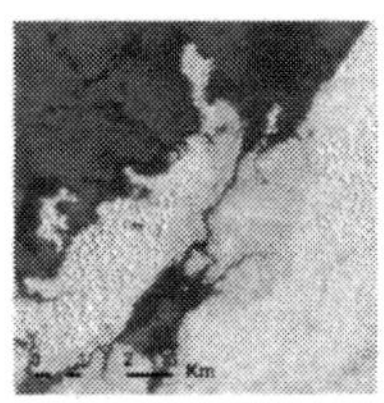
23. Siargao island, Philipines

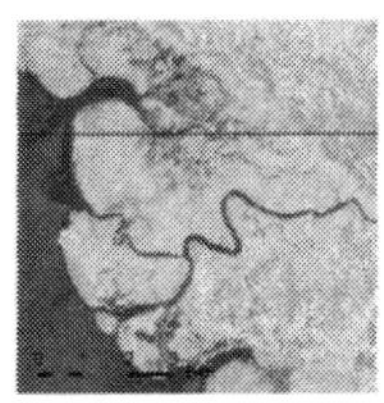
24. Santa Margarita, Philipines

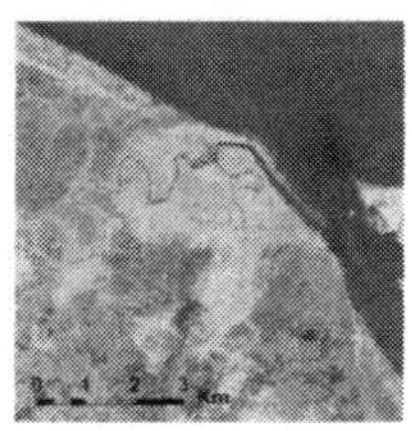
25. Abulug Cagayan, Philipines

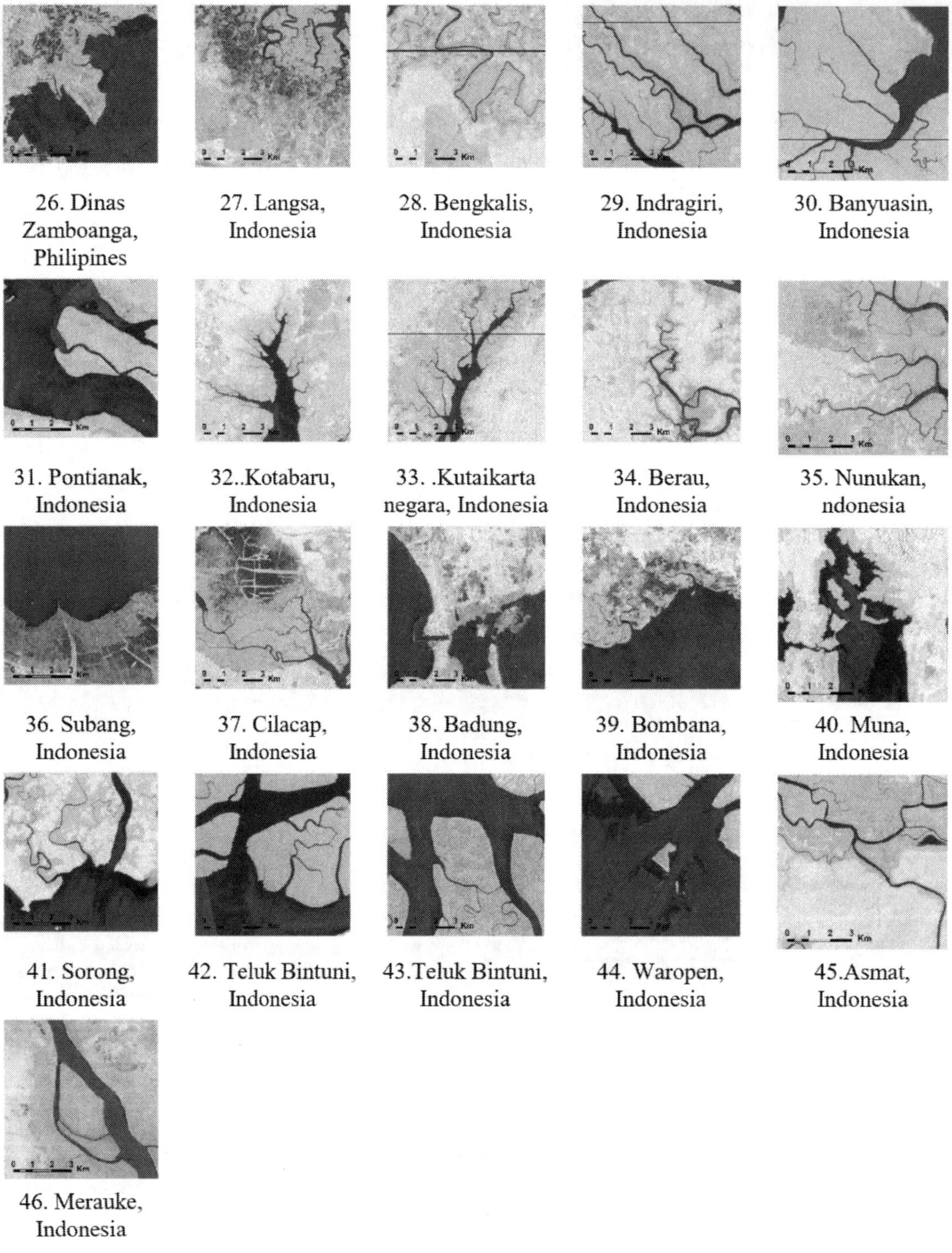

26. Dinas Zamboanga, Philipines | 27. Langsa, Indonesia | 28. Bengkalis, Indonesia | 29. Indragiri, Indonesia | 30. Banyuasin, Indonesia

31. Pontianak, Indonesia | 32..Kotabaru, Indonesia | 33. .Kutaikarta negara, Indonesia | 34. Berau, Indonesia | 35. Nunukan, ndonesia

36. Subang, Indonesia | 37. Cilacap, Indonesia | 38. Badung, Indonesia | 39. Bombana, Indonesia | 40. Muna, Indonesia

41. Sorong, Indonesia | 42. Teluk Bintuni, Indonesia | 43.Teluk Bintuni, Indonesia | 44. Waropen, Indonesia | 45.Asmat, Indonesia

46. Merauke, Indonesia

Figure 3. Visualization of mangrove forest on ALOS PALSAR composite R=HH, G=HV, B=HH/HV.

We collected ROI to know characteristics of mangrove forest based ALOS-PALSAR. ROI spread overall Southeast Asia region from **(1)** Lay Taung, Myanmar until **(46)** Merauke Papua, Indonesia. After that calculate mean and standard deviation from each ROI. Mean and standard deviation value for each ROI can see in the figure 4. In figure 4, we can see characterization of backscatter coefficient of mangrove forest on polarization of HH and HV. The "HH" modes indicates the combination of a horizontally (H) polarized transmitted signal and a horizontally (H) polarized received signal. Similarly, "HV" stands for the combination of a horizontally polarized (H) transmitted signal and a vertically polarized (V) received signal. Based on figure 4, characterization of mangrove forest for average of HH value around -10.88 dB to -6.65 dB and average of HV values around -16.49 dB to -13.26 dB. On polarization of HH, the highest backscattering value is mangrove forest in Preak Piphot River, Cambodia (18), Thái Thủy Thai Binh, Vietnam (21), and Vạn Ninh, tp. Móng Cái Quảng Ninh Vietnam (22) dan whereas the lowest backscattering value is mangrove forest in Thailand area. On polarization of HV, the highest backscattering value is mangrove forest Preak Piphot River, Cambodia (18), Sorong (41) and Teluk Bintuni (32) Indonesia whereas the lowest backscattering value is mangrove forest in Subang, Indonesia (36), Giao Thiện, Giao Thuỷ Nam Định, Vietnam (20) and Puyu Mueng Satun, Thailand (8).

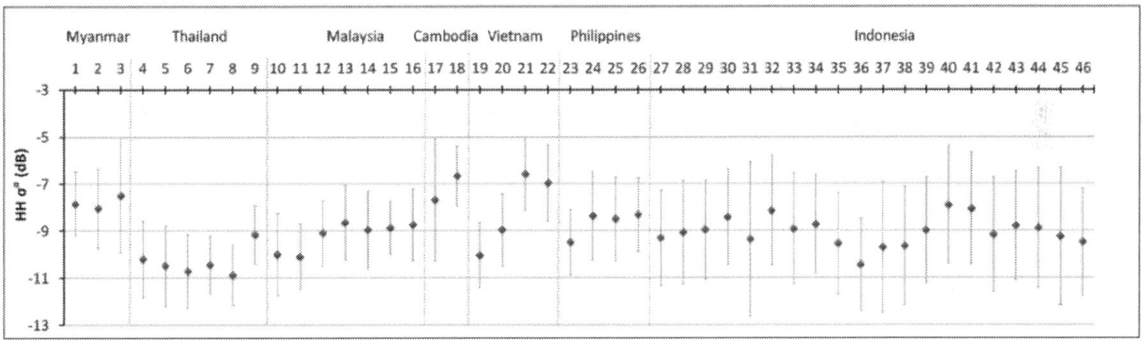

(a)

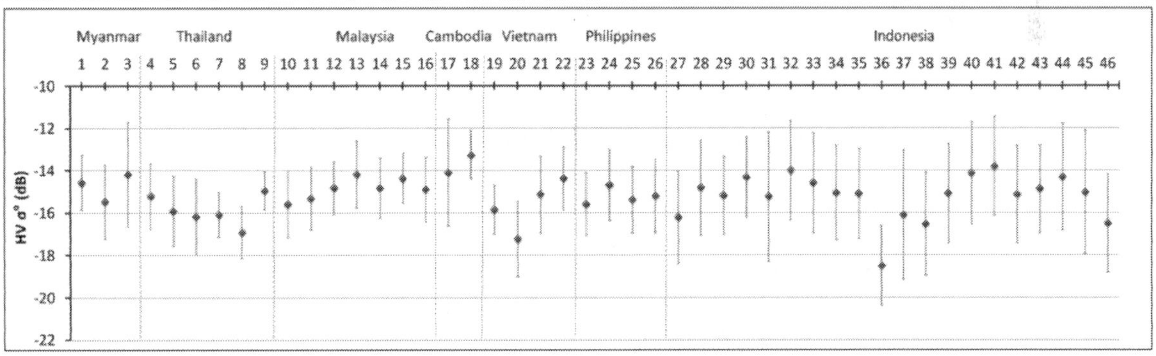

(b)

Figure 4. Backscatter characteristics of mangrove forest based on ALOS PALSAR polarization on (a) HH and (b) HV.

The sensitivity of the backscatter to mangrove forest parameters and the saturation level is rather sites-dependent, since mangrove forest structure influences the relative contribution of the scattering mechanism [10,13]. In addition, the individual contribution to the total mangrove forest backscatter is also dependent on environmental conditions (i.e. weather conditions, moisture conditions, and weather dynamics) and also topography and tidal height [14] which can affect the dielectric properties of the vegetation and ground surface. The interactions between the radiation and the plant's internal properties (e.g. moisture content influencing the dielectric constant of a material and cell structure) and external components (e.g. size, geometry and orientation of leaves, trunks, branches, and aerials or stilt roots) also result in a specific backscatter signal (e.g. radar calibration and orthorectification) [15,16].

Scatter graphs for visualization both HH and HV value can be seen in figure 5. Based on figure 5 correlation of HH and HV value not exactly linear. Generally, the sensitivity of the backscatter intensity to the biophysical parameters of land surface vegetation is examined in different combinations of polarized modes [17]. The HV polarization has more the stable relationship revealed between the SAR backscatter and above-ground forest biomass rather than HH polarization [18] and HV is much less influenced by soil and vegetation moisture [19]. Based on visual interpretation from scatter graphs in figure 5, backscatter value of mangrove forest on polarization HH and HV can be classified on three classes. First class which has higher value of HH and HV, second class which middles values of HH and HV and third class which has lower value of HH and HV. If compared with previous research [19] about characteristics of mangrove types based on ALOS PALSAR 50m-resolution in 2008, the results are similar.

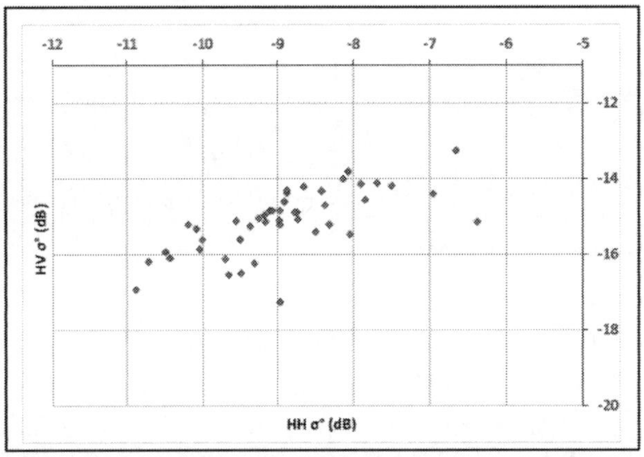

Figure 5. Scatter graph on HH and HV backscatter value of mangrove forest.

3.2 Spatial distribution of mangrove forest in Southeast Asia

We classify mangrove forest based on backscattering characteristics of mangrove forest and also based on topography data. On this case the method that was used is rule based classification method. Parameter that was used on rule based classification is HH, HV and topography data derived from SRTM 90. The threshold based on lowest and highest of HH and HV value on study area (figure 4) and average of topography data derived from SRTM 90m. Based on characterization, we found the lowest HH value of -10.88 to -6.65 dB and HV value of -16.49 to -13.26 dB also average of height of topography is 9.92 m. Based on that criteria we have found spatial distribution of mangrove forest based on ALOS PALSAR. Spatial distribution of mangrove forest in Southeast Asia can be seen on figure 6. Basically for getting mapping of mangrove forest on regional areas with high accuration we suggest classifying smaller areas by smaller areas. This condition because of mangrove forest has unique characteristics on each area. However for classification of smaller areas by smaller areas on whole Southeast Asia need big effort and time.

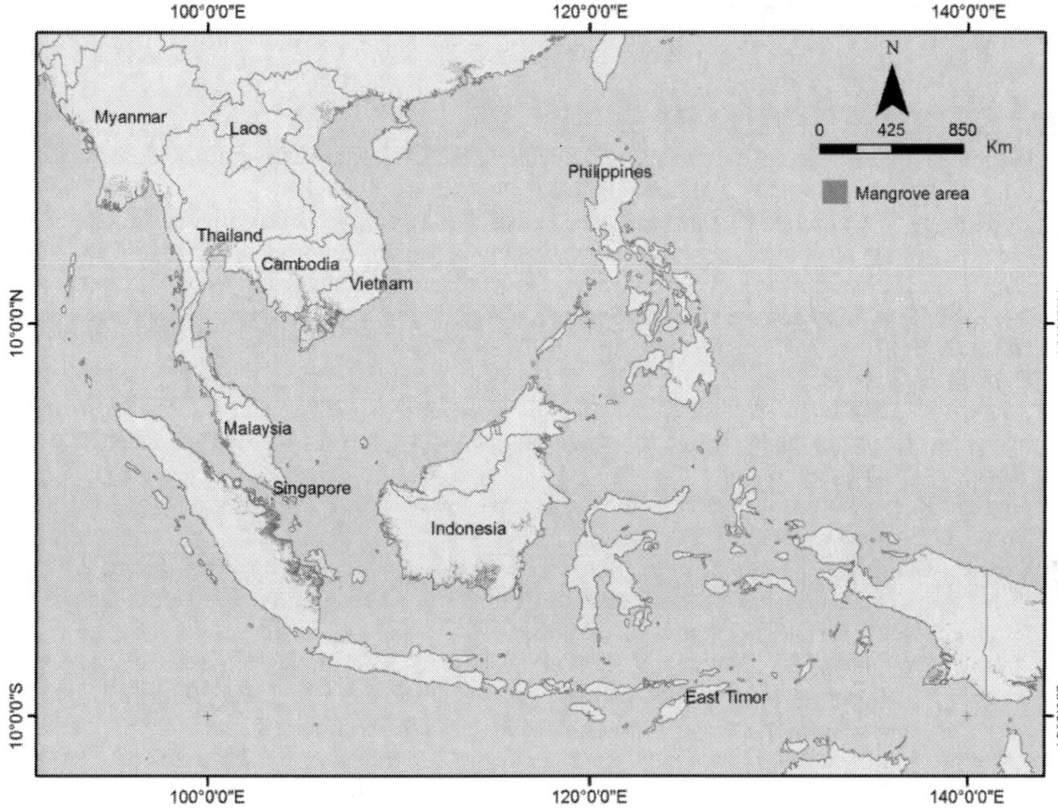

Figure 6. Spatial distribution of mangrove forest in Southeast Asia.

4. Conclusion

We showed backscattering characteristics of mangrove forest in Southeast Asia based on ALOS PALSAR mosaic 25m-resolution with polarization on HH and HV. We found characteristics of mangrove forest on HH value around -10.88 dB to -6.65 dB and on HV value around -16.49 dB to -13.26 dB. Also we found backscatter value of mangrove forest on polarization HH and HV can be classified on three classes. First class which has higher value of HH and HV, second class which has middles values of HH and HV and third class which has lower value of HH and HV. On polarization of HH which the highest backscattering value is mangrove forest in Preak Piphot River Cambodia, Thái Thủy Thai Binh Vietnam, and Vạn Ninh tp. Móng Cái Quảng Ninh Vietnam whereas the lowest backscattering value is mangrove forest in Thailand area. On polarization of HV which the highest backscattering value is mangrove forest Preak Piphot River Cambodia, Sorong and Teluk Bintuni Indonesia whereas the lowest backscattering value is mangrove forest in Subang Indonesia, Giao Thiện Giao Thuỷ Nam Định Vietnam and Puyu Mueng Satun Thailand. Based on characterization, we create a rule criteria for classification of mangrove areas. Finally we found spatial distribution of mangrove forest based on ALOS PALSAR 25m-resolution in Southeast Asia. Actually mangrove forest has unique characteristics on each area, needed some specific criteria to identify mangrove with the highest accuration result, for next step we will verify and validate our result and also for enhancement our results we try to find more rule criteria for classification of mangrove forest on regional area.

Acknowledgement
This study is carried out as a JAXA SAFE activity. The author would like to thank JAXA for their support.

References

[1] Giesen, W., Wulffraat, S., Zieren, M., and Scholten, L., 2007, Mangrove guidebook for south east asia, ISBN: 974-7946-85-8, FAO and Wetlands International, Bangkok.
[2] Nellemann, C., Corcoran, E., Duarte, C. M., Valdes, L., De Young, C., and Fonseca, L., 2009, Blue Carbon, A rapid response assessment. Norway; United Nations Environment Programme (80 pp).
[3] Daniel, C. D., Kauffman, J. B., Murdiyarso, D., Kurnianto, S., Sthidham, M., and Kanninen, M., 2011, Mangroves among the most carbon-rich forests in the tropics, Nature Geosciences Letters, NGEO1123 (5pp).
[4] Daniel, M., Purbopuspito, J., Kauffman, J. B., Warren, M. W., Sasmito, S. D., Donato, D. C., Manuri, S., Krisnawati, H., Taberima, S., and Kurnianto, S., 2015, The potential of Indonesia mangrove forest for global climate change mitigation, Nature Climate Change Letters, DOI: 10.1038/NCLIMATE2734
[5] Giesen, W. & S. Wulffraat (1998) - Indonesian mangroves part I: Plant diversity and vegetation. Tropical Biodiversity, 5(2):11-23.
[6] Giri, C., Ochieng, E., Tieszen, L. L., Zhu, Z., Singh, A., Loveland, T., Masek, J., and Duke, N., 2011, Status and distribution of mangrove forests of the world using earth observation satellite data, Global Ecology and Biogeography, 20, 154-159.
[7] Takeuchi, W., Tien, D. V., Phuong, V. T., Van, A. N., and Oo, K. S., 2011, Above ground biomass mapping of mangrove forest in Vietnam by ALOS-PALSAR. ISBN 978-1-4577-1351-4, http://ieeexplore.ieee.org/ stamp/ stamp.jsp?arnumber=06087044&tag=1
[8] Lucas, R. M., Mitchell, A., Rosenqvist, A., Proisy, C., Melius, A., and Ticehurst, C., 2007, The potential of L-band SAR for quantifying mangrove characteristics and change: Case studies from the tropics. Aqua. Conserve., 17, 245-264.
[9] Tsomon, R., Tateishi, R., and Tetuko, J.S.S., 2002, A Method to estimate forest biomass and its application to monitor Mongolian Taiga using JERS-1 SAR, International Journal of Remote Sensing, 23, 4971-4978.
[10] Lucas, R. M., Moghaddam, M., and Cronin, N., 2004, Microwave scattering from mixed-species forest, Queensland, Australia. IEEE Transaction on Geoscience and Remote Sensing, 42, 2142-2159.
[11] Shimada, M., Takahiro, O., 2010, Generating large-scale high quality SAR mosaic datasets: Application to PALSAR data for global monitoring, IEEE Journal of selected topics in applied earth observations and remote sensing, Vol. 3, No. 4 DOI: 10.1109/JSTARS.2010.2077619
[12] Shimada, M., Isoguchi, O., Tadono, T., and Isono, K., 2009, PALSAR radiometric and geometric calibration, IEEE Transactions on Geoscience and Remote Sensing, 47, 3915-3932.
[13] Imhoff, M. L., 1995, A Theoretical analysis of effect of forest structure on synthetic aperture radar backscatter and the remote sensing of biomass, IEEE Transaction on Geosciences and Remote Sensing, 33, 2739-2759.
[14] Darmawan, S., Takeuchi, W., Vetrita, Y., Wikantika, K., Sari, D. K., 2015, Impact of topography and tidal height on ALOS PALSAR polarimetric measurements to estimate aboveground biomass of mangrove forest in Indonesia, Journal of Sensor, Article ID 641798.
[15] Van Zyl, J. J., 1993, The effect of topography on radar scattering from vegetated areas, IEEE Transactions on Geoscience and Remote Sensing, Vol. 31, No. 1.
[16] Kuenzer, C., Bluemel, A., Gebhardt, S., Tuan Vo Quoc and Dech, S., 2011, Remote Sensing of Mangrove Ecosystems: A Review, Remote Sensing, ISSN 2072-4292, www.mdpi.com/journal/ remotesensing.
[17] Suzuki, R., Kim, Y., Ishii, R., 2013, Sensitivity of the backscatter intensity of ALOS/PALSAR to the above-ground biomass and other biophysical parameters of boreal forest in Alaska, Polar Science, 7, 100-112.

[18] Peregon, A., and Yamagata, Y., 2013, The use of ALOS/PALSAR backscatter to estimate above-ground forest biomass: A case study in Western Siberia, Remote Sensing of Environment 137, 139–146.
[19] Sandberg, G., Ulander, L. M. H., Fransson, J. E. S., Holmgres, J., & Le Toan, T. (2011). Land P-band backscatter intensity for biomass retrival in hemiboreal forest. Remote Sensing of Environment, 115, 2874–2886.
[20] Darmawan, S., Takeuchi, W., Vetrita, Y., Wikantika, K., and Sari, D., K., 2014, Characterization of mangrove forest types based on ALOS-PALSAR in overall Indonesian archipelago, IOP Conf. Ser.: Earth Environ. Sci. 20 012051.

Characterization of mangrove species using ALOS-2 PALSAR in Hai Phong city, Vietnam

Tien Dat Pham[1*], Kunihiko Yoshino[2]

[1]Graduate school of Systems and Information Engineering, University of Tsukuba, 1-1-1 Tennodai, Tsukuba city, Ibaraki, Japan 305-8573
[2]Faculty of Engineering, Information and Systems, University of Tsukuba, 1-1-1 Tennodai, Tsukuba city, Ibaraki, Japan 305-8573

[*]Email: dat6784@gmail.com

Abstract. Hai Phong city is located in the Northern coast of Vietnam where the mangroves are distributed between zone I and zone II among the four mangrove zones in Vietnam. This city is vulnerable to rising sea levels associated with climate change and tropical cyclones, which are forecasted to become more severe due to the impact of climate change. In the past, mangrove forests in this city have decreased markedly because of over expansion of shrimp farming. Thus, identification of mangrove forests is important to monitor and support mangrove conservation and management in the coastal zone. The objectives of this research were to identify the locations of mangrove and characterize mangrove species in Hai Phong using HH and HV backscatters of the Advanced Land Observing Satellite 2 (ALOS-2) with enhanced Phased Array L-band Synthetic Aperture Radar (PALSAR). Image segmentation was used to identify the locations of mangrove forests. Moreover, Geographic Information System (GIS) was applied to update current status of mangrove species in 2015. The results showed that the means of HH and HV backscatter coefficients of *K. obovata* are lower than *S. caseolaris*. *K. obovata* has HH value around -13.9 dB until -10.3 dB and HV value around -20.6 dB until -16.2 dB. Higher HH values between about -14.9 dB and -6.8 dB and HV values between roughly -20.6 dB and -14.3 dB have observed by *S. caseolaris*. The total area of mangrove forest in Hai Phong in the year 2015 was around 4084 hectares, of which *S. caseolaris* occupied over 68% and mixed mangrove species was approximately 25.6%. This research indicates the potential for the use of L-band ALOS-2 PALSAR in characterizing mangrove forest species in the coastal zone.

1. Introduction

Mangrove forests are productive and the most important ecosystems which appear in the inter-tidal zones along the coast in most tropical and semi-tropical climates. They play an important role in protecting land from erosion, mitigate the effects of tropical cyclones [1], and reduce damage from the effects of tsunamis [2]. Moreover, their ecosystems could stabilize coastlines, clean water, and act as highly efficient carbon sinks in the tropics [3]. For centuries, mangroves have contributed significantly to the socio-economic lives of coastal dwellers since they could provide building materials, charcoal, tannin, food, honey, herbal medicines, and other forest products [4].

Nevertheless, mangroves are under severe threat due to high population growth, aquaculture expansion and migration into coastal areas. Globally more than 3.6 million hectares of mangroves have been lost since 1980. Asia has suffered the greatest loss of 1.9 million hectares [5]. The mangrove areas in Vietnam have decreased dramatically over the last 50 years [6]. In northern coast of Vietnam, throughout the periods 1964-1997, mangrove area decreased by 17,094 ha. This loss of mangrove forests in the Northern coast of Vietnam due to over expansion of shrimp aquaculture [7]. Therefore, it is necessary to monitor mangroves, and map their locations in order to support coastal zone management and planning programs.

Satellite remote sensing data can be used for large areas and over time and thus represents an alternative tool for monitoring mangrove forests. Throughout the world, many researchers have utilized various satellite remote sensing data for mapping mangrove forests including optical imagery [8-12] and SAR data [9, 13]. However, in Vietnam few studies have used remotely sensed data to characterize mangrove forest types. There is a lack of available and updated data on the current status of mangrove area in Vietnam [14]. Mangrove mapping and change detection have been done only in some specific regions of Vietnam such as Ca Mau Peninsula in the South [15-17], Hai Phong and Quang Ninh in the North [18-20].

The Advanced Land Observing Satellite 2 (ALOS-2) was successfully launched in May 2014 equipped with an enhanced Phased Array L-band Synthetic Aperture Radar (PALSAR-2). ALOS-2 PALSAR provides higher resolution mode and wider area observation compared to ALOS PALSAR. As a result, disaster and environmental monitoring at local, regional and global scales will continue by ALOS-2 [21]. The objectives of this study are to map the locations of mangrove forest types and to investigate characteristics of mangrove species in Hai Phong using ALOS-2 PALSAR. In this study, an object-based method was used for ALOS-2 PALSAR image processing. In addition, Geographic Information System (GIS) was also utilized update information of different mangrove species.

2. Materials and methods

2.1. Materials

2.1.1. ALOS-2 PALSAR. ALOS-2 PALSAR imagery level 2.1 with high sensitive polarimetric mode was used to map the locations of mangrove forest types and characterize distribution of dominant mangrove species in Hai Phong. Data acquisition is shown in Table 1.

Table 1. Acquired satellite remote sensing data.

Satellite sensor	Date of acquisition	Level of process	Pixel Spacing	Polarizations
ALOS-2 PALSAR	2015/7/31	2.1	6.25 m	L band (HH, HV)

Source: JAXA, Japan

2.1.2. Field survey data. The field data was collected during several field trips carried out during the period of this research (September 2014, July & August 2015). The foundation for these field trips was the collection of land cover information in particular mangrove forest types. Five coastal districts (Fig. 1) were visited in Hai Phong and we collected ground-truth points (GTPs) using GPS (Global Positioning Systems) to characterize dominant mangrove species which are present in Hai Phong. In total, 157 GTPs were collected during field survey, of which 57 locations were recorded for mangrove species.

2.2. Methods

2.2.1. Study site. This study was conducted in mangrove forests of Hai Phong city, which is located between 20^{0}30'to 20^{0}01' N latitude and 106^{0}23' to 107^{0}08' E Longitude. This city belong to the Northern coast of Vietnam which is about 120 km from the capital Hanoi. The length of sea coast of Hai Phong is about 125 km including the length of coast surrounding the offshore islands. A vast majority of mangrove forests in the North is distributed in Hai Phong and surrounding provinces [18]. Mangrove forests in this city are under severe threats due to over shrimp aquaculture [7]. Five coastal districts were visited during this research (Fig. 1). Regarding to mangrove species, about 57 different locations were recorded during July and August 2015.

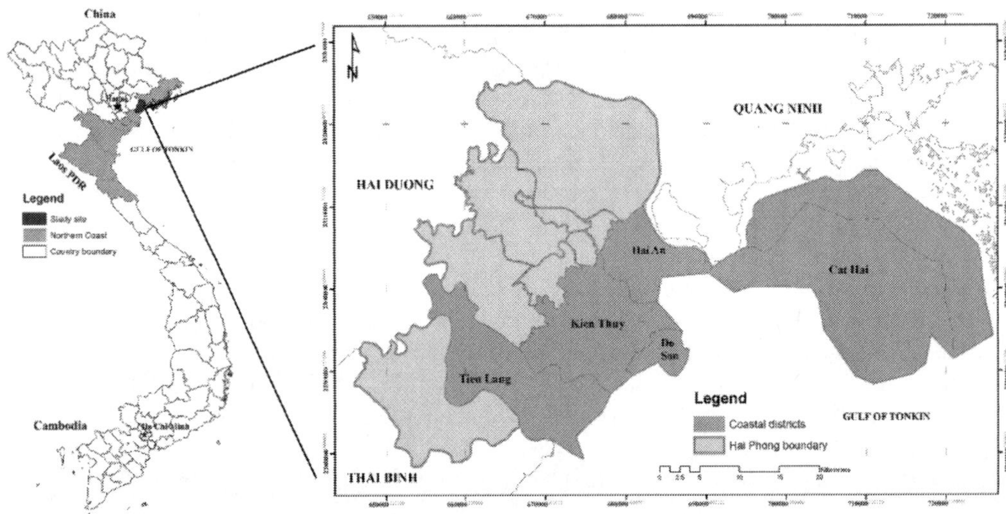

Figure 1. Location map of the study area.

2.2.2. ALOS-2 PALSAR imagery processing. ALOS-2 PALSAR data level 2.1 was acquired from the Remote Sensing Technology Centre (RESTEC) of Japan. The DN (Digital Number) was converted to normalized radar sigma-zero by the equation 1

$$\sigma^{0}\,[\mathrm{dB}] = 10.\,\log_{10}(DN)^2 + CF \qquad \text{(Eq. 1)}$$

Where σ^{0}: backscattering coefficient,
DN: digital number of the amplitude image
CF (Calibration Factor) = -83 dB for both HH and HV polarizations [22]. The CF used to process ALOS-2 PALSAR is similar to ALOS PALSAR [23].

The digital number of each pixel was transformed into backscattering sigma naught (σ^{0}) in decibel (dB) after applying Forst filters with moving windows of 5 x 5 in order to reduce speckle noise of SAR data.

ALOS-2 PALSAR imagery was classified using a supervised training method by applying maximum livelihood algorithms. The ENVI 5.2 software was also employed for SAR imagery processing. The details of the whole image processing of ALOS-2 PALSAR for 2015 could be similar to image processing for ALOS PALSAR which can be found in [19].

Figure 2 shows an example of image segmentation created from image composite of ALOS-2 PALSAR (RGB = HH, HV, and HH/HV) for land-cover types of Hai Phong city in 2015.

2.2.3. Characterization of mangrove species. In order to characterize mangrove species, a total of 57 sampling plots with size 30 m x 30 m were randomly selected in the study site during field survey in July and August 2015. The species composition of each plot were recorded using GPS. In fact, two dominant mangrove species including: *Kandelia obovata* and *Sonneratia caseolaris* were recorded in the study area. All information on mangrove species which are presented in Hai Phong during the year 2015 was updated and stored in ArcGIS 10.3.1 geodatabase.

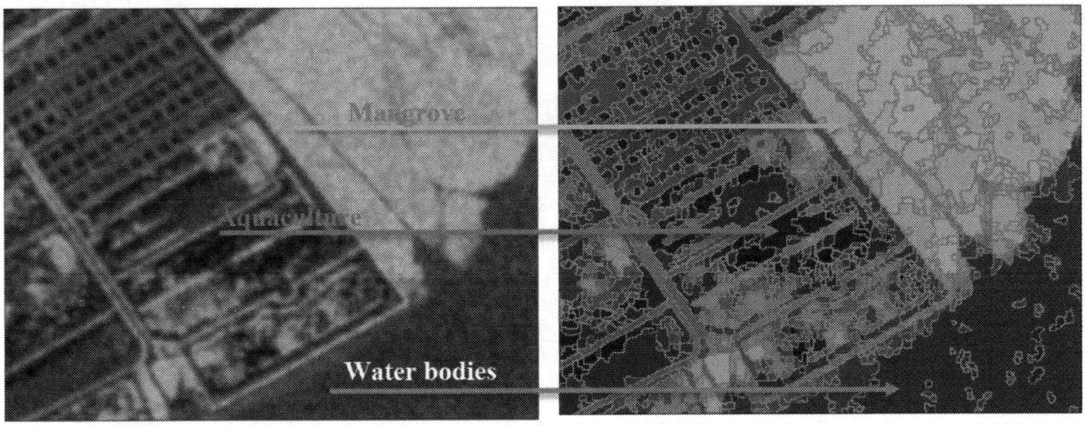

ALOS-2 PALSAR composite (R = HH, G=HV, and B = HH/HV)

Image segmentation of land cover types (Mangrove, Aquaculture, and Water bodies)

Figure 2. Image segmentation of land cover types of Hai Phong in 2015.

3. Results and discussion

3.1. Spatial distribution of mangrove forests

Figure 3 displays spatial distribution of mangrove forest types in Hai Phong in the year 2015. Mangroves are presented mostly on the exterior of sea dykes or in the river mouths of Hai Phong. It is quite clear because mangrove could defend against tropical cyclones during stormy reason [7].

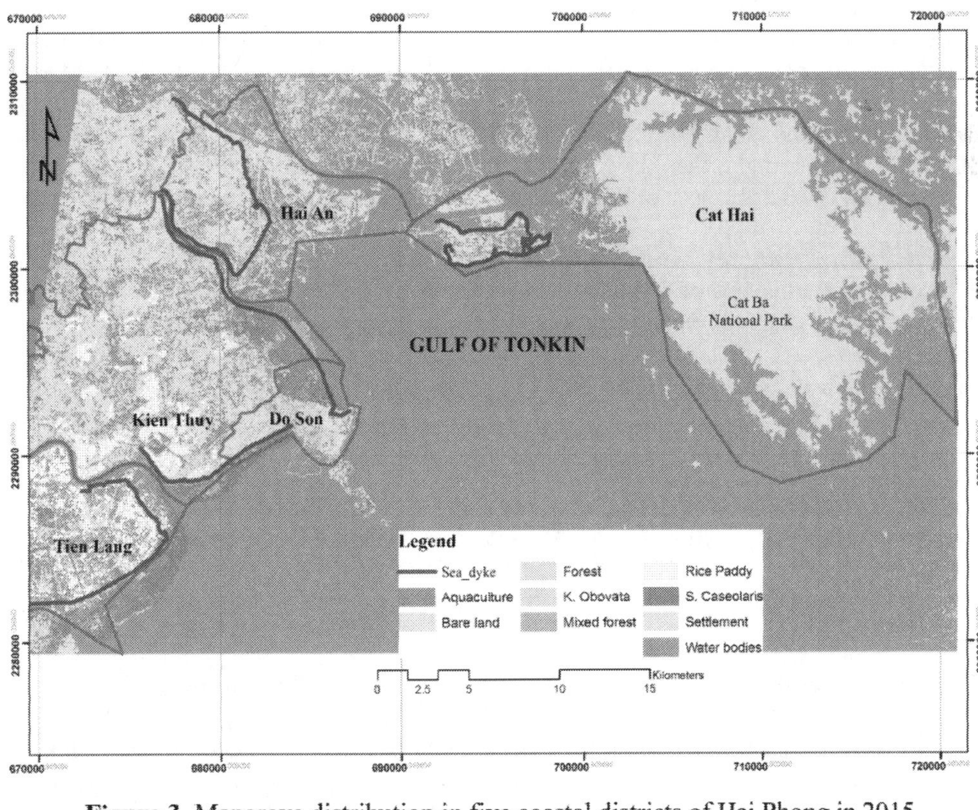

Figure 3. Mangrove distribution in five coastal districts of Hai Phong in 2015.

Mangrove forests appear in tidal regions of 5 coastal districts of Hai Phong city. Based on the image segmentation from ALOS-2 PALSAR composite, it can be identified different mangrove forest types among other land cover types.

In order to identify the locations of mangrove forests, we collected Ground Truth Points (GTPs) for different land cover types. Then each GTP was extracted HH, HV backscatters values. Finally, all GTPs of each land cover type were calculated min, max, mean and standard deviation for HH and HV backscatters values. Table 2 shows the characterization of backscatter coefficients (HH, HV) for land-cover types of Hai Phong in 2015.

Table 2. Backscatter coefficients of land-cover types of Hai Phong in 2015.

	HH			**HV**		
	Max	Min	Mean	Max	Min	Mean
Mangrove	-7.1	-13.9	-10.9	-12.1	-20.5	-16.8
Water bodies	-14.6	-23.4	-20.8	-22.8	-30.1	-26.8
Settlement	2.3	-13.5	-8.4	-1.5	-24.1	-15.4
Forest	-7.8	-19.2	-11.5	-11.9	-20.0	-15.1
Bare land	5.8	-24.3	-12.8	-7.4	-29.3	-20.8
Aquaculture	-10.7	-27.0	-20.5	-18.6	-31.2	-26.4
Rice Paddy	-14.3	-22.5	-17.8	-21.8	-28.6	-24.5

It can be seen from table 2 that HH and HV backscattering coefficients of water bodies and aquaculture are lowest among various land cover types, followed by the corresponding figures of rice paddy. The means of backscattering coefficients are about -20 dB at HH and roughly -26 dB at HV for water bodies and aquaculture. In contrast, the means of backscattering coefficients of settlement are lowest among seven land cover types. These numbers are approximately -8.4 dB at HH and around -15.4 dB at HV. Interestingly, the means of backscattering coefficients of forest and mangrove are almost similar at HH whilst the mean of HV for forest cover type is slightly higher than that of mangrove.

3.2. Characteristics of mangrove species in Hai Phong

We compared characterizations of dominant mangrove species in Hai Phong based on backscattering coefficients values at HH and HV polarizations. The result is shown in table 3.

Table 3. Backscatter coefficients of mangrove species of Hai Phong in 2015.

	HH			**HV**		
	Max	Min	Mean	Max	Min	Mean
K. obovata	-10.3	-13.9	-12.0	-16.2	-20.6	-18.2
S. caseolaris	-6.8	-14.9	-9.8	-14.3	-20.6	-16.4
Mixed species	-7.8	-15.3	-11.4	-13.4	-22.7	-17.9

The result indicates that the means of HH and HV backscatter coefficients of *K. obovata* are lower than *S. caseolaris*. In fact, *K. obovata* has HH value around -13.9 dB until -10.3 dB and HV value around -20.6 dB until -16.2 dB. Higher HH values between about -14.9 dB and -6.8 dB and HV values between roughly -20.6 dB and -14.3 dB have observed by *S. caseolaris*. These numbers are slightly lower than those of mangrove forest types in Indonesia due to the significant differences in mangrove species. In

Indonesia, three dominant species accounted for majority of mangrove forests are *Rhizophora*, *Bruguiera* and *Ceriops* [24]. Figure 4 depicts three kinds of mangrove forests in the study sites.

K. obovata *S. caseolaris* Mixed mangrove forest

Source: Survey, July 2015

Figure 4. Mangrove forests communities in the study area.

3.3. Mangrove area statistics of Hai Phong in 2015.

Table 4 indicates the area in hectares and in percentage of each mangrove forest type. As table 4 showed, the largest area is occupied by *S. caseolaris*. The mangrove forest area covered by *S. caseolaris* is approximately 2781 hectares accounting for about over 68% while the area of *K. obovata* is just around 256 hectares. The total mangrove forests area of Hai Phong in the year 2015 is about 4084 hectares.

Table 4. The mangrove forest area in hectare and percentage of Hai Phong in the year 2015.

Mangrove type	Area (ha)	Area (%)
K. obovata	256.05	6.27
S. caseolaris	2781.07	68.09
Mixed species	1046.83	25.64
Total	4083.95	100

It is quite clear because *S. caseolaris* could grow well in the high level of tidal due to their pneumatophores while *K. obovata* could not survive from high tidal inundation. Therefore, *K. obovata* often plants together with *S. caseolaris* [4]. From 2001 to 2013, mangrove forests increased slightly in Hai Phong then continued its upward trend in 2015 thanks to community-based forest management in cooperation with local authorities [25].

4. Conclusion

This research demonstrates that L-band ALOS-2 PALSAR data could be used to characterize the spatial distribution of mangrove species in the coastal zone. The results of this research indicated that *K. obovata* has HH value around -13.9 dB until -10.3 dB and HV value around -20.6 dB until -16.2 dB. Higher HH values between about -14.9 dB and -6.8 dB and HV values between roughly -20.6 dB and -14.3 dB have observed by *S. caseolaris*. The total area of mangrove forest in Hai Phong during the year 2015 is around 4084 hectares, of which *S. caseolaris* occupied over 68% and mixed mangrove species is approximately 25.6%. This research indicates the potential for the use of ALOS-2 PALSAR in characterizing mangrove forest species in the coastal zones. For future works, we will estimate above ground biomass and carbon storage of mangrove species using ALOS-2 PALSAR and field survey measurements.

Acknowledgements

The authors would like to thank CARES (Center for Agricultural Researches and Ecological Studies), Vietnam National University of Agriculture (VNUA), Vietnam for logistical support during fieldwork. We are thankful to MEXT (Ministry of Education, Culture, Sports, Science, and Technology) for financial support to the author whilst studying at the University of Tsukuba, Japan.

References

[1] Y. Mazda, M. Magi, M. Kogo, P.N. Hong, 1997, Mangroves as a coastal protection from waves in the Tong King delta, Vietnam, *Mangroves and Salt Marshes*, **1**, 127-135.

[2] F. Danielsen, M.K. Sørensen, M.F. Olwig, V. Selvam, et al., 2005, The Asian Tsunami: A Protective Role for Coastal Vegetation, *Science*, **310**, 643.

[3] A. Komiyama, J.E. Ong, S. Poungparn, 2008, Allometry, biomass, and productivity of mangrove forests: A review, *Aquatic Botany*, **89**, 128-137.

[4] P.N. Hong, H.T. San, 1993, *Mangroves of Vietnam*, in: IUCN, Bangkok, Thailand, pp. 173 pages.

[5] FAO, 2007, *The World's mangroves 1980-2005*. A thematic study prepared in the framework of the Global Forest Resources Assessment 2005, in: FAO Forestry Paper, Food and Agriculture Organization of the United Nations, Rome, pp. 89.

[6] L.X. Tuan, Y. Munekage, Q.T.Q. Dao, N.H. Tho, P.T.A. Dao, 2003, Environmental Management in Mangrove Areas *Environmental Informatics Archives*, **1**, 38-52.

[7] P. Tien Dat, K. Yoshino, 2016, Impacts of mangrove management systems on mangrove changes in the Northern Coast of Vietnam, *Tropics*, **24,** 141-151.

[8] B.G. Long, T.D. Skewes, 1996, A Technique for Mapping Mangroves with Landsat TM Satellite Data and Geographic Information System, *Estuarine, Coastal and Shelf Science*, **43**, 373-381.

[9] V. Pasqualini, J. Iltis, N. Dessay, M. Lointier, O. Guelorget, L. Polidori, 1999, Mangrove mapping in North-Western Madagascar using SPOT-XS and SIR-C radar data, *Hydrobiologia*, **413**, 127-133.

[10] G. Conchedda, L. Durieux, P. Mayaux, 2008, An object-based method for mapping and change analysis in mangrove ecosystems, *ISPRS Journal of Photogrammetry and Remote Sensing*, **63**, 578-589.

[11] S. Nandy, S. Kushwaha, 2011, Study on the utility of IRS 1D LISS-III data and the classification techniques for mapping of Sunderban mangroves, *Journal of Coastal Conservation*, **15**, 123-137.

[12] J.B. Long, C. Giri, 2011, Mapping the Philippines' Mangrove Forests Using Landsat Imagery, *Sensors*, **11**, 2972-2981.

[13] R.M. Lucas, A.L. Mitchell, A. Rosenqvist, C. Proisy, A. Melius, C. Ticehurst, 2007, The potential of L-band SAR for quantifying mangrove characteristics and change: case studies from the tropics, *Aquatic Conservation: Marine and Freshwater Ecosystems*, **17**, 245-264.

[14] R. McNally, A. McEwin, T. Holland, 2011, The Potential for Mangrove Carbon Projects in Vietnam.

[15] T.N.K.D. Binh, Vromant, N., Hung, N.T., Hens, L., Boon, E.K., 2005, Land cover changes between 1968 and 2003 in Cai Nuoc, Ca Mau Peninsula, Vietnam, *Environment, Development and Sustainability*, **7**, 519-536.

[16] V. Tran Thi, A. Tien Thi Xuan, H. Phan Nguyen, F. Dahdouh-Guebas, N. Koedam, 2014, Application of remote sensing and GIS for detection of long-term mangrove shoreline changes in Mui Ca Mau, Vietnam, *Biogeosciences*, **11**, 3781-3795.

[17] S. Nguyen-Thanh, C. Chi-Farn, C. Ni-Bin, C. Cheng-Ru, C. Ly-Yu, T. Bui-Xuan, 2015, Mangrove Mapping and Change Detection in Ca Mau Peninsula, Vietnam, Using Landsat Data

and Object-Based Image Analysis, *Selected Topics in Applied Earth Observations and Remote Sensing, IEEE Journal of*, **8**, 503-510.

[18] P.T. Dat, K. Yoshino, 2011, Monitoring Mangrove Forest using Multi-temporal Satellite Data in the Northern Coast of Vietnam, in: The 32nd Asian Conference on Remote Sensing, Taipei, Taiwan

[19] P. Tien Dat, K. Yoshino, 2012, Mangrove analysis using ALOS imagery in Hai Phong City, Vietnam, 85250U-85250U.

[20] T.D. Pham, K. Yoshino, 2015, Mangrove Mapping and Change Detection Using Multi-temporal Landsat imagery in Hai Phong city, Vietnam, in: International Symposium on Cartography in Internet and Ubiquitous Environments 2015, The University of Tokyo, Japan.

[21] A. Rosenqvist, M. Shimada, S. Suzuki, F. Ohgushi, T. Tadono, M. Watanabe, K. Tsuzuku, T. Watanabe, S. Kamijo, E. Aoki, 2014, Operational performance of the ALOS global systematic acquisition strategy and observation plans for ALOS-2 PALSAR-2, *Remote Sensing of Environment*, **155**, 3-12.

[22] JAXA, 2014, ALOS-2/PALSAR-2 Level 1.1/1.5/2.1/3.1 CEOS SAR Product in, Japan Aerospace Exploration Agency pp. 233.

[23] M. Shimada, O. Isoguchi, T. Tadono, K. Isono, 2009, PALSAR Radiometric and Geometric Calibration, *Geoscience and Remote Sensing, IEEE Transactions on*, **47**, 3915-3932.

[24] S.Darmawan, W. Takeuchi, Y. Vetrita, G. Winarso, K. Wikantika, D.K. Sari, 2014, Characterization of mangrove forest types based on ALOS-PALSAR in overall Indonesian archipelago, *IOP Conference Series: Earth and Environmental Science*, **20**, 012051.

[25] Dat P.T, K. Yoshino, 2013. Comparing Mangrove Forest Management in Hai Phong city, Vietnam towards Sustainable Aquaculture. *Procedia Environmental Sciences*, **17**, 109-118.

An investigation of age and yield of fresh fruit bunches of oil palm based on ALOS PALSAR 2

S Darmawan [1,3], W Takeuchi [1], A Haryati [2], R Najib A M [2] and M Na'aim [2]

[1] Institute of Industrial Sciences, University of Tokyo, Komaba, Meguro-ku, Tokyo, 153-8505 Japan.
[2] Felda Agriculutral Services Sdn. Bhd (FASSB), Platinum Park No 11, Persiaran KLCC, 50088, Kualalumpur Malaysia.
[3] Institut Teknologi Nasional, PHH Mustofa No 23 Bandung 40124, West Java Indonesia.

Email : soni@iis.u-tokyo.ac.jp and soni_darmawan@yahoo.com

Abstract. The objective on this study is to investigate age and yield of FFB of oil palms based on ALOS PALSAR 2. Study areas in oil palm plantations areas of Jerantut, Pahang Malaysia. Methodology consists collecting of ALOS PALSAR 2 and tabular data on the study area, processing of ALOS PALSAR 2 including of converting digital numbers to normalize radar cross sections (NRCS), topography correction and filtering, making of regions of interest according to areas of age and yield of FFB of oil palms and making of relationship analysis between backscatter value of HH, HV and age and yield of FFB of oil palm. The results have showed relationship between HH, HV and age of oil palm which R^2 of 0.63 for HH and 0.42 for HV that indicated increasing of age of oil palm as increasing of HH and HV value. Also relationship between HH, HV and yield of FFB of oil palm which R^2 of 0.26 for HH and 0.15 for HV, that indicated increasing of yield of FFB as decreasing of HH and HV value.

1. Introduction

Oil palm is the most productive oil seed and is becoming an increasingly important agriculture in the world which is around 85 % of the crude oil originates come from Malaysia and Indonesia [1]. High yield and low production costs of palm oil are the reasons driving commercial plantation companies to cultivate oil palm trees on a large scale [2]. Since the 1990s, palm plantations are in expansion in humid tropical areas, and land under oil palms increased to 12.6 million hectares in 2010 with annual palm oil production exceeding 32 million tons suggests that global demand for palm oil will be about 62–63 million tons in 2015 [3]. In the case of Malaysia, first oil palm plantations established in 1917, the oil palm area has expanded continuously, and recent increases in the demand for food, chemicals and biofuels have led to a surge in land conversion rates [4,5,6,7]. Oil palm plantations in Malaysia covered approximately 5.08 million ha in 2012, an increase of 1.6 per cents against 5.00 million hectares recorded in 2011 [4,8]. The state of Sabah is the largest producer of oil palm in Malaysia, with 1.44 million ha of oil palm plantations or 28.4 per cent of the total oil palm planted area in Malaysia, followed by Sarawak with 1.08 million ha (21.2 per cent) [4,8]. On the oil palms cultivation needs humid equatorial conditions to thrive, and conditions in Southeast Asia are ideal [9]. Seasonal droughts at higher tropical latitudes greatly reduce yields [10]. Also age of the oil palms is one of the important factors influencing the production of fruit bunches [11,12]. According to [11,13] age data on oil palm have the potential to be used in precision farming. Organization or management of zones based on such data is effective for maximization of palm oil yield, which is one of the most important variables affecting profit [11]. According to [11] collecting age data on oil palm trees is time consuming and costly, especially on a large or regional scale.

Remote sensing has been widely proven to be essential in monitoring and mapping highly threatened on land use land cover including forest and plantation area. Although sensors in the optical range of the electromagnetic spectrum have received the greatest attention but optical remote sensing is the degradation of the imagery through haze which is a very common phenomenon in Malaysia and other countries in South East Asia [12]. The launch of the Japanese Space Exploration Agency's (JAXA) Advanced Land Observing Satellite (ALOS) Phased Arrayed L-band SAR (PALSAR) represented a milestone in the global observation, characterization, mapping and monitoring of land use land cover including forest and plantation area largely, because these provide more information on the three dimensional structure and biomass of woody vegetation. As data can be days or night regardless of weather conditions, land use land cover including forest and plantation area can be observed more frequently, even in regions with prevalent cloud cover. ALOS-PALSAR penetrates through the foliage and interact primarily with the woody components of vegetation. Horizontally transmitted waves are either depolarized through volume scattering by branches in the canopy, with a proportion of vertically polarized microwaves returning to the sensor, or penetrate through the canopy and interact with the trunks, returning primarily through double bounce scattering as a horizontally polarized wave [13]. Longer L-band (e.g. L-band 15-20 cm) microwaves have a greater likelihood of penetrating the foliage and small branches at the upper canopies of the forest, and interacting with woody trunks and larger branch components as well as the underlying surface [14,15]. It has been proven that radar is sensitive to the structure of the canopy. The received backscatter intensity represented in the image is a composition of interactions with the crown, the trunk and the ground surface. Using fully polarimetric SAR it is possible to derive a relationship between backscatter, texture and crop status. If we consider an oil palm as crops this is very interesting. It would help to derive certain patterns for different growth stages of the oil palm. This is very relevant information for optimized and sustainable oil palm plantation management [16].

On wholes our study is to investigate utilizing ALOS PALSAR 2 for oil palm plantation management including mapping, estimating of biomass and yield producing but now just started on the preliminary study. On this paper we focus to investigate age and yield of fruit fresh bunches (FFB) of oil palms based on ALOS PALSAR 2. Study areas in oil palm plantations areas of Jerantut, Pahang Malaysia.

2. Methodology

Methodology consists collecting of ALOS PALSAR 2 and tabular data on the study area, processing of ALOS PALSAR 2 including of converting digital numbers to normalize radar cross sections, topography correction and filtering, making of regions of interest in according to areas of age and yield of FFB of oil palms and making of relationships analysis between backscatter value of HH, HV and age and yield of FFB of oil palm. Generally the methodology on this study can be seen on figure 1.

2.1 Data collection

We used ALOS PALSAR 2 (Figure 2) that was launched on May 24, 2014. ALOS PALSAR L-band with 1.2 GHz center frequency and 14 or 28 MHz bandwidths and dual polarization HH (horizontal transmitting and horizontal receiving) and HV (horizontal transmitting and vertical receiving) developed by Japan Aerospace Exploration Agency (JAXA). We collected ALOS PALSAR 2 Fine Beam Double

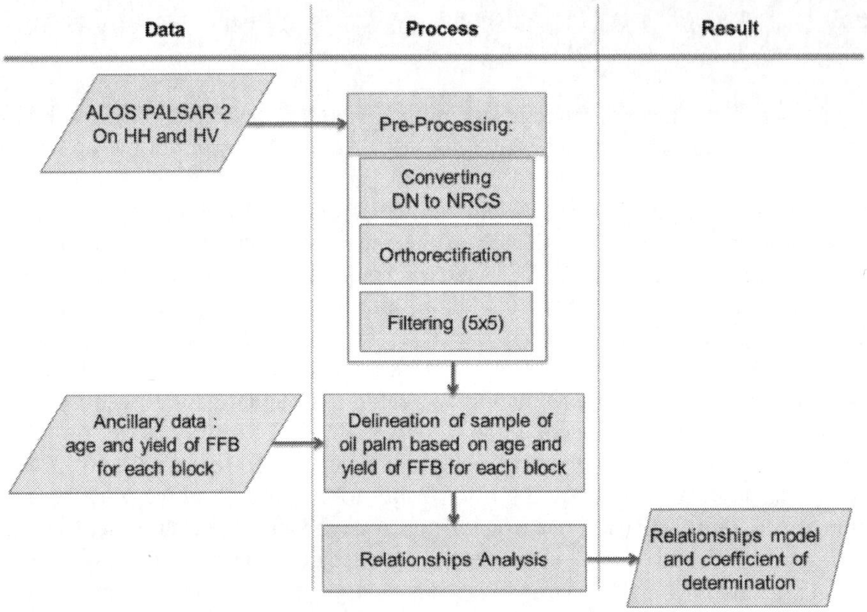

Figure 1. Methodology on this study.

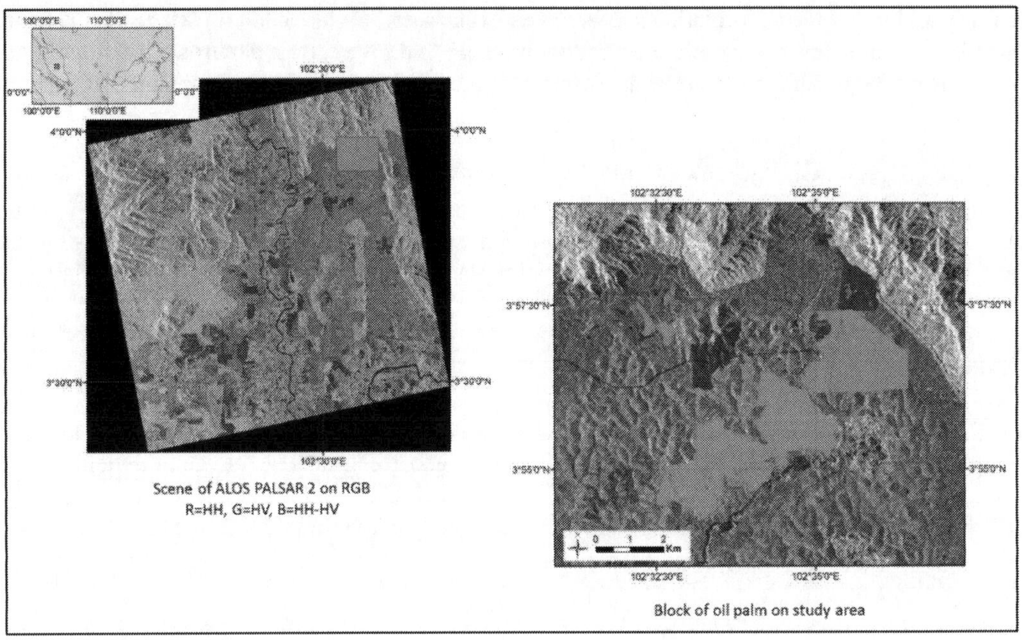

Figure 2. Data used on the study area.

Polarization (FBD) product level 1.5 that was acquired on February 28, 2015. FBD product level 1.5 has spatial resolution of 10 meters and swath of 70 km which multi-look data on slant range from map projection amplitude data with range and azimuth compressed. For this study we also collected tabular data about age and yield of fruit fresh bunches (FFB) on the study area. Age and yield of FFB based on each block. We collect age and yield of FFB with totally on 12 blocks in study area but for this case we have limitation because we collect age of oil palm majority on 11 to 21 years old.

2.2 Pre processing

Preprocessing focuses converting of digital number (DN) value of HH (DN_{HH}) and HV (DN_{HV}) to a normalized radar cross section (NRCS) in decibel (dB), (i.e., $\sigma°_{HH}$ and $\sigma°_{HV}$) by the following equations [17]:

$$\sigma°_{HH} = 10 \text{ x } \log_{10}(DN^2_{HH}) - CF \quad (1)$$
$$\sigma°_{HV} = 10 \text{ x } \log_{10}(DN^2_{HV}) - CF \quad (2)$$

Where, $\sigma°$ is backscattering coefficient and CF is the calibration factor. The CF is dependent on the processing date, in this study CF is equal to -83.0 both for HH and HV.

For reducing multiplicative speckle noise and topography effect on ALOS PALSAR 2 imaging we used frost enhancement filtering [18] with windows size 5x5 and orthorectification. Since lack of detail topography data, on this study we used Shuttle Radar Topography Mission (SRTM) data for processing of orthorectification.

2.3 Region of interest (ROI) and relationships analysis

We create ROI on ALOS PALSAR 2 with considering of age and yield of FFB region. We have totally 12 of ROI. Each ROI has information of age and yield of FFB of oil palm. On ALOS PALSAR 2 imaging data contain response from characteristics of oil palm. Backscattering response of oil palm area depends on structure and density of oil palm. All oil palm stand and even for that matter various structure and density have different backscattering and texture pattern in various wavelengths. Thus the relationships between age and yield of FFB and remotely sensed data are unique. For analyzing these relationships, Pearson's correlation coefficient is used. If the coefficient is close to 1, it means there is a strong relationship between them. In this study, the first variable is age, yield of FFB and HH, HV value derived from ALOS PALSAR 2. The coefficient of determination (R^2) is indication of the regression model. The R^2 value will show the percentage of variation as the regression models variables.

3. Result and discussion

On this study we have some variables such as age, yield of FFB of oil palms derived from field surveys and ancillary data and HH, HV value derived from ALOS PALSAR 2. We integrate each other to know regression models and coefficient determination. Based on regression models and coefficient determination we can get information about relationships both each other.

Firstly we investigate relationship between age and yield of FFB on the study area. We collect information of age and yield of FFB from 12 blocks in study areas. The limitation on this study area is variety of age of oil palms majority from 11 to 21 years old. Scatter plot between age and yield of FFB can be seen on figure (3). Based on figure 3 has indicated that increasing of age of oil palm as decreasing of yield of FFB. On this case we use linear analysis because we start from 11 to 21 years old. According to [11,12] commonly oil palm trees have an economical life pan up to 25 years old which begin at the time a tree reaches maturity at about 3 years, with production reaching a peak between 6 and 10 years [11, 13]. Yield reach a peak at around 10 years and decrease after that [11,17].

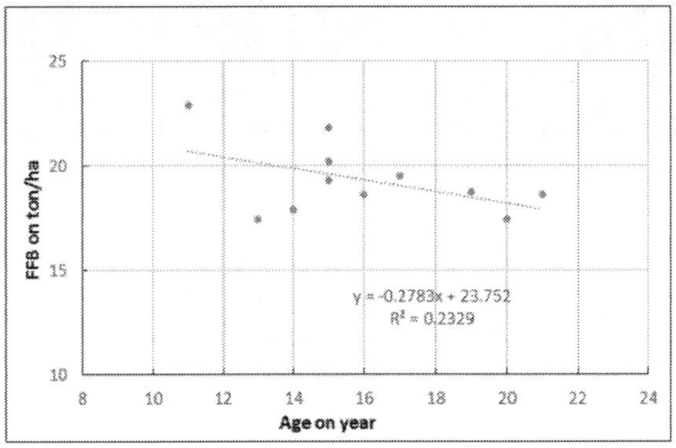

Figure 3. Relationship between age and yield of FFB of oil palm based on 11 years old to 21 years old.

Secondly we investigate relationship between age and HH, HV derived from ALOS PALSAR 2 on the study area. Relationship between HH, HV and age of oil palm can be seen on figure 4. Based on figure 4 relationship between HH, HV and age of oil palms is moderate, the coefficient of determination (R^2) is 0.63 for HH and 0.42 for HV. Figure 4 can be indicated increasing of age of oil palms as increasing of HH and HV value. We hypothesized oil palms growing up that cause higher of trunks, more leafs, more branches and larger of canopy. The condition will add more double and volume backscattering value that overall will add HH and HV backscattering value. According to [11] height of oil palms has strong correlation with age, the coefficient of determination values is 0.90 and canopy of oil palm has strong correlation with age that larger of canopy as older of oil palm, the coefficient of determination value is 0.88 [18] also higher of HH and HV value indicated of higher of above ground biomass of oil palm [19].

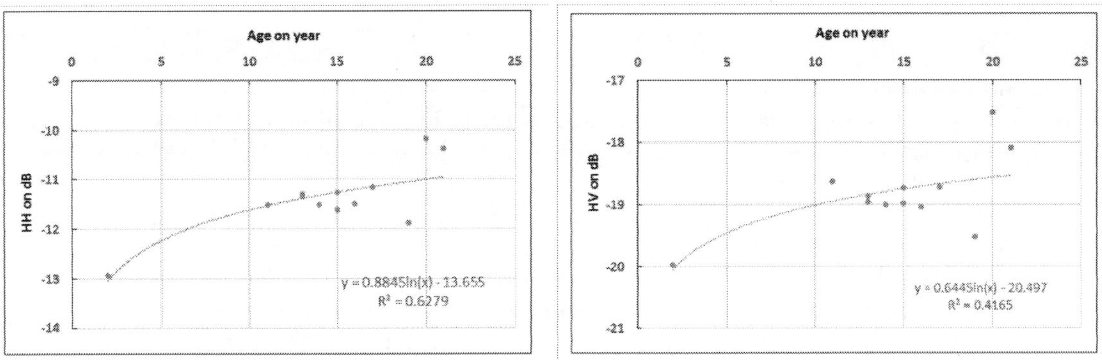

(a) (b)

Figure 4. Relationship between HH and age of oil palm (a) and HV and age of oil palm (b).

Thirdly we investigate relationship between yield of FFB and HH, HV derived from ALOS PALSAR 2 on the study area. Relationship between HH, HV and yield of FFB of oil palm can be seen of figure 5. Based on figure 5 relationship between HH, HV and yield of FFB of oil palms are not strong. The coefficient of determination (R^2) is 0.26 for HH and 0.15 for HV, that was indicated increasing of yield as decreasing of HH and HV value. For this case we also use yield of FFB derived from age of oil palms from 11 to 21 years old. We believe if use more yield of FFB derived from variety of age of oil palms will get more better results. Our hypothesized of age has significant correlation with HH, HV derived from ALOS PALSAR and also age has significant correlation with FFB of oil palm because of that HH, HV has significant correlation with FFB of oil palm. However backscatter values derived from ALOS PALSAR depend on wavelengths, polarization, incident angle and temporal data, environment (e.g. moisture, landscape) [20, 21] and structure of oil palm (e.g. size, geometry and orientation of leaves, trunks, branches, and aerials or stilt roots) [22, 23].

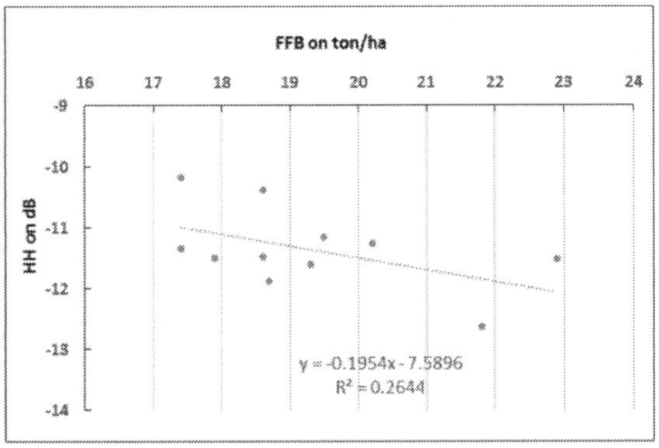

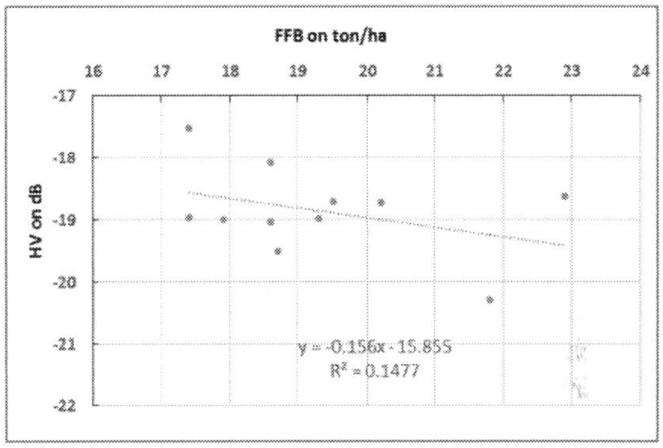

(a) (b)

Figure 4. Relationship between HH and yield of FFB of oil palm (a) and HV and yield of FFB of oil palm (b), FFB derived from oil palm from 11 to 21 years old.

4. Conclusion

The study has showed an investigation of age and yield of FFB of oil palms based on ALOS PALSAR 2. Based on age of oil palms from 11 to 21 years old we found increasing of age of oil palms as decreasing of yield of FFB. Also we found relationship between HH, HV derived from ALOS PALSAR 2 and age of oil palm is moderate. R^2 is 0.63 for HH and 0.42 for HV that be indicated increasing of age of oil palm as increasing of HH and HV value. We hypothesized oil palms growing up that cause higher of trunks, more leafs, more branches and larger of canopy. The condition will add more double and volume backscattering value that overall will add HH and HV backscattering value. Finally we found relationship between HH, HV derived from ALOS PALSAR 2 and yield of FFB derived from age of oil palm from 11 to 21 years old is not strong, R^2 of 0.26 for HH and 0.15 for HV that has indicated increasing of yield as decreasing of HH and HV value. We believe utilization of ALOS PALSAR 2 is promising for monitoring and management of oil palm plantation, for future works we will collect more ALOS PALSAR and samples of oil palm for mapping and creating a model to estimate biomass, age and FFB in regional area also considering effect of rainfall and topography on oil palm management.

Conflict of interest

The author declare that there is no conflict of interests regarding the publication of this paper.

Acknowledgement

This study is carried out as supporting for JAXA SAFE activity. The author would like to thanks to JAXA-SAFE and FELDA Malaysia as executor on JAXA-SAFE activity.

References

[1] Mc Donald P A, Carlson K, Curran L M, and Ponette-Gonzalez, 2013, A NASA satellite data used to study the impact of oil palm expansion across Indonesian Borneo, *The Earth Observer* 25(5) 12-16.

[2] Wahid M B, S N A Abdullah, and I E Henson, 2005, Oil Palm – Achievements and Potential, *Plant Production Science* 8: 288–297.

[3] H Aholoukpè, B Dubos, A Flori, P Deleporte, G Amadji, J L Chotte, D Blavet, 2013, Estimating aboveground biomass of oil palm: Allometric equations for estimating frond biomass, *Forest Ecology and Management* 292, 122–129.

[4] L Khoon Kho and M R Jepsen, 2015, Carbon stock of oil palm plantations and tropical forests in Malaysia: A review, *Singapore Journal of Tropical Geography* 36, 249–266.

[5] Corley R H V, 2009, How much palm oil do we need ?, *Environmental Science & Policy* 12 (2), 134–9.

[6] Basri W M, Abdullah S N A, Henson I E, 2005, Oil Palm, *Plant Production Science* 8 (3), 288–97.

[7] Danielsen F, Beukema H, Burgess N D, Parish F, Brühl C A, Donald P F, Murdiyarso D, Phalan B, Reijnders L, Struebig M, Fitzherbert E B, 2009, Biofuel plantations on forested lands: double jeopardy for biodiversity and climate, *Conservation Biology* 23 (2), 348–58.

[8] Malaysian Palm Oil Board, 2012, *Malaysian Oil Palm Statistics 2012*. Malaysian Palm Oil Board, Selangor.

[9] Sheil, D, Casson A, Meijaard E, van Nordwijk, M Gaskell J, Sunderland-Groves J, Wertz K and Kanninen M, 2009, The impacts and opportunities of oil palm in Southeast Asia: What do we know and what do we need to know ?, *Occasional paper* No. 51. CIFOR, Bogor, Indonesia.

[10] Basiron Y, 2007, Palm oil production through sustainable plantations, *European Journal of Lipid Science and Technology* 109: 289–295.

[11] K P Tan , K D Kanniah and A P Cracknell, 2013, Use of UK-DMC 2 and ALOS PALSAR for studying the age of oil palm trees in southern peninsular Malaysia, *International Journal of Remote Sensing*, 34:20, 7424-7446, DOI:10.1080/01431161.2013.822601.

[12] Van Gelder J W, 2004, Greasy Palms: European Buyers of Indonesia Palm Oil, http://www.foe.co.uk/resource/reports/greasy_palms_buyers.pdf

[13] Mc Morrow J, 2001, Linear Regression Modelling for the Estimation of Oil Palm Age from Landsat TM, *International Journal of Remote Sensing,* 22: 2243–2264.

[12] Wielaard N, 2011, Impact of oil palm plantations on peatland conversion in Sarawak 2005-2010 *Report SarVision,* 14 pages http://www.wetlands.org/Portals/0/publications/Report/Malaysia%20Sarvision.pdf

[13] Lucas R M, Mitchell A, Rosenqvist A, Proisy C, Melius A, and Ticehurst C, 2007, The potential of L-band SAR for quantifying mangrove characteristics and change: Case studies from the tropics. *Aqua. Conserve.*, 17, 245-264.

[14] Tsomon R, Tateishi R, and Tetuko J S S, 2002, A Method to estimate forest biomass and its application to monitor Mongolian Taiga using JERS-1 SAR, *International Journal of Remote Sensing,* 23, 4971-4978.

[15] Lucas R M, Moghaddam M, and Cronin N, 2004, Microwave scattering from mixed-species forest, Queensland Australia, *IEEE Transaction on Geoscience and Remote Sensing*, 42, 2142-2159.

[16] Christine P, 2014, Mapping palm oil expansion using SAR to study the impact on the CO2 cycle, *IOP Conf. Series: Earth and Environmental Science*, 20, 012012 doi:10.1088/1755-1315/20/1/012012.

[17] Corley R H V, B S Gray, 1976, *Oil Palm Research, Development in Crop Science I: Yield and Yield Component*. Amsterdam: Elsevier.

[18] Chemura A, 2012, Determining Oil Palm Age from High Resolution Satellite Imagery, *MSc*

Thesis, International Institute for Geo-Information Science and Earth Observation (ITC), The Netherlands.

[19] Vena S, A R M Shariff, L Nordin, B Pradhan, 2012, Estimation of above ground biomass of oil palm trees by PALSAR, *IEEE Colloquium on Humanities, Science & Engineering Research* (CHUSER), December 3-4, Kota Kinabalu Sabah, Malaysia.

[20] Lucas R, Armston J, Fairfax R, Fensham R, Accad A, Carreiras J, Kelley J, Bunting P, Clewley D, Bray S, Metcalfe D, Dwyer J, Bowen M, Eyre T, Laidlaw M, and Shimada M, 2010, An Evaluation of the ALOS PALSAR L-Band Backscatter—Above Ground Biomass Relationship Queensland, Australia: Impacts of Surface Moisture Condition and Vegetation Structure, *IEEE Journal of selected topics in applied earth observations and remote sensing*, vol. 3, no. 4, 576-593.

[21] Darmawan S, Takeuchi W, Vetrita Y, Wikantika K, Sari D K, 2015, Impact of topography and tidal height on ALOS PALSAR polarimetric measurements to estimate aboveground biomass of mangrove forest in Indonesia, *Journal of Sensor*, Article ID 641798.

[22] Morel A C, Saatchi S S, Malhi Y, Berry N J, Banin L, Burslem B, Nilus R, and Ong R C, 2011, Estimating aboveground biomass in forest and oil palm plantation in Sabah, Malaysian Borneo using ALOS PALSAR data, *Forest Ecology and Management,* 262, 1786–1798.

[23] Darmawan S, Takeuchi W, Vetrita Y, Wikantika K, and Sari D, K, 2014, Characterization of mangrove forest types based on ALOS-PALSAR in overall Indonesian archipelago, *IOP Conf. Ser.: Earth Environ. Sci.* 20 012051.

Yield estimation comparison of oil palm based on plant density coefficient variation index using spot-6 imagery in part of Riau

H. A. Setyowati, S. H. Murti B S, and W. Widyatmanti

Departement of Cartography and Remote Sensing, Faculty of Geography, Universitas Gadjah Mada, Sekip Utara, Sleman, Yogyakarta 55281
Corresponding author **heratania.a.s@mail.ugm.ac.id**

Abstract. Oil palm plantations consist of diverse plant density level that influence the appearance of soil surface or commonly in remote sensing terms called as soil background. Choosing the right density coefficient of vegetation transformation can decrease the noise of soil background for estimating oil palm yield. This research aims 1) to examine the accuracy of SPOT-6 to identify the oil palm l plant growth level and to estimate their yield 2) to know the variation of oil palm yield based on SAVI index vegetation using different density coefficient; and 3) to determine the best density coefficient to estimate the yield of oil palm.

This research was held in part of Air Molek, Indragiri Hulu Regency, Riau, one of the largest oil palm plantations in Indonesia. This research method utilises SAVI transformation with density coefficient L-0, L-0.5, and L-1, and regression statistics analysis. The land-cover primary data is derived from SPOT-6 imagery archived in 13rd June 2013. The field survey was conducted in the same month of image's acquisition time and 120 sample areas were taken during that time. Two steps of regression analyses were applied to see the correlation between, first, vegetation index value and oil palm plant; and second, oil palm plant, vegetation index values, and oil palm yield from field observation. These steps produced a model to estimate the oil palm yield based on the index values of yield and vegetation, and the productivity estimation. The result shows that SPOT-6 imagery has 96% accuracy level which is considered high for identifying the oil palm variation. The R value for L-0 density coefficient is 0.8, for L-0.5 is 0.81 whereas for L-1 is 0.82. The best plant's density coefficient for estimating oil palm yield/yield is L-0 with yield estimation accuracy of 83.33%.

1. Introduction

Indonesia is listed as the fourth biggest population in the world [16] and will reach about 300 million people in 2035 [3]. Indonesian economic level is mostly elevated by industrial sectors including oil palm plantation, either owned by private, community, or Government Company (BUMN). Data from Ministry of Agriculture Data Centre (2013) stated that the yield centres of oil palm plantation in 2011 were Riau, North Sumatera, Central Kalimantan, South Sumatera and Jambi Provinces. Since 2006, Indonesia had exceeded Malaysia in producing oil palm, leading it to the biggest oil palm producer country in the world [8]. Big Private Plantation (PBS) dominates the total area of oil palm. Riau is the most oil palm contributor province in Indonesia [11].

Oil palm plant or in scientific language named as Elaeis guineensis jacq is a palm plant family originating from Nigeria, West Africa and also from Brazil, South America [5]. East coast of Sumatra and Aceh is the first oil palm plantations in Indonesia [7] and subsequently spread to South Sumatra, West Sumatra, Riau, Bengkulu, Jambi, Lampung, West Java, West Kalimantan, Central Kalimantan, East Kalimantan, Maluku, Sulawesi, and Papua [20]. From many parts of oil palm, its fruit arranged in bunches or called TBS (Fresh Fruit Bunches) has the highest economic value compared to others [2].

Oil palm plantation management activities in the form of certain spacing, density levels, and the age of the plants will affect the appearance of the soil background. This soil background condition affects the reflection recorded on satellite imagery therefore a correction to minimize those effect is needed. Researches on oil palm commonly has been using sensors from such as AIRSAR [14] , ALOS PALSAR [12], LIDAR [1], and UK-DMC 2 [22]. This current study uses SPOT-6 imagery as the primary data which is the next generation of SPOT-1 to 5 which has high vegetation object sensitivity. The addition of the blue band which is sensitive to atmospheric disturbances, can increase the variety of transformations used in this study. The use of vegetation index transformation to compress the effect on the soil background image on SPOT-6 is more considered to improve the generated spectral values. SAVI [22] is one of vegetation index that can minimize the influence of soil background using different levels of vegetation densities. In conjunction with oil palm yield parameter, applying SAVI on satellite imagery is expected to produce a better estimation of oil palm yield. To examine this vegetation index in oil palm mapping thus this research aims 1) to examine the accuracy for identifying the oil palm plant and to estimate its yield; 2) to know the variation in every yield results based on SAVI vegetation index with each density coefficient; and 3) to determine the best density coefficient and result for to estimate the yield of oil palm-based imagery SPOT-6.

2. Study Area

This study was carried out in Indragiri Hulu, Riau province, precisely on the oil palm plantations belongs to Tunggal Perkasa Plantations (TPP) Company, a private company engaged in agriculture industry which is a subsidiary of PT. Astra Agro Lestari. This company's main yields are CPO (Crude Oil palm) and kernel (Palm Kernel Oil). There are 23 departments which are divided into 447 blocks of plantations (Figure 1).

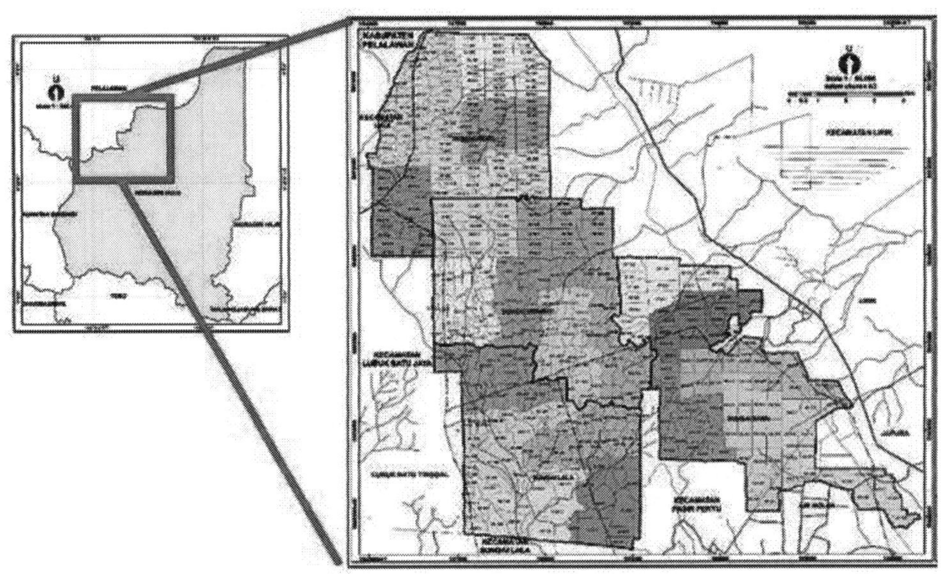

Figure 1. Area of study.

3. Materials and Methods

3.1. Data

This study utilizes SPOT-6 multispectral imagery with 6 meters spatial resolution recorded on 13-06-2013, provided by Indonesian National Institute of Aeronautics and Space (LAPAN). Other data were also used i.e. Map of Indonesian Topographic (RBI) Scale 1: 50,000 Sheet Kelayang Simpang 0915-41, Sheet 0915-42 Air Molek, Sheet 0915-43 Ukui One, and Sheet 0915-44 Base Kapau; and block plantation maps used for the area boundaries extraction and basic map information. Yearly data of planting growth is also used to determine sample yield areas.

3.2. Methods

3.2.1 ToA radiometric correction

LAPAN has provided SPOT-6 data guidance to do image pre-processing. Top of Atmosphere (ToA) radiometric correction were applied to convert digital number (DN) into at sensor radiance and then to at sensor reflectance, using below formulas:

$$\text{Radiance: } L\lambda = \frac{DNi}{Gi} + Bi \quad (1)$$

Description:
$L\lambda$: Radian Band-I, DNI: Digital Number Band I, Gi: Gain Band-i, Bi: Bias Band-i (for satellite SPOT, bias value of each Band = 0).

After *at sensor radiance* correction, the next step performed *at reflectance sensor* correction. The process is as follows:

$$\text{Reflectance: } \rho\lambda = \frac{\pi x L\lambda x d^2}{ESUN\lambda x cos\theta} \quad (2)$$

Description:
Reflectance Toa, $L\lambda$ = Radian Band-I, = Irradiance Sun, = angle Zenith Solar, d^2 = Ratio Distance Earth Sun, d^2 = 1-0.0168 * cos ((Julian Day / 365 * 360) * (π / 180))) 2; Zenith Angle of the Sun: (a) = cos (900 - angle of elevation). This study uses SPOT-6 imagery obtained in June 13, therefore, the value of Julian day is 164.

The results of calculations for radiometric correction process was performed using formulas as follows:

$$\rho\lambda\ Band\ 1 = \frac{3.141593\ x\left(\frac{B1}{9.529}\right)x\ 1.000023}{1540.494123\ x\ 0.793251276} \quad (3)$$

$$\rho\lambda\ Band\ 2 = \frac{3.141593\ x\left(\frac{B2}{10.4416}\right)x\ 1.000023}{1826.087443\ x\ 0.793251276} \quad (4)$$

$$\rho\lambda\ Band\ 3 = \frac{3.141593\ x\left(\frac{B3}{14.0063}\right)x\ 1.000023}{1982.671954\ x\ 0.793251276} \quad (5)$$

$$\rho\lambda \; Band\ 4 = \frac{3.141593\ x\left(\frac{B4}{8.5175}\right)x\ 1.000023}{1094.747446\ x\ 0.793251276} \tag{6}$$

Value of L factors are various depended on the vegetation density levels. For high vegetation cover, the L value is 0.0, the medium one is 0.5, and the low one is 1.0 [19]. The existence of object above the ground will contribute as a background to the vegetation reflection, so their effect on the reflection is recorded on the imagery digital data value (DN).

$$\boldsymbol{SAVI} : \frac{(\boldsymbol{Near\ Infra\ Red - Red})(\boldsymbol{1+L})}{(\boldsymbol{Near\ Infra\ Red + Red + L})}$$

(7)

The use of SAVI transformation is expected to reduce that effect on the objects background. In this case, oil palm plants are planted with a wide enough space between them so that the reflection of the soil will greatly affect the vegetation reflection, thus SAVI is needed.

3.2.2 Regression Analysis

Regression analysis used in this study was the linear regression between the variables of the age of the oil palm plant and its vegetation index values in. Then, multiple regression analysis was applied to yield estimation model, using age of plant as x and yield y variables as to create the image of oil palm productivity, using following formula:

a. Linear Regression

$$Y = a + bx \tag{8}$$

Description :

Y = the dependent variable (dependent) (year of planting);

a, b = regression coefficient;

x = the independent variable (independent) (transformation of vegetation index)

b. Multiple Linear Regression

$$Y = a + bx + cx2 \tag{9}$$

Description :

Y = the dependent variable (dependent) (yield);

a, b, c, .. n = regression coefficient;

x = the independent variable (independent) (transformation of vegetation index and age)

3.3Accuracy Test

The accuracy of the estimation model of yield was conducted based on the parameters that influenced the yield and value of vegetation index calculated using the formula used in the study belong to [21]. This formula uses the difference between the deviation of the yield estimation results and the real yield. The result show a good value if its deviation is close to 0 and a higher degree of accuracy if the percentage is approaching 100%. To check the accuracy of the interpretation of the results this study use confusion matrix. Formula test the accuracy of the model is as follows:

$$\text{Deviation} = \text{Yield Estimated} - \text{Real Yield} \tag{10}$$

$$\text{Accuracy (\%)} = 100\% - (\frac{(Deviation\ Value)}{(Real\ Yield)} x100\%) \tag{11}$$

4. Research analysis procedures

The stages of data processing analysis are illustrated in **Figure 2.**

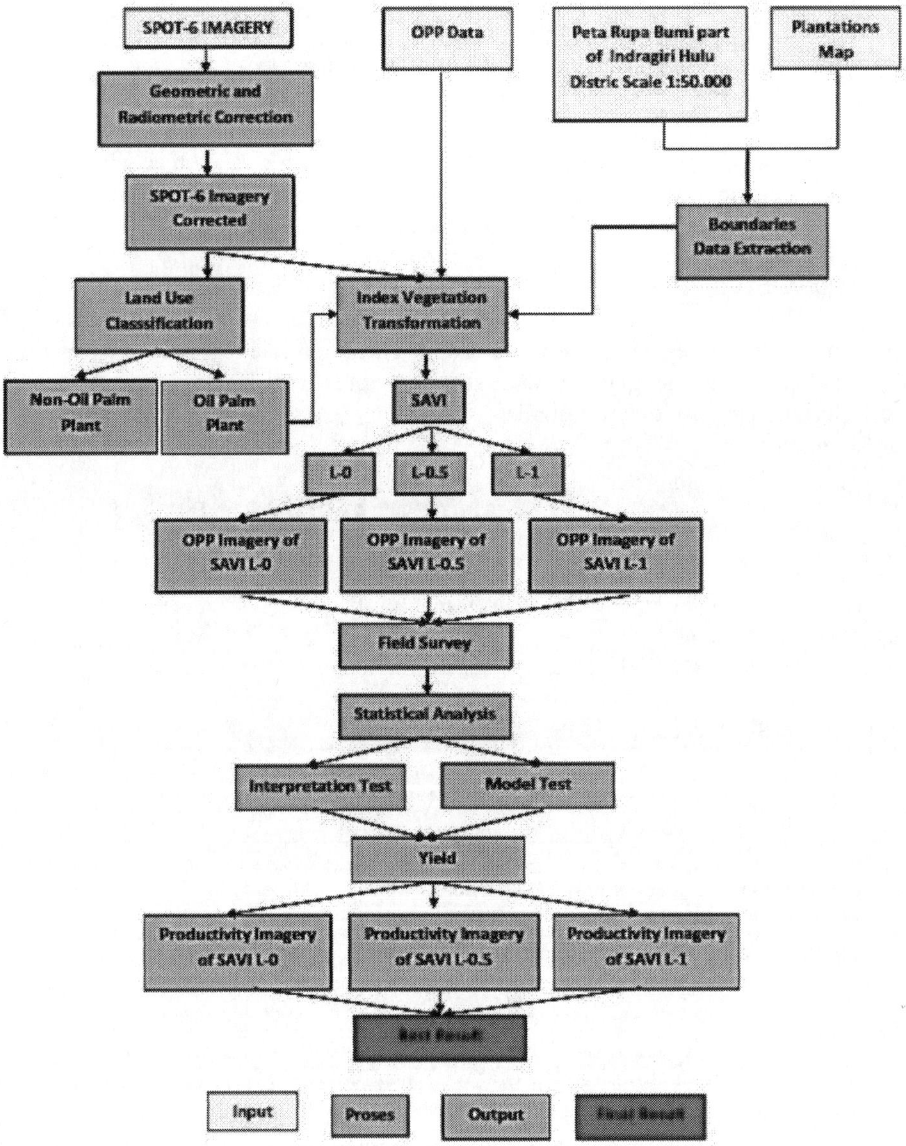

Figure 2. Flowchart of research.

5. Result, Analysis, and Discussion

5.1 *Pre-processing*

The satellite imagery of study area shows that almost 100% of plantation area was covered by both thin and thick haze. The TOA Radiometric correction applied to reduce the influence of these hazes on the digital imagery in order to reveal the real object value in the field. The combination of

IOP Conf. Series: Earth and Environmental Science **37** (2016) 012038 doi:10.1088/1755-1315/37/1/012038

blue channel in SPOT-6 is also considered to be able to reduce the haze effects. The display of SPOT-6 image sharpening with equalization (Equalization Enhancement) is shown in Figure 3.

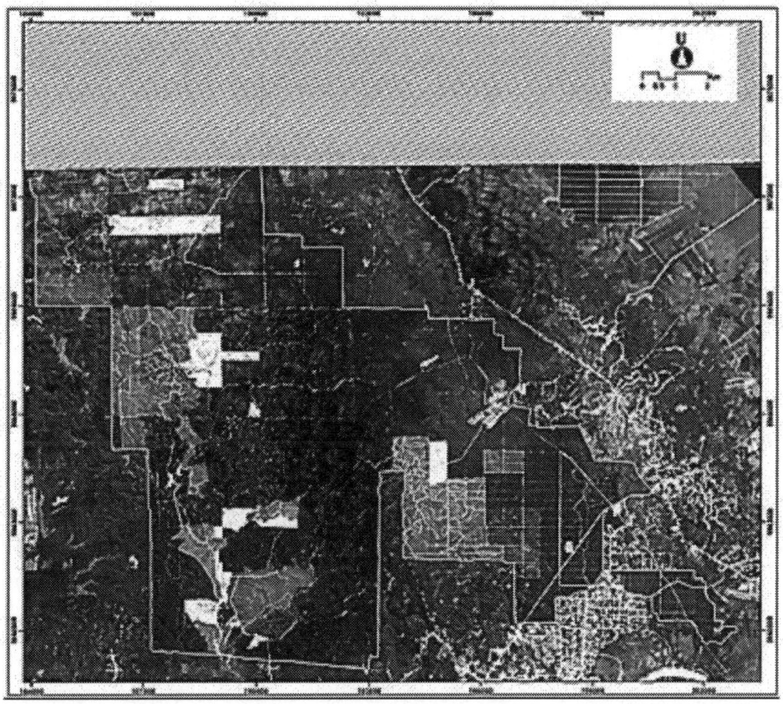

Figure 3. The appearance of the condition image acquired SPOT-6.

Figure 4 presents the image comparison between the one that has not been corrected radiometrically and the one that has. The appearance of hazes over the vegetation is reduced and the appearance of objects appears sharper. Those differences can be seen on the pixel values resulted from TOA correction (see the explanation in white box in **Figure 4**).

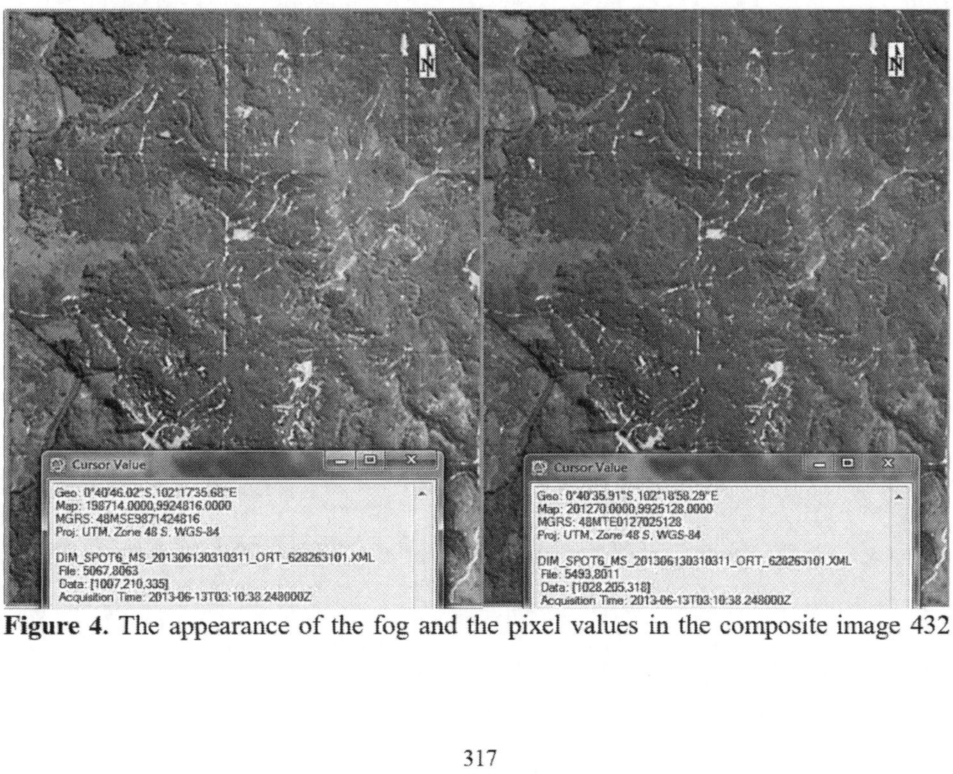

Figure 4. The appearance of the fog and the pixel values in the composite image 432

before the correction (left) and a reduction in the visual area of the fog and the pixel value after correction at reflectance sensor (right)

5.2. Vegetation Index Transformation

The correlation analysis between the vegetation index, year of planting, and SAVI transformation result using L-0 presents the lowest value. This low correlation is caused by plantation management that may influence the digital quality of satellite imagery. The spacing system, age of plant, regular midrib pruning or trimming to reduce evaporation and to open the canopy cover so the sunlight can penetrate the ground, cause the canopy conditions from different year of planting or ages looked the same. This finding support previous studies [10], [19], and [22] which identified the similar result in mapping oil palm. SAVI Transformation produces a planting year regression formula as follow:

FORMULA OPP SAVI L-0 Y = 1892.62338654888 + (144.342393790123) SAVIL-0 (12)

FORMULA OPP SAVI L-0.5 Y = 1920.37592846425 + (176.312417272302) SAVIL-0.5 (13)

FORMULA OPP SAVI L-1 Y = 1924.41864322699 + (200.974132796737) SAVIL-1 (14)

SAVI transformation has an inverse relationship between values of oil palm productivity and vegetation index but has a proportional one between vegetation index and the age of plant. This finding is in the contrary with pervious study [13] who could not find any relationship between vegetation index value and the yield of seasonal crops (in the case of tobacco). In general, the comparison of the correlation between vegetation index value, planting year, and productivity in this study are illustrated in **Figure 5**.

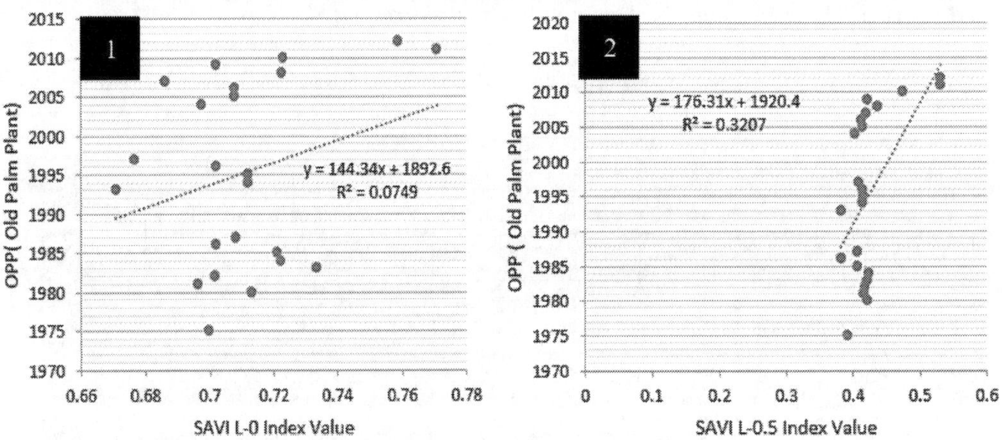

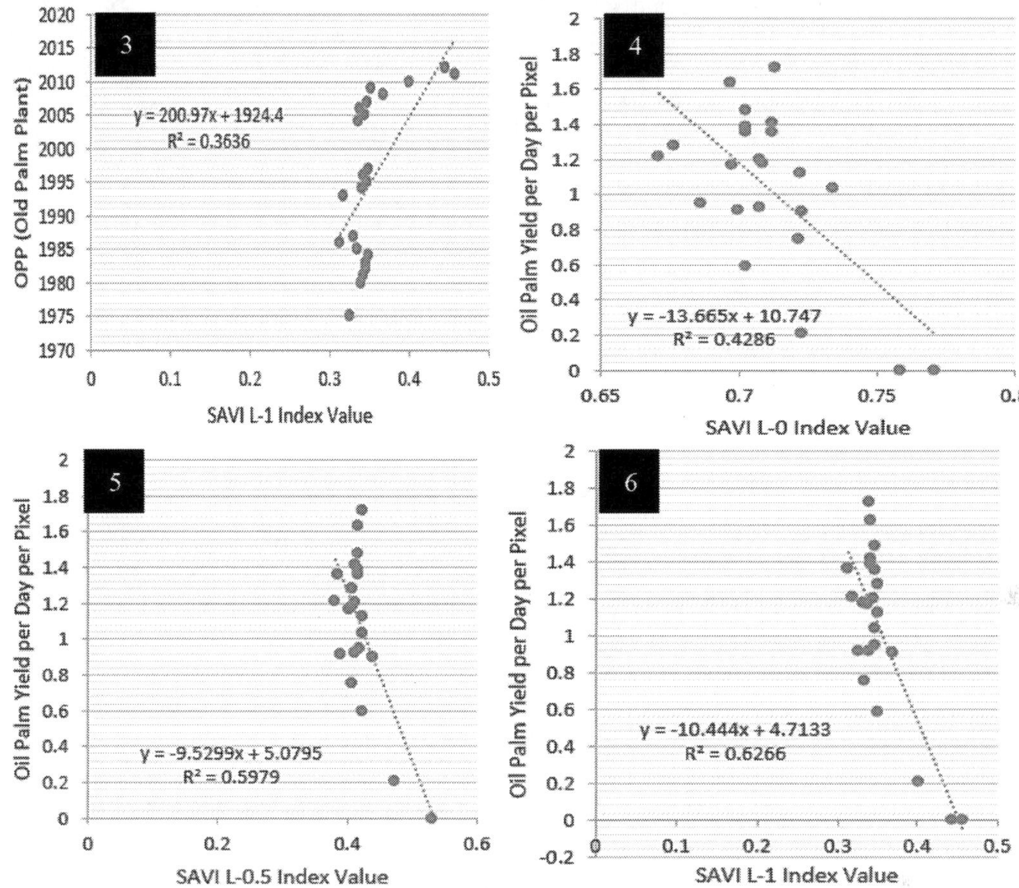

Figure 5. (1, 2, and 3) Correlation of vegetation index and opp (year of planting) and (4, 5, and 6) correlation of vegetation index and productivity.

5.3. Multispectral classification

Purwadhi [17] mentions the area of oil palm plantations is very suitable when it is located on tahe dry-field, bushes, bare-land, or convertible yield forest (HPK), but also has settlement (APL) nearby. This corresponds to the land use function in study area, after identifying the appearance of oil palm plantations and land cover. There are eight classes of land cover which are artificial lakes, settlements, bushes, bareland, replanting area, old and young oil palm area.

Multispectral classification is used to distinguish different objects in oil palm plants using maximum likelihood method, followed by majority/minority classification filters. The accuracy test shows that multispectral classification accuracy with the actual conditions in the field (overall accuracy) is 96%. This finding support previous study [18] that was conducting research on the disease detection and mapping of oil palm plant using Quickbird imagery to detect the disease and also the pattern of basal stem rot disease in North Sumatra. The use of six kinds of vegetation index that is applied to the visible band (λ: 450-695 nm) and near infrared (λ: 760-890 nm) is used to identify the plants affected by the disease. Accuracy segmentation palm trees infected by 84% and the mapping accuracy of 96%. Meanwhile the overall accuracy of test results for classification of old oil palm crops or crop yield, reaches 99.46%.

5.4. Modeling Estimation of Oil Palm Yield

Table 1 shows the values of R and R^2 of the regression analysis result which presents the magnitude of the correlation between afore mentioned variables and also the direction of relationship. Variables which have a strong relationship and influence each other, present a strong values regression. Variables which have strong influence or relationship but combined with variables which have a weak relationship and affect each other variables will reduce the value of regression.

Table 1. Correlation and regression on transformation SAVI.

	Vegetation Index	*OPP and Vegetation Index (R^2)*	R	*Yields and Vegetation Index (R^2)*	R	*OPP, Yields, and Vegetation Index (R^2)*	*R*
1	SAVI L-0	**0.273752**	**0.5232131**	**-0.65469**	**0.80912916**	**0.655937**	**0.80989938**
2	SAVI L-0.5	**0.566327**	**0.752547**	**-0.77327**	**0.87935772**	**0.656756**	**0.81040484**
3	SAVI L-1	**0.603008**	**0.7765359**	**-0.79158**	**0.88970782**	**0.667054149**	**0.81673383**

Oil palm crop yield model was built using three variables: the age of the oil palm [9], the vegetation index, and yield itself. The model was developed using two stages, first: regression between the years of planting the vegetation index which was applied to the vegetation index transformation image to produce a new image that illustrated the distribution of age; second: regression analysis between all the variables that produced new images that depict the distribution of productivity. This stage of modeling produces oil palm yield estimation formula as follows:

FORMULA SAVI L-0 Y = 47.8866175050089 + (- 0.0196233740165436) OPP + (-10.83260743103) SAVIL-0 (16)

FORMULA SAVI L-0.5 Y = 27.447058489428 + (- 0.011647479092411) OPP + (-7.4763101487276) SAVIL-0.5 (17)

FORMULA SAVI L-1 Y = 23.9219264018003 + (- 0.0099815208365571) OPP + (-8.43758981829296) SAVIL-1 (18)

5.5. Yields Estimation Based on Vegetation Index Transformation

Yields estimation model produces an image that provides the value of crop productivity of oil palm per pixel in one year. This image is able to show the total value of yield within one year by summing all pixel values. **Figure 6 (1, 2, 3)** show a comparison of transformation SAVI productivity image using L-0, L-0.5, and L-1 respectively.

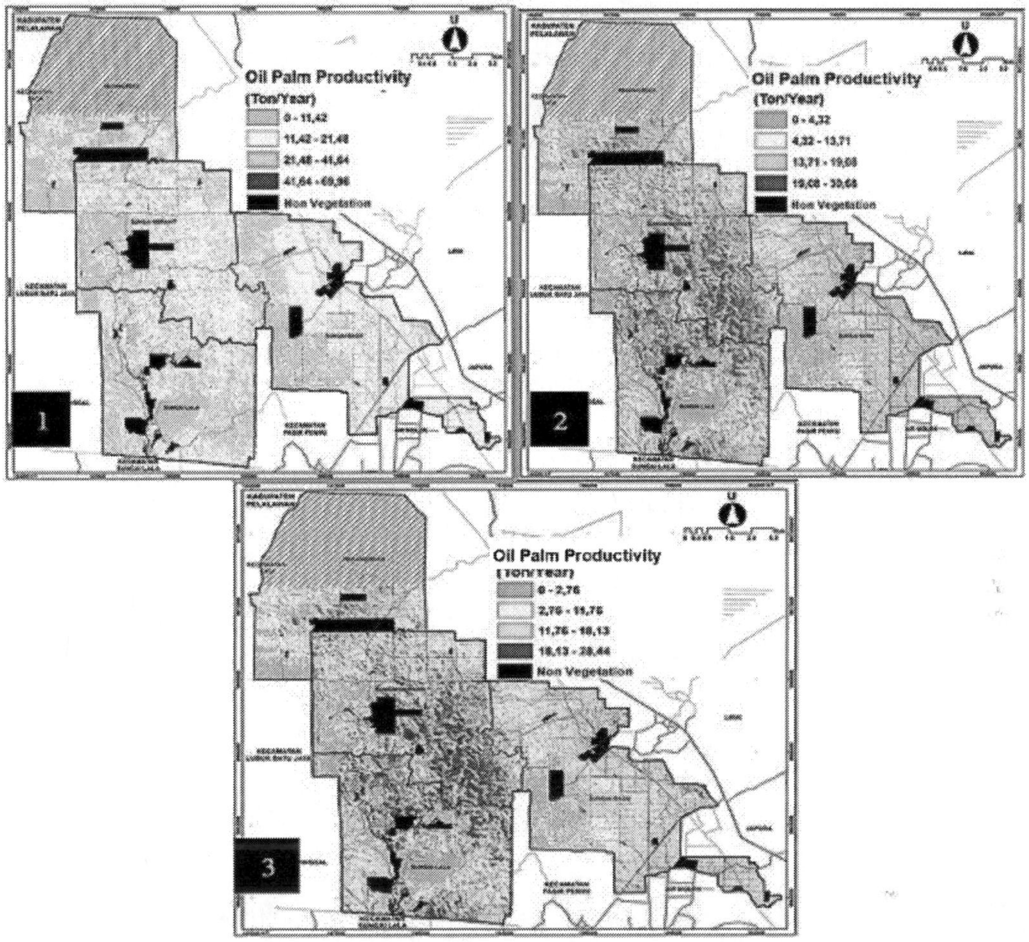

Figure 6. The appearance of the final result image transformation productivity SAVI L-0 (1), SAVI L-0.5 (2), and SAVI L-1 (3).

The result of yield estimation analysis using SAVI transformation L-0 has the greatest value of accuracy of 83.33%. This result is related to the density of vegetation cover coefficient which indicates the high density plantations level of oil palm. Transformation SAVI L-0.5 and L-1 have smaller values with accuracy of 79.36% and 78.10%, respectively. These differences show that the moderate and low density coefficients are not suitable for oil palm plantations because the reflection of atmospheric disturbances background from the ground is still very high. Thus, SAVI transformation with high density coefficient is suitable for the oil palm plantation area imagery analysis because it can manage the variety of vegetation cover density conditions.

5.6. Accuracy of Interpretation and Estimation

A total of 120 samples of plantation blocks were determined using stratified random sampling. From that number, more samples were taken in the blocks that consist of more variety of years of planting. Overall classification accuracy test for old palms trees or plants reaches 99.46% when it applied using image analysis of 432 false colors composite. Table 2 shows the accuracy test result of interpretation and of model accuracy estimation for oil palm yield which shows variations ranging from 78 to 83%. This accuracy test result is not considered high based on agronomists expertize field for the classes of plants [6].

Table 2. Results of yield estimation accuracy on SAVI transformation.

	Vegetation Index	*Yield Estimated (ton/year)*	Real Yield *(ton/year)*	*Deviation (ton/year)*	Percent	*Percent of Deviation*	*Accuracy (%)*
1	**SAVI L-0**	**151421.5**	**181702.59**	**-30281.085**	**100**	**-16.67**	**83.33**
2	**SAVI L-0.5**	**144200.1**	**181702.59**	**-37502.485**	**100**	**-20.64**	**79.36**
3	**SAVI L-1**	**141918.3**	**181702.59**	**-39784.285**	**100**	**-21.90**	**78.10**

Those three transformations show that the transformation of the SAVI is capable to detect the density of vegetation as also indicated [10]. Height cover coefficient provides the best transformation result in oil palm crop yield estimation. It was also shown in the field that oil palm plants is the majority plants there. However to maintain the ecological functions of the soil and the environment, and soil moisture, other plants were also planted (*Legumes muconna, ferns nefrolephis,* flower plants like *Turnera subulata, Euphorbia heterophylla, Cassia tora,* and *Antigonon leptopus*). The natural fertilization containing stem from pruned the oil palm plant, empty fruit bunches, and etc. will cause the soil changing into different chemical condition and backgrounds.

6. Conclusion

SPOT-6 satellite imagery has a very high accuracy to identify the oil palm vegetation, reaching value of 99.46%. The results show the value of the yield estimation for the SAVI transformation coefficients with low coverage (L-1) is 141,918.3 tons, for a medium coverage (L-0.5) is 144,200.1 tons, and high coverage (L-0) is 151,241.5 tons. When compared with the data field yield which is 181,702.6 tons, the accuracy of yield estimation from SAVI transformation of L-1 is 78.10%; SAVI L-0.5, 79.36% and SAVI L-0, 83.33%. The estimation result of the oil palm yield is best when appliying the transformation of SAVI L-0 with high coverage coefficient.

Acknowledgments

The author would like to thank the Administrator PT. Tunggal Perkasa Plantations, Mr. Sumarno who have given permission the author to study there and all the staff who have helped this research activities. The gratitude also goes to the ground staffs who have helped the author in doing field work, as well as Mr. Hadi Purnomo Aulia as a mentor in PT. Tunggal Perkasa Plantations, and PT. Astra Agro Lestari.

Head of Data Processing LAPAN, Mr. Drs. Kustiyo, M.Sc. who has given the author to access to the image of SPOT-6 in this study, Mrs. Sukentyas Estuti Siwi, S.Si., M.Sc. and Mrs. Diannovita, S.Si., as a mentor in LAPAN, and the entire staffs who have been helpful in assisting the author are also very much appreciated.

References

[1] Arno', J., Sanz, R., Escola`, A., Masip, J., Valle`s, J M., Palacı'n, J., Llorens, J., Rosell-Polo, J R 2012 Leaf area index estimation in vineyards using a ground-based LiDAR scanner. Precision Agriculture. (2013) 14: 290-306

[2] Astra Agro Niaga 1995 Guidelines Basic Brevet-1 Crop Plant Engineering. Jakarta: PT. Astra Agro Lestari Tbk

[3] Central Statistics Association 2013 Indonesian Population Projection 2010-2035.Jakarta: BPS Jakarta-Indonesia

[4] Central Statistics Association 2014 Regional Statistics Indragiri Hulu 2014 ISBN 979 484 6503 Riau: BPS (Central Statistics Association) Indragiri Hulu.

[5] Corley, R.H.V., Tinker, P.B. 2003 The Oil palm (fourth edition) USA: Blackwell Science Ltd-Blackwell Publishing Company.

[6] Danoedoro, Projo. 2012 Introduction to Remote Sensing Digital Yogyakarta: Yogyakarta

ANDI pp 243 260

[7] Fauzi, Yan., Widyastuti, YE, Setyawibawa, I., and Hartono, R 2002 Oil palm: Cultivation, Utilization and Waste, Business Analysis and Marketing Jakarta: Penebar Organization.

[8] Hasan, Fadhil. Oil palm Industry Reflection 2013 and 2014 2014 The prospect of the Indonesian Oil palm Association www.gapki.co.id. last accessed 7 Oktober 2014

[9] Idrus, Erwin. 1998 Digital Data Analysis SPOT Multispectral for Estimating Oil Palm's Shareable Productivity in Oil Palm Plantations Sebrang, North Sumatra, Thesis, Yogyakarta: Gadjah Mada University, Faculty of Geography.

[10] Imanishi, Junichi., Morimoto, Yukihiro., Imanishi, Ayumi., Sugimoto, Kayoko., Isoda, Kazutaka.2007.The independent detection of Drought stress and leaf density using resolution hyperspectral data. International Consortium of Landscape and Landscape Engineering and Springer 2007. Ecol Eng (2007) 3: 55-65

[11] Maulidar, Indri. 2014 Ministry of Trade: Europe Respond Positively Indonesian CPO.www.tempo.co last access October 7, 2014

[12] Morel, Alexandra C., Sassan S. Saatchi, Yadvinder Malhi, Nicholas J. Berry, David Burslem,Reuben Nilus, Robert C. Ong, 2011 Estimating Aboveground Biomass in Forest and Oil Palm Plantation in Sabah, Malaysian Borneo Using ALOS PALSAR Data. International Journal of Forest Ecology and Management: Elsevier. 262 (2011) 1786–1798.

[13] Murti BS, Heru Sigit 1997 Estimated Yield of Tobacco Leaf Processing Integration Based on Landsat Thematic Mapper with Geographic Information Systems, Thesis, Yogyakarta: Faculty of Geography, University of GadjahMada.

[14] Nordin, L., Shahruddin, A., and Mariamni, H. 2002. Application of AIRSAR Data to Oil palm Tree characterization. MACRES Bulletin. ISSN: 1511-7748. Kuala Lumpur.

[15] Data Center and Remote Sensing Technology, PUSTEKDATA LAPAN. (2014), Radiometric Correction Data Satellite Spot-6 (http://pustekdata.lapan.go.id/) last access October 12, 2014

[16] Purnomo, Hendaru. 2014 State of the World's Most residents, RI 4 Great Sign. http://finance.detik.com/ October 7, 2014

[17] Purwadhi, Florentina Sri Hardiyanti .; Haryani, Nanik Suryo .; Siwi, Sukentyas Estuti. 2010 Spatial Information Distribution and Potential of Oil palm Plantations of Remote Sensing Data in the Province of South Sumatra .. News sensing. Pekayon, East Jakarta: Remote Sensing Data Center, LAPAN. Vol.IX ISSN 1412-4564.

[18] Santoso, Heri., Gunawan, Totok., Jatmiko, Retnadi Heru., Darmosarkoro, Witjaksana., Minasny, Budiman 2010 Mapping and Identifying Basal Stem Rot Disease in Oil palms in North Sumatra with Quickbird Imagery, International Journal of Remote Sensing , Springer. Precision Agric (2011) 12: 233-248

[19] Sasaki, Takesi., Imanishi, Junichi., Ioki, Keiko., Morimoto, Yukihiro., Kitada, Katsunori. 2008 Estimation of leaf area index and canopy openness in broadleaved forest using an airborne laser scanner in comparison with high- resolution near-infrared digital photography. International Consortium of Landscape and Landscape Engineering and Springer 2008. Ecol Eng (2008) 4: 47-55

[20] Sunarko 2007 Practical Oil palm Cultivation and Processing, Jakarta: Agromedia Library.

[21] Susetyo, Imam 2012 Modeling Crop Yield Estimates Rubber (Hevea brasiliensis Muell.-Arg) Based Satellite imagery, genetic potentials, and Terrain Data Unit (Study Gardens Getas, PTPN IX Salatiga, Semarang regency, Central Java), Thesis, Yogyakarta : Faculty of Geography, University of Gadjah Mada.

[22] Tan Kian Pang 2011 Use of UK-DMC for Identifying Age of Oil palm Trees in Southern Peninsular Malaysia. International Journal of Remote Sensing.

A study on the oil palm fresh fruit bunch (FFB) ripeness detection by using Hue, Saturation and Intensity (HSI) approach

Muhammad Kashfi Shabdin[2], Abdul Rashid Mohamed Shariff[1,*], Mohd Nazrul Azlan Johari[1], Nor Kamilah Saat[2], Zulkifly Abbas[2]
[1]Department of Biological and Agricultural Engineering, Faculty of Engineering, Universiti Putra Malaysia, Serdang, Selangor, Malaysia
[2]Department of Physics, Faculty of Science, Universiti Putra Malaysia, Serdang, Selangor, Malaysia

*Email: rashidpls@upm.edu.my

Abstract. To increase the quality of palm oil means to accurately grade the oil palm fresh fruit bunches (FFB) for processing. In this paper, HSI color model was used to determine the relationship between FFB's color with the underipe and ripe category so that the grading system could be accurately done. From the analysis manipulation, a formula was generated and applied to the data obtained. The by linear regression in the data shows an average success rate at 45% accuracy for oil palm ripeness detection. Artificial Neural Network (ANN) however return a better accuracy result for both underipe and ripe categories which are 60% and 80% respectively. This yield an overall accuracy of 70%. This can be increased more by improving the grading system.

1. Introduction
Quality is the most important factor in the oil palm production because high quality products were believed to be the key to success in today's competitive market. In oil palm productions, quality is always determining by its texture, shape and color. These features of oil palm Fresh Fruit Bunches (FFB) is always been observed by the human's vision which leads to inconsistencies and inaccurate. Besides, this job is tiresome and time consuming. These faults lead research on methods to make this system computerized using machine vision based technologies by using the imaging technique.

Oil palm fruit as shown in figure 1 is one of the major agricultural product exports by Malaysia. Palm oil has become the ingredient in the making of margarine, candles, soaps, domestic frying oil and snack foods [1]. Oil palm FFB is very mutual in Malaysia. Grading of oil palm fruit conventionally observed by human's vision but today, many types of research has been carried out to find the correlation between oil content in the oil palm fruit against the color of the oil palm fruit [2].

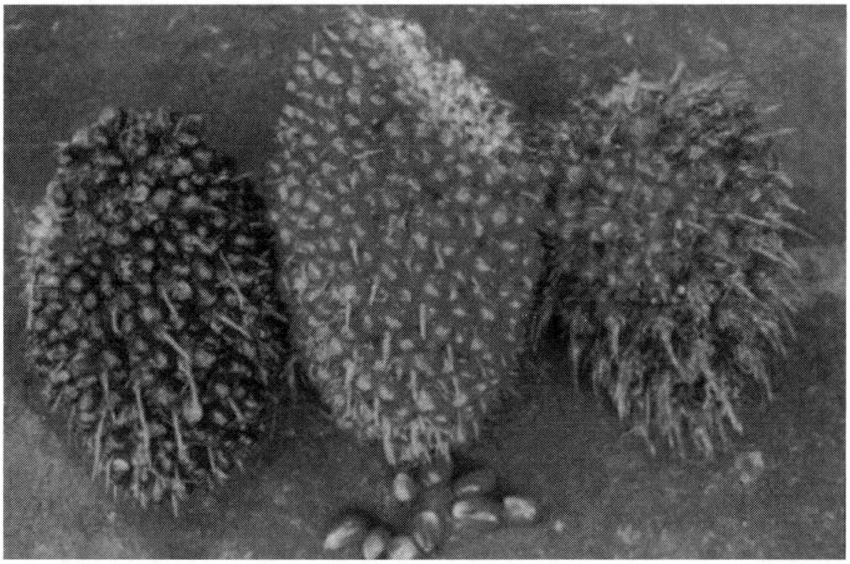

Figure 1. Three categories of oil palm: left is underipe, the center is ripe, right is overipe.

In this research, oil palm is utilized as sample for testing the improved fresh fruit bunch (FFB) grading system because this machine is fabricated especially to assist the conventional oil palm grading system. This machine uses the imaging technique in getting important information regarding the feature of the FFBs. With a closed environment and balanced illumination, a good image could be captured thus analyzed. Imaging technique can utilize features and color of the fruit. Therefore, in this study, colors (HSI) were chosen as the main parameter because the fabrication of the machine already utilized all the features and not the colors of the FFBs. The scope of work will only cover the two main category of oil palm which is underipe and ripe. This is because these two categories are hard to differentiate between each other and are the main categories that are in need to be segregated. Overipe category, however, is easier to differentiate when compared with ripe. Overipe is not in concern because it will be processed along with ripe fruit but underipe has to be segregated because processing the fruit will be costly and yield low quantity palm oil.

Color provides valuable information in estimating the ripeness of the oil palm FFB. Color is one of the most significant criteria related to fruit ripeness identification [2]. Light reflected from an object determines the color of the object thus these variations provide a base for image processing and analysis. Red, Green and Blue (RGB) are the main components in the color code. The other color is the additive of these three main colors. Although a human's perception of color is a subjective process, the physical nature of color still can be distinct by various experiments and results which explained by the frequent response of the brain to the stimuli when in contact with lights, naturally produced when incoming light reacts with several types of Cone cells in the eye.

According to [3], when a beam of light comprising of many wavelengths approach the eye, its response is not given by the sum of the responses that would be produced by each one of the component wavelengths acting alone. Although there are many different distributions of wavelengths that can be observed, our characterization of the appearances of these distributions can be expressed using a relatively small number of parameters. That is, there are many different combinations of wavelengths which appear to produce the same visual color.

Even though a number of different methods can be used to characterize a combination of wavelengths, it turns out that all of them use either three or four parameters since this small number of parameters is related to the way the eye recognizes color. The simplest triplet of parameters is called hue, saturation and intensity.

If a beam of light is break up into each of its component wavelengths and if the intensity of each component is plot as a function of wavelength, then, the hue is the peak of this plot – the wavelength (or relatively insignificant band of wavelengths) which has the extreme intensity. The hue is generally the single word that can be used to describe a composite color. Hue values range from about 440 nm for violet, 450 nm for blue, up to about 700 nm for red light. The names associated with different hues follow the spectral decomposition of a rainbow: red, orange, yellow, green, blue, and violet. These descriptive colors are associated with ranges of wavelengths rather than with unique values, and some people can see colors outside of this conventional range of wavelengths (ultraviolet with a wavelength shorter than violet or infra-red with a wavelength longer than red).

The saturation of a beam of light is related to the width of the plot of intensity vs. wavelength described above. A completely saturated beam would have only one wavelength and would be called monochromatic, which a completely unsaturated beam would contain all wavelengths in equal fraction and would appear white. An absolutely saturated beam therefore, has a very narrow intensity distribution function (possibly consisting of only one non-zero value in the limit), which a completely unsaturated beam has a very wide distribution function, possibly consisting of a constant value over most or all of the visible spectrum.

The intensity is related to the strength of the light beam. Intensity is very difficult to specify because the apparent brightness and the actual brightness can differ significantly. The intensity is related to the total power in the light beam as measured by some objective instrument (such as a photographic light meter), but the perceived brightness of a light (or lightness of a surface) is strongly influenced by lots of other factors and cannot always be specified objectively.

These parameters are often not independent of each other. For example, the intensity and hue of a standard light bulb are related through the black-body relationships – decreasing the output intensity of a black body also shifts the hue towards longer wavelengths.

The hue, saturation and brightness of a light beam are often specified using a three-dimensional color tree, as shown in Figure 2. The vertical axis of the tree specifies the intensity of the beam, from nothing at the bottom which is black, through gray to some maximum value at the top equivalent of the brightest possible white. At each level of the tree (which corresponds to a given lightness or brightness), we draw a circle whose circumference shows the various pure, fully saturated, monochromatic colors of the rainbow in wavelength order from red to violet. The points on a radius line from the center of the tree to some point on the circumference represent different unsaturated colors formed by mixing some amount of white from the center of the tree with some amount of the color at the endpoint of the line.

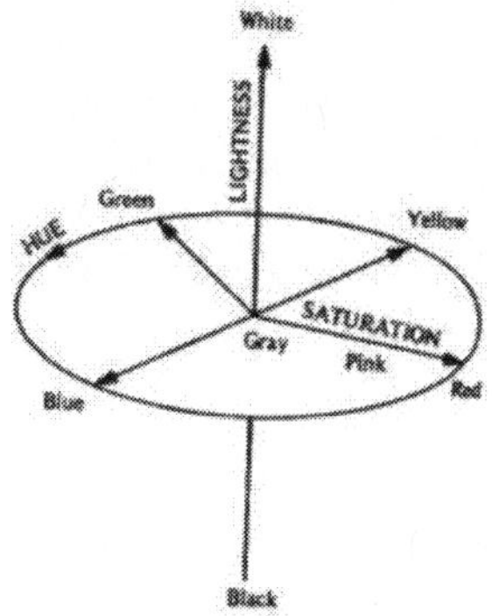

Figure 2. The HSI representation of color [3].

Whereas in oil palm application, shape and pattern cannot be a typical guide for one to estimate the ripeness of FFBs. This is because the fruit may vary in the shape and pattern but share equal quality. In order to overcome this, the grading device should be able to analyze the color of the fruit therefore measures the density of the color by using RGB model. It was believed that color changes resulting from the chemical reactions in the fruit texture can be related to the fruit maturity [4]. This application would decrease the usage of the human grader and minimize the time consume for processing purposes.

Latest development and application of the basic concepts and technologies associated with image capturing and processing were reviewed. There was an approach done to classify potato chips by using pattern recognition from color digital images which include a) image acquisition, b) preprocessing, c) segmentation, d) feature extraction and e) classification [5]. There were also some of the sensor systems which have been explored for machine vision based FFB grading were using optical RGB cameras [6][7][8][9][10]. In [6], the harvesting machine was designed by using the RGB color intensity where they can segregate between different FFB categories. The charged couple device (CCD) camera was used in this system. Hue, Saturation and Intensity (HSI) were used in [7] to predict the best time for harvesting the FFB and by using this color model with multivariate discriminant analysis, they found that the average misclassification for the vision system at 8%. In [8], visible and near infrared wavelength is used to differentiate the old wheat plant which is infected by the yellow stripe, leaf and stem rust. They claimed to found that the sequential application of Anthocyanin Reflectance index can separate the healthy and infected plant under the laboratory conditions. In [10], the experiment was done by using the RGB color model to differentiate between different categories. The data was then trained by using supervised learning Hebb technique and graded using fuzzy logic. The results showed that the automated grading by using the RGB produced an average of 49% success rate while the neuro-fuzzy logic used achieved an accuracy level of 73.3%. Therefore, the motivation here comes to analyzed different categories of FFB by using HSI color model in order to correctly classify whether the FFB is good to be processed or not.

2. Methodology

In this study, the main concern is the two category which is underipe and ripe. This is because they are the main class that determines the quality and quantity of the oil. Therefore, the data collection and the data analysis will only revive on these two categories. Today, the milling industries have difficulties in defining underipe and ripe category but has less problem in detecting between ripe and overipe category. This is because from the physical appearance both ripe and overipe can be differentiated by the naked eye but the underipe and ripe category is difficult to distinguish.

2.1 Data collection

Malaysian Palm Oil Board (MPOB) has established standards to correctly classify the oil palm FFB which is : 'underipe' with 1 – 9% loose fruits from the bunch, and 'ripe' with 10 – 50% loose fruits from the bunch [2]. In this study, images were captured by using 60 samples of fresh fruit bunches (FFB) that already been categorized according to the mill graders and inspectors before the data collection being carried out.

There are few types of oil palm in Malaysia, but the most planted are Elaeis guineensis (subsps: Nigrescens) which acquired from Pertubuhan Peladang Salak Tinggi, Sepang, Malaysia. The FFB was placed in the FFB grading machine and images were taken from the side of the FFB. The area covered was about 100cm x 100cm. Digital images were captured from overall 30 bunches under each category (underipe and ripe) using a GigE camera. Readings taken was then saved for further analysis.

2.2 Color feature extraction

This paper investigated on oil palm FFB grading using hue saturation and intensity (HSI). The image was taken in the grading system with the closed environment so that no ambient light would disturb the quality of the image. Figure 3 shows the image before and after background extraction. ENVI classic was used to extract the background from the image and then the image was uploaded in the MATLAB software to extract the RGB values. The HSI values were then extracted from the RGB values. The HSI values are then analyzed by using WEKA 3.6 software.

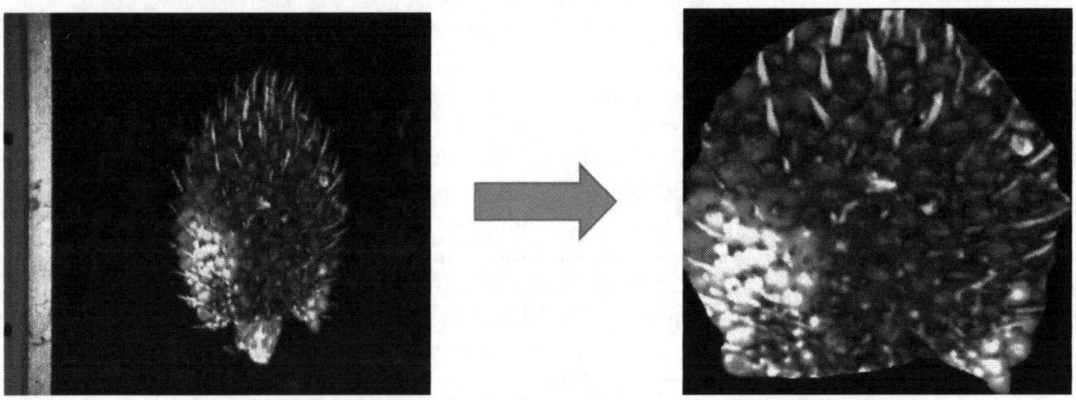

Figure 3. FFB image after background extraction using ENVI 5.1.

3. Results & Discussion

Results are divided into two parts, which are the analysis by using WEKA software and artificial neural network analysis. The WEKA analysis shows different results compared to the ANN analysis. This is perhaps the methods of running the analysis is different from each other. WEKA software uses cross validation to validate all data but ANN analysis uses training and testing data to validate all sample.

3.1 Analysis by WEKA software.

The analysis of the data in the WEKA 3.6 software uses cross validation to determine the correlation coefficient of the data set. Figure 4 shows the correlation coefficient for the category and the HSI. From the correlation, we can indicate that the accuracy for the underipe and the ripe category by using the HSI model is an average of 45%. The mean absolute error is 38% and the root means squared error is 44% showing that the data is averagely correlated but can be improved.

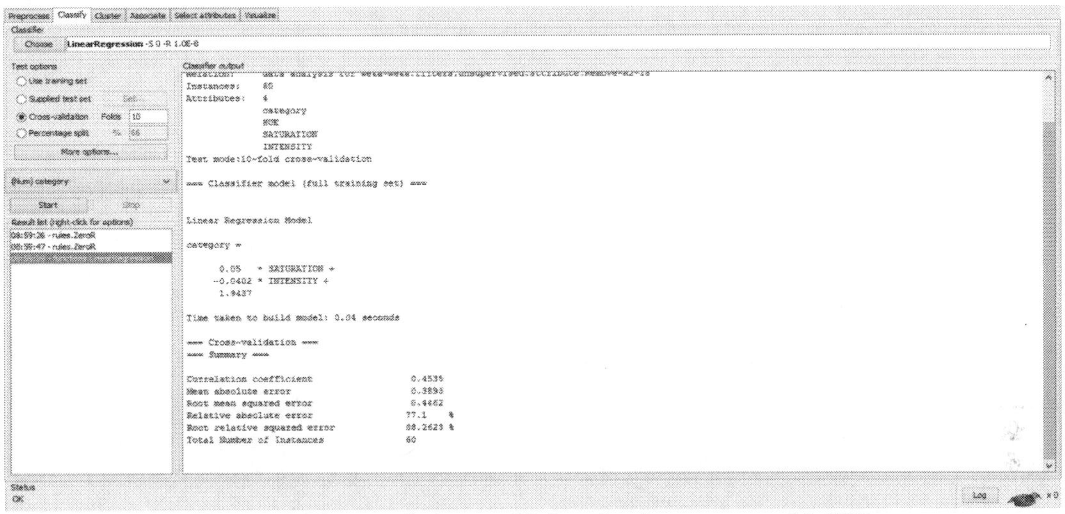

Figure 4. The correlation coefficient for the category with HSI.

Table 1 shows the difference in accuracy shown by different type of combination for the HSI data. Oil palm category was taken as the independent variable whereas HSI as dependent variables. For combination number one, the category is taken into cross validation with Hue and produce 17% accuracy, which is the lowest accuracy compared to other combinations. The highest average accuracy obtained was for combination number five, which is category taken with saturation and intensity. The accuracy was 45%. Combination number five also shows the lowest mean absolute error and root mean squared error which is 38% and 44% respectively. Combination number six shows the highest mean absolute error for 52% and for root mean squared error, combination number one and six shows the highest at 53%.

Table 1. The difference in accuracy shown by different HSI combination.

number	combination	accuracy	mean absolute error	root mean squared error
1	category, H	17%	51%	53%
2	category, S	25%	45%	48%
3	category, I	34%	51%	52%
4	category, H,S	18%	46%	50%
5	category, S,I	45%	38%	44%
6	category, H,I	20%	52%	53%

Figure 5 shows the difference in accuracy by different combination as explained in Table 1. The accuracy for the Hue is lower than when independent category variable is combined with saturation and intensity. This situation happens perhaps because of the lightness that is a shift to a longer distribution wavelengths. This cause it to be a non – monochromatic and unsaturated beam as mentioned by [3].

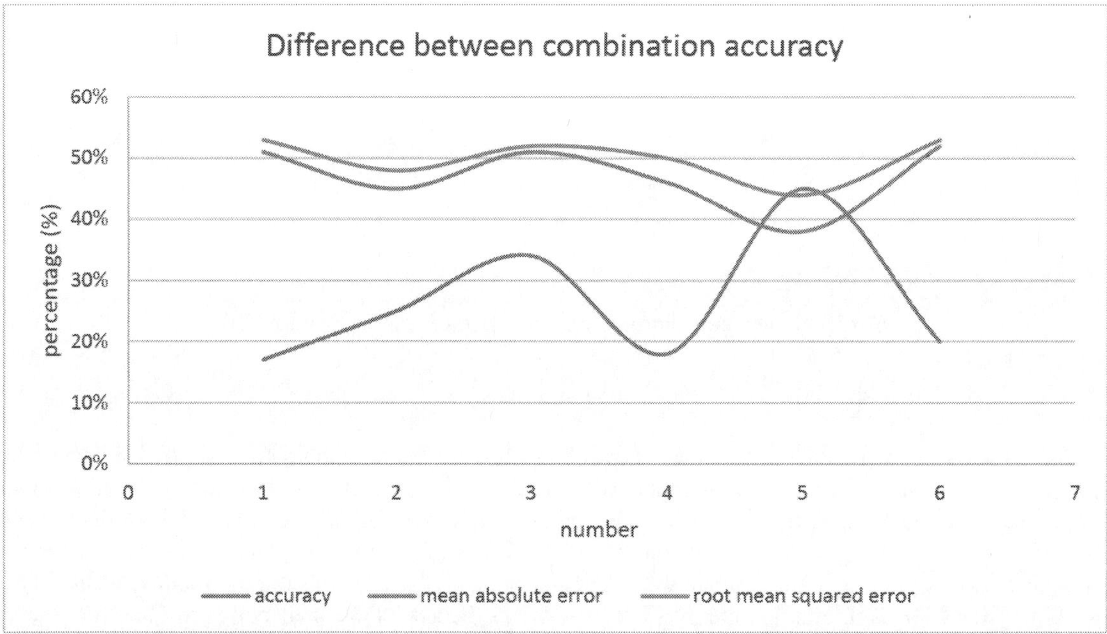

Figure 5. The accuracy, mean absolute error and root mean squared error of the HSI combination.

3.2 Analysis by artificial neural network (ANN).

The artificial neural network allows the user to input the training data and testing data. Training data are used to train the system to recognize the sample features so that it can be tested accurately. Figure 6 shows the data obtained when the system is trained. the R value for the training data is 0.8 and this shows a near correlation between the sample and the output. Validation and test result however, shows better R value which is 0.99 which means that the sample has a high correlation with the output. From the regression, the R value for all correlation is 0.85. This suggests that this neural network can be used for the training and testing purposes.

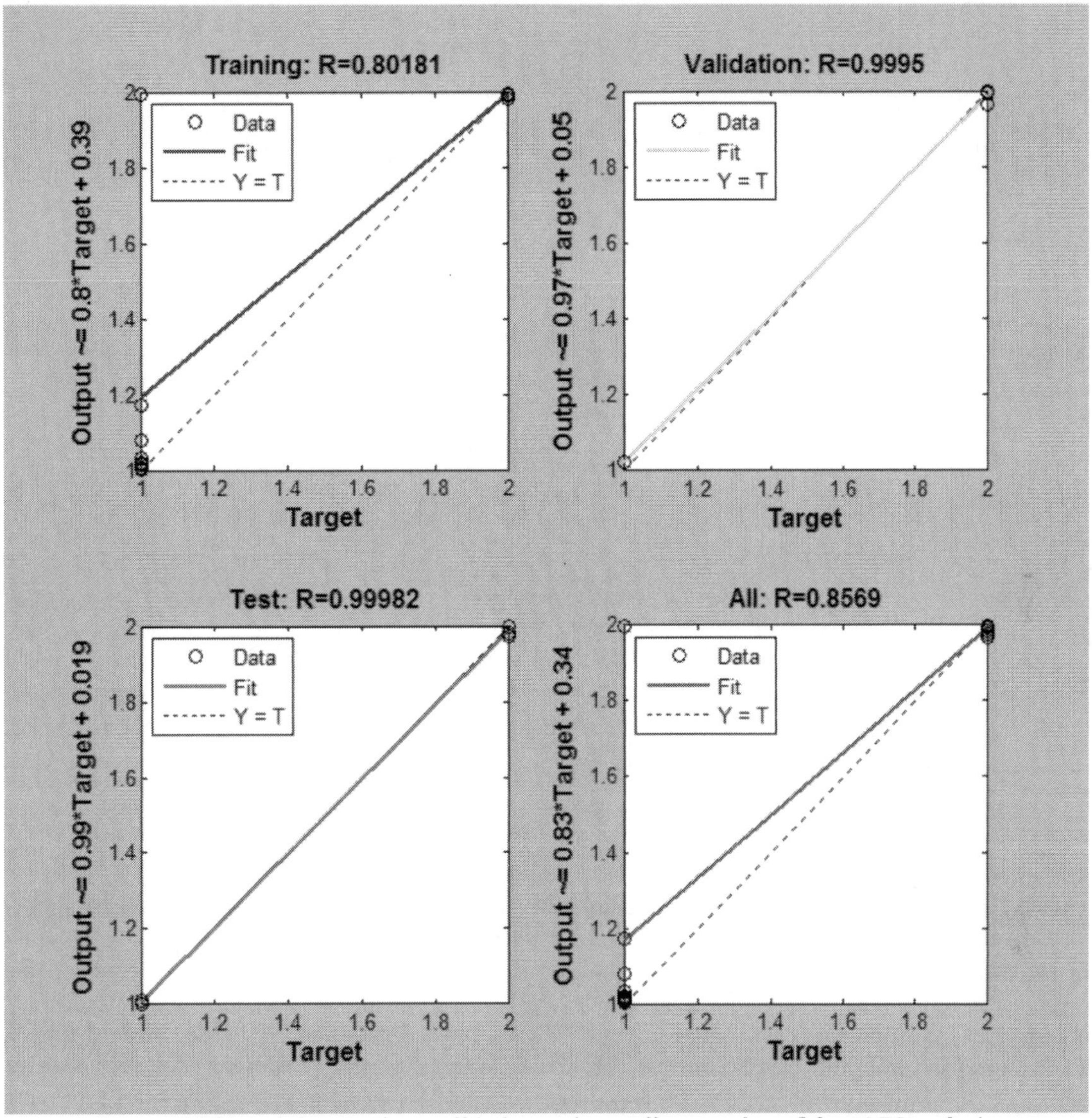

Figure 6. The training, testing, validation and overall regression of the ANN analysis.

After few times of training the data to fit in the equation, the testing of the data is carried out so that the accuracy and efficiency of the system can be tested. The results are shown in Figure 7. Hue, Saturation and Intensity was plotted on the same axis while the result is on the secondary axis. The yellow dot shows the system categorizing the FFB into their correct classification while the red dot shows incorrect classification. For the ANN system to work, earlier it was set into the training system that for any result with number one, that means underipe, whereas if the result comes out number two, that means ripe. 20 samples were taken into account for testing where sample number one – ten was underipe and 11 – 20 was ripe. The classification shows that the accuracy for underipe is 60% because of four incorrect classifications and for ripe, the accuracy is 80% because of two incorrect classifications. This makes the overall accuracy of 70% for the ANN system.

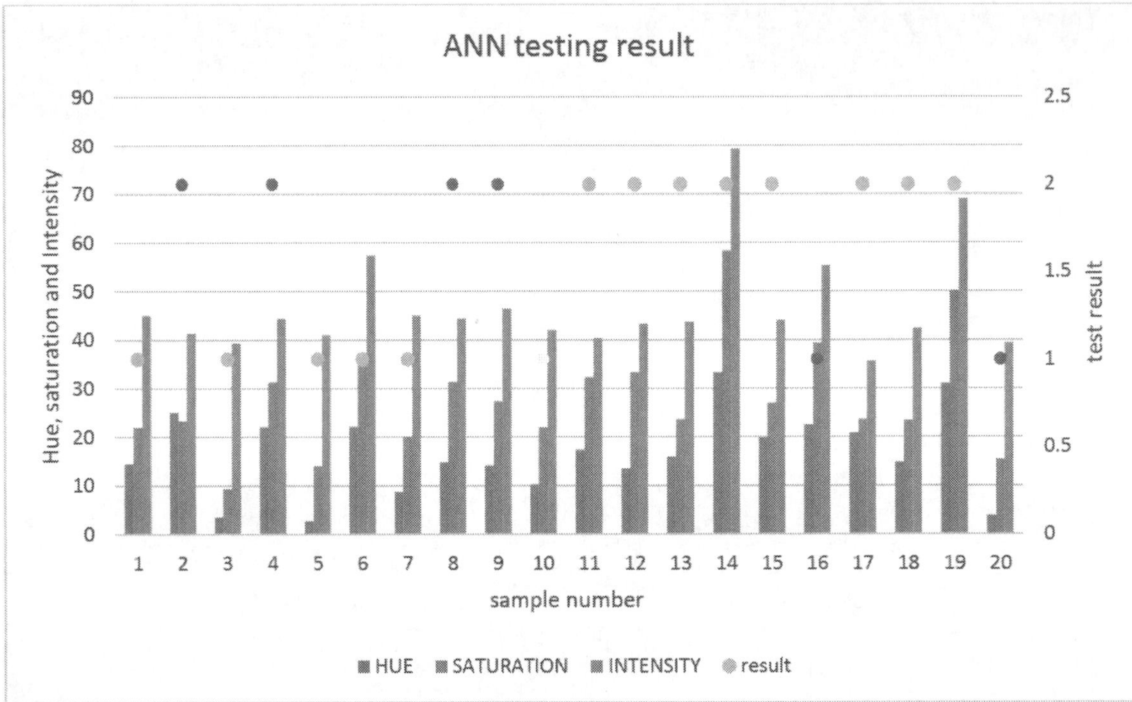

Figure 7. The testing result of the ANN analysis.

4. Conclusion

This study evaluated the applicability of a new GigE camera with HSI color model in determining the oil palm underipe and ripe maturity in a closed environment. The incident beam from the illumination in the closed environment is in low intensity thus not enough causes the beam to be unsaturated and non – monochromatic. This affects the image of the FFBs consequently produced the highest accuracy obtained for the analysis is at only an average of 45%. This average result was obtained by applying the linear regression from the WEKA software. Artificial Neural Network (ANN) however brings about higher accuracy in oil palm FFB ripeness detection. The underipe and ripe categories were taken into account and yield an acceptable accuracy result of 60% and 80% respectively. Those result then yield an overall accuracy of 70%. ANN shows more reliable results than WEKA and it could be utilised with the HSI measurement for oil palm. Furthermore, this could be a great analysis if the sample condition and the illumination in the system can be increased to a higher level but this would introduced noise into the system. Further research and analysis will combine the automated grading system that can be tested with a larger data set and thus more accurate analysis.

5. Acknowledgement

The authors would like to thank warm support and cooperation received from Pertubuhan Peladang Salak Tinggi, Sepang, Selangor.

6. References

[1] N. Fadilah, J. Mohamad-Saleh, Z. A. Halim, H. Ibrahim, and S. S. S. Ali, "Intelligent color vision system for ripeness classification of oil palm fresh fruit bunch," *Sensors (Switzerland)*, vol. 12, pp. 14179–14195, 2012.

[2] M. S. M. Alfatni, a R. M. Shariff, H. Z. M. Shafri, O. M. B. Saaed, and O. M. Eshanta, "Oil palm fruit bunch grading system using red, green and blue digital number," *Journal of Applied Sciences*, vol. 8. pp. 1444–1452, 2008.

[3] Judah Levine, "Light and Color. Topic 45. Hue, Saturation and Intensity." [Online]. Available: http://www.colorado.edu/physics/phys1230/phys1230_fa01/topic45.html. [Accessed: 22-Mar-2016].

[4] F.-R. A. Thomas S.Y. Choong, Saad Abbas, Abdul Rashid Shariff, Rohaya Halim, Mohd Halim S. Ismail, Robiah Yunus, Salmiaton Ali, "Digital Image Processing of Palm Oil Fruits," *Int. J. Food Eng.*, vol. 2, no. 2, pp. 1556–3758, 2006.

[5] J. M. A. F. P Edreschi, D. Mery, F. Mendoza, "Classification of Potato Chips using Pattern Recognition," *J. Food Sci.*, vol. 69, no. 6, pp. E264–E270, 2004.

[6] Wan Ishak Wan Ismail, Mohd Zohadie Bardaie, Abdul Malik Abdul Halim, "Optical properties for mechanical harvesting of oil palm FFB," *J. Oil Palm Res.*, vol. 12, no. 2, pp. 38–45, 2000.

[7] Albdullah, M.Z., Guan, L.C., Mohamed, A.M.D., Noor, M.A.M., "Color vision system for ripeness inspection of oil palm Elaeis Guineensis.," *J. Food Process. Preserv.*, vol. 26, pp. 213–235, 2002.

[8] Devadas, R., Lamb, D.W., Simpfendorfer, S., Backhouse, D., "Evaluating ten spectral vegetation indices for identifying rust infection in individual wheat leaves," *Precis. Agric.*, vol. 10, pp. 459–470, 2009.

[9] A. Alfatni, M.S.M., Shariff, A.R.M., Shafri, H.Z.M., Saaed, O.M.B., Eshanta, O.M. and M., "Automated oil palm fruit bunch grading system using density of color (RGB).," *Proc. 7th Saudi Eng. Conf. King Saud Univ. Riyadh*, pp. 299–313, 2007.

[10] S. Jamil, N., Mohamed, A., Adbullah, "Automated grading of palm oil fresh fruit bunches (FFB) using neuro-fuzzy technique.," *Proc. 2009 IEEE Int. Conf. Soft Comput. Pattern Recognition, Malacca, Malaysia,* pp. 245–249, 2009.

IOP Conf. Series: Earth and Environmental Science 37 (2016) 012040 doi:10.1088/1755-1315/37/1/012040

Detecting nutrients deficiencies of oil palm trees using remotely sensed data

Faradina Marzukhi, Aina Liyana Elahami, Sharifah Norashikin Bohari
Universiti Teknologi MARA Perlis, Malaysia

Email: faradinamarzukhi.fm@gmail.com

Abstract. Oil palm plantation management involve crucial role for the farmers. The remote sensing imagery has widely used nowadays in order to monitor oil palm tree in plantation. To pact with the problem, the use of vegetation indices analysis on satellite image on plantation will examine the ability of spectral data in determining the greenness of the trees. Vegetation Indices are used for estimating the crops and vegetation variables by using visible and near-infrared region (NIR) from the electromagnetic spectrum. The healthy tree will display very low reflectance and transmitted in visible region and very high reflectance transmitted in NIR. The chlorophyll absorption in reflectance and normalizes pigment chlorophyll vegetation indexes will show a loss of chlorophyll pigment compared to healthy oil palm trees. Besides, pH. value and soil nutrient will be examined to determine their effect towards the trees. In addition, the laboratory test sample is done to analyse the pH. value and major nutrient status of nitrogen (N), phosphorus (P) and potassium (K) together with their relationship with the remotely sensed data.

1. Introduction

Oil palm has been planted in large scale of area and it requires on time monitoring. In order to overcome the problem, satellite imagery has being used to identify the unhealthy tree at the plantation. The result of this project will reduce the time constraint for monitoring and help to control the plantation. It is important for planning to build an integrated information system in order to access accurate and up-to-date data which is used in the activities of research including planning, management, controlling and also in overcoming unpredictable situation. This study will help the farmers in decision making. Besides, the large area of oil palm plantation considers many criteria and implementation in order to get the large productivity of crops or fruit bunches. As to maximize the efficiency of management in oil palm plantation, geospatial analysis is implemented. This is because, the useful information from the remote sensing data can be integrated in plantation especially in oil palm. Farm manager can obtain details and specific crops together with soils over the large area to assist decision making. If the plantation is managed in accordance with the right decision and rule by the farmer, it will affect the productivity improvement of oil palm fruits bunches. The criteria that must be considered are including the land selection, planting material, technical management, harvesting and environment. If all the criteria in the oil pam plantation can be managed and integrated with each other, production of fruits can be obtained. Therefore, the next precaution and caution can be taken in order to get the huge profit in plantation production. Thus, the result of analysis can be used to maintain the sustainable analysis.

2. Study area

This study was carried out at Felcra Kampung Kurnia Gelong Gajah. It is situated in Ayer Tawar, Perak, Malaysia. There are three phases of the oil palm plantation in Felcra Kampung Kurnia but the study area for this project is on the Phase 3 and its geographical coordinates are 4° 23' 00' North, 100° 47' 29" East. The area of oil palm plantation is about 100 acre and it is been planted since year 2001. The oil palm plantation had been conducted and supervised by Felcra Berhad Perak.

Figure 1. Study area (Ayer Tawar, Perak).

3. Methodology

There are two method that are being used for this project. The methods are image processing on remote sensing and soil test for palm trees ground data. The following steps are conducted in order to identify the condition of unhealthy and healthy oil.

3.1 Soil sampling

Nutrient content (N.P.K) and pH value are examined from five sample of soils at the plantation area. The soil test is conducting in the laboratory. It is being done for the purpose of estimating and analysing the plant-available such as method and chemical solution that are used, tested or verified for the correlation of exact nutrient. Then, the regression equation is constructing to get the nutrient content (N.P.K) on soil tested. Apart from that, the soil pH estimation is important to know either the oil palm trees growth suitability for neutral, acidic or alkaline.

3.2 Image processing

Image processing is the important step in the remote sensing technique. The processing of SPOT-5 image is done through certain tools in the ERDAS imagine 2014 software. The tool includes reprojection, subset, and vegetation index of NDVI and SAVI. The processing is done for determination of accessing, interpreting and analysing the result.

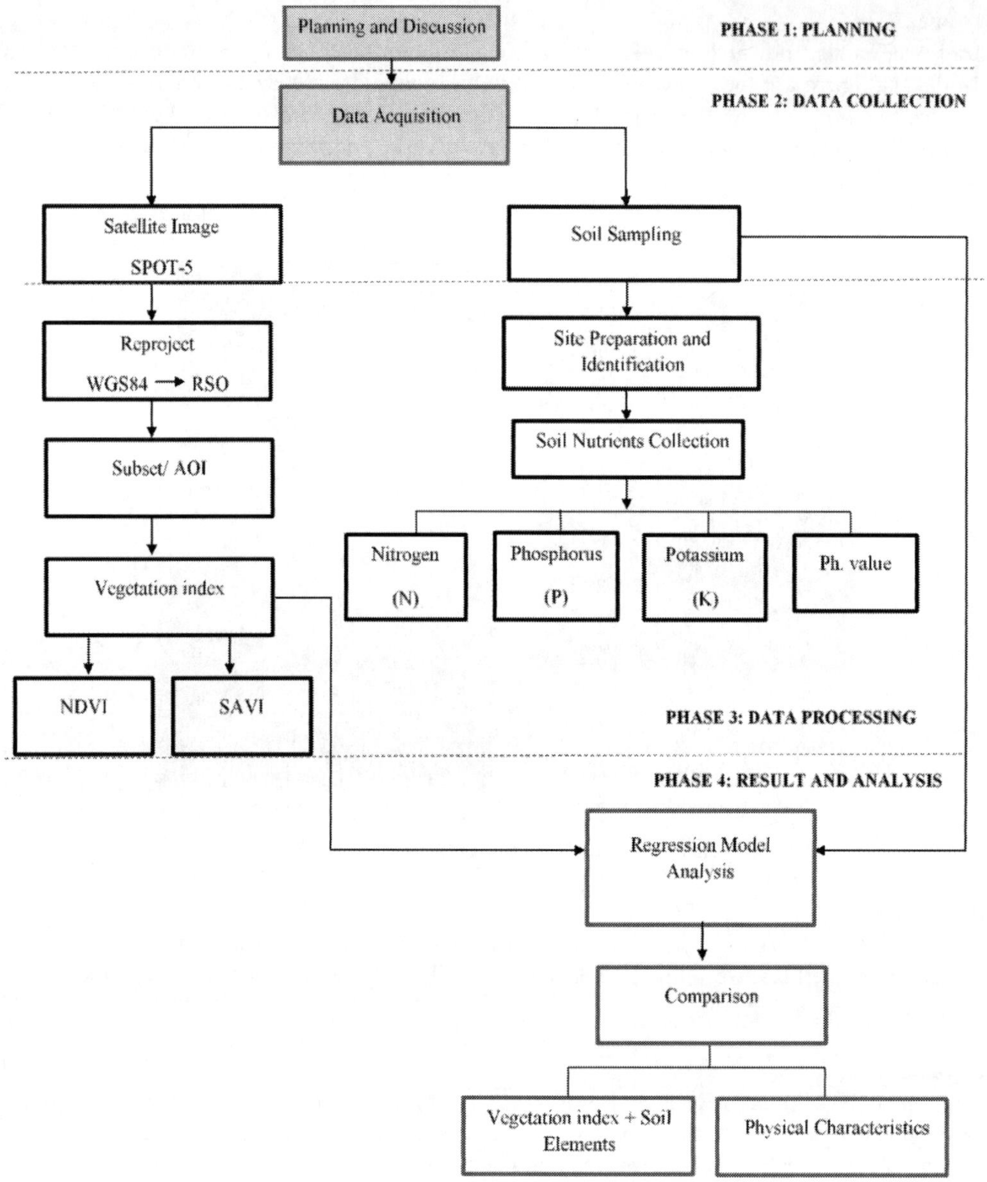

Figure 2. Flowchart of methodology.

4. Results and discussion

The result and analysis of the project study is based on the image processing using ERDAS Imagine software, soil analysis of nutrient contents including nitrogen, phosphorus, potassium and pH value. Apart from that, the relationship of NDVI and SAVI are described by regression model towards the soil analysis.

4.1 Analysis on plot area

There are five plot area in this study area. Before the soil sampling was collected, the observation is made on the physical botanical on the oil palm trees on the plantation. The healthy and unhealthy plot area was selected for identifying the nutrients content. The result of the plot sample is stated on table below.

Table 1. Identification on the plot area.

Plot sample	**Coordinate location of plot area (m)**		**Area (m^2)**	**Physical condition of the plot area**
	x	*y*		
1	311554.58	485331.00	261605.45	Healthy
2	312203.02	485689.58	461572.07	Healthy
3	311884.60	484837.14	385592.61	Unhealthy
4	312531.23	485258.92	160433.07	Unhealthy
5	311796.43	485933.77	600754.37	Healthy

4.2 Analysis on nutrients content

The soil nutrient testing has be done for five composite sample on the site location. There are ten soil samples for each plot are collected and mixed into single sample. The soil sample have been labelled and analyzed in the soil laboratory.

Table 2. Result of soil sampling.

Plot	**Element**						pH. Value
	Total Nitrogen (N) %	Status	Phosphorus (P) mg/kg	Status	Potassium (K) meq/100g	Status	
1	0.25	Very High	197	High	0.75	Very High	5.8
2	0.20	Very High	40	Very Low	0.98	Very High	5.4
3	0.17	Moderate	7	Very Low	0.20	Low	5.0
4	0.14	Moderate	81	Very Low	0.19	Low	4.8
5	0.21	Very High	180	Low	0.65	Very High	5.0

Based on the five samples analyses of soil stated on Table 2, the highest percentage of total nitrogen in soil for sample on the Plot 1, which is 0.25 %, while the lowest is in sample on the Plot 4, which is 0.14%. By referring to Goh, K.J. [3] on the soil nutrient, it is suggested that the oil palm plantation specifically above nine years by the optimum composition of nitrogen needed in soil for oil palm tree should less or same 0.12 %. Thus, any percentages that below than that level is consider as low and the value that higher than the level is consider as high. In this case, the Plot 4 has the lowest nitrogen content which is 0.14% and Plot 1 has the very high of nitrogen content which is 0.25%. The value of nitrogen is optimum. E.G.Ilori, P.N.Okonjo, [2] said that all plant need nitrogen for growth and almost all the nitrogen in the soil is in an organic form. By referring to the five samples analyses of soil, the highest content of available phosphorus is on Plot 1, which is 197 mg/ kg followed by Plot 5 is 187 mg/kg. By referring to Goh, K.J. [3], the available phosphorus in soil for oil palm tree should above 60mg/kg. The value that indicate above than 60 mg/kg is consider as high. Any value that below than 25 mg/kg is consider as low. Based on this study, the lowest of available phosphorus content is on Plot 3, which is 7 mg/kg. Plot 3 has a lot of different of soil analyzed, compared to other plots. The low content of phosphorus may be caused by many factors. Ian Rankine and T.H. Fairhurst [4] said that phosphorus deficient in plants may be underdeveloped with short fronds, and the palm trunk may have a noticeable pyramid shape. Based on the five sample analyses of soil, the highest content of potassium is on Plot 1, which is 0.98 meq/100g and followed by Plot 1 which is 0.75 meq/100g. Meanwhile, the lowest of phosphorus content is on Plot 4, which is 0.19 meq/100g. Both soil potassium status in Plot 3 and 4 has slightly difference by 0.1 meq/100g. As referred to Goh, K.J. [3], the study of the soil that indicated as has high potassium content is above 0.30 cmol/ kg and the value below than 0.20 cmol/ kg is point to as low of potassium content. Therefore, any value that above than 0.30 cmol/ kg is consider as high and contrary. The nutrient of potassium can affects fruit bunch size, fruit bunch, and also prevention of disease disruption. Furthermore, the potassium deficiency is usually on peat and sandy soils. Besides, the low acid in soil also cause potassium deficiencies.

4.3 Analysis on soil properties

The soil properties that has been analyse is pH value of the soil. The pH. value indicate the acidic or alkali status of the soil sample. The chart below show the pH value varies on the five plot. Soil pH plays a great role because a lot of change either on biological nor chemically activities in the soil. E.G.Ilori, P.N.Okonjo [2] said that the pH will affect the availability intake of other nutrients. Based on the result from the five plot of soil analyses, the Plot 3 and 4 has the lowest of pH value and its equivalent to nutrients of nitrogen, phosphorus and potassium (N.P.K) that available in soil. All the pH value of soil sample has range between 5.8 to 4.8 and the mean of the entire area is about 5.2 and had proven that the soil is slightly acid. In addition, Plot 3 and Plot 5 has the same pH value, which is 5.0. By referring to the study in Pathumthani province in Thailand by Ukrit Sahapatsombut [9] and Nigerian Institute for Oil Palm Research in Benin City, Nigeria, both study have proven that the suitable pH value for oil palm plantation is ranges from 4 to 6. It can be planted in acid soil. Based on the study of soil fertility on oil palm by Dr. E Mutert [6], in Southeast Asia, the oil palm are grown on acid soil. In the meantime, the range of pH value in soil in the study area is 5.2 and indicated that the pH of the soil on the study area is acidic.

4.4 Analysis on nutrients content

The vegetation index is for detecting the greenness of the tree. Image SPOT-5 is used during this project. It is because the resolution of SPOT-5 is higher than the Landsat imagery. In addition, SPOT-5 has two spectral band that is important in NDVI analysis which are Band 3 and Band 4. Band 3 is red and band 4 is near infrared (NIR). Generally, the healthy vegetation will absorb most of the visible light that falls on it and reflect a large portion of the near infrared. Sim Wright [7] said that the unhealthy vegetation reflect more visible light less near- infrared light. It can be concluded that the healthy oil palm has the high of NDVI value and the unhealthy oil palm tree has the lowest NDVI value. NDVI is calculate from the formula that is stated below:

$$NDVI = (NIR - RED) / (NIR + RED) \quad (1)$$

The RED stand for spectral measurement of Band 3 and NIR is spectral measurement in band 4. The NDVI value varies from range -1.0 to +1.0. Desmond Ofosu Anim,Amos Tiereyangn Kabo-bah,Philip Nti Nkrumah, Raphinos Tackmore Murava [1] said that the positive value show different type of vegetation classes, where near zero and negative value show non vegetation classes such water, snow, built-up area and barren area. Based on this project, the lowest of NDVI value is 0.30 as there is only oil palm plantation spot. The non-vegetation is distinguished by subset the plot area.

Figure 3. NDVI process on the plot area.

Table 3. The result of NDVI.

NDVI	Min	Max	Median	Mean
Plot 1	0.5000	0.67816	0.65032	**0.649**
Plot 2	0.45833	0.68000	0.64796	**0.645**
Plot 3	0.21818	0.34375	0.31285	**0.312**
Plot 4	0.28000	0.0.3813	0.30707	**0.306**
Plot 5	0.56204	0.67251	0.30705	**0.645**

The NDVI of vegetation indices is process through ERDAS Imagine 2014 software. By using the reflectance of spectral, the green vegetation of the oil palm can be detected. Table 3, shows the NDVI value of the five plots. The value of NDVI range from 0.30 to 0.65. These variety of analysis show the classification of the unhealthy and healthy oil palm tree on the plantation. The healthy tree has low red light reflectance and high near infrared reflectance. Therefore, the value of NDVI is analysed as high. Based on the study of Veena Shashikant, Abdul Rashid Mohamed Shariff, Laili Nordin and Biswajeet Pradhan [10], the unhealthy tree has high red light reflectance and lower near infrared reflectance so the result of the NDVI value is low. It also indicated that the lowest of NDVI value are on the Plot 3 and Plot 4. It was predictable because of the observation on the physical of the trees for ground data collection contain most of the oil palm trees that nearly dead and damages.

4.5 Result of SAVI

SAVI analysis (Equation 2) was developed to minimize soil influences on canopy spectral by incorporation a soil adjustment factor L into the dominator of normalized differences vegetation index (NDVI) of (Equation 1). For optimal adjustment of the soil effect, L factor should inversely with the amount of vegetation present.) SAVI equation created by Ji.Qi, A.Chehbouni, A.R.huete, [5] is shown as below:

$$SAVI = NIR - RED / NIR + RED + L \times (1 + L) \quad (2)$$

Figure 4. Result of SAVI analysis on the plot area.

Table 4. The result of SAVI.

SAVI	Min	Max	Median	Mean
Plot 1	0.89701	1.0143	0.97262	**0.972**
Plot 2	0.68512	1.0171	0.96911	**0.964**
Plot 3	0.90492	1.0148	0.97491	**0.975**
Plot 4	0.87197	1.0029	0.95843	**0.958**
Plot 5	0.8400	1.0058	0.96826	**0.965**

The value of NDVI range from 0.30 to 0.65. These variety of analysis show the classification of the unhealthy and healthy oil palm tree on the plantation. As Veena Shashikant, Abdul Rashid Mohamed Shariff,Laili Nordin, Biswajeet Pradhan [10] have mentioned that the healthy tree has low red light reflectance and high near infrared reflectance. Thus, the value of NDVI is analysed as high. It also indicated that the lowest of NDVI value are on the Plot 3 and Plot 4. The average of the SAVI value is 0.966 and it show that the value is not much differences. Based on the SAVI analysis the result do not as same as the soil analysis. Based on the SAVI analysis, the result do not as predicted towards the soil analysis.

4.6 Comparison of NDVI and SAVI analysis

The relationship of both data from soil test analysis and vegetation indices is done by regression model. The purpose of the regression model analysis is to predict the value of the dependent variable for individual and estimating the effect of some explanatory variable on the dependent. The regression algorithm is examined to get the R^2. The R^2 shows the correlation of NDVI, SAVI and the nutrients analysis. The relationship can be seen on the simple linear of the regression model analysis.

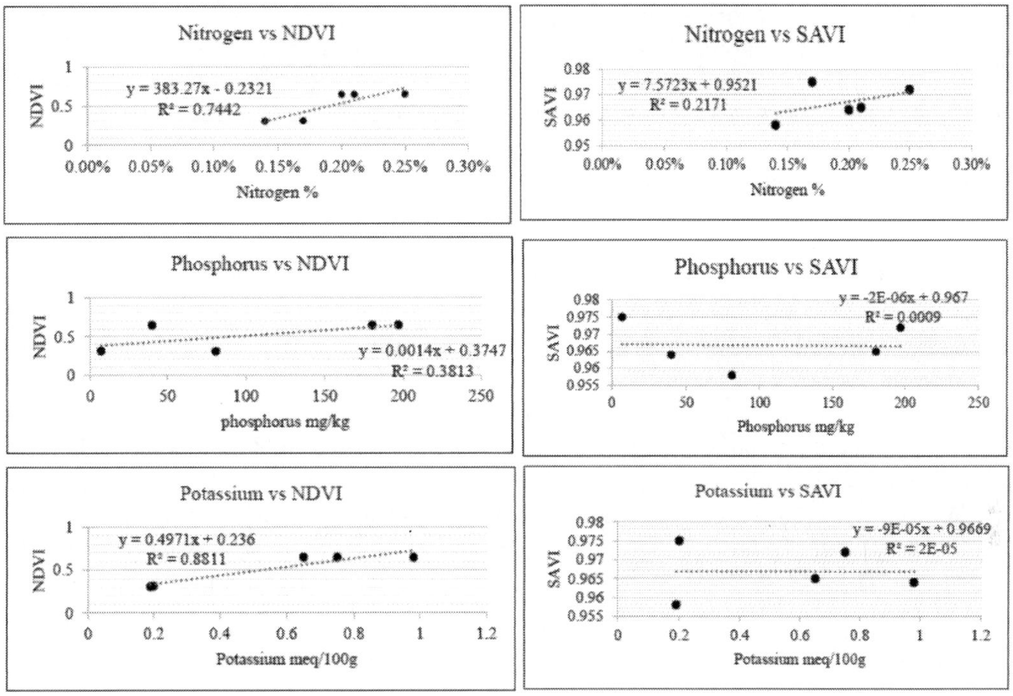

Figure 5. The correlation relationship of NDVI and SAVI towards nutrients analysis.

4.7 Comparison on the vegetation indices and soil analysis

After done the correlation analysis of both technique of NDVI and SAVI towards the soil properties, the vegetation indices show the positive and negative correlation can be compared in order to evaluate which of the vegetation indices has the strong correlation relationship. The R^2 value are listed and compared according to the nutrients element.

Table 5. Comparison of both correlation relationship of vegetation indices towards soil properties.

Correlations element	**R^2 value**	
	NDVI	SAVI
Nitrogen (N)	0.7442	0.2171
Phosphorus (P)	0.3813	0.0009
Potassium (K)	0.8811	0.012
pH. value	0.4789	0.1969

By referring to Table 5, the correlation of NDVI show positive relationship compared to the SAVI. On the NDVI, the value indicate exact as the physical condition on the plot area. The correlation of SAVI towards the soil properties indicate negative correlation relationship whereby the element range from 0.100 to 0.2171. Moreover the correlation of R^2 on Potassium element cannot be identified. Therefore, the Normalized Differences Vegetation Indices (NDVI) has a strong correlation compared to Soil Adjusted Vegetation Indices (SAVI) towards element in soil nutrients.

4.8 Comparison on the physical characteristics
After done the soil analysis and vegetation indices on the plot area, the physical characteristics of the oil palm tree towards the nutrient deficiencies that related is identified. There are many physical characteristics of oil palm tree that can be diagnosed nutrient deficient such as trunk, leaf, fruits and frond, but the easiest to spot is on its leaves. Beside soil analysis, Timothy, K. Broschat [8] said that the leaves of the oil palm tree also can resemble the soil profile. Nitrogen (N) is the most important element in plant including oil palm. Based on the oil palm tree in the Plot 3, the leave indicated lacking on the nitrogen composite. It is because the leaves and petiole of the oil palm tree show yellowish colour and the frond is widely open. Based on the research study in Florida by Timothy, K. Broschat [8] regarding the nutrient deficiencies on palm tree, the nitrogen deficiencies started on the oldest leaves turn into light green then slowly spread to young leaves and in a time will becoming discoloured. In addition, golden-yellow brown petiole, rachis and crown shaft colour on the oil palm tree is cause by nitrogen deficiencies. The nitrogen deficiencies is cause by insufficiencies N in the soil. The lack of nitrogen deficiencies on the Plot 3 is proven, whereby the nitrogen in soil is lowest than other plots. In long period, the sizes of leaves will becoming much smaller and trunks tapering to a point as a result.

Figure 6. Oil palm tree spot on the plot 3 area.

There are no specified study that relate to the phosphorus (P) deficiencies as the positively react to the nitrogen and potassium. Based on the study in TNAU Agritech Horticulture, the symptom on the lack of phosphorus is high frequency of premature dryness of older leaves occurs and the oil palm tree that has enough phosphorus will increase the fruit productivity and number of female inflorescences. On the Plot 4, there are few of oil palm tree that shown low quality of fruit bunch production and the leaves indicate dryness as show in brown colour on the most frond. Referred to the soil analysis, the Plot 3 has the lowest phosphorus element and it can be conclude that the oil palm tree on the area is lacking on phosphorus content.

Figure 7. Leaves dryness indicate in brown colour. Fruit bunches is unusually small.

Potassium (K) deficiencies is usually nutritional symptom in palm and always show on the oldest leaves. The symptom show on the older leaves mottled with yellowish spot that are translucent when viewed. Apart from that, Timothy, K. Broschat [8] also said that the early symptoms appear as translucent yellow or orange spotting on the leaves and may be accompanied by necrotic spotting as well. Those symptoms are obviously shown on the area of Plot 3 and 4.

Figure 8. Symptom in lacking potassium on leaves.

5. Conclusion

According to the analysis of correlation relation acquired, both technique of soil sampling and NDVI are correlated to each other in finding the nutrient deficiencies of oil palm tree and this method can give a good identification of vegetation area. The regression model shown the correlation relationship for each nutrient elements with the vegetation indices. Based on the comparison of the analysis on soil data, it shows that the Vegetation Indices of NDVI gives better information compared to SAVI in order to find the nutrients deficiencies. Others than that, for the pH value, does not have strong correlation compared to the nutrients content towards the vegetation indices.

6. Acknowledgement

The author is grateful to Malaysia Remote Sensing Agency (MRSA) on SPOT-5 image provided.

7. References

[1] Desmond Ofosu Anim,Amos Tiereyangn Kabo-bah,Philip Nti Nkrumah,Raphinos Tackmore Murava . (2013). Evaluation of NDVI Using SPOT-5 Satellite Data for Northen Ghana. Environmental Management and Sustainable Development, 167-182.

[2] E.G.Ilori, P.N.Okonjo. (2014). Assestment of Soil Nutrient status of a Oil Palm Plantation. *Agriculture Journal*, 127-131. Food and Agriculture Organization of the United Nations.

[3] Goh, K.J. (2004). Fertilizer Recommendation Systems for Oil Palm Estimating the Fertilizer Rates. 1-37. Haifa Chemicals ltd. (2014). How to avoid common nutrient deficiencies in oil palm.

[4] Ian Rankine and T.H. Fairhurst. (1999). Management of Phosphorus, Potassium and Magnesium. Better Crop International, 1-15.

[5] Ji.Qi, A.Chehbouni, A.R.huete. (1994). A Modified Soil Adjusted Vegetation Indices. Remote Sensing Environment, 112-126.

[6] Mutert, D. E. (1999). Suitability of Soils for Oil Palm in Southeast Asia. Better Crops International, Vol. 13, No. 1.

[7] Sim Wright. (1999). Retrieved from Normalized Difference Vegetation Index analysis on forestry and crop management. Timothy, K. Broschat. (2005). Nutrient Deficiencies of Landscape and Field-Grown Palms in Florida. Environmental Horticulture Department University of Florida, 1-10.

[8] Timothy, K. Broschat. (2005). Nutrient Deficiencies of Landscape and Field-Grown Palms in Florida. Environmental Horticulture Department University of Florida, 1 -10.

[9] Ukrit Sahapatsombut, Rapeepun Rahong and Samai Jai-in. (n.d.). Oil Palm Plantation in Acid Soil at Rangsit field. Thailand.

[10] Veena Shashikant, Abdul Rashid Mohamed Shariff, Laili Nordin, Biswajeet Pradhan. (2012). NDVI of oil palm trees by LANDSAT-5 Imagery. The 3rd Asian Conference on Remote Sensing, (pp. 1-4). Selangor.

Development of young oil palm tree recognition using Haar-based rectangular windows

S Daliman[1], S A R Abu-Bakar[1], S H Md Nor Azam[2]

[1]Computer Vision, Video and Image Processing (CvviP) Research Lab, Department of Electronics and Computer Engineering, Faculty of Electrical Engineering, Universiti Teknologi Malaysia, 81310 Skudai, Johor, MALAYSIA.

[2]Sime Darby Research Sdn. Bhd., Jalan Pulau Carey, 42960 Pulau Carey, Selangor, MALAYSIA.

Email: [1]shaparas@utm.my

Abstract. This paper presents development of Haar-based rectangular windows for recognition of young oil palm tree based on WorldView-2 imagery data. Haar-based rectangular windows or also known as Haar-like rectangular features have been popular in face recognition as used in Viola-Jones object detection framework. Similar to face recognition, the oil palm tree recognition would also need a suitable Haar-based rectangular windows that best suit to the characteristics of oil palm tree. A set of seven Haar-based rectangular windows have been designed to better match specifically the young oil palm tree as the crown size is much smaller compared to the matured ones. Determination of features for oil palm tree is an essential task to ensure a high successful rate of correct oil palm tree detection. Furthermore, features that reflects the identification of oil palm tree indicate distinctiveness between an oil palm tree and other objects in the image such as buildings, roads and drainage. These features will be trained using support vector machine (SVM) to model the oil palm tree for classifying the testing set and subimages of WorldView-2 imagery data. The resulting classification of young oil palm tree with sensitivity of 98.58% and accuracy of 92.73% shows a promising result that it can be used for intention of developing automatic young oil palm tree counting.

1. Introduction

Malaysia has a vast area of oil palm plantation. The current standard of monitoring the number of oil palm trees have been either manually counting by deploying human worker at the plantation itself or by manually counting from the given airborne images. With high resolution of 2m in WorldView-2 images, these images can be further enhanced by using Haar-based rectangular windows to extract information beneficial for farm monitoring purposes, such as tree detection. The purpose of detecting oil palm tree is to locate or count the oil palm trees based on satellite imagery of WorldView-2. This information is the key factor for oil palm plantation management and monitoring in each plantation area, where yield prediction can be obtained by knowing the numbers of oil palm tree.

Haar-based rectangular features also known as Haar-like rectangular features was firstly introduced by [1]. The name of Haar is intuitively similar to Haar wavelets and these Haar-based rectangular features have been used for face recognition in Viola-Jones object detection framework as the first real-time face recognition [2].

Viola-Jones algorithm have becoming popular for face detection since then. Haar-based rectangular features particularly Viola-Jones algorithm has been widely experimented for other object recognition and detection as well. Recently, Haar-based rectangular features or also known as Haar-based rectangular windows have also been used for pedestrian tracking [3], stop sign detection [4] and mostly for other features in faces, such as eyes and mouth detection. Thus, this study will use the Haar-based rectangular windows for features extraction of oil palm tree.

2. Feature extraction and oil palm tree recognition

2.1. Haar-based rectangular windows

Haar-based rectangular windows used in [2] are based on Haar wavelets which are single wavelength square waves. In two dimensions image, a square wave is a pair of adjacent rectangles with one-sided light pixels and another one sided having dark pixels. The rectangles used for face detection are not truly Haar wavelets. The rectangle combinations used in face recognition is designed for better suit to visual recognition tasks. Thus, these rectangles are usually referred to as Haar-based rectangular windows, or Haar-like features, rather than wavelets.

The presence of a Haar feature is determined by subtracting the average dark-region pixel value from the average light-region pixel value. If the difference is above a threshold which was set during learning algorithm, that feature is said to be present [2]. This binary determination is implemented to identify as face or not a face. Haar features can be described as a simple and inexpensive image features based on intensity differences between rectangle-based regions that share similar shapes to the Haar wavelets [5], which will be defined as:

$$F_{Haar} = E\,(R_{white}) - E\,(R_{black}) \tag{1}$$

where R_{white} refers to the light region pixels of an image and R_{black} refers to the dark region pixels of an image. However, some images of the same object may have different monotonic illumination changes due to the light differences. Thus, Haar features as defined in Equation 1 have to be normalized as follows:

$$F_{Haar} = \frac{E(R_{white}) - E(R_{black})}{\sqrt{\left|E(R_{white_\mu})^2 - E(R_{black_\mu})^2\right|}} \tag{2}$$

In addition to the application of Haar-based rectangular window, an integral image computation need to be done for fast feature calculation. The Haar-based rectangular features can be computed very fast and in constant time for any size by means of two auxiliary images [6]. The auxiliary image is the Summed Area Table (SAT) which is defined as the sum of the pixels of the upright rectangle ranging from the top left corner at (0,0) to the bottom right corner at (x,y) as in Figure 1a

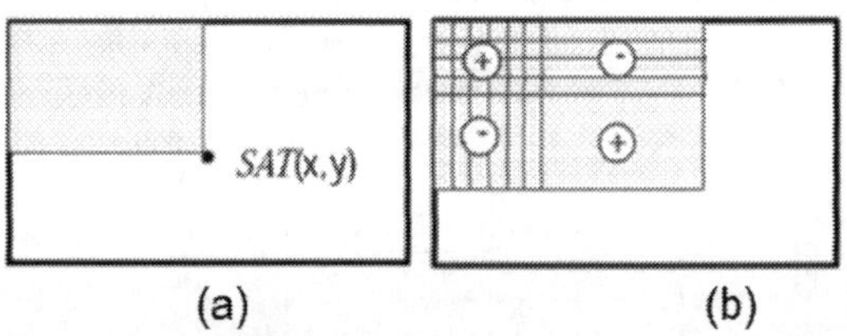

Figure 1. Summed Area Table for fast computation of Haar features [6].

Summed Area Table, SAT(x,y) can be stated as:

$$SAT(x,y)=\sum_{x'\leq x,y'\leq y} I(x',y') \tag{3}$$

SAT(x,y) can be calculated with one pass over all pixels from left to right and top to bottom by modifying Equation 3.25 into:

$$SAT(x,y)=SAT(x,y-1)+SAT(x-1,y)+I(x,y)-SAT(x-1,y-1) \tag{4}$$

where

$$SAT(-1,y)=SAT(x,-1)=0 \tag{5}$$

Based on Equation 5, the sum of pixels in Haar-based rectangular window can be computed using four lookups table as in Figure 1b and can be defined as:

$$RecSum(r)=SAT(x-1,y-1)+SAT(x+w-1,y+h-1)-SAT(x-1,y+h-1)-SAT(x+w-1,y-1) \tag{6}$$

where h and w refers the height and width of the Haar-based rectangular window, respectively. For computing the Haar-based rectangular windows on young oil palm tree and non-oil palm tree database, Equation 6 will be implemented for calculating SAT of oil palm tree and non-oil palm tree features.

Similar to face recognition, the oil palm tree recognition would also need a suitable Haar-based rectangular windows that best suit to the characteristics of oil palm tree. Evaluation of Haar feature based on Equation 2 will be considered as WorldView-2 images tend to have different illumination changes. A set of seven Haar-based rectangular windows, namely as hf1, hf2, hf3, hf4, hf5, hf6 and hf7 have been designed to better match specifically the young oil palm tree as the crown size is much smaller compared to the matured ones. Illustrations of Haar-based rectangular windows used for oil palm tree recognition are depicted in Figure 2, while Figure 3 shows samples of young oil palm tree with each of Haar-based rectangular windows.

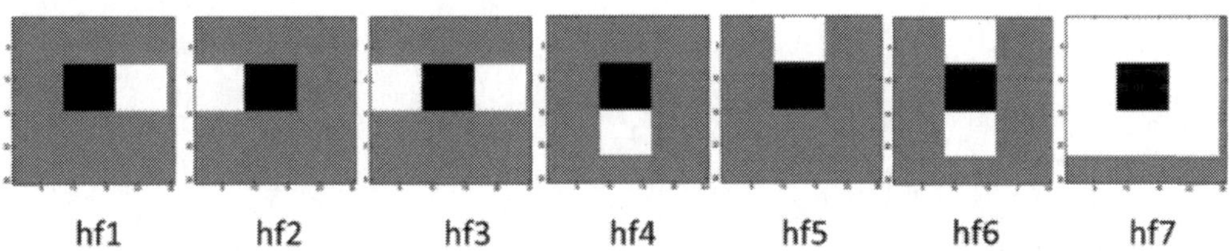

Figure 2. Haar-based rectangular windows for oil palm tree recognition.

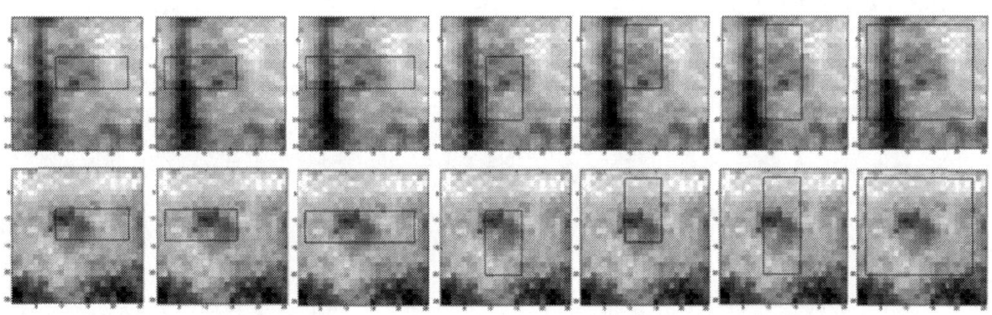

Figure 3. Samples of young oil palm tree with Haar-based rectangular windows.

The Haar-based rectangular windows are also designed due to fact that the young oil palm tree in WorldView-2 image has a significantly low contrast compared to its surrounding area. As shown in Figure 4, these seven Haar-based rectangular windows aim at characterizing several distinct descriptors of young oil palm tree in a 25x25 pixels window. This is because some of the young oil palm trees are located near to the drainage or land banks.

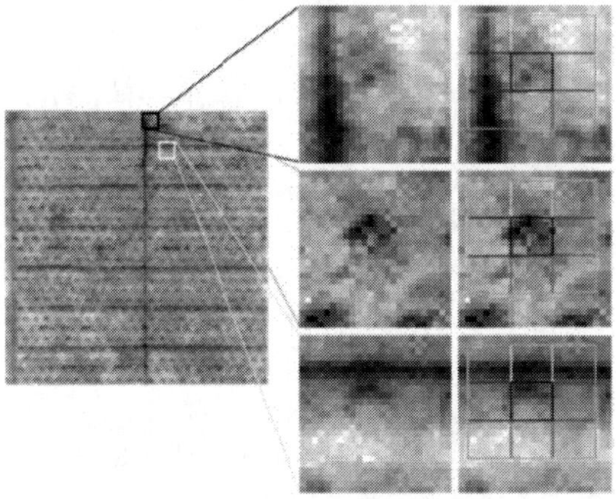

Figure 4. Seven Haar-based rectangular windows on different characteristics of young oil palm tree.

2.2. Support vector machine

The support vector machine (SVM) classifier will be considered in this study. This optimal separating hyperplane of support vectors [7] can be described based in Figure 5 below.

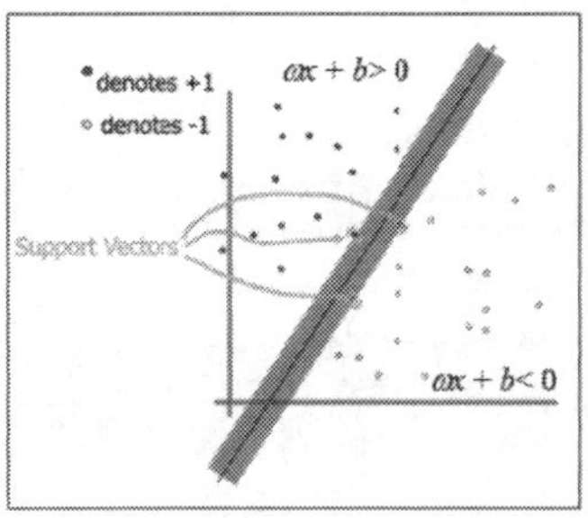

Figure 5. The optimal separating hyperplane of support vectors [7].

The foundations of support vector machines have been developed by [8]. The support vectors as illustrated in Figure 5 are the data points that the margin pushes up against in creating the maximum margin linear classifier [7], which in this case to classify the data into two categories of +1 and -1. This is the simplest kind of SVM that is also known as linear SVM. The hyperplane, $\langle w,x\rangle+b$ in Figure 5 can be defined as [9]

$$\langle w,x\rangle+b=0 \tag{7}$$

The margin, ρ can be denoted as

$$\begin{aligned}\rho(\omega, b) &= \min_{x^i:y^i=-1} d(\omega, b; x^i) + \min_{x^i:y^i=1} d(\omega, b; x^i) \\ &= \min_{x^i:y^i=-1} \frac{|\langle\omega, x^i\rangle + b|}{\|\omega\|} + \min_{x^i:y^i=1} \frac{|\langle\omega, x^i\rangle + b|}{\|\omega\|} \\ &= \frac{1}{\|\omega\|}\left(\min_{x^i:y^i=-1}|\langle\omega, x^i\rangle + b| + \min_{x^i:y^i=1}|\langle\omega, x^i\rangle + b|\right) \\ &= \frac{2}{\|\omega\|} \end{aligned} \tag{8}$$

The linear SVM needs to be developed in assessing issues linear constraints by optimizing a quadratic function. By constructing a dual problem that involves Lagrange multiplier, α_i, the classifying function will have the form of [7]

$$f(x) = \sum \alpha_i\, y_i {x_i}^T x_j + b \tag{9}$$

The linear classifier is based on the dot product between vectors as in Equation 10 as follows

$$K(x_i, x_j) = {x_i}^T x_j \tag{10}$$

In solving the non-linear classification, every data point needs to be mapped into high-dimensional space through transformation as follows

$$\Phi: x \rightarrow \varphi(x) \tag{11}$$

and the dot product in Equation 10 will becomes the kernel function as follows

$$K(x_i, x_j) = \varphi(x_i)^T \varphi(x_j) \tag{12}$$

The kernel function is a function that corresponds to an inner product of expanded feature [7]. Several different types of kernel function such as linear, polynomial and Gaussian kernel functions can be generated by modifying the Equation 12. The linear kernel function is defined as:

$$K(x_i, x_j) = (x_i)^T (x_j) \tag{13}$$

and the polynomial of power, p kernel function is denoted as

$$K(x_i, x_j) = \left(1 + {x_i}^T x_j\right)^p \tag{14}$$

The Gaussian kernel function or also known as radial-basis function network [7] is defined as:

$$K(x_i, x_j) = exp\left(-\frac{\|x_i - x_j\|^2}{2\sigma^2}\right) \tag{15}$$

The classifying function in Equation 9 will be modified for the kernel functions as follows:

$$f(x) = \sum \alpha_i\, y_i K(x_i x_j) + b \tag{16}$$

In this study, the kernel function of linear has been used for the young oil palm tree recognition.

IOP Conf. Series: Earth and Environmental Science **37** (2016) 012041 doi:10.1088/1755-1315/37/1/012041

3. Methodology

3.1. Study area and data collection

A database of 561 young oil palm trees and 214 matured oil palm trees has been selected from the WorldView-2 imagery data. The coordinate of each oil palm tree is recorded based on the center of the tree crown visible in the WorldView-2 image. The distribution of selected young and matured oil palm tree is depicted as in the following Figure 6, while Figure 7 depicted the example of a true colour WorldView-2 image of a matured and a young oil palm tree.

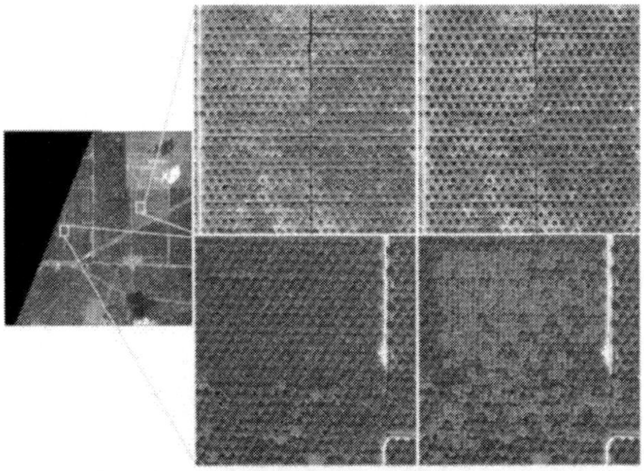

Figure 6. Distribution locations of young oil palm trees (blue) and matured oil palm trees (green).

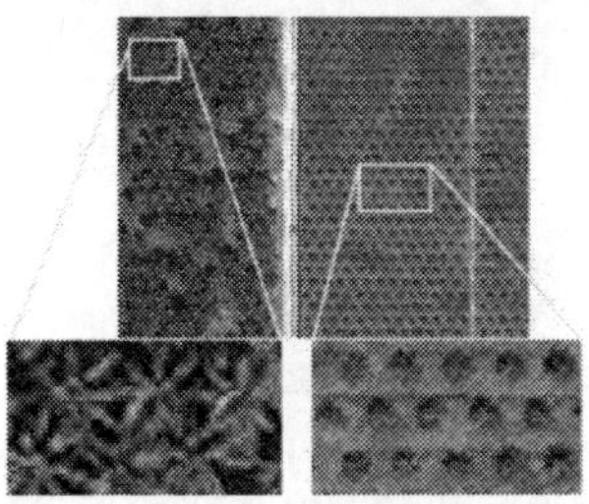

Figure 7. Samples of matured and young oil palm trees in true colour of WorldView-2 imagery data.

By comparing matured and young oil palm tree as in Figure 7, the difference in ages of oil palm tree is clearly visible. The crown for matured oil palm tree looks like a flower petals, whilst the crown for the young oil palm tree looks darker and circular-shaped with its size is much smaller. These features would be essential value for features extraction of young oil palm tree recognition.

3.2. Calculation of oil palm tree features

Features for young oil palm trees are computed using the seven Haar-based rectangular windows. Location of the seven Haar-based rectangular windows are illustrated individually in Figure 8 for sample of young oil palm tree.

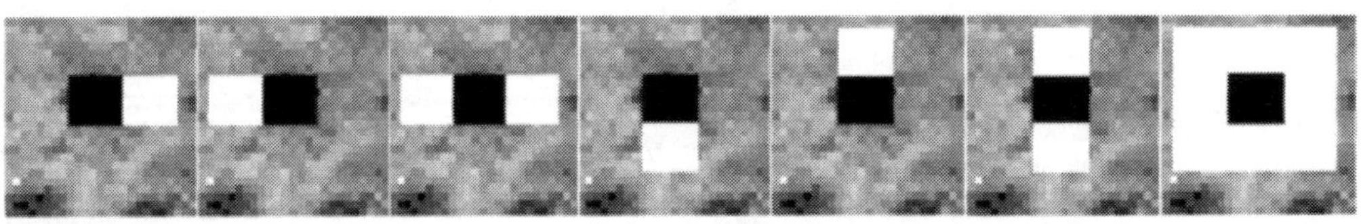

Figure 8. Seven Haar-based rectangular windows individually on sample of young oil palm tree.

4. Results and discussion

A database of 561 young oil palm trees, 214 matured oil palm trees, and also 761 points of non-oil palm tree objects have been selected from the WorldView-2 imagery data. These data were divided into two parts: one for training and another one for testing set as shown in Table 1.

Table 1. Distribution of training and testing set for oil palm tree classification.

Oil Palm	Train Set	Test Set	Total Points
Young Oil Palm Tree	280	281	561
Matured Oil Palm Tree	107	107	214
Total Oil Palm Tree	387	388	775
Non-Oil Palm Objects	**Train Set**	**Test Set**	**Total Points**
Water body	14	15	29
Roads	41	41	82
Buildings	57	57	114
Open Area/Field	74	75	149
Ground/Background of oil palm tree	193	194	387
Total Non-Oil Palm Tree Objects	379	382	761

The performance of each features class were evaluated by performing classification on the testing set. Each features set were measured by classification accuracy and sensitivity which defined as [10]:

$$\textit{Classification accuracy} = \frac{C}{T} * 100\% \tag{17}$$

$$\textit{Sensitivity} = \frac{P}{Tp} * 100\% \tag{18}$$

where C represents a total of correctly classified data, T is a total of test data, P representing a total of correctly classified oil palm tree and Tp representing a total of tested oil palm tree. The classification accuracy is used to represent overall classification performance while sensitivity represents the capability of classification system to identify the oil palm tree. The latter indicator is also used to represent the capability of oil palm tree recognition model to be developed. Sensitivity is important to ensure high capability of oil palm tree recognition model to identify the oil palm tree correctly and accurately. High sensitivity ensures good performance in automated oil palm tree counting.

Features of oil palm tree were computed based on Haar-based rectangular windows by implementing hf1, hf2, hf3, hf4, hf5, hf6 and hf7 windows on the database of oil palm and non-oil palm tree objects. Classification of young, matured and non-oil palm tree using Haar-based rectangular windows have been implemented using SVM. Results of classification for band 1, band 2, band 3 and band 4 of WorldView-2 imagery data are shown in Table 2.

Based on the accuracy and sensitivity results for each features derived Haar-based rectangular windows, the best features was obtained using band 4 of WorldView-2 image with 98.58% and 98.13% sensitivity for young and matured oil palm tree respectively, and overall accuracy of 92.73%. Higher accuracy and better sensitivity response from Haar-based rectangular windows may result from the selection of Haar features that match with the specifications of the oil palm tree especially the young oil palm tree with the highest sensitivity performance.

Table 2. Results of oil palm tree classification using Haar-based rectangular windows for band 1, 2, 3 and 4 of WorldView 2 image.

Class of Classification	Band 1			Band 2			Band 3			Band 4		
	True Positive	False Positive	Sensitivity	True Positive	False Positive	Sensitivity	True Positive	False Positive	Sensitivity	True Positive	False Positive	Sensitivity
Young Oil Palm Tree	277	4	98.58%	279	2	99.29%	275	6	97.86%	277	4	98.58%
Matured Oil Palm Tree	103	4	96.26%	101	6	94.39%	104	3	97.20%	105	2	98.13%
Water body	15	0	100%	15	0	100%	15	0	100%	15	0	100%
Roads	41	0	100%	41	0	100%	41	0	100%	41	0	100%
Buildings	57	0	100%	57	0	100%	57	0	100%	56	1	98.25%
Open Area/Field	73	2	97.33%	73	2	97.33%	73	2	97.33%	73	2	97.33%
Ground/ Background of oil palm tree	131	63	67.53%	132	62	68.04%	137	57	70.62%	147	47	75.77%
Accuracy	697	73	90.52%	698	72	90.65%	702	68	91.17%	714	56	92.73%

The oil palm tree detection has been implemented by using oil palm tree recognition model based on Haar-based rectangular windows. Two tested 100x100 pixels subimage for young oil palm tree have been used to determine the effectiveness of the model. Distance of horizontal and vertical between location of the first three detected oil palm tree have been implemented to identify the existence of the tree, as the locations of young oil palm tree can be predicted based on the first three locations of oil palm tree detected as shown in Figure 9 and 10. In Figure 9, all predicted locations were detected leading to 100% accuracy of young oil palm tree detected. Furthermore, Figure 10 shows the ability of the young oil palm tree recognition model to achieve 100% accuracy of detected tree and signify the location of non-oil palm tree object as indicated. This proves that oil palm tree recognition model gives high performance and promising result for identifying and counting young oil palm tree.

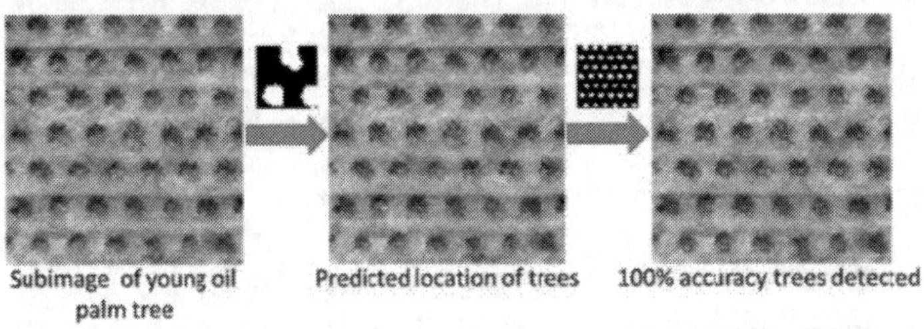

Figure 9. Young oil palm tree detection with 100% accuracy.

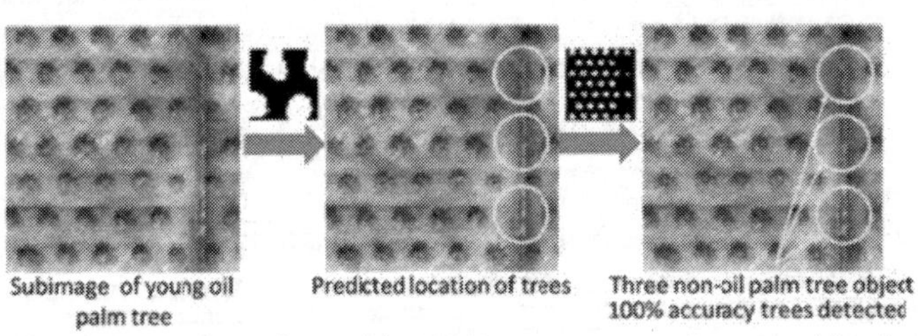

Figure 10. Young oil palm tree detection with 100% accuracy and succeed in detecting non-oil palm tree object.

5. Conclusions

The classification of young, matured and non-oil palm tree have been performed. As a result, it is found that features obtained from Haar-based rectangular windows has achieved 92.73% overall accuracy with 98.58% sensitivity to young oil palm tree classification. Implementation of oil palm tree detection using oil palm tree recognition model based on Haar-based rectangular windows also have shown promising result for identifying and counting the young oil palm tree. Thus, it can be concluded that oil palm tree features derived from Haar-based rectangular windows is the most suitable for oil palm tree recognition model.

6. References

[1] Papageorgiou C P, Oren M and Poggio T 1998 A general framework for object detection *IEEE Sixth International Conference on Computer Vision* pp 555-562

[2] Viola P and Jones M 2001 Rapid object detection using a boosted cascade of simple features *Proceedings of IEEE Computer Society Conference on Computer Vision and Pattern Recognition CVPR 2001* vol 1 pp 511-518

[3] Viola P, Jones M J and Snow D 2005. Detecting pedestrians using patterns of motion and appearance *International Journal of Computer Vision*, 63(2), pp 153-161

[4] Brkic, K., 2010. An overview of traffic sign detection methods. *Department of Electronics, Microelectronics, Computer and Intelligent Systems Faculty of Electrical Engineering and Computing Unska*, 3, p 10000

[5] Gerónimo D 2009 Haar-Like Features and Integral Image Representation *Universitat Autónoma de Barcelona*

[6] Lienhart R and Maydt J 2002 An extended set of Haar-like features for rapid object detection *Proceedings of IEEE International Conference on Image Processing* vol 1 pp 900-903

[7] Tan M 2004 Support vector machine & its applications *The University of British*, Columbia

[8] Vapnik V N 1995 *The nature of statistical learning theory* New York: Springer-Verlag

[9] Gunn S R 1998 *Support Vector Machines for Classification and Regression* Technical Report, Image Speech and Intelligent Systems Research Group, University of Southampton

[10] Liang H and Hartimo I 1998 A heart sound feature extraction algorithm based on wavelet decomposition and reconstruction Proceedings of IEEE 20th Annual International Conference of the Engineering in Medicine and Biology Society vol 3 pp 1539-1542

Acknowledgments

The authors would like to thank Sime Darby for providing the WorldView-2 imagery data which operated by DigitalGlobe. Authors would also like to express appreciation to the members of CvviP for their support and valuable comments and ideas. Authors also acknowledge the Ministry of Higher Education Malaysia and Universiti Teknologi Malaysia (through research grant Q.J130000.2527.11H22) for providing financial support in completing this study.

Quantification of terrestrial laser scanner (TLS) elevation accuracy in oil palm plantation for IFSAR improvement

N A Muhadi[1], A F Abdullah[1] and M S M Kassim[1]

[1] Department of Biological and Agricultural Engineering, Faculty of Engineering, Universiti Putra Malaysia, 43400 Serdang, Selangor, Malaysia

tiramuhadi@gmail.com

Abstract. In order to ensure the oil palm productivity is high, plantation site should be chosen wisely. Slope is one of the essential factors that need to be taken into consideration when doing a site selection. High quality of plantation area map with elevation information is needed for decision-making especially when dealing with hilly and steep area. Therefore, accurate digital elevation models (DEMs) are required. This research aims to increase the accuracy of Interferometric Synthetic Aperture Radar (IFSAR) by integrating Terrestrial Laser Scanner (TLS) to generate DEMs. However, the focus of this paper is to evaluate the z-value accuracy of TLS data and Real-Time Kinematic GPS (RTK-GPS) as a reference. Besides, this paper studied the importance of filtering process in developing an accurate DEMs. From this study, it has been concluded that the differences of z-values between TLS and IFSAR were small if the points were located on route and when TLS data has been filtered. This paper also concludes that laser scanner (TLS) should be set up on the route to reduce elevation error.

1. Introduction

Site selection is one of the essential factors that affect yield productivity of oil palm plantation. It is required during the early phase to determine potential plantation area and throughout the life cycle of oil palm plantation. Climate condition, soil type and slope are some factors that need to be considered in choosing a good site for better yield productivity and plantation management. A good map of the plantation area together with elevation information is required in order to evaluate and manage the plantation wisely. In order to do that, Digital elevation models (DEMs) are used to produce the map of plantation area. Thus, it is crucial to obtain DEM data that are accurate. Light Detecting and Ranging (LIDAR) and Interferometric Synthetic Aperture Radar (IFSAR) are the most commonly used sources for creating DEMs. Airborne LIDAR data has been used in many applications such as in individual tree identification in forest area [1] [2] [3], flood modelling in urban area [4] [5] [6], detection of roads for road network [7] and hydrologic modelling [8]. Airborne LIDAR has the capability to collect three-dimensional information very densely over a large area. It provides high accuracy of DEMs and this is the reason why LIDAR data is costly compared to other data sources. Due to its price, airborne LIDAR data is found to be not economical to be used in agricultural sector. Therefore, an alternative approach to generate DEMs is introduced.

Terrestrial LIDAR data or can also be known as terrestrial laser scanning (TLS) is a technique that provides detailed information for small-scale area. The use of TLS in many applications has been done successfully previously. It has been applied in recognition of building structures [9], recording historical building [10], forestry management [11] and in agricultural sector [12] [13]. In agricultural sector, oil palm plantation normally covers over large areas, so multiple scans should be done to cover all areas. TLS dataset contains millions of data points, which take longer time to process even for one scanned data. Hence, it is not practical to use TLS data alone in oil palm plantation. IFSAR data has become the ultimate choice in using space borne data due to its ability to operate in day and night and during unwanted weather such as haze, rain and fog. Furthermore, the main reason why users are prefers to purchase IFSAR data is because of its affordable price with acceptable accuracy. For wide area coverage, IFSAR has been proved to be more economical compared to other data sources including airborne LIDAR [14].

Thus, the combination of IFSAR data and TLS data is proposed to overcome this problem. The main aim of this on-going research work is to develop a new method or algorithm to combine both airborne IFSAR and terrestrial LIDAR specifically for development of DEMs in oil palm plantation. However, this paper focuses on the evaluation of elevation value of TLS points and selection of TLS position in oil palm plantation. Besides, this paper also studied the importance of filtering process while using TLS data.

2. Methodology

The study was carried out in oil palm plantation in UPM using FARO laser scanner Focus 3D and RTK-GPS. The point clouds data were collected using FARO laser scanner. FARO laser scanner uses laser technology to acquire millions of three-dimensional points in a few minutes. It sends infrared laser beam to the centre of the rotating mirror before projecting it outward from the scanner to the surrounding. Once the infrared light is in contact with the object, it is then reflected back to the scanner. This FARO laser scanner Focus 3D can provide detailed scanning between 0.6 m up to 130 m. The distance accuracy of Faro laser scanner is up to 2 mm based on the specifications given by manufacturer. Nevertheless, this measurement accuracy is actually based on several factors such as the distance and incidence angle of scanner, air temperature gradients and wind. This study used Trimble R6 receiver and applying RTK positioning technique which manufacturer claimed the given error is 15 mm vertically. The base position of FARO laser scanner station was determined using Real Time Kinematic (RTK-GPS) to geo-reference the data. Figure 1 shows the mounted FARO laser scanner during field data collection.

Figure 1. The mounted FARO laser scanner.

After all data were collected, pre-processing and post-processing steps were conducted in FARO laser scanner software called SCENE software version 5.1.2. The scanning project is then being processed in SCENE software for point clouds registration. In order to get the ground values, the scan points need to be filtered. Several algorithms have been developed for filtering process. However, Abdullah et. al. [15] stated that not all algorithms are competent to produce good DEM data. The filtering algorithms should be chosen based on their ability to produce desired result. In this study, the scan points were filtered using TerraScan software. TerraScan is a commercial software that has been commonly used to remove non-ground points. The results were then exported to ArcGIS for further analysis. To evaluate the accuracy of scan points, 50 random points were collected using RTK-GPS as a reference point within the same area. These random points would be used to compare the elevation value or also known as z-values from both RTK-GPS and FARO laser scanner.

3. Results and discussions

This study was carried out to evaluate the elevation value of TLS. The scanner station was set up at longitude 414834.66 m and latitude 329855.23 m which located on the field route. For the first part, this study used raw scan points to compare the z-values from TLS and RTK data. Figure 2(a) shows map of the study area that contains the whole TLS scan points in preliminary study. It is clearly seen that the area within 10 m radius from the scanner position contained more scan points compared to the area outside the range.

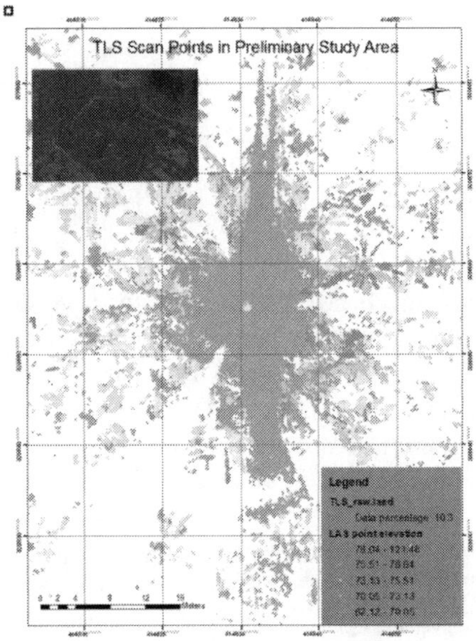

Figure 2(a). Map of TLS scan points in study area.

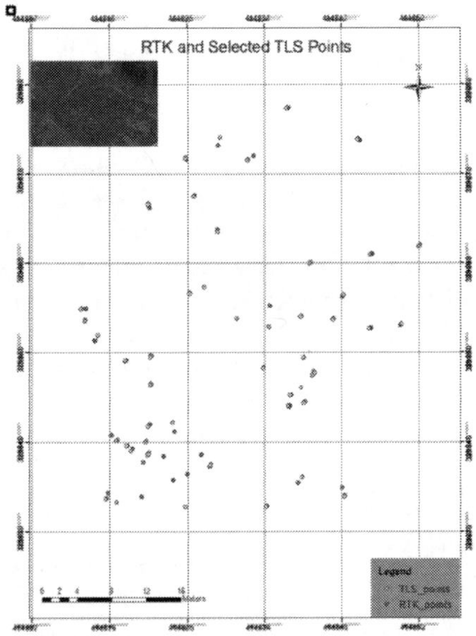

Figure 2(b). Map of RTK and selected TLS points.

In order to evaluate the TLS values, 50 points were measured randomly inside the study area using GPS-RTK. Therefore, only 50 points at the same spot were selected from the whole TLS points so that the z-values from both data could be evaluated. Figure 2(b) shows the map of RTK and selected TLS points in the study area. When RTK points were measured in dense area, the total of TLS scan points in that area was abundant. The laser could hit the surrounding thoroughly. Therefore, there were TLS points that available on the same coordinates. However, if the RTK points were measured in area with fewer points, there were possibility of having no TLS points on the same coordinate. Thus, the nearest TLS point from RTK data was selected to compare the elevation value. The range of distance between RTK and TLS points was within 0 to 1 meter. The difference between z-values from both TLS and RTK points were plotted in figure 3.

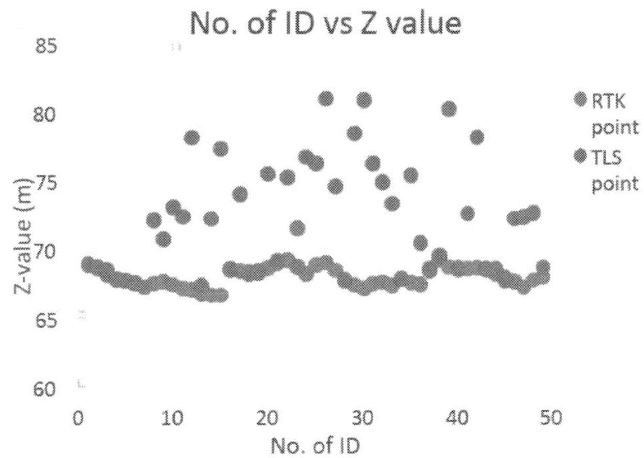

Figure 3. The difference of z-values between RTK and TLS points.

Based on the graph, it could be seen that there were some points that had huge differences of z-values. It was confirmed that most of these points were representing trees rather than the ground because the points were located inside the plantation area. The distance between the points and the scanner position that were far from each other could also affect the z-value of TLS points. This situation might happen because laser could not reach that far away area. Thus, laser could not hit the ground as it was being blocked by the canopy or neighbouring trees. Figure 4 shows the location of both RTK and selected TLS points.

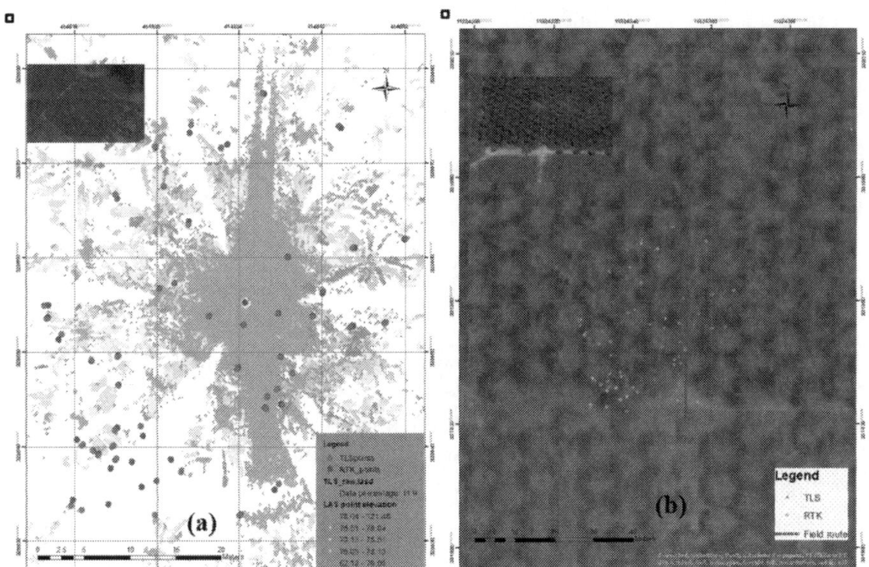

Figure 4. Map of RTK and selected TLS points in study area. (a) Selected TLS and RTK points with the whole TLS data (b) Location of the points on Earth.

The ground points could be obtained easily when the points were dense and in the field route area. However, the ground points were barely obtained in less dense area and in tree planted area. This was because laser from the scanner could not hit the ground as the distance was too far from the scanner. Besides, the laser might only hit the canopy of trees due to the instrument limitation. To quantify the TLS z-values, root mean square error (RMSE) for each point was calculated. Table 1 shows that only 21 points had obtained RMSE values less than 0.1. Most of these points were located on the field route with abundant scan points available.

Table 1. RMSE for TLS dataset

	RTK	TLS	RMSE		RTK	TLS	RMSE
1	68.924	68.870	0.008	**26**	69.076	80.980	1.683
2	68.742	68.650	0.013	**27**	68.455	74.590	0.868
3	68.144	68.500	0.050	**28**	67.681	67.800	0.017
4	67.895	67.810	0.012	**29**	67.364	78.470	1.571
5	67.717	67.700	0.002	**30**	67.141	80.890	1.944
6	67.625	67.500	0.018	**31**	67.462	76.270	1.246
7	67.246	67.260	0.002	**32**	67.601	74.870	1.028
8	67.526	72.160	0.655	**33**	67.362	73.310	0.841
9	67.675	70.730	0.432	**34**	67.885	67.820	0.009
10	67.384	73.070	0.804	**35**	67.566	75.360	1.102
11	67.157	72.390	0.740	**36**	67.423	70.470	0.431
12	67.064	78.200	1.575	**37**	68.570	68.440	0.018
13	66.751	67.340	0.083	**38**	69.247	69.560	0.044
14	66.642	72.250	0.793	**39**	68.703	80.220	1.629
15	66.656	77.340	1.511	**40**	68.643	68.490	0.022
16	68.594	68.480	0.016	**41**	68.581	72.600	0.568
17	68.422	74.020	0.792	**42**	68.653	78.160	1.344
18	68.383	68.230	0.022	**43**	68.547	68.520	0.004
19	68.465	68.290	0.025	**44**	68.265	68.570	0.043
20	68.651	75.510	0.970	**45**	67.915	67.700	0.030
21	68.962	69.150	0.027	**46**	67.648	72.250	0.651
22	69.239	75.220	0.846	**47**	67.269	72.330	0.716
23	71.548	68.760	0.394	**48**	67.733	72.650	0.695
24	68.234	76.740	1.203	**49**	68.029	68.720	0.098
25	68.881	76.290	1.048	**50**	68.252	73.980	0.810

3.1. Filtering process

The covered area of raw TLS data after processed from SCENE software was huge. The process is done to reduce noise and to limit the scanned so that it will cover the interested area only. The subject of this study was the ground points of the area. Nevertheless, the raw data contained millions of points including unnecessary object points. The process of extracting ground points from mixture of ground and objects is called as filtering. Therefore, filtering process was carried out to overcome this problem. Moreover, filtering could also help in reducing storage space and processing time. Thus, it could improve data processing performance.

Figure 5 shows TLS points after removing object points and the original TLS data. The ground points had been extracted from the raw TLS data using TerraScan software. Initially, there were 26,465,412 total points before filtering and the points reduced to 1,819,190 points after filtering process. In order to classify the position of each point, the data was segmented into three sections, section A, B and C as shown in figure 6. Section B contained 1,622,974 points that were located on the field route. On contrary, section A and C both consisted of 178,541 and 17,675 points that were located on the left and right sides of the field route respectively. More than half of the remaining points were located in section B. It is clearly shown that most of the points that had been removed were located in the plantation areas. This is mainly because lots of scan points in those areas were classified as object points, thus resulted in removal of points.

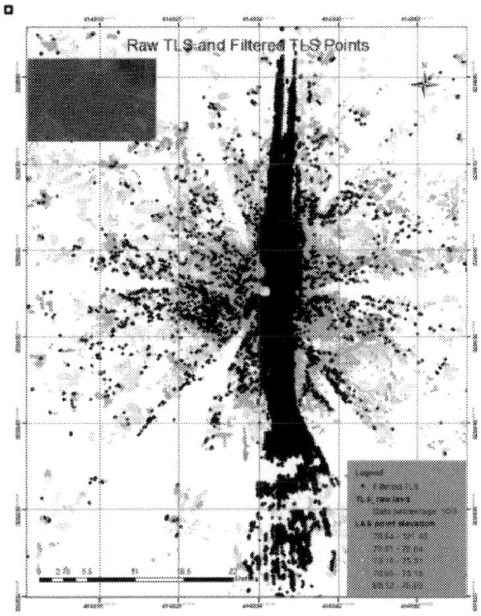

Figure 5. Filtered and raw TLS points of the area.

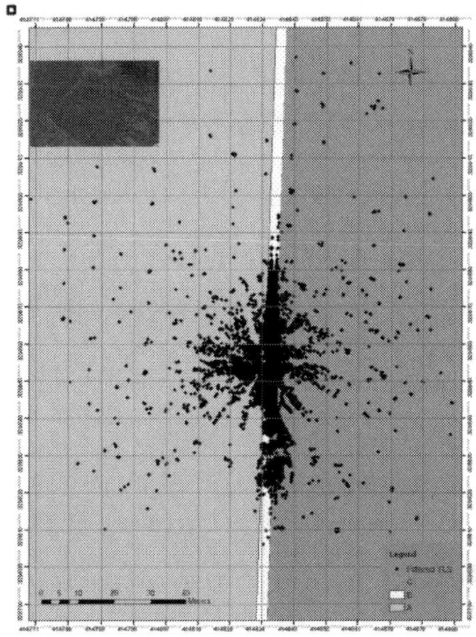

Figure 6. TLS points were divided into three sections; A, B and C.

50 RTK points from the first part were used again to determine the difference of z-values between RTK and filtered TLS points. Based on the same coordinate between those points, the elevation values were compared. If there were no TLS point available on the same coordinate, the nearest point would be selected within 1 meter range. As a result of filtering process, only 31 TLS points had met the requirement. The points had reduced from 50 to 31 because several points nearby RTK data had been removed during the process. Figure 7 shows the map of RTK and selected TLS points after filtering process in the study area.

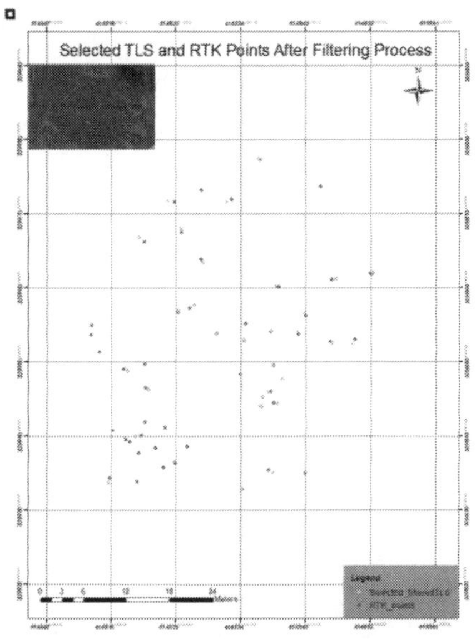

Figure 7. Map of selected TLS points after filtering process and RTK points.

The difference between z-values from both TLS and RTK data were plotted in figure 8. The figure clearly showed that there was no significant difference between Z value of RTK and TLS points after filtering process. The difference between those values was less than 1 meter.

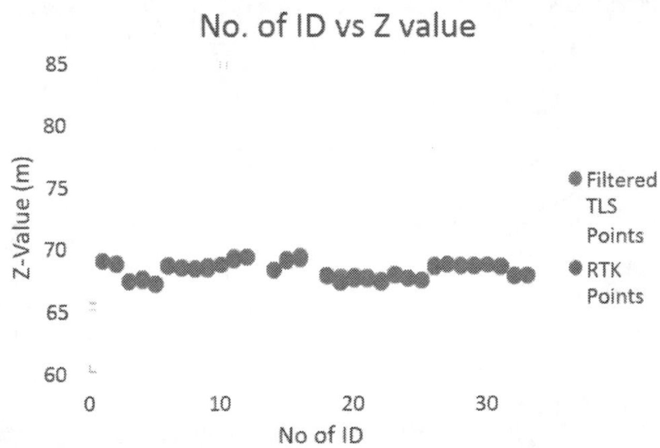

Figure 8. The difference between TLS and RTK data after filtering process.

The RMSE values were also calculated to measure how much error between TLS and RTK points. The result showed that all 31 points were having small RMSE with no point obtained RMSE value more than 0.1. Table 2 represents the RMSE value for each of the TLS points.

Table 2. RMSE of TLS data after filtering process.

	RTK	TLS	RMSE		RTK	TLS	RMSE
1	68.924	68.870	0.009	**16**	67.681	67.780	0.017
2	68.742	68.650	0.016	**17**	67.141	67.610	0.082
3	67.246	67.260	0.002	**18**	67.462	67.700	0.041
4	67.384	67.570	0.032	**19**	67.601	67.470	0.023
5	67.064	67.140	0.013	**20**	67.362	67.140	0.039
6	68.594	68.480	0.020	**21**	67.885	67.780	0.018
7	68.422	68.350	0.013	**22**	67.566	67.540	0.005
8	68.383	68.220	0.028	**23**	67.423	67.380	0.007
9	68.465	68.260	0.036	**24**	68.57	68.440	0.023
10	68.651	68.660	0.002	**25**	68.703	68.700	0.001
11	68.962	69.240	0.048	**26**	68.643	68.480	0.028
12	69.239	69.240	0.000	**27**	68.581	68.480	0.018
13	68.234	68.180	0.009	**28**	68.653	68.660	0.001
14	68.881	69.110	0.040	**29**	68.547	68.420	0.022
15	69.076	69.280	0.036	**30**	67.648	67.760	0.019
				31	67.733	67.760	0.005

4. Conclusion

This study showed that the z-values from terrestrial laser scanning (TLS) are totally reliable because the difference between z-value from TLS and RTK is small especially when the point is located in canopy-free area. Thus, this study proposes that laser scanner should be located on route and not placed in canopy area to reduce elevation errors. Another finding of this study is when using TLS data, filtering is a critical stage before doing further analysis. This paper showed that the RMSE value decreased significantly after the filtering process. In addition, filtering process could reduce the size of data which results in reducing the storage space and processing time.

References

[1] Warner T a, Lee J Y, McGraw J B, Hill D a and Leckie D G 1999 Delineation and identification of individual trees in the Eastern Deciduous Forest *Automated Interpretation of High Spatial Resolution Digital Imagery for Forestry, Int. Forum* pp 81–91

[2] Samal A, Brandle J R and Zhang D 2006 Texture as the basis for individual tree identification *Inform Sciences* **176(5)** 565–76

[3] Reitberger J, Heurich M, Krzystek P and Stilla U 2007 Single Tree Detection in Forest Areas With High-Density Lidar Data *Int. Arch Photogram Remote Sens Spatial Inform Sci* 36((3/W49B)) pp 139–144

[4] Haile A and Rientjes T 2005 Effects of LiDAR DEM resolution in flood modelling: A model sensitivity study for the city of Tegucia- galpa, Honduras *Proc. of ISPRS Workshop Laser Scanning 2005* pp 168–73

[5] Sarker L R, Nichol J and Mubin A 2013 *Developments in Multidimensional Spatial Data Models* ed Abdul Rahman A et al (Berlin, Heidelberg: Springer Berlin Heidelberg) pp 69-83

[6] Abdullah, A. F., Vojinovic, Z., Price, R. K., & Aziz, N. A. A. (2012). Improved methodology for processing raw LiDAR data to support urban flood modelling accounting for elevated roads and bridges. *Journal of Hydroinformatics* **14(2)** 253–69

[7] Li Y, Yong B, Wu H, An R and Xu H 2015 Road detection from airborne LiDAR point clouds adaptive for variability of intensity data *Optik* **126(23)** 4292–98

[8] Jones K L, Poole G C, O'Daniel S J, Mertes L a K and Stanford J a 2008 Surface hydrology of low-relief landscapes: Assessing surface water flow impedance using LIDAR-derived digital elevation models *Remote Sense of Environ* **112(11)** 4148–58

[9] Vosselman G, Gorte B G H, Sithole G and Rabbani T 2004 Recognising strucutre in laser scanner point clouds *Int. Arch Photogram Remote Sens Spatial Inform Sci* vol XXXVI pp 33–8

[10] Vozikis G, Haring A, Evangelos Vozikis E, Kraus K 2004 Laser Scanning: A New Method for Recording and Documentation in Archaeology *Proc. of FIG Working week 2004 "The Olympic Spirit in Surveying"* (Athens Greece)

[11] Srinivasan S, Popescu S C, Eriksson M, Sheridan R D and Ku N-W 2014 Multi-temporal terrestrial laser scanning for modeling tree biomass change. *Forest Ecol and Manage* **318** 304–17

[12] Barneveld R J, Seeger M and Maalen-Johansen I 2013 Assessment of terrestrial laser scanning technology for obtaining high-resolution DEMs of soils *Earth Surf Proc Landforms* **38(1)** 90–4

[13] Ehlert D and Heisig M 2013 Sources of angle-dependent errors in terrestrial laser scanner-based crop stand measurement *Comput Electron Agric* **93** 10–6

[14] Dowman I 2004 Integration Of Lidar And Ifsar For Mapping *Proc. of XXth ISPRS Congress 12-23 July 2004* (Istanbul, Turkey Commission 2)

[15] Abdullah A F, Rahman A and Vojinovic Z 2009 Lidar algorithms for urban flood application: Revew on current algorithms and filters test *Int. Arch Photogram Remote Sens Spatial Inform Sci* XXXVIII (Part 3) pp 30-6

Geospatial application for the identification and monitoring of rubber smallholders in the Malaysian state of Negeri Sembilan

Mohd Hafiz Mohd Hazir[1] and Tuan Mohamad Tuan Muda[2]

[1,2]Extension and Development Division, Malaysian Rubber Board, 14th Floor, Bangunan Getah Asli (Tower), Jalan Ampang, 50450, Kuala Lumpur, Malaysia

E-mail: mohdhafiz@lgm.gov.my

Abstract. The Malaysian rubber industry, especially in the upstream sector, is much dependent on smallholders to produce latex or cup lumps. Identification and monitoring of rubber smallholders are essential tasks when it comes to the Malaysian rubber industry's sustainability. The authorised agencies who support the rubber smallholders can do better planning, arranging, and managing. This paper introduces a method of calculating the total number of smallholders as well as identifying the location of their planted rubber area. The scope of this study only focused on land owners as rubber smallholders in the selected study area of Negeri Sembilan. The land use map provided by the Department of Agriculture Malaysia gave information on distribution of rubber area in Malaysia, while the cadastral map from the Department of Survey and Mapping Malaysia was specifically used for identifying land owners of each rubber parcel or rubber lot. Both data were analyzed and processed with ArcGIS software to extract the information, and the results were then compared to the Malaysian Rubber Board smallholders database.

1. Introduction

Malaysia is a well-known producer of raw material commodities. Today, the largest commodity sector in Malaysia is oil palm, followed by rubber. Malaysia's rubber industry was established in 1876, when the British introduced the rubber tree in Malaya [1]. A great deal of research has been conducted by various institutions and organizations to ensure the sustainability of the rubber industry — this involved upstream and downstream study [2]. The Rubber Research Institute of Malaysia (RRIM) is already known as the leading rubber research institution in Malaysia. In 1998, the organization was rebranded by combining with other rubber agencies, and it was eventually recognized as the Malaysian Rubber Board (MRB). The board became a pioneer — and custodian organization — in the Malaysian rubber industry activities operated under the Ministry of Plantation Industry and Commodity (MPIC).

The rubber industry in Malaysia, especially the upstream sector, is highly dependent on smallholders [3]. Rubber smallholders covered about 93% of total cup lumps and latex production in 2014. As reported by MRB in 2015, the total Malaysian rubber area in 2014 was 1,065,630 hectares; smallholders owned 985,510 hectares (Figure 1). Rubber smallholders can be divided into two categories: rubber land owners and rubber tapper workers [4]. However, throughout this paper, the term "rubber smallholders" will specifically refer to rubber land owners. The term smallholder refers to the owner of land less than 40 hectares; on average, each rubber smallholder owns about 2 to 4 hectares of rubber area [5]. The monitoring of rubber area is crucial for estimations of national rubber productivity. Rubber area in Malaysia has been decreasing recently because rubber landowners and plantation companies have been moving to other types of commodities. The real estate industry and the development of rural areas have also affected the total land area planted with rubber [6].

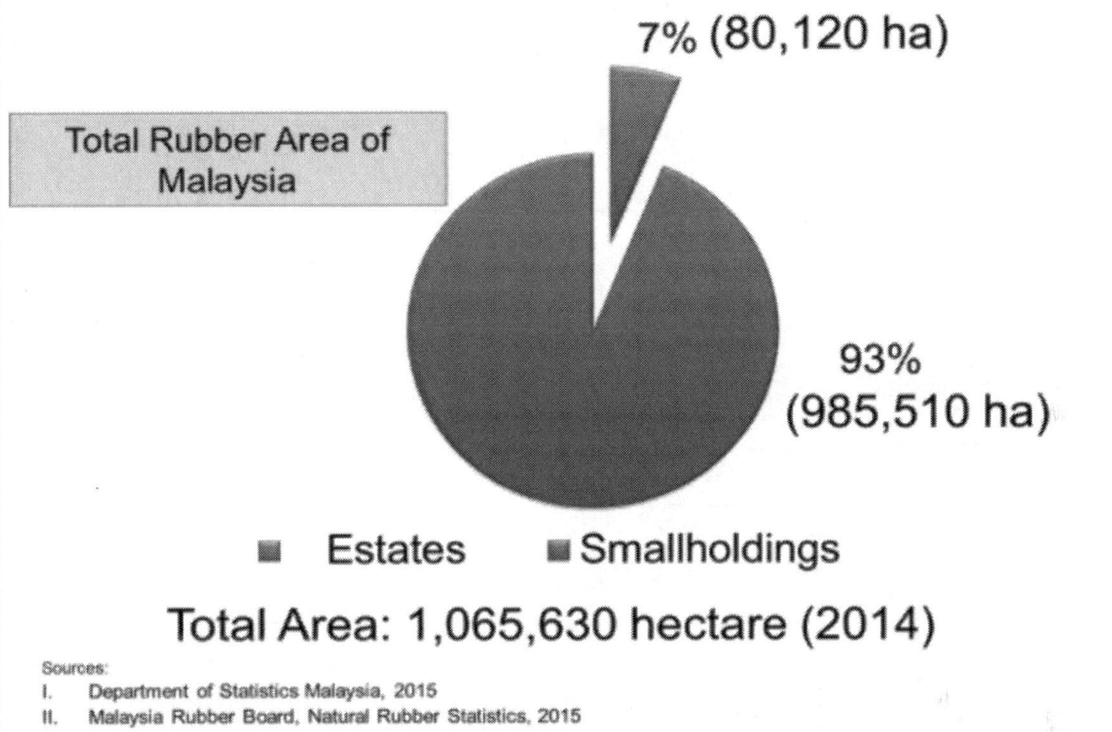

Figure 1. Total rubber area in Malaysia.

Therefore, methods are needed to determine the total number of smallholders, calculate smallholders' rubber areas, and identify the location of smallholders' rubber areas. The MRB has already taken proactive action by collecting smallholders' information through the Permit Autoriti Transaksi Getah (PAT-G) database. The purpose of PAT-G is to ensure that all transactions to buy and sell cup-lumps and latex are conducted legally. Smallholders also need to have a PAT-G card to make use of the Insentif Pengeluaran Getah (IPG) program. This program was initiated by the Malaysian government to help rubber smallholders during times when the cup - lump and latex prices are low by giving them an interest in the rubber commodity.

MRB also published the nationwide Independent Rubber Smallholder's Survey in 2012. The objective of this survey was to determine how well technology created by MRB has been disseminated to the rubber smallholders. The survey was done for the whole country. The survey also allowed MRB to collect invaluable information on smallholders, and that information was developed into a very useful database for MRB. The information was quite detailed, but addressed only the independent rubber smallholders. The definition of 'independent rubber smallholder' is any smallholder who plants rubber without being attached to any scheme or agency. Independent rubber smallholders have their own land, and are self-managed.

The objective of this study is to introduce a practical method for indicating, calculating, and identifying rubber smallholder's numbers and locations using a Geospatial application. The scope of this study is limited to the rubber land owners who have been granted full land title from the

Malaysian Land Department and who are recorded by Malaysian Mapping and Survey Department (JUPEM). The results can then be compared with the rubber smallholders database that has already been established by MRB.

2. Materials and methods

The materials used in this study were provided by other government agencies, under the belief that sharing information should be practiced among agencies in order to promote the optimization of government resources. The study area was chosen based on the availability of the data, resources, and manpower. The proposed method used in this study was the most practical and effective method for obtaining the rubber smallholders' information. Figure 2 shows the overall method used in this study. All the pre-processing and analysis of the data in this study was carried out using ArcGIS software version 10.3, developed by ESRI.

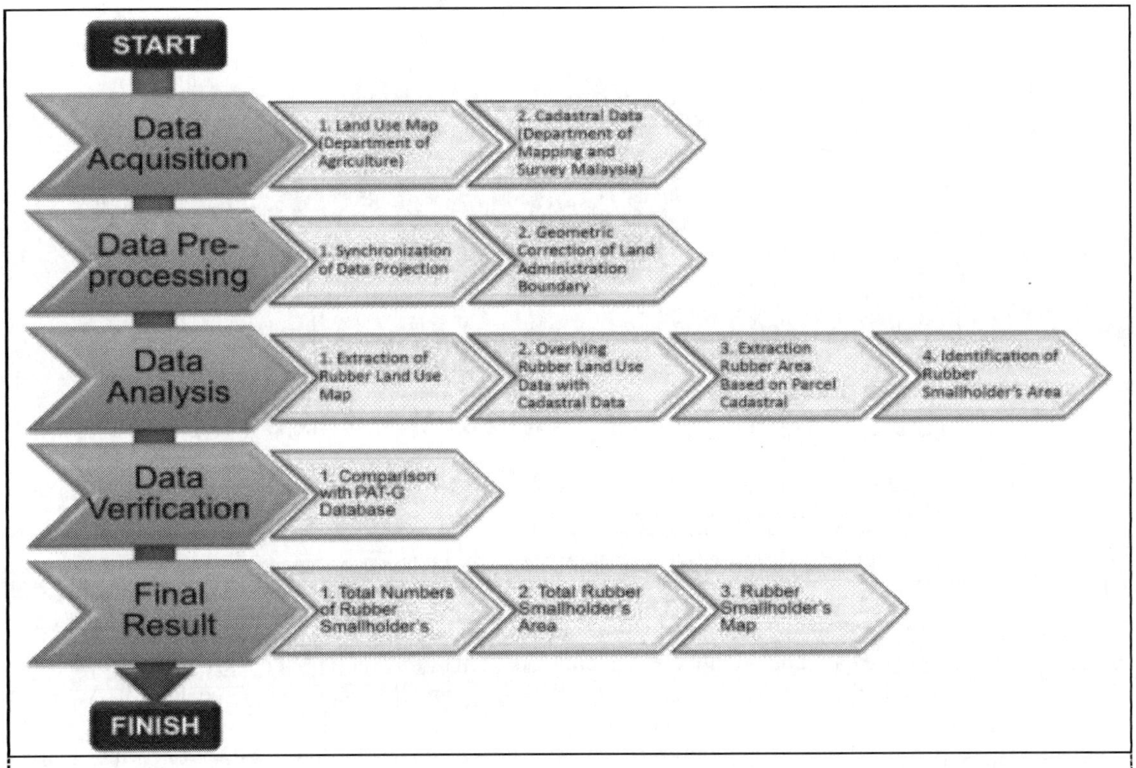

Figure 2. The overall method.

2.1. *Study area*

The selected study area was the Malaysian state of Negeri Sembilan, located on the western coast of the peninsula (Figure 3). Negeri Sembilan has quite a number of rubber smallholders within its seven districts consisting of Seremban, Jelebu, Kuala Pilah, Jempol, Tampin, Rembau and Port Dickson. Jempol is the largest district, followed by Jelebu and the capital, Seremban. Negeri Sembilan's economy was once dominated by agriculture, but manufacturing has quickly become a major contributor to the state's income today [7].

The total area of Negeri Sembilan is about 669, 506 hectares with the total rubber-producing area being about 20% of that. In the year 2010, the Department of Agriculture (DOA) reported approximately 158,836 hectares were planted with rubber trees. The Malaysian Rubber Board (MRB) and the Department of Statistics (DOS) also reported that in the year 2013, the total area used for rubber cultivation in Negeri Sembilan was 108,522 hectares. Based on MRB Independent

Smallholders Surveys in the year 2012, the total number of rubber smallholders in Negeri Sembilan was 21,593 smallholders. The Rubber Industry Smallholders Development Authority (RISDA) established the statistic of the total number of Negeri Sembilan rubber smallholders to be 25,548 smalholders. This study was piloted partly by the Malaysian Rubber Board Scientific and Economic Advisory Council (SEAC) project where the objective is to identify rubber-producing areas using satellite imagery. The coordinates of the Negeri Sembilan axis are 3°17.100' N, 101° 56.022' E and 2°24.948' N, 102°35.772' E.

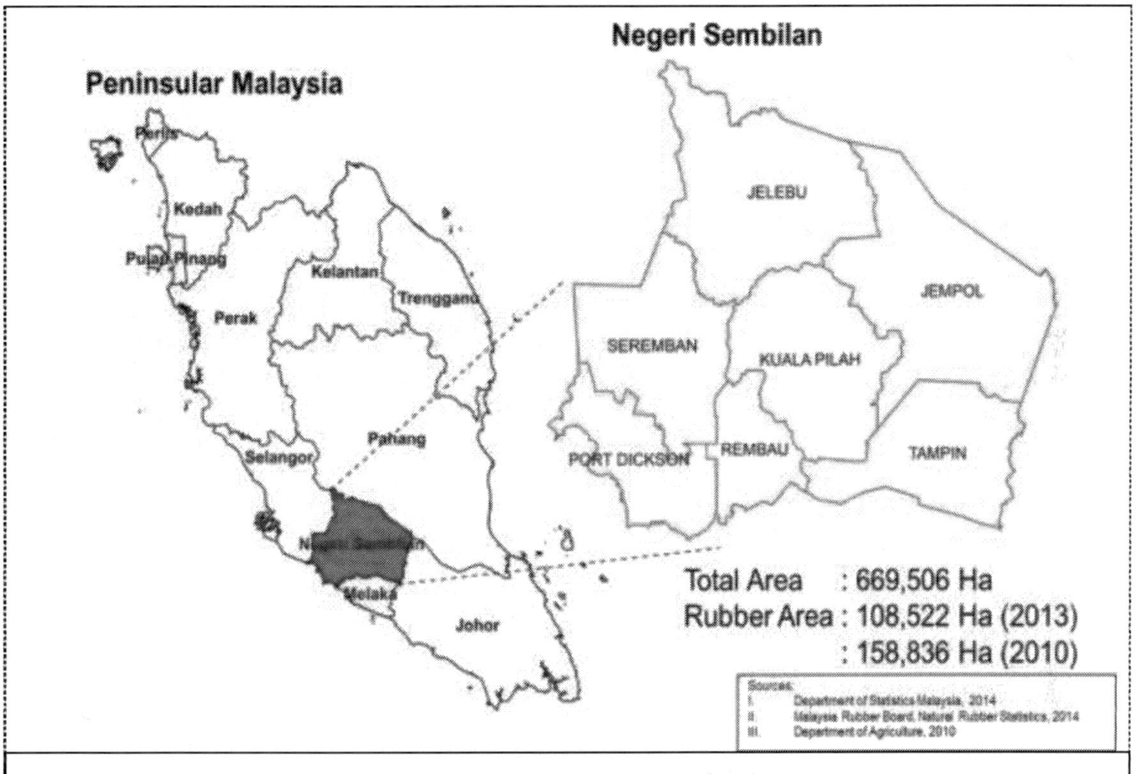

Figure 3. Negeri Sembilan, Malaysia.

2.2. Data acquisition

Two important sources of data were used in this study. The first source of data was the Land Use Map produced by the Department of Agriculture Malaysia (DOA), which shared the map through the Government to Government (G2G) system. The Land Use Map was in shapefile format and a projected coordinate system of RSO Peninsular Malaysia (GDM2000). It contained all the information on land use, such as commodity plantation, fruit plantation, vegetable farm, road, highway, and mining area (Figure 4). The second source of data was the digital cadastral map created by the Department of Survey and Mapping Malaysia and known as the National Digital Cadastral Database (NDCDB) (Figure 5). The map shows the final parcels of land based on titles, and the layout, boundary and stone of surveyed lots in Cassini Geocentric projection. It can also be applied for in shapefile format. The NDCDB data were successfully acquired from Malaysian Centre Geospatial Data Infrastructure (MaCGDI), a Geospatial government agency that promotes sharing information bases.

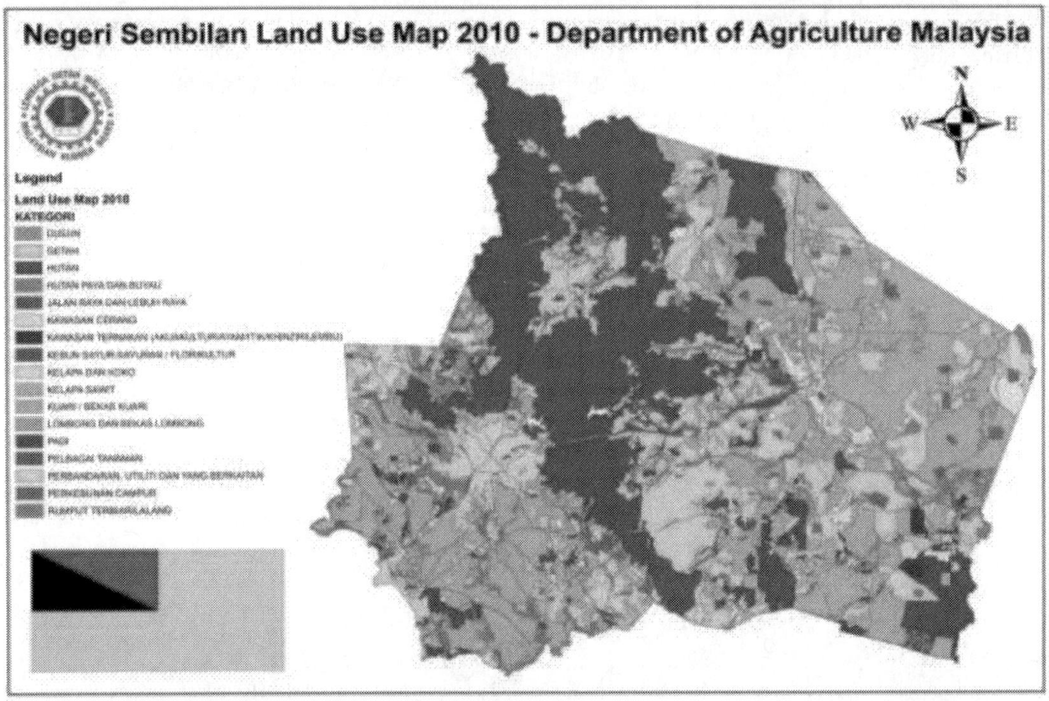

Figure 4. Negeri Sembilan land use map.

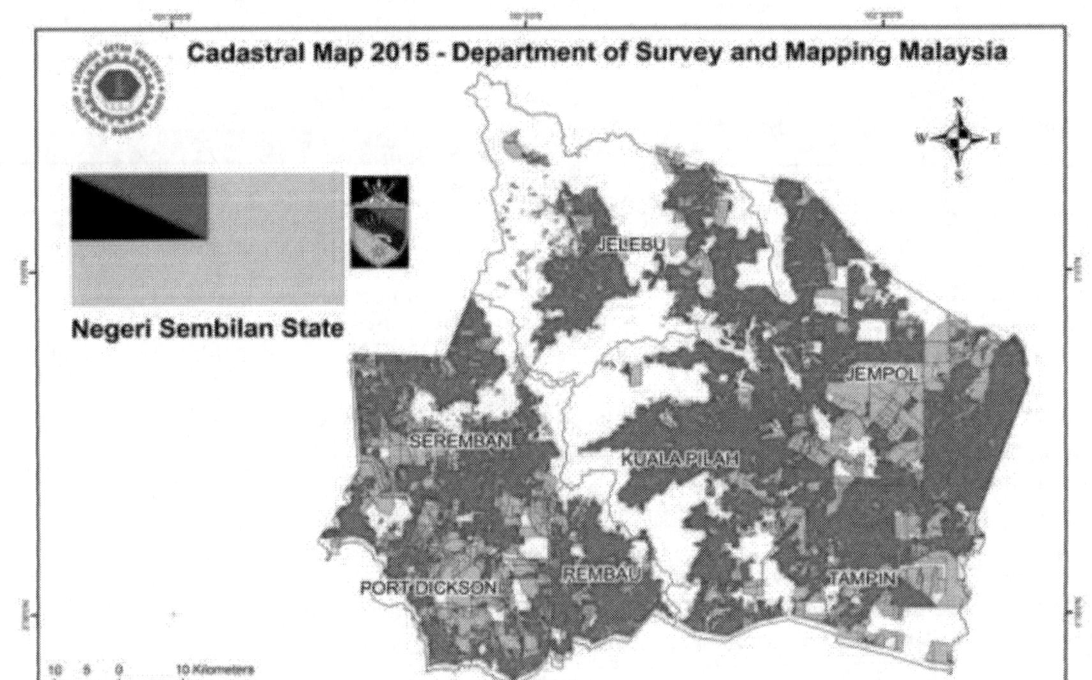

Figure 5. Negeri Sembilan National Digital Cadastral Database (NDCDB).

2.3. Data pre-processing

The Land Use Map and the National Digital Cadastral Database (NDCDB) were not in the same coordinate projection system. Therefore, a pre-processing task was performed to overlay the two maps on top of each other. The chosen projection was World Geodetic System 1984 (WGS84) and both data were converted to this projection. The rubber land use map was extracted from original Land Use Map as this study is only interested in the rubber area.

2.4. Data analysis

The analysis of the data was started by overlaying Rubber Land Use with NDCDB data. The next step was to apply the selection by location and choose the 'intersect' option. The idea was that any NDCDB data intersecting with rubber area for Rubber Land Use would be selected. After that, the selected data were exported into a new file and generated Rubber Area Based on Land Parcel. The next step was calculating the total number parcels containing rubber area, which can be divided into two categories: namely the rubbers smallholder's area and plantation area. The definition of rubber smallholders is less than 40 hectares of land planted with rubber. A land parcel area greater than 40 hectares will be acknowledged as plantation area. The result of this analysis was the total number of rubber smallholders based on total parcel cadastral lot covered with rubber area less than 40 hectares.

2.5. Data verification

The generated rubber smallholder's data from the analysis was compared with PAT-G database. The comparison method used was the 'join tool' in ArcGIS software. Both data sets must have a Unique Parcel Identifier (UPI) to provide basic information. The results show how many rubber smallholders are already registered under PAT-G and how many have not yet registered. The same process was carried out in comparison with the MRB Independent Rubber Smallholders Survey 2012 database. The estimation of the total number of rubber smallholders was based on the calculation of NDCDB lots. The PAT-G database is more likely to include rubber smallholders with mature tree stages, which already have cup-lumps or latex to sell. This database provides the latest information that MRB has on rubber smallholders. Meanwhile, the MRB Independent Rubber Smallholders Survey 2012 database provides information on immature and mature rubber area.

3. Results and discussion

The results of this study highly depend on the accuracy of the Rubber Land Use data provided by the Department of Agriculture (DOA). The extracted rubber area information from the original Land Use Map is shown in Figure 6. This data is from the year 2010. Even though this data does not represent the latest rubber land use, the method used in this study can be applied to more current data when it becomes available. Figure 7 shows the overlay data between NDCDB cadastral data and the Rubber Land Use Map.

The important task was to extract only the lots and parcels that had rubber area. Figure 8 shows the results of this extraction. The total rubber land parcels measured 145,938 hectares. As this study was only interested in smallholders, only the parcels and lots smaller than 40 hectares were selected. The total land parcels or lots smaller than 40 hectares stood at 140,516, as shown in Figure 9. The study assumed that each parcel or lot represents one smallholder. This means the total number of rubber smallholders at Negeri Sembilan was around 140,516 smallholders.

The final step was verifying the result. A comparison was made with the MRB smallholders' database. The first database was MRB Independent Smallholders Surveys 2012. The total number of Negeri Sembilan rubber smallholders from this database stood at 22,729. This database was focused on independent rubber smallholders and did not cover the entire information on rubber smallholders. This data was merged based on the Unique Parcel Identifier (UPI) information. The result shows that only 7,041 entries were merged (Figure 10). The figure may be lower because of several factors. Not all independent rubber smallholder databases were successfully converted into the UPI format as it depends on the availability of information during the surveying task.

A comparison was also made with the Negeri Sembilan Permit Autoriti Transaksi Getah (PAT-G) database, which is the latest available smallholder database. The total records of rubber smallholders owning a rubber land parcel or lot according to the PAT-G database stood at 14,688. The result shows that only 2,929 records could be merged with the analysis results of the rubber land parcel (Figure 10) as it depends on how well this database was converted into the UPI format. This format should essentially feature detailed information on the state code, district code, sub-district code, section code, and lot or parcel number. The PAT-G database mostly focuses on the rubber smallholder owners who have a mature rubber tree as they need to register to sell their cup-lump or latex.

The estimated total number of Negeri Sembilan rubber smallholders based on the analysis result was quite high compared to the available Malaysian Rubber Board (MRB) database. This is mainly because it totally depends on how well the rubber land use map was produced. The changes in the latest Negeri Sembilan's land use for the year 2010 may also have an impact on the result as the MRB database was developed in the year 2012 and 2015. The MRB method of developing the database was used as a survey method for the Independent Rubber Smallholders Survey 2012 and for voluntarily registering for the PAT-G database. Even though it is compulsory to register with PAT-G and bear a PAT-G card, ultimately it depends on the rubber smallholder's decision and acceptance.

The NDCDB cadastral database also changes from time to time. The number of parcels or lots can increase because of reasons such as registration of new parcel lots, approval of new full land title, and separated parcels. Availability of a high-accuracy and latest land use map was important for this kind of study. It is the main factor to obtain the accurate estimated total number of rubber smallholders. The rubber land use map can be produced by aerial mapping and ground data collection. The Department of Agriculture (DOA) created the land use map based on aerial imagery specifically based on satellite imagery analysis. The resolution of the imagery should be as high as possible to capture the rubber smallholder's area. The MRB's vision is to go into as much detail as possible for acquiring rubber smallholders' information using the Geospatial application.

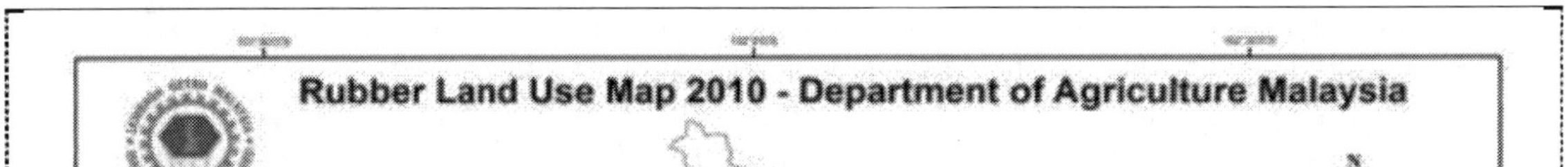

Figure 6. Negeri Sembilan rubber land use map.

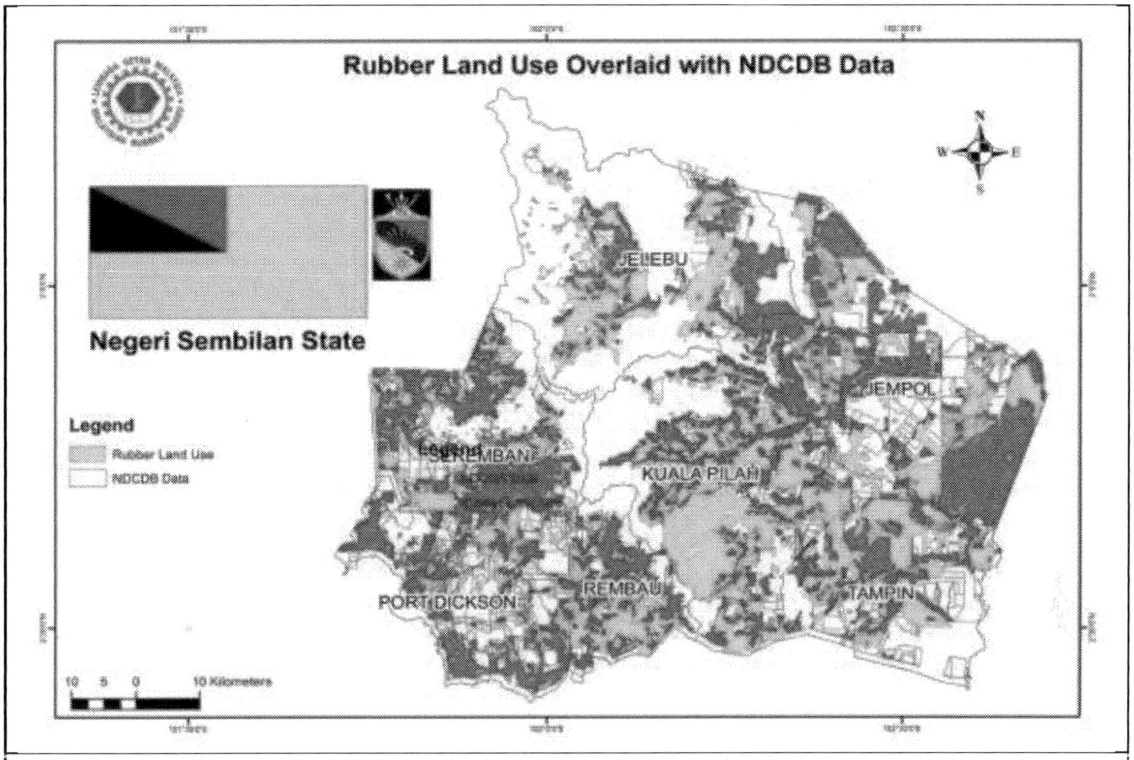

Figure 7. Rubber land use overlaid with NDCDB data.

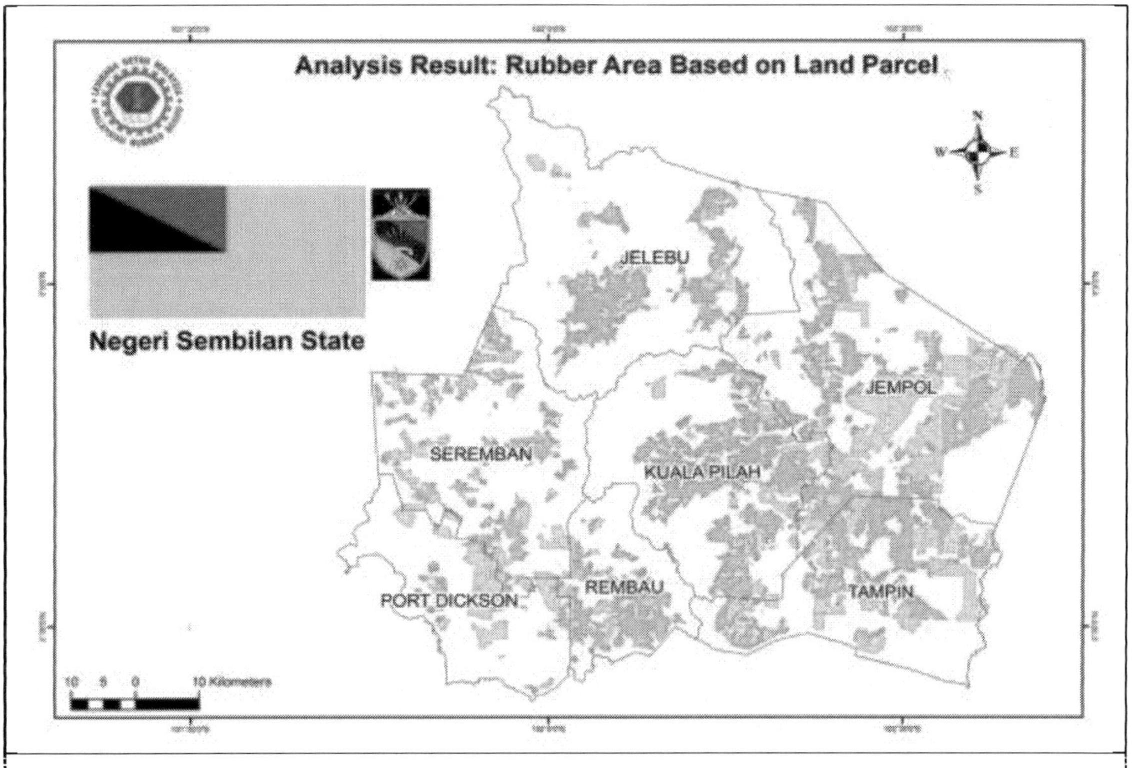

Figure 8. Analysis results of rubber area based on land parcel.

IOP Conf. Series: Earth and Environmental Science **37** (2016) 012043 doi:10.1088/1755-1315/37/1/012043

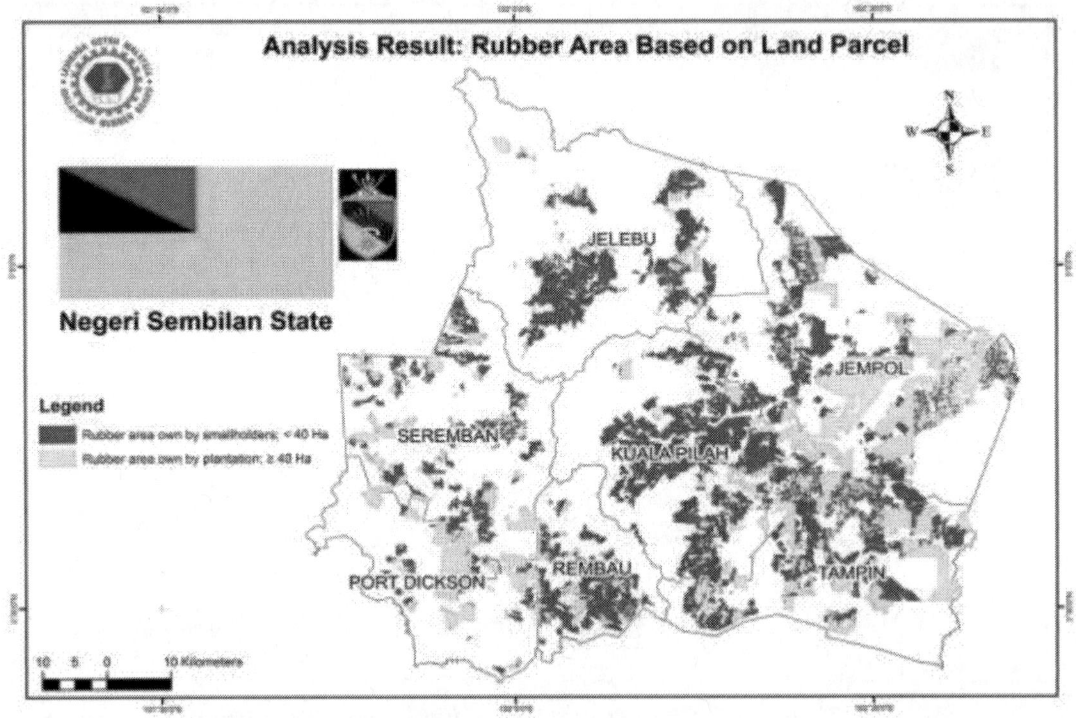

Figure 9. Analysis results of rubber area based on hectarage.

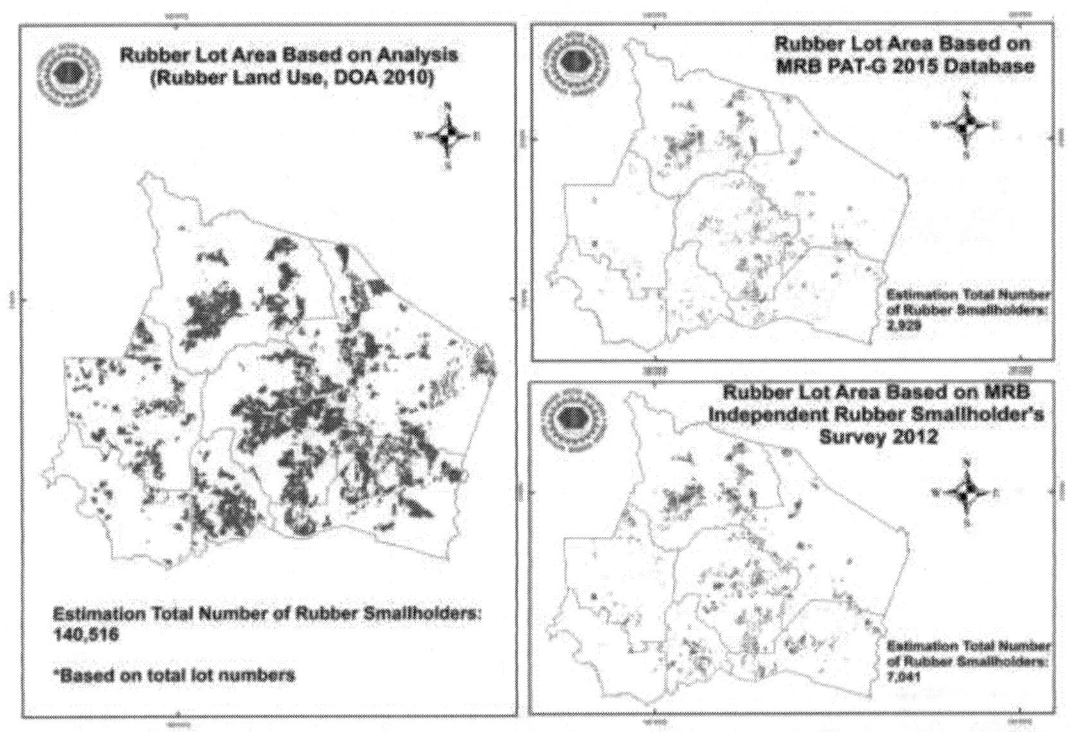

Figure 10. Verification of analysis results with MRB database.

4. Conclusions

The estimated total number of Negeri Sembilan rubber smallholders based on the analysis results was 140,516. This number assumes that each NDCDB lot was owned by one rubber smallholder. The accuracy of this study and its results were highly dependent on the accuracy of the Land Use Map provided by the Department of Agriculture (DOA) and availability of NDCDB final title land parcel from the Department of Survey and Mapping Malaysia (JUPEM). This study successfully presented a method to estimate the total number of rubber smallholders based on the available data.

Acknowledgement

This study was partly conducted under the Scientific and Economic Advisory Council project code of S14EST0522 and was supported fully by the Malaysian Rubber Board. The researchers would like to thank the Malaysian Centre Geospatial Data Infrastructure (MaCGDI), Department of Survey and Mapping Malaysia (JUPEM), Department of Agriculture (DOA), Federal Department of Town and Country Planning, and the Malaysian Remote Sensing Agency for sharing data and information. Our special thanks to the Extension and Development Division staff and others who contributed to this study.

References

[1] Drabble J H 2015 Rubber in Malaya 1876–1922: The genesis of the industry *Oxford University Press*, Kuala Lumpur and Singapore, 1973. Pp. xx, 256

[2] Azmi I M and Alavi R 2013 Patents and the practice of open science among government research institutes in Malaysia: The case of Malaysian Rubber Board *World Patent Information*, 35(3), pp.235-242.

[3] Theriault V 2014 New Directions for Smallholder Agriculture *Journal of Economic Issues*, 48(4).

[4] Fox J and Castella J C 2013 Expansion of rubber (Hevea brasiliensis) in Mainland Southeast Asia: what are the prospects for smallholders? *The Journal of Peasant Studies*, 40(1), pp.155-170.

[5] Stubbs R 1983 Malaysia's Rubber Smallholding Industry: Crisis and the Search for Stability *Pacific Affairs*, 56(1), pp.84-105.

[6] Tan-Soo J S Adnan N Ahmad I Pattanayak S K and Vincent J R 2016 Econometric evidence on forest ecosystem services: deforestation and flooding in Malaysia *Environmental and Resource Economics*, 63(1), pp.25-44.

[7] "Negeri Sembilan" Portal Rasmi Kerajaan Negeri Sembilan *Negeri Sembilan State Government*, 24 Feb. 2016. Web. 24 Feb. 2016.

Agriculture land suitability analysis evaluation based multi criteria and GIS approach

Goma Bedawi Ahmed[1, 2*], Abdul Rashid M. Shariff [1], Siva Kumar Balasundram [3], Ahmad Fikri bin Abdullah[4]
[1]Geospatial Information Science Research Centre (GISRC), University Putra Malaysia
[2]Arab Center for Desert Research and Development of Desert Communities Morzok- Libya
[3]Department of Agriculture Technology, University Putra Malaysia
[4]Department of Biological and Agricultural Engineering, University Putra Malaysia

E-mail: joma762001@gmail.com

Abstract. Land suitability evaluation (LSE) is a valuable tool for land use planning in major countries of the world as well as in Malaysia. However, previous LSE studies have been conducted with the use of biophysical and ecological datasets for the design of equally important socio–economic variables. Therefore, this research has been conducted at the sub national level to estimate suitable agricultural land for rubber crops in Seremban, Malaysia by application of physical variables in combination with widely employed biophysical and ecological variables. The objective of this study has been to provide an up–to date GIS-based agricultural land suitability evaluation (ALSE) for determining suitable agricultural land for Rubber crops in Malaysia. Biophysical and ecological factors were assumed to influence agricultural land use were assembled and the weights of their respective contributions to land suitability for agricultural uses were assessed using an analytic hierarchical process. The result of this study found Senawang, Mambau, Sandakan and Rantau as the most suitable areas for cultivating Rubber; whereas, Nilai and Labu are moderately suitable for growing rubber. Lenggeng, Mantin and Pantai are not suitable for growing rubber as the study foresaw potential environmental degradation of these locations from agricultural intensification. While this study could be useful in assessing the potential agricultural yields and potential environmental degradation in the study area, it could also help to estimate the potential conversion of agricultural land to non–agricultural uses.

1. Introduction

The growing population numbers, acutely in developing countries, has intensified the pressure on both natural and agricultural resources. To meet the economic demands of the growing world population, an increased economic return is required. Both population increases and the process of urbanization have increased the pressure on agricultural resources [1]. This increased pressure on the available land resources may result in land degradation [2]. Dependable and accurate land evaluation is, therefore, indispensable to the decision-making processes involved in developing land use policies that will support sustainable rural development. If self-sufficiency in agricultural production is to be achieved in developing and transitional countries, land evaluation techniques will be required to develop models for predicting the land's suitability for different types of agriculture [3].

Multi-criteria evaluation processes are already used in some regional planning processes since they aim at ''estimating the potential of land for alternative land uses; among which, the agricultural land use may be the most important area where it is applied'' [4]. This method could play a key role in future land-use planning [4, 5]. Agricultural land suitability classification based on indigenous knowledge is vital to land use planning. The systematic assessment of land and water potential aims to identify and put into practice future alternative land uses that will best meet the needs of the people while at the same time safeguarding resources for the future [6].

The land evaluation method is the systematic assessment of land potential to find out the most suitable area for cultivating some specific crop. Theoretically, the potential of land suitability for agricultural use is determined by an evaluation process of the climate, soil, water resources and topographical, as well as the environmental components under the criteria given and the understanding of the local biophysical restraints [7]. The use of GIS Multi-Criteria Decision Making (MCDM) methods allows the user to derive knowledge from different sources in order to support land use planning and management [7, 8]. One multi-attribute technique that has been incorporated into the GIS-based land use suitability procedure is the Analytical Hierarchy Process (AHP) [9, 10]. MCDM methods, such as the AHP method, have been successfully applied to land evaluation techniques [11]. These methods, which aim to allow for a transparent decision-making base, are, however, only rarely used in developing and transitional countries such as Malaysia. We have used a GIS-based MCDM land suitability analysis method to classify the study area (Seremban District) with respect to the potential for Rubber cultivation. We assumed that this goal could effectively help agricultural insurance through the identification and separation of land-based capabilities with regards to environmental, Biophysical and Socio-economical potential.

2. Multi-criteria decision making MCDM approaches

Multi-criteria decision making MCDM approaches were developed in the 1960s in order to assist decision makers in incorporating numerous options, reflecting the opinions of concerned parties into a potential or retrospective framework. This framework is ''primarily concerned with how to combine the information from several criteria to form a single index of evaluation'' [10]. They were designed to define the relationship between data input and data output. The integration of the GIS and MCDM methods provides powerful spatial analysis functions[12, 13]. In the MCDM approach, GIS is best suited for handling a wide range of data criteria at multi-spatial, multi-temporal and multi-scale from different sources for a time-efficient and cost-effective analysis. Therefore, there is growing interest in incorporating the GIS capability with MCDM processes[14].

The Analytical Hierarchy Process is a well-known multi-criteria technique that has been incorporated into GIS-based suitability procedures [10]. For the classification of land suitability within our case study area in Seremban, we utilized the AHP's ability to incorporate different types of input data, and the pairwise comparison method for comparing two parameters, simultaneously. The application of the AHP process involves several steps in order to rank Criteria or factors to the set of suitable criteria. This is usually achieved by domain and experts' opinions: The consistency of the overall set of pairwise comparisons is assessed using its Consistency Ratio CR).

One of the most important factors affecting the land suitability classification for cultivation is soil properties. The soil properties' criteria consist of soil texture, surface stoniness, soil depth, pH, EC, soil phosphorus, Potassium and organic matter.

3. Materials and Methods

3.1. Study area

Seremban city is the largest district and the capital of the Negeri Sembilan State. The study area is located between longitudes (101o 45' 0" E and 102o 6' 0" E) and latitudes (3o 0' 0" N and 2o 30' 0" N). The city occupies a total land area of approximately 935.78 sq. km. Seremban City includes the districts of Seremban, Setul, Labu, Rasah, Ampangan, Rantau, Pantai and Lenggeng (Fig. 1). Seremban is located at about 20 kilometers from Putrajaya, which is considered the national capital of Malaysia, and at about 67 kilometers from Kuala Lumpur, the economic center of Malaysia.

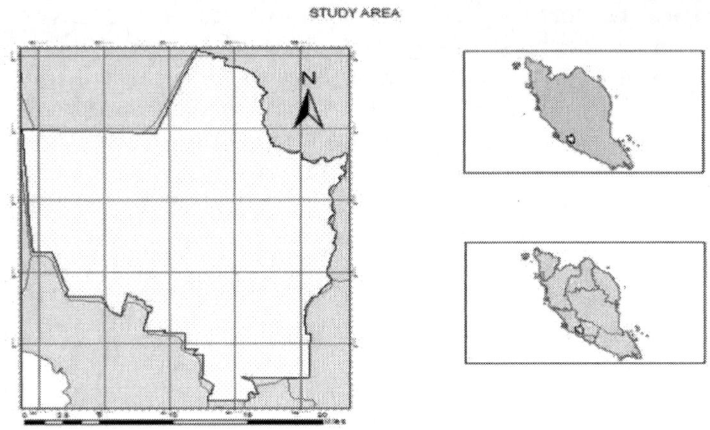

Figure 1. Location of the study area, Seremban district.

3.2 Methodology

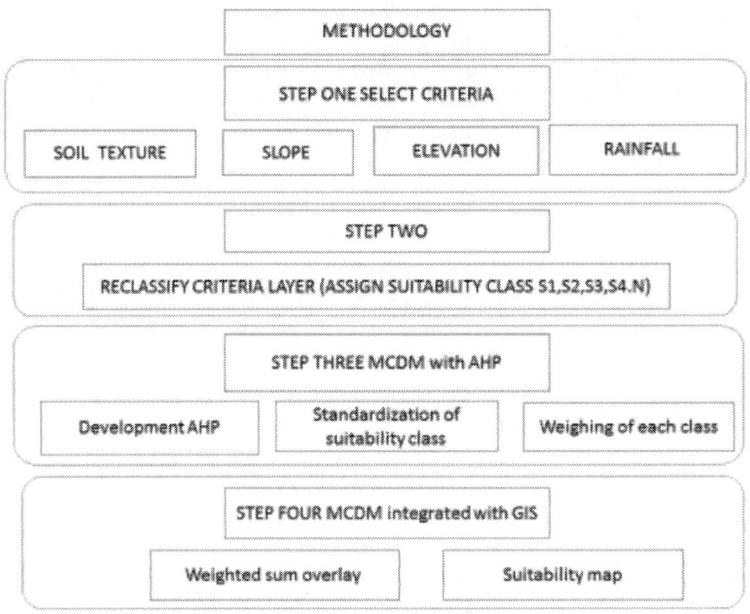

Figure 2. Flowchart of methodology.

3.3 Data collection and preparation using GIS

For the Data Preparation for the Spatial Analysis, geographical, climatic, proximity and socioeconomic factors that influence land suitability for agricultural uses were aggregated in this study [15]. All these factors constitute the criterion maps; and, the maps were projected to the same scale, boundary extent, resolution and spatial reference before they were standardized to a cell size of 300 m and UTM Zone 47N projection.

3.3.1Soil

Landform data for the study area were derived from the 1:50 000 topographic map of Peninsular Malaysia (JUPEM, 2010). The soil characteristic was acquired from the data supplied by the Malaysian department of Agriculture (2010). The original landform and soil data were obtained in vector format and were converted into 300m raster data after they were clipped to the boundary of the study area.

3.3.2 Slope

The slope data were derived from the 1:50 000 contour data using the method described by (Deng, 2011). The slope dataset was re-classified based on the NPP classification sequence to represent different suitability situations.

3.3.3 Climatic factor

Climate variables were aggregated based on the annual averages. Kriging interpolation was used to determine the spatial approximations and to calculate the value of the variables for each grid.

4. Suitability Classification

This study used the four levels of (Table 2) [highly suitable (S1), moderately suitable (S2), marginally suitable (S3) and unsuitable (N)] suitability classes commonly used by the Food and Agricultural Organization (FAO, 1976)[16]. A complex decision problem was decomposed into its constituent criteria. The criteria were, therefore, prioritized according to their relative importance within each level.

Table 1. Classes of suitability *Source[16].*

Code	Class	Description
S1	Highly suitable	land having no significant limitation for agricultural productivity
S2	Moderately suitable	land having some limitations that are severe for sustained productivity
S3	Marginally suitable	land with major limitations for sustained agricultural productivity
N	Unsuitable	land with extreme limitations for sustained agricultural productivity

4.1. Generation of the Criterion Maps

The criterion maps were classified into four classes (Table 1). Raster layers have numerical values 4, 3, 2 or 1, which represent S1 (highly suitable), S2 (moderately suitable), S3 (marginally suitable) and N (not suitable).

4.2. Standardization of the criteria

The process of setting the relative importance of each criterion is known as the standardization of the criteria [10]. In this process, scales of 0 to 1, 0 to 10 or 0 to 100 (etc.) are normally used for the criteria standardization. A pairwise comparison technique is typically used for rating and standardizing the ordinal values. In order to compare the criteria with each other, all values needed to be transformed to the same unit of measurement scale (from 0 to 1).

Table 2. Scales for the pairwise AHP comparisons[10] .

Intensity of importance	Description
1	Equal importance
3	Moderate importance
5	Strong or essential importance
7	Very strong or demonstrated importance
9	Extreme importance
2,4,6,8	Intermediate values
Reciprocals	Values for inverse comparison

4.3. Selection of the Criteria

Four criteria were selected for evaluating the agricultural land suitability in the study area (Table 3). These criteria were selected based on an extensive literature review of potential factors affecting agricultural land use and a review of the recommendations of the Malaysian National Physical Plan[15] .

Table 3. Weighting matrix for the main criteria and sub criteria for rubber.

No.	Criteria	Weight
1	Soil texture	0.845
2	Rainfall index	0.049
3	Elevation	0.059
4	Slope	0.048
Total	-	1

4.4. Weighing of the criteria

Criterion weights are the weights assigned to the objective and attribute maps. Deriving weights for the selected map criteria (land characteristic map layers) is a fundamental requirement for applying the AHP method. For determining the relative importance of the criteria, the pairwise comparison matrix using Saaty's nine-point weighing scale was applied (Table 2).

Table 4. Rubber requirement.

Agro–suit classes	S1	S2	S3	N
Annual Rainfall (mm/yr)	2500 - 4000	>4000	2000 – 5000	<1500
Dry month	<1	<1	2	>2
Land drainage	Well	Somewhat excessive Moderately well	Somewhat excessive	Very poor, poor Excess drain
Texture	SaL , L, SaCL, SiL, CL, Si, SicL	Lsa, SaC	C, SiC	Gravel, sand, massive clay, peat
Soil depth	>200	130 – 120	80 – 130	<80
Slope %	< 10	10 – 37	37 – 47	>47

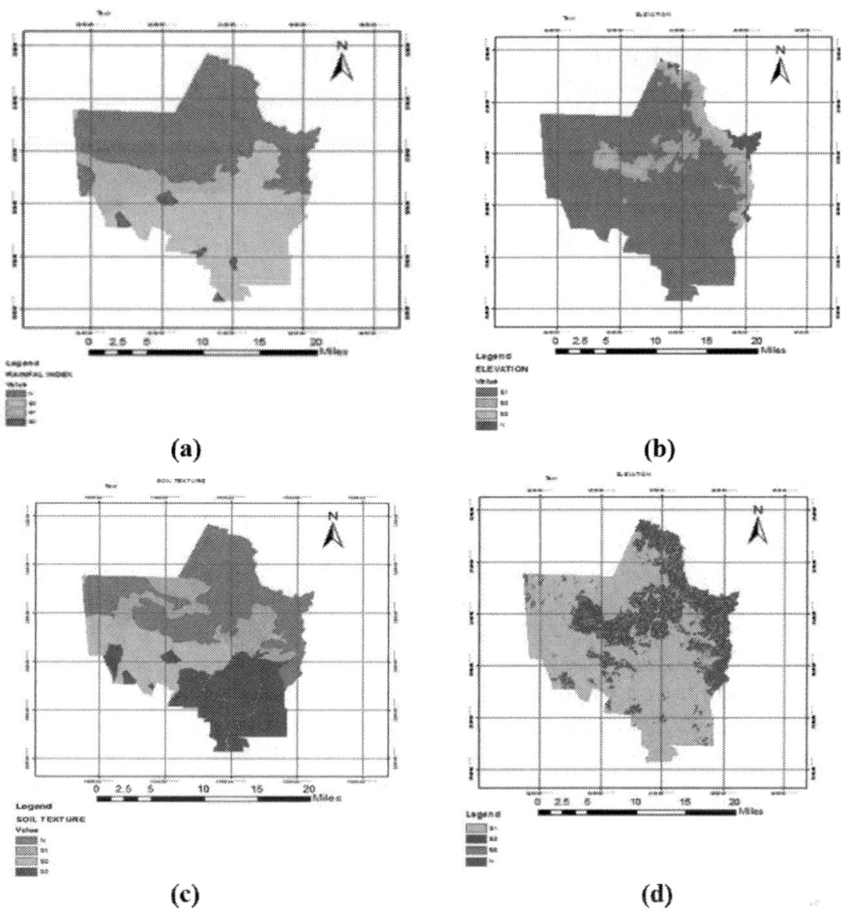

Figure 3. (a) Rainfall (b) Elevation (c) Soil texture (d) Slope.

4.5. Overlaying map layers

The weighted overlay is a technique for applying a common scale of values to diverse and dissimilar input data to create an integrated analysis. After weighing of the criteria, regarding their importance for the land suitability analysis, all the criteria maps were overlaid using a suitability index.

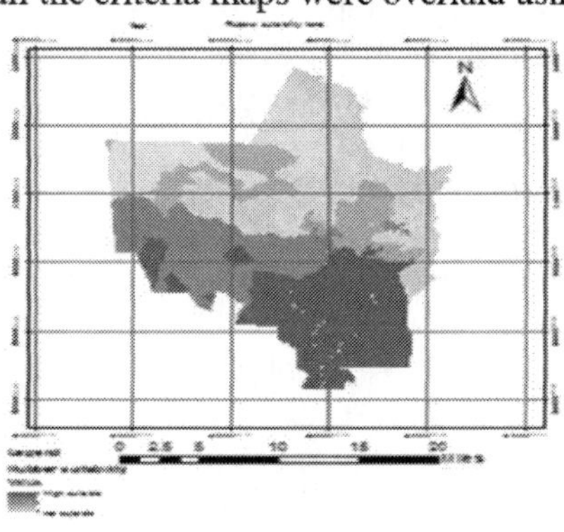

(e)

Figure 4. (e) Rubber land suitability in the study area.

5. Results & Discussion

The weight was derived for each of the independent variables as obtained from the analytical hierarchical process (AHP) presented in Table 3 and Figure 4. From the table, Soil texture was the most important factor affecting agricultural land suitability; whereas, slope had less weight, and elevation and rainfall had an equal effect on rubber in the study area.

The North-western and South-western (Nilai, Labu, Mambau and Rantau) parts of the district are the most suitable locations for cultivating rubber; whereas, the central parts (Mantin, Labu) have been found to be less suitable for rubber. The North-western and South-western parts (Lenggeng, Pantai and Senawang) of the district are not suitable for rubber cultivation (Figure 4). This shows the potential of such areas being over exploited for agricultural purposes and the impending environmental implications of this practice. This study showed the potentially sustainable agricultural land use in the study area after the exclusion of locations that are prone to topsoil erosion. The analysis of the current agricultural land use in Selangor, Malaysia indicated that land is mostly being used for what it is exactly suited for. However, there exists some challenges in the future agricultural land use in the district. For instance, the land that has been shown as less suitable and not suitable for rubber cultivation might be suitable for different crops, such as oil palm, in the Seremban district.

6. Conclusion

This study has revealed the potential agricultural land in Seremban, Malaysia. Unfortunately, available evidence showed that some of these agricultural lands are being taken up by non – agricultural uses and this has a potential of negatively affecting the country's economics. On the other hand, the scenario could create a situation where the country may be achieving its food security at greater economical and environmental costs. For crops to be matched with the biophysical conditions, the biophysical variables (climatic, geomorphological, number of rainy days, average temperature, and relative humidity) of the study area were collected to enable the interpretation of the climatic variables with reference to their suitability for specific crop production. The climatic adaptability of crops forms the basis of defining the crop-climatic requirements.

References

[1] S. A. Abdullah and A. A. Hezri, "From forest landscape to agricultural landscape in the developing tropical country of Malaysia: pattern, process, and their significance on policy," *Environmental Management,* vol. 42, pp. 907-917, 2008.

[2] A. P. D. Aguiar, G. Câmara, and M. I. S. Escada, "Spatial statistical analysis of land-use determinants in the Brazilian Amazonia: Exploring intra-regional heterogeneity," *Ecological modelling,* vol. 209, pp. 169-188, 2007.

[3] E. Attual and J. Fisher, "Land Suitability Assessment for Pineapple Production in the Akwapim South District, Ghana:. A GIS-MultiCriteria Approach," *Ghana Journal of Geography,* vol. 2, pp. 47-84, 2014.

[4] J. Malczewski, *GIS and multicriteria decision analysis*: John Wiley & Sons, 1999.

[5] J. Malczewski, "GIS-based land-use suitability analysis: a critical overview," *Progress in planning,* vol. 62, pp. 3-65, 2004.

[6] O. Marinoni, "Implementation of the analytical hierarchy process with VBA in ArcGIS," *Computers & Geosciences,* vol. 30, pp. 637-646, 2004.

[7] A. MashayekhanA and A. S. MahinyB, "A Multi-Criteria Evaluation approach to Delineation of Suitable Areas for Planting Trees (Case Study: Juglans regia in Gharnaveh Watershed of Golestan Province)," *Journal of Rangeland Science,* vol. 1, 2011.

[8] V. Meyer, D. Haase, and S. Scheuer, "A multicriteria flood risk assessment and mapping approach," 2009.

[9] H. Pourkhabbaz, S. Javanmardi, and H. Faraji Sabokbar, "Suitability Analysis for Determining Potential Agricultural Land Use by the Multi-Criteria Decision Making Models SAW and VIKOR-AHP (Case study: Takestan-Qazvin Plain)," *Journal of Agricultural Science and Technology,* vol. 16, pp. 1005-1016, 2014.

[10] T. L. Saaty, "The Analytic (Hierarchy) Process, New York, St," *Louis ua,* 1980.

[11] C. HUYNH VAN, "MULTI-CRITERIA LAND SUITABILITY EVALUATION FOR SELECTED CROPS USING GIS AT COMMUNITY LEVEL IN CENTRAL VIETNAM," 2008.

[12] H. Van Chuong, "MULTICRITERIA LAND SUITABILITY EVALUATION FOR CROPS USING GIS AT COMMUNITY LEVEL IN CENTRAL VIETNAM."

[13] H. Voogd, *Multicriteria evaluation for urban and regional planning*: Taylor & Francis, 1983.

[14] F. SarmadianA, A. KeshavarziA, B. RajabpourA, and S. AskariA, "Application of MCDM method in fuzzy modeling of land suitability evaluation," in *19th World Congress of Soil Science*, 2010.

[15] M. M. S. MALAYSIA, "Department of Statistics, Malaysia," 2014.

[16] A. FAO., *A framework for land evaluation*: Food and agriculture organization of the United Nations, 1976.

Agricultural water demand, water quality and crop suitability in Souk-Alkhamis Al-Khums, Libya

Mohamed Ali Abunnour*, **Noorazuan Bin Md. Hashim and Mokhtar Bin Jaafar**
School for Social, Development and Environmental Studies, Faculty of Social Sciences and Humanities, Universiti Kebangsaan Malaysia, 43600 UKM, Bangi, Selangor

*Email: Mohamed79ali11@gmail.com

Abstract. Water scarcity, unequal population distribution and agricultural activities increased in the coastal plains, and the probability of seawater intrusion with ground water. According to this, the quantitative and qualitative deterioration of underground water quality has become a potential for the occurrence, in addition to the decline in agricultural production in the study area. This paper aims to discover the use of ground water for irrigation in agriculture and their suitability and compatibility for agricultural. On the other hand, the quality is determines by the cultivated crops. 16 random samples of regular groundwater are collected and analyzed chemically. Questionnaires are also distributed randomly on regular basis to farmers.

1. Introduction

World over, agricultural water is used to grow fresh produce such as crops, fruits and vegetables as well as raising livestock for our diet. The United States Geological Survey (USGS) states that water used for agricultural irrigation accounts for nearly 65 percent of the world's freshwater withdrawals excluding thermoelectric power. Even though water demand in an irrigated area depends on the region's climate, agricultural water demand is a serious issue particularly at the coastal area of Souk-Alkhamis Al-Khums, Libya. Generally, one of the critical problems that hinder sustainable development in Libya is the lack of renewable water resources [1]. This study on agricultural water demand and crop suitability in Souk-Alkhamis Al-Khums a historical city and one of the oldest agricultural sites intend to examine the region's water resources and types of crops suitable for cultivation. This is therefore due to over-exploitation of fossil groundwater resources to meet irrigation demands that has already affected the northern aquifers of the country.

Incidentally, in Libya, problems caused by water scarcity and excessive groundwater mining are uneven population distribution and intensive agricultural activities in the coastal plains, seawater intrusion, water quality deterioration, poor crop yields and fragmentation of agricultural land holdings were among others found to be the principal issues in the water sector[1]. In the recent years, the water resources are increasingly becoming a scarce form of natural resources and a judicious and sustainable way of utilization is a priority today. On the basis of that, the present investigation would be carried out to ascertain the quality and quantity of the available agricultural water demand and the crops that should be encouraged for cultivation in the area. This would help agriculturalists to improve productivity by cultivating suitable crops that can boost the country's entire economy.

The groundwater quality study provides an insight for various usage of the water beyond agriculture. By evaluating quality, a conclusion can be drawn on how the water resources can be best utilized for optimum capacity [2]. In view of the fact that water is an important and indispensable commodity for all human beings as well as the plants and animals, history has shown that human civilizations originated around water resources such as rivers, lakes and springs. However, in Libya, due to seasons, the water demand increases from dry to semi-dry environments because it is scarce, limited and

IOP Conf. Series: Earth and Environmental Science **37** (2016) 012045 doi:10.1088/1755-1315/37/1/012045

irregular due to its geographical distribution. Intrinsically, most people devise means of preserving and securing water resources to ensure water availability for contemporary and future demands.
All over the world, most countries and the international organizations are concerned with water and have developed laws and programs to enhance and organize water uses in various fields including agricultural activities. [3] States that for five thousand years bed irrigation system was used in Egypt and a dam was built on the Nile River, which was one of the oldest dams built by humans on earth. As time passes by, new methods of water resource investment strategies were developed. Modern machineries are now use in pulling out the ground water to meet the required water demand and supply to the plants due to insufficient rainfall. This continues to increase drawing of ground water prevalence to irrigate agriculture in the large areas of the world including in dry and semi-dry areas including Souk-Al khamis Alkhums.
Consequently, the spread of irrigation agriculture and overuse of agricultural water affect the region's water quality thereby reflecting low crop productivity. In fact, the situation is further compounded by environmental issues such as drought and desertification which comes from the hinterland areas of the country. According to the General Authority for Water (1992), they confirm that ground water contribution in Libya forms 98% of the country's total water resources. It was estimated that water resources required would amount up-to 1679 million m^3 in 2000. This value was also expected to reach 4735 million m^3 in 2025. Hence, this research seeks to find out the relationship between agricultural water efficiency and crops cultivation suitability in the study area.

1.1 The location
The area geographically lies between north latitudes 32°25'N and 32°40'N, and between east longitudes 14° 10'E and 14° 25'E' with an estimated land mass of about 456.96 km2. In fact, this region is one of the oldest agricultural areas in Libya with the best arable soils for agricultural activities. However, the region has experience remarkable changes in the various living areas very quickly.

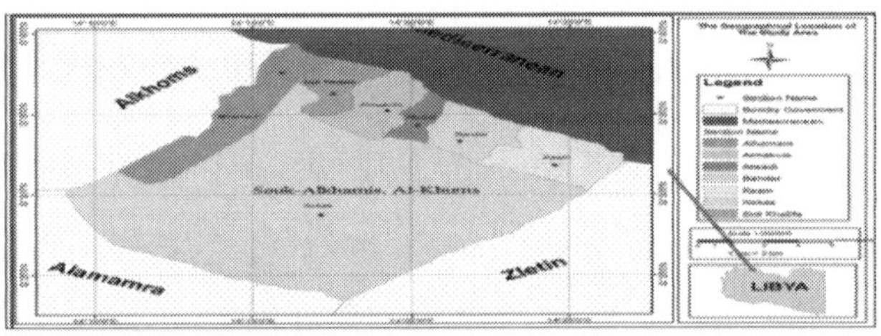

Figure 1. The geographical location of study area.

1.2 Problem statement
Water is essential to all life forms. Uneven population distribution and intensive agricultural activities in the coastal plains of Libya including Souk-Al Khamis is characterized by high population densities, intensive commercial, industrial and agricultural activities that cause the gap between water demands and supply wider [1]. The author added that the imbalance between supply and demand is expected to grow wider in the future, especially for the northern basins. Therefore, the scarcity of agricultural water remains a major issue of concern to most countries particularly countries such as Libya that has a large dry or semi-dry climate. As a result of increased demand for food, the study area is experiencing rapid expansion in cultivated land areas. This requires an urgent need to improve efficiency in the use of water resources due to the area's growing population. With an annual rainfall of 189.1 mm/annum [4], Souk-AlKhamis Al-khums is strategically located on the coastal strip area. However, the town experience seawater intrusion that affects groundwater quality leading to poor crop yields in the study area.
In the last three decades, the study area's population increased from 28, 647 in 1973 to about 80, 557 in 2005. This population increase requires a balanced increase in demand for food hence leading to an

increased in irrigated agricultural land as well as drilling of wells for water supply to crop production. Hence, the uneven population distribution has brought intensive agricultural activities to a more densely populated area and caused agricultural water scarcity in the coastal plains of Souk-AlKhamis Al-khums. Furthermore, the possibility of the emergence of certain contaminants in groundwater as a result of poor sanitation networks, as most of the population uses black wells to drain the water for domestic purposes. The fact is that the proportion of dissolved salts in the groundwater exceeded almost 4,300 mg/L [5].

In Libya, before the discovery of oil in 1959, the agricultural land was actually very small compared to what it is today. However, the state invested a significant part of the oil revenues in the development of agriculture, particularly increasing agricultural lands as well as increasing groundwater sources. In addition, there were increases in drilling of wells for agricultural operations and a change in irrigation methods for agricultural operations. What should be noted here is that irrigation methods in the period before 1959 were limited to the surface irrigation method and farmers continued to use this method until 1980. At present, there are various methods that include sprinkler irrigation and drip irrigation methods.

Finally, it is worthy to note that most population lives in the northern parts of the country, which resulted in the shrinkage of agricultural areas. On the other hand, there is an increase in the demand for ground water to supplement the inadequate rainfall for agricultural activities. Therefore, the deterioration of groundwater in terms of quantity and quality would lead to deterioration in agricultural production as well as degrading agricultural soil. Thus, the increase in the irrigation "ground" water salinity levels over the years in many places in Libya has led to soil salinization and, as a result, irrigation water demand increases and crop yields decreased [1]. This study intends to investigate further and determine the extent of available groundwater suitable for agricultural use so as devise means of its efficient use.

1.3 Research questions

To achieve the above objectives, the following constitutes the research questions to be pursued in this research:

- How does farmers' demographic and socio-economic characteristic affect water demand and crop production?
- What are the types of agricultural crops grown and is it fit or suitable with the current water condition?
- What are the major sources of agricultural water used and is it compatible for agricultural purposes with crops grown in terms of quantity and quality and how people perceive on the agricultural water quality?
- How to enhance the agricultural water efficiency to help the farmers in improving their agricultural technology and crop productions?

1.4 Aim and objectives of the study

The major aim of this study is to examine the agricultural water demand and its suitability for crop production in Souk-Alkhamis Al-Khums town of Libya. However, to overcome the above problem, the study intended to pursue the following specific operational objectives:

- To identify farmers' demographic and socio-economic status in Souk-Alkhamis Al-Khums, Libya;
- To evaluate the various agricultural crops in the area and determine their water and soil requirements;
- To assess the quality and quantity of water resources in Souk-Alkhamis Al-Khums Town, Libya so as to determine the most appropriate crops to be grown in the region;
- To suggest best practices for agricultural water use efficiency and recommend suitable crops to be cultivated in the region.

IOP Conf. Series: Earth and Environmental Science 37 (2016) 012045 doi:10.1088/1755-1315/37/1/012045

1.5 Research hypotheses

The study intends to attain the following hypotheses null in achieving the above objectives:

- Farmers' demographic and socio-economic status is not the homogeneous in Souk-Alkhamis Al-Khums, Libya.
- The agricultural crops in the study area may not necessarily have the same water and soil requirements.
- The actual and perception on water quality and water resources quantity do not differ by geographical location in the study area.
- Agricultural water use efficiently may not depend on the types of crops in a given area.

1.6 Conceptual study:

- Groundwater
  Groundwater is the water located beneath the earth's surface in soil pore spaces and in the fractures of rock formations. A unit of rock or an unconsolidated deposit is called an aquifer when it can yield a usable quantity of water.
- Agriculture
  Agriculture, also called farming or husbandry, is the cultivation of animals, plants, fungi, and other life forms of food, fiber, biofuel, medicinally and other products used to sustain and enhance human life.
- Soil
  Soil is the mixture of minerals, organic matter, gases, liquids and a myriad of organisms that can support plant life. It is a natural body that exists as part of the biosphere and it performs four important functions: it is a medium for plant growth; it is a means of water storage, supply and purification; it is a modifier of the atmosphere; and it is a habitat for organisms that take part in the decomposition and the creation of a habitat for other organisms.

2. Literature review

The literature review provides a general overview of past studies and the current states of knowledge on the agricultural water use, conservation and efficiency on the first part. It is followed by water demand, crop suitability and management as well as models related to agricultural water demand. The agricultural water demand equations and finally, it examines literature on agricultural production and water shortage in Libya among others.

2.1 Agriculture water use conservation and efficiency

The agricultural water is used worldwide to grow fresh produce and sustain livestock for human use. As [1] revealed, this water also makes it possible to grow fruits and vegetables and even raise livestock for our diet. In support of this, the United States Geological Survey (USGS) states that water use for irrigation accounts for nearly 65 percent of the world's freshwater withdrawals excluding thermoelectric power. The author elaborates further that there is 330 million acres of land used for agricultural purposes in the United States that produces an abundance of food and other products. Hence, water is used effectively and safely, production and crop yield are positively affected for the benefit of mankind. However, a decrease in applied water can cause production and yield to decrease. As such the author added that management strategies are the most important way to improve agricultural water use and maintain optimal production and yield of crops. This can only be achieved if the management strategies improving water use efficiency without decreasing yield are strictly implemented. For instance, improve irrigation scheduling and crop specific irrigation management are some of the strategies that can be followed to achieve maximum harvest in Libya.

Furthermore, the author shows that water efficiency can improve the reliability of existing supplies and reduce vulnerability to drought and other water-supply constraints. In addition, improvements in water efficiency can result in reductions of both consumptive and non-consumptive water use in a given area. [6] Explains that reductions in consumptive use provide additional water supply that can become available for other uses, even though, there are also compelling reasons to seek reductions in non-consumptive use. Finally, the author assumed that water used efficiently can improve the timing for agricultural productions and maximize the amount of water left in the natural environment, providing benefits to downstream water quality, the environment, recreation and even upstream use [6].

2.2 Water demand, crop suitability and management

Past studies revealed that water is indispensable for farming and farming in turn uses the vast majority of all water withdrawn for human use. Food production needs to be increased in the coming decades to support a growing world population [7]. He further elaborates that the scarcity of freshwater resources proportionally increases the intensity of potential for political conflicts over water within and between countries in the world. [8], views water as public goods which means that it cannot be sold in the market while [8], describes it as the resource that sustains all life on earth and is a key element of sustainable development. The author further stressed that in Pakistan, studies revealed that the available water resources are becoming fully committed and the irrigation expansion option will be increasingly difficult to pursue because many river basins have already been developed to their maximum capacity. Moreover, groundwater resources play a significant role in the hydrologic cycle and a region's water balance since they act as a buffer, providing significant amounts of water during years of low precipitation or storing a significant part of runoff on the occasion of high precipitation. This water stored in the ground forms major source for irrigation activities in many countries including Libya. However, in view of meagre recharge in areas of low rainfall, aquifers are destined for ultimate depletion unless prudent and very conservative development is carried out. He further argued that the exploitation of aquifers in regions of low natural recharge and scarce surface water is to a large extent controlled by the low rainfall and the relatively small amounts of runoff, which is often associated with high erosion when the vegetative cover is poor and flash floods are the rule. Hence, the amount of erosion depends considerably on the type and state of the land surface [10].

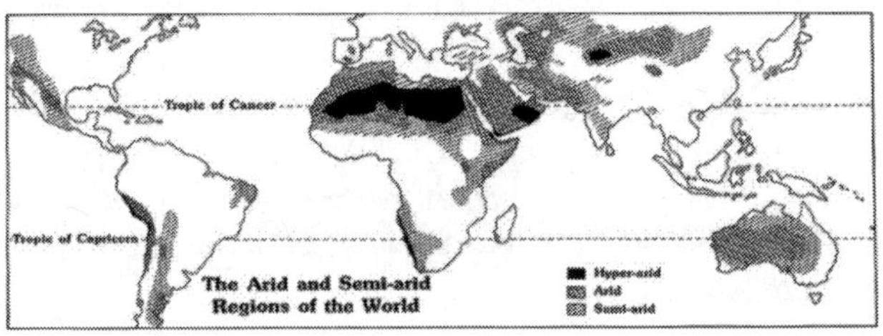

Figure 2. The arid and semi-arid regions of the world.

2.3 Factors affecting water demand

A report from the Food and Agricultural organisation reveals that factors affecting water demand are all anthropogenic by nature. These factors include; population, where its growth rate and changes in consumption patterns directly affect demand for goods and services, and the water associated with their production, processing and delivery. Secondly, it was observed that human pressure on water resources increases as their income grow which applies not just for household water demand for domestic but also to municipal demand and most importantly the growth in demand for industrial and agricultural products [7]. Moreover, the author added that increasing incomes lead to a rise in the per capita demand for food in which people diversify their diets (meat and dairy products) which also requires more water than a diet based on staple crop products (cereals or root crops). And lastly, urbanisation is also another factor that affects water demand. The authors explains that urbanization affects food consumption for instance in cities where supermarkets, restaurants and convenience foods that demand more water usage become more important.

2.4 Models related to agricultural water demand

The Agriculture Water Demand Model (AWDM) was developed in the Okanagan Watershed. The model was developed in response to rapid population growth, drought conditions from climate change, and the overall increased demand for water which indeed, reflects the current situation of Libya. Therefore, the author asserts that AWDM helps to understand current agricultural water use and helps to fulfil the commitment under the "Living Water Smart - BC Water Plan" to reserve water for agricultural lands. The Model calculates water use on a property-by-property basis, and sums each property to obtain a total water demand for the entire basin or each sub-basin which provide current and future agricultural water demand. Hence according to him, the type of crop, type of irrigation system, soil texture and climate data are used to calculate the water demand. However, the model is based on a Geographic Information System (GIS) database that contains information on cropping, irrigation system type, soil texture and climate [11].

2.4 Agricultural production and water shortage

According to [12], in Libya agricultural production is dominated by crop production where most of the activities are limited to a long narrow strip along the Mediterranean coast, low mountains and scattered oases in the desert. They elaborate that agricultural production account for 5 percent of the Gross Domestic Product (GDP) and occupy about 13percent of the total labour force of the country. Libya'z vast dry areas of land constraint the entire region's agricultural activities. However, the total arable land is approximately 2,170,000 ha which is 355,000 ha are permanent crops while the permanent pasture and rangeland occupy about 13,300,000 ha and the forest and woodland about 320,000 ha. In view of this therefore, majority of lands are been cultivated despite the need for crop production as stressed by the Food and Agricultural Organisation.

2.5 Research design

The findings of this research rely upon the combination of quantitative and qualitative frameworks (See Fig. 3 Framework for the research approach) to understand the socio cultural behaviors and the natural factors influencing levels of agricultural water demand, quality and crop suitability within the study area.

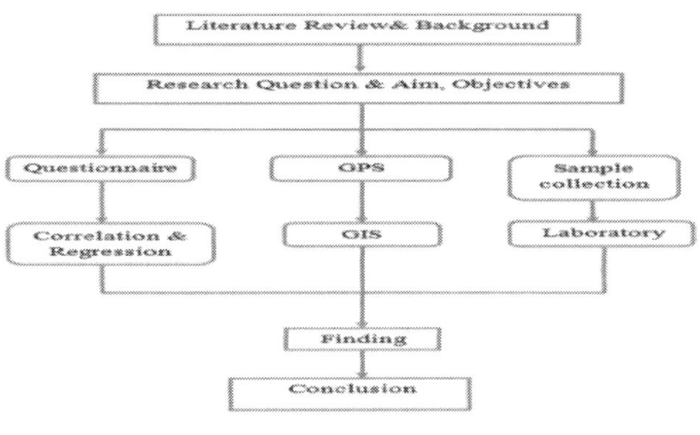

Figure 3. Conceptual framework.

3. Research methodology

This research will be consisting of four basic stages:

3.1 Data collection stage

In this stage, we will rely on books, journals, thesis and scientific researches to collect all the information and data related to the topic of study.

3.2 Sample of study

Due to the favorable climate, soil for agricultural activity and the availability of groundwater, the study area is of the same nature of the agricultural areas in Libya.It could be argued that it will be collecting groundwater samples from agricultural wells, including 5% of the agricultural wells, estimated at about between (800-1000) wells.

3.3 Field study

There is much information and data needed to go to field work to collect them, via following steps:

- Conducting personal interviews with peasants and specialists in the field of agriculture and groundwater.
- Using (GPS) the Global Positioning System to determine sites the agricultural wells in the study area.
- Design a questionnaire which contains series questions that related to the subject of study.

3.4 Data analysis stage

This stage will involve the following steps:

- Analyse the samples of water in special laboratories.
- All the samples will be submitted for analysis in special laboratories, to detect pollutants which pollute the wells
- Use (SPSS) to analyze the data.
- Use (GIS) Geographic Information System to produce maps which will contribute to determine groundwater current status.

3.5 Data collection and analysis

Objective	Data collection	Data analysis
Inventory and mapping of groundwater demand identify the best agricultural crops for cultivation	Field measurement, GPS	GIS approach
Factors leading to the deterioration of groundwater.	Questionnaire Water samples	SPSS Laboratory analysis
Recommendation	Literature review, questionnaire	SPSS

Table 1. data collection and analysis.

4. Expected results

The expected results evolve from the current study are based on the data in the region. Firstly, deterioration in the quality of groundwater might cause a decline in agricultural production in the area of study. In fact, this kind of decline is located at the lower levels of deterioration as it heads south and moves away from the sea. Secondly, the agricultural areas more likely show a notable increase in the region in comparison with the previous conditions. In point of fact, such increase contributes widely to the demand for water. Thirdly, there is a possibility that some cultivated crops are not in line with the quality of water used for irrigation.

5. Reference

[1] CEDARE 2014. Libya Water Sector M&E Rapid Assessment Report", Monitoring & Evaluation for Water in North Africa (MEWINA) Project,. Water Resources Management Program, C.

[2] Manjunath H.et al.T.2014.groundwater quality for agriculture in wessuvarnamukhi river basin, karnataka,india.Departmentof Geology, Bangalore University, Bangalore, Karnataka, India

[3] Fathi Ibrahim Masoud, The basics of agricultural irrigation, house new publications, Alexandria, Egypt, 1976.

[4] Jomaa AlMalyan.T.2005. The Relationship between Water Exploitation and Available Water Resource in Wadi Agricultural.(Unpublished Master degree) Al- Mergib University)

[5] General Authority for Water in Libya.T.2002. Branch Central Region, information on the region's groundwater and surface wells and dams .

[6] Mustafa Al Qadi -T 2002.The Excessive Exploitation of Underground Water in Zawia and the Surrounding Area and its Effect on its Quality (1977- 2002) Unpublished Master Degree 7th April University

[7] FAO. 2012. Coping with Water Scarcity an Action Framework for Agriculture and Food Security. FAO Water Report.

[8] Lawgali, F. F. 2008. Forecasting Water Demand for Agricultural, Industrial and Domestic Use in Libya. Dundee, Scotland, University of Abertay Dundee.

[9] hatti, A. M., Suttinon, P. And Nasu, S. 2009. Agriculture Water Demand Management in Pakistan: A Review and Perspective. Society for Social Management Systems (SSMS-2009)

[10] Pereira, L. S., Cordery, I. & Lacovides, I. 2009. Coping with Water Scarcity- Addressing the Challenges. Paris: Springer

[11] Van Der Gulik, T., Neilsen, D., Fretwell, R. & Tam, S. 2013. Agriculture Water Demand Model, Report for Metro Vancouver. Regional Adaptation Collaborative – Natural Resources Canada

[12] Al-Idrissi, M., Sbeita, A., Jebriel, A., Zintani, A., Shreidi, A. & Ghawawi, H. 1996. Libya: Country Report to the Fao International Technical Conference on Plant Genetic Resources. Libya Country Report.

Linear spectral unmixing to monitor crop growth in typical organic and inorganic amended arid soil

El Battay, A[1*], H. Mahmoudi[2]

[1]Arabina Gulf University, Department of Geoinformatics, College of Postgraduate Studies, P.O.Box: 26671, Kingdom of Bahrain, www.agu.edu.bh
[2] International Center for Biosaline Agriculture, Dubai, UAE

[*]Email: alieb@agu.edu.bh

Abstract. The soils of the GCC countries are dominantly sandy which is typical of arid regions such as the Arabian Peninsula. Such soils are low in nutrients and have a poor water holding capacity associated with a high infiltration rate. Soil amendments may rehabilitate these soils by restoring essential soil properties and hence enable site revegetation and revitalization for crop production, especially in a region where food security is a priority. In this study, two inorganic amendments; AustraHort and Zeoplant pellet, and one organic locally produced compost were tested as soil amendments at the experimental field of the International Center for Biosaline Agriculture in Dubai, UAE. The main objective is to assess the remote sensing ability to monitor crop growth, for instance Okra (*Abelmoschus esculentus*), having these amendments, as background with the soil. Three biomass spectral vegetation indices were used namely; NDVI, TDVI and SAVI. Pure spectral signatures of the soil and the three amendments were collected, using a field spectroradiometer, in addition to the spectral signatures of Okra in two growing stages (vegetative and flowering) in the field with a mixed F.O.V of the plant and amended soil during March and May 2015. The spectral signatures were all collected using the FieldSpec® HandHeld 2 (HH2) in the spectral range 325 nm – 1075 nm over 12 plots. A set of 4 plots were assigned for each of the three amendments as follow: three replicates of a 1.5 by 1.5 meter plot with 3kg/m^2 of each amendment and 54 plants, one plot as control and all plots were given irrigation treatments at 100% based on ETc. Spectra collected over the plots were inversed in the range of 400-900 nm via a Linear Mixture Model using pure soil and amendments spectral signatures as reference. Field pictures were used to determine the vegetation fraction (in term of area of the F.O.V). Hence, the Okra spectral signatures were isolated for all plots with the three types of amendments. The three vegetation indices were then calculated and compared in the vegetation fraction of the entire F.O.V. The key outcome of this analysis was that a considerable bias was induced when using a mixed F.O.V. In fact, the compost as an organic soil amendment containing dead vegetation affects similarly the sensitivity of the three vegetation indices in the vegetative stage of Okra compared to AustraHort and Zeoplant pellet amended plots. However, the TDVI is very sensitive to vegetation presence even with unmixed crop spectra. AustraHort amendment led to better status of Okra both in March and May with values of 0.19 and 0.28 respectively. Bias induced by some soil amendments can be misleading when upscaling spectral information to satellite imagery with low spectral and spatial resolutions. The results obtained are encouraging for further use of spectral information for crop monitoring in soil containing amendments.

1. Introduction

The strength of remote sensing in agricultural applications has been refining ever since the emergence of this field. From satellite based, airborne imagery or terrestrial imaging systems, myriad of research and publications have come to assert an eminent contribution in all aspects and stages of agricultural practices. Climate change impact and increasing global demand of food and other commodities called for innovations in agricultural research especially to improve water and nutrients use efficiencies which are major limiting factors in crop production in the Arabian Peninsula (AP). The soils of AP in general and the GCC countries in specific are coarse textured and have low fertility status. High temperature and poor soils necessitate daily irrigation for crop production. Improving the water use efficiency through innovative technologies such as climate smart agriculture may be an option. In this regard, a number of organic and inorganic amendments have been used by various researchers to improve water and nutrient use efficiencies. However, little has been done in this respect on U.A.E, and hence this subject has formed the focus of this study, involving two inorganic amendments, namely AustraHort and Zeoplant Pellet, and compost as an organic amendments. The remote sensing component of this study targeted to integrate close range sensing techniques to assess the impact of using soil amendments on crop production. The aim was to propose a time and cost effective method to measure vital bio-physical parameters in a perspective of large scale farming practice. This study contributes to build the preliminary and necessary in field remote sensing knowledge and subsequently the models that fits the studied soils, used soil amendments and crops. It is a crucial step before any remote sensing upscaling to use satellite imagery, airborne or even UAVs based earth observation imagery. Proposing to farmers a way to improve the soil capabilities and water efficiency is as important as giving them the tools to monitor the state of the crop during various stages of the phenological cycle. While soil amendments is proposed in this study to reach a better use of water in poor soils in U.A.E, the remote sensing component paves the way to an intelligent tracking of the state of the crops using less day to day resources and avoiding long and fastidious field work especially in arid zones and harsh weather conditions.

Myriad of spectral vegetation indices are used to assess various aspects of vegetation in general and agriculture in specific [1, 2, 3, 4]. Many of these indices relies on the spectral information collected in the near infrared and visible bands and the most popular, and not necessary the most effective, is the normalized difference vegetation index (NDVI) [5]. The NDVI has been used for numerous application including leaf area index [6], biomass and yield [7], evapotranspiration [8], Chlorophyll content estimation; [9, 10], Nitrogen content [11, 12], etc. Other reflectance indices are also evaluated and used to estimate other parameters such as growth, productivity and water status, *inert alia*, Enhanced Vegetation Index, the Normalized Difference Water Index, Simple Ratio Index, Atmospherically Resistant Vegetation Index, Moisture Stress Index, Normalized Difference Infrared Index, etc. are few example of such indices. In this study, three vegetation indices were selected, namely, NDVI (as one of the most spread vegetation indices), TDVI (Transformed Difference Vegetation Index) and SAVI (Soil Adjusted Vegetation Index). The three indices rely on the spectral information collected in the near infrared and visible portions of the electromagnetic spectrum, for which equivalent multispectral bands are available in most Erath Observation Satellites, including U.A.E national E.O satellites program (e.g. Dubaisat-1, DubaiSat-2 and coming Khalifa Sat).

2. Material and methods

The plots trial was conducted at ICBA station using sandy soil dominant in UAE. Two inorganic amendments; AustaHort and zeoplant pellet, and a locally produced compost were used (Figure 1). These three amendments are referred to respectively in the text and figures as A1, A3 and A5. Standard fertilizer rates were used (N= 150 kg/ha, P_2O_5= 300 kg/ha and K_2O= 150 kg/ha) as a first dose applied two weeks after germination. The second dose of fertilizer (N= 150 kg/ha, and K_2O= 75 kg/ha) was applied three weeks later. Treatments were triplicated in Randomized Complete Block Design (RCBD). The plot size was 2.25 m^2 (1.5m by 1.5m) and the distance between plants is 0.25 meter between lines and 0.30 meter between plants. At sowing two seeds per hole were placed and after germination seedling were thinned to 1 plant. A set of 4 plots were assigned for each of the three amendments as follow: three replicate of a 1.5 by 1.5 meter plots with 3kg/m^2 of amendments and 54 plants, one plot as control and all plots were given irrigation treatments at 100% based on ETc.

Figure 1. The three used amendments and plots preparation at ICBA.

It is clear visually that these three amendments are very different in their colour as the Zeoplant pellet is the lighter, the compost is the darker, while the AustraHort is just in between. The three amendments and their mixtures with soil at a rate of 3kg/m^2 were used to determine water retention at 1/10 and 15 bar using pressure membrane apparatus (Table 1).

Table 1. Water retention at 0.1 and 15 bars by pure amendments, native sand soil of UAE and their mixtures at 3 kg/m^2.

Amendment/ Soil mixture	1/10 bar water retention (%)	15 bars water retention (%)
Pure Amendments		
AustraHort	48.20	40.60
Zeoplant pellet	27.48	22.40
Compost	75.50	57.11
Soil:amendment mixtures		
AustraHort @ 3.0 kg /m^2	4.40	1.27
Zeoplant pellet @ 3.0 kg /m^2	3.55	2.39
Compost @ 3.0 kg /m^2	3.32	2.90
Native UAE sand soil (ICBA)	5.59	1.72

Table 1 and figure 2 show the distribution of water retention at 0.1 and 15 bars pressure. Among the amendments the compost shows the highest water retention relative to other amendments, which is expected from an organic material. Among the inorganic amendments, AustraHort is showing better water retention than zeoplant pellet.

Figure 2 depicts a general overview of the methodology adopted to achieve the objectives set using close range remote sensing. For this, pure spectral signatures were collected in the range of 400 to 900 nm of the selected U.A.E soil and the three soil amendments mixed with soil at the rate of 3kg/m^2 used in this study. Furthermore, mixed spectral signatures of crop, soil and amendments were collected at two growing stages (vegetative and flowering) of Okra (*Abelmoschus esculentus*). The spectral signatures were all collected in a zenith angle [13,14] of the sun-target-sensor geometry near Nadir ($\theta s= \theta v$ = 90°), Field of View (FOV) of 25° and constant height of 0.5 meter for soil and amendments measurements and 1.5 meter for vegetation measurements in plots. White and dark current calibration were performed for each single spectra to avoid discrepancies related to slight solar zenith angle variation in time of collection. The FieldSpec® HandHeld 2 (HH2) spectroradiometer (figure 3) was used to collect spectral signature in the 325 nm – 1075 nm spectral range. With a wavelength range of 325 nm – 1075 nm, an accuracy of ±1 nm and a resolution of <3 nm at 700 nm. The software used with the Handheld 2 is R3 for accurate data visualization. Integration time in the field was set to 534 ms, because of the harsh field conditions and the big number of spectra to collect while the sun is in the zenith. Table 2 presents the different dates of spectral signatures collection at ICBA experimental farm.

The pure collected spectral signatures of soil were used (1) to compare the spectral behaviour of different amendments vs soil between 400-900 nm. (2) to inverse the spectra and isolate the crop signature from the mixed spectral signatures by using a basic Linear Mixture Model [15] (equation 1) and knowing the fraction of the FOV occupied by vegetation collected, and (3) to calculate three spectral vegetation indices, namely; Normalized Difference Vegetation Index (NDVI), Transformed Difference Vegetation Index (TDVI) and Soil Adjusted Vegetation Index (SAVI). Moreover, a combination of the soil and amendments spectral signatures using the same linear spectral mixture model was performed to assess the effect of that mixture on vegetation indices, in this case NDVI. Table 3 present the three used vegetation indices.

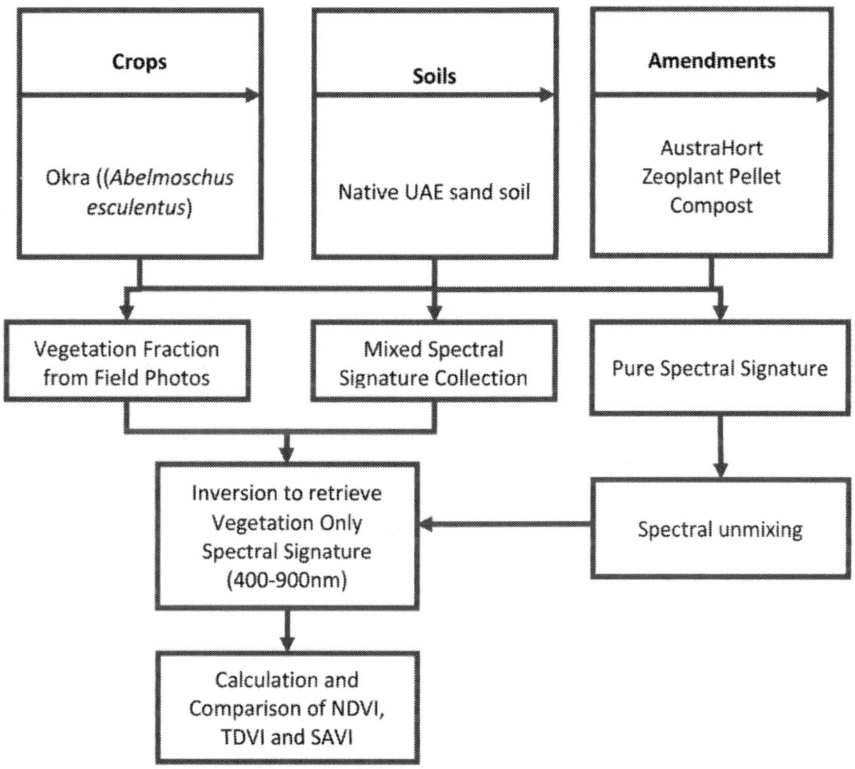

Figure 2. General overview of Remote Sensing Methodology.

$$R_i = \sum_{j=1}^{n} f_{ij} r_j + \varepsilon_i \qquad (1)$$

where:

R_i is the resulting reflectance in wavelength I for a given FOV (value from 0 to 100%)
r_j is the reflectance of any given pure surface element "*j*" within the FOV (value from 0 to 100%)
f_{ij} is the fraction of area that the surface element "*j*" occupied in the FOV in the wavelength *I* (Values from 0 to 1)
ε_i is an element of error related to all intrinsic and extrinsic factors

Table 2. Spectral Vegetation Indices used in the study.

Normalised Difference Vegetation Index	$NDVI = \frac{R_{850} - R_{680}}{R_{850} + R_{680}}$	Most widely used vegetation index. Difference between the high chlorophyll absorption and the reflection of leaf cellular structure
Transformed Difference Vegetation Index	$TDVI = 1.58\left(R_{850} - R_{680} / \sqrt{R_{850}^2 + R_{680} + 0.5}\right)$	Similarly to NDVI, the TDVI used difference between the high chlorophyll absorption and the reflection of leaf cellular structure but is more robust in case of sparse and low density vegetation
Soil Adjusted Vegetation Indices	$SAVI = \frac{1.5 \times (R_{850} - R_{680})}{(R_{850} + R_{680} + 0.5)}$	Help to reduce the effects from background soils, Coefficient L is considered as equal to 0.5 in this study accordingly with the density of crop.

Table 3. List of fieldwork visits and dates.

Period	**location**	**Purpose**
26-29 January 2015	ICBA-UAE	Collection of spectral signatures of pure amendments, .soil, and bare plots experiments
15-19 March 2015	ICBA-UAE	Collection of spectral signatures at field of plots experiments with vegetation (Okra) before blossom stage
12-14 May 2015	ICBA-UAE	Collection of spectral signatures at field of plots experiments with vegetation (Okra) after blossom stage

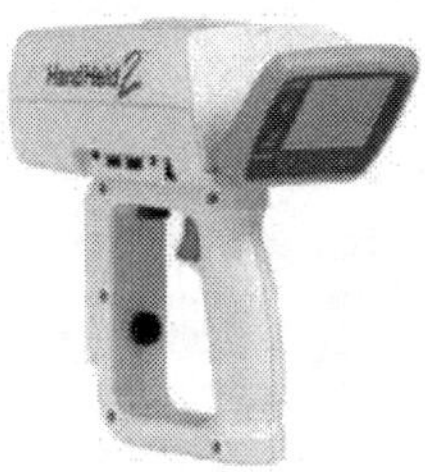

Figure 3. The FieldSpec® HandHeld 2 used to collect spectral signatures.

3. Results and discussion

As described above, visually the three used amendments show clearly a significant distinction of the colour ton. In fact, figure 4 shows that the Zeoplant Pellet (A3) which is the clearer one has a distinctive spectral signature with relatively high reflectance values, in the range 400-900nm compared to AustraHort (A1) and Compost (A5). Figure 5 presents the spectral signatures of the native U.A.E soil which present a typical sand signature. All the signature of amendments and soil represent the normal trend of soils spectral signatures. It is clear from figure 5(b) that the mixture rate of 3kg/m^2 applied of amendments to the soil does not affect significantly its reflectance.

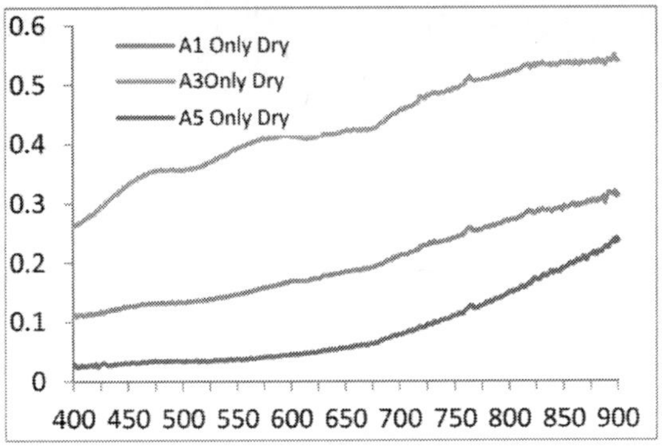

Figure 4. Spectral Signature of the three used soil amendments using in dry conditions.

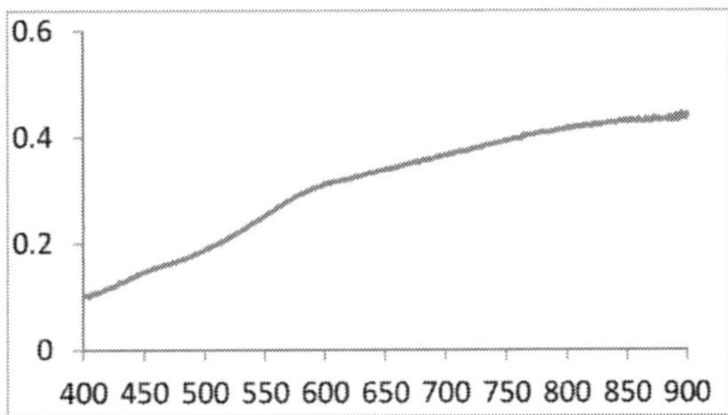

a)

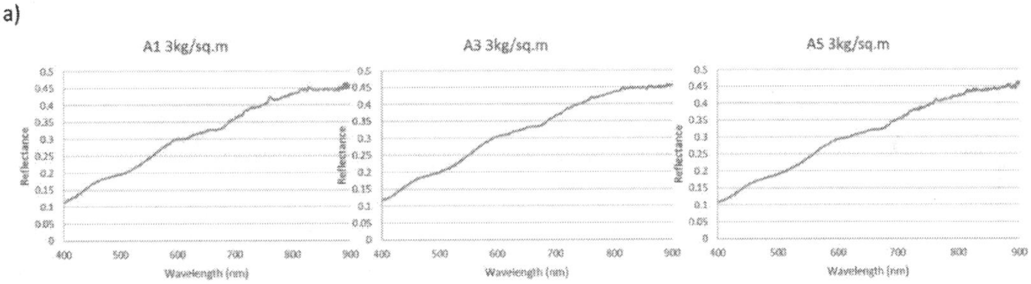

b)

Figure 5. Spectral Signatures (reflectance from 400-900 nm) of (a) UAE used soils and (b) mixture of the three amendments with the soil at 3 kg/sq.m.

As explained above, before using spectral vegetation indices to infer any relation between them and vegetation status, it is primordial to understand the soil amendments impact on such indices even without the presence of any vegetation (on bare soil). The following figure 6, depicts the theoretical effect of a mixture of A5 Amendment (compost) with UAE soil on the Normalized Difference Vegetation Index (NDVI). A5 amendment was selected because it is a compost and contains remains of dead vegetation. To realise this curve, the LMM was used to combine gradually, in term of surface, the A5 curve presented in figure 4 and the soil curve in figure 5. While no crop is present, the NDVI values are very high, starting from 20% of Compost mixture with soil, and this bias induced can be misleading when upscaling spectral information to satellite imagery with low spectral and spatial resolutions.

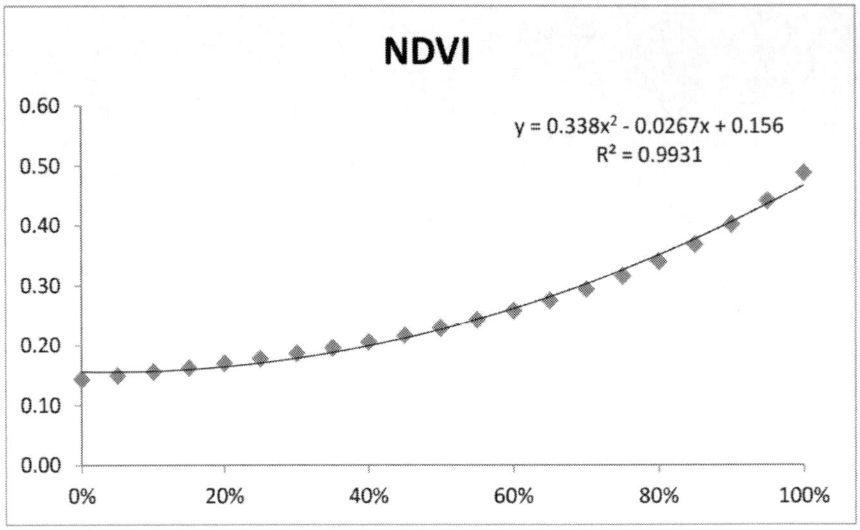

Figure 6. Effect of A5 mixture with UAE soil (passing from 0% to 100%; X axis) on NDVI values.

Unmixing collected field spectra to get only the vegetation spectral response is necessary. For this purpose, spectral signature of bare soil and amendments collected on January 2015 is needed as reference. In fact, this spectral signature, for each plot, is giving the spectral response of the soil and amendments only, while the subsequent spectral signatures collected in March and May 2015 contain additional information on vegetation as well. Figure 7 shows the FOV of the HH2 spectroradiometer which includes, vegetation, soil, amendments, irrigation pipes and shadow of vegetation.

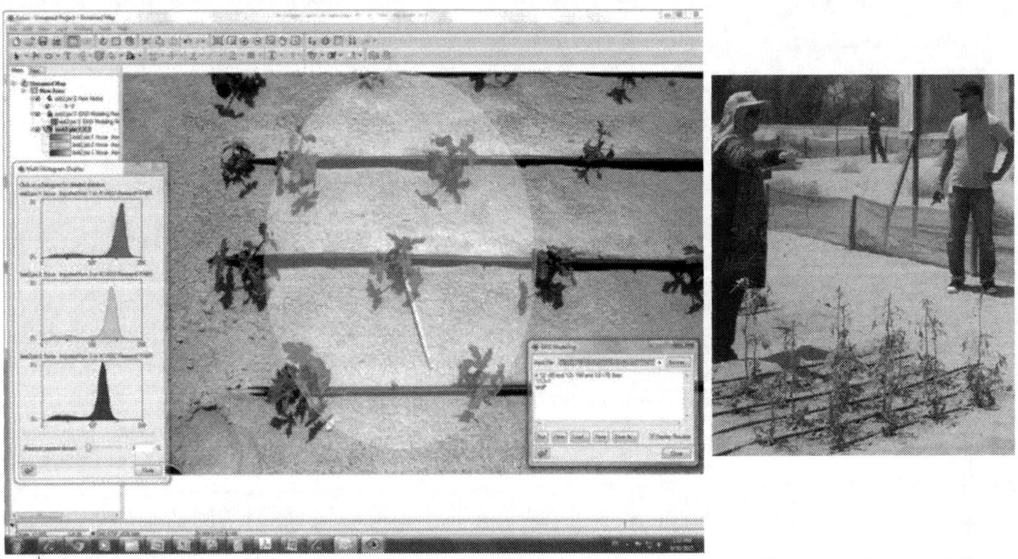

Figure 7. Isolation of vegetation (illustrated in red color) using a simple Histogram Thresholding technique via Geomatica 2015.

Figure 8 present an example of different spectral signatures collected in the field with various height values of the spectroradiometer (HH2). It is clear that plant signature (yellow) and the soil one (red) are mixing when got together; blue and green curves. The idea is how to retrieve from a mixed signature such as the one shown in figure (24), the vegetation one. As mentioned earlier, using the linear mixture model presented in equation 1 above, and knowing the reference spectral signature of soil and amendment for each single plot an endmembring is possible to extract only vegetation signature. A nominal surface of vegetation of 30% of the FOV is used for this purpose, based on analysis of field photos. Figure (9) shows the same signatures presented in the previous figure (10) but with the retrieved vegetation one (in green colour).

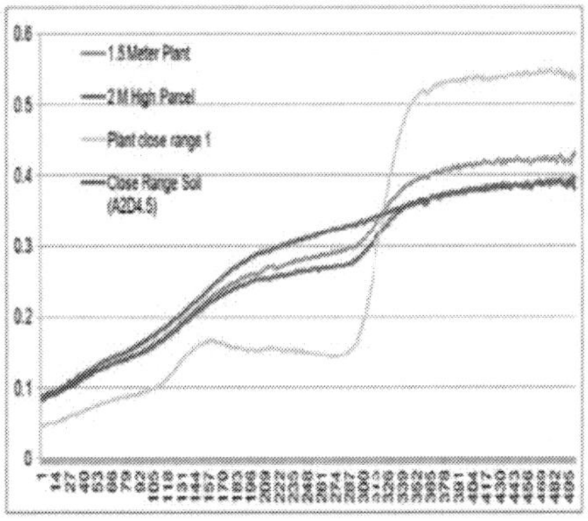

Figure 8. Example of spectral mixture of soil, amendments and vegetation.

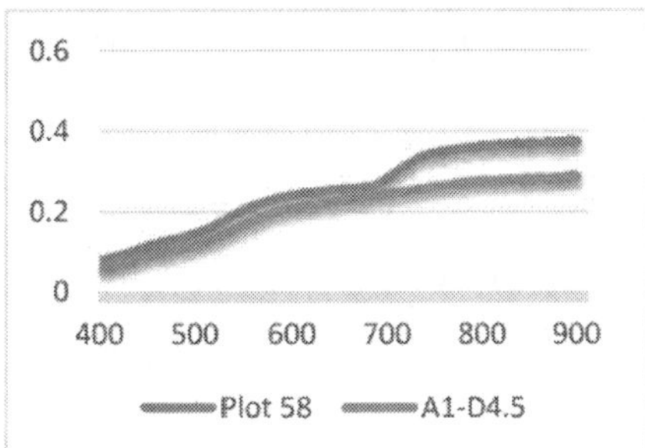

Figure 9. Mixed soil, amendment and vegetation spectral signature aside reference Soil and Amendment spectral signature.

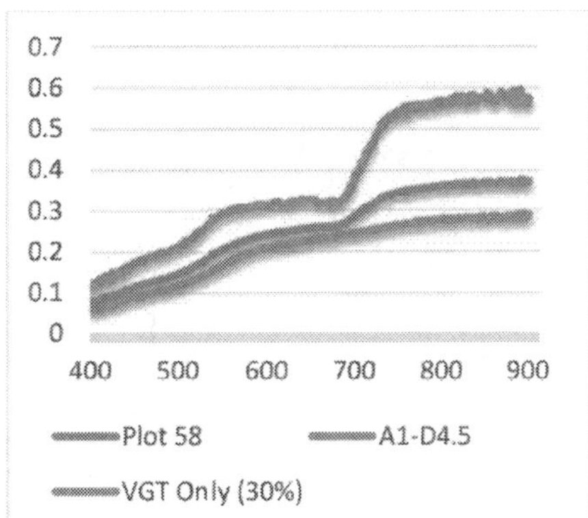

Figure 10. Vegetation spectral signature retrieved using the LMM from the mixed signature (blue) and the reference one (orange).

The same approach is applied to all plots for March and May 2015 (ICBA-UAE) field spectral signatures to retrieve only the vegetation one, see figure 11 (a and b)

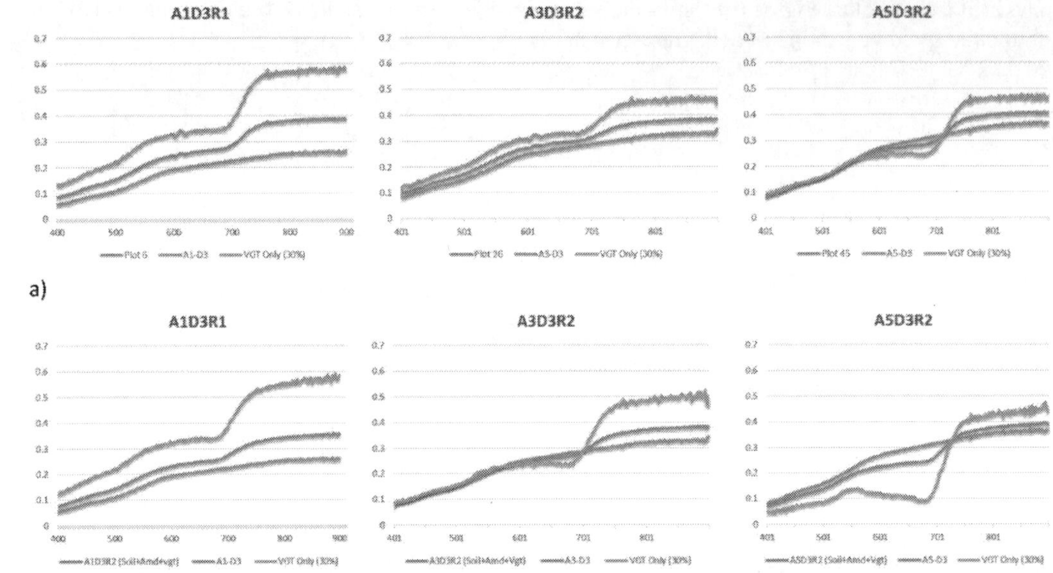

Figure 11. Unmixing of Vegetation spectral signatures from field collected spectra in (a) March 2015 and (b) May 2015 at ICBA-UAE.

Tables 4 and 5 presents the three vegetation indices (NDVI, TDVI and SAVI) calculated for the crop only during March and May 2015. These values are obtained from crop spectral signature extracted using the LMM as presented in figure 11.

Table 4. Crop only Vegetation indices values during March 2015.

	680nm	850 nm	NDVI	TDVI	SAVI
Control	0.25	0.37	0.19	0.19	0.15
A1 Replica 1	0.30	0.36	0.10	0.11	0.09
A1 Replica 2	0.27	0.38	0.17	0.19	0.15
A1 Replica 3	0.29	0.38	0.13	0.14	0.11
A3 Replica 1	0.31	0.39	0.12	0.13	0.10
A3 Replica 2	0.30	0.38	0.12	0.13	0.10
A3 Replica 3	0.31	0.39	0.11	0.13	0.10
A4 Replica 1	0.28	0.41	0.18	0.20	0.15
A4 Replica 2	0.28	0.42	0.20	0.23	0.18
A4 Replica 3	0.29	0.41	0.16	0.18	0.14

Table 5. Crop only Vegetation indices values during May 2015.

	680nm	850 nm	NDVI	TDVI	SAVI
A1 Replica 1	0.26	0.36	0.17	0.17	0.14
A1 Replica 2	0.53	0.75	0.17	0.28	0.19
A1 Replica 3	0.26	0.35	0.16	0.16	0.13
A3 Replica 1	0.24	0.35	0.20	0.20	0.16
A3 Replica 2	0.24	0.37	0.21	0.22	0.18
A3 Replica 3	0.27	0.38	0.17	0.18	0.14
A4 Replica 1	0.26	0.37	0.18	0.19	0.15
A4 Replica 2	0.22	0.36	0.23	0.23	0.19
A4 Replica 3	0.24	0.39	0.23	0.24	0.19
A1 Replica 1	0.25	0.38	0.21	0.22	0.18

4. Conclusion

It was possible in this study to unmix plot collected spectral signatures using a simple LMM to obtain Okra (Abelmoschus esculentus) spectral information in the range of 400-500nm. For this purpose, separate spectral signature of soil, amendments and soil mixed with amendments at a rate of 3kg/m^2 were used as reference. The fraction of vegetation if a mixed FOV was obtained by simple thresholding classification of field photos using Geomatica 2015. The key outcome of this analysis was the considerable bias was induced when using a mixed F.O.V. In fact, the compost as an organic soil amendments containing dead vegetation affect similarly the sensitivity of the three Vegetation indices in the vegetative stage of Okra compared to Austrahort and Zeoplant pellet amended plots. However, the TDVI is very sensitive to vegetation presence even with unmixed crop spectra. Austrahort amendment led to better status of Okra both in March and May with values of 0.19 and 0.28 respectively. Bias induced by some soil amendments can be misleading when upscaling spectral information to satellite imagery with low spectral and spatial resolutions. The results obtained are encouraging for further use of spectral information for crop monitoring in soil containing amendments. The next step to this study is to collect the spectral signatures on pure vegetation samples to try to correlate the spectral vegetation indices with the crop characteristics. Hence, if achieved successfully, spectral crop model(s) will be generated to quantify spectrally, using close range remote sensing data, crop characteristics such as chlorophyll content/absorption which is linked in its turn to the status of the crop.

References

[1] Bannari A Morin D Bonn F and Huete A R. A review of vegetation indices. Remote Sensing Reviews, 13 (1995) p95-120

[2] Elvidge CD and Chen Z K. Comparison of broad-band and narrow-band red and near-infrared vegetation indexes. Remote Sensing of environment 54 (1995) p38-48

[3] Gamon J A Peñuelas J and Field C B. A narrow-waveband spectral index that tracks diurnal changes in photosynthetic efficiency. Remote Sens. Environ., vol. 41, no. 1 (1992) p35–44

[4] Chen, J.. Evaluation of vegetation indices and modified simple ratio for boreal applications. Can. J. Remote Sens. 22 (1996) p229–242

[5] Gao, B C. NDWI - a normalized difference water index for remote sensing of vegetation liquid water from space. Remote sensing of environment 58 (1996) p257-266

[6] Hasegawa K Matsuyama H Tsuzuki H and Sweda T. Improving the estimation of leaf area index by using remotely sensed NDVI with BRDF signatures. Remote Sensing of Environment 114 (2010). p514-519

[7] Moriondo M Maselli F and Bindi M. A simple model of regional wheat yield based on NDVI data. European Journal of Agronomy 26, (2007) 266-274

[8] Hunsaker D Fitzgerald G French A Clarke T Ottman M and Pinter Jr O. Wheat irrigation management using multispectral crop coefficients. I. Crop evapotranspiration prediction. Trans. ASABE 50, (2007) p2017-2033

[9] Jones C L Weckler P Maness N O Jayasekara R Stone and Chrz D. Remote sensing to estimate chlorophyll concentration in spinach using multi-spectral plant reflectance. Transactions of the ASABE 50, (2007) p2267-2273

[10] Haboudane D Miller J R Tremblay N Zarco-Tejada P J and Dextraze L. Integrated narrow-band vegetation indices for prediction of crop chlorophyll content for application to precision agriculture. Remote Sens. Environ., 81 (2002) p416– 426

[11] Cabrera-Bosquet L Molero G Stellacci A Bort J Nogues S Araus J. NDVI as a potential tool for predicting biomass, plant nitrogen content and growth in wheat genotypes subjected to different water and nitrogen conditions. Cereal Research Communications 39 (2011) p147-159

[12] Strachan I B Pattey E and Boisvert J B. Impact of nitrogen and environmental conditions on corn as detected by hyperspectral reflectance. Remote Sens. Environ. 80 (2002) p213-224

[13] Latifovic R Cihlar J and Chen J. A Comparison of BRDF Models for the Normalization of Satellite Optical Data to a Standard Sun-Target-Sensor Geometry. IEEE transactions on geoscience and remote sensing, vol. 41, no. 8, august (2003), pp 1889-1898

[14] Liang S Strahler A H Barnsley M J. Borel C C Gerstl S Diner D. J, Prata A J and Walthall C. Multiangular remote sensing past, present and future. Remote Sensing. Review, vol. 18, (2000) p83-102

[15] Fortin J P Bernier M El Battay A and Gauthier Y. Estimation of Surface Variables at the Sub-pixel Level for Use as Input to Climate and Hydrological Models. Vegetation 2000, Belgirate, Italy, 3-6 April (2000), 59 pages.

Acknowledgment
This study is made possible by the support of the Arabian Gulf University (kingdom of Bahrain) and the International Center for Biosaline Agriculture (Dubai, UAE). It is also made possible by the support of the American People through the United States Agency for International Development (USAID.). It is a part of the joint research project "Improving Agricultural Soil Properties Using Soil Amendments to Enhance Water and Nutrient Use Efficiency for Crop Production in Dry Lands and Assessing These Efficiencies Via Remote Sensing Techniques" between AGU and the International Center for Biosaline Agriculture (ICBA) under the Further Advancing the Blue Revolution Initiative (FABRI), USAID-DAI Contract No. AID-EPP-00-04-00023. The contribution of PCI Geomatica for providing image processing capabilities via its Educational Alliance with AGU is also acknowledged.

Spatial variability of soil nutrient in paddy plantation: Sites FELCRA Seberang Perak

H Kamarudin[1], N A Adnan[2], M R Mispan[2], A.Athirah.A[2]
[1] [2]Centre of Studies . Surveying Science and Geomatics, Faculty of Architecture, Planning and Surveying, University Teknologi MARA, 40450 Shah Alam, Malaysia.
[2]Agrobiodiversity & Environment Research Center, MARDI, Persiaran MARDI-UPM, 43400 Serdang Selangor, Malaysia

Email: hasliana@gmail.com

Abstract: The conventional methods currently used for rice cultivation for Malaysia unable to give maximum yield although the yield production of paddy is increasing. This is due to the conversional method unable to included soil properties as one of their parameter in agriculture management. Soil properties are vary spatially in farm scale due to differences in topography, parent material, vegetation or land management and soil characteristics as well as plantation productivity can varies significantly over small spatial scales. Knowledge of spatial variability in soil fertility is important for site specific nutrient management. Analysis of spatial variability of soil nutrient of nitrogen (N), phosphorus (P) and potassium (K) were conducted in this study with the aid of GIS (i.e ArcGIS) and statistical softwares. In this study different temporal and depths of soil nutrient were extracted on the field and further analysis of N,P,K content were analysed in the chemical laboratory and using spatially technique in GIS sofware. The result indicated that for the Seberang Perak site of 58 hactares area, N and K are met minimum requirements nutrient content as outlines by the MARDI for paddy cultivation. However, P indicated poor condition in the study area; therefore the soil needs further attention and treatment.

KEY WORDS: spatial variability, interpolation technique, GIS, paddy plantation

1. Introduction

1.1Scenario paddy industry in world

Paddy is a major source of atmospheric methane, yet vast amounts of paddy is continue to be grown in order to meet increasing global food demand. Countries in the East Asia region such as China, Japan, Taiwan and Korea are an important rice growing region of the world (Maclean et al. 2002)with China is higher than other countries for rough paddy production in 6.23 (t/ha) which is 20% for cultivation area and 31.8 % for paddy production in the world reported by (Maclean et al. 2002) and requires large field area compared to the other plantations and estimated are about 90% of paddy cultivations produced in Asia. It indicates that Asia's people need rice as staple food such as Malaysian people.

1.2 Overview paddy production in Malaysia

In Malaysia, paddy is one commodity that is given priority in the national food security the agenda. To ensure that food production is assured, many efforts have been taken by the government to increase the production of commodity (Firdaus et al. 2014). Every year, thousands of tones of rice are produced to fulfill demand of 29.34 million Malaysians(MOA 2012). However, the country's rice supply is still insufficient. (MOA 2011)Paddy by season which indicated that paddy production is increasing every year in Malaysia from 1980 to 2010. In 2010, Malaysia experienced 30% shortage of rice supply and caused Malaysia government importing 700,000 tones of rice from other countries (FELCRA 2013).

To solve this problem, the government initiated a goal for Malaysia to produce and supply 100% of the rice needs in 2015.
The farmers also had problem with soil and element of fertilizer which is individual farms management usually practicing conventional method of fertilizer distribution. However, conventional methods unable to generate maximum yield of rice cultivation (Liu et al. 2014). Among the issues related tosoil are fertility problem and the fertilizer is not uniformly and equally distributed in paddy field area. Nutrient levels suitable for paddy cultivation are required to increase rice seedling growth and to achieve targeted production levels. Since soil nutrient levels vary across paddy regions, it will cause less paddy production. Therefore, soil treatment needs to be done. However, this approach leads to unnecessary financial costs and can have adverse effects on the environment waterways and weed pests. In addition, crop production will be affected. Soil fertility assessments commonly rely on a random soil sampling protocol to obtain an average fertility. Thus, spatial variability is ignored and, consequently, some parts may receive excessive fertilizer while other parts may suffer nutrient deficiency (Liu et al. 2014).The farmers should monitor their paddy plant by own since it is private individual lots of paddy. The problem normally facing by the farmers is related to soil fertility and unequal fertilizer distribution which normally associated with less paddy production. Many factors caused different in paddy yield, however, this research will only focuses on soil fertility condition and spatial fertilizer distribution.

1.3 Cultivation paddy in Malaysia
Paddy cultivation in Malaysia has two growing seasons. The season 1 (off-season) is between March to July which normally associated with drought condition for the Malaysia's weather. Farmers have challenges to accept risks environment like drought due to water requirement is affected in paddy cultivation. During this season, the main requirement for paddy plantation is water supply that is pumping from the river. The season 2 (main season) is between August to February. In this season, of the paddy field normally receives alarge amount of rainfall which is a major of water supply for paddy plantation (MOA 2012).Figure 1 shows the growth phase of paddy cultivation in Malaysia. The average time for paddy cultivation is 90 days. Therefore, in this study, 2 different conditions and period of soil sampling were acquired for soil nutrient. The first sampling is before paddy cultivation which is land preparation stage by the farmer. In this period, the land is cleared, loosened and flattened before planting the rice seedlings and before fertilization stage. Secondly, soil sampling before the harvesting time approximately 100 days of paddy cultivations with soil fertilization has taken place (Figure 1). The first session of soil sampling was using an auger to carry out during the first preparation works of land. In this period, the land is cleared, loosened and flattened before planting the rice seedlings. The purpose of this soil sampling is to examine the soil fertility status of the land taken before the fertilizing process takes place. Soil sample was done systematically with one sample for each lot. One lot of total study area is equivalent to 1.03 hectares. For every lot soil sampling was taken at two different depths of 0-20cm and 20-40cm as normally practiced by the MARDI for soil sampling activity to extract soil chemical properties. After that process, appropriate soil treatment will able to be suggested.

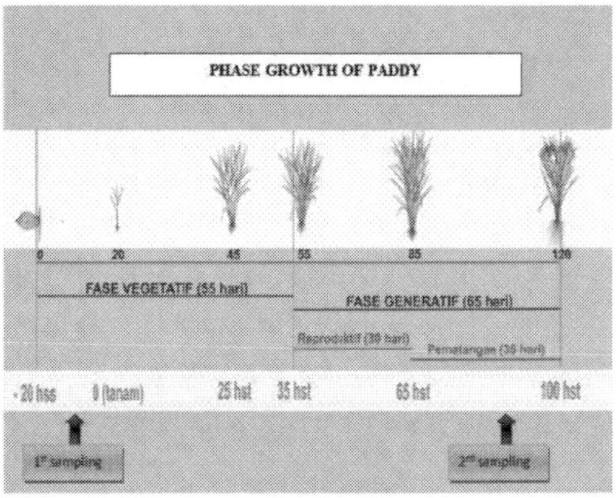

Figure 1. Phase growth of paddy

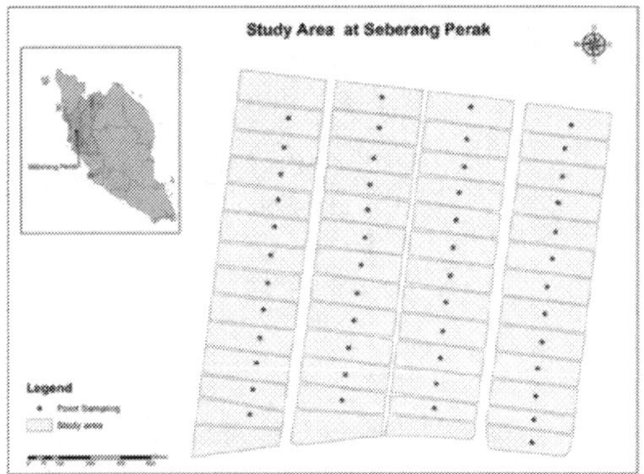

Figure 2. Site location of soil sampling Felcra Seberang Perak

2. Study area

2.1 Site of Felcra Seberang Perak, Perak

FELCRA is big organization estate plantation which established in Malaysia such as the FELCRA Seberang Perak. Total granary area in Seberang Perak is about 16,437ha (MOA 2011). Figure 2 shows the study area of FELCRA Seberang Perak. The farms management is managed by the FELCRA Berhad. The study area is also research study site that conducted by Malaysia Agriculture Research Development Institute (MARDI) and Kawasan Pembangunan Pertanian Bersepadu (IADA) agencies at the small plots area Felcra Seberang Perak with estimated total area is 1.03 hectares. FELCRA Seberang Perak is one of plantation estate managed by FELCRA. This area is totally having a systematic management by FELCRA and it's proven in 2013, total rice production in the fields FELCRA is 45,951.95 tones(FELCRA 2013). Figure 4 is paddy plots field with estimated total area is 58 hectares. It is under MARDI Research Station authority. The fields have facilities of irrigation water from Perak River to canal and direct discharge of water to tertiary drains.

2.2 Soil sampling and laboratory analysis requirements

The soil cores of each point for every plot were collected that cover area of approximately 58 ha. Core sampler and augerwas used to collect topsoil samples at depth 0-20cm with the soil surface cleared from coarse rice straw debris. Soil samples were collected inside the box rather than at grid intersection in order to reflect the entire field. After activity of soil sampling is done in the field the samples were taken to soil chemist laboratory and marking it using a systematic numbering. Analysis of nitrogen using a method kjedahl(Zarina Zakaria, Alina Rahayu Mohamed 2013), then phosphorus (P) and potassium (K) will be derived (Figure 3).

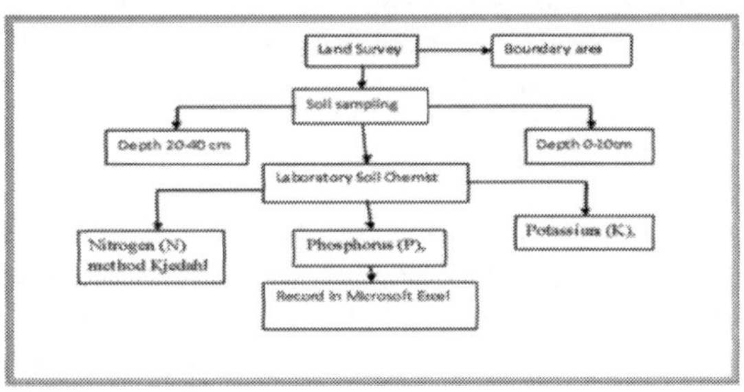

Figure 3. Flow chart of sampling and laboratory works.

According to (Eltaib1 et al. 2002)the distribution of soil chemical properties, especially the amount of nitrogen (N), available phosphorus (P), and converted Potassium (K). It is important for farm management practices and to assess the impact agriculture on the environment. The range of the concentration for the soil parameter is classified according to the Department of Agriculture (DoA) as shown in Table 1.

Table 1. Soil classification of nutrient content; N, P, K (Source: DoA, 1997)

Nutrients	N (g/kg)	P (mg/kg)	K (cmol$^{(+)}$/kg)
Very High	> 10	>45	>1.40
High	6-10	25-45	0.80-1.40
Moderate	3-6	10-25	0.45-0.80
Low	1-3	3-10	0.14-0.45
Very Low	<1	<3	<0.14

Table 2. Optimum soils chemicals properties value for paddy requirement

Chemical Property	Optimum requirement
Total Nitrogen (%)	0.2-0.3
Available P (mg/kg)	>40
Exchangeable K cmol$^{(+)}$/kg	>0.1

In this study GIS analysis was used to derive nutrient content of N, P, and K classification. In comparison of soil nutrients this research will adopt optimum values of soil nutrients for paddy planting as recommended by Malaysian Agriculture Research and Development Institute (MARDI) (Table 2).The levels for each variable on the spatial maps were set based on the standard range for paddy soil, recommended by MARDI.

2.4 Geostatistical analysis in soil management of paddy plantations

This study will use Geographic Information System (GIS) as mapping technique and and SPSS as geostatistical analysis method. According to(Xu et al. 2013) techniques of geostatistics are used to perform traditional statistical analysis and spatial structural analysis with ArcGIS, geostatistical software GS+ and statistical software SPSS in order to obtain the knowledge of characteristics of distribution and spatial variability of soil nutrients. Software ArcGIS geostatistical analyst is an extension that provides capability for surface modeling using deterministic and geostatistical methods. (Qiu et al. 2011) also mentioned geostatistics coupled with geographic information system (GIS) technologies has substantial potential to discern spatial patterns. (Qiang et al. 2011) stated that descriptive analysis and Kolmogorov-Smimov (KS) test were carried out on SPSS17.0 and calculation of experimental semivariogram, model fitting and map drawing were completed on ArcGIS 9.3 software. (Aishah et al. 2010)Geostatistical analyses were applied to examine the within-field spatial variability using semivariograms and kriged maps (Junusi & Hasmadi 2009) in the previous research is descriptive statistic and variation of NPK nutrient status in soil was analyzed using SPSS tool. GIS software was used in producing map to show the spatial distribution of the N, P, and K content. The use of spatial information for managing plant nutrients focuses on using maps and map soil to determine appropriate application of fertilizer can be added or subtracted to correct the deficiencies, saving input costs, and make the field more uniform.

3. Result and Discussion

3.1 Temporal soil nutrient spatial mapping

The data consists of 49 sampling points in Seberang Perak site. Variables of Nitrogen (N) which is measured in term of gram/kilogram (g/kg), phosphorus (P) is measured in as part per million (ppm or mg/kg) and potassium (K) is measured in cmol (+)/kg. Figure 6 illustrates spatial analysis distribution of variable N, P, K for two different times sampling which are before and after fertilization. The results of variables N, P, K were analyzed according to Table 2 that shows the range of the optimum soil chemicals properties values for paddy requirement is classified according to the MARDI guidelines. Table 3 shows the mean value for 1st cycle samples of N, P, K on depth of 0-20cm with the results of 1.783g/kg, 9.938mg/kg, 0.399cmol(+)/kg respectively. For the 2nd cycle sample on depth 0-20cm in the study area were 2.0485g/kg, 12.868mg/kg, 0.7649cmol (+)/kg respectively.

Table 4. Summary of N, P, K content on depth 20-40 in two (2) circle sampling at study area Seberang Perak

Variables	N(g/kg)		P(mg/kg)		K(cmol +/kg)	
Circle sampling	1st	2nd	1st	2nd	1st	2nd
Standard deviation(δ)	0.1598	0.1572	0.5051	0.908	0.0360	0.0951
Mean(μ)	1.1271	1.5461	2.5213	5.702	0.3126	0.6149
Coefficient of variation (%) δ/μ*100	14.18	10.17	20.03	15.92	11.52	15.47

Table 3. Summary of N, P, K content on depth 0-20cm in two (2)circle sampling at study area Seberang Perak

Variables	N(g/kg)		P(mg/kg)		K(cmol +/kg)	
Circle sampling	1st	2nd	1st	2nd	1st	2nd
Standard deviation(δ)	0.308	0.309	2.204	1.968	0.065	0.264
Mean (μ)	1.783	2.0485	9.938	12.868	0.399	0.7649
Coefficient of variation (%)δ/μ*100	17.27	15.08	22.17	15.29	16.29	34.51

However these mean value for 1st circle shows low contents of N, P, K in the soil compared to the range reported by DOA (refer Table 2). In the 2nd session cycle sample indicated values of N and K higher and achieved optimum range 2.0485g/kg and 0.7649 cmol(+)/kg as compared to variable P of 1.9678mg/kg which is very low. The optimum range as stipulated by the MARDI guidelines for variable P is 40mg/kg.

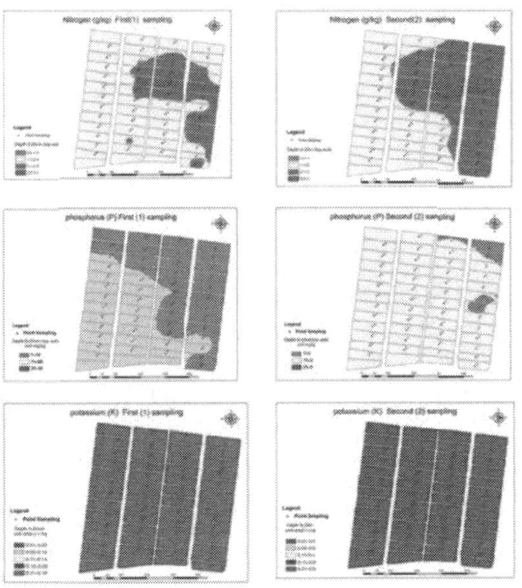

Figure 4. Spatial analysis distribution of N, P,K values on depth 0-20cm in Seberang Perak

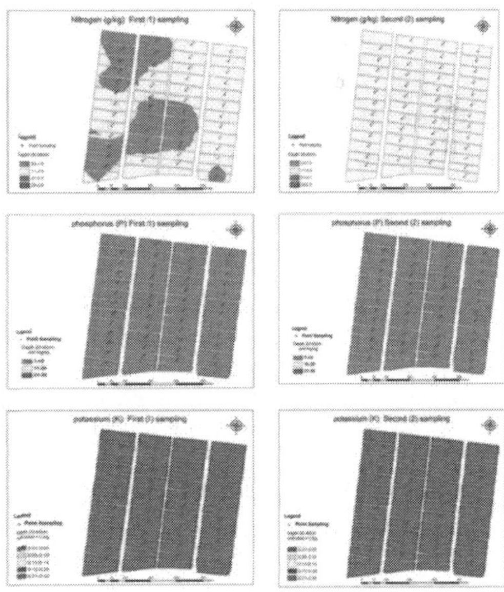

Figure 5. Spatial analysis distribution of N, P, K values on depth 20-40cm in Seberang Perak

Table 4 shows the mean value of 1st circle sample of N, P, K on depth of 20-40cm of 1.1271g/kg, 2.5213mg/kg and 0.3126cmol (+)/kg respectively. For the 2nd cycle sample variable of N, P, K on depth of 20-40cm were 1.5461g/kg, 5.705mg/kg and 0.6149cmol (+)/kg respectively. The results indicated that the values of N, P, and K in two different depths are not much difference to each other's (Figure 7). In term of soil sample depth, (Aminuddin et al. 2003) reported assessment of paddy soil degradation has significant different result for two different layers of depth. The studies had proven soil sampling at two different soil layers which is in the top soil (0-20cm) has more reflected the soil degradation and the sub soil (20-40cm) which represent the subsoil indicated least affected soil degradation.

4. Conclusions

The spatial mapping of the N, P, and K soil nutrient indicated varies result for both condition of before and after fertilization as well as two different depths. The result derived on the study site was compared to the optimum values of soil chemical properties recommended by MARDI. The results showed that the study needs further treatment of soil fertilization of N, P, and K distribution. Furthermore, the result revealed that the variable P are very minimal with mean value range of 9.938mg/kg to 12.868mg/kg and 2.5213mg/kg to 5.702mg/kg in two sampling cycles respectively compared to minimum requirements of 40mg/kg. Therefore, further, nutrient treatment of value of phosphorus (P) is needed for this study area of the Felcra Seberang Perak. It is hope that with the geospatial analysis as conducted in this study, the plantation management such as paddy plantation able to makes better decision making which regard to soil nutrient fertilization and management in order to minimizing cost and maximizing paddy production.

Acknowledgements

Acknowledgements personally to Mr. Ab. Rahim Jais, Mr. Czahari B. Mohamad, Mdms Aimi Athirah to support this paper.

6.0 References

Aishah, a. W. et al., 2010. Spatial variability of selected chemical characteristics of paddy soils in sawah sempadan, selangor, Malaysia. *Malaysian Journal of Soil Science*, 14(1), pp.27–39.

Aminuddin et al., 2003. an Assesment of Paddy Soil Degradation and Its Impact on Sustainable Rice Production.Pdf. *Malaysian Journal of Soil Science*, Vol. 7(3947990), pp.45–56.

Eltaib1, Ss.M., 3, Mohd Musa Hanafi, 4, A.R.M.S. & Wayayok, A., 2002. Spatial variability of N , P , and K in rice field in sawah sempadan Malaysia. *Songklanakarin J. Sci. Technol*, 24, pp.322–328.

FELCRA, 2013. *FELCRA BERHAD | laporan tahunan 2013*,

Firdaus, R.B.R. et al., 2014. Penghidupan petani padi dalam mendepani cabaran perubahan iklim: peranan intervensi kerajaan menerusi skim subsidi harga padi (sshp). *Kajian Malaysia*, 32(2), pp.73–92.

Junusi, R. & Hasmadi, ismail mohd, 2009. Determining and Mapping Soil Nutrient Content Using Geostatistical Technique in a Durian Orchard in Malaysia. *journal of agricultural science*, 1(1), pp.86–91.

Liu, Z. et al., 2014. Geoderma A simple assessment on spatial variability of rice yield and selected soil chemical properties of paddy fi elds in South China. *Geoderma*, 235-236, pp.39–47.

Maclean, J.L. et al., 2002. Rice Almanac :Source Book for the Most Important Economic Activity on Earth. In p. 5.

MOA, 2011. *Kementerian Pertanian & Industri Asas Tani Malaysia : Perangkaan Agromakanan 2011*,

MOA, K.P. dan I.A.T.M., 2012. *Perangkaan Pertanian Malaysia. Agricultural Handbook. Kementerian Pertanian dan Industri Asas tani Malaysia.*, www.moa.gov.my.

Qiang, L. et al., 2011. Study on Spatial Distribution of Soil Available Microelement in Qujing Tobacco Farming Area, China. *Procedia Environmental Sciences*, 10, pp.185–191.

Qiu, W., Curtin, D. & Beare, M., 2011. Spatial Variability of available nutrients & soil carbon under arable cropping in Canterbury. , pp.1–7.

Xu, Y. et al., 2013. Geostatistical Analysis of Soil Nutrients Based on GIS and Geostatistics in the Typical Plain and Hilly-Ground Area of Zhongxiang, Hubei Province. *Open Journal of Soil ...*, 2013(September), pp.218–224.

Zarina Zakaria, Alina Rahayu Mohamed, N.H.M.S. and S.N.A.M., 2013. Total Nitrogen content from earthworm(Eisenia foetida) using the Kjeldahl method. *IIUM Engineering Journal*, 14(1), pp.43–51.

IOP Conf. Series: Earth and Environmental Science 37 (2016) 012048 doi:10.1088/1755-1315/37/1/012048

Monitoring spatial and temporal variations of the rice backscatter coefficient ($\sigma°$) at different phenological stages in Sungai Burong and Sawah Sempadan, Kuala Selangor

Siti Aishah Mohd Rasit[1], Abdul Rashid Mohammed Shariff[2*], Ahmad Fikri Abdullah[2], Aimrun Wayayok[2]

[1]Department of Civil Engineering, Faculty of Engineering, Universiti Putra Malaysia, 43400 UPM Serdang, Selangor, Malaysia
[2]Department of Biological & Agriculture, Faculty of Engineering, Universiti Putra Malaysia, 43400 UPM Serdang, Selangor, Malaysia

[*]Email address: rashidsnml@gmail.com

Abstract. Monitoring rice growth and yield estimation using optical remote sensing data constitutes a big challenge largely due to cloud conditions that are typical of tropical regions. Using Radar remote sensing data helps because it overcome the cloud issue and as well distinguishes the behaviour of the radar backscattering of rice crops specifically. This study indicated the temporal change of rice backscatter ($\sigma°$) at two different growth stages using HH polarimetric Radarsat-2. The aims of this study are: (1) to identify crop with different life spans based on the backscatter coefficient's values from a single polarisation for understanding the backscatter characteristic of rice over the entire growth cycle, and (2) to understand the advantages and limitations using the RADARSAT-2, C band with HH polarisation. The values of backscattering coefficients have been related to the Malaysia rice crop calendar to get the information of the growth status. The result shows strong backscatter coefficient values on the 21st of May that referred to the reproductive-maturity of rice in the Sawah Sempadan area, and out of season for the Sungai Burong area. While for the August 1st imagery, the result shows weak backscatter values which refers to early vegetative and vegetative-reproductive. The values of backscattering coefficient are found to be much less for early vegetation compare to mature rice crop. In this paper, we have also performed a classification of a rice field using Landsat 8 OLI.

1. Introduction

Advanced satellite technology has in turn spawned remote sensing technology, allowing observation of the earth from space using a wide range of sensors. Agriculture, natural disasters, and ecological system monitoring and forecasting has become a priority field in remote sensing based space observation. Remote sensing data, both optical and radar, have the capability of earth surface observation such as forecasting, monitoring changes, detecting object, predicting and estimating. The Synthetic Aperture Radar (SAR) has become an indispensable tool to understand vegetation's problem and characteristics. The biophysical and quantitative approach of SAR-based technology is very powerful for rice vegetation observation.

Oryza sativa (rice) is one of the important crops in developing countries since rice is a global staple food for humans supplying a large fraction of the needs for energy-rich materials and generally a significant proportion of the intake of other nutrients as well. Because of the growth of the population increasing day by day, there is a high demand for the rice production to meet the demand. Rice becomes the concerning crop in the Group of Earth Observation (GEO) Global Agriculture Monitoring Initiative (GEOGLAM) in improving the crop yield assessment and forecasting at a global level to monitor agriculture and support crop prediction in order to better plan for food availability and cope with the world's demand. Rice crop monitoring is important for a sustainable resource and food security. The purpose of GEO GLAM is to improve food security information using Earth Observation (EO) at national, regional, and global scales [1].

Monitoring at every stage of rice's phenological stages is an important trait that determines the grain production of rice, which is mainly affected by certain factors, including tillering, plant height, and panicle morphology. Other than these, other factors such as pest occurrence, weather conditions, water supply, fertilizer, and pesticide practices also affect the rice production. It is a must to have timely monitoring work on the rice plant development. However, timely monitoring and the high accuracy of information is a challenge in remote sensing based on rice agriculture monitoring and observation. With the high resolution of spatial and temporal polarimetric data such as Radarsat-2 along with the support of statistical data from the agricultural department, the issue will be solved.

Polarimetric SAR is a very powerful tool for extraction of information for identification and classification of different natural features, as each polarisation is sensitive to different surface characteristics and properties. Furthermore, SAR data will overcome the cloud, and weather problems as well as distinguish the behaviour of radar backscattering of rice crops, specifically because at these tropical regions, heavy clouds and rainfall make it difficult to acquire optical remote sensing data [3].

The relationship between radar backscattering coefficients and rice crop parameters using European Remote Sensing (ERS-1) satellite C-band SAR VV polarisation data was the first result gained in the year 1995 [6]. Using the ERS-1 satellite, C-band SAR, a simulation on the Monte Carlo rice canopy was applied in the theoretical model that described the rice canopy backscatter mechanism [7].

A previous study, used RADARSAT-1 HH polarisation data to monitor rice crops and retrieve rice growth parameters [3, 9] However, for single polarisation, such as ERS-1/2 VV and RADARSAT-1 HH data, limited information was provided. When ENVISAT was launched, various polarisation mode data were investigated [2, 4, 10, and 12] With the successful launch of RADARSAT-2, multi-temporal and quad-polarisation data have become available for crop monitoring. Classification for rice field in differentiating between rice and non-rice has been also carried out using RADARSAT, ENVISAT, ALOS and ERS images based on the SAR polarisation, [2,3,4,7,6,9,10,7, and11].

2. Materials and method

2.1. Study area

Tanjung Karang in Kuala Selangor is a famous rice granary area in Selangor, Peninsula Malaysia. In this study, we selected two test sites, Sawah Sempadan and Sungai Burong, located approximately around latitudes $3^0$27'41,61"N and $3^0$30'99" N, and longitudes $101^0$12'38.30"E and $101^0$9'29.90"E respectively. The rice growing area is the fourth-largest area of rice cultivation land in Malaysia and is under monitoring by the Integrated Agriculture Development Area (IADA), Barat Laut Selangor. The topography of Tanjong Karang is a coastal area characterised by a relatively flat landscape as shown in Figure 1.

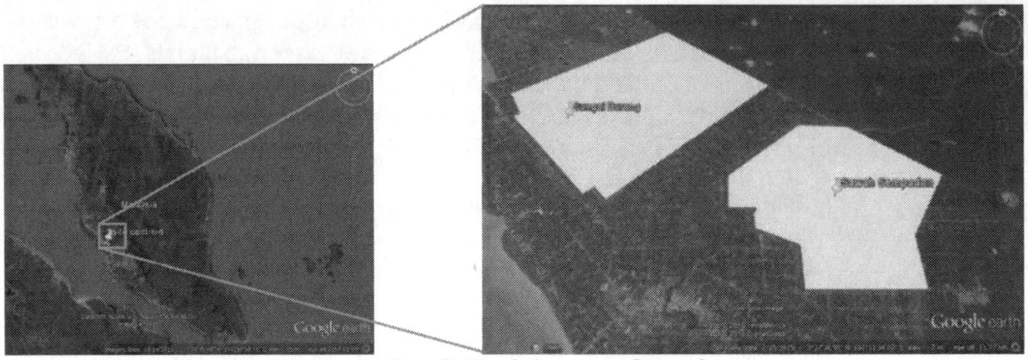

Figure 1. Google Earth images of test sites.

2.2. Dataset description

Two types of datasets had been used, Radarsat-2 (Figure 2(a)) and Landsat 8 OLI (Figure 2(b)). The Radarsat-2 data was HH polarisation with a Fine Wide beam mode, acquired on the 21st of May and 1st of August 2014 respectively covering the rice growing season in the study area. Both Radarsat-2 images were 12.5m resolutions. Similarly, Landsat 8 OLI with 30m resolutions were used for retrieving the LULC of the granary area.

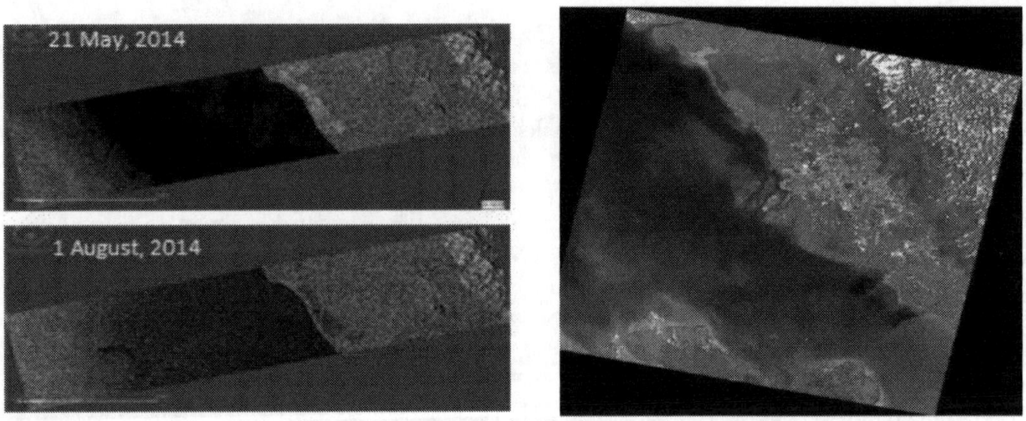

Figure 2. Visualisations of the datasets used: (a) Radarsat-2 images. (b) Landsat-8 OLI.

2.3. Malaysian rice crop calendar

In monitoring paddy rice, it is significant to know the phases of the paddy rice growth. Target dates or months for satellite images play an important role in the developing stage like discriminating between paddy rice varieties [7]. The growth calendar stipulates the classification of rice growing stages in the month as shown in Table 1. The timing of sowing, planting, growing and harvesting follows the growth calendar. Rice's cultivation practices in Malaysia is twice a year based on double cropping. This is because the climate factors in the equatorial regions have wet and dry seasons continuously. It is important to identify the time and weather conditions to determine the growth period before planting. Weather conditions must be suitable for every growth stage.

In the Rice Growth Outlook, weather involves precipitation, solar radiation and temperature. Excellent weather condition will produce a good harvest whiles, bad weather will produce a poor harvest (weather disaster, growth injury, and other damages such as disease and insects)[9]. In Tanjung Karang, rice growth stages can be divided into three stages: i) planting-early vegetative (January-February, July-August), ii) vegetative-reproductive (March-May, August-October) and iii) reproductive-maturity (June-July, October-December). Rice planting in Tanjong Karang follows the schedule as in Table 2 for every season.

Table 1. Temporal differences in rice crop growth stages in Malaysia, MADA, 2015.

Days after Planting	Description	Paddy Rice Stages	Growth Phases Based on Malaysian Crop Calendar
0-20	Pre-flood, Germination	Germination	Planting - Early Vegetative
20-30	Tilter initiation	Vegetative	
30-40	Early Tilter, mid-tiltering	phase	Vegetative - Reproductive
40-55	Panicle initiation	Reproductive	
55-70	Booting, heading, flowering		Reproductive - Maturity
70-115	Milky, dough, yellow, ripe, mature	Ripening	

Table 2. Malaysian yearly rice crop calendar at Sungai Burong and Sawah Sempadan, MOA, 2014.

Region	J1	J2	F1	F2	M1	M2	A1	A2	M1	M2	J1	J2	J1	J2	A1	A2	S1	S2	O1	O2	N1	N2
Sawah Sempadan	1	1	2	2	2	2	3	3					1	1	2	2	2	3	3			
Sungai Burong			1	1	2	2	2	2	3	3					1	1	2	2	2	2	3	3

1 = *Planting - Early Vegetative*, 2 = *Vegetative - Reproductive*, 3 = *Reproductive - Maturity, and Blank = Out of Season*

2.4. SAR data processing.

Image processing was performed using NEST 5.1 (Next ESA SAR Toolbox), an open source software for exploiting ESA and other satellite data. The data processing in NEST can be divided into four parts; sigma nought calibration, geometric correction and co-registration, multi-temporal speckle filtering and terrain correction. Figure 3 below shows the workflow of the processing. The methods of Radarsat-2 data processing in this study were employed as follows;

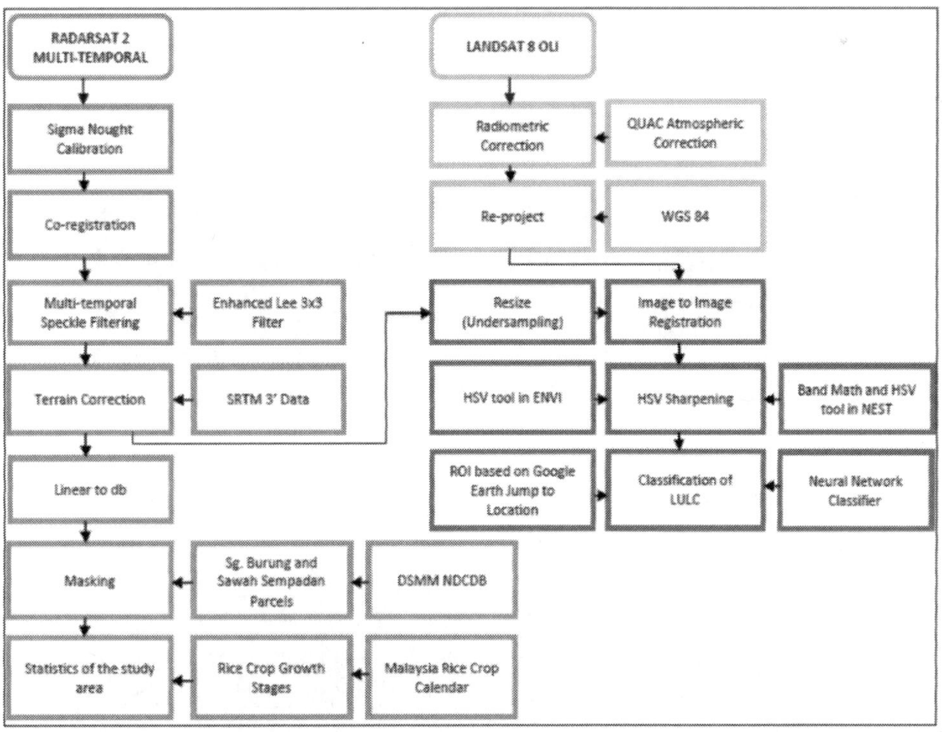

Figure 3. An outline of the flow of process in monitoring crop rice growth stages.

2.4.1 Graph Builder

Developed to automate the processing in NEST, a network of processing chains has been designed to allow for batch processing on the stack of Radarsat-2 multi-temporal images, which included the sigma nought calibration (from amplitude to normalised radar backscatter coefficients). The backscatter coefficients (σ^{o}) were calculated using the following calibration equation (1):

$$\sigma^{\circ} = 10 * \log \frac{(DN^{2}_{jk} + A_{\circ})}{A_j} + 10 * \log(sin I_j)i \qquad (1)$$

where DN is the digital number of the Radarsat-2 image, A_o and A_j are the automatic gain control factors and I_j is the local incidence angle of each pixel across the range direction.

2.4.2 Satellite orbit parameters

With the satellite orbit parameters in the images header files, geometric correction and automatic co-registration performed on both Radarsat-2 images, it produced high positional accuracy. The RMSE for the co-registration was set up at 0.5 m.

2.4.3 Filtering technique

An Enhanced Lee filter with a 3x3 kernel was used for multi-temporal speckle filtering of the radar imageries to reduce the noise and small patches over the paddy parcels prior to data analysis.

2.4.4 Terrain correction

Doppler range radar terrain correction was performed on the multi-temporal images using 3' SRTM which showed an obvious difference between the area with a higher elevation and the flat area (Sungai Burong and Sawah Sempadan) located near the coastal region of Kuala Selangor.

2.4.6 *Linear to dB*

The conversion of radar backscatter coefficients (σ^{o}) from linear to dB from pixel intensity values to physical quantities is shown below (Figure 4).

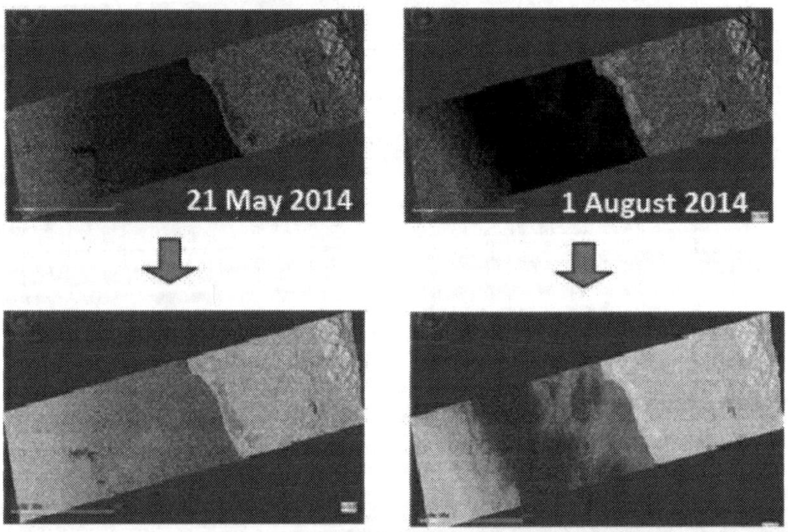

Figure 4. Conversion of Radarsat-2 images' radar backscatter coefficients (σ^{o}) from linear to dB

2.5. Determining the characteristics of radar backscatter in rice crop fields

Basic statistics were calculated for paddy rice parcels in the Sungai Burong and Sawah Sempadan areas based on the land parcel data retrieved from the Department of Survey and Mapping Malaysia NDCDB (National Digital Cadastral Database). The extracted backscatter coefficient values were achieved by masking Radarsat-2 multi-temporal images using ESRI shapefile vector data of the rice field parcels. From the NDCDB vector data obtained from DSMM, 62 parcels were chosen for Sawah Sempadan and 24 parcels were chosen for Sungai Burong. These parcels are chosen randomly without any specific requirements being used to entail the selection.

2.6 Classification of LULC to monitor the distribution of crop rice in the study area

The ESRI shapefile of the land parcels and Radarsat-2 multi-temporal images were exported to Google Earth .kml file. These datasets were overlaid with Google Earth to identify the land cover types such as paddy rice, water bodies, other vegetation, urban areas and others. Google Earth ENVI Jump to Location was also used to verify the land cover types to be used as ROIs. For the rice field, the classes were identified according to their growth stages specified in the Rice Crop Calendar: i) planting-early vegetative, ii) vegetative-reproductive, and iii) reproductive-maturity. It must be noted, however, that these stages were assigned based on a mere visual interpretation of the Landsat 8 OLI image collected on 21st of May 2014. This image was fused with Radarsat-2 images also taken on the 21st of May 2014.

Due to a heavy percentage (>50%) of cloud cover during the period when both Radarsat-2 datasets were used, the 21st of May 2014 and 1st of August 2014, a proper data fusion of Landsat 8 OLI and Radarsat-2 datasets on these dates for LULC classification using the Neural Network classifier and accurate delineation of the growth stages of the rice were difficult to be achieved. Even if the classification was performed, the accuracy of the LULC classification result would have been low due to the misclassification of several classes that were affected by the cloud cover. The purpose of the LULC classification using SAR-multispectral datasets in this study was also not to monitor the growth stages of the rice crops but rather to analyse the distribution of rice crops in the study areas. It was to analyse visually, whether it was fit with the location of the Sungai Burong and Sawah Sempadan parcels that were used as masks to generate the statistical data, prominent in the analysis of this study.

3. Result and analysis

3.1 Experimental lot

From the NDCDB vector data obtained from DSMM, 62 parcels were chosen for Sawah Sempadan and 24 parcels were chosen for Sungai Burong (Figure 5). These parcels were chosen randomly without any specific requirements being used to entail the selection. The characteristics of σ° in the rice crop fields involving the correlation between the strength of the radar σ° and the growth stages of the rice crops, were reviewed to provide an analysis of their growth in terms of the visual perspective.

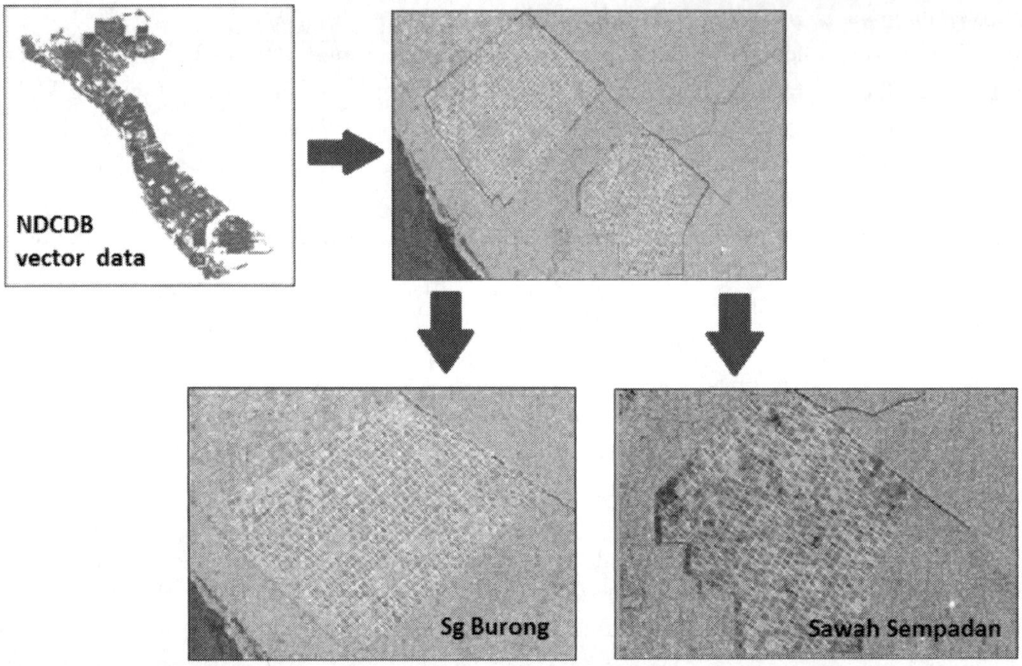

Figure 5. Land parcels of Sungai Burong and Sawah Sempadan lots obtained from DSMM NDCD.

3.2. Visual Interpretation

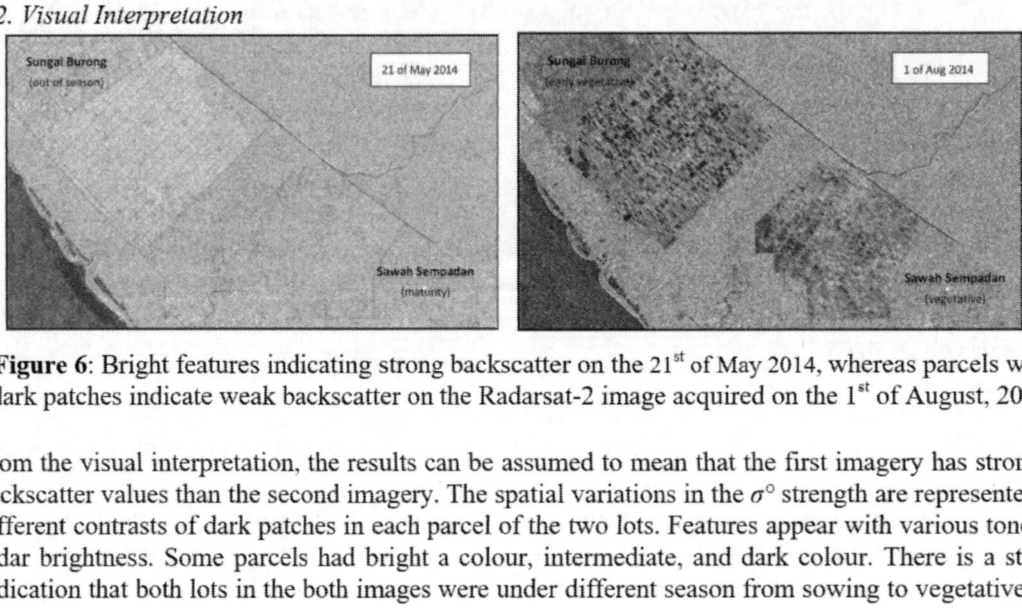

Figure 6: Bright features indicating strong backscatter on the 21st of May 2014, whereas parcels with dark patches indicate weak backscatter on the Radarsat-2 image acquired on the 1st of August, 2014.

From the visual interpretation, the results can be assumed to mean that the first imagery has strongest backscatter values than the second imagery. The spatial variations in the σ° strength are represented by different contrasts of dark patches in each parcel of the two lots. Features appear with various tones of radar brightness. Some parcels had bright a colour, intermediate, and dark colour. There is a strong indication that both lots in the both images were under different season from sowing to vegetative and the vegetative to the reproductive periods. This situation can give the assumption that, the cultural practice is different in every parcel in both lots. The rice tree in these parcels may experience a difference in the growth of the rice crop in any of these stages: germination, tiller initiation, early tillering, mid-tillering or panicle initiation. The rices fields show up as brighter and brighter patches in the growth of the rice crops, [8, 15]. This is due to an increase in the intensity of the backscattered microwaves.

Based on the agriculture statistics from the Department of Agriculture, Kuala Selangor, around seven types of rice varieties were planted at Sawah Sempadan and Sungai Burong for both seasons, which were, MR 220CL1 & CL2, MR219, MR253, MR263, MR269, and MR325. Different varieties of rice will have different maturity period and life cycles. For example, the rice type of MR253, MR263 and MR269, have less than the 120-day maturity period. Other characterisation such as the shape of the plant and height, resistance to disease, and capacity of grain production are also different. This type of rice is the highest quality and gives a high yield [13]. The rice variety also is one of the important factors that influence the rice yield.

In the other cases, very bright parcels or targets appeared due to the corner reflector or double bounce effect, where the radar pulse bounced off the horizontal ground towards the target, and was then reflected from one vertical surface of the target back to the sensor [5]. Built-up areas and many man-made features usually appear as bright patches in a radar image due to the corner reflector effect. Some of the areas in Sawah Sempadan and Sungai Burong were built with houses, community places and shop lots. So in these cases, buildings reflected most of the radar energy back to the sensor.

Besides, surface roughness and moisture content of bare soil will affect the backscatter values. The brightness of areas covered by bare soil may vary from very dark to very bright depending on its roughness and moisture content [5]. Typically, rough bare soil appears bright in the image. For similar soil roughness, the surface with a higher moisture content will appear brighter. For flooded area, it will appear dark because all of radar energy specularly reflected away. The familiarity of ground conditions of the areas imaged, makes us difficult to interpret. In Radar, there is a rule which is, the higher the backscattered intensity, the rougher is the surface being imaged. Some of the flat surface such as road and water bodies will appear dark because radar the incident radar pulses are specularly reflected away [5]. All this situation resulting the images appear in various tones of brightness and darkness.

3.3. Average backscatter coefficient in Sungai Burong and Sawah Sempadan

From the two growth periods, the $\sigma°$ on the 21st of May (from 55 to 115 days after transplanting) was the highest as compared to the backscatter coefficient on the 1st of August (Figure 7). At Sawah Sempadan, on the 21st of May, during the reproductive-maturity period, the mean $\sigma°$ was -7.80 dB, showing a strong backscatter between -33.91 dB (minimum) to 22.37 dB (maximum). The mean $\sigma°$ at Sungai Burong followed the same trend with the mean $\sigma°$ on the 21st of May being -6.42 dB at the range between -32.88 dB (minimum) to 22.06 dB (maximum).

Around the early-vegetative to vegetative reproductive period (from 0 until 55 days after transplanting), the $\sigma°$ also change with the decreasing trend. The HH $\sigma°$ polarisation at Sawah Sempadan on the1st of August shows the mean of -10.51 dB, ranging from -36.26 dB (minimum) to 22.06 dB (maximum). The $\sigma°$ at Sungai Burong followed the same pattern with the mean $\sigma°$ at -11.45 dB, ranging from -39.37 dB to 21.70 dB. These $\sigma°$ temporal variations from the HH polarisation Radarsat-2 datasets can be used as the function of the rice crop growth period which refers to the months of the year.

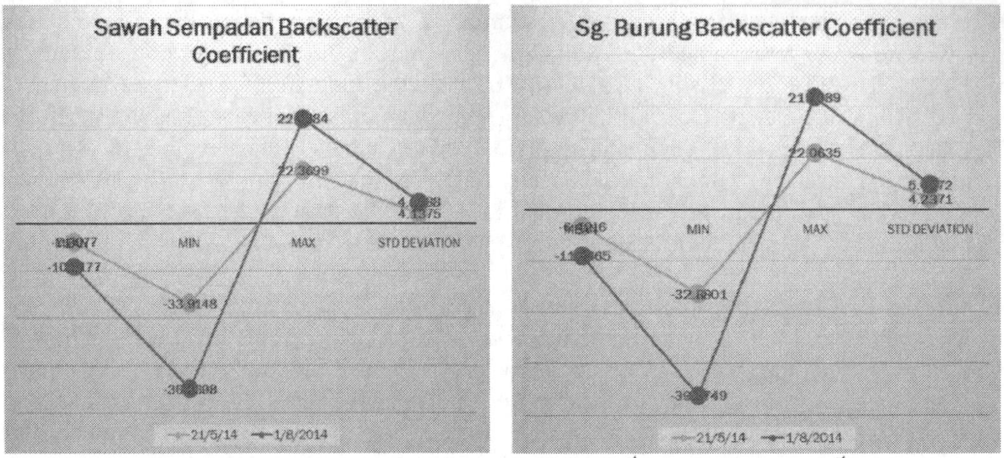

Figure 7. The average backscatter coefficient value on the 21st of May and the 1st of August, 2014 in Sawah Sempadan and Sungai Burong.

3.4. Spatial and temporal variation

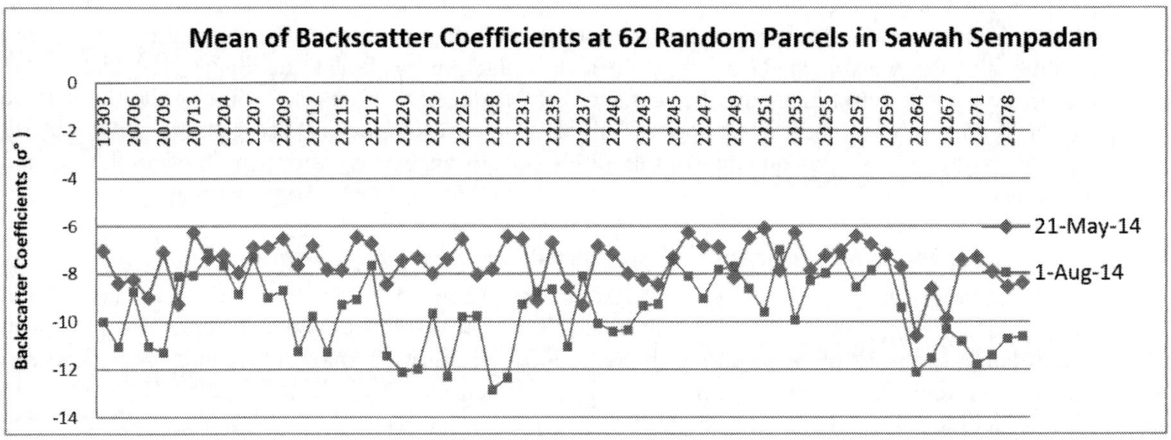

(a)

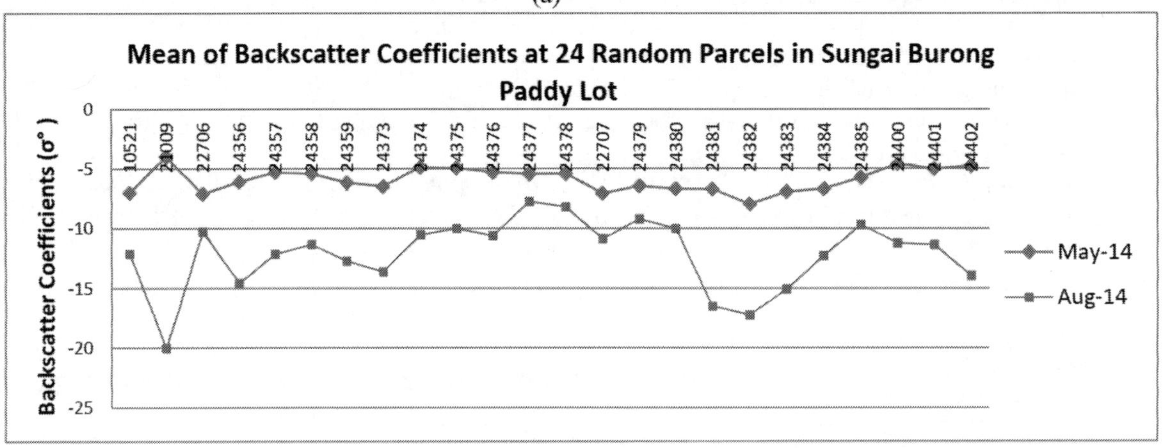

(b)

Figure 8:Backscatter coefficients at selected parcels in (a) Sawah Sempadan and (b) Sungai Burong for both images.

Both Figures 8 (a) and (b) above show the pattern of $\sigma°$ at each parcel in the two different growth periods. The computed statistics reveal that the growth phases in each parcel within the Sawah Sempadan and Sungai Burong lots were different, although they are categorised into the same growth period based on the Malaysian Rice Crop Calendar. Each parcel exhibited different $\sigma°$ values in terms of their mean values, either lower or higher than one another. The differences in their mean values however were not too subtle, having a mean $\sigma°$ ranging from -6 dB to -11 dB and -7 dB to -13 dB on May 21st and the 1st of August respectively, for both lots. However, in a certain parcel such as parcel no. 22 in Sawah Sempadan it shows high maximum values of $\sigma°$ at both months (May and August) at 13.30 and 12.88 dB, respectively. Other parcels have maximum $\sigma°$ values between -0.5 dB to 7 dB only for the observed period. This may be due to the existence of other features within that parcel such as dwellings which significantly change the $\sigma°$ at that particular parcel.

The statistics for the sample parcels selected from Sungai Burong show the same $\sigma°$ pattern. The mean $\sigma°$ for the 24 sample parcels was between -4 dB to -7 dB and -7 dB to -20 dB for the 21st of May and the 1st of August respectively. The range (maximum/minimum) on the 21st of May and the 1st of August at the Sungai Burong sample parcels were -22 dB to 12 dB and -33 dB to 13 dB respectively. The significant difference is observed at lot 21009 with minimum/maximum $\sigma°$ is -4.37 dB/-3.61 dB, and -21.54 dB/-18.52 dB in May and August respectively. The low $\sigma°$ values of this parcel as compared to the others may be caused by other features in the parcel with low radar $\sigma°$ such as flooded water or another feature (s) with high a moisture content.

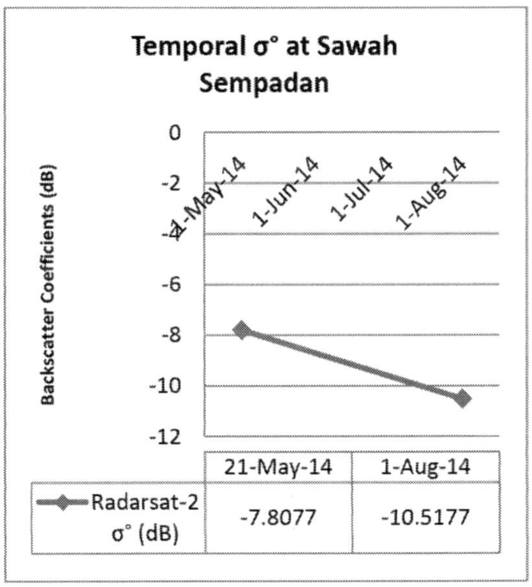

(a)

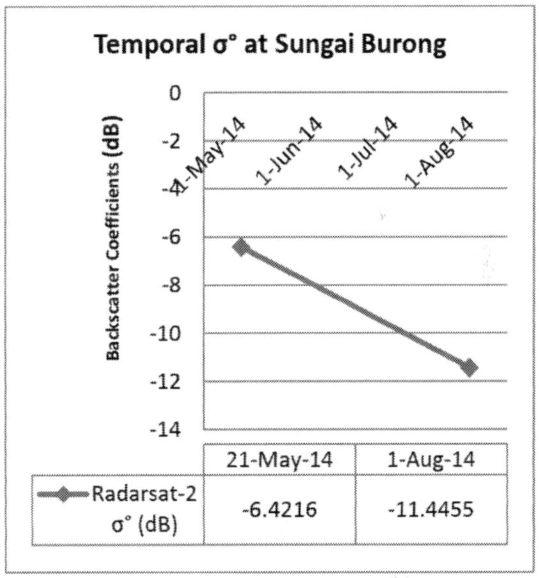

(b)

Figure 9 (a),(b), and **Table 3.** Variations of the rice backscatter coefficients at both lots corresponding to the age of the crop based on the Malaysian rice crop calendar broadcasted by Radarsat-2 image acquisition.

Area	Date	Growth Stages	Age (days)	Radarsat-2 $\sigma°$ (dB)	Backscatter strength
Sawah Sempadan	21st of May 2014	Out of season	-	-7.8077	High
	1st of August 2014	Vegetative-reproductive	45-75	-10.5177	Low
Sungai Burong	21st of May 2014	Reproductive-maturity	75-90	-6.4216	High
	1st of August 2014	Planting-early vegetative	0-30	-11.4455	Low

Figure 9 and Table 3 have indicated the relationship between the radar $\sigma°$ and the growth stages (refer to the days in the month when the images were acquired) on both rice field lots which can strongly differentiate the $\sigma°$ in the two growth stages. The first date of the image (21st of May) results in a strong $\sigma°$. This strong backscatter suggests that both lots were in their reproductive-maturity period of the rice crop growth. This situation occurred due to the rice crop height temporal variations. During the reproductive-maturity stage, the rice plant height increases rapidly, and the difference in the cultivation time of rice crops at Sungai Burong and Sawah Sempadan may have caused the difference in the rice crop height, increases in the number of leaves, grain filling, and thus caused the double bouncing between plants to occur, causing the image on the 21st of May to appear brighter (strong backscatter).

The bright features of the rice crop may have appeared in the image due to the corner reflector or double bounce effects where the radar pulses bounced of the horizontal grounds towards the targets and then, were reflected from one vertical surface of the targets back to the sensor [2]. However, the values of $\sigma°$ in Sawah Sempadan were still high even though there wass no vegetative growth at the time (out of season) because of the bare soil of rice field hawing rough surface with low moisture content.

On the 1st of August, the $\sigma°$ was relatively weak during the planting to early vegetative and vegetative to reproductive periods. However, in Sawah Sempadan it was quite high compared to Sungai Burong because of the different volumetric scattering, and the number of leaves and panicle were less at this age. Furthermore, in the early stages, the moisture content of the crop and the soil moisture of the rice field is really high. In the optical range, early vegetation will appear greenish, indicating a healthy rice plant because of the high moisture content and nutrient level. Different from the microwave range, it will appear dark because of the energy being absorbed by the moisture or water content in the soil and rice plant. Daily weather conditions also cause changes in moisture content of the rice crop and the soil that will affect the strength of the $\sigma°$ values.

3.5. Distribution of rice crop in the study area

Figure 10 shows the result of the LULC classification using Landsat 8 OLI and Radarsat-2 data fusion of the study area to show the distribution of rice crop covering the area. The areas marked by the white line polygon vector are the study areas of Sungai Burong and Sawah Sempadan. This map was created only for verifying the types of vegetative features covering the Sungai Burong and Sawah Sempadan land parcels. The majority of the LULC in this area wass covered by rice crops. As we can see, the location of Sungai Burong and Sawah Sempadan are near the coastal areas.

To sum up, if an exact period of Landsat 8 OLI datasets are obtained during the day/period when Radarsat-2 images are acquired with cloud cover less than 10%, a proper visual analysis and classification map can be produced to monitor the rice crop growth stages using either the fused or combination of both SAR and multispectral datasets. Using the fused Radarsat-2 and Landsat 8 OLI datasets with different dates, the different growth periods as used in a small part of this study subjected to the analysis of the LULC classification were very limited, so as to show only the distribution of rice crops in the study area.

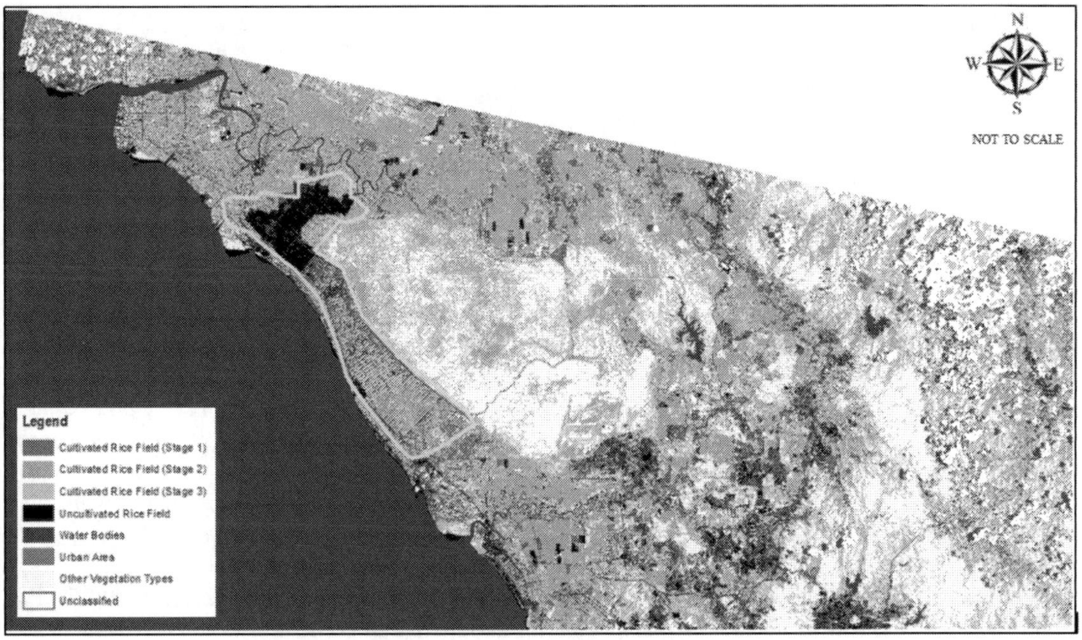

Figure 10. Distribution of crop rice in the study area.

4. Conclusion

The values of the backscatter coefficients, which were related to the Malaysian rice crop calendar, easily differentiate the rice growth stages in the rice area. Results indicated, strong backscatter coefficient values on the 21st of May that referred to the reproductive-maturity rice in the Sawah Sempadan area and out of season for the Sungai Burong area. Whilst for the 1st of August imagery, the results show weak backscatter values, which were referred to as early vegetative and vegetative-reproductive. The values of the backscattering coefficient were much lower for early vegetation compared to mature rice crops due to temporal changes. Referring to the factor affecting the σ° values, moisture content and surface roughness of the target determine the values of σ° and strength of the scattering. The different contrasts of parcels in the same area and date showed spatial variations that were caused from different cultural practices amongst the farmers. This indicator can help to monitor the farmers' activities in order to ensure a high rice yield production. The limitation of this study is that it has only focused on the rice growth stages (age), the next study will examine various parameters such as biomass, moisture content etc., which can be correlated with the ground data.

5. Recommendations

As for this study, monitoring of the growth stages of the rice crop growth was performed by using the Multi-Temporal Radarsat-2 dataset using only one observation and only using single polarisation (HH). In the near future, another polarisation could be used to see the different results of monitoring the rice growth stages In conjunction with using the SAR image, more parameters could be extracted and used for other purposes, such as rice plant height, LAI, biomass, moisture content etc. Apart from that, the backscatter coefficient from the image could be correlated with ground data to actual condition and get better accuracy.

Aknowledgement

This study was supported by research collaboration under SOAR-JECAM Project (Food Security Malaysia (GEO-GLAM Asia Rice, No: 5281). SOAR-JECAM Project referred to Science and Operational Applications Research and development program. Crop Assessment and Monitoring Initiative was a part of SOAR program in developing techniques using RADARSAT-2 and ALOS-2 satellite data. We thank our colleagues from Canadian Space Agency (CSA), Canada, and Japan Aerospace Exploration Agency (JAXA), Japan, who provided data, knowledge, and expertise that greatly assisted the research.

References

[1] Shin-ichi Sobue. 2014. *Asia Rice Crop Team Activity for GEO GLAM to Improve Food Security Information using Earth Observation at National/Regional/Global Scale.* Report Paper of Remote Sensing Technology Center of Asia Rice Crop Team.

[2]Bouvet, A., Le Toan, T., Lam-Dao, N., (Feb. 2009). *Monitoring of the rice cropping system in the Mekong delta using ENVISAT/ASAR dual polarisation data. IEEE Trans. Geosci. Remote Sens.*, vol. 47, no. 2, pp. 517–526.

[3]Chakraborty, M., Manjunath, K. R., Panigrahy, S., Kundu, N., and Parihar, J. S., (Aug. 2005). *Rice crop parameter retrieval using multi-temporal, multiincidence angle RADARSAT SAR data. ISPRS J. Photogramm. Remote Sens.*, vol. 59, no. 5, pp. 310–322.

[4]Chen, J. S., Lin, H., and Pei, Z. Y. 2007. *Application of ENVISAT ASAR data in mapping rice crop growth in Southern China. IEEE Geosci. Remote Sens. Lett.*, vol. 4, no. 3, pp. 431–435.

[5]CRISP. 2001. Interpreting SAR Images. Retrieved on May 25, 2015 at http://www.crisp.nus.edu.sg/~research/ tutorial/ sar_ int.htm

[6]Kurosu, T., Fujita, M., Chiba, K. 1995. *Monitoring of rice crop growth from space using the ERS1 C-band SAR, IEEE Trans. Geosci. Remote Sens.*, vol. 33, no. 4, pp. 1092–1096.

[7]Le Toan, T., Ribbes, F., Wang, L. F., Floury, N., Ding, K. H. 1997. *Rice crop mapping and monitoring using ERS-1 data based on experiment and modeling results. IEEE Trans. Geosci. Remote Sens.*, vol. 35, no. 1, pp. 41–56.

[8]Ribbes, F. and Toan, T.L. 1999. Rice field *mapping and monitoring with RADARSAT data. International Journal of Remote Sensing*, 20: 745-765.

[9] Shao, Y., Fan, X. T., Liu, H., Xiao, J. H., Ross, S., Brisco, B., Brown, R., and Staples, G. 2001. *Rice monitoring and production estimation using multitemporal RADARSAT, Remote Sens. Environ.*, vol. 76, no. 3, pp. 310–325.

[10]Stankiewicz, K. A. 2006. *The efficiency of crop recognition on ENVISAT ASAR images in two growing seasons. IEEE Trans. Geosci. Remote Sens.*, vol. 44, no. 4, pp. 806–814.

[11]Wu, F., Wang, C., Zhang, H., Zhang, B., and Tang, Y. 2011. *Rice Crop Monitoring in South China With RADARSAT-2 Quad-Polarisation SAR Data. IEEE Geoscience And Remote Sensing Letters*, VOL. 8, NO. 2, pp. 196-200.

[12]Yang, S. B., Shen, S. H., Li, B. B., Le Toan, T., and He, W. 2008. Rice mapping and monitoring using ENVISAT A SAR data, *IEEE Geosci. Remote Sens. Lett.*, vol. 5, no. 1, pp. 108–112, Jan. 2008.

[13] DOA. Laporan Hasil Padi. 2014. Department of Agriculture (DOA), Kuala Selangor.

Smart facility application: exploiting space technology for smart city solution

A A A Termizi[1], N Ahmad[1], M F Omar[1], N A Wahap[1*], D Zainal[1], N M Ismail[2],

[1] Agensi Angkasa Negara (ANGKASA), Pusat Angkasa Negara, 42700 Banting, Selangor, Malaysia
[2] Geoinfo Services Sdn. Bhd. 31, Jalan Bandar 2, Taman Melawati, 53100 Kuala Lumpur, Wilayah Persekutuan Kuala Lumpur, Malaysia

E-mail: ammar@angkasa.gov.my

Abstract. Facilities and amenities management is amongst the core functionalities of local government. Considering the vast area that local government has to manage, a smart solution is extremely inevitable to solve issues such as inefficient maintenance of public parks, drainage system and so forth. Therefore, this paper aims to offer a smart city solution which exploits the benefit of space technology. This proposed solution is one of the modules developed in Spatial Smart City Service Delivery Engine (SSC SDE) Project undertaken by Agensi Angkasa Negara (ANGKASA). Various levels of local government have been chosen to understand real issues faced by them. Based on this data, a Smart Facility application has been developed with the aim to enhance the service delivery by the local government hence improving citizens' satisfaction. Since this project is still in progress, this paper will merely discussing the concept of this application.

1. Introduction

Facility management is not a new discipline altogether but it is a modern concept of maintenance. In the past, and still in practice, maintenance of a facility or a precinct is disintegrated from one to another. A company or a government agency normally has multiple contractors that are doing various maintenance tasks such as cleaning services, landscaping services, mechanical and electrical services, civil services and many others. However, lately, many organisations are moving towards facility management concept where one facility management division is doing all those maintenance work.

Maintaining a facility or a precinct is not an easy task especially if it involve the public at large. In this urbanised society, every little hiccup will cause real damage to the organisation's image as it became viral in the social media. Thus, organisation such as local government will have to improve the way they manage their facility to provide better services to the public. This is where space technology could be exploited as part of smart city solution to manage massive area under the local government management.

Therefore, the objective of this paper is to explore the idea of utilizing space technology application in smart city initiative to improve facility management in local government. This paper focusses its discussion based upon the Malaysian context and environment although it does not limit its discussion by taking examples from abroad. In addition, data and information in this paper are merely an extrapolation of data obtained from selected local governments. Furthermore, this paper is more interested to share the overall concept of the solution and does not intend to go into details of the proposed solution as it is still under development. However, certain proposed working solution will be shared where necessary.

Space technology as defined by [1], is any technology that is designed, intended, deployed, or put into use in outer space or any celestial body. This includes satellites such as earth observation satellites, navigation satellites, communication satellites and others. These satellites provide avenues for many applications in various fields from government sector to private sector. Both sectors will consequently benefit the public at large. For instance, the application of navigation satellite technology in mobile application such as Waze and Google Maps has tremendously reduce travelling time taken by users. Another good example of the utilization of space technology is in the management of vast areas of forestry. The authority body could use images provided by earth observation satellites to quality of stock, mitigate potential problems and plan for the future of forestry managed. These examples of the utilization of space technology in various sectors shows that such technology could possibly be applied in smart city initiatives.

Smart city initiative or solution is relatively a new idea in Malaysia, although it has been around in many places worldwide. Cities such as Barcelona in Spain, Amsterdam in the Netherland, Toyota City in Japan and Beijing Haidian District in China are famous to be smart although their definition of smart might be different from one to another. Smart city market size is rapidly expanding and it is predicted by ABI Research that it will reach USD 39.5 billion by 2018 and Frost and Sullivan further predicted that it will reach up to total 3.3 trillion by 2025 [2]. In Malaysia, there are also initiatives on smart city development including Cyberjaya and Iskandar Malaysia [3] [4].

2. User requirement study

In this project, facilities is defined to be any facilities and amenities managed by the local government. In Malaysia, local government is the tertiary level government under federal government and state government. There are many facilities and amenities managed by the local government. These include public halls, roads, public parks and many others. In order to understand the real problem faced by local government in managing these facilities and amenities, site surveys and interviews in selected local government has been conducted. Six local governments in various locations were selected in this project. These include urban, suburban and rural areas. This will provide better picture of the problems at various levels.

Interviews and surveys in selected local governments concluded that the main issues or problems encountered by them and based on complaints by the public are waste collection problem, bad condition roads, blocked drainage system and long grass in public parks and places. Somehow, these are common problems in most local government in Malaysia. Therefore, this paper intends to propose applications of space technology to solve part of these problems and issues hence developing smarter city.

3. Smart Facility Management

Facility management is a discipline that involves preventive maintenance and corrective maintenance. In the proposed Smart Facility Management application, both aspects are integrated in the system. This is in line with the best practises of facility management. The preventive aspect of this application is where the supervisors of local government schedule the working plan of each

IOP Conf. Series: Earth and Environmental Science **37** (2016) 012049 doi:10.1088/1755-1315/37/1/012049

contractor or local government worker. For instance, contractors or local government workers are schedule on certain date of every month to mow the lawn in the public parks and clean the drains. This is the routine schedule for every specified period of time, however, there are unforeseen events where this routine is not sufficient. This is where the corrective aspects is critical to complement and complete the maintenance activity. Complaints from users or public will provide useful information for the corrective aspect of facility management. For instance, complaint from the public about overgrown lawn or blockages in drains will provide input for supervisors to arrange or if required reschedule the working plan of contractors or local government workers.
In addition, considering the large precinct and various types of facilities managed by local government, locating the exact location of the problem occurred is critical to improve the efficiency of the service. It is always a problem for local government to navigate to the complained area as most of the time location given by the public is ambiguous. It is time-consuming for the officer or maintenance worker to navigate or search for the point of problem. However, with space technology playing a key role, this navigation and searching time could be massively reduced. This Smart Facility application is linked to a Smart Complaint application where the public could pin point the exact location of the problem and attach a photo taken from their smart phone. Therefore, when the manager who receive the complaint could direct the officer or maintenance worker to the closest location if not exact location of the problem hence improving the efficiency of the service.

Hence, Geo-location data from Global Navigation Satellites and images from Remote Sensing Satellites is indeed very useful to manage facilities and related maintenance problems. Space technology will be integrated with spatial technology as well as information technology to provide the proposed smart city solution. The following sub-sections will explain further about some of the concept of the Smart Facility Application.

3.1. Work flow
Work flow of the system is designed in such a way it caters both preventive maintenance and corrective maintenance as explained in the previous paragraph. There are three main actors in the work flow. The first actor is the manager who is responsible to schedule maintenance works, monitor works and manage complaints. The manager could be the top management of local government or the directors of related departments. The second actor is the one who lodges complaints about problems occurred in the facilities and precincts under local government management. This actor includes local citizen or the public at large as well as the officer in-charge within the local government. The final actor is one who executes maintenance tasks. This is normally the contractors and the local government maintenance workers. The following work flow diagram shows the overall work flow between the three main actors.

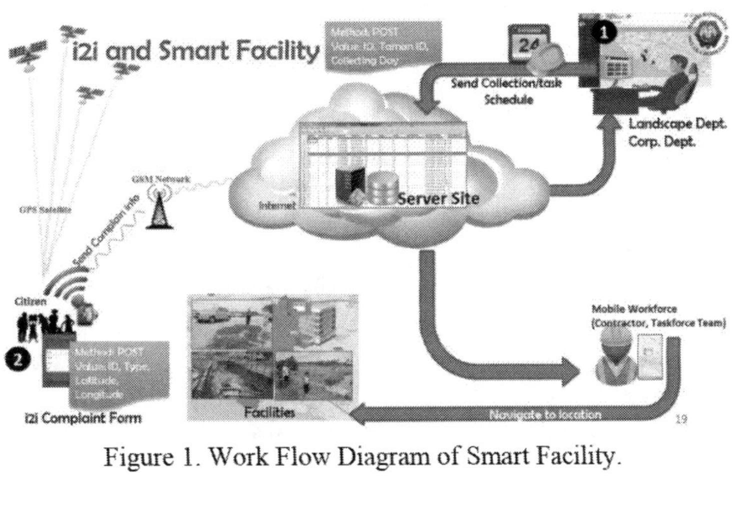

Figure 1. Work Flow Diagram of Smart Facility.

Firstly, the manager will schedule maintenance workers and contractors daily work over a period of time. For instance, they will be scheduled to do routine condition assessment for every facility, mow the lawn and clean the public hall. This is the preventive procedure of the facility maintenance. This task schedule will be uploaded to the cloud so that relevant access the schedule.

Secondly, complaints from the public and local government officer will be uploaded to the cloud through GSM network. A mobile application is developed to be used by the public to lodge complaints. The mobile application is developed in such a way it could read the GPS coordinate given by the navigation satellite. This is a critical information for the manager in order to give clear direction to the contractors and maintenance workers.

Finally, contractors and maintenance workers will access the cloud daily to get information about the updated daily work schedule which comes with GPS coordinate of the task locations. After task is completed, contractors and maintenance workers will report the task completed including photos of the work done.

The Smart Facility application will provide local government useful information instantly and anywhere there is internet connection. In addition, savings could be made as managers of local government could monitor and supervise contractors and maintenance workers remotely. Apart from that, paperless monthly report could be generated for payment procedure and reporting exercise.

3.2. Web application

The Smart Facility application is developed in both web platform and mobile platform. This application is to be accessed by the manager, the public and the maintenance worker. The access level is differentiated depending on the type of user or actor. In addition, the interface will be customised in accordance to user. The following diagram is an example of the screen capture of the web application developed.

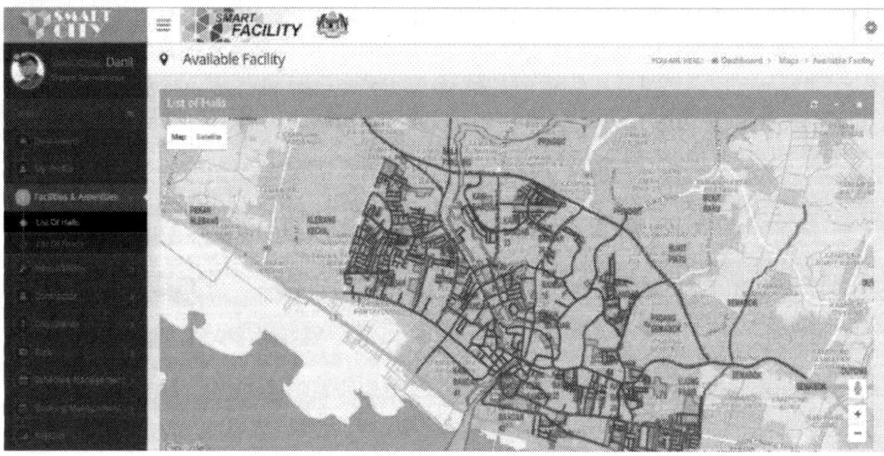

Figure 2. Viewing available maintained facilities.

Since this project is still in progress, the interface and design is not final yet. The project team is working hard to improve it to maka it as user friendly as possible. However, these screen shots provide a good idea how the application will appear finally.

4. Further enhancement
In the era of Internet of Things (IoT), almost everything becomes interconnected and automated. The use of sensors in many application is common nowadays. Sensor could reduce human intervention hence could reduce significant amount operating cost using manpower. Thus, the utilisation of sensor to provide autonomous corrective input instead public complaints to notify related authority is a smarter solution to improve the system. This sensor should come with Global Navigation Satellite System (GNSS) module to provide exact location of the problem arising. These notifications and locations can then be viewed on remote sensing satellite maps to give better graphical overview to the relevant officer. The diagram below briefly explains the system.

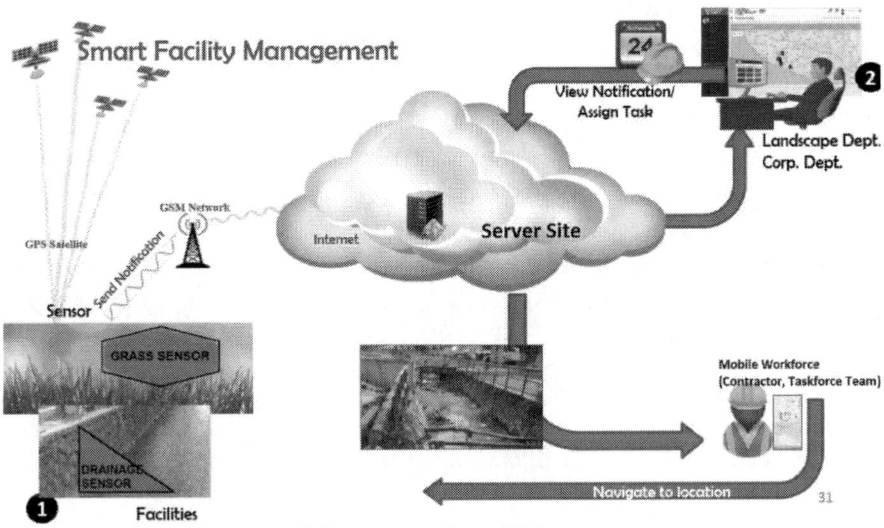

Figure 3. Enhancement by utilising sensors.

5. Challenges
There are many challenges in implementing a smart city solution including enormous precinct managed local government, contractual issues and multiple owners involved. The challenges mentioned are not exhaustive but these are some of the key challenges realized along the development of the project and based on local literature available.

Firstly, according to the current demarcation of local government, the geographical area covered by each local government is massive. The size of area managed by local government became a big challenge because many factors have to be considered in each decision making including huge population, geo-politics, infrastructure facilities, socio-economy and physical and many others [5]. In addition, the bigger the size of area the bigger the cost to implement the solution if sensors were to be included.

Secondly, another challenges foreseen in implementing this solution is contractual issues. Local government has many different contracts where each is unique from one to another. This uniqueness makes it difficult to design a generic application that suits all contracts. In addition, certain local government are reluctant to share the extracts of the contract as it is the property of them. This causes more challenges to understand the scope of the contract made that is indeed needed to design a working and realistic application.

Finally, the difficulty foreseen in this project is there are multiple owners or departments in charge of certain facility. For instance, maintenance of the court in a hall is maintained by one department while another department maintains the lighting system and another department in charge of the drainage system. This causes difficulty in terms of providing maintenace directive as it could involve multiple departments. These challenges require the team project to study thoroughly the local government procedure thus propose the best possible solution.

6. Conclusion

In summary, space technology is indeed a beneficial technology to be exploited to improve local government service delivery. The main improvement would be faster navigation to problem area, user friendly public complaints and efficient reporting of maintenance workers. The utilisation of the space technology enabled application is part of smart city solution proposed by Agensi Angkasa Negara. In addition, further improvement with the integration of sensors could reduce human intervention hence result in higher maintenance efficiency. Furthermore, challenges from procedural and organisational factor provide room for the application to be more dynamics in a sense it can be used in multiple situations.

References

[1] Mineiro M 2011, *Space Technology Export Controls and International Cooperation in Outer Space* (Springer Science & Business)
[2] Yanrong K Whyte J et al 2014 *Comparative Study of Smart Cities in Europe and China*, EU-China Smart and Green City Cooperation, EU-China Policy Dialogues Support Facility II
[3] Angelidou M 2015 Smart City Strategy Cyberjaya (Malaysia) retrieved from http://www.urenio.org/2015/02/09/smart-city-strategy-cyberjaya-malaysia/
[4] Jaffar M 2015 Revitalizing and Revamping Overstretched City System And Infrastructure, IRDA
[5] Ngah K 2009 *The future challenges of local authorities in Malaysia* (Centre for Policy Research and International Studies, CenPRIS, USM) Working paper No. 111/09

A planning support system to optimize approval of private housing development projects

M Q Hussnain[1], K Wakil[1], A Waheed[1] and A Tahir[2]
[1] Department of Urban and Regional Planning, National Institute of Transportation, National University of Sciences and Technology, Islamabad, Pakistan
[2] Institute of Geographical Information Systems, National University of Sciences and Technology, Islamabad, Pakistan

E-mails: plannerqadeer.urp@nit.nust.edu.pk
khydijawakeel.urp@nit.nust.edu.pk
waheedabdul@nit.nust.edu.pk
ali.tahir@igis.nust.edu.pk

Abstract. Out of 182 million population of Pakistan, 38% reside in urban areas having an average growth rate of 1.6%, raising the urban housing demand significantly. Poor state response to fulfil the housing needs has resulted in a mushroom growth of private housing schemes (PHS) over the years. Consequently, only in five major cities of Punjab, there are 383 legal and 150 illegal private housing development projects against 120 public sector housing schemes. A major factor behind the cancerous growth of unapproved PHS is the prolonged and delayed approval process in concerned approval authorities requiring 13 months on average. Currently, manual and paper-based approaches are used for vetting and for granting the permission which is highly subjective and non-transparent. This study aims to design a flexible planning support system (PSS) to optimize the vetting process of PHS projects under any development authority in Pakistan by reducing time and cost required for site and documents investigations. Relying on the review of regulatory documents and interviews with professional planners and land developers, this study describes the structure of a PSS developed using open-source geo-spatial tools such as OpenGeo Suite, PHP, and PostgreSQL. It highlights the development of a Knowledge Module (based on regulatory documents) containing equations related to scheme type, size (area), location, access road, components of layout plan, planning standards and other related approval checks. Furthermore, it presents the architecture of the database module and system data requirements categorized as base datasets (built-in part of PSS) and input datasets (related to the housing project under approval). It is practically demonstrated that developing a customized PSS to optimize PHS approval process in Pakistan is achievable with geospatial technology. With the provision of such a system, the approval process for private housing schemes not only becomes quicker and user-friendly but also transparent.

1. Introduction

Housing is a basic human need. With the average growth rate of 1.6%, population in Pakistan has reached to 185 million out of which 38% reside in urban areas [1]. In 2009, total housing demand was 8 million which will grow to 13 million by 2025 with an annual incremental need of 1.1 million [2]. In the absence of adequate public sector response, there has been a mushroom growth of private housing development projects. As compared to 120 public sector housing schemes (covering 9,462 acres) in five major urban areas of Punjab (largest province of Pakistan), there has been 383 private housing development projects (covering 21,987 acres) [3]. Additionally, over 150 illegal private housing schemes exist within jurisdictions of development authorities. It is estimated for Lahore that 75% of residential land development is done by informal/unplanned sector.

A major factor behind the cancerous growth of unapproved private housing schemes is the delays in the approval process. While the hindrances in the approval process include lack of awareness in developers about planning regulations, confusion about jurisdictions of development authorities (DAs') and lack of coordination among planning agencies. The approval of PHS or land subdivision is governed by zoning regulations and private housing scheme rules of particular development control authorities. According to the officials, the average time taken for approval has been 13 months which affects the whole project in terms of cost and development.

The approval of PHS is governed under a set of rules/regulations of concerned development control authorities or Town Municipal Administration (TMA). Applicable regulations for private housing development project vary from one area to another with the change in number and weightage of factors under consideration. On the other hand, in the case of Islamabad (the capital), they vary for various development zones while in the case of Lahore (capital of Punjab province); they differ on the basis of area of housing project. In most of the cases, any private housing development project has to be in line with the latest development plan, zoning regulations and housing scheme rules for that particular area. Currently, a manual and paper based approach is used for vetting and for granting the permission which is highly subjective and non-transparent [4], [5].

This study explains the structure of a PSS which has been developed using open source set of tools (OpenGeo Suite[1], PHP, and PostgreSQL) to improve the efficiency of PHS approval process by reducing time and cost required for site and documents investigations. The system development follows the ideas advocated by an earlier school of thought [6]–[8]. It is believed that the developed PSS will provide answers to the following questions to speed up the decision process:

- Under which authority's jurisdiction, the housing project location falls?
- Are there any zoning regulations for this location? If yes, in which zone this project falls?
- Based on location and size (area), what are the applicable zoning regulations and housing scheme rules?
- Analyze the housing scheme layout plan and provide the following information:
    - percentage of various land uses
    - Schedule of plots
    - Schedule of street widths
- Whether the division of land uses, plot sizes and street widths are in line with permissible rules? If not, identify problematic areas.

The remainder of the paper is organized as follows; Section 2 outlines the scope of the research project while the related literature is presented in Section 3. The detailed legislations and the local government rules are described in Section 4 while Section 5 explains the data requirements, system flow diagram and the implementation details of the prototype. Finally, conclusions and directions to future work are given in Section 6.

2. Scope

In terms of scope, the knowledge module of PSS will be based on the review of documents from all development authorities of Punjab and Islamabad Capital Territory. It will make it geographically usable for the whole of Punjab (most populated province of Pakistan accommodating 56% of total population) as well as for capital territory. Furthermore, the system has been designed such that user will be able to add/ define the parameters for any other part of the country to use it elsewhere.

[1] http://boundlessgeo.com/solutions/opengeo-suite

The first stage in the formulation of this PSS included the development of knowledge module by extracting rules from regulatory documents. It contained equations related to scheme type, size (area), location, access road, components of layout plan, planning standards and other related approval checks. At the second stage, the architecture of database module was designed where data requirements were enlisted and categorized as base datasets (built in part of PSS) and input datasets (related to the housing project under approval). The *base datasets* included jurisdiction boundaries of DAs, zonal limits, existing regional road network and future roads proposed in approved development plan while the *input datasets* comprised of the boundary of the housing project, plots distribution, road network within proposed project and location of proposed civic facilities. Each of the input datasets possessed a predefined set of attributes. Figure 1 illustrates the generalized flow chart, placement, and interaction between the developed PSS components.

3. Related work

Deciding about the approval of a private housing project is very subjective and relatively complex in nature. Owing to the uniqueness of location, design, physical surroundings, designer vision, relevant regulatory body etc., each approval case requires dealing with a considerable level of uncertainty. Around the globe, planning support systems have been frequently used in such circumstances to aid planners by combining GIS, urban models, and visualization tools [9]–[14]

Anjum [3], Larkham [15] and Yaakup et al. [16] have acknowledged the complexity of procedures undertaken by planning authorities during planning approval process. They highlighted that the process of analysing the suitability of planning applications involves many stages of decision-making which faces delays, inconsistent decision making owing to personal judgment, and lack of transparency.

In 2007, Yaakup et al. proposed a planning support system for local authorities in Malaysia to facilitate development plan implementation, improved service delivery, and enhanced planning approval process [16]. From the review of literature from advanced world, it was concluded that a spatial decision support tailored to the Pakistani context will serve the purpose.

Ballatore et al. (2011) have articulated a comprehensive comparison of open source geospatial technologies for web mapping where they have reflected good marks for PostGIS, OpenLayers and ExtJS [17]. Additionally, in another similar study, an open source web architecture is discussed which is built on PostgreSQL, OpenStreetMap and MapServer[18]. We use above mentioned open source geospatial technologies which are widely used and tested in various applications.

4. Understanding private housing schemes approval process

4.1. Review of relevant regulation

To develop the desired PSS, it was important to review relevant rules and regulations from an operational perspective. Following documents have been reviewed to extract rules governing approval of private housing schemes.

- Punjab Private Housing Schemes and Land Subdivision Rules 2010
- LDA Private Housing Schemes Rules 2014
- Islamabad Capital Territory Zoning Regulations 2005
- Capital Development Authority (CDA) Housing Scheme Regulation for Zone, 2, 4 and 5

Table 1 provides a summary of key attribute areas and corresponding rules contributing to the approval of private housing scheme in above-mentioned areas [19]–[21].

Table 1. Rules governing approval of private housing schemes in Pakistan.

Attribute	Rule Definition
Type	• Land subdivision (up to 100 kanal) • Farm housing scheme • Private housing scheme (more than 100 kanal)

Size
For LDA; Area <40 kanal
40 – 99 kanal
100- 300 kanals
300- 500 kanals
>500 kanals
For RDA; Min area is 100 kanal

Access Road
Have frontage on Access Road, RDA needs min 60 ft

Location
Not in Flood zone
Not notified for acquisition by government
Fall in area marked as residential in approved development plan

Layout Plan
Compulsory components:
Scale of scheme plan is 1:1200 to 1:2400
Scale of location plan is 1:4800
Size and dimensions of each plot
Plot number
Corner plots should be chamfered; up to 1 kanal by 5X5 and >1 kanal by 10X10
Right of way of roads
Dimensions of scheme boundary
Boundary of the mouzas or a revenue estate
Number and boundary of each Khasra in a mouza
Land use breakup in area and percentage for

- Residential
- Commercial (Com)
- Open Spaces (OS)
- Roads
- Graveyard (GY)
- Public buildings (PB)
- Other land uses

Schedule of plots with area and size
Area of existing graveyard, if any
Location of pumping station, overhead water tank, and tube well if any
Aggregate of various categories of plots to be mortgaged

Planning standards

Land use	**Land subdivision**		**Farm housing scheme**	**Private housing scheme**		
	< 40k	*40-99 k*		*<300 k*	*300-500 k*	*>500 k*
OS	-	Min 5%	Min5%	Min7%	Min 7%	>=7%
GY	-	-	Min2%	Min2%	Min 2%	Min 2%
Com	-	Max 1%	Max5%	Max5%	Max5%	Max10%
PB	-	Min2%	Min2%	2-10%	3-10%	4-10%
Plot size	2k max	2k max	4k min	2k max	2k max	2k max
Street	Min 30ft	Min 30ft	Min30ft	Min30ft	Min30ft	Min30ft
Solid waste mgmt.	-	Min5m	Min10m for every 500 kanal	Min10m for every 500 kanal	Min10m for every 500 kanal	Min10m for every 500 kanal
Major Road	Min40ft	Min40ft	Min40ft up to 300k and 60ft above 300k	Min 100ft	Min 100ft	Min 150ft
Society Office	-	Min 5m	Min 10m	Min 10m	Min 10m	Min 1k

Other Checks
If the project site intersects any road proposed in approved development plan, then accommodate in plan
If the project site intersects a high tension electricity line, provide green strip
10% of the plots shall be allocated for 3-5 marla
10% of the remaining plots shall be allocated for multistory flats or apartments

4.2. Semi-structured interviews with professionals and developers
Considering the importance of stakeholders in the approval process of a private housing scheme, views of both professional planners and developers have been noted by means of semi-structured interviews. This helped in understanding current practices and ground realities, and identifying loopholes and time-consuming stages. It has been ensured that their comments and suggestions are also incorporated in the design of the system architecture.

5. PSS system description
In terms of data requirements, the system requires seven different vector layers including jurisdiction boundaries of DAs, land-use zones, regional road network, the boundary of the project site, plot subdivisions, street network within proposed project and location of proposed civic facilities. Table 2 provides detailed characteristics of each of the data layer containing layer type (polygon, polyline or point) and associated attributes. Additionally, comments have been added for further explanation.

Table 2. Data requirements in the proposed PSS.

	Dataset	Type	Attribute	Comments
1	Jurisdiction boundaries of Development Authorities	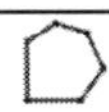	Name of authority	Polygonal demarcation of jurisdictions of development control authorities
2	Zones		Zone number/ names	Zonal polygons obtained from land-use zoning map
3	Regional road network		Road class, width, name	Network of Existing roads + road proposed in approved future development plan of the area
4	Location of housing project		Size	Shapefile containing boundary of the project site
5	Plot distribution		Plot size, land use, height, address	Shapefile containing land subdivided into plots such that each polygon of plot contains its attributes
6	Road network within proposed project		Width, Type (major road, internal road etc.), name	Centre line road network of the proposed project
7	Location of proposed civic facilities		Use, Name	Shapefile containing point location of proposed civic facilities including schools, health facilities, parks, commercial center etc.

Overall workflow and interaction of components in proposed PSS is given in Figure 1. The system uses four data layers as base datasets. These layers (control area of DA, zoning plan and both existing and proposed regional road network) have been made built-in part of the system. Four other layers are needed as input datasets and these are related to the PHS under consideration. Rules derived from regulations have been made part of knowledge module in the form of built-in equations.

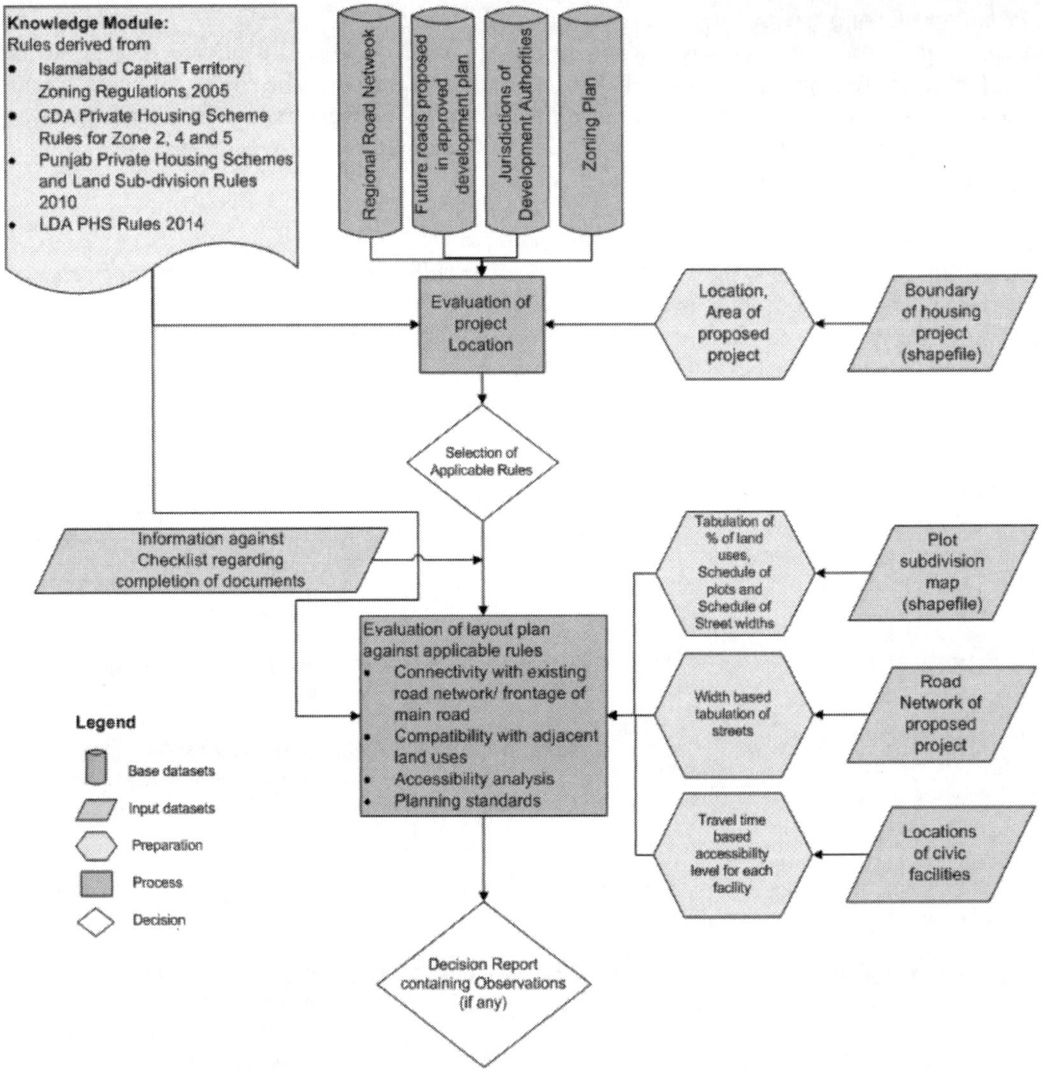

Figure 1. Generalized workflow and interaction between PSS components.

In terms of user interface, a combination of OpenLayers, GeoExt and ExtJS has been used. ExJS is a JavaScript framework for building cross-platform web applications leveraging HTML5 features while GeoExt provides a bridge between OpenLayers and ExtJS. Figure 2 reflects the arrangement of various tools, layer tree, map window and legend panel of the system which is currently displaying the control area of DA, alignment of proposed roads and location of the project site. Figure 3 illustrates the land-use breakup analysis of a loaded PHS layout map while Figure 4 shows the information on key decision indicators for the PHS project under consideration based on its location.

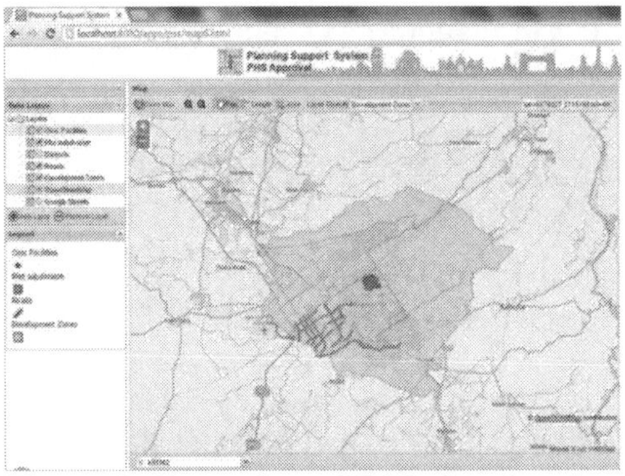

Figure 2. Screenshot of PSS showing the jurisdiction of development control authority and future planned roads.

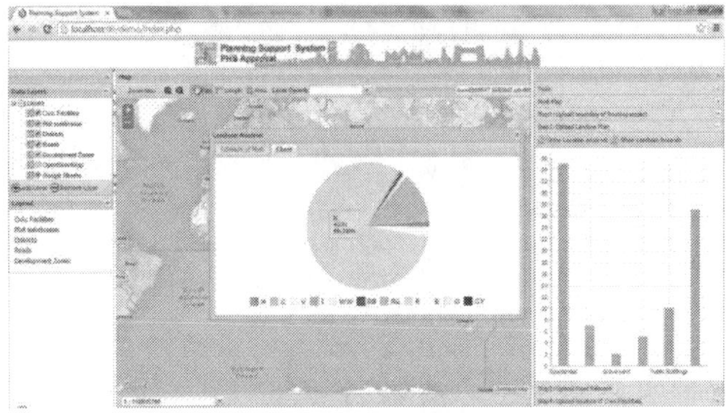

Figure 3. Snapshot reflecting land-use breakup of the proposed PHS derived from attributes of loaded Shapefile.

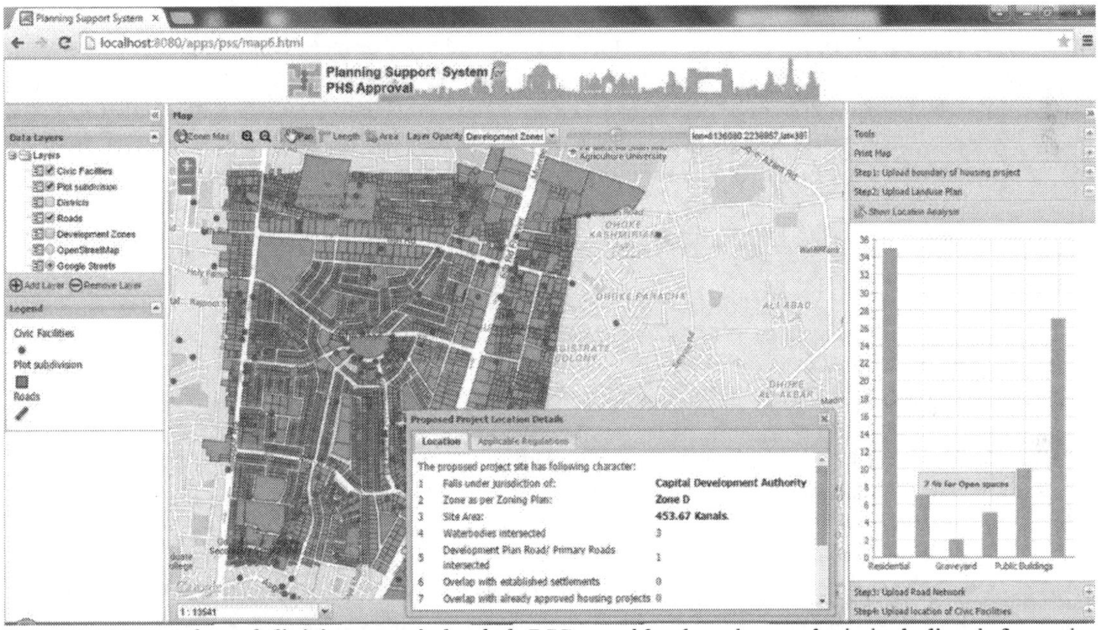

Figure 4. Once plot subdivision map is loaded, PSS provides location analysis including information about associated development authority, zone, site area, intersection with existing water bodies, overlap with already approved PHS etc.

6. Conclusions and future work

This study proves that developing a customized planning support system to achieve time and cost efficiency in private housing schemes approval process in Pakistan is very much achievable with open source tools. In doing so, the approval process for private housing schemes not only becomes quicker and user-friendly but also becomes transparent. However, availability of adequate human resources and existing institutional setup are few of the biggest challenge in the implementation of such PSS in concerned development authorities and TMA's. Wider utilization of the system is recommended, however, awareness and capacity building is direly needed to remove the reluctance of professional planners working in development authorities in Pakistan.

Since the PHS regulations are subject to changes based on area and time, the knowledge module of the PSS has to be updated. Further work is needed to develop and configure graphical user interfaces to add/update equations in knowledge module. Similarly, work can be done to enhance reporting module so that advisory reports containing suggestions for improvements in PHS layout can be produced.

References

[1] The World Bank, "World Development Indicators," 2014. [Online]. Available: http://data.worldbank.org/country/pakistan. [Accessed: 11-Sep-2015].

[2] S. S. Hasnain, "Housing Finance Review 2005-2011," 2014.

[3] G. A. Anjum, "Assessment of urban land development and management practices in five cities of Punjab," Lahore, 2008.

[4] M. Q. Hussnain, "A Study on the Utilization of Information and Communication Technology in Preparation and Implementation of Development Plans: a Case Study of Rawalpindi," University of Engineering and Technology, Lahore, 2013.

[5] M. Q. Hussnain, G. A. Anjum, K. Wakil, and P. H. T. D. Tharanga, "Improving Efficiency in Data Collection for Urban Development Plans through Information and Communication Technology," in *International Conference on Town Planning and Urban Management (ICTPUM)*, 2014.

[6] B. Harris and M. Batty, "Locational Models, Geographic Information and Planning Support Systems," *J. Plan. Educ. Res.*, vol. 12, no. 3, pp. 184–198, Apr. 1993.

[7] R. E. Klosterman, "Planning Support Systems: A New Perspective on Computer-Aided Planning," *J. Plan. Educ. Res.*, vol. 17, no. 1, pp. 45–54, Sep. 1997.

[8] S. Geertman, "Potentials for planning support: a planning-conceptual approach," *Environ. Plan. B Plan. Des.*, vol. 33, no. 6, pp. 863–880, 2006.

[9] H. Timmermans, Ed., *Decision Support Systems in Urban Planning*. London: E & FN SPON, 1997.

[10] D. J. . Saarloos, T. A. Arentze, A. W. J. Borgers, and H. J. P. Timmermans, "Towards a Local Planning Support System, Introducing the MASQUE Framework," *Design & Decision Support Systems in Urban Planning*. Avegoor, The Netherlands, pp. 210–224, 2002.

[11] S. Geertman and J. Stillwell, "Planning support systems: An inventory of current practice," *Comput. Environ. Urban Syst.*, vol. 28, no. 4, pp. 291–310, 2004.

[12] R. K. Brail and R. E. Klosterman, *Planning Support Systems for Cities and Regions*. Cambridge: Lincoln Institute of Land Policy, 2008.

[13] S. Geertman, F. Toppen, and J. Stillwell, Eds., *Planning Support Systems for Sustainable Urban Development*. Springer, 2013.

[14] S. Geertman, J. J. Ferreira, R. Goodspeed, and J. Stillwell, Eds., *Planning Support Systems and Smart Cities*. Springer, 2015.

[15] P. J. Larkham, "The concept of delay in development control," *Plan. Outlook*, vol. 33, no. 2, pp. 101–107, 1990.

[16] A. Yaakup, S. Sulaiman, F. Johar, and M. Che Ngah, "Computerised Planning Approval System For A Local Authority In Malaysia," *10th Int. Conf. Comput. Urban Plan. Urban Manag.*, no. September, p. 15, 2007.

[17] A. Ballatore, A. Tahir, G. McArdle, and M. Bertolotto, "A comparison of open source geospatial technologies for web mapping," *Int. J. Web Eng. Technol.*, vol. 6, no. 4, pp. 354–374, 2011.

[18] G. McArdle, A. Ballatore, A. Tahir, and M. Bertolotto, "An open-source web architecture for adaptive location-based services," in *14th International Symposium on Spatial Data Handling (SDH), at the Joint International Conference on Theory, Data Handling and Modelling in GeoSpatial Information Science Hong Kong, 26-28 May, 2010*, 2010.

[19] Lahore Development Authority, "LDA Private Housing Scheme Rules." Housing, Urban Development & Public Health Engineering Department, Lahore, 2014.

[20] Punjab Local Government, "Punjab Private Housing Scheme and Land Sub-division Rules 2010." The Punjab Weekly Gazatte, 2010.

[21] Capital Development Authority, "Islamabad Capital Territory Zoning Regulation 2005." Islamabad, pp. 1–10, 2005.

Development of online Spatial Multi-Criteria Decision Analyst application: Case study on determining area suitability for school location in Surabaya, Indonesia

Daniel HaryPrasetyo[1,2], JamilahMuhamad[1], RosmadiFauzi[1]

[1]Department of Geography
University of Malaya,
50603 Kuala Lumpur, MALAYSIA

[2]Information Technology Department
University of Surabaya
Raya Kalirungkut Surabaya, Indonesia

Abstract. A decision sometimes needs to consider many aspects and to be judged by many people. Presenting a case of finding a suitable location for a new school this research proposes seven factors which its emphasis will be differently sorted by various people perspective. Each factor comes in a form of a multi-polygon layer that valued from 0 to 9, representing the suitability value of the certain aspect in the whole city area. The Public and some expert will judge by using the pair-wise comparison of those aspects. This research will provide web GIS application that will use by the public and the expert in this justification process and analyzing the result.

1. Introduction

Surabaya is an expanded old city and grows up reaching its form like today. Starting in the colonial era, Surabaya citizens lived in the district area. Surabaya populated area has expanded to the south, east, and west area. This expanded population is not accompanied by new public school development, especially the high school level. Figure 1 shows the distribution of school in secondary and high-level school respectively, overlaying the distribution of population areas. By a simple interpretation, it can be seen that the city has a lack of public high schools. It also shows that there exists a scarcity of educational institutions at the higher levels, making it common for students to travel much longer distances as they move to higher levels of study [6].

Surabaya city government need to build some new schools in an optimal location to increase the service and accessibility of education of the town. They have proposed some analysis factor that corresponded with school location. Those factors are Administration factor, Accessibility factor, Population factor, Public transportation factor, Land use factor, Student flow factor, and Public Preferences factor. Figure 2 shows these factors with a graduate colour presentation from green that indicates the most suitable location to red that indicates the least one.

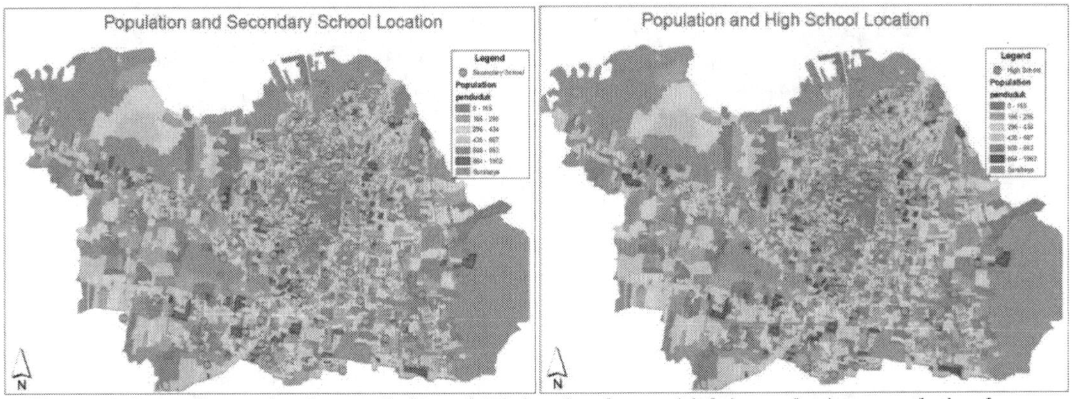

Figure 1. Secondary school map (left) and High school map (right) overlaying population layer.

Administration factor is a factor that represents the number of school in the sub-districts and its neighbourhood. The government want the schools spread over the sub-district equally. Thus, the sub-district that has no school and it surrounding sub-district also have no school will get high priority. Sub-district that already has school and it's surrounding has school will get low priority. This priority coloured from red to green to represent low to high priority.
Accessibility factor is a factor that calculated from school location and street network using Service Area analysis. An area that already has good accessibility to school is an area with low value because it does not need more school around. In contrast, an area with low accessibility will have high priority.
Population factor is a factor that calculated based on population number. High priority location is location with a high-density population.
Transportation factor is a factor that calculated by the number of public transportation in the area. An area with much public transportation routes founded will have high priority in placing new school.
Land use factor is a factor that based on the usability of the area. This factor wants to make the school area are close to vegetation and public facility area, and not in and in a distance from military, graveyard, and industrial zone. Thus, military, graveyard, and industrial zone, and also areas surround it was marked as low priority.
Student flow factor is a factor that calculated from the density of student passing through the area in their journey from their home to their school. An area that was passed though by the crowded student will have low priority for the new school location. A new school in a quiet area is expected to spread the crowd. Therefore, the quiet area marked as high priority.

Public preferences factor is a factor that excavates citizen preferences in choosing school location. A new school location will be better if it lies in the location that loved by the public. A survey for collecting public preferences has been conducted with a Web GIS application at http://participatorygis.net. This survey got public preferences by retrieving this information: a. Are they prefer the school located near their home, what is accepted distance, and how strong this preferences. b. Are they prefer the school located near their workplace, what is accepted distance, and how strong this preferences. c. Are they prefer the school located in some place along from their home to their workplace, and how strong this preferences.

This work wants to combine those seven factors with optimal weight setting. The optimal weight will be obtained from Multi-Criteria Decision Analyst method.

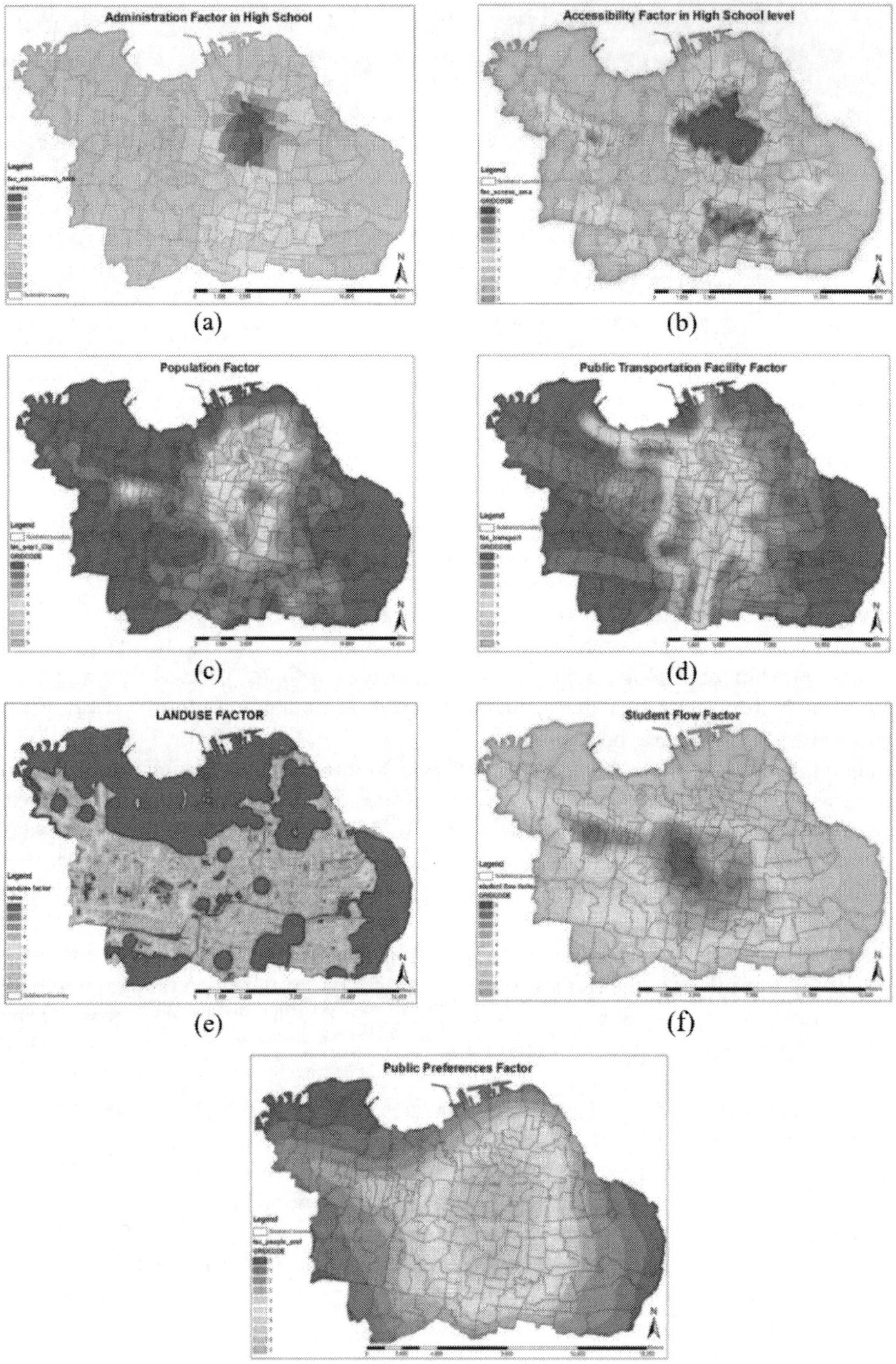

Figure 2. The seven factors. (a) Administration factor, (b) Accessibility factor, (c) Population factor, (d) Transportation factor, (e) Land Use factor, (f) Student flow factor, (g) Public Preferences factor.

2. GIS-MCDA

Since the 1990s, the combination of MCDA and GIS has been promoted for use in solving spatial problems in urban planning, forest conservation and site determination [1]. In land-use suitability assessment, for example, alternative options and preferences for land use can be determined using MCDA, through the identification of desirable objectives, related attributes, and criteria, such as social, economic, environmental and ecological factors. Some successful work of combining GIS and MCDA can be seen in the work of Jessy Paquette and John Lowry about assessing flood hazard [7], BachtiarFezizadeh and Thomas Blaschke about landslide analyzing [8], Gbanie and colleagues about modelling landfill location [9], and Hamadouche and colleagues about finding best preservation area [10].

Quantitative analysis is necessary for MCDA, including scoring, ranking and weighting. A single conclusion is produced after the quantitative analysis. Phua and Minowa work explain that the main steps in MCDA for land-use suitability assessments include defining the objectives and the corresponding judgment criteria, analyzing the criteria, assessing the standards, quantitatively analyzing the criteria for the assessment units and aggregating the judgments. GIS-MCDA is a process that transforms and combines geographical data and value judgments (the decision-maker's preferences) to obtain information for the decision making. Different GIS-MCDA techniques exist depending on the operations needed to obtain the final valuation of the solution alternatives [2]. Vega and colleagues mention that in some cases, these transformations may be simple arithmetic operations as in the weighted linear summation methods, but in others, as for example for the ideal/reference point or the outranking methods, more complex operations is needed.

Great work of Chakhar and Martel make classification of GIS and MCDA integration. There are three types of GIS and MCDA integration. Those are Indirect, Build In, and Full integration [3]. In an indirect GIS-MCDA integration mode, integration of GIS software and stand-alone MCDA software is made possible by the use of an intermediate system. The Intermediate system permits to reformulate and restructure the data obtained from the overlay analysis which is performed through the GIS into a form that is convenient to the MCDA software. The other parameters required for the analysis are introduced directly via the MCDA software interface. The results of the analysis (made in the MCDA part) may be visualized by using the presentation capabilities of the MCDA package, or feedback to the GIS part, via the intermediate system, for display and for further manipulation. A built-in GIS-MCDA integration mode In this mode, a particular MCDA model is directly added to the GIS software. The MCDA model develops its own database independently but still integrated. The use of the interface of the GIS part alone increases the interactivity of the system. Here, there is no need for an intermediate system because the MCDA model is reformulated in such a way that the exchange of data and analysis results between the two parts is performed directly. Full GIS-MCDA integration mode, the third mode, yields itself to a fully GIS-MCDA integrated system that has a unique interface and a unique database. Here, the MCDA model is stimulated directly from the GIS with standard spatial operation tools. The GIS database supports both the geographical and descriptive data and support the parameters required for the multi-criteria evaluation techniques. The single theme of graphical interface makes the system fully integrated.

There is some method experimented to get a better decision. However, AHP is the popular one. AHP is becoming one of the most common methods of coupling decision-making processes and geospatial analysis [4]. AHP is the simplest decision-making method to be prototyped. It means the formula is easy to implement. However, it provides accurate results in geospatial analysis. AHP has become popular due to its straightforward implementation and positive results. These characteristics served as the basis for this research to develop a simple but functional geospatial AHP-based decision-making framework which serves as a mechanism that ranks inputs and produces output maps input ranking–output maps mechanism facilitating decision making at different levels. AHP method, cited from work by Karnatak and colleagues [5] about finding best site for biodiversity conservation, in any decision making process the decision maker has to identify the goal or ultimate objective of decision problem.

In AHP the decision maker initiates to identify the objective and decision problem. The process continues to the choice phase. The main objective of this phase is to select one particular alternative from a set of known options. Therefore, the first step is to list all the alternatives. It continues to define the criteria that will be used to judge the alternatives and develop a decision hierarchy. The hierarchy represents the structure of the decision problem and forms the basis of the comparisons. The next step in the AHP process is pair-wise comparison alternatives, i.e., for each criterion; the decision maker compares all the alternatives pair-wise. The decision maker can make numerical or verbal judgments. In the verbal mode, statements are selected varying from 'equally prefer' to 'greatly prefer'. In the numerical method, the decision maker selects a score on a scale of one to nine.

3. Methodology

There is some option for implementing Web GIS application. Some of them are by spatial services server like Mapserver or Geoserver. Both can directly connect with map files and send data to the client request and show the map in client's web browser. But they need a particular handle in the server. We must own the server by ourselves or we have to pay more hosting service that includes Mapserver or Geoserver, a service that not exist in the common hosting provider. Another way that can be hosted in common hosting service is to store the map in a common database like MySQL or Postgres and using Javascript based tool for showing the map in client's browser. This work is using this second way. Factors Layers in previous section comes in a form of a shapefile with detail resolution. In average it has size about 600 Kbytes per factor. It is big enough for taking time in the process of showing the map in the browser. In the MCDA process, the map will be simplified to 20% of the real size to supply fast respond to the user. Thus, what will show in the MCDA web application is not the actual map, but still have a good representation of the real condition. By referencing classification of Chakhar and Martel, this work becomes Indirect GIS-MCDA since the MCDA process does not connect with the original spatial data, but it come with a facility to show the simplifying map for better user understanding.

There are three big steps in this work: Data Preparation, Judgement process, and Result Calculation. Figure 3 showsthese steps.

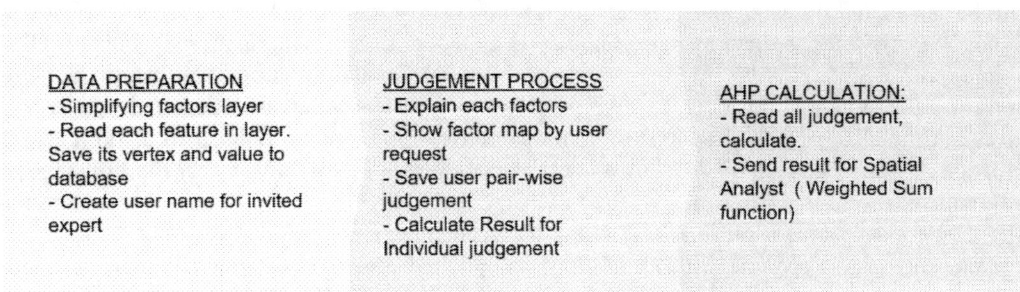

Figure 3. Processes in the GIS-MCDA.

3.1 Data preparation

This step use three tools: ArcGIS 9.3, Visual Basic 6.0 and MySQL. ArGISis used to simplifying the polygon factors. It uses 'simplify polygon' function from Generalization tool package. The simplified polygon then read by a custom application made from Visual Basic 6.0 equipped with ShapefileReader library. It read polygon by polygon in each shapefile and put the vertex data to MySQL database. The Database structure is shown in Figure 4.

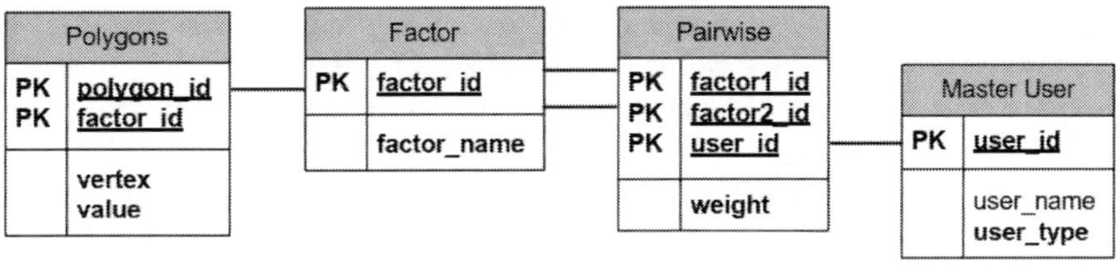

Figure 4. Database Structure.

The database has four tables. Those are table Factor, table Polygons, table Master_user, and table Pairwise. Table Factor stores list of factor layer. Table Polygons store a collection of the polygon that constructs factor layer. Table Master User saves user data that will give judgment in MCDA process. Table pairwise is saving the pairwise judging from the user. There are two types of user, Public and Expert. The public user data will generate automatically when they start using the MCDA application. In contrast, user data for the expert user need to be prepared. The User Id and User Name field have to be filled first; then User ID will be informed to the expert before they are using the application. They have to use this ID in the login page. At this step, data preparation process is done.

3.2 Judgement process

The judgement process uses web-based application. The application builds using PHP language, above MySQL database, andOpenLayers library for presenting the map. The application has three pages. The first page is a login page. In this page, the user has to choose whether they are will login as Public or as Expert. If the user selects an expert door, they must type their gifted ID. The second page is an introduction page. It explains all factors that will be judged. The user also can see a presentation of each Factor on the map, as an overlay layer in the Google map. Figure 5 shows this second page.

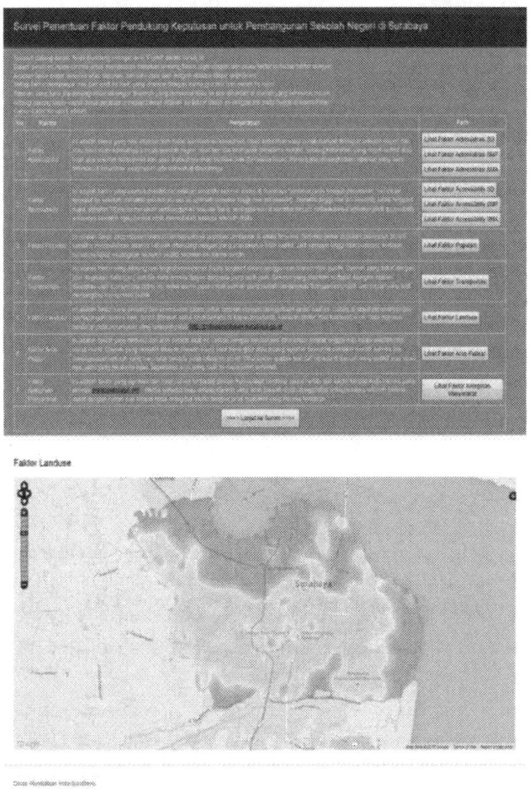

Figure 5. Introduction page.

The third page is the pairwise judgement page. The user has to decide each factor whether it have influence in choosing the location as absolute strong, very strong , strong, slightly strong , or the same strong with another factor. There are 21 pairs of seven factors. It means user have to make judgement 21 times with different factor pair. After the user makes decision for all pair combination, the user can take a look the result for their judgement. The user input will be calculated in AHP method. This step is the first calculation in this work. The second calculation will be used in the combining process of all respondent's judgement. The calculation method will explain in the next section. This calculation will generate the weight for each factor. These weights will be used to define the transparency of factor layers while it is drawn in the browser. Each layer, using the polygons value, will display with graduate colour from red to yellow to green colour. All factor layers will overlay with different transparency value. This step will make a city area that coloured by seven different layers. Green area indicates that area have dominant green coloured polygons from all layers while red area has dominant green coloured polygons. Figure 6 shows this third page.

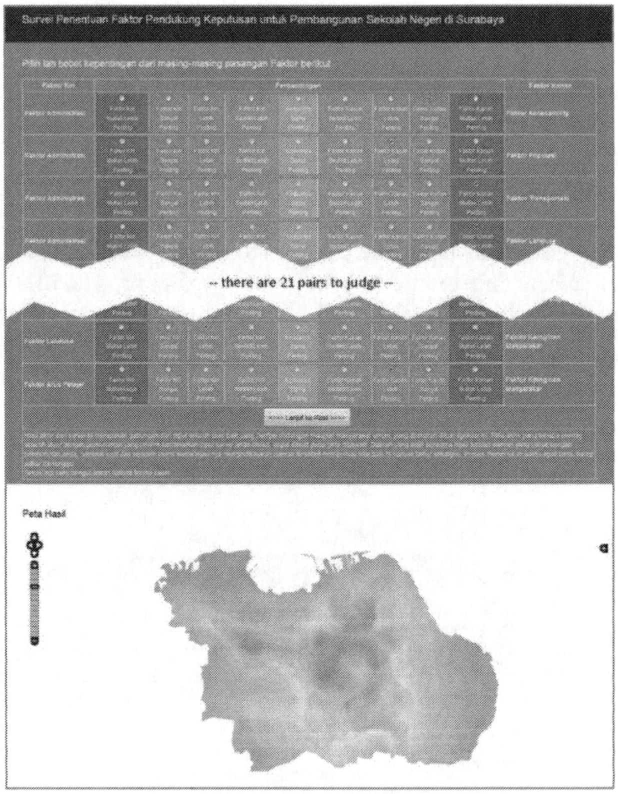

Figure 6. Pairwise judgement page.

3.3 AHP calculation

AHP Calculation step will do after all respondents done the pairwise judgement. Not all result will be brought to the final calculation. Only the credible result that passed the consistency ratio will be used in the next stage. All valid respondent result then will be combined with geometric mean formula. This process uses visual basic 6.0 application for read from MySQL database and calculates the result and display the calculation and result in Microsoft Excel. Below is the algorithm for this process, explained along with Figure 7 that show the example of a respondent calculation and Figure 8 that shows the equation of consistency ratio.

- *Looping for all respondent judgement*
  - *For every judgment, make a comparison matrix (see Figure 7 range A1:H8)*
  - *Sum each column. (see Figure 7 range A10:H10)*

IOP Conf. Series: Earth and Environmental Science **37** (2016) 012051 doi:10.1088/1755-1315/37/1/012051

- *Make second matrix by Normalizing relative weight by divide each cell in the first matrix with the sum of column (see Figure 7 rangeA12:H19)*
- *Calculate Average of every value in a row in the second matrix (see Figure 7 range J13:J19). This will become weight result if Consistency* Ratio (CR) is accepted.
- *Multiply above average with the corresponded Sum (see Figure 7 range H13:H19), Sum the result to get λmax (see Figure 7 cell H20)*
- *Calculate Consistency Index CI using λmax , then calculate Consistency Ratio (CR) using CI and Random Consistency Index (RI), which for n=7 , RI=1.32.If CR < 0.2 the result is accepted and use for calculate all weighted mean.*

- *Combine all accepted respondent result by using Geometric Mean formula.*

Geometric Mean formula will produce the final weight of each layer. Next, the last step, is bringing the final weight of each factor in the ArcGIS 9.3. The real factor layers, not the simplified ones, will use this weight in the ArcGIS Weighted Sum function.

	A	B	C	D	E	F	G	H	I	J	K
1		adm	acc	popul	trans	landuse	flow	pref			
2	adm	1	0.2	0.33	5	3	5	7			
3	acc	5	1	1	3	5	5	7			
4	popul	3	1	1	7	7	7	9			
5	trans	0.2	0.33	0.143	1	1	3	5			
6	landuse	0.33	0.2	0.143	1	1	0.33	3			
7	flow	0.2	0.2	0.143	0.33	3	1	3			
8	pref	0.143	0.143	0.111	0.2	0.33	0.33	1			
9											
10	SUM	9.873	3.073	2.87	17.53	20.33	21.66	35			
11											
12		adm	acc	popul	trans	landuse	flow	pref		W	W*SUM
13	adm	0.101286	0.065083	0.114983	0.285225	0.147565	0.23084	0.2		**0.163569**	1.614916
14	acc	0.506432	0.325415	0.348432	0.171135	0.245942	0.23084	0.2		**0.289742**	0.890378
15	popul	0.303859	0.325415	0.348432	0.399315	0.344319	0.323176	0.257143		**0.328808**	0.94368
16	trans	0.020257	0.107387	0.049826	0.057045	0.049188	0.138504	0.142857		**0.080724**	1.415084
17	landuse	0.033424	0.065083	0.049826	0.057045	0.049188	0.015235	0.085714		**0.050788**	1.032521
18	flow	0.020257	0.065083	0.049826	0.018825	0.147565	0.046168	0.085714		**0.06192**	1.341182
19	pref	0.014484	0.046534	0.038676	0.011409	0.016232	0.015235	0.028571		**0.024449**	0.855712
20										λ Max=	**8.093473**
21										CI=	**0.182246**
22								RI(7)=1.32		CR=	**0.138065**

Figure 7. Example of the single calculation.

$$CR = \frac{CI}{RI} \qquad CI = \frac{\lambda_{max} - n}{n-1}$$

Random Consistency Index (RI)

n	1	2	3	4	5	6	7	8	9	10
RI	0	0	0.58	0.9	1.12	1.24	1.32	1.41	1.45	1.49

Figure 8. Formula for calculate Consistency Ratio (CR).

4. Result and discussion

The web-based application build in this work is still active and still collect Public opinion at http://ristek.net until now. The Experts comes from the government, scientist, and professional. This research subject is having immediate concern within two departments in Surabaya city, the Educational Department and Planning and Development Department. According to that, the first type of expert respondents consist of the head of those two departments, and it's subordinate that concern with this subject. Some Scientist from Surabaya Tenth November Institute of Technology from Urban Development Department also invited to join as MCDA respondent. Some professional who support Surabaya city government and have a concern in Urban Planning work are also encouraged. They fill in by themselves in the online application independently. The expert respondent result shows that they prefer school location mainly have to be placed in the area with having high population. The next factor with high priority after population factor is the administrative factor, accessibility factor, and transportation factor. Figure 9 shows the final weight of expert judgement result and the final reliability area map.

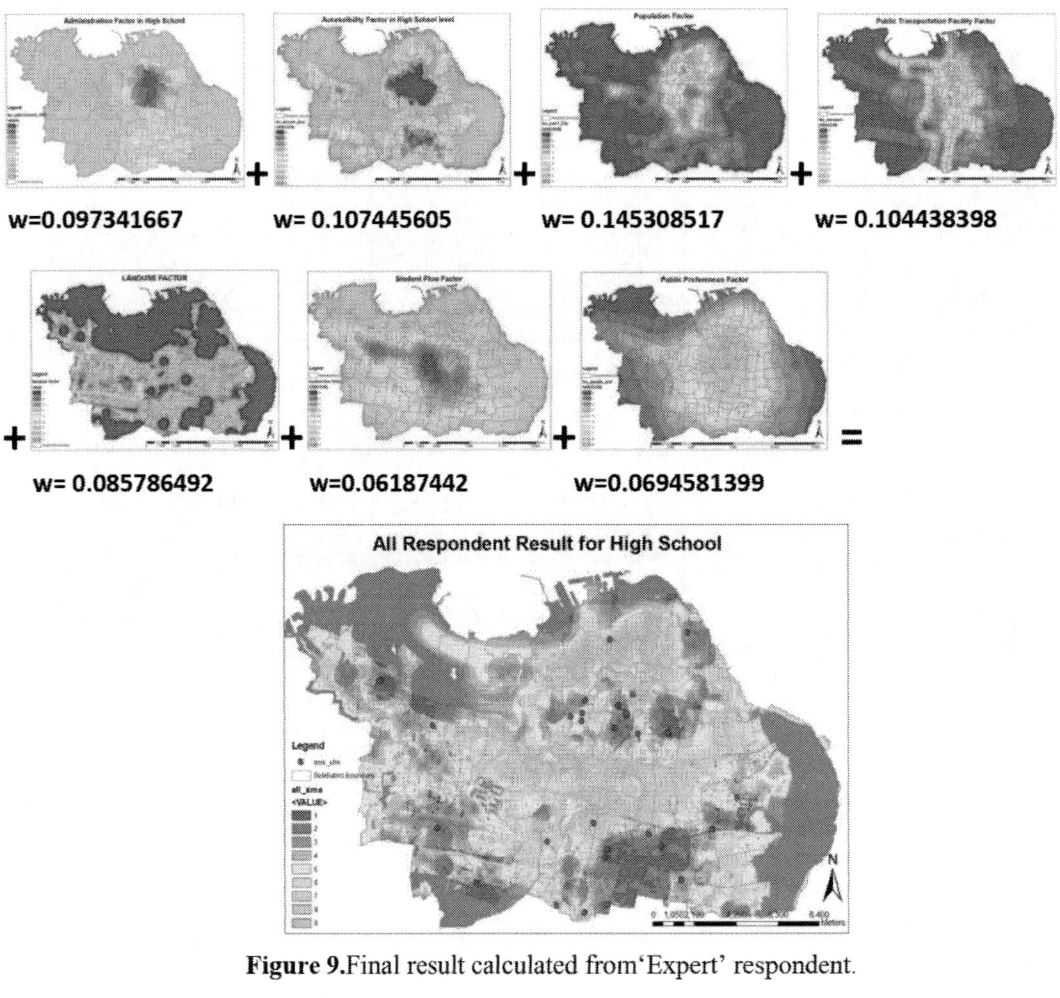

Figure 9. Final result calculated from'Expert' respondent.

From the final weight discovered by AHP calculation, it can be seen that Expert are choosing population factor as the main factor. The weight have significant gap from the second position. In the next position there is the second group of factors with slight different value; those are transportation factor, accessibility factor, and administration factor. The rest are factors that not considered as primary factors, those are land use factor, student flow factor, and public preferences factor.

5. Conclusion

This outcome then informs back to the Expert respondent to get their opinion about the research. All respondent can accept the result and appreciate the application. The web-based Application as a product of this research has been proven can be used to find a suitable location for a new school by using multi-criteria decision analyst. Some other research that has to be decided by multiple factors and by many respondents which have to work remotely can duplicate the model proposed in this research for their tool.

References

[1] Mui-How P and Mitsuhiro M. 2005. A GIS-based multi-criteria decision making approach to forest conservation planning at a landscape scale: a case study in the Kinabalu Area, Sabah, Malaysia. *Landscape and Urban Planning*, **71**, 207–222.
[2] Rafael SV, Dolores R, Pablo D, and Manuel R. 2011. A multi-criteria GIS based procedure to solve a network competitive location problem. *Applied Geography*, **31**, 282–291.
[3] Chakhar S and Martel JM. 2003. Enhancing Geographical Information Systems Capabilities with Multi-Criteria Evaluation Functions. *Journal of Geographic Information and Decision Analysis*, **7(2)**, 47 – 71.
[4] Stitch B, Holland JH., Noberga RA, and O'Hara C . 2011. Using multi-criteria decision making to highlight stakeholders' values in the corridor planning process. *Journal of Transport and Land Use*, **4(3)**, 105–118.
[5] Karnatak HC, Saran S, Bhatia K, and Roy PS. 2007. Multicriteria Spatial Decision Analysis in Web GIS Environment. *Geoinformatica*, **11**, 407–429.
[6] Prasetyo D., Muhamad J, and Fauzi R. 2011. Supporting needy student in transportation: a population based school bus routing in spatial environment. *International Conference on Social Science and Humanity. IACSIT Press, Singapore*.
[7] Paquet J and Lowry J. 2013 Flood hazard modelling and risk assessment in the Nadi River Basin, Fiji, using GIS and MCDA.*The South Pacific Journal of Natural and Applied Sciences*, **30(1)**, 33 – 43.
[8] Feizizadeh B and Blaschke T. 2013 GIS-multicriteria decision analysis for landslide susceptibility mapping: comparing three methods for the Urmia lake basin, Iran.*Natural Hazards*,**65(3)**, 2105-2128
[9] Gbanie, S. P., Tengbe, P. B., Momoh, J. S., Medo, J., &Kabba, V. T. S. 2013. Modelling landfill location using geographic information systems (GIS) and multi-criteria decision analysis (MCDA): case study Bo, Southern Sierra Leone. *Applied Geography*, **36**, 3-12.
[10] Hamadouche, M. A., Mederbal, K., Kouri, L., Regagba, Z., Fekir, Y., &Anteur, D. 2014. GIS-based multicriteria analysis: an approach to select priority areas for preservation in the Ahaggar National Park, Algeria. *Arabian Journal of Geosciences*, **7(2)**, 419-434.

IOP Conf. Series: Earth and Environmental Science **37** (2016) 012052 doi:10.1088/1755-1315/37/1/012052

Rethinking urban space in cities – A study of parks in Hyderabad, India

B. Shrinagesh
Associate Professor, Department of Geography, Osmania University, Hyderabad, India

Kalpana Markandey
Professor, Department of Geography, Osmania University, Hyderabad, India

E-mail: geosri74@gmail.com

Abstract. Urban areas being economically diversified attract large streams of migrants making for a burgeoning population. This is more prevalent in the developing countries. The concomitants of this are high density, heavy traffic movement and increased pollution levels. To reduce the stressful life of city dwellers it is important to have open spaces, where one can pursue leisure time activities a few removes from clutter. A public space is a space that is generally open and accessible to people. Roads, public parks, libraries etc, are typically considered public space. The term 'public space' is also often misconstrued to mean other things such as 'gathering place', which is an element of the larger concept of social space. Hyderabad, the historical city is the capital of Telangana, India and extends from longitude 78°23' to 78°33'E and latitude of 17°17' to 17°31'N. It is the second largest city in terms of area and fifth largest in terms of population. It is one of the fastest growing cities in India. There is a huge influx of people from other states in search of better opportunities. The main objectives of the study are; to study the sprawl and changing demographic structure of the city of Hyderabad, to study the accessibility of parks, to study the need for the emergence of a local public sphere. The data base will be mainly on secondary data collected from various government sources. A primary survey will be conducted based on a structured questionnaire. GIS and other mapping techniques will be applied to analyse the data.

1. Introduction

A place space is a social space that is generally open and accessible to people. Roads, public parks, libraries etc, are typically considered public space. The term 'public space' is also often misconstrued to mean other things such as 'gathering place', which is an element of the larger concept of social space. To a limited extent, government buildings which are open to the public, such as public libraries are public spaces, although they tend to have restricted areas and greater limits upon use. Although not considered public space, privately owned buildings or property visible from sidewalks and public thoroughfares may affect the public visual landscape, for example, by outdoor advertising. Recently, the concept of Shared space has been advanced to enhance the experience of pedestrians in public space jointly used by automobiles and other vehicles.

Public space has also become something of a touchstone for critical theory in relation to philosophy, (urban) geography, visual art, cultural studies, social studies and urban design. One of the earliest examples of public spaces is common. For example, no fees or paid tickets are required for entry. Non-government-owned malls are examples of 'private space' with the appearance of being 'public space'. The characteristics of a public space include: Promotes human contact and social activities, Is safe, welcoming, and accommodating for all users, Has design and architectural features that are visually interesting, Promotes community involvement, Reflects the local culture or history, Relates well to bordering uses, Is well maintained, Has a unique or special character. Public spaces are regarded as democratic because everybody can use them: places that, rhetorically at least, allows 'community' to exist and flourish. Public space is 'our open-air living room, our outdoor leisure centre' [1], important to the health and well-being of residents of all ages. Drummond states that, the space appears to the whole community; similarly Pumiao says the place that all residents have the right to use. In other words, public spaces link people with a space. Public spaces are always shared and used by the whole community. Claiming social space and being seen in public becomes a way for social groups to legitimate their right to belong in society. Yet because they can be used by everyone, public spaces are frequently considered contested spaces; places where the opposition, confrontation, resistance and subversion can be played out over 'the right to space' [2]. How space is understood and used may also depend on individual and group characteristics [3]. The public spaces are imbued with power relations: particular social groups can be encouraged, tolerated, regulated, and sometimes excluded from public space depending on the degree to which they might be deemed 'in' or 'out of place'. For instance, acts as loitering, drinking, skateboarding or even ' hanging out' in public may be constructed as inappropriate because of particular social representations about what sorts of groups and activities should be seen out in public[4, 5].

Objectives: Parks provide places for people to experience nature, engage in physical activity, and relax. We studied how the city has been expanding and how the communities use public, urban parks and how parks contribute to physical activity.

Methodology: In 10 public parks, we used a survey method to document the number, gender, age group, and activity level of park users who visit daily on an average. We conducted face-to-face interviews in either English or Telugu (Native Language) with both park users (n = 100). Only persons over 18 years of age were eligible. At parks, respondents were recruited by field staff between observations (7:30 AM–11:30 AM and 5:30 PM–7:30 PM). Participants were selected from the busiest and least-busy target areas, and half in each target area was selected because they were sedentary, and half because they were active. We interviewed 100 park users for all the parks at an average of 10 nos. Per park. A questionnaire was formulated and designed to suit the needs of the study. SPSS software was used to generate the tables by using various variables, such as age, gender, religion, proximity to the park, and perceptions of park safety, park characteristics, and performance of park staff. To predict whether residents used the park once each week or more often versus less than once per week, we fit the data to bivariate correlation, Crosstab methods and frequency methods.. We analyzed the frequency of weekly leisure exercise as the number of the times per week a person exercises. Later, Arc GIS is used to generate maps.

2. Study area

Hyderabad, the historic, beautiful city is the capital of Telangana, India and extends from longitude 78°23' to 78°33'E and latitude of 17°17' to 17°31'N. It is the second largest city in terms of area and fifth largest in terms of population. It is one of the fastest growing cities in India. There is a huge influx of people from other states in search of better opportunities. The study area shown in Figure 1 covers about 217 km^2 and encompasses the administrative districts in the integrated town and city planning scheme which are greatly influenced by urban sprawl.

Source:GHMC

Figure 1. Map showing study area.

Public parks may have an important role to play in facilitating physical activity. They provide places for individuals to walk or jog, and many have specific facilities for sports, exercise, and other vigorous activities In other words, parks can play a role in facilitating physical activity, but do not necessarily do so; indeed, parks also provide opportunities for people to engage in sedentary behavior. Information on who uses public parks and what they do there can elucidate the current and potential contribution of parks to physical activity. Features other than size may influence park use, including accessibility, availability, and quality of amenities. Use is also likely a reflection of individual preferences, as well as age, exercise habits, and race/ethnicity. Other important characteristics include the surrounding land use and availability of organized events that draw people to the park. In a review article, Godbey et al emphasized the need to include objective measures of physical activity when studying parks. In this study, we used several methods, including direct observation, to examine how 10 parks in the City of Hyderabad were used, and how much physical and other activities occurs in them. We also explored how services might be changed to better serve residents. Table 1 gives a gist of parks and urban bio-diversity of the city of Hyderabad. There are a total of 49 parks and one reserved forest. The size of the parks, range from 1 acre to more than 300 acres.

IOP Conf. Series: Earth and Environmental Science **37** (2016) 012052 doi:10.1088/1755-1315/37/1/012052

Table 1. List of parks and urban biodiversity.

Sl. No.	Resource	No.	Area in acres
1	Major parks	49	266.91
2	Colony parks in East Zone	65	27.36
3	Tree parks in East Zone	29	11.60
4	Traffic islands and central media in East Zone	5	0.43
5	Colony parks in South Zone	115	46.83
6	Tree parks in South Zone	122	43.04
7	Traffic islands and central media in South Zone	30	8.39
8	Colony parks in Central Zone	212	83.31
9	Tree parks in Central Zone	22	59.14
10	Traffic islands and central media in Central Zone	59	16.29
11	Colony parks in West Zone	58	30.88
12	Tree parks in West Zone	46	38.74
13	Traffic islands and central media in West Zone	22	11.51
14	Colony parks in North Zone	71	22.52
15	Tree parks in North Zone	69	23.74
16	Traffic islands and central media in North Zone	43	3.71
17	Fountains	85	-
18	Cement pots for plants	4778	-
19	Native trees for Urban Landscape	127	-

Source: GHMC

A total of 266.91 acres of land is under major parks which are acting as the lung space for the city residents. Apart from major parks we find that colony park, three parks and traffic parks. These parks are divided according to municipality zone wise and category wise. Table 2 shows types of parks such as Major park, City level park, Theme park and Reserved Forest.

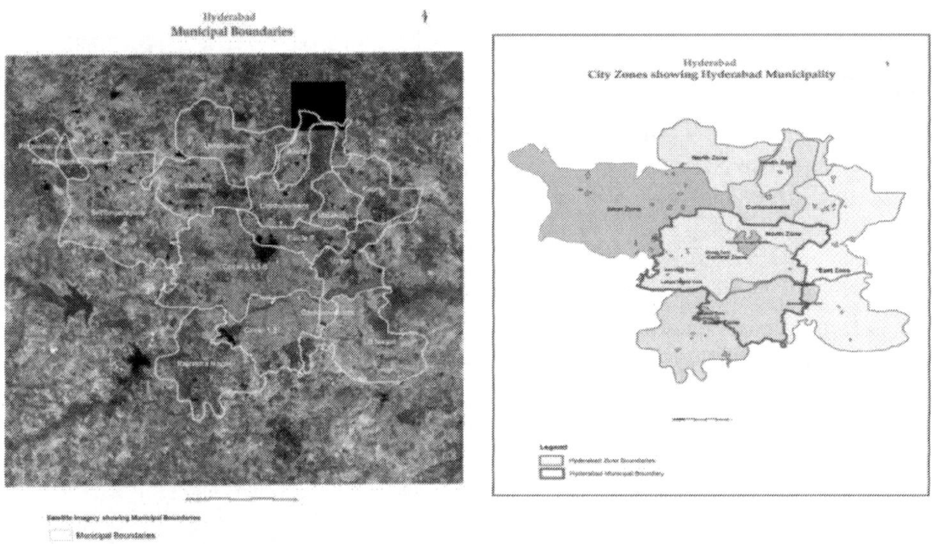

(a) **(b)**

Figure 2. (a) Satellite imagery of zones. (b) City zones of Hyderabad.

Table 2. Selected parks for the field study.

S.No.	Zone Name	Circle No.	Name of the Park	Category	Area In Acres
1	East	1	Dr. A. S. Rao Nagar Park	Major Park	2.00
2	East	1	Sainikpuri'E' Park	Major Park	2.00
3	South	5	Imliban Park	City Level Park	9.00
4	South	5	Musi Landscape Garden	Major Park	3.50
5	Central	7	Cha Cha Nehru Park	City Level Park	12.75
6	Central	8	Feroz Gandhi Park	Major Park	0.60
7	Central	9	Indira Park	City Level Park	69.40
8	West	11	Gulmohar Park	Major Park	3.20
9	West	14	Kukatpally Phase – VII Park	Major Park	2.10
10	West	10	KBR Park	Reserve Forest	360.0

Source: GHMC

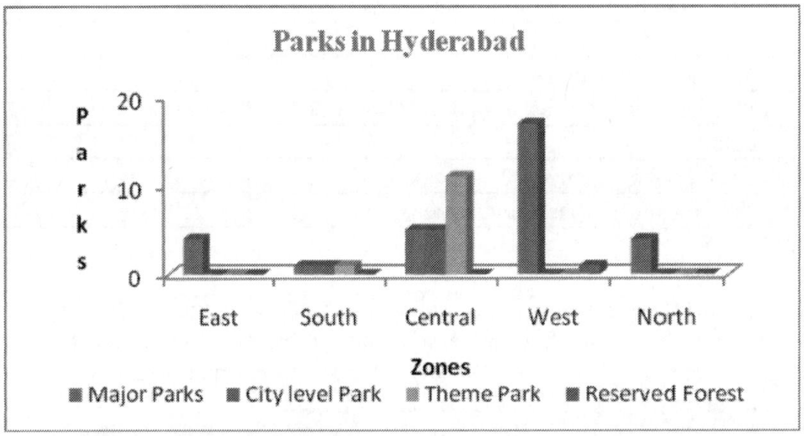

Figure 3. Types of parks in Hyderabad.

3. Results & discussion

All 10 parks were public, urban, parks, and each had a building with an office. All had, Park staff, Garden with flowers walking path and play area. Seven had Tracks, 4 had skate courts, and 6 had picnic areas. Two parks had running tracks, but only one (KBR Park) was accessible. Some parks provided programming, such as health related, religious programmes.

The variables from Table 3 shows that 65 percent of people come to park daily and alone with less than 1km of distance. Whereas 32 percent of people have said that they come with others. These others comprise mainly, children and friends. Only 3 percent have stated that they come only once in a week and the distance travelled is more than 5 km. About 97 percent of people are daily users.

Table 3. Details on visitors to the park.

You come alone or with others			How often you use this park		Total
			Daily	Weekly	
Alone	Distance to park	0-1/2 km	16	0	16
		1/2 - 1 km	22	0	22
		More than 1 km	21	0	21
		More than 5 km	6	1	7
	Total		65	1	66
With Others	Distance to park	0-1/2 km	8	0	8
		1/2 - 1 km	9	0	9
		More than 1 km	12	0	12
		More than 5 km	3	2	5
	Total		32	2	34

The utility of parks is mostly for health upkeeping. A total of 39 percent of people have said that they visit the park for walking as a daily routine, similarly 30 percent of people have agreed that they come daily to park as part of exercise or jogging. About 12 percent have people come to do yoga, while 11 percent of people just come to relax in the park.

Table 4. Purpose of visit to the park.

		Frequency	Percent	Valid Percent	Cumulative Percent
	Bike	2	2.0	2.0	2.0
	Bring Children	2	2.0	2.0	4.0
	Skating	4	4.0	4.0	8.0
	Exercise/jog	**30**	**30.0**	**30.0**	**38.0**
	Relax	12	12.0	12.0	50.0
	Walking	**39**	**39.0**	**39.0**	**89.0**
	Yoga	11	11.0	11.0	100.0
	Total	100	100.0	100.0	

Table 5. Space particulars in the park.

		Count	**Percent**
Enough Space in the park Space	**Enough**	55	55%
	Too Crowded	45	45%
Overall		100	100%
Excluded		0	
Total		100	

When asked about the space for their activities, about 55 percent of people have agreed that there is enough space, while 45 percent of people have said that they do not much space and it is too crowded. Table 5 explains the peoples perception towards the space. This clearly shows that the urban spaces are congested and need for more parks with enough space is required.

Any person visiting a particular place means he has a choice of his own. When asked about the priorities for a visit to a particular park, 46 percent of people have said that they come to park due to availability of Paths(walking), which correlates with their purpose of visit. About 25 percent of people have stated that the cleanliness is their priority and that's why they visit the those parks.

Table 6. Priorities for visit to the park.

		Frequency	Percent	Valid Percent	Cumulative Percent
Valid	Available Park Staff	2	2.0	2.0	2.0
	Benches	2	2.0	2.0	4.0
	Cleaning/Litter Removal	25	25.0	25.0	29.0
	Gardens and Flowers	1	1.0	1.0	30.0
	Paths	46	46.0	46.0	76.0
	Lighting	1	1.0	1.0	77.0
	Recreation	6	6.0	6.0	83.0
	Security	4	4.0	4.0	87.0
	Track	3	3.0	3.0	90.0
	14	8	8.0	8.0	98.0
	15	2	2.0	2.0	100.0
	Total	100	100.0	100.0	

4. Conclusion

The city of Hyderabad, though growing at a faster rate compared to the other cities in India, it has its own urban space problems. The findings suggest that urban spaces should be designed such a way, that all people have a park within at least 1 km of their residence. Data shows that more people used specific areas when they were provided organized activities, suggesting that increasing the availability of structured, supervised activities will also likely to increase park use; however, only 9 percent of all observations found areas supervised, suggesting that greater attention should be paid to staffing. Perceptions of safety may affect the use of recreational areas, but they did not predict park use in this study. Our analysis, however, was restricted to 8 parks, mostly in low-income, minority neighborhoods. A larger sample of parks with greater variation might provide different results.

The primary finding that residential proximity to a park was the most robust predictor of both park use and self-reported leisure exercise in urban, minority communities should be noted by urban planners and officials responsible for ensuring safe and healthy neighborhoods. Facilitating larger numbers of people being physically active is critical for improving overall population health.

5. References

[1] Lipton, S. (2002) The Value of Public Space, Foreword, York: CABE Space.

[2] Matthews, H., Limb, M. and Taylor, M. (2000) 'The street as thirdspace: class, gender and public space', in S. Holloway and G. Valentine (eds) Children's Geographies: Living, Playing, Learning and Transforming Everyday Worlds, London: Routledge, pp 63-79..

[3] Mitchell, D. (1995) 'The end of public space? People's park, definitions of the public, and democracy', Annals of the Association of American Geographers, 85, pp 108-33.

[4] Valentine, G (1996) 'Children should be seen and not heard', Area, 17, pp 205-20.

[5] Wilson, E. (1991) The Sphinx in the City: Urban Life, the Control of Disorder and Women, Berkeley, CA: University of California Press

[6] Gobster P. Managing urban parks for a racially and ethnically diverse clientele. Leisure Sci.2002;24:143–159.

[7] Godbey GC, Caldwell LL, Floyd M, Payne LL. Contributions of leisure studies and recreation and park management research to the active living agenda. Am J Prev Med. 2005;28(suppl 2):150–158.

[8] Godbey GC, Graefe A, James, SW. The Benefits of Local Recreation and Park Services: a Nationwide Study of the Perceptions of the American Public. Ashburn, Va: National Recreation and Park Association; 1992.

[9] Mertes J, Hall J. Park, Recreation, Open Space and Greenway Guidelines. Ashburn, Va: National Recreation and Park Association; 1996

[10] Walker C. The public value of urban parks. In: Beyond Recreation: a Broader View of Urban Parks. Washington, DC: The Urban Institute; 2004.

GIS modelling for new landfill sites: critical review of employed criteria and methods of selection criteria

Sohaib K. M. Abujayyab*[1], Mohd Sanusi S. Ahamad[1], Ahmad Shukri Yahya[1], Mohammed J.K. Bashir[2], Hamidi Abdul Aziz[1&3]
[1]School of Civil Engineering, Engineering Campus, Universiti Sains Malaysia, 14300 Nibong Tebal, P. Pinang, Malaysia
[2]Faculty of Engineering and Green Technology (FEGT), University Tunku Abdul Rahman, 31900 Kampar, Perak, Malaysia.
[3]Solid Waste Management Cluster, Science and Engineering Research Centre, Engineering Campus, Universiti Sains Malaysia, 14300 Nibong Tebal, Penang, Malaysia.Corresponding author E-mail: s.jayyab@hotmail.com

Abstract. Policy makers and the public are increasingly concerned with the determination of landfill-siting input criteria (DLSIC) in landfill modelling procedures as an area of research. Thus, its procedures are complicated and decision makers are increasingly pressured. These procedures can be considerably develop in order to reduce the negative effect of landfill locations on the environment, economy, and society. In this review article, literature related to the developments of 64 models and their procedures in the past 18 years (from 1997 to 2014) were comprehensively survey. DLSIC are determined through a conventional method. The frequency of criterion usage reflects the limitation of Conventional method for DLSIC. Moreover, some of these studies utilize unrelated criteria that are time-consuming, costly, arduous, and fruitless. Potential improvement in Geographic information systems GIS modelling parameter for landfill sites via utilizing multivariate analysis (MVA) instead of Conventional method (CM) through for DLSIC (e.g., Input variables, Accuracy, objectivity, reliability of criteria, time consumption, cost and comprehensiveness) were emphasize. It can be conclude that expenses can be reduce by implementing MVA in DLSIC for landfill modelling using geographic information systems (GIS) based on the corresponding significant level. Moreover, the determined criteria can be accurate, satisfying, sufficient, and free of bias from experts and human error.

1. Introduction

The process of searching for a new municipal waste disposal site is time-consuming [1]. The related procedures are extremely complex because its involve combination of several knowledge from diverse interesting areas, and numerous gatherings are in charge of or influenced by the outcomes [2][3–5]. To properly identify and select appropriate landfill sites, systematic procedures must be adopted and followed carefully [6]. [7] stated that the determination of suitable landfill locations is a decision that requires extensive land evaluation. Therefore, the process of landfill siting considers environmental, ecological, and technical parameters [8,9]. The selected locations must also fulfil the necessities of existing legislative guidelines and reduce health and environmental expenses in the meantime [10]. Furthermore, design considerations, area availability and prospects of development. Similarly, [11] concluded that Geographic information systems (GIS) modelling for new landfill sites must take operational and economic issues into account. In addition; insurance and social considerations and political opposition explained through the phenomenon "not in my back yard (NIMBY)" [12–15]. On the other hand, GIS modelling aims to find suitable spatial patterns for upcoming landuse based on exact constraints, preferences or predictions of some activity. The spatial nature of modelling of new landfill sites problem involves the usage of GIS to expand the traditional frameworks in order to concern equally spatial and attributes data. According to McHarg's map layering theory [16], these frameworks constructed to reinforced performance and managing geographic data to improve their part in landfill GIS modeling. Additionally, the synergy between multiple criteria decision analysis (MCDA) techniques and previous frameworks come from the improvement of multicriteria spatial decision support systems (MC-SDSS). The aim of synergizing is to rating the suitable sites based on their significance in sustaining modelling goals. The research of MC-SDSS concentrations on improving software packages and decision techniques, which lead to facilitate the assessment process to select the new landfill sites.

Recent last two decades, a noteworthy number of articles conducting for landfill GIS modelling. It demonstrates the benefits resulting through the synergy of MCDA and GIS. Development of framework is mutual practice between these publications, which intention to modelling the landfill sites. These frameworks delivers consistent stages for handling conflicting objectives among the diverse involved crowds that react with decision making procedure, as well as in the meantime to deliver compromises between the conflicting aims. In addition, it permits the analyzer to consider the stakeholders preferences. These frameworks contain coherent sequence of stages. It can be discriminate through four stages. First, determining objective trees and identifying maps of constraint and factor criterion. Second, execute standardization procedures on the maps. Third, in choice stage, specific decision rule implemented to the set of the criteria in order to generated map of suitability index and classify it. Forth, to define the robustness of the modelling result, sensitivity analysis were implemented [17]. Consequently, in implementation stage, trustiness evaluated for the new sites through assessment of public opposition. On the off chance that the chosen site does not fulfil the expectations, then all prior four stage require to recurrence till ensures minimizes environmental consequences and achieving public acceptance.

In addition, policy makers and the public are increasingly aware and concerned with the environmental problem related to landfill-siting criteria. Thus, landfilling procedures must consider numerous of the different related criteria and variables [18] involved in the selection of a proper site for healthy urban landfills [19]. Consequently, GIS modelling framework for landfill site are complicated and decision-makers are increasingly pressured [2,5]. However, the number of considered criteria for landfill modelling increases, which may result in the adoption of unnecessary criteria that will be high cost, time-consuming, and wasting resources without reaping any benefit. In contrast; it potential to disregarded some significant criteria, or consider some criteria as more important than it's real important due to vague and subjective estimates. Consequently decision-makers may make poor siting decisions and inappropriate landfill sites might be selected [18]. Moreover, these problems can have negative and far-reaching effects on the environment, economy and ecology situations.

Recently, various methods have been widely employed to determine a list of related criteria for landfill GIS modelling framework without testing their suitability. Conventional method (CM) to selecting the significant criteria such as guidelines, regulation requirements, previous literature and expert interviews were used recently.

This article aim to review the employed criteria and methods of selection criteria concerning landfill GIS modelling to clarifying its problems in this filed and propose guidelines for future research. In this paper, (1) an extensive survey of GIS modelling scientific published literature and critically reviews to determine the list of the used criteria and criterion usage scores. (2) demonstration and evaluation for the employed methods for selecting GIS criteria which implemented formerly in past 18 years for landfill GIS modelling were performed, and (3) investigate the potential improvement of the implementing spatial multivariate analysis SMVA method over the covenantal method CM to determine the landfill siting input criteria (DLSIC).

2. Methodology

A numerous number of publications consider the application of MCDA for landfill GIS modelling issue [2–4,6,14,15,20–35]. This is owing to the necessity for developing new frameworks as well as, the fast development in computer science, dramatically contributed to the enlargement of MCDA application. Published articles related to frameworks have been reviewer over the years from 1997 to 2004. An exploration to select the relevant articles was executed via variety of scientific journals publishers such as; ScienceDirect (www.sciencedirect.com), SpringerLink (link.springer.com), ASCE (www.asce.org), SAGE (online.sagepub.com), Taylor and Francis (www.tandfonline.com), as well as Scopus database (www.scopus.com). The seeking was made through the following terms; multi criteria and GIS, landfill modeling, landfill siting, mapping and analysis, selection of Municipal Solid Waste (MSW) landfill, suitability analysis, landfilling, and MCDA. Additionally, their references list were explore hastily to select further related papers in the same scope. An extensive survey was conduct to extract the common criteria employed for GIS modelling for landfill sites as highlighted earlier in the literature. Moreover, statistical summary were perform to determine the criterion usage scores alongside all the selected articles. In addition, another literature survey was accompanies to summarize the considered method adopted for selection criteria for landfill siting.

3. Results and discussion

3.1. Articles analysis

From the literature, 75 published studies have been select; these studies and models commonly focus on developing and improving the frameworks that can identify the most suitable location for landfill sites. It summarizes the progress of main published research, concerning landfill GIS modelling documented in literature since 1997. Nearly 85% of articles being released after 2008, (data extracted from Scopus and existing articles), which reflecting the growing interest and awareness. [7] concerning the benefits of GIS in this research scope as shown in **Figure 1** and the increasing interest of Asian countries in this kind of research as demonstrated in **Figure 2**.

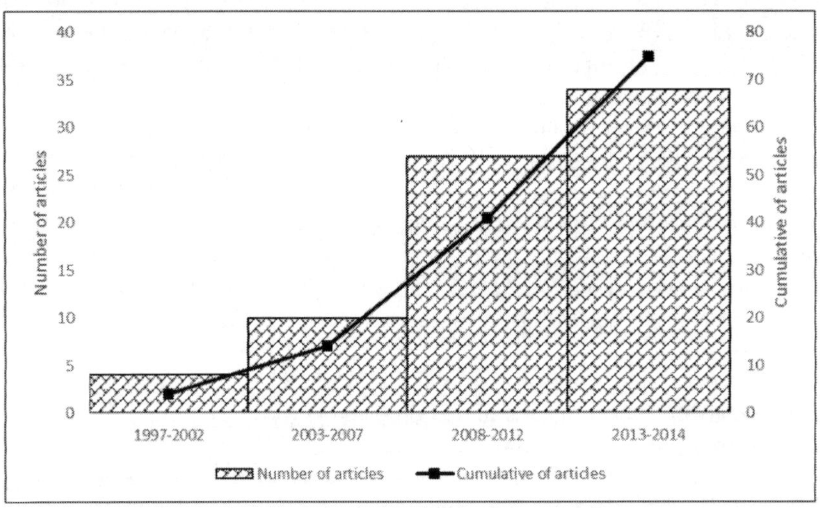

Figure 1. Recent published works concerning the development of landfill GIS modelling.

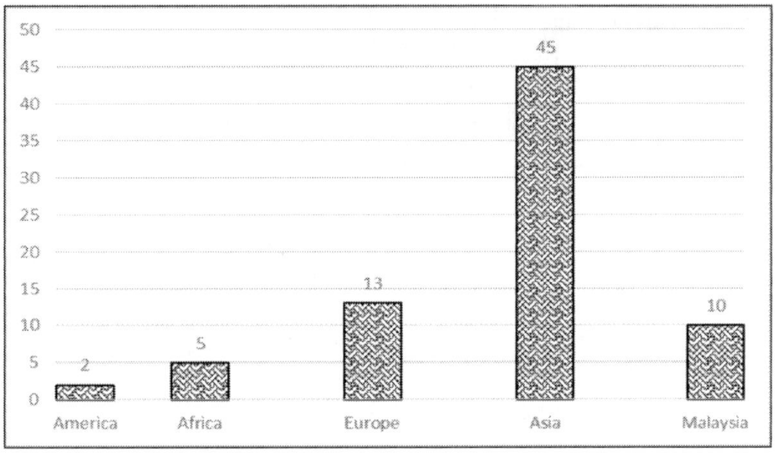

Figure 2. Published article based on different location.

3.2. Input criteria of landfill GIS modelling

Even though the effectual of landfill GIS modelling mainly influenced by the adequacy of evaluation criteria through the entire framework, huge amount of criteria were implemented via legislation and decision makers [17]. Decision makers have been accomplished wide work to reach to this amount of criteria, which targeting to minimize the social, environmental, and economic expenses. Additionally fulfil government regulations requirements as well as public acceptance [2]. Under the heading 'Input criteria of landfill GIS modelling', Table 1, 2 and 3 summarizes the number of criteria that have been implemented to evaluate appropriate sites via GIS modelling. The wide investigation of the literature released that 80 criterion have been executed until present to improving landfill GIS modelling frameworks. The number of criteria is maximized as a result of the stress and the pressure from increased awareness of researchers, the public opposition, decision makers, and the government regulations with regard to landfill modelling [3,7].

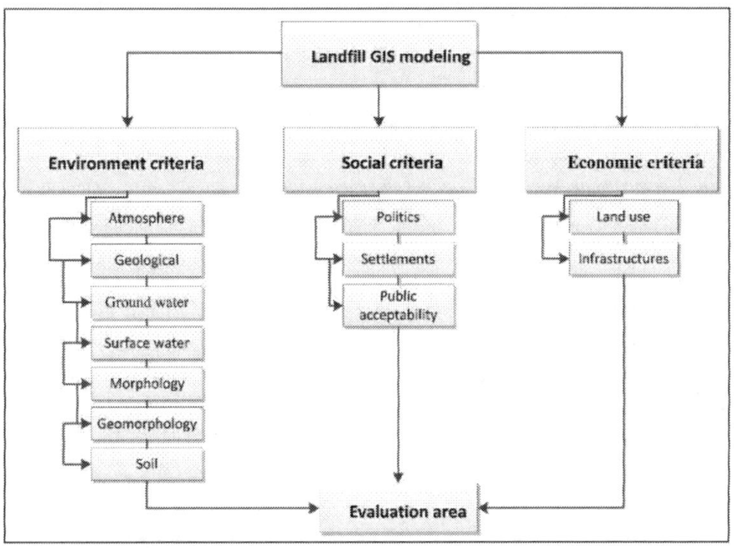

Figure 3. Objective and sub-objectives decision tree for Landfill GIS modelling.

To deliver a coherent discussion concerning the applied criteria, it have been gathered into three major groups **Figure 3**, which establishing a three-level decision tree, which reflect the sustainable objectives (environment, economy and social) and for further and easier understanding of the massive number of criteria. These groups explained more general considerations (objectives) than that should be satisfied through the GIS modelling. In addition, twelve sub-groups that deliver additional definitions in evaluation process were conduct (atmosphere, geological, ground water, surface water, morphology, soil, geomorphology, politics, settlements, public acceptability, land use, and infrastructures). **Figure 4** shows the classification of the decision objectives and sub-objectives used in the previous work.

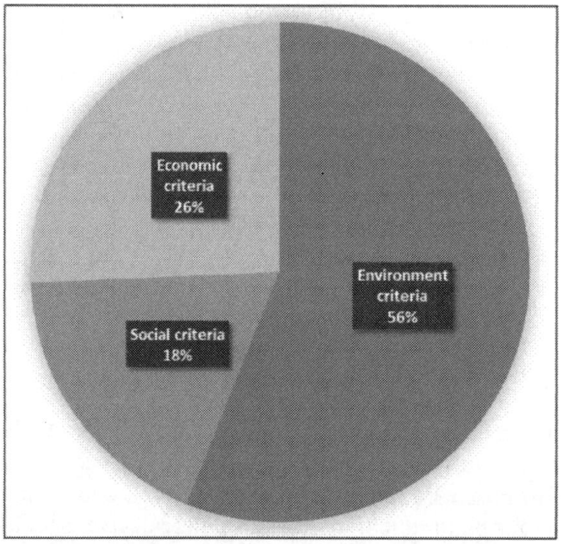

Figure 4. Percentage of criteria frequency in landfill modelling.

On the other hand, the sub-criteria were statistically accounted the usage score (based on 75 reviewed articles) to generate primary knowledge regarding the used criteria and its usage frequency, which is presented in **Figure 4**. The percentage for the frequency of sustainable criteria application is considerably high. The sub-criteria were group into three categories as per this percentage. **Figure 4** indicates that the environmental criteria constitute 56% of the total criteria. This proportion reflects the stress related to environmental issues during landfill modelling operation. The economic and social aspects accounted for 26% and 18%, respectively. These percentages emphasize that many articles focused on environmental criteria in general and either ignored or barely considered the social and economic aspects and vice versa [9,36–39]. These criteria reflect the limitation of CM for DLSIC, and it is not necessary to be appropriate for landfill modelling. On the other hand, all the adopted 80th sub-decision criteria through the three sections were discussed briefly and separately in three sections, as well as the criterion score of usage in the previous articles, In the following sections;

3.2.1. Environment criteria

Matters relevant to environment safeguard were exemplified in evaluation via verity necessary criteria, (Meteorology, Geological, Ground water, Surface water, morphology, soil, Geomorphology and Sensitive zones criteria) according to Malaysian guideline [40]. The environment criteria represent the first part in the sustainable development and planning, it embodied the analysis via implementing variety of criteria, these criteria play substantial roll to select the landfill sites. The goal of environment criteria is to protect the surrounding environment against any water, soil, air pollution resource [2]. Moreover, avoiding any degradation and minimize the risk in the desire area. These sensitive sites should exclude from the evaluation zone and placing the landfills a sufficient distance far away from the sensitive areas. By providing the attention to environmental criteria, it is easily to explain the consideration of the exacting guidelines through local and universal regulation. **Table 1** illustrates the summary of employment criterion in the environment section. The fact that 39 criteria implemented in the literature were relate to environmental section.

Table 1. Summary of the employment criterion in the Environment section.

Objective	Criteria	Sub-criteria	Reference
Environment	Meteorology/ Atmosphere	Evaporation	[12]
		Climatic regimes	[4]
		Humidity	[41]
		Precipitation	[42]
		Temperature	[43]
		Air pollution	[44]
	Geological	Landslides	[45]
		Geological formation/lithology	[46]
		Earthquake zones	[34]
		Faults	[11]
	Ground water	Aquifers	[2]
		Groundwater quality	[47]
		Groundwater direction	[12]
		Hydraulic conductivity of groundwater	[1]
		Groundwater depth	[39]
	Surface water	Lakes/ponds	[48]
		Rivers/streams	[7]
		Springs	[28]
		Flood plains	[21]
		Wetlands	[22]
		Dam	[8]
	Morphology	Elevation	[42]
		Land aspects	[49]
		Slope	[10]
	Soil	Soil depth	[28]
		Soil type	[38]
		Soil permeability	[50]
		Mining and land resources	[51]
		Land contamination	[48]
	Geomorphology	Geomorphology (karst, erosion)	[51]
		Snow and glacial period	[10]
		Drillings	[20]
		Geothermal fields	[2]
		Gorges and caves	[13]
	Sensitive zones	Petrochemical storage plants	[11]
		Protected forest	[52]
		Protected areas	[15]
		Cultivation and natural cover	[19]
		Sensitive ecosystems	[13]

Figure 5. Frequency of criterion usage score in environment sectionclearly shown the criterion importance score of usage, were the five most significant landfill modelling criteria are distance from rivers, which referred to 54 articles in the literature, while distance from lakes (45 articles), slope (41 articles), elevation (30 articles) and soil (25 articles), which is imposed from the national/international guidelines [40]. In these studies, 10 out of 39 criteria were applied only once, including humidity, landslides, groundwater directions, the hydraulic conductivity of groundwater, snow and glacial periods, drillings, land contamination, geothermal fields, gorges and caves, petrochemical storage plants, climatic regimes, and air pollution [40]. And it is not enforced by national/international regulations [42], While the remained criteria referees frequently in 2 to 24 articles and it varied in existing in the guidelines or not. The small score does not essentially mean the criterion is insignificant. Moreover, the restricted usage can be possibly due to the some responsible causes such as deficiency of infrastructure, data availability or natural feature).

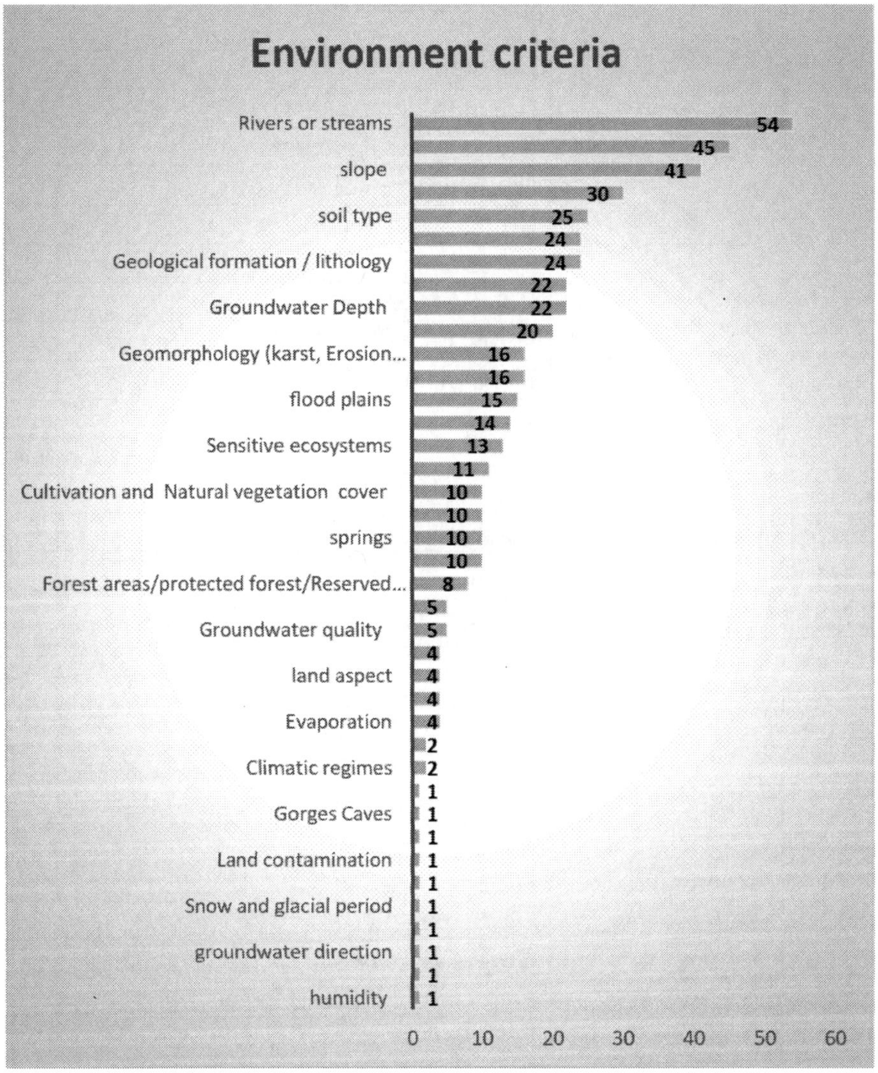

Figure 5. Frequency of criterion usage score in environment section.

3.2.2. Social criteria

The significance of the social criteria lies in the public acceptability or social consensus and political opposition that can arise from establishment of a new landfill site. The population do not want landfill sites to be located near their residential areas because it consider as obnoxious locations. this fact expressed through the phenomenon “not in my back yard (NIMBY)” [12–15], which in another term call locally unacceptable land uses (LULUs), particularly in heavily inhabited zones. During the past, NIMBY phenomenon has been mightily relevant to GIS modelling plans and landfill planning. These NIMBY controversies need the involvement of individuals personify different competence areas, social values and political agendas to handle the geographic decision issues [2]. The different preferences participants (technical experts, Decision makers and stakeholders) improved the understanding of the spatial issues (evaluate situations, needs assessment, clarifying goals, define criteria). As a result, the difficulty were decrease and the decisions made further thoughtful, precede to an efficient landfill GIS modelling and hence reducing the danger of occur public opposition. Thus, several criteria used in the literature such as; Sufficient distance from danger of odor and insects were taken from (residential areas, Archaeological, tourism, and cultural sites, sports sites, education, historical sites, recreational sites, Natural parks, theaters, Hospital, schools, universities, military areas, airports. Furthermore, dense population areas, Villages/rural areas, Visible travel distance from roads, Visibility from residential areas, Wind direction, Wind speed, District boundaries and National borders) to get over resident conflict. **Table 2** delivers a summary of employment criterion in current survey from collected works.

Table 2. Summary of the employment criterion in the Social section.

Objective	Criteria	Sub-criteria	Reference
Social	Politics	District boundaries	[38]
		National borders	[41]
	Settlements	Proximity to dense population	[53]
		Villages/rural areas	[14]
		Population of urban centers	[14]
	Public acceptability	Visible travel distance from roads	[24]
		Visibility from residential areas	[10]
		Wind direction	[28]
		Wind speed	[42]
		Urban/residential areas	[54]
		Archaeological, tourism, and cultural sites	[6]
		Sports education	[55]
		Tourism services	
		Natural parks	[45]
		Theaters	[32]
		Hospital/public health safety	[2]
		Schools/universities	[32]
		Military areas	[56]
		Airport/helipad/airfield	[57]

This social section referred to 19 criteria in the literature. **Figure 6.** Frequency of criterion usage score in social section obviously displayed the criterion importance score of usage. The five most significant landfill modelling criteria are distance from residential areas, which referred to 32 articles in the literature. While distance from urban centers population and airports (27 articles), distance from villages/Rural areas (20 articles), distance from archaeological/tourism/cultural sites (18 articles), and Wind direction (14 articles). which is forced by national/international guidelines [40]. In these studies, 7 of the total 19 criteria were applied only once or twice, including Proximity to dense population, distant from theaters, distant from schools/ university, distant from district boundary, wind speed, distant from sports-Education sites and distant from services for Tourism sites [2]. And it is not enforced by national/international regulations [40]. While the remained criteria referees frequently in 5 to 1 articles and it varied in existing in the guidelines or not.

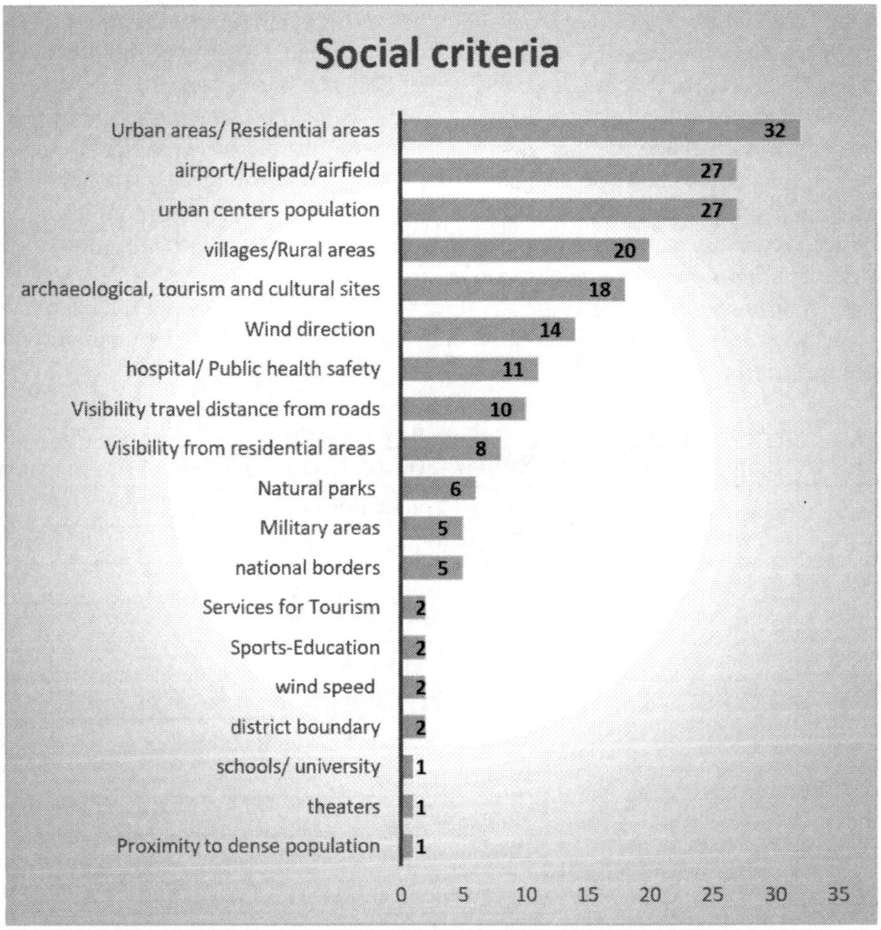

Figure 6. Frequency of criterion usage score in social section.

3.2.3. Economic criteria

The aim of economic criteria is to minimize the final cost of modelling landfill sites and to guarantee suitability of landfill sites as well as effective services process. Furthermore, targeting to sidestep in value deterioration of property, tourism industry, future use of the land and industrial areas. Therefore, landfills must take a least space from the private properties or some land use type. The accessibility of landfill locations is a factor that essential to assessed throughout the GIS modelling process to guarantee garbage trucks sufficient accessibility to landfill locations to reduce waste transference and collection expenses from the origin source to landfill sites, which mean in another way minimization of travel cost, in addition to reduce the complementary infrastructure building requirements [58]. The nearness of the landfill sites to the local and federal streets networks wanted, although landfill sites preferable to be place far away from the major transportation network. Additionally; an fitting planning procedure must guarantee the conservation of infrastructures, water facilities and public utilities, thus should take an exclusion zones from it [58]. Moreover, to ensure the compliance with current landfills to avoid of services area conflicting. As shown in Table 3, there are 22 criteria were implemented in landfill GIS modelling in the previous work.

Table 3. Summary of the employment criterion in the economic section provides the criterion importance score of usage, were the five most significant landfill modelling criteria are road accessibility, which referred to 52 articles in the previous works, while Land use (43 articles), Ground infrastructures (18 articles), Industrial areas (16 articles), and Municipal and local wells (12 articles), which is forced by national/international guidelines [40]. In these articles, 5 of the total 19 criteria were utilized just one time communication infrastructures, ownership, salinization zones, grazing and public utilities. Furthermore, it is unenforced by any national/international regulations [40]. While the remained criteria referees frequently in 12 to 3 articles and it varied in existing in the guidelines or not.

Table 3. Summary of the employment criterion in the economic section.

Objective	Criteria	Sub-criteria	Reference
Economic	Land use	Land use/cover	[59]
		Coastlines	[48]
		Ownership	[48]
		Salinization zones	[58]
		Industrial areas	[60]
		Compliance with current landfills	[36]
		land cost	[37]
		Future use of the land	[54]
		Special disposal location (Abandoned mines, quarries)	[10]
		Deserted fields	[13]
		Grazing	[55]
		Availability of cover materials	[51]
		Recycling/ferrous exploitation	[15]
	Infrastructures	Public utilities	[2]
		Ground infrastructures	[61]
		Communication infrastructures	[11]
		Railways	[14]
		Road accessibility to the site	[62]
		Water facilities	[58]
		Municipal and local wells	[63]
		Waste production centres (transfer station)	[39]
		Capacity of landfill sites	[25]

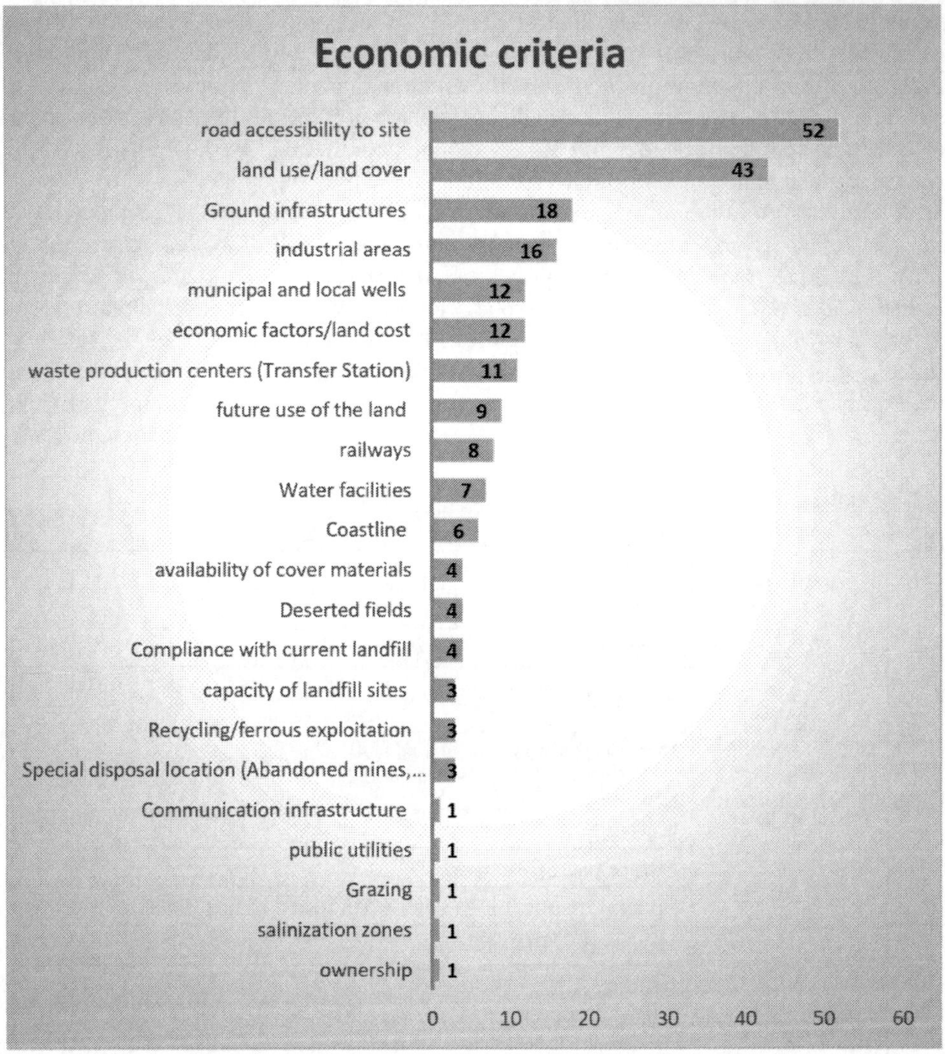

Figure 7. Frequency of criterion usage score in economic section.

3.3. Use of CM to determine the input criteria

The survey results of adopted method for selection criteria for landfill-siting are depicted in Fig. 8 based on the survey of literature review above, the most common method of landfill-siting criteria selection is to follow guideline and regulation requirements. **Figure *8*** suggests that 46 studies selected criteria according to this method; 26 extracted criteria from previous literature; 11 conducted expert interviews; and 6 statistically analyzed survey questionnaires. The significant level was determined as per the weighting of the criteria extracted from these questionnaires. However, four studies did not describe the criteria selection processes they used, and other works justify their disregard of some criteria based on data availability.

Some studies [2,10,12,28,42] apply various comprehensive models that consist of many criteria determined without testing (either attribute or spatial testing) or pre-processing. This orientation complicates landfill modelling procedures and renders them time-consuming because of the high amount of spatial data that must be purchased, collected, prepared, and processed for landfill location selection [12,24,47,59,62,64]. The weighting process is challenging, because face-to-face interviews must be conducted [59], and all of the criteria must be explained to each expert or specialist. Nonetheless, ignoring any of these criteria may result in the selection of inappropriate landfill locations and in miscalculations. These outcomes can negatively affect the environment, the economy, and ecology [18]. Unfortunately, some studies discarded some important criteria [1,14,55], which may result in poor siting decisions. Instead, these studies utilize unrelated criteria that are time-consuming, costly, arduous, and fruitless. In addition, some criteria (variables) may be accompanies by other criteria, for instance, modelling land use can satisfy the objective and can act as an alternative to many criteria, such as schools and utilities, especially within the constructed area. Thus, the limited data do not affect the accuracy of the result, thereby minimizing the expense attributed to additional unnecessary data.

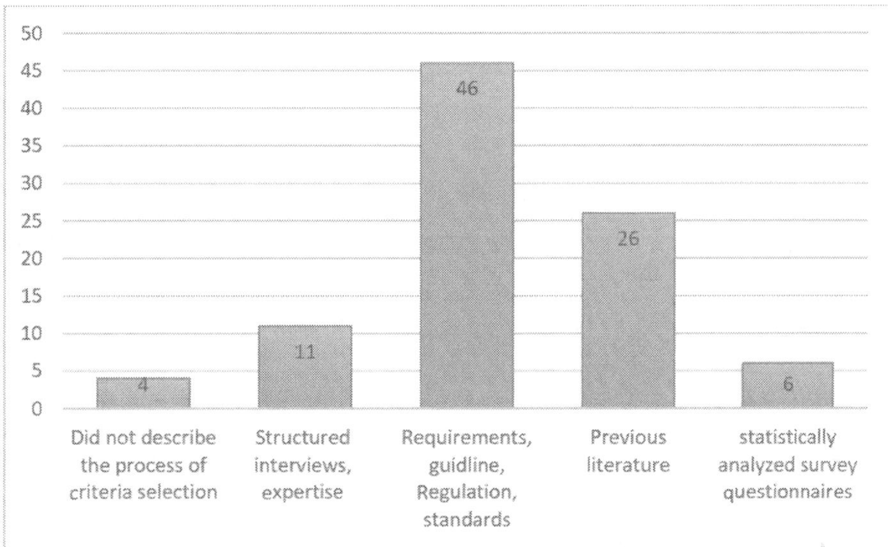

Figure 8. Popular method adopted for landfill-siting criteria selection.

Aforementioned, the conventional method for selection landfill siting criteria are inaccurate, unsatisfactory, insufficient, time-consuming, and unacceptable. Therefore, there is an urgent need for a new method that can avoid the previous limitation and guarantee reasonable result, and reduce environmental influences as well as at meantime rise probability to override opposition of public.

The methods can improved through multivariate analysis (MVA) to determine the landfill siting input criteria (DLSIC). It select the criteria according to their significant levels for each criterion based on the spatial measurements extracted from the input maps of existing landfills [28], which lead to enhances the accuracy and efficiency [65–69]. This method helps to manage the budget for GIS projects, which are short on fund support given the high cost of project operation and design. The collection and manipulation of unnecessary data waste budgeted resources because these data are unrelated to either projects or models.

3.4. Potential improvement through MVA determination

To investigate the potential improvement of MVA method for DLSIC instead of CM, these two methods were compare. **Table** *4* summarizes their effect on results of the model in different aspects, such as input variables (data collection, processing, and modeling), accuracy, reliability of criteria validation, and the comprehensiveness of the criteria.

Table 4. Potential improvement in GIS modelling parameter for landfill sites via utilizing MVA instead of CM through for DLSIC.

Parameter	CM determination	MVA determination	Reference
Input variables (Data collection, processing and modelling)	• Both related and unrelated criteria are inputted; as a result, landfill modelling is particularly time-consuming, arduous, and costly.	• Significant spatial criteria alone are inputted, thereby reducing modelling time considerably. In the process, costs are typically lower as well.	[1,10,24, 31,49,70]
Accuracy and objectivity	• CM determines criteria based on expert knowledge. • Simple determination based on either public or expert opinion. • Subject to bias from experts and human error.	• MVA is a scientific and statistic determination method that shifts based on the spatial measurements obtained from the real Geo-database. It is also highly objective.	[6,7,37,48,71]
Reliability of criteria validation	• The level of validation reliability of the landfill modelling model criteria is low.	• Confidence can be enhance to high level by adapting significant criteria.	[6] [38][34]
Comprehensiveness of the criteria	• Comprehensiveness levels vary intermittently among the different models.	• Comprehensiveness can be improve by testing most of the criteria prior to determining the significant ones.	[2,28,42]

4. Conclusions

A comprehensively surveys literature conduct related to the developments of 75 landfill modelling models and their procedures in the past 18 years (from 1997 to 2014) reflects the growing awareness of this research scope. Approximately 80 criteria used in previous works to evaluate land for landfill modelling given different importance levels. In addition, the different criteria lists derived from previous models are justified by data availability, which are relate to the application area and are suitable for landfill policies. The varied frequency in the input criteria of previous models reflects an imbalance in determination of criteria in landfill modelling and limitation. This imbalance induces miscalculations and various problems. Moreover, this result signifies that dissimilarities in the input criteria increase the cost of search procedures for new landfill sites. The DLSIC methods in landfill modelling were develop based on previous literature, expert interviews, guidelines and regulation requirements, and questionnaires. The use of MVA determination instead of conventional determination in landfill modelling could be progressed from a costly, less scientific, and less efficient approach to a highly scientific technique that is very beneficial. These benefits are induce by the different DLSIC procedures and are as follows: (1) time-scale reduction, which implicitly lowers cost in most cases; (2) a decline in the number of steps of overall procedures to eliminate the need for expert interviews and data collection, manipulation, modeling, and weighting; and (3) the increased effectiveness of overall procedures. These benefits suggest that the final models obtained from MVA for DLSIC procedures may be more economically competitive than those generated using conventional techniques. MVA determination technology recently has been applies for DLSIC in landfill modelling. Nonetheless, no research has been conduct on the DLSIC in landfill modelling using GIS based on spatial measurements obtained through MVA. Based on this review and to the best of our knowledge. The same is true with respect to research on site selection for GIS spatial modelling and planning fields. Thus, future studies should aim to understand the procedures of low-cost DLSIC in landfill modelling using GIS and to demonstrate MVA effectively. Furthermore, none of the previous studies tested the significant levels of landfill modelling criteria using MVA techniques, to the best of our knowledge.

Acknowledgements

The authors graciously acknowledge the funding support from the University Sains Malaysia (USM) through the USM Fellowship and University Sains Malaysia (RU Grant 814219). In addition, the authors graciously acknowledge the Ministry of Higher Education for their funding allocation under (FRGS grant 6071258). This work is also funded by Universiti Sains Malaysia under Iconic grant scheme (Grant no. 1001/CKT/870023) for research associated with the Solid Waste Management Cluster, Engineering Campus, Universiti Sains Malaysia.

References

[1] W. Guiqin, Q. Li, C. Lijun, Landfill site selection using spatial information technologies and AHP: A case study in Beijing, China, J. Environ. Manage. 90 (2009) 2414–2421. doi:http://dx.doi.org/10.1016/j.jenvman.2008.12.008.

[2] O.E. Demesouka, A.P. Vavatsikos, K.P. Anagnostopoulos, Suitability analysis for siting MSW landfills and its multicriteria spatial decision support system: method, implementation and case study, Waste Manag. 33 (2013) 1190–1206. doi:10.1016/j.wasman.2013.01.030.

[3] P. V Gorsevski, K.R. Donevska, C.D. Mitrovski, J.P. Frizado, Integrating multi-criteria evaluation techniques with geographic information systems for landfill site selection: a case study using ordered weighted average, Waste Manag. 32 (2012) 287–296.

[4] M. Sharifi, M. Hadidi, E. Vessali, P. Mosstafakhani, K. Taheri, Integrating multi-criteria decision analysis for a GIS-based hazardous waste landfill sitting in Kurdistan Province , western Iran, Waste Manag. 29 (2009) 2740–2758. doi:10.1016/j.wasman.2009.04.010.

[5] M.I. Yesilnacar, M.L. Süzen, B.Ş. Kaya, V. Doyuran, Municipal solid waste landfill site selection for the city of Şanliurfa-Turkey: an example using MCDA integrated with GIS, Int. J. Digit. Earth. 5 (2011) 147–164. doi:10.1080/17538947.2011.583993.

[6] A. Babalola, I. Busu, Selection of landfill sites for solid waste treatment in Damaturu Town-using GIS techniques, J. Environ. Prot. (Irvine,. Calif). 2 (2011) 1–10.

[7] H. Effat, M. Hegazy, Mapping potential landfill sites for North Sinai cities using spatial multicriteria evaluation, Egypt. J. Remote Sens. Sp. Sci. 15 (2012) 125–133.

[8] M. Uyan, MSW landfill site selection by combining AHP with GIS for Konya, Turkey, Environ. Earth Sci. 71 (2014) 1629–1639. doi:10.1007/s12665-013-2567-9.

[9] M. Zamorano, E. Molero, Á. Hurtado, A. Grindlay, Á. Ramos, Evaluation of a municipal landfill site in Southern Spain with GIS-aided methodology, J. Hazard. Mater. 160 (2008) 473–481. doi:http://dx.doi.org/10.1016/j.jhazmat.2008.03.023.

[10] A. Nazari, M.M. Salarirad, A.A. Bazzazi, Landfill site selection by decision-making tools based on fuzzy multi-attribute decision-making method, Environ. Earth Sci. 65 (2012) 1631–1642.

[11] O.B. Delgado, M. Mendoza, E.L. Granados, D. Geneletti, Analysis of land suitability for the siting of inter-municipal landfills in the Cuitzeo Lake Basin, Mexico, Waste Manag. 28 (2008) 1137–1146. doi:http://dx.doi.org/10.1016/j.wasman.2007.07.002.

[12] M. Khamehchiyan, M. Nikoudel, M. Boroumandi, Identification of hazardous waste landfill site: a case study from Zanjan province, Iran, Environ. Earth Sci. 64 (2011) 1763–1776. doi:10.1007/s12665-011-1023-y.

[13] T.D. Kontos, D.P. Komilis, C.P. Halvadakis, Siting MSW landfills on Lesvos island with a

GIS-based methodology, Waste Manag. Res. 21 (2003) 262–277.

[14] B. Şener, M.L. Süzen, V. Doyuran, Landfill site selection by using geographic information systems, Environ. Geol. 49 (2006) 376–388.

[15] T.Z. Vasiljević, Z. Srdjević, R. Bajčetić, M.V. Miloradov, GIS and the analytic hierarchy process for regional landfill site selection in transitional countries: a case study from Serbia, Environ. Manage. 49 (2012) 445–458.

[16] B. Yang, M.-H. Li, S. Li, Design-with-Nature for Multifunctional Landscapes: Environmental Benefits and Social Barriers in Community Development, Int. J. Environ. Res. Public Health. 10 (2013) 5433–5458. doi:10.3390/ijerph10115433.

[17] J. Malczewski, GIS and multicriteria decision analysis, John Wiley & Sons, 1999.

[18] M.H. Ghobadi, R. Babazadeh, V. Bagheri, Siting MSW landfills by combining AHP with GIS in Hamedan province, western Iran, Environ. Earth Sci. 70 (2013) 1823–1840. doi:10.1007/s12665-013-2271-9.

[19] H. Shahabi, S. Keihanfard, B. Bin Ahmad, M.J.T. Amiri, Evaluating Boolean, AHP and WLC methods for the selection of waste landfill sites using GIS and satellite images, Environ. Earth Sci. 71 (2014) 4221–4233.

[20] D. Alexakis, A. Sarris, Integrated GIS and remote sensing analysis for landfill sitting in Western Crete, Greece, Environ. Earth Sci. (2013) 1–16. doi:10.1007/s12665-013-2966-y.

[21] M. Banar, B.M. Kose, A. Ozkan, I.P. Acar, Choosing a municipal landfill site by analytic network process, Environ. Geol. 52 (2007) 747–751.

[22] H. Basaiaoclu, E. Celenk, M.A. Mariulo, N. Usul, SELECTION OF WASTE DISPOSAL SITES USING GIS1, JAWRA J. Am. Water Resour. Assoc. 33 (1997) 455–464. doi:10.1111/j.1752-1688.1997.tb03524.x.

[23] Chua, E.J.M. Sahid, Leong, Sustainable municipal solid waste management and GHG abatement in Malaysia, ST-4 Green Energy Manag. 04 (2011) 1–8.

[24] M. Eskandari, M. Homaee, S. Mahmoodi, E. Pazira, Integrating GIS and AHP for Municipal Solid Waste Landfill Site Selection, J. Basic Appl. Sci. Res. 3 (2013) 588–595.

[25] S.K. Jung, An integrative approach for municipal solid waste (MSW) landfill selection: A case study of O'ahu, Hawai'i, University of Hawai'i at Manoa, 2011. http://search.proquest.com/docview/962454521?accountid=14645.

[26] J. Jusoh, S. Samsudin, Improving MSW landfilling system of developing countries: an assessment of current conditions and situations in Kedah, Malaysia, Int. J. Manag. Stud. 14 (2007) 17–33.

[27] Y. Meng, Evaluating Web-based Public Participatory GIS for Multicriteria Site Selection Analysis: A Case Study in Canmore Alberta, The University of Western Ontario (Canada), 2010. http://search.proquest.com/docview/868837007?accountid=14645.

[28] M. Moeinaddini, N. Khorasani, A. Danehkar, A.A. Darvishsefat, M. Zienalyan, Siting MSW landfill using weighted linear combination and analytical hierarchy process (AHP) methodology in GIS environment (case study: Karaj)., Waste Manag. 30 (2010) 912–20. doi:10.1016/j.wasman.2010.01.015.

[29] B. Nas, T. Cay, F. Iscan, A. Berktay, Selection of MSW landfill site for Konya, Turkey using GIS and multi-criteria evaluation, Environ. Monit. Assess. 160 (2010) 491–500. doi:10.1007/s10661-008-0713-8.

[30] B. Pradhan, L. Billa, GIS modeling for selection of a transfer station site for residential solid waste separation and recycling, Pertanika J. Sci. Technol. 21 (2013) 477–488.

[31] M.O. Saeed, H.A. Aziz, M. Sanusi, A review of Municipal solid waste landfill siting techniques , Glob. J. Environ. Sci. Technol. 2 (2012) 1–10.

[32] M.A. Shaikh, Using GIS in Solid Waste Management Planning: A case study for Aurangabad, India, Linkopings University, 2006.

[33] Z. Tang, Selecting optimal residential locations using fuzzy GIS modeling, University of North Texas, 2006. http://search.proquest.com/docview/305295586?accountid=14645.

[34] iulia fontanine valentina mănoiu romulus costache, remus prăvălie, iulian mitof, Using gis techniques for assessing waste landfill placement suitability. case study: prahova county, romania, Geogr. Tech. 8 (2013) 47–56.
[35] G. Yal, H. Akgün, Landfill site selection utilizing TOPSIS methodology and clay liner geotechnical characterization: a case study for Ankara, Turkey, Bull. Eng. Geol. Environ. 73 (2014) 369–388. doi:10.1007/s10064-013-0562-8.
[36] S.Z. Ahmad, M. Sanusi, M.S. Yusoff, Spatial effect of new municipal solid waste landfill siting using different guidelines, Waste Manag. Res. 32 (2013) 24–33.
[37] H.-Y.K. Lin, Jehng-Jung, Enhanced spatial model for landfill siting analysis, J. Environ. Eng. 125 (1999) 845–851.
[38] Mahamid, S. Thawaba, Multi criteria and landfill site selection using GIS: a case study from Palestine, Open Environ. Eng. J. 3 (2010) 33–41.
[39] N.S. Romali, W.F.W. Ishak, N.B. Mokhtar, M.A.A. Samah, Solid Waste Management Development Of AHP Model for Application of Landfill Sites Selection in Kuantan Pahang, Malaysia, in: Int. Symp. Anal. Hierarchy Process (ISAHP 2013), Malaysia, 2013.
[40] M. Ministry of Housing and Local Government, Criteria for Siting Sanitary Landfills: National Strategic Plan for Solid Waste Management, Ministry of Housing and Local Government, Malaysia, Kuala Lumpur, Malaysia:, 2005.
[41] N. Al-Ansari, R. Pusch, S. Knutsson, Suggested landfill sites for hazardous waste in Iraq, Nat. Sci. 5 (2013) 463–477.
[42] K. Paul, A. Dutta, Krishna, A Comprehensive Study on Landfill Site Selection for Kolkata City, J. Air Waste Manage. Assoc. 64 (2014) 846–861.
[43] M.I. Yesilnacar, H. Cetin, Site selection for hazardous wastes: A case study from the GAP area, Turkey, Eng. Geol. 81 (2005) 371–388. doi:10.1016/j.enggeo.2005.07.012.
[44] L. Abdullah, N. Naim, A.F. Wahab, Determination of weight for landfill-siting criteria under conflicting bifuzzy preference relation, J Sustain Sci Manag. 6 (2011) 139–147.
[45] Milevski, Dimitrovska, Markoski, Svemir, Optimal landfill site selection as prevention of diseases , Tradit. Mod. Med. Geogr. Disaster Med. 1 (2012) 100–106.
[46] A.R.M. Dadras M. Farjad. B, Integration of GIS and multi-criteria decision analysis for urban solid waste management and site selection landfill in Bandar Abbas city, south of Iran, Int. Geoinformatics Res. Dev. J. 1 (2010) 14–27.
[47] Vatalis, Manoliadis, A two-level multicriteria DSS for landfill site selection using GIS: case study in western Macedonia, Greece, J. Geogr. Inf. Decis. Anal. 6 (2002) 49–56.
[48] A.F. Abdalqader, Landfills needs assessment in Gaza strip and sites selection using GIS, The Islamic University of Gaza, 2011.
[49] S.P. Gbanie, P.B. Tengbe, J.S. Momoh, J. Medo, V.T.S. Kabba, Modelling landfill location using Geographic Information Systems (GIS) and Multi-Criteria Decision Analysis (MCDA): Case study Bo, Southern Sierra Leone, Appl. Geogr. 36 (2013) 3–12. doi:10.1016/j.apgeog.2012.06.013.
[50] G.J. Lunkapis, N. Ahmad, A.R.M. Shariff, S. Mansor, GIS as Decision Support Tool for Landfills Sitting, University Putra Malaysia and GIS Support Unit, Town and Regional Planning, Sabah , 2002.
[51] N.H. Moghaddas, H.H. Namaghi, Hazardous waste landfill site selection in Khorasan Razavi Province, Northeastern Iran, Arab. J. Geosci. 4 (2011) 103–113.
[52] M.N.A. Azman, M. Sanusi, Z.A. Hamid, C.P. Gomez, K. Anuar, N.D. Hilmi, et al., The Selection of IBS Precast Manufacturing Plant in Malaysia Using GIS, Malaysian Constr. Res. J. 10 (1) (2012) 1–180.
[53] P. V. Gorsevski, K.R. Donevska, C.D. Mitrovski, J.P. Frizado, Integrating multi-criteria evaluation techniques with geographic information systems for landfill site selection: A case study using ordered weighted average, Waste Manag. 32 (2012) 287–296. doi:10.1016/j.wasman.2011.09.023.

[54] H. Houshyar, Location and Capacity Assessment, Disposal Sites, Municipal Waste, Using a Hybrid Model of AHP and Boolean, Case Study, City Bukan, Sci. Park. I (2014).

[55] Despotakis, Economopoulos, A GIS model for landfill sitting, Glob. NEST J. 9 (2007) 29–34.

[56] Zeinhom El Alfy, Rasha Elhadary, Ahmed Elashry, Integrating GIS and MCDM to Deal with Landfill Site Selection, Int. J. Eng. Technol. IJET. 10 (2010) 33–40.

[57] S.Z. Ahmad, M. Sanusi, M.S. Yusoff, Spatial Comparison of New Landfill Sites Using Criteria Derived From Different Guidelines, Res. J. 29 (2013) 29–39.

[58] O.E. Demesouka, a. P. Vavatsikos, K.P. Anagnostopoulos, Suitability analysis for siting MSW landfills and its multicriteria spatial decision support system: Method, implementation and case study, Waste Manag. 33 (2013) 1190–1206. doi:10.1016/j.wasman.2013.01.030.

[59] W. Hussin, Modeling landfill suitability based on multi-criteria decision making method , Interdiscip. Themes J. 2 (2010) 20–30.

[60] G. Biotto, S. Silvestri, L. Gobbo, E. Furlan, S. Valenti, R. Rosselli, GIS, multi-criteria and multi-factor spatial analysis for the probability assessment of the existence of illegal landfills, Int. J. Geogr. Inf. Sci. 23 (2009) 1233–1244. doi:10.1080/13658810802112128.

[61] M.S. OzeairAbessi, Site selection of a hazardous waste landfill using GIS technique and priority processing, a power plant waste in Qazvin Province case example, Environ. Sci. 6 (2009) 121–134.

[62] W.Z.W. Jaafar, How GIS can be a useful tool to deal with landfill site selection , Int. Symp. Geoinformatics Spat. Infrastruct. Dev. Earth Allied Sci. . 2 (2008) 535–541.

[63] B. Nas, T. Cay, F. Iscan, A. Berktay, Selection of MSW landfill site for Konya, Turkey using GIS and multi-criteria evaluation, Environ. Monit. Assess. 160 (2010) 491–500.

[64] H. Ersoy, F. Bulut, Spatial and multi-criteria decision analysis-based methodology for landfill site selection in growing urban regions, Waste Manag. Res. 27 (2009) 489–500.

[65] V. Akbari, M.A. Rajabi, S.H. Chavoshi, R. Shams, Landfill site selection by combining GIS and fuzzy multi criteria decision analysis, case study: Bandar Abbas, Iran, World Appl. Sci. J. 3 (2008) 39–47.

[66] R. Daneshvar, Customizing ArcMap interface to generate a user-friendly landfill site selection GIS tool, J. Solid Waste Technol. Manag. 31 (2005) 1–12. http://search.proquest.com/docview/305051124?accountid=14645.

[67] T.A. Ebistu, A.S. Minale, Solid waste dumping site suitability analysis using geographic information system (GIS) and remote sensing for Bahir Dar Town, North Western Ethiopia, African J. Environ. Sci. Technol. 7 (2013) 976–989.

[68] J. Malczewski, GIS-based multicriteria decision analysis: a survey of the literature, Int. J. Geogr. Inf. Sci. 20 (2006) 703–726.

[69] A.A. Mornya, Identification of potential landfill sites by using GIS and multicriteria method in Batam, Indonesia, 3rd Int. Grad. Conf. Eng. Sci. Humanit. (2010).

[70] R. Greene, Addressing Accessibility Challenges of GIS-based Multiple-Criteria Decision Analysis for Integrated Land Management: Case study in the Humber region of Newfoundland and Labrador, Canada, Memorial University of Newfoundland (Canada), 2010. http://search.proquest.com/docview/902448551?accountid=14645.

[71] A. Afzali, J.M. V Samani, M. Rashid, Municipal landfill site selection for Isfahan City by use of fuzzy logic and analytic hierarchy process, Iranian J. Environ. Health Sci. Eng. 8 (2011) 273–284.

One map policy (OMP) implementation strategy to accelerate mapping of regional spatial planing (RTRW) in Indonesia

Fuad Hasyim, Habib Subagio, and Mulyanto Darmawan

Center of Mapping for Spatial Planning and Atlas, Geospatial Information Agency

Email : fuad.hasyim@big.go.id

Abstract. A preparation of spatial planning documents require basic geospatial information and thematic accuracies. Recently these issues become important because spatial planning maps are impartial attachment of the regional act draft on spatial planning (PERDA). The needs of geospatial information in the preparation of spatial planning maps preparation can be divided into two major groups: (i). basic geospatial information (IGD), consist of of Indonesia Topographic maps (RBI), coastal and marine environmental maps (LPI), and geodetic control network and (ii). Thematic Geospatial Information (IGT). Currently, mostly local goverment in Indonesia have not finished their regulation draft on spatial planning due to some constrain including technical aspect. Some constrain in mapping of spatial planning are as follows: the availability of large scale of basic geospatial information, the availability of mapping guidelines, and human resources. Ideal conditions to be achieved for spatial planning maps are: (i) the availability of updated geospatial information in accordance with the scale needed for spatial planning maps, (ii) the guideline of mapping for spatial planning to support local government in completion their PERDA , and (iii) capacity building of local goverment human resources to completed spatial planning maps. The OMP strategies formulated to achieve these conditions are: (i) accelerating of IGD at scale of 1:50,000, 1: 25,000 and 1: 5,000, (ii) to accelerate mapping and integration of Thematic Geospatial Information (IGT) through stocktaking availability and mapping guidelines, (iii) the development of mapping guidelines and dissemination of spatial utilization and (iv) training of human resource on mapping technology.

1. Introduction

In general surveying and mapping activities will produce data and information about condition and potential of resources in any region. Data and information are very important for suistanable development of country. The increasing need of data from surveying and mapping activities absolutely requires guidance, policies, strategies and programs.

Based on Republic of Indonesia (RoI) law No 26 of 2007 about Spatial Planning [1], mandated that spatial planning map must be drawn up on a national, provinces, and region levels is spatial structure and pattern. In Indonesia, a spatial planning is a product of regional regulation that would be depicted on map. Geospatial information (GI) is not only used for visualization of spatial planning in form of a map, but GI has been intensived also used for inventory, monitor changes and analysis of various kind of spatial aspect to support spatial planning activities. Under The RoI Law No 4 of 2011 about Geospatial Information [2] defined that, Geospatial is data about location of area, object or an event that is naturally or artificially occuring in, on, and above the earth. The data has been processed through certain analysis method is namely geospatial information and this GI could be used as a tool in policy formulation, decision-making and/ or implementation of activities related to spatial aspects.

A preparation of spatial planning documents would require accurate and reliable of basic geospatial information (IGD) and thematic geospatial information (IGT) This is important because map would be part of attachment of regional act draft on spatial planning (PERDA). Text description written in the PERDA should illustrated spatially in a map. Therefore, the aspect of precision and accuracy is important. The need for geospatial information for spatial planning divided into two major groups:

- Basic Geospatial Information (IGD), the Topographic map of Indonesia, at scale of 1: 250,000, 1: 50,000 and 1: 25,000.
- Thematic Geospatial Information (IGT), thematic data required for spatial planning is shown at Table 1 [3].

Table 1. The need for IGT in spatial planning.

IGT types	Scale
IGT climatology, include: • rainfall map	• province: 1: 250,000 • district: 1: 50,000 • city: 1: 25,000
IGT topography, include: • elevation map • slope map	• province: 1: 250,000 • district: 1: 50,000 • city: 1: 25,000
IGT geology and soil, include: • geological map • soil type map	• province: 1: 250,000 • district: 1: 50,000 • city: 1: 25,000
IGT minerals, include: • potential excavation class C map • potential mining map	• province: 1: 250,000 • district: 1: 50,000 • city: 1: 25,000
IGT hydrology, include: • surface water map • groundwater map	• province: 1: 250,000 • district: 1: 50,000 • city: 1: 25,000
IG natural disaster, include: • tsunami disaster risk map • volcano disaster risk map • landslide disaster risk map • flood disaster risk map • earthquake disaster risk map	• province: 1: 250,000 • district: 1: 50,000 • city: 1: 25,000
IGT land cover and land use map	• province: 1: 250,000 • district: 1: 50,000 • city: 1: 25,000
IGT integration evaluation of land resources • land capability map • land suitability map	• province: 1: 250,000 • district: 1: 50,000 • city: 1: 25,000
IGT infrastructure system	• province: 1: 250,000

• transportation system map • energy/ electricity system map • telecommunications system map • water resources map	• district: 1: 50,000 • city: 1: 25,000
IGT physical development policy • forestry region map • mining region map	• province: 1: 250,000 • district: 1: 50,000 • city: 1: 25,000

Source: the ministry of public works regulation No 20 of 2007 about technical guidelines for physical aspects and environmental analysis, economic and socio-cultural in the preparation of the spatial planning map.

Availability of IGD and IGT as mentioned above are a prerequisite for ideal conditions. Maximum utilization of geospatial information should involve the accuracy and precision of IGD and completeness of IGT analysis to obtain staisfied spatial planning pattern (landuse plan and structure plan). Implementation of mapping of spatial planning in field has not optimally well because of several conditions. It has been affected for such as the data availability, guidelines, human resources, and as well as technology indeed. The solution to these conditions required a comprehensive strategy that would includes stages of the program that meet the needs for mapping of spatial planning.

2. Analysis of the spatial planning maps condition

To view conditions of spatial planning maps at regional and national levels, it can be grouped into three aspects: (i) the completion progress of spatial planning regulation, (ii) the availability of geospatial information including IGD and IGT, along with guidelines (iii) the trans boundary distribution of human resources.

2.1. The completion progress of spatial planning regulations

Completion of planning regulations refers to the RoI law No 26 of 2007 about Spatial Planning, article 78 paragraph 4 states that: "all the regulations of the district/ city on spatial planning should be arranged or adapted in 3 (three) years since of this law (2007)". The statement urges spatial planning has to be arranged or adapted at least for two years. The table below shows the progress of the legislative drafting on the spatial planning [4].

Table 2. Status of the spatial planning legislation at provincial district and city.

Status	Province	District	City
Not finished	9	70	9
Already finished	25	329	84
Total	34	399	93
Percentage of completion	73.5 %	82.5 %	90.3 %

Source: www.penataanruang.net (updated May 26, 2015).

According to table 2 can be seen that completion of the spatial plan for the province only 73.5%, 82.5% for the district and 90.3% for the city. A fact that could be drawn, until now there are 9 provinces, 70 districts and 9 cities that have not completed their spatial planning legislation draft yet. Of course more efforts are needed to assist the completion of spatial planning regulation.

2.2. Availability of geospatial information and guidelines

Table 3 shows the availability of topographic map for supporting spatial planning work. The availablity of IGD, in form of the topographic map scale of 1: 250,000 has been completed in 2013. However, topographic map scale of 1: 50,000 available at arround 72% of national territory in 2014. Topographic map scale of 1: 25,000 are available only for Java, Bali, Nusa Tenggara and Sulawesi. The other island such as Sumatra, Central Kalimantan, East Kalimantan, Papua, as well as Banten province have been pushed up to finished topographic maps at scale of 1: 25,000. In general, the availability of topographic maps for small scale (below 1:25,000) spatial planning maps is quite sufficient, however for large scale (above 1:25,000) of spatial planning map the availability of its map limited. The table below shows the availability of the topographic maps at various scales [5].

Table 3. Availability of topographic maps

Scale	Number of sheets			Percentage
	National coverage	Already available	Not yet available	
1 : 5,000	379,012	539	378,473	0.14%
1: 10,000	91,547	1,074	90,473	1.17%
1 : 25,000	13,020	3,894	9,126	29.91%
1 : 50,000	3,899	2,837	1,062	72.76%
1 : 250,000	309	309	0	100%

Source: Center for Topographic Mapping and Toponymy, Geospatial Information Agency (BIG).

The RoI Law No 4 of 2014 on Geospatial Information has mandated that the preparation of IGT is the responsibility of the Ministry / Institution, while Geospatial Information Agency (BIG) would be able to organize IGT that has not been conducted by Ministry / Institution instead of BIG. The preparation of IGT for spatial planning has encountered several problem especially in terms of coordination and technical system indeed. Issues relating to the implementation of IGT to support spatial planning may be grouped into four majors:

- The existency and distribution of the IGT for spatial planning still remain unclear that would influence on the speed of completion of spatial planning maps nationally.
- The availability of IGT that could be used for spatial planning still limited. This conditionn actually can be a trigger for starting point for the completion of IGT nationally.
- Limited guidelines document for spatial plan mapping. The guidelines will accelerate the completion of the spatial planning mapping.
- Lack of action plan in the context of accelerating the implementation of IGT involving all Ministry/ Institution.

The roadmap of topographic maps completion at scales of 1: 25,000 and 1: 50,000 have been completed. This means that the need for spatial plan mapping at these scale area already supported. The problems that still need to be resolved is how to coordinate the completion of IGT, concerning the appointment of custodianship, guidelines and make the completion for spatial plan IGT become mainstream so that it becomes a target for the entire Institution.

2.3. *The trans boundary distribution of human resources*

Geospatial Information Law in article 57 mentioned that BIG has a mandate support management of IGT. A fostering Geospatial Information to users can be done through: socialization of geospatial information and its utilization; technical education and training. In order to develop these approachment, BIG has socialized technical guidance for mapping of spatial planning, assistance and supervision of spatial planning maps, and develop human resources through geospatial education and training.

Completion of spatial planning maps are often constrained by the limitations of the survey and mapping human resources. Many regions have good data, the adequate budget, but not supported by the availability of personnel who capable to conduct about the survey and mapping. Therefore trans boundary distribution of human resources in the field of surveying and mapping is very important. Each district / city should have an adequate human resources in surveying and mapping. In addition to the trans boundary distribution of human resources, it is also related to the technological adaption aspects. Every region should be supported by an adequate survey and mapping equipment. Human resources development policy mentioned that BIG has a big role through education and training, technical consulting, and coaching mapping surveyors.

3. Ideal Conditions

Having these considerations, some ideal conditions should be pursued through programs implemented by Ministry/ Institution. The ideal situation are as follows:

- The availability of geospatial information that is accurate and reliable. The availability of geospatial information as baseline data in the preparation of spatial maps are needed in a variety of scales to the new conditions, considering the spatial planning document that could be implemented for a long time (20 years). IGD and IGT are important for spatial planning at the moment at scale of 1: 50,000 and 1: 25,000. As for the detail spatial planning required IGD at scale of 1: 5,000.
- Availability of IGT from multisector that should be integrated with spatial planning data. Spatial planning is a multidisciplinary activity which requires a lot of data, both physical and socio-cultural environment. The data should be available in accurate and appropriate scale to support spatial planning.
- Availability of documents or guidelines about IGD and IGT should be prepare for spatial planning. When such custodianship cannot be able to organize data, the data should be supported and prepared by other sector, with the guidelines that have been set by its own custodian of data it is effective to answer the availability of IGD and IGT nationally.
- The availability of human resources that could be transfer among region and technology that could support for spatial plan mapping in all regions of Indonesia. This is obviously very necessary because the maps of spatial plans are prepared by all the administrative regions in Indonesia.

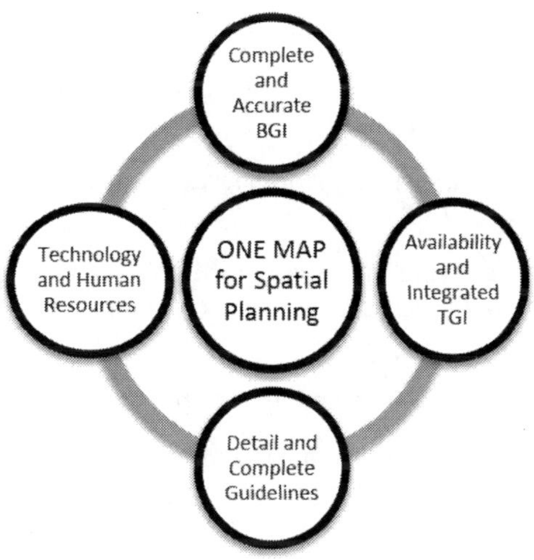

Figure 1. Ideal conditions diagram of geospatial information implementation.

Four conditions mentioned above are prerequisite that will have a direct impact on accelerating the completion of PERDA. These conditions encourage the completion of regulation of spatial plans which are equipped with IGD and IGT within the framework of "one map policy".

4. One map policy implementation strategy

One map policy implementation strategy intens to identify problems that have been outlined to be solved systematically. Along with this strategy, there are five clusters based on the analysis of the issue:

4.1. To accelerate completion of IGD for spatial planning map

- Formulate priorities of IGD completion to the specified region by proritize area that have not completed the spatial plan regulation and region which will revise spatial plan regulations on a scale of 1: 50,000 or 1: 25,000.
- Formulate priorities of IGD completion scale of 1: 5,000 for the mapping of detailed spatial plan.
- Encouraging the use of available IGD to accelerate the completion of spatial plan map through technical consultation and dissemination mechanism.

4.2. To acceleration mapping and integration of IGT

- Inventory availability of IGT based locations nationwide, as well as collecting existing IGT that can directly ready be accessed by local governments.
- Provide alternative solutions to the lack of availability of IGT for spatial planning through guidelines issued by the related Ministry/ Institution. This would encourage the use of IGT for spatial planning analysis.

4.3. To develop guidelines for mapping

- Develop an inventory of the existence guidelines and priorities of the guidelines drawn up by BIG and Ministry/ Institution relates to spatial planning.
- Prioritize guidance on the preparation of IGD alternative when the topographic maps has not available or not been updated, such as preparing high resolution satellite imagery with the standards of orthorectification that was issued by BIG.
- Disseminating of the existing guidelines for preparation of spatial planning.

4.4. Human resource development on mapping

- To conduct an employees inventory in the expertize of mapping in Indonesia.
- To socialize the requirement post for mapping surveyor and estimate the needs of human resources in spatial planning program.
- To organizes training on surveys and mapping to the government, local governments and the private sector.
- To strengthen national surveying and mapping industry and professional certification in the field of geospatial information, through the certification of the experts on geospatial information.

4.5. Coaching and consulting for spatial planning maps

- Build a team to speed up of spatial planning map program that should coordinated with BIG.
- Provide technical consulting and mapping clinic to the local government, with regional that have not completed spatial planning regulation as a priority.
- Provide technical guidance on mapping to the provincial government so that the approval process of spatial plans in each province goes well. It would become an attention to geospatial information aspect for spatial planning.

5. Conclusion

The need of geospatial information in the preparation of spatial planning that could be divided into two major groups: (i). Basic Geospatial Information (IGD), in the form Topographic maps Indonesia (RBI), (ii). Thematic Geospatial Information (IGT) include basic thematic and infrastructure network.

Obstacles faced by local governments in spatial plan mapping are: the availability of geospatial information; the availability of guidelines; and the trans-boundary resources of human resources and technology.

Ideal conditions to be achieved for spatial planning are: (i) the availability of geospatial information that is updated, (ii) the existence of guidelines for mapping that allow non-custodian to suport a thematic geospatial information according to their needs, and (iii) an adequate human resources in each region for mapping of the spatial plan mapping.

A formulated strategies to achieve ideal conditions for mapping of spatial planning are as follows : (i) to acceleratete the implementation of IGD on a scale of 1: 50,000, 1: 25,000 and 1: 5,000; (ii) to accelerate mapping and integration of IGT through providing a guidelines for IGT mapping; (iii) to develop mapping guidelines and dissemination of spatial utilization; and (iv) human resource development through training and technical guidance.

References

[1] National Spatial Planning Coordinating Board. Republic of Indonesia law No 26 of 2007 about Spatial Planning. Accessed on March 4, 2016 www.bkprn.org.

[2] Geospatial Information Agency. Republic of Indonesia law No 4 of 2011 about Geospatial Information. Accessed on March 2, 2016 www.big.go.id.

[3] Ministry of Public Works. Regulation of the Minister of Public Works No. 20 / PRT / M / 2007 on Technical Guidelines for Physical Aspects and Environmental Analysis, Economic and Socio-Cultural in the preparation of the Spatial Planning. Accessed on March 4, 2016 www.birohukum.pu.go.id.

[4] Ministry of Public Works. 2015. Status RTRW Bylaw provincial, district and city. Accessed on March 4, 2016 www.penataanruang.net.

[5] Center for Topographic Mapping and Toponimy, Geospatial Information Agency. Availability Index Topographic Map of Indonesia. 2015.

Landscape analysis of urban growth patterns in Seremban, Malaysia, using spatio-temporal data

Maher M Aburas[1, 2], Sabrina H Abdullah[1*], Mohammad F Ramli[1], Zulfa H As'shari[1]

[1] Faculty of Environmental Studies, University Putra Malaysia, 43400 Malaysia.
[2] The High Institution for Engineering Vocations Almajori, Benghazi, Libya.

E-mail: yuekming@upm.edu.my.

Abstract. Urban growth is one of the major issues that have played a significant role in destroying the ecosystem in recent years. Landscape analysis is an important technique widely used to evaluate urban growth patterns. In this study, four land-use maps from 1984, 1990, 2000, and 2010 have been used to analyze an urban landscape. The values of a built-up area were initially computed using a geographic information system environment based on the spatial gradient approach. Mathematical matrices were then used to determine the amount of change in urban patches in each direction. Results of the number of patches, landscape shape index, aggregation index, and total edges confirmed that the urban patches in Seremban, Malaysia, have become more dispersed from 2000 to 2010. The urban patches have also become more continuous, especially in the north-western part of Seremban as a result of the urban development in the Nilai District. These results indicate the necessity to create new policies in the city to protect the sustainability of the land use of Seremban.

1. Introduction

Urban growth is a global issue that leads planners and decision makers to be concerned about its future effects on the ecosystem [1]. This complex issue includes large-scale changes in the land cover and use at the local and regional levels [2-4]. Geographic information systems (GISs) and remote sensing (RS) have been used in recent years for several applications, including the detection of land-use changes [5]. Land-use maps extracted from satellite images refer to the significant data commonly used in urban studies [6, 7]. The number of land-use classes in each map is determined on the basis of image resolution. An increase in satellite image resolution should correspond to an increase in the number of land-use classes in each map [8].

Spatio-temporal data, such as satellite images and land-use maps, are widely used to identify urban growth patterns using landscape analysis [3] as they provide a strong visualization of phenomena in different periods. They can also provide a statistical representation to analyze, measure, and identify urban growth in specific areas [9]. Landscape analysis is a powerful technique used to quantify urban dynamics spatially and temporally [3]. Landscape metrics are extensively used to explain the structures of urban built-up areas [10] and to describe the interrelationship of intra- and inter-land uses and the driving forces of urban growth [11, 12].

* Corresponding author. Tel: +1-603-89466735; Fax No: +1-603-89467468

In this study, the gradient approach has been used to understand the urban growth trends in Seremban, the capital of the Negeri Sembilan State, Malaysia, in consideration of the temporal land-use maps in the periods of 1984, 1990, 2000, and 2010. This study has mainly aimed to quantify urban dynamics using temporal RS data through landscape analysis.

2. Methodology

2.1. Study area

Seremban City is the largest district in the Negeri Sembilan State. The study area is located between longitudes of 101° 45' 0" E and 102° 6' 0" E and latitudes of 3° 0' 0" N and 2° 30' 0" N). The city occupies a total land area of approximately 935.78 km². Seremban City is composed of the districts of Seremban, Setul, Labu, Rasah, Ampangan, Rantau, Pantai, and Lenggeng (Fig. 1). This city is located approximately 20 km from Putrajaya, which is the national capital of Malaysia, and approximately 67 km from Kuala Lumpur, the economic center of Malaysia.

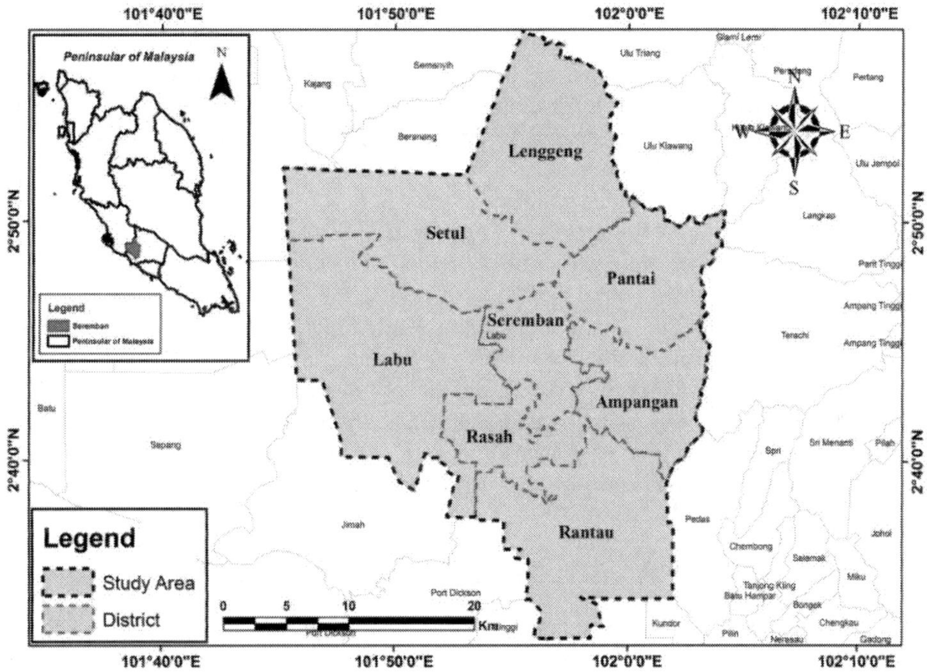

Figure 1. Location of the study area.

2.2. Materials and method

This study utilized certified land-use maps from the Department of Agriculture of Malaysia from 1984, 1990, 2000, and 2010. These land-use maps were extracted from Spot 2, 4, and 5 images. All spot images were registered and geo-corrected using ground control points through the global positioning system (GPS). Image-enhancement techniques were used to classify all the images with accuracy. The supervised classification method was used to classify all clipped images to extract land-use categories. Field data were collected using GPS to assess the classification accuracy by comparing the classified images with the GPS points from the field for each type of land use. The accuracy assessment values reached the acceptable values of the kappa index, and thus the image classification was acceptable. According to the Anderson scheme, the acceptable kappa index value for the accuracy assessment is more than 0.85 [8].

The accuracy assessment method was applied to the land-use maps of 2000 and 2010 to ensure that the classification accuracy of the land-use maps generated by the Malaysian Department of Agriculture reached acceptable values. This study assumed that if the values of the overall accuracy and kappa coefficient of the land-use maps of 2000 and 2010 were acceptable, then all the land use maps used in this study would also be acceptable. High-resolution Google maps were used to conduct the accuracy assessment of the 2000 and 2010 maps. Random samples of each class were used on the basis of the distribution and density of each class in the study area. This method was easily conducted by using the create random point tool in the ArcGIS 10.2 software. The sample number of each class was identified on the basis of the stratified random sample. The minimum number of samples of each class should be 50, except for water, which could be less [13]. This study randomly selected 340 points for verification. Fifty-five errors were identified. The total accuracy values of the land-use maps of 2000 and 2010 were 92% and 91%, respectively. The kappa coefficients of the land-use maps of 2000 and 2010 were 0.90 and 0.89, respectively. Thus, the classification of the land-use maps met the study requirements.

The land-use maps were reclassified into two types of land use; namely, built-up and non-built-up areas, to comply with the general objective of this study. The built-up areas included residential, commercial, services, industrial, transportation, communications, utilities, mixed urban or built-up land, and other urban or built-up land. The non-built-up areas included other types of land use, such as water bodies, agricultural lands, forests, and open areas. All categories of built-up areas were combined according to the Anderson scheme. The main reason for classifying land-use maps to be built-up and non-built-up area classes was that all of the types of spatial analysis in this study were applied to the built-up area to evaluate and measure the urban growth patterns in Seremban [3, 4, 9, 14]. The spatial analysis applied to the category of built-up area is explained as follows.

Built-up area analysis based on the gradient approach: The study area was divided into four zones based on spatial directions. Each zone was further divided into concentric circles of incrementing radii of 3 km from the central business district (*CBD*) to understand the process of changes based on spatial directions and the gradient from the center to the edge. Each direction was divided into nine concentric circles from the *CBD* to explain the urban dynamics at the local level (Fig. 2). This approach can help planners and decision makers determine the causal factors of urban growth, and it can also help in visualizing the forms of urban expansion by using multiple landscape matrices (Table 1). All landscape matrices were employed by using the Fragstats software. All the built-up areas in each direction and circle were clipped and converted into the ASCII format in ArcGIS 10.2 to use them in the Fragstats environment.

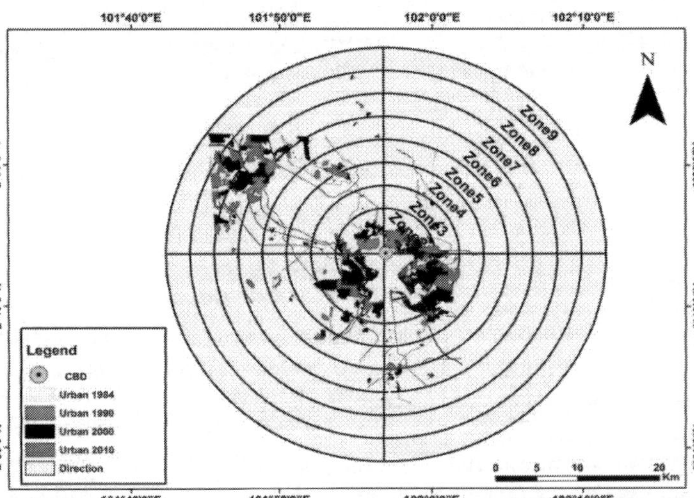

Figure 2. Diagram of the approach used to evaluate urban growth in Seremban according to gradient analysis.

Table 1. Landscape metrics used to analyze the status of the overall urban dynamics.

Landscape metrics	Equation	Description
Number of patches (*NP*)	$N = n_i$	Range $NP \geq 1$ NP = number of patches of the corresponding patch category. n_i = number of patches of a particular category.
Landscape Shape Index (*LSI*)	$LSI = e_i/min\ e_i$	$LSI > 1$, without limit e_i = total length of edge of class i in terms of the number of cell surfaces. $min\ e_i$= minimum total length of edge of class i in terms of the number of cell surfaces.
Aggregation Index (*AI*)	$AI = \left[\frac{g_{ii}}{max_{g_{ii}}}\right](100)$	$1 \leq AI \leq 100$ g_{ii} = number of like adjacencies (joins) among pixels of patch type (class) i according to the single count method. $max_{g_{ii}}$ = maximum number of like adjacencies (joins) among pixels of patch type class i according to the single count method.
Total Edge (*TE*)	$TE = \sum_{k}^{m} = 1e_{ik,}$	$e_{ik,}$ total length (m) of edge in landscape involving patch type (class) i; it includes landscape boundary and background segments involving patch type i.

3. Results and discussion

3.1. Landscape analysis

3.1.1. Number of urban patches(NP)

NP refers to the level of fragmentation of built-up areas [9]. Figure 3 shows the direction and gradient wise number of built-up patches during the different time periods. A major increase in the *NP* from 1984 to 2010 in all zones and in all gradients indicates that the landscape became more fragmented after 1990. The *NP* values decreased near the *CBD*; this decrease explains the process of aggregation of the built-up patches. The northwestern (NW) and southwestern directions illustrate a major increase. A significant increase occurred in built-up patches in most gradients, especially along the edges of the NW direction. The occurrence of urban sprawl is indicated.

3.1.2. Landscape shape index (LSI)

LSI is a simple method used to determine the aggregation and disaggregation of built-up areas [3]. Figure 4 presents lower values after 1990 at the *CBD*, thus indicating simpler shapes and more compact built-up patches. Fringes indicate increased *LSI* values. Fringes refer to complex shapes with fragmented growth, as highlighted by the *NP* in 2000 and 2010 and the development of complex patches. The zones without *LSI* values indicate no built-up area.

3.1.3. Aggregation index (AI)

AI is widely used to measure the aggregation and dispersion of built-up area patches, and it ranges from 0 to 100 [9]. Figure 5 shows the results of the *AI*. The NW zone was generally more dispersed than other directions because of new development projects, especially in the Nilai District, which considered new business areas in the NW part of Seremban [15]. The northeastern part of Seremban was less affected by urban development, and urban patches presented more aggregation near the *CBD*. By contrast, the zones near the edge were devoid of urban patches because they are considered highlands and are inappropriate for urban development[15].

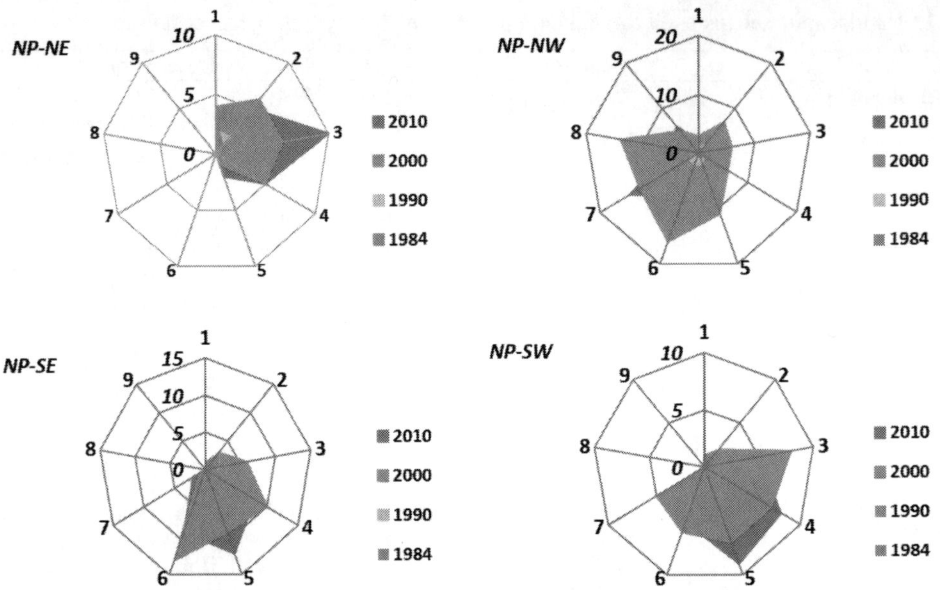

Figure 3. Number of patches metric– direction-wise/circle-wise.

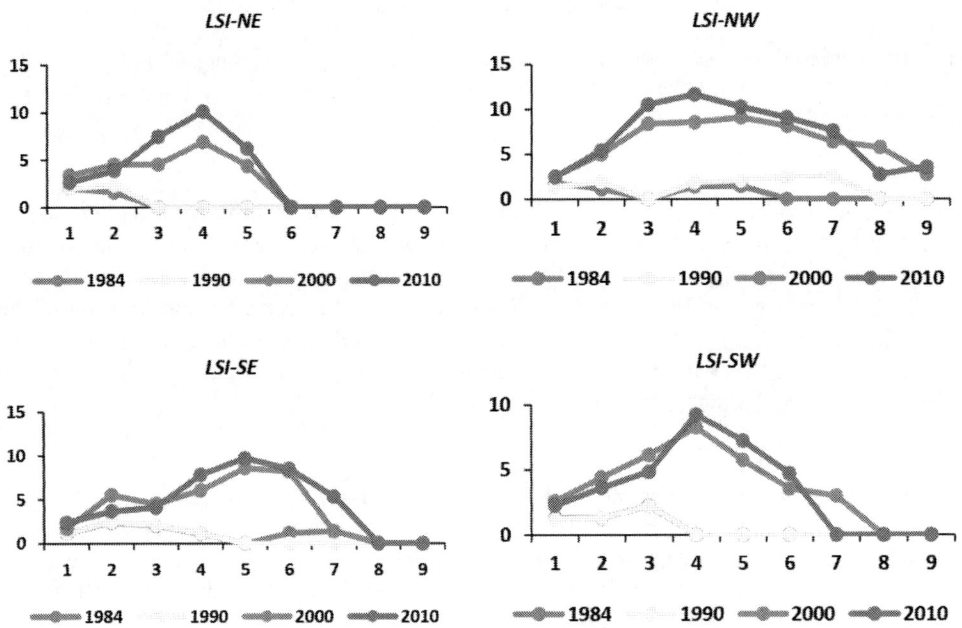

Figure 4. Landscape shape index– direction-wise/circle-wise.

3.1.4. Total edge (TE)

TE is the absolute measure of the total length (perimeter) of the built-up areas of a particular patch in a landscape in meters. Large values of *TE* indicate large, continuous urban patches. *TE* values are greater than or equal to zero. Figure 6 illustrates that in 1984 and 1990, the edges were smaller, and thus the urban patches were discontinuous because the landscape was fragmented. In 2000 and 2010, the results of the *TE* show larger edges. These results indicate that urban patches were ubiquitous and continuous in 2000 and 2010. The *TE* findings imply continuous urban growth and the breadth of its area in most zones of Seremban.

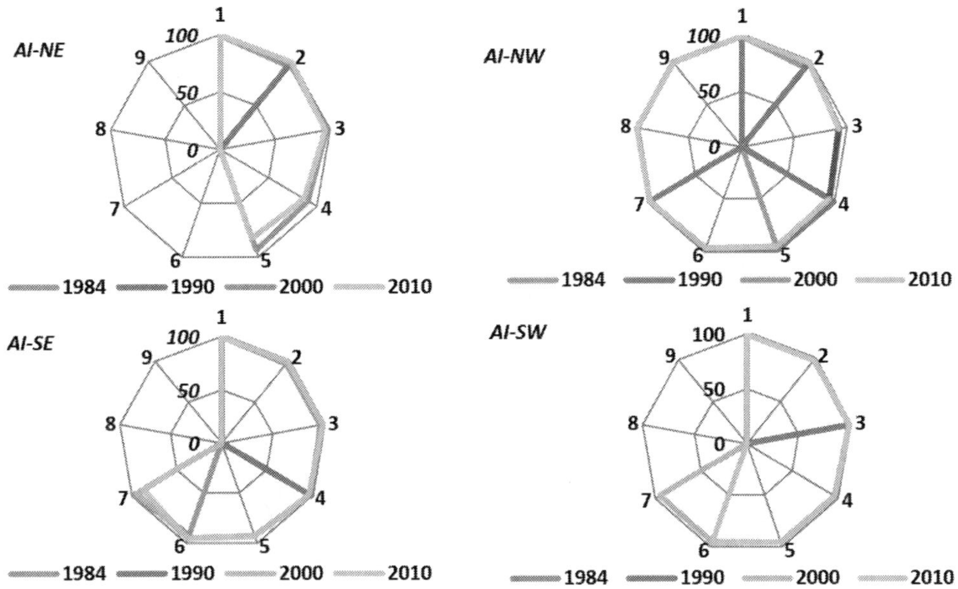

Figure 5. Aggregation Index– direction-wise/circle-wise.

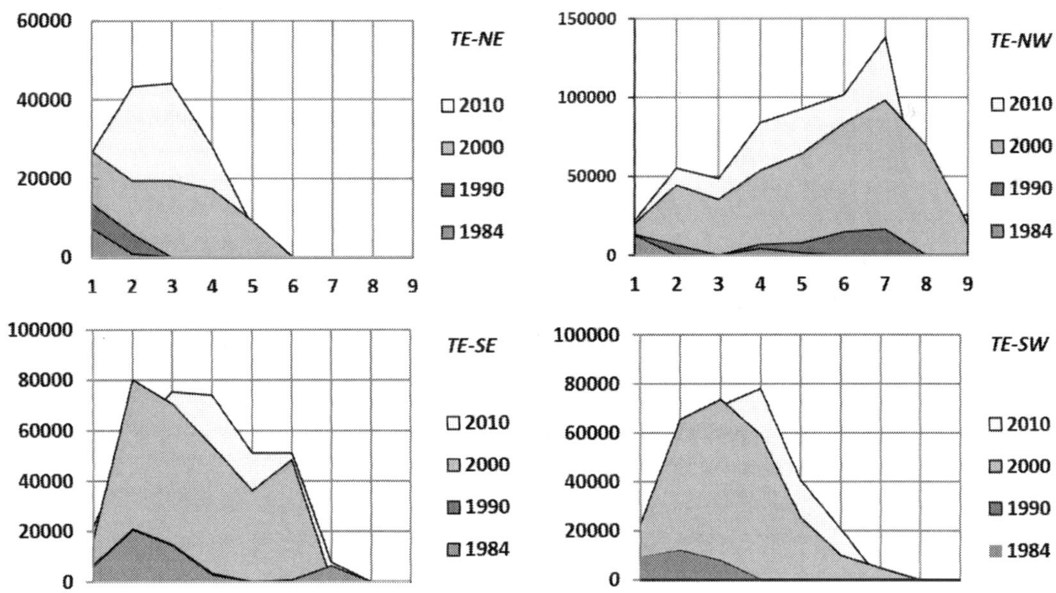

Figure 6. Total Edge metric – direction-wise/circle-wise.

4. Conclusion

Landscape analysis has recently become an important indicator used to evaluate urban growth patterns. This analysis can effectively help monitor urban dynamics and support planners and decision makers in creating sustainable urban strategies. The results of the landscape analysis confirmed that Seremban City is facing expansibility and fragmented and continuous urban patches. The results also showed that the *NW* direction of Seremban has a significant increase in built-up patches in most gradients, and this increase explains the process of urban dynamics in Seremban. This study has demonstrated the potential of *RS* and *GIS* data in providing an accurate means, supplemented with landscape analysis and gradient approach, to understand urban development trends.

Acknowledgments The study presented here is part of the research project funded by the University Putra Malaysia under the research grant scheme.

5. References

[1] Bihamta, N., et al., Using the SLEUTH Urban Growth Model to Simulate Future Urban Expansion of the Isfahan Metropolitan Area, Iran. Journal of the Indian Society of Remote Sensing, 2014: p. 1-8.
[2] Bhatta, B., S. Saraswati, and D. Bandyopadhyay, Quantifying the degree-of-freedom, degree-of-sprawl, and degree-of-goodness of urban growth from remote sensing data. Applied Geography, 2010. 30(1): p. 96-111.
[3] Aithal, B.H. and D.D. Sanna, Insights to urban dynamics through landscape spatial pattern analysis. International Journal of Applied Earth Observation and Geoinformation, 2012. 18: p. 329-343.
[4] Ren, P., et al., Spatial Expansion and Sprawl Quantitative Analysis of Mountain City Built-Up Area, in Geo-Informatics in Resource Management and Sustainable Ecosystem. 2013, Springer. p. 166-176.
[5] Maasikamäe, S., H. Hass, and E. Jürgenson. The Impact of Uncontrolled Development on the Use of Arable Land. in Proceedings: The Fifth International Scientific Conference Rural Development 2011, 24-25 November, 2011, Akademija. 2011.
[6] Uy, P.D. and N. Nakagoshi, Application of land suitability analysis and landscape ecology to urban greenspace planning in Hanoi, Vietnam. Urban Forestry & Urban Greening, 2008. 7(1): p. 25-40.
[7] Dewan, A.M. and R.J. Corner, Spatiotemporal Analysis of Urban Growth, Sprawl and Structure, in Dhaka Megacity. 2014, Springer. p. 99-121.
[8] Anderson, J.R., A land use and land cover classification system for use with remote sensor data. Vol. 964. 1976: US Government Printing Office.
[9] Ramachandra, T., H. Bharath, and M. Sowmyashree, Analysis Of Spatial Patterns Of Urbanisation Using Geoinformatics And Spatial Metrics. Theoretical and Empirical Researches in Urban Management, 2013. 8(4): p. 5-24.
[10] Herold, M., J. Scepan, and K.C. Clarke, The use of remote sensing and landscape metrics to describe structures and changes in urban land uses. Environment and Planning A, 2002. 34(8): p. 1443-1458.
[11] Deng, J.S., et al., Spatio-temporal dynamics and evolution of land use change and landscape pattern in response to rapid urbanization. Landscape and Urban Planning, 2009. 92(3): p. 187-198.
[12] Setturu, B., et al. Landscape Dynamics through Spatial Metrics. in Proceedings of India GeoSpatial Conference, Epicentre, Gurgaon, India. 2012.
[13] Congalton, R.G., A review of assessing the accuracy of classifications of remotely sensed data. Remote sensing of environment, 1991. 37(1): p. 35-46.
[14] Al-sharif, A.A., et al. Quantitative analysis of urban sprawl in Tripoli using Pearson's Chi-Square statistics and urban expansion intensity index. in IOP Conference Series: Earth and Environmental Science. 2014. IOP Publishing.
[15] Seremban, L.C.o., Local plan for Seremban area. Report, 2015. Volume I: Part B.

Modeling urban expansion in Yangon, Myanmar using Landsat time-series and stereo GeoEye Images

Tanakorn Sritarapipat, Wataru Takeuchi

Institute of Industrial Science, The University of Tokyo, 4-6-1 Komaba, Meguro-ku, Tokyo, 153-8505, Japan

tanakorn@iis.u-tokyo.ac.jp

Abstract. This research proposed a methodology to model the urban expansion based dynamic statistical model using Landsat and GeoEye Images. Landsat Time-Series from 1978 to 2010 have been applied to extract land covers from the past to the present. Stereo GeoEye Images have been employed to obtain the height of the building. The class translation was obtained by observing land cover from the past to the present. The height of the building can be used to detect the center of the urban area (mainly commercial area). It was assumed that the class translation and the distance of multi-centers of the urban area also the distance of the roads affect the urban growth. The urban expansion model based on the dynamic statistical model was defined to refer to three factors; (1) the class translation, (2) the distance of the multi-centers of the urban areas, and (3) the distance from the roads. Estimation and prediction of urban expansion by using our model were formulated and expressed in this research. The experimental area was set up in Yangon, Myanmar. Since it is the major of country's economic with more than five million population and the urban areas have rapidly increased. The experimental results indicated that our model of urban expansion estimated urban growth in both estimation and prediction steps in efficiency.

1. Introduction

Urban growth is rather a complex process. Since urban expansion causes from many factors such as human behaviors, population rates, the economic states, the policies of the government and so on. Remote sensing technology provides the physical information that can directly observe the urban areas from the past to the present, and can detect the center of the urban areas. Hence, it can lead us to understand the mechanism of urban expanding and create the model. Urban expansion model can be taken an employment to understand how urban expands and predict urban areas in the future. This information can be used to support the urban development and management.

Urban expansion model has been widely introduced. Urban land-use model based on spatial interaction model was developed by Lowry [1]. The statistical model was used for introducing urban expansion model [2]. An Urban growth model based on automata cellular was proposed by Batty [3]. Moreover, by using multi-agent-based model, the residential distribution estimation was developed [4].

This work introduced a methodology to model the urban expansion based dynamic statistical model using Landsat and GeoEye Images in Yangon, Myanmar. Yangon is the major of country's economic with more than five million population, and the urban areas have rapidly increased [5].

2. Methodology

2.1. Factors to indicate urban expansion
Firstly, we defined that urban expansion is related to (1) the class translation, (2) the distance from the multi-centers of the urban areas and (3) the distance from the roads. For the first factor, we assumed that the number of new urban areas can be identified following the statistical of the class translation. For the second and third factors, the locations of new urban areas can be calculated regarding the distance from the multi-centers of the urban area and the distance from the roads. We assumed that the center of the urban area is the group of the high-rise buildings such as department store, hotel, office, school and so on. We supposed that each area depends on the nearest center of the urban area. The flowchart of our methodology to model urban expansion is displayed in figure 1.

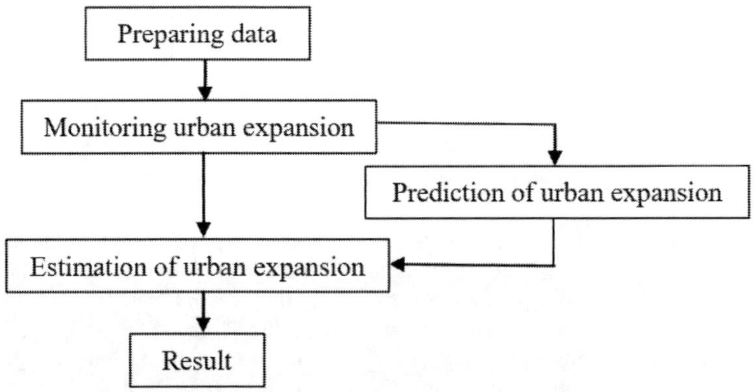

Figure 1. The flowchart of our methodology.

2.2. Preparing data for urban expansion model
In this research, four multi-spectral Landsat images with a 30 m.-resolution and 1,000x1,000 pixels from 1978 to 2009 were used to provide land cover change. Stereo GeoEye images with a 0.5 m.-resolution were employed to obtain the height of the building. The locations of the roads were provided by ICUS (International Center for Urban Safety Engineering, The University of Tokyo, Japan). The details of satellite dataset are shown in table 1 and the images are displayed in figure 2.

Table 1. The details of Satellite images.

No.	Satellite and Sensor	Bands	Resolution	Acquired Date
1	Landsat -3 MMS	4	60m. x 60m.	1978-11-22
2	Landsat -4 TM	7	30m. x 30m.	1990-11-12
3	Landsat -7 ETM	8	30m. x 30m.	2000-11-7
4	Landsat -5 TM	7	30m. x 30m.	2009-11-08
5	GeoEye Stereo-mode	3	0.5m x 0.5m.	2013-11-08 2013-11-16

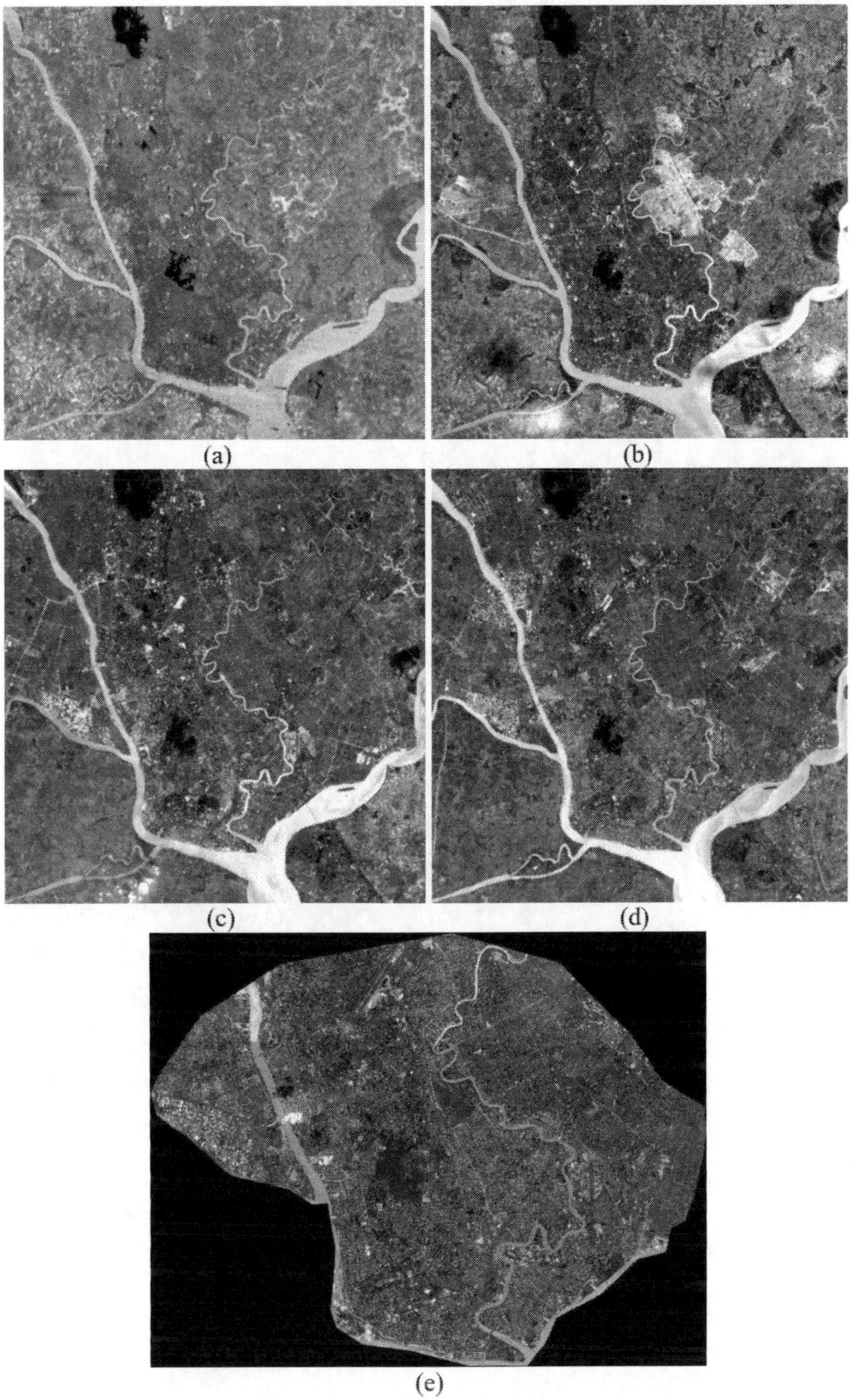

Figure 2. (a) Landsat image in 1978 (False color), (b) Landsat image in 1990, (c) Landsat image in 2000, (d) Landsat image in 2009, (e) GeoEye image.

For obtaining the land cover change, the Landsat images were classified into three classes of (1) urban, (2) vegetation, and (3) water by using Mahalanobis distance method (supervised classification). Unfortunately, there is still some noise in the classification results. To improve the classification results, the rule of urban expansion and removing cloud effect were applied. The classification results are shown in figure 3 (a) and 8 (a, c, e).

For detecting the multi-centers of the urban area, the stereo GeoEye images were extracted to obtain the height of the building (figure 3 (b)). Then, the heights of the buildings were separated into two classes of (1) low and (2) high buildings by manual thresholding. After that, the locations of the high buildings were grouped into seven classes by using K-means (Unsupervised classification). The center positions of seven classes were defined as the multi-centers of the urban area (figure 3 (c)). The region of the multi-center areas is displayed in figure 3 (d).

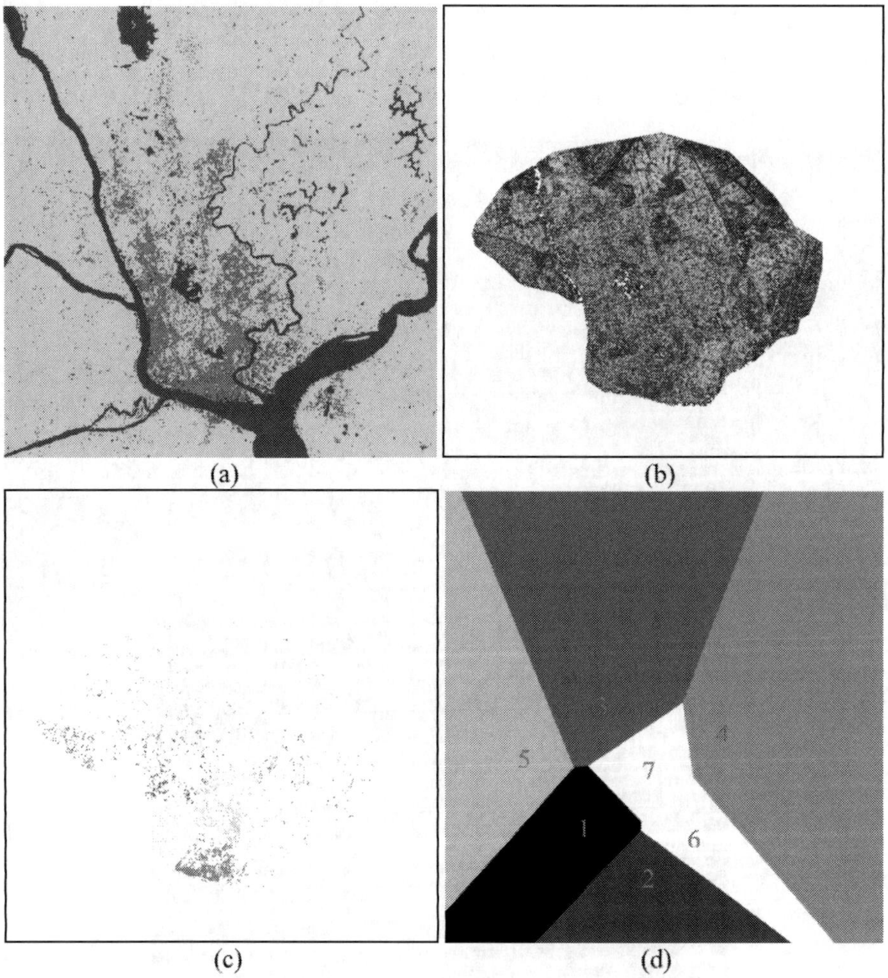

Figure 3. (a) the land cover result in 1978 (b) the heights of the buildings (c) the classification result of the multi-centers, and (d) the region of multi-centers.

2.3. Monitoring urban expansion

For the statistical of class translation, we received the number of class translation by observing the land cover areas from the past to the present. By combining with the multi-centers of the urban area, the examples of the number of class translation in the region of the center #1 with time variation are illustrated in table 2 and figure 4.

Table 2. The number of class translation in the region of the center #1 from 1978 to 2009.

	1978-1990	1990-2000	2000-2009
urban→urban	8,410	10,942	12,780
vegetation→urban	2,123	1,774	2,208
water→urban	409	64	69

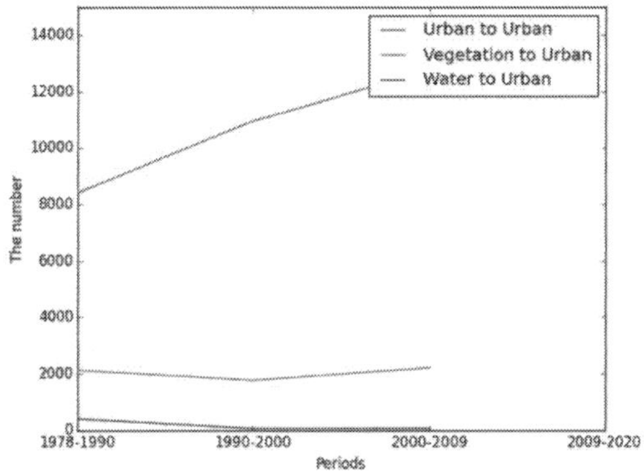

Figure 4. The number of class translation in the region of the center #1 from 1978 to 2009.

For the distance from the multi-centers of the urban area, we observed the urban growth to obtain the histogram of urban expansion affected by the distance from the multi-centers. The examples of the histograms of the urban expansions among the varied distance from the center of the region #1 with time variation are depicted in figure 5.

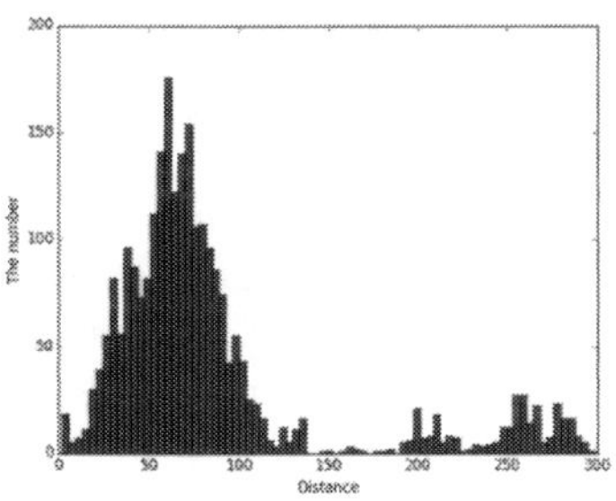

(a)

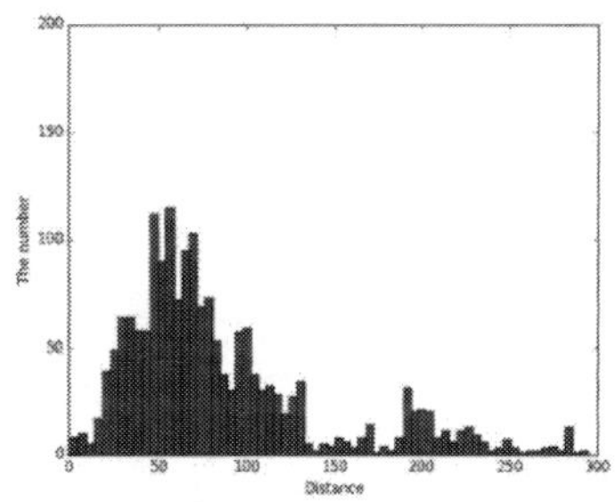

(b)

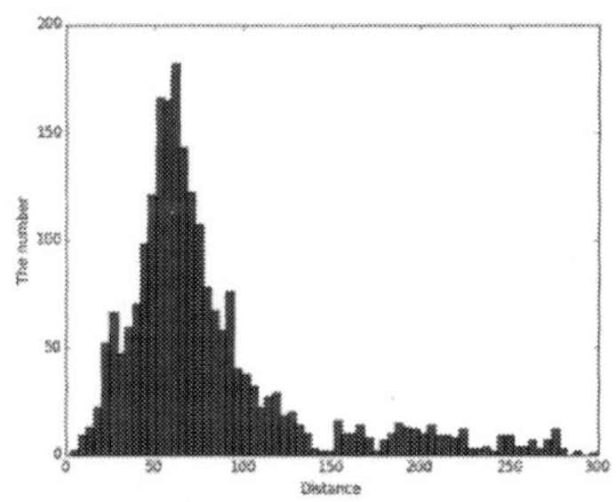

(c)

Figure 5. The histograms of the urban expansions among the varied distance from the center of the region #1 in the years of (a) 1978-1990, (b) 1990-2000, (c) 2000-2009.

Table 3. Means and variances of the distance from the center that was effect to urban expansion in the region #1 from 1978 to 2009.

	1978-1990	1990-2000	2000-2009
Mean	87.12	90.23	87.51
Variance	4411	4260	4608

We investigated that the urban expanded by beginning from the close distance from the center of the urban area to the far distance. The distributions of the histograms seem as Gaussian distribution.

For the distance from the roads, the histogram of the urban expansion affected by the distance from the roads was provided by observing urban growth. The histogram of the distance from the roads is shown in figure 6. The mean and variance of the distance from the roads by urban expansion are 1.42 and 4.34, respectively.

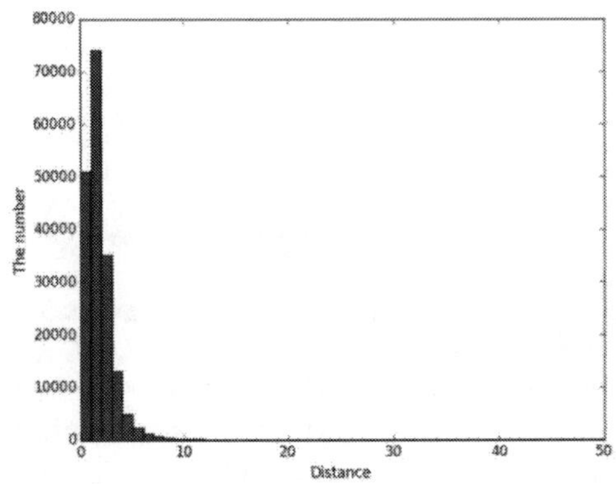

Figure 6. The histogram of the urban expansion among the varied distance from the roads.

We found that the urban grew along the roads, and the distribution appears as Gaussian distribution.

2.4. Estimation of urban expansion

The probabilities of class translation with time variations were defined as Markov chain. The probabilities of the distance from the center and the roads were assumed as Gaussian distribution. By integrating the probabilities, the estimation of our model by using maximum likelihood estimator was expressed in equation 1.

Maximize *The Probability of Class translation* + *The probability of the distance from the multi-centers* + *The probability of distance from the roads* (1)

Since the Probability of Class translation and the probability of the distance from the multi-centers and the distance from the roads are independent, we separated them into two parts with equation 2 and 3.

Maximize *The Probability of Class translation* (2)

Maximize *The probability of the distance from the multi-centers* + *The probability of distance from the roads* (3)

$$\text{Maximize} \prod_{i=1,j=1}^{1000,1000} [\beta_1 \frac{1}{\sigma_{center}\sqrt{2\pi}} \exp(-\frac{(disCenter(\mathrm{i},\mathrm{j}) - \mu_{center})^2}{2\sigma_{center}^2}) \times \beta_2 \frac{1}{\sigma_{road}\sqrt{2\pi}} \exp(-\frac{(disRoad(\mathrm{i},\mathrm{j}) - \mu_{road})^2}{2\sigma_{road}^2})] \quad (4)$$

Since some parts are insignificant, and some parameters are constants, we converted equation 4 into equation 5.

$$\text{Minimize} \sum_{i=1,j=1}^{1000,1000} [\lambda_1 \frac{(disCenter(\mathrm{i},\mathrm{j}) - \mu_{center})^2}{\sigma_{center}^2} + \lambda_2 \frac{(disRoad(\mathrm{i},\mathrm{j}) - \mu_{road})^2}{\sigma_{road}^2}] \quad (5)$$

i and j are the locations of the images in x-axis and y-axis. λ_1 and λ_2 are required to assign. In our experiment, we found that $\lambda_1 = 0.3$ and $\lambda_2 = 0.7$ that provided a high accuracy. By using equation 2, the number of urban expansion was obtained. By using equation 5, the locations of urban expansion were provided. Then, the classification result in 1978 was set as the initial land cover image. We used the initial land cover image with observed parameters as input for our model to estimate the land cover images in 1990. Next, we used the estimated land cover in 1990 with observed parameters as input for our model to estimate the land cover image in 2000. We repeated the same step to estimate land cover image in 2009.

2.5. Prediction of urban expansion

Since the probability of the class translation and mean and variance of the distance from the center of the urban area in the future could not be observed. They are required to estimate. We used the previous information with polynomial regression to calculate the parameters used for prediction in the future. In this research, we estimated the urban areas in 2020 as the future time. The example of estimated parameter of the number of class translation to predict the urban areas in 2020 is shown in figure 7.

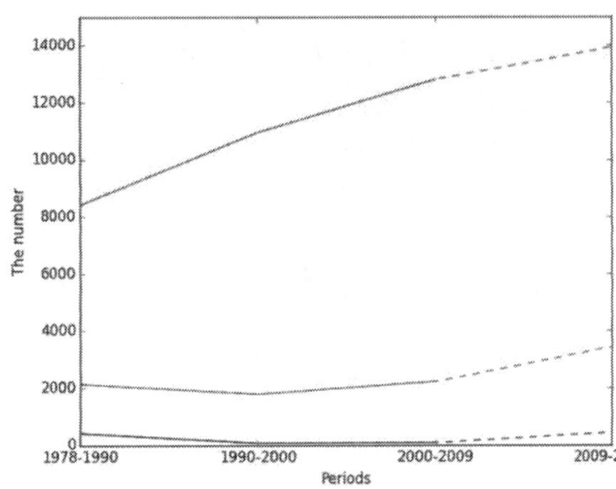

Figure 7. The estimated number of class translation in the region of the center #1 from 1978 to 2020.

IOP Conf. Series: Earth and Environmental Science **37** (2016) 012056 doi:10.1088/1755-1315/37/1/012056

After that, we used the estimated land cover in 2009 with estimated parameters as input for our model to estimate the land cover image in 2020.

3. Results and discussion

3.1. Estimation of urban expansion

The classification results in the section of 2.2 preparing data for urban expansion were defined as original land cover images. We compared the estimated land cover images by using our model (figure 9 (b, d, f)) with the original land cover images (figure 9 (a, c, e)). For the accuracy, two classes with urban and non-urban (vegetation and water) were used to calculate the accurate result. The accuracy of original versus estimated land cover images in 1990, 2000, and 2009 are expressed in table 4.

Table 4. The accuracy of original versus estimated urban areas in 1990, 2000, and 2009.

Year	Accuracy (%)	True positive (%)	True negative (%)
1990	93.51	64.95	96.52
2000	88.35	61.37	93.26
2009	83.72	63.68	89.62

3.2. Prediction of urban expansion

The Landsat image on November 25, 2015 (figure 8) was selected and classified to be a land cover image as an unseen land cover image (figure 10 (a)). We compared the predicted land cover image in 2020 (figure 10 (b)) with the unseen land cover image in 2015. The accuracy is 81.24% with true positive of 70.86% and false positive of 85.64%.

Figure 8. Landsat image in 2015

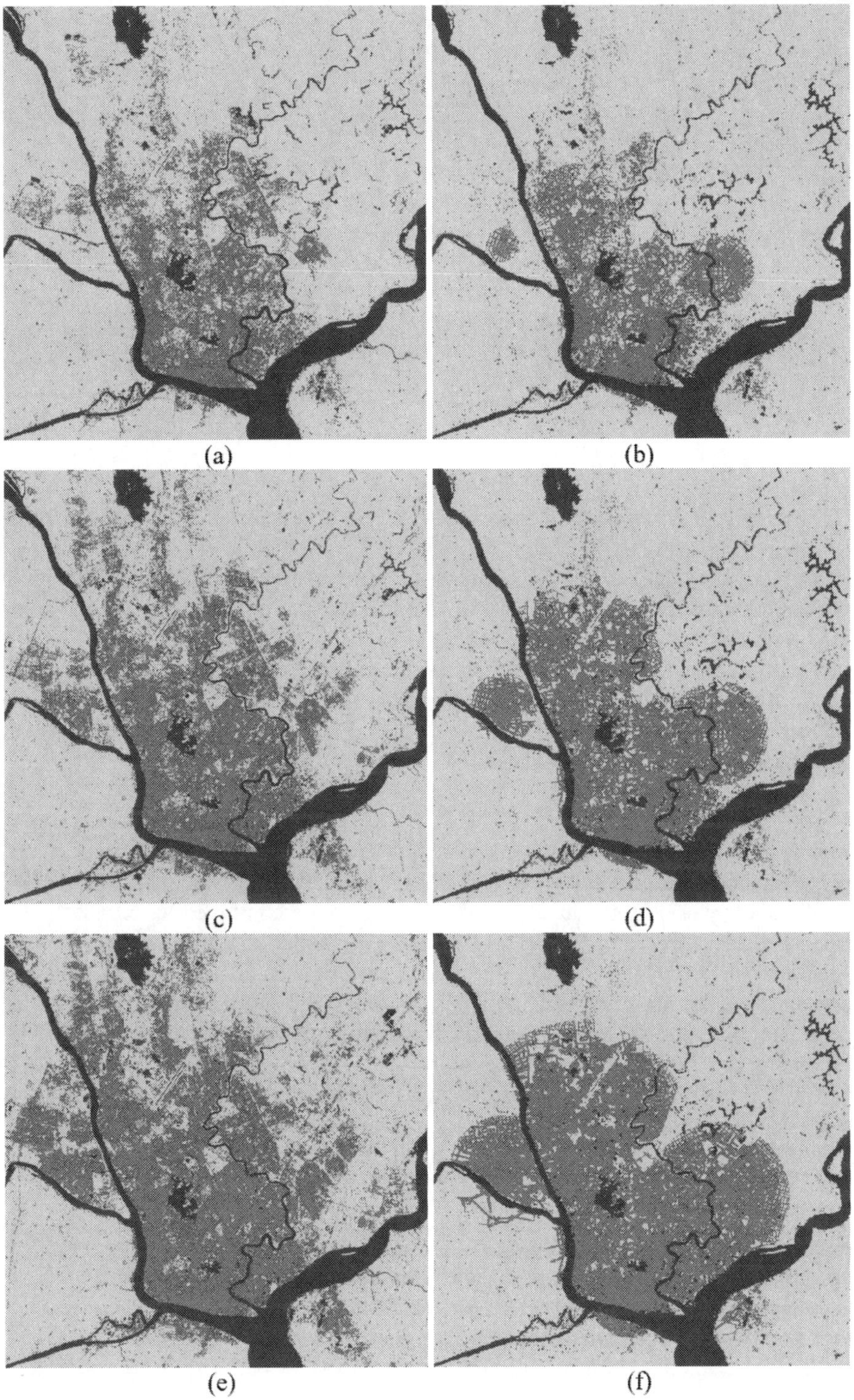

Figure 9. (a) original land cover in 1990, (b) estimated land cover in 1990, (c) original land cover in 2000, (d) estimated land cover in 2000, (e) original land cover in 2009, and (f) estimated land cover in 2009

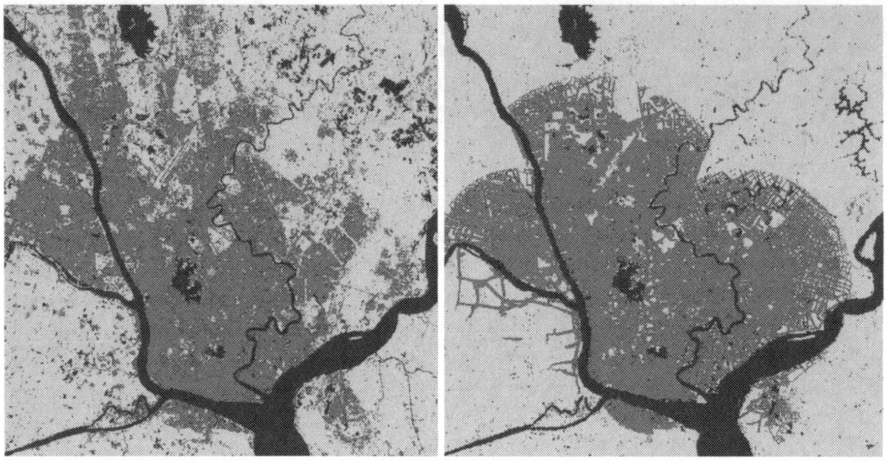

(a) (b)

Figure 10. (a) original land cover in 2015, and (b) estimated land cover in 2020

In the experimental results, our model of urban expansion estimated the urban growth in the years of 1990, 2000, 2009 with an average accuracy of 88.53% and also predicted the urban expansion in the future (in the year of 2020) with an accuracy of 81.24%.

Our model relies on the distance from the multi-centers and estimated urban areas grow up from the closest distance from the multi-centers of the urban area. However, In particular, some urban areas grow up far from the centers. As a result, our method cannot estimate the urban areas that grow up far from the multi-centers.

4. Conclusion

This research introduced a methodology to model urban expansion using Landsat time-series and stereo GeoEye images in Yangon, Myanmar. Multi-spectral Landsat images from 1978-2009 were used to provide land cover change. Stereo GeoEye images were employed to extract the height of the building that can detect the center of the urban area. The model of urban expansion was defined to refer three factors of class translation, the distance from the multi-centers, and the distance from the roads. Based on the dynamic statistical model, Estimation and prediction of urban expansion were formulated. In the experimental results, our method estimated urban areas from 1990 to 2009 and predicted urban areas in 2020 with an accuracy.

For future work, the other factors such as the direction of urban expansion, an elevation should be included to obtain a higher accuracy. The significant structure such as transportation, the bridge would be related to our model to understand the practical mechanism of urban expansion. Since some parameters are required to specify, the optimization method could be employed.

References

[1] Lowry I S 1964 *A model of metropolis Santa Monica*, CA (Rand Corporation)
[2] Sklar F H and Costanza R 1991 *The development of dynamic spatial models for landscape ecology* (A review and prognosis) pp 239-288
[3] Batty M 1997 Cellular automata and urban form (A primer) *Journal of the American Planning Association* 63(3) pp 264-274
[4] Benenson I 1998 Multiagent simulations of residential dynamics in the city *Computers Environment and Urban Systems* 22(1) 25-42
[5] Morley I 2013 Rangoon *Cities* (31) 601–614

Socio-economic impact of Trans-Siberian railway after the collapse of Soviet Union by integrated spatial data analysis

Seina Uchida[1], Wataru Takeuchi[1], Kiichiro Hatoyama[2], Yuri Mazurov[3]

[1]Institute of Industrial Science, The University of Tokyo, Meguro 4-6-1, Tokyo 153-8505, Japan
[2]Graduate School if Civil Engineering, The University of Tokyo, Japan
[3]Department of Geography, Moscow State University, Russia

seina@iis.u-tokyo.ac.jp

Abstract: How Russian cities have stood up again after the collapse of Soviet Union will be discussed in this paper. In order to know how the cities has managed the difficult period after the change of social system, transition of urban area, population, and nighttime light is searched. Although Far East will not stop as one of the most important area with abundant resources, overpopulation in towns and depopulation in countryside is going on. By searching the present situation, this research also aims to predict the future of Far East and Russia. First of all, Landsat data from 1987 to 2015 is collected over Moscow, Vladivostok, Novosibirsk, Tynda, and Blagoveshchensk and urban area is calculated by land cover classification. Secondly, population and retail turnover data are collected from year books in Russia. Thirdly, gross regional product (GRP) is estimated by nighttime light images from DMSP-OLS and VIIRS DNB dataset. In addition, these data are compared and difference of development stage after the collapse of Soviet Union between the unstable era (1990s-2000) and development era (2000-) will be discussed. It is expected that these analysis will give us useful information about Russian strategy for the future.

1. Introduction

Trans-Siberian railway (TSR) is the longest railway in the world. It is constructed in 1916 and has been used for 100 years. TSR plays an important role not only for passenger transport, but also for cargo transport. Since its history is very long, there are so many researches about this railway. Mori, T. [1] studied about resent use of TSR and guessed future use. However, there are very few research which focused on the cities along the railway. Although Rishat, M. [2] mentioned about development of Novosibirsk, which is along TSR, its analysis is limited in that it does not based on enough quantitative analysis. We can say that this study is unique in that it aims to detect the changes of the cities along TSR by using statistical number and information extracted from satellite images.

This study aims to know how the cities along TSR has restored from the huge impact of collapse of Soviet Union. TSR has succeeded in improving Russian economy and developing cities along it. Cities along TSR has equally developed before the collapse of Soviet Union, but after that, the way to recover from the damage is different to city to city. This research also tried to know the difference of the cities.

2. Methodology

2.1 Flow chart of this study

Flow chart of this study is shown in Figure 1. Population was collected in the Year Books in library of Moscow State University. For calculating urban area, Landsat images were used over 1985 to 2015. Each city was masked according to its shapefile and supervised classification was conducted. For nighttime light, VIIRS monthly image of August 2014 was used. After collecting these data, these data were compared.

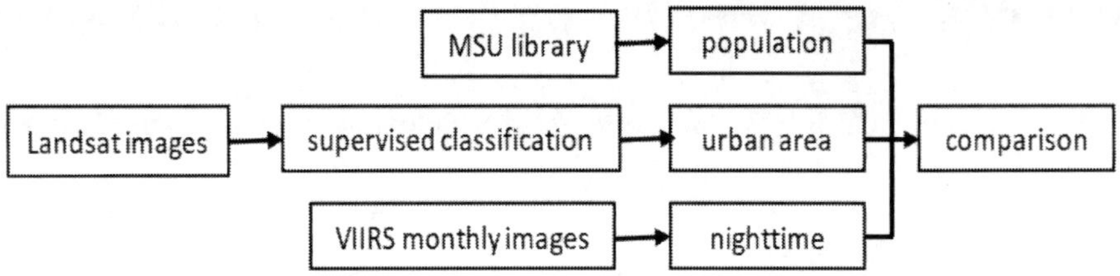

Figure 1. Flow chart of this study about data sources and analytical process.

2.2 Test site

Six cities shown in Figure 2 were selected. Tynda is not along TSR, but all of the other cities are along TSR. Moscow, Ekaterinburg, Novosibirsk, Blagoveshchensk, and Vladivostok were chosen because these were big, developed cities and were along TSR. Tynda was selected as a comparison target.

Figure 2. Test site of this study along TSR.

2.3 Data used in this study

Population data were collected by some books in the library of Moscow State University. Population of each six cities were collected for about 100 years. Urban area was calculated by Landsat images. Images were gained from 1990 to 2015. Urban area, water, vegetation, and follow land were chosen as region of interest and supervised classification was conducted. Since Tynda is a very small city and even inside administrative district of Tynda, there is forest, which produces more error in classification. That's why only urban area around the station is cut and classification was conducted only in the square region in the case of Tynda.

Nighttime light was gained by NPP-VIIRS (http://ngdc.noaa.gov/eog/viirs.html). Data of August in 2014 were used because in winter, reflects of snow increase the amount of light and it will decrease the research quality. GADM shapefile was used to cut the cities according to administrative definition. Total light amount was calculated for each city and it was divided by area of each city. However, in the case of Tynda, since the administrative area was too large and it was difficult to compare, the same area which was used when classification was chosen and nighttime light amount in the square area was calculated.

3. Results and discussion

3.1 Urban area changes by Landsat

Figure 3 and 4 shows the change of urban area. The area of the first year of observation is set one and rate is shown in this figure. Tynda is the only one city whose urban area is decreased since 1990. We can see that all the cities along TSR has expanded urban area. Especially in Ekaterinburg and Vladivostok, urban area did not change before 2010 so much, but after 2010 urban area started to increase greatly.

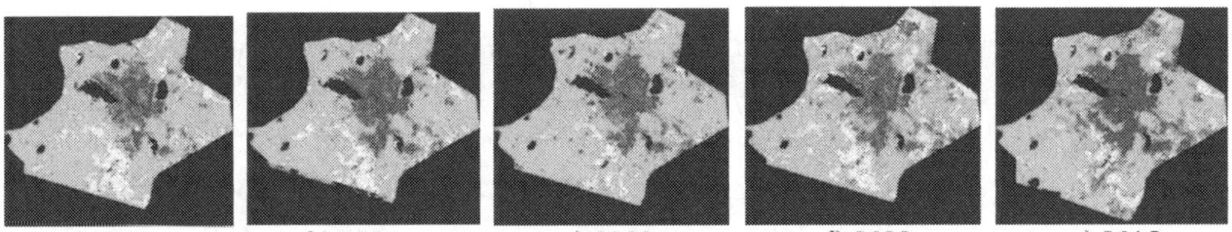

a) 1991 b) 1994 c) 2000 d) 2009 e) 2015

Figure 3. Classified result of Ekaterinburg by using Landsat images. 10km

■ (red) urban area ■ (green) vegetation (yellow) fallow land ■ (blue) outside the

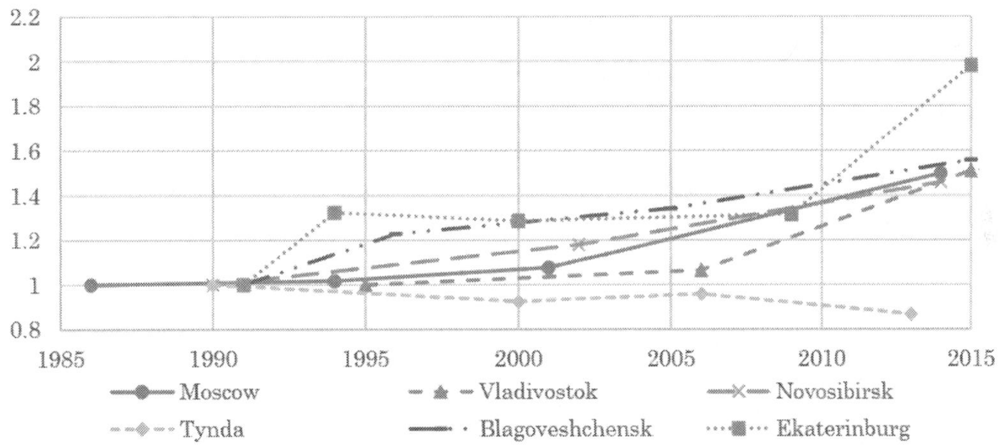

Figure 4. Urban area change of six target cities shown with the value of the first observation year set as one in each city.

3.2 Nighttime light at 2014

Table 1 shows that the total amount of light is much higher in Moscow than in other cities. According to Figure 5, nighttime light is concentrated especially at the cross point of railway. This area is the place where station is located, so we can assume that city has developed around railway station.

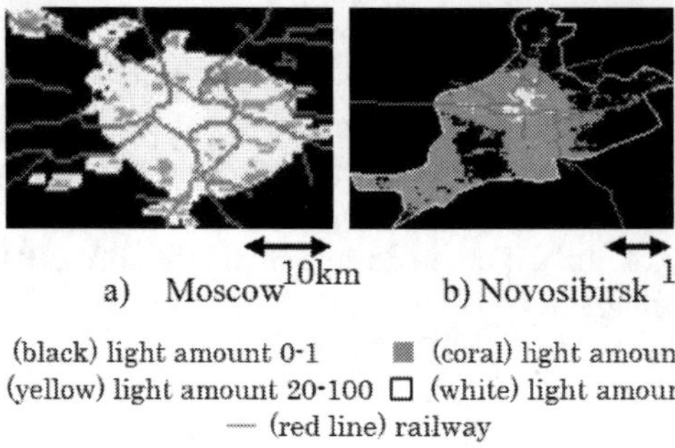

Figure 5. Nighttime image in August 2014 taken by VIIRS in (a) Moscow and (b) Novosibirsk.

Table 1. Sum of Nighttime light in August 2014 by VIIRS dataset. Unit is [nanoWatts/cm2/sr].

	Moscow	Ekaterinburg	Novosibirsk	Tynda	Blagoveshchensk	Vladivostok
Sum of nighttime light	101907	17550	22960	1795	3373	6739

3.3 Population changes

According to Figure 6, population of cities along TSR has increased greatly after construction of TSR. However, there was an unstable era after the collapse of Soviet Union, during which population has stopped increasing. Recently, from around 2010, population has started increasing again. However, in the case of Tynda, which is not along TSR, population has increased greatly only in 1970s and 1980s. It might be because government's policy to attract workers to Far East. Right after the collapse of Soviet Union, which means after this policy has over, Tynda population starts to decrease.

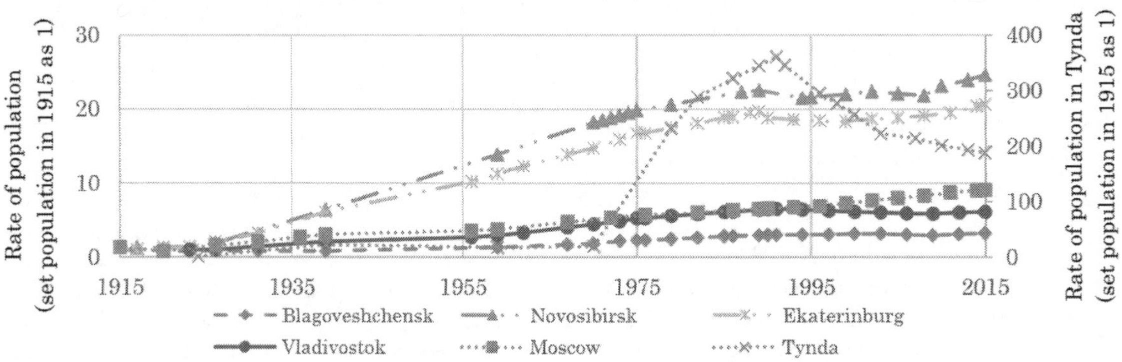

Figure 6. Population transition of six cities (set population in 1915 as one).

3.4 Discussions

Figure 7 shows population and nighttime light has strong relation. It can be a good reason for high value of nighttime light in Moscow in Table 1. Moscow is the most populated city in Russia, so its nighttime light is also very bright.

Figure 8 shows that in Moscow and Novosibirsk, nighttime light assumed by urban area is higher than other cities, and in Vladivostok and Ekaterinburg, nighttime light assumed by urban area is lower than other cities. This fact means that in Moscow and Novosibirsk, the city is more developed to vertical. Moscow is the largest, and Novosibirsk is the third largest city (second is Saint Petersburg) in Russia, and there might be many skyscraper in these cities.

Figure 8 also shows that the plot of Tynda is on the approximate curve. Figure 5 and 6 shows that although the population of Tynda has decreased greatly, urban area does not change so much. It might be because even though the number of people has decreased and there are some houses which are not used anymore, satellite image cannot tell them from houses which is still in use. From this fact, it can be said that what lit in Tynda does not houses but some factories or commercial buildings. These kind of buildings will not stop to light even if the number of people working there decreases.

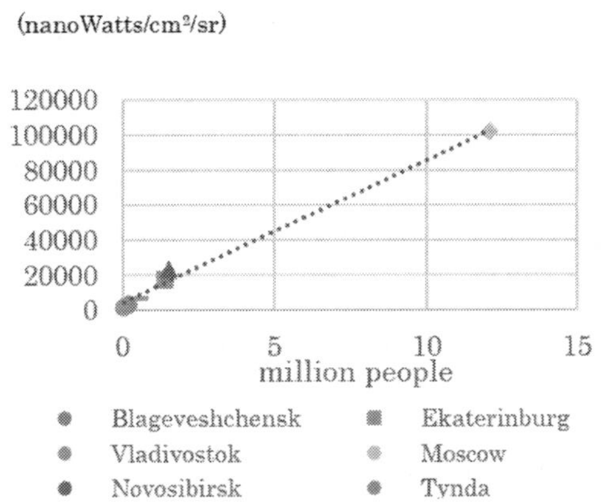

Figure 7. Population and nighttime light of each city.

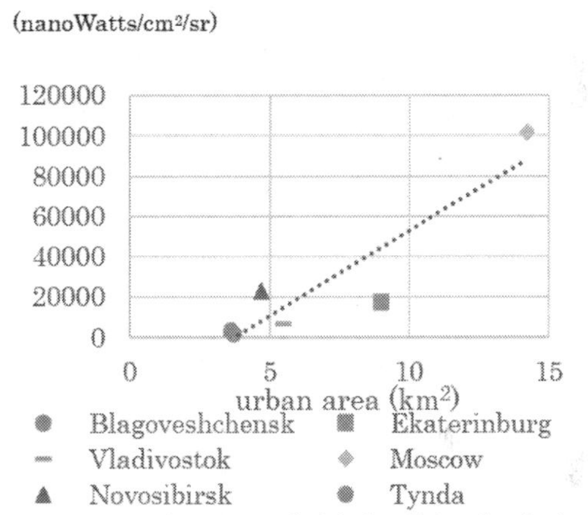

Figure 8. Urban area and nighttime light of each city.

Figure 5 shows that Ekaterinburg and Vladivostok has greatly increased its population after 2010. In the case of Vladivostok, there was APEC conference held in 2012, so Russian government made some policies to construct infrastructure and made some roads. We can guess that the same kind of thing have happened also in Ekaterinburg. That may be the reason why in these cities nighttime light amount is smaller than expected from urban area. Since the great development of these cities has happened just recently, there were not that much people living in the city. According to Figure 7, population and nighttime light amount has liner relation, which leads the result in Figure 8.

4. Conclusion and future works

In conclusion, we can say that TSR has played an important role in increasing the number of people in the cities along TSR. However, because of collapse of Soviet Union, population transition pattern after 1990 differs from city to city. But Russia has gradually stood up against this difficulty and after around 2010, each city along TSR has turned to increase its population. TSR has also effected urban area along it. Urban area of the cities along TSR has increased after 1990, but urban area of Tynda, which is not along TSR, has not increased. Although all the five cities along TSR has increased its urban area after 1990, transition pattern differs from city to city. In those cities, Ekaterinburg and Vladivostok has greatly increased urban area after 2010. Seeing that other cities has increased gradually, it might be the result of some policy. In the case of Vladivostok, APEC conference was held in 2012, and government has invested a lot of money to construct infrastructure. The same kind of thing might have happened in Ekaterinburg. And this result might have affected the relation of urban area and nighttime light. And about Tynda, concerning population has decreased, urban area has not decreased so much, and nighttime light is not that small, people might be working in factories or commercial buildings. In order to know the effect of TSR more exactly, more cities which is not along TSR need to be searched.

Acknowledgements
I would like to thank JSPS for supporting the Russian and Japanese student exchange program managed by the cooperation of the natural science and civil engineering.

References

[1] Mori, T. (2006). Second spot light of Siberian land bridge: basics of global logistics. 92-93.
[2] Rishat, M. 2003. A study on new cities development related to Trans-Siberian railway development on the sample of Novosibirsk city (1893-1940). Technical papers of annual meeting Architectural Institute of Japan, 481-482
[3] Federal State Statistics Service. (n.d.). Retrieved March 3, 2016, from http://www.gks.ru/wps/wcm/connect/rosstat_main/rosstat/en/figures/transport/
[4] Knoema world data atlas. (n.d.). Retrieved March 3, 2016, from http://knoema.com/atlas/russian-federation
[5] Shi, K. et al. (2014). Evaluating the Ability of NPP-VIIRS Nighttime Light Data to Estimate the Gross Domestic Product and the Electric Power Consumption of China at Multiple Scales: A Comparison with DMSP-OLS Data. *Remote Sensing, 6.* 1705-1724.
[6] Welch, R. (1980). Monitoring urban population and energy utilization patterns from satellite data. *Remote Sensing of Environment 9.* 1-9.

Tank Update System: A novel asset mapping approach for verifying and updating lakes using Google Maps

Bhaskar Reddy Pulsani
Telangana State Remote Sensing Applications Centre (TRAC), Hyderabad, India.

E-mail: pulsani@yahoo.co.in

Abstract. Mission Kakatiya is one of prestigious programs of Telangana state government under which restoration of tank across ten districts is being implemented. As part of the program, government plans to restore about 9,000 lakes. Therefore, to have a comprehensive list of lakes existing in Telangana state, Samagra Tank Survey was carried out. Data collected in this survey contained about 45,000 tanks. Since the mode of collection of data was not in a standard format and was made using excel, a web interface was created to fill the gaps and to standardise the data. A new approach for spatially identifying the lakes through Google maps was successfully implemented by developing a web interface. This approach is less common since it implements the nature of asset mapping for the lakes of Telangana state and shows the advantages of using online mapping applications such as Google maps in identifying and cross checking already existing lakes on it.

1. Introduction

After the division of the state of Andhra Pradesh and formation of new state called Telangana on 2nd June 2014, the Government of Telangana has come up with a new programme "Mission Kakatiya" for restoration of tanks across the ten districts of Telangana. The name was given as a tribute to the Kakatiya dynasty rulers who had built a large number of irrigation tanks in Telangana state. "Mission Kakatiya" is one of prestigious flagship programmes of the government which went with the famous tagline - "Mana Vooru - Mana Cheruvu (Our Village - Our Lakes)". Under the program, the government plans to restore about 45,000 tanks in the next five years. It will also be addressing the issues where tanks have been neglected and encroached.

Irrigation & CAD department of Telangana state is responsible for planning, designing, construction and maintenance of major, medium and minor dams [1]. The government of Telangana has instructed I & CAD department to take up Samagra Tank Survey during October 2014. As part of this exercise mandal level officers have gathered information on lakes existing in their jurisdiction and submitted them to the Chief Engineer of I & CAD. Initial estimates from the raw data indicate that there are about 45,000 tanks in the State. Since the data received from different mandal level officers was not in a standard format, there was an urgent need to compile the data and place it in a proper format. In this effort Telangana State Remote Sensing Applications Centre (TRAC) took initiative in compiling the data. After examining the data it was found that there were numerous missing fields where data was incomplete. Therefore, to standardize the data a Web based Tank Update System was designed and developed for inserting, updating and deleting the attributes of the surveyed data.

The approach taken up for the current study implements the nature of asset mapping which is defined as mapping of organized attributes of a community [3]. Asset mapping has been used by John McKnight and John Kretzmann for community development. It involves strengthening a community after analysing the mapped assets [2]. Most of the applications of asset mapping have been in the field of community development. The other major implementation has been in the field of crowd sourcing. Openstreetmap is one of renowned examples where users from all around the world can contribute free geographic data by digitising on the portals provided by it. iD and Potlatch are two famous editors which provide the tools needed for creating, manipulating and updating free geographic data. Both the editors use aerial imagery of Bing maps as background for tracing [5]. Delineation of water bodies forms one component of the various components existing on these interfaces. The current study implements this approach for identifying and capturing information of lakes using Google Maps. A few examples of identifying some archaeological sites were put forward by Fragen to bring forth the advantages of using Google Earth for reconnaissance survey. He suggests that even though Google Earth survey through remote sensing maps could never replace ground survey, it provides systematic aerial reconnaissance for planning and pinpointing sites before going into the field [6]. The current study uses this kind of approach in identifying the already existing lakes on Google maps which also acts as the background map for retrieving the coordinates of these lakes. ASP.NET, Google Maps and PostgreSQL database were used to build the application.

2. Data sets

Raw data collected as part of Samagra Tank Survey by the Junior Engineers (JE) was taken as datasets. Figure 1: Format for collecting data through Excel shows a format for data collection defined in an excel sheet. It contains twenty fields which need to be collected for each tank by all the JE's.

Since, the format for collection of data was given as an excel sheet, the collected data had numerous errors and the standards were not followed while entering. To be able to implement the web interface for update by Junior Engineers (JE), the data was edited and was standardised at mandal level. A total of 45,000 records were submitted by the JE's at the end of the survey. When all records were clubbed together, 3,000 records that were not in the standard format were discarded. The clubbed 42,000 records then were imported to the database for further processing. Two tables were created in the database - Tank Table and Login Table. Tank table was created with the defined fields of the survey

S.No	Name of the Tank	Village	Mandal	Constituency	District	Local Name of the Source if any	Type of Source 1. Minor Irrigation 2. Anicuts / Diversion Schemes 3. Small PR Tank 4. Check Dam 5. Percolation Tank 6. Private Kunta 7. Any Water Bodies created by any dept.
1	2	3	4	5	6	7	8
1	Marri Kunta	Chagal	Stn.Ghar	Ghanpur(Stati	Warangal	Marri Kunta	Minor Irrigation
2	Garla Kunta	Chinnapi	Stn.Ghar	Ghanpur(Stati	Warangal	Garla Kunta	Minor Irrigation
3	Puttalamma cheru	Ghanpu	Stn.Ghar	Ghanpur(Stati	Warangal	Puttalamma cheru	Minor Irrigation
4	Nagula cheru	Ippagud	Stn.Ghar	Ghanpur(Stati	Warangal	Nagula cheru	Minor Irrigation
5	Koya cheru	Lingamp	Stn.Ghar	Ghanpur(Stati	Warangal	Koya cheru	Minor Irrigation
6	Oora cheru	Malkapu	Stn.Ghar	Ghanpur(Stati	Warangal	Oora cheru	Minor Irrigation
7	Pedda cheru	Namilgo	Stn.Ghar	Ghanpur(Stati	Warangal	Pedda cheru	Minor Irrigation
8	Venkatadri cheru	Pallagutt	Stn.Ghar	Ghanpur(Stati	Warangal	Venkatadri cheru	Minor Irrigation

S.No	Longitude	Latitude	Basin	Sub Basin	Ayacut (In Acres)	Direct Ayacut / Indirect Ayacut	Available in rolls of the Department (Yes / No)	Approximate length of Bund (In Mtrs)	Weir existing (Yes / No)	Sluice Existing (Yes / No)	Photo along Team Members Enclosed or Not	Remarks
1	9	10	11	12	13	14	15	16	17	18	19	20
1	79°19	17°54	Krish	K12 (	127	Direc	yes	220	yes	yes		Repairs Required
2	79°24	17°53	Krish	K12 (	105	Direc	yes	600	yes	yes		Repairs Required
3	79°21	17°50	Krish	K12 (	105	Direc	yes	607	yes	yes		Repairs Required
4	79°22	17°46	Krish	K12 (	139	Direc	yes	1000	yes	yes		Repairs Required
5	79°14	17°56	Krish	K12 (	133	Direc	yes	1000	yes	yes		Repairs Required
6	79°17	17°56	Krish	K12 (	264	Direc	yes	1050	yes	yes		Repairs Required
7	79°24	17°50	Krish	K12 (	202	Direc	yes	900	yes	yes		Repairs Required
8	79°54	17°19	Krish	K12 (	132	Direc	yes	1000	yes	yes		Repairs Required

Figure 1: Format for collecting data through Excel.

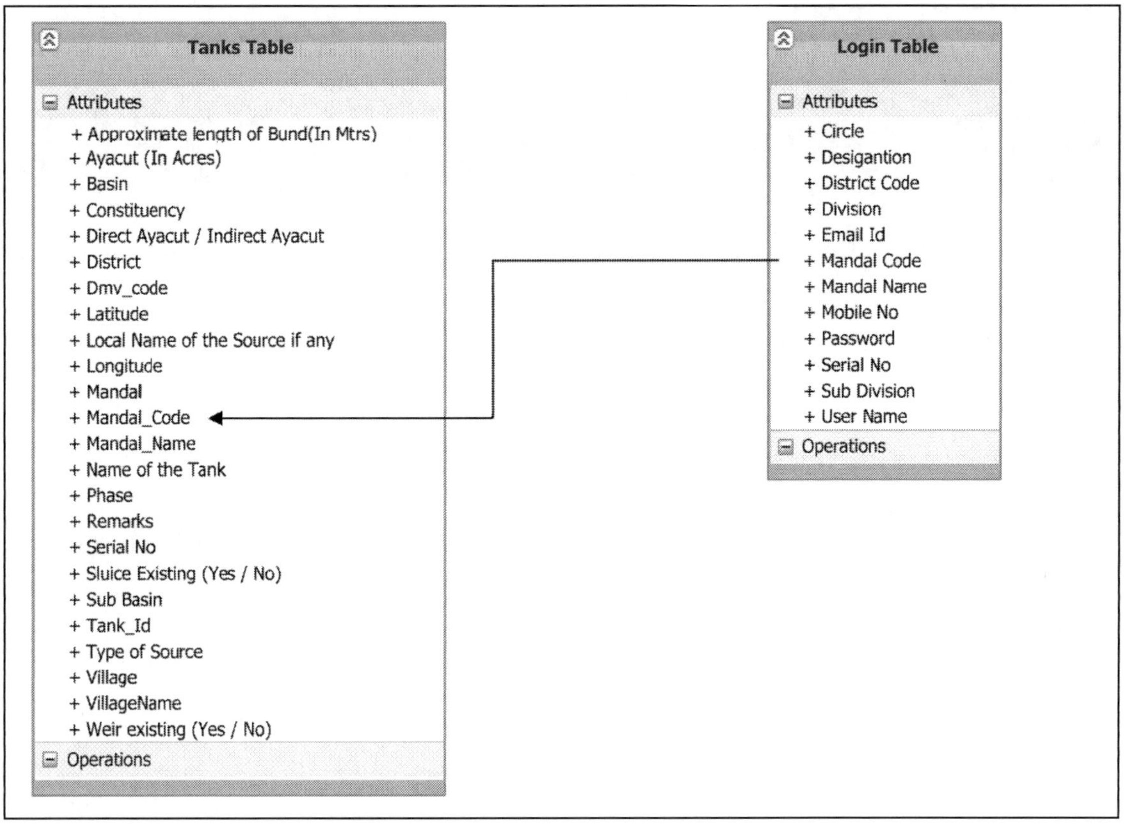

Figure 2: Shows the database table structure.

with some additional fields. These fields are used to link tank table to the login table. Figure 2: Shows the database table structure shows the structure of the table in the database where mandal code of the login table is used as an alternate key in the tank table. This key is used to query all the records from the tank table for that particular query of the mandal.

3. Design and application development

The web application for Tank Update System was developed using ASP.NET, JavaScript and Google Maps API. Junior engineers (JE) from I & CAD department are responsible for preparing plans and estimates for special repairs, additions and alterations of works/schemes. Since the preliminary work of collecting information for the survey was assigned to them, the application was built to have a login interface for each JE. Mandal name as login name was provided for each JE to login and to update

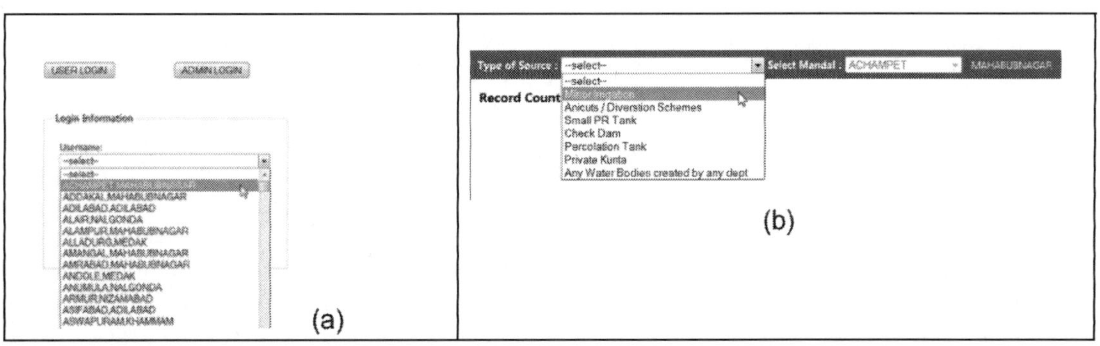

(c)

Figure 3: (a) Mandal names as user names for login; (b) Dropdown for selecting the type of tank; (c) Display of records in grid view after selecting type of tank.

information collected by them during Samagra Tank Survey. Figure 3: (a) Mandal names as user names for login; (b) Dropdown for selecting the type of tank; (c) Display of records in grid view after selecting type of tank. (a) shows a snapshot of the login interface provided for these JE's. For 472 mandals existing in the state of Telangana, 472 logins as mandal name were created and the password were distributed to them by the irrigation department.

Once the user logs into the applications, a drop down is provided for selecting the tanks based on categories. Figure 3: (a) Mandal names as user names for login; (b) Dropdown for selecting the type of tank; (c) Display of records in grid view after selecting type of tank. (b) shows a list of the existing categories under which the tanks fall. To demonstrate the capability of the application, let's suppose a Minor Irrigation tank from the list is selected. This would brings up a grid view which displays all tanks that come under this category for that particular mandal like shown in Figure 3: (a) Mandal names as user names for login; (b) Dropdown for selecting the type of tank; (c) Display of records in grid view after selecting type of tank.(c). Towards the left of Figure 3: (a) Mandal names as user names for login; (b) Dropdown for selecting the type of tank; (c) Display of records in grid view after selecting type of tank.(c), an "Edit Details" button is provided for editing data displayed in this grid view. A click on this edit button navigates the user to the edit interface of the tank.

The edit interface pre-populates the form text boxes with already existing data from the database for that particular record of edit button. The form contains some editable and non editable information. Fields with non editable information like Serial No., District, Constituency, Mandal and Village are greyed out and disable. This edit interface also provides the user with a novel way of collecting information from Google maps which is show towards the right side of the form in Figure 4: Edit interface of individual tanks with Google maps. Let's suppose a JE from Achampet has logged into the application. When the user navigates to the page shown in Figure 4: Edit interface of individual tanks with Google maps a marker at Achampet is displayed on Google maps. If any JE from any mandal logs into the application, a marker for the mandal he has logged into is displayed on Google maps. This way, the interface for identification of tanks existing in mandals or villages is made user friendly.

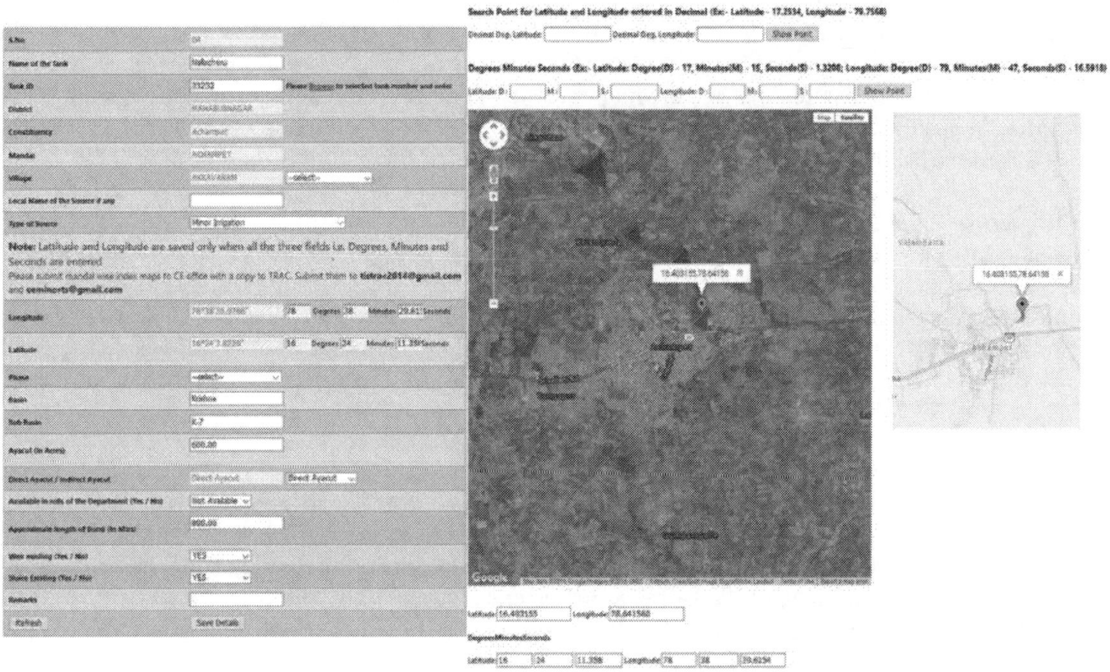

Figure 4: Edit interface of individual tanks with Google maps.

To spatially identify a tank two ways of collecting spatial information was provided. One way of collecting this information was by using markers in Google maps. The developed application has a provision to retrieve coordinates by moving the marker on the map. While moving this marker, the coordinates are updated dynamically and are displayed in the text boxes of the form for latitude and longitude columns. In Figure 4: Edit interface of individual tanks with Google maps a marker is placed at a tank that could be identified with sky blue colour towards the right. When this marker is moved, like explained it populates the latitude and longitude columns with coordinate information. The application has provision to switch between Map and Satellite view. In Figure 4: Edit interface of individual tanks with Google maps one can notice that only one tank with sky blue colour could be identified on the map view where as on the satellite view dark colour patches for water bodies could also be identified. In addition to the tanks that have already been identified by Google Maps, the satellite view provides us with clear view of tanks which have not yet been identified. This way a preliminary survey could be carried out to obtain all the tanks that are visible on Google Maps and its satellite view to save time and effort in searching them on the ground. The other way of retrieving coordinates was by using GPS unit. As part of the survey, JE's were provided with a GPS unit for obtaining the coordinates of tanks. By entering the GPS coordinates in the provided text boxes placed right above the Google maps satellite view window and by clicking on "Show Point" button, a marker is placed in the map for these coordinates. If the coordinates are in decimal degrees, they are automatically converted to degree-minutes-seconds and are then populated into the latitude and longitude section of the form. These two ways of spatially identifying the tanks was made as user friendly as possible for the JE's. Hence, the developed application integrates both management of information and spatial tagging of this information.

4. Results and discussion

After building the application, it was deployed on IIS 7.5 server and the access was given to all the JE's. A progress page was created to monitor the activities of JE's for the records that were being updated and for new records that were being entered. While the irrigation personnel have not much expertise in handling web based technologies, the response shown by them in updating the data on the web was overwhelmingly positive.

Edit interface in Figure 4 has been the unique feature of the application. Its capability 1) to identify already existing tanks, 2) a satellite view to identify new tanks 3) the asset mapping approach of capturing geographic coordinates of these tanks through markers and 4) a provision to enter coordinates for the identified tanks through GPS has certainly reduces time and effort, and the amount of survey needed for establishing the database of the tanks. At the initial level of establishing the tank database, more than half of the tanks database could spatially be identified and created this way using Google Maps.

5. Conclusion

The purpose of the project was to provide a user friendly web interface to streamline the data obtained from tank survey and to spatially tag this data using Google maps. The kind of asset mapping approach implemented in this article is a novel way of retrieving already existing information on Google Maps which is shown to have been implemented successfully. Further it also helps in cross checking the collected information with established Google maps. The approach lowers the costs of survey by helping the junior engineers in identifying already existing lakes on Google maps and for identifying new tanks using its satellite view.

Acknowledgements

The author is thankful to Director Admin Sri Rufus Dattam and Director Technical Sri D.V.J.Sastry of Telangana State Remote Sensing Application Centre (TRAC) for giving permission to carry out this work. Very special thanks to my mentor, Sri. Y.M Gopal Krishna (Deputed to TRAC from I & CAD Department) for all the help received in providing the data. Thanks to him for reviewing and also in giving valuable suggestions while developing the web interface.

References

[1] Irrigation & CAD Department, Government of Telangana. [Online]. Available: http://irrigation.telangana.gov.in/icad/home

[2] John L. McKnight and John P. Kretzmann "Building Communities from the Inside Out: A Path Toward Finding and Mobilizing a Community's Assets," ACTA Publications, Chicago, 1993.

[3] Tony Fuller, Denyse Guy, Carolyn Pletsch, "Asset Mapping, A Handbook", Canadian Community Economic Development Network, 2002, Available: https://ccednet-rcdec.ca/en/toolbox/asset-mapping-handbook

[4] OSM Foundation, The OpenStreetMap Project 2011 [Online], http://www.openstreetmap.org

[5] OpenStreetMap. (n.d.). In Wikipedia. Retrieved May 14, 2015, from http://en.wikipedia.org/wiki/OpenStreetMap

[6] Fagan B. "Ancient Lives: An Introduction to Archaeology and Prehistory", 2011, Routledge publication

Implementation of spatial smart waste management system in Malaysia

M F Omar[1*], A A A Termizi[1], D Zainal[1], N A Wahap[1], N M Ismail[2], N Ahmad[1]

[1] Agensi Angkasa Negara (ANGKASA), Pusat Angkasa Negara, 42700 Banting, Selangor, Malaysia
[2] Geoinfo Services Sdn. Bhd. 31, Jalan Bandar 2, Taman Melawati, 53100 Kuala Lumpur, Wilayah Persekutuan Kuala Lumpur, Malaysia

E-mail: farid@angkasa.gov.my

Abstract. One of the challenges to innovate and create an IoT-enabled solution is in monitoring and management of the environment. Waste collection utilizing the Internet of Things (IoT) with the technology of smart wireless sensors will able to gather fill-level data from waste containers hence providing a waste monitoring solution that brings up savings in waste collection costs. One of the challenges to the local authority is how to monitor the works of contractor effective and efficiently in waste management. This paper will propose to the local authority the implementation of smart waste management in Malaysia to improve the city management and to provide better services to the public towards smart city applications.

1. Introduction

One of the challenges to innovate and create an IoT-enabled solution is in monitoring and management of the environment. Effective management of solid wastes is one of the prerequisites for Malaysia to achieve a developed country status. It is generally recognized that there is a strong relationship between the effective management of solid wastes and a good quality of life and a healthy environment. The attractiveness of the country to foreign visitors and investments is very much influenced by a clean and healthy environment. By 2030, almost two-thirds of the world's population will be living in cities. This requires the development of sustainable solutions for urban living. These trends support the development of Smart City concepts, which are intended to improve living in urban areas by using innovative technologies. The "Internet of Things" provides new opportunities for making cities smarter.

National Space Agency Malaysia (ANGKASA) under Ministry Of Science Technology, Innovation Malaysia (MOSTI) taking an initiative to developed Smart Waste Management System (SWMS). SWMS is one of the modules developed in Spatial Smart City Service Delivery Engine (SSCSDE) Project funded by Mosti under Technofund grant. The Objective of the project is to develop a spatial service delivery engine that are capable of exploiting the smart (geospatial) technologies to support city operations which includes managing, using, and exploring spatial data with scientific analysis in all possible phases of the decision making process in the organization.

The reason and idea for choosing SWMS as one of the development module is to monitor the environment by enforcement. Local authority always have difficulty to monitor the job done by waste collection services provider. Contractor doing waste collection services seems doing collection not follow the schedule as per agreed in contract. Hence SWMS will act as automatic enforcement system to penalize the contractor who break the contract. This solution not only help local authority to monitor the contractor, it's also brings information of garbage container status and report automatically when it's full. This valuable information can be used by local authority for monitoring purpose and enforcement. For waste collection contractor this information provide alerts on the status of containers in terms of fill levels in order to optimize collection routes based on historical and real-time information hence save operation cost. For joint management body who in charge of the waste chamber in resident area can know the status of waste of each chamber and keep it clean. And for citizen there is no more waste issue complaint. The proposed SWMS had been operated as a pilot project in two places for pilot study area which is Cyberjaya in Sepang District and Bandar Saujana Putra, Taman Langat Murni in Kuala Langat District in Selangor, Malaysia.

The rest of this paper is organized as follow. Section 2 describes the case study for SWMS, Section 3 details the architecture of the SWMS consist of conceptual system design, conceptual workflow, web application and hardware structure. Section 4 elaborate further spatial elements of SWMS. Section 5 show the implementation of the SWMS in Sepang and Kuala Langat District and the operation result. Finally, some concluding remarks and direction for future work are given in Section 6.

2. Case Study for Smart Waste Management System

2.1. Enevo

Enevo one is a comprehensive logistics solution that saves time, money and the environment. It uses wireless sensors to measure and forecast the fill-level of waste containers and generates smart collection plans using the most efficient schedules and routes. The solution provides up to 50% in direct cost savings. Receive automatically generated schedules and optimised routes which take into account an extensive set of parameters future fill level projections, truck availability, traffic information, road restrictions, container and content types the vehicle can collect etc. New schedules and routes are planned not only looking at the current situation, but considering the future outlook as well. Enevo using WSN and Ultrasound sensor for measuring the fill level of waste container [1].

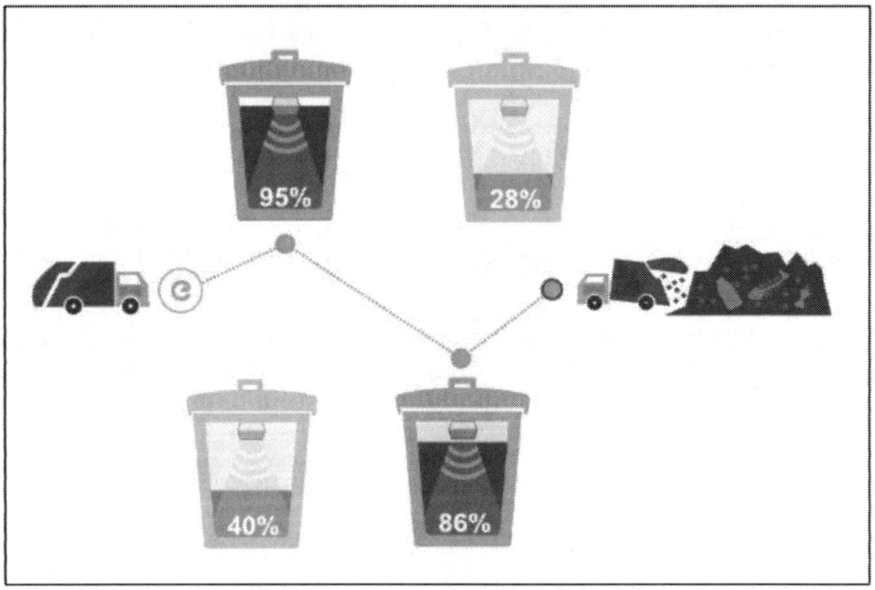

Figure 1. Enevo system architecture.

2.2. SENSdumpster filling level monitoring

SENSdumpster is an innovative device that monitors the filling level of a dumpster. SENSdumpster works with any kind of waste and fits to any type of dumpster. It makes use of a volumetric sensor that provides information of the level of filing of a dumpsters. SENSdumpster can be applied to any kind of waste, and a temperature sensor is able to detect significant temperature increases that may result in fire. SENSdumpster is a fully ZigBee wireless device that can be integrated to any ZigBee network deployment within the smart-city [2].

Figure 2. SENSdumpster wireless device to monitor filling level of a dumpster.

3. Smart Waste Management System Architecture

The SWMS can help local authority in decision making for future waste collection schedule. The waste collection regularity can be monitored, analysed, and rescheduled based on the waste level in garbage containers. The collection schedule for area with high regularity of full garbage container can reschedule to be more often. By using sensor technology in waste collection can help to reduce the operation time for waste collection process. This sensor can detect the level of garbage in garbage bin.

3.1. Conceptual system design

GSM modules are used for communication for SWMS. GSM can be used to transmit data from the sensor to the local server. The sensors need to be equipped with the GSM module including the SIM card and need to subscribe mobile data packets. The coverage network depends to the providers like Celcom, Maxis, Digi, Red One, U mobile Altel and Tunetalk. Figure 3 show the concept GSM for Smart Waste Management.

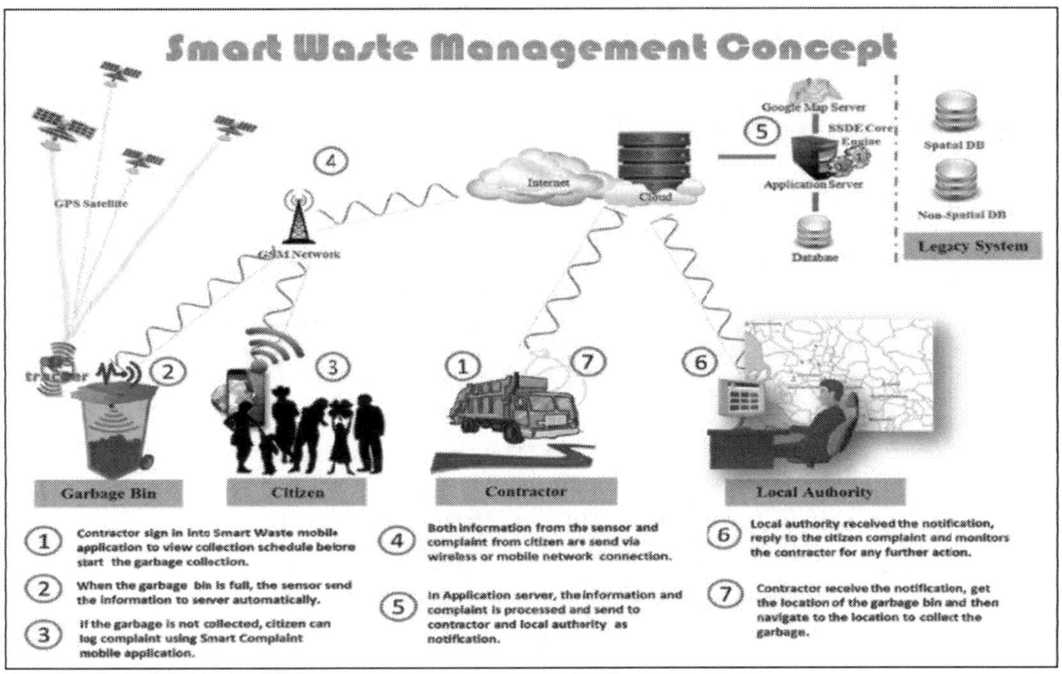

Figure 3. SWMS concept by GSM communication.

3.2. Conceptual workflow

There are three main actors in the work flow of SWMS consist of local authority, waste collection contractor and citizen. The first actor is the local authority who is responsible to add, register and view data sensor. The local authority also responsible to register, update and view contractor moreover the first actor will view and update waste schedule plus view and response to the complaints. The second actor is the waste collection contractor who is responsible to pick up waste based on the indicator appear on the map. This actor will response each of the status of bin in their job scope. The final actor is citizen who will give the feedback by fill in the complaint form online. Figure 4. Shows the overall work flow diagram of SWMS between the three main actors and Table 1 summarised function of each actor.

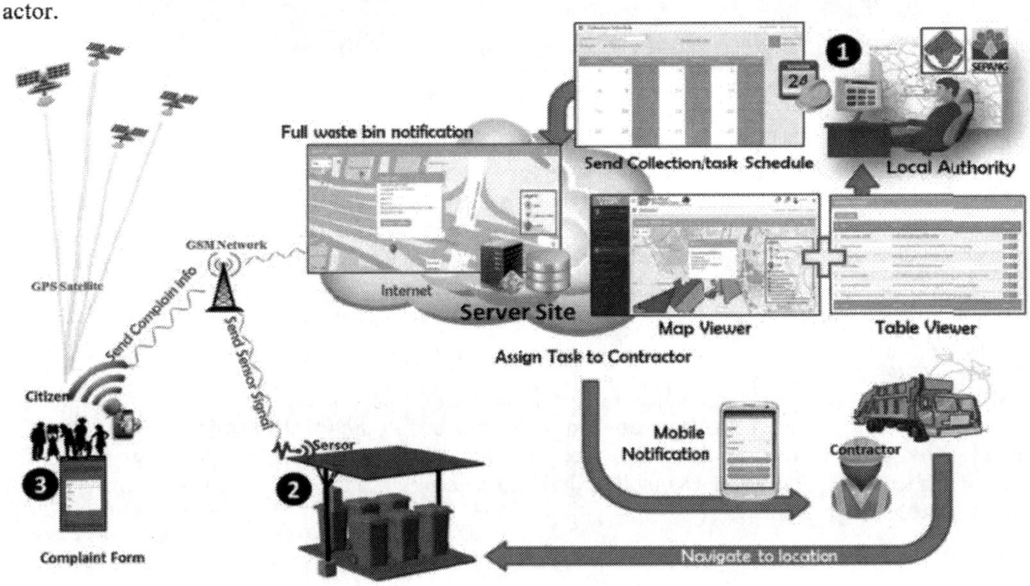

Figure 4. Work flow diagram of SWMS.

Table 1. Summary of the function of each actor.

No.	Module & Sub module	Functions
1	Web module 1. Sensor Management	• View data from sensor • Register new sensor • Delete sensor • Assign contractor -> Notify contractor
	2. Contractor Management	• View contractor list • Register new contractor • Update contractor information • Delete contractor
	3. Waste Collecting Schedule Management	• View waste collecting schedule • Update waste collecting schedule • Delete waste collecting schedule
	4. Complaint Management	• View complaint list • Create task • Assign enforcer -> Notify enforcer -> Verify complaint • Assign contractor -> Notify contractor -> Valid complaint • Response to the complaint
2	Mobile_apps_module	

	1. Complaint	• File a complaint • View response of their complaint
3	Back end module 1. Smart Waste API	• Manage request from Web and Mobile app module -> read and write smart waste data

3.3 Web Application

The web application for SWMS consist of web browser, mobile browser and mobile apps as shown in Figure 5. Interface response towards the device resolution. The method used to design and develop Core Smart City application is called Device Agnostic Solution. This is to address the increasing number of people browsing information over the web with difference devices, viewer resolutions and browsers. Device Agnostic Solution is suitable for most devices (ideally all devices) and be responsive nevertheless. These responsive designs which strictly obey the device-agnostic approach, without any doubt present the content in the best possible manner that is appropriate for the device on the screen.

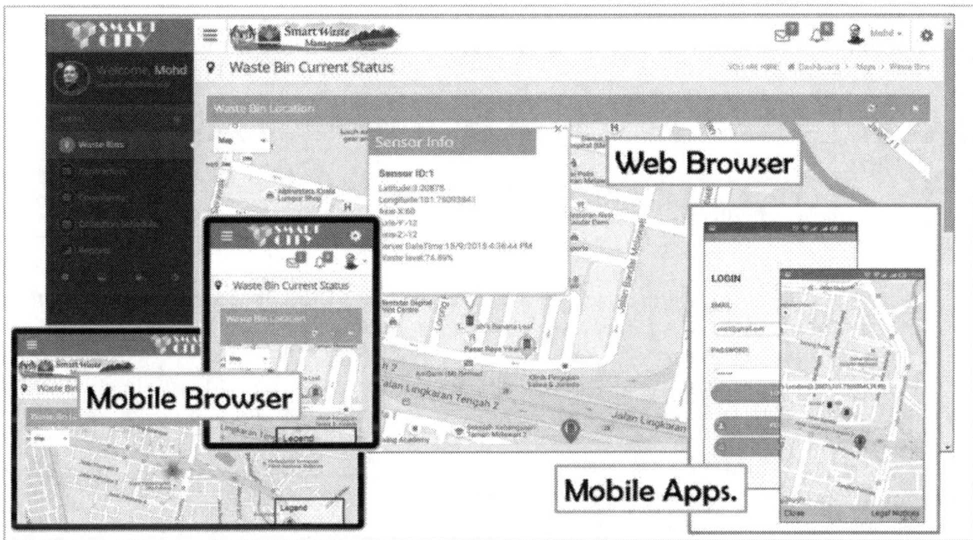

Figure 5. Web application for SWMS.

3.4 Hardware

The hardware involve during the development as shown in Figure 6. Three type of platform is been used which is waspmote [3], myrio and arduino. The platform powered by solar panel.

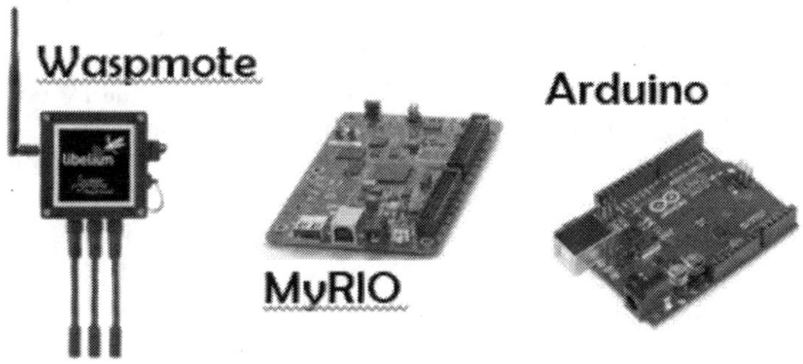

Figure 6. Hardware used for SWMS.

4. Smart Waste Management System Spatial Elements

The scope of work for proposed SWMS GIS spatial data is to provide geo-location of authority boundary area, contractor maintenance are and waste bin location. Figure 7 summarised the methodology stages of spatial data process.

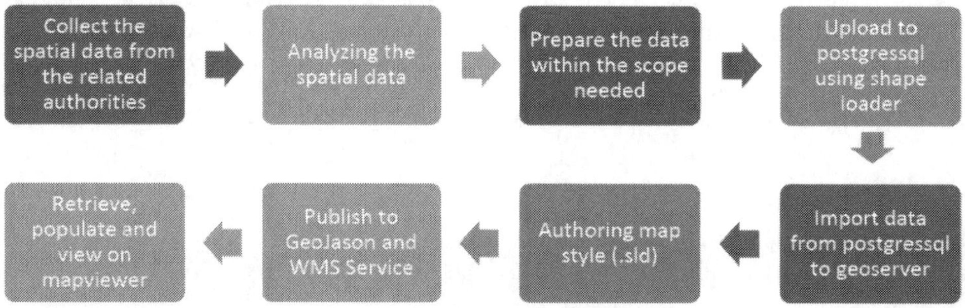

Figure 7. Methodology for spatial data process.

i. Collect the Spatial Data: The data will be gather by meeting with related authorities, which is Majlis Perbandaran or Majlis Bandaraya for each selected area according to their Smart City GIS System proposed category.
ii. Analysing the Spatial Data: Once receive the spatial data, the data will be analyse using certain GIS software's to make sure all data is in good condition and search for any issues related to the data.
iii. Solve the Issues Occurred: After finish analysing and acquire the outcome results, some recommendations and solutions are suggested to overcome the issues.
iv. Prepare the Data within the Scope Needed: After the data is corrected, the spatial data will be prepared according to each category of Smart City GIS System for each selected authorities.

5. System Implementation

5.1. Ultrasonic wireless sensor

Ultrasonic sensor measures the distance between the sensor and the waste inside the container and sends its information through the cellular network directly to the server. In the server the data is processed and transformed into useful information. The filling percentage of the container is estimated based on the latest measurements produced by the sensor. The information is been processed and the result is appear on the web application. There is criteria of sensor selection for ultrasound sensor. The sensor can continuously measure the fill level, the sensor work with any type of bin materials and the sensor able to measure without physical contact with materials. Figure 8 shows how the sensor works.

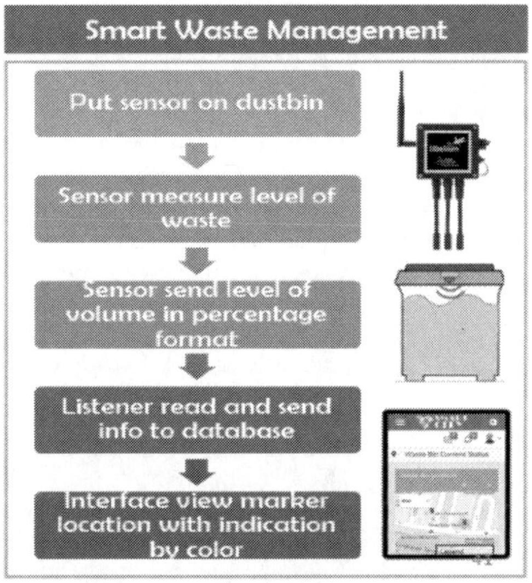

Figure 8. SWMS implementation.

There are four level indicator for waste bin level:
1. Green: less than 50%
2. Yellow: less than 100% and more than 50%
3. Red: more than 100%

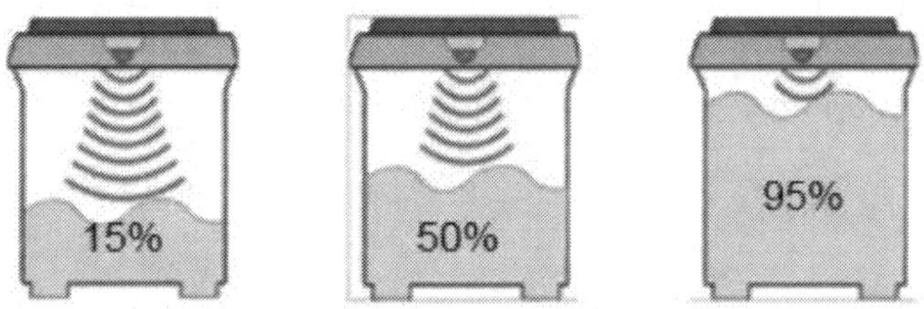

Figure 9. Ultrasonic fill level sensor.

A typical concept for Smart Waste Management has being concluded that the system will use ultrasound sensor and GSM network as tool to measure and transmit sensor data to the server. It will take time to find a total solution for SWM instrumental, in Malaysia does not have a ready-made instrument that can use in the system. There are two vendor has being selected that are Libelium and Visi. Libelium is the Internet of Things platform provider based in Zaragoza, Spain. Libelium delivers a powerful, modular, easy to program open source sensor platform for the Internet of Things enabling system integrators to implement reliable Smart Cities and M2M solutions with minimum time to market.

5.2 Pilot area

MP Sepang and MD Kuala Langat was selected as pilot test for SWM. Site visit was carried out to obtain the current condition and information regarding the waste container that includes detailed information such as location, type, size, materials, the capacity and the surrounding area. Discussion with Local Authority has conducted with MP Sepang and MD Kuala Langat respectively. Local Authority has suggested to plant level sensor on waste container MGB 660 litre type. Because this MGB 660 litre are commonly used in Apartment area and also withstand rugged conditions, light and inexpensive. Figure 9 and 10 shown the area that had been chosen for the pilot study of Smart City GIS System on Smart Waste Management are located at the area within MPS and MDKL authority.

Figure 10. District of Sepang, Majlis Perbandaran Sepang [4].

Figure 11. District of Kuala Langat, Majlis Daerah Kuala Langat [5].

6. Conclusion and Future Work

The objective of SWMS is to value-add existing Solid Waste Management process for waste management under local authorities, to help in decision making for waste management process, to ensure the contractors follow the work procedure, and to improved waste collection services delivery which comes under the responsibility of local authority. In this study, MPS and MDKL waste management monitoring process was reviewed. By manipulating geospatial technology and intelligence sensor such as ultrasonic sensors via IoT technology, SWMS application has provided the waste collection operator with a platform to work in real-time mode to improve the service by optimizing operation time and cost. SWMS application as a tools for local authority to monitor waste collection operator to ensure waste collection services is deliver accordingly as per contract. The impact of SWMS to resident, residents associations and joint management body is direct when the environment is now clean and give a healthier life.

Nevertheless, the proposed SWMS requires more maintenance cost than the existing system. There is a need to make it a sustainable which is the development cost is acceptable to implement in local authority. The most important issue is how to deliver to local authority with competitive price and less maintenance cost. With the current system, the implementation only limited for apartment and condominium which the sensor device put on top of the big garbage bin inside of apartment waste chamber. In the future, small IoT gadget for waste monitoring can be develop and put inside the waste chamber in front of each terrace house and bungalow to widening the implementation to the citizen.

Acknowledgement

The authors would like to thank to Ministry of Science, Technology, and Innovation Malaysia (MOSTI) for funding the project. Thank to Sepang Municipal Council and Kuala Langat Municipal Council (MPS & MDKL) for the support given in providing information of the existing solid waste management process and platform to conduct the integration of modules and verification for the SWMS prototype. Thank to Cyber HeightsVilla, Mutiara Ville Cyberjaya, Bandar Saujana Putra Joint Manegement Body and Taman Langat Murni Residents Association to allowed installation of sensor in their residents area.

References

[1] Enevo. (2015). Enevo ONe Waste Collection for Smart Cities.
[2] Sensor, S. (2014). Dumpster Smart lling level monitoring.
[3] Libelium. (2015b). Smart Metering 2.0 Technical Guide, *4.9*. Retrieved from http://www.gedigitalenergy.com/smartmetering/catalog/SGM3000_Residential.htm#sgm3000tab5
[4] Portal Rasmi Majlis Perbandaran Sepang; (n.d.-b). Retrieved May 7, 2015, from http://www.mpsepang.gov.my/peta-daerah-sepang
[5] Portal Rasmi Pejabat Daerah / Tanah Kuala Langat Peta - Peta Daerah Kuala Langat. (n.d.). Retrieved May 7, 2015, from http://www.selangor.gov.my/kualalangat.php/pages/view/186

Nomenclature

SWMS:	Smart Waste Management System.
SSCSDE:	Spatial Smart City Service Delivery Engine.
GSM:	Global System Mobile Communication.
ANGKASA:	National Space Agency of Malaysia.
MOSTI:	Ministry of Science, Technology & Innovation Malaysia.
MPS:	Sepang Municipal Council.
MDKL:	Kuala Langat Municipal Council.
MGB:	Mobile Garbage bin
GIS:	Geographic Information System
M2M:	Machine to Machine
IoT:	Internet of Thing

IOP Conf. Series: Earth and Environmental Science **37** (2016) 012060 doi:10.1088/1755-1315/37/1/012060

Regulating outdoor advertisement boards; employing spatial decision support system to control urban visual pollution

K Wakil[1], MQ Hussnain[1], A Tahir[2] and M A Naeem[1]
[1] Department of Urban and Regional Planning, National Institute of Transportation, National University of Sciences and Technology, Islamabad, Pakistan
[2] Institute of Geographical Information Systems, National University of Sciences and Technology, Islamabad, Pakistan

E-mails: khydijawakeel.urp@nit.nust.edu.pk
plannerqadeer.urp@nit.nust.edu.pk
ali.tahir@igis.nust.edu.pk
asghar.naeem@nit.nust.edu.pk

Abstract. Unmanaged placement, size, location, structure and contents of outdoor advertisement boards have resulted in severe urban visual pollution and deterioration of the socio-physical living environment in urban centres of Pakistan. As per the regulatory instruments, the approval decision for a new advertisement installation is supposed to be based on the locational density of existing boards and their proximity or remoteness to certain land-uses. In cities, where regulatory tools for the control of advertisement boards exist, responsible authorities are handicapped in effective implementation due to the absence of geospatial analysis capacity. This study presents the development of a spatial decision support system (SDSS) for regularization of advertisement boards in terms of their location and placement. The knowledge module of the proposed SDSS is based on provisions and restrictions prescribed in regulatory documents. While the user interface allows visualization and scenario evaluation to understand if the new board will affect existing linear density on a particular road and if it violates any buffer restrictions around a particular land use. Technically the structure of the proposed SDSS is a web-based solution which includes open geospatial tools such as OpenGeo Suite, GeoExt, PostgreSQL, and PHP. It uses three key data sets including road network, locations of existing billboards and building parcels with land use information to perform the analysis. Locational suitability has been calculated using pairwise comparison through analytical hierarchy process (AHP) and weighted linear combination (WLC). Our results indicate that open geospatial tools can be helpful in developing an SDSS which can assist solving space related iterative decision challenges on outdoor advertisements. Employing such a system will result in effective implementation of regulations resulting in visual harmony and aesthetic improvement in urban communities.

1. Billboards and hoardings; challenges and opportunities for local governments

Major cities in Pakistan are losing their aesthetics because of the high concentration of billboards and hoardings, their improper locations, inappropriate size, bad heights, unsuitable colors and faulty construction [1]. These advertisement structures on busy roads and urban centers are not only disturbing urban visual fabrics, deteriorating the social and physical living environment, causing nuisance for surrounding land uses and road users, but also result in physical losses and causalities under storm conditions [2]–[6]. On the other hand, hoarding and billboards have emerged as a lucrative revenue source for controlling authorities. Only from Karachi (the largest city of the country) city government earned more than Rupees 825 million rupees from billboard in year 2012-13. On the other hand, Clifton Cantonment (a cantonment town in Karachi) generated Rupees 147 million for the same year [7]–[10].

Throughout Pakistan, regulatory instruments for controlling and management of outdoor advertisement boards are at a very nascent stage[11]. Only few large cities such as Karachi, Lahore and Faisalabad follow regulations to manage billboards but with prime focus on revenue generation [12], [13]. Only two years back, in 2013, Punjab government devised the very first provincial level policy - Punjab Outdoor Advertisement and Signboards Policy - requiring the local authorities formulate their bylaws on the subject [14]. Response from local governments has been slow but gradually they are formulating legal instruments talking about the location, size, type and density of outdoor advertisements boards.

All of the regulatory instruments advise to base the approval decision for a new advertisement installation on linear density of existing billboards and their proximity or remoteness from certain land-uses. However, such decisions are mostly driven by political pressure or revenue potential rather than urban visual harmony. Other than Karachi, there appears no use of spatial information in deciding location of advertisement boards in any other city. In case of Karachi, City Government has been using diagrammatic illustrations (not to scale) to demarcate and allocate sites for hoardings of various sizes (see Figure 1). In both cases, responsible authorities remain handicapped in effective implementation of regulations due to the absence of geospatial analysis capacity.

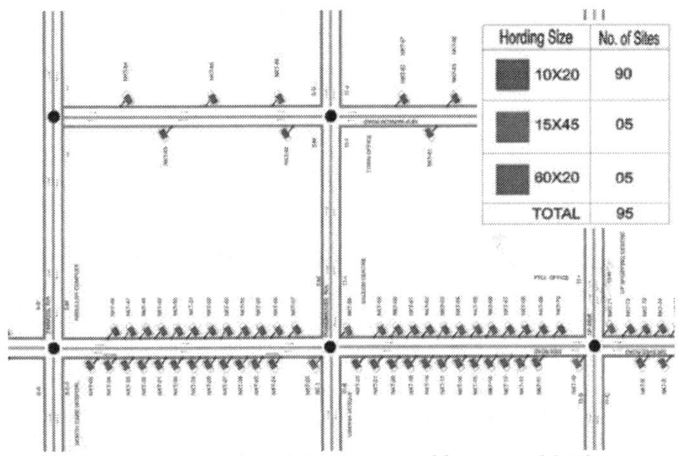

Figure 1. Sample of diagram used by Karachi City Government to allocate sites for billboards of various sizes on a primary road.

Intermittent and irregular control results in mushroom growth of boards under corporate pressure. From time to time, political leaders pushes authorities to start 'war-on-billboards' causing immense stress on local authorities as well as advertisers [4], [15], [16]. In a nutshell, there exists no decision support system which can be used to identify the problematic locations because of boards, extent of compliance with the policies and rules and identification of optimal locations for new installments from advertisers and planning authorities' point of view.

This study targets at solving decision challenges in billboards' location by conceptualizing and developing an SDSS based on open source tools including OpenGeo Suite, PHP and GeoExt. For effective functioning, the SDSS must have the following capabilities:

1. Visualization: to visualize the existing and proposed sites of advertisement boards in relation with the legal requirements of location and distance in residential areas or on roads
2. Scenario evaluation : to evaluate the impact of change in permissible linear density of boards and to evaluate the impact of change in buffer restriction around a particular landuse
3. Generation of reports
4. Iteration: to provide user friendly interfaces where user can define and redefine parameters

2. Related work

The related work provides evidences that spatial technologies have been in practice to identify best locations for advertisements. Haidu et al. have analyzed the potential of GIS implementation for optimizing advertisements. They believed that effectiveness of advertisement is linked with geo-demographic, socio-economic and educational characteristics and the infrastructure of the targeted area. Hence, they used spatial analysis to locate certain categories of potential clients. They employed a series of factors which would attract or repel various advertisement activities including street network pattern, advertisement prohibited areas, population density, and regulatory bindings to identify and grade advertising area [17].

Luke et al. have used GIS to explain locations and characteristics of tobacco billboards in St Louis. They used GIS based analysis to measure the extent to which tobacco companies were placing billboards in close proximity to minority neighborhoods and schools [18]. Similarly, Hillier et al. have studied the clustering of unhealthy outdoor advertisements around child serving institutions and have emphasized the need of regulatory provisions to keep outdoor advertisement boards' locations away for such institutions [19].

In Nigeria, private companies have been using GIS to keep updated inventory of advertisement boards for outdoor advertisement industry clients [20]. While in USA, Europe and Canada, Software as a Service (SaaS) has been in practice where service provider make use of demographic data, traffic flow, elevation and exposure data to help advertisers decide locations and return on investment for their outdoor advertisements [21]. Additionally, open source geospatial technologies have gained momentum in the last decade or so. In a very interesting study, several open source technologies are discussed which can be applied in many domains [22]. In another study, the authors conducted an online survey and asked the users of these geospatial technologies about their experience while using these technologies [23]. Some of the technologies which scored higher were used in our study such as PostgreSQL, OpenLayers, ExtJS and GeoExt. These well-known open source geospatial technologies have also been used in different applications such as map personalization in the context of location-based services [24].

However, most of the literature focuses on advertisers' perspective where GIS is used to make location of advertisement more and more profitable. While our study focuses more on outdoor advertisement's management from local authorities stand point by making use of an SDSS. Controlling authorities are largely concerned about checking if the new installation request complies with the policy requirements in spatial terms or who will be affected if a change is made in the threshold of linear density of boards on roads or a buffer restriction around a particular land-use.

3. Methodology

This study has followed a four step methodology to reach the desired objective. First, it reviews the relevant regulations to enlist all the prescribed spatial provisions and restrictions in the form of knowledge module of the system. Secondly, database module has been developed by arranging relevant data layers with desired attributes in a geo-database. Thirdly, site suitability analysis has been performed using analytical hierarchy process with three criteria layers. Finally, a web application has been developed using open source tools to provide users with a friendly interface to perform iterative actions.

4. Results and discussions

4.1. Review of relevant regulations

This study has reviewed all the available regulatory documents which are in practice in various urban areas of Pakistan and include the following:

i. Regulations for outdoor Advertisements and Billboards by Parks and Horticulture Authority (PHA)
ii. Punjab Local Government Ordinance 2001
iii. City District Government Faisalabad Advertisement Bye-laws 2013 [25]
iv. Punjab Outdoor Advertisements and Sign Boards Policy 2008 by Punjab Housing and Urban Development Department (PHUDD)[26]

v. SOP-Hoarding in Cantonment Areas, July 2011
vi. Military Lands and Cantt Hoarding Policy 2012 [27]
vii. Advertisement, Signage Bye-laws for Karachi 2003 [28]

As per these regulatory instruments, the approval decision for a new advertisement installation is supposed to be based on two key factors:

a. The linear density of existing billboards; spacing between two boards vary from 150 to 250 feet based on their location (within city, intercity road, highway) and size (20X60 feet or 30X90 feet) [25], [26]
b. The proximity or remoteness of a billboard/ hoarding from certain land-uses; prevailing regulations in Pakistan have specified locations prohibited for placing advertisements such as residential buildings, notified areas of heritage importance and road medians [25]–[28]. International practices provide more comprehensive criteria related to placement near parks, educational and religious buildings, monuments and cultural sites etc. [17], [29]

4.2. Preparation of geo-data

Primarily, three data sets have been used in this study and have been managed using PostgreSQL database with PostGIS as a spatial extension. These include building parcels along with land-use information, road network and location of existing billboards. Table 1 provides details about type, associated attributes and data format of each data layers.

Table 1. Datasets used in geo-database.

Sr. #	Data set (name & snapshot)	Type	Attribute	Data format
1	Parcel Land-use 	Polygon	Commercial, residential, religious, historical, educational, health institutions	Vector
2	Road network 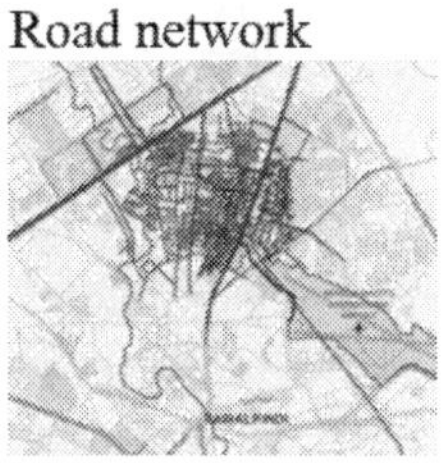	Polyline	Length, width, type	Vector
3	Location of boards 	Points	Height, size, number, type (traditional, mechanical, electrical), display surfaces	Vector

4.3. Site Suitability Analysis

Site suitability analysis has been used to identify suitable locations for siting new bill boards. Locational suitability has been calculated using pairwise comparison through AHP and WLC where distances from roads, existing billboards and educational institutions have been taken into consideration. For this purpose Quantum GIS, loaded with Easy AHP – a QGIS python plugin, has been used. Figure 2 explains the step by step process involved in finding suitable sites for new installations using AHP and WLC. From step 1 to step 3, criteria map layers are illustrated; step 4 shows the pairwise comparison for AHP; step 5 depicts the weight assignment for WLC while step 6 shows the identified resultant sites.

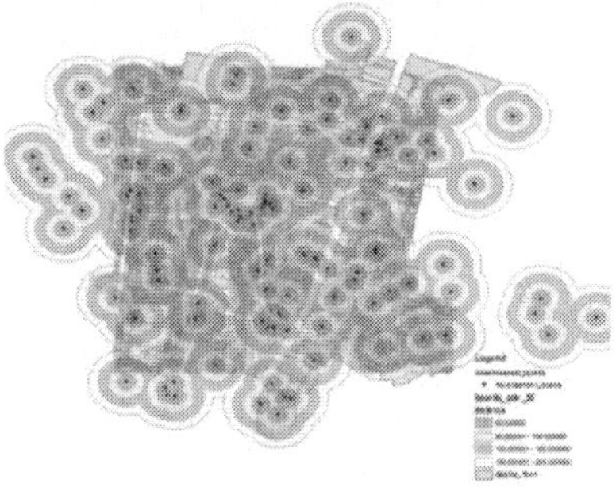

Step 1: Buffers based on distance from existing bill boards

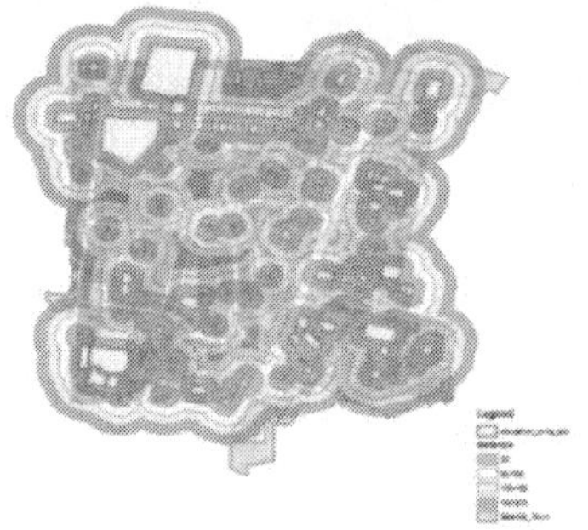

Step 2: Distance from education institutions

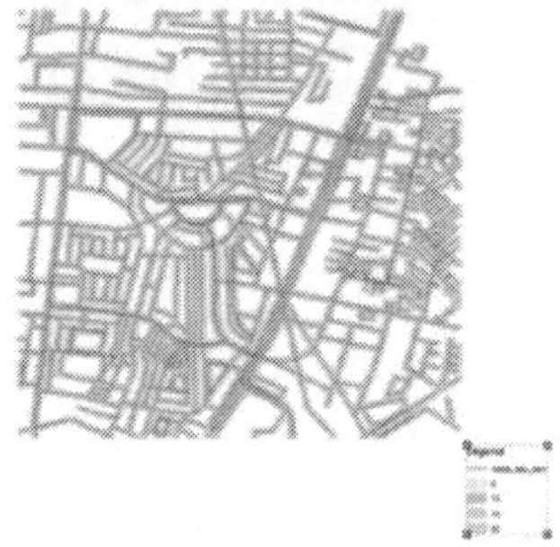

Step 3: Distance from roads

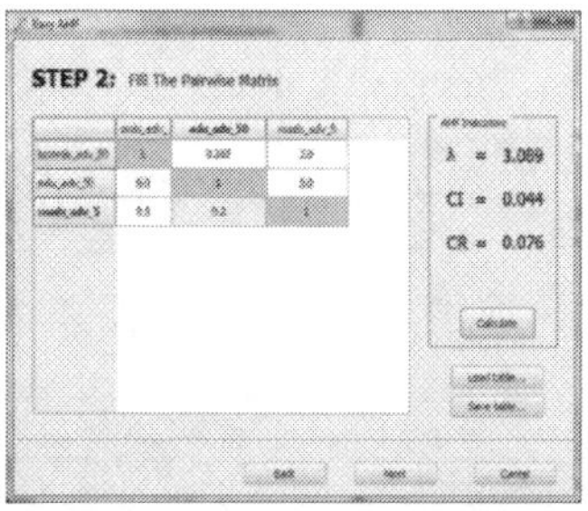

Step 4: Assigning order of importance to layers

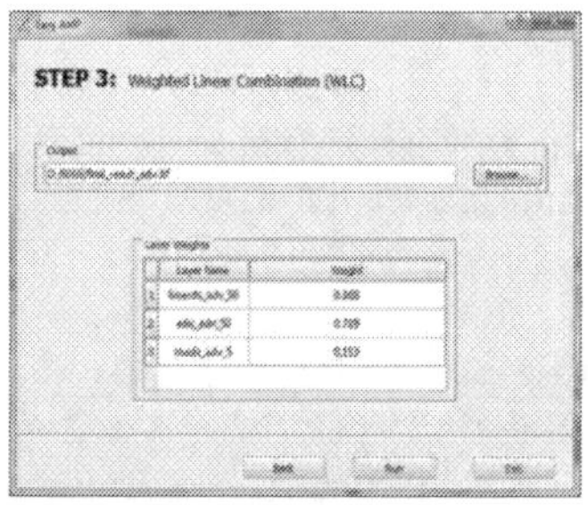

Step 5: Weighted linear combination of data layers

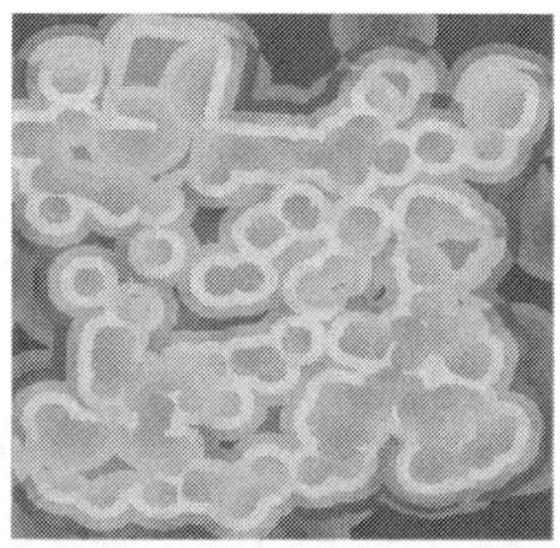

Step 6: Suitable sites are given in green color result

Figure 2. Finding suitable sites for new installation using AHP and WLC.

4.4. Web Application Development

Finally, an OpenLayers based web application has been developed to display datasets and resultant raster of AHP process. OpenLayers allows users to display data layers while using OpenStreetMap and Google Street Map as background. The raster layer showing suitable areas can be adjusted for opacity via slider tool. Additionally users can view the linear density of existing advertisement boards along a selected road. Figure 3 illustrates the final shape of web SDSS application where individual map layers are shown to the right which can be turned on and off while corresponding legend appears to the left of Web interface. Based on the location of existing billboards, users can see the linear density along any selected road.

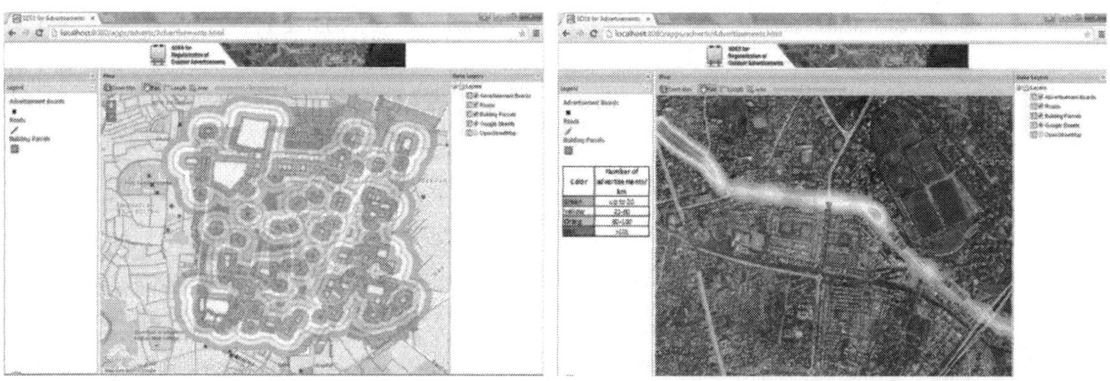

Figure 3. Snapshots of the user interface of web application.

5. Conclusions

This study reflects that open geospatial tools can be helpful in developing an SDSS which can assist solving space related iterative decision challenges on outdoor advertisements. It has been demonstrated that, in the light of existing outdoor advertisement regulations, such system can work with minimum need of datasets i.e. spatial road inventory, location of existing boards and land-use parcels while AHP and WLC can be used for spatial multi-criteria decision making for site suitability analysis. Although the main structure of the system has been established under this study, considerable amount of programming related effort is required to strengthen the graphical user interface. In this exercise, all criterion maps have been developed in QGIS and pairwise comparisons have been performed using Easy-AHP. However, for a robust SDSS on-the-fly calculations are desired. Efforts are required to further improve the system in order to perform these tasks in an online environment with user given inputs. Employing such a system will result in effective implementation of regulations resulting in visual harmony and aesthetic improvement in urban communities.

Bibliography

[1] K. Wakil, "A Study of Regulating Advertisement Boards in Lahore," University of Engineering and Technology, Lahore, 2012.
[2] Daily Times, "Daily Times," *Daily Times*, Lahore, 25-Mar-2006.
[3] R. S. Hassan, "Billboard plays havoc with life, property," *Dawn*, Karachi, 15-Feb-2006.
[4] A. Qudoos, "CM Punjab orders to remove big and ugly billboards," *Seen Report*, Lahore, 03-Dec-2008.
[5] B. Ali, "But nobody's listening," *Pakistan Today*, Karachi, 12-Jul-2012.
[6] K. Wakil, M. Q. Hussnain, A. Waheed, and A. M. Naeem, "Contextual Review of Outdoor Advertisements: Impacts and Regulatory Practices," *Sci. Int.*, vol. 1, pp. 531–535, 2016.
[7] Dawn, "Hoardings: money for officials, threat to people," *Dawn*, Karachi, 27-Jun-2007.
[8] F. Ilyas, "When billboards make money and trees don't," *The Express Tribune*, Karachi, 01-Aug-2013.
[9] A. R. Alam, "Pawning the family jewels – in Lahore," *Lahore Nama*, 2008. [Online]. Available: https://lahorenama.wordpress.com/2008/03/31/pawning-the-family-jewels-in-lahore/#more-98. [Accessed: 09-Dec-2015].
[10] M. Maher, "Skin City: The story of Karachi's billboards," *The Express Tribune*, Karachi, 08-Mar-2014.
[11] A. R. Alam, "PHA and the regulation of billboards," *Daily Times*, Lahore, 28-May-2005.
[12] A. R. Alam, "Towards proper regulation of outdoor advertising," *The News*, Lahore, 04-Aug-2008.
[13] K. Wakil and M. Q. Hussnain, "Regulatory Mechanisms for Outdoor Advertisement Boards in Lahore," in *International Conference on Town Planning and Urban Management (ICTPUM 2014)*, 2014.
[14] K. Hasnain, "Regulations on outdoor advertising," *Dawn*, Lahore, 07-Mar-2013.
[15] Dawn, "PHA's war on billboards - without any rules," *Dawn Newspaper*, Lahore, 07-Dec-2008.
[16] S. Hasan, "Scramble to remove 'dangerous' billboards as SC deadline looms," *Dawn*, Karachi, 22-Aug-2015.
[17] I. Haidu, A. I. Crăciun, C. Lupău, and C. Şerban, "Using GIS in Order to Detect the Optimal Areas for Dedicated Advertising," *Geogr. Tech.*, no. 1 / 2009, pp. 40–49, Jan. 2009.
[18] D. Luke, E. Esmundo, and Y. Bloom, "Smoke signs: patterns of tobacco billboard advertising in a metropolitan region," *Tob. Control*, vol. 9, no. 1, pp. 16–23, Mar. 2000.
[19] A. Hillier, B. L. Cole, T. E. Smith, A. K. Yancey, J. D. Williams, S. A. Grier, and W. J. McCarthy, "Clustering of unhealthy outdoor advertisements around child-serving institutions: a comparison of three cities," *Health Place*, vol. 15, no. 4, pp. 935–945, 2009.
[20] Spatial Technologies Ltd, "GIS Solutionf for Outdoor Advertisement in Nigeria," 2011. [Online]. Available: http://www.spatialtechnologiesltd.com/outdoor_advertiser.php. [Accessed: 12-Dec-2015].
[21] L. Malloy, "Using Location-Based Information to Get More From Your Out-of-Home Advertising," *AdSmart Blog*, 2014. [Online]. Available: http://www.intermap.com/adsmart-

blog/2014/09/get-more-from-out-of-home-advertising. [Accessed: 12-Dec-2015].

[22] J. Sanz and M. Montesinos, "Current Panorama of the FOSS4G Ecosystem," *Eur. J. Informatics Prof.*, vol. X, no. 2, pp. 43–51, 2009.

[23] A. Ballatore, A. Tahir, G. McArdle, and M. Bertolotto, "A comparison of open source geospatial technologies for web mapping," *Int. J. Web Eng. Technol.*, vol. 6, no. 4, pp. 354–374, 2011.

[24] G. McArdle, A. Ballatore, A. Tahir, and M. Bertolotto, "An open-source web architecture for adaptive location-based services," in *14th International Symposium on Spatial Data Handling (SDH), at the Joint International Conference on Theory, Data Handling and Modelling in GeoSpatial Information Science Hong Kong, 26-28 May, 2010*, 2010.

[25] City District Government, "City District Government Faisalabad Advertisement Byelaws, 2013," 2013. [Online]. Available: http://pcgip.urbanunit.gov.pk/docs/Final Advertisement Byelaws 2013 Faisalabad [MEHRAN] without tc.pdf. [Accessed: 12-Dec-2015].

[26] PHUDD, "Punjab Outdoor Advertisements and Sign Boards Policy." Government of Punjab, Lahore, 2008.

[27] Military Lands and Cantonments, "Military Lands and Cantt Hoarding Policy." Military Lands and Cantonments, Karachi, 2012.

[28] The City Government, "Advertisement, Signage By-laws Karachi," *The City Government Laws*. Karachi, 2003.

[29] Auckland City Council, "Auckland City Council Bylaws for Billboards 2007." Auckland, 2007.

A comparison of pixel-based and object-based approaches for land use land cover classification in semi-arid areas, Sudan

H E Adam[1,*], E Csaplovics[2] and M E Elhaja[1]

[1] Faculty of Natural Resources and Environmental Studies (FNATRES). University of Kordofan, Elobeid, Sudan.
[2] Institute of Photogrammetry and Remote Sensing (IPF), Technical University of Dresden, D-01062 Dresden, Germany

* corresponding author's address: hassan_adam@hotmail.com

Abstract. This paper deals with the comparison between application of pixel-based and object-based approaches in land use land cover classification in semi-arid areas in Sudan. The second aim is to assess the accuracy of classification for each approach. The study was conducted in the gum arabic belt in North Kordofan State, which is affected by modifications in conditions and composition of vegetation cover trends. The study used ASTER L1B registered radiance at the sensor image acquired on (19.10.2010). The image was radiometrically corrected by using ENVI-FLAASH software. Subset with an area of (40880) ha was created. The image classification (pixel-based and object-based) and accuracy assessment were conducted. Total number of (47) GCPs were surveyed and used in accuracy assessment using ERDAS 9.1. Image segmentation process was implemented using Definiens eCognition 7.1 software. Segmentation level 4 of scale parameter 25 was selected for classification based on colour and form homogeneity. Land use land cover classes were derived by classification using the nearest neighbor classifier with membership functions (fuzzy logic) for each class. The land use land cover distribution in the area for forest dominated by *Acacia senegal is (20%) and* for residential area is (1.50%) for the two methods of classification. While for bare and farm land, grass and bush land and mixed woodland classes are (6.69% and 1.63%), (18.62% and 15.16%) and (53% and 61%) for pixel based and object based methods, respectively. The overall accuracy and Kappa statistic of the classification produced by the pixel-based and object-based were (72.92%, and 54.17%) and (0.6259 and 0.3810), respectively. The pixel based approach performed slightly better than the object-based approach in land use land cover classification in the semi-arid land in gum Arabic belt.

1. Introduction

Land cover refers to the surface cover on the ground, whether vegetation, urban infrastructure, water, bare soil or other. Identifying, delineating and mapping of land cover are important for global monitoring studies, resource management and planning activities [2] and [12]. Remote sensing techniques help in developing areas in studying the changes in vegetation cover. Also geographic information system (GIS) is very powerful tool when applied to earth sciences and land use studies [5]. Land use land cover classification of vast areas using traditional methods is a time consuming and expensive process. Remote sensing offers a quick and efficient approach to the classification and mapping of land use land cover in a basis for future planning. Hard classification methods assume that each pixel represents a homogeneous area on the ground and showing only one land cover type [19]. But in fact, the spatial structure and variation of land cover can cause numerous 'mixed pixels' in remotely sensed imagery also in semi-arid areas (e.g. gum arabic belt). The soft classification methods like sub-pixel and fuzzy classification are an alternative to the standard 'hard' classifier and providing the user with a measure of the degree (termed membership grade) to which the given pixel belongs to some or all of the candidate classes [12] and [20]. Object based image analysis is observed and used in the past few years. Its application has been extended in various fields, especially for forest mapping and Land use land cover classification [15] and [17]. Object-based, has emerged, and has generally had better success with narrow band and high spatial resolution data [26]. An object based image analysis (OBIA) approach has been proposed as an alternative analysis framework that can mitigate the deficiency associated with the pixel-based approach [6]. OBIA approaches for analyzing remotely sensed data have been established and investigated since the 1970s [6].The classification process, in this case, begins with a segmentation of neighboring pixels into homogenous units or objects [3]. Object-oriented methods of image classification have become more popular in recent years due to the availability of software (eCognition®).

The gum arabic belt contains large areas of land with little land resources information. The belt is dominated with *Acacia senegal* trees (*Hashab*). [9] reviewed many studies conducted in Sudan using various types of satellite imageries to identify the features of land use land cover changes. In Sudan most land use land cover classifications were derived from remotely sensed data using pixel based methods of classification mainly by maximum likelihood classification (hard classification) as the commonly used technique for classification. Therefore, the study aims to compare classification of land use land cover using pixel based and object based approaches in semi arid areas in Sudan, which can be performed and provides the ground cover information for future land management in the gum arabic belt in Kordofan region. Secondly, to achieve accuracy assessment for the two classification approaches.

2. Material and methods

2.1. Study area

The selected study area is located in North Kordofan State, which is one of the most vulnerable areas in the country concerned with desertification processes. It is located in central Sudan between latitudes 9° 30' and 16° 24' N and longitudes 27° to 32° E. Kordofan, which lies largely within the arid zone (figure 1). The area is located in gum arabic belt in North Kordofan State. The rainfall in the state is concentrated during a few summer months (high seasonality) and to relatively few occasions (high intensity). The mean annual temperature varies between 28° and 30°C.

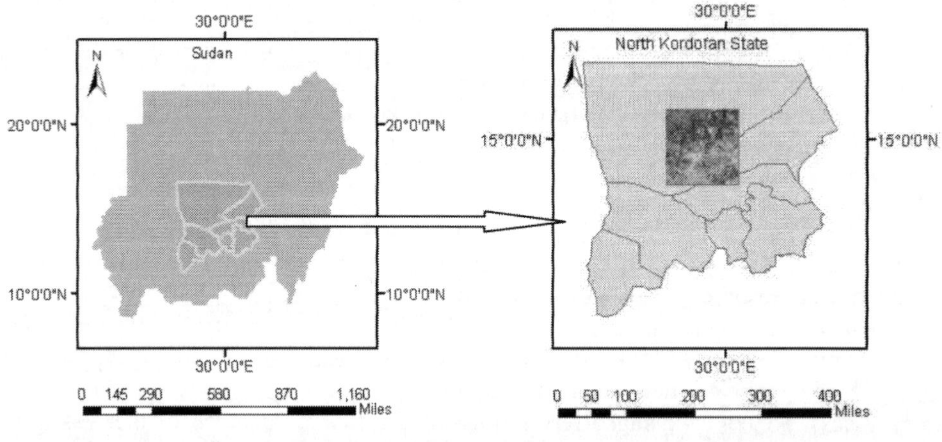

Figure 1. Location of the study area.

2.2. Material and methods

An image of ASTER L1B registered radiance at the sensor on acquired on (19.10.2010) was used in the study. ASTER image was geo-referenced to the WGS84 datum and Universal Transverse Mercator (UTM) projection [1]. The image was radiometrically corrected by using ENVI-FLAASH software (Fast Line-of-sight- Atmospheric Analysis of Spectral Hypercube) module in ENVI 4.5. Then a subset covering an area of (40880 ha) was created from ASTER image. Field data was carried for the investigation of ground control points (GCPs). The coordinates for each point were recorded using hand-held GARMIN eTrex Venture HC Global Positioning System (GPS). Information on land use and cover was recorded. Total number of (47) GCPs were surveyed and used in accuracy assessment using the calculation of a confusion matrix [8]. The segmentation and classification were conducted on three levels to be incorporated into the hierarchical image object network. Classification accuracy was assessed by comparing the automated technique with the field inventory data using Kappa coefficient. The image classification (pixel-based and object-based) and accuracy assessment were conducted (figure 2). Image segmentation process was implemented using Definiens eCognition 7.1 software. The process involved multi-resolution segmentation and classification. The pixel-based classification using maximum likelihood method was applied for the same subset image. The comparison and analysis of these two classification approaches used three spectral bands of visible/near Infrared (VNIR).

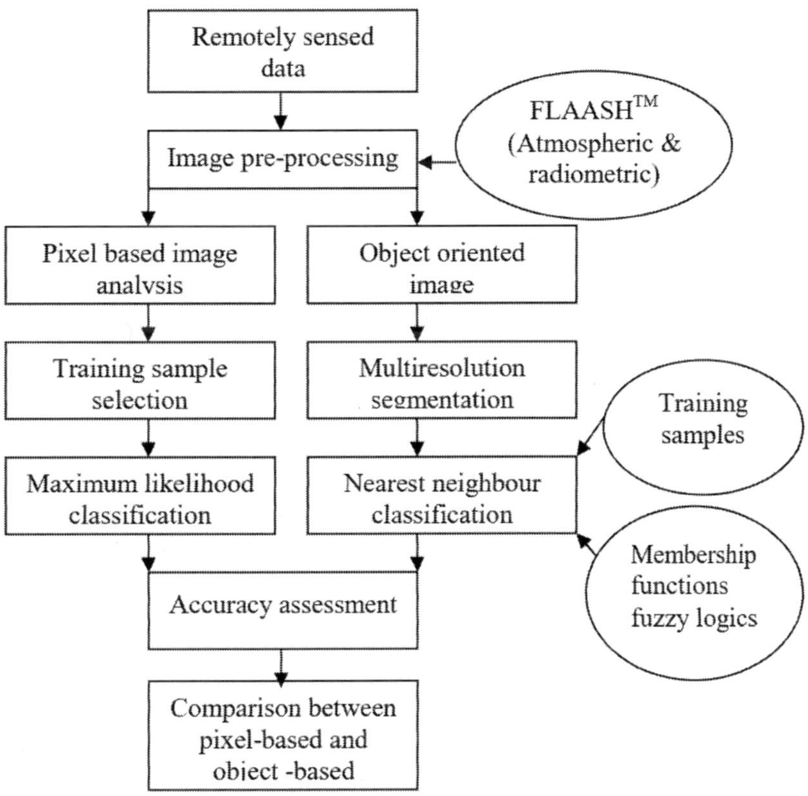

Figure 2. Flowchart for image processing and classification using object-based and pixel-based approaches.

3. Results

The comparison of both techniques was based on a visual analysis of the respective thematic maps outputs and on evaluation of the corresponding accuracy assessment measures (overall, producer's and user's accuracies, kappa coefficient). The classification of the land use land cover classes are indentified into five classes, which are bare and farm land, grass and bush land, forest dominated by *Hashab*, mixed woodland and residential areas.

3.1. Object-based classification

The multi-resolution segmentation technique was used to build up a hierarchical network of image objects that allowed the definition of relations between neighboring objects of different size [4]. Multi-resolution segmentation is a bottom-up segmentation strategy assembling objects to create larger objects and is based on the chosen scale, colour and shape parameters. Each parameter was weighted from 0 to 1. The scale parameter is the most important determining factor of multi-resolution segmentation [24].

Four levels of scales (200, 100, 50 and 25) combined with shape factor 0.7 and compactness 0.7 was tested and one optimum set of scale parameter was selected for the study (table 1). Within each segmentation level, an image object is not only linked to its neighbours, but also to its super-object and its sub-object providing useful context information for classification analysis [14]. In this study, segmentation level 4 of scale parameter 25 was selected for classification as it gave the best result in visualization of land use land cover classes based on colour and form homogeneity. The segmentation parameter 25 produced homogeneous segments for the optimal separation compared with the other levels.

Table 1. Segmentation parameters and criteria.

Segmentation level	Scale parameter	Shape factor	Compactness	Number of objects
Level 1	200	0.2	0.7	1311
Level 2	100	0.2	0.7	4935
Level 3	50	0.2	0.7	18917
Level 4	25	0.7	0.7	79614

For each individual image object, more than 150 object properties are created, which are also known as object features are provided [10]. Each object class has characteristic features enabling it to be separated from other object. The appropriate application of object features in OBIA for class description is based on previous user experience and knowledge or it can utilize feature reduction algorithms [25].

Table 2. Feature categories used for separability and threshold.

Type	Feature
Customized	NDVI, SAVI, RVI, Ratio (IR/R), (IR-G), (IR-R), (R-G)
Layer Value	Brightness, Max. Diff., Mean (L1, L2, L3), Stddev (L1, L2, L3)
Texture After Haralick	GLCM* Homogeneity (all dir.) (L1, L2, L3, all dir)
	GLCM Mean (all dir.) (L1, L2, L3, all dir)
	GLCM Contrast (all dir.) (L1, L2, L3, all dir)
	GLCM Dissimilarity (all dir.) (L1, L2, L3, all dir)

* Gray Level Co-occurrence Matrix
Stddev = Standard deviation

The feature extraction and analysis were applied based on 32 selected features to identify the characteristics of each feature of the objects determining gland use land cover classes, and to find the suitable threshold values for class separation and distinguishing among the given classes (table 2). To achieve this measurement and to identify the characteristic features for the individual object classes for the classification, separability and threshold calculation were used [22], [7] and [21].

[7] used the Bhattacharyya distance (*B*) for calculating class separability and feature selection (equation 1) for mapping urban areas. [22] developed a tool called SEaTH (Separability and Thresholds), designed to determine suitable features and their threshold values in the Definiens software.

On the basis of these representative training data for each land use land cover class, the probability distribution for each class was estimated and used to calculate the separability between two land use land cover classes.

The study distinguished between the two classes C1 and C2 followed by using a threshold of separation (*T*). The thresholds were determined by Bayes' rule as following equations using [21]:

$$B = \frac{1}{8}(m_1 - m_2)^2 \frac{2}{s_1^2 + s_2^2} + \frac{1}{2}\ln\left[\frac{s_1^2 + s_2^2}{2s_1 s_2}\right] \tag{1}$$

$$J = \left[2(1 - e^{-B})\right]^{1/2} \tag{2}$$

$$T = \frac{1}{\sigma_{c_1}^1 - \sigma_{c_2}^2}\left[m_{c_2}\sigma_{c_1}^2 - m_{c_1}\sigma_{c_2}^2 \pm \sigma_{c_1}\sqrt{(m_{c_1} - m_{c_2})^2 + 2A(\sigma_{c_1}^2 - \sigma_{c_2}^2)}\right] \tag{3}$$

Where m_i and σ_i^2, i = 1, 2, are the mean and the variance, respectively, for the two feature distributions.

The Jeffries–Matusita (*J*) distance between a pair of probability functions is the measure of the average distance between the two class density functions [23]. The range of (*B*) falls in half-closed interval [0, ∞]. This range is transformed into the closed interval [0, 2] by using a simple transformation of Jeffries Matusita distance measure (*J*). When calculating the separability between the two classes, the features which have high Jeffries-Matusita value (J = 2), is the optimum features for separation between the two classes.

The results of all possible combinations for the separation of the land use land cover classes are presented in table (3). Separation between two land use land cover classes, only two features were selected when the value of (*J*) is greater than 1.7, otherwise more than two features were selected when the (*J*) value is less than 1.7 in order to increase the classification quality and permit best separation between these two classes. Therefore all objects in bare and farm land class could easily to be separated from those in the forest dominated by *Hashab* trees when the (*J*) value for feature brightness is 1.91 and 1.85 for mean layer 1. In other words, all objects with thresholds (*T*) for feature brightness greater than 1.91653 should be assigned to bare and farm land class, and all objects with thresholds (*T*) for feature mean layer 1 greater than 1.85754 also should be assigned to bare and farm land class. All other combinations between the other are land use land cover classes (table 3).

Finally the land use land cover classes were derived by classification using the nearest neighbor classifier with the defined membership functions (fuzzy logic) for each class. Each class was classified separately. Firstly bare and farm land class was separated using two features then forest dominated by *Hashab* (using three features), grass and bush land (using four features), mixed woodland (using one feature) and the residential areas (using one feature) were classified (table 4).

IOP Conf. Series: Earth and Environmental Science **37** (2016) 012061 doi:10.1088/1755-1315/37/1/012061

3.2. Pixel-based classification
Digital pixel-based methods aim to offer objective and repeatable procedures for identifying land use land cover classes. This approach is referred to pixel-based methods given that each pixel is compared from one date to another independently from its' neighbours. The maximum likelihood technique was selected for supervised classification and for comparison with the object-based classification. The distribution of the pixel based classification for the bare and farm land, grass and bush land, forest dominated by *Hashab,* mixed woodland and residential area classes are found as 18.62%, 53.16%, 20.02%, 6.695 and 1.50, respectively (table 4).

Table 3. Object features used for the separation of classes using seperability (*J*) and threshold (T).

No.	Classes separation	Feature	Seperability (J)	Direction	Threshold (T)
1	Bare and farm land from forest dominated by Hashab	Brightness	1.91653	<	3248.64
		Mean Layer 1	1.85754	<	2009.00
2	Bare and farm land from grass and bush	Brightness	1.91275	<	3251.27
		Man Layer 1	1.76075	<	2046.27
3	Bare and farm land from mixed woodland	Brightness	1.9998	<	2997.40
		Mean Layer 1	1.99976	<	1795.90
4	Forest dominated by Hashab form grass and bush	Brightness	1.6921	>	1325.10
		Ratio	1.0404	<	1.6442
		Mean Layer 1	0.85241	>	1321.12
		Mean Layer 2	0.75248	>	1423.1
5	Forest dominated by Hashab from mixed woodland	Brightness	1.16844	<	2411.30
		Ratio	0.82175	<	0.31251
		Mean Layer 1	1.0000	>	1470.99
		Mean Layer 2	0.95889	>	2066.00
6	Grass and farm from mixed woodland	NDVI	0.97539	>	0.24453

Table 4. Land use land cover classes derived from pixel-based and object- based classification.

	Area (ha)	%	Area (ha)	%
Class name	MLC1		OBC2	
Bare and farm land	7611.8	18.62	6197.6	15.16
Grass and bush land	21731.8	53.16	24970.0	61.08
Forest dominated by Hashab	8183.9	20.02	8428.5	20.60
Mixed woodland	2736.4	6.69	667.5	1.63
Residential area	615.2	1.50	615.2	1.50

[1] Maximum likelihood classification
[2] Object-based classification

IOP Conf. Series: Earth and Environmental Science **37** (2016) 012061 doi:10.1088/1755-1315/37/1/012061

The visual interpretation and comparison of results of mixed-based versus object-based classification of the respective image revealed some differences between the land use land cover classes (figure 3).

3.3. Accuracy assessment

The overall accuracy of the classification produced by the pixel-based method was 72.92%, with Kappa statistic of 0.6259, whereas the overall accuracy and Kappa statistic of the classification produced by the object-based method were 54.17% and 0.3810, respectively (table 5 and table 6).

Table 5. Classification accuracy assessment of pixel-based approach.

Class name	Reference totals	Classified totals	Number correct	Producers Accuracy (%)	Users Accuracy (%)
Bare and farm land	14	11	10	71.43	90.91
Grass and bush land	15	24	14	93.33	58.33
Forest dominated by *Hashab*	13	8	6	46.15	75.00
Mixed woodland	2	1	1	50.00	100.00
Residential area	4	4	4	100.00	100.00
Total	48	48	35		

Overall classification accuracy = 72.92%
Overall Kappa Statistics = 0.6259

Table 6. Classification accuracy assessment of object-based approach.

Class name	Reference totals	Classified totals	Number correct	Producers Accuracy (%)	Users Accuracy (%)
Bare and farm land	14	11	10	71.43	90.91
Grass and bush land	15	5	1	06.67	20.00
Forest dominated by *Hashab*	13	27	10	76.92	37.04
Mixed woodland	2	1	1	50.00	100.00
Residential area	4	4	4	100.00	100.00
Total	48	48	25		

Overall classification accuracy = 54.17%
Overall Kappa Statistics = 0.3810

4. Discussion

Quantitative results from this study give a good overview for understanding the differences in performance of the two approaches for the same data. Considerable variability in the performance of these methods was observed. The pixel-based approach creates individual pixels or group of pixels while the object-based approach creates multi-pixel features. The bare and farm land as well as grass and bush land classes are more or less classified with the same area of coverage by the two approaches. The discrimination of mixed woodland class is confused with other classes in object-based classification due to the similarly of pixel-based spectral information between the classes. The most significant misclassification between the two classification approaches occurred in the mixed woodland and grass and bush land classes. The mixed woodland class seems to be underestimated while the grass and bush land class seems to be overestimated in object-based classification compared with the pixel-based approach. This is due to medium resolution of the multi-spectral bands of the ASTER image (i.e. 15 m) which is not be sufficient for classification of heterogeneous nature of the land use land cover classes in the study area (mixed pixels problem).

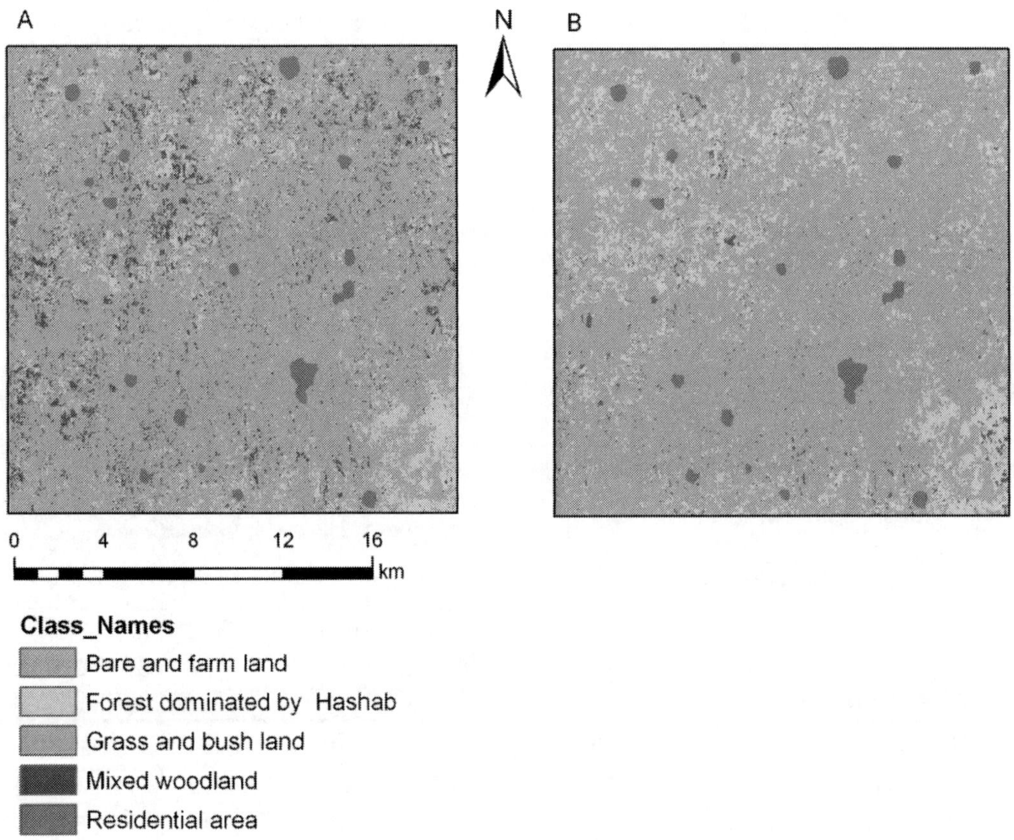

Figure 3. Results of land use land cover classification for (A) pixel- based and (B) object-based approaches.

Grouping of pixels to objects in the object-oriented classification method decreases the variance within the same land cover type by averaging the pixels within the objects, which prevents the significant *salt* and *pepper* effect in pixel-based classification [18].Therefore, the object-oriented classification approach is becoming more suitable for the needs of mapping when dealing with the high resolution imagery.

The object-based method misclassifies pixels especially in spectrally heterogeneous classes like mixed woodland and *Hashab* forests. However, the pixel-based approach generally performs better since most of the land use land cover classes selected for classification is highly homogeneous. Object based largely depends on the experimental objects used, which the successful results attributed to the careful selection of training objects [27].

The pixel-based approach is still efficient when used for land cover classification in semi-arid areas based on medium resolution imagery. Unlike pixel-based classification, the object-oriented approaches produce an output which is composed of segments that can be easily exported to GIS for creating and updating the information [13]. In addition to that object-based analysis combines spectral and spatial information as well as the texture and context information in the image [11]. The mixed pixel problem associated with conventional medium spatial and spectral resolution satellite imagery of semi-arid lands can be overcome by moving from conventional crisp per-pixel classification to soft spectral object-based approaches when using very high resolution.

Acknowledgement
I thank the chair of remote sensing at the Institute of Photogrammetry and Remote Sensing (IPF) in Technical University of Dresden, Germany for providing us with ERDAS 9.3 eCognition software for remote sensing data analysis during my research visit in 2012.

References

[1] Abrams M 2000 The advanced space borne thermal emission and reflection radiometer (ASTER) data products for the high spatial resolution imager on NASA's Terra platform. Int. *J of Remote Sensing,* **21** pp 847-860

[2] Aspinalls R J and Hill M j 2008 *Land Use Change, Science Policy and Management* (New York) CRC Press Taylor and Francis Group

[3] Baatz M, Benz U, Dehghani S and Heynen M 2004 *eCognition User Guide 4* (Munich) *Definiens imagine GmbH* (Germany)

[4] Baatz M, Heynen M, Hofmann P, Lingenfelder I, Mimler M, Schäpe A, Weber M and Willhauck G 2002 *eCognition User Guide* (München) Definiens imaging GmbH (Germany)

[5] Barredo J I and Sendra J B 1998 Multi-criteria evaluation methods for ordinal data in a GIS environment *J. Geographical Systems* pp 313-327

[6] Blaschke T 2010 Object based image analysis for remote sensing *ISPRS Int. J of Photogrammetry and Remote Sensing* **65** 1, pp 2–16

[7] Carleer A P and Wolff E 2006 Urban land cover multi-level region-based classification if VHR data by selection reverent features *Int. J of Remote Sensing,* **27** 6 pp 1035-1051

[8] Congalton R G and Green K 1999 *Assessing the Accuracy of Remotely Sensed Data, Principles and {ractices* (Lewis Publishers)

[9] Dafalla M S 2007 *Mapping and Assessment of Land Use/Land Cover Using Remote Sensing and GIS in North Kordofan State, Sudan PhD thesis* Technical University of Dresden (Germany)

[10] Definiens 2007 *Definiens Imaging Developer 7. eCognition User Guide* (München) Definiens AG. (Germany)

[11] Flanders D Mryka H and Joan P 2003 Preliminary evaluation of eCognition object based software for cut block delineation and feature extraction *Canadian J of Remote Sensing,* **20**, pp 441-452

[12] Foody G M and Atkinson P M 2002 *Uncertainty in remote sensing and GIS* (London) John Wiley and Sons

[13] Geneletti D and Gorte B G H 2003 A method for object-oriented land cover classification combining TM data and aerial photographs. *International J of Remote Sensing* **24** pp 1237-1286

[14] Hall D K., Key J, Casey K A, Riggs G A and Cavalieri D J 2004 Sea ice surface temperature product from the Moderate-Resolution Imaging Spectroradiometer (MODIS) *IEEE Transactions on Geoscience and Remote Sensing,* **42** 5 pp 1076-1087

[15] Huth J Kuenzer C Wehrmann T Gebhardt S Tuan V and Dech S 2012 Land cover and land use classification with TWOPAC: towards automated processing for pixel- and object-based image classification. *J. Remote Sens.* **4** pp 2530-2553 doi:10.3390/rs4092530

[16] Key J Yang P Baum B and Nasiri S 2002 Parameterization of shortwave ice cloud optical

properties for various particle habits. *J of Geophysical Research* **107** pp 4181-4191

[17] Kim S, Lee W Kwak D, Biging G S, Gong P, Lee J and Cho H 2011 Forest cover classification by optimal segmentation of high resolution satellite imagery. *Sensors* **11** PP 1943-1958 doi:10.3390/s110201943

[18] Laliberte A S, Rango A, Havstad K M, Paris J F, Beck R F, McNeely R and Gonzalez A L 2004 Object-oriented image analysis for mapping shrub encroachment from 1937 to 2003 in southern New Mexico *Remote Sensing of Environment* **93** 1 pp 198-210

[19] Markham B L and Townsend R G 1981 Land cover classification accuracy as a function of sensor spatial resolution. *Proceedings of the 15th International Remote Sensing of the Environment*; 11–15 May 1981. *Ann Arbor,* (University of Michigan Press: USA) pp1075-1090

[20] Mather P M 2004 *Computer Processing of Remotely Sensed Images* (3rd ed) John Wiley and Sons, Ltd (USA)

[21] Nussbaum S and Menz G 2008 *Object-Based Image Analysis and Treaty Verification, New Approached in Remote Sensing-Applied to Nuclear Facilities* (Iran) *Springer Science-Business Media B.V.D*

[22] Nussbaum S, Niemeyer I and Canty M J 2006 SEATH- A new tool for automated feature extraction in the context of object based image analysis *Proc. 1st Int Conf on Object-based Image Analysis (OBIA 2006),* , 4-5 July 2006 Salzburg *ISPRS vo.* XXXVI – 4/C4224 *of Remote Sensing* **22** pp 305-1334

[23] Richards J A and X Jia 2006 *Remote Sensing Digital Image Analysis (4th ed.): An Introduction* (Heidelberg) Verlag (Berlin,: Geramny)

[24] Su W, Li J, Chen Y, Liu Z, Zhang J, Low T M, Suppiah I and Hashim S A M 2008 Textural and local spatial statistics for the object-oriented classification of urban areas using high resolution imagery *Int. J. of Remote Sensing* **29** pp 3105-3117

[25] Tarantino E and Figorito B 2012 Mapping rural areas with widespread plastic covered vineyards using true color aerial data *Remote Sens*png **4** pp 1913-1928 doi:10.3390/rs4071913

[26] Willhauck G 2000 Comparison of object oriented classification techniques and standard image analysis for the use of change detection between SPOT multispectral satellite images and aerial photos, *XIX Congress of the Int. Society for Photogrammetry and remote Sensing (ISPRS).* Geomatics Information and Trading Center (GITC) (Amsterdam: The Netherlands) pp 214-221

[27] Yang X (Eds.) 2011 *Urban Remote Sensing: Monitoring, Synthesis and Modeling in the Urban Environment* (Wiley-Blackwell) p 408

Land cover classification of Landsat 8 satellite data based on Fuzzy Logic approach

Afirah Taufik[1] and Sharifah Sakinah Syed Ahmad[2]

Faculty of Information and Communication Technology, Universiti Teknikal Malaysia Melaka, Hang Tuah Jaya, 76100 Durian Tunggal,Melaka, Malaysia[1].
Department of Industrial Computing, Faculty of Information and Communication Technology, Universiti Teknikal Malaysia Melaka, Hang Tuah Jaya, 76100 Durian Tunggal, Melaka, Malaysia[2].

Email: afirahtaufik@gmail.com

Abstract. The aim of this paper is to propose a method to classify the land covers of a satellite image based on fuzzy rule-based system approach. The study uses bands in Landsat 8 and other indices, such as Normalized Difference Water Index (NDWI), Normalized difference built-up index (NDBI) and Normalized Difference Vegetation Index (NDVI) as input for the fuzzy inference system. The selected three indices represent our main three classes called water, built-up land, and vegetation. The combination of the original multispectral bands and selected indices provide more information about the image. The parameter selection of fuzzy membership is performed by using a supervised method known as ANFIS (Adaptive neuro fuzzy inference system) training. The fuzzy system is tested for the classification on the land cover image that covers Klang Valley area. The results showed that the fuzzy system approach is effective and can be explored and implemented for other areas of Landsat data.

1. Introduction

The classification of Landsat 8 satellite data using fuzzy logic is a method to extract features of a land cover into certain classes. Landsat 8 contains multispectral bands which are invisible and infrared bands. The combination of certain bands is able to classify the features of the classes such as water, vegetation and built-up land. The indices that are used are input for fuzzy inference system are Normalized Difference Vegetation Index (NDVI), Normalized Difference Water Index (NDWI) and Normalized Difference Built-up Index (NDBI).

The aim of this paper is to propose a method to classify the land covers of a satellite image based on fuzzy rule-based system approach. The purpose of the image classification system is to assign each input of the indices to certain classes according to fuzzy logic approach. Fuzzy logic has been used widely to make a decision with incomplete or uncertain data. The combination of the neural networks and fuzzy theoretic approaches, are formed into Adaptive neuro fuzzy inference system (ANFIS) by attaching the fuzzy inference system to the framework of adaptive networks, The parameter selection of fuzzy membership function using supervised method ANFIS training is important to achieve the best performance for the fuzzy inference system. The fuzzy system is tested for the classification on the land cover image that covers the Klang Valley area. The results are based on the fuzzy rules-based used as detection for a land cover classes and based on the accuracy assessment for the classified pixels.

2. Related works

Vegetation index can be used as an indicator to quantify the greenness of plants within the satellite data. There are several vegetation indices, but the most frequently used index is the Normalized Difference Vegetation Index (NDVI) [1]. NDWI is sensitive to changes in liquid water content of vegetation canopies. NDWI is less sensitive to atmospheric effects than NDVI, NDWI does not remove the background soil reflectance effects completely as NDVI [2]. Urban spatial areas have expanded in an accelerated speed during the last five decades, and rates of urban population growth are higher than the overall growth in most countries because urban areas are the locus of economic activity and transportation nodes [3]. The paper proposes a technique to extract urban built-up land features from Landsat Thematic Mapper (TM) and Enhanced Thematic Mapper Plus (ETM) imagery taking two cities in the South Eastern China as examples. The study selected three indices, Normalized Difference Built-up Index (NDBI), Modified Normalized Difference Water Index (MNDWI), and Soil Adjusted Vegetation Index (SAVI) to represent three major urban land-use classes, built-up land, open water body, and vegetation, respectively [4]. Fuzzy set theory [5] provides useful concepts and tools to deal with imprecise information. Partial membership allows the information about more complex situations, such as cover mixture or intermediate conditions, can be better represented and utilized. However, works in remote sensing image analysis using fuzzy sets are rather scarce. The proposed ANFIS can construct an input-output mapping based on both human knowledge (in the form of fuzzy if-then rules) and stipulated input-output data pairs. In the simulation, the ANFIS architecture is employed to model nonlinear functions, identify nonlinear components on line in a control system, and predict a chaotic time series, all yielding remarkable results [6]. In this study, object-based image analysis is used for image classification. The main problems in high resolution image classification are the uncertainties in the position of object borders in satellite images and also multiplex resemblance of the segments to different classes. In order to solve these problems, fuzzy logic is used for image classification, since it provides the possibility of image analysis using multiple parameters without requiring inclusion of certain thresholds in the class of assignment process [7] .

3. Materials and methodology

Figure 1 shows a framework of the land cover classification study. The details of materials and methodology are explained in the next section.

3.1. Data

As the initial stage, the study was conducted with information collection of satellite image at Klang, Selangor. This study area is situated between longitudes 101° 16' to 102° 00' E and scope 3° 05' to 3° 22' N. The satellite information was procured from USGS Landsat 8 information that was utilized as a part of this study. The information was chosen by considering the high calibre of information gained by the Landsat satellite. The multispectral bands Landsat 8 data contains visible and infrared wavelength. The data were acquired on 24th March 2014. By utilizing an unearthly subset as a part of ENVI programming, the chosen region of interest (ROI) was picked as training pixels. The DAT record contains 7 groups with 349 pixels × 329 pixels.

3.2. Pre-processing

The Landsat 8 digital data of study area was read into MATLAB. Image processing was carried out to produce better CIR (color-infrared) composite using a decorrelation stretch. Band combinations were performed many times/several times in order to obtain visual of interpretation of classes and other information. The image processing system can display three band combinations by assigning three different colors to each band. The images before and after processed are shown in Figure 2. The shown images adopt band 5, 4 and 3.

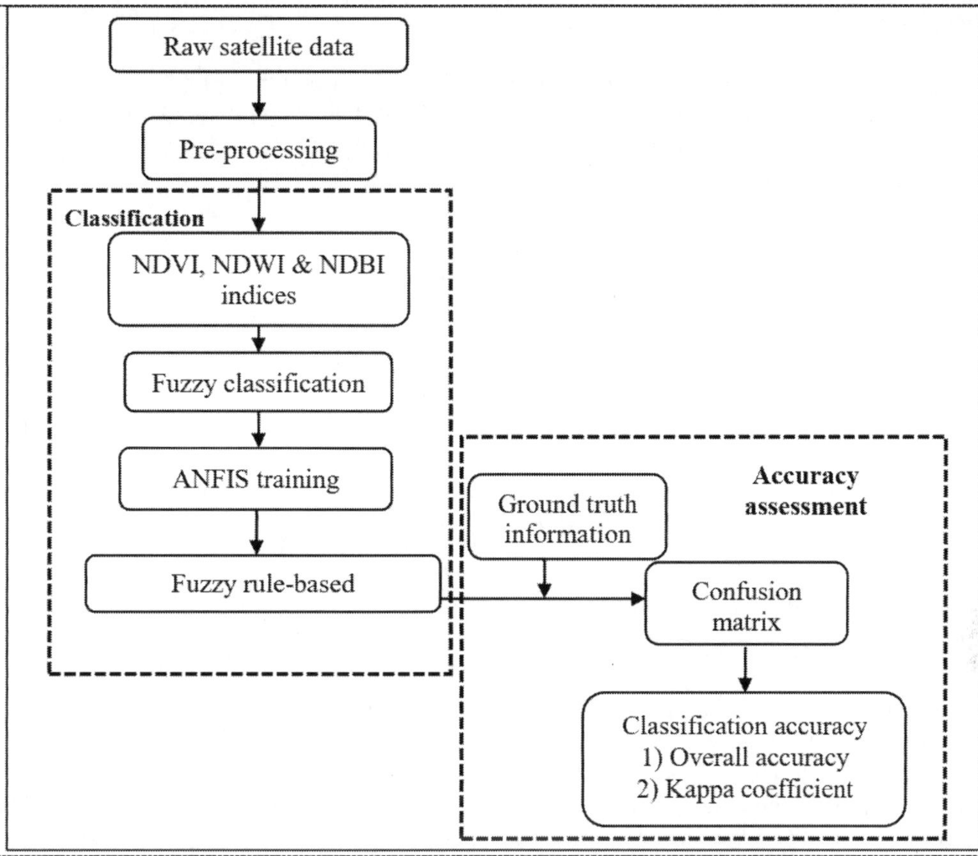

Figure 1. The framework of classification experiment.

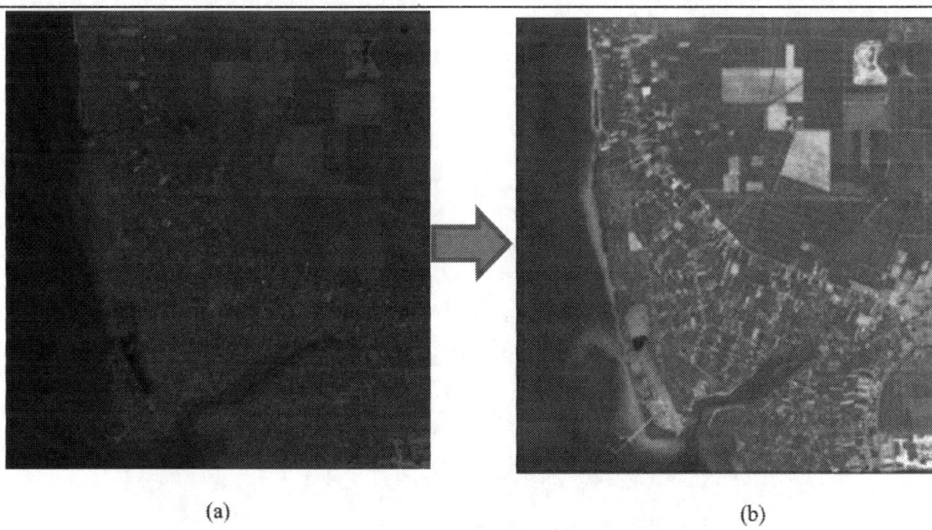

Figure 2. The process before (a) and after (b) pre-processing.

3.3. Classification

In this study, the layer of the multispectral bands consist of NIR (near-infrared), SWIR (short-infrared), Red and Green are used. The layer of the bands 5-4-3 (NIR, Red and Green) are extracted and mapped into red, green and blue (RGB color). The RGB color is a standard color of infrared (CIR) image. The NDVI, NDWI and NDBI were determined by utilizing the formula (1), (2) and (3):

$$NDVI = \frac{NIR - Red}{NIR + Red} \tag{1}$$

$$NDWI = \frac{Green - NIR}{Greeen + NIR} \tag{2}$$

$$NDBI = \frac{NIR - SWIR}{NIR + SWIR} \tag{3}$$

3.3.1. ANFIS modelling

Fuzzy inference systems are the fuzzy rule based systems which consists of a fuzzification interface, rule base, database, decision making unit, and defuzzification interface [8]. ANFIS is the supervised training method to determine the selection of membership function for the three inputs based on the output (ground truth information). Figure 3 shows the fuzzy modelling by using ANFIS training to determine the selection of parameter of membership function of the inputs for the fuzzy inference system. The data are divided into training and testing data. The three inputs are NDVI, NDWI and NDBI. The output data is also a vital part where the ground truth information is needed to classify the satellite image data. Output data consists of three classes based on the ground truth information, water, built-up land and vegetation. The left-nodes are input and the right-nodes are output. Each node performs a particular function from input node as well as asset of parameters to the node. There are 27 rules are used for the land cover classification by using fuzzy inference system. The fuzzy if-then rules for three classes are discussed in the next section.

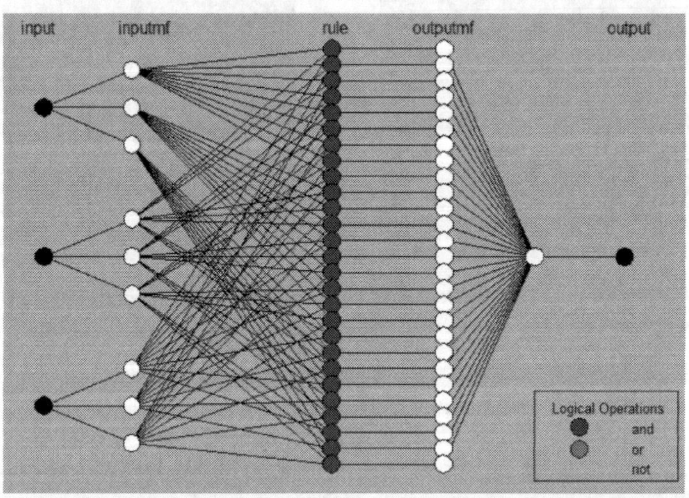

Figure 3. ANFIS model structure.

3.4 Accuracy assessment

The classified image was compared with information of ground truth data in order to check the accuracy. The user, producer and overall accuracy were carried out to measure the classification accuracy. Producer's accuracy is where the individual class accuracy can be acquired by dividing the sum of correctly classified pixels. Meanwhile, the user's accuracy is a measurement where the individual class acquired from the classified pixels in same group [8]. The number of misclassified pixels into another class were recorded. The overall accuracy (4) was calculated according to the confusion matrix that obtained for user's accuracy and producer's accuracy. The description of overall accuracy is shown below:

$$Overall\ accuracy = \frac{Total\ number\ of\ correct\ classified}{Total\ number\ of\ pixels} X\ 100 \tag{4}$$

Kappa coefficient (5) is another measurement to measure the training pixels with the ground truth data information. The Kappa values are +1.0 to -1.0. The positive values show high accuracy and the value of zero shows no correlation in the classification.

$$K = \frac{n\sum_{i=1}^{p} x_{ii} - \sum_{i=1}^{p}(x_{i+} \times x_{+i})}{n^2 - \sum_{i=1}^{p}(x_{i+} \times x_{+i})} \tag{5}$$

Where,
n = total number of training pixels
p = number of classes
$\sum x_{ii}$ = total number elements of confusion matrix
$\sum x_{i+}$ = sum of row i
$\sum x_{+i}$ = sum of column i

The results of accuracy assessment for a classification based on the confusion matrices were obtained and recorded in the next section.

4. Results and discussion

4.1 Accuracy assessment results.

The results of this study are acquired based on the accuracy assessment and the fuzzy rule based. The classification process to satellite data must be quantitatively determined their accuracy of classification. By using the ANFIS training to determine the selection of parameter membership function in fuzzy inference system, the three classes output were classified. The pixels that have been classified using fuzzy classification were compared with the ground truth information. In order to evaluate the fuzzy inference system for land cover classification, an area (349 pixels × 329 pixels) of image is classified based on three input indices: NDVI, NDWI and NDBI. The results of classification showed the color mapping and labelling of 3 classes. As we can observe in Figure 3, the images are grouped into 3 categories, water, built-up land and vegetation. Table 1 shows the related confusion matrices assessment and classification results (overall accuracy and kappa coefficient). Based on the results, the total pixels are correctly classified is very high rather than misclassified pixels for every category class. The misclassified pixels for the vegetation class are 228 and 9 pixels, built-up land and water, respectively.

The misclassified pixels for the built- up land class are in the other two classes; vegetation and water, 387 and 168, respectively. The misclassified pixels for water class into the other class; built-up land, is 60 pixels. The user's accuracy and producer's accuracy were obtained from the submission of every row and column from the confusion matrix table. The correctly classified pixels of every row and column are divided with the sum of total pixels for every row and column. The percentage for overall accuracy that obtained from formula (4) is 97.6%, where the result is highly accurate. Kappa coefficient is 0.956 and it is noteworthy that the value is the positive range. This result is classified as high accuracy for kappa coefficient. The final accuracy is very high and exceeds 90 % for every classes.

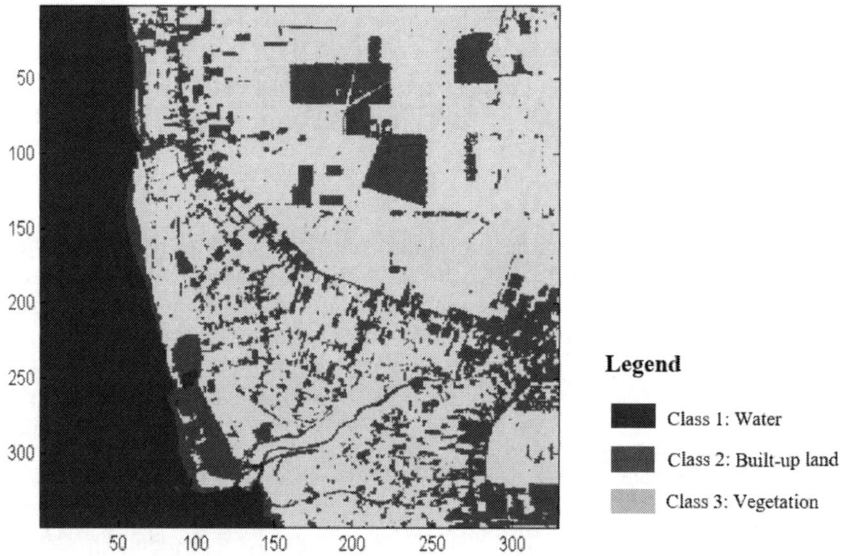

Figure 4. Color mapping and labelling.

Table 1. Confusion matrices for fuzzy classification of Klang image dataset.

Class	Vegetation	Built-up land	Water	Total	User's Accuracy (%)
Vegetation	22,559	228	9	22,796	98.9%
Built-up land	387	5,875	168	6,430	91.3%
Water	0	60	8,928	8,988	99.3%
Total	22,946	6,163	9,105	38,274	
Producer's Accuracy (%)	98.3%	95.3%	98%		

Overall accuracy = 97.6%
Kappa coefficient = 0.956

Defuzzification of the fuzzy output function into the crisp output is one of the processes in fuzzy logic. Based on the fuzzy if-then rules, there are three rules are used for a class detection from Takagi-Sugeno. The three classes are vegetation, water and built-up land. The rules for vegetation shows that the membership function for NDVI is high, and another two membership functions for two inputs are low and medium; NDWI and NDBI, respectively.

The rules for water shows that the membership function for NDWI is high, and another two membership functions for two inputs are medium and low; NDVI and NDBI, respectively. The rules for built-up land shows that the membership function for NDBI is high, and another two membership functions for two inputs are low for NDVI and NDWI.

If **NDVI** is *high* and **NDWI** is *low* and **NDBI** is *medium* then class is **Vegetation.**
If **NDVI** is *medium* and **NDWI** is *high* and **NDBI** is *low* then class is **Water.**
If **NDVI** is *low* and **NDWI** is *low* and **NDBI** is *high* then class is **Urban.**

5. Conclusion

The results that presented in this study demonstrate the effectiveness of fuzzy approach for a land cover classification by using ANFIS modelling. The parameter selection of fuzzy membership function using supervised method ANFIS training is important to achieve the best performance for the fuzzy inference system. The fuzzy rule-based are used for class detection. The supervised Adaptive neuro that is formed with fuzzy inference system is effective and can be explored and implemented for other areas of the Landsat data.

6. References

[1] J. W. Rouse, R. H. Haas, J. A. Schell, and D. W. Deering, "Monitoring the vernal advancement and retrogradation (green wave effect) of natural vegetation," 1973.
[2] B. C. Gao, "NDWI - A normalized difference water index for remote sensing of vegetation liquid water from space," *Remote Sens. Environ.*, vol. 58, no. 3, pp. 257–266, 1996.
[3] "The Dynamics of Global Urban Expansion 31," pp. 31–48, 2000.
[4] H. Xu, "Extraction of urban built-up land features from Landsat imagery using a thematic-oriented index combination technique," *Photogramm. Eng. Remote Sensing*, vol. 73, no. 12, pp. 1381–1391, 2007.
[5] L. A. Zadeh, "Fuzzy Sets-Information and Control." , vol. 8, pp. 338-353, 1965.
[6] J. S. R. Jang, "ANFIS: Adaptive-Network-Based Fuzzy Inference System," *IEEE Trans. Syst. Man Cybern.*, vol. 23, no. 3, pp. 665–685, 1993.
[7] S. Jabari and Y. Zhang, "Very High Resolution Satellite Image Classification Using Fuzzy Rule-Based Systems," *Algorithms*, vol. 6, no. 4, pp. 762–781, Nov. 2013.
[8] S. N. Sivanandam, S. Sumathi, and S. N. Deepa, *Introduction to Fuzzy Logic using MATLAB*. 2007.
[9] R. G. Congalton, "A review of assessing the accuracy of classifications of remotely sensed data," *Remote Sens. Environ.*, vol. 37, no. 1, pp. 35–46, 1991.

Acknowledgements
This research is supported by Universiti Teknikal Malaysia Melaka and the Ministry of Education Malaysia under grant no PJP/2015/FTMK (4B)/S01433.

Temporal assessment on land use land cover of Somalia after the effect of the civil war using remote sensing

Abdinur Abdulle, Adhwa Amir Tan, Biswajeet Pradhan* and Saleh Abdullahi

Department of Civil Engineering, Faculty of Engineering, University Putra Malaysia, 43400, UPM, Serdang, Selangor, Malaysia

Corresponding Address: Biswajeet Pradhan, Tel: +603-89466383,
Fax. +603-86567129; Email: biswajeet24@gmail.com or biswajeet@lycos.com

Abstract. The aim of this study is to analyse land use and cover changes for the studied area during 1992-2015 and particularly evaluate the effect of civil war on these changes. Three Landsat images were used; Landsat 4 (1992), Landsat 7 (2000) and Landsat 8 (2015). Assessment of changes has been applied through three supervised classification algorithms, support vector machine, minimum classifier, and mahalanobis classifier. The result shows that SVM is providing highest overall accuracy of 98.5% for the years 2000 and 2015 with kappa coefficient of 0.9803 in year 2015. The change detection result show that the higher changes is between year 1992-2000 where vegetation land cover has dropped down to 11.1% and undeveloped area has increased to 11.4%. Whereas for year 2000-2015, higher changes belongs to build up area by 3.30% while undeveloped area and vegetation land cover keep decreasing by 2.64% and 1.93% respectively.

1 Introduction

Land use land cover (LULC) plays an important role in land management, natural recourse department, economic activities, and effect of war zone; it is also a necessary task for environmental monitoring and management [1]. It is essential for the authorities to have an up-to-date land use land cover map since the world is continuously reshaping and changing [2]. LULC maps provides the quantity of changes to the landscape, as well as where the changes took place. It also offers crucial information on the state environment, development trends, shift of vegetation areas and diversity of the desert and others [3]. Generally, this can be accomplished by utilizing the temporal ability of the earth observation satellite systems [4]. Change detection is the process of identifying differences in the state of an object or phenomenon by observing it at different times [5].

Remote sensing (RS) is a tool that able to capture the earth system since the first earth orbiting satellite was launched into space. RS imageries have been used to detect and monitor land cover changes at various scales [6, 7]. A range of algorithms generated for change detection including, pre-classification and post-classification comparison, image geometric correction and etc. [8]. A variety of sensors can produce several images for the same area at different spectral, temporal and spatial resolutions. These capabilities help to solve various issues regarding agricultural, forestry, geology and environmental management. The digital change detection is effected by spatial, spectral, radiometric and temporal constraints since many change detection techniques are available. Selection of suitable method or algorithm is important for land use and cover changes for obtaining a better outcomes [5].

The objectives of this research are to assess the LULC changes of the study area in the period of 1992-2015, and identifying the difference between natural and the man-made changes especially occurred during civil war. The main target feature is the urban areas and the abrupt increase of population. Hence, the pattern changes to the city of Mogadishu can be found out. The land covers features for the study area has been divided into four types as follow:

i. Desert: A barren area which is covered with sand,
ii. Vegetation: The plant of a region,
iii. Built up area: Areas that has been developed by the human, and
iv. Undeveloped area: Bare soil or the land that has not been used.

The scope of this research is to compare three supervised classification method;1) Minimum Distance Classifier (MINDP), 2) Mahalanobis Distance Classifier (MAHADP) and 3) Support Vector Machine (SVM), in order to get better classification result and evaluate the overall accuracy for change detection in the study area. Landsat imagery has been utilized for this research with spatial resolution of 30m.

2 Study area and dataset

2.1 Study area

The study area, Mogadishu is in coastal Banaadir region on the Indian Ocean and the capital city of Somalia. It is the largest and the fastest growing city in the country with the area coverage approximately 91 km square (35 sq. mi). The city coordinates is 02°02′N 45°21′E and 2.033°′N 45.350°′E. The city has administered an important port for many centuries. The population of the city has been estimated 2.12 million at 2015 [9]. Figure 1 shows the geographical location of Mogadishu City, Somalia. Mogadishu has a relatively dry climate with the average temperature of 27°C and irregular rainfall. There are 47 wet days annually, which are associated with a 12% daily probability of rainfall. The city has an average of 3,066 hours of sunshine per year, with 8.4 hours of sunlight per day. Mean daylight hours and minutes per day are 8h 24'. The annual percentage of sunny versus cloudy daylight hours is 70% and 30%, respectively. Average sun altitude at solar noon on the 21st day of the month is 75.

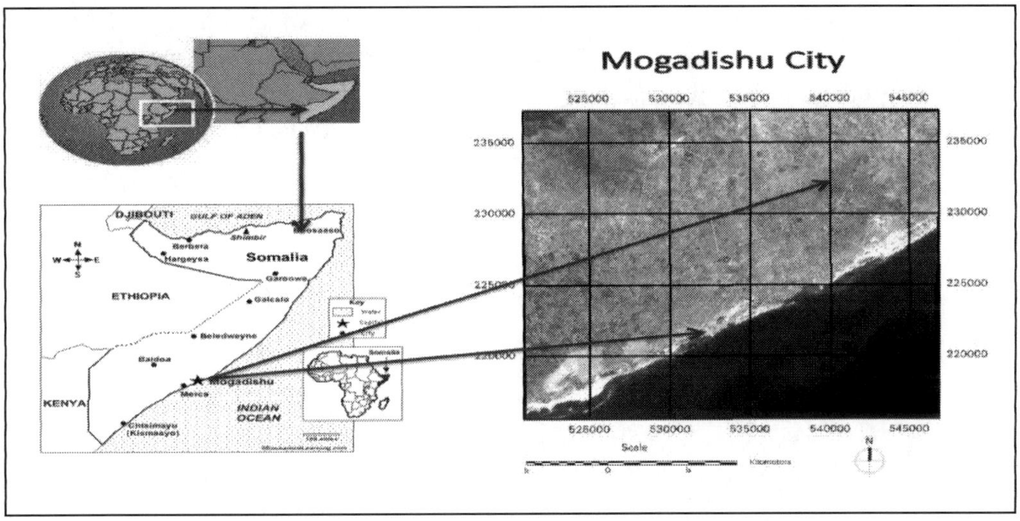

Figure 1. Map and the Landsat 8 image scene covering the study area.

2.2. Dataset

In this research, three Landsat images were used; Landsat 4 (1992), Landsat 7 (2000) and Landsat 8 (2015). Images of the study area were downloaded from the USGS Earth Explorer, especially the periods of 19 December 1992, 02 February 2000 and 18 January 2015. The details of the images are shown in Table 1. The selection of the dates of the datasets is based on the objectives of the research which to monitor the changes of the land cover during the period. Solid correlation of LULC is needed in a period of time in order to obtain these changes.

Landsat imagery is selected due to high quality images and multispectral band that allows extraction of valuable information. With the moderate spatial resolution (30 m), it has provided many layers of data across the visible and invisible light spectrum. These data can be manipulated to reveal what the Earth's surface looks like, including what types of vegetation are present or how a natural disaster has impacted an area [10].

Table 1. Properties of Landsat data for the research.

Details/Data	Landsat 4 (TM)	Landsat 7 (ETM+)	Landsat 8 (OLI & TIRS)
Acquisition date	12 December 1992	2 February 2000	18 January 2015
Source	USGS (Earth Explorer)	USGS (Earth Explorer)	USGS (Earth Explorer)
Spatial Resolution	30 m Band 1 to 5, 7 120 m (30 m) Band 6	15 m Band 8 (Pan) 30 m Band 1 to 5, 7 60m (30 m) Band 6	15 m Band 8 (Pan) 30 m Band 1 to 7, 9 60m (30 m) Band 10 and 11
Spectral Resolution	7 spectral bands	8 spectral bands	11 spectral bands

3 Methodology

3.1 Pre-processing phase

As shown in figure 2, the methodology for the research has been divided in to four phases. First is the pre-processing phase where the datasets has been gone through atmospheric and geometric correction due to atmospheric effect. Correction of the dataset in terms of the spectra information shall provide efficient input for the next step processing. Quick Atmosphere Correction (QUAC) is an atmospheric correction method for multispectral and hyperspectral imagery that works with the visible and near-infrared through shortwave infrared (VNIR-SWIR) wavelength range [11]. This followed by the geometric correction process. With this process, the images were rectified and able to be directly compared using the location wise. ROI has been established by subset the study area.

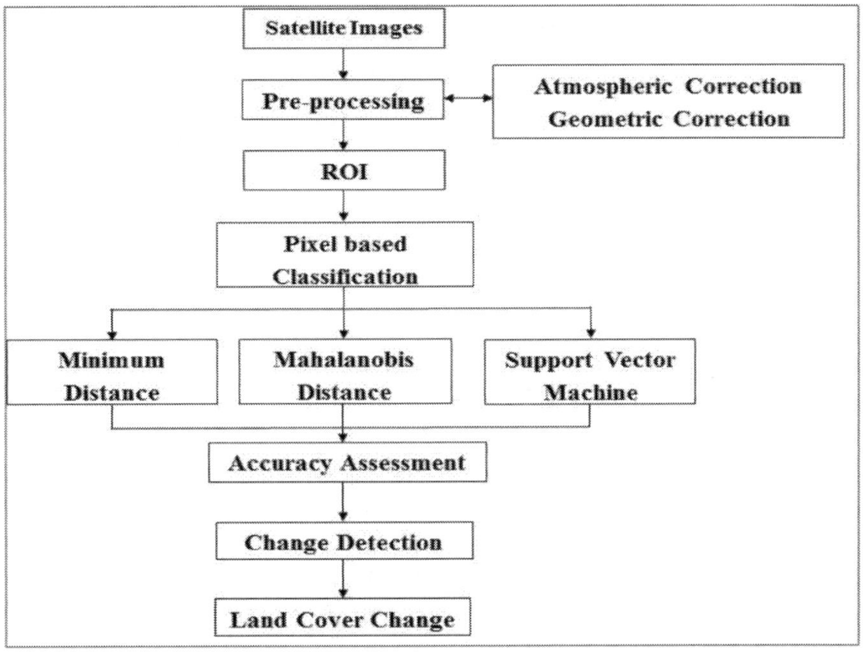

Figure 2. Methodological flowchart of the study.

3.2 Data classification phase

Next phase of methodology deals with image classification methods. Land cover features for the classification is desert, vegetation, built up area and undeveloped area. The supervised image classification based on the pixel information was selected in order to accomplish the classes. Samples of the features were collected from the images. In this case, the selection of the pixel based classification is due to the moderate resolution of the Landsat imagery.

Three different classification methods were applied; Minimum Distance Classifier (MINDP), Mahalanobis Distance Classifier (MAHADP) and Support Vector Machine (SVM). Based on the review papers, LULC usually have been classified by using SVM and Maximum Likelihood classifier (MLC) where its gives best result in terms of overall accuracy and kappa coefficient. This research focused on evaluation and comparison of these three methods to examine their aftermath. The result of the best classification will be used for the change detection analysis. This selection is based on the result of accuracy assessment.

3.3 Analysis

Research analysis phase is focusing on the accuracy assessment and the change detection findings. Accuracy assessment is the difference between the sample area and the training area. It is involves with the statistical estimation that obtained from the classification output and the independent references in the image for the probability error measurement of the classified map [12, 13]. This is to ensure the classification result is accurate and reliable. For each feature, sample areas and the training area need to be established in order to perform this assessment. Accuracy assessment has been applied for the classified map for year 1992, 2000 and 2015.

Change detection is based on the classification image differences between the initial state and the final state images [14-16]. The subtracted information from the initial and final image is the changes of the features properties. The result of changes detection will be deliberated in result section by comparing the each result based on the classification result. The first comparison is between years 1992 and 2000 and second is between 2000 and 2015. The main reason of the two comparisons is to know what is the trend or pattern of the changes. The total of the change detection will be reviewed in the discussion section.

4 Result

4.1 Image classification

Classification results were evaluated based on two main parameters. First is about the overall accuracy of the classification result; second is the Kappa coefficient for the measure of association between two categorical variables. Land cover maps based on the classification methods on the MINDP, MAHADP and SVM has been produced. Figure 3 shows that SVM method has producing high accuracy result for all images. For the year 2000 and 2015, the overall accuracy level is 98.5% while for year 1992 is 97.2%. For the MINDP classifier, the highest overall accuracy was recorded for year 2015 with 95.4%, followed by 93% for image of year 1992 and 92.4% for image of year 2000. Meanwhile for MAHADP, the highest overall accuracy is for year 2015 with 96.9%. Next is the image of year 2000 with 94.5% and 91.6% for image of year 1992.

Table 2 shows the overall accuracy and kappa coefficient of all image classification methods. The kappa coefficient ranking level is the same as the pattern in overall accuracy. From this phase, only the land cover map based on the SVM method remains for further details analysis. Figure 4 illustrates the SVM classification map for the year 1992, 2000 and 2015. It shows four main land cover classes; desert in grey, built up area in red, undeveloped area in yellow and vegetation in green. Table 3 represents classification results of the land cover features.

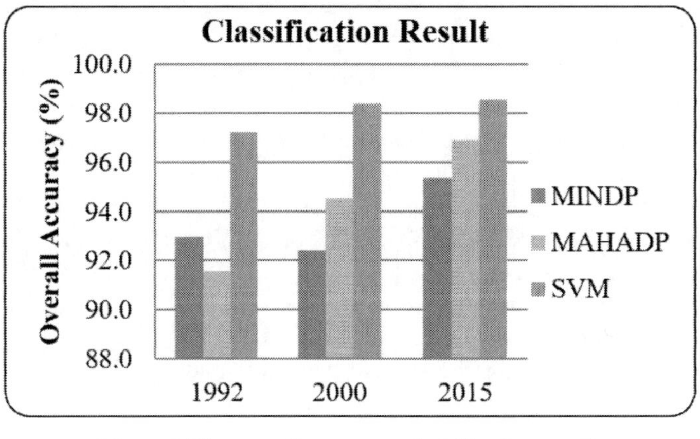

Figure 3. The overall accuracy classification result.

Table 2. Compilation of the overall accuracy result and kappa coefficient.

	1992		2000		2015	
	Overall Accuracy	Kappa	Overall Accuracy	Kappa	Overall Accuracy	Kappa
MINDP	93.0	0.9053	92.4	0.8957	95.4	0.9378
MAHADP	91.6	0.8868	94.5	0.9254	96.9	0.9583
SVM	97.2	0.9625	98.5	0.9798	98.5	0.9803

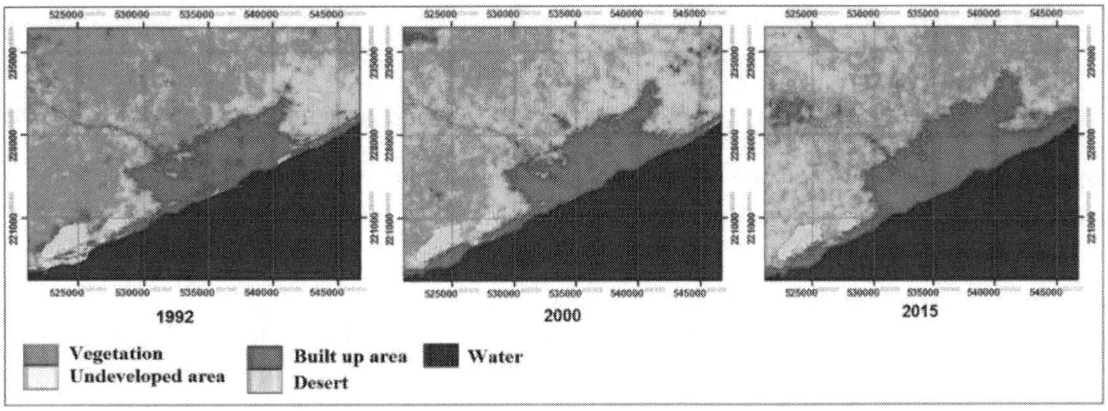

Figure 4. Land Cover Map based on SVM classification for year 1992, 2000 and 2015.

Table 3. Land Cover Features Classification Result.

	1992	2000	2015
Desert	2.27%	1.47%	1.25%
Undeveloped Area	17.20%	28.56%	25.91%
Built Up Area	12.75%	13.59%	16.89%
Vegetation	38.00%	26.88%	24.95%
Water	29.79%	29.51%	30.99%

4.2 Area of changes

Evaluation on the each feature area shows that there is lot of change happen to the Mogadishu city and its surrounding. Figure 5 illustrates the changes of the built up area that is expanding from 1992 to 2015. From the visual interpretation, the most growth of built up area is in east and north-east side of the city. In addition, in city centre also several new built up area can be observed. The vegetation land cover has displayed much of changes from a dense vegetation land cover to scatter vegetation area. The undeveloped area is rapidly spreading and desert area is reduced significantly. Water changes have been excluded due to insignificant analysis for this research. Further analysis on what caused the changes is explained in discussion section.

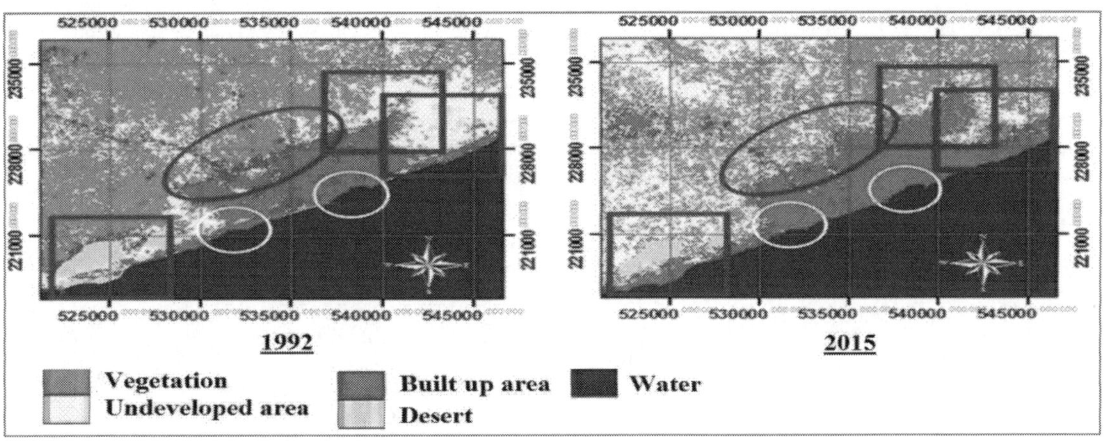

Figure 5. Land Cover changes for between year 1992 and 2015.

IOP Conf. Series: Earth and Environmental Science **37** (2016) 012063 doi:10.1088/1755-1315/37/1/012063

5 Discussion

In order to have better understanding to the changes, two comparisons has been implemented between year 1992 and 2000, and between year 2000 and 2015. Change detection for the first time period (1992-2000) has revealed that the most growth is for undeveloped area with increase of 11.36%. Vegetation land cover and desert were decreased 11.12% and 0.80% respectively. In contrast, built up area has increased 0.84%.

Meanwhile in 2000 to 2015 time period, the highest rate of change is for built up area with the growth rate of 3.30%. Other land cover areas had reduction of 2.64% for undeveloped area, 1.93% for vegetation and 0.22% for desert. In overall, desert area had slow and vegetation had high reduction rate during selected period of time. Undeveloped zones were expanded at the initial years, but then reduced in the subsequent years. The only land cover that keeps increasing was the built up area. Based on these two comparisons, the most changes rate happen to Mogadishu city is between years 1992 to 2000. Figure 6 depicts the change detection rate for both comparison years. Table 4 presents the rate of change detection for the comparison result and the overall changes between years 1992 to 2015.

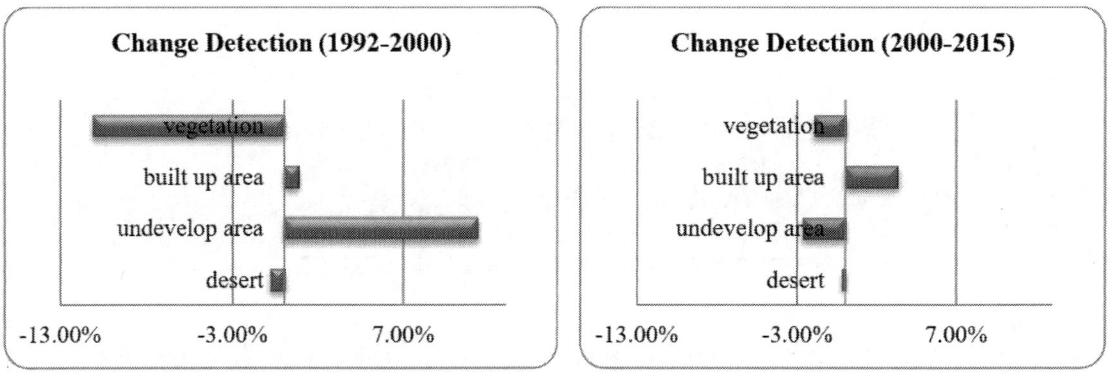

Figure 6. Change Detection rate for year 1992-2000 and 2000-2015.

Table 4. Changes rate based on the comparison result.

Changes	1992-2000	2000-2015	1992-2015
Desert	-0.80%	-0.22%	-1.01%
Undeveloped area	11.36%	-2.64%	8.71%
Built up area	0.84%	3.30%	4.14%
Vegetation	-11.12%	-1.93%	-13.04%

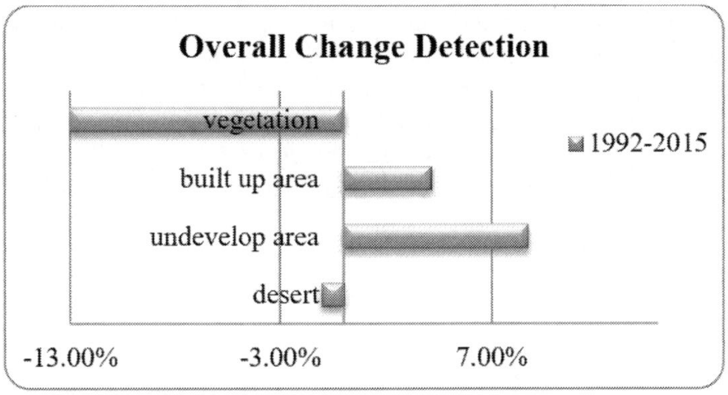

Figure 7. Overall Land Cover Changes for year 1992-2015.

The overall changes for each land cover type indicate that the built up area and undeveloped area has increased through the years. However, the land cover type for vegetation and desert is decreasing. Figure 7 displays the overall changes from year 1992 to 2015. Based on these findings, there are two types of changes that happen in the Mogadishu city; natural and man-made changes. Obviously the built up area growth is due to rapid growth of the settlement area after the war. The war incident has changed the topography of city where the location of the settlement area is moving out circuit from the city centre. The growth of the population over the years after the war has also increased the built up area. Figure 8 represents the change of the built up area for first (1992-2000) and second comparison (2000-2015). Figure 9 indicates reduction of desert area in both comparison stages. These finding show that the desert area is decreased because of the desert greening activities that has occurred at the north-east of the city. In addition, for the south-west desert, natural plant has over take some of the area.

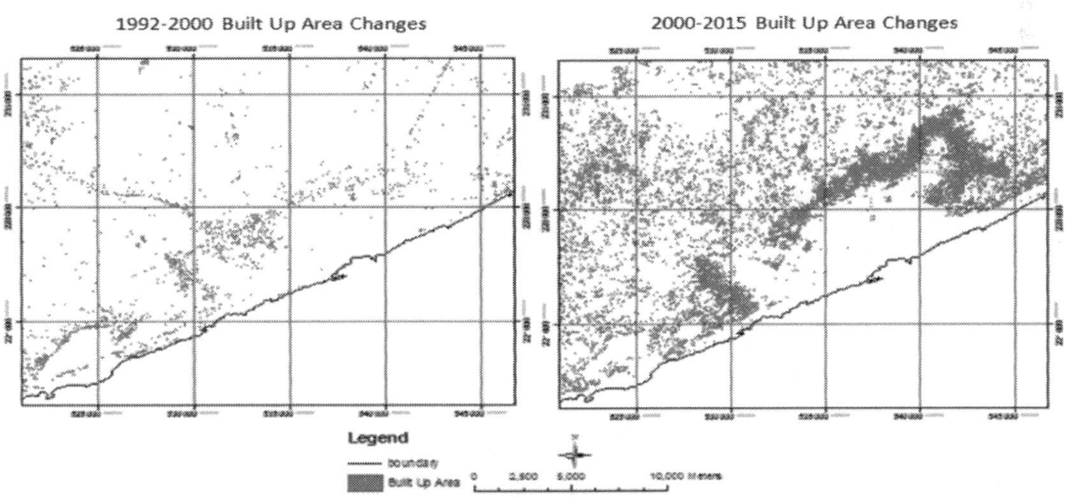

Figure 8. Growth of the Built up Area in Mogadishu City and surroundings.

The first comparison identified that a widening land cover has turn out to be undeveloped area. But in the second comparison, only small area has changed as built up and vegetation area. This change has been shown in Figure 10 where the proportional of undeveloped area changes in first comparison contrasted to second comparison. For the vegetation area, the land cover has reduced from year 1992 to 2015. Figure 11 shows the change details for the vegetation area. It is relevant to determine that the change between undeveloped area and the vegetation area is correlated. Back in year 1992, 1999, 2004, 2005, 2008, 2010 and 2012, historical data has shown that drought event has surged the Horn of Africa (Ethiopia, Somalia, and Kenya) [17]. Drought event has affected the vegetation area as it become dry and deforestation. This has the most signification contribution of land cover changes for the natural features.

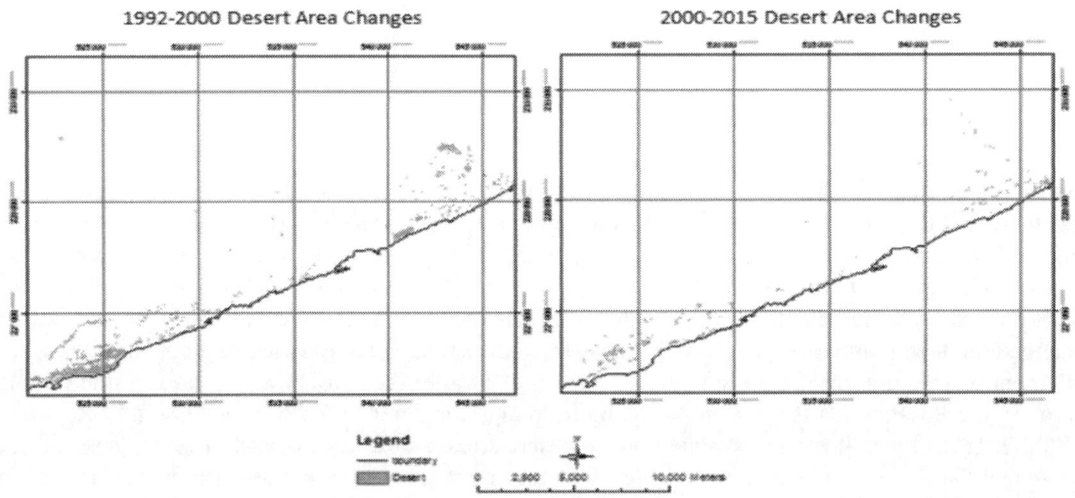

Figure 9. Decrease Desert Area over the period 1992-2015.

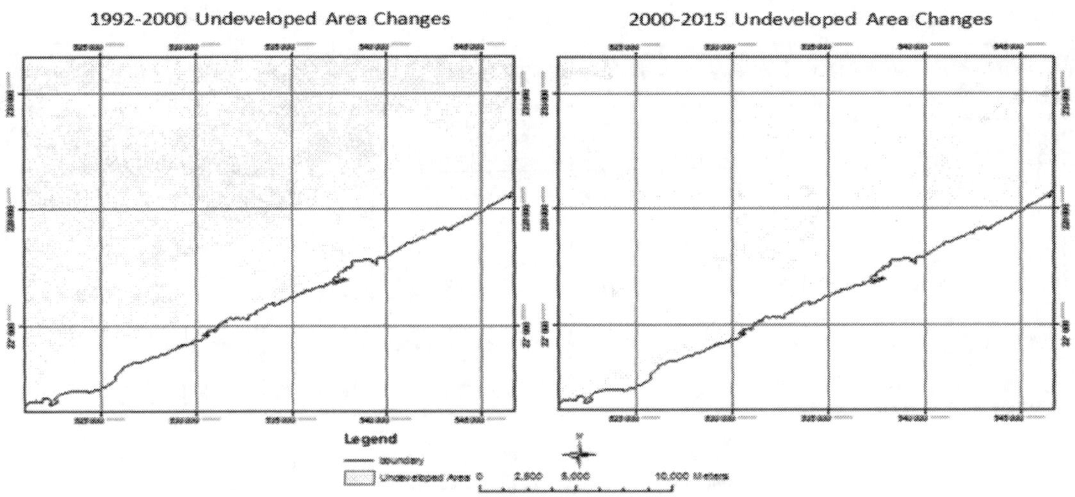

Figure 10. Increase of the Undeveloped Area over the period 1992-2015.

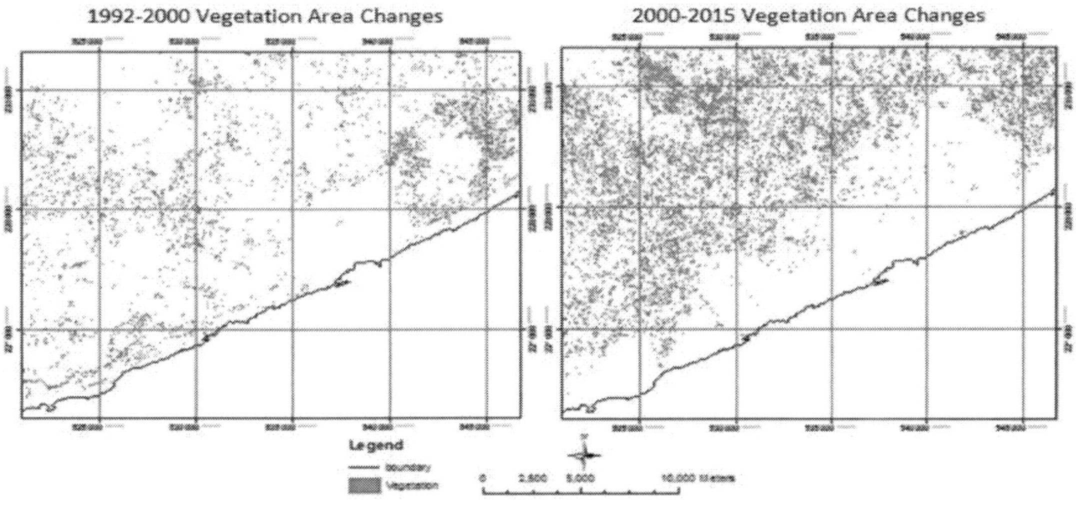

Figure 11. Vegetation Area shirking over the period 1992-2015.

Population Analysis: The study conducted above was to estimate the population of the city of Mogadishu, in order to better manage and supply sufficient care and need for the people of Mogadishu. Figure 12 represents the information about the population growth of the Mogadishu city. Based on the study Mogadishu population was estimated in 1990 around 779,000, while in 2015 it was estimated around 2,120,000 and it was concluded that Mogadishu remains the largest and the fastest growing city in the entire country. Population of year 2000 estimated at 1.2 million while in 2015 the population was doubled around 2.12 million.

Year	Population
1950	47,000
1955	66,000
1960	94,000
1965	134,000
1970	191,000
1975	271,000
1980	385,000
1985	548,000
1990	779,000
1995	965,000
2000	1,219,000
2005	1,552,000
2010	1,955,000
2015	2,120 ,000

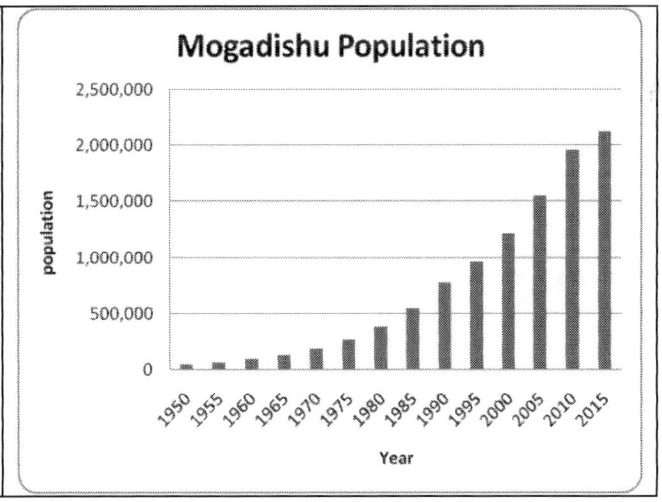

Figure 12. Mogadishu population over the period of time.

6 Conclusion

Land use land cover analysis is a unique method in describing the natural and man-made features over the area. This process has produced better fitting in locating the changes of the area and what is happening to the Mogadishu city in Somalia. Time series data and the image classification have indicated the changes that have occurred. SVM method has produced high accuracy result and high relationship correlation on the features for each image that have exploited.

The war incident has affected the city centre growth at the beginning. Major reconstruction area in the city centre and the abundant building has forced the people to migrate outside the area in order to start new life after the war. Major finding of the research shows that the growth of the downtown area at north-east of the city is due to space limitation in the city centre and the reconstruction.

The pattern of the comparison has revealed that the area of the natural and man-made changes is possible to distinguish. This is because most of the area that near to the city is unrelated to farming activities. The drought event has caused lots of changes to the vegetation area and the undeveloped area. If this situation is unable to be properly managed, deforestation rate will increased over the period of time.

For the future work, extension classification can be done through object based and reflectance (spectral) classification. Details analysis for the features can be prepared by using the data fusion, hyperspectral and high resolution data so that the characteristic of the urban is able to be carry out.

Acknowledgement
The Landsat data used in this study were downloaded from US Geological Surveyweb sites.

References

[1] Abdullahi S and B Pradhan 2015, Sustainable brownfields land use change modeling using GIS-Based weights-of-evidence approach. *Appl Spat Anal Polic*: p. 1-18.

[2] Abdullahi S, A R b Mahmud, and B Pradhan 2014, Spatial modelling of site suitability assessment for hospitals using geographical information system-based multicriteria approach at Qazvin city, Iran. *Geocarto International*. **29**(2): p. 164-184.

[3] Yao K, M O Idrees, B Pradhan, and S M Nabeel 2015, Detecting Land Use Land Cover Change Impacted by Civil Crisis in Ivory Coast using Remote Sensing and GIS. *Asian J Appl Sci*. **3**(05).

[4] Al-sharif A A, B Pradhan, H Z M Shafri, and S Mansor 2013, Spatio-temporal analysis of urban and population growths in Tripoli using remotely sensed data and GIS. *Indian J Sci Tech*. **6**(8): p. 5134-5142.

[5] Singh A 1989, Review article digital change detection techniques using remotely-sensed data. *Int J Remote Sens*. **10**(6): p. 989-1003.

[6] Treitz P and J Rogan 2004, Remote sensing for mapping and monitoring land-cover and land-use change—an introduction. *Prog Plann*. **61**(4): p. 269-279.

[7] Abdullahi S, B Pradhan, and M N Jebur 2015, GIS-based sustainable city compactness assessment using integration of MCDM, Bayes theorem and RADAR technology. *Geocarto Int*. **30**(4): p. 365-387.

[8] Hamedianfar A and H Z Shafri 2014, Development of fuzzy rule-based parameters for urban object-oriented classification using very high resolution imagery. *Geocarto Int*. **29**(3): p. 268-292.

[9] Cox W, *Demographia World Urban Areas*, in *10th Annual Edition (Belleville, IL: Demographia. com, 2014)*2012. p. 20.

[10] Miller H, N Sexton, and L Koontz, *Landsat imagery: a unique resource*, 2011, US Geological Survey.

[11] Guide E U s, *ENVI on-line software user's manual*, in *ITT Visual Information Solutions*2008.

[12] Foody G M 2002, Status of land cover classification accuracy assessment. *Remote Sens Envir*. **80**(1): p. 185-201.

[13] Tehrany M S, B Pradhan, and M N Jebuv 2014, A comparative assessment between object and pixel-based classification approaches for land use/land cover mapping using SPOT 5 imagery. *Geocarto Int.* **29**(4): p. 351-369.

[14] Al-Doski J, S B Mansor, and H Z M Shafri 2013, Image Classification in Remote Sensing. *Envir Earth Sci.* **3**(10): p. 141-147.

[15] Coppin P, I Jonckheere, K Nackaerts, B Muys, and E Lambin 2004, Review ArticleDigital change detection methods in ecosystem monitoring: a review. *Int J Remote Sens.* **25**(9): p. 1565-1596.

[16] Radke R J, S Andra, O Al-Kofahi, and B Roysam 2005, Image change detection algorithms: a systematic survey. *Image Process, IEEE Trans.* **14**(3): p. 294-307.

[17] Masih I, S Maskey, F Mussá, and P Trambauer 2014, A review of droughts on the African continent: a geospatial and long-term perspective. *Hydrol Earth Syst Sci.* **18**(9): p. 3635-3649.

Spatial assessment of land surface temperature and land use/land cover in Langkawi Island

Suzana Binti Abu Bakar, Biswajeet Pradhan*, Usman Salihu Lay and Saleh Abdullahi

Department of Civil Engineering, Faculty of Engineering, University Putra Malaysia, 43400, UPM, Serdang, Selangor, Malaysia

Corresponding Address: Biswajeet Pradhan, Tel: +603-89466383,
Fax. +603-86567129; Email: biswajeet24@gmail.com or biswajeet@lycos.com

Abstract. This study investigates the relationship between Land Surface Temperature and Land Use/Land Cover in Langkawi Island by using Normalized Difference Vegetation Index (NDVI), Normalized Difference Build-Up Index (NDBI) and Modified Normalized Difference Water Index (MNDWI) qualitatively by using Landsat 7 ETM+ and Landsat 8 (OLI/TIRS) over the period 2002 and 2015. Pixel-based classifiers Maximum Likelihood (MLC) and Support Vector Machine (SVM), has been performed to prepare the Land Use/ Land Cover map (LU/LC) and the result shows that Support Vector Machine (SVM) achieved maximum accuracy with 90% and 90.46% compared to Maximum Likelihood (MLC) classifier with 86.62% and 86.98% respectively. The result revealed that as the impervious surface (built-up /roads) increases, the surface temperature of the area increased. However, land surface temperature decreased in the vegetated areas. Based from the linear regression between LST and NDVI, NDBI and MNDWI, these indices can be used as an indicator to monitor the impact of Land Use/Land Cover on Land Surface Temperature.

1 Introduction

Rapid urbanization has caused significant changes in Land Use/ Land Cover [1, 2], and this process especially affect urban heat island [3-5]. As cities grow, changes occur in the physical landscape, roads, building and other infrastructures thus replacing open land and vegetation [6-8]. Previous researches has shown that integration of remote sensing and geographic information system (GIS) can been used as a tool to estimate the impact of land surface temperature with the rapid urban growth development [9].

Human activity is massively altering the Earth's vegetative cover where it directly contributes to climate change through a variety of processes. These include the growth or degradation of surface vegetation and changes in the land surface, which affect regional and global climate by producing changes in the surface energy budgets. This concern leads to an increasing research effort to understand the complex interactions between both urban and rural land surfaces and their atmospheric boundary layers. It also leads to efforts towards developing techniques for monitoring spatial and temporal land surface climate processes [10, 11].

Land Surface temperature can be obtained from direct ground measurement [12]. With the availability of satellite remote sensing, thermal band retrieved from the satellite can be used to estimate surface temperature. Several researches have been carried out to estimate land surface temperature by using split-window algorithms in Landsat 8 using band 10 and 11 [13]. The work demonstrates that linear regression method (LST vs NDVI) has negative correlation since the district consists of more vegetative cover in hilly regions. Raja et al. [14] investigates the spatial analysis of land surface temperature in Dhaka area by using the linear correlation between LST and NDVI and LST and NDBI which indicates the positives relationship with NDBI due to the urban growth. Yusuf et al. [6] assessed the spatio-temporal of Urban Heat Island effects in Kuala Lumpur metropolitan city by using Landsat Images. This work demonstrates the use of mono-window algorithm using band 6 ETM+. The results revealed that remote sensing satellite image can be used to estimate the LU/LC changes and urban heat island around Kuala Lumpur metropolitan city. Chen et al. [15] stated that appropriated indices such as Normalized Difference Vegetation Index (NDVI), Normalized Difference Build-Up Index (NDBI) and Modified Normalized Difference Water Index (MNDWI) can be used to classify land use/ land cover.

The current study aims to estimate the relationship between land surface temperature and LU/LC in Langkawi Island, Malaysia. Therefore, the objectives of this study are (1) to estimate land surface temperature for the study area during years 2002 to 2015; (2) to generate LU/LC map using pixel based approach and finally (3) to analysis the relationship between LST and LU/LC using linear regression between LST and NDVI, NDBI and MNDWI.

2 Study area and data sets

2.1 Study area

The study area is in Langkawi Island, Kedah, Malaysia. Langkawi is located between longitude 99° 51′ 03′′E and latitude 06° 18′ 19′′N. The study area is characterized by a variety of agricultural fields, residential areas, and water body. Langkawi and its 99 islands were declared as a Geopark by the United Nations Educational, Scientific and Cultural Organization (UNESCO) in 2007.

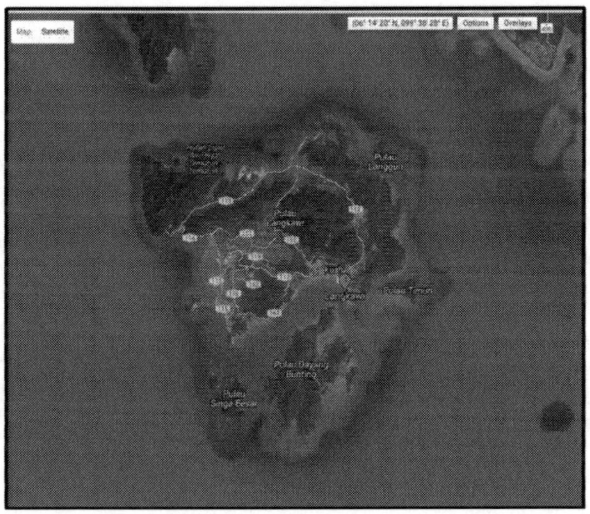

Figure 1. Study area showing the site location on Google Map.

2.2 Data used

In this study LANDSAT 8 (OLI) and Landsat ETM+ imagery was utilized. The multispectral Landsat 8 image was acquired on January 13, 2015, local time overpasses was GMT 03:34:01, while Landsat 7 ETM+ was acquired on January 17, 2002, local time overpasses was GMT 03:22:46. Both data were downloaded from Earth Resources Observation and Science Center of the US Geological Survey, where the data were geometrically and radiometrically corrected. The Landsat 8 image consists of 11 spectral bands, 30m except 15m for panchromatic band and 100m for TIRS band. The bands are Coastal (0.43–0.45 μm), Blue (0.00–0.51μm), Green (0.53–0.59 μm), Red (0.64–0.67 μm), Near-IR (0.85–0.88 μm), SWIR 1 (1.57–1.65 μm), SWIR 2 (2.11–2.29μm), Panchromatic (0.50 – 0.68 μm), Cirrus (1.36 – 1.38 μm). TIRS 1 (10.6 – 11.19 μm) and TIRS 2 (11.5 – 12.51 μm) both are at 100m but have been resample to 30m to match with multispectral data. Landsat 7 ETM+ consists of 8 spectral bands, 30m except 15m for panchromatic band and 60m for TIR band. The bands are Blue (0.45 – 0.52 µm), Green (0.52 – 0.60 µm), Red (0.63 – 0.69 µm), Near-IR (0.77 – 0.90µm), SWIR 1 (1.55 – 1.75µm), and SWIR 2 (2.09 – 2.35µm), Panchromatic (0.52 – 0.90 µm). TIR band (10.40 – 12.50 µm) is 60m but has been resampled to 30m to match with multispectral data.

3 Methodology

Figure 2 and 3 illustrate the data processing workflow which starts with preprocessing of Landsat 7 ETM+ and Landsat 8 respectively. This is followed by image preprocessing, image pan-sharpening, image classification, accuracy assessment for LU/LC, employing indices, land surface temperature estimation, linear correlation and finally result and discussion.

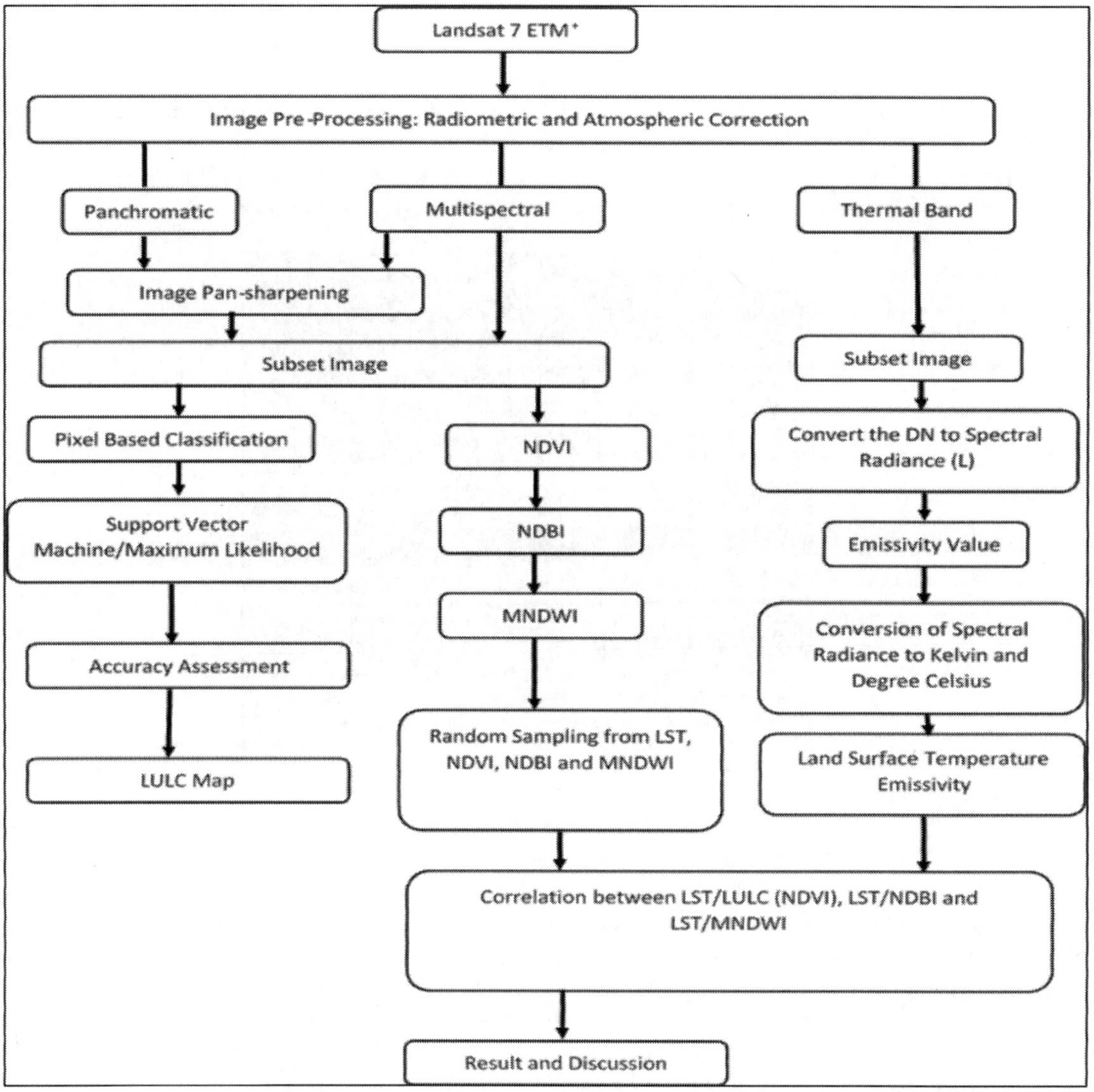

Figure 2. Data processing workflow for Landsat 7 ETM+.

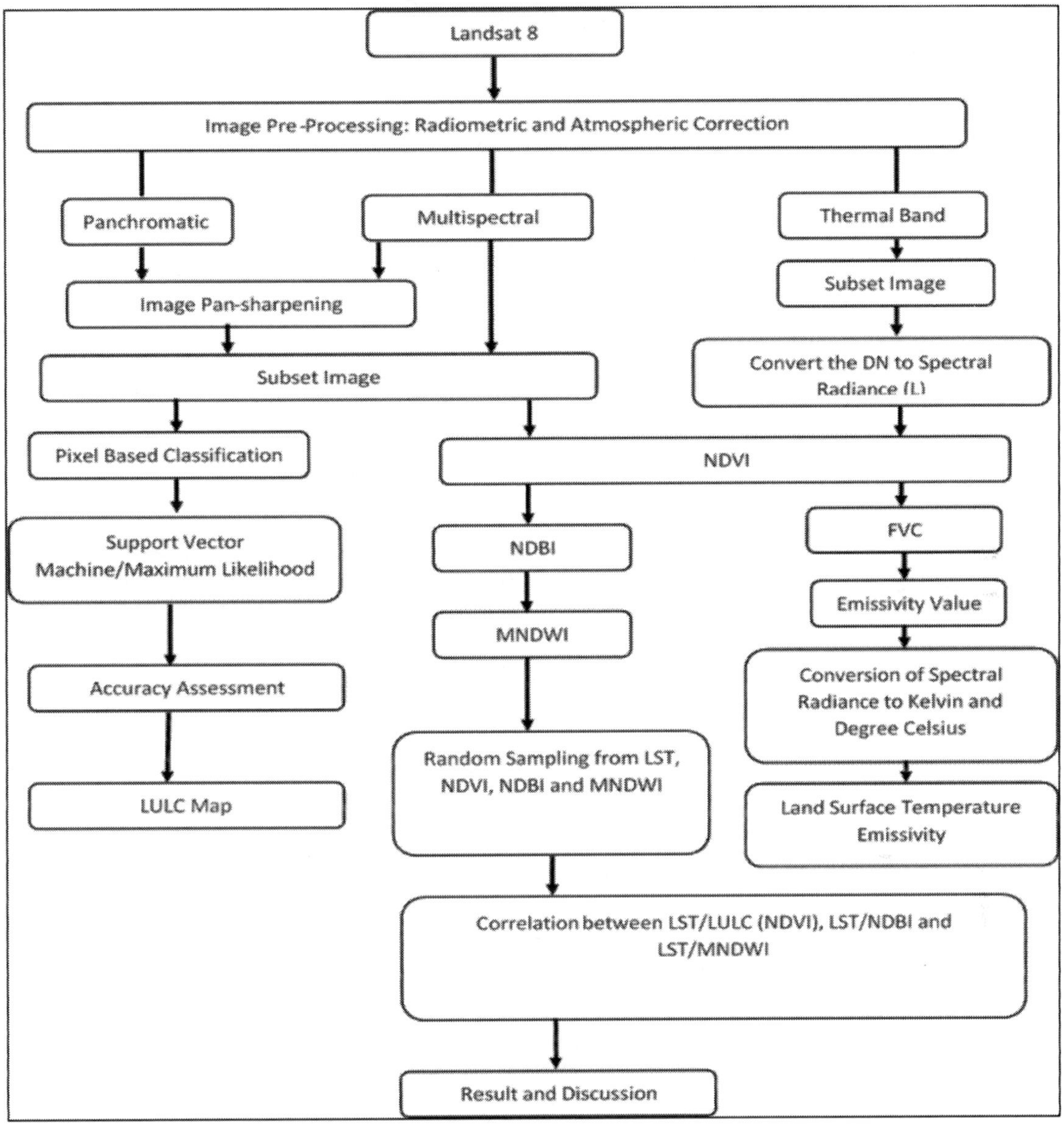

Figure 3. Data processing workflow for Landsat 8.

3.1 Land use / land cover classification

First, the Landsat 7 ETM+ and Landsat 8 data were corrected for radiometric and atmospheric correction (Internal Average Reflectance) in ENVI 5.1. After that both Landsat imagery with spatial resolution of 30m were pan-sharpened using Gram-Schmidt method with the 15m panchromatic band to improve the spatial resolution for better LU/LC classification. The study area was then subset from the preprocessed datasets. By using pixel-based classification methods such as Maximum Likelihood and Support Vector Machine classifiers, the LU/LC map were produced for the year 2002 and 2015. Marino et al. [16] stated that pixel-based classification methods extract information from images based on pixels which rely on prior knowledge of the diverse land classes in the image, which can be known through field surveys, laboratory measurements, or from any existing source of information. The knowledge of these features was used to select training pixels for classification into seven different land cover classes which are forest, built-up/roads, water, grassland, vegetation, cultivated land, and bare land.

3.2 Computation of land surface temperature from landsat 7 ETM+
The atmospheric correction of thermal band 61 has been performed using ENVI 5.1 Software. The purpose of atmospheric correction is to remove the atmospheric contributions from thermal infrared data.

i) Conversion of Digital Number to Spectral Radiance: According to Amiri et al. [17] and Coll et al. [18], the method has been taken from USGS Landsat User handbook:

$$L_\lambda = \frac{LMAX_\lambda - LMIN_\lambda}{QCALMAX - QCALMIN} \times (QCAL - QCALMIN) + LMIN_\lambda \quad (1)$$

Where L_λ is cell value as radiance in W/(m^2*sr*µm), $LMAX_\lambda$ is spectral radiance that is scaled to QCALMAX in W/(m^2*sr*µm) and $LMIN_\lambda$ is Spectral radiance that is scaled to QCALMIN in W/(m^2 *sr*µm) .QCALMAX is a maximum quantized calibrated pixel value (corresponding to LMAX) in DN: 255 and QCALMIN is a minimum quantized calibrated pixel value (corresponding to LMIN) in DN: 1.

Table 1. $LMAX_\lambda$ and $LMIN_\lambda$ of Band 61 (Landsat 7 ETM+).

$LMAX_\lambda$	17.040
$LMIN_\lambda$	0.000

ii) Conversion of Spectral Radiance (Lλ) to Kelvin with emissivity value:

$$T_B = \frac{K_2}{ln\left(\frac{K_1 \times £}{L_\lambda} + 1\right)} \quad (2)$$

Where T_B is value at satellite brightness temperature (K), L λ is a cell value as radiance in W/ (m^2 *sr*µm). **£** is emissivity value (typically 0.95) for Landsat TM and ETM+ and K1 and K2 are band specific thermal conversion constant from the metadata of band 61(low gain).

Table 2. K_1 and K_2 Constant value of Band 61 (Landsat 7 ETM+).

Thermal Constant	Band 61
K_1	666.09
K_2	1282.71

iii) Conversion of Kelvin to Degree Celsius:

$$T_C = T_B - 273.15 \quad (3)$$

Where T_B is value at satellite brightness temperature (K) and T_C is temperature in Degree Celsius.

3.3 Computation of Land Surface Temperature from Landsat 8

The atmospheric correction of thermal band 10 has been performed using ENVI 5.1 Software in order to remove the atmospheric contributions from thermal infrared data.

i) Top of Atmospheric Spectral Radiance:

$$L_\lambda = ML \times Qcal + A_L \tag{4}$$

Where L_λ is top of Atmospheric Radiance in watts/ (m^2*srad*μm), ML is band specific multiplicative rescaling factor (band 10), Qcal is band 10 image. A_L is band specific additive rescaling factor band 10.

Table 3. Qcal and A_L of Band 10 (Landsat 8).

Qcal	A_L
0.00033420	0.1

ii) Normalized Difference Vegetation Index (NDVI) and Fractional Vegetation Cover (FVC): In this study NDVI has been used as an index to estimate vegetation cover and Fractional Vegetation cover has been calculated to calculate the amount of vegetation distribution on the flat background.

$$NDVI = \frac{NIR-RED}{NIR+RED} \tag{5}$$

$$f_v = \frac{NDVI_{(x,y)}-NDVI_{min}}{NDVI_{max}-NDVI_{min}} \tag{6}$$

iii) Surface Emissivity

Sobrino et al. [19] stated that surface emissivity can be obtained from the surface mixture of bare soil and vegetation as in the equation 7 as below:

$$£ = £_v \,.\, f_v + £_s\,(1 - f_v) \tag{7}$$

Where f_v is fractional vegetation cover and emissivity of vegetation ($£_v$) is 0.985 and emissivity of soil ($£_s$) is 0.96.

iv) Conversion of Spectral Radiance (L_λ) to Kelvin with emissivity value: Similar to equation (2), T_B is value at satellite brightness temperature (K), L λ is a cell value as radiance in W/ (m^2 *sr*μm). £ is emissivity value (typically 0.945) and K1 and K2 are band specific thermal conversion constant from the metadata of band 10.

Table 4. K_1 and K_2 Constant value of Band 10 (Landsat 8).

Thermal Constant	Band 10
K_1	774.8853
K_2	1321.0789

v) Rescale of Land Surface Temperature: In order to make a comparison of the data at different times, the parameter of LST were used to rescale the data using minimum and maximum values as in equation 8, 9 [20]:

$$LST\ 2002 = \frac{LST - LST_{min}}{LST_{max} - LST_{min}} \tag{8}$$

$$LST\ 2015 = \frac{LST - LST_{min}}{LST_{max} - LST_{min}} \tag{9}$$

3.4 Indices

In order to find the relationship between land surface temperature, the NDVI, NDBI and MNDWI has been employed on Landsat 7 ETM+ and Landsat 8 based on the equation (5). The normalized difference vegetation index (NDVI) (Rouse et al., 1974) has been extensively used to monitor and identify vegetation and land cover changes. NDVI is expressed by the following formula based on Near-infrared and Red bands.

The Normalized Difference Build-Up Index has been developed by Zha et al. (2003) in order to identify the built-up areas. NDBI is expressed by the following formula based on Mid-infrared and Near-infrared bands:

$$NDBI = \frac{MID - NIR}{MID + NIR} \tag{10}$$

Xu [21] developed the Modified Normalized Difference Water Index (MNDWI) in order to identify the water purity degree and the matter content of water due its dynamic range. MNDWI is expressed by the following formula based on Green and Mid-infrared bands:

$$MNDWI = \frac{GREEN - MIR}{GREEN + MIR} \tag{11}$$

3.5 Accuracy assessment

Finally accuracy assessment was performed to evaluate the quality of classification output [22]. Error matrix [23, 24] based on assessment of the overall accuracy, producer's accuracy, user's accuracy, and kappa coefficient [25, 26] was utilized to evaluate the pixel-based classification output for LU/LC classification.

4 Results and discussion

4.1 Land surface temperature emissivity for Landsat 7 ETM+ and Landsat 8

Figure 4 shows the Land Surface Temperature Emissivity extracted from Landsat 7 ETM+ of the year 2002 and Landsat 8 of the year 2015. From the result, it can be concluded that the minimum temperature of the year 2002 was 14 degree Celsius and maximum temperature was 32 degree Celsius. However, land surface temperature emissivity of the year 2015 shown a minimum temperature of 15 degree Celsius and maximum temperature of 34 degree Celsius.

IOP Conf. Series: Earth and Environmental Science **37** (2016) 012064 doi:10.1088/1755-1315/37/1/012064

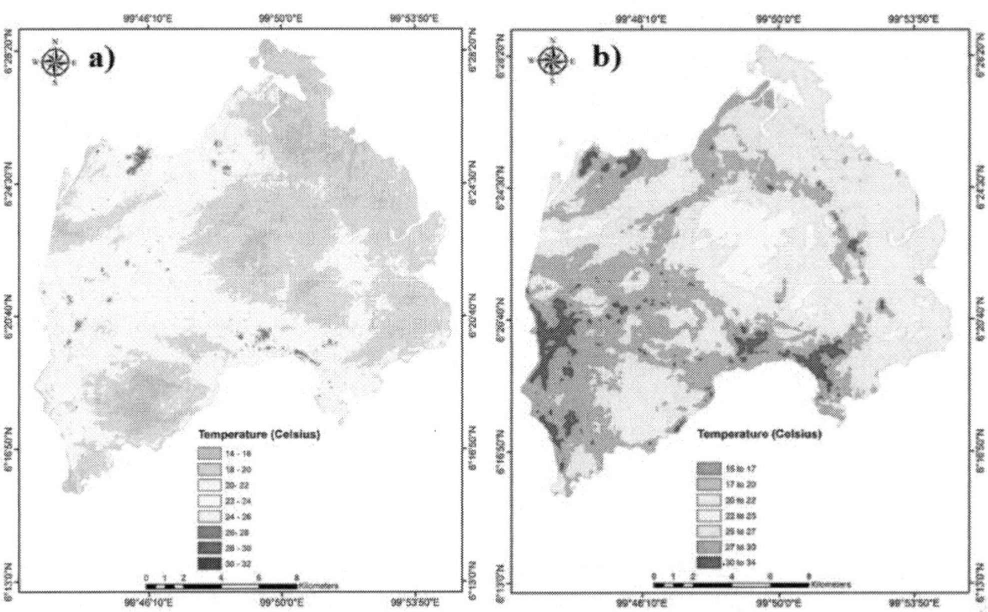

Figure 4. Land Surface Temperature Emissivity outputs of: (a) Landsat 7 ETM +, (b) Landsat 8.

4.2 Pixel-based classification results for land use / land cover classification 2002

The Landsat 7 ETM+ has been classified using Maximum Likelihood (MLC) and Support Vector Machine (SVM) classifier. From the result, it can be observed that water and vegetation classes have been mixed with forest class. The SVM classifier (90%) with kappa coefficient 0.735 provided the highest result, followed by the MLC (86.62%) with kappa coefficient 0.746.

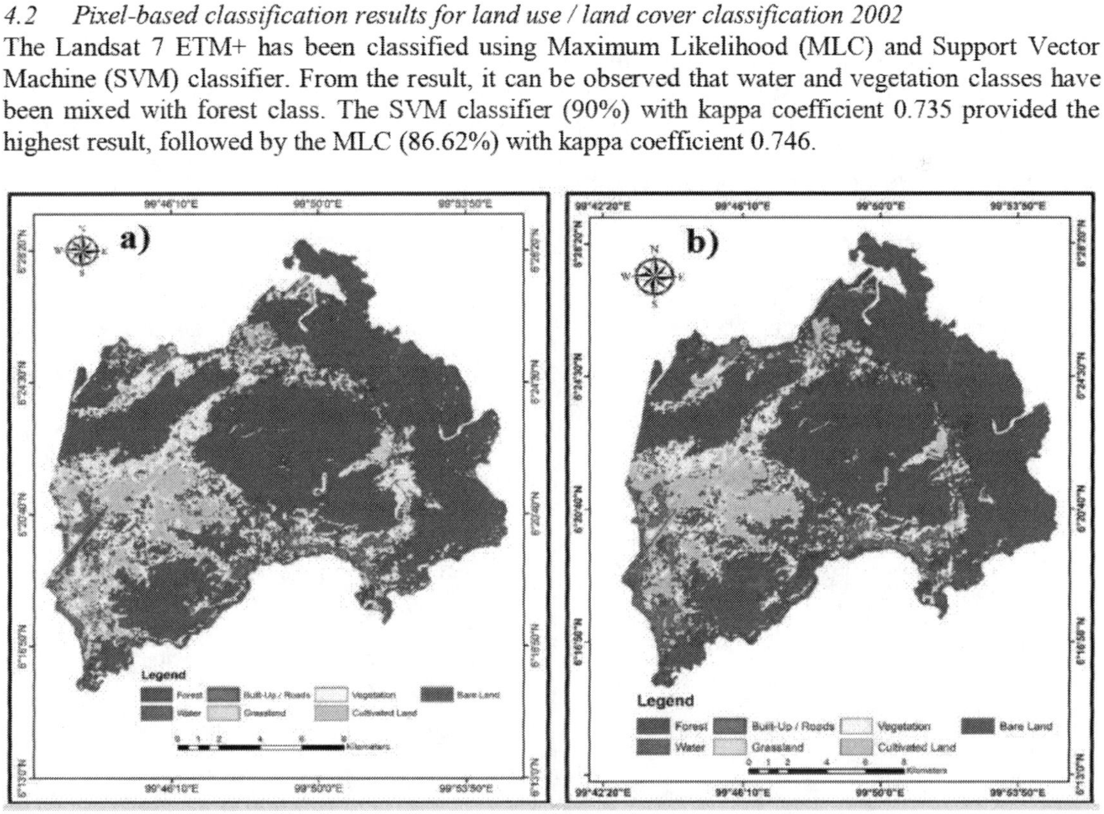

Figure 5. Classification outputs of Landsat 7 ETM + imagery by applying: (a) MLC, (b) SVM.

Table 5. Overall classification accuracy of Landsat 7 ETM+.

Landsat 7 ETM+: Maximum Likelihood Overall Accuracy = (7914/9136) 86.6243%; Kappa Coefficient = 0.7464							
	Forest	Water	Built-Up/Road	Grassland	Vegetation	Cultivated Land	Bare Land
Forest	5935	0	41	4	166	0	0
Water	28	156	20	0	9	0	0
Built-Up/Road	6	7	612	5	15	1	0
Grassland	2	4	79	19	100	1	0
Vegetation	83	5	60	22	719	9	1
Cultivated Land	0	0	14	0	402	301	18
Bare Land	0	0	14	1	49	56	172
Total	6054	172	840	51	1460	368	191
Landsat 7 ETM+: Support Vector Machine Overall Accuracy = (7220/8028) 90 %; Kappa Coefficient = 0.7335							
	Forest	Water	Built-Up/Road	Grassland	Vegetation	Cultivated Land	Bare Land
Forest	6031	3	20	0	458	0	0
Water	0	156	0	0	0	0	0
Built- Up/Road	0	2	397	18	0	0	0
Grassland	0	0	2	4	0	0	0
Vegetation	23	0	23	33	237	0	6
Cultivated Land	0	0	35	6	75	359	79
Bare Land	0	0	23	2	0	0	36
Total	6054	161	500	63	770	359	121

4.3 Pixel-based classification results for land use / land cover classification 2015

The Landsat 8 has been classified using Maximum Likelihood (MLC) and Support Vector Machine (SVM) classifier. From the observation, it is obvious that vegetation class has been mixed up with forest class. The SVM classifier (90.4574 %) with kappa coefficient 0.8068 achieved a better result compared to MLC (86.9779 %) with kappa coefficient 0.8045.

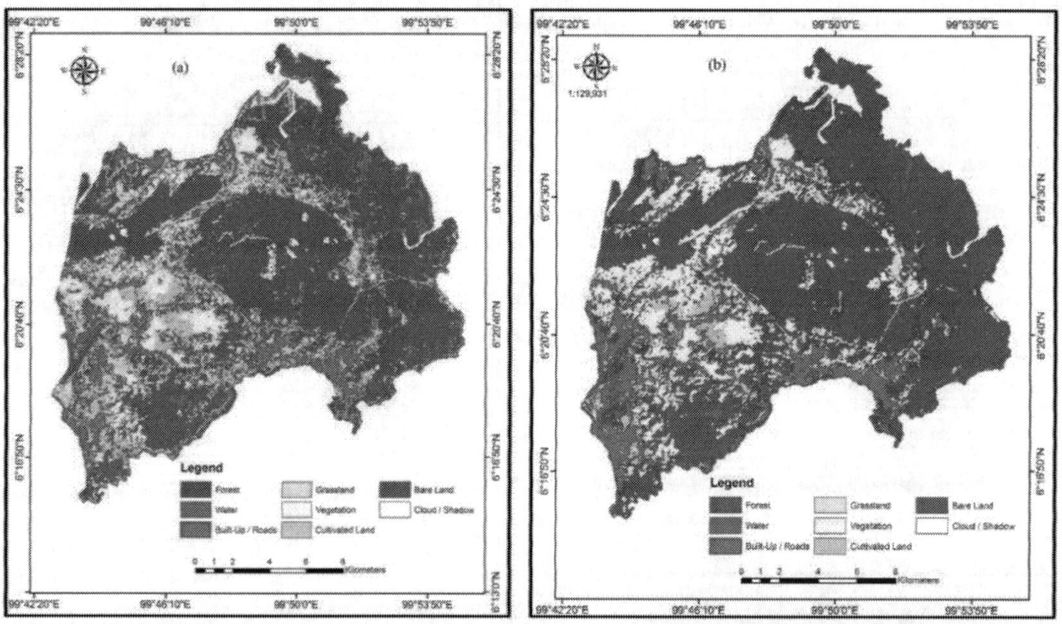

Figure 6. Classification outputs of Landsat 8 imagery by applying: (a) MLC, (b) SVM.

Table 6. Overall classification accuracy of Landsat 8.

Landsat 8: Maximum Likelihood Overall Accuracy = (4956/5698) 86.9779 ; Kappa Coefficient = 0.8045								
	Forest	Water	Built-Up/Road	Grassland	Vegetation	Cultivated Land	Bare Land	Cloud/ Shadow
Forest	2713	0	0	0	350	0	0	35
Water	0	148	12	0	0	0	0	69
Built-Up/Road	1	1	838	2	0	0	0	12
Grassland	23	0	3	105	63	0	27	0
Vegetation	76	0	7	11	868	0	1	0
Cultivated Land	0	0	7	0	21	158	0	0
Bare Land	0	0	1	2	0	2	82	0
Cloud/Shadow	2	0	11	3	0	0	0	44
Total	2815	149	879	123	1302	160	110	160
Landsat 8: Support Vector Machine Overall Accuracy = (8247/9117) 90.4574 %; Kappa Coefficient = 0.8068								
	Forest	Water	Built-Up/Road	Grassland	Vegetation	Cultivated Land	Bare Land	Cloud/ Shadow
Forest	6007	1	0	3	560	0	0	2
Water	0	155	25	0	0	0	0	1
Built- Up/Road	0	14	787	12	3	83	23	0
Grassland	0	2	0	24	30	0	24	0
Vegetation	46	0	0	12	867	0	0	0
Cultivated Land	0	0	0	0	0	285	0	0
Bare Land	0	0	3	0	0	0	84	0
Cloud/Shadow	1	0	25	0	0	0	0	38
Total	6054	172	840	51	1460	368	131	41

4.3.1 Area changes of LU/LC between year 2002 and year 2015

Table 7 shows a statistical analysis of LU/LC area changes from year 2002 and 2015. From the observation, it can be concluded that there was drastic change in built-up/roads, forest. Built-Up/ Roads has been increased by percentage from 4.746% to 8.297% and forest has been decreased from 80.927% to 78.04% respectively. Total vegetation and cultivated land from the year 2002 with 11.421% has been decreased by 9.527% corresponding to the loss of 61.782 km^2 to 51.537 km^2 respectively. The rapid changes have been detected due to tourism factor and the conversion of vegetated and cultivated land to water due to soil digging activities.

Table 7. Area changes of LU/LC in year 2002 and 2015.

Year 2002			**Year 2015**		
Types	**Total Area (km^2)**	**Percentage (%)**	**Types**	**Total Area (km^2)**	**Percentage (%)**
Forest	198.036	80.92	Forest	182.472	78.04
Water	11.025	2.03	Water	11.262	2.08
Built-Up/Road	25.674	4.74	Built-Up/Road	44.884	8.29
Grassland	1.067	0.19	Grassland	6.338	1.17
Vegetation	37.916	7.09	Vegetation	46.935	8.67
Cultivated Land	23.865	4.412	Cultivated Land	4.601	0.85
Bare Land	3.628	0.67	Bare Land	1.357	0.25

IOP Conf. Series: Earth and Environmental Science **37** (2016) 012064 doi:10.1088/1755-1315/37/1/012064

4.4 Indices result of the years 2002 and 2015

From the observation of the year 2002, the minimum and maximum values of NDVI were -0.694147 and 0.450689, respectively. For NDBI, the minimum value was -0.489924 and maximum value was 0.468670. For MNDWI, the minimum and maximum values of the year 2002 were -0.460173 and 0.745609 respectively. In contrast, the observations of the year 2015 shown that, the minimum and maximum values of NDVI were -0.948714 and 0.582615, respectively. For NDBI, the minimum value was -1.00 and maximum value was 0.684669. For MNDWI, the minimum and maximum values of the year 2015 were -0.635346 and 1.000 respectively.

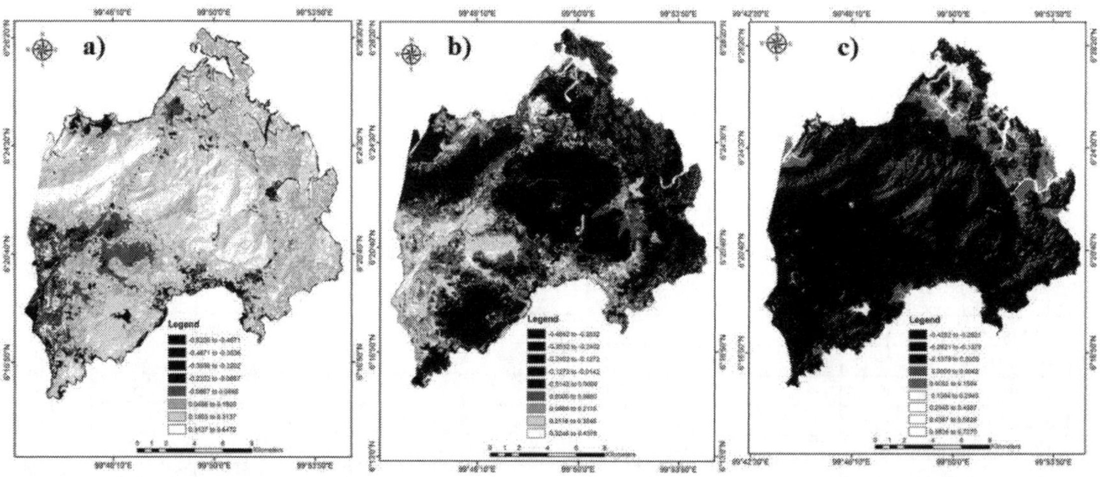

Figure 7. Indices outputs of Landsat 7 ETM + imagery by applying: (a) Normalized Difference Vegetation Index, (b) Normalized Difference Build-Up Index and (c) Modified Normalized Difference Water Index.

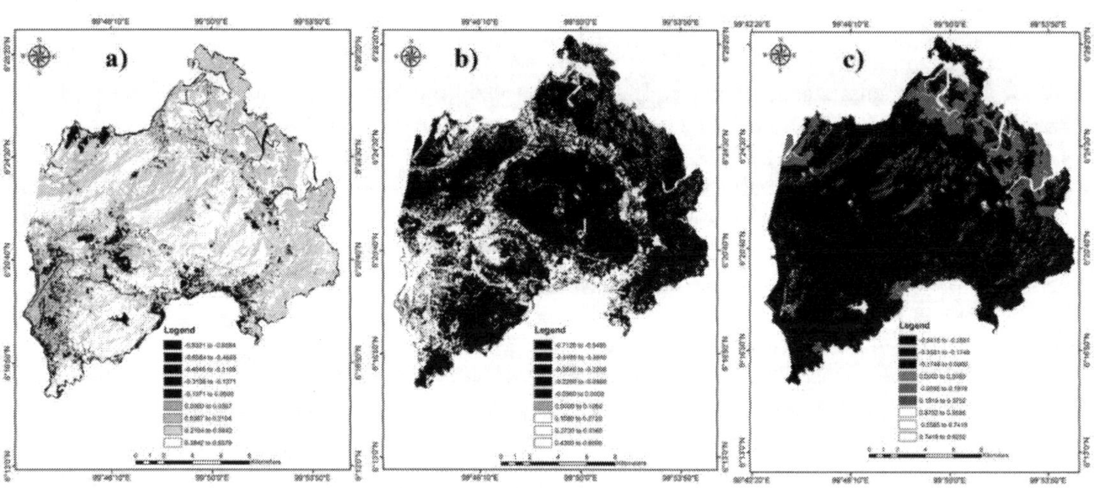

Figure 8. Indices outputs of Landsat 8 imagery by applying: (a) Normalized Difference Vegetation Index, (b) Normalized Difference Build-Up Index and (c) Modified Normalized Difference Water Index.

4.5 The relationship between Land Surface Temperature (LST) and land use/land cover 2002

4.5.1 Land Surface Temperature (LST) and Normalized Difference Vegetation Index (NDVI)
Table 8 shows a summary of the linear regression between Land Surface Temperature 2002 with Normalized Difference Vegetation Index (NDVI). 20 random samplings have been done in order to see the relationship. From the observation, it was obvious that the coefficients value between LST vs NDVI was good in forest with R^2 0.548. However, the coefficient value for water, built-up / roads, grassland, cultivated land, vegetation and bare land were small due to presence of water on the which lead to the negative values. The correlation between LST and NDVI of the year 2002 was $R^2 = 0.1524$.

Table 8. LST vs NDVI 2002.

LU/LC	**Equations**	**R^2**
Forest	y = 16.353 × NDVI + 15.459	$R^2 = 0.548$
Water	y = -3.1366 ×NDVI + 19.136	$R^2 = 0.0835$
Built-Up/Road	y = 0.9858 × NDVI + 25.251	$R^2 = 0.0018$
Grassland	y = 3.0676 × NDVI + 23.214	$R^2 = 0.0161$
Cultivated Land	y = 9.9899 × NDVI+ 25.492	$R^2 = 0.101$
Vegetation	y = 11.076 × NDVI+ 18.778	$R^2 = 0.1021$
Bare Land	y = -5.2356 × NDVI+ 22.999	$R^2 = 0.0582$
LST vs NDVI	**y = -2.9335× 21.672** **R^2 = 0.1524**	

4.5.2 Land Surface Temperature (LST) and Normalized Difference Build- Up Index (NDBI)
Twenty random samplings have been performed on LST and NDBI data of the year 2002. From the observation, it is obvious that the correlation between LST and NDBI were significantly positive with R^2 0.8137 as can be seen from the Figure 9 below:

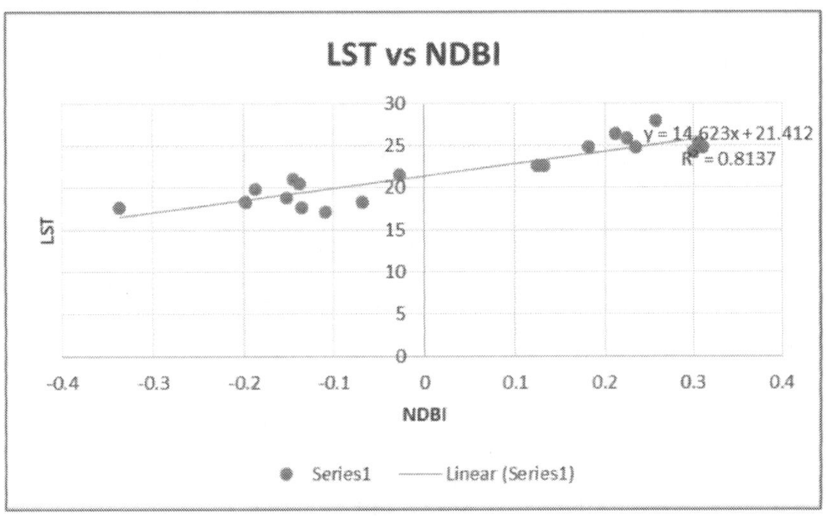

Figure 9. LST vs NDBI .

4.5.3 Land Surface Temperature (LST) and Modified Normalized Difference Water Index (MNDWI)
Figure 10 shows that LST vs MNDWI has significantly negative correlation from 20 random samplings that has been made on the LST and MNDWI data of the year 2002. From the observation, it shown that LST and MNDWI have negative correlation implied that pure water decrease the surface temperature and polluted water increase the surface temperature.

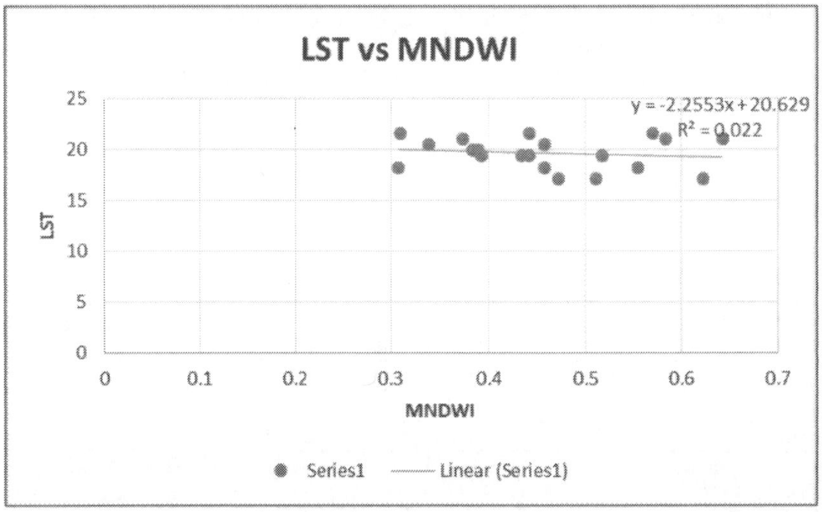

Figure 10. LST vs MNDWI.

4.6 The relationship between Land Surface Temperature (LST) and Land Use/Land Cover 2015

4.6.1 Land Surface Temperature (LST) and Normalized Difference Vegetation Index (NDVI)
Table 9 shows a summary of the linear regression between Land Surface Temperature 2015 with Normalized Difference Vegetation Index (NDVI). 20 random samplings have been done in order to see the relationship. From the observation, it was obvious that the coefficients value between LST vs NDVI was good in forest, grassland and vegetation with R^2 = 0.548, R^2 = 0.2655 and R^2 = 0.3406 respectively. However, the coefficient value for water, built-up/roads, cultivated land, and bare land were small. The correlation between LST and NDVI of the year 2015 was R^2 = 0.055.

Table 9. LST vs NDVI 2015.

LU/LC	Equations	R^2
Forest	$y = 6.0237 \times NDVI + 23.741$	$R^2 = 0.4395$
Water	$y = -0.578 \times NDVI + 26.097$	$R^2 = 0.0646$
Built-Up/Road	$y = -3.3429 \times NDVI + 29.318$	$R^2 = 0.1151$
Grassland	$y = 6.4735 \times NDVI + 27.334$	$R^2 = 0.2655$
Cultivated Land	$y = 2.9025 \times NDVI + 28.519$	$R^2 = 0.1092$
Vegetation	$y = 3.2486 \times NDVI + 25.42$	$R^2 = 0.3406$
Bare Land	$y = -5.5912 \times NDVI + 30.289$	$R^2 = 0.0684$
LST vs NDVI	**$y = -1.0973 \times 28.195$ $R^2 = 0.055$**	

4.6.2 Land Surface Temperature (LST) and Normalized Difference Build- Up Index (NDBI)

Figure 11 shows the significant positive correlation between LST and NDBI of the year 2015 with 20 random samplings of each class. Built-Up/ Roads indicates positives value while others features indicates negatives values. However, bare land/cultivated land (no Vegetation) also resulting in positive values in NDBI imagery. From the observation, LST would increase with the density of urban built-up, bare land and cultivated land (no vegetation) where the R^2 is 0.8419.

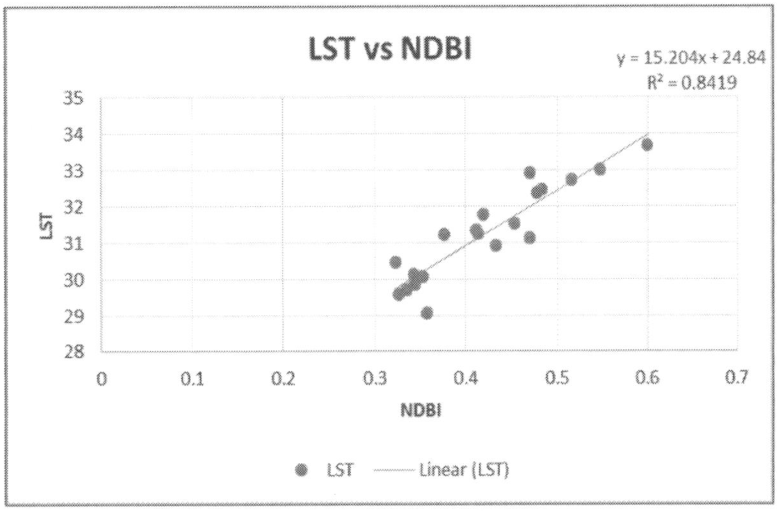

Figure 11. LST vs NDBI.

4.6.3 Land Surface Temperature (LST) and Modified Normalized Difference Water Index (MNDWI)

Figure 12 shows that LST vs MNDWI has significantly negative correlation from 20 random samplings that has been made on the LST and MNDWI data of the year 2015. From the observation, this index enhanced the water features in a MNDWI image. Water will indicates positives values while other features such as built- up area/roads, forest, vegetation, bare land, cultivated land and will have negative value. The correlation between LST and MNDWI is negatives with $R^2 = 0.1584$.

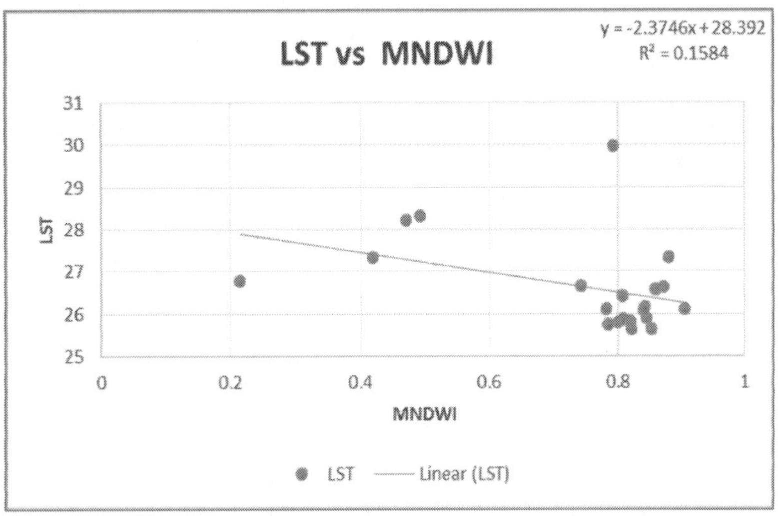

Figure 12 LST vs MNDWI.

4.7 Comparison of LST datasets (2002 and 2015)

Figure 13 shows the comparison of the dataset between the year 2002 and 2015. The rescale factor ranges between 0 and 1. From the observation, it was obvious that the temperature increase from 2002 to 2015 from all classes. The main reason that may support this result is due to the rapid urbanization from the tourism sector in Langkawi Island and the conversion of the cultivated land and vegetation area to water due to soil digging activities and residential areas.

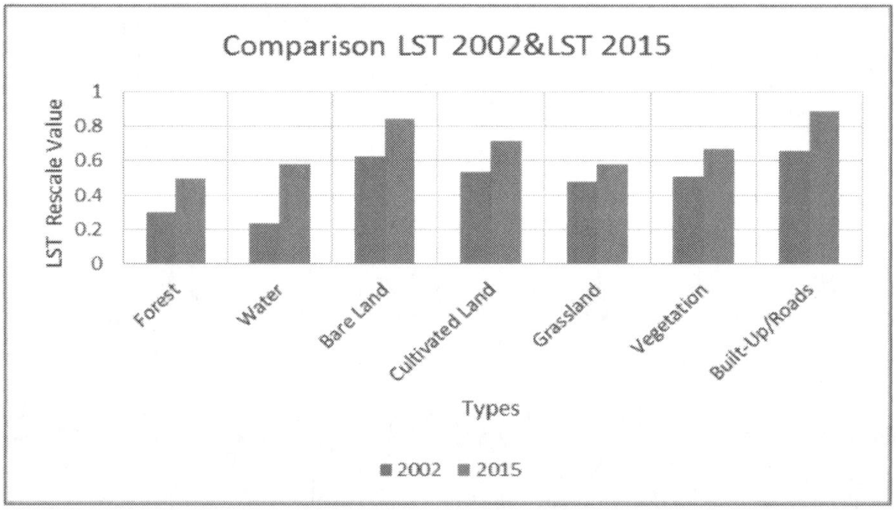

Figure 13. Comparison dataset of years 2002 and 2015.

5 Conclusion

In this study, pixel-based classifiers were used to classify seven land cover types: forest, built-up/roads, water, grassland, vegetation, cultivated land, and bare land. The analysis indicates that Support Vector Machine (SVM) is more efficient in mapping land cover features compared to Maximum Likelihood classifier. NDVI has been used widely as an indicator of vegetation abundance to estimate LST in studies of urban heat islands. However, for other LU/LC classes, it is not effective index to correlate with LST. By employing NDBI and MNDWI positive relationship were found on NDBI and LST. In addition, this process indicated that, land surface temperature increases in correlation to significant increase of impervious surface on the area. While MNDWI shows a negative correlation due to unpolluted water that significantly help to reduce the temperature. In conclusion, the correlations between land surface temperature with NDVI, NDBI and MNDWI can be used as an indicator in monitoring urban thermal environment.

Acknowledgement

The authors sincere thank the Department of Town and Country Planning in Kuala Lumpur for providing the land use map of the study area. Landsat 7 ETM+ and Landsat 8 were downloaded from the Earth Resources Observation and Science Center of the US Geological Survey (USGS). Not forgotten, our lecturer Assoc. Prof. Dr. Biswajeet Pradhan and Maher Ibrahim Sameen for providing guidance and advice.

References

[1] Sun Q, Z Wu, and J Tan 2012, The relationship between land surface temperature and land use/land cover in Guangzhou, China. *Envir Earth Sci*. **65**(6): p. 1687-1694.

[2] Abdullahi S and B Pradhan 2015, Sustainable brownfields land use change modeling using GIS-Based weights-of-evidence approach. *Appl Spat Anal Pplic*: p. 1-18.

[3] Gallo K P and T W Owen 1999, Satellite-based adjustments for the urban heat island temperature bias. *J Appl Meteorol*. **38**(6): p. 806-813.

[4] Streutker D R 2003, Satellite-measured growth of the urban heat island of Houston, Texas. *Remote Sens Envir*. **85**(3): p. 282-289.

[5] Weng Q 2003, Fractal analysis of satellite-detected urban heat island effect. *Photogramm Eng Remote Sens*. **69**(5): p. 555-566.

[6] Yusuf Y A, B Pradhan, and M O Idrees 2014, Spatio-temporal Assessment of Urban Heat Island Effects in Kuala Lumpur Metropolitan City Using Landsat Images. *J Indian Soc Remote Sens*. **42**(4): p. 829-837.

[7] Abdullahi S, B Pradhan, and M N Jebur 2015, GIS-based sustainable city compactness assessment using integration of MCDM, Bayes theorem and RADAR technology. *Geocarto Int*. **30**(4): p. 365-387.

[8] Abdullahi S, B Pradhan, S Mansor, and A R M Shariff 2015, GIS-based modeling for the spatial measurement and evaluation of mixed land use development for a compact city. *GISci Remote Sens*. **52**(1): p. 18-39.

[9] Liu H and Q Weng 2008, Seasonal variations in the relationship between landscape pattern and land surface temperature in Indianapolis, USA. *Envir Monitor Assess*. **144**(1-3): p. 199-219.

[10] Hall F G, K F Huemmrich, S J Goetz, P J Sellers, and J E Nickeson 1992, Satellite remote sensing of surface energy balance: Success, failures, and unresolved issues in FIFE. *J Geophy Res Atm (1984-2012)*. **97**(D17): p. 19061-19089.

[11] Sellers P, Y Mintz, Y e a Sud, and A Dalcher 1986, A simple biosphere model (SiB) for use within general circulation models. *J Atm Sci*. **43**(6): p. 505-531.

[12] Hussain A, P Bhalla, and S Palria 2014, Remote Sensing Based Analysis of the Role of Land Use/Land Cover on Surface Temperature and Temporal Changes in Temperature; a Case Study of Ajmer District, Rajasthan. *ISPRS-Int Arch Photogramm, Remote Sens Spat Info Sci*. **1**: p. 1447-1454.

[13] Rajeshwari A and N Mani 2014, Estimation of Land Surface Temperature of Dindigul District using Landsat 8 Data. *Int J Res Eng Tech*. **3**(05): p. 122-126.

[14] Raja D R 2012, Spatial Analysis of Land Surface Temperature in Dhaka Metropolitan Area. *J Bangladesh Institut Plann ISSN*. **2075**: p. 9363.

[15] Chen X-L, H-M Zhao, P-X Li, and Z-Y Yin 2006, Remote sensing image-based analysis of the relationship between urban heat island and land use/cover changes. *Remote Sens Rev*. **104**(2): p. 133-146.

[16] Marino C M, C Panigada, and L Busetto. *Airborne hyperspectral remote sensing applications in urban areas: Asbestos concrete sheeting identification and mapping*. in *Remote Sensing and Data Fusion over Urban Areas, IEEE/ISPRS Joint Workshop 2001*. 2001. IEEE.

[17] Amiri R, Q Weng, A Alimohammadi, and S K Alavipanah 2009, Spatial–temporal dynamics of land surface temperature in relation to fractional vegetation cover and land use/cover in the Tabriz urban area, Iran. *Remote Sens Envir*. **113**(12): p. 2606-2617.

[18] Coll C, J M Galve, J M Sanchez, and V Caselles 2010, Validation of Landsat-7/ETM+ thermal-band calibration and atmospheric correction with ground-based measurements. *Geosci Remote Sen, IEEE Transactions on*. **48**(1): p. 547-555.

[19] Sobrino J, N Raissouni, and Z-L Li 2001, A comparative study of land surface emissivity retrieval from NOAA data. *Remote Sens Envir*. **75**(2): p. 256-266.

[20] Fathian F, A Prasad, Z Dehghan, and S Eslamian 2015, Influence of land use/land cover change on land surface temperature using RS and GIS techniques. *Int J Hydrol Sci Tech.* **5**(3): p. 195-207.

[21] Han-qiu X 2005, A Study on Information Extraction of Water Body with the Modified Normalized Difference Water Index (MNDWI)[J]. *J Remote Sens.* **5**: p. 589-595.

[22] Bhatta B, *Remote sensing and GIS*2008: Oxford University Press, USA.

[23] Foody G M 2002, Status of land cover classification accuracy assessment. *Remote Sens Envir.* **80**(1): p. 185-201.

[24] Nichol J and M S Wong 2008, Habitat mapping in rugged terrain using multispectral IKONOS images. *Photogramm Eng Remote Sens.* **74**(11): p. 1325-1334.

[25] Congalton R G and K Green 1993, A practical look at the sources of confusion in error matrix generation. *Photogramm Eng Remote Sens.* **59**(5): p. 641-644.

[26] Hamedianfar A and H Z Shafri 2014, Development of fuzzy rule-based parameters for urban object-oriented classification using very high resolution imagery. *Geocarto Int.* **29**(3): p. 268-292.

Land use change and soil loss risk assessment by using geographical information system (GIS): A case study of lower part of Perak River

Fasihah Mohd Yusof*, Nor Rohaizah Jamil , Nyvee Inthano a/p Cha Laew, Norfadilah Aini, Latifah Abd Manaf

Faculty of Environmental Studies, Universiti Putra Malaysia (UPM), Maalysia

*Email: fasihah.mohd.yusof@gmail.com

Abstract. The developing mode of the nation enhance more land area being exploited to generate economy income. Objectives of this study were to analyse the land use changes from year 2010 to 2013 and soil erosion potential rate for year 2013 of lower part of Perak river basin. All of the spatial analysis work were carried out in the GIS environment using the ArcGIS version 9.3 software. Land use maps were obtained from Department of Agriculture and been digitized accordingly. The total area was 2914.91 km^2 and land use categories were clustered into various classes. Based on land use change analysis, oil palm plantation recorded some increment from year 2010 to 2013. While, area of forest depleting from 95.54km^2 to 86.01 km^2 indicating that the forest area were being exploited and shifted to other land use type. In the other hand, the rubber plantation decrease due to land conversion into palm oil plantation. Urban area showed some increment in coverage proving the current blooming number of population occurs rapidly. In context of cleared land, 2013 recorded higher coverage of cleared land compared to the year 2010 which recorded a shifting from 8.89km^2 in 2010 to 21.24 km^2 in 2013. By adopting the RUSLE model, in 2013, the soil erosion potential was categorised as very low (0-1 tons/ha/year) with some soil erosion hotspot spotted within the study area. The soil erosion range from very low to extreme class. A very low soil erosion potential class (0-1 ton/ha/yr) recorded the majority of 61% (1765.60 km^2) of total area. The extreme classes (>100 ton/ha/yr) recorded about 18% (536.19km^2) of the total area. According to the result, it can be concluded that the middle part of study area experience low to severe classes of potential soil erosion.

1. Introduction

Recently, the over-utilization of land use rises because of the demand from rapidly increasing number of population and plantation along the river basin. These factors speed up the occurrence of soil erosion along the river which indirectly will contribute toward water quality status declining and alteration of river hydro-morphology.

Land use activity such as agriculture sectors overly speed up the soil erosion process and lead to the deteriorating of the river water quality resulted from the high transportation of the suspended sediment loads [1]. This is due to the transported sediment from the surrounding activities along the river will be deposited at the bottom of the river. River system degradation is expected to alter the river morphology as the river bank and depth of the river will be affected [2]. Consequently, suitability of a habitat for the various biodiversity will decrease because of the river bottom becoming shallower [3]. For instance, many previous research had proven that land use change can deteriorate

IOP Conf. Series: Earth and Environmental Science **37** (2016) 012065 doi:10.1088/1755-1315/37/1/012065

many environmental aspects such as environmental services [4], carbon and nitrogen sequestrations [5], water yield [6], carbon stocks [7] and aquatic life [8]. As soil erosion becoming a severe environmental threat mainly towards the river water conservation, the risk assessment of soil erosion triggered by water is important in designing new and effective water and soil management policies [9].

Thus, aims of this study are to investigate the land use changes pattern, soil erosion risk analysis and its potential impact towards the river ecosystems in Perak River, by adopting geospatial analysis in GIS environment. The river ecosystem will reflects the health of the river and its suitability for the biotic components such as fish, terrapins, and others. The sequence of land use maps and soil maps were sourced from the Department of Agriculture and been digitized to determine the land use change percentage and soil loss risk categories.

2. Methodology

2.1 Study area

The study was carried out in the lower part of Perak River basin located at the north of Peninsular Malaysia. The study area is located at the Perak River which is the second longest river in Peninsular Malaysia after the Pahang River. It covered 70% of the state land and flows over 400km in 15,000 km^2. Various activities carried out along the Perak River such as sand mining, development of residential area, industrial area, plantation, aquaculture and others.

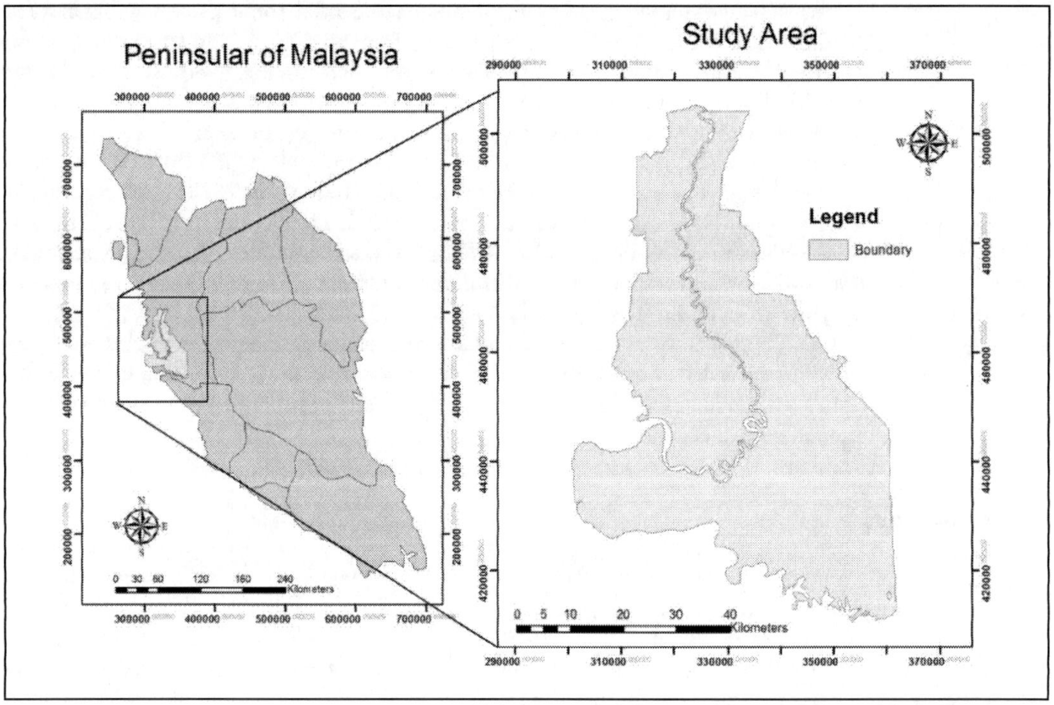

Figure 1. Study area.

In terms of ecosystem conservation, Perak River is an important habitat for endangered species of river terrapin or *Batagur affinis*. This top 25 most threaten turtle species also has been listed as a critically endangered species according to International Union for Conservation of Nature (IUCN) Red List 2009 and Convention of International Trade of Endangered Species of Wildlife Flora and Fauna (CITES) Appendix 1. In conjunction of that, Bota Kanan Tuntung Centre, the conservation centre for terrapin was established by the Perak Wildlife and National Parks Department in 1967 to nurture the sustainability of river terrapin population in Perak River. Thus, it is vital to ensure the health of this river must be the concern of many not only to preserve the precious endangered species of river terrapin, but also to sustain other aquatic organism and any environment services that Perak River would offer.

2.2 ArcGis 9.3

By using the ArcGIS version 9.3, the land use change and soil loss risk analysis were conducted based on the 2010 and 2013 land use maps and soil map of current years. The digitizing process included the soil reconnaissance map and the land use map for each of the year of 2010 and 2013.

2.3 Dataset

Primary data used in this research were the coordinate obtained from ground truthing activity. Secondary dataset employed in this research were reconnaissance soil map and land use maps of Perak River of year 2010 and 2013. The map was published by the Management and Conservation of Land Resources, Department of Agriculture Malaysia (DOAM).

2.3.1 Data acquisition

There are primary and secondary data involve in the study.

2.3.2 Primary data

Site verification is performed to compare the real observation with the output which represent by the map [10]. Site verification is by determined the coordinates using GPS and observed the current situation at the study area. Observation at the study area are proved by captured some pictures.

2.3.3 Secondary data

The topography map sourced from Department of Survey and Mapping Malaysia (JUPEM) was used as the references of study area. The land use maps of 2010 and 2013 were sourced from Department of Agriculture (DOAM). Land use maps were provided in the image format (.jpeg) of 1:300 000 in scale. The rainfall data was obtained from the Department of Irrigation (DID). The rainfall data of year 2010 and 2013 are obtained to determine the annual rainfall of the study area.

3. Results & discussion

3.1 Land use

The land use map of Perak lower part river basin from year of 2010 and 2013 have been used as it required to perform the land use change analysis from year 2010 to 2013 which encompasses the total area of 2914.91 km^2.

Table 1. Land use type of Perak River lower part basin. The type of land use has been categorised into ten different types.

No	Land Use Type	Area (km^2)	
		2010	**2013**
1	Others Plantation/Crops	311.15	261.19
2	Cleared Land	8.89	21.24
3	Forest	95.54	86.01
4	Oil Palm	2061.34	2115.74
5	Mining Area	18.45	16.44
6	Paddy Field	228.21	232.7
7	Others Economic Activities	9.67	9.9
8	Rubber	78.18	77.1
9	Urban Area	52.25	59.6
10	Water Bodies	51.24	34.99
	TOTAL	**2914.91**	**2914.91**

3.1.1 Land use change of 2010

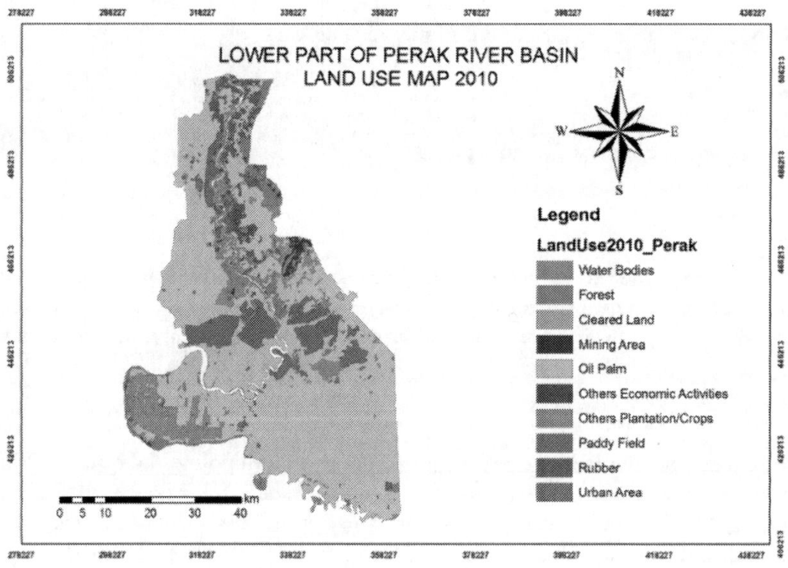

Figure 2. Land use map of lower part of Perak River basin of year 2010.

Based on land use map of year 2010 (Figure 2.) and land use area (Table 1.), the dominant land use in study area was oil palm plantation covering about 70.72 % (2061.34km^2) of the study area. The second highest of land use area was Other Plantation/Crops which cover about 10.67% (311.15km^2). The third highest reading was the paddy field which contribute about 7.83% (228.21km^2). Forest only cover about 3.28% (95.54km^2) of the total area. The other land use type were rubber plantation and urban area which contribute about 2.68% (78.18 km^2) and 1.79% (52.25km^2) respectively.

3.1.2 Land use change of 2013

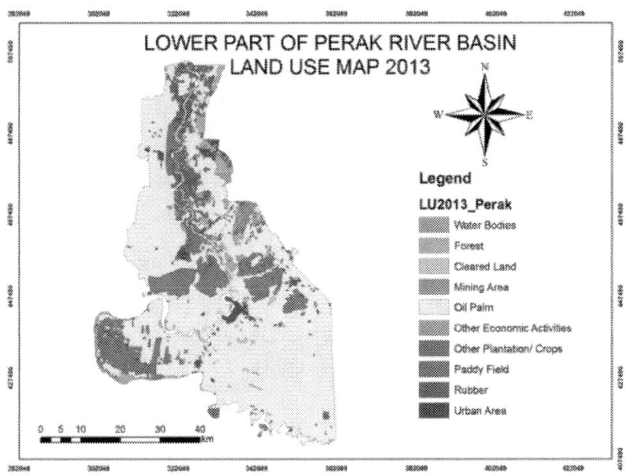

Figure 3. Land use map of lower part of Perak River basin of year 2013.

Based on the Figure 3, the land use of year 2013 shows that the oil palm plantation recorded the highest percentage of land use type covering an area of 72.58% (2115.74km^2). The second highest was the others plantation/crops which recorded for about 8.96% (261.19km^2).These classes of land use represent the mixed horticulture, scrubs and others plantation or crop except the oil palm, paddy and rubber plantation. Some of the area within the study area had been converted to the others plantation/crops land use class. The paddy field is the third highest of the land use type which is 7.98% (232.7km^2). The others land use type or class such as rubber plantation, urban area, forests, water bodies, cleared land, other economic activities and mining area donated approximately about 10% to the total study area. The 2013 land use map of 2013 sourced from DOA represent the most recent land use status in lower part of Perak river basin.

3.2 Land use change

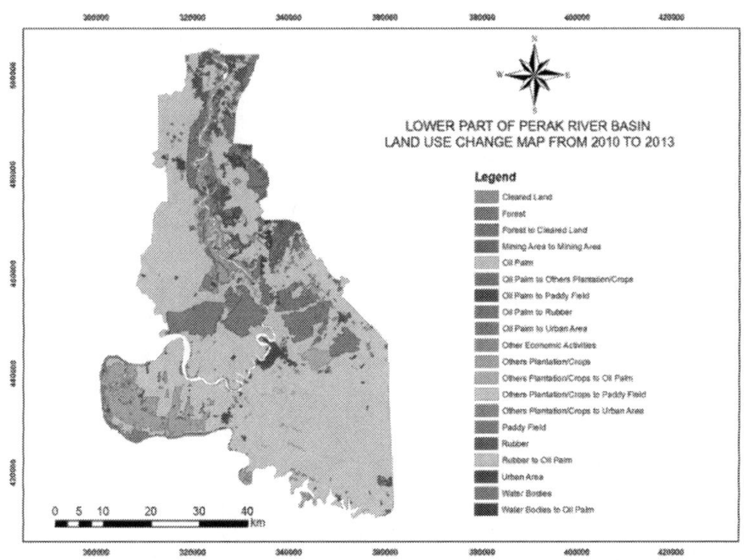

Figure 4. Land use change from year 2010 to 2013

Table 2. Area of land use change. Depending to various factor, land use classes can be shifted from one to another land use classes.

No.	Land Use Change	Area (km^2)
1	Urban Area	50.56
2	Other Economic Activities	10.41
3	Cleared Land	6.21
4	Others Plantation/Crops	260.47
5	Others Plantation/Crops to Urban Area	8.57
6	Others Plantation/Crops to Paddy Field	3.60
7	Forest to Cleared Land	6.79
8	Water Bodies	31.89
9	Mining Area to Mining Area	19.89
10	Oil Palm to Urban Area	14.09
11	Oil Palm to Others Plantation/Crops	22.74
12	Oil Palm	2008.54
13	Oil Palm to Paddy Field	6.25
14	Others Plantation/Crops to Oil Palm	24.91
15	Paddy Field	223.71
16	Rubber to Oil Palm	43.61
17	Rubber	63.34
18	Forest	89.50
19	Oil Palm to Rubber	6.71
20	Water Bodies to Oil Palm	13.04

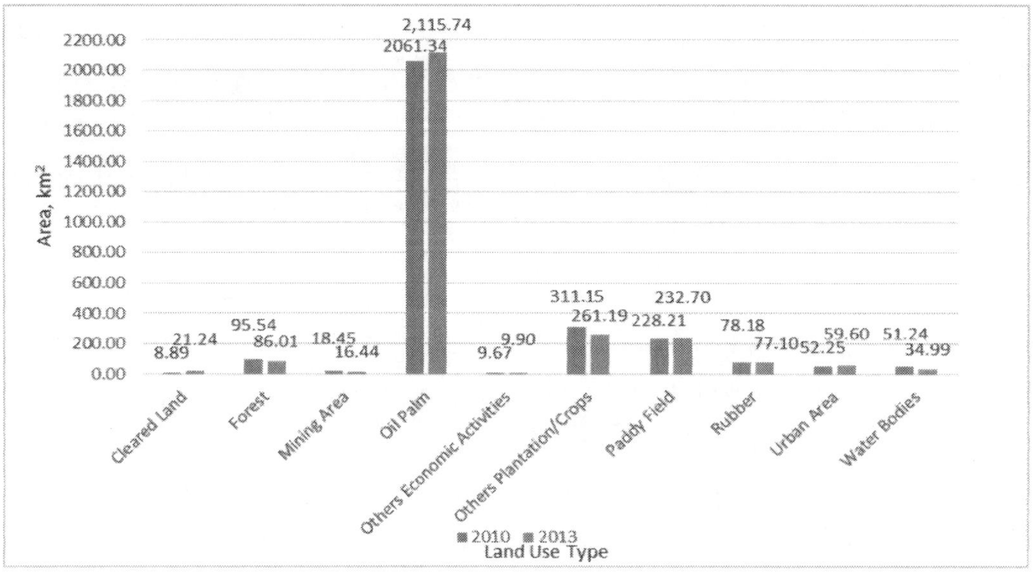

Figure 5. Land use change type against area of land use change.

IOP Conf. Series: Earth and Environmental Science 37 (2016) 012065 doi:10.1088/1755-1315/37/1/012065

The above figure shows the land use change from year 2010 to 2013. Depending to various factor, land use classes can be shifted from one to another land use classes (Table 2). The overall land use change from 2010 to 2013 as displayed in Figure 4. Based on result, oil palm plantation recorded some increment from year 2010 to 2013. While, area of forest depleting from 95.54km^2 to 86.01 km^2 indicating that the forest area were being exploited and shifted to other land use type. In the other hand, the rubber plantation decrease due to land conversion into palm oil plantation. Urban area showed some increment in coverage proving the current blooming number of population occurs rapidly. In context of cleared land, 2013 recorded higher coverage of cleared land compared to the year 2010 which recorded a shifting from 8.89km^2 in 2010 to 21.24km^2 in 2013. The others land use also undergoes changes from year of 2010 to 2013.

3.3 Soil erosion analysis

The soil erosion analysis is performed to determine the hotspot of soil erosion in study area. There are four factors that considered during performed the soil erosion analysis such as soil type, management practices, crop system, annual rainfall and elevation [11].

Models such as USLE [12] and its revised version RUSLE [13] are widely used method used for estimating the annually soil erosion potential rate as this method employ lesser field data instead of any other more complex data input [14]. The RUSLE model was employed in this research in order to perform the soil erosion analysis.

The RUSLE model are as below:

$$A = LS \times R \times K \times C \times P \quad (1)$$

A = Annual Soil Erosion Rate, ton/ha/yr
R = Annual rainfall erosivity factor, MJmm/ha/h/yr
K = Soil erosivity factor Mg h/MJ/mm
LS = Slope length and steepness factor
C = cover management factor
P = Conservation practise factor

3.3.1 Soil type

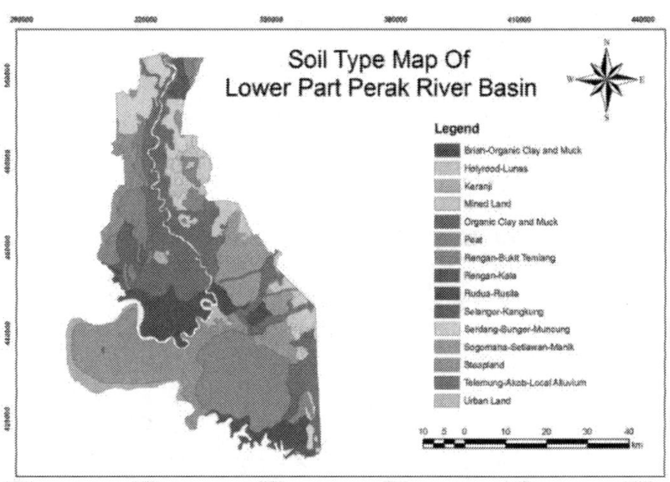

Figure 6. K-Value map of lower part of Perak River basin.

The soil type within the study area are determined according to the soil reconnaissance map sourced from the DOA. A variety of soil type are found in the study area. The soil type found in the study area are briah-organic clay and muck, holyrood-lunas, peat, mined land, rengam-kala, sogomana-setiawan-manik, rudua–rusila, Selangor, kangkung, organic clay and muck, kranji, rengan-bukit temiang, serdang-bungor-muncung, steepland, telemung-akob-local alluvium and urban land. The k-value factor were analysed according to the value obtained from Department of Agriculture.

3.3.2 P-factor map

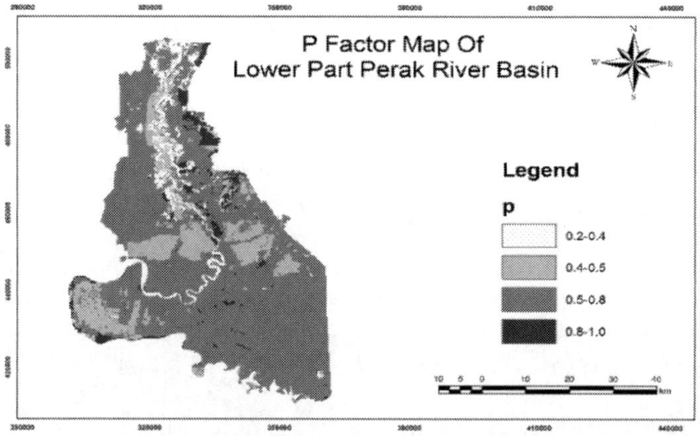

Figure 7. The P-Value map of lower Part Perak River basin.

From the map above (Figure 7), the P factor value are in the range of 0.2 to 1.0. The lower the P value, the more effective of adopted conservation practice to reduce the soil erosion. According [15], P value reflects the amount of erosion happened when any practice is employed compared to what would occur without that particular practice.

3.3.3 C-factor map

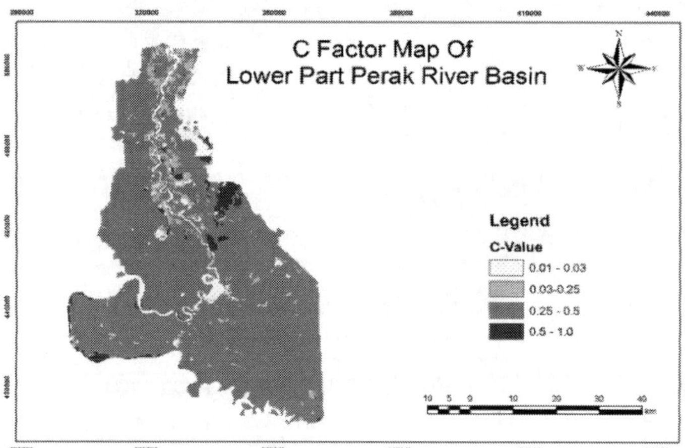

Figure 8. The C-Value map of lower part Perak River basin.

Figure 8 shows the Cover Management Factor of the study area. This factor are determine by referring to the land use type within the study area. It indicates the effect of conservation plan towards the annual soil erosion potential rate [16]. The cover management factor is referring to the ratio of soil erosion with the vegetation cover over that land use type. The vegetation cover is different for the land use type where the protected land have higher C value compared to bare land. The C value were sourced from Department of Agriculture.

3.3.4 R-factor map

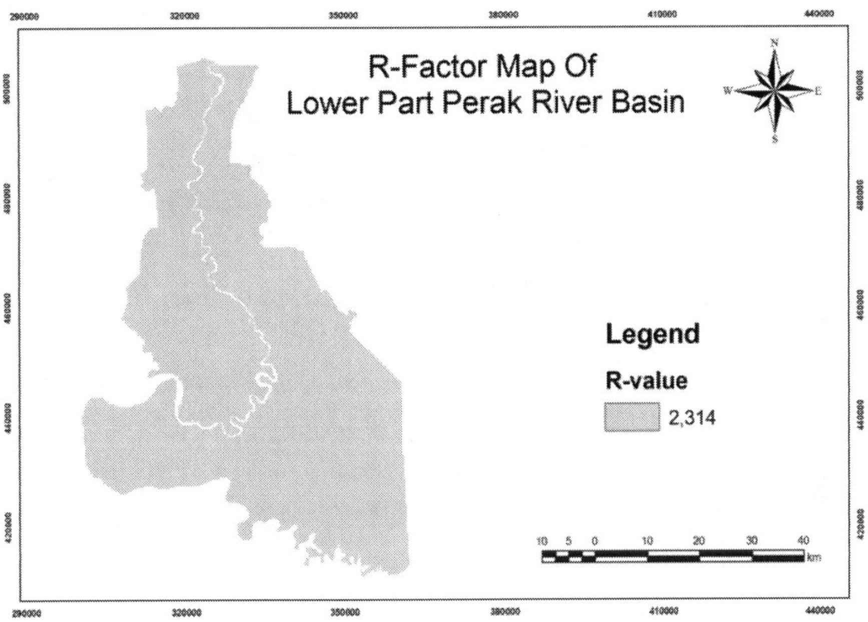

Figure 9. R-Value map of lower part Perak River basin.

The annual rainfall factor is referring to the erosion potential of the study area related to the rainfall intensity and the potential rain ability to create the erosion. For this study, the annual rainfall capacity recorded for the year 2013 was 2314 mm (Figure 9). This will represent the total annual rainfall capacity for the study area. The rainfall data is obtained from Department of Irrigation and Drainage.

3.3.5 Slope factor map

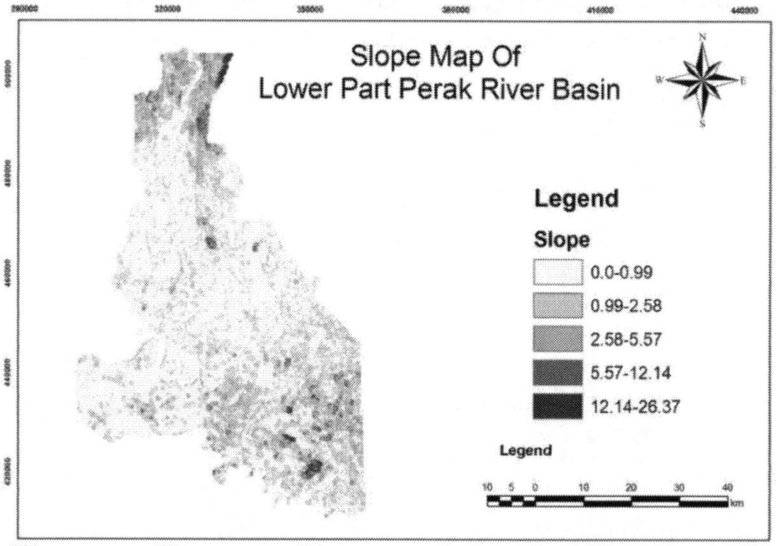

Figure 10. Slope map of lower part Perak River basin.

LS factor derivation was obtained from the Survey and Mapping Department (JUPEM). Figure 12 shows the slope map of the study area. Dominantly, the study area have the slope rate range from 0.00 to 0.09. Higher slope area was observed at the south part of study area. The highest elevation found in the study area is about 600 m from the mean sea level. However, majority of the area lies under 100 m from the mean sea level.

3.3.6 Soil erosion map

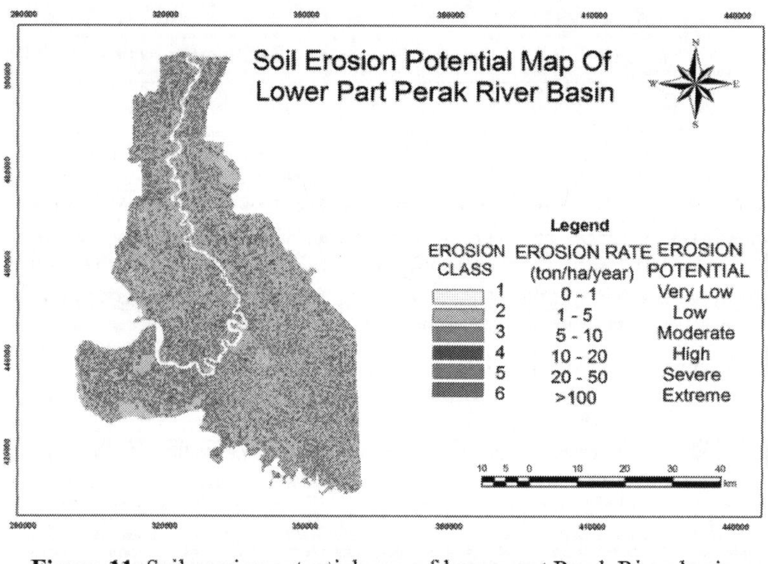

Figure 11. Soil erosion potential map of lower part Perak River basin.

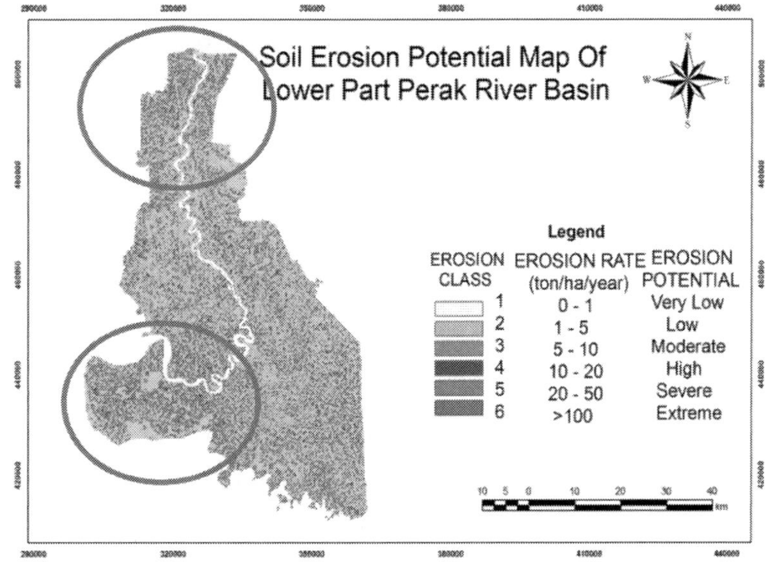

Figure 12. Soil erosion potential hotspot (circle) map of lower part Perak River basin.

Table 3. Soil erosion potential. Erosion potentials are categorised into 6 different classes.

Erosion Class	Erosion Rate (ton/ha/yr)	Soil Erosion Potential	Area (km^2)
1	0-1	Very Low	1765.60
2	1-5	Low	94.61
3	5-10	Moderate	102.53
4	10-20	High	160.05
5	20-50	Severe	255.03
6	>100	Extreme	536.19
		TOTAL	**2914.00**

Figure 12 and Table 3 shows the soil erosion potential within the study area. The soil erosion range from very low to extreme class. The soil erosion potential for very low (0-1 ton/ha/yr) classes recorded the majority of 61% (1765.60 km^2) of total area. The extreme classes (>100 ton/ha/yr) recorded about 18% (536.19km^2) of the total area. It can be concluded that the middle part of study area experienced low to severe classes of potential soil erosion.

4. Conclusion

According to the study, the land use in lower part of Perak river basin was dominantly been covered by oil palm plantation for both year of 2010 and 2013. In addition of that, most of the study area were exploited and converted from forest area to agriculture sector especially into oil palm and rubber plantation. The forest area in the study area decreases from 2010 to 2013. The soil erosion range from very low to extreme class. The soil erosion potential for very low (0-1 ton/ha/yr) classes recorded the majority of 61% (1765.60 km^2) of total area. The extreme classes (>100 ton/ha/yr) recorded about 18% (536.19km^2) of the total area. It can be conclude that the middle part of study area suffer low to severe classes of potential soil erosion.

References

[1] Ahearn, D. S., Sheibley, R. W., Dahlgren, R. a., Anderson, M., Johnson, J., & Tate, K. W. (2005). Land use and land cover influence on water quality in the last free-flowing river draining the western Sierra Nevada, California. Journal of Hydrology, 313(3-4), 234–247. http://doi.org/10.1016/j.jhydrol.2005.02.038

[2] Allan, J. D. (2004). Landscapes and Riverscapes: The Influence of Land Use on Stream Ecosystems. *Annual Review of Ecology, Evolution, and Systematics*, *35*(1), 257–284. http://doi.org/10.1146/annurev.ecolsys.35.120202.110122

[3] Aiken, R.R., & Leigh, C.H. (1985). On the Declining Fauna of Peninsular Malaysia in the Post Colonial Period. *Ambio*, (1491), 15-22.

[4] Gao, J., Li, F., Gao, H., Zhou, C., & Zhang, X. (2016). The impact of land-use change on water-related ecosystem services: a study of the Guishui River Basin, Beijing, China. *Journal of Cleaner Production*, 1–8. http://doi.org/10.1016/j.jclepro.2016.01.049

[5] Deng, L., Wang, G., Liu, G., & Shangguan, Z. (2016). Effects of age and land-use changes on soil carbon and nitrogen sequestrations following cropland abandonment on the Loess Plateau, China. *Ecological Engineering*, *90*, 105–112. http://doi.org/10.1016/j.ecoleng.2016.01.086

[6] Jia, X., Fu, B., Feng, X., Hou, G., Liu, Y., & Wang, X. (2014). The tradeoff and synergy between ecosystem services in the Grain-for-Green areas in Northern Shaanxi, China. *Ecological Indicators*, *43*, 103–111. http://doi.org/10.1016/j.ecolind.2014.02.028

[7] Tao, Y., Li, F., Wang, R., & Zhao, D. (2015). Effects of land use and cover change on terrestrial carbon stocks in urbanized areas: A study from Changzhou, China. *Journal of Cleaner Production*, *103*, 651–657. http://doi.org/10.1016/j.jclepro.2014.07.055

[8] Elosegi, A., Díez, J., & Mutz, M. (2010). Effects of hydromorphological integrity on biodiversity and functioning of river ecosystems. *Hydrobiologia*, *657*(1), 199–215. http://doi.org/10.1007/s10750-009-0083-4

[9] Xu, L., Xu, X., & Meng, X. (2013). Risk assessment of soil erosion in different rainfall scenarios by RUSLE model coupled with Information Diffusion Model: A case study of Bohai Rim, China. *Catena*, *100*, 74–82. http://doi.org/10.1016/j.catena.2012.08.012

[10] Dawn C. Parker, Steven M. Manson, Marco A. Janssen, Matthew J. Hoffmann & Peter Deadman (2003) Multi-Agent Systems for the Simulation of Land-Use and Land-Cover Change: A Review, Annals of the Association of American Geographers, 93:2, 314-337, DOI: 10.1111/1467-8306.9302004

[11] Stone, R. P., & Ontario. Ministry of Agriculture, Food and Rural Affairs. (2000). Universal Soil Loss Equation, USLE. Ministry of Agriculture, Food and Rural Affairs.

[12] Wischmeier, W.H., Smith, D.D., 1978. Predicting rainfall-erosion losses — aguide to conservation planning. Agriculture Handbook. N°537. USDA Forest Service, Washington, D.C.

[13] Renard, K.G., Foster, G.R., Weesies, G.A., Mc Cool, D.K., Yoder, D.C., 1997. Predicting Soil Erosion by Water: a Guide to Conservation Planning with the Revised Universal Soil Loss Equation (RUSLE). Agriculture Handbook N° 703. Natural Resources Conser- vation Service.Washington DC, USDA.

[14] Fernández, C., & Vega, J. A. (2016). Geoderma Evaluation of RUSLE and PESERA models for predicting soil erosion losses in the fi rst year after wild fi re in NW Spain. *Geoderma*, *273*, 64–72. http://doi.org/10.1016/j.geoderma.2016.03.016

[15] Treoh, Frederick R & Hobbs, J. Arthur (James Arthur), 1914-& Donahue, Roy Luther, 1908- (1999). Soil and water conservation : productivity and environmental protection (3rd ed). Prentice Hall, Upper Saddle River, N.J

[16] Ramli, M. F., Mei, C. S., Yusoff, M. K., & Makmom, A. (2004). Preliminary Assessment of Erosion Hazard using Open Source, (September), 12–14.

Analysis of land cover change and rainfall on the global land surface water coverage database for 1987-2015

X Li and W Takeuchi

Institute of Industrial Science, The University of Tokyo, Ce-506, 6-1, Komaba 4-chome, Meguro, Tokyo 153-8505, Japan

lixi@iis.u-tokyo.ac.jp

Abstract. In this paper, taking into account population density of the world, major river basin was delineated continent wise all over the world using HYDRO1k data. Then, monthly rainfall change from the year 1981 to 2014 and daily LSWC (Land surface water coverage) change from 1987 to 2015 based on each major river basin was computed and compared with each other. A good agreement was found between LSWC pattern and rainfall pattern, showing a seasonal variation characteristic in each year. However, it could be seen that rainfall is not the only factor that bring about change in LSWC. Also, it was found that the change of urban area is very strong. Especially in Yangtze basin, from 2000 to 2012, the urban changed from 0.07% to 0.83%. Moreover, the proportion of cropland increased significantly, especially in Ganges basin increased by 57.64%, grew to nearly 70% from 1992 to 2012. Besides, the trend of consistent growth was showed both in cropland and LSWC. It is indicated that the widespread expansion of cropland may bring about LSWC increasing.

1. Introduction

Flooding is known as frequent and most devastating event worldwide. Flooding analysis and estimate has received great attention of planners and decision makers in government and enterprises in order to master accurate information and mitigate catastrophic effects [1]. Since flood events are dynamic processes, high temporal resolution data is a desirable condition for flooding studies [2]. In previous research, global daily LSWC database derived from passive microwave radiometers including SSM/I, AMSR-E, WindSAT, AMSR2 during 1978-2015 has been made. and indicated its availability and potential in flood events detection [3-5] since LSWC could be associated to relative increase of both water and soil moisture [6].

Land cover change is known to influence both surface water hydrology and soil hydraulic properties by altering the hydrological characteristics of the land surface and modifying the patterns and rates of water flow [7]. In addition, rainfall also seriously influences open water and soil moisture and thus plays an important role in flooding researches. Ohana-Levi has modeled the effects of land cover change on rainfall runoff relationships in a Semiarid, Eastern Mediterranean watershed, which was found a strong relationship between vegetation cover and the runoff volume. Moreover, the land cover changes with most pronounced effects on runoff volumes were related to urbanization and vegetation removal [8]. Panahi discussed the effect of land use/cover changes on the floods of the Madarsu basin of northeastern Iran which was found that the discharge rate of 2003 flood was about 10 times larger than that of the 1964 flood, since the direct effect of the land use/cover change from the stable forests and rangelands to the unstable agricultural lands on the both soil moisture retention capacity and run off rate [9]. Ferrazzoli carried out analysis of the effect of rain and flooding events on AMSR-E Signatures of La Plata Basin, Argentina, which has been found that the amount of the effect and the correlation between variables are dependent on the properties of the areas surrounding the stations [10]. Hydrological response to land cover change and human activities in arid regions using a geographic information system and remote sensing has been discussed by Mahmoud which has been indicated that changes in land cover are predicted to result in an annual increase in irrigated cropland and dramatic decline in forest area in the study area over the next few decades [11].

However, most of this field of research is aimed at a certain watershed or a region until now. Therefore, the objective of this study is to make clear rainfall pattern of each major river basin worldwide and analysis the effects of land cover change and rainfall on the global LSWC database during 1987-2015 derived from passive microwave remote sensing.

2. Methodology

2.1. The flowchart of this study

Figure 1 shows the framework of this research. Firstly, taking into account population density of the world, global major river basins were derived from HYDRO1k data. Secondly, combining with major river basins, CRU_TS precipitation dataset from 1981 to 2014 was integrated to obtain the rainfall pattern of each river basin in order to compare with LSWC pattern of each river basin. Thirdly, from land cover map of 1992 (UMD_LC), 2000 (GLC2000) and 2012 (BU_LC), cropland, forest, urban and water body were masked out respectively and the percent area change of each land cover type in each river basin was calculated to explore its relationship with LSWC change.

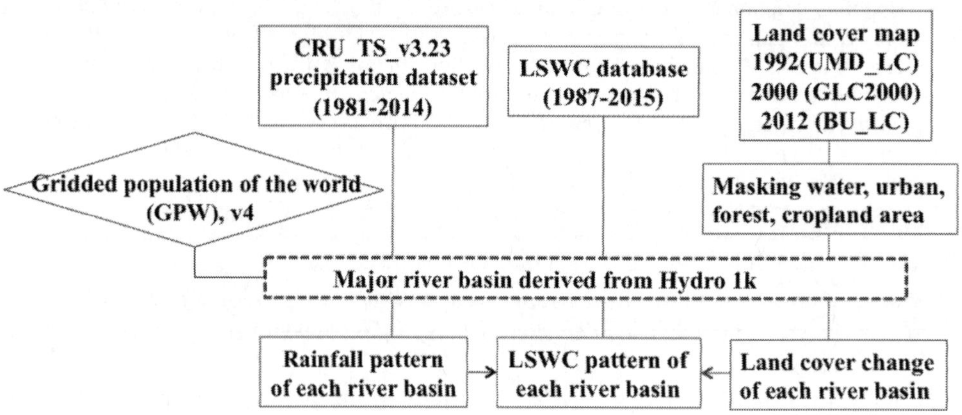

Figure1. The flowchart of this study.

2.2. River basin delineation using HYDRO1k

HYDRO1k is a geographic database developed to provide comprehensive and consistent global coverage of topographically derived data sets, which provides a standard suite of geo-referenced data sets at a resolution of 1 km. The HYDRO1k data sets are being developed on a continent-by-continent basis, for all landmasses of the globe without Antarctica and Greenland.

Vector river basin data from HYDRO1k was imported into ArcGIS 10.2.2 for processing. After converting the projection from Lambert-Azimuthal Equal Area to Geographic Lat/Lon in WGS84, the first level of basins were highlighted and selected based on the Pfafstetter Coding System [12]. Thereafter by combining it with population density data of the world, derived from gridded population of the world (GPW) which provides estimates of population density for the years 2000, 2005, 2010, 2015 based on counts consistent with national censuses and population registers with respect to relative spatial distribution [13-14], a subdivision for the first level of river basin were carried out in densely populated or river rich areas.

2.3. Rainfall pattern analysis by CRU_TS_v3.23

The gridded CRU_TS (time-series) v3.23 dataset, produced by the Climatic Research Unit (CRU) at the University of East Anglia, provides month-by-month variations in climate over the period from 1901 to 2014 with 0.5×0.5 degree grids [15-16]. Firstly the same coordinate system (Geographic Lat/Lon in WGS84) was defined to CRU_TS precipitation data as projection of river basin in order to combine rainfall dataset with river basin. Then rainfall pattern of each major river basin was computed by integrating long time series of data from 1980 to 2014 in order to compare with the LSWC pattern of the same river basin.

2.4. Land cover change analysis

The land cover maps from year of 1992 (UMD_LC), 2000 (GLC2000) and 2012 (BU_LC) were used to discuss and explore the relationship between land cover change and LSWC change [17-19]. The specification of three land cover products is shown in Table 1 [20-20]. Based on three kind of classification schemes and conversion legends as shown in Table 2 [22], urban area, forest, cropland area and water body for the years 1992, 2000 and 2012 was masked out respectively. Finally the land area change of each land cover type in each major river basin was calculated.

Table 1. Characteristics of the three global land cover data sets assessed in this study.

Global land cover data set	Classification scheme	Sensor	Date	Number of classes	Resolution
UMD_LC	Simplified IGBP	AVHRR	1992	14	1km
GLC2000	FAO LCCS	SPOT-4	2000	23	1km
BU_LC	IGBP	MODIS	2012	17	1km

Table 2. Conversion legends table of three land cover maps.

Target legend	1992 (UMD-LC)	2000 (GLC2000)	2012 (BU-LC)
Cropland	Class 11	Class16-18	Class 12, 14
Forest	Class1-6	Class1-6, 9-10	Class 1-5, 8
Urban	Class 13	Class 22	Class 13
Water	Class 0	Class 20	Class 0

3. Results and discussion

3.1. River basin delineation using HYDRO1k

Global map of gridded population density in 2015 was prepared and the first level river basins all over the world were derived from HYDRO1k data set. Figure 2 shows the gridded population density of the world in 2015 overlaid by the first level of river basins wherein each pixel value represents persons per square kilometer. It can be seen that the population density in Asia and some area in Europe and Africa is very high, especially that in India, Bangladesh and China is much high.

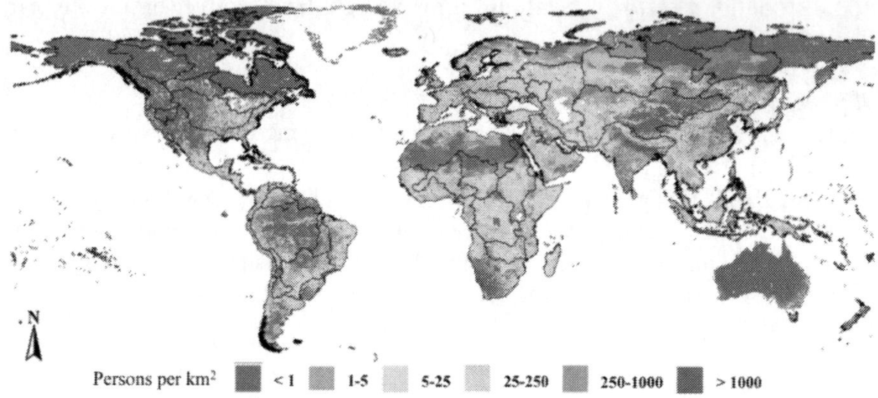

Figure 2. Gridded population density of the world in 2015 overlaid by the first level of river basins, the pixel values represent persons per square kilometer.

Since flooding causes more damage in densely populated areas. Therefore, taking into account the river distribution and population density we carried out a subdivision for the first level river basin in densely populated or rich basin area and finally got the global major river basins with ID as shown in Figure 3. Table 2 shows the list of several significant major river basins and the corresponding ID of each continent.

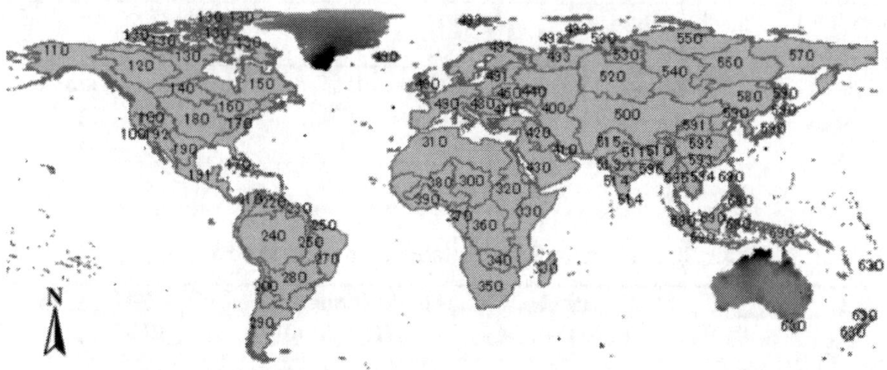

Figure 3. Major river basin with ID of the world derived from HYDRO 1k data set.

Table 3. Selected several major river basins and the corresponding ID by continents.

Continent	ID	River basin	ID	River basin
North America	120	Mackenzie	160	St Lawrence
	140	Saskatchewan/Nelson	180	Mississippi
South America	240	Amazon Basin	280	La Plata Basin
Africa	320	Nile	360	Congo
	340	Zambezi	380	Niger
Europe	420	Tigris/Euphrates	480	Danube
Asian	510	Brahmaputra Basin	580	Amur Basin
	511	Ganges	591	Huang He Basin
	515	Indus Basin	592	Yangtze Basin
	520	Ob	594	Mekong Basin
	540	Yenisey	596	Irrawaddy Basin

3.2. Comparison between rainfall pattern and LSWC pattern

In order to make clear the relationship between rainfall and LSWC, the monthly precipitation change of all river basins worldwide was computed by integrating long-term monthly CRU precipitation dataset from 1981 to 2014. Three major river basins, which are Brahmaputra (510), Ganges (511) and Yangtze (592), from Asian were chosen for further analysis. In addition, in previous research, we built long-term LSWC database in time series from 1987 to 2015 by combining SSM/I, AMSR-E, WindSat and AMSR2 [3-5]. LSWC has been proved a good performance in large-scale flooding detection. Then, daily LSWC change from 1987 to 2015 based on each river basin was also computed.

Figure 4 shows the corresponding relationship between precipitation and LSWC. It can be seen that the LSWC pattern basically coincides with the rainfall pattern, showing a seasonal variation characteristic within each year. Moreover in the long term, the trend of continued growth or decreasing in rainfall is not clear whereas LSWC showed an increasing trend according to the trend line trough 1987 to 2015. Compared to previous years, increase in average LSWC value after 2000 was 2% for Brahmaputra, 1% for Ganges and 1% for Yangtze than before 2000. It can be indicated that rainfall is not the only factor that makes the change of LSWC.

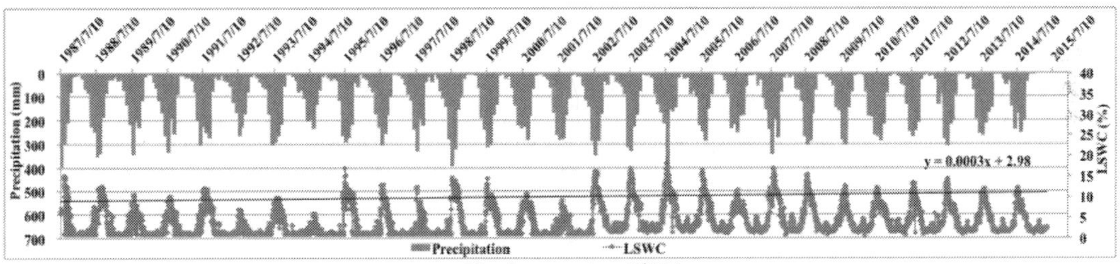

a. Brahmaputra (510)

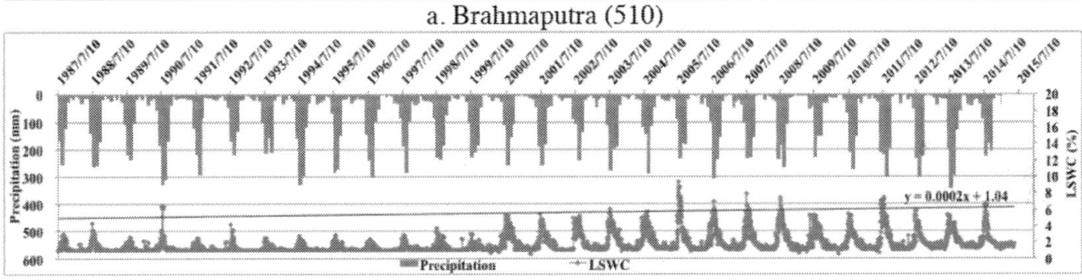

b. Ganges (511)

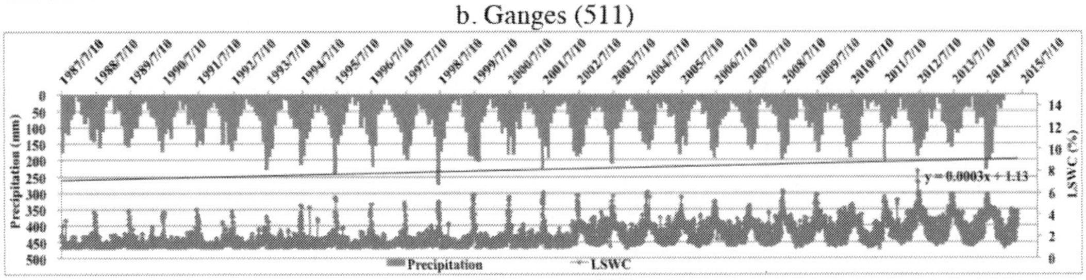

c. Yangtze (592)

Figure 4. Compare of the long-term precipitation and LSWC of each river basin.

3.3. Land cover change analysis

To further explore factors affecting LSWC, land cover change has been discussed. Figure 5 shows the change in cropland, forest, urban and water body area among the years 1992, 2000 and 2012 for the Brahmaputra (510), Ganges (511) and Yangtze basin (592). We saw that the proportion of forest area showed a decreasing trend and no much change observed for the Yangtze basin. Moreover a strong increase in urban area was also found in each basin. Especially in Yangtze basin, from 2000 to 2012, the urban region increased from 0.07% to 0.83%. It can also be seen that from 1992 to 2000 and then to 2012, the proportion of croplands increased significantly. Brahmaputra basin area increased by 13.78%, Yangtze basin increased by 22.01%, especially Ganges basin increased by 57.64%, grew to nearly 70%.

By comparing cropland with LSWC change from 1992, 2000 to 2012 in each basin as shown in Table 4, we can find that the cropland presented consistent growth situation along with LSWC. According to the relative change in Brahmaputra basin, the croplands increased by 8% and 6%, while LSWC increased by 4% and 3%. It is hypothesized that since during sowing season irrigation causes cropland surfaces to be covered by water, it could have led it to be detected as an inundated area. Moreover, since the increase of human population and standards of living demand more harvest and production from the earth resources, increased irrigation frequency or sowing time in one year can be anticipated. Thus it is expected that the widespread expansion of cropland may bring about LSWC increasing. In addition, it can also be inferred that the rapid expansion of urban and reduction of forest area may be factors affecting LSWC change.

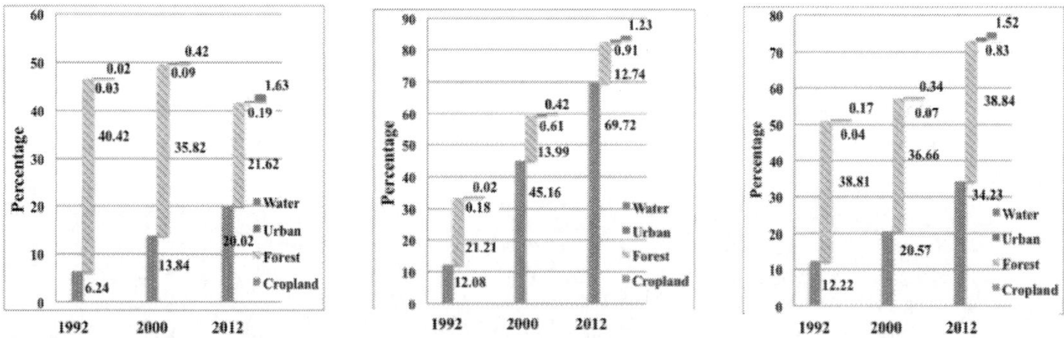

a. Brahmaputra (510) b. Ganges (511) c. Yangtze (592)

Figure 5. Land area change of different land cover type for 1992, 2000 and 2012 in three river basins.

Table 4. Compare of cropland and LSWC change from 1992, 2000 to 2012 in three basins.

River basin	Category	Percentage (%)				
		1992	2000	2012	Relative change	
					1992-2000	2000-2012
Brahmaputra	Cropland	6	14	20	8	6
	LSWC	7	11	14	4	3
Ganges	Cropland	12	45	70	33	25
	LSWC	4	5	6	1	1
Yangtze	Cropland	12	21	34	9	13
	LSWC	4	6	8	2	2

4. Conclusion and future work

In this paper, firstly, taking into account population density of the world we delineated major river basin continent wise all over the world using Hydro 1k data. Secondly, by comparing precipitation monthly change with LSWC daily change of each river basin, we found that LSWC pattern basically coincide with rainfall pattern, showing a seasonal variation characteristic in each year. However rainfall is not the only factor that brings about the change in LSWC. Finally, it was found that the change in urban was very strong in Yangtze basin, from 2000 to 2012, changed from 0.07% to 0.83%. Moreover, the proportion of cropland increased significantly, especially Ganges basin increased by 57.64%, grew to nearly 70%. In addition the trend of consistent growth was showed both by cropland and LSWC. It is expected that the widespread expansion of cropland may bring about LSWC increasing.

For the future work, we advance to analyse all the river basin worldwide in order to find regulation among rainfall, land cover change and LSWC. Moreover, we wish to conduct spatial analysis of land cover change and combine factors of social economy to explore its influence on LSWC change.

Nomenclature

LSWC	Land Surface Water Coverage
SSM/I	Special Sensor Microwave Imager
AMSR-E	Advanced Microwave Scanning Radiometer - Earth Observing System
AMSR2	Advanced Microwave Scanning Radiometer - 2
CRU_TS	Climatic Research Unit - Time Series
UMD_LC	University of Maryland - Land Cover product
GLD2000	Global Land Cover 2000
BU_LC	Boston University – Land Cover product
IGBP	International Geosphere Biosphere Programme
FAO LCCS	Land Cover Classification System of Food and Agricultural Organizations
AVHRR	Advanced Very High Resolution Radiometer
SPOT- 4	Satellite Pour l'Observation de la Terre - 4
MODIS	Moderate Resolution Imaging Spectroradiometer

References

[1] Brivio P A, Colombo R, Maggi M and Tomasoni R 2002 Integration of remote sensing data and GIS for accurate mapping of flooded areas *Int. J. Remote Sens.* **23** 429–41

[2] Singh Y, Ferrazzoli P and Rahmoune R 2013 Flood monitoring using microwave passive remote sensing (AMSR-E) in part of the Brahmaputra basin, India *Int. J. Remote Sens.* **34** 4967–85

[3] LI X and Takeuchi W 2015 Estimation of land surface water coverage with PALSAR and AMSR-E for flood detection on global scale *The International Symposium on Remote Sensing (IRSR), Tainan, Taiwan*

[4] LI X and Takeuchi W 2015 Development of large scale flooding detection method by integrating historical global record using AMSR-E/AMSR2 with PALSAR *IEEE IGARSS, Milan, Italy*. 822–5

[5] LI X and Takeuchi W 2015 Global flood detection and spatio-temporal analysis from 2002-2015 *The 36th Asian Conference of Remote Sensing (ACRS), Manila, Philippines*

[6] Temimi M, Leconte R, Brissette F and Chaouch N 2007 Flood and soil wetness monitoring over the Mackenzie River Basin using AMSR-E 37 GHz brightness temperature *J. Hydrol.* **333** 317–28

[7] Savary S, Rousseau A N and Quilbé R 2009 Assessing the effects of historical land cover changes on runoff and low flows using remote sensing and hydrological modeling *J. Hydrol. Eng.* **14** 575–87

[8] Ohana-Levi N, Karnieli A, Egozi R, Givati A and Peeters A 2015 Modeling the Effects of Land-Cover Change on Rainfall-Runoff Relationships in a Semiarid, Eastern Mediterranean Watershed *Adv. Meteorol.* 16

[9] Panahi A, Alijani B and Mohammadi H 2010 The Effect of the Land Use/Cover Changes on the Floods of the Madarsu Basin of Northeastern Iran *J. Water Resour. Prot.* **2** 373–9

[10] Ferrazzoli P, Rahmoune R, Moccia F, Grings F, Salvia M, Barber M, Douna V, Karszenbaum H, Soldano A, Goniadzki D, Parmuchi G, Montenegro C, Kandus P and Borro M 2010 The effect of rain and flooding events on AMSR-E signatures of la plata basin, argentina *IEEE J. Sel. Top. Appl. Earth Obs. Remote Sens.* **3** 81–90

[11] Mahmoud S H and Alazba A A 2015 Hydrological response to land cover changes and human activities in arid regions using a geographic information system and remote sensing *PLoS One* **10** 1–19

[12] Pfafstetter O, 1989, "Classification of hydrographic basins: coding methodology", unpublished manuscript, DNOS, Rio de Janeiro; translated by J.P. Verdin, U.S. Bureau of Reclamation, Brasilia, Brazil, September 5, 1991.

[13] Tobler W, Deichmann U, Gottsegen J and Maloy K 1997 World population in a grid of spherical quadrilaterals. *Int. J. Popul. Geogr.* **3** 203–25

[14] Deichmann U, Balk D and Yetman G 2001 Transforming Population Data for Interdisciplinary Usages: From census to grid Palisades, NY: NASA Socioeconomic Data and Applications Center (SEDAC), CIESIN, Columbia University. 20pp

[15] Hulme M 1992 A 1951-80 global land precipitation climatology for the evaluation of general circulation models *Clim. Dyn.* **7** 57–72

[16] Hulme M, Osborn T J and Johns T C 1998 Precipitation sensitivity to global warming: Comparison of observations with HadCM2 simulations *Geophys. Res. Lett.* **25** 3379

[17] Loveland T R, Reed B C, Brown J F, Ohlen D O, Zhu Z, Yang L and Merchant J W 2000 Development of a global land cover characteristics database and IGBP DISCover from 1 km AVHRR data *Int. J. Remote Sens.* **21** 1303–30

[18] Friedl M, McIver D, Hodges J C, Zhang X, Muchoney D, Strahler A, Woodcock C, Gopal S, Schneider A, Cooper A, Baccini A, Gao F and Schaaf C 2002 Global land cover mapping from MODIS: algorithms and early results *Remote Sens. Environ.* **83** 287–302

[19] Bartholome E and Belward A S 2005 GLC2000: A new approach to global land cover mapping from Earth observation data *Int. J. Remote Sens.* **26** 1959–1977

[20] Zhang N and Tateishi R 2013 Integrated Use of Existing Global Land Cover Datasets for Producing a New Global Land Cover Dataset with a Higher Accuracy: A Case Study in Eurasia *Advances in Remote Sens.* **2** 365–72

[21] Song X P, Huang C, Feng M, Sexton J O, Channan S and Townshend J R 2014 Integrating global land cover products for improved forest cover characterization: an application in North America *Int. J. Digit. Earth* **7** 709-24

[22] Bai Y, Feng M, Jiang H, Wang J, Zhu Y and Liu Y 2014 Assessing consistency of five global land cover data sets in China *Remote Sens.* **6** 8739–59

Application of PALSAR-2 remote sensing data for structural geology and topographic mapping in Kelantan river basin, Malaysia

Amin Beiranvand Pour[1], Mazlan Hashim

Geoscience and Digital Earth Centre (Geo-DEC), Universiti Teknologi Malaysia

Abstract. Natural hazards of geological origin are one of major problem during heavy monsoons rainfall in Kelantan state, peninsular Malaysia. Several landslides occur in this region are obviously connected to geological and topographical features, every year. Satellite synthetic aperture radar (SAR) data are particularly applicable for detection of geological structural and topographical features in tropical conditions. In this study, Phased Array type L-band Synthetic Aperture Radar (PALSAR-2), remote sensing data were used to identify high potential risk and susceptible zones for landslide in the Kelantan river basin. Adaptive Local Sigma filter was selected and applied to accomplish speckle reduction and preserving both edges and features in PALSAR-2 fine mode observation images. Different polarization images were integrated to enhance geological structures. Additionally, directional filters were applied to the PALSAR-2 Local Sigma resultant image for edge enhancement and detailed identification of linear features. Several faults, drainage patterns and lithological contact layers were identified at regional scale. In order to assess the results, fieldwork and GPS survey were conducted in the landslide affected zones in the Kelantan river basin. Results demonstrate the most of the landslides were associated with N-S, NNW-SSE and NE-SW trending faults, angulate drainage pattern and metamorphic and Quaternary units. Consequently, geologic structural map were produced for Kelantan river basin using recent PALSAR-2 data, which could be broadly applicable for landslide hazard assessment and delineation of high potential risk and susceptible areas. Landslide mitigation programmes could be conducted in the landslide recurrence regions for reducing catastrophes leading to economic losses and death.

1. Introduction

The state of Kelantan is located in north-eastern corner of Peninsular Malaysia (Figure 1). Kelantan river is the major river in the region. It appears at the convergence of the Galas river and Lebir river near Kuala Kari and meanders over the coastal plain until it finally degrades into the South China Sea. Kelantan river basin covers 923 km^2, which is about 85% of the

[1] Corresponding Author. Tel: +607 -5530666; Fax: +607- 5531174; Email address: beiranvand.amin80@gmail.com; a.beiranvand@utm.my; mazlanhashim@utm.my.

Kelantan state's surface area. It is composed of flat slope to moderately sloping areas in northern part and steep scraps and high slopes in the southern part of the river basin [1]. A wide variety of rocks consisting of igneous, sedimentary and metamorphic rocks could be seen in the state of Kelantan, which are distributed in a north-south trend. Typically, four types of rocks are classified in the region, including granitic rocks, sedimentary/metasedimentary rocks, extrusive rocks (volcanic rocks) and unconsolidated sediments (Figure 2). Localised geological features comprise folding, faulting and jointing in the sedimentary rocks and faulting and jointing in the granitic rocks.

Figure 1. Location of the Kelantan state in Peninsular Malaysia.

Yearly, several landslides occur during heavy monsoon rainfall in Kelantan river basin, peninsular Malaysia, which are obviously connected to geological structures and topographical features of the region. In recent years especially, there have been many severe flooding events (in the year 2005, 2006, 2007, 2008, 2009 and 2014) which have led to significant damage to livestock, agricultural produce, homes and businesses in the Kelantan river basin [2]. The problem stems from the inappropriate use of lands that are vulnerable to erosion, quick water runoff and slope failure. Recent challenge is to identify high potential risk and susceptible zones for natural hazards of geological origin in the Kelantan river basin using advanced remote sensing technology. In this study, the recently launched ALOS-2-Phased Array type L-band Synthetic Aperture Radar-2 (PALSAR-2), remote sensing data were used to identify high potential risk and susceptible zones for geological origin hazards in the Kelantan river basin at regional and district scales.

2. Materials

In this investigation, a ScanSAR mode dual polarization (level 3.1) and two Fine mode dual polarization (level 3.1) PALSAR-2 scenes were obtained from ALOS-2 data distribution consortium online system Remote Sensing Technology Center of Japan (RESTEC) (http://www.restec.or.jp/english/index.html) and PASCO Corporation (http://en.alos-pasco.com; https://satpf.jp/) for comprehensive analysis of major geological structures and detailed characterizations of lineaments in the state of Kelantan. In particular, L-band microwave from PALSAR-2 has ability to penetrate vegetation due to relatively long wavelengths (about 24 cm), making the data particularly useful for geological structural mapping in tropical environments [3,4]. The wavelength of the L-band is relatively long among microwaves (C-band: about 6 cm and X- band: about 3 cm), allowing it to travel all the way down to the ground through vegetation. Not only can information be, obtained about vegetation but information of the ground surface can be obtained as well [5,6]. The data were processed using the ENVI (Environment for Visualizing Images) version 5.1 software package.

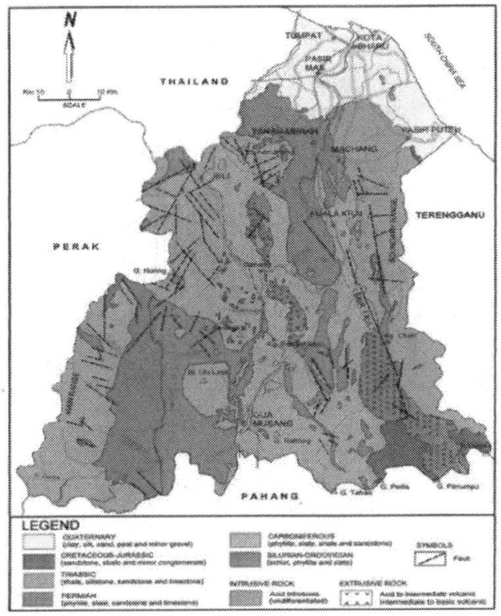

Figure 2. Geologic map of the Kelantan state [2].

3. Methods

In this study, the adaptive Local Sigma filter was selected and applied to accomplish speckle reduction and preserving both edges and features. The Local Sigma filter uses the local standard deviation computed for the filter box to determine valid pixels within the filter window. It replaces the pixel being filtered with the mean calculated using only the valid pixels within the filter box [7]. Directional filtering technique was applied to the Local Sigma resultant image for detailed lineament extraction and edge enhancement. Directional filter is used for producing artificial effects suggesting tectonically controlled linear features. It is a straightforward method for extracting edges in the spatial domain that approximates the first derivative between two adjacent pixels. The algorithm produces the first difference of the image input in the horizontal, vertical, and diagonal directions [8]. As a result, many additional edges of diverse orientations are enhanced. Edge enhancement is performed by convolving the original data with a weighted mask or kernel. Directional filters were used to enhance specific linear trends in the Local Sigma resultant images. Four principal Directional filters: N-S, E-W, NE-SW, and NW-SE with a 7*7 kernel size were applied. Directional filter angles were adjusted as N-S: 0°, E-W: 90°, NE-SW: 45°, and NW-SE: 135°. North (up) is zero degrees and the other angles are measured in the counterclockwise direction. A Global Positioning System (GPS) survey was carried out using a Garmin® MONTANA® 650 to provide accurate locations for landslide affected zones. Additionally, image processing results were compared with the geological and general topography maps of Kelantan state (1:100,000 scale).

4. Results and discussion

Figure 3 shows ScanSAR HV polarization image that is superimposed by general topography map of the Kelantan state. It is evident that the morphology of any area is largely controlled by rock type and structure. High elevation areas (500-100 m and <1000 m) in the Kelantan state are mountainous areas associated with Main Range granites (in the west) and Boundary Range granite (in the east), which are detected as grey to cream colour in ScanSAR image. Hilly, plain and coastal areas with altitude between 500 to 50 m are associated with sedimentary rocks, which are manifested pink to yellow colour in ScanSAR image.

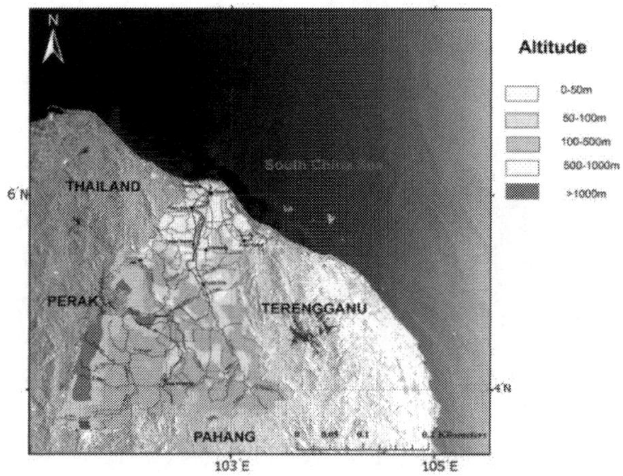

Figure 3. ScanSAR HV polarization image of the Kelantan state superimposed by general topography map.

Figure 4 shows the results of RGB colour-composite for N-S, NE-SW, and NW-SE (R: 0°, G: 45°, B: 135°) directional filters. Major change in deformation style is obvious from the west to the east in Figure 6. Structural analysis reveals four distinct parts from the west to the east, including (ii) western part of the scene by ductile fabrics; (ii) western of the Bentong-Raub Suture Zone (BRSZ) affected mainly by brittle deformation; (iii) ductile-brittle deformation between the BRSZ and Lebir Fault Zone; and (iv) brittle-ductile fabrics between Lebir Fault Zone and eastern coastal line. Lineament occurrence in Figure 4 is mainly linked to the N-S trending of the BRSZ and Lebir Fault Zone. Generally, major faults are strike-slip with both dextral and sinistral movements, which trend N-S and NW-SE. NW-SE trending strike-slip faults moved sinistrally in the Lebir Fault Zone. The sinistral movement along the Lebir Fault Zone is responsible to the formation of folding and reverse faulting adjacent to the fault and surrounding area. The collision zone and compressional structures appear clearly in the west of the BRSZ in Main Range granites (Figure 4). Deformation in this region shows the shortening zone oriented parallel to the BRSZ. Several faults, joints and fractures represent brittle deformation events in the region that mostly strike NW-SE. Generally, most of the short lineaments are clustered in the collision zone. Ductile deformation in the western margin of the image (Figure 4) includes upright asymmetrical mega folds with axial surfaces oriented W-E. The deformed area zone between the BRSZ and Lebir Fault Zone represents faults system and folded area. Brittle-ductile fabrics in the eastern part of image between

Lebir Fault Zone and eastern coastal line illustrate curved shear zone that occupied by several N-S and NW-SE striking faults, fractures and joints.

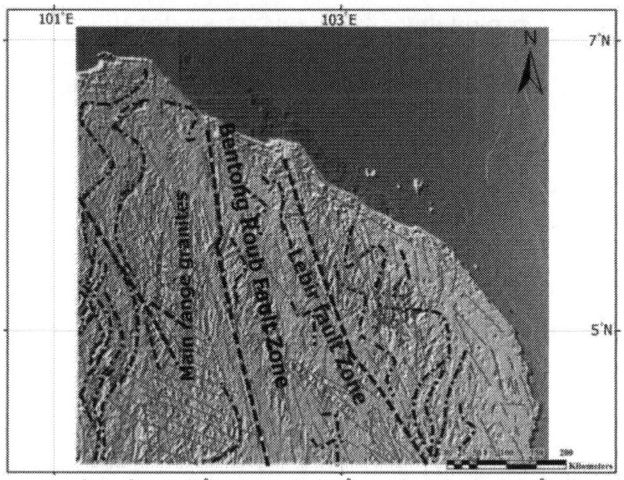

Figure 4. RGB colour-composite of N-S (0°), NE-SW (45°), and NW-SE (135°) directional filters of northern part of Peninsular Malaysia.

Figure 5 shows geologic structural map for the Kelantan state. It is produced based on the integration of directional filter images for fine PALSAR-2 scenes of northern and southern parts of Kelantan. The most important feature in the map is the presence of river systems, drainage lines patterns and fault zones. The N-S, NE-SW and NNE-SSW lineament trends are commonly dominant in the image map (Figure 5). Additionally, few short NW-SE trending lineaments are detected in the western and eastern parts of the study area. The dominant lineaments tend to run in the N-S direction. Pattern of the lineament map suggest that some faults belong to the Lebir Fault Zone are detected in south-eastern part of the image. They are N-S and NE-SW striking system distributed in the south eastern part of Kelantan. On the contrary, pattern of the lineament map in north-western part of the image map displays the occurrence of BRSZ fault zone, which contains lineaments in N-S and NNE-SSW direction. The rivers in the study area are structurally controlled. The drainage of the Kelantan river basin shows dendritic, sub-dendritic and rectangular pattern in most part of the region. Surface drainage patterns are controlled by lithology and geological structures. Dendritic and sub-dendritic patterns with a large number of tributaries are typical of drainage in areas of impermeable crystalline rock such as gneiss. It is the most common drainage pattern and is characteristic of essentially flat-lying and/or relatively homogeneous rocks and impervious soils and lack of structural control. Rectangular pattern is usually caused by jointing or faulting of the underlying bedrocks. It is usually associated with massive, intrusive igneous and metamorphic rocks. Therefore, the area with dendritic and sub-dendritic patterns is subjected to hydro-geological hazards such as flooding because of low infiltration runoff. Rectangular drainage pattern is susceptible zone that could be affected by landslide easily due to slope of the land, litho-structural conditions and speed of runoff. Results of fieldwork verified that the most of the landslides were associated with N-S, NNW-SSE and NE-SW trending faults, angulate drainage pattern and metamorphic and Quaternary units.

5. Conclusions

The results presented in this study demonstrate the importance and advantages of the use recently launched PALSAR-2 data for geological structural mapping in the Kelantan state, north of Peninsular Malaysia. The state of Kelantan is located between two significant fault zones, including the Bentong-Raub Suture Zone and Lebir Fault Zone. Most of the landslide affected areas were associated with N-S, NNW-SSE and NE-SW trending faults, angulate drainage pattern and metamorphic and Quaternary units. Results of this investigation could be considered for governmental or private construction projects in the landslide recurrence regions for reducing catastrophes leading to economic losses and death.

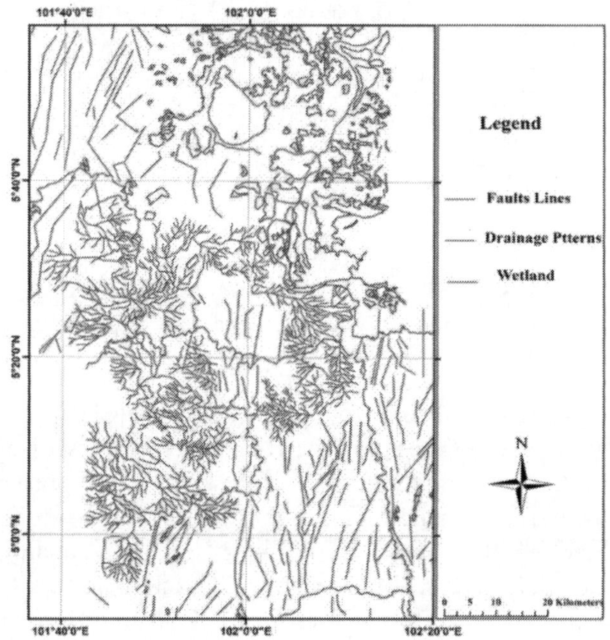

Figure 5. Geologic structural map for the Kelantan state derived from directional filtering images.

Acknowledgements

This study was conducted as a part of TRGS grant (Vote no: R.J130000.7809.4L837), Ministry of Higher Education (MOHE) Malaysia. We are thankful to the Universiti Teknologi Malaysia for providing the facilities for this investigation.

References

[1] Pradhan B 2009. Flood susceptible mapping and risk area delineation using logistic regression, GIS and remote sensing. Journal of Hydrology 9, No: 2.

[2] Nazaruddin DA, Fadilah, NSM, Zulkharnain Z, Omar, SAS, Ibrahim, MKM, 2014. Geological Studies to Support the Tourism Site: A Case Study in the *Rafflesia* Trail, Near Kampung Jedip, Lojing Highlands, Kelantan, Malaysia. International journal of Geosciences, 835-851.

[3] Pour BA, Hashim M, Marghany M, 2014. Exploration of gold mineralization in a tropical region using Earth Observing-1 (EO1) and JERS-1 SAR data: a case study from Bau gold field, Sarawak, Malaysia. Arabian Journal of Geosciences 7(6), 2393-2406.

[4] Pour BA, Hashim M, 2014. Structural geology mapping using PALSAR data in the Bau gold mining district, Sarawak, Malaysia. Advances in Space Research, 54(4), 644-654.
[5] Pour AB, Hashim M, 2015. Structural mapping using PALSAR data in the Central Gold Belt Peninsular Malaysia, Ore Geology Reviews 64, 13-22.
[6] Pour AB, Hashim, M, 2015. Integrating PALSAR and ASTER data for mineral deposits exploration in tropical environments: a case study from Central Belt, Peninsular Malaysia. International Journal of Image and Data Fusion 6(2), 170-188.
[7] Eliason EM and McEwen AS 1990. Adaptive Box Filters for Removal of Random Noise from Digital Images. Photogrammetric Engineering & Remote Sensing, April, 1990, V56 No. 4, p. 453.
[8] Jensen JR 2005. Introductory Digital Image Processing: A remote sensing perspective. Pearson Prentice Hall, Upper Saddle River NJ 07458. Third edition, p: 276-287.

Mapping of post-event earthquake induced landslides in Sg. Mesilou using LiDAR

Habibah Hanan Mat Yusoff[1], Khamarrul Azahari Razak[1,2], Florence Yuen[1], Afifi Harun[3], Jasmi Talib[4], Zakaria Mohamad[5], Zamri Ramli[6], Razain Abd Razab[7]

[1]UTM RAZAK School of Engineering and Advanced Technology, Universiti Teknologi Malaysia, 54100 Jalan Sultan Yahya Petra, Kuala Lumpur, Malaysia; Tel: +6019-3649495; Fax: +603-26934844; khamarrul.kl@utm.my
[2]Disaster Preparedness and Prevention Center, Malaysia-Japan International Institute of Technology, Universiti Teknologi Malaysia, 54100 Jalan Sultan Yahya Petra, Kuala Lumpur, Malaysia;
[3]Faculty of Architecture, Planning & Surveying, Universiti Teknologi MARA,
[4]Faculty of Geosciences and Petroleum Engineering, Universiti Teknologi Petronas,
[5]Southeast Asia Disaster Prevention and Research Initiative (SEADPRI), Universiti Kebangsaan Malaysia, 43600 UKM Bangi, Selangor
[6] Technical Services Section, Minerals and Geoscience Department Malaysia, Ministry of Natural Resources and Environment, Malaysia, Malaysia
[7]BUMITOUCHplmc Sdn. Bhd., Suite 21, Level 21, Tower Block, Bangunan PERKIM, Jalan Ipoh, Kuala Lumpur,

Abstract. Earthquake is a common natural disaster in active tectonic regions. The disaster can induce cascading disasters such as debris flow, mudflow and reactivated old landslides. M 6.0 Ranau earthquake dated on June 05, 2015 coupling with intense and prolonged rainfall caused several mass movements such as debris flow, deep-seated and shallow landslides in Mesilou, Sabah. This study aims at providing a better insight into the use of advanced LiDAR mapping technology for recognizing landslide induced by earthquakes particularly in a vegetated terrain, assessing post event hazard and analyzing its distribution for hazard zonation. We developed the landslide inventory using LiDAR-derived visual analysis method and validated in the field. A landslide inventory map improved with the support of LiDAR derivative data. Finally, landslide inventory was analysed by emphasizing its distribution and density in such a way that it provides clues of risky zone as a result of debris flow. We recommend that mitigation action and risk reduction should be taken place at a transport zone of the channel compared to other zones. This study indicates that modern airborne LiDAR can be a good complementary tool for improving landslide inventory in a complex environment, and an effective tool for rapid regional hazard and risk assessment in the tropics.

1. Introduction

Earthquakes induce other major natural disasters such as landslide, tsunami, and volcanic eruption. These cascading hazards often resulted in large damages to infrastructure and even caused more injuries and casualties than the earthquake [1]. Earthquake-induced landslide is one of the common disasters that occurs in a tectonically active region and its complete inventory either at national or global scales remain elusive and even more challenging in the anthropogenically tropical region. For examples, M 7 earthquakes in Nepal triggered 5,600 landslides [2]. The Ms 8.0 Wenchuan earthquake, induced 257 landslide lakes which were distributed along the fault rupture zone and river channels [3]. The 1994 Northridge earthquake triggered more than 10,000 landslides in the hills around Los Angeles [4]. On May 18, 1980, a magnitude-5.1 earthquake triggered the collapse of the north flank of Mount St. Helens, resulting in the largest landslide ever recorded [4]. The number, type, and size of landslides varies depending on magnitude of the earthquake, type and depth of the ruptured fault.

The type, activity and process of earthquake-induced landslides have similarities and differences from the previous studies of the prominent earthquakes. The recent Nepal earthquakes induced new and reactivated landslides [2]. The most common types of earthquake-induced landslides during the Wenchuan earthquake were rock falls, rock avalanches, and rock debris slides in the area [5]. Deep-seated large landslides were moderately common and abundant in the vicinity of the fault rupture [5]. Landslides caused by the Chi-Chi earthquake were new and retrogressive landslides, which the landslides were not reactivated by the historical landslides [6]. These studies show that large earthquakes can reactivate old, dormant landslides and cause new landslides with different landslide types, mechanisms and activities. To understand the characteristics of earthquake-induced landslides, it is crucial to carefully map, identify and characterize landslides induced by earthquakes using appropriate landslide mapping techniques resulting with sufficient accuracy in a relatively short time.

Landslide mapping for earthquake-induced landslide consists of field mapping, remote sensing technique with the support of local- and expert knowledge. A combination of these techniques can produce a reliable landslide map. Passive and active remote sensing such as optical satellite images and LiDAR (Light Detection and Ranging) images, respectively have been widely used to map and identify landslide induced by earthquakes [7]. In Sabah, M 6.0 earthquake that ruptured on 5th June 2015 is the second large earthquake after 1991 earthquake in Ranau. The earthquake induced rock avalanche occurred on the summit of Mount. Kinabalu caused 18 fatalities [8]. After intense and prolonged rainfall, aftershocks, and tremors, a devastated debris flow occurred on 15th June 2015 in Sungai Mesilou, which caused a house, homestays and roads damages along the Sungai Mesilou area and indirectly affect the socio-economic activity of the local community [9].

2. Study area

The study area is located in Ranau district and it is situated in a well-known touristic area with attractive remarks such as the first World Heritage Site in Malaysia, Mount Kinabalu. Several touristic places and infrastructures in the vicinity of Kg. Mersilau are Mount Kinabalu Golf Club, Mesilou Nature Resort, and Desa Dairy Farm. The study area is situated between 500 to 2,000 m above mean sea level on the southern flank of Mount Kinabalu (4,101m) as shown in Figure 1. Since the study area is located along the main channel, therefore the geomorphology is dominated by floodplain, river terraces and plateau. The study area is described as plains surrounded by mountains. Dusun is the main tribe who resides in Mesilou, who are mostly work as farmers, which makes the major land cover is plantation area aside from recreational and tourism area [10].

The geology of the upper and lower part of the study area consists of igneous rock and sedimentary rock, respectively. The oldest igneous rock of the study area is gabbro and ultramafic, followed by granodiorite. The oldest sedimentary rock is Trusmadi Formation, Crocker Formation, and Pinosuk Gravel. Trusmadi Formation is described as strongly folded and faulted grey and dark grey argilite, slate, siltstone and sandstone with volcanics, whereas Crocker Formation is referred as strongly folded and faulted sandstone, silstone, red and grey shale, mudstone and argilite. Pinosuk Gravel is poorly consolidated unsorted gravel up to boulder size in a sandy to clayey matrix [11]. The dominant lithology in the study area is sedimentary rock, specifically Pinosuk Gravel as shown in Figure 1. The main active faults present in the study area are Mensaban Fault and Lobo-Lobo Fault segments. Mensaban Fault is a normal fault trending northwest-southest and west-east. Lobo-Lobo Fault is left-lateral strike slip fault, which trends N20°E [12].

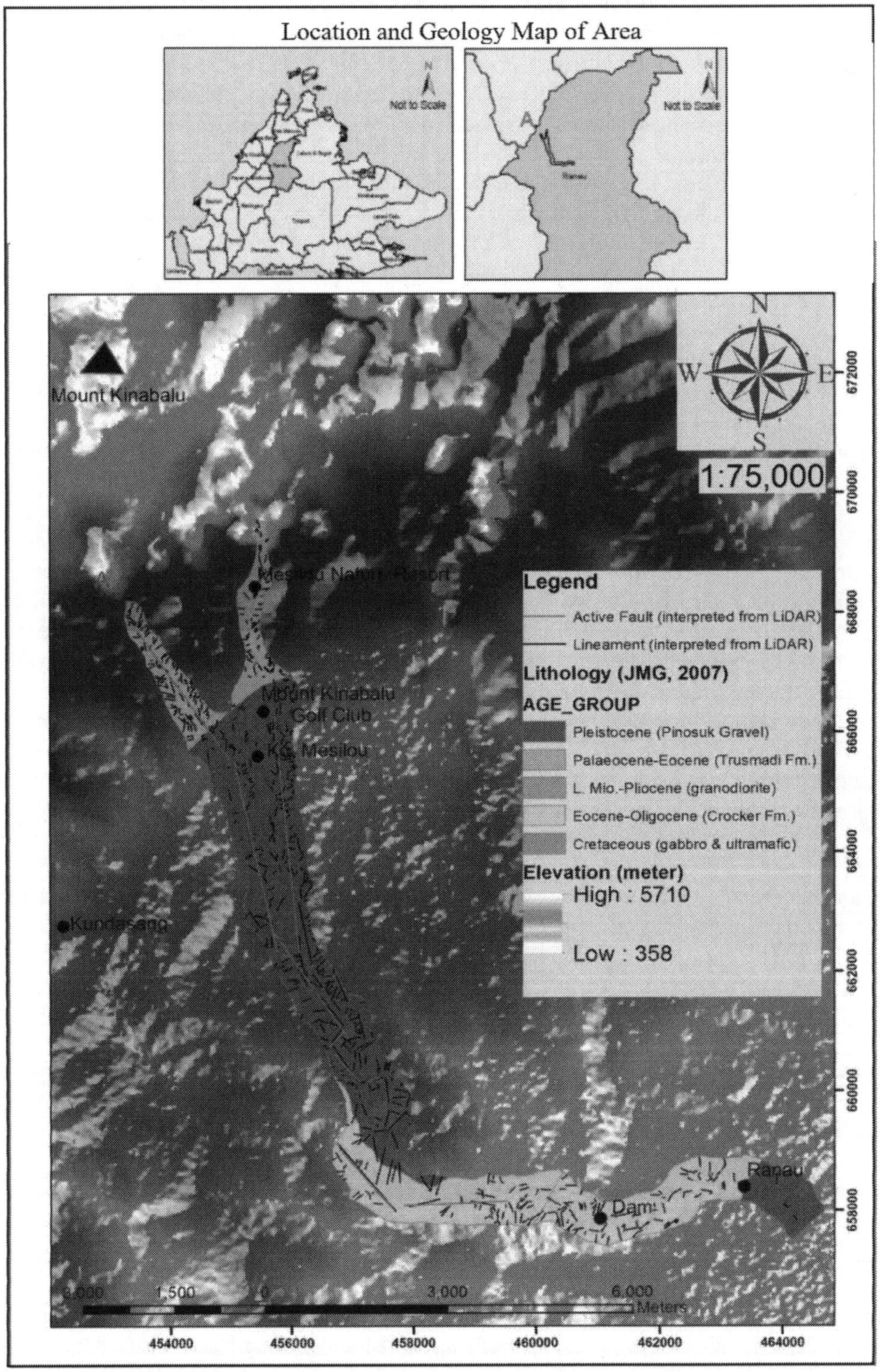

Figure 1. Location and geology map of the study area.

Mass movement investigation has been done in Mesilou area by monitoring Jalan Cinta Mata off Jalan Kundasang-Ranau because the road experienced constant damages of tension cracks, settling, and lateral movements [13]. Subsurface investigation and instrumentation indicated that the rate of ground movement is 21mm per week at maximum [13]. This high reading is caused by the massive creep movements of the post-glacial deposits and ancient mudflow [13]. Studies have been done on the origin of the huge granite boulders or Pinosuk Gravel lithology. The boulders make the dominant panorama in Mesilou. The earliest finding defines Pinosuk Gravel as piedmont fan from glaciation of Mount Kinabalu [14]. Later, the interpretation was improved by emphasizing the mechanism that brought down the boulder is mudflow by observing at the different sizes of boulders [15], [16]. Moreover, sources of boulders deposition were originated from four main channels, which are located at Desa Cattle, Mesilou Golf Club, Pusat Latihan Dakwah, and plantation area [10]. The latest work on Pinosuk Gravel is a proposal to gazette the area that are covered with this deposits as a geoheritage place due to its remarkable geological origin and history [10].

Previous landslide mapping has been done in Kundasang to identify the causal factors of the ground instability and creeping mass movement in the area. The extensive and detailed landslide mapping using conventional methods resulted in a landslide complex of Kundasang [17]. However, the study area did not include Mesilou. The previous studies in Mesilou area revealed that most of the research activities were focused on debris flow in Sungai Mesilou and none of the landslide mapping has been done yet in the study area except landslide monitoring at Kampung Mesilou by JMG. Landslide has been occurring in the area on 2008, 2010, and 2013 with one recorded fatality [18]. After earthquake and debris flow events on June 2015, Mesilou becomes attention because of the high disaster impact.

The main objective of this paper is to implement post event hazard and risk assessment by mapping landslide spatially and analysing the landslide density for hazard zonation utilizing LiDAR technology. The scope of works include pre-processing LiDAR datasets for georectification, generating LiDAR-derived images for image visualization, interpreting landslides in a systematic GIS environment, preparing basemaps for field verification, and finally analysing landslide inventory for supporting landslide hazard and risk assessment.

3. Methods

3.1. Airborne LiDAR campaign

The airborne LiDAR data was captured on August 2015 using RIEGL system mounted in a helicopter. The acquisation and processing of raw airborne LiDAR data was carried out by BUMITOUCHplmc Sdn Bhd. It is worth to mention that this airborne LiDAR 3D point clouds recorded up to 120 points per square meter; the highest point density reported over the disaster area in Malaysia. The processed LiDAR data consists of classified 3D point clouds, 0.25 m digital terrain model (DTM), 0.25 digital surface model (DTM), and 0.07 m orthophoto of the debris flow area. These datasets are very high resolution and the best quality datasets.

3.2 LiDAR processing for visualization

LiDAR-derived datasets are re-processed for visualization in two-dimensional and three-dimensional models using several open-source and commercial softwares. One of the common LiDAR-derivatives for visualization is a default hillshade (azimuth: 315, altitude: 45), which is based on solar illumination from a single direction. Hillshade image is overlaid with slope and curvature images in ArcGIS software for a better visualization and interpretation. In slope image, steep slope indicates scarp area and flatter slope can be indicated as debris area. In curvature image, concave slope indicates denudation and scarp area and convex slope indicates debris or accumulation zone.

A new mode of visualization is topographic openness generated from SAGA-GIS software that visualizes the topographic dominance or enclosure from digital elevation models (DEM). Positive openness signifies convex topography and negative openness shows concave topography [19]. Another new technique of visualizing a LiDAR-derived DTM is known as a color composite image, which can be produced using ILWIS software [20].This technique produces one color DTM into a combination of three shades of color (red, green, and blue), indicating the solar illumination of the image in three directions compared to the conventional hillshade. A 3D orthophoto is also used to identify recent landslide and observe the morphological- and disrupted drainage coupling with geoindicators in a forested terrain. These visualization techniques generate 2-D outputs for landslide interpretation. Another possibility of visualization is 3-D visualization generated from ILWIS by creating stereo pair images from DTM. The images are used to visualize landslide in 3D by highlighting the depth of landslide to differentiate between deep-seated and shallow landslides using 3-D glasses. These types of different visual are helpful to visualize landslide features for accurate interpretation.

3.3 Landslide interpretation using LiDAR-derived datasets

Landslide interpretation becomes easier with the help of image visualization and high resolution data. All the generated images are overlaid and the images are used alternately in GIS software for landslide interpretation. The landslide interpretation process was guided by a group of senior geologists and geomorphologists, who have been working in the area for several years and familiar with the landslide processes and activities in the study area.

In this study, we developed a landslide geodatabase consists of landslide ID, body, type, activity, speed of movement, and certainty. Body attribute consists of scarp area, accumulation area and undifferentiated. Landslide types attribute are divided into deep-seated, shallow, flow and rockfall and subdivided into translational, rotational, and complex landslide. Activity attribute contains active, stable, dormant, and stable. Speed of movement attribute consists of slow and fast, whereas certainty attribute has two options, which are certain or uncertain.

The first step of landslide interpretation is to identify geomorphology, geology, land covers, land uses, past landslide events, and topography of the study area to understand regional landslide process, causal and conditioning factors. Then, semi-circular or crown shape is delineated as a scarp. The shape varies depending on the type and activity of the landslide. Fresh, active, retrogressive landslide usually has semi-circular shape with tension crack at the scarp. Dormant and old landslide shows a perfect semi-circular shape. In certain cases for complex landslide, the scarp has incomplete semi-circular shape because it is overlapped and erased by the recent scarps. After recognizing the scarp, the next step is to identify the landslide accumulation area, which commonly signifies as hummocky and bowl-shaped feature. In non-forested and developed areas, the accumulation area is not clearly seen because the displaced material is often removed after the event, which makes delineating accumulation area is quite challenging.

3.4 Field observation

The field work was carried out after interpreting landslides for almost two weeks, the selected landslides were chosen based on the certainty attribute to be verified on the field. Fieldwork basemaps in A3 size were prepared by incorporating orthophoto and color composite overlaid with interpreted landslides. Field verification was accompanied by local expert from the Minerals and Geoscience Department Malaysia (JMG) and other senior geologists.

Landslide interpretation using remote sensing images is done before and after the field mapping to improve the landslide boundary mapping. Landslide validation through field mapping is useful to understand and visualize landslide in real correlated to remote sensing images. In addition, geological and geotechnical information, which are investigated during field mapping, can be a complement to landslide inventory. Local and expert knowledge is essential for landslide mapping because these people witnessed the event. From these sources of information, a reliable landslide map can be produced. This map is important as an input for hazard and risk assessment. In addition, the map also can be utilized for post-event mitigation to identify possible avalanches and landslides.

3.5 Landslide inventory analysis
The interpreted, verified, and improved landslides are systematically stored in a GIS geodatabase environment. A landslide distribution, density and multi-temporal analysis can be derived from the database. Remarkably, the landslide inventory maps can be further utilized for analysing landslide susceptibility, assessing landslide hazard and associated risk. Landslide density map is produced by converting the interpreted landslides from polygon to point shape file. Then, the point density is classified into five classes (high, medium, low, very low and nil density) by applying 1 km^2 buffer.

Another important analysis is analysing multi temporal landslide from satellite images before and after the debris flow. Worldview image in 2011 is acquired to compare with the orthophoto to observe the condition of the disconnected road before and after the disaster. SPOT image in 2010 also used to compare multi temporal landslides. The selected area is located in Kampung Mesilou, where the landslide occurred in March 2008, March 2010, and July 2013 [18]. These observation are important for land planners and engineers to identify element-at-risks before planning and construction.

4. Results
LiDAR data covers 3m to 4m river banks along Sungai Mesilou. The upstream of study area is started at the Mesilou Nature Park and ended at Ranau town. The length of the selected Sungai Mesilou section is approximately 7 km. The main finding of this study is the first earthquake-induced landslide inventory created from LiDAR-derived landslide interpretation supported with field verification. The inventory is further analysed to generate landslide density map and relate with element-at-risk at the study area by using multi temporal images.

4.1. Image visualization from LiDAR-derived datasets
The initial step before landslide interpretation is image visualization, which consists of hillshade image (Figure 2a), color composite image (Figure 2b), and topographic openness image (Figure 2c). These layers are used alternately to visualize and identify landslide morphology such as semi-circular scarp, concave-convex, and hummocky topography. These images have their own advantages and disadvantages. A color composite image has better visualization than hillshade image because it brightens the shaded area in various perspectives. Topographic openness images emboss the concavity of the topography, whereas landslide scarp can be easily identified as shown in Figure 2. Visualizing landslide in various ways is an added value that only can be obtained from LiDAR datasets, which significantly helps the landslide interpreters.

4.2. Landslide interpretation diagnostic features
Landslide interpretation was carried out by recognizing the geomorphologic signatures to deduce type of landslides. Coupling with color composite DTM images, an orthophoto can be used to identify landslides by observing vegetation pattern, river course, and anthropogenic features. Tree density and pattern such as at the scarp and accumulation area are quite distinctive. Scarp area usually has tall and greener trees, whereas accumulation area shorter and less greener vegetation that indicating of re-growing process. Fresh landslide can easily be seen in the orthophoto as a brown patch. For high resolution orthophoto, tilted and backtilted trees can be seen as well. Thus, there are multiple and

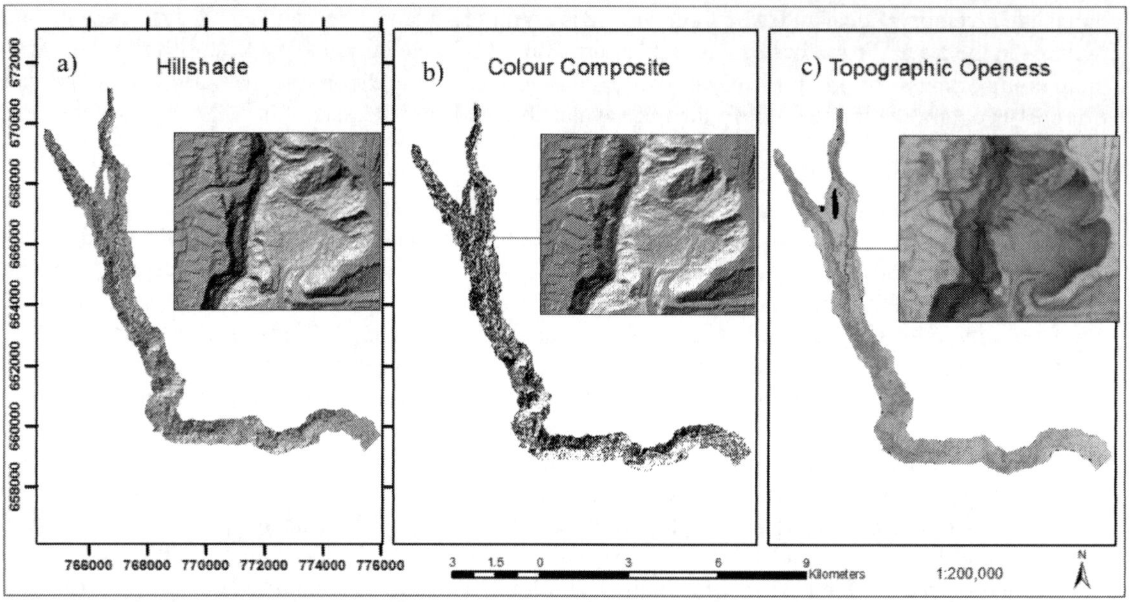

Figure 2. LiDAR-derived image visualization consists of (a) hillshade (b) color composite (c) topographic openness of debris flow channel in Mersilau, Sabah.

various methods to visualize and interpret landslide using LiDAR-derived datasets and the approach varies depending on the availability and resolution of the data. The diagnostic features as mentioned in Figure 3 is used not only to identify landslides but most importantly to infer landslide process, activity, type for further analysis. For examples, figure 3a is an earthquake-induced landslide at the Golf Club because this landslide presents after the M 6.0 earthquake. The type of landslide is rotational and retrogressive landslide characterized by semi-circular scarp, step-like slope, and hummocky deposition part. Backtilted trees can be seen on the orthophoto. Figure 3b is an active landslide at Kampung Mesilou. It is characterized by concave-convex morphology. Figure 3c is landslide complex and deep-seated rotational due to existing of multiple series of landslide events. Disrupted plantation terrace can be seen on 3c(i). The recent scarp is induced by the 5 June earthquake. Based on figure 3d, there is complex landslide in a forested terrain at Kampung Mesilou. Multiple scarps presents showing that the landslide is active. The recent scarp has the semi-circular shape but the older scarp does not have the perfect semi-circular shape. The multiple scarps shows that the landslide is active. The deposition part has hummocky and flow feature. The identified signatures are later compiled for field verification.

4.3. Field observation

The topography of the upper of study area is quite mountainous with steep terrain and the lower part is relatively flat and low-lying terrain. In general, the upper part of Sungai Mesilou is a straight river and the lower part of the river is a meandering river due to the presence of floodplain in the lower part compared to upper part of the river. The straight channel are related to the presence of active faults along and across the upper part of the river. The other dominant geomorphology of the study area are terraces, which can be observed along the river, and plateau at the right side of the river. The observation was made along the channel of the debris flow that occurred on 15th June 2015 in Sungai Mesilou. The event caused a house, homestays and roads damages along the Sungai Mesilou area. Based on field observation from the source zone, transport zone, and deposition zone, the rock boulders that filling up the channel are huge and the sizes are getting bigger towards deposition zone as shown in Figure 6. Another important observation is the recent debris flow event was not the first event from the evidence of older layer of rock boulders along the channel. The similar observation has been made at upstream of

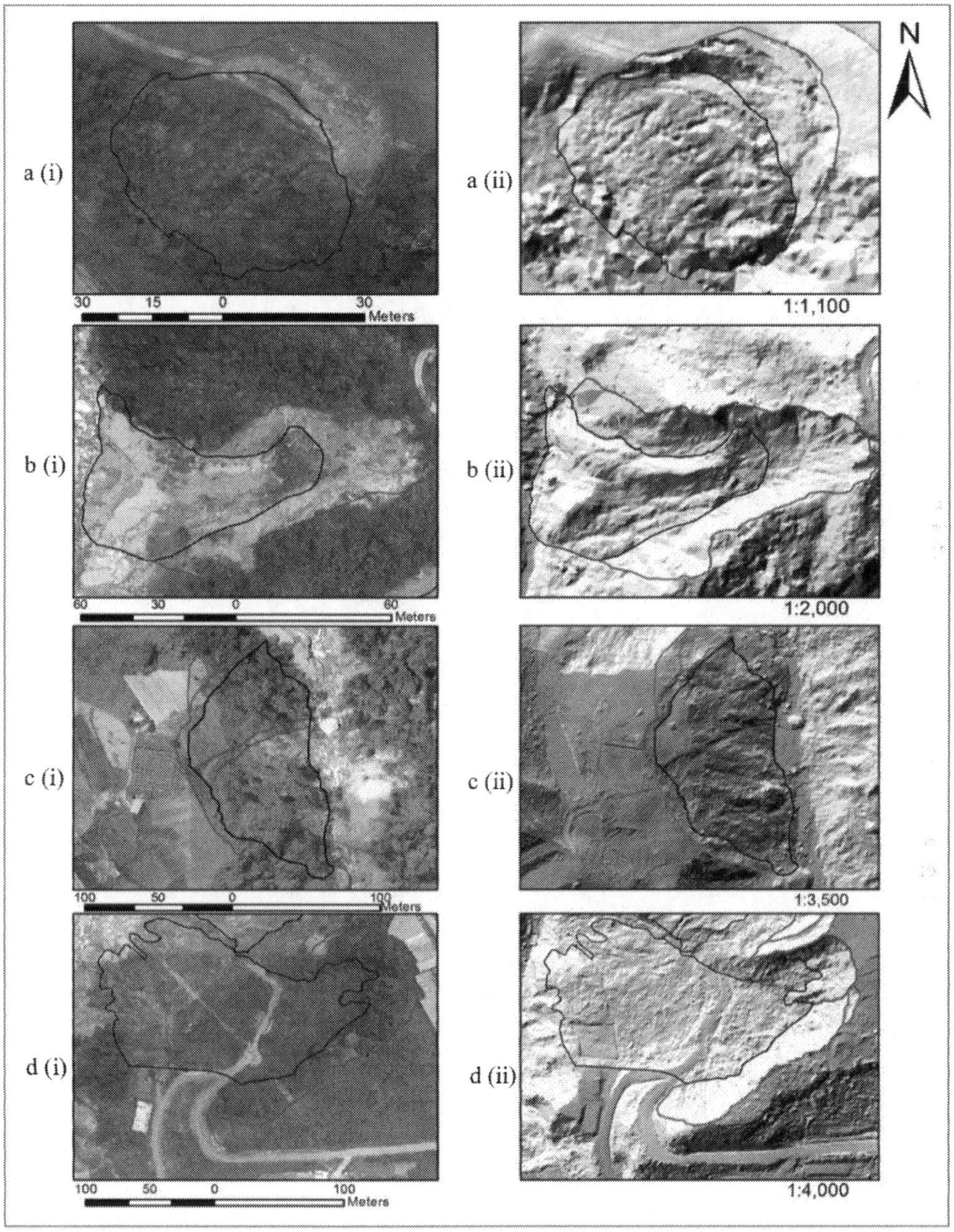

Figure 3. Landslide diagnostic features along Sungai Mesilou. 0.07 m orthophoto (left) and color composite DTM (right) are used to recognize landslide geomorphology. The landslide parts are scarp (red line) and accumulation zone (black line).

Sungai Mesilou as different sizes of granite boulders and weathered granite choked the main river [21]. The author added that the boulders are the deposits from Mount Kinabalu's summit carried down by the river over thousands of years [21]. This observation reveals the possibility of future debris flow event in Sungai Mesilou and in general, the possibility of landslides to occur again at the same location in future. Thus, producing a landslide map of debris flow area is important to identify new, reactivated and historical landslides that were induced after the earthquake and debris flow event. About 70% of the interpreted landslides were verified in the field. The landslide boundary, type, activity and process are improved after getting input from local and expert knowledge and visualizing on the field as shown in Figure 4 and 5.

Figure 4. Field verification at Kampung Mesilou looking at landslides (see Figure 3b and 3d (right) and debris flow channel (left).

Figure 5. Field verification of a landslide (Figure 3c) indicating straight line of scarp by observing vegetation pattern and sudden drop of plantation area.

Figure 6. Debris flow observation at the source area (A), transport area (B), and deposition area (C).

4.4. Landslide inventory analysis

A final landslide inventory is systematically stored in ArcGIS geodatabase system and overlaid with multi temporal satellite images to observe the land use changes and the impact of the disaster. Along the debris flow channel, five disconnected roads and bridges are observed and the event significantly stripped off the vegetation along the channel as shown in Figure 7. Another impact of landslides towards land use changes is landslides at Kampung Mesilou. Figure 8 shows images of before- and after the landslides event. The landslides damages the roads and plantation area and caused road to be re-routed. Multi temporal images reveal the important of landslide inventory using LiDAR for land-use planning to avoid future damages and monetary loss. In addition, landslide inventory also can generate landslide density map to identify the distribution and intensity of landslides. Based on Figure 9, the landslide has high density at the transport zone, low to medium density at source zone, and no density at deposition zone. This finding correlated with the understanding of river incision, which it usually highly incises the sediment at the transport zone and deposits the sediment at the deposition zone.

5. Discussion

Landslide mapping using both techniques of remote sensing technique and field mapping is an efficient way of creating and updating an event-based landslide inventory. Landslide interpretation technique using different methods of landslide image visualization is helpful compared to the common hillshade image. High resolution LiDAR also helps to interpret landslides accurately by observing smaller geomorphologic features and differentiating source and deposition zone compared to 30m Aster GDEM and 10m IFSAR images. Interpreting landslides in a digital form is more efficient than done in a printed material because landslides and their specification can be visualized in multi scales and interactively. Thus, LiDAR is the best and superior technology for landslide recognition.

Landslide interpretation using LiDAR-derived datasets made possible to map and characterize recent landslide induced by earthquake, dormant, and historical landslides. The datasets show not only morphology features but other small-scale geodynamic features such as disrupted drainages and cracked- and displaced roads. This result indicates the possibility to develop other methods of detecting landslides in an objective manner and consider the vegetation anomaly. Bringing LiDAR into the field as a primary field material is quite challenging. This study proved that the LiDAR data was practically beneficial and can be directly assisted the field investigators Interpreting landslide using LiDAR is easier than verifying on the field because of different scales and physical constraints on the field such as forested area, plantation, and buildings. Identifying boundary for large landslide is quite difficult compared to small and new landslide. However, this problem can be avoided by viewing the landslides

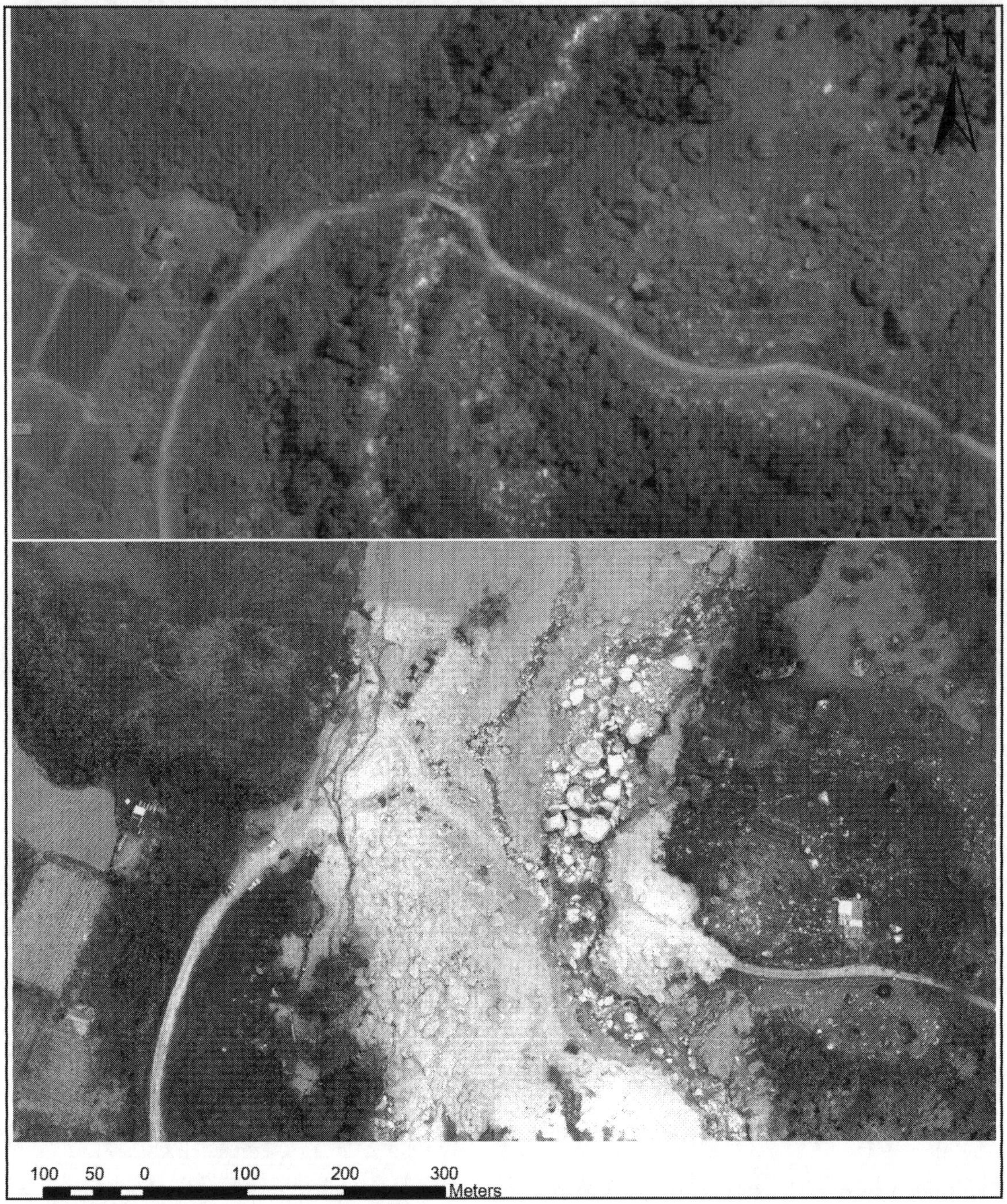

Figure 7. Multi temporal images disconnected roads before debris flow from Worldview image in 2011 (top) and after debris flow from orthophoto in 2015 (below). 5m of road was displaced after the debris flow event.

IOP Conf. Series: Earth and Environmental Science 37 (2016) 012068 doi:10.1088/1755-1315/37/1/012068

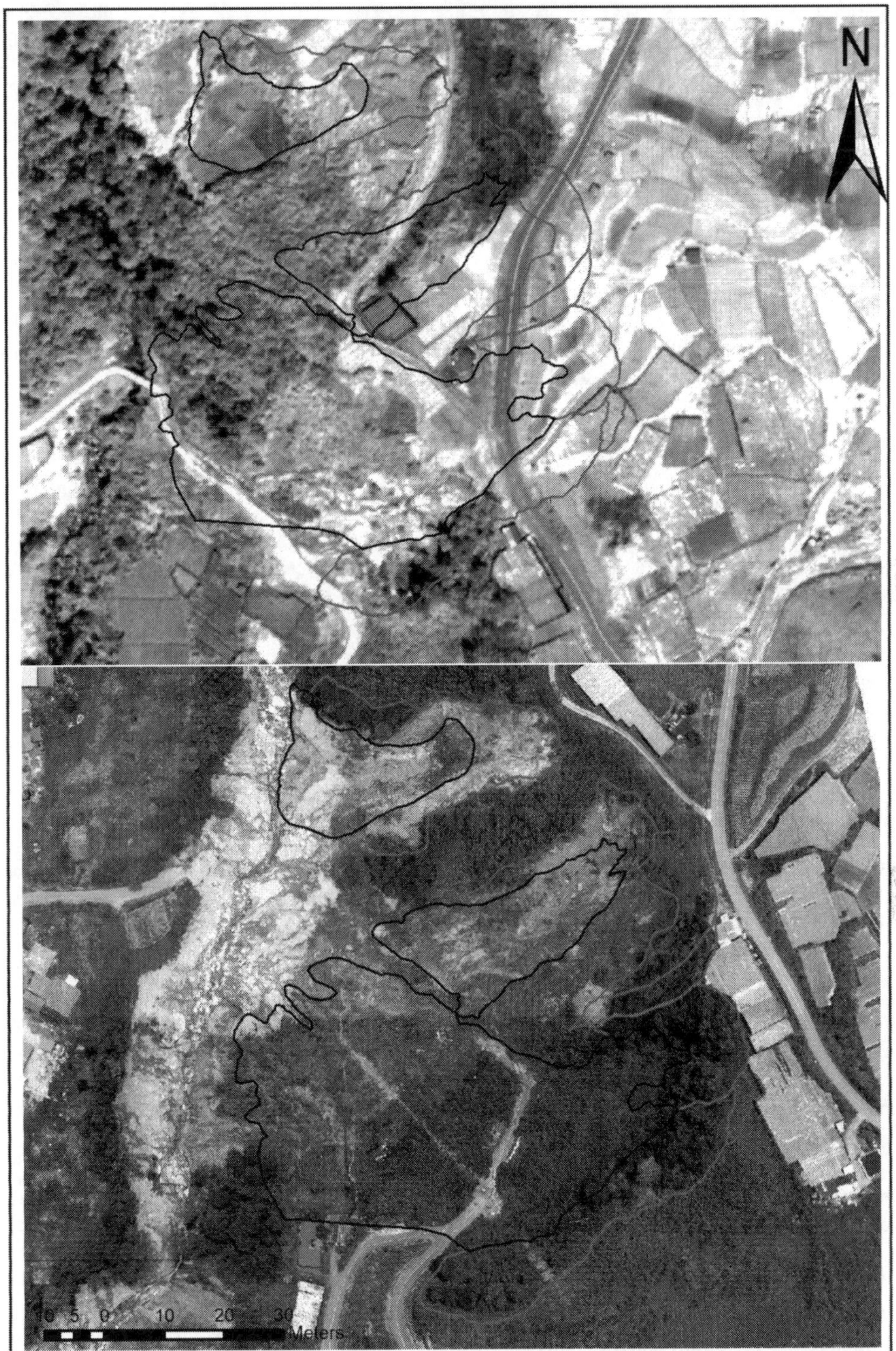

Figure 8. Scale 1: 1,400. Multi temporal images of SPOT (2010) and orthophoto (2015) of landslides in Kampung Mesilou. A very distinct and fast landscape changes within 5 years.

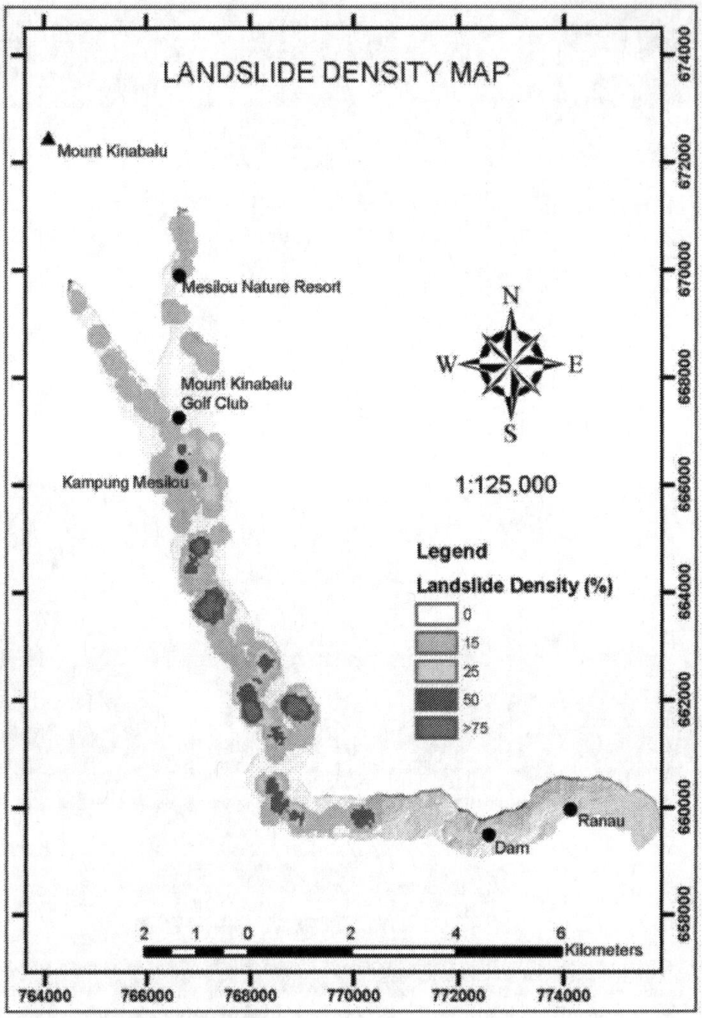

Figure 9. Landslide density map as a result of LiDAR-derived data validated in the field.

at distant and high place. Going to the field with experts and locals are quite helpful in guiding and understanding the past landslide events and regional geology.

Based on landslide density map, high density of landslide is concentrated in a transport zone of debris flow channel compared to source and deposition zones. The density map also reveals that the high landslide density are located in active river incision areas, fault zone, and characterized by Pinosuk Gravel lithology, which describes that the landslide activity along the Sungai Mesilou is mainly controlled by geological factors. Disconnected roads and damages houses can be seen at the transport zones compared to deposition zone. As a result of field observation, we found that transport zone is In terms of hazard and risk assessment, transport zone should be put in red zone for mitigation actions because there is possibility for debris flow to occur again at the same channel.

6. Conclusion

This paper presents an inventory analysis of landslides induced by earthquake in a tectonically active region in Kundasang (Sabah, Malaysia). It is a new landslide record and compiled based on the recent data relatively captured after the 6.0 Mw earthquake on June 5, 2016. The landslide inventory was largely developed based on high resolution airborne LiDAR data coupled with field investigation.

Multi-scale LiDAR data have revolutionized the analysis and assessment of hillslope-, and tectonic geomorphology in a tropical environment. An intensive 3D point cloud- and image processing with stringent analysis routine made the LiDAR technology and massive geospatial data worth to invest and obtain. Landslide diagnostic feature is one of the indicators to landslide identification but interpretation still depends on the study area, geology, and topography. Field observation becomes easier with the help of LiDAR because it directs the field investigators into the accurate location with relatively detailed characteristics of mapped landslides and their precursors at different working scales. A combination of the aforementioned methods produce the first reported landslide inventory after the 5th June earthquake by taken several considerations, which are the map utilized the best resolution for LiDAR, recorded landslide causal factor, and the level of hazard, as indicated in the field is very high because of the existing debris and rock boulders. The possibility of constructing new roads, bridges and other infrastructures is very challenging as all the infrastructures, e.g. disconnected roads should be carefully investigated and monitored. Landuse planning for the surrounding area along Sungai Mesilou also should be taken into consideration. These mitigation actions show the usefulness of landslide inventory map helps hazard and risk assessment.

Acknowledgement

This research is financially supported by the Research University Grant (GUP-05H66), Universiti Teknologi Malaysia (UTM) and Fundamental Research Grant Scheme (FRGS-4F480), Ministry of Higher Education Malaysia. The author would like to thank BUMITOUCHplmc Sdn. Bhd for sharing the best LiDAR quality data for research purpose. The author also would like to convey appreciation to geologists from JMG for the knowledge sharing and all researchers from Cascading Geo-Hazard Research Initiative team for the helps throughout the Scientific Expedition UTM @ Kundasang Sabah 2015, November 21-23, 2015.

References

[1] Fagel MJ. *Principles of Emergency Management: Hazard Specific Issues and Mitigation Strategies*. CRC Press; 2011.

[2] Alex Densmore, Tom Dijkstra, Colm Jordan, Vanessa Banks, Nick Rosser, John Rees GJ and JW. *UPDATED (30 June) landslide inventory following 25 April and 12 May Nepal earthquakes*. June, 30 2015. http://ewf.nerc.ac.uk/2015/06/30/updated-30-june-landslide-inventory-following-25-april-and-12-may-nepal-earthquakes/.

[3] Cui P, Zhu Y yan, Han Y shun, Chen X qing, Zhuang J qi. *The 12 May Wenchuan earthquake-induced landslide lakes: Distribution and preliminary risk evaluation*. Landslides 2009;6:209–23. doi:10.1007/s10346-009-0160-9.

[4] Diane Noserale TL. *Earthquakes that Trigger Other Natural Hazards* 2006:16. http://www.usgs.gov/homepage/science_features/docs/plw1906/chain_reaction.pdf.

[5] Gorum T, Fan X, van Westen CJ, Huang RQ, Xu Q, Tang C, et al. *Distribution pattern of earthquake-induced landslides triggered by the 12 May 2008 Wenchuan earthquake*. Geomorphology 2011;133:152–67. doi:10.1016/j.geomorph.2010.12.030.

[6] Chigira M, Wang WN, Furuya T, Kamai T. *Geological causes and geomorphological precursors of the Tsaoling landslide triggered by the 1999 Chi-Chi earthquake, Taiwan*. Eng Geol 2003;68:259–73. doi:10.1016/S0013-7952(02)00232-6.

[7] Mat Yusoff HH, Razak KA. *Different Methods of Landslide Mapping in Cameron Highlands*.

Malaysian J Remote Sens GIS 2015;4:85–91.
[8] USGS. *M6.0 - 14km WNW of Ranau, Malaysia 2015.* http://earthquake.usgs.gov/earthquakes/eventpage/us20002m5s#general_summary.
[9] Goh R. *A house collapsed into Sungai Mersilau as it widened following heavy rain*. NST 2015. http://www.nst.com.my/news/2015/09/house-collapsed-sungai-mersilau-it-widened-following-heavy-rain?m=1.
[10] Sarman M, Komoo I. *Kundasang-Ranau : Dataran Warisan Ais Gunung*. Geol. Soc. Malaysia Annu. Geol. Conf., 2000.
[11] JMG. *Geological Terrain Mapping of the Kundasang Area, Sabah*. 2003.
[12] Tjia HD. *Kundasang (Sabah) at the intersection of regional fault zones of Quaternary age*. Geol Soc Malaysia 2007;53:59–66.
[13] Liew SS, Gue SS. *Massive Creep Movements of Post-Glacial Deposits in Kundasang Areas*. GSM-IEM Forum "Engineering Geol. Geotech. Slopes," 2001, p. 13.
[14] Collenette P. *The geology and mineral resources of the Jesselton-Kinabalu area, North Borneo*. Geological Survey of Malaysia, Borneo Region Memoir; 1958.
[15] Tjia HD. *Accretion tectonics in Sabah: Kinabalu Suture and East Sabah accreted terrane*. Bull Geol Soc Malaysia 1988;22:237–51.
[16] Kong LT. *Bouldery Mudftow Deposit at Ranau, Sabah, East Malaysia*. 1970:139–46.
[17] Komoo I, Lim C-S. *The Kundasang Landslides Complex : A detailed mapping of the Kundasang National Secondary School area*. Bull Geol Soc Malaysia 2003;46:387–92.
[18] JMG. *Laporan Awal Kejadian Tanah Runtuh di Kg. Mesilau, Kundasang, Ranau*. 2013.
[19] Yokoyama R, Shirasawa M, Pike RJ. *Visualizing Topography by Openness : A New Application of Image Processing to Digital Elevation Models*. Photogramm Eng Remote Sens 2002;68:257–65.
[20] Razak KA, Straatsma MW, van Westen CJ, Malet JP, de Jong SM. *Airborne laser scanning of forested landslides characterization: Terrain model quality and visualization*. Geomorphology 2011;126:186–200. doi:10.1016/j.geomorph.2010.11.003.
[21] Hall R, Cottam MA, Suggate SM, Tongkul F, Sperber C, Batt G. *The geology of Mount Kinabalu*. Sabah Park Publ 2009:76.

Estimation of Peak Ground Acceleration (PGA) for Peninsular Malaysia using geospatial approach

Amir Nouri Manafizad, Biswajeet Pradhan* **and Saleh Abdullahi**

Department of Civil Engineering, Faculty of Engineering, University Putra Malaysia, 43400, UPM, Serdang, Selangor, Malaysia

Corresponding Address: Biswajeet Pradhan, Tel: +603-89466383,
Fax. +603-86567129; Email: biswajeet24@gmail.com or biswajeet@lycos.com

Abstract. Among the various types of natural disasters, earthquake is considered as one of the most destructive events which impose a great amount of human fatalities and economic losses. Visualization of earthquake events and estimation of peak ground motions provides a strong tool for scientists and authorities to predict and mitigate the aftereffects of earthquakes. In addition it is useful for some businesses like insurance companies to evaluate the amount of investing risk. Although Peninsular Malaysian is situated in the stable part of Sunda plate, it is seismically influenced by very active earthquake sources of Sumatra's fault and subduction zones. This study modelled the seismic zones and estimates maximum credible earthquake (MCE) based on classified data for period 1900 to 2014. The deterministic approach was implemented for the analysis. Attenuation equations were used for two zones. Results show that, the PGA produced from subduction zone is from 2-64 (gal) and from the fault zone varies from 1-191(gal). In addition, the PGA generated from fault zone is more critical than subduction zone for selected seismic model.

1 Introduction

Natural hazards' management is one of the most important applications of remote sensing and GIS technologies [1-3]. GIS as the technology of gathering, combining, analysis and manifesting spatial data [4], has a very significant role in improving the quality of studies related to natural hazards and possibility of prediction, management and mitigation of their disastrous consequences. Collecting and classification of historical data about a disaster makes it possible to have a relatively realistic estimation of its behavior and rhythm of occurrence [5]. Especially in developed countries, by utilization of modern communication technologies, GIS are used in the real-time manner for managing natural disasters.

Furthermore, modelling of calamities and evaluating of their risks provide a very strong tool for decision-makers to find the best options for land use and development of infrastructures. These models are beneficial for some businesses like insurance companies to evaluate and estimate the amount of risk in investments.

Among natural disasters, earthquake is considered as one of the most destructive and devastating phenomenon which seriously threatens the life of people, properties and infrastructures. This fact makes it a necessity to investigate various regions of the world to gain a scientific understanding about the mechanism of occurrence and recurrence of it. There are nearly daily reports about earthquakes in different parts of the world, but until now there is no way to predict the exact location and intensity of the next event. As earthquakes imposes a great deal of damages to societies, seismologists and engineers concentrate on studies to find out the rhythm of occurrence of seismic events and seismic hazard analysis to apply the results in construction standards and codes in order to mitigate the rate of losses related to earthquakes.

An earthquake is an amount of energy which is released because of crash and displacement of rocks under the surface of the ground. Although the rupture is usually originated tens of kilometres under the ground level, at the surface of the earth; earthquakes manifest themselves by shaking and displacing the ground [6]. In case the epicentre is located at offshore, the earthquake at seabed shows itself as a tsunami. The occurrence of earthquake can intensify happening of other natural disasters like landslides and occasionally volcanic activities [2].

It has been recognized that urban areas located at large distances from tectonic plate margins (broadly categorised as low- to moderate-seismicity regions), may also be affected by earthquake tremors [7]. This is explained as the "Bowl of Jelly" phenomenon which was exemplified in occurrence of a huge devastation in Mexico City in year 1985 resulted from the long period component of the shear wave of a distant earthquake (approximately 350 km far from the city). The mechanism of this type of energy transfer is illustrated in Figure 1.

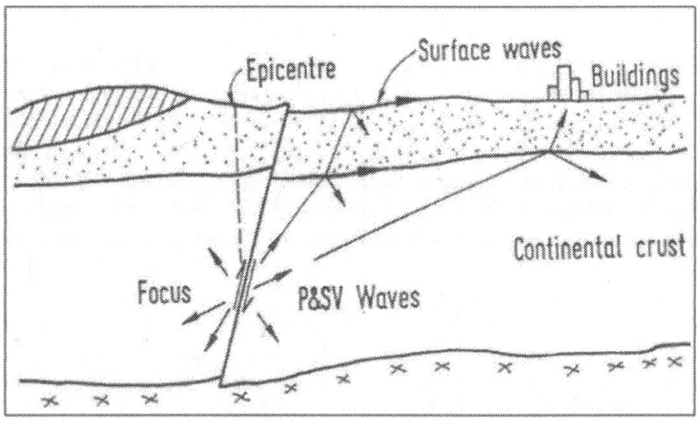

Figure 1. Schematic diagram for far-field effects of earthquakes [8].

Although Peninsular Malaysia is located in the stable Sunda Shelf with low to medium seismic activity level, the seismic risk, in terms of damage potential (to structures, loss of lives, assets, properties, businesses and services, as well as its psychological and cultural impacts), cannot be ignored since there is not a national code for building and structures for seismic loads.

Tremors due to Sumatra earthquakes have been reported several times [9]. Recent great magnitude earthquakes (Mw=9.1) which occurred in this area have resulted in extreme damages and death of lots of people. Especially 2004 Aceh earthquake was notably devastating, and was one of the largest to be recorded in the history. Fortunately, the rupture was not directed towards Malaysian territories, thus the level of ground shake on Malaysian grounds was insignificant [10]. Based on the findings of Bolt & Dreger [11] the directivity of rupture can intensify the ground acceleration more than 10 times. According to MMD [12], between years 1977 and 2014 more than 100 tremors have been felt in Peninsular Malaysia.

These findings besides to considering the rapid development of civic texture and construction of high-rise buildings and various infrastructures during recent decades make the necessity to study seismic events in this area. These studies usually are based on two stages. The first stage is estimation of peak ground movements using proper attenuation equations which is the subject of this research. The second stage (known as "Microzonation"), focuses on investigation of the impact of the type of local soil and other geological and geomorphological specifications of a site. The main aim of this stage is to achieve a very accurate estimation about the amount of earthquake waves' intensity that reaches to the surface of the ground and influences constructions and infrastructures.

Although there have been many attempts to investigate the issue of calculating PGA in this region during recent years, it is required to deal with it from different viewpoints and implementing various methods. Furthermore these types of research need to be updated for recently happened seismic events. Therefore, the objectives of this study are; to visualize important attributes of seismic events in the region; and to calculate PGA and prepare hazard maps based on deterministic method by using proper attenuation equations for two effective zones and evaluating the results.

2 Literature review

Many researches have been conducted by utilizing of GIS for manifesting spatial features combined with their various attributes as well as data integration platform [3, 13, 14]. Relating to seismicity, several application and tools for the analysis of geotechnical data and hazard mapping exist in a GIS platform. Furthermore, most GIS communicate easily with external modelling and simulation programs that are essential for seismic analysis. The results of an external analysis can be used by GIS as both graphic and non-graphic data for further interpretation and analysis. With these wide areas of application, GIS play a unique role for hazard preparedness and management.

Generally hazard mapping analysis is implemented in two macro and micro levels. Macro level which is the subject of this research, reports some general seismic characteristics of zones and sites like PGA, without considering the detailed specifications of the site under study. For this level the application of GIS is compiling the seismic data containing earthquake catalogues and maps of seismic sources which are compiled with site map and analysis.

Seismic micro-zonation is subdividing a region into smaller areas having different potential for hazardous earthquake effects. The earthquake effects depend on ground geomorphological attributes consisting of geological, geomorphology and geotechnical information. The parameters of geology and geomorphology, soil coverage/thickness, and rock outcrop/depth are some of the important geomorphological attributes. Other attributes are the earthquake parameters, which are estimated by hazard analysis and effects of local soil for a hazard (local site response for an earthquake). The Peak Ground Acceleration (PGA), amplification/site response, predominant frequency, liquefaction and landslide due to earthquakes are some of the important seismological attributes. These datasets are compiled and analysed to generate thematic maps based on assigning proper weightages to attributes. The Earth's outermost part (called the lithosphere) is not one continuous shell, but consists of several large (and some small) stable rock slabs adjoining each other which called tectonic plates. The way these plates interact with each other, provides a geological model that properly explains the occurrence of the majority of earthquakes. Each plate extends to a depth of about 100-200 km and includes the Earth's outermost rigid rocky layer, called the crust. The moving tectonic plates of the Earth's surface also provide an explanation of the various mechanisms of most significant earthquakes [11]. The collision of neighbouring plates that results in subduction of one plate beneath the other is the cause of fracturing of regional crustal rocks. Figure 2 provides a schematic global view of the arrangement of these plates.

The seismologic events occurring in these boundary regions are named "Inter-plate Earthquakes". The most devastating events all around the world that happened in shallow depths such as Chile, Peru, the eastern Caribbean, Central America, Southern Mexico, California, Southern Alaska, the Aleutians the Kuriles, Japan, Taiwan, the Philippines, Indonesia, New Zealand, the Alpine-Caucasian-Himalayan belt are of plate-edge type. The other sources to generate earthquakes are "Strike-Slip" or Transform Faults along them plates slide past each other.

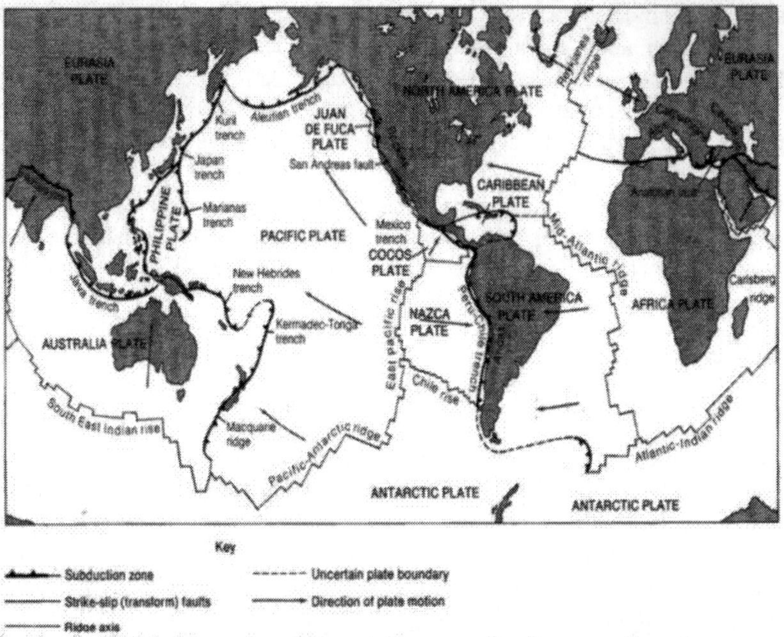

Figure 2. The major tectonic plates, mid-oceanic ridges, trenches and transform faults (Source: Bruce A. Bolt, Nuclear Explosions And Earthquakes: 1976).

However the plate tectonics theory explains the great part of seismic activities, some destructive earthquakes happen very far from plate boundaries within continents which are named" Intra-plate earthquakes". The general concept of how the earthquake sources interrelate with each other is demonstrated in Figure 3.

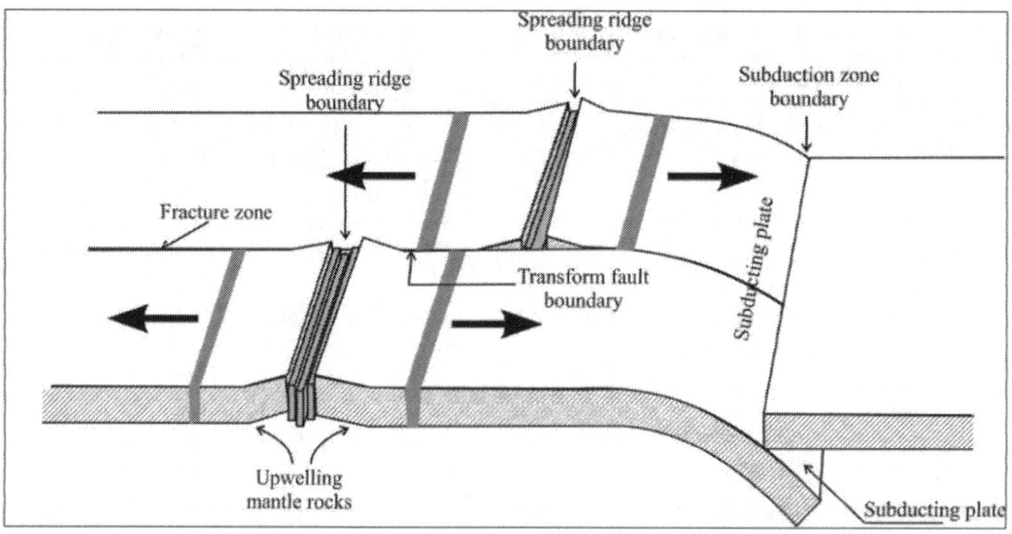

Figure 3. Interrelationship among spreading ridge, subduction zone, and transform fault plate boundaries [6].

Southeast Asia is a region of variable seismic hazard, ranging from high seismic hazard associated with the subduction process beneath the Indonesian and Philippine archipelagos to moderately low seismic hazard across a large stable region that contains the Peninsular Malaysia [15]. The concise introduction to the tectonic setting of the region is presented in Figure 4. It can be seen that three tectonic plates, Indian-Australian, Eurasian and Pacific plates are meeting each other in this region and the amount and the direction of convergence have been displayed as well. In addition to illustration of the tectonic texture of the region, the extent of active volcanoes as another source of earthquakes and the more earthquake-prone part of Malaysia are presented.

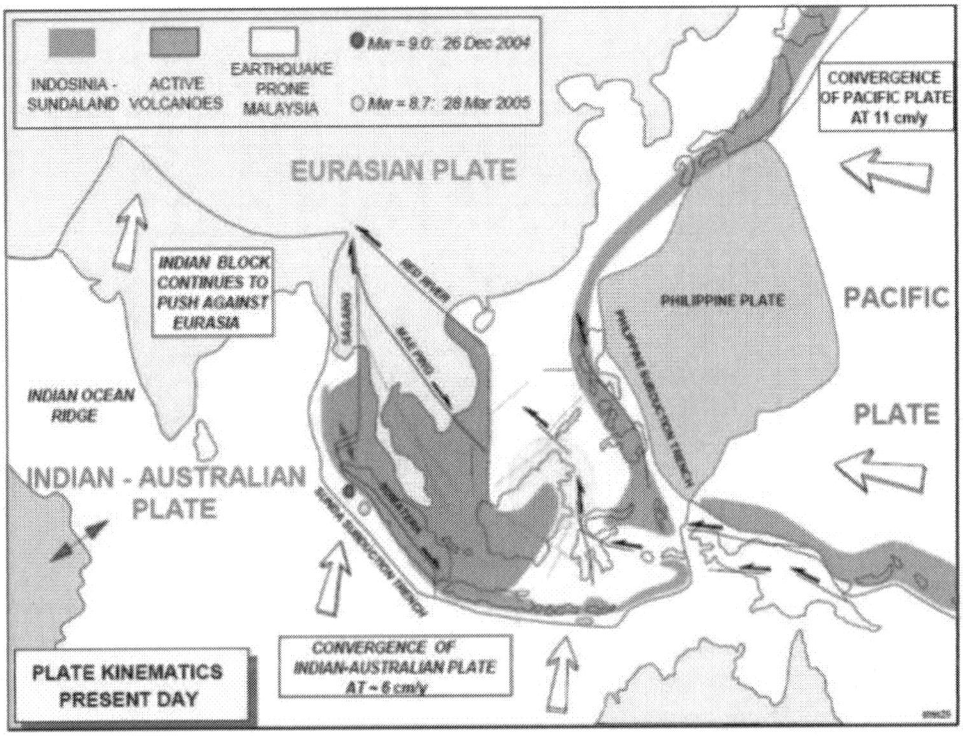

Figure 4. Earthquake-prone region of Malaysia [16].

In general, Malaysia is taken into account as a low- seismic activity country because of the fact it is situated on the stable part of Sunda crust and the nearest active source is located at least 350 km far from it. Until recently (before the occurrence of 2004 Sumatran giant earthquake), it was widely believed that West Malaysia will not be shacked because of activity of seismic sources inside the territory. According to Rosaidi [17] no earthquake had been originated from the area, although the flooding of the Kenyir Dam in Terengganu during 1984-1987 created some seismic activity (maximum magnitude was 4.6 on the Richter scale).

Considering the findings of Marto et al., [18] there have been records of more than 32 events that were originated from local sources during recent years. Although the magnitudes of these earthquakes are very small and there is not an official report and catalogue to refer, it seems inevitable to consider the reactivation phenomenon of inactive faults because of reformation in the core of the Sunda-land. According to Jeffery Chiang, the head of the group working on designing national building code in Malaysia [18], near field earthquake especially within the vicinity of 80km-long Bentong Fault should be given considerable attention.

Although East Malaysia and especially Sabah is tectonically more active and it is named earthquake-prone area of Malaysia, Peninsular has experienced some earthquakes as well. According to MMD [12], Peninsular Malaysia has felt tremors from local origin earthquakes (e.g. Bukit Tinggi Earthquakes, Kuala Pilah Earthquakes, Manjung Earthquake, Jerantut Earthquake and Terengganu Earthquake) and some of these local events had caused considerable damages to poorly built or badly designed building in the area. Table 1 depicts the maximum intensity and frequency of recorded earthquakes in various states of Malaysia.

Table 1. Earthquake Intensity recorded in Malaysia [12].

State	Frequency	Maximum intensity (Modified Mercalli Scale)
Peninsular Malaysia (1909-2010)		
Perlis	3	V
Kedah	18	V
Penang	41	VI
Perak	24	VI
Selangor	52	VI
Negari Sembilan	14	V
Melaka	19	V
Johor	32	VI
Pahang	35	III
Terengganu	2	IV
Kelantan	3	IV
Kuala Lumpur/Putrajaya	38	VI
Sabah (1897-2010)		
Sabah	41	VII
Sarawak (1874-2010)		
Sarawak	17	VI

Based on historical records, the major part of earthquakes that influenced Peninsular Malaysia is originated from two external sources: Sumatran subduction zone and Sumatran strike slip fault. Only seismic source zones within a radius of 500 km from the site are considered. Those outside of this radius may not significantly influence the peak ground acceleration [19]. Large earthquakes that originated from these two active areas did create considerably ground motion over western part of West Malaysia [17].

3 Data and methodology

This study was conducted to estimate the peak ground acceleration in Peninsular Malaysia by using earthquake catalogue of the region from the year 1900 to 2014, and for an approximately 500 km distance from every side of this region which covers the longitude of 90°E to 110°E and latitude of 10°S to 10°N. The required data was collected from NEIC (USGS, 2014). There are more than 16000 recorded events for the mentioned time period. The catalogue contains more than 10 various types of magnitude scales; so the first step is unification of data scales by using equations of relation proposed by Adnan et al. [9].

The methodology followed for (Deterministic Seismic Hazard Analysis) DSHA described as four steps and illustrated in Figure 5.

Data was shortened for records less than Mw=4, because this range of data has no impact for estimation of PGA. As the seismic sources were identified previously, the zonation model was generated (Figure 6) based on the model of Petersen et al. [20].

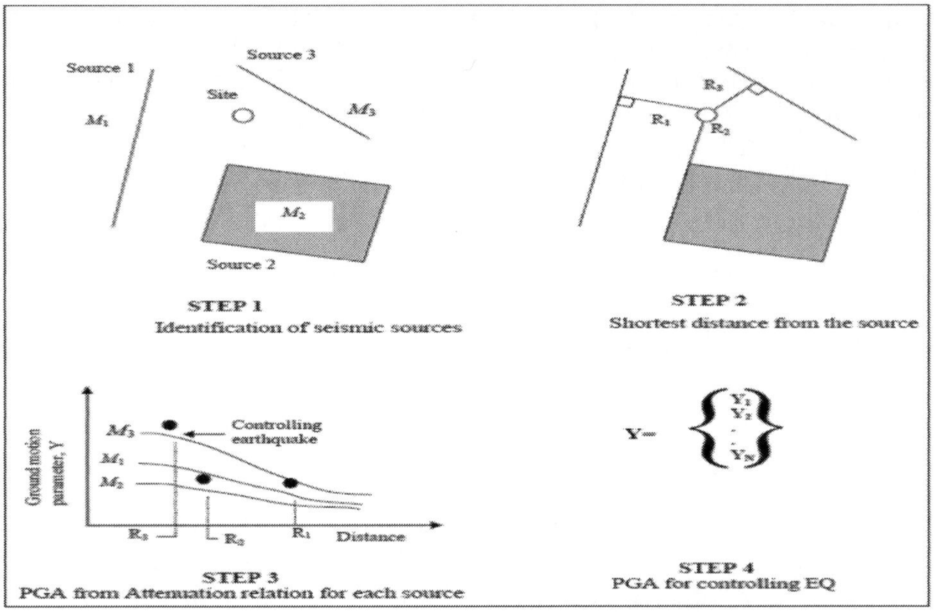

Figure 5. Different steps for deterministic seismic hazard analysis [6].

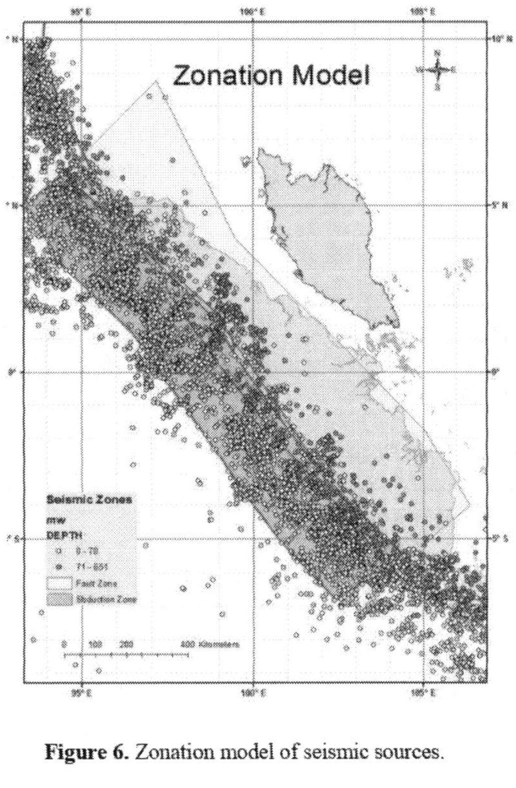

Figure 6. Zonation model of seismic sources.

MCE was estimated in the way proposed by Gupta [21] by using an increment of 0.5 to the maximum recorded magnitude. Attenuation equation of Sadigh et al. [22] which has been developed for distances more than 200 kilometres by Petersen et al. [20], was implemented for calculation of PGA for subduction zone:

$$\ln y_{\text{Sadigh et al.}}(M,R) = C_1^* + C_2 M + C_3^* \ln\left[R + e^{C_4^* - \frac{C_2}{C_3^*}M}\right] + C_5 Z_{SS} + C_8 Z_t + C_9 H$$

Equation (1)

$$\ln y_{\text{Petersen}}(M,R) = \ln y_{\text{Sadigh et al.}}(M,R) + [-0.0038*(R-200)]$$

Equation (2)

For Fault zone, the equation of Campbell [23] was implemented:

$$\ln Y = c_1 + f_1(M_W) + f_2(M_W, r_{rup}) + f_3(r_{rup})$$

$$f_1(M_W) = c_2 M_W + c_3(8.5 - M_W)^2,$$

$$f_2(M_W, r_{rup}) = c_4 \ln R + (c_5 + c_6 M_W) r_{rup},$$

$$R = \sqrt{r_{rup}^2 + [c_7 \exp(c_8 M_W)]^2},$$

$$f_3(r_{rup}) = \begin{cases} 0 & \text{for } r_{rup} \leqslant r_1 \\ c_7(\ln r_{rup} - \ln r_1) & \text{for } r_1 < r_{rup} \leqslant r_2 \\ c_7(\ln r_{rup} - \ln r_1) + \\ \quad c_8(\ln r_{rup} - \ln r_2) & \text{for } r_{rup} > r_2 \end{cases}.$$

Equation (3)

Eventually rest of the analysis was done by using software ArcGIS 10.2.2.

4 Results and discussion

The findings of this work are based on the earthquake data collected from NEIC (USGS, 2014) which was accessible for years 1900 – 2014. Distribution of data which consists of 16150 recorded events for the region is summarized in Table 2.

Table 2. Number of earthquakes reported for study area from year 1900 to 2014.

MW	Mw<4	4≤Mw<5	5≤Mw<6	6≤Mw<7	≤Mw<8	Mw≥8	Total
EVENTS	4598	8339	2826	327	55	5	16150

Data distribution based on zonation model for fault zone and subduction zone are depicted in Figure 7 and Figure 8 respectively.

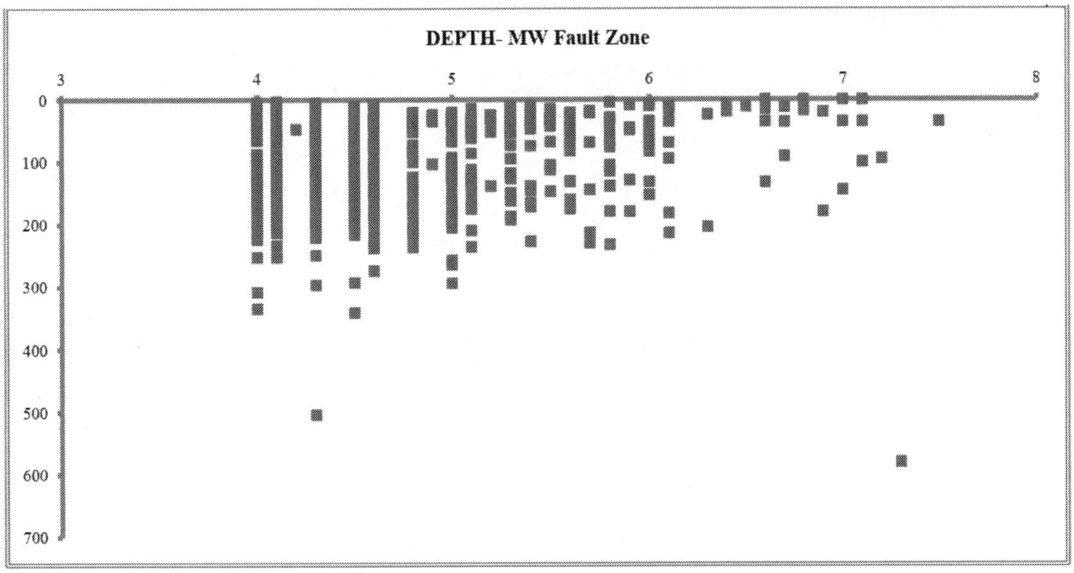

Figure 7. Distribution of events based on depth for Fault Zone.

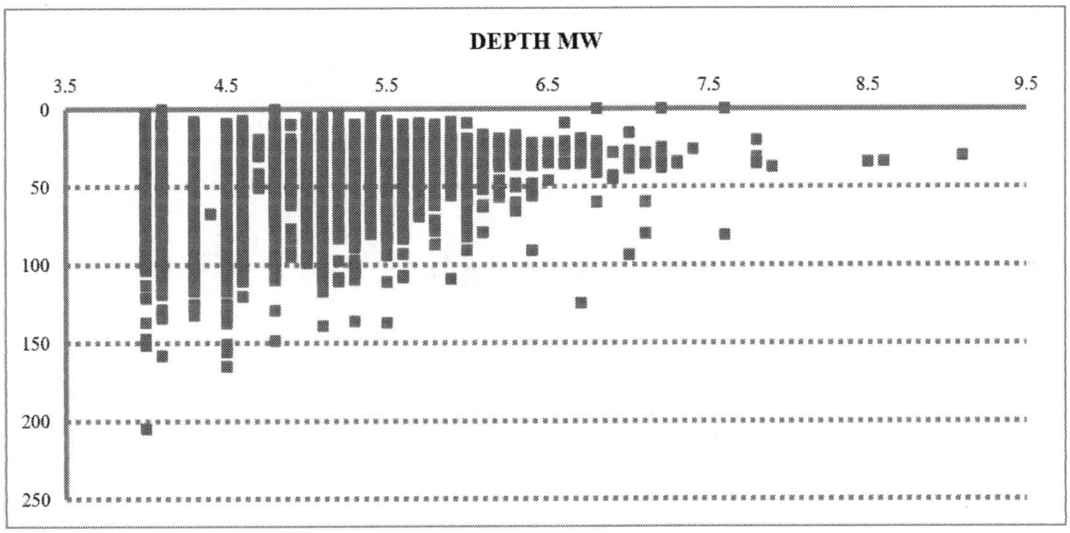

Figure 8. Distribution of events based on depth for Subduction Zone.

A comparison between two zones reveals that events in fault zone happen in depths deeper than subduction zone. Events smaller than Mw=4, because of having a very subtle effect in analysis are intentionally removed from report. The events in fault zone have occurred between 0 – 200 kilometres depth for Mw=4 to Mw=6, and for greater magnitudes the shallow events (lesser than 50 km) are more than deep events. The maximum magnitude recorded for this zone is Mw=7.5. On the other hand, for the subduction zone the majority of events are happened in shallow depths (lesser than 50 km) for whole range of magnitudes. Although for magnitudes greater than Mw=6 the events are restrained to 100 km, for greater magnitudes (Mw>7.5) events have happened within depths lesser than 35 km. The maximum magnitude recorded for this zone is Mw=9.1.

The results of PGA analysis for MCE= 9.6 for Subduction zone and MCE=8 for Fault zone are presented in Figure 9 and Figure 10 respectively.

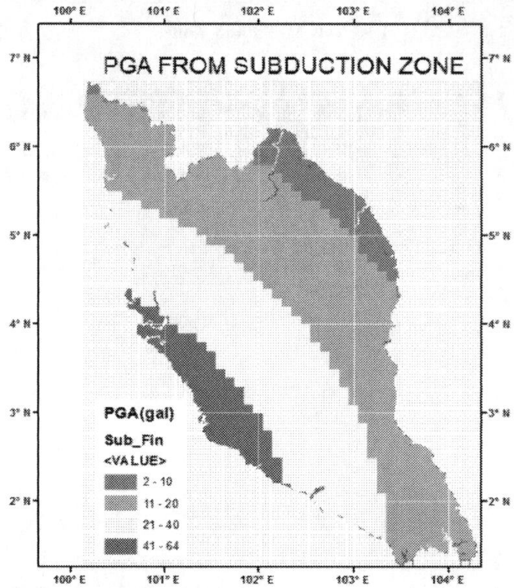

Figure 9. PGA for Subduction Zone using deterministic approach.

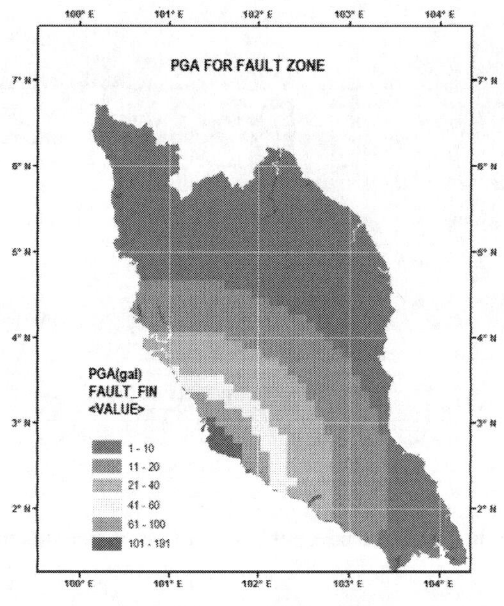

Figure 10. PGA for Fault Zone using deterministic approach.

There are some published results for estimation of PGA for Peninsular Malaysia which have been conducted using deterministic approach. Among them Adnan et al. [9] have published a result for deterministic approach and divided the PGA map of Peninsular into two zones, i.e. the zone for range between 30 and 50 gals on the east side of Peninsular Malaysia and the zone between 50 and 70 gals on the west side. The difference between results of that research and this work might relates to the zonation model of seismic sources, the estimation of MCE for each zone, and updating the earthquake catalogue as well.

5 Conclusion

GIS provides strong tools to visualize various attributes of seismic data and its analysis capabilities make it possible to estimate the PGA for a region accurately. Results from this research show the PGA from subduction zone fluctuates from 2-64 (gal) and from the fault zone varies from 1-191(gal). The results show the PGA generated from fault zone is more critical than subduction zone for selected seismic model. Finally, it can be concluded that, the amount of PGA for Peninsular Malaysia varies between 1 to 191 gals as per results of deterministic analysis.

References

[1] Tehrany M S, B Pradhan, and M N Jebur 2014, Flood susceptibility mapping using a novel ensemble weights-of-evidence and support vector machine models in GIS. *J Hydrol.* **512**: p. 332-343.

[2] Umar Z, B Pradhan, A Ahmad, M N Jebur, and M S Tehrany 2014, Earthquake induced landslide susceptibility mapping using an integrated ensemble frequency ratio and logistic regression models in West Sumatera Province, Indonesia. *Catena.* **118**: p. 124-135.

[3] Youssef A M, B Pradhan, H R Pourghasemi, and S Abdullahi 2015, Landslide susceptibility assessment at Wadi Jawrah Basin, Jizan region, Saudi Arabia using two bivariate models in GIS. *Geosci J*: p. 1-21.

[4] Abdullahi S, B Pradhan, S Mansor, and A R M Shariff 2015, GIS-based modeling for the spatial measurement and evaluation of mixed land use development for a compact city. *GISci Remote Sens.* **52**(1): p. 18-39.

[5] Youssef A M, B Pradhan, and A M Hassan 2011, Flash flood risk estimation along the St. Katherine road, southern Sinai, Egypt using GIS based morphometry and satellite imagery. *Envir Earth Sci.* **62**(3): p. 611-623.

[6] Kramer S L, *Geotechnical earthquake engineering*. Vol. 80. 1996: Prentice Hall Upper Saddle River, NJ.

[7] Megawati K, Lam, N. T. K., Chandler, A. M., & Pan, T.-C., *Cities without a seismic code I: Hazaed assessment*, in *13th World Conference on Earthquake Engineering*2004: Vancouver, B.C., Canada.

[8] Balendra T and Z Li 2008, Seismic hazard of Singapore and Malaysia. *Electron J Struct Eng*, ((special issue, Earthquake engineering in the low and moderate seismic regions of Southeast Asia and Australia)): p. 57-63.

[9] Adnan A, H Hendriyawan, A Marto, and M Irsyam 2005, Seismic hazard assessment for Peninsular Malaysia using gumbel distribution method. *J Tech B*, (42B): p. 57-73.

[10] AULOV A and M S LIEW, *Seismic Hazard Analysis (SHA) Considerations for South China Sea Territories of Malaysia*, 2013.

[11] Bolt B A, & Dreger, D., *Nature of Earthquakes and seismic hazards*, in *Designing for Earthquakes A MANUAL FOR ARTICHECTS, FEMA* 2006.

[12] MMD., *The effective ness of public awareness campaigns on Earthquake and Tsunami hazard in Malaysia*, 2011.

[13] Abdullahi S, A R b Mahmud, and B Pradhan 2014, Spatial modelling of site suitability assessment for hospitals using geographical information system-based multicriteria approach at Qazvin city, Iran. *Geocarto Int.* **29**(2): p. 164-184.

[14] Abdullahi S, B Pradhan, and M N Jebur 2015, GIS-based sustainable city compactness assessment using integration of MCDM, Bayes theorem and RADAR technology. *Geocarto Int.* **30**(4): p. 365-387.

[15] Anderson J L, et al. 2007, ACC/AHA 2007 guidelines for the management of patients with unstable angina/non–ST-elevation myocardial infarction. *J Am Coll Cardiol.* **50**(7): p. e1-e157.

[16] Tjia J, et al. 2008, Medicare beneficiaries and free prescription drug samples: a national survey. *J Gen I Med.* **23**(6): p. 709-714.

[17] Mohd Rosaidi C. *Earthquake Monitoring In Malaysia*. in *Seismic Risk Seminar on 25th September*. 2001.

[18] Marto A, C Tan, F Kasim, and N Mohd Yunus. *Seismic Impact in Peninsular Malaysia*. in *Proceedings of the 5th International Geotechnical Symposium-Incheon, Seoul, Korea*. 2013.

[19] Irsyam M, et al. 2008, Proposed seismic hazard maps of Sumatra and Java islands and microzonation study of Jakarta city, Indonesia. *J Earth Syst Sci.* **117**(2): p. 865-878.

[20] Petersen M D, et al. 2004, Probabilistic seismic hazard analysis for Sumatra, Indonesia and across the Southern Malaysian Peninsula. *Tectonophysics*. **390**(1): p. 141-158.

[21] Gupta I 2002, The state of the art in seismic hazard analysis. *ISET J Earthquake Tech*, (428): p. 311-346.

[22] Sadigh K, C-Y Chang, J Egan, F Makdisi, and R Youngs 1997, Attenuation relationships for shallow crustal earthquakes based on California strong motion data. *Seismol Res Lett.* **68**(1): p. 180-189.

[23] Campbell K W 2003, Prediction of strong ground motion using the hybrid empirical method and its use in the development of ground-motion (attenuation) relations in eastern North America. *Bull Seismol Soc Am.* **93**(3): p. 1012-1033.

Spatial relationship between earthquakes, hot-springs and faults in Odisha, India

Biswajeet Pradhan[1*], Ratiranjan Jena[2]

[1]Department of Civil Engineering, Geospatial Information Science Research Centre (GISRC), Faculty of Engineering, University Putra Malaysia, 43400, UPM, Serdang, Selangor, Malaysia, Email: biswajeet24@gmail.com or biswajeet@lycos.com

[2]SEOCS, IIT Bhubaneswar, Samantapuri Swostik Plaza, Bhubaneswar

Abstract. Odisha is famous for Mineral rich Eastern-Ghat mobile belt, hot springs and cultural Heritage. The hot springs are known for centuries and are used by public as a place for worship. Odisha falls under the II and III seismic zones in India. Most of the seismicity in Odisha is due to motion along some active normal faults along the Mahanadi Graben. Therefore, it is necessary to identify the active faults and understand spatial distribution of seismic activity in Odisha. It is also important to understand the Earthquakes and their relation with the Geology of Odisha and understand the neo-tectonic activity. There are 7 major hot springs found along the North Odisha Boundary Fault and Mahanadi Shear Zone. The hot water percolates deep into the Earth through porous and permeable fractured rocks along the fault. Depth of source for most of the hot springs in Odisha must be some few feets to few meters; however most of these observations are not based on scientific geophysical data. Therefore, spatial relationship between thermal springs, earthquakes, and geology of Odisha may provide better understanding of the hot-spring setting. By using the earthquake and fault data, the sense of motion along faults can be easily interpreted. All these information can explain the spatial distribution and inter-relation between hot-springs, faults and earthquakes in Odisha.

1 Introduction

Odisha has experienced very few moderate to large earthquakes. Some events with magnitudes in excess of 5.0 have originated in the Bay of Bengal off the coast of the state. The Mahanadi Graben is a narrow elongated belt covering parts of Odisha, Chatishgarh and Madhya Pradesh along Mahanadi river valley with a NW-SE orientation [1, 2]. It meets the Narmada-son river valley on the North West in Madhya Pradesh and lies hidden beneath the coastal alluvium in Odisha [1]. The Graben lies between two important fault zones, North Odisha boundary fault [3] on the north and the Mahanadi shear zone [2] on the south.

The important sediment filled depressions are IB Basin, Talcher basin, Jajpur-chandbali basin which are lying on the northern part and the Athgarh and Salepur basin on the southern part of Mahanadi Graben. The faults have shown evidence of movement during the Holocene epoch. The Brahmani Fault in the vicinity of Bonaigarh is among them. The Mahanadi also flows through a Graben structure [3]. Several deep-seated faults are situated beneath the Mahanadi Graben. However, it must be stated that proximity to faults does not necessarily translate into a higher hazard as compared to areas located further away, as damage from earthquakes depends on numerous factors such as subsurface geology [4].

The geo-tectonic setup of India, the fault map zones and the epicentre of seismicity indicate about the anomalous nature of different types of geothermal distributions [5, 6]. GSI [7] had identified 340 Hot-water springs which are characterized by Organic activity. A total of about 10,000 MW could be generated from Himalaya, Naga Lushai, Andaman- Nicobar Islands, West Coast of Cambay Graben, Aravalli, Son-Narmada-Tapti, Godavari and Mahanadi, South Indian Cratonic geothermal provinces in India. However, the geothermal reservoirs in India are low to medium enthalpy type where surface temperatures vary from 80 to 105°C thus used to convert their geothermal energy into electricity. All the Hot-Springs of India were classified on the basis of their geo-tectonic setup and grouped into six Geothermal Provinces as follows: (a) Himalayan Province-Tertiary Organic belt with Tertiary magmatism; (b) Areas of Faulted blocks Aravalli belt, Naga-Lushi, West coast regions and Son-Narmada lineament; (c) Volcanic arc-Andaman and Nicobar arc; (d) Deep sedimentary basin of Tertiary age such as Cambay basin in Gujarat; (e) V. Radioactive Province Surajkund, Hazaribag, Jharkhand; and (f) Cratonic province peninsular India.

The Geology of Peninsular India mainly comprises Precambrian assemblages with Jurassic to Quaternary sedimentary rocks and volcanic of Cretaceous-Eocene [8]. The on-shore and off-shore regions of East coast are mainly younger sequences with prospects of hydrocarbons. The Coastal areas have undergone repeated cycles of sea level changes during Quaternary. The Coastal risk along Peninsular Coast has been worked out prior to the advent of cyclone of 2004. Later Inter-mitten seismic activity along parts of East coast is reported around the following places from reports during the period.

1) Off. Odisha coast, May. 2014.
2) Off. Visakhapatnam, Jan., 1996.
3) Off. Pondicherry, Dec., 2001.

The shallow depth origin and possible rejuvenation of faults are either on shore or off-shore. Evidences of uplift and subsidence are reported. Furthermore, there are evidences of disturbances in the top layers of sediments along parts of East Coast by way of displacements in the layers as seen in the sub-bottom profiles. There are also rise-like structures. Presence of hot springs and field evidence showing displacement on a small scale are noticeable. There are evidences of coarse sized sand and pebbly rich aggregates of sediments on the deep sea floor at around 2500m depth beyond slope as observed at off Pennar. Such exotic sediments are possibly seisomogenic turbidities (GSI unpublished sea bed survey reports 1983-2000). Thus, it is prudent to infer seismic activity, neo-tectonism and upper continental crust along East Coast at places [9, 10]. Therefore, the current study aims; to understand the spatial distribution and relation between Earthquakes, Hot-springs and Faults in Odisha; to identify active Faults and understand spatial distribution of Seismic activity; and to interpret the sense of motion along faults by using the Earthquake and Fault data.

2 Study area and dataset

The seismic hazard map of India was updated in 2000 by the Bureau of Indian Standards (BIS). There are no major changes in the zones in Odisha with the exception of the merging of Zones I and II in the 1984 BIS map. Districts that lie in the Mahanadi river valley lie in Zone III, and within Odisha this zone stretches from Jharsuguda along the border with Chhatisgarh in a south-easterly direction towards the urban centres of Bhubaneswar and Cuttack on the Mahanadi Delta (Figure 1).

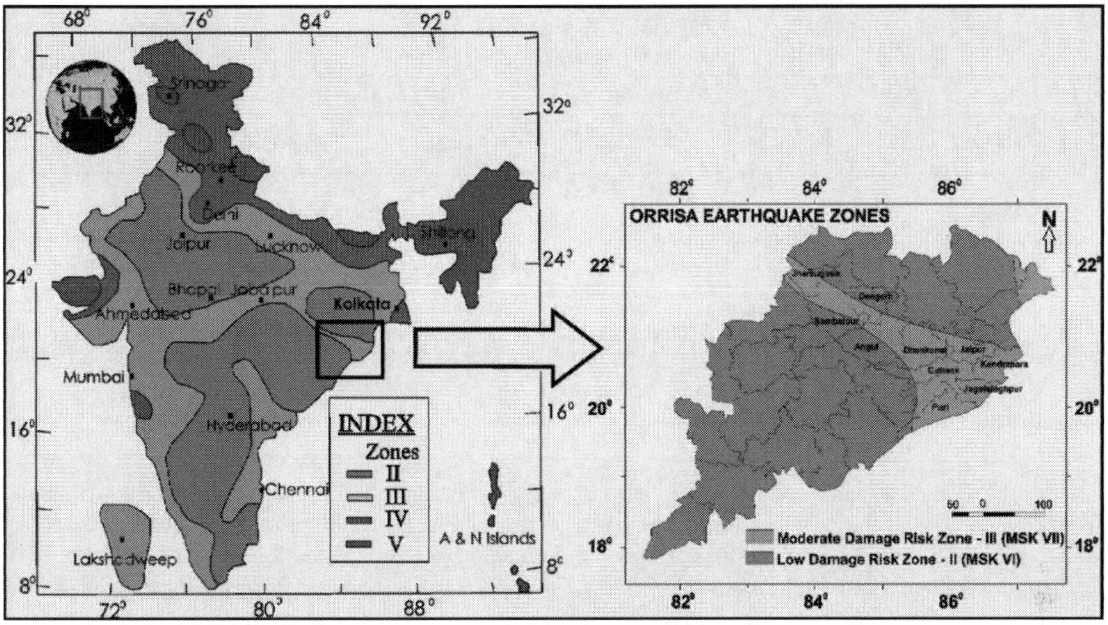

Figure 1. Map and the Mahanadi Graben covering the study area.

In this research, three earthquake data were used to prepare a map where all the seismic and active regions can be interpreted especially the periods of 1676 to 2014. The details of the images are shown in Table 1. The selection of the datasets is based on the objectives of the research and literature which explains the spatial distribution of earthquakes [11, 12]. Data of hot spring temperature and location has taken from the geological map of Odisha prepared by GSI. These data can be manipulated to reveal the spatial distribution of earthquakes, hot spring and faults in Odisha and to explain neo-tectonism including their relation with each other [12-14].

Table 1. Earthquake Data (ASC Website).

No	Year	Magnitude	Intensity	Location	Latitude	Longitude	Damage/Loss of Life
1	1676	4.0	IV	Balasore Area, Odisha	21.500N	86.900E	No
2	1837	5.0	VI	Rambha- Paluru Area, Odisha	19.500N	85.100E	No
3	1858	4.9	V	Balasore- Chandipur Area, Odisha	21.500N	87.000E	No
4	1860	4.9	V	Karantola area, Odisha	19.400N	84.900E	No
5	1891	4.9	V	Near Palmyras Point, Odisha	20.800N	87.000E	No
6	1963	6.0	VII	Bijakuli Banei Area, Odisha	21.700N	84.900E	Damage of Departmental Building and crop loss
7	1979	4.7	V	Dublabera Majhgaon Area, Odisha	22.100N	84.900E	No
8	1982	5.2	VI	Bay of Bengal	18.510N	86.310E	Damage of cottages in Village

9	1982	4.7	V	Khajuripada Banigochha Area, Odisha	20.390N	84.420E	No
10	1985	5.4	VI	Bay of Bengal	18.367N	88.188E	Less Damage
11	1995	4.6	V	Laimura Deogarh Area, Odisha	21.671N	84.565E	No
12	1995	4.7	V	Kasijodi Nuakoat Area,Odisha	21.780N	85.327E	No
13	2001	4.7	V	Kanakjora- Sundargarh Area,Odisha	22.240N	83.918E	No
14	2014	5.9	VII	Bay of Bengal	18.201N	88.019 E	Less Damage

3 Methodology

Literature survey for historical and recent earthquake was conducted to prepare the map of earthquake locations, faults and thermal springs. This step was done for better interpretation and understands the spatial distribution; identifying historical zones and recent seismicity. Modelling of the cross section of Mahanadi Graben creatively design the Gondwana sediments and their relation with normal faults. The Relation with seismic zonation map and earthquake distribution is also important to describe the percentage of hazardous area. Preparing the model explains about the Relation between faults and thermal springs (Figure 2).

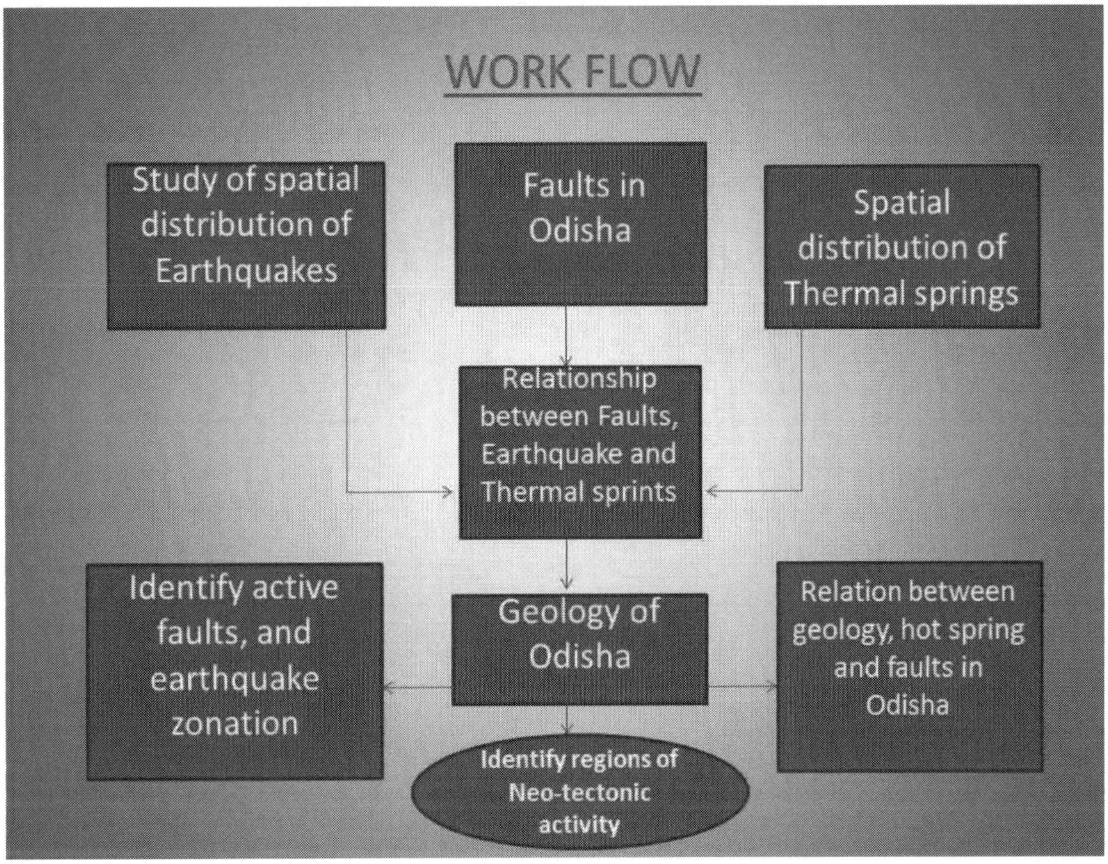

Figure 2. Overall methodological flowchart

Research analysis phase was focusing on the arguments and change detection of the findings. Accuracy of analysis is depends on the study area and data reliability. It was involved with the statistical estimation that obtained from the magnitude vs. year output and the independent references for the probability error measurement of the prepared map. This was to ensure the prepared map and the models are accurate and reliable. For each feature, the data gives the accurate analysis which needs to be established in order to perform good result. The subtracted information from the prepared map explains the features properties. The relation between earthquakes, Hot Springs and faults deliberated in result and discussion section by comparing the features with each other. The details of the analysis will be viewed in the result and discussion section.

4 Result & Discussion

The spatial distribution of earthquakes in India characterized with very high magnitude earthquake zones in Himalayan region, Gujarat and some other places. But earthquakes happened till now in Odisha are generally with low magnitude of 2.5 to 5.5. The districts which are present in Mahanadi Graben are going under II seismic zone. The recent earthquake was during 2003 in Sundergarh area and at a distance of 150 km from the Odisha coast with a magnitude of 5.9. This explains the active movement of normal faults in different regions of Mahanadi Graben. Therefore, it gives the information about neo-tectonic movements in different regions of Mahanadi Graben.

The magnitude vs. year graph explains the (Figure 3);

- Distribution of Earthquakes with Magnitude
- Equal magnitude of earthquakes occurred in same area but year is different and no earthquakes of magnitude higher than 6 and lower than 3.

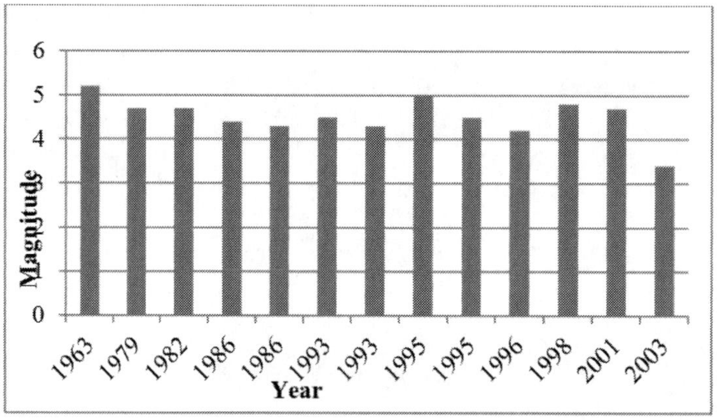

Figure 3. Earthquakes Distribution with year.

There are many regions in India where earthquakes are happening continuously. But some places where very low magnitude earthquakes are happening. Odisha is such a place where the Magnitudes of earthquakes are generally 3 to 5.5, thus making these earthquakes non-dangerous. The 3D model explains the origin of hot spring is at a depth of 15-20 feet from the surface (Figure 4 and 5).

IOP Conf. Series: Earth and Environmental Science **37** (2016) 012070 doi:10.1088/1755-1315/37/1/012070

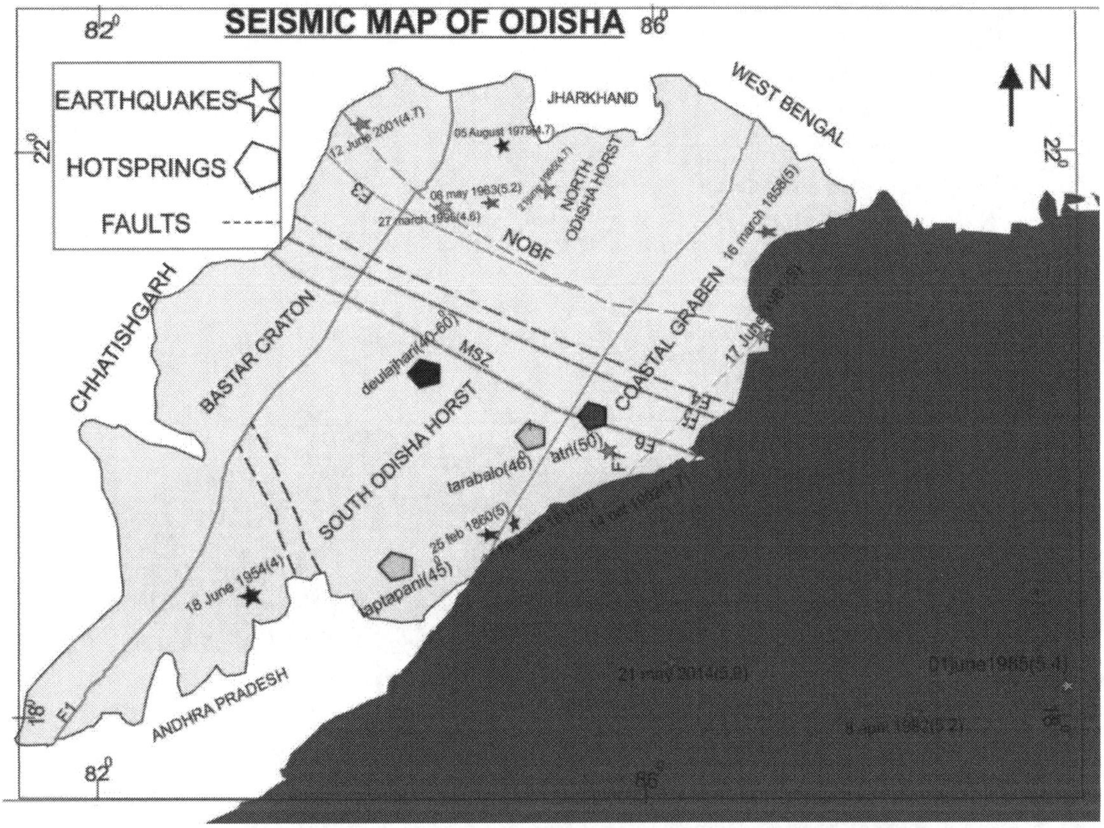

Figure 4. Map of Odisha shows distribution of earthquakes, hot springs and faults.

Hot springs are the sites that discharge hot ground water. The springs usually emerge from the deep faults or fissures of earth along which the ground water comes out.

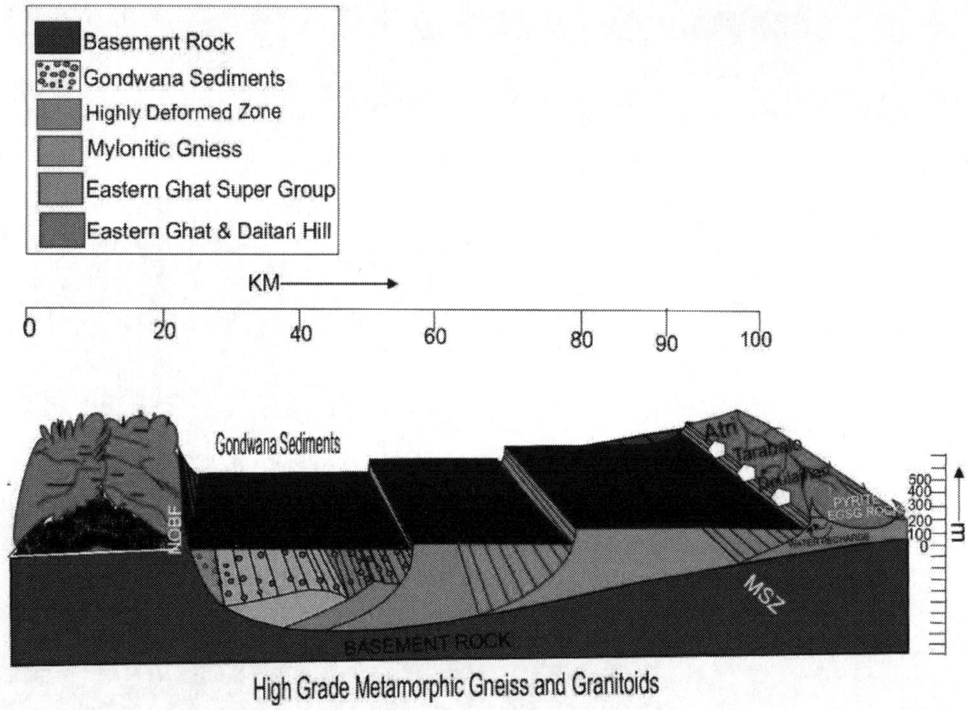

Figure 5. Model of Mahanadi Graben with Faults and Hot Springs [15].

The high temperature of hot spring water is because of the geothermal energy, exothermic reactions and disintegration of radioactive elements. Here the basement rock is high grade metamorphic gneiss and Granitoids. The cyclic movement of water occurs. Here the water infiltrate into the porous and permeable rocks and come in contact with the country rocks. Reaction takes place where the rock contains pyrite and pyrohtite. Reaction of water occurs with pyrite and pyrohtite which finally gives basic composition of the thermal spring water.

The area is covered with vegetation. The basement rock type of Mahanadi Graben is generally gneiss and the basin is totally covered with Gondwana sediments. There are two main faults present as boundaries known as MSZ and NOBF.

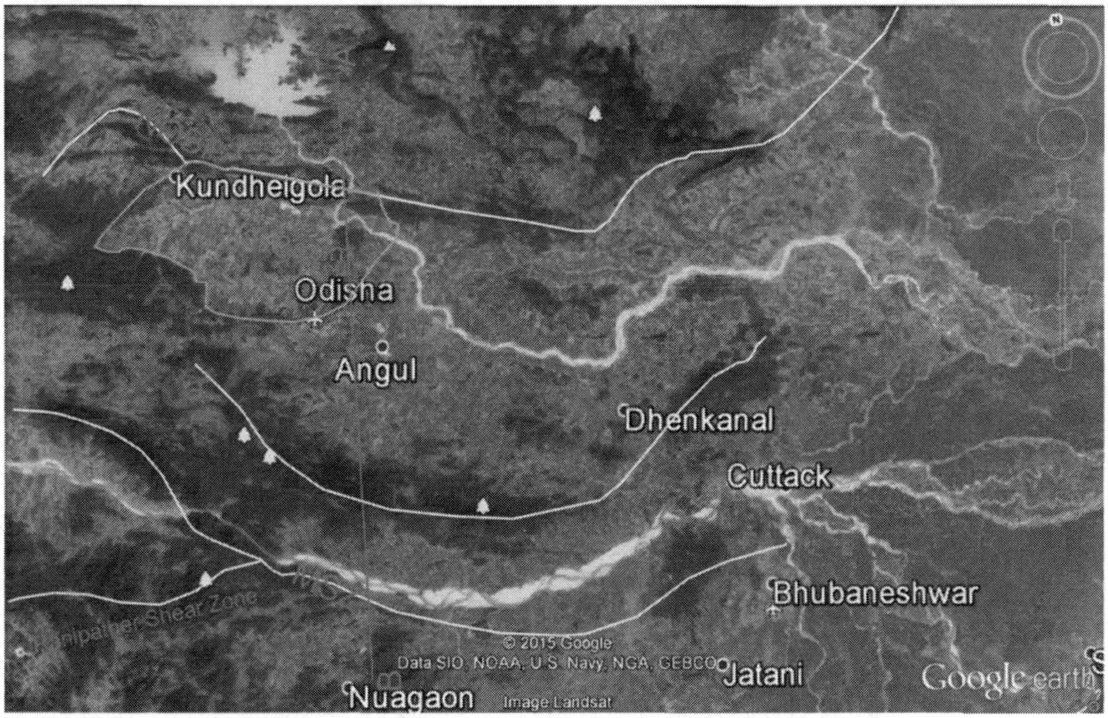

Figure 6. Identification of faults and cross section along AB.

Neo-tectonics is the motions and deformations of Earth's crust that are current or recent in geological time. Geologists refer to the corresponding time-frame as the Neo-tectonic period, and to the preceding time as the Palaeo-tectonic period. Neo-tectonics defining the field as recent tectonic movements occurred in the upper part of Tertiary and in the Quaternary. Active tectonic movements that are expected to occur within a future time span of concern to society. Any fault that moved say the last 35,000 years (10,000 years) is considered as active faults. Active tectonics involved dynamic process operating in the present day regimes. Combining active tectonics and records of reported seismicity new generation maps or models of seismo-tectonics could be framed combing the inputs of geology, geophysics and seismic micro zonation maps. Pertaining the east coasts, for the last 200 years there are some historical records of low magnitude earthquakes. The Extension of the continental crust is well postulated together with active faults or those rejuvenated along pre-existing faults. Therefore, all the earthquakes happened till now in different regions of Mahanadi Graben are due to neo tectonic activity. Neo-tectonic activity is mainly found in central part of the basin because of high thickened and deformed sediments. These loose sediments are also responsible to produce high magnitude earthquakes because it amplifies the earthquake waves.

The focal mechanism of an earthquake describes the deformation in the source region that generates the seismic waves. In the case of Mahanadi Graben fault-related event it refers to the orientation of the fault plane that slipped and the slip vector and is also known as a fault-plane solution. Focal mechanisms are derived from various methods, including first motion analysis and a solution of the moment tensor for the earthquake, which itself is estimated by an analysis of observed seismic waveforms. The focal mechanism can be derived from observing the pattern of "first motions", that is, whether the first arriving P waves break up or down. This method was used before waveforms were recorded and analyzed digitally and this method is still used for earthquakes too small for easy moment tensor solution. Focal mechanisms are now mainly derived using semi-automatic analysis of the recorded waveforms. Seismologists refer to the direction of slip in an earthquake and the orientation of the fault on which it occurs as the focal mechanism.

This information was used from seismograms to calculate the focal mechanism and typically display it on maps as a "beach ball" symbol. This symbol is the projection on a horizontal plane of the lower half of an imaginary, spherical shell (focal sphere) surrounding the earthquake source. A line is scribed where the fault plane intersects the shell. The stress-field orientation at the time of rupture governs the direction of slip on the fault plane, and the beach ball also depicts this stress orientation. These focal mechanisms explain that all the faults that are present in Mahanadi Graben are generally normal faults. For a double-couple source mechanism (or only shear motion on the fault plane), the compression first-motions should lie only in the quadrant containing the tension axis, and the dilatation first-motions should lie only in the quadrant containing the pressure axis.

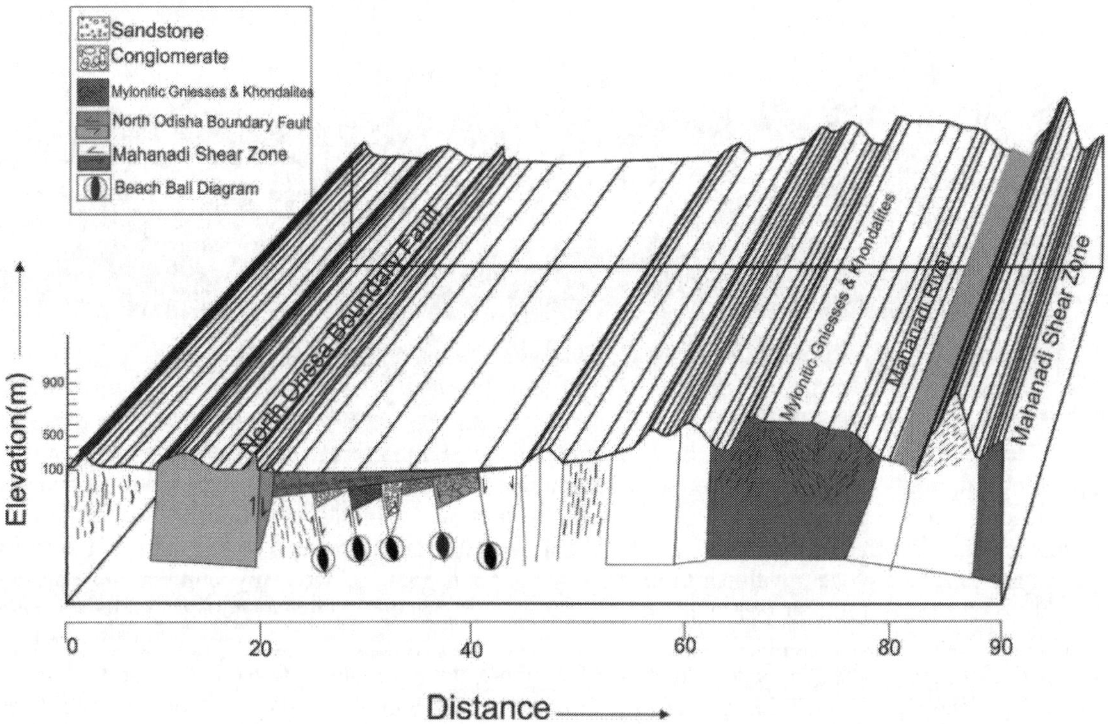

Figure 7. Model of Mahanadi Graben along the cross section [15].

A hot water spring, also known as a thermal spring, is a natural discharge of hot water from the earth. Such springs normally occur in areas where underground water passes through hot igneous or metamorphic rocks along faults. They can form pools, geysers or fumaroles .The spatial distribution of hot water springs and faults explains that the hot springs are the result of the non-active faults.

The relationship is scale-dependent in time and space. Therefore a correlation between location of hot springs and present day seismic activity is not necessary but the faults were indeed active at some point in history. Therefore, in Odisha the NOBF and MSZ faults were active but not today. The other normal faults between these boundary faults must be active because there is no hot springs now and earthquakes are the result of these active faults.

Table 2. Temperature of all the Hot Springs in Odisha.

Sl. No	Name of the Hot Springs	Temp. (C°)
1	Taptapani	30°- 40°
2	Deulajhari	40°- 62°
3	Atri	55°
4	Tarabalo	67°
5	Bankhol	40°- 50°
6	Badaberena	45°
7	Magarmuhan	55°
8	Boden	32°

Water generating from thermal spring is heated by geothermal gradient i.e. heat from earth's interior. The water percolating deep down into the crust through structural breaks are heated up, while coming in contacts with the host rocks. The source of heat in India for hot springs is generally regional heat flow. Five heat flow zones in India subcontinent are reported where heat flow values are ranges between 30 and 468mW/m2. Odisha falls in zone II (100-180mW/m2 and zone III (70- 100mW/m2) on the basis of heat flow values.

Water in hot springs comes from a depth that is when the porous and permeable rocks get recharged and water passes through the rock up to the basement rock of gneiss. Then water can't pass through it further. Therefore, by cycling process the water get back to the surface through the faults. Therefore, the depth at which the cycling process starts is called as the depth of hot spring. By using the geothermal gradient and the depth dataset it can be determine that the water is coming from a specific depth of 13-15 feet from the surface of the Mahanadi.

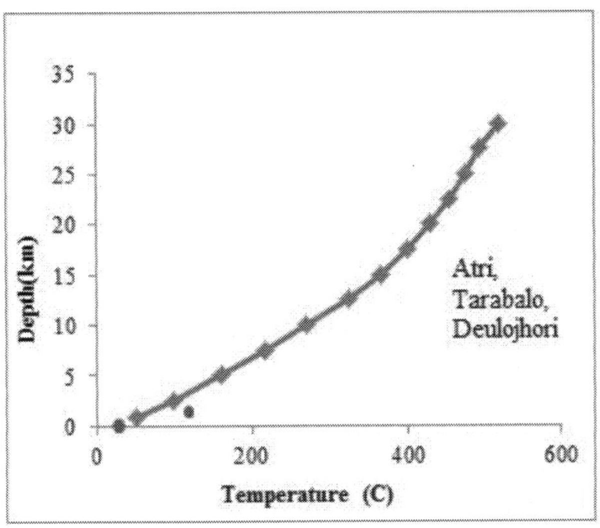

Figure 8. Depth of the Hot Springs.

5 Conclusion

Mahanadi Graben is an active tectonic area where earthquakes are continuously happening. However, these are of small magnitude and not considered as hazardous. The Eastern Coastal areas along the Peninsula witness moderate events as evidenced by the recent reports. The study of outcrop of the whole Graben by using remote sensing techniques for identifying the Hot Springs and faults have great potential for research. Earthquakes related science as an attempt towards progresses of at least the vulnerability criteria could be a worthwhile endeavour with the aid of all modern tools of geology, geomorphology, geodesy, geo physics- geo chemistry and remote sensing. As the only visible marking available is active faults and neo-tectonics for generating new class of seismic hazard zonation maps. The search for hidden faults through geology centred techniques along with geophysics and geochemistry analogies should become a programme of national endeavour Organisations.

Acknowledgement

The Earthquake Data used in this study were downloaded from ASC Website.

References

[1] Kumar T S, R Mahendra, S Nayak, K Radhakrishnan, and K Sahu 2010, Coastal vulnerability assessment for Orissa State, east coast of India. *Journal of Coastal Research*: p. 523-534.
[2] Mahala S, P Singh, M Das, and S Acharya 2012, Genesis of thermal springs of Odisha, India. *International Journal of Earth Sciences and Engineering*. **5**(6): p. 1572-1577.
[3] Mahalik N K 2013, Coastal Tract of Odisha, Resources and Environment Geology. *Journal of the Geological Society of India*. **81**(1): p. 1421-1439.
[4] Rajendran C and K Rajendran 2001, Characteristics of deformation and past seismicity associated with the 1819 Kutch earthquake, northwestern India. *Bulletin of the Seismological Society of America*. **91**(3): p. 407-426.
[5] Roy S and R Rao 1996, Regional heat flow and the perspective for the origin of hot springs in the Indian shield. *Visesa Prakasana-Bharatiya Bhuvaijñanika Sarveksana*, (45): p. 39-40.
[6] Sharma R S, *Cratons of the Indian Shield*2009: Springer.
[7] Mahadevan T M 1994, Deep continental structure of India: a review. *GSI Publications*. **1**(1).
[8] Rao B R and P S Rao 1984, Historical seismicity of peninsular India. *Bulletin of the Seismological Society of America*. **74**(6): p. 2519-2533.
[9] Verma G, M Verma, and R Tripathi 2013, Dynamic Analysis of Hirakud Dam Due to Seismic Forces. *Global Journal of Researches In Engineering*. **13**(1).
[10] Valdiya K 1976, Himalayan transverse faults and folds and their parallelism with subsurface structures of north Indian plains. *Tectonophysics*. **32**(3): p. 353-386.
[11] Chopra A K 2012, Earthquake analysis of arch dams: factors to be considered. *Journal of Structural Engineering*. **138**(2): p. 205-214.
[12] DAY A L 1939, The hot-spring problem. *Geological Society of America Bulletin*. **50**(3): p. 317-336.
[13] Khattri K, A Rogers, D Perkins, and S Algermissen 1984, A seismic hazard map of India and adjacent areas. *Tectonophysics*. **108**(1-2): p. 93111-108134.

[14] Guha S K 1986, Status of exploration for geothermal resources in India. *Geothermics*. **15**(5): p. 665-675.

[15] Kumar P S, R Menon, and G Reddy 2007, The role of radiogenic heat production in the thermal evolution of a Proterozoic granulite-facies orogenic belt: Eastern Ghats, Indian Shield. *Earth and Planetary Science Letters*. **254**(1): p. 39-54.

8th IGRSM International Conference and Exhibition on Remote Sensing & GIS (IGRSM 2016) IOP Publishing
IOP Conf. Series: Earth and Environmental Science **37** (2016) 012071 doi:10.1088/1755-1315/37/1/012071

Spatial modulation of the hydrological risk at Praia, Cape Verde

José Lino Silva[1], Ana Cristina Neves[2] and João Garrott Marques Negreiros[3]

[1] UniCV - Palmarejo, Praia, Cape Verde
[2] University of Saint Joseph, Macao, China
[3] University of Saint Joseph, Macao, China

E-mail: joslino.silva@gmail.com

Abstract. Hydrology modeling became a relevant topic for the Cidade da Praia, Cabo Verde, Africa, due to negative impact risk to local population and its assets. The modeling via Geographical Information Systems (GIS) can help the decision-making process of space occupation and characterization for this type of risk. Under the municipalities of Praia, the phenomenon of flash flood is common, causing soil erosion and landslide. This constitutes a risk for the local habitat, particularly in districts with a lack of strong human infrastructures. To simulate, analyze and generate risk maps using GIS to help this county governance authorities for decision-making, thus, becomes the main aim of this article.

1. Introduction

The county of Praia in Cape Verde features a large morphological diversity and some adverse weather conditions [1]. Environmental vulnerability and hydrological risk have always existed in this county, due to its own territorial morphology derived from the local climate and aggravated by human intervention upon the physical environment. On the other hand, the economic and social changes on the Praia county over several decades and the accelerated pace of the economic growth in recent years have led to strong pressures on the environment, leaving behind a negative imprint on the territories, which sometimes requires huge financial costs to reverse the situation [2].

GIS have the advantage of allowing simulations and develop spatial support systems in the sense of finding a balance in the development of human activities. For example, rainfall takes place with increased frequency in Cape Verde in an intensive and irregular manner. Between the months of September and October, there are records of, at least, one high intensity rain event, causing extensive damage and endangering the lives and safety of the population.

This writing goals the characterization of hydrological risk (flash flood) that the municipality of Praia (133,000 inhabitants) is subject to, especially in urban areas, a phenomenon that frequently occurs in regions featuring this dry tropical climate [3]. With this purpose in mind, we intend to produce a set of information that allows to (A) Model the physical environment of Praia county and determine the associated hydrological risks; (B) Provide the mapping of hydrological risk areas based on a surface runoff model; (C) Analyze and quantify the various factors of soil occupation risks associated with climatic, environmental and social aspects; (D) Highlight the importance of GIS in the planning urban areas at Praia.

This research paper is divided into seven sections. Starting with this introduction and followed by an overview of hydrological risk, the third section recalls the fundamental concepts of digital terrain models (DTM) and the theoretical references to the current drainage models (D8, NGFLOW and SCS). The following two sections include an overview of the county of Praia and the hydrological methodology used for the realization of rainfall simulation, analysis and generation of risk maps (section 6). The last chapter expresses some recommendations to the local government of Cape Verde.

2. Hydrologic risk

Territory disordering reflects on the potential danger of flooding, caused by the surface runoff increased and on the dry cargo effect (destruction of vegetation, increased erosion of slopes, soil sealing). However, for a correct assessment of the hydrological risk, it is essential to know the existing rainfall regime in the area. It is known that precipitation is a natural event, which has great variability in terms of its distribution in the planet's surface. This is a phenomenon that feeds the hydrological cycle and is the major factor for surface runoff, infiltration, evaporation, aquifer recharge and alike. Thus, precipitation is an essential element in studies of infrastructures planning and projection. Certainly, measuring rainfall data is crucial for identifying fields with conditions for agriculture, design of water resources, environmental assessment and quantification of soil erosion [4].

Within arid regions, the phenomenon of flash flooding can be particularly dangerous for several reasons. Although they are rare, these storms can discharge a large amount of water in a short time. Secondly, these rains usually fall in low permeability soils such as clay by greatly increasing the amount of runoff, overloading the rivers and drainage channels. Sometimes these regions do not have adequate infrastructural conditions to divert water such as manholes, underground holding tanks and retention basins, either due to lack of population, poverty or because the residents still ignore the real risk of flooding. Moreover, the lack of regular rain water to clean water channels can cause flooding because of a large amount of accumulated debris throughout the dry period which are then carried by floods.

Generally, the hydrological risk approach can be broken down into four stages, according to [5]: (A) Analysis of the region morphology by considering their historical values; (B) Survey on the changes and introduced by man (it can serve to counteract the effects or minimize the likely impacts); (C) Survey of existing vulnerabilities due to the presence of man (urban areas) or due to the economic lacking infrastructures; (D) Presentation of the risks/dangers chart and their respective spatial weighting. Note that this last step requires a thorough analysis, taking into account the evaluation of the intensity of risks, vulnerabilities and environmental conditions.

3. Digital model of lifting (DML) and drainage network (D8, NGFlow and SCS)

Digital Elevation Model (DEM) is a major key element in any hydrological phenomena study. Undoubtedly, the topography of the terrain influences the hydrological flow. However, the advent of GIS based on DEM has facilitated the hydrological modeling at different scales of watershed areas [6]. Figure 1 depicts the DEM in the southern region of the island of Santiago (municipality of Praia), where water spatial distribution in the basin requires the use of spatial data, particular at the borders of the river basins and sub-basins, slopes and drainage channels. Certainly, all these topographic attributes are determined by the DEM.

With regard to the dynamics of hydrological processes and movement of soil, slope is another significant factor. The definition of a slope, [4] considers the formula $\varphi = \Delta Z \frac{1}{100X}$, where φ stands for slope, ΔZ for the altitude variation and x equals the distance between the center of any GIS cells. Thus, it is possible to determine, at the pixel level, the current direction of the water that will take, allowing the generation of an image with flow directions. These channels are identified as cell lines, whose flow accumulation exceeds a specific number of cells. By computing the number of cells above a particular threshold setup for the drainage network, one can determine the accumulated flow in that cell [7]. The algorithm that describes the flow direction became known as the Deterministic Algorithm 8 or D8 [8]. This methodology is based on the fact that water is able to move in eight possible directions, as shown in Figure 2.

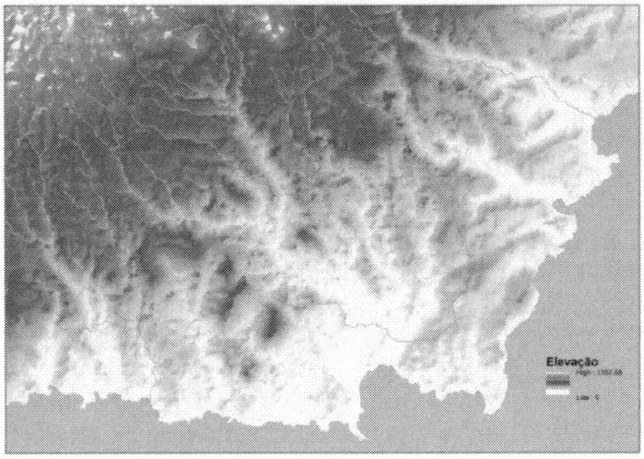

Figure 1. DEM of Praia County.

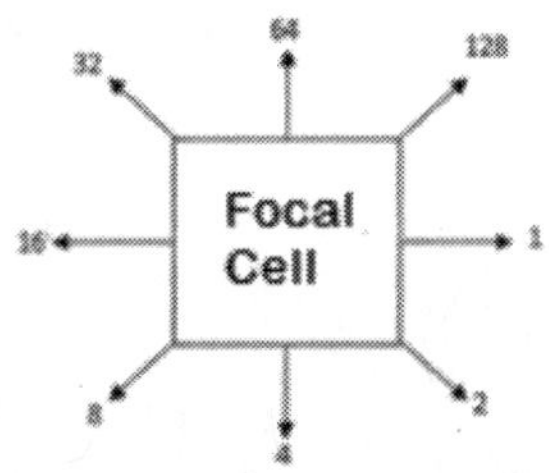

Figure 2. Directions codification of the flow at the cell level.

During the creation of the drainage system with GIS, the first step consists in filling small depressions existence on the model (fill sinks) that arise for geographical data entry errors. The correction of these depressions is an important step for the generation of the river system within the DEM. Thereafter, one delimits the contribution basins that are identified as a set of all cells that flow to a certain target cell from which drain lines are defined (Figure 3).

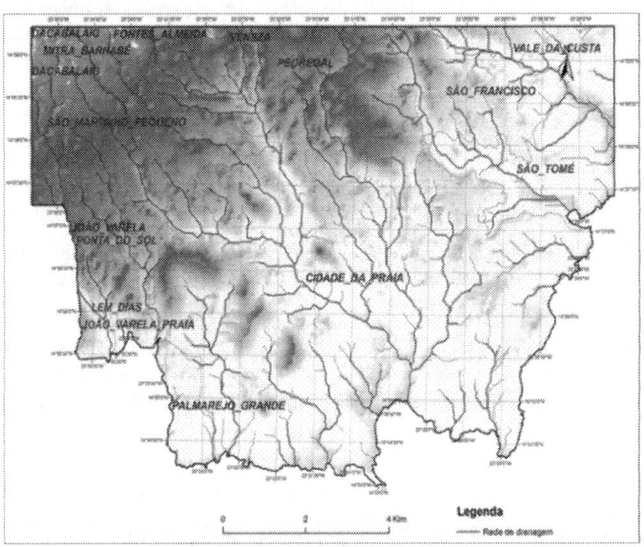

Figure 3: Draining network generated from a DTM.

In turn, the NGFlow model is based on the DEM as a spatial representation of weather stations, providing the precipitation data records and rainfall time series [9]. In situations where dams can be found, an increase of timely information is needed about the volume, length and release of water. To the input data, one adds the estimated water balance based on the precipitation data, soil storage capacity and potential evaporation. Therefore, the excess flow is formed by water that does not evaporate or infiltrates the soil. In order to calculate the excess runoff flow, we took into consideration equation 1. The left side of the formula describes the water absorption by the soil in a given time interval. Note that the absorption at t time corresponds to the absorption occurred in the previous time, plus the amount of precipitation minus evaporation. The right side of the equation describes the flow. While the soil capacity to absorb water is not met ($w(t) \leq w'$), the flow rate is zero. At the moment that the soil infiltration capacity is reached ($w(t) > w'$), it is then possible to determine the flow rate that corresponds to the water surplus (not absorbed by the soil in a given time interval).

$$\frac{W(T)}{\ddot{A}} = \frac{W(t-1)}{\ddot{A}} + P(t) - E(t) \quad S(t) = \frac{(W(t) - W')}{\ddot{A}} \tag{1}$$

Under this equation, $S(t)$ stands for the flow rate at a given t time of the simulation, $P(t)$ indicates the precipitation, $E(t)$ represents the evaporation; $W(t)$ means the soil capacity to absorb water while W specifies the water holding capacity of the soil where $\ddot{A}$ denotes the considered time interval.

The SCS-CN (Soil Conservation Service - Curve Number) method is widely used in hydrology for its simplicity and quality of results in estimating the direct surface runoff from a precipitation event [10]. In this model, the main elements that determine the surface runoff volume (effective precipitation) are the retention of rain on the ground depressions and infiltration. This effective precipitation estimate considers three variables: (A) precipitation in the time interval; (B) soil characteristics and moisture which defines the retention potential; (C) loss on early rainfall. As expected, the flow rate has an influence on the hydrological dynamics of soil due to its action on the surface, on the mass movement and on the flow time or concentration. The evaluation of this speed can be carried out by the Soil Conservation Service (Table 1).

Table 1. Speed in m/s based on the slope and soil occupation (Source: Soil Conservation Service, 1972).

Slope in %	Forests	Natural grasslands	Slope in %	Almost bare grounds
0-4	0.3048	0.4572	0-2	0.6096
4-8	0.6096	0.9144	2-4	0.6096
8-12	0.9144	1.2192	4-6	0.6096
12-15	1.0668	1.3716	6-10	0.9144
			10-12	1.2192
			12-15	1.52

The estimated speed computation shown in the previous table can, hence, be performed by equation 2 (S represents the slope or gradient). In turn, the coefficient K is estimated taking into account Table 2.

$$V = K \times S^{0.5} \tag{2}$$

Table 2. K coefficient value of the flow speed assessment (Source: Soil Conservation Service, 1972).

Types of areas	**Speed**
Forest with a lot of foliage on the soil	0.076
Area with little cultivation	0.152
Meadow-grasses	0.213
Cultivated land	0.274
Bare ground	0.305
Flowlines	0.457
Paved surface	0.61

The time of flow path (the water flow distance from one location to another) is also an important variable in hydrological modeling. According to NRC [11], the flow time between two points in a water basin is determined on the basis of the principle of equation 3 (*Tp*: flow path time; *n*: roughness coefficient; *l*: stream length; $P_{24}^{0.5}$: 24 hours average of rainfall over the last two years; *S*: slope or gradient).

$$Tp = \frac{0{,}007x(nl)^{0.8}}{P_{24}^{0.5} \times S^{0.4}} \tag{3}$$

The determination of the roughness coefficient presumes the assessment of soil conditions and occupation. To this end, the NRSC has set a coefficient, complying with these factors as it is described in Table 3.

Table 3. Index of roughness, according to NRSC [8].

Hard surfaces (concrete, asphalt...)	0.011
Fallow soil	0.05
Cultivated land	
With coverage < a 20%	0.06
With coverage > a 20%	0.17
Formation of meadow-grassland	
Grassland	0.15
Dense grassland	0.24
Bermuda grass	0.41
Forest	
Relatively high density	0.8
Low density	0.4

The calculation of the discharge assumes this rational method for determining the specific discharge of each pixel in the runoff basin. It can be computed with the SCS estimate as described in equation 4 (*Tp*: time; *Q*: discharge peak for each cell (m^3/s); *S*: pixel area (ha); *I*: precipitation average intensity (mm/min); α: flow coefficient).

$$Q = 0.167 \times S \times I \times \alpha \tag{4}$$

By definition, flow rate equals the volume of a fluid flowing through a given section of a free channel per time unit (a free channel can be a river or a pipe, for instance). Thus, one can determine the flow rate as $F = A \times V$, in which A stands for the area in question and V for the speed of flow expressed in m³/sec. In order to measure this speed rate of a water course, one typically uses the windlass (an apparatus provided with a propeller and a rotation meter). This measurement is universally used to determine the flow of a natural watercourse and consists in determining the cross-sectional area and the average speed in any section.

Finally, the surface runoff in rivers and channels is represented mathematically by two differential equations that describe mass or volume preservation and the quantity of flow movement, named the equations of Saint Venant [12] where Q equals volumetric flow rate, A is the area of the wet section, X stands for the distance in the longitudinal direction, t denotes time and qL represents the input or output flow per unit of width.

$$\frac{\partial Q}{\partial X} + \frac{\partial A}{\partial T} = q_L \tag{5}$$

4. The city of Praia

The archipelago of Cape Verde is located in the Atlantic Ocean, approximately 450 km from the West Coast of Africa, between the latitude of 14°48'00 and 17°12'13 " North and the meridians 22°23'59 " and 25°20'40 " West . This archipelago consists of 10 islands and 8 islets, organized into two groups, according to each geographic location: Windward and Leeward. The archipelago has a total land area of 4033 km^2 whose corresponding coast line extent is 1,020 km.

Cape Verde's climate is part of an extensive range of arid and semi-arid climates in the heart of the Sahelian region (situated between the subtropical high pressures of the Atlantic and the equatorial low pressures). This track of African territory (Senegal, Mauritania, Mali, Burkina Faso, Chad, Sudan, Ethiopia, Erithrea, Djibouti and Somalia) has a precipitation regime, ranging from 150-300 mm per year.

The county of Praia is located in the south of the Santiago island and covers an area of 102 km^2 (about 10% of the surface of the island of Santiago). It is bordered to the North by the mountain of Antonia Peak and to the South by the limits of the maritime coastline. Administratively, it has borders with the municipalities of São Domingos and Ribeira Grande de Santiago.

Towards morphology, the elevations of the islands of Santo Antão, St. Nicholas, Santiago and Fogo reach an altitude above 1,300 m. Yet, the highest altitude corresponds to Pico on Fogo Island, with 2,829 m, followed by the Crown Top on the island of Santo Antão with 1,985 m. The Antonia peak in Santiago reaches 1,395 m and Monte Gordo, in St. Nicholas, presents 1,304 m. Considering the geomorphological characteristics of Praia, there was a sharp erosion lately and, along with the poor vegetation cover, it creates a positive situation for dynamic erosion (Figure 4). The existing valleys in the area are relatively well embedded between the highlands (achadas), among which some relevant elevations stand out, such as the Hill of Vacas with 392m, the Ilhéu or Hill of São Filipe with 274 m, Red Hill with 195 m, Hill of Gonçalo Afonso with 235 m and the Hill of Ilhéu with 259 m.

The drainage networks extend from the highlands of the North (Hill of Rui Vaz) to the South of the county, flowing into the Atlantic. As it can be observed, there is a main basin where there is a set of flow networks that converge downstream. It turns out that the urban area of the municipality is located at the crossroads of these drainage networks (Figure 5). However, the existing security infrastructure cannot always withstand the water discharges coming from upstream areas.

IOP Conf. Series: Earth and Environmental Science **37** (2016) 012071 doi:10.1088/1755-1315/37/1/012071

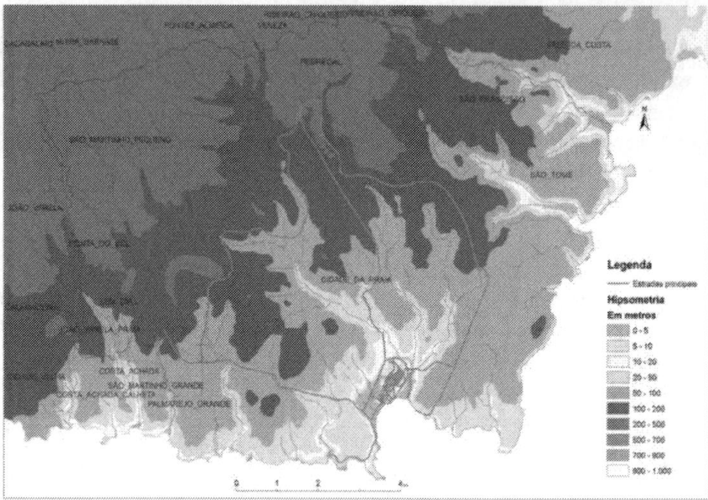

Figure 4: Hypsometric southern map of Santiago Island.

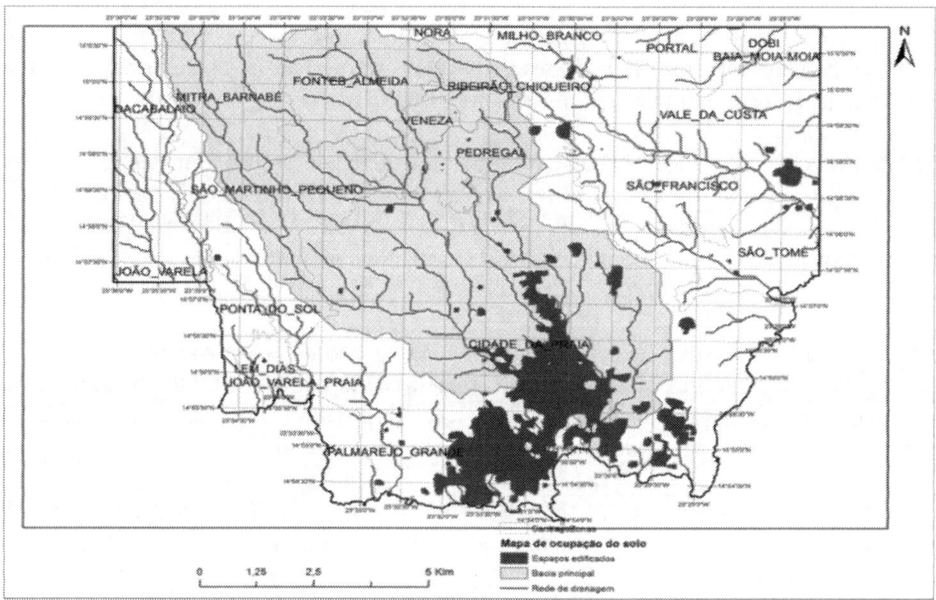

Figure 5: Drain main basin, edification areas and network drain.

The average annual rainfall in the Cape Verde archipelago is no more than 300mm in the lowlands (less than 400m high). According to [13], the precipitation of downward trend has been recorded since the 60s, reflected negatively on the conditions of farming and water supply. The areas under the influence of hot and dry air masses that blow from the eastern quadrant during the Harmattan (December/January) are the driest, corresponding to an average rainfall of only 150 mm. In spots located more than 500m above sea level and exposed to the trade winds, rainfall can exceed 700 mm. In relation to water balance, it is considered that about 20% of the precipitation water is lost by surface runoff, 13% goes towards aquifers recharging and 67% will disappear by evaporation [13].

However, extreme precipitation in the form of flash floods is a phenomenon that occurs in all latitudes, including arid regions and desert climates. This phenomenon has been a reason for concern to the scientific community due to its increasing frequency as a result of recent climate changes. These effects of extreme precipitation are sometimes more serious due to the existing lack of vegetation cover that could reduce the impact on the ground. Moreover, notwithstanding the existence of side protection dikes, certain cross-sectional passages of circulation in the city streams do not hold the sufficient height to allow the flow of water when intensive precipitation occurs (Figure 6).

Figure 6: Superficial draining at Safende riverside.

5. The hydrological methodology

The present hydrological risk simulation is based on a model with three main components: (A) Flow module, in which considers the various layers of soil types/occupation and of precipitation; (B) Flow path time module, in which the DEM and flow rate are considered; (C) Discharge module, which results from the interaction of the fallen precipitation and the pedagogical and geomorphological processes characteristics. As expected, the different layers interact together, having an influence on the runoff dynamics. For this research, the water discharge at the end of the drainage basin was simulated. The graphical user interface provided by the Model Builder is significant, as most of the functions are already integrated within ArcGIS (Figure 7). The first component is the flow. Based on the SCS procedure, several parameters are taken in consideration such as the Curve Member, the initial infiltration and the potential one. These elements play a key role in the final activity when determining the flow coefficient that is obtained by the ratio between the total precipitation and the flow amount. This calculation presupposes the raster format associated to all layers: soil type, foregoing moisture of soil (amount of moisture before precipitation), land use, precipitation and DEM. The intermediate results obtained are the curve number, initial infiltration and maximum potential of moisture retention.

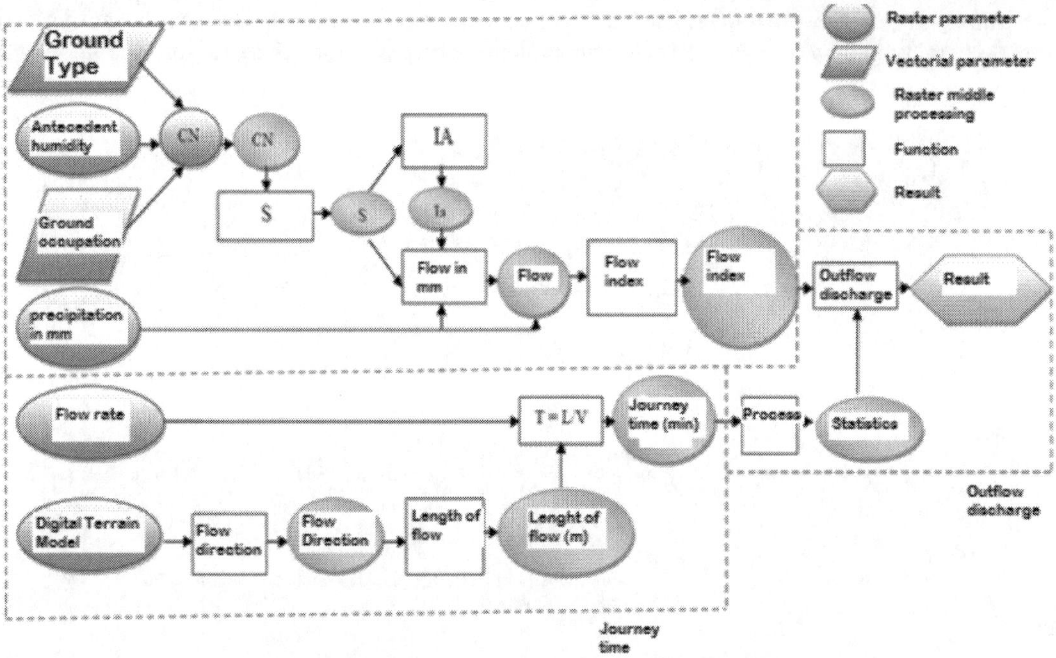

Figure 7: Simulation model for the hydrologic flow.

For framing soil types, it was considered the different soil groups according to the FAO classification [14]. The map of Figure 8 shows the predominant soil types of the county of Praia. As demonstrated, there is a predominance of the Xerossol class that is distributed throughout the northern and northwestern regions of the county of Praia. These soils display a reddish coloring of great thickness. Next, in terms of expressiveness, there are the Cambisol soils, which are little developed soils from non-limestones formations. These soils are usually associated with rocky outcrops, often being characterized by a high proportion of stony elements. One can also identify lithosols belonging to the group of non-climatic erosion mineral soils, which are young, little evolved and associated with outcrops of hard consolidated rocks and basalts. There is still a small portion of undefined soils, which occupies the Plateau neighborhood of Praia.

The foregoing humidity is a parameter that can be modified, depending on the time acclimatization: condition I for dry soils, condition II for moderate soils and condition III for wet soils. Under the land layer drawn from Landsat TM5 images (Figure 9), one can see a predominance of bare ground and rocks in over 80% of the county of Praia's area. However, there is some scattered green area, consisting of bushes and acacia trees, a species that was introduced in the last 40 years, framed in the national afforestation program.

The discharge makes up the final phase of the flow process. This index takes place at the pixel level (see equation 4 and 5) and, subsequently, it is followed by the accumulation of the discharges, resulting on the flow output. As expected, the total flow in the drainage basin is obtained by the sum of discharges. Finally, we proceed to the discharge values reclassification by the equidistance of the flow path time. In this case, we selected a temporal equidistance of every 30 minutes.

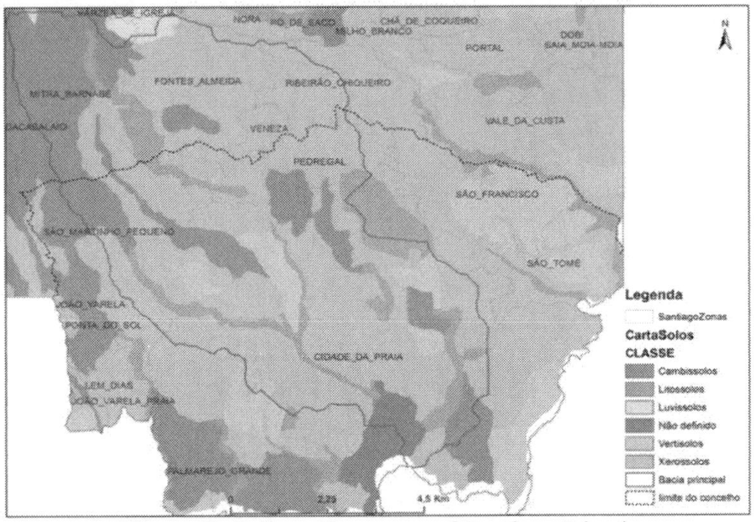

Figure 8: Soil southern map of Santiago Island.

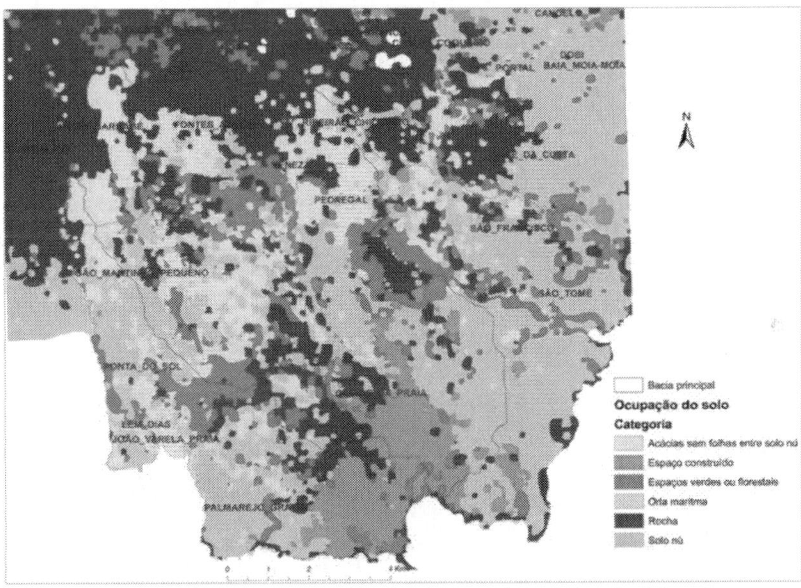

Figure 9: Soil map occupation in the southern island of Santiago.

6. The county of Praia: simulation findings

The precipitation data was obtained from the sensor 3B42 of the TRMM satellite. The historical rainfall records indicate some variation of rainfall occurring in the county. As a rule, higher areas receive a greater amount of rainfall as shown in Figure 10 [4]. The central region of the island matches with the areas of higher altitudes whilst lower precipitation rates are situated along the coast. As a reference, the county of Praia presents an annual variation between 73-163 mm of rain.

IOP Conf. Series: Earth and Environmental Science **37** (2016) 012071 doi:10.1088/1755-1315/37/1/012071

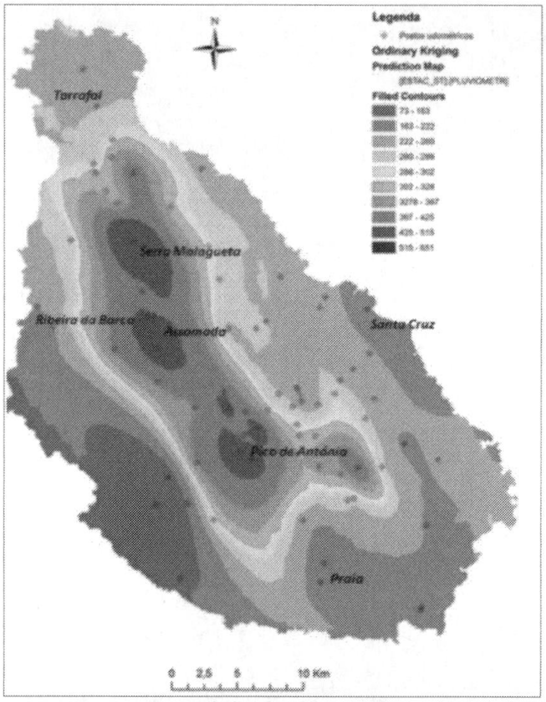

Figure 10: Average precipitation map of Santiago Island [4].

It is particularly notable that the TRMM satellite estimates precipitation in a resolution of 0.25 x 0.25, corresponding to an area of 2,162 km^2. Given the size of Santiago, approximately 900Km^2, this resolution covers 2.4 times the island of Santiago. In order to carry out this study, it was considered the rainfall data occurred on October 22, 2010 (Figures 11 and 12).

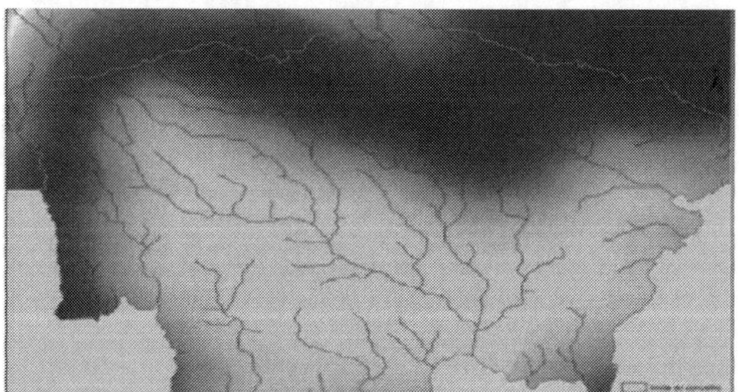

Figure 11: Precipitation map on October 22, 2010.

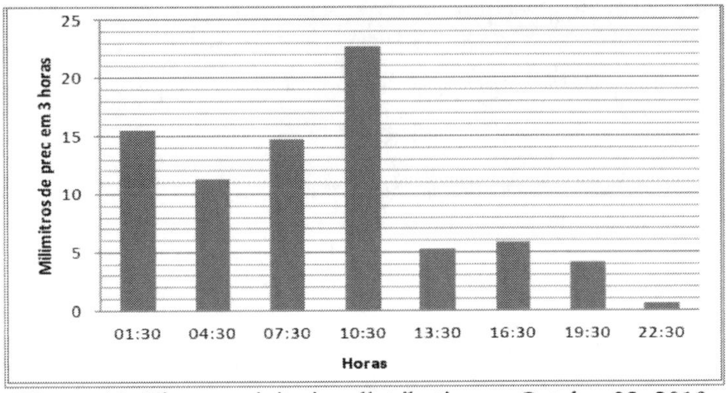

Figure 12: Time precipitation distribution on October 22, 2010.

By definition, the potential infiltration presents the soil's capacity to retain water. Figure 13 illustrates the sectors with greater retention capacity, according to the existence of some dispersion towards the infiltration potential level. These differences are linked to the pedagogical features, degree of vegetation cover, soil occupation and land use. As one might expect, the built-up areas, rocky zones and bare ground spaces are generally of lower potential retention, implying a higher degree of surface runoff.

The flow path time is another significant variable as far as it influences the flow volume in a specified section of the basin, either in volume or in terms of duration. The approach used for determining the flow time is based on [15] formula. In a simplified manner, it defines the flow time as the ratio between the flow length and the average speed over the entire area of the basin. Subsequently, the stream length is used to estimate the flow time for its relationship with the average velocity of the basin flow. Afterwards, this flow time is reclassified every 30 minutes and applied to the basin for the delimitation purposes of the different patterns. Thus, spaces are shown after having been reclassified according to the flow path time. Figure 15 characterizes the flow time of surface runoff generated from the ratio between the watercourses length in different parts of the basin and the average speed recorded across the basin. Certainly, the flow time is shorter as it moves towards downstream.

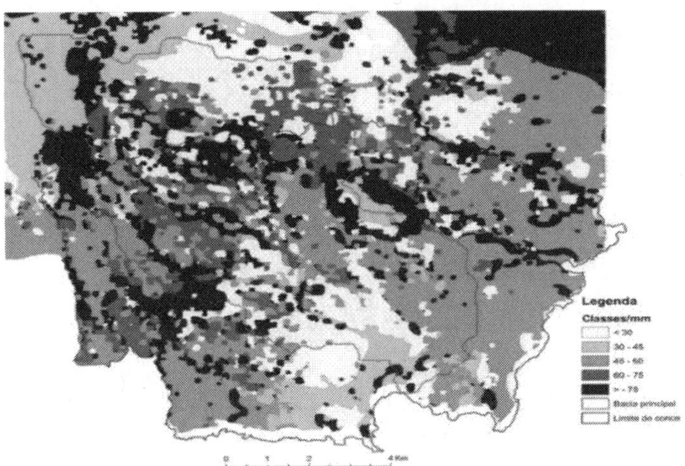

Figure 13: Potential infiltration map in mm at Praia County.

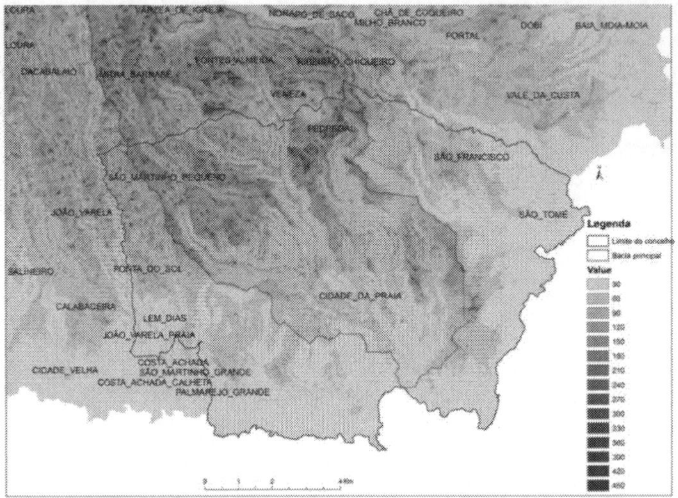

Figure 14: Time map of flow route.

The flow speed involves the *K* coefficient and the relief slope of the basin. After an analysis of the basin conditions for determining this constant, we chose the value 0.305, resulting from the fact that most of the space within the drain basin is almost a bare soil. It is noted that the average flow velocity is 1.9 m^{-1}, with the minimum speed of 0.053 m^{-1} and a maximum of 6.49 m^{-1}. As expected, higher speeds occur in the areas of greater slope, which are well represented in regions further north of the county, coinciding with the most mountainous areas.

The discharge is the last element obtained from the overall model. This is obtained by the interaction of rainfall parameters in mm/min, flow time and infiltration potential (soil capacity to retain water). From this latter factor, we may obtain the flow coefficient (effective precipitation). Subsequently, the precipitation that was initially in mm per hour is now converted to mm per minute in order to make the discharge reclassification in every 30 minutes easier. It should also be noted that the discharge is initially obtained at the level of the pixel and reclassified in areas depending on the flow path time by ArcGIS 10 Spatial Analyst (Zonal Statistics function).

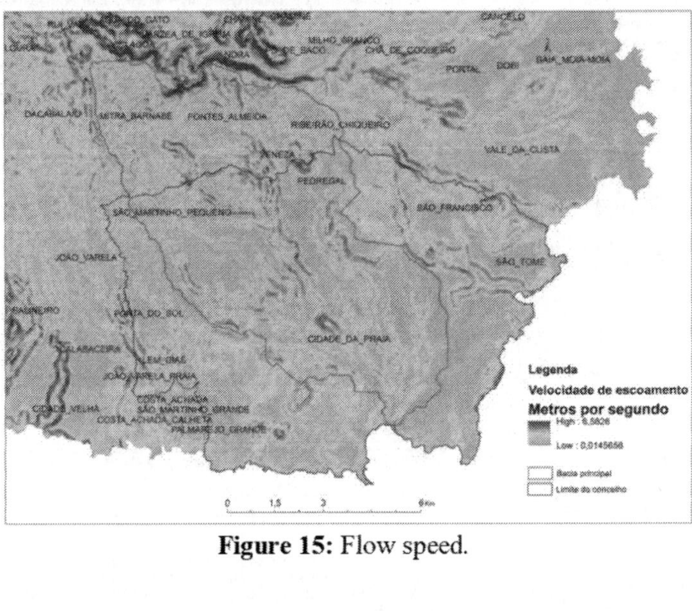

Figure 15: Flow speed.

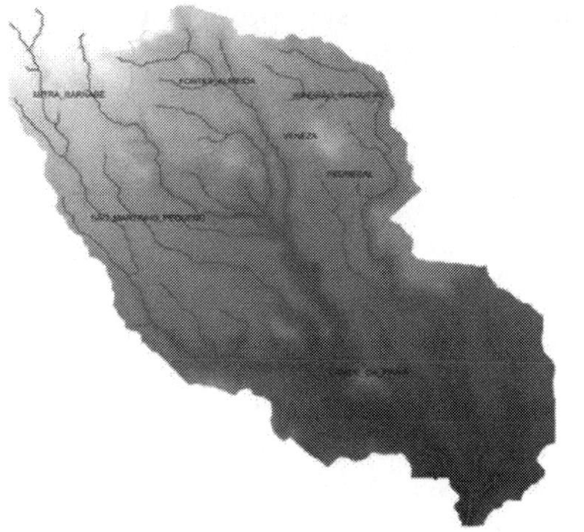

Figure 16: Basin flow at Praia county and the water volume drained in m^3 during 450 minutes.

Figure 16 shows the discharge result from the precipitation data. This allows the water volume evaluation in cubic meters that crosses throughout the final section at intervals of 30 minutes for 8 hours. It is demonstrated that 76% of the precipitation volume stems from the flow coefficient that passed through the final part of the basin in 120 minutes and 99% crossed in 300 minutes.

Finally, Figure 17 shows the flow to the main basin, already reclassified at an equidistance of 30 minutes in terms of flow time. There is, however, some inconsistence of the results for the sectors with lower discharge that occurs further within the basin region, supposedly associated with areas of low permeability and rapid surface runoff.

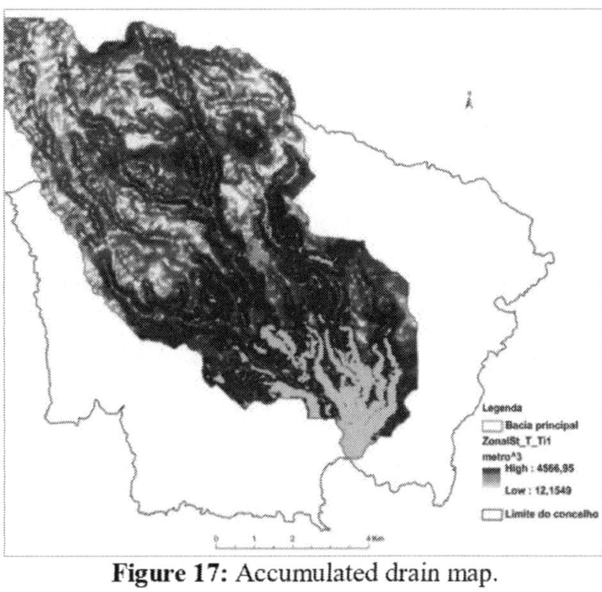

Figure 17: Accumulated drain map.

7. Conclusion

In the county of Praia, there is a large population growth living under a hydrological risk area, especially along the water streams. The morphology and configuration of the surface runoff network in the surrounding areas of this county increase that risk mainly due to the confluence of several basins that flow into the urban area. Indeed, the areas of greatest hydrological risks in the county of Praia are in the lowland areas (along the flow streams). Despite the existence of lateral protection dikes in these streams, there is water overflow in some zones (particularly in Vila Nova creek) because of the large amount of material carried from the slopes and the highland areas of Fontes Almeida, St. Martinho and Mitra Barnabé. Moreover, there is a certain shortage of infrastructures for torrential correction in several ramps of the city and along the side protection dikes delimiting the rivers, where one can find a high density of soil occupation, especially in the settlements of San Pedro, Safende, Calabaceira, Vila Nova and Fazenda.

In this sense, it is recommended interventions in these risk areas in order to protect households that are at the edge of these flood lines, despite the financial costs. The intervention on these slopes becomes a priority in order to change the arrangement of the surface runoff lines. This intervention may consist of afforestation and construction of grooves to prevent excessive transport of soil debris toward the low lying areas of the city during precipitation.

Certainly, GIS solutions can play an essential tool for the surveillance and prevention of hydrological phenomena in this municipality. GIS adds a set of spatial information, facilitating the identification of existing vulnerabilities. For example, in this project we tried to define a model to characterize the hydrological risk flexible enough to operate at different scales or to be applied in other geographical areas with the necessary adjustments. Municipal authorities are, therefore, in a position to setup a warning and risk prevention system, as well.

8. References

[1] Pedrosa & Perreira (2005). Diagnóstico dos factores condicionantes das susceptibilidade face ao risco de inundação urbano no concelho de Matosinhos. Territorium.

[2] Nascimento, J. (2009). La croissance et le système de gestion et de planification de la ville de Praia. Universite de Rouen: Tese de Doutoramento.

[3] Amaral, I. (1964). Santiago de Cabo Verde. A Terra e os Homens. Lisboa: Memórias da Junta de Investigação do Ultramar.

[4] Silva, J., Monteiro, P., Negreiros, J.: Spatial Modulation of Rainfall in Santiago, Cabo Verde, with Geostatistical Analyst, International Association for the Scientific Knowledge - InterTIC International Conference, Pedro Acosta &Pedro Cravo (Eds), ISBN 978-989-06-04, pp. 272-276, Oviedo, Spain (Dec 10).

[5] Lelièvre, M. A. (2009). L'approche hydrogéomorphologique pour la cartographie des zones à risque d'inondation dans les vallées de petites et moyennes tailles: un exemple commenté pour la vallée de la rivière-au-renard. Géorisques IV Geohazardes. Georisques IV Geohazardes.

[6] Gesller, P. (1996). Modeling soil-landscape and ecosystem proprieties using terrain attributes. Australian National University.

[7] Ferraz, B. F., Milde, L. & Mortatti, J. (1999). Hydrologic models attached to Geographic Information Systems: a case study. Revista de de Ciência & Tecnologia.

[8] Jones, R. (2002). Algorithms for using a DEM for mapping catchment areas of stream sediment samples. Computers & Geosciences.

[9] Santana, I. D. (2010). Modelo hidrológico NGFLOW aplicado à bacia hidrográfica do Alto Aroguaia.

[10] Jain, K. & Mishra, K. (2006). Evaluation of AMC-dependent SCS-CN-based models using watershed characteristics. Water Resources Management.

[11] Natural Resources Conservation Service (2010). Hidrology National Engeenering Handbook. USA: Unite States Department of Agriculture.

[12] Tucci, C. (2002). Hidrologia: ciência e aplicação. Editora UFRGS/ABRH.

[13] INMG (2003). Instituto Nacional de Meteorologia e Geofísica. Clima de Cabo Verde.

[14] UNESCO (1988). Soil map of the world: revised legend. World Soil Resources Report 60. Roma, Itália.

[15] Domnita, M., Ionut, A., Cracium, Haidu, I. & Magyari, S. (2010). Geographical Information System module for deriving the flash flood hydrograph in mountainous areas. 4th European computing conference.

Assessment of drought impacts on vegetation health: a case study in Kedah

M Othman[1], Z H Ash'aari[1], F M Muharam[2], W N A Sulaiman[1], H Hamisan[1], N D Mohamad[1] and N H Othman[1]

[1] Department of Environmental Sciences, Faculty of Environmental Studies, Universiti Putra Malaysia, 43400 Serdang Selangor, Malaysia.
[2] Faculty of Agriculture, Universiti Putra Malaysia, 43400 Serdang Selangor, Malaysia.

Email: zulfa@upm.edu.my

Abstract. Prolonged drought in the early of 2014 has caused Malaysia to experience water supply shortage which directly affects both health and growth of vegetation. Thus this study aims to assess the risk vegetation areas that were impacted during 2014's drought by integrating the Standardized Precipitation Index (SPI) and Normalized Differentiation Vegetation Index (NDVI) methods. These two methods were able to assess the risk areas for the vegetation by measuring its health and classifying them according to its severity while considering the rainfall reduction at the specific time and location. The results obtained from this study shows that the central and north west of Kedah was vulnerable to the occurrence of drought. Kedah was more impacted by the dry event during the northeast monsoon. This study is significant as a fundamental input for further research and as an alternative approach by the application of space technology.

1. Introduction

In recent years, the increasing numbers of extreme weather had caused lots of damages to many sectors such as agricultures, fisheries and societies' development. One of the main concerns affecting Malaysia is the occurrence of drought events which lately happens with high intensity for a long duration of time. In early 2014, Malaysia has been hit by a prolonged drought. Malaysia is in the midst of three-month-long heat waves that has effected on the surrounding environment and communities [1] [2] [3]. In addition, the effects of the El Niño phenomenon was felt in Peninsular Malaysia at the end of 2014 and early 2015 [4]. During this period, vegetation will be more stress as it lack of water supply to store the moisture which usually reduces the length of the growing seasons.

Drought has been a big issue which can impact the society and economy as well. Therefore, the natural resources management in every country need rapid and continuous information about the changes of the variables. In a large scale, drought can cause death of animals, plants and even men when there is a shortfall of precipitation in a long time period that led to water shortage problems. Thus, monitoring and managing of this occurrence need a proper quantification economically and environmentally. The impacts of drought can be economic, social and environmental depending on how the surroundings reflect to the drought itself. There is a complex link between the drought events where the impact from the drought could be measure in the large scale and when experiencing a physical drought. Drought can affect its target either directly or indirectly. The direct impacts of drought are forest fires, rangeland, reduced crop, reduced in water level, damage to fish habitat, and increased wildlife and livestock mortality rates. On the other hand, indirect impacts occurs as a consequence of direct impacts, for instance, a reduction in crop production and farmer's income, increase in market value of goods, migration, and others. In term of economic impact, Malaysia will face declination of 30% GDP if the drought continues for a long period after March [2]. In addition, the situation that happened in Malaysia during the prolonged drought impacted the oil palm production and followed by a decrease in the crude percentage if any long stretch of the hot weather continues a few months more. This is a challenging situation as water is a very important component to the palm oil growth [2].

There is no direct definition for droughts [4] and it is hard to measure as there are various ways and indices that can be used to define drought. Generally, drought can be classified into four groups that are meteorological drought, agriculture drought, hydrological drought and socio-economic drought. The first three types of drought are in respect of measuring the drought as a physical phenomenon and the last is about the supply and demand, in term of the impact of water shortage throughout the socio-economic system.

Table 1. Types of drought classified by the National Drought Mitigation Centre, University of Nebraska-Lincoln.

Types of drought	Explanation
Meteorological	Defined as related with the total of precipitations. It is comparing the precipitations at certain places and at certain period of times off the average precipitation for the places ("normal" or average amount). Thus, meteorological drought is regional-specific.
Hydrological	Hydrological drought is associated with the effect of the precipitation deficiency on surface or subsurface of water sources. Usually this type of drought is identified after meteorological drought.
Agriculture	This drought happens when there is lack of water for crops that links with several of characteristics of meteorological or hydrological droughts to agriculture impact such soil water deficits, reduced groundwater or water reservoir, precipitation deficiency and differences between actual and potential evapotranspiration.
Socio-economic	It is implies all the element of meteorological, hydrological and agricultural drought in term of the demand and supply aspect of economic goods. It happens when the demand of the economic goods exceeds it supply due to the drought event.

There are several indices that measure the distribution of precipitation in a period of time such as the Palmer Drought Severity Index (PDSI), Standardized Precipitation Index (SPI), Percent of Normal, Surface Water Supply Index (SWSI) and few more indices. Each of the indices works differently depending on the need that arises. PDSI was developed to measure the departure of the moisture supply [6] where the parameters needed are the precipitation and the temperature data. It is believed to be more complex than the Standardized Precipitation Index. However, PDSI held an advantage since it takes into account the basic effect of global warming through potential evapotranspiration. The index is roughly ranging from -4.0 to +4.0 and some researchers suggested that PDSI is suitable for agriculture but does not accurately represent the hydrological droughts [7]. This index however has many other problems related to the calibration and spatial comparability [8] [9] [10].

Standardized Precipitation Index or better known as SPI is an index that is simple and easy to be calculated and statistically relevant. This index is developed by Mckee and others in 1993 to understand that the rainfall deficit has different impact of groundwater, reservoir storage, soil moisture, snowpack and stream flow. A historical monthly precipitation data of 30 years is needed in order to calculate the SPI [11]. The SPI was designed to quantify the precipitation deficit for multiple time scales. These time scales reflect the impact of drought to the availability of different water sources. Soil moisture condition responds to the precipitation anomalies on a relatively short scale while groundwater, stream flow and reservoir storage reflect the longer term of precipitation anomalies. Thus, the SPI was calculated for 3-, 6-, 12-, 24- and 48- time scales originally [7]. A drought event occurs any time the SPI is continuously negative and reaches an intensity of -1.0 or less. The event ends when the SPI becomes positive and a wet event take place [7]. Each drought event therefore has a duration defined by its beginning and end, and the intensity for each month that the event continues. The positive sum of the SPI for all the months within a drought event can be defined as the drought's "magnitude". In this study, SPI is chosen as the index to measure the drought condition in Kedah based on ability and the suitability of the index.

The advancement of the technology to detect a slow disaster like drought is a very helpful solution and initiative that provide a meaningful data to deal with this kind of event. Therefore, with the help of environmental space sensors, drought can be detected 4-6 weeks earlier and delineated more accurately [12]. Geographic information system (GIS) and remote sensing (RS) have played an important role in studying different types of hazards [13]. Remote sensing techniques also provide variations of resolution for the image needed to be used. Satellite derived drought indicators calculated from satellite-derived surface parameters have been widely used to study drought such NDVI and VCI. The Normalized Difference Vegetation Index (NDVI) is a numerical indicator that uses the visible and near-infrared bands of the electromagnetic spectrum, and is adopted to analyse remote sensing measurements and assess whether the target being observed contains live green vegetation or not. The wavelength of visible lights are absorbed and reflected by the green plants on the Earth surface [14]. Generally, healthy vegetation will absorb most of the visible light that falls on it, and reflects a large portion of the near-infrared light. Unhealthy or sparse vegetation reflects more visible light and less near-infrared light. Bare soils on the other hand reflect moderately in both the red and infrared portion of the electromagnetic spectrum [15].

NDVI is a good indicator of green biomass, leaf area index and patterns of production [16] [17] and it is the most commonly used vegetation indices in the field previously mentioned. Theoretically, NDVI values are represented as a ratio ranging from -1 to 1. A low NDVI value usually shows the areas of barren rock, sand or snow (0.1 or less) while a moderately value of NDVI shows sparse vegetation like shrubs and grasslands (0.2 to 0.5). Dense vegetation like temperate and tropical forest will reflect a high NDVI value which approximately 0.6 to 0.9.

Several studies have been devoted towards drought with the aid of satellite-derived information. The relationship between NDVI and rainfall is known to vary spatially due to the effects of variation in properties such as vegetation type and soil background [18] [19], with the sensitivity of the NDVI values to fluctuations in rainfall, therefore they are varying regionally. There was a study related to the early detection of drought in East Asia was done previously where a standard NDVI and up-to-date NDVI were calculated to derive different NDVI image, of which the intensity and agricultural area damaged by drought was detected [20]. The difference image was used to create a drought risk maps. The study was successful in detecting and monitoring drought effects in agriculture. Previous research was able to utilize NDVI time series and gridded precipitation estimates at different spatial resolutions to investigate temporal and spatial patterns of vegetation greenness and was also able to explore the relationship between rainfall and vegetation dynamics. [21].

To fill this gap, this study aims to assess the vegetation health of Kedah in year 2014 by combining two types of drought assessment which are the meteorological part using SPI and the agricultural part using the NDVI. Integration of both indices was used to identify the exact vegetation areas that are at risk which help authorities in future plan and mitigation measures.

2. Methodology

2.1. Study area

Kedah is located at the northwestern part of Peninsular Malaysia which is 6.1184° N and 100.3685° E. It comprises of 12 districts with a total of geographical area of 9427 square kilometers and a total of over 2,000,000 populations. The landscape is dominated by paddy field and Kedah is the main rice producer of the country. There are dense rainforests on the highland, mangrove forest on the west coast of the state and practically two third of the state is forest area and some of it is reserved for legal lodging activities while few of it are opened to the public as a forest recreational park.

Based on Koppen-Geiger climate classification system, Kedah experienced a tropical climate and received a significantly rainfall that classified in the Af group (equatorial rainforest and fully humid) [22]. Even in a dry season the state is still experiencing rainfall. As other peninsular state of Malaysia, Kedah face the changes of the wind flow pattern that can be distinguished as Southwest (SW) monsoon, Northeast (NE) monsoon and inter-monsoon seasons. Because of some local topographic features, the rainfall distribution pattern in Kedah is affected and there was a previous study done showed that during NE monsoon, Kedah was one of the driest regions [23] in Malaysia. Figure 1 shows the rainfall station used in the study and information was obtained from the Department of Irrigation and Drainage Malaysia (DID).

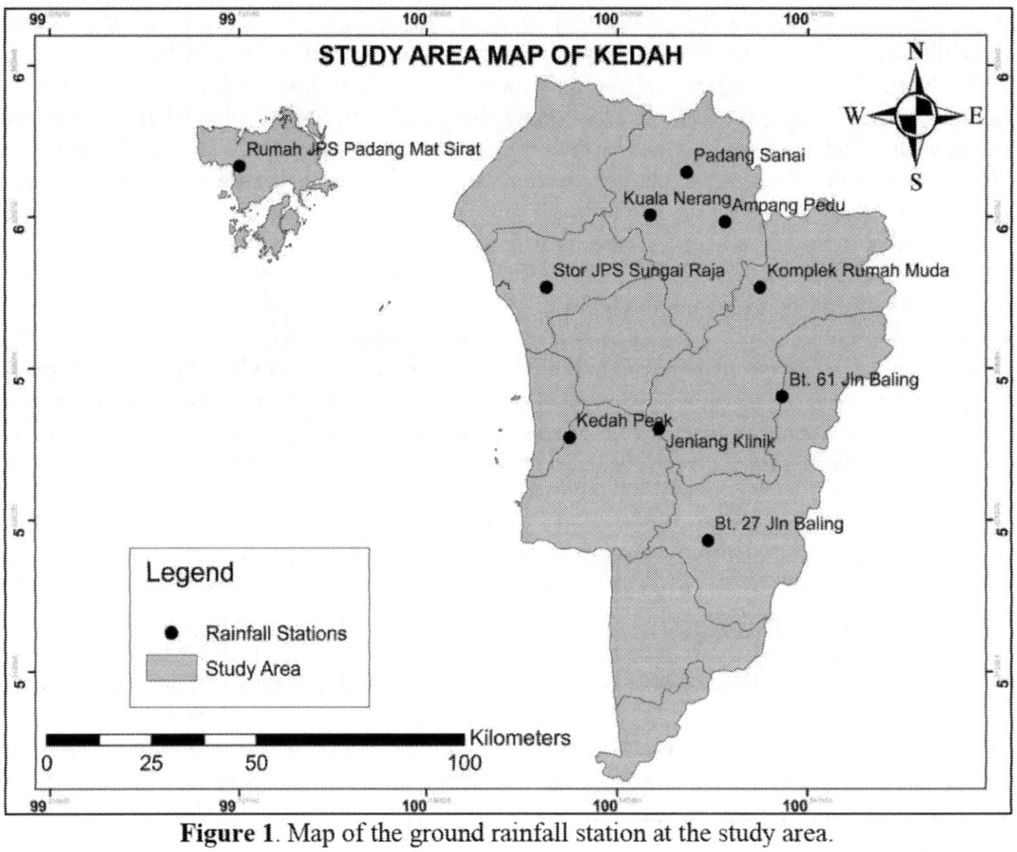

Figure 1. Map of the ground rainfall station at the study area.

2.2. Data acquisition

2.2.1 Meteorological data. Daily rainfall data for the period of 30 years from 1984 to 2014 for 10 stations was obtained from the DID Malaysia. Only ground rainfall station with 30 years of rainfall data was chosen in this study.

2.2.1 Satellite image. Landsat 8 OLI and TIRS data were obtained from the U.S. Geological Survey Earth Resources Observation and Science (EROS) Centre Landsat archive. The Landsat archive consists processed data of Level 1 Terrain corrected (L1T) level defined in the GeoTIFF file format in the Universal Transverse Mercator (UTM) map projection. Landsat 8 satellite has OLI and TIRS sensor on board to capture earth's surface with 16 days temporal resolution. Two images from February and July 2014 was selected from Landsat8 and the selection was based on the amount of the cloud coverage of the image and its consistency with the monsoon seasons that occurred in Malaysia.

2.3. Data processing

2.3.1. Meteorological data. The data has undergone statistical treatment due to some daily missing values. Even though the SPI program can be run with missing data however; it will affect the confidence level of the result obtained, depending on the distribution of the missing data [24]. Based on the SPI classification system a drought event will ends when a positive values start to appear, as negative values indicate dry spell while positive values indicate wet spell [7]. Interpreting a 3-month SPI provides a seasonal estimation of precipitation and reflects short-term and medium-term moisture conditions [24]. The SPI values are very helpful in obtaining the time and areas of interest of a dry spell occurrence. Interpolation method was then used to predict the values for areas that were far and not representable by the existing ground rainfall stations. Inverse Distance Weight (IDW) was chosen to predict the values by using the collected data points which was the monthly average of the SPI values.

2.3.2. Satellite image data. The image used in this study was corrected geometrically and atmospherically. The atmospheric correction was done to remove the cloud covers. Next procedure was to derive the NDVI values from the image by transforming the raw satellite data into NDVI values ranging from -1 to +1.

2.3.3. Vegetation health map. The last step in this study was to produce the drought vulnerability map by using the combination the SPI and the NDVI map. Co-kriging method was used due to its ability to take advantage of the covariance between two or more regionalized variables that were related and appropriate when the main attribute of interest is sparse, but related secondary information (seismic) is abundant. Ordinary co-kriging was similar to simple co-kriging in which the mean was assumed constant, but it differs in terms of the mean was estimated locally within each set of neighborhood control points, rather than being specified globally.

3. Results and discussion

3.1. Temporal variation of SPI -3

Drought risk area has been identified using SPI in Kedah for the year of 2014 by utilizing 30 years of historical rainfall data that was interpolated into the map for the identification process. SPI during the year of 2014 has been presented into two ways which were temporal and spatial pattern. Based on figure 2, the drought intensities can be determined by observing the values following the classification system by Mckee and others in 1993 [7]. Due to high humidity in Malaysia climate [25], the modified classification system as shown in table 2 was referred during the process of classifying the SPI result and all stations showed consistent variation with the driest peak on March and the wettest peak on December. A stockpile of values less than –1.0 were grouped together from January to April which indicate severe drought event while there were abundance of values that was more than +1.0 mostly from May to July and October to December that indicates severe wet seasons. Kedah Peak station have the lowest value of -1.56 which makes it the driest regions and Kuala Nerang station recorded the highest value of +2.39 in December which make it the wettest region among the entire part of the study area.

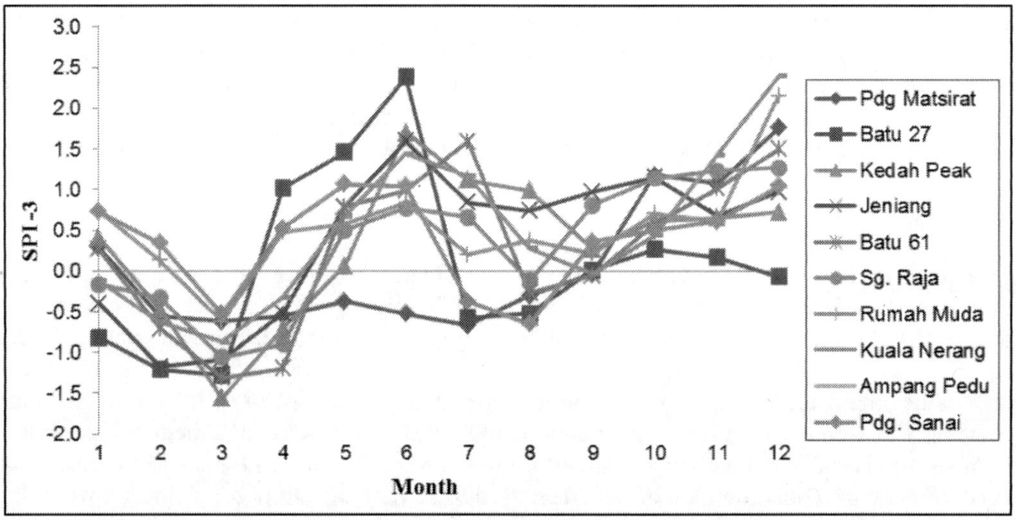

Figure 2. The monthly variation of the SPI -3 in 2014.

Table 2. Modified SPI classification system.

SPI value	Condition
More than 1.0	Severely wet
0.5 to 1.0	Moderately wet
0 to 0.49	Normal
-0.01 to -0.99	Moderately dry
Less than -1.0	Severely dry

3.2. Spatial variation of SPI -3 during Southwest (SW) monsoon

The interpolation of the SPI-3 variations for the study area in 2014 during SW monsoon season was shown in figure 3. SW monsoon was the driest season throughout the country and it experienced monthly minimum rainfall from late May to September [26] and it is a dry period with occasional short dry spells [27]. Result shows that Langkawi was the area with the most recurrent event of dry period compared to the other area in Kedah. Langkawi also experienced minimum rainfall received over the SW monsoon. Observations on the whole part of the study area shows that it experienced a wet season during May and June while certain area faces dry condition during July. The intensity of the dry period that occurs in July was high at the northern and southern part of Kedah. During August, the entire state of Kedah was in a moderate dry condition while the area represented by the Kedah Peak station recorded a moderately wet condition. Generally, the study area was not so affected by the SW monsoon and it was proven by a previous research where SW monsoon caused rainfall reduction on the western part of peninsular since all the rainfall indices that are tested were higher than other parts of the Peninsular Malaysia [23]. Thus, the study area which is located in the northwest region of peninsular was considered to be the wettest region during this season.

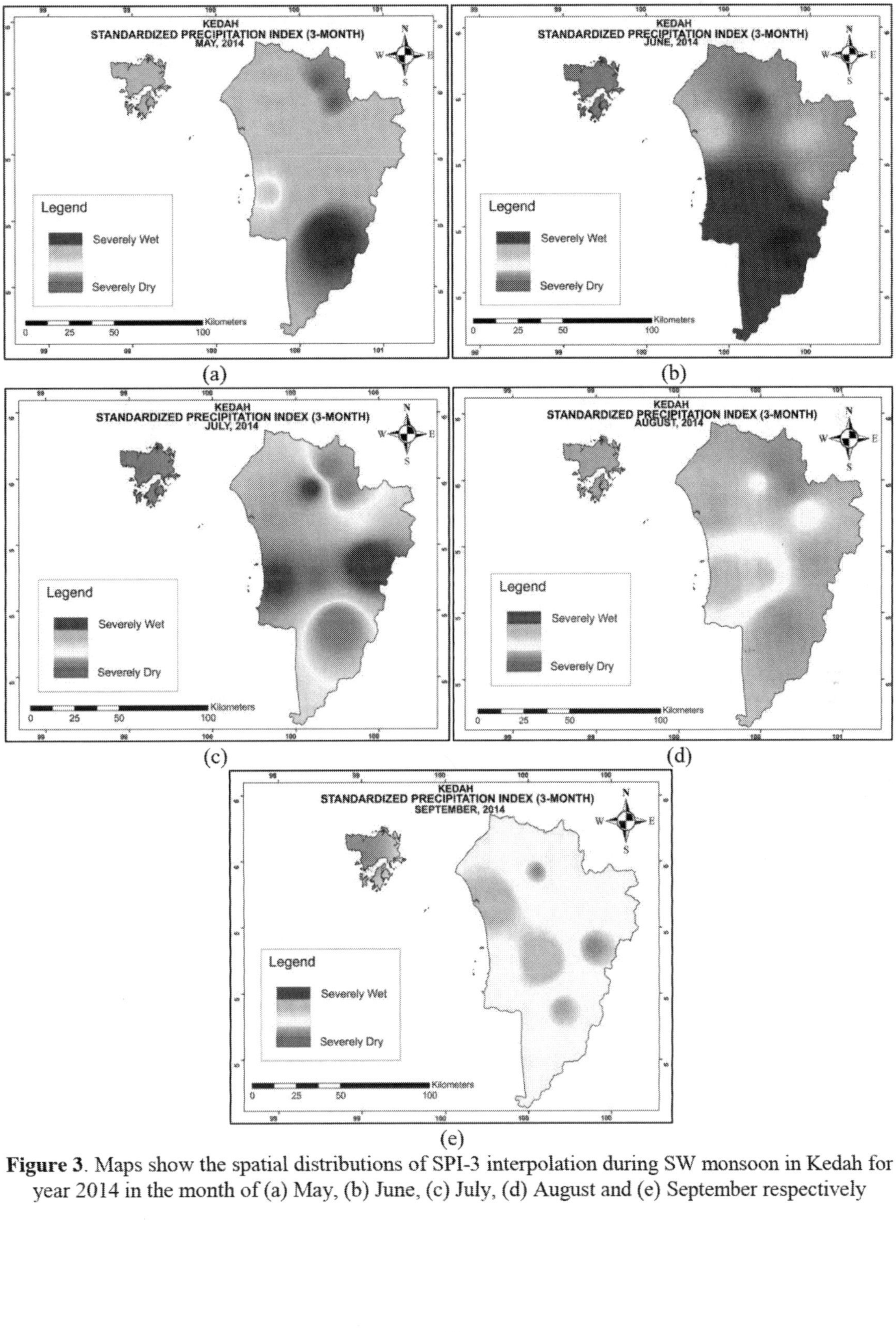

(a) (b) (c) (d) (e)

Figure 3. Maps show the spatial distributions of SPI-3 interpolation during SW monsoon in Kedah for year 2014 in the month of (a) May, (b) June, (c) July, (d) August and (e) September respectively

IOP Conf. Series: Earth and Environmental Science **37** (2016) 012072 doi:10.1088/1755-1315/37/1/012072

3.3. Spatial variation of SPI -3 during Northeast (NE) monsoon

The observation made during NE monsoon shows an increasing trend of dry areas in the first three months; January, February and March. NE monsoon occurred from November to March which was the major rainy season with a maximum rainfall received in Malaysia [26]. It is usually known as the wettest period every year. However according to the maps in the figure 4 (c), (d) and (e), the study area experienced a dry condition during those three months and the phenomenon was proven by a previous finding done in 2009, indicated that the northwest region could be classified as the driest area during NE monsoon [23]. This might be due to the existence of Titiwangsa range. This is also consistent with the Drought Report of Kedah that stated, Kedah had experienced a prolonged drought from January to March and faced a great loss in the rice production during this period [28].

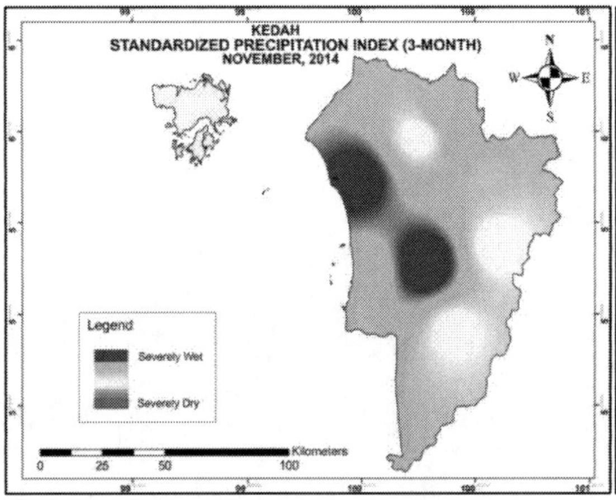

(a)

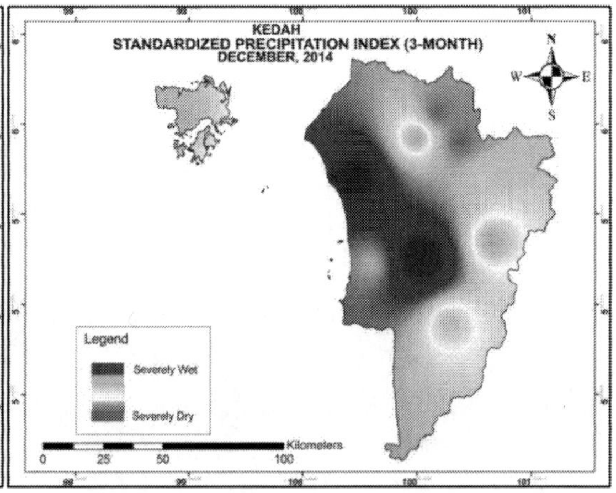

(b)

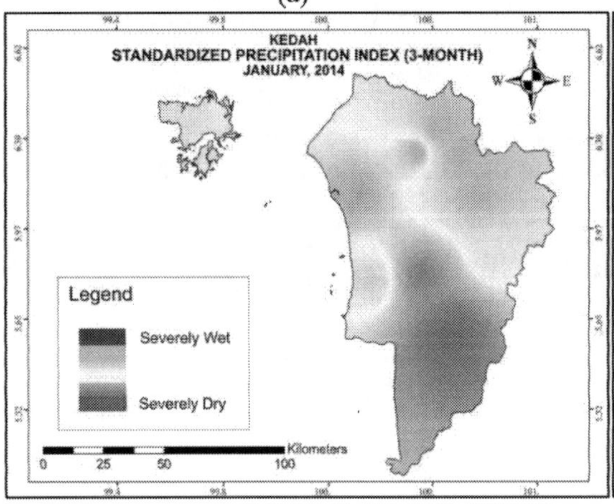

(c)

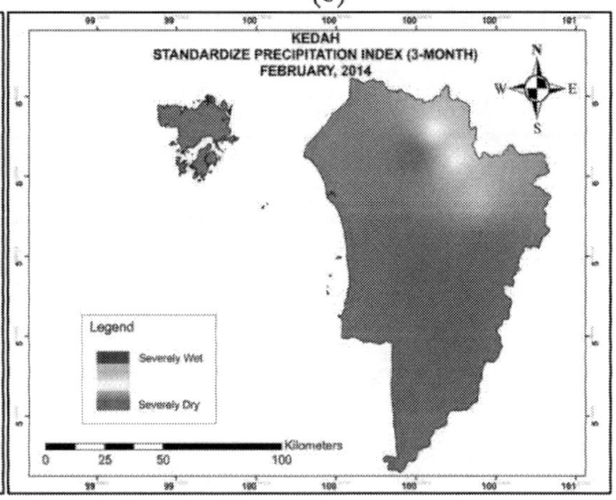

(d)

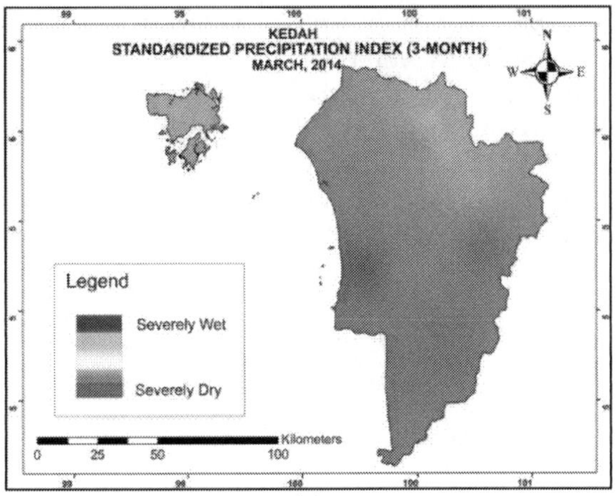

(e)

Figure 4. Maps show the spatial distribution of SPI-3 interpolation during NE monsoon in Kedah for year 2014 in the month of (a) November, (b) December, (c) January, (d) February and (e) March respectively.

3.4. Spatial pattern of vegetation health

Figure 5 (a) and (b) depicts the spatial pattern of NDVI of February and July respectively in 2014. Based on figure 5 (a) the west part of the study area shows red colour that indicate low NDVI values which means less vegetation or greenness. On the other hand, in figure 5 (b) the north part of the study area were dominated by low NDVI values and this might be caused by the problem of cloud coverage on that area that was treated before processing. From figure 5 (a) and (b) we can interpret that compare to July, February records low greenness value which indicate the deteriorating health of the vegetation in that particular area. The finding is in line with the report from DOA that recorded from January until March in 2014, where the yield encountered losses due to the drought event that last during those three months [28]. The deterioration of vegetation health during this period might be contributed by the insufficient amount of rainfall received.

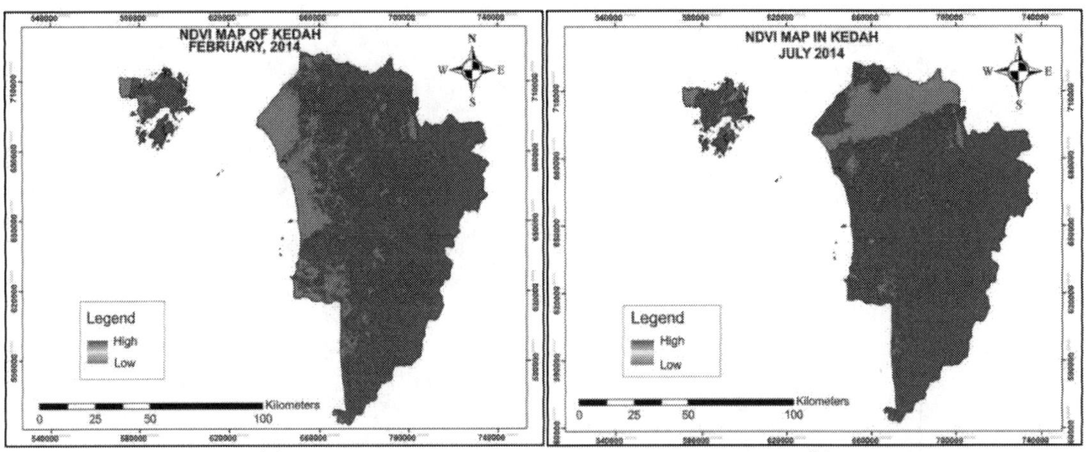

(a) (b)

Figure 5. The spatial pattern of NDVI in Kedah for (a) NE monsoon and (b) SW monsoon.

The result of SPI and NDVI were consistent with each other in term of areas affected by the rainfall reduction and that indicate the rainfall variation can be applied to detect the drought event in the study area. Severity map shows the level of the drought intensities and its strength for a certain period of time. Figure 6 (a) is the severity of drought in February which represents the condition during NE monsoon. Meanwhile figure 6 (b) is the condition during the SW monsoon. Condition during NE monsoon was much severe compared to the condition during the SW monsoon. This result was in line with the SPI result obtained in figure 3 and 4 that shows dry condition was much more intense during the NE monsoon.

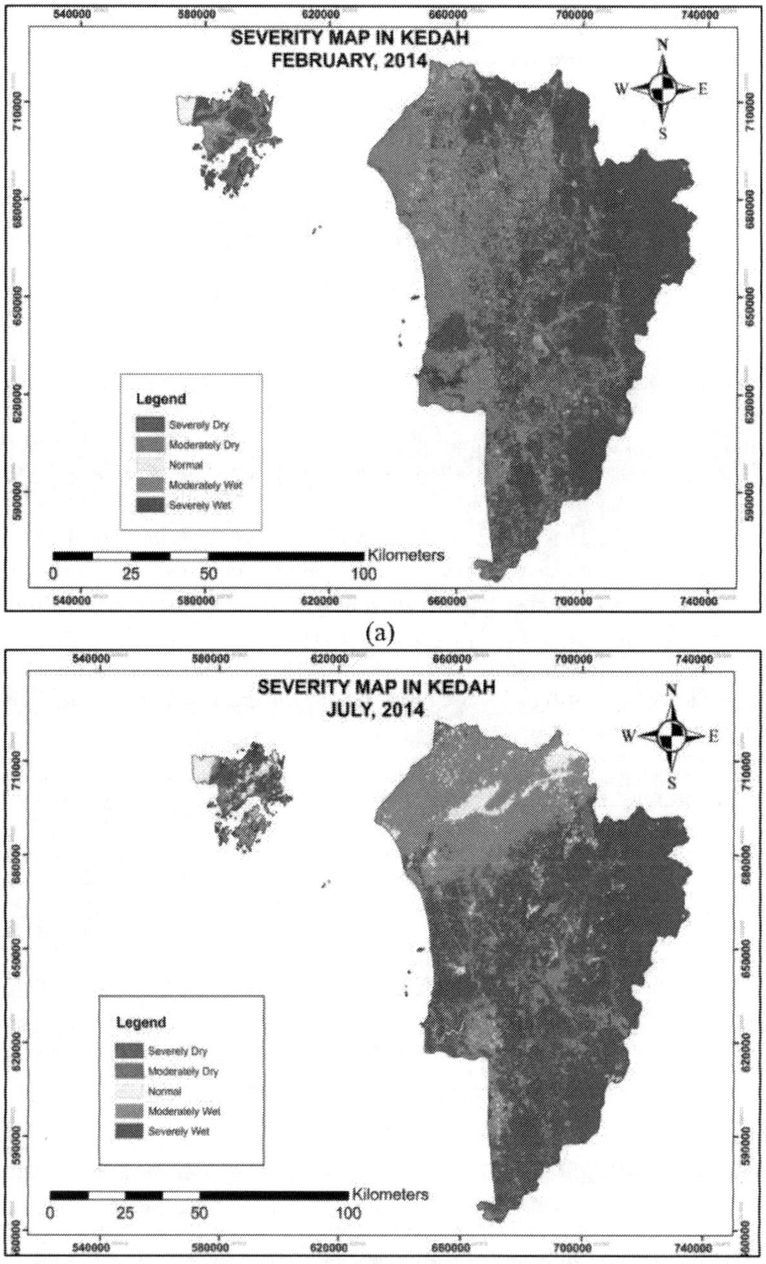

(b)

Figure 6. The severity map of Kedah during (a) NE monsoon and (b) SW monsoon.

3.5. Vulnerability map for vegetation in Kedah

In this study, the application of SPI identified the area that faced drought while the NDVI result showed the area that record the health of the vegetation and the combination of these two indices showed vegetation health that was affected by rainfall reduction or occurrence of drought event. Referring to figure 7, the northern part of the map that governed by the unhealthy vegetation condition was at Kuala Nerang. Predominantly, the villager's economic source for this area is derived from the paddy and mango plantation. Based on the previous SPI result in figure 3 and 4, Kuala Nerang mainly experienced moderate rainfall during the early SW monsoon and ranged between moderately to severely dry during the NE monsoon. Another unhealthy vegetation area was at Pendang which was also a paddy production area. Langkawi Island was one of the areas that face risk of moderately dry condition resulting in unhealthy vegetation in that area.

Since the lack of water can affect the soil moisture and the reservoir storage, scientific analysis can help to point out the exact area that need attention and at the same time improve the level of society's awareness prior to this intense issue. In this regard, early planning and action could be taken as a proper step in connection with the water shortage problem.

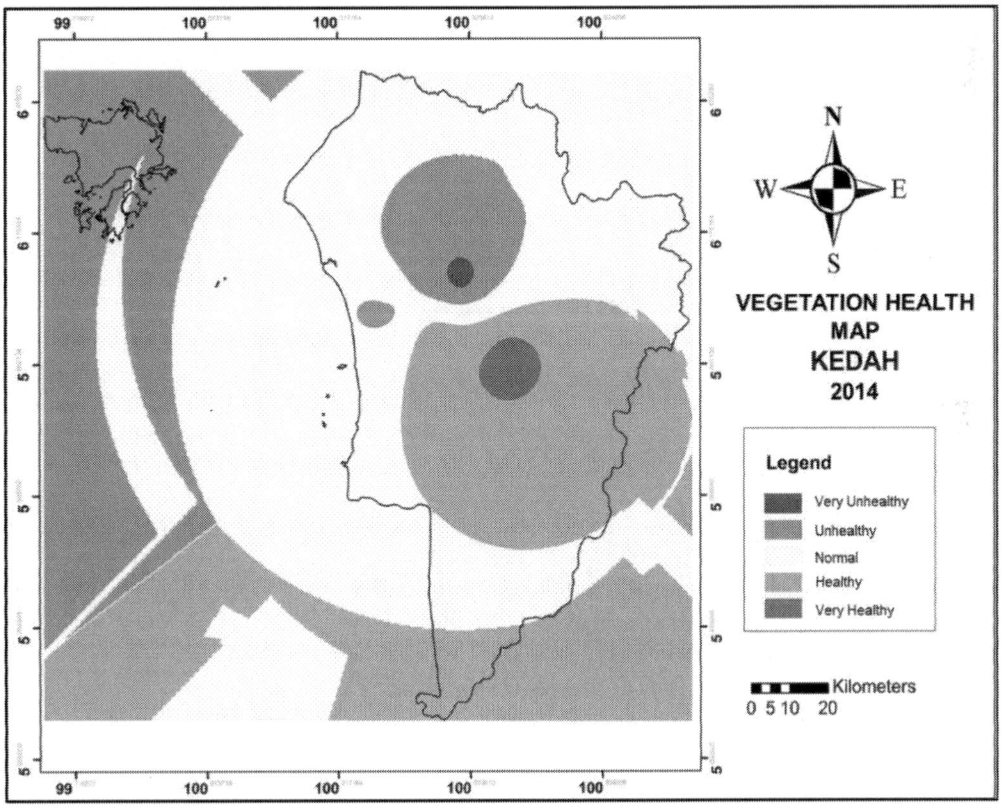

Figure 7. Vegetation health map in Kedah for year 2014.

4. Summary

Kedah was affected by the two main monsoons in Malaysia which were the NE monsoon that brings dry condition to Kedah especially at Kedah Peak station and SW monsoon that shows result in a wet condition throughout the state. Generally, Kedah's weather fluctuates all along the year. Kuala Nerang, Pendang and Langkawi Island were the areas with its vegetation's health affected by the rain reduction. This study was done to help the authorities and farmer to take action based on drought event that occurs in 2014 resulting in economic losses to the state. The result has shown areas that were more prominent to the drought occurrence and by identifying the specific area, extra precaution will be easier. The main limitation in this study was due to the rainfall station that was distributed sparsely. Resulting in some of the area was not exactly represented by its exact condition which led to errors in the vegetation health map.

The overall outcome showed that the risk identification process can be done by integrating many methods and sources with the benefit of minimizing time and cost taken for the process. The finding of this study can help the authorities and even the farmers in preparing the mitigation measures. The findings of this study will definitely contribute in identifying the risk area of Malaysia which is in the process of researching at the moment.

Acknowledgements
The author would like to acknowledge the Department of Irrigation and Drainage Malaysia for providing the daily rainfall data and also to the Ministry of Education Malaysia for the funding of this research (Grant reference code: FRGS/1/2014/STWN01/UPM/02/1).

References

[1] Leister E 2014 Drought Worsen in Malaysia and Thailand *Accuweather* Retrieved from http://www.accuweather.com
[2] Chong P K and Pakiam R 2014 Water Rationing Begins Outside Malaysia Capital Amid Drought *BloombergBusiness* Retrieved from http://www.blommberg.com
[3] Lim C Y 2014 Oil Palm Industry Braces for Drought *The Star Online* Retrieved from http://www.thestar.com.my
[4] Bernama 2014 Effect of El-Nino in the peninsular at the end and beginning of the year *Sinar Online* Retrieved from www.sinarharian.com.my
[5] Wilhite D A and Glantz M H 1985 Understanding the drought phenomenon: The role of definitions *Water International* **10** (3) 111-120
[6] Palmer W C 1965 Meteorological Drought Research Paper *US Weather Bureau Washington DC* **45** 58
[7] McKee T B, Doesken N J and Kleist J 1993 The relationship of drought frequency and duration to time scale *Proceedings of the Eighth Conference on Applied Climatology* 179–184.
[8] Guttman N B, Wallis J R and Hosking J R M 1992. Spatial comparability of the Palmer Drought Severity Index *Water Resoures Bulletin* **28** 1111–19
[9] Karl T R 1983 Some spatial characteristics of drought duration in the United States *Journal Climate and Applied Meteorology* **22** 1356–66
[10] Alley W M 1984 The Palmer Drought Severity Index: Limitations and assumptions. *J. of Climate and Applied Meteorology* **23** 1100–09
[11] Guttmann N B 1998 Comparing the Palmer drought index and the standardized precipitation index *J. of American Water Resources Association* **34** 113–121
[12] Kogan F N 1997 Global Drought Watch from Space *Bull. Amer. Meteor. Soc.* **78** 621-636
[13] Price, R., & Price, K. (2009). Remote Sensing Imagery in Agriculture: Image Conversion to NDVI 1–4 Retrieved from http://www.ksre.ksu.edu/bookstore/Item.aspx?catId=381 &pubId=2959
[14] Aggarwal S 2004 Principles of remote sensing *Satellite Remote Sensing and GIS Applications in Agricultural Meteorology* 23–38
[15] Holme A M, Burnside D G and Mitchell A A 1987 The development of a system for monitoring trend in range condition in the arid shrublands of Western Australia *Australian Rangeland Journal* **9** 14-20
[16] Thenkabail P S, Gamage M S D N, Smakhtin V U, 2004 The use of remote sensing data for drought assessment and monitoring in Southwest Asia *Int. Water. Manage. Inst. Colombo, Sri Lanka Research report* **85**

[17] Wang Q, Tenhunen J, Falge E, Bernhofer C, Granier A and Vesala T 2004 Simulation and scaling of temporal variation in gross primary production for coniferous and deciduous temperate forests *Global Change Biology* **10** 37

[18] Li B, Tao S and Dawson R W 2002 Relation between AVHRR NDVI and ecoclimatic parameters in China *Int. J. Remote Sensing* **23** 989-999

[19] Nicholson S E and Farrar T J 1994 The influence of soil type on the relationships between NDVI,rainfall, and soil moisture in semiarid Botswana .I. NDVI response to rainfall *Remote Sensing of the Environment* **2** 107–121

[20] Song X, Saito G, Kodama M and Sawada H 2004 Early detection system of drought in East Asia using NDVI from NOAA/AVHRR data *Int. J. Remote Sensing* **20** 3105-11

[21] Hermann S M, Anyamba A and Tucker C J 2005 Recent trends in vegetation dynamics in the African Sahel and their relationship to climate *Global Environmental Change* **15** 394 – 404

[22] Kottek M, Grieser J, Beck C, Rudolf B and Rubel F 2006 World map of the Köppen-Geiger climate classification updated *Meteorologische Zeitschrift* **15** (3) 259–263

[23] Deni S M, Suhaila J, Wan Zin W Z and Jemain A A 2009 Spatial trends of dry spells over Peninsular Malaysia during monsoon seasons *Theoretical and Applied Climatology* **99** 3-4 357–371

[24] World Meteorological Organisation 2012 Standardized Precipitation Index User Guide **24**

[25] Fredolin T T, Juneng L, Salimun E, Sei K M, Le L J and Muhammad H 2012 Climate change and variability over Malaysia : Gaps in science and research information. *Sains Malaysiana* **41** (11) 1355–66

[26] Wong C L, Venneker R, Uhlenbrook S, Jamil A B M and Zhou Y 2009 Variability of rainfall in Peninsular Malaysia *Hydrology and Earth System Sciences Discussions* **6** (4) 5471–5503

[27] Othman M, Ash'aari Z H and Mohamad N D 2015 Long-term daily rainfall pattern recognition: Application of principal component analysis *Procedia Environmental Sciences* **30** 127-132

[28] Department of Agriculture Malaysia 2014 Kedah drought disaster *Rice crop damage report*

IOP Conf. Series: Earth and Environmental Science 37 (2016) 012073 doi:10.1088/1755-1315/37/1/012073

Developing hydrological model for water quality in Iraq marshes zone using Landsat-TM

Maged Marghany*[1], Hashim Ali Hasab[2], Shattri Mansor[1], Abdul Rashid Bin Mohamed Shariff[1]

[1]Geospatial Information Science Research Centre, Faculty of Engineering University Putra Malaysia, 43400 UPM, Serdang, Selangor, Malaysia
[2] Ministry of Higher Education & Scientific Research, Foundation of Technical Education, Baghdad, Iraq
*Email: magedupm@hotmail.com

Abstract. The Mesopotamia marshlands constitute the largest wetland ecosystem in the Middle East and Western Eurasia. These wetlands are located at the confluence of the Tigris and Euphrates Rivers in southern Iraq. However, there are series reductions in the wetland zones because of neighbor countries, i.e. Turkey, Syria built dams upstream of Tigris and Euphrates Rivers. In addition, the first Gulf war of the 1980s had damaged majority of the marches resources. In fact,the marshes had been reduced in size to less than 7% since 1973 and had deteriorated in water quality parameters. The study integrates Hydrological Model of RMA-2 with Geographic Information System, and remote sensing techniques to map the water quality in the marshlands south of Iraq. This study shows that RMA-2 shows the two dimensional water flow pattern and water quality quantities in the marshlands. It can be said that the integration between Hydrological Model of RMA-2, Geographic Information System, and remote sensing techniques can be used to monitor water quality in the marshlands south of Iraq.

1. Introduction

In the last century, there are several critical environmental problems have been raised up such as pollution, coastal and soil erosions [1-6]. Man-made is beyond all the environmental crises due to war, wrong behaviors and managements. It is not forgotten the three Gulf wars of 1980s , 1990, and 2003 which lead to several damages for environmental specially in the Arabian Gulf and specially in Iraq. For instance, during the 1991 Gulf War, roughly one million oil tones blackened the Arabian Gulf (Figure 1).

In Iraq, there are other problems have raised up in last decades such as water quality. American-Iraq war in 2003 has been affecting the water quality of rivers, streams, lakes and marshes in Iraq [7]. Further, inadequately treated sewage, poor land use practices, industrial waste waters discharges excessive use of fertilizers, and a lack of integrated watershed management are other factors impact water quality in Iraq [8]. The effects of these problems threaten ecosystems, endanger public health risks, and intensify erosion and sedimentation, leading to land and water resources degradation. Many of these negative effects arise from environmentally destructive development, a lack of information on the situation regarding water quality and poor public awareness and education on the protection of water resources [9].

However, clean water is essential to human survival as well as to aquatic life. Much surface water is used for irrigation, with lesser amounts for municipal, industrial, and recreational purposes: only 6% of all inland water is used for domestic consumption. An estimated 75% of the population of developing nations lacks adequate sanitary facilities, and solid waste is commonly dumped into the nearest body of flowing water. Pathogens such as bacteria, viruses and parasites make these waste materials among the world's most dangerous environmental pollutants, waterborne diseases are estimated to cause about 25,000 deaths daily worldwide [10].

Figure 1. Oil spill blackened the Arabian Gulf.

Remote sensing techniques play major roles for monitoring and mapping, water quality [11]. With advances using remote sensing for data acquisition and the integrating finite element numerical model with the spatial capabilities of GIS and the spatial and temporal capabilities of remote sensing applications could provide a powerful tool for management and assessment to surface water quality problems in the marshes zone of southern Iraq. This study is very important for monitoring and assessing the water quality in the wetland south of Iraq. The benefits of this study are restoration, rehabilitation of the wetlands as well as the environmental and water quality improving in the marshes in southern of Iraq. This research work mainly images of LANDSAT (TM) were selected to identify the spatial changes in the marsh zones and to support the catchment delineation marshes.

2. Study Area

The Mesopotamian Marshlands of Southeastern Iraq represents one of the largest wetland ecosystems in all of Asia and covered more than 15,000 km^2 and is formed by the confluence of the Tigris and Euphrates rivers. But in 2003 the marshes had been reduced in size to less than 7% of their 1973 levels (8,926 km^2 within Iraq)[12]. The major marshes in Iraq such as Al Huwazia marsh is located east of Tigris River by approximate area 3000 km^2 and the central marshes are located in a triangular area between the Tigris and the Euphrates by approximate area 2800 km^2 and Abu-Ziriq is located north of Nasiriyah city on the western side of the former Central Marsh with area 2000 km^2 then Hammer marsh is south of the Euphrates River, extend from near Al Nasiriyah in the west to Al Basrah on the Shat Al Arab in the east with area 4500 km^2, this study will focus on theses the marshes only (Figure 2).

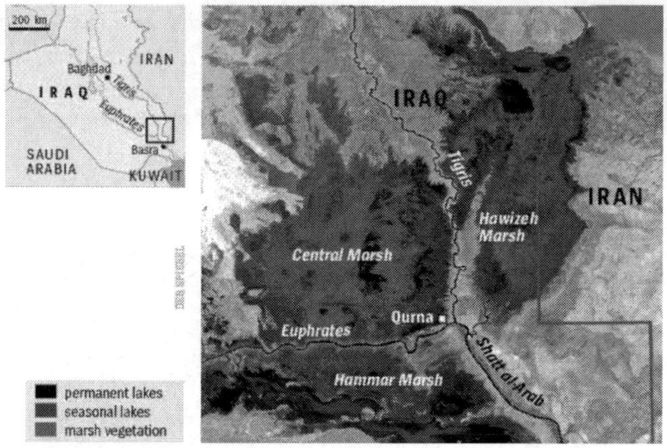

Figure 2. Mesopotamian marshlands of southeastern Iraq.

3. Data Collection and Model

Data collection are required, such as a hydrological data (inflow and outflow) of all marshes and topographical data as the marsh boundaries, area, surface water elevation, depth water, geographical location of sources, pollution stations and geographical location for sampling stations as well as the satellite images such as Landsat-7 TM, ETM that resolution 30 m^2 and seven bands. The finite element model (FEM) is the dominant discretization technique in structural mechanics. The basic concept in the physical interpretation of the FEM is the subdivision of the mathematical model into disjoint (non-overlapping) components of simple geometry called finite elements or elements for short. The response of each element is expressed in terms of a finite number of degrees of freedom characterized as the value of an unknown function, or functions, as a set of nodal points.

The response of the mathematical model is then considered to be approximated by that of the discrete model obtained by connecting or assembling the collection of all elements. The disconnection-assembly concept occurs naturally when examining many artificial and natural systems. It can be applied to a fluid as the water to determine water depth and water flow analysis as well as for assessing and monitoring water quality when integrate with remote sensing techniques and Geographic information system applications.

4. Results and Discussion

In order to fulfill the overall aim and objectives of the study, the results and analyses will achieve based on the pre-processing and processing of the input datasets. The obtaining results as the water depth, velocity patterns as well as water quality parameter distribution, such as (TDS, TSS, Salinity) for four seasons (seasonally) in marshes of southern Iraq. Figures 3 and 4 show an exponential relationship between the march zones and elevations. The maximum elevation of 9 m coincides with an area of 2,500 km^2. Clearly the storage volume increase exponentially with elevation of 9 m. This can prove that there are exponentially relationships between area, storage volume and elevation of march zones. This study confirms the work done by Dunia Frontier Consultants [13].

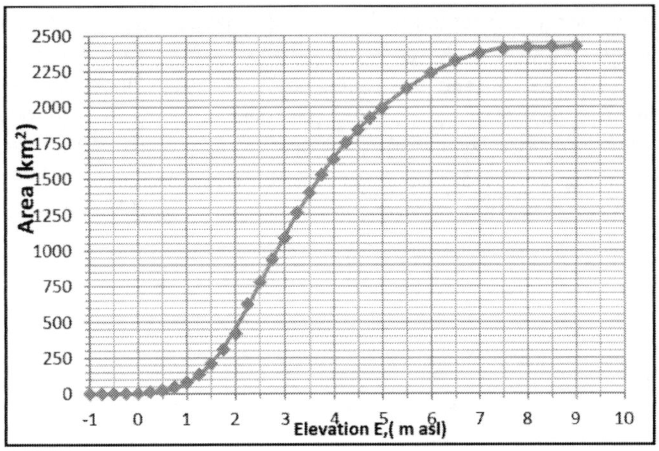

Figure 3. Area elevation curve of Al Qurna marshes.

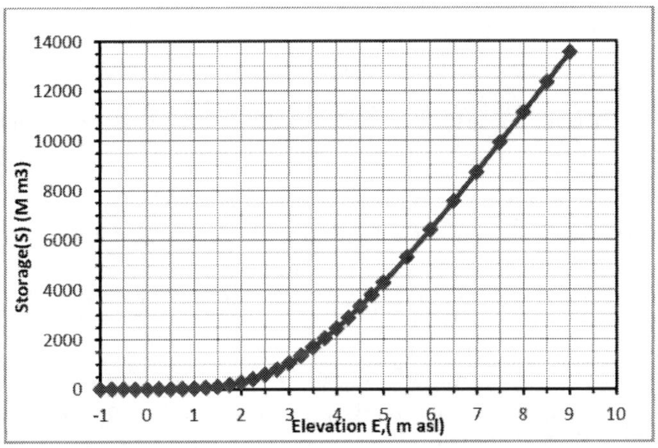

Figure 4. Storage elevation curve of Al Qurna marshes.

Figure 5 shows the rainfall spatial variations in the maximum amount of 300 mm. Clearly the maximum range of rainfall amount matches with high concentration of water drainage zones which are closed to the Arabian Gulf. Further, the heavy dense concentration of march zones are existed along the drainage concentration in both river banks of Tigris and Euphrates rivers (Figures 2 and 5). In fact, march zones are required much amount of water to grow up and to keep on living.

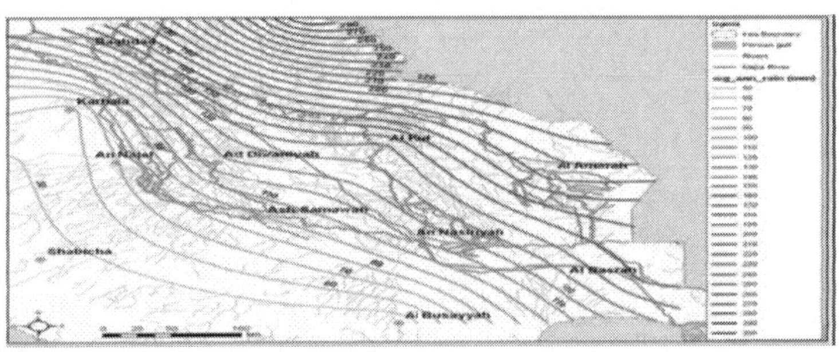

Figure 5. Rainfall distribution in the study area (mm).

In the last few years, however, water levels in Iraq's rivers have rapidly decreased to less than a third of their normal capacity. Water levels may fall further in the coming years due to declining precipitation, gradual desertification, and upstream water use and damming. The long-term average annual precipitation in Iraq is equal to 216mm per year with high variability across time and governorate. In 2011, 507.4 mm of water fell in Suleimaniyah while only 65.3 mm fell in Basrah. Lakes, reservoirs, and minor rivers are also experiencing diminished levels of water. If present conditions remain unchanged, Iraq will experience a water shortage of over 33 million cubic meters a year by 2015 [14].

5. Conclusion

This study demonstrate method for water quality distribution in Iraq. The integration between LANDSAT satellite data and finite model was used to model volume of water quality variations. The study shows that the storage volume increase exponentially with elevation of 9 m. This can prove that there are exponentially relationships between area, storage volume and elevation of march zones. It can be said that the heavy dense of march zones are existed with area of high water storage which can be identified using LANDSAT-TM data.

References

[1] Marghany,M. (2013). Three Dimensional Coastline Deformation from Insar Envisat Satellite Data. In Beniamino Murgante, Sanjay Misra, Maurizio Carlini, Carmelo M. Torre, Hong-Quang Nguyen, David Taniar, Bernady O. Apduhan, and Osvaldo Gervasi Computational Science and Its Applications – ICCSA 2013, 7972, 2013, pp 599-610.

[2] Marghany,M. (2013). Genetic Algorithm for Oil Spill Automatic Detection from Envisat Satellite Data. In Beniamino Murgante, Sanjay Misra, Maurizio Carlini, Carmelo M. Torre, Hong-Quang Nguyen, David Taniar, Bernady O. Apduhan, and Osvaldo Gervasi. Computational Science and Its Applications – ICCSA 2013, 7972, pp 587-598.

[3] Marghany,M. (2012). DEM Reconstruction of Coastal Geomorphology from DINSAR. In Murgante B. et al. (eds.): Lecture Notes in Computer Science (ICCSA 2012), Part III, LNCS 7335, pp. 435–446.

[4] Marghany,M. (2012).Three-Dimensional Coastal Front Visualization from RADARSAT-1SAR Satellite Data. In Murgante B. et al. (eds.): Lecture Notes in Computer Science (ICCSA 2012), Part III, LNCS 7335, pp. 447–456

[5] Marghany,M. (2014). Simulation of three dimensional of coastal erosion using differential interferometry synthetic aperture radar. Global nest Journal 16(1): 80-86.

[6] Marghany,M. (2013). DInSAR technique for three-dimensional coastal spit simulation from Radarsat-1 fine mode data. Acta Geophysica. vol. 61, no. 2, Apr. 2013, pp. 478-493.

[7] Brooks, K. N, P.F. Ffolliott, H.M. Gregersen, and L.F. Debano. 1997. Hydrology and the Management of Watersheds, 2nd Ed. Iowa State University Press, Ames, Iowa.

[8] Haith, D. A., and Tubbs, L. J. (1981). Watershed loading functions for nonpoint sources. J.

[9] Envir. Eng. Div., ASCE, 107(1), 121–137. Environmental Protection Agency, Washington, D.C.

[10] World Bank. (2000). World Development Report 2000/2001: Attacking Poverty.

[11] McCutcheon, Steve C. (1990). Water Quality Modeling – Volume 1 – Transport and Surface Exchange inRivers, Boca Raton: CRC Press Inc.

[12] Yang, M. D., Kuo, J. T and Yeh F. Y.(2000). Application of Remote Sensing and GIS in Water Quality Simulation and Calibration. 4th International Conference on Integrating GIS and Environmental Modelling (GIS/EM4): Problems, Prospects and Research Needs.

[13] Zacharias, I., Gianni, A., (2008). "Hydrodynamic and dispersion modeling as a tool for restoration of coastal ecosystems. Application to a re-flooded lagoon." Environmental Modelling & Software. 23: pp.751-767.

[14] Dunia Frontier Consultants (2013). Water and Sewage Sectors in Iraq: Sector Report-February 2013[www.duniafrontier.com].

[15] UN (2014). Water in Iraq Factsheet. http://www.jauiraq.org/documents/1866/Water-Factsheet.pdf.

Real-time Web GIS to monitor marine water quality using wave glider

Siti Maneesa Amiruddin

GIS Assistant Manager, Alam Sekitar Malaysia Sdn Bhd (ASMA)

E-mail: sitimaneesa@enviromalaysia.com.my

Abstract. In the past decade, Malaysia has experienced unprecedented economic development and associated socioeconomic changes. As environmentalists anticipate these changes could have negative impacts on the marine and coastal environment, a comprehensive, continuous and long term marine water quality monitoring programme needs to be strengthened to reflect the government's aggressive mind-set of enhancing its authority in protection, preservation, management and enrichment of vast resources of the ocean. Wave Glider, an autonomous, unmanned marine vehicle provides continuous ocean monitoring at all times and is durable in any weather condition. Geographic Information System (GIS) technology is ideally suited as a tool for the presentation of data derived from continuous monitoring of locations, and used to support and deliver information to environmental managers and the public. Combined with GeoEvent Processor, an extension from ArcGIS for Server, it extends the Web GIS capabilities in providing real-time data from the monitoring activities. Therefore, there is a growing need of Web GIS for easy and fast dissemination, sharing, displaying and processing of spatial information which in turn helps in decision making for various natural resources based applications.

1. Introduction

Alam Sekitar Malaysia Sdn Bhd, or better known as ASMA, has been at the forefront of the nation's environmental monitoring programme since 1993. With a concession spanning the past 20 years with the Department of Environment Malaysia (DOE) for monitoring the nation's air and river water quality, the ocean is the largest and the next frontier for ASMA. Riding on the experience gained from conducting the nation's comprehensive Island Marine Monitoring Programme in 2008, ASMA is also looking forward to provide more value through its latest technology acquisition, the Wave Glider.

Wave Glider is a state-of-the-art marine technology in harvesting oceanic data. It is an autonomous marine surface vehicle that operates only on renewable energy. Wave energy is converted into forward propulsion whilst solar energy powers communication and on- board sensors. Its operation is zero carbon footprint, although a vessel is required for launch and recovery at sea. The Wave Glider is ruggedly built to withstand long durations at sea and harsh oceanic conditions. The speed can go up to 3 knots (kts) depending on the wave intensity and can communicate via cell, satellite and Wi-Fi (depending on the location). Embedded on the platform is a command and control computer that is capable of on-board processing and communication. The units also complements other marine technologies or vessels in providing relevant and vital data for making informed decisions during pollution or exceedance events. Each unit is equipped with on-board sensors to collect 22 in-situ parameters [1]. The unit has an on-board data logger for data safekeeping before new data overwrites the older data when the storage has reached its maximum capacity of approximately 10 gigabytes. The Wave Glider unit is in operation for 24 hours a day, 7 days a week, 365 days per year. Data is collected once every 30 minutes and transmitted once every hour via Iridium satellite to the Wave Glider Management System (WGMS) web. Figure 1 illustrates the 3D layout for the unit. The sensor layout of the unit is reflected in Figure 2.

Figure 1. 3-Dimensional layout of the Wave Glider.

1. Schematic	2. Sensor Detail
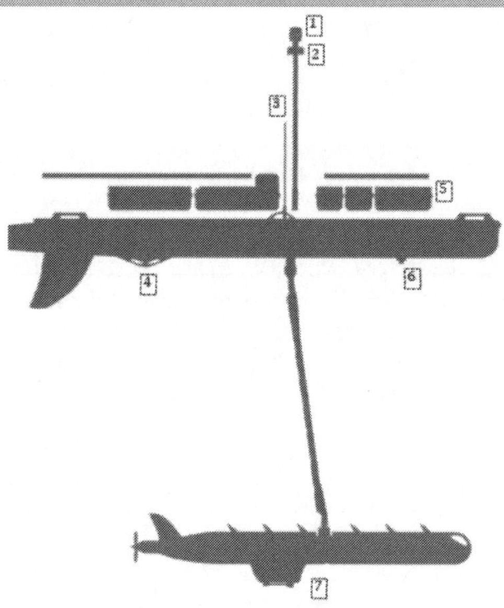	1. Air Temperature, Wind Speed & Direction, Barometric Pressure 2. Wave Height, Wave Direction, Wave Period 3. AIS (Automatic Identification System) 4. Current Direction, Current Speed, 5. Turbidity, pH, Oil – Crude, Oil – Fine, Chlorophyll in vivo, CDOM/FDOM, and Fluorescein dye 6. Water Speed 7. Water Temperature, Dissolved Oxygen, Electrical Conductivity, Salinity, Density

Figure 2. Sensor layout of the Wave Glider.

2. Objectives

The objectives of this project are:

- To continuously monitor marine water quality in real-time basis and provide the data wirelessly to the Web GIS application,
- To function as an early warning system that alerts to any unexpected changes in water quality along its transect routes for indication of exceedance or pollution events through Web GIS,
- To provide long term status and trends data used to determine the changes in water quality conditions and to determine the impact of natural influencing factors (i.e. climate and ocean) on water quality year to year variability via Web GIS Dashboard.

3. Project Area

Round-the-island mission at Bidong Island, Terengganu (5°37'2.57"N, 103° 3'47.17"E), one square kilometre (1km^2) in area and is situated off the coast of Terengganu, Malaysia in the South China Sea. It is the northeast side of Kuala Terengganu city and Merang town. To get to the island, it takes about 40 – 50 min by boat [2].

4. Methodology

4.1 Mission planning

Mission planning was carried out in order to achieve the best possible methodology to execute the mission and meet the desired uptime of collected data. Mission planning began by determining the best possible route for the glider to fulfil the monitoring objectives. The first step in determining the route is by carrying out an Operational Risk Assessment (ORA).

An ORA describes the environmental, political and safety conditions that act as a recommendation of the best practicable route for the units, to ensure the efficiency and reliability of the mission and ultimately to meet the monitoring objectives. Intended coordinates or waypoints of the mission is pre-configured and acknowledged by the satellite. This enables autopilot mode of the operations although dedicated pilots assigned to each unit 24/7 are able to interject the autopilot mode and manoeuvre whenever deemed as necessary. After the necessary interjection, the unit is able to return to its autopilot mode, following the pre-configured waypoints.

4.2 Launch and recovery

A small vessel is required with minimum requirements of 8 m long and installed with a hoisting capability on-board (crane/davit) with minimum Safe Working Load of 500 kg. The small vessel shall bring the Wave Glider off shore up to a point of fulfilling the minimum required depth for it to operate, which in this case is either 8 or 10 m. After a Wave Glider is launched, it will make its way to its first pre-configured coordinate/waypoint and begin data collection (see Figures 3 and 4).

4.3 Piloting

The Wave Glider unit is equipped with physical security features, remote safety watch and alarm to secure against vandalism and theft.

i. Real-time transmission of redundant Global Positioning System (GPS) reports from sensors in real-time
ii. Low visibility
iii. Localized, on-unit labelling
iv. Automatic Identification System (AIS) to autonomously avoid approaching vessels with transponders

The Wave Glider is remotely traced and controlled by dedicated personnel whom are positioned at the Environmental Data Centre (EDC). The WGMS portrays mapping for piloting of the units, where pilots send commands to the units through satellite for instructions and to receive notifications such as the physical or software condition of the unit and to monitor real-time data as the units transmit them. A screenshot of the web-based piloting interface is illustrated in Figure 5.

Figure 3. On-boat preparation before the launch of the Wave Glider.

Figure 4. The Wave Glider is ready for launch and begin mission.

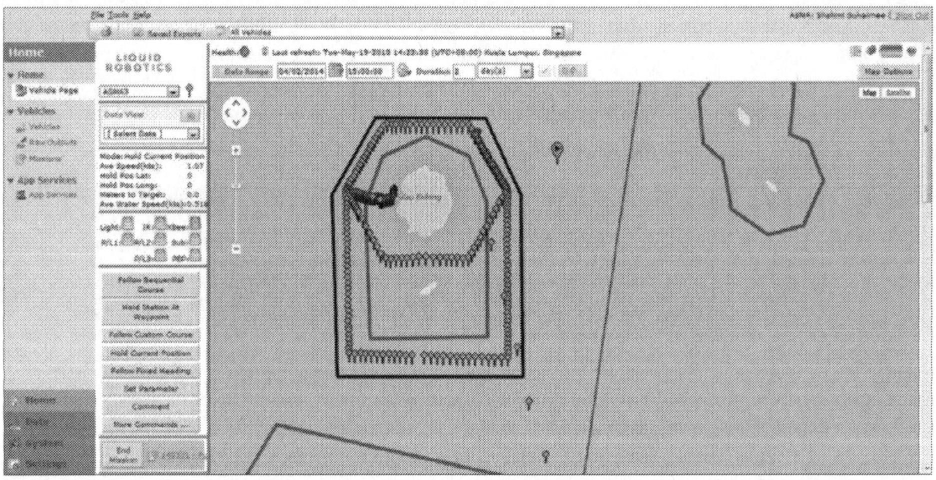

Figure 5. Web based piloting interface that can be viewed on a tablet, laptop and etc. Primary piloting will be at EDC through wireless application.

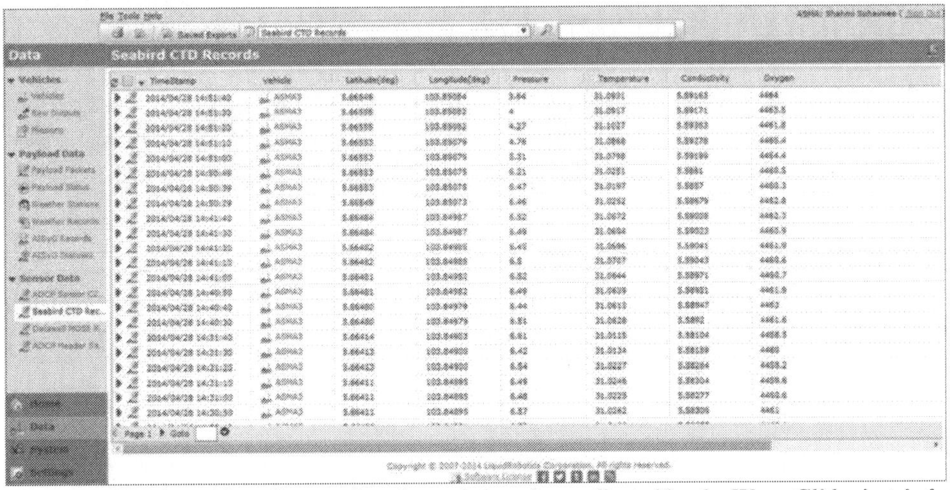

Figure 6. Web based interface to view the real-time data collected by the Wave Glider in tabular format.

4.4 Building Real Time Web Applications Using GeoEvent Processor

Over the past few years interactive mapping or Internet GIS has rapidly developed resulting in the migration of some GIS functionality. By using GeoEvent Processor, it extends the capabilities of ArcGIS for Server by bringing live data from Wave Glider and issuing alerts when specified conditions occur—all in real time. GeoEvent Extension can:

- Stream (push) event data to the FlexViewer via WebSockets,
- Direct event data into feature services hosted on ArcGIS for Server so that maps created will represent the most up-to-date information occurring in the real world,
- View the latest feature status using FlexViewer,
- Filter GeoEvents using spatial or attribute conditions to focus on the most required event data,
- Enable GeoFence to avoided areas during the transection,
- Archive event data within featured services or tables,
- Enrich incoming events with data from a secondary feature service or system file [3].

ArcGIS GeoEvent Manager is used to create and configure service components, as well as design, publish, and manage GeoEvent Services. Input Connectors, Output Connectors and published GeoEvent Services are created in this page. Getting the real-time data into Web Application is done by polling the CSV file from WGMS web. The real-time data is viewed via FlexViewer developed using Flex application builder. ArcGIS Viewer for Flex provides a smart, intuitive framework for viewing and interacting with maps. It is a configurable web application that allows user to easily build user own custom mapping application in just a few minutes, with no programming required. Figure 7 illustrates the process from data collection stage until delivering the information through Web GIS and Mobile Application.

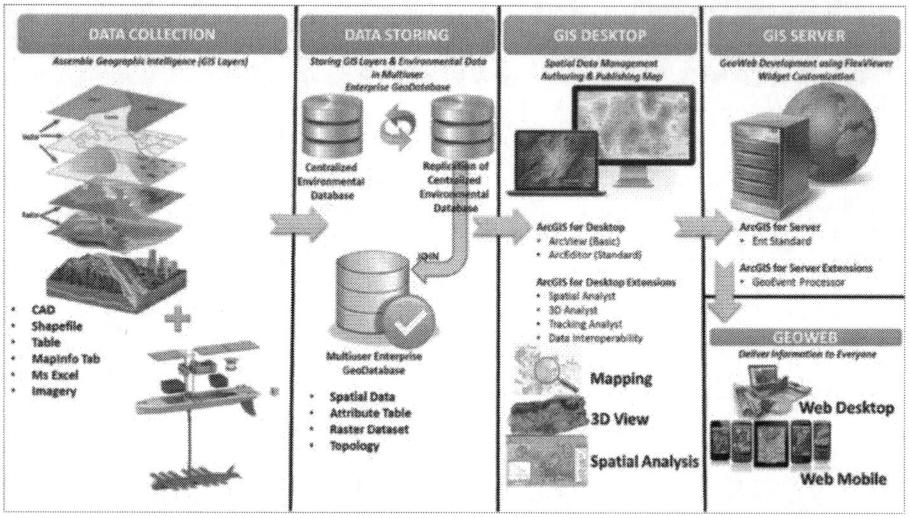

Figure 7. Process in Building Real Time Web Applications Using GeoEvent Processor

This Web GIS has different analysis tools to help the users understand marine water concerns in various areas. This Marine Water Quality Database (MWQDB) contains more than twenty (20) data layers representing various marine water quality themes namely Dissolved Oxygen (DO), Temperature (Temp), pH, Conductivity (Cond) and meteorology data such as Wind Speed, Wind Direction and other parameters. Using the classification capabilities of ArcGIS, the quality of marine water is represented through symbols to show variations on a map that allow visual differentiation, enabling users to easily view the parameters that are exceeding their safe limits, along with their percentage of exceedance. This helps anyone to easily see areas where water quality is exceeding safety limits and by what amount according to the Malaysian standards and best practices. The users can see the results of these calculations as color-coded ranges on a map and easily understand the water quality levels at that particular location.

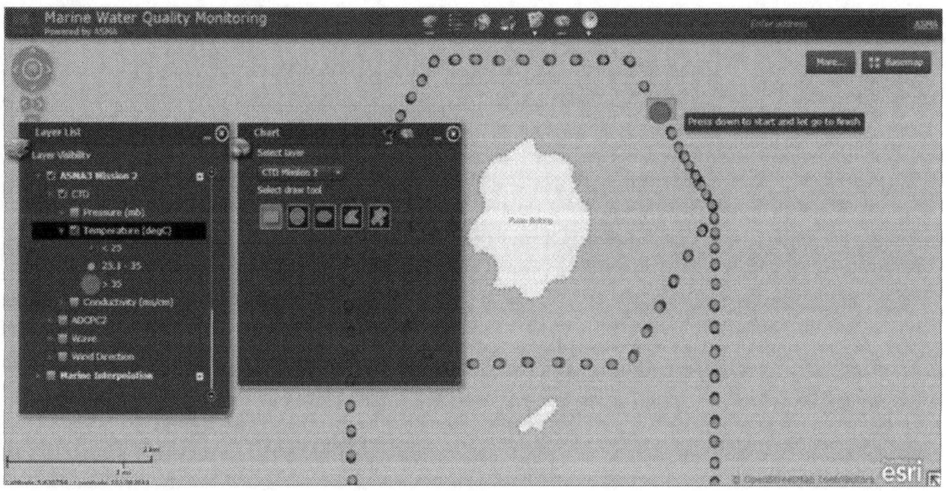

Figure 8. Colour coded ranges and instantaneous colour schemes for each monitoring locations.

It is easy to understand marine conditions by looking at the dashboard, which includes charts, statistics, summary tables and maps. Users can get access to the updated data in areas where water quality exceeds safety parameters and find complete sample details, parameter variation trends and annual summary of marine water quality status. Water quality changes over time are depicted on intuitive, multilevel, customizable charts that allow users to track marine water quality by years, months or even weeks. They can review and compare the water quality at one or more sites as average annually, monthly or weekly values. Users are also able to identify threshold-violating samples through instantaneous colour schemes and timely automatic alerts on the map.

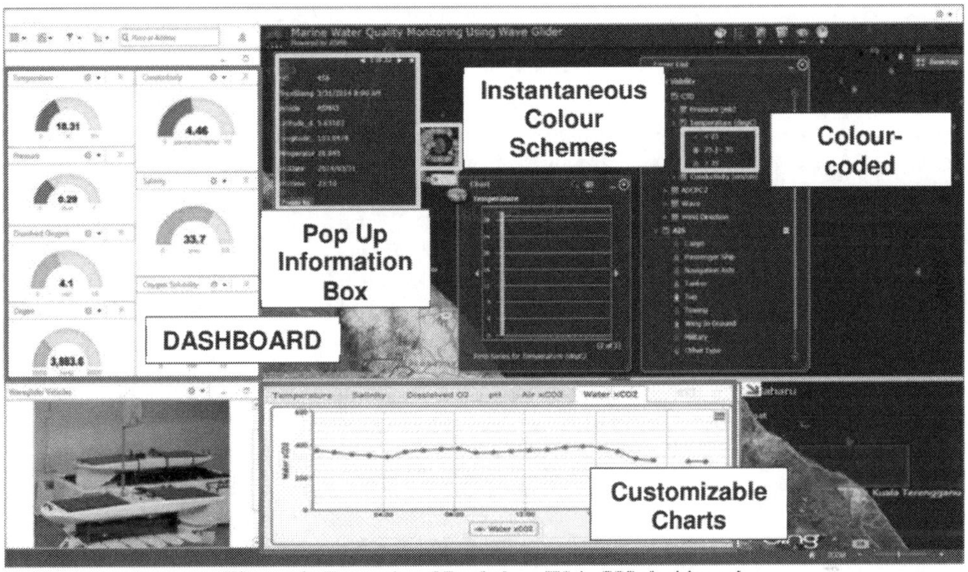

Figure 9. Example of Real-time Web GIS dashboard.

5. Conclusion

The Wave Glider can provide real-time, autonomous, zero-carbon-footprint derived oceanography data - in order for decision and policy makers alike to make more well-informed decisions and even to assist in early warning detection of natural disasters. This platform has many tools to help the users understand marine water condition in various areas. Continuously moving sensors measure temporal and spatial characteristics of surface gradients in water and therefore Wave Glider is a critical feature that ensures corrective actions and response to be taken before potential issues arise in regard of marine water quality.

By using GeoEvent Processor to incorporate the real-time data into Web GIS, it is easier for anyone to analyse the marine water quality data. It shows areas with poor water quality and draws attention to the possible source of environmental violations. Furthermore, by manipulating the different environment map layers, users can make correlations between at-risk water samples and findings with the neighbouring environment. The Web GIS application improves the efficiency dissemination of information to the stakeholders particularly the decision makers. This web helps the staff in managing and analysing relevant data, as well as publishing this information as maps, charts, and reports.

The benefits of having the real-time Web GIS for this monitoring project are:

- Reducing monitoring costs,
- Improving the speed of decision making by supporting the decision-makers with real time information,
- Ability to access by everyone and everywhere over the internet,

- Reducing time and minimizing effort to reach data,
- Ability to produce reports based on user specified parameters.
- GIS classification enables the users to distinguish marine water safety on color-coded maps.

6. References

[1] Liquid Robotics – Wave Glider Technology. Retrieved from http://www.liquidr.com/prodserv/wg/environmental.html
[2] Lamvi Dao, Rosli Mohamad, and Ghani Ibrahim. Pulau Bidong. Retrieved from http://pulaubidong.org/pulau-bidong-island/
[3] ESRI. What is ArcGIS GeoEvent Extension for Server? Retrieved from http://server.arcgis.com/en/geoevent-extension/latest/get-started/what-is-arcgis-geoevent-extension-for-server.htm

Acknowledgments

Author wishes to acknowledge the assistance and encouragement from colleagues in ASMA, especially the Chief Operating Officer, Shamuddin bin Sulaiman and Assoc. Prof. Sr. Mohamad Nor bin Said for the support and guidance.

A feed-forward Hopfield neural network algorithm (FHNNA) with a colour satellite image for water quality mapping

Ahmed Asal Kzar[*1], **M. Z. Mat Jafri**[*], **Lim Hwee San**[*],**Ali A. Al-Zuky**[2], **Kussay N. Mutter**[*3], **Anwar Hassan Al-Saleh**[4]

[*]School of Physics, Universiti Sains Malaysia, 11800 Penang, Malaysia.
[1]Physics Department, Faculty of science, Kufa University, Najaf, Iraq.
[2]Physics Department, College of science, Al-Mustansiriya University, Baghdad, Iraq.
[3]Physics Department, College of Education, Al-Mustansiriya University, Baghdad, Iraq,
[4]Department of Computer science, College of science, Al-Mustansiriya University, Baghdad, Iraq.

[*1] ahmedassal77@yahoo.com, (The corresponding author).

Abstract. There are many techniques that have been given for water quality problem, but the remote sensing techniques have proven their success, especially when the artificial neural networks are used as mathematical models with these techniques. Hopfield neural network is one type of artificial neural networks which is common, fast, simple, and efficient, but it when it deals with images that have more than two colours such as remote sensing images. This work has attempted to solve this problem via modifying the network that deals with colour remote sensing images for water quality mapping. A Feed-forward Hopfield Neural Network Algorithm (FHNNA) was modified and used with a satellite colour image from type of Thailand earth observation system (THEOS) for TSS mapping in the Penang strait, Malaysia, through the classification of TSS concentrations. The new algorithm is based essentially on three modifications: using HNN as feed-forward network, considering the weights of bitplanes, and non-self-architecture or zero diagonal of weight matrix, in addition, it depends on a validation data. The achieved map was colour-coded for visual interpretation. The efficiency of the new algorithm has found out by the higher correlation coefficient (R=0.979) and the lower root mean square error ($RMSE$=4.301) between the validation data that were divided into two groups. One used for the algorithm and the other used for validating the results. The comparison was with the minimum distance classifier. Therefore, TSS mapping of polluted water in Penang strait, Malaysia, can be performed using FHNNA with remote sensing technique (THEOS). It is a new and useful application of HNN, so it is a new model with remote sensing techniques for water quality mapping which is considered important environmental problem.

1. Introduction

Water is essential to humans, animals and plants. especially for human being, it has important in many different ways like for food production, drinking, domestic and industrial activities [1]. Water is one of the most important natural resources and the lifeblood for sustaining economic development in any country [2]. Water pollution can come from both point and non-point sources. Point sources can be traced to a single source, such as a pipe or ditch. Non-point sources are diffused and typically associated with a watershed and its response to the water movement, human activities, land use and management, and natural influences. Industrial, agriculture and urban areas are anthropogenic sources of point and non-point pollutants. Water pollutants that deteriorate water quality affect many estuarine and freshwater ecosystems on earth [3]. Water quality changes in surface water bodies can create various health problems for living creatures [4]. To solve water pollution issues, we must identify the causes and detect their locations, beginning with major pollution sources that impact places of significant interest or importance.

Penang strait is used as our study area. The area was chosen because of its urgent need for environmental and ecosystem protection. The area is centrally located between the two sections of Penang state, with people living in both areas. Thus, human activities and natural influences continue to increase pollution levels. We must apply monitoring techniques, analyse water bodies and determine pollution sources to solve pollution problems. All of these requirements can be satisfied via remote sensing techniques [5]. The use of artificial neural networks (ANNs) is an important technique in remote sensing applications. ANNs help to provide several useful processes, such as recognition, classification, enhancement, analysis, estimation and prediction. Many studies have been carried out related to water quality. In this study, we focus on water quality mapping. Among several types of ANNs, the Hopfield neural network (HNN) is found suitable for solving optimization problems. In this study, we modify the modified Hopfield neural network (MHNN) [6] for TSS mapping over Penang Strait, Malaysia. The drawback associated with using an HNN with high information, multi-colour images is that information is reduced when these images are converted into white and black, which is required by HNN. Therefore, this drawback is a challenge for colour images, such as satellite images, where each pixel is considered a significant piece of useful information used to interpret a certain phenomenon. To use HNN with colour images, a modified Hopfield neural network (MHNN) has been presented by [6]. This algorithm focuses on the pixel depth of colour images, such as colour satellite images.

In this study, the MHNN technique has been developed for mapping total suspended solids (TSS), Depending on the modification technique, each band of the adaptive HNN image was sliced into binary bitplanes [6-8]. Each band in the colour image is converted to eight bitplanes by converting each band value in the pixel from decimal to binary, then, to a bipolar system, which is convenient for HNN. The new algorithm is named Feed-forward Hopfield Neural Network Algorithm (FHNNA).

2. FHNNA

The FHNNA has two phases: Learning phase and Converging phase. The modification was with both of these phases.

2.1. Learning phase

This phase depend totally on the network vector, called the known vector, and the learning weight. The adopted vector of this network consists of three elements (three bipolar numbers), which are extracted from one band. The sample vectors which are considered known vectors are used for initializing the learning weights according to equation (1). This means the weight will be a matrix of nine elements. The weight initializing occurs in the learning phase of this neural network, and these weights are stored in a look-up table to be ready for the converging phase of the HNN with the unknown vectors, which also consist of three elements from the image data. The weight equation is called the Hebb rule [6, 9, 10].

$$W_{ij} = \sum_{i,j=1}^{3} V_i V_j \qquad \text{.........(1)}$$

Then, it is multiplied by the majority description of the known vector $mdkv(V_j)$:

$$W_{ij} = \sum_{i,j=1}^{3} V_i V_j \; mdkv(V_j) \qquad \text{.....(2)}$$

Where V_j is the known vector and V_i is the transpose of V_j.

$$mdkv(v_j) = sgn(\textstyle\sum_{j=1}^{3} v_j) \qquad \text{.....(3)}$$

In addition, the initialized weight has been determined with zero-diagonal architecture (non-self-connection architecture). The majority description is very important in this network for avoiding similar weights produced by orthogonality phenomenon. Table 1 shows the possible cases of the adopted vector (known or unknown) and its weights, with the correction by $mdkv(V_j)$.

Table 1: All the possible vectors cases with their weight matrices.

Index (decimal)	Vector states (binary)	Vector states (bipolar)	Learning Weight states	majority description md(v)	Corrected Weight states
0	0 0 0	-1 -1 -1	$\begin{bmatrix} 0 & 1 & 1 \\ 1 & 0 & 1 \\ 1 & 1 & 0 \end{bmatrix}$	-1	$\begin{bmatrix} 0 & -1 & -1 \\ -1 & 0 & -1 \\ -1 & -1 & 0 \end{bmatrix}$
1	0 0 1	-1 -1 +1	$\begin{bmatrix} 0 & 1 & -1 \\ 1 & 0 & -1 \\ -1 & -1 & 0 \end{bmatrix}$	-1	$\begin{bmatrix} 0 & -1 & 1 \\ -1 & 0 & 1 \\ 1 & 1 & 0 \end{bmatrix}$
2	0 1 0	-1 +1 -1	$\begin{bmatrix} 0 & -1 & 1 \\ -1 & 0 & -1 \\ 1 & -1 & 0 \end{bmatrix}$	-1	$\begin{bmatrix} 0 & 1 & -1 \\ 1 & 0 & 1 \\ -1 & 1 & 0 \end{bmatrix}$
3	0 1 1	-1 +1 +1	$\begin{bmatrix} 0 & -1 & -1 \\ -1 & 0 & 1 \\ -1 & 1 & 0 \end{bmatrix}$	+1	$\begin{bmatrix} 0 & -1 & -1 \\ -1 & 0 & 1 \\ -1 & 1 & 0 \end{bmatrix}$
4	1 0 0	+1 -1 -1	$\begin{bmatrix} 0 & -1 & -1 \\ -1 & 0 & 1 \\ -1 & 1 & 0 \end{bmatrix}$	-1	$\begin{bmatrix} 0 & 1 & 1 \\ 1 & 0 & -1 \\ 1 & -1 & 0 \end{bmatrix}$
5	1 0 1	+1 -1 +1	$\begin{bmatrix} 0 & -1 & 1 \\ -1 & 0 & -1 \\ 1 & -1 & 0 \end{bmatrix}$	+1	$\begin{bmatrix} 0 & -1 & 1 \\ -1 & 0 & -1 \\ 1 & -1 & 0 \end{bmatrix}$
6	1 1 0	+1 +1 -1	$\begin{bmatrix} 0 & 1 & -1 \\ 1 & 0 & -1 \\ -1 & -1 & 0 \end{bmatrix}$	+1	$\begin{bmatrix} 0 & 1 & -1 \\ 1 & 0 & -1 \\ -1 & -1 & 0 \end{bmatrix}$
7	1 1 1	+1 +1 +1	$\begin{bmatrix} 0 & 1 & 1 \\ 1 & 0 & 1 \\ 1 & 1 & 0 \end{bmatrix}$	+1	$\begin{bmatrix} 0 & 1 & 1 \\ 1 & 0 & 1 \\ 1 & 1 & 0 \end{bmatrix}$

2.2. Converging phase

The convergence phase of this network is represented by the energy function equation. It provides the value that denotes the amount of convergence between the known vector (from the sample), which is represented by the learning weight, and the unknown vector (from the image). The energy function equation of the Hopfield neural network is given below [11]:

$$E = -\frac{1}{2}\sum_{i}^{n}\sum_{j}^{n} v_i\, w_{ij} v_j \; + \delta\, v_j \qquad \ldots\ldots\ldots (4)$$

Where E = energy function, n = number of vector elements, v_j= unknown vector = v_i^T or v_i= v_j^T, Wij = learning weight of the known vector which is the counterpart of the unknown vector v_j, where this weight from the output of neuron i to the input of neuron j and δ = the limiting value, which equals zero in the Hopfield neural network because it is a single layer.
Because HNN here in this study is single layer, the δ will be equal to zero, then the energy function equation becomes:

$$E = -\frac{1}{2}\sum_{i}^{n}\sum_{j}^{n} v_i\, w_{ij} v_j \qquad \ldots\ldots\ldots (5)$$

In the FHNNA, the energy function is modified by considering weights for the bitplanes (Wb) to keep the energy function level of each of the eight bitplanes. Furthermore, the energy function equation must be multiplied by the majority description of the unknown vector, $mduv(v_j)$, to avoid producing the reverse result for the energy function [12]. For this, the equation for E becomes:

$$E = -\frac{1}{2}\sum_{i}^{n}\sum_{j}^{n} v_i\, w_{ij} v_j \, * Wb * mduv(v_j) \quad \dots (6)$$

$$mduv(v_j) = sgn(\textstyle\sum_{j=1}^{3} v_j) \quad \dots\dots\dots (7)$$

The weight of a bitplane (*Wb*) is represented by the following equation:

$$Wb = 2^{L-1} \quad \dots\dots(8)$$

Where *Wb* is the weight of a bitplane in the binary system and L is the bitplane order [5], so that these weights will be equal when the binary system is converted to the bipolar system, which is suitable for the HNN.

Table 2: Bitplane orders with their weights in the binary system.

L	1	2	3	4	5	6	7	8
Wb	1	2	4	8	16	32	64	128

According to the energy function value, we can know how similar to, or different from, the unknown vector the known vector is, according to the sequence of energy values (-3, -1, +1, +3). The investigation of the energy function value in the case of a non-self connection architecture (weight of zero on the diagonal) for all states between the unknown vectors and the weights of the known vectors are 64 states of energy function values, this number comes from all cases that produced from the 8 possible cases weights of the known vectors and the 8 possible cases of the unknown vectors, as shown in table 3.

Table 3: All possible energy function values of the FHNNA.

W \ V	1	2	3	4	5	6	7	8
1	-3	1	1	-1	1	-1	-1	3
2	1	-3	1	-1	1	-1	3	-1
3	1	1	-3	-1	1	3	-1	-1
4	-1	-1	-1	-3	3	1	1	1
5	1	1	1	3	-3	-1	-1	-1
6	-1	-1	3	1	-1	-3	1	1
7	-1	3	-1	1	-1	1	-3	1
8	3	-1	-1	1	-1	1	1	-3

3. The study area

The Penang Strait, Malaysia is the study area of this work. The image used for the study is located between latitudes 5° 19' N and 5° 26' N and longitudes 100° 17' E and 100° 24' E. Penang state is divided into two parts. The first is an island and the second part is a coastal strip on the mainland, which is called Wellesley Province (Seberang Perai) [13]. Penang Island has an equatorial climate, which is uniform throughout the year [6]. The area has a warm, humid climate. The temperature ranges from 25 °C (night time) to 33 °C (day time) [13, 14]. The average annual relative humidity varies between 70% and 90%. The total annual rainfall can be as high as 624 cm, and the average annual rainfall is approximately 267 cm.

Penang state is faced by monsoon winds. During this period, the weather conditions become rainy at night and sunny during the day. Figure 1 shows the study area.

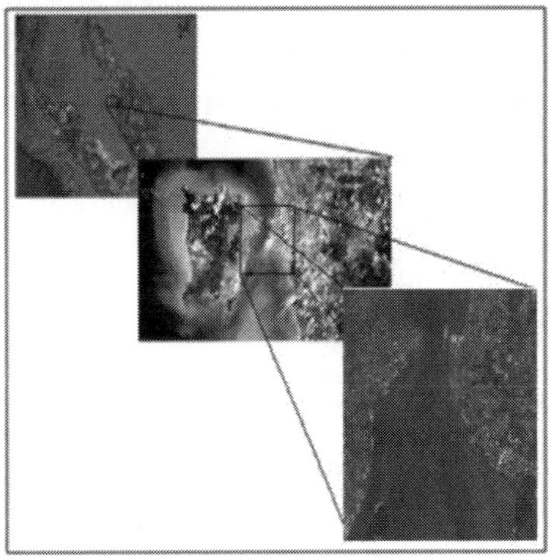

Figure 1: Study area: Penang strait.

4. The used data

The data in this study is divided into two essential parts: the remote sensing image (THEOS) and the validation data, which related with samples that took simultaneously with capturing the satellite image.

4.1. The satellite image

In this study, the used remote sensing (satellite) image is Thailand earth observation system (THEOS). There are technical specifications for this type that make it able to achieve several applications; Figure 2 shows THEOS [15]. Table 4 demonstrates these specifications [16], and Table 5 gives characteristics about THEOS satellite orbit parameters.

Figure 2: The used satellite: Thailand Earth Observation System (THEOS).

Table 4: Technical specifications of THEOS.

	Panchromatic	Multispectral
Wavelength (nm)	450–900	Blue: 450–520 Green: 530–600 Red: 620–690 Near Infrared: 770–900
Resolution (m)	2	15
Swath width (km)	22	90
Pixels	12,000	6000

Table 5: Characteristics of THEOS.

Operator	GISTDA
Mission type	Earth Observation
Launch date	1 October 2008 06:37 GMT
Carrier rocket	Dnepr
COSPAR ID	2008-049A
Mass	715 kg
Power	840 W
Orbital elements	
Regime	LEO
Inclination	98.78°
Apoapsis	826 km (513 mi)
Periapsis	825 km (513 mi)
Orbital period	101.4 minutes

4.2. Validation data

The location for sample collection (longitudes and latitudes) in the study area were determined using a handheld global positioning system (GPS). Samples from these locations are collected simultaneously with satellite (THEOS on 29-1-2010) image acquisition. The samples were analysed in the lab to measure the validation data, represented by TSS concentrations in milligrams per litre (mg/L) for each location. The raw satellite imagery and sampling locations used for validation data and the TSS concentrations with indices are shown in Figure 3.

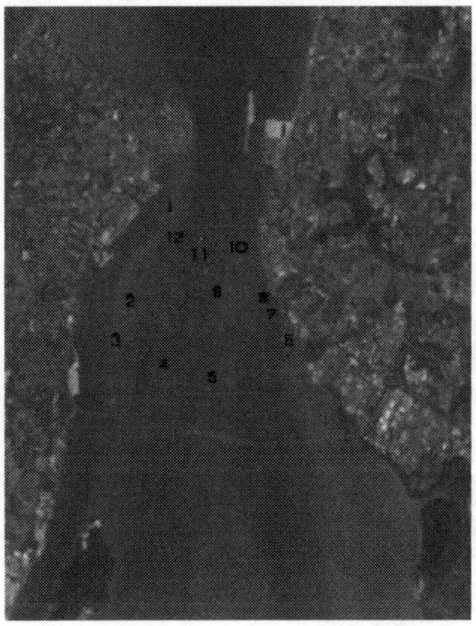

a

Sample index	Longitude	Latitude	Sample value (mg/L)
1	100d20'20.6002"E	5d23'56.0900"N	75
2	100d19'45.9098"E	5d22'45.8500"N	68
3	100d19'40.1200"E	5d22'16.9400"N	70
4	100d20'22.2601"E	5d22'01.2400"N	59
5	100d21'01.9199"E	5d21'50.5000"N	50
6	100d22'17.1098"E	5d22'15.2800"N	98
7	100d22'11.3300"E	5d22'40.0700"N	91
8	100d22'03.0698"E	5d22'56.6000"N	101
9	100d21'25.0600"E	5d22'58.2000"N	71
10	100d21'33.3202"E	5d23'21.3800"N	62
11	100d20'59.4398"E	5d23'13.1200"N	50
12	100d20'09.8700"E	5d23'37.0800"N	71

B

Figure 3: THEOS and validation data: a- Raw satellite images and sampling indices. b- Samples locations and their concentrations values.

5. Methodology

It is important to mention the FHNNA architecture, that is demonstrated in the Figure 4 [6]. This architecture consists of three neurons, where each node is connected with every other neuron but not with itself, as a result we used architecture called non-self-architecture (zero diagonal weight matrixes). The possible number of vectors is 2^3, the small number of neurons gives FHNNA the ability for computing faster than other size, in addition to ability of dealing with a big number of patterns.

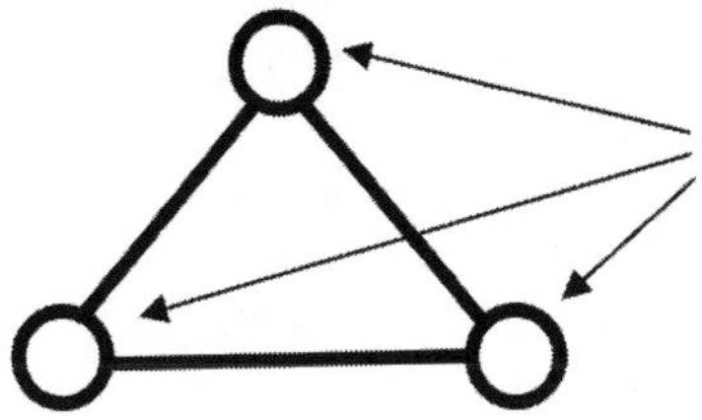

Figure 4: The architecture of FHNNA.

Modification of the FHNNA was conducted via the following contributions:

5.1. The validation

The validation is very important in studies like that related with water quality, where the validation data (real data) representing TSS concentrations are divided into two equal groups. The first group is used to classify samples (source samples) through the converted red, green, and blue values of these samples locations for using in the FHNNA. The second group is used as test samples by finding out their position in the produced classes. The band values are converted from decimal to binary representation, then to a bipolar representation as known vectors. They produce the learning weights in the learning phase of the FHNNA. The weights will then combine with the unknown vectors from the image bands by the energy function value to test the corresponding or nearest class. The concentration values will not be used in the proposed algorithm, but instead for determining accuracy through calculating the R and $RMSE$ between the concentrations from the produced classes and those are found out in the produced class locations. Previous studies [6] have used only two samples. In this study, twelve real samples were used. They give a better range of samples, which is ideal for satellite image classification success and algorithm accuracy. Matching then occurred between the image pixels vectors (unknown vectors) and the counterparts weights of the samples vectors (known vectors). This matching is an important step in providing the energy function value.

5.2. FHNNA as a feed-forward associative memory

In the new algorithm, HNN is represented as a feed-forward network, which acts in one direction. It is not recurrent because this artificial neural network has been determined as ideal for classification purposes. Two types of convergence methods exist [12, 17, 18]:

Type one: Detection method – used to detect the corresponding vector by calculating the value of the energy function. This method is related to a feed-forward net (non-iterative net) and is typically used for pattern classification.
Type two: Generating method – used to generate the lost vector based on the energy function value calculation. This method is a recurrent net (iterative net) and is typically used for several applications such as enhancement or restoring purposes.

We have adopted a detection method using satellite (THEOS) images to classify pollutants concentrations and contribute to water quality mapping. FHNNA in converging phase as detection method (without iterations) provides two advantages: (1) The network will not face local minimum problems, which produce errors in the results; and (2) It will increase algorithm speed, i.e., reduces the algorithm run time. In addition, because [6] used an HNN generating method for dealing and optimizing colour images and we used the HNN for dealing and classifying colour images, a comparison will be made between the minimum distance classifiers, which have been used to classify satellite images in many previous studies [19, 20].

5.3. Considering weights of bitplane in the energy function

The binary number value in the bitplane increases as the order increases from lowest to highest. But Without *Wb*, the energy function value of a bitplane will be based on the nearest or similar value of the energy function of another bitplane, without considering the binary number level of the bitplane. Therefore, each bitplane has a differing effect on the summation value of the eight energy functions. The selected sample (class) for image pixels, as a result of classification, is the sample which gives the lowest summation of energy functions because the Hopfield neural network denotes the corresponding or nearest class. The *Wb* in the FHNNA is very important for correcting the results. This variable avoids error in the summation value of energy functions values.

Thus, we have modified the HNN, a simple network, to be suitable with modern techniques (satellite imagery) to analyse an important topic, water pollution. Where, each image band has been contributed in the algorithm depending on its ratio as shown in the table 6. These ratios gave the best results.

Table 6: The depended bands ratios in the used image.

Band	Ratio
Red	0.42
Green	0.38
Blue	0.20

The methodology of this work is demonstrated clearly in the following figure:

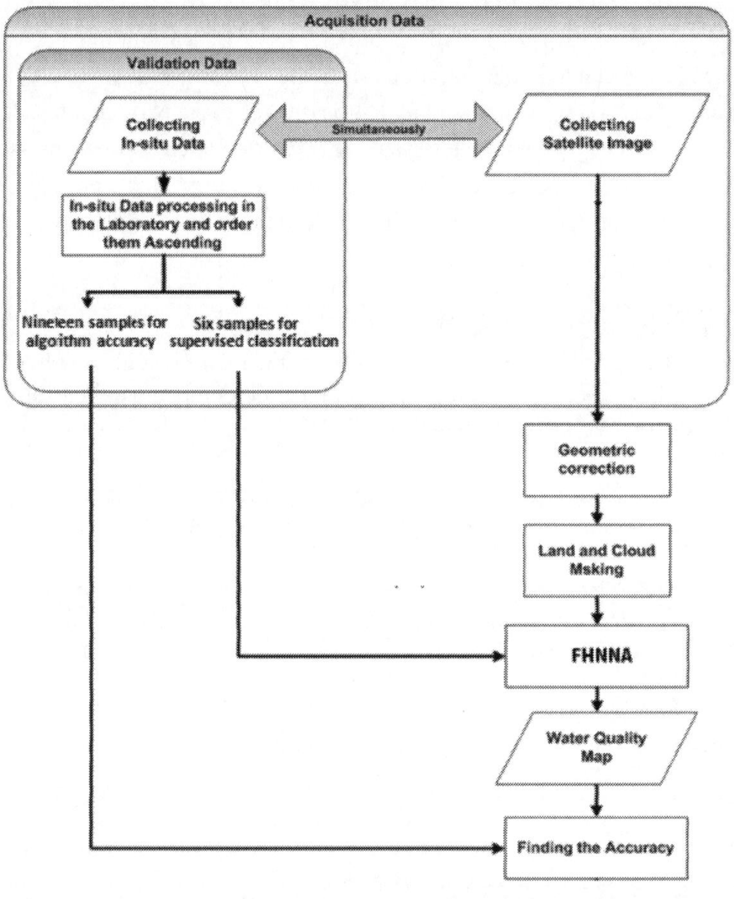

Figure 5: The work methodology.

Figure 5 shows the methodology of using FHNNA with remote sensing image for water quality mapping, but the main steps of the new algorithm (FHNNA) are explained below:

FHNNA:

Step 1– Collect samples to be validation data (in situ data) from the study area, the collection is simultaneously with the acquisition of the satellite imagery, fixing the position of these data via a handheld global positioning system (GPS).

Step 2– Determine the pollutant concentration values of the collected samples in the lab.

Step 3- Order the pollutant concentration values in ascending order. Then, separate them into two groups. The first group includes values that have even indexes, which are used as classes for supervised classification. The second group includes odd values, which are used to validate the algorithm accuracy.

Step 4– Use the satellite image and give the non-water pixels a black colour (Red=0, Green=0, Blue=0).

Step 5- **Learning phase:** For water pixels, analyse each pixel of the RGB bands and multiply each band by its ratio through table 6, then for each band, do:

Step 6- Convert each digital number from decimal to binary representation (eight bits), i.e., replacing the band with eight binary bitplanes (eight binary images). Each pixel in a bitplane will be 0 or 1. Note that for small binary values (e.g., 3 is equal to binary 11), the remaining six bits (bitplane) to the left will have zero values (e.g., 00000011).

Step 7– Convert the bitplane numbers from binary to bipolar representation by consider each (0) to be (-1) and each (1) to be (+1), or via the following equation:

$$N_{bipolar}(x,y) = 2 * N_{binary}(x,y) - 1 \quad ...(9)$$

Step 8- Choose 3x3 pixel x 8 bitplane sample sizes for each image band, which represent the samples for supervised classification. The locations of these samples are the exact locations (longitudes and latitudes) of collected samples (real data) from the study area.

Step 9- Initialize 1x3 vectors by dividing each sample size in each band in each bitplane by the 1x3 size (i.e., each chosen sample size will have 3 vectors) to be included in the known (learning) vector.

Step 10- For each known vector, find the learning weight via equation (2) and depending on the vector information. These weights are given sequentially symbols and saved in the lookup table:

$$W_{ij}^{sbpc} = \sum_{i}^{n}\sum_{j}^{n} V_i^{sbpc}.V_j^{sbpc}\, md(V_j^{sbpc}) \quad ....(10)$$

Where $\boldsymbol{s}$ is number of samples in band $\boldsymbol{b}$ in bitplane $\boldsymbol{p}$ of number of known vector $\boldsymbol{c}$.

Step 11- **Converging phase:** Starting from the beginning of the water pixels in image $\boldsymbol{I}$, for each band $\boldsymbol{b}$ and each bitplane $\boldsymbol{p}$, obtain a 3x3 sample size, then divided these samples into three 1x3 vectors, which will be unknown vectors: V_j^{Ibpc}.

Step 12- Find the energy function values for each vector in the first sample and its counterpart vector from the image, as shown in:

$$E^{bpc} = -\frac{1}{2}\sum_i^n \sum_j^n V_i^{Ibpc}\, W_{ij}^{sbpc}\, V_j^{Ibpc} * Wb^p * md(V_j^{Ibpc}) \quad ....(11)$$

Step 13- Calculate the Energy Function summation values for the eight bitplanes of each band. Then, find the *SEF* value for the three bands (Red, Green, and Blue) related convergence of the first sample.

$$SEF = E^r + E^g + E^b \quad ....(12)$$

Where E^r, E^g and E^b are the Energy Function summation values for the red, green and blue bands, respectively.

Step 14- Repeat steps (12) and (13) for all other samples.

Step 15- Find the minimum summation value (SEF) from all summations values. Then, note the image size of the sample (class) that produced this minimum value.

Step 16- After producing the classified image, validate the results through identifying the class values that cover the location of the second group values (e.g., each value that represents its class and the value from the second group it covers). Then, find the correlation coefficient, *R*, and *RMSE* between them.

6. Results and discussion

The results obtained by applying the proposed algorithm to the adaptive satellite image are shown in Figure 6.

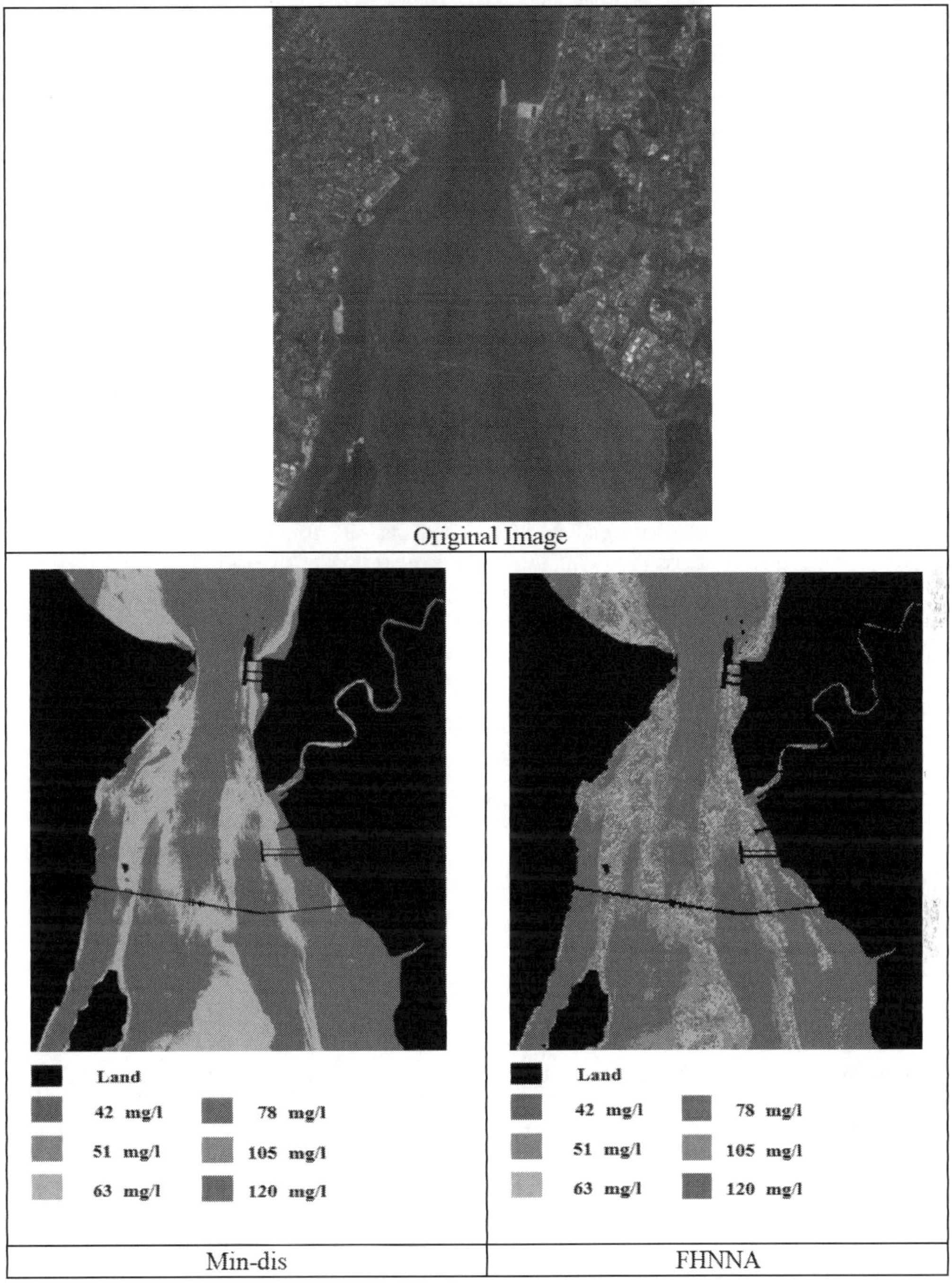

Figure 6: TSS mapping using the FHNNA compared with a Min-dis classifier using the THEOS 29-1-2010 image.

The results are visually clear for evaluating FHNNA and its comparison with Min-dis, where the latter gives a classification but is not synoptic for all classes. There are classes that have not been classified, especially with the two classes of TSS=78 mg/l and TSS=105 mg/l. This outcome does not give a completed mapping for the TSS. When we look at the FHNNA map, we find that all of the classes have been classified, which gives a completed mapping with respect to TSS. With respect to a visual comparison, more accuracy can be given by the tables (Tables 7 and 8).

Table 7: The accuracies of the produced classes by FHNNA and Min-dis.

Min-dis		**FHNNA**	
Classification Samples (mg/l)	**Test samples (mg/l)**	**Classification Samples (mg/l)**	**Test samples (mg/l)**
50	50	50	50
68	62	68	62
68	70	71	70
98	71	75	71
98	91	98	91
98	101	98	101

Table 7 denotes the concentration values of the classes produced via each method. Although the majority of the test samples are located in the correct classes, some were inaccurately classified, particularly within the Min-dis results. This proves that the proposed algorithm is better than Min-dis. The FHNNA is capable of high quality TSS mapping (classification). Regarding the efficiency of the proposed algorithm, we calculated the correlation coefficient (*R*) and the root mean square error (*RMSE*) between the first and the second groups of validation data (real data). Table 8 displays the accuracy results.

Table 8: Accuracy Results.

	R	*RMSE*
FHNNA	0.979	4.301
Min-dis	0.846	11.740

This table provides evidence that FHNNA yields a more accurate result, and this result is the first that proves the FHNNA efficiency with the remote sensing image by the band ratios method, where the different ratios for the bands refer to the physical meaning of the water surface phenomenon, which gives these bands values and has a relationship with the reflectance of the water surface and the conditions above the studied area.

7. Conclusion

This study provides a brief overview of TSS mapping at the Penang Strait, Malaysia. Satellite imagery can be used to provide information for effective planning management. Applying the FHNNA to THEOS data for TSS mapping in the study area produced reliable, high quality results. Our results exhibit successful evidence of applying the new algorithm to the water quality mapping of colour images, which is considered a new application of HNN. In addition, FHNNA is considered a new model with remote sensing techniques for water quality that is critical environmental problem.

8. Acknowledgements

This project was supported by the University Sains Malaysia. We would like to express our thanks and appreciation to all technical staff who contributed to this project.

9. References

[1] Mekonnen, *et al*, Spatial Runoff Estimation and Mapping of Potential Water Harvesting Sites: A GIS and Remote Sensing Perspective, Northwest Ethiopia, in Landscape Dynamics, Soils and Hydrological Processes in Varied Climates. 2016, Springer. p. 565-584.

[2] Bhatti, A.M., Modelling and monitoring of suspended matter in surface waters using remotely sensed data. 2008.

[3] Malthus, T. *et al*, First derivative indices for the remote sensing of inland water quality using high spectral resolution reflectance. Environment International, 1995. 21(2): p. 221-232.

[4] Alaguraja, P., et al., Remote sensing and GIS approach for the water pollution and management in Tiruchirappli Taluk, Tamil Nadu, India. International Journal of Environmental Sciences, 2010. 1(1): p. 66.

[5] Kzar, A.A., et al. Modified Hopfield Neural Network Algorithm (MHNNA) for TSS mapping in Penang strait, Malaysia. in Signal and Image Processing Applications (ICSIPA), 2013 IEEE International Conference on. 2013: IEEE.

[6] Mutter, K.N., Developing Hopfield Neural Network For Multiresolution Image Recognition, in school of physics. 2010, usm.

[7] Gonzalez, R.C., *et al*, Digital image processing. Prentice Hall, 2002: p. 299-300.

[8] Milan, S., *et al*, Image processing, analysis, and machine vision. Photocopy Edition, Posts & Telecom Press, Peking, 2002.

[9] Fausett, L., Fundamentals of neural networks: architectures, algorithms, and applications. 1994: Prentice-Hall, Inc.

[10] Samarasinghe, S., Neural networks for applied sciences and engineering: from fundamentals to complex pattern recognition. 2006: CRC Press.

[11] Neshat, M., et al., Using Fuzzy Hopfield Neural Network for Diagnosis of the Hepatitis Disease. Turkish Journal of Engineering, 2014. 2: p. 63-75.

[12] Kareem, E.I.A., *et al*, Multi-Connect Architecture (MCA) Associative Memory: A Modified Hopfield Neural Network. Intelligent Automation & Soft Computing, 2012. 18(3): p. 279-296.

[13] Daraigan, S.G., et al. Multispectral absorption algorithm for retrieving TSS concentrations in water. in Geoscience and Remote Sensing Symposium, 2007. IGARSS 2007. IEEE International. 2007: IEEE.

[14] San, L.H., *et al*, Aerial Photogrammetry Method for Water Quality Monitoring Using Digital Camera. Open Environmental Sciences, 2009. 3: p. 20-25.

[15] Yu, G., *et al*, Image compression systems on board satellites. Acta Astronautica, 2009. 64(9): p. 988-1005.

[16] Sritarapipat, T., *et al*, Fusion and registration of THEOS multispectral and panchromatic images. International Journal of Remote Sensing, 2014. 35(13): p. 5120-5147.

[17] Fausett, L.V. *et al*, Fundamentals of neural networks: architectures, algorithms, and applications. Vol. 40. 1994: Prentice-Hall Englewood Cliffs.

[18] Kzar, A.A., et al., A Modified Hopfield Neural Network Algorithm (MHNNA) Using ALOS Image for Water Quality Mapping. International Journal of Environmental Research and Public Health, 2015. 13(1): p. 92.

[19] Rashed, T., et al., Revealing the anatomy of cities through spectral mixture analysis of multispectral satellite imagery: a case study of the Greater Cairo region, Egypt. Geocarto International, 2001. 16(4): p. 7-18.

[20] Yu, B., et al., Object-based spatial cluster analysis of urban landscape pattern using nighttime light satellite images: A case study of China. International Journal of Geographical Information Science, 2014. 28(11): p. 2328-2355.

IOP Conf. Series: Earth and Environmental Science **37** (2016) 012076 doi:10.1088/1755-1315/37/1/012076

The application of remote sensing techniques for air pollution analysis and climate change on Indian subcontinent

S. N. Palve[1*], Dr.P.D.Nemade [2], Dr.S.D.Ghude[3]

[1]Research scholar DYPIET Pimpri, University of Pune, India,
[2]S.B. Patil C.O.E. Indapur, Pune,
[3]Indian Institute of Tropical Meteorology, Pune

[*]Email: sandippalve@hotmail.com

Abstract. India is home to an extraordinary variety of climatic regions, ranging from tropical in the south to temperate and alpine in the Himalayan north, where elevated regions receive sustained winter snowfall. The subcontinent is characterized by high levels of air pollution due to intensively developing industries and mass fuel consumption for domestic purposes. The main tropospheric pollutants (O_3, NO_2, CO, formaldehyde (HCHO) and SO_2) and two major greenhouse gases (tropospheric O_3 and methane (CH_4)) and important parameters of aerosols, which play a key role in climate change and affecting on the overall well-being of subcontinent residents. In light of considering these facts this paper aims to investigate possible impact of air pollutants over the climate change on Indian subcontinent. Satellite derived column aerosol optical depth (AOD) is a cost effective way to monitor and study aerosols distribution and effects over a long time period. AOD is found to be increasing rapidly since 2000 in summer season that may cause adverse effect to the agricultural crops and also to the human health. Increased aerosol loading may likely affect the rainfall which is responsible for the observed drought conditions over the Indian subcontinent. Carbon monoxide is emitted into the atmosphere by biomass burning activities and India is the second largest contributor of CO emissions in Asia. The MOPITT CO retrievals at 850 hPa show large CO emission from the IG region. The development of convective activity associated with the ASM leads to large scale vertical transport of the boundary layer CO from the Indian region into the upper troposphere. TCO over the Indian subcontinent during 2007 has a systematic and gradual variation, spatial as well as temporal. Higher amount of TCO in the northern latitudes and simultaneous lower TCO at near equatorial latitudes indicates depletion of ozone near the equator and accumulation at higher latitudes within the subcontinent. In addition, changes in stratospheric ozone and atmospheric abundances of aerosols alter the energy balance of the climate system.

1. Introduction

Major sources of energy in India include coal, petroleum, biomass, nuclear power and hydropower. Combustion of these fuels is the main source of pollution in the natural environment. India has a large variation in climate from region to region, due to its Geographical nature. Extremes of climate and weather events are increasingly being recognized as key aspects of climate change and they adversely affect humans and ecosystems, with serious socioeconomic consequences [1].The complexity of the climate system is enormous. The thermal gradient caused by the lower tropospheric warming and corresponding cooling over the upper troposphere largely regulates the atmospheric general circulation patterns. In the changing climatic scenario, the recent sharp rising trend in the surface air temperature over the Indian landmass is attributed to the large-scale declining trend in Indian summer monsoon rainfall [2]. Not only the temperature gradient but also the pressure gradients across the Indian landmass plays a major role in controlling the climate system in and around the Indian subcontinent [3].Anthropogenic changes in the atmosphere can have a profound impact on the climate and the consequences are assumed to occur at present [4]. Climate change may result in an increase in the frequency of floods in many regions. However, natural changes in the Earth's climate might have been augmented by the enhanced greenhouse effect, caused by manmade changes in the Earth's environment. Greenhouse gases and aerosols are the two most important elements that affect the radiation balance of the earth's atmosphere. It is well established now that the anthropogenic emissions of greenhouse gases (most importantly CO_2) into the atmosphere cause global warming of the troposphere [1].

In this paper we highlighted and put together role of air pollutant on climate change over the Indian subcontinent. Aerosols play a crucial role in the climate of the Earth–atmosphere system by means of their direct and indirect impacts, and are considered to be one of the largest uncertain components of the global climate system [4]. The overall composition of aerosol density is likely to affect the large-scale heating and pressure gradients in the atmosphere, which can amend the circulation pattern, cloud dynamics, precipitation and so on [5]. CO is another important atmospheric pollutant for a number of reasons. The major source of CO emission in India is traditional bio-fuel use, which is almost 50% of the total Indian CO emission [6]. Carbon monoxide can be measured from space and can be used to identify sources of air pollution. The industrialization of eastern Asia has influenced the chemical climate of India through transport processes and thus through the Indian monsoon [7]. The small belt of ozone around the globe filters the ultraviolet radiations reaching the ground and thereby protects the whole biosphere from harmful effect [8].

2. Aerosols mechanism and its influence on climate

In the monsoonal countries like India and China, aerosol problem is ever becoming acute due to increased loading of atmospheric pollutants from anthropogenic as well as natural sources [9].They interact with various monsoonal activities, for example, energy balance, cloud formation processes and so on [10] and perturb their normal occurrences. The magnitude and sign of aerosol radiative impact depend on physical, chemical and radiative properties of these aerosols. Aerosols are classified into natural and anthropogenic, according to their origin. Sea salt, dust, natural sulphates, etc., are naturally occurring aerosols, whereas soot, industrial sulphates, black carbon, etc., are of anthropogenic origin. Soot is an absorbing aerosol whereas dust and organic matter are partly absorbing. Interactions between dust aerosols and clouds in the upper troposphere and lower stratosphere can produce a significant impact on atmospheric radiation budget, and hence on global and regional climates.

Carbonaceous aerosols have received much attention recently because of their potential role in regional climate change [11].Black carbon is a light-absorbing aerosol, that is, the by-product of incomplete combustion of carbonaceous fuel. Because of its absorptive nature, Black Carbon accounts directly for the reduction in incoming short-wave solar radiation at the Earth's surface, leading to heating of the atmosphere and thus possibly changing the temperature structure in the troposphere, which in turn affects the cloud microphysical properties and thereby rainfall mechanisms [11].Global warming produced by greenhouse gases is partly suppressed by aerosols; they therefore have a substantial role in the radiation budget and climate. The radiative effects of aerosols on the Earth's atmospheric system are governed by the quantity of aerosols in the atmosphere, their vertical distribution, size distribution and single scattering albedo, and the reflectivity of the underlying surface. Climate change is one of the most burning issues globally; aerosols have great potential to bring out changes in climatic conditions at regional and global scales [12].

2.1 MODIS AOD data

Moderate Resolution Imaging Spectroradiometer (MODIS), one of the first passive satellite spectroradiometer designed to observe aerosols from an altitude of about 700 km.MODIS measures reflected solar radiance and terrestrial emissions in 36 wavelength bands between 0.405 and 14.385 μm, with resolutions between 250 and 1 km. It was launched onboard the Terra satellite in late 1999 and onboard the AQUA satellite in May 2002. The data obtained from MODIS Aura has a spatial resolution of 1.0° (latitude) × 1.0° (longitude). Collection 5 (MOD08_M3-005) of MODIS-Terra level-3 monthly aerosol data was used. The MODIS AOD available from Giovanni (http://disc.sci.gsfc.nasa.gov/giovanni) is used in order to analyze spatial variations of aerosol loading. MODIS with its 2330 km viewing swath provides almost daily global coverage. The MODIS AOT uncertainty is ±0.05 ± 0.15 × (AOT) over the land [13].

2.2 AOD profile over Subcontinent

The dominance of the monsoon climate over the Indian subcontinent controls the seasonal highs and lows of aerosol concentration. Due to the climatic characteristics, during the winter and summer dry seasons (prior to the onset of the monsoon season), the concentration of aerosol particles remains high in the atmosphere. Hence the aerosol properties and their effects on weather and the climate of Indian Subcontinent are certainly different from rest of the world.AOD distribution during pre-monsoon affects cloud formation and hence rainfall distribution which is found to be prominent in last 4 years. Complex interaction between aerosols, clouds, climate and vegetation needs to be addressed in the light of harmful effects observed in several studies by different workers [14]. Organizations (WMO) recognize dust as a major component of atmospheric aerosol, which is an essential climatic variable to study. Mineral dust emitted from arid and semi-arid regions plays an important role in climate and also cause changes in cloud properties, such as the number, concentration, and size of cloud droplets, which can alter both cloud albedo and cloud lifetime. The phenomenon termed the 'indirect aerosol effect'. [10], using aerosol forcing derived from atmospheric brown clouds field experiments, suggested that aerosol induced cooling decreases surface evaporation and reduces the north south surface temperature gradient over the Indian Ocean, leading to a weakened monsoon circulation. [5] Found that an abundant amount of dust aerosols from the Thar Desert and the Middle East deserts are transported into northern India, during the summer season (April through early June).Forced by the prevailing wind against the steep topography of the Himalayas, the dust aerosols pile up against the foothills and spread over the Indo-Gangetic Plain (IGP). The airborne dust particles become even more absorbing when transported over megacities of the IGP and coated by fine black carbon aerosols from local emissions [15].

The Indo-Gangetic Plain is an aerosol "super hotspot", hosting the world's highest population density and concentration of coal-firing industrial plants. AOD is found to be very high (>.6) in Ganga basin with increasing aerosol concentration at an alarming rate in eastern part of basin shown in Figure1.Southern parts of India show a much cleaner environment with AOD less than 0.4. The spatial gradient of AOD is showing increase from southern part of Indian subcontinent to northern part up to Himalaya. The central India shows moderate AOD values. Input of aerosols (mostly coarser fraction) from Thar Desert and dry season during pre-monsoon cause high AOD in the Ganga basin.AOD at eastern part of subcontinent increased during summer 2014 shown in Figure2 indicates high level of concentration over IG region. The concentration of AOD is in variable nature over the subcontinent shown in Figure3, frequency distribution curve of AOD. Indo Gangetic plane and eastern part of continent shows high values of AOD as compared to rest of the world. The impact of increasing aerosol loading over Indian subcontinent during summer season affecting monsoonal rainfall distribution which is responsible for serious drought conditions and adverse effect to the agricultural crop production and also to the human health.

IOP Conf. Series: Earth and Environmental Science 37 (2016) 012076 doi:10.1088/1755-1315/37/1/012076

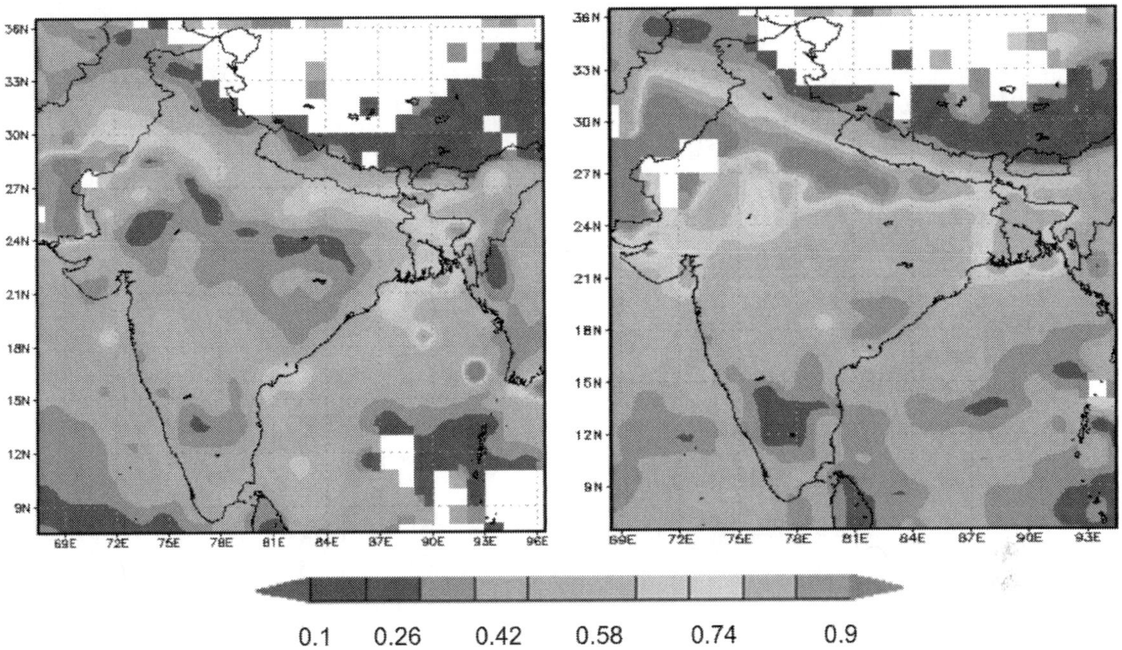

Figure 1. MODIS AOD over the Indian Subcontinent during summer 2000 and 2004.

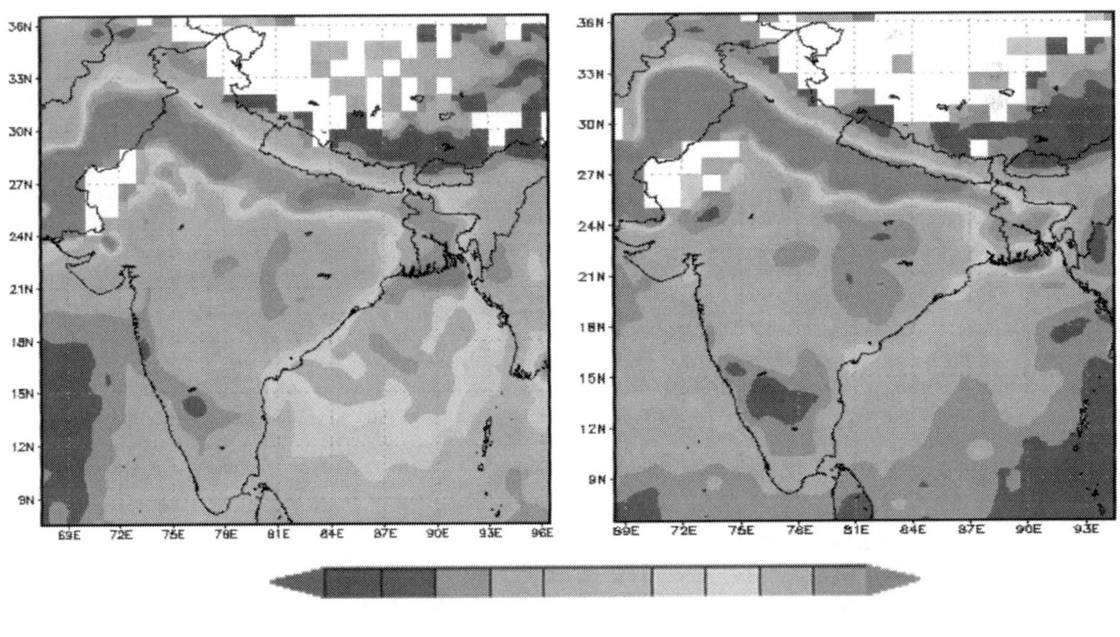

Figure 2. MODIS AOD over the Indian Subcontinent during summer 2008 and 2014.

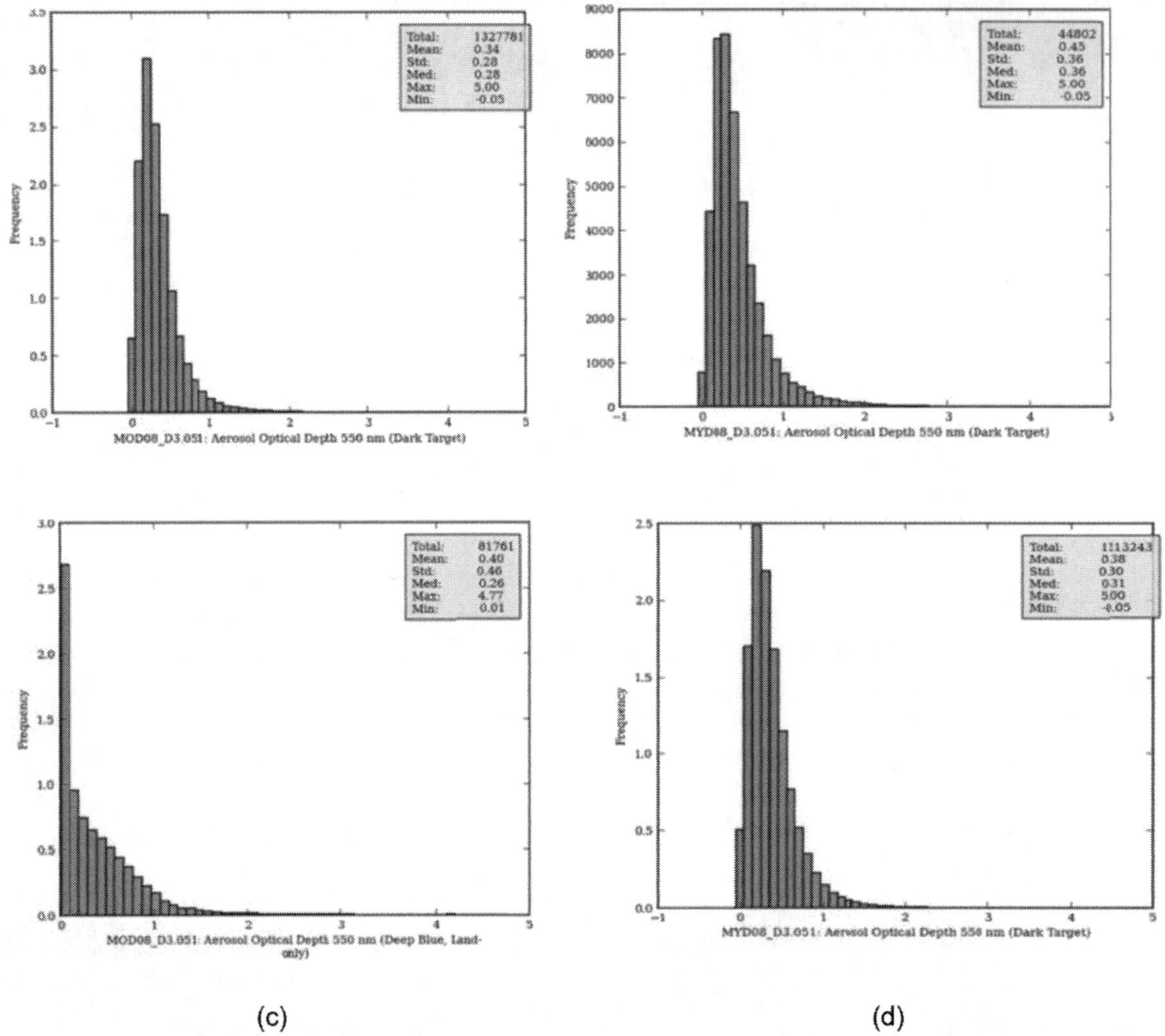

Figure 3. MODIS frequency distribution curve of AOD over the Indian Subcontinent during summer 2000, 2004, 2008 and 2014 respectively.

3. Carbon monoxide over the tropopause

CO contributes to air pollution, acts as a precursor to tropospheric ozone and is a sink for OH radicals. The photochemical oxidization of CO and hydrocarbon in the presence of NO_x is a major source of tropospheric ozone at both regional and global scales [8].CO is produced at the Earth's surface by incomplete combustion processes related to industry, traffic or biomass burning [16]. The main sources of CO are from anthropogenic and natural direct emissions and from the oxidation of methane and other gases. Dominant emissions are from road traffic, fossil fuel and biomass burning together with smaller contributions from vegetation and the oceans. High levels of CO pollution are found around the world, and they result from different types of biomass burning in different locations [17].

Emissions from biomass burning account for about one-quarter of the CO released to the atmosphere, with an average of around 600 Mt (600×106 tonnes) CO per year. CO is an ideal tracer for monitoring the air pollution sources and transports due to its moderately long lifetime (weeks to months) and inhomogeneous distribution in the troposphere [18]. The estimation of CO profiles and total CO

IOP Conf. Series: Earth and Environmental Science **37** (2016) 012076 doi:10.1088/1755-1315/37/1/012076

column has been identified as a very important objective in order to improve our understanding of the global climate system.

3.1 Retrieval of atmospheric CO profile from MOPITT

The MOPITT instrument [19] is onboard the Terra platform and measures tropospheric CO with nadir sounding. MOPITT is a correlation radiometer for estimating vertical profiles of atmospheric CO through the thermal radiance received in the 4.7 μm spectral region. The horizontal resolution of MOPITT CO data (version 3) is 22 km×22 km. The super-observations give around 8000 daily vertical profiles that are retrieved on seven pressure levels (surface, 850, 700,500, 350, 250 and 150 hPa). This sensor was initially designed to carry out measurements of both CO and CH_4, but, currently, there are no plans for dissemination of CH_4 data.

The global MOPITT retrieved CO data have been used for estimating CO distributions, atmospheric transport, sources, and sinks. MOPITT is a correlation radiometer, which means that the CO profiles are retrieved using an optimal estimate of the maximum likelihood solution. With this technique, the retrieved CO profiles depend not only on MOPITT radiance measurements, but also on the a priori CO profile and the averaging kernels which are also averaged and are taken into account in the assimilation system.

CO mixing ratios over the Indo Gangetic region for the month of May, averaged for the 8-year period 2000–2007shown in Figure 4. It is clearly observed that CO mixing ratios over the northern part of India is much higher compared to other part of India.CO mixing ratio higher values of 185ppb to 220 ppb are observed over the Indo Gangetic region due to rapid growth of Industrialisation, urbanisation and growing population. Besides strong seasonal variations as shown by previous studies [20], CO has also been shown to have large interannual variability in the troposphere [21]. Despite of these higher values, CO mixing ratios over the central part of India observed by MOPITT at 850 hPa level is low due to less industrial activities and low biomass burning.

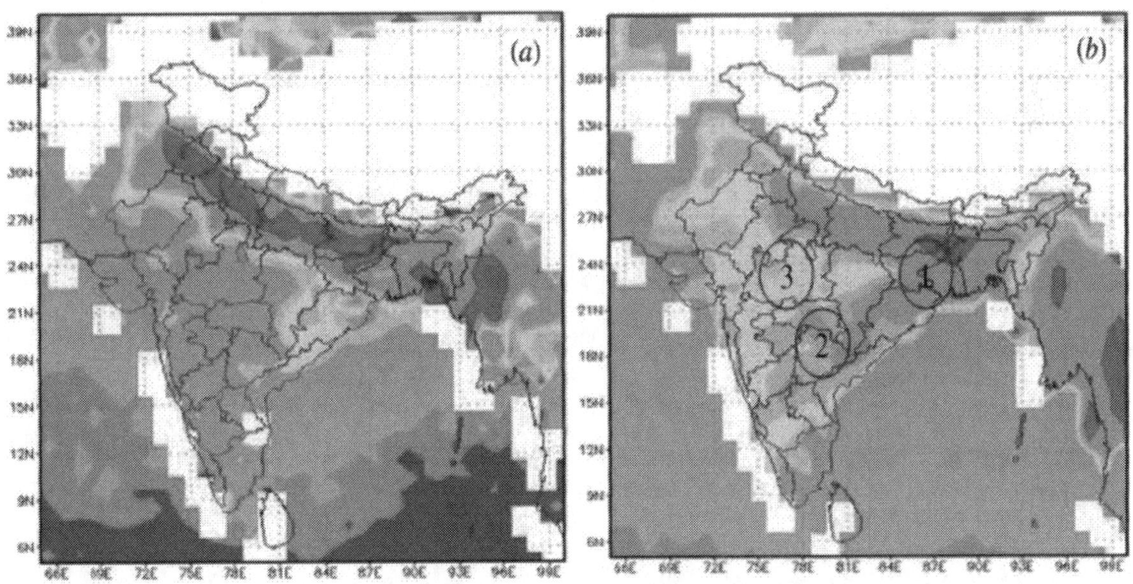

Figure 4. MOPITT carbon monoxide (CO) mixing ratios (ppb) at 850 hPa over the Indian subcontinent (a) averaged during May 2000–2007, (b) averaged for 2000–2007.

IOP Conf. Series: Earth and Environmental Science **37** (2016) 012076 doi:10.1088/1755-1315/37/1/012076

4. Atmospheric ozone and climate

Atmospheric ozone is a strong oxidizing agent. Together with water vapor, as a precursor of the hydroxyl radical, ozone has a strong influence on the oxidizing power of the atmosphere and hence on the rate at which many natural and anthropogenic compounds are eliminated from the atmosphere [22]. Tropospheric ozone impacts air quality and human health, atmospheric radiative forcing and ecosystem productivity with resulting impacts on food and climate. It also modifies the "oxidizing capacity" of the troposphere, impacting the lifetime and radiative forcing of methane. The small belt of ozone around the globe filters the ultraviolet radiations reaching the ground and thereby protects the whole biosphere from harmful effect [8]. The ozone located in the vicinity of tropopause where temperature is considerably lower than near the surface is the largest contributor to the 'greenhouse' warming [23]. The presence of ozone in the Earth's atmosphere protects the human biological life system from harmful solar UV-B radiation and also plays a pivotal role in controlling the thermal structure of the stratosphere [24]. Changes in UV irradiance can influence the structure of the middle atmosphere through modification of photochemical dissociation rates; with associated effects on ozone [22].Solar UV radiation, apart from its dependence on stratospheric ozone, is closely associated with tropospheric ozone abundance [8]. The Indian region has indicated a possible rise in tropospheric ozone, which could be related to increasing industrialization, transportation growth, and urbanization.

4.1 Measurement of Total Column Ozone (TCO) by using TOMS

TCO values were measured by Total Ozone Mapping Spectrometer (TOMS) carried on board NASA Earth Observing System, Aura Ozone Monitoring Instrument (OMI).OMI is a nadir-viewing spectrometer that measures the solar irradiance and Earth backscattered radiance from 270 to 500 nm with a spectral resolution of approximately 0.5 nm [25]. It provides near global coverage with a nadir pixel size of 13 km×24 km in the UV-2 channel used to retrieve TCO.Monthly average TCO data were retrieved from NASA's website ftp://toms.gsfc.nasa.gov/pub/. It is observed that TCO over the Indian subcontinent remains nearly equal to 260 ± 20 DU during the winter 2007and 2008 shown in Figure5.Spatial gradients in TCO observed during this period, gradually increases from a low value near the equator to a maximum at high latitude. This observed latitudinal variation pattern of TCO over the Indian subcontinent may be due to the geographical and meteorological characteristics of the region which strongly influence the transportation as well as the production and loss of TCOshown in Figure 6 during monsoon period.

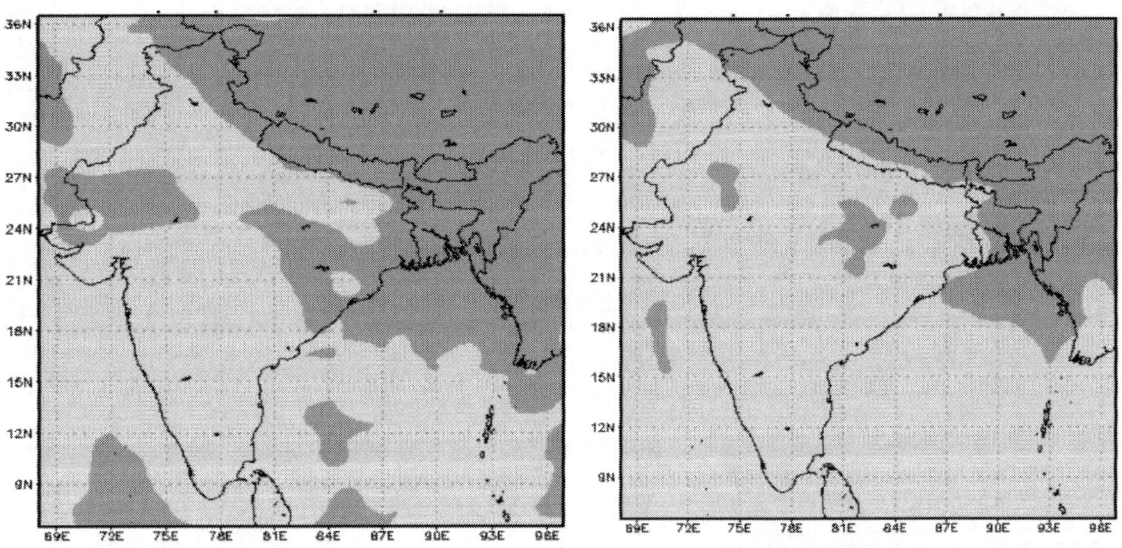

Figure 5. Latitudinal variation of total column ozone (TCO) over the Indian subcontinent for winter 2007 and 2008.

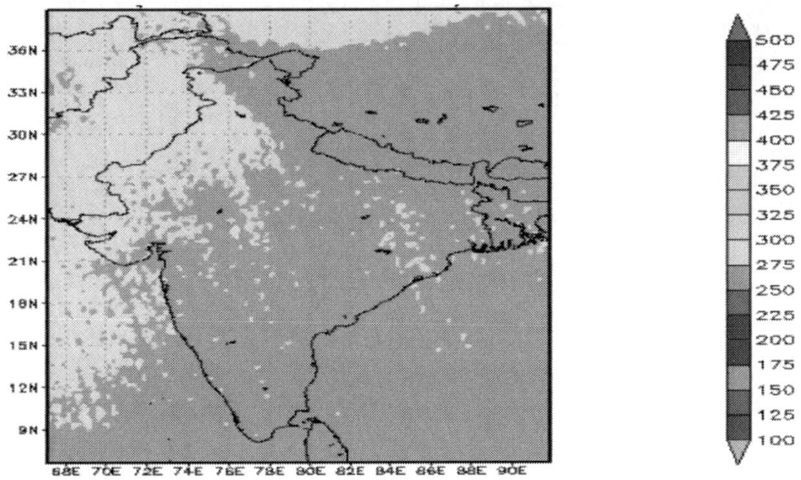

Figure 6. Total column ozone (TCO) over the Indian subcontinent during July 2014.

5. Conclusions

These study summaries the following aspects which influence the climate change over the subcontinent due to increased level of air pollution. AOD is found to be very high in the Indo-Gangetic region with increasing aerosol concentration in eastern part India. AOD is increasing from southern part of Indian subcontinent to northern part up to Himalaya. The increasing aerosol loading over the Indian subcontinent affects the rainfall distribution which is responsible for agricultural production and human health. MOPITT CO daytime retrievals provide sufficient information about the vertical and horizontal transport of CO and the deep convective activities during the monsoon period. Higher values of CO mixing ratios over the central part of India are observed low due to less industrial activities and low biomass burning. Spatial gradients in TCO during winter period, gradually increases from a low value near the equator to a maximum at high latitude. The variation pattern of TCO over the Indian subcontinent may be due to the geographical and meteorological characteristics of the region which strongly influence the transportation as well as the production and loss of TCO. The geographical diversity of the Indian Peninsula may be one of the major causes of spatial variability of total ozone concentration.

Alknowledgement
The authors are thankful to the Indian Institute of Tropical Meteorology for their valuable guidance and encouragement during the work.

Note: Following abbreviations have been used in this paper.

O_3	Ozone
NO_2	Nitrogen Dioxide
CO	Carbon Monoxide
HCHO	Formaldehyde
SO_2	Sulphur Dioxide
CH_4	Methane
AOD	Aerosol Optical Depth
AOT	Aerosol Optical Thickness
ASM	Asian Summer Monsoon

TCO	Total Column Ozone
MODIS	Moderate Resolution Imaging Spectroradiometer
MOPITT	Measurement of Pollution In the Troposphere
TOMS	Total Ozone Mapping Spectrometer
OMI	Ozone Monitoring Instrument
WMO	World Meteorological Organization
IGP	Indo-Gangetic Plain

References

[1] Kondratyev, K.Y. and Varotsos, C.A., 1995, Atmospheric ozone variability in the context of global change. *International Journal of Remote Sensing*, 16, pp. 1851–1881.

[2] Singh, R.P., Prasad, R.K., Chauhan, S.S. and Singh, S., 2005, impact of growing urbanization and air pollution on the regional climate over India. *International Association for Urban Climate Newsletter*, 14, pp. 5–10.

[3] Patil,S.D. and Revadekar, J.V., 2009, Extremes in total ozone content over northern India. *International Journal of Remote Sensing*, 30, pp. 2389–2397.

[4] Craknell, A.P. and Varotsos, C.A., 2007, The IPCC Fourth Assessment Report and fiftieth anniversary of Sputnik. *Environmental Science and Pollution Research*, 14, pp. 384-387.

[5] Lau, K. M. and Kim, K. M., 2006, Observational relationships between aerosol and Asian monsoon rainfall and circulation. *Geophysical Research Letter*, 33, L21810, Doi: 10.1029/2006GL027546.

[6] Dalvi, M., Beig, G., Patil, U., Kaginalkar, A; Sharma, C. and Mitra, A.P., 2006, A GIS based methodology for gridding large scale emission inventories: application to carbon monoxide emissions over Indian region. *Atmospheric Environment*, 40, pp. 2995-3007.

[7] Krishnamurti, T.N., Chakraborty, A; Martin, A., Lau, W.K; Kim, K.M; Sud, Y. and Walker, G; 2009, Impact of Arabian Sea pollution on Bay of Bengal winter monsoon rains. *Journal of Geophysical Research*, 114, D 06213.

[8] Kondratyev, K.Y. and Varotsos, C.A., 1996, Global total ozone dynamics: impact on surface solar ultraviolet radiation variability and ecosystems. *Environmental Science and Pollution Research*, 3, pp. 205–209.

[9] Lau, K. M., Kim, K. M., Hsu, C.N. and Holben, B.N., 2009, Possible influence of Air Pollution, dust- and sandstorms on the Indian monsoon.*WMO Bulletin* 58 (1), pp.22-30.

[10] Ramanathan, V. and Ramana, M., 2005, persistent widespread and strongly Absorbing haze over the Himalayan foothills and the Indo-Gangetic Plain. *Pure and Applied Physics*, 162, pp. 1609–1626.

[11] Menon, S., Hanse, J., Nazarenko, L. and Luo, Y., 2002, Climate effects of black carbon aerosols in china and India. *Science*, 297, pp. 2250–2253.

[12] Ghude, S.D., Kulkarni, P.S., Kulkarni, S.H., Fadnavis, S. and Van Der, R. J., 2011, Temporal variation of urban NOx concentration in India during the past decade as observed from space. *International Journal of Remote Sensing*, 32, pp. 849–861.

[13] Remer, L., Kaufmen, Y., et al., 2005, The MODIS aerosol algorithm, product and Validation. *Journal of Atmospheric Sciences*, 62, pp. 947-973.

[14] Kaufman, Y.J., Boucher, O., Chin, M., Remer, L.A. and Takemura, T; 2005, Aerosol anthropogenic component estimated from satellite data. *Geophysical Research Letter*, 32, L17804.

[15] Prasad, A.K. and Singh, R.P., 2007, changes in aerosol parameters during major dust

storms events (2001-2005) over the Indo-Gangetic Plain using AERONET and MODIS data. *Journal of Geophysical Research* 112, D 09208.

[16] Hollay, T., Levy, H. and Kasibatla, P; 2000, Global distribution of carbon monoxide. *Journal of Geophysical Research*, 105, pp. 12123-12147.

[17] Allen, D. J., Kasibhatla, P., Thompson, A. M., Rood, R.B., Doddridge, B.G., Pickering, K. E., Hudson, R. D., and Lin, S.J., 1996, Transport induced Interannual variability of carbon monoxide using a chemistry and transport model. *Journal of Geophysical Research*, 101, pp. 28655–28670.

[18] Leliveld, J., Crutzen, P.J; Ramanathan, V., Andreae, M.O., et al; 2001, The Indian Ocean experiment: widespread air pollution from South and South- East Asia. *Science*, 291, pp. 1031-1036.

[19] Drummond, J. R., 1992, Measurement of Pollution in the troposphere, in the use of EOS for Studies of *Atmospheric physics*, edited by J.C. Gille and G. Visconti, pp. 77-101.

[20] Liu, J., Drummond, J.R., Li, Q., Gille, J. C. and Ziskin, D. C., 2005, satellite mapping of carbon monoxide emission from forest fires in Northwest America using MOPITT measurement. *Remote Sensing of Environment*, 95, pp. 502–516.

[21] Edward, D. P., Emmons, L.K; Gille, J.C., Chu, A., Attie, J.L., Giglio, L; Wood, S.W; Haywood, J; Deeter, M. N; Massie, S.T; Ziskin, D.C. and Drummond. J.R; 2006, Satellite observed Pollution from southern hemisphere biomass burning. *Journal of Geophysical Research*, 111, p. D 14312.

[22] Braeeeur, G. and Granier. G., 1992, Mount Pinatubo aerosols, chlorofluro carbons and Ozone depletion. *Science*, 257, pp. 1239-1242.

[23] Lacis, A. A., Wuebbles, D. J. and Logan, J.A., 1990, radiative forcing by changes in the Vertical distribution of ozone. *Journal of Geophysical Research* 95, pp. 9971-9981.

[24] Ghude, S.D., Jain, S.L., Arya, B.C., Beig, G., Ahammed, Y.N., Kumar, A. and Tyagi, B., 2009b, Ozone in ambient air at tropical mega city, Delhi: trends and cumulative ozone exposure indices. *Journal of Atmospheric Chemistry*, 60,pp.237–252.

[25] Levelt, P. F., Van, G, H, J; Dobber, M., R., et al; 2006, The Ozone monitoring instrument. *IEEE Transactions on Geosciences and Remote sensing*, 44, pp.1093-1101.

Analysis of air quality and nighttime light for Indian urban regions

Prakhar Misra and Wataru Takeuchi

Institute of Industrial Science, The University of Tokyo, Meguro 4-6-1, Tokyo 153-8505, Japan

mprakhar@iis.u-tokyo.ac.jp

Abstract. Indian urban regions suffer severe air pollution issues. A 2014 study by WHO highlighted that out of 20 cities globally with worst air quality, 13 lie in India. Although insufficient ground monitoring data and incomplete air pollution source characterization impedes putting policy measures to tackle this issue, remote sensing and GIS can overcome this hurdle to some extent. To find out how much of this hazard is due to economic growth, past researches have tried to make use of socio-economic growth indicators like GDP, population or urban area to establish its correlation with air quality in urban centres. Since nightlight has been found to correlate well with economic conditions at national and city level, an attempt has been made to analyse it with air quality levels to find regions with high contribution of anthropogenic emissions. Nighttime light activity was observed through DayNight Band (DNB) of VIIRS sensor while the air quality levels were obtained for ANG and AOD (using MODIS sensor) and SO_2 and NO_2 (using OMI sensor). We have classified Indian landmass into 4 air-quality and DNB classes: LowLight-HighPollution, HighLight-HighPollution, LowLight-LowPollution and HighLight-LowPollution for each air quality species using June 2014 data. It was found that around half of urban regions show high AOD and ANG values. On the other hand almost all urban regions exhibit high SO_2 and NO_2 values.

1. Introduction

Indian cities have some of the worst air pollution levels in the world, as was pointed out by World Health Organization (WHO) in 2014 [1]. In absence of sufficient temporal and spatial data from ground monitoring sensors, satellite imagery can be used to study pollution thanks to their long running data legacy. Socio-economic growth is one of the drivers of pollution, however it varies with composition of economy [2] or social development indicators like literacy, population, etc. [3]. In a previous study [4], satellite data derived NO_2 and aerosol optical depth levels for 16 Indian cites were found to have high correlation with population (0.75) and population density (0.56), respectively. As a new approach and in absence of province level socio-economic data, studies are trying to analyze socio-economic growth by using nightlight data [5]. High correlations of nightlight with gross domestic product (GDP) (0.88) and motor-vehicle count(0.91) have been reported at national level by Katayama and Takeuchi [6] on a global scale. For India, province or city level data of gross regional product (GRP) is not available barring few economic centers. Moreover, in developing countries it is often common for a lot of economic activities to be also centered around informal or unregulated industries whose estimates do not show up in national or state-level GDP values [7]. Such a study that incorporates informal economic sector too by using nighttime light as a proxy for economic activity to study air quality and human activity is being carried out for the first time for India and its urban centers.

The objective is to analyze air quality species for Indian landmass and urban regions on the basis of economic activity. The specific aim is to classify air quality species: angstrom exponent (ANG), aerosol optical depth (AOD), SO_2 and NO_2 (collectively referred to as AQ hence) and nighttime light data into four classes based on histogram deduced threshold values. Thus, the percentage of pixels falling within each class shall reveal which AQ species is most related with economic activity induced human emissions.

2. Data and methodology

2.1. Data-sources

2.1.1. Air quality data. To study ANG and AOD levels, daily 'MOD04L2' Level-2 data at 10km×10km resolution obtained from Moderate Resolution Imaging Spectroradiometer (MODIS) sensor onboard NASA's EOS Terra satellite for June 2014 was used. 'Deep_Blue_Aerosol_Optical_Depth_550_Land dataset was used for AOD images and 'Deep_Blue_Angstrom_Exponent_Land' dataset was used for ANG images. To study SO_2 and NO_2 column concentration levels for June 2014, 'OMSO2e' and 'OMNO2d' Level-3 Ozone Monitoring Instrument (OMI) datasets from NASA's EOS Aura were used respectively. This data has a spatial resolution of approximately 13km×24km. A mean monthly composite for all AQ species images was prepared using daily values.

2.1.2. Nighttime light data. To study nightlight radiance, monthly composite of day-night band (DNB) product [8] of Visible Infrared Imaging Radiometer Suite (VIIRS) dataset for June 2014 was used. It consists of light only from persistent sources. This data has not been filtered for forest fires, volcano activity, northern lights or any other activity that may generate light from natural sources. Its spatial resolution is approximately 0.75km.

2.1.3. Land cover map. MODIS based global land cover map having a 0.5km resolution, developed by Broxton et al. [9], based on 10 years (2001-2010) of Collection 5.1 'MCD12Q1' land cover type data was used.

2.2. Flowchart of this study

Figure 1 lays down the flowchart of data selection, processing and analysis to obtain results.

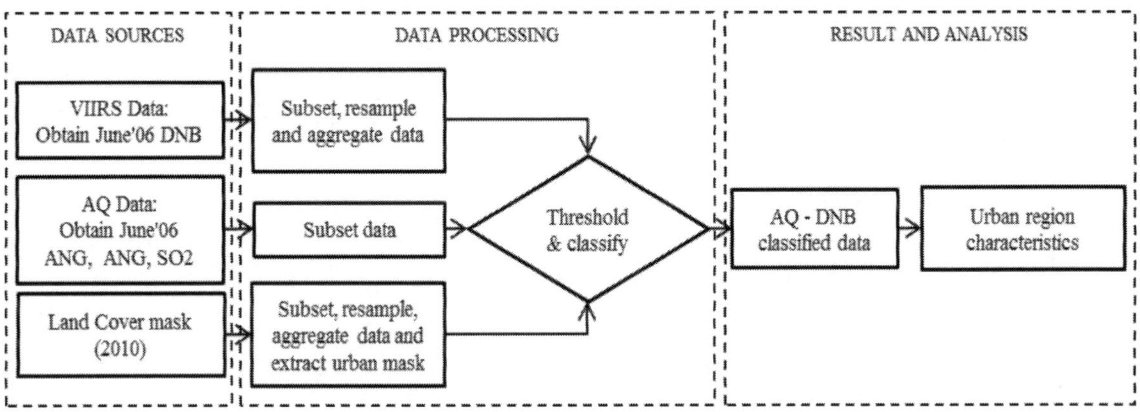

Figure 1. Flowchart of study outlining data sources, processing and analysis.

2.3. Classification by economic activity and air quality

Since economic activity levels can be judged fairly from the nightlight observed from space, first the Indian administrative region was subset from all the data sources using the shapefiles developed by Hijmans [10]. Since any forest fires were not reported during June 2014, DNB image was used without any corrections. Further as DNB monthly composites have higher resolution, it was downscaled to the coarser resolution of MODIS and OMI by mean aggregation, henceforth referred to as DNB_{MODIS} and DNB_{OMI} respectively. Downscaling factor used were 20 and 60 respectively.

As the objective is to classify regions by socio-economic activity and air quality, we needed to first bin the data into low values and high values. For this binned frequency histograms for each AQ parameter and DNB map values were plotted. Histograms were then thresholded either at valley or peak of the trend plot. Through these thresholds it is expected non-urban urban regions can be identified. Then using the threshold values for each AQ parameter and nightlight values (DNB), the region was classified as: a) LowLight-HighPollution (LLHP), b) HighLight-HighPollution (HLHP), c) LowLight-LowPollution (LLLP) and d) HighLight-LowPollution (HLLP). It is show in Figure 2 for clarity.

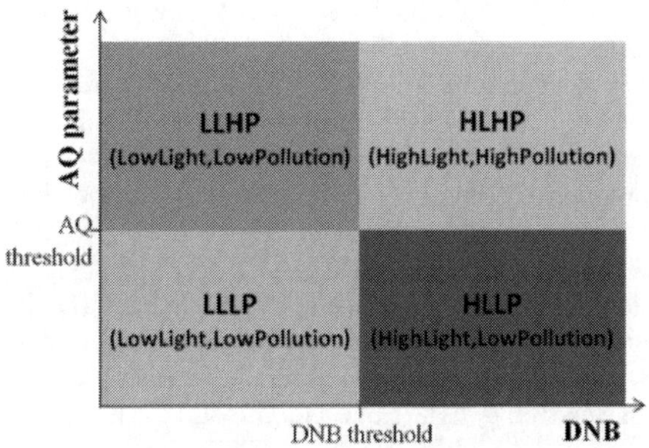

Figure 2. Classification scheme followed based on high and low AQ parameter values and DNB values. The 4 classes: LLHP, HLHP, LLLP, HLLP are colour coded as per their respective quadrant colour.

The HL regions signify locations of non-agricultural economic activity. They may not necessary be urban regions but may include factories or industrial areas often situated outside urban limits. Similarly LL regions include mountainous, forest, deserts and other regions with no or very less human activities. Since all villages do not share the same economic development characteristics they may fall in a LL or a HL region.

In the next part of processing ‘urban and built-up’ class map was prepared from the Land cover map data. Since the class mask was at higher resolution than MODIS and OMI, it was downscaled using a kernel block of size 20 and 60 respectively. Urban regions in various Indian towns and small cities are often not as large as a MODIS or OMI pixel. To preserve those urban regions in the downgraded urban cover mask, if at least ¼ pixels in the kernel block belonged to urban class, the whole block was marked as urban. Using this urban mask, percentage of urban region was classified to find about their characteristics.

3. Results and discussion

After subsetting data for Indian boundary region, DNB image was downscaled to resolution of MODIS and OMI respectively. To obtain classification thresholds following histogram were generated as shown in Figure 3.

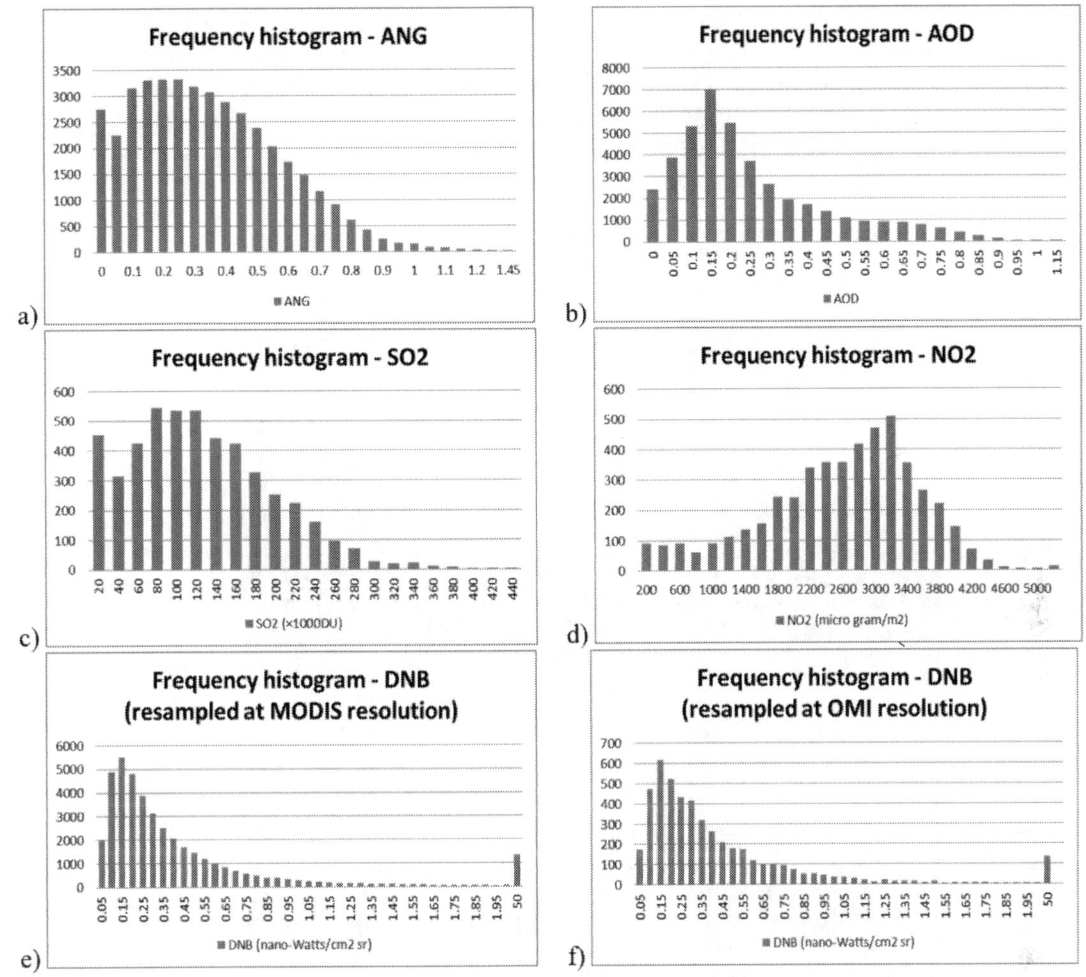

Figure 3. Threshold values for all the image datasets. a) ANG threshold at 0.2, b) AOD threshold at 0.25, c) SO_2 threshold at 100E3 DU, d) NO_2 threshold at 2000 microgram/m^2, e) DNB_{MODIS} threshold at 0.3 nanoWatt/cm^{2}sr, f) DNB_{OMI} threshold at 0.3 nanoWatt/cm^{2}sr.

The histogram plot were not found to be bimodal, which otherwise would have led to easier identification of thresholds. Instead others inflection point values of the plot were checked for their suitability as threshold values. For ANG, SO_2 and NO_2 the first valley comes quite early in the plot (at 0.05, 40(×1000 DU) and 800 microgram/m^2 respectively). However considering they are physically too low to serve as threshold, the next inflection points in the histogram plot were chosen as thresholds. DNB_{MODIS} showed sharper trends in lower mid-ranges of 0.15 to 0.75 nanoWatt/cm^{2}sr compared to DNB_{OMI}. This was expected since DNB_{OMI} was obtained by degrading the original DNB image to a greater extent than DNB_{MODIS}. Threshold of 0.3 nanoWatt/cm^{2}sr was chosen because a lower inflection value from the plot could not distinguish the human inhabitance of regions based on light. For determining AOD threshold, inflection values were adjudged to be either too high or too low compared to their physical suitability. So, we tried to find an approximate AOD equivalent of particulate matter (PM2.5 and PM10) based WHO 24-hour mean air-quality guidelines (25μg/m^3 and 50μg/m^3 respectively) [11]. By using AOD and PM regression relationships developed for the cities of New Delhi [12] and Agra [13] approximate corresponding AOD values were obtained. Based on those values, 0.25 (no dimension) was adjudged as a suitable AOD threshold. Thus, thresholds were chosen as 0.2 for ANG, 0.25 for AOD, 100(×1000 DU) SO_2, 2000 microgram/m^2 for NO_2 and 0.3

nanoWatt/cm^{2}sr for both DNB_{MODIS} and DNB_{OMI}. Most of the frequency histograms in Figure 3 follow a left skewed shaped curve, implying that the bulk of pixels have low values. However for NO_2, the bell shaped curve is right skewed. Thus it can be inferred that higher NO_2 values are more common than lower values. Using the combination of threshold values of AQ parameters and DNB (as inferred from the plots in Figure 3) and following the classification scheme mentioned earlier, classified maps for India were developed. These maps are shown in Figure 4.

a) c)
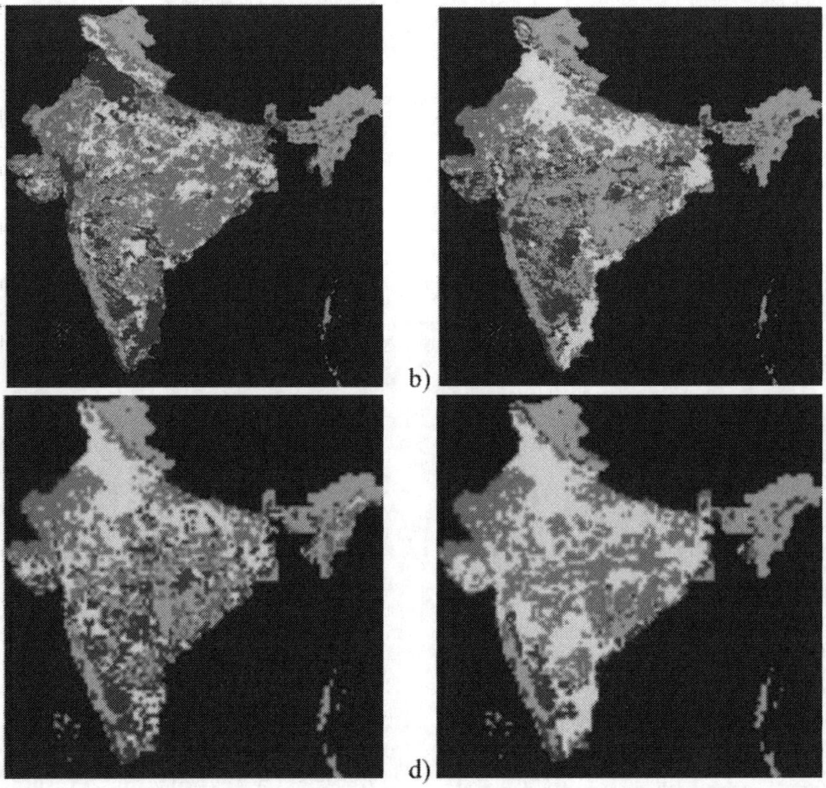

Figure 4. Figure 4 AQ parameter (a-ANG, b-AOD, c-SO_2, d-NO_2) and nightlight (DNB) classified map of India. Legend used is same as Figure 2:

From classification maps in Figure 4, it was found that most of LLHP and HLHP regions lie in mainland northern regions. The eastern coast of southern India shows dominance of HLHP region while the western region shows existence of LLLP regions. A region in northern western edge of India shows HLHP for AOD, SO_2 and NO_2 yet shows HLLP for ANG values.

Since low ANG values point to the existence of mineral dust, it can be deduced that those regions are highly industrialized and face severe pollution not only from man-made sources (as can be seen from NO_2 and SO_2 maps) but also a high amount of mineral dust. The small islands of HLHP within the HLLP belt of ANG-DNB classification map are actually cities of Ludhiana, Amritsar and New Delhi. Amongst cities, they tend to have some of the highest ANG levels in the country [4].

To explore which classes do the urban regions belong to, these classification map were overlaid with the land cover map. The comparison of urban and non-urban pixel values is presented in Figure 5.

IOP Conf. Series: Earth and Environmental Science **37** (2016) 012077 doi:10.1088/1755-1315/37/1/012077

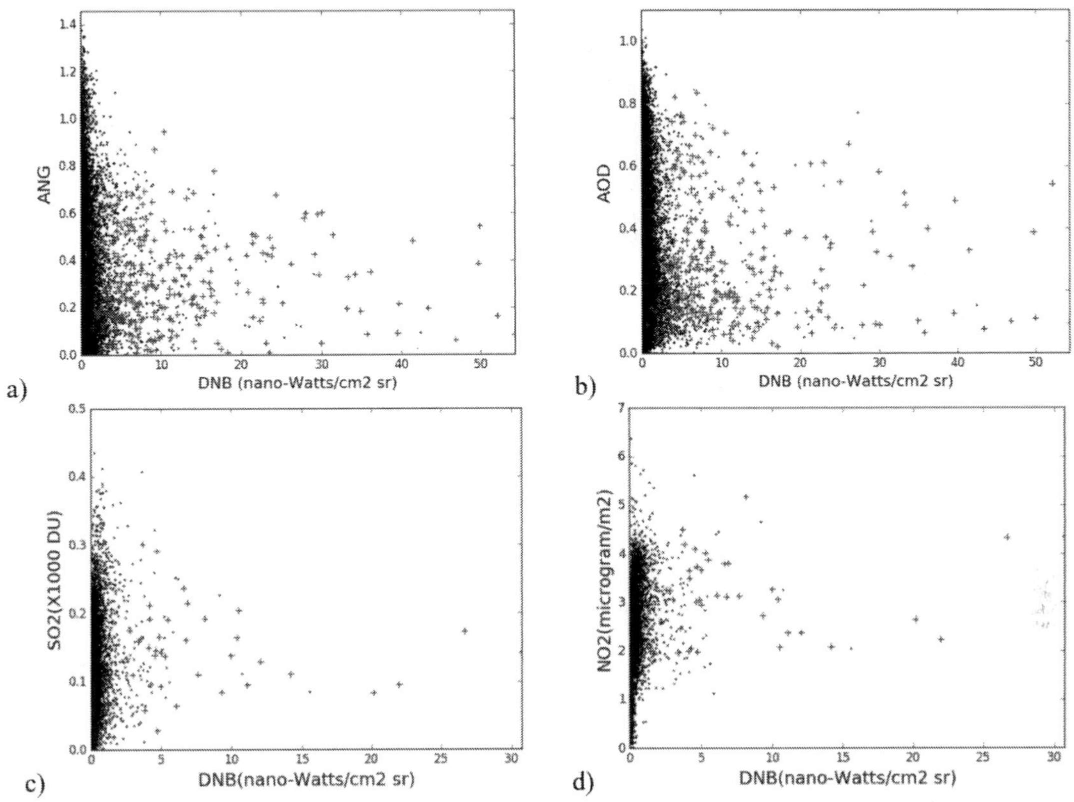

Figure 5. Comparison of urban and non-urban pixels on the basis of AQ parameter (a-ANG, b-AOD, c- SO_2, d-NO_2) and nightlight (DNB). Urban pixels are represented by + while non-urban pixels are represented by •.

The urban pixels can easily be spotted due to their high DNB values. However considering the threshold of DNB, which is set at 0.3, there are considerable locations which despite not being urban or built-up still have high DNB and AQ values. They are likely to be the suburban areas around the cities. In our analysis, urban pixels represented only 0.7% of all pixels which is considerably less than the reported urban area (2.35% of Indian land mass) for the year 2008 by the Department of Land Resources [14]. Thus there is under-representation of urban regions in the land cover map. Also identity of original urban mask pixels has diminished due to resampling and downgrading.

In Table 1, classification result of urban regions is presented. As discussed earlier only high nightlight (HL) region pixels were identified as urban.

Table 1. Classification percentages of different classes for urban regions.

Urban AQ & DNB map	LLHP	HLHP	LLLP	HLLP
ANG & DNB	0.00%	54.51%	0.00%	45.49%
AOD & DNB	0.00%	54.51%	0.00%	45.49%
SO_2 & DNB	0.00%	73.68%	0.00%	26.32%
NO_2& DNB	0.00%	89.47%	0.00%	10.53%

Incidentally the percentage of pixels identified as HLHP and HLLP is exactly same for ANG-DNB and AOD-DNB maps. It remains to be checked if indeed they are also the same pixels. The problem of SO_2 is comparatively much higher in HL regions while high NO_2 pollution seems to exist almost only in HL regions. Thus, almost half of the urban regions have relatively high aerosol levels. In contrast a much greater number of urban regions have higher values of SO_2 and NO_2. This implies that SO_2 and NO_2 emission sources (e.g. vehicle emissions, biomass burning) are almost present in all urban regions. Whereas about half of the urban regions have greater (or stronger) number of aerosol sources that are not present in other urban regions. This implies the almost all urban areas are similar in respect to SO_2 and NO_2 emissions sources but about one half of the urban regions are different from the other half in terms of aerosol sources. This is an important result to check the source of aerosol in urban regions. Also it can also be used to predict disease outbreaks likely to be born out different polluting species.

As a next step, a more recent land cover map should be used to generate urban cover mask to analyze this in detail. If sufficient urban pixels can be detected then thresholding could be done on the urban pixels to assess classification of urban region. Also combining the result with a population density raster can inform the number of people at health-risk due to anthropogenic pollution.

To study a long term trend, workflow of this analysis can be adapted to the Defense Meteorological Satellite Program's Operational Linescan System (DMSP-OLS) datasets. It has a long-running legacy and can be used with data MODIS and OMI data which is available since 2001. As a final step if this analysis is performed for multiple periods using DMSP-OLS and VIIRS-DNB datasets, how cities transition from one AQ-DNB class to another can be studied.

4. Conclusion

In this study, Indian landmass was classified according into regions with 4 different levels of air quality and economic activity by thresholding their histograms into 'high' and 'low'. By using urban land cover mask this process was also applied to urban areas where it was found all urban area lie in high light regions. NO_2 levels was found to be significantly high for about 90% urban area while SO_2 levels was found to be high for about 74% urban area. ANG and AOD classification exhibited a similar behavior with same percentage of pixels being identified as high (about 55%). Using the same methodology, we wish to process DMSP-OLS data and study how classification behavior of urban regions has changed in the past 15 years. This can help locate regions where air quality has changed under influence of its economic activities

Nomenclature of abbreviations used

WHO	World Health Organization
GDP	Gross Domestic Product
GRP	Gross Regional Product
AOD	Aerosol Optical Density; used to study the density of particulate matter in environment
ANG	Angstorm Exponent; used to study the size of particulate matter in air. Larger values indicate smaller size of particulate matter particles
SO_2	Sulphur Dioxide
NO_2	Nitrogen Dioxide
AQ	Air Quality species considered in this study, namely:
EOS	Earth Observation Satellite
MODIS	Moderate Resolution Imaging Spectroradiometer satellite sensor
OMI	Ozone Monitoring Instrument satellite sensor
VIIRS	Visible Infrared Imaging Radiometer Suite
DNB	Day Night Band
DNB_{MODIS}	DNB image resampled to MODIS resolution
DNB_{OMI}	DNB image resampled to OMI resolution
LLHP	LowLight-HighPollution class
HLHP	HighLight-HighPollution class

LLLP	LowLight-LowPollution class
HLLP	HighLight-LowPollution class
DMSP-OLS	Defense Meteorological Satellite Program's Operational Linescan System
DU	Dobson Unit - measurement of the columnar density of a trace gas in the Earth's atmosphere

References

[1] WHO 2014 Ambient (outdoor) air pollution in cities database 2014 *Public Heal. Environ. Soc. Determ. Heal. (PHE), WHO*
[2] Dinda S, Coondoo D and Pal M 2000 Air quality and economic growth : an empirical study *Ecol. Econ.* **34** 409–23
[3] Holian M J 2014 The Effect of Social and Economic Development on Air Pollution in Indian Cities *Environ. Urban. ASIA* **5** 1–15
[4] Misra P and Takeuchi W 2015 Analysis of Air Quality in Indian Cities using Remote Sensing and Economic Growth Parameters *36th Asian Conference of Remote Sensing* (Manila)
[5] Doll C N H, Muller J-P and Morley J G 2006 Mapping regional economic activity from night-time light satellite imagery *Ecol. Econ.* **57** 75–92
[6] Katayama N and Takeuchi W 2014 Comparison between Nighttime Light and Socioeconomic Indicators an an International Scale Using VIIRS Day-Night Band *35th Asian Conference of Remote Sensing,* (Nay Pyi Taw)
[7] Ghosh T, Powell R L, Elvidge C D, Baugh K E, Sutton P C and Anderson S 2010 Shedding Light on the Global Distribution of Economic Activity *Open Geogr. J.* **3** 148–61
[8] Baugh K, Hsu F-C, Elvidge C D and Zhizhin M 2013 Nighttime Lights Compositing Using the VIIRS Day-Night Band: Preliminary Results *Proc. Asia-Pacific Adv. Netw.* **35** 70
[9] Broxton P D, Zeng X, Sulla-Menashe D and Troch P A 2014 A global land cover climatology using MODIS data *J. Appl. Meteorol. Climatol.* **53** 1593–605
[10] Hijmans R 2009 Global Administrative Areas
[11] World Health Organization 2006 WHO Air quality guidelines for particulate matter, ozone, nitrogen dioxide and sulfur dioxide: global update 2005: summary of risk assessment *Geneva World Heal. Organ.* 1–22
[12] Kumar N, Chu A and Foster A 2007 An empirical relationship between PM2.5 and aerosol optical depth in Delhi Metropolitan *Atmos. Environ.* **41** 4492–503
[13] Chitranshi S, Sharma S P and Dey S 2014 Satellite-based estimates of outdoor particulate pollution (PM10) for Agra City in northern India *Air Qual. Atmos. Heal.*
[14] Department of Land Resources 2013 *National Land Utilisation Policy*

IOP Conf. Series: Earth and Environmental Science **37** (2016) 012078 doi:10.1088/1755-1315/37/1/012078

Evaluation of groundwater pollution risk (GPR) from agricultural activities using DRASTIC model and GIS

Sabrina Mohd Ariffin, Mohamed Azwan Mohamed Zawawi, Hasfalina Che Man
Biological and Agricultural Department, Faculty of Engineering, Universiti Putra Malaysia, 43000 UPM Serdang, Selangor

E-mail: bienasabrina@gmail.com

Abstract. Groundwater Pollution risk (GPR) map which utilized groundwater quality is important in order to prevent the groundwater contaminant concentration due to the agricultural activities. DRASTIC model and GIS application are two important tools that had been used for accessing and predicting the quality of groundwater. These supplementary tools are calculating, visualizing, and presenting the GPR by using DRASTIC index for each hydrogeologic factor through ArcGIS software. This study was covered approximately Selangor basin area where the GPR has been defined. There are four categories of agricultural activities in the Selangor basin which are animal husbandary areas, horticultural lands, short term crops and tree, palm and other permanent crops. The map showed that the "low" zones of GPR occupied 56% of the east side of the Selangor basin, 34% of the west side of the Selangor basin exposed to "medium" zones of GPR and the "high" zones of GPR covered 10% at the north side and the south to the west side of the Selangor basin. As a particular, for agricultural activities which is 52% of Selangor basin area, the "low", ''medium" and "high" zones of GPR was occupied as 42%, 43% and 15% respectively. Based on four categories of agricultural landuse, GPR map validated by nitrate distribution map, shows that the 99% of the variation in nitrate distribution zones are explained by GPR zones. In conclusion, groundwater pollution risk was affected by agricultural activities.

1. Introduction

Recently groundwater has become a major source of water for many industries and municipalities and also for agriculture activities [2], [5]. Most of the groundwater is clean, but groundwater can become polluted or contaminated if no further protection is being considered [12], [22]. The movement of contaminants that infiltrate into soil and groundwater occurs through wastes disposals, discharge of effluent from industries or release of chemicals through agriculture activities [13, 17]. Agricultural activities have been identified as one of the non-point sources of pollutants to groundwater and soil. They affect groundwater quality in many ways. For example, it can become polluted when people use too much fertilizer or pesticides on their crops area. Agricultural pollutants such as nitrate and heavy metals from fertilizer will dissolve in the surface water such as river or irrigation water and then infiltrate through the soil and groundwater zone [1].

Agricultural activities for example paddy field, palm oil plantation and large scale farming have high risk to contaminate the groundwater such as by nitrate and heavy metals leaching generated from fertilizer used in agricultural lands [14]. The fertilizer which contains nitrogen, phosphorus and potassium will be more soluble in water. Solute transport movement is due to groundwater flow. [5] have reported growing incidences of nitrate pollution and increase of nitrate concentration in groundwater in intensive agricultural areas. Nitrogen rich fertilizers is a major source which makes the level of nitrate concentrations increased [7]. Leaching of various pollutants from agricultural land through the unsaturated zone and groundwater zone gives rise to contamination in these zones. This process varies from one location to another [6]. According to Malaysian Environmental Quality Report 2006 by Department of Environment (DOE), Ministry of Natural Resources and Environment

Malaysia showed that in nitrate sample collected from 12 monitoring wells in agricultural land use area exceeds the National Guidelines for Raw Drinking Water Quality.

Many technical methods have been created and applied by scientist and other researchers from previous study to monitor issue of groundwater vulnerability to contamination. There were several methods and approaches for groundwater vulnerability assessment through model and mapping such as DRASTIC [3], GOD [10], AVI [23] and SINTACS [8]. These models can be classified as conventional methods. According to [24], conventional methods can differentiate the level of vulnerability at wide scales where the existence of various lithologies. The DRASTIC model is a famous method developed by U.S. Environmental Protection Agency (EPA). It is a standardized system which is employs numerical ranking system that assign to relative weights to various hydrogeological parameters affecting the level of groundwater vulnerability.

Nowadays with the advancement and development in technology, most scientist and researchers were using the help of technology for planning, monitoring and decision making. Geographic Information System (GIS) is a technology that is very useful for agricultural scientists and other researchers such as [2], [4], [15] and [20] to locate and explore groundwater. Using this technology, the techniques were widely used in groundwater pollution potential mapping. The main benefit of GIS mapping is the unification of data layers and modification of data parameters that applied in vulnerability classification [25].

In this study, landuse, water quality and hydrogeological factors data from Malaysian government agencies were acquired for Selangor basin area and together assembled in a GIS environment by using ArcGIS software and DRASTIC system to achieve three main objectives as follows:

i. To calibrate hydrogeological parameters of DRASTIC model of Selangor basin.
ii. To determine Groundwater Pollution Risk (GPR) zones based on agricultural activities of Selangor basin.
iii. To correlate the significance of GPR zones and nitrate distribution zones on agricultural area.

2. Methodology

2.1 The study area

The selected study area is Selangor basin where it is located at the northern part of the state of Selangor (Figure 1). The basin is lies within latitudes 3° 9' 57.6" - 3° 43' 48" North (350969-412414 RSO meters) and longitude 101° 6' 14.4" - 101° 49' 44.4" East (344864-426160 RSO meters). The area of basin is approximately 2900 km^2 which is covering about a one third of the state of Selangor. The basin is covering four districts in Selangor state which is Kuala Selangor, Hulu Selangor, a west part of Gombak and a north part of Petaling.

Sungai Selangor is the major river in Selangor basin. It starts at elevation approximately 700 m to 7000 m above mean sea level from the west part of Banjaran Titiwangsa between Pahang state and Selangor state boundary. The river flows from the northeast of Hulu Selangor toward the west of Kuala Selangor and finally ends at Straits of Melaka as far as about 110 km. The main tributaries that relate to this river are Sungai Buloh, Sungai Belatak, Sungai Kerling, Sungai Batang Kali, Sungai Rening, Sungai Kul and Sungai Gumut.

Throughout the year, Selangor basin has characterized with tropical climate. The climate of the basin is normally warm but sometimes hot and sunny days, and cool in evenings. Temperatures typically range from 23°C to 34°C. Humidity levels are generally 80% or higher and annual rainfall often exceeds 2,000mm. Selangor basin is also rich with a variety of agricultural activities which is covering about a half of the basin. Most of the agricultural activities are converging at Kuala Selangor and at west side of Hulu Selangor.

IOP Conf. Series: Earth and Environmental Science 37 (2016) 012078 doi:10.1088/1755-1315/37/1/012078

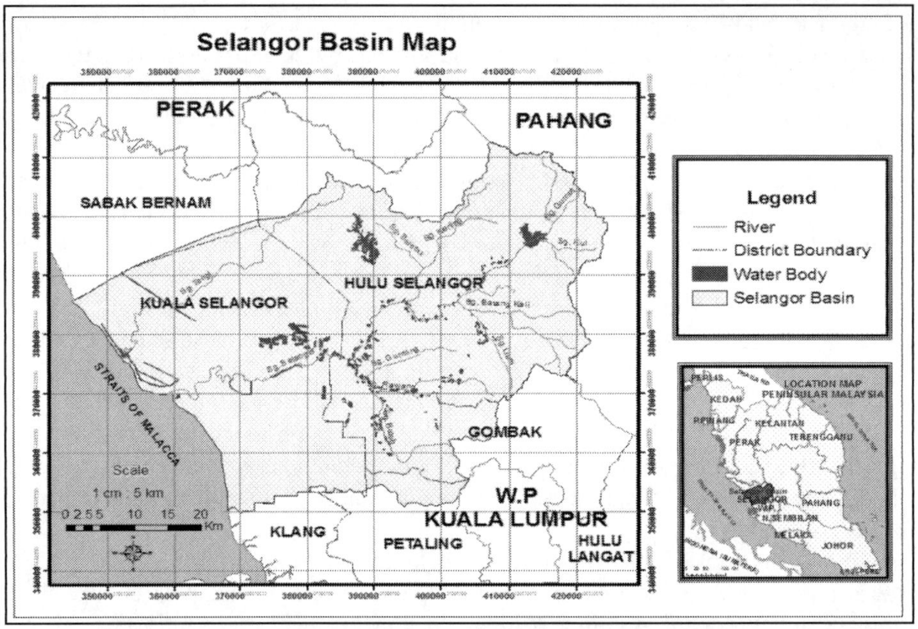

Figure 1. Selangor basin base map.

2.2 DRASTIC model

In this study, the DRASTIC model will be used to evaluate the potential for groundwater contaminant in the selected agricultural area. DRASTIC is an empirical model developed by U.S. Environmental Protection Agency (EPA) which is widely used for evaluating relative groundwater pollution susceptibility by the use of hydro-geological factors [3]. It can be used to assist planners, managers and administrators in the task of evaluating the relative vulnerability of an area to groundwater contamination from various sources of contamination. It is an acronym for the 7 most important hydro-geological factors affecting groundwater pollution, which are D (depth to water table), R (net recharge), A (aquifer media), S (soil media), T (topography), I (impact of vadose zone media), and C (hydraulic conductivity of aquifer).

[1], [4], [19] and [21] have demonstrated the successful use of this model to develop a groundwater contamination risk map using GIS techniques. The outcome of the project will be mapped as Groundwater Pollution Risk (GPR) zones which could be used to estimate the groundwater pollution risk due to agricultural activities. It is necessary to provide guidelines for groundwater protection. This study is also to employ Information Technology to the field of agriculture. GIS is used for data capture, data management and map presentation design.

Evaluation of DRASTIC index is based on 3 significant parts which are ranges, ratings and weights that have been devised using the DRASTIC factors. The DRASTIC factors are measurable parameters where it is generally available from miscellaneous of sources. The use of DRASTIC involves rating for each of the individual parameters at the certain location and multiplying the rating by a relative importance or weight constant. The formula of DRASTIC or known as DRASTIC index to evaluate the total of score (Rating x Weight) for each range in every parameter on certain area is (Equation 1):

$$DrDw + RrRw + ArAw + SrSw + TrTw + IrIw + CrCw = GPR \qquad (1)$$

When DRASTIC index has been evaluated, it is possible to determine zones which are more tend to be affected to groundwater pollution relative to one another. The higher of the DRASTIC index, the greater of the groundwater pollution risk [3]. The DRASTIC index gives just a comparative evaluation tool and it is not designed to provide actual answer.

2.3 GIS data and analysis

The systematic procedure and planning are needed in preparation of 7 DRASTIC parameter thematic maps to produce groundwater pollution risk map. Table 1 shows the sources of data based on each of DRASTIC parameter.

Table 1. List of sources for preparing DRASTIC parameters.

Government agencies	D	R	A	S	T	I	C
Department of Survey and Mapping Malaysia (JUPEM)	√	√			√	√	
Malaysian Meteorological Department (MET)		√					
Minerals and Geoscience Department Malaysia (JMG)	√		√			√	√
Department of Agriculture Malaysia (DOA)				√			

All the data has been processed by using GIS software which is ArcGIS 9.3. The main task is to bring all the appropriate data together into a GIS database. Basically, all the available spatial data will be converting into GIS platform. The data was added together and assemble in digital form with right coordinate system to make sure the spatial component will overlay correctly.

2.4 Depth to water analysis

In process to design D parameter layer, the elevation data has been generated from topographic map by DEM. Then, water level data (point layer) based on well location has been interpolated using IDW method. Next, the difference between elevation layer and water level layer has been computed to produce the depth to water layer because both of data are measure from mean sea level. The map was reclassified based on range in DRASTIC system and has been assigned with rating and weight to generate D parameter layer.

2.5 Recharge analysis

Recharge layer was generated from calculation of total percentage of slope layer and difference of rainfall and evapotranspiration distribution layer based on general water balance equations which is (equation 2). Calculation has been constructed through spatial analyst.

$$\text{Recharge (\%)} = \text{Slope (\%)} + [\text{Rainfall (\%)} - \text{Evapotranspiration (\%)}] \quad (2)$$

Slope layer in percentage are generated from DEM while Slope layer in percentage are generated from DEM while total of rainfall data and evapotranspiration data distribution layer are generated from difference of interpolation of both data using IDW method and raster calculator. The data was obtained from rainfall station in the whole of peninsular Malaysia due to there are no average evapotranspiration data in Selangor basin area and only two principal stations that near to Selangor basin included Lapangan Terbang Sultan Abdul Aziz Shah (Subang) and Petaling Jaya. The recharge layer has been reclassified based on range in DRASTIC system and has been assigned with rating and weight to generate R parameter layer.

2.6 Aquifer media analysis

Based on lithology in attribute table of geology map, its characteristics have been reclassified based on DRASTIC range for aquifer media factor. Type of aquifer media was added in the same attribute table and then the layer properties of the map have been edited to produce aquifer media layer. Each range has been assigned with rating and weight according to the permeability of each aquifer media based on DRASTIC system to generate A parameter layer.

2.7 Soil media analysis
The soil media has been determined based on the percentage of clay, silt and sand data in soil map attribute table using soil textural classification chart (STC). Then, the class of soil media was added into attribute table of soil map based on DRASTIC ranges. The layer properties of the map have been edited to produce soil media layer and it has been assigned with rating and weight to generate S parameter layer.

2.8 Topographic (slope) analysis
Topographic slope was designed by extracted elevation data from topography layer by using digital elevation model (DEM). Then, slope layer in percentage value has been generated and the range has been reclassified based on DRASTIC ranges. Then the layer has been assigned with rating and weight to generate T parameter layer.

2.9 Impact of vadose zone analysis
The impact of vadose zone media is identified based on well logs report and depth to water layer. Once the media has been set based on DRASTIC range, the data was added into attribute table of depth to water layer. Then, the layer properties of the map have been edited to produce impact of vadose zone layer and it has been assigned with rating and weight to generate I parameter layer.

2.10 Hydraulic conductivity analysis
Due to the lack of hydraulic conductivity of aquifer data, the value has been taken from previous research based on type of aquifer media (Table 2). The value was added into attribute table of aquifer media map and layer properties of the map was edited to produce hydraulic conductivity layer and it has been assigned with rating and weight to generate C parameter layer.

Table 2. Literature review for hydraulic conductivity value.

Aquifer media	Hydraulic conductivity (m/s)	References
Igneous rock	1.6×10^{-6} (values range from 8×10^{-9} to 3×10^{-4}	[9]
Shale	2.2×10^{-6} (values range from 1.5×10^{-6} to 3.3×10^{-6})	[21]
Bedded Sandstone, Limestone and Shale	7.1×10^{-5} (values of K range from 2.4×10^{-6} to 1.4×10^{-4})	[16]
Limestone	5.0×10^{-3} (values range from 2.5×10^{-2} to 9.3×10^{-5})	[11]
Alluvium	1.9×10^{-4} (values of C range from $7. \times 10^{-6}$ to 3.7×10^{-4})	[19]

2.11 Evaluation of GPR through map integration
The groundwater pollution risk (GPR) layer has been prepared using 7 DRASTIC parameters layers in GIS environment. After that, the layers were converted between raster to vector and vector to grid based on DRASTIC index (R x W) in attribute table for each layer with pixel size 250 m x 250 m. The final stage involves combination of all DRASTIC thematic layers based on DRASTIC index by using raster calculator in spatial analyst to evaluate GPR. This system will evaluate product of rating multiplied by weight of these seven parameters. The outcome which is GPR layer in Selangor basin area has been generated with new range. The range of GPR index is within 40 to 176. Based on DRASTIC index, the range was classified into 3 classes which are low (40 – 119), medium (120 – 159) and high (160 – 176). Figure 2 shows the summary of DRASTIC system process where all the 7 DRASTIC parameter layers were overlaid using GIS platform in format raster (grid). Grid represents the layer with pixels which has been set with 250m x 250m size. Every pixels for each the 7 layers have their own value based on (R x W). The pixels values of the 7 layers have been computed to evaluate the new value of pixel in 1 layer which is new layer of GPR.

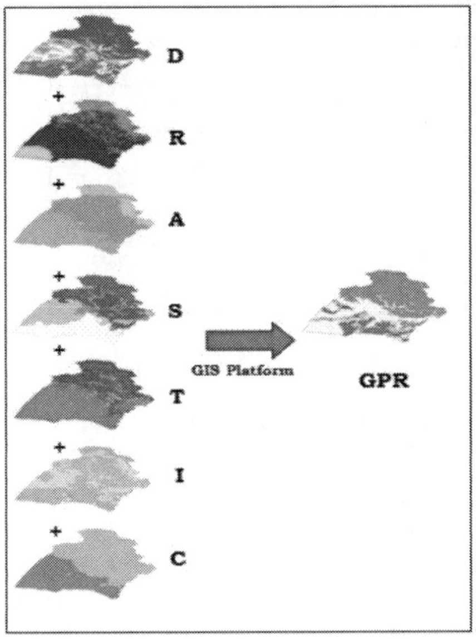

Figure 2. Summary of DRASTIC parameter evaluation.

3. Results and discussion

3.1 Depth to water (D)

Based on Figure 3, the west part of Selangor basin is the shallowest location and mostly occupied on coastal plain area. The east side of the basin, the depth to water are more than 31 m from ground level and mostly covering district of Hulu Selangor and Gombak where the area are dominated by mountainous and hilly area. In general, the higher ground surface elevation, the deeper water level and the highest value of (R X W) index, indicates the highest of groundwater pollution risk where the ground surface is close to groundwater (Figure 4).

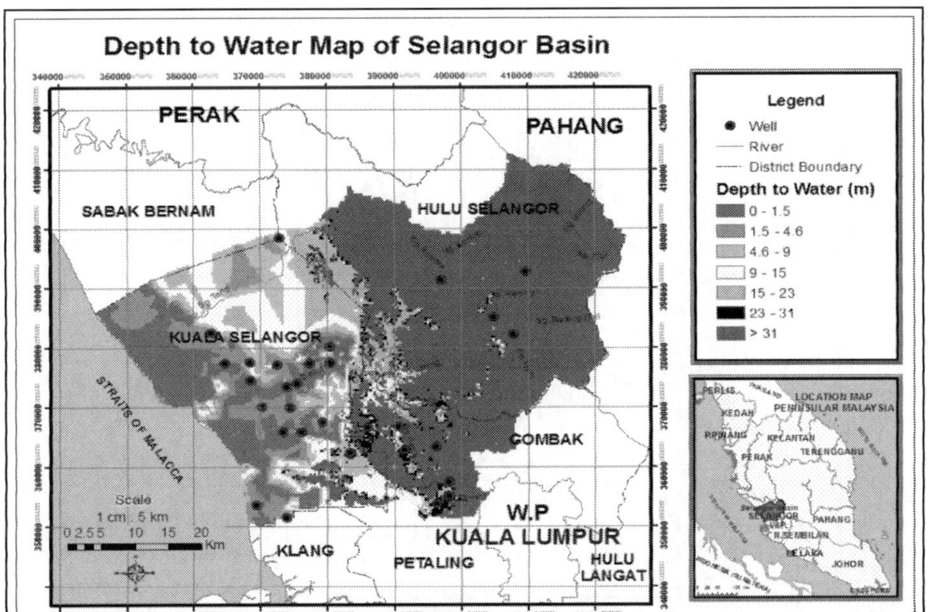

Figure 3. Depth to water map.

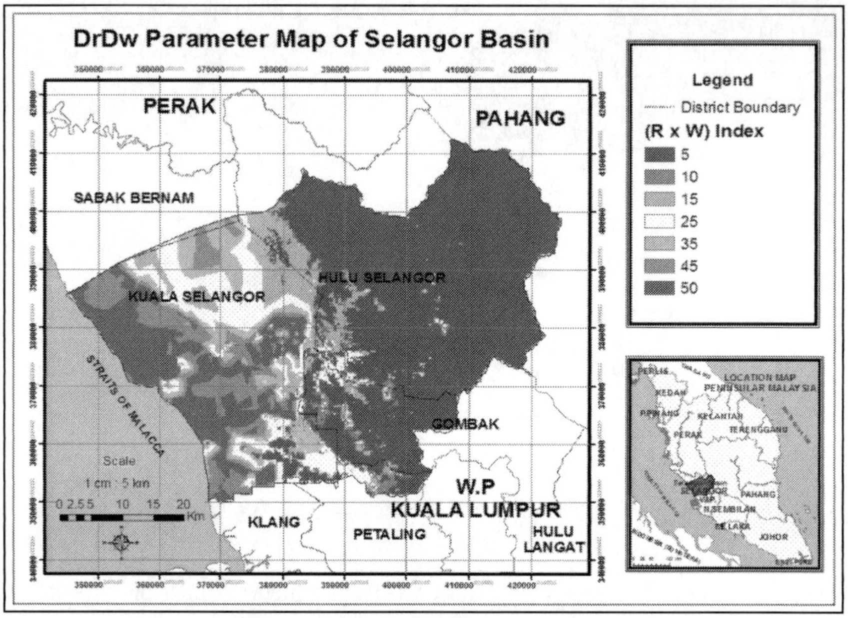

Figure 4. Map of DrDw parameter.

3.2 Recharge (R)

Figure 5 shows that more than half of Selangor basin area has a highest recharge value which is more than 250 mm. The low slope contributes to the more amounts of rainfall that will infiltrate into the aquifer. At the northwest part of Selangor basin has the moderate recharge value due to the annual rainfall with low value and influenced by monsoon (seasonal revising wind). In general, the steeper slopes, the lower of recharges value, the lesser groundwater pollution risk. Based on Figure 6, the highest of (R x W) index indicating the highest of groundwater pollution risk. Increases of recharge value are parallel to increases of (R x W) index value.

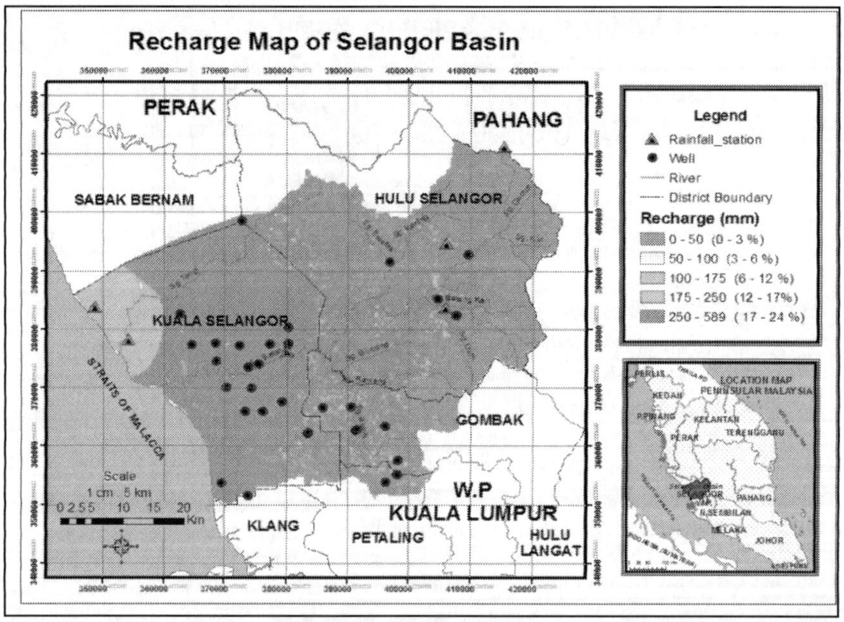

Figure 5. Recharge map.

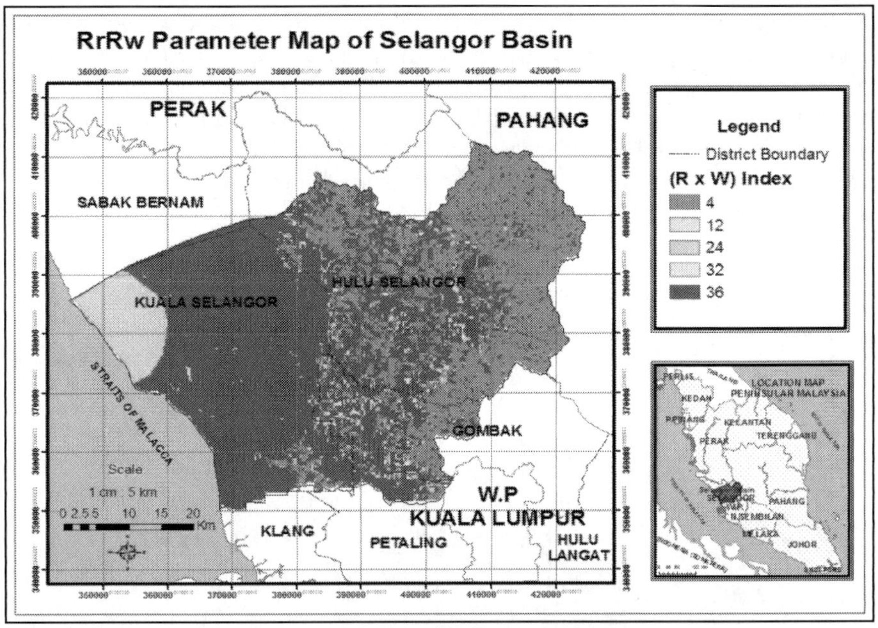

Figure 6. Map of RrRw parameter.

3.3 Aquifer media (A)

Figure 7 shows about two-thirds of Kuala Selangor district is covered with alluvium media and the centre part of the basin are occupied by bedded sandstone, limestone and shale. At the east side of the basin which is mountainous area are dominated by igneous rock and a little part of the basin are covered with a massive limestone and shale. Alluvium medium is unconsolidated mixtures of sand to gravel sized particle which contain varying amounts of fine materials. The geology age of alluvium in the basin is quaternary alluvium and it includes stream alluvium and coastal alluvium (gravel, sand and silt) and floodplain alluvium (sand, silt, clay and peat) [3].

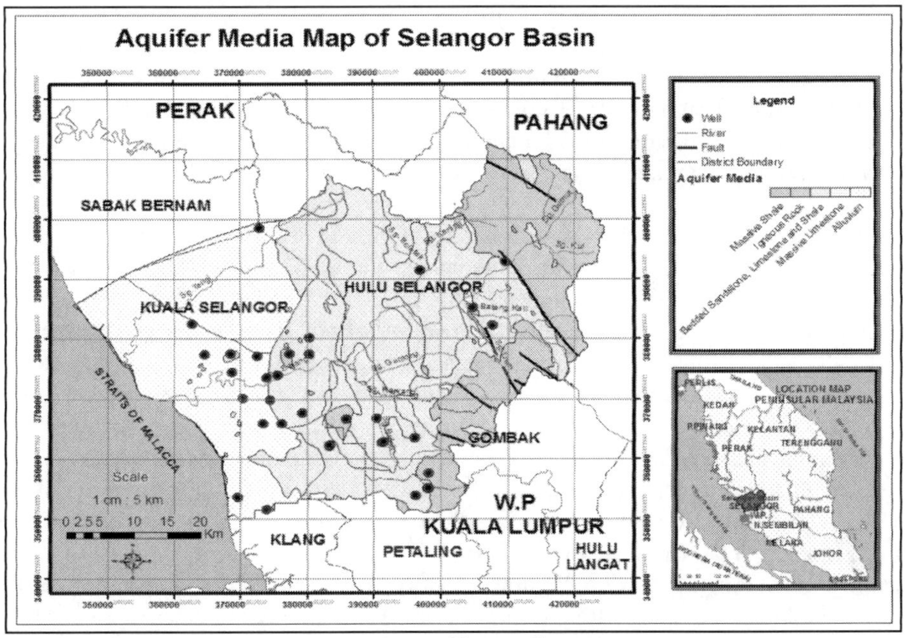

Figure 7. Aquifer media map.

Generally, the larger grain size, the higher permeability and the higher of groundwater pollution risk. Therefore, based on Figure 8, the alluvium media indicates the highest value of (R x W) index. Massive shale has a lower (R x W) index value because it is thick bedded shales, clay or claystone which normally yield low amounts of water from fractures and have a low pollution risk. The other aquifer medium in the Selangor basin is consolidated rock and the level of groundwater pollution risk is controlled by degree of fracturing. Hence, bedded sandstone, limestone and shale media and massive limestone media have a same value of (R x W) index.

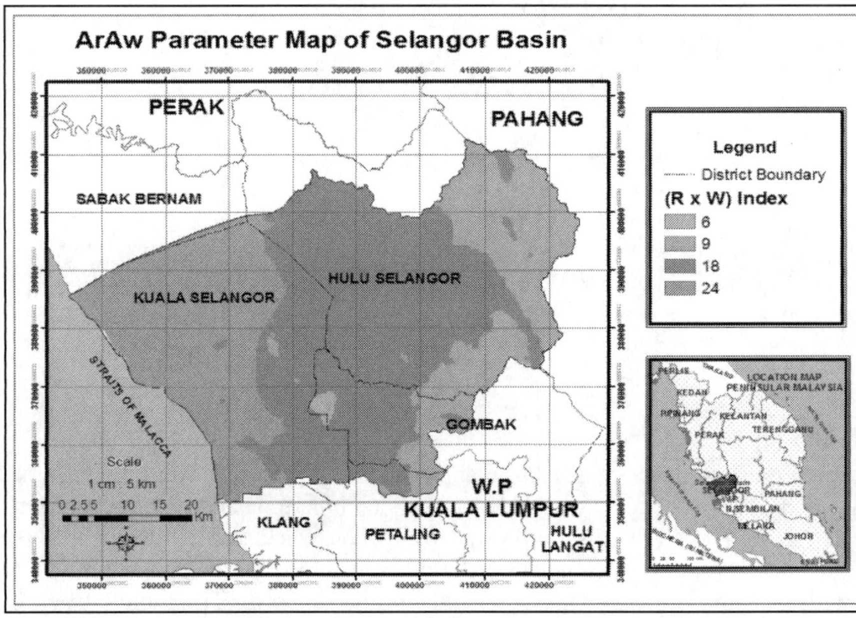

Figure 8. Map of ArAw parameter.

3.4 Soil media (S)

Generally, the type of soil media is influenced by topography. Figure 9 shows the east part of Selangor basin which is mountainous area where the soil layer is almost thin or absent because of rock formation within the mountain is more significant. The west part of the basin which is lowland area mainly was underlain by peat and clay. Peat media are covered by swamps while clay is covered by plantation such as paddy and palm oil. The centre part of the basin is dominated by sandy clay but mostly at rivers area is underlain by sand and loam media and there is a little part is underlain by gravel.

Highest (R x W) index value represents the higher of groundwater pollution risk (Figure 10). There are 2 pairs of soil media type that share the same value of (R x W) index. The first pair is gravel and thin or absent media which is have the highest value of (R x W) index. If soil layer is not present or very thin, the pollution risk is higher because attenuation of pollutants is not productive. The size of particles normally is larger than 2 mm. Gravel soils is include a mixture of sand, silt, clay and gravel particles. The second pairs is sandy clay loam and sandy clay which is have the second lower value of (R x W) index but pollution risk still high due to the higher percentage of sand. Peat and clay have (R x W) index value in the middle within the other media. Peat contains organic matter which may be significant for pollutants attenuation but they are relatively permeable, thus pollution risk is high [3]. Clay has low permeability but it can have high pollution risk based on secondary vertical permeability created by the cracking of the media when it is drying [3]. The rest part of the basin that covered by sand which have second higher of (R x W) index value due its size of particles is smaller than gravel and loam have the lowest value of (R X W) index because of the lower percentage of clay and silt so that the groundwater pollution risk is low.

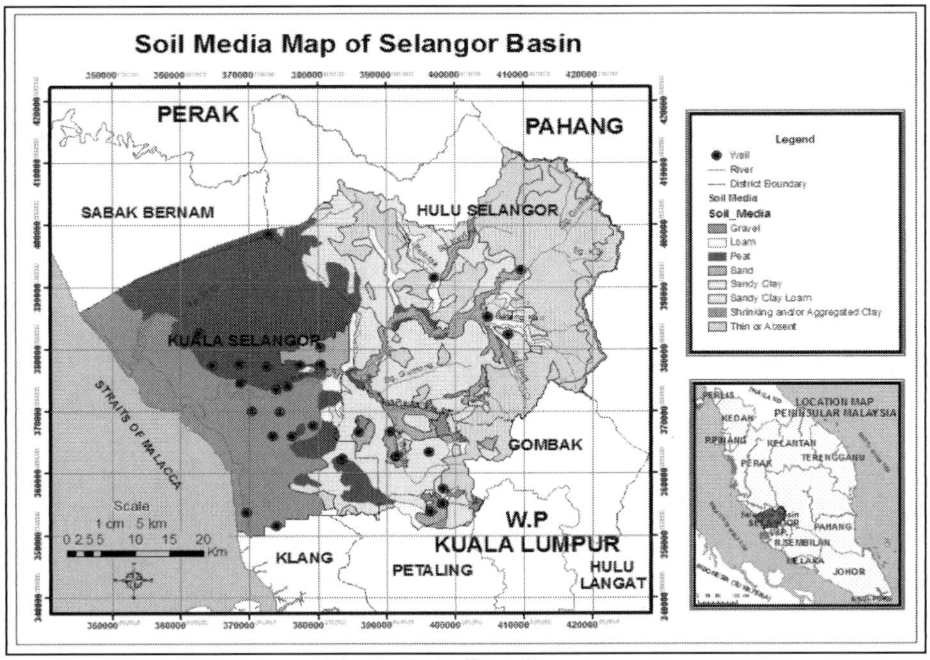

Figure 9. Soil media map.

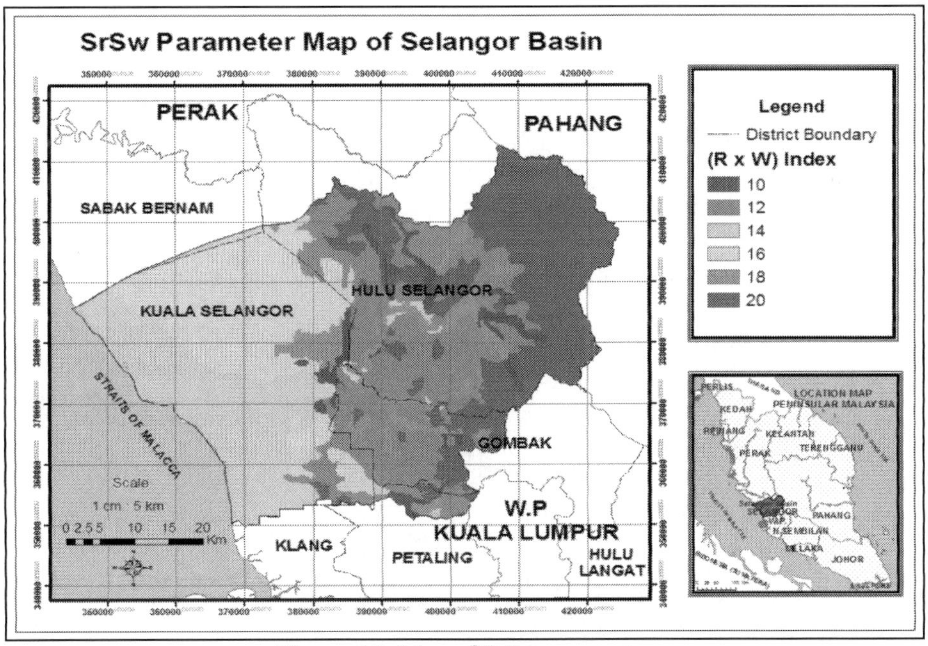

Figure 10. Map of SrSw parameter.

3.5 Topographic slope (T)

Figure 11 shows the west side of the basin which is Kuala Selangor district and near to coastal plain is underlain by range of slope area which is 0 – 2% is the lowest slope and its known as lowland area. The highest slope which is more than 18% and its known as mountainous and hilly area was underlain at the east part of the basin. Only a little part mostly at rivers area is underlain by the lower slope to the middle slope between 0 - 12% which is at rivers area. In general, the slope of land is decreasing from east to west of the basin until it reaches the sea.

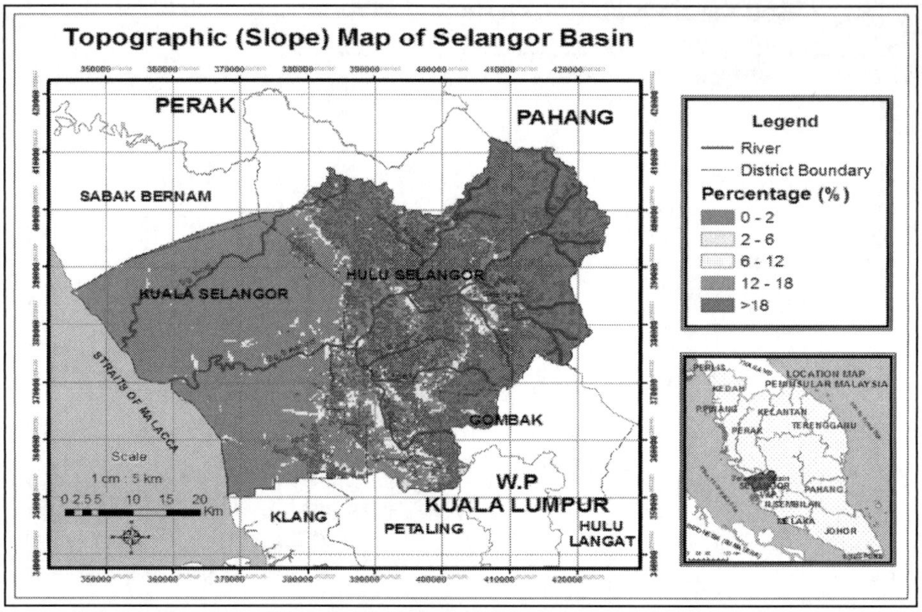

Figure 11. Topographic slope map.

The lowest slope which is covering the west side of the basin was assigned with the highest value of (R x W) index due to its ability for pollutants to infiltrate into groundwater because it is remain on the surface for long enough to infiltrate (Figure 12). It will contribute to the higher pollution risk. The highest slope which is occupied at the east part of the basin have the lowest value of (R x W) index because it affords a high runoff capacity and therefore a lesser probability of pollutants infiltration and a subsequent a lower groundwater pollution risk. However steep slopes are more conducive to rapid erosion and pollution of surface water [3].

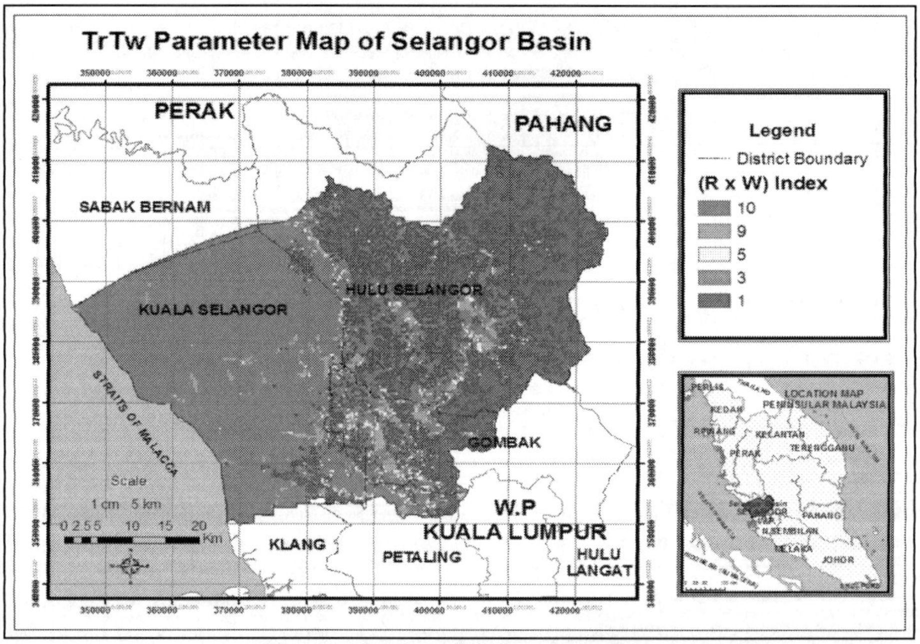

Figure 12. Map of TrTw parameter.

3.6 Impact of vadose zone (I)

Figure 13 shows the area division of 5 classes of media is almost equal within the Selangor basin started from the east which are metamorphic/igneous, bedded sandstone, limestone and shale, shale, sand and gravel with significant silt and clay and silt/clay. Figure 14 shows that the value of (R x W) index become 3 classes only. Silt/clay and shale has a lower value of (R x W) index. Deposits of silt and clay particles act as a barrier to restrict the movement of pollutants.

The shrinking clay and higher silt concentrations increase the pollution risk while the high clay contents provide a low pollution risk. Shale is thick bedded clay rock which may be fractured. Pollution risk is low but can be high with the degree of fracturing. Bedded limestone, sandstone, shale is typically thin bedded sequences of sedimentary rocks which contain primary porosity but the level of pollution risk is controlled by degree of fracturing within the media [3]. Sand and gravel with significant silt and clay is unconsolidated mixtures of fine materials includes sand and gravel. These deposits have lower permeability due to the high concentration of clay, thus the pollution risk is lower. Both of the media have the higher value of (R x W) index compare to the other media in the Selangor basin. Metamorphic/igneous rock is consolidated rock which is contain primary porosity but not significant. The movement of pollutants is through the fractures.

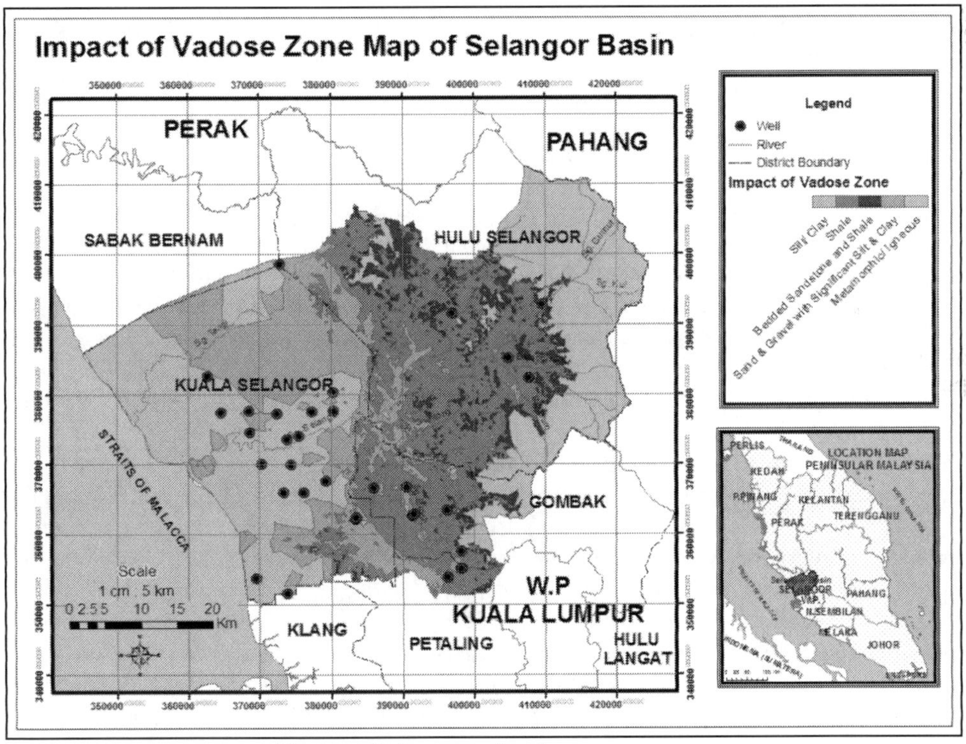

Figure 13. Impact of vadose zone map.

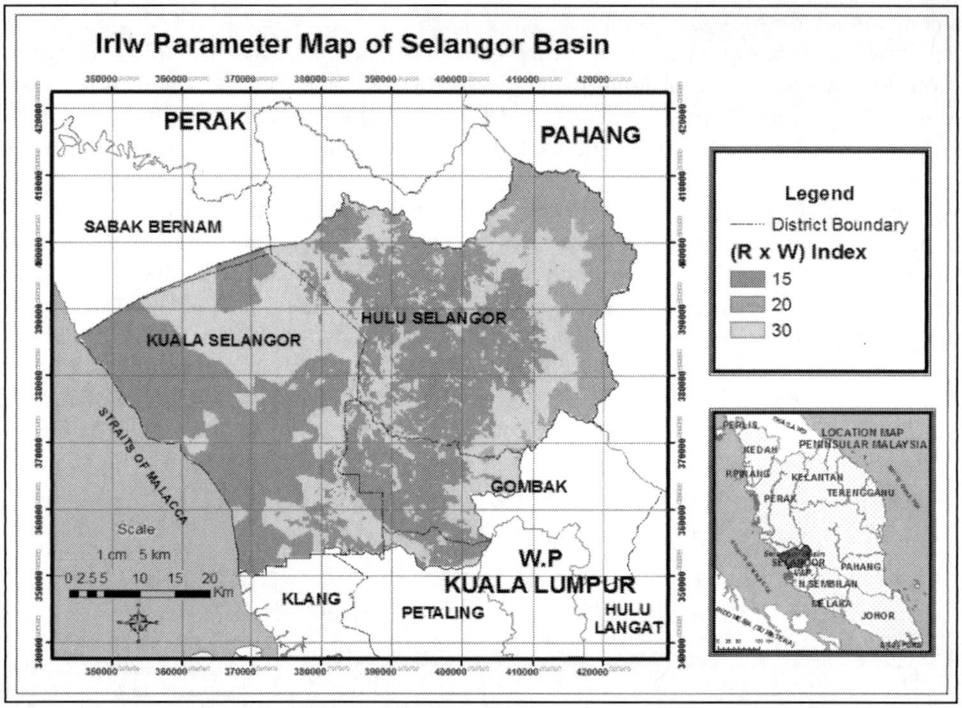

Figure 14. Map of IrIw parameter.

3.7 Hydraulic conductivity (C)

In this study, hydraulic conductivity has been determined based on the type of aquifer media. According to Figure 15, the limestone has the highest of hydraulic conductivity and alluvium media at a second place. Hydraulic conductivity for igneous rock is the lowest and shale is the second lower. Bedded limestone, sandstone and shale have moderate hydraulic conductivity. Higher or lower of hydraulic conductivity is depends on a consequences of intergranular porosity, fracturing or bedding planes. Based on Figure 16, the highest value of (R x W) index indicates the highest of groundwater pollution risk.

Consolidated limestone which typically contains fewer bedding planes than bedded limestone, sandstone and shale sequences [3]. Pollution risk is highly affected by the degree of fracturing and the amount of solution cavities in the limestone. Therefore, limestone get the highest value of (R x W) index for highest hydraulic conductivity. The sequences of bedded limestone, sandstone and shale contain primary porosity and pollution risk is influenced by degree of fracturing [3]. For that reason, hydraulic conductivity for these sedimentary rocks is quite low and have the second lower of (R x W) index value. Igneous rock and shale have same value of (R x W) index due to the value of hydraulic conductivity below than 4.72 x 105 m/s. Thick bedded shales and clays yield only a small amount of water from fractures while igneous rock is consolidated bedrock which is contains a little or no primary porosity and it yield water only from fractures within the bedrock [3]. Therefore, pollution risk is controlled by the degree of fracturing. Alluvium with 1.9 x 10 4 m/s of hydraulic conductivity has the moderate value of (R x W) index within the others. It is unconsolidated mixtures of sand to gravel sized particles which is containing varying amounts of fine materials [3]. Thus, lower amounts of fine materials in aluvium will increases the level of groundwater pollution risk.

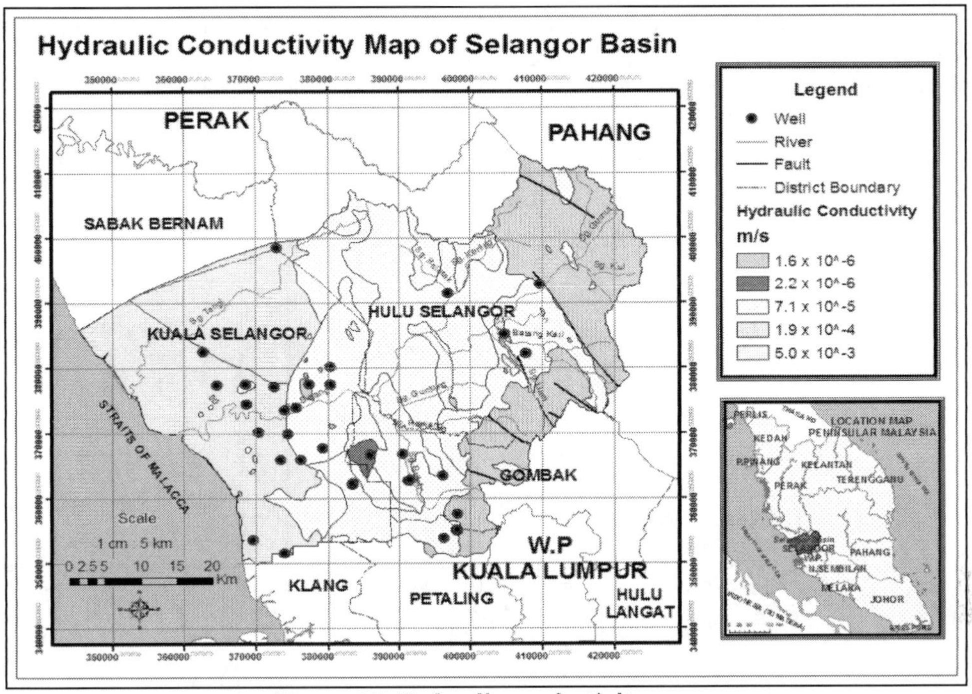

Figure 15. Hydraulic conductivity map.

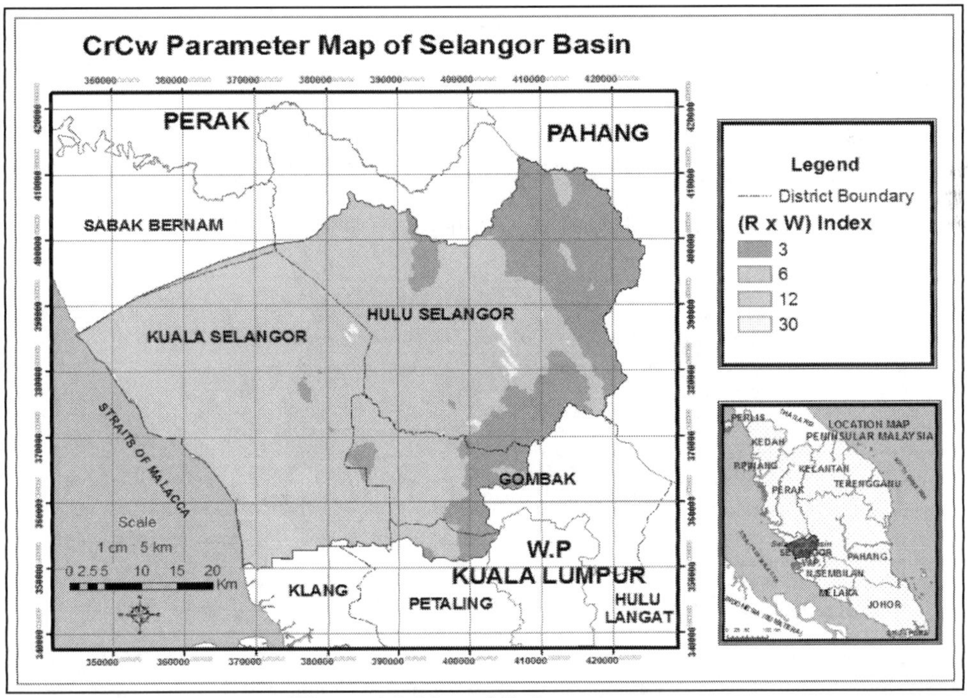

Figure 16. Map of CrCw parameter.

Based on Figure 17, predominantly area in Kuala Selangor district was covered with medium range of GPR. Only at the northeast side and the south to the west side of the district was covered with high range of GPR. The low range of GPR was covering a little part of the southwest side of the district. Furthermore, the whole of Hulu Selangor and a part of Gombak district was dominated by low range

GPR zones. Generally, the index range value of GPR zone is become higher starting from the mountainous area and followed by hilly area and lowland area until it reaches at coastal plain. Overall, the Selangor basin area was occupied by low, medium and high of GPR ranges is about 56%, 34% and 10% respectively (Figure 18).

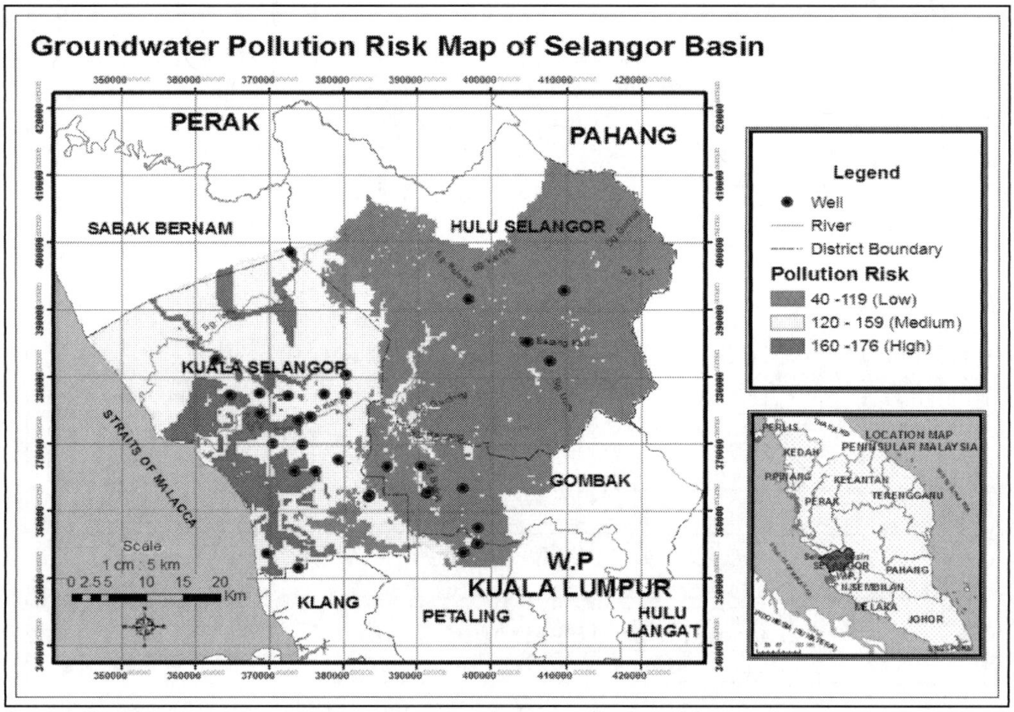

Figure 17. Groundwater pollution risk.

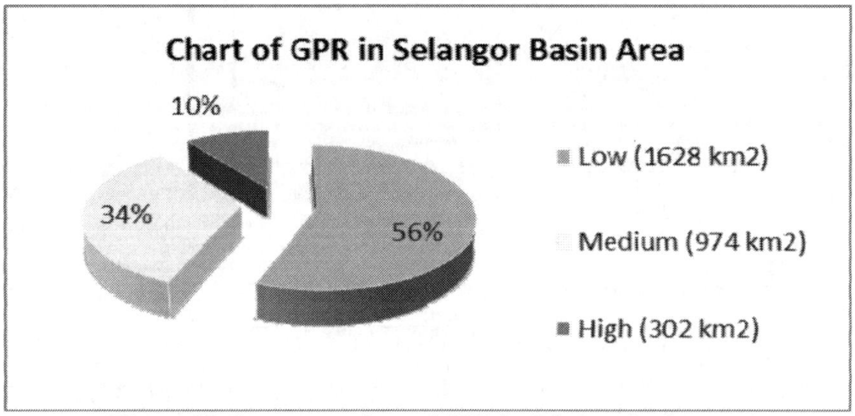

Figure 18. Percentage chart of GPR in Selangor basin area.

Based on landuse map of Selangor basin, the agricultural landuse layer was extracted from the map through database query. There are 4 categories of agricultural landuse which are animal husbandary areas, horticultural lands, short term crops and tree, palm and other permanent crops. Then, GPR layer of Selangor basin was through the clipping process using analysis tools in ArcGIS based on agricultural landuse layer. Figure 19 shows the agricultural activities on GPR zones. On the whole, low, medium and high GPR ranges that covering agricultural area is about 42%, 43% and 15% respectively (Figure 20).

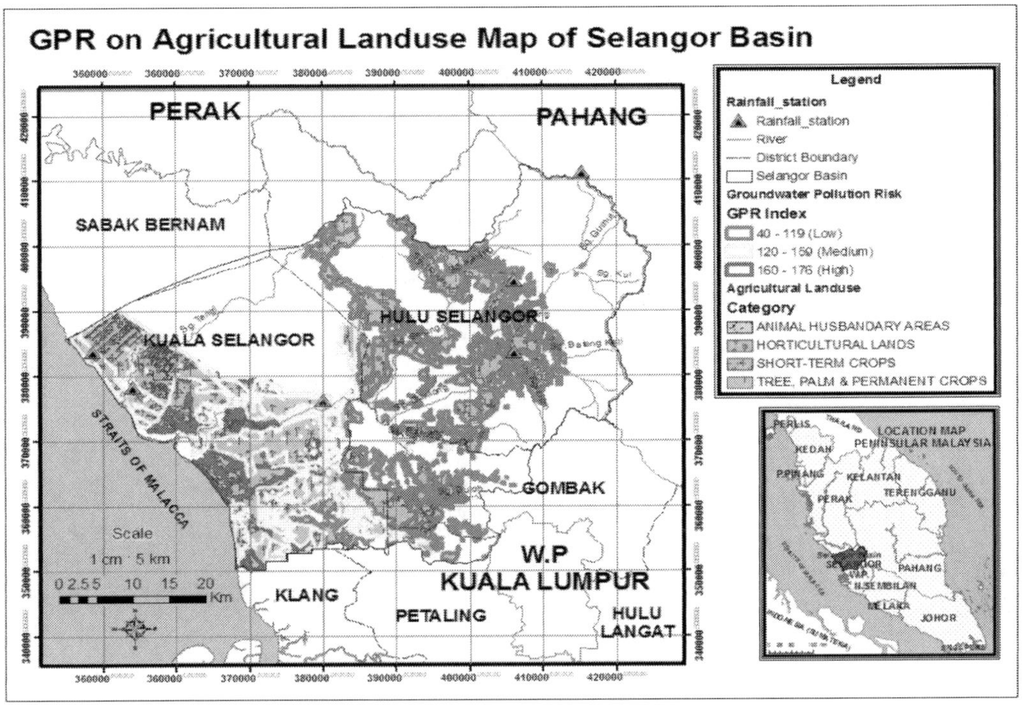

Figure 19. GPR on agricultural landuse.

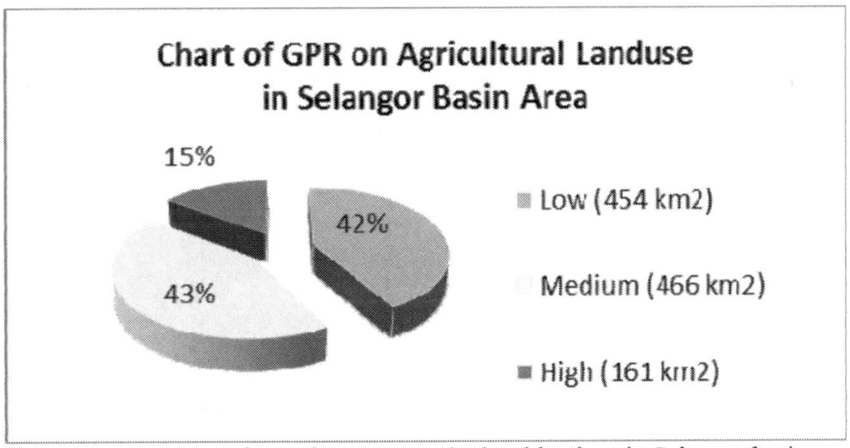

Figure 20. Percentage chart of GPR on agriculural landuse in Selangor basin area.

3.8 Significance of GPR and nitrate distribution zones based on agricultural landuse

Same to water level and rainfall data, the nitrate distribution layer was generated by using nitrate values that obtain from water quality data in JMG report for each well in Selangor basin through IDW interpolation method. According to national guidelines for raw drinking water quality, benchmark for nitrate is 10 mg/l. Therefore, the range of nitrate distribution was classified into 3 classes which are low (0-5 mg/l), medium (5-10 mg/l) and high (>10 mg/l). After that, the nitrate distribution layer of Selangor basin was through the clipping process using analysis tools in ArcGIS based on agricultural landuse layer. Based on Figure 21, Selangor basin area was dominated with low range of nitrate level. Generally, the Selangor basin area was occupied by low, medium and high nitrate level ranges is about 90%, 7% and 3% respectively (Figure 22).

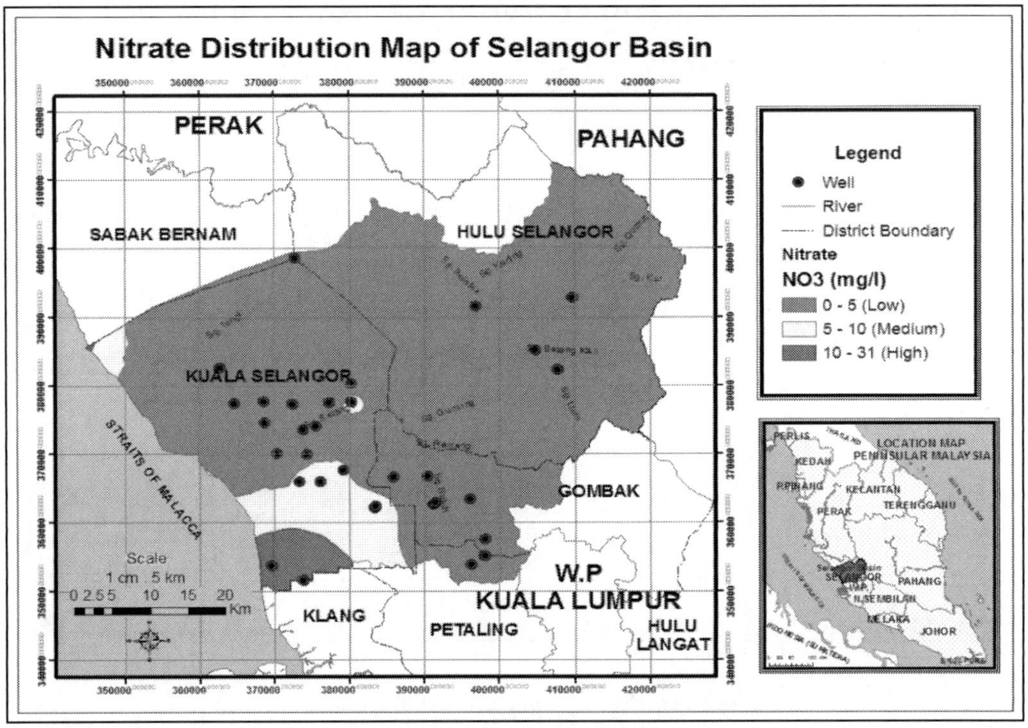

Figure 21. Nitrate distribution.

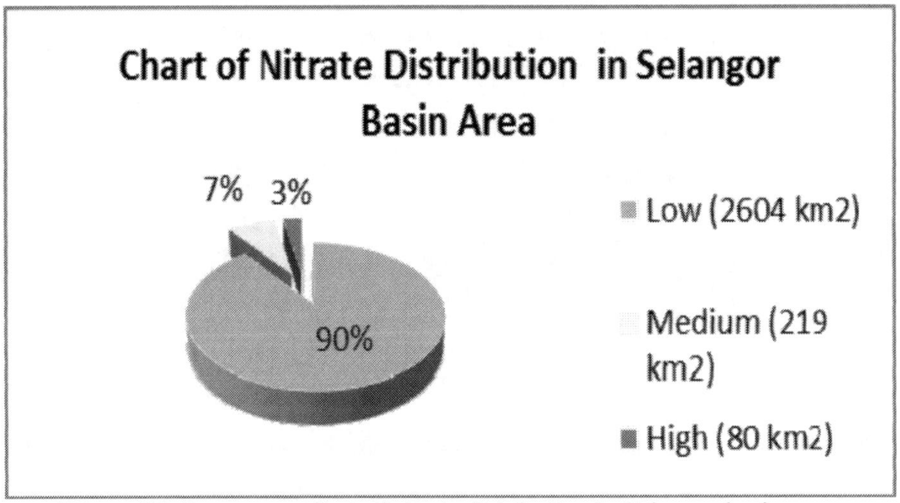

Figure 22. Percentage chart of nitrate distribution in Selangor basin area.

Figure 23 shows the nitrate distribution based on agricultural landuse. The high nitrate level range was underlain by categories of tree, palm and other permanent crops while the medium and low nitrate level range mostly was underlain by all the categories of agricultural activities which are tree, palm and other permanent crops, short term crops, animal husbandary areas and horticultural lands. By and large, the agricultural landuse area in Selangor basin was occupied by low, medium and high nitrate level ranges is about 78%, 16% and 6% respectively (Figure 24).

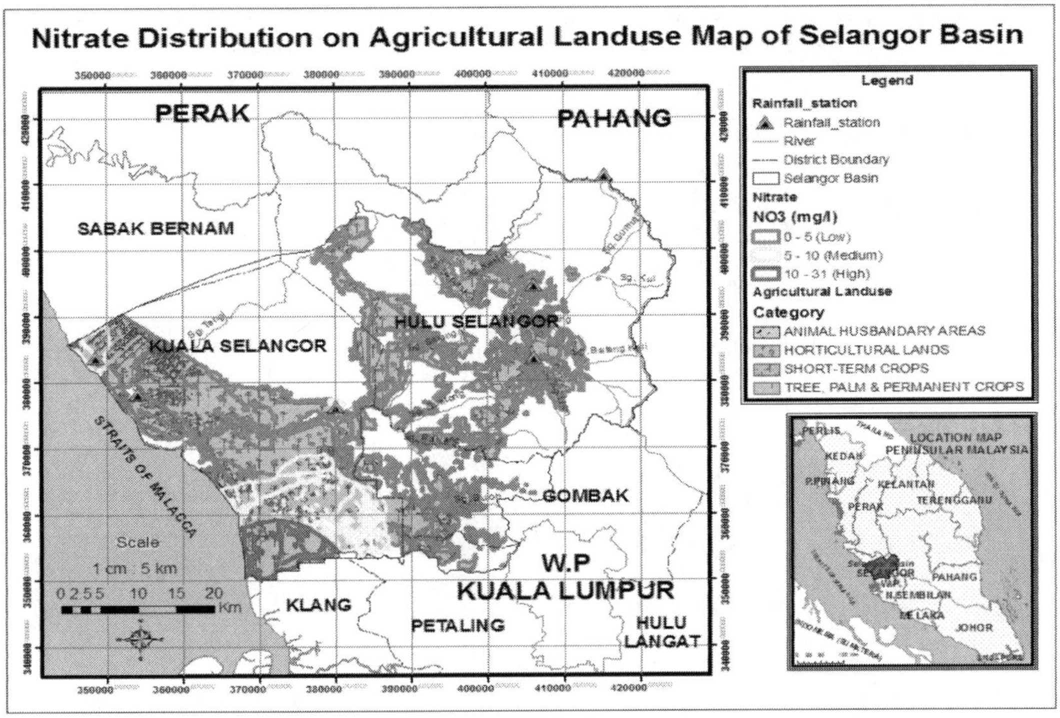

Figure 23. Nitrate distribution on agricultural landuse.

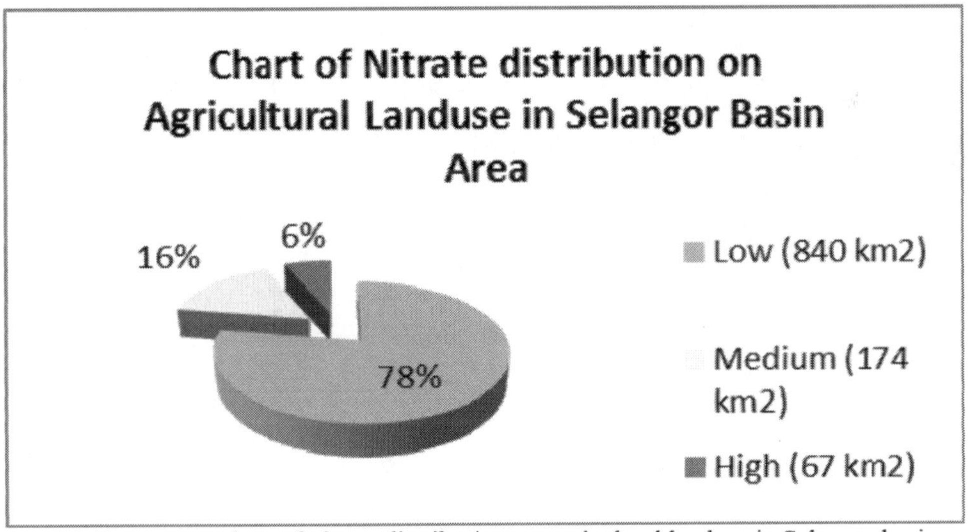

Figure 24. Percentage chart of nitrate distribution on agriculural landuse in Selangor basin area.

GPR zones were validated by correlate with nitrate distribution layer. Validation process was analysed by apply Pearson correlation coefficient tests through Micosoft Excel software with add in XL stat software. The pie charts below show the high, medium and low area percentages for GPR zones and nitrate distribution on agricultural landuse (Figure 25). Category of tree, palm and other permanent crops has the highest percentage for all pie charts which is range from 80% to 95%. The rest categories only have percentage 0% to 11%. Furthermore, the percentages division of 3 level areas on 4 categories of agricultural landuse for both layers zones is almost precisely with each other.

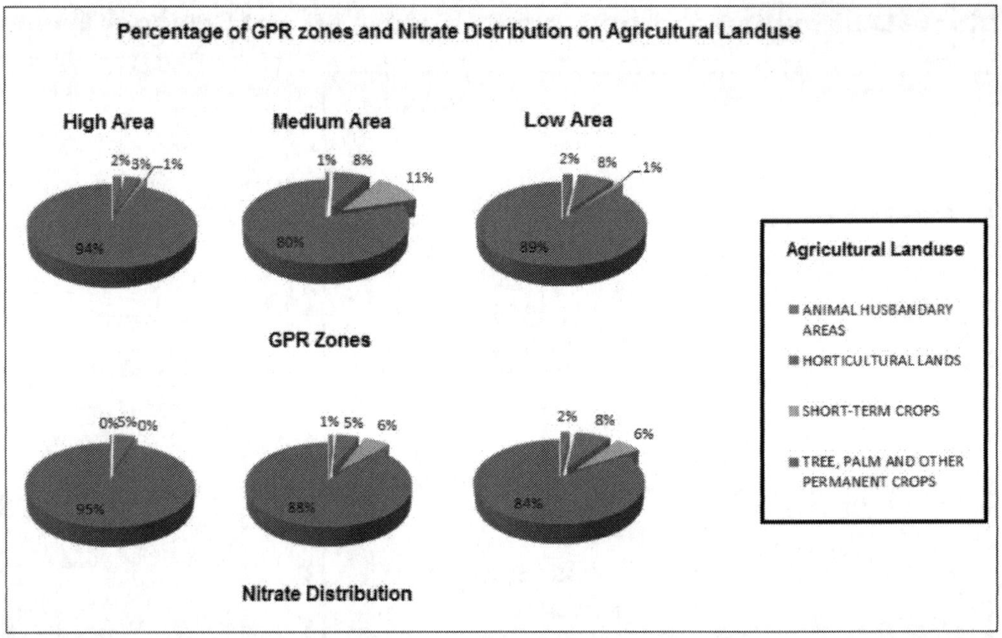

Figure 25. Pie Charts of percentages for each level area of GPR zones and nitrate distribution on agricultural landuse.

Based on Figure 26 to Figure 28, the scatter plots for high, medium and low area shows the positive slope relationship. Therefore, GPR zones and nitrate distribution have high positive correlation that indicates both layers increase together. Besides that, the correlation tests results indicate that the strength of relationships between the 2 maps layer for these 3 level regions is very high which are for high area (r = 1.0), medium area (r = 0.998) and low area (r = 0.998) and that the correlation coefficient is highly significantly different from zero (P < 0.05). Also, we can say that 99% of the variation in nitrate distribution zones is explained by GPR zones.

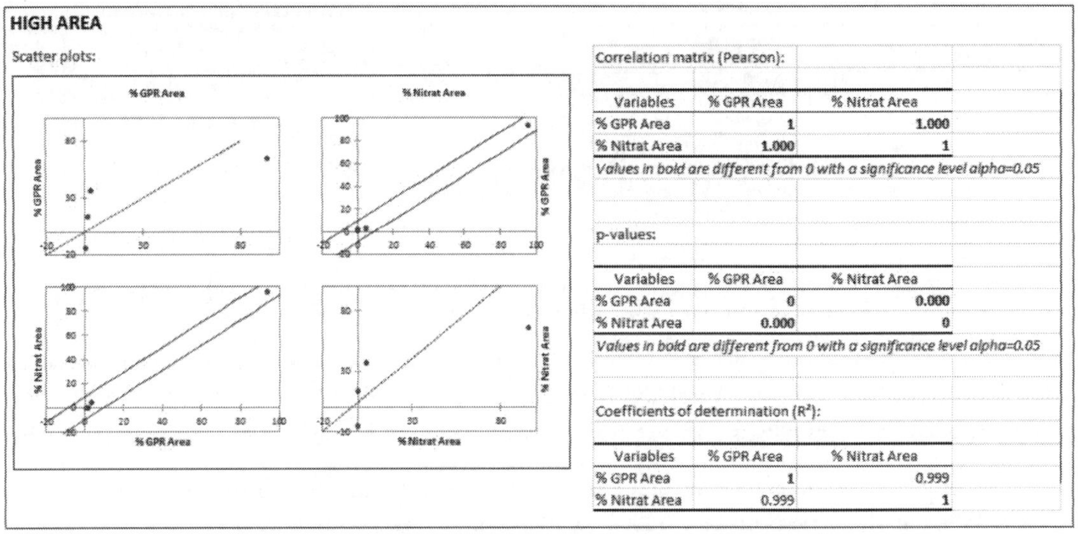

Figure 26. Pearson correlation coefficient tests for high area.

IOP Conf. Series: Earth and Environmental Science 37 (2016) 012078 doi:10.1088/1755-1315/37/1/012078

MEDIUM AREA

Scatter plots:

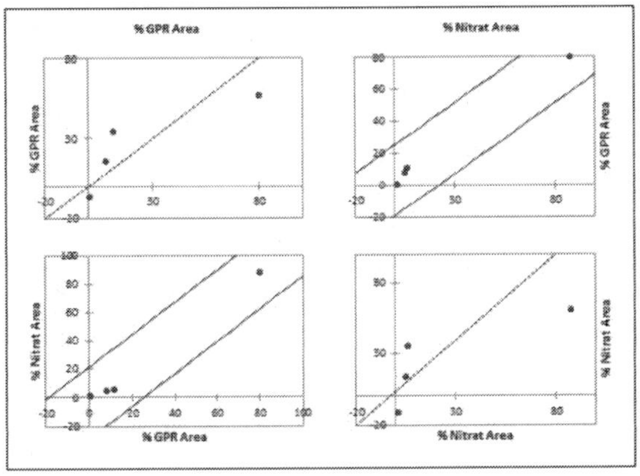

Correlation matrix (Pearson):

Variables	% GPR Area	% Nitrat Area
% GPR Area	1	**0.998**
% Nitrat Area	**0.998**	1

Values in bold are different from 0 with a significance level alpha=0.05

p-values:

Variables	% GPR Area	% Nitrat Area
% GPR Area	0	**0.002**
% Nitrat Area	**0.002**	0

Values in bold are different from 0 with a significance level alpha=0.05

Coefficients of determination (R^2):

Variables	% GPR Area	% Nitrat Area
% GPR Area	1	0.995
% Nitrat Area	0.995	1

Figure 27. Pearson correlation coefficient tests for medium area.

LOW AREA

Scatter plots:

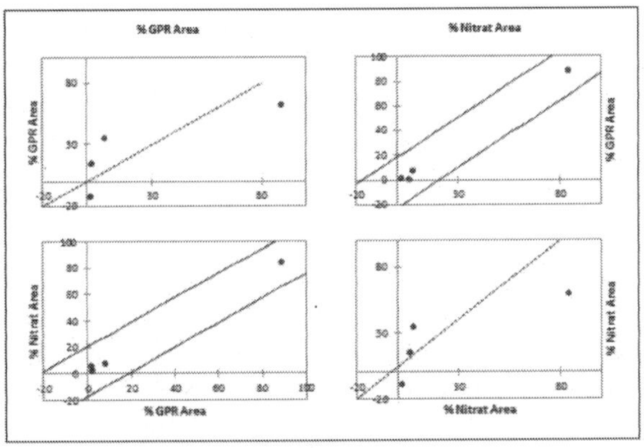

Correlation matrix (Pearson):

Variables	% GPR Area	% Nitrat Area
% GPR Area	1	**0.998**
% Nitrat Area	**0.998**	1

Values in bold are different from 0 with a significance level alpha=0.05

p-values:

Variables	% GPR Area	% Nitrat Area
% GPR Area	0	**0.002**
% Nitrat Area	**0.002**	0

Values in bold are different from 0 with a significance level alpha=0.05

Coefficients of determination (R^2):

Variables	% GPR Area	% Nitrat Area
% GPR Area	1	0.997
% Nitrat Area	0.997	1

Figure 28. Pearson correlation coefficient tests for low area.

4. Conclusion

As demonstrated in this study, seven hydrogeological parameters of DRASTIC were successfully shown through the maps produced by using ArcGIS software. GIS tasks which are spatial database building, spatial data analysis, geographic data modeling and visualization and representation of data has been applied and demonstrated for evaluation of groundwater pollution risk in the Selangor basin area.

Groundwater pollution risk are depends to the environment and hydrogeologic factors. Each DRASTIC factor have own characteristics and play an important role in determining the level of pollution risk. Integration of seven grid layers of hydrogeological parameters through DRASTIC index overlay has been successfully evaluated to construct GPR map. With the aid of agricultural landuse map, GPR based on agricultural activities map was successfully generated.

Application of fertilizers is a main contributor in present of nitrate in groundwater. Accuracy and significance of GPR classification based on agricultural activities has been validated with nitrate distribution layer through Pearson correlation coefficients tests. Significant of GPR zones and nitrate distribution are highly correlated.

5. References

[1] Ahmed, AA., *Using generic and pesticide DRASTIC GIS based models for vulnerability assessment of the quarternary aquifer at Sohag. Egypt. Hydrogeol. J.* doi:**10**.1007/s10040-009-0433-3. 2009

[2] Ali, E., Nezar, H. and Mustafa, K., *Gis-based evaluation of groundwater vulnerability in the Russeifa area, Jordan. Revista Mexicana de Ciencias Geologicas*, v. **23**. num. 3, p. 277-287, 2006.

[3] Aller, L., Bennet, T., Lehr, J.H. and Petty, R.J., *'DRASTIC: A standardized system for evaluating groundwater pollution potential using hydrogeologic setting'*; EPA/**600**/2-85/018, R.S.Kerr Environmental Research Laboratory, US EPA, Ada, Oklahoma, 1985.

[4] Atiqur, R., *A GIS based DRASTIC model for assessing groundwater vulnerability in shallow aquifer in Aligarh, India. Applied Geography*, **28** (2008) 32-35, 2008.

[5] Azwan, M. M. Z., Kamil, M. Y., Hazilia, H. and Suhani, N., *Nitrate-nitrogen concentration variation in groundwater flow in a paddy field. Journal-The Institutions of Engineers, Malaysia*, Vol. **71**, No. 4, December 2010.

[6] Baalousha, H., *Vulnerability assessment for the Gaza Strip, Palestine using DRASTIC. Env Geol* **50**:405-414, 2006.

[7] Babiker, I.S., Mohamed, M.A.A., Terao, H., Kato, K. and Ohta, K., *Assessment of groundwater contamination by nitrate leaching from intensive vegetable cultivation using GIS. Environment International* **29**:1009-1017, 2004.

[8] Civita, M. and De Maio, M., *Valutazione e cartografia automatic della vulnerabilit_ degli acquiferi all'inquinamento con il systema parametrico SINTACS R5 [Evaluation and automatic cartography of aquifer vulnerability using the parametric system SINTACS R5]*. Pitagora Editrice, Bologna, 2000.

[9] Davis, S.N., *Porosity and permeabiliy of natural materials. Flow through porous media, ed.* R.J.M. De Wiest, *Academic Press, New York*, **54**-89, 1969.

[10] Foster, S.S.D., *Fundamental concepts in aquifer vulnerability, pollution risk and protection strategy*. In: Van Duijevenboden W, Van Waegeningh HG (eds) Vulnerability of soil and groundwater to pollutants, vol **38**. TNO Committee on Hydrogeological Research, Proceedings and Information, The Hague, pp 69–86, 1987.

[11] Kreft, A., A. Lenda, A. Turek, A. Zuber, and K. Czauderna, *Determination of effective porosities by the two-well pulse method, Isot. Tech. Groundwater Hydrol*., Proc. Symp., **2:** 295-312, 1974.

[12] Kumar, S., Thirumalaivasan, D. and Radhakrisnan, N., *GIS based assessment of groundwater vulnerability using DRASTIC model, Arab J Sci Eng* **39**: 207-216, 2014.

[13] Lihe, Y., Zhang, E., Wang, X., Wenninger, J., Dong, J., Guo, L. and Huang, J., *A GIS-based DRASTIC model for assessing groundwater vulnerability in the Ordos, Plateau, China, Environ Earth Sci,* **69** (1): 171-185, 2013.

[14] Majumdar, D., *Methane and nitrous oxide emission from irrigated rice fields: proposed mitigation strategies. Current Science*, **84**(10), 1317-1326, 2003.

[15] Neshat, A., Pradhan, B. and Dadras, M., *Groundwater vulnerability assessment using an improved DRASTIC method in GIS. Resources, Conservation and Recycling*, **86**: 74-86, 2014.

[16] Oakes, D.B., and D.J. Edworthy, *Field measurements of dispersion coefficients in the United Kingdom, in Ground Water Quality, Measurement, Prediction, and Protection*, pp. **327**-340, Water Research Center, Reading, England, 1977.

[17] Ramachandra, T. V., *Soil and Groundwater Pollution from Agricultural Activities*. (pp **13**-14). Capital Publishing Company (New Delhi, Kolkata, Bangalore), 2006.

[18] Ratnakar, D., Singh, V. S., Negi, B. C., Subhash, C. and Ananda Rao V., *Geomorphological and geophysical approach for locating favorable groundwater zones in granitic terrain, Andhra Pradesh, India. Journal of Environmental Management*, **88**(4): 1373-1383, 2007

[19] Robson, S.G., *Feasibility of Digital water quality modelling illustrated by application at Barstow, California*, U.S. *Geol. Surv. Water Resourc. Invest.*, **46**-73, 1974.

[20] Sener, E., Sener, S. and Davraz, A., *Assessment of aquifer vulnerability based on GIS and DRASTIC methods: A case study of the senirkent-uloborlu basin (isparta, turkey). Hydrogeology Journal*, **17**(8), 2023-2035, 2009.

[21] Smith, R.T. and Ritzi Jr., R.W., *Designing a nitrate monitoring program in a heterogeneous carbonate aquifer, Ground Water*, **31**(4), 576-584, 1993.

[22] The Groundwater Foundation. 5561 S. 48th St., Ste. 215 Lincoln, NE 68516 USA *http://www.groundwater.org/ kc/whatis.html* (accessed 1 Nov 2012), 1985

[23] Van Stempvoort D., Ewert L. and Wassenaar, L., *Aquifer vulnerability index (AVI): a GIS compatible method for groundwater vulnerability mapping. Can Water Resour J* **18**:25-37, 1993.

[24] Vias, J. M., Andreo, B., Perles, M. J. and Carraso, F., *A comparative study of four schemes for groundwater vulnerability mapping in a diffuse flow carbonate aquifer under Mediterranean climatic conditions. Env Geol* **47**:586-595, 2005.

[25] Wang, Y., Merkel, B. J., Li, Y., Ye, H., Fu, S. and Ihm, D., *Vulnerability of Groundwater in Quarternary Aquifers to Organic Contaminants: a case study in Wuhan City, China. Env Geol* **53**:479-484, 2007.

Diachronic analysis of the occupation of the steppe area of the department of Sidi Bel Abbes (Western Algeria)

B Hellal [1], A Ayache [2], N Ayad [2] and T Hellal [2]

[(1)] Laboratory of geomatics and sustainable development.

University Ibn Khaldoun. Tiaret (Algeria)

[(2)] Faculty of Natural Sciences and Life

University Djillali Liabes. Sidi Bel Abbes (Algeria)

E-mail : hellal_b@yahoo.com ; ayache.abbassia@yahoo.fr ; helayad@yahoo.fr ;

hellaltijania@yahoo.com

Abstract
Modes of occupation of the soil of the steppe area of the department of Sidi Bel Abbes (Western Algeria) know lots of mutations during the period 1987/2013; compromising the future of pastoral activity. This dissection based on supervised classification TSAVI values (Transformed Soil Adjusted Vegetation Index) using images of remote sensing of average spatial resolution of type Landsat-TM 5 and 8. The determination of the state of occupation of the ground and validation of remote sensing map shows that the status of the halophytic/psammophytic steppes and the Matorrals are detected in 38.38 % and 55,71 % of cases, respectively. On the other hand, the steppes chamaephytic mark -9,81 % regression only, agricultural land -24,51 %, and -46,24 % dense vegetation are correctly mapped. The sensing medium resolution is therefore, in the light of these figures, a management tool of the steppe field relevant and effective, which, in addition, allows enriching the field for a proper plan for the fight against desertification.

1. Introduction
The steppe area of the department of Sidi Bel Abbes (Western Algeria) is subjected to very high spatial transformations. These transformations are ecological, climatic, sociological and professional [1]. This has increased the degradation of plant cover of the steppe and promoted the event of desertification throughout the steppe space [2]. The study of changes in the steppe vegetation cover is undertaken by remote sensing. The availability of digital data from previous years allows to quantitatively and qualitatively assess land use patterns. The dynamics of the degradation is followed by the application of the index transformed vegetation adjusted to the ground (TSAVI) [3]. The data used to design maps of land use are satellite scenes "Landsat 5 and 8" TM (Tematic Mapper) from March 1987 and 2013. They thus help to establish a more complete diagnosis of the evolution the steppe area of the department of Sidi Bel Abbes for a period of 26 years.

2. Ecological characteristics of the study area
The steppe zone of the Department of Sidi Bel Abbes is between 34° 30' and 34° N and 1° and 0° East. It is located nearly 300 km from the coast of western Algeria (figure 1).

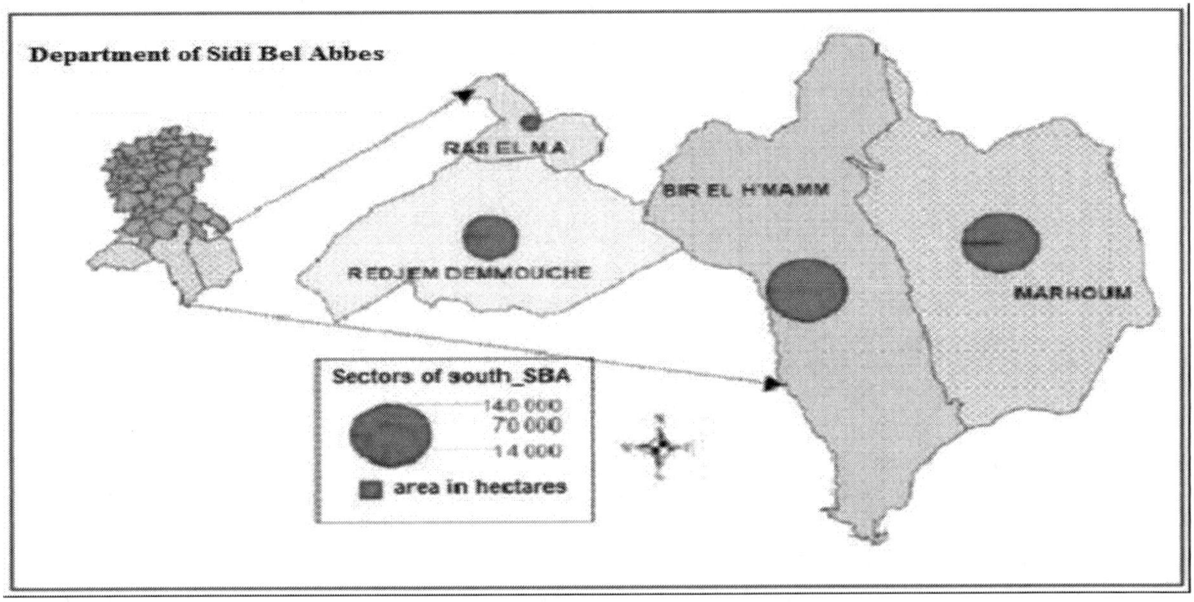

Figure 1. Location of the steppe area south of the department of Sidi Bel Abbes (Algeria).

It covers an area of 3531.18 km² or 38.58% of the area of the Department of Sidi Bel Abbes. It is located at an average altitude of 1200 m. They have four municipalities with Ras Elma (143.52 km²), Redjem demmouche (765.84 km²), Marhoum (1253.13 km²) and Bir El' Hmmam (1368.69 km²). Annual rainfalls of 125.1 mm to over 200 mm. Maximum temperatures (M) are between 34.5°C and 35°C [4]. The region is moving in the arid bioclimatic moderate cold winter [5]. The predominant brown limestone soils are characterized by a depth of 30-40 cm and a sandy clay texture clay surface and at depth. They rest on a limestone crust more or less indurated. Calcimagnesic xeric soils are also represented in the study area. The glaze of the south of the region is characterized by calcareous brown soil crusting xeric to less than 30 cm deep.

3. Materials and methods

The study is based on two satellite images Landsat TM5 and Landsat-March 1987 TM8 March 2013. They are geo-referenced and projected into the conical Lambert system (geographic coordinates system applied in Algeria). The pixel size on the photographs is 30 m.

3.1. Pretreatments of satellite scenes

Satellites scenes have undergone digital pretreatments using appropriate software (ENVI, version 4.7). This allows among others to perform geometric and atmospheric corrections to improve the colored composition of the three selected channels. This step makes it possible to reduce to a maximum of artifacts related to the conditions of shooting.

The image of Landsat TM5: TM1 (blue channel), TM3 (red channel) and TM4 (near infrared channel).
The image of Landsat TM8: TM2 (blue channel), TM4 (red channel) and TM5 (near infrared channel).
The radiometric corrections (atmospheric) is supported on the physical model based radiative transfer "MODTRAN". This second step corrects the wavelengths in the visible through near infrared and shortwave infrared up to 3 m [6]. It makes the two satellite images from different dates directly comparable [7].

Georeferencing of satellite images is a correction operation in a raw image, geometric distortions due to variations in orientation of the observation platform. This correction can bring the geometry of the raw image to a map geometry.

3.2. Treatment index transformed vegetation adjusted to the floor (TSAVI)
The index converted adjusted vegetation soil (TSAVI) is based on the predetermination of right soil from the spectral band of the sensor. It is expressed by the following formula [8]:

$$TSAVI = a\ (PIR - a \times R - b)/[R + a \times (PIR - b) + 0{,}08\ (1 + a2)]$$

with :

(PIR) : near infrared channel; **(R)** : infrared channel ; **(a)** : the slope of the soil ; **(b)** : the intercept of the straight soil, values range from (- 1 et + 1) respectively to absence and presence of vegetation.
The clustering is proposed to define the land use classes on thematic bases. The results are then compared with control zones using a confusion matrix [9]. The land cover classes are determined from different values of TSAVI calculated on two satellite scenes.
The evaluation of the performance of the classification structure is based on the calculation of the percentage and number of pixels correctly classified within each test polygon of the control zones. The confusion matrix of the estimation on the two finally selected pictures will illustrate this performance.

4. Results

4.1. Index of TSAVI
The results of the processing of satellite images are shown in figures 2 and 3. The analysis identified seven main types of land use in the study area: (Class 1: wetland and urban area, Class 2: ground uncovered, Class 3: halophytic and psammophytic steppes, class 4: chamaephytic steppes, class 5: grain farming, class 6: matorral and maquis and class 7: dense vegetation).

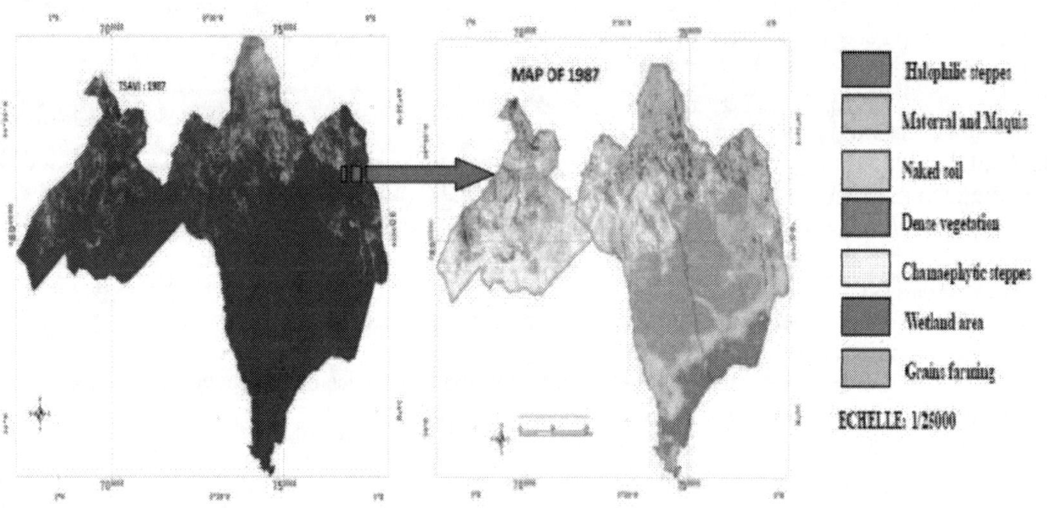

Figure 2. Images of index of TSAVI of the study area (from year 1987).

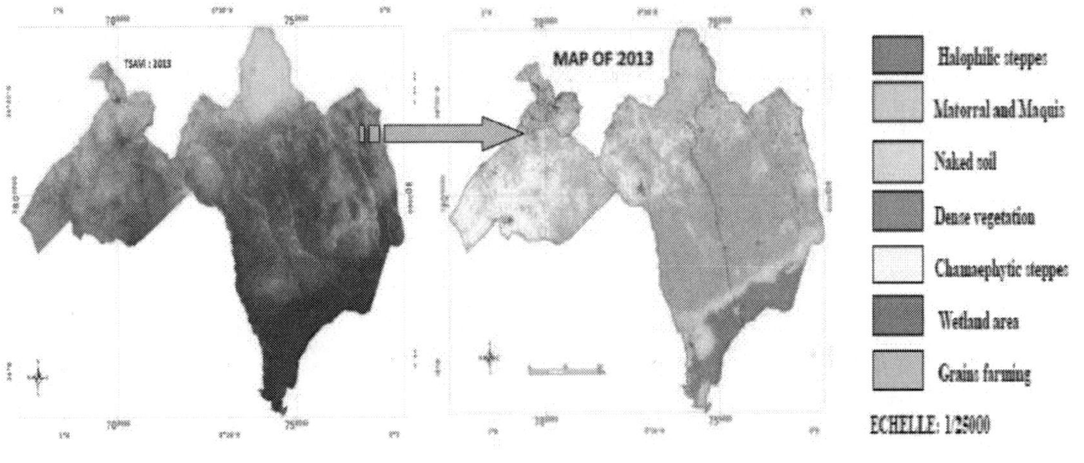

Figure 3. Images of index of TSAVI of the study area (from year 2013).

4.2. Change in vegetation cover by municipality

Figure 4 shows the status of land use in the municipalities of Ras Elma and Marhoum for the years 1987 and 2013. The changes between the two years are shown on the same figure (figure 4).

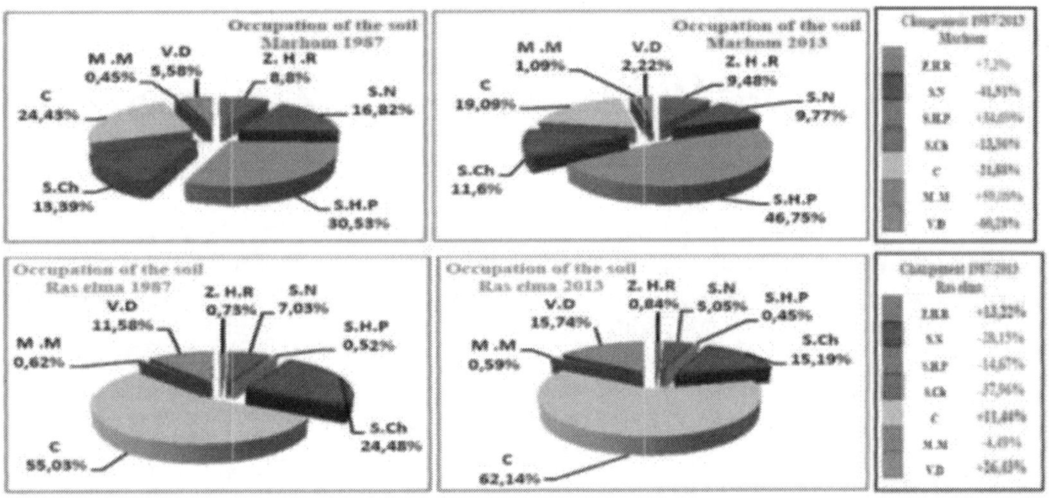

Figure 4. Percentage of land cover and change (municipalities: Ras Elma and Marhoum period 1987/2013).

*NB : (**Z.N.R** : wetland and urban area). (**S.N** : ground uncovered). (**S.H.P** : halophytic and psammophytic steppes). (**S.Ch** : chamaephytic steppes). (**C** : grain farming), (**M.M** : matorral and maquis). (**V.D** : dense vegetation: forests / growing vegetables /cultivation of fruit trees/ replant trees).*

The municipality of Marhoum recorded notable increases halophytic and psammophytic steppes, matorral and maquis couple with 34.69% and 59.08% respectively. The other classes fell to -41.91%

for the ground uncovered and -60.28% for dense vegetation. The Municipality of Ras Elma was characterized by an increase of 26.43% in irrigated crops and orchards. Cereal crops increased by 11.44%. Steppe rangelands and ground uncovered have declined by against because of land clearing, farming and put into crops.

The state of land use and changes between 1987 and 2013 in the municipalities of Bir El'Hmmam and Redjem Demouche are given in figure 5.

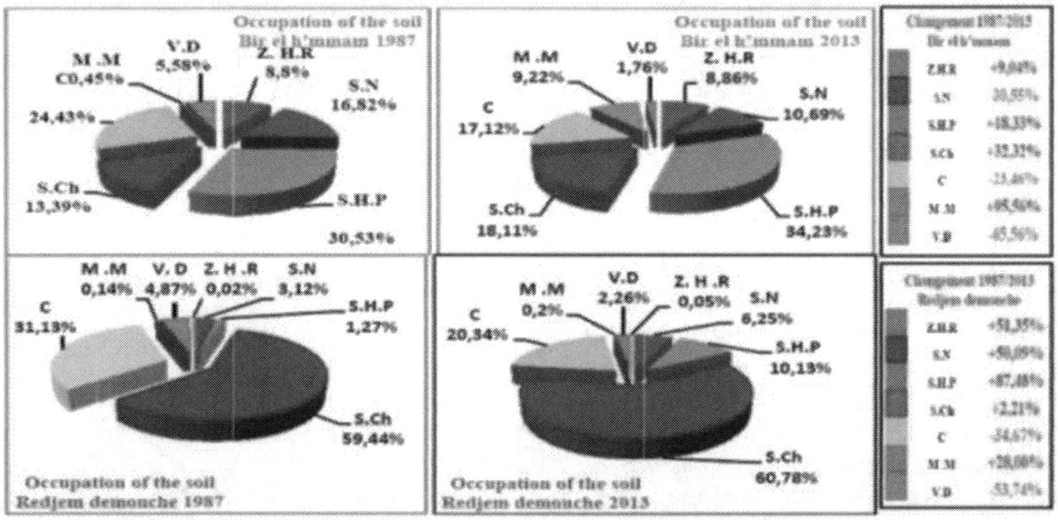

Figure 5. Percentage of land cover and change (municipalities: Bir El'Hmmam and Redjem demouche period 1987/2013).

*NB : (**Z.N.R** : wetland and urban area). (**S.N** : ground uncovered). (**S.H.P** : halophytic and psammophytic steppes). (**S.Ch** : chamaephytic steppes). (**C** : grain farming), (**M.M** : matorral and maquis). (**V.D** : dense vegetation: forests / growing vegetables /cultivation of fruit trees/ replant trees).*

The forest areas in the municipality of Bir El'Hmmam experiencing a remarkable extension of 95.66% by the reforestation work. The degradation of alfa grass and siltation of land favored the installation of halophytic steppes and psammophytic. She recorded by against a regression in agriculture with a percentage of -23.46%. The municipality of Redjem Demouche is significantly affected by the phenomenon of silting. Diachronic analysis of land use reveals a change in the natural area with increased halophytic and psammophytic steppes (38.38%), an extension of the matorral and maquis (55.71%), a reduction cereal crops (-24.51%) and the ground uncovered (-27.10%). The area occupied by the dense vegetation and chamaephytic steppes have undergone a significant change respectively in the order of (-46.24%) and (-9.81%).

4.3. Land use map

Halophytics, psammophytics and steppe sagebrush (*Artemisia herba-alba.*Asso) remain the two main forms of land use in the study area (figure 6). They occupy, in 2013, respectively 32.07% and 24.94% of the total area. The dense vegetation covering 4.83% in 1987. This class consists of forest areas dominated by the Aleppo pine (*Pinus halepensis*) and holm oak forests (*Quercus ilex*).

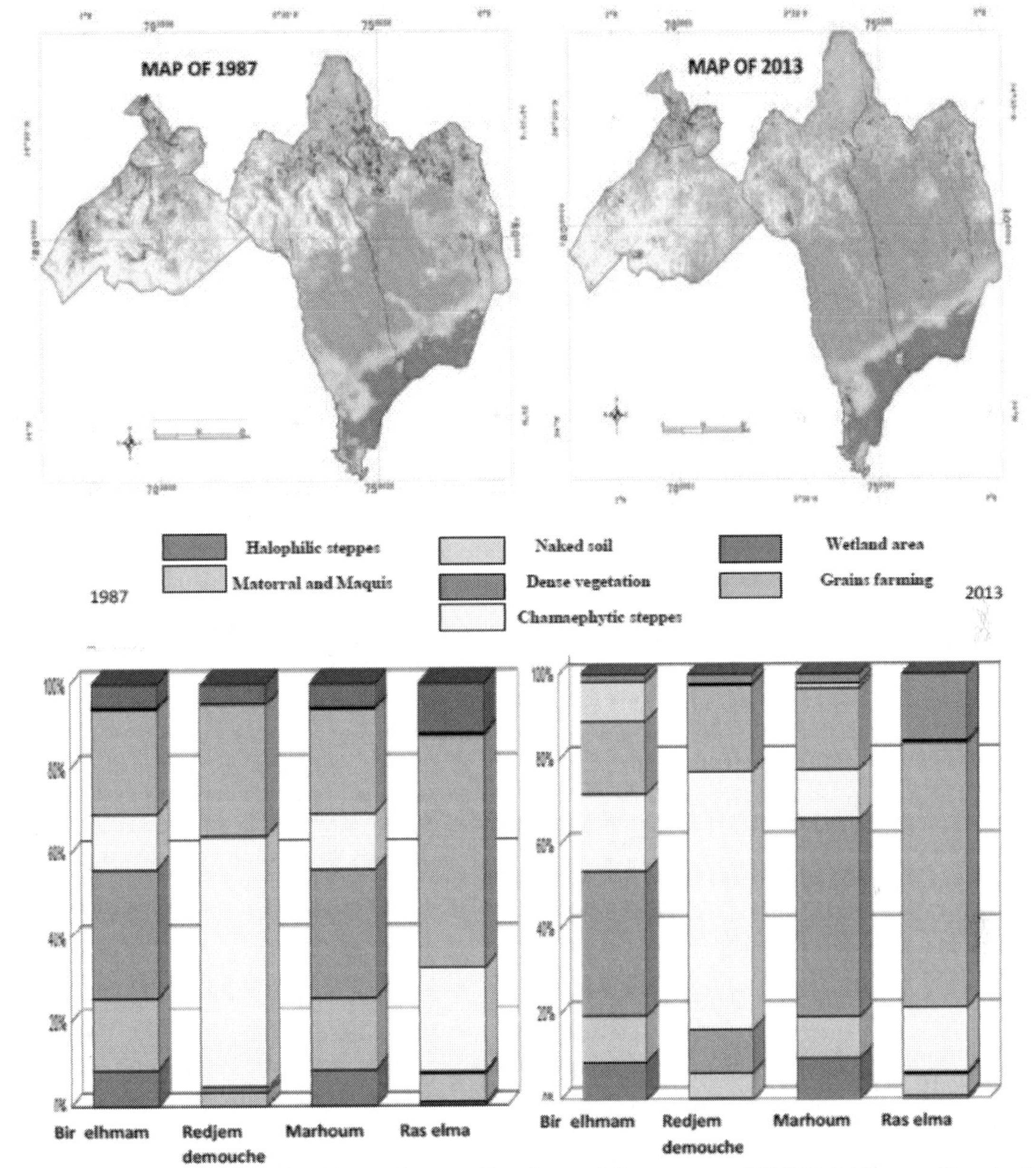

Figure 6. Use maps of soil of the study area (Southern of department of Sidi Bel Abbes – Algeria -).

4.4. Change analysis

The diachronic analysis of land use between the two dates in 1987 and 2013 reveals a change in the nature area with a salt-tolerant and increased psammophytic steppes (38.38%), an extension of matorrals and maquis (55.71 %), reduced grain farming areas (-24.51%) in favor of such crops as arboriculture and market gardening. Bare soil by recorded against a decrease (-27.10%) in favor of halophytic and psammophytic steppes and grain farming areas. The area occupied by dense vegetation and chamaephytic steppes respectively suffered a significant change in the order of (-46.24%) and (-9.81%).

4. Discussion

The space steppe has undergone changes following irregular rainfall, overgrazing and land clearing for agricultural purposes. Livestock has increased during the last 23 years. The increasing availability of

cleared areas and abandoned by farmers is causing the regression of vegetation. These areas have become unproductive and converted rangelands. Forestry training at the rate of 2% of the total area of the study site, are frequented by livestock. Forest grazing remains an almost continuous activity throughout the year, the herd finds appreciable fodder units, estimated at 150 UF / ha for Aleppo pine forests and 250 UF / ha in the lower formations of holm oak [10-11]. The pastoral admitted steppe zone should not exceed 0.5 sheep equivalent per hectare. So according to recent agricultural statistics on the area, the pastoral care in the course of the municipality Marhoum is 0.53 sheep equivalents per hectare. It is respectively 0.63 and 1.53 sheep per hectare equivalent to municipalities of Redjem demmouche and Ras Elma [12]. The agricultural areas are reused as rangelands. The yields of these agricultural lands are still very low and closely dependent on the distribution of annual rainfall [5]. The land cleared for agriculture are abandoned when drought lasts for several successive years.
The cultivation of fodder has increased significantly. It is explained by the high demand of sheep for forage supplement [13]. The strong use of land in crops has resulted in the decrease in the rate of organic matter in the surface horizon offering a favorable particle structure to water and wind erosion. Almost all cereal production is processed into feed due to the reduction in the number of plant species eaten by herbivores such as sagebrush, alfa and Esparto. Southern department of Sidi Bel Abbes (Western Algeria) is essentially agropastoral where space for vegetable crops seems to be affected by fluctuations in rainfall rates below average and therefore recorded low yields. Degradation chamaephytic steppe sagebrush (*Artemisia herba-alba* Asso.) and rising halophytic / psammophytic steppes on silted lands are favored by sedentary livestock causing overgrazing. This situation has become irreversible with a persistent imbalance between supply and demand of steppe rangelands [14-15-16]. Some spaces sagebrush are colonized by other plant species as *Salsola vermiculata, Noaea mucronata, Peganum harmala, Stipa parviflora, Atractylis serratuloides* and *Circium syriacum*.
In the extreme south of the department of Sidi Bel Abbes (Western Algeria) the soil is mostly covered by the particles of sand and silt. The winds from the south warm and dry, very common in autumn and summer accentuate the intensity of wind erosion that usually affects the soil of southern steppe of southern Oran [17]. The rise of salts following a high evaporation leads to dissociation of the soil particles, making them very susceptible to wind erosion and therefore unsuitable for any cereal crops [18].

5. Conclusion

The study of changes in the steppe south of the department of Sidi Bel Abbes (western Algeria), from the satellite scenes from 1987 and 2013, revealed various changes of vegetation. It appears from this study that different types of land use were reported cases of progression and regression during this period. Steppe and agricultural areas have suffered constant pressure imposed by the weather, overgrazing and land clearing. The dominance of livestock in the region and the almost permanent presence of livestock are mainly due to the settlement of the nomadic population. The appearance of desertification is encouraged by signs of clearing very fragile soil and climatic conditions binding for all cereal crops.

6. References

[1] Bouabdellah H 1991 Degradation of the vegetation cover of the Steppic area South Oranaise: the case of aricha. Thesis Magister, Geography, University of Oran (in French).
[2] Hadeid M 1998 Effects of settlement of nomads in sparse area in strengthening desertification of the South Oran steppe. In Proceedings of the Seminar "Drylands : retrospectives, Issues and Strategies," Adrar on 25, 26 and April 27, Edition CRSTRA, Algiers (in French).
[3] Baret F, Guyot G and Major D 1989 TSAVI vegetation index which minimizes soil brightness effects on LAI or APAR estimation, in 12th Canadian symposium on remote sensing and IGARSS'90, Vancouver, Canada, 10-14 July.
[4] ONM 2010 Reports of climatic data of the wilaya of Sidi Bel Abbes. Leaves of 1979-2010 monthly statements. National Meteorological Office, Sidi Bel Abbes (in French).

[5] Hellal T, Hellal B, Baghdadi A, Ayad N and Bensmira Z 2014 Impact of rainfall on the cultivation of cereals in Southern steppe prefecture of Sidi Bel'Abbes (Western Algeria). *European Scientific Journal,*. Vol 10 N° **17**. (in French).

[6] Haddouche I, Gacemi M and Bouhanifia K 2011 Spatial analysis of forest regeneration after fire in the forest "Fergoug" in Mascara, Algeria. *Wood magazine and Tropical Forests* **307**: **23-31**.

[7] Benhanifia K, Bekrada A and Smith S 2002 Multitemporal remote sensing and geographic information systems: a useful tool for detecting and mapping forest changes. *Surveying and Land Information Science*, **62 (3)**: **171-178**.

[8] Gilabert MA, González-Piqueras J, García-Haro FJ and Meliá J 2002 A generalized soil-adjusted vegetation index. *Remote Sensing of Environment*, N° **82**: **303-310**.

[9] Lacombe JP 2001-2002 Introduction to software ENVI 3.4. INP, ENSAT

[10] Mederbal K 1995 Understanding of the vegetation transformation mechanisms: phytoecological approach by aerospace remote sensing and dendroecology *Pinus halepensis* Mill. in western Algeria. State Doctorate, University of Aix-Marseille III (in French).

[11] Labani A 1999 Analysis of the dynamics of land use and eco-development perspectives: Case of the municipality of Ain-El-Hadjar (Saida, Algeria). Sidi Bel Abbes: Magister, University Djillali Liabes. (in French).

[12] DSA 2010. Technical Assessment and Agricultural Statistics of the wilaya of Sidi Bel Abbes. Report of the Directorate of Agricultural Services, Algiers (in French).

[13] Armand R 2009 Research of surface states and their dynamics for different soil practices. Development of runoff indicator. Doctorate Thesis of Strasbourg University (in French).

[14] Barbero M, Loisel R and Quézel P 1990 Changes and disturbances of forest ecosystems caused by human activities in the western part of the Mediterranean basin. *Vegetatio* **87**: **151-173**.

[15] Ayad N, Hellal B and Maatoug M 2007 Dynamics of populations of *Artemisia herba-alba* Asso in the southern steppe Oran (western Algeria) *Rev. Secheresse*, **18 (3)**, **193-98** (in French).

[16] Ayache A, Hellal B, Ayad N, Benhanifia K and Gacemi MA 2013 Diachronic analysis of the steppe land cover of the department of Sidi Bel-Abbes using the remote sensing (Western Algeria). *International journal of geomatics and geosciences*, Vol **6** Issue **2**.

[17] Benabadji N and Bouazza M 2000 Some climatic changes in the southwest of Oran (western Algeria). *Revolution Energy Renovation*, **3**, **117-125** (in French).

[18] Saidi D, Le Bissonnais Y, Duval O, Daoud Y and Halitim A 2004 Effect of exchangeable sodium and salt concentration on the physical properties of soils of the plain Cheliff (Algeria). *Soils Research and Management*, **11 (2)**, **81- 92** (in French).

Application of GIS technologies to monitor secondary radioactive contamination in the Delegen mountain massif

O. Alipbeki[1], G. Kabzhanova[1], G. Kurmanova[2], Ch.Alipbekova[2]
[1]Kazakhstan Gharysh Sapary, Astana, Kazakhstan
[2]S.Seifullin Kazakh AgroTechnical University, Astana, Kazakhstan

E-mail: kurmanova_gul@mail.ru

Abstract. The territory of the Degelen mountain massif is located within territory of the former Semipalatinsk nuclear test site and it is an area of ecological disaster. Currently there is a process of secondary radioactive contamination that is caused by geodynamic processes activated at the Degelen array, violation of underground hydrological cycles and as a consequence, water seepage into the tunnels. One of the methods of monitoring of geodynamic processes is the modern technology of geographic information systems (GIS), methods of satellite radar interferometry and high accuracy satellite navigation system in conjunction with radioecological methods. This paper discusses on the creation of a GIS-project for the Degelen array, facilitated by quality geospatial analysis of the situation and simulation of the phenomena, in order to maximize an objective assessment of the radiation situation in this protected area.

1. Introduction

During over 40 years of confrontation of nuclear powers, 456 nuclear tests were conducted in the territory of the former Semipalatinsk nuclear test site (SNTS). In these tests, 616 nuclear and thermonuclear devices were detonated, including underground (131 in the wells and 209 in the tunnels). Although the tunnels were created with a stemming complex with hermetically sealed elements of high strength with damping devices and gas blocking, explosions of high enough power (up to 150 kilotons of TNT equivalent (TE) contributed to rock strata deformation, cracks, activation of geodynamic changes [1]. These processes are typical for the Degelen mountain massif (DMM). Given that the main aquifer located at a depth of 10 to 30 m from the surface, geodynamic changes as a result of explosions contributed to underground hydrological cycles disturbance.

According to the literature data [2, 3], there are about two dozen tunnels with the manifestation of radioactive water in the DMM SNTS. Water from some tunnels flow into local rivers, which in turn are in the basin collecting waters of Shagan river, falling into the Irtysh river. Water manifesting tunnels of SNTS are sources of the secondary pollution by long-lived radioactive isotopes, which is extremely dangerous for human and animal health. According to Polyakov et al. [4], the underground water of DMM is located in an open fracture (intensive weathering) of rocks and partly in unconsolidated sediments of the Quaternary age. This increased radioactivity is identified in the south, east, north and northwest massif underground water. In general, the condition of water resources of the Degelen massif appears unsatisfactory.

Figure 1. View of the Degelen mountain massif.

The effectiveness of monitoring the secondary pollution of the DMM due to modern geodynamic processes depends on the methods of control. Using modern technology of geographic information systems (GIS), satellite radar interferometry and high accuracy satellite navigation systems contribute to the overall analysis of the current situation and elaborate recommendations to overcome the effects of pollution of the environment. Integration of GIS, remote sensing and modeling of processes applied for the environmental protection fields is an important tool to the management and decision-making [5]. One of the research goals in SNTS is a creation of expanded geographic information platform of the DMM for environmental monitoring and control the secondary pollution of the environment based on GIS technologies and ERS data.

2. Methodology

For development of a geoinformation system of the DMM and the surrounding regions, digitization of individual thematic layers of the DMM SNTS was conducted using ArcGIS. The development of a geodatabase and spatial models were done using the ArcGIS' 3D Analyst and Spatial Analyst applications. In order to create a DMM spatial model, electro-optical space images from ALOS and Spot-5, as well as radar images from COSMOSkyMed were used.

Elaboration of a geoinformation system of any object requires inter-summand and mutual complements of each component. The structure of the geoinformation system for the DMM contains the following interrelated components: attribute database, which contains the narrative components and information about the DMM; mapping information database, which is a complex of electronic thematic maps for the study area; block data analysis and processing, which are formed by means of problem solving, statistical data processing, mathematical modeling tools.

The DMM is located in the south-western part of SNTS with lowlands of mountains, dissected by valleys and decays. The DMM represents spatial object with a complex structure, which is the basis of the developed system of low mountains. Therefore, the using of such a powerful and convenient means of integrating spatially-distributed data as GIS is justified and promising.

3. Results & discussion

Direct use of DMM geoinformation system provides information support for monitoring the effects of nuclear explosions on the mountain massif territory. To fulfill the main task of developing a geographic information system - monitoring of the ecological balance of the DMM territory - a graphical representation of information form of plans and schemes applied to data on topography, elevation, geology, human-induced changes, etc. The creation of digital maps on the basis of GIS technologies according to actual data of remote sensing reduces the complexity of creation of informative materials, and increases efficiency and effectiveness of measures for management of ecological safety area. The GIS structure of the DMM and its subsystems, core modules and data relationships are shown in Figure 2.

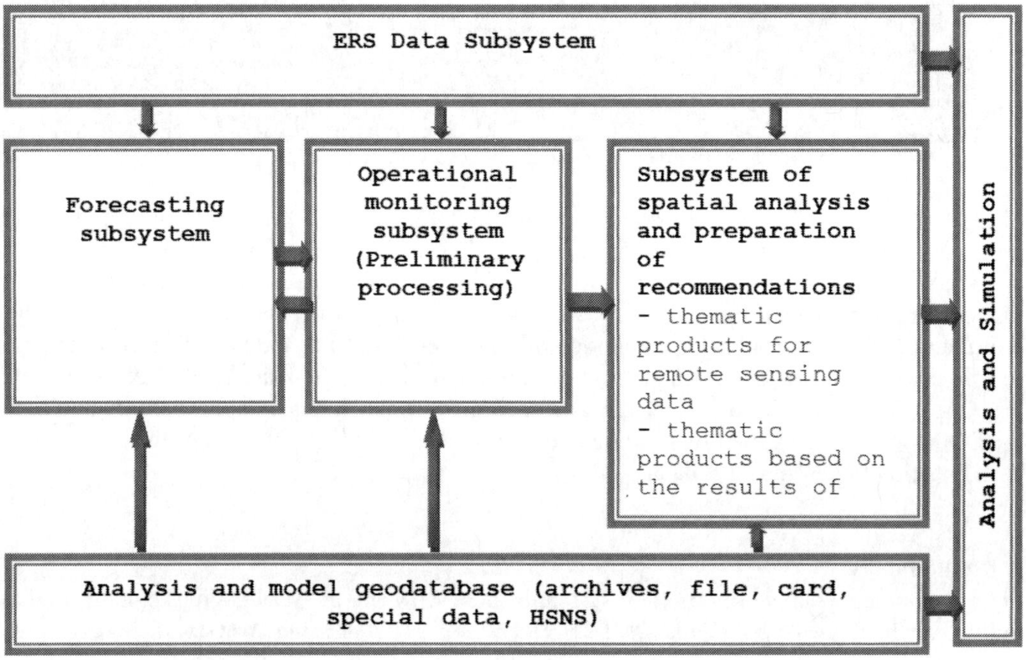

Figure 2. Structure of GIS monitoring of the DMM.

As a rule, at the initial stage of creation of the GIS, the inventory of the district topographical data is produced through the creation of different classes of geodatabase or sets of classes. The geodatabase of the DMM as vector model of the district consists of six directories and 18 thematic layers with attributive information. Figure 3 presents all the thematic layers and directories of the DMM GIS.

In order to identify environmental trends in the DMM, we need to know about the relationship of drainage tunnels and hydrological features of the study area. The directory has been created as "Hydrology", which contains thematic layers of "rivers_full", "river_dry", "dumb" and "spring". The mask "rivers_full" can be traced to some rivers, such as Karabulak, Uzynbulak, Baitlesov and Toktakushik waters fall from the radioactive tunnels. Visual representation of hydrological posts in the form of thematic layers contributes analysis of possible radioactive contamination areas. The formation of a vector layer "shtolny_water" in the directory "Shtolny" displays of all water-manifested tunnels of the Degelen massif. The attribute information shows the exposure dose of radiation, reflected by the intensity of the manifestations of water (Figure 4).

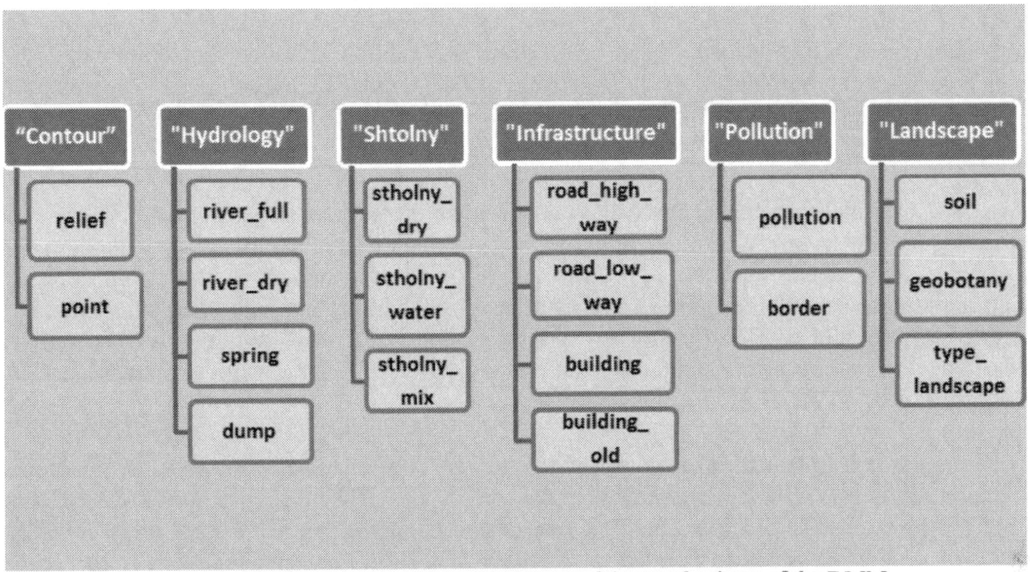

Figure 3. Directory and thematic layers of the geodatabase of the DMM.

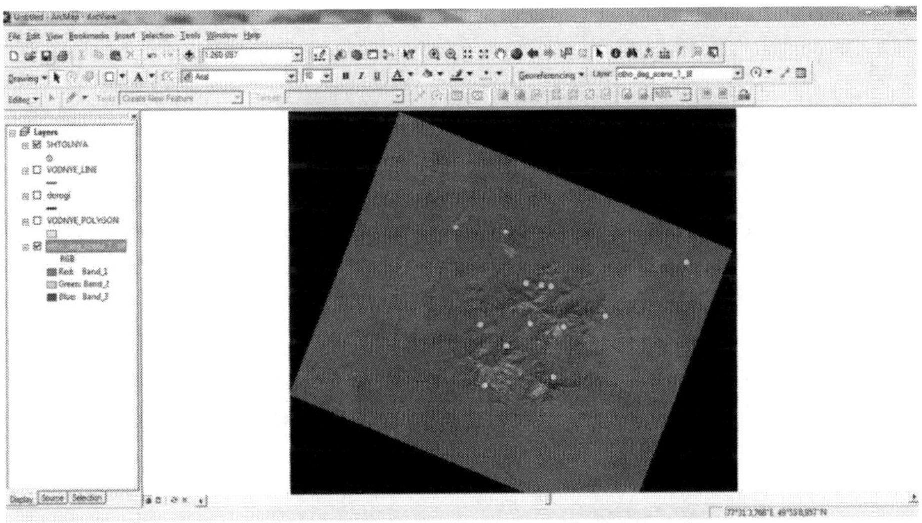

Figure 4. Location of water-manifested tunnels.

This vector layer represents only 12 tunnels with signs of water seepage. Attribute sampling was formed from the layer "schtolny", which indicates the location of the 180 tunnels. For a comprehensive assessment of the environmental situation in the DMM for the emergence of secondary contamination risk by radioactive substances, as well as to improve the accuracy of solving problems of long-term forecasting of multivariate object situations, this requires clustering of data in the DMM GIS. The main blocks of the DMM GIS (block of hydrometeorological data, geological unit performance, block of physical and geographical factors, block of the socio-economic indicators, and unit of the environmental performance) are composed a number of constituent components. Moreover, the indicators of each block (cluster) must be ranked by character and optional historical aspect.

Hence, the block of environmental performances is considered background contamination, maximum permissible concentrations of harmful substances, the potential contamination and others. The meteorological data block is formed by data-term and annual rainfall averages, temperature and wind data, which have an indirect impact on the ecosystem of the DMM. The hydrological data area of the DMM includes the length and mesh of the river system, presence and depth of groundwater, as well as indicators of pollution.

The geological data includes exogenously and endogenously-geological and geological changes, and also indicators of modern geodynamic changes of the DMM defined on the basis of radar pictures. The physiographic aspect is very important in the creation of the GIS for the mountain territory as the key characteristic of intensity of processes. The relief defines zones of accumulation and redistribution of polluting substances and soil characteristics causing buffer action and resistance of the territory pollution. The landscape characteristic is given on the basis of genetic types of landscapes and possible scenarios of development the denudation processes.

The socio-economic infrastructure block includes infrastructures of nearby settlements, population, transportation routes, the results of recent medical inspections, etc. The design of the DMM GIS does not apply for completeness of the accounting factors and communications. Nevertheless, it gives a reasonable picture of the ecological situation in DMM. In connection with specifics of the studied area, conditions of environment, and also degree of technogenic loading, the GIS can be supplemented also with other clusters.

The creation of the DMM GIS uses the most common software and hardware implementation for realization and establishing standard structure of the mountain areas GIS database, which ensures optimum control of the manifestations of radioactive contamination that includes sequences and content thematic layers, structure of thematic clusters. The results of the geospatial analysis of the situation of the DMM generates three-dimensional model of the DMM in order to maximize the objective assessment of the radiation situation in this protected area. On the basis of archival optical-electronic space images (stereo pair), a digital three-dimensional model of the DMM was created in different scale. The GIS system is necessary, especially for high-precision of forecasting development processes proceeding in the highlands (the analysis of geomechanical changes – subsidence of the soil, shift of rocks; the analysis of a geodynamic situation – change of crust, a motion, etc.), which gives the advance to prevent the occurrence of environmental problems. The integrated three-dimensional digital model of the investigated mountain massif at 1:50 ,000 scale is shown in Figure 5. Lighter areas is subjected to the destruction of rocks, where intensive geodynamic processes are possible.

Figure 5. Three-dimensional digital model of the DMM at scale of 1:50,000.

The developed GIS was used to study modern activity geodynamic processes of the DMM using radar interferometry methods and high-accuracy satellite navigation systems. In the north-eastern part of the DMM, during the observation period, displacement (shifts) of the earth's surface, size of 3.1 cm was detected, which indicates a relatively high activity of geodynamic processes in this area [6].

Based on GIS modeling and forecasting, the situations that led to increased environmental hazards in the DMM include increase of groundwater levels, violation of radioactivity balance, secondary pollution processes of the territory and changes in the nature of the biosphere.

4. Conclusion

The DMM GIS allows us to give environmental assessment of the territory. The spatial geodatabase actualized by the DMM GIS provides a more detailed spatial analysis of the migration and transformation of artificial radionuclides in the DMM. The prepared vector and attribute basis of the DMM GIS allows for subsequent creation of geoproducts with high added value as a set of indicators can reflect more realistically the anthropogenic influence, in order to develop recommendations for the reduction of social and ecological intensity of the territory.

References

[1] Matuschenko A.M., Logatchev V.A., Maltsev A.A et al. Semipalatinsk radiation phenomenology of underground nuclear explosions, the residual contamination of the area. Report at the conference. Antwerp (Belgium). - 1993.

[2] Akhmetov M.A., Artemyev O.I., Ptitskaya L.D., Sinyaev V.A Radiation monitoring streams and problems of rehabilitation of upland array Delegen Semipalatinsk test site.Vestnik NNC RK, 2000 - Issue 3. S. 23-28.

[3] Aydarhanova G.S Ecological assessment of influence of underground nuclear testing on the environment massif Delegen: diss. ... Candidate of biol. Sciences. Almaty, 1998. - 130c.

[4] Polyakov L.E And others. The radiological state of groundwater and surface water south-eastern part of the territory of the Semipalatinsk test site and the border areas. Vestnik NNC RK, 2006 - Issue 2. - S.106-111.

[5] G.DROJ: GIS and Remote Sensing in Environmental Management. J.of Environmental Protection and Ecology, 1 (13), 361, 2012.

[6] Alipbeki O.A., Kabzhanova G.R., Makhmetova A.A., Maksut J.J Characteristics of geodynamic processes in the Degelen mountain massif by applying radar interferometry method. Bulletin of the National Nuclear Center of the Republic of Kazakhstan. Issue 4 (64).- 2015.-S.79-84.